Plant Hazard Analysis and Safety Instrumentation Systems

Plant Hazard Analysis and Safety Instrumentation Systems

Swapan Basu

ELSEVIER

AMSTERDAM • BOSTON • HEIDELBERG • LONDON
NEW YORK • OXFORD • PARIS • SAN DIEGO
SAN FRANCISCO • SINGAPORE • SYDNEY • TOKYO

Academic Press is an imprint of Elsevier

Academic Press is an imprint of Elsevier
125 London Wall, London EC2Y 5AS, United Kingdom
525 B Street, Suite 1800, San Diego, CA 92101-4495, United States
50 Hampshire Street, 5th Floor, Cambridge, MA 02139, United States
The Boulevard, Langford Lane, Kidlington, Oxford OX5 1GB, United Kingdom

Library of Congress Cataloging-in-Publication Data

A catalog record for this book is available from the Library of Congress

British Library Cataloguing-in-Publication Data

A catalogue record for this book is available from the British Library

ISBN: 978-0-12-803763-8

For information on all Academic Press publications visit our website at https://www.elsevier.com/

Publisher: Joe Hayton
Acquisition Editor: Fiona Geraghty
Editorial Project Manager: Maria Convey
Production Project Manager: Lisa Jones
Designer: Inês Cruz

Typeset by TNQ Books and Journals

Dedicated to Gurudeb I trust
and
to all my gurus in various spheres of my life
and
my loving wife

Contents

Foreword

There is no facility in the world which is free from hazards and risks. There have been a number of catastrophes in the world (e.g., Bhupal in India). After these, people realized that there should have been a systematic approach toward safety. Each facility is different from others; naturally hazard analyses will be varying in nature. Since in the modern era of high competition, people depend a lot on electronic instrumentation and control. Keeping these in mind, IEC came out with IEC 61508 and 61511with systematic approaches toward safety instrumented systems (SISs) and safety integrity level (SIL).

Hazard analysis and SIS is a vast subject to cover all sectors of industrial facilities. It is indeed a good idea that the author has felt to put forward his rich industrial experience in a single book to cover various hazard analysis methods, calculations of SILs for developing SIS.

There are a few good books on different topics, but it is fact there is no single book to cover both hazard analysis methods as well as detailed SIS. The chapters this book covers including risk assignment fundamentals, hazard analysis methods, SIL calculations to develop SIS, and effects of SIS on various components staring from sensors, fieldbus, logic solver, and final elements are really worth praising. The book is well balanced to cover basic mathematics as well as sector-wise industrial applications.

The author has rich industrial experience in India and abroad, and is still in touch with modern developments. I truly feel that the book, which is developed based on the author's experience, would be extremely helpful to the practicing engineers as well as for fresh engineers. Also, I am glad to know that this book has been selected by IChemE in their series of publications.

I wish all the success for the effort of the author for the book.

Professor S.K. Sen
BE, PhD(LOND.) DIC(LOND.), FIE, FNAE, SM CSI,
Fellow, Imperial College,
London

Preface

Risks and hazards are a part of everyday life. And it is our natural tendency to make decisions after calculating risk for the action. Looking at a long distant traffic light, a driver controls the speed and applies brake at the correct time and place to stop the car within stop line without any jerking. However, the person may not be safe for long! In the area of technology, too, there are hazards and risks everywhere; nothing is totally free from hazards or risk. In all of industrial facilities, process plants, civil/architectural construction sites, transportation systems, aerospace, space craft — even for a financial decisions there will be hazards in different forms. Accordingly, there will be associated risks in different forms and in varying nature. Therefore, it is obvious that hazard analysis is an absolutely necessity in all spheres of life.

The book has two parts; one part is mainly related to hazard analysis methods applicable for all disciplines of engineering. The other part is based around recent IEC standards for life cycle analysis for safety instrumented systems (SISs). The book covers all major hazard analysis methods and safety instrumentation applicable for all industrial plants. The book enjoys *granted permission from* ***IEC Switzerland*** to utilize some figures from their standards IEC 61508 and 61511; the premier international standards for E/E/PEs for SIS and life cycle methods (duly acknowledged). In view of the total coverage of the book, international society ***IChemE*** has included the book in their series of technical books.

There are a number of books available on hazard analysis as well as quite a few books on SIS. However, there are hardly any technical books where people get the tastes of both in a single book. Also, the majority books are with theoretical approach with huge mathematical details, or only practical details without any basic approach. This book endeavors to act as a balancing means between two extreme lines of thinking, giving a comprehensive approach toward hazard analysis as well as safety instrumented systems. The book takes into account basic mathematics, but more emphasis has been put on physical explanations. Hazard analysis covers both qualitative type, word-based systems, and quantitative systems (including human errors analysis) have also been covered comprehensively with automated processes for hazard analyses. Hazard analysis part will help instrumentation practicing engineers in participating in various hazard analysis processes. For each case, both mathematical approaches as well as physical explanations have been put forward. Step-by-step SIS determination from various failure modes, reliability, and safety integrity level (SIL) calculations and selections from various approaches, for example, risk matrix, risk graph, etc. help to arrive at correct and appropriate SIL calculation. The book comprehensively discusses each and every stage in different phases, that is, all the aspects of life cycle process mentioned in IEC 61508 and 61511 for E/E/PEs, with respect to continuous process and manufacturing industries. The book also covers safety systems of batch process in line with ISA 88. It also includes an alarm as an independent protection layer (IPL), area classification including explosion protection, fire and gas system, process shutdown (PSD) and emergency shutdown

(ESD). Finally, a discussion on SIS would be incomplete unless certification methods, FMEDA, proof testing, and partial stroke testing, etc. are not covered.

Detailed discussions at component level starting from sensors, safety field bus, safe logic solvers, and final control elements are a unique addition to the book. The book also covers the security aspects of various networks, viz. firewalls and zone-conduit to name a few, which is also another unique feature of the book. The book also covers application of SIS in various plants covering fossil fuel power stations, nuclear power stations, oil and gas sector such as upstream, midstream, refinery, and petrochemicals. There have been supplementary data and information on statistical approaches, embedded controls, and cost impact and life cycle cost analysis to take care of advancement in technology and systematic approach toward the problem.

The book is primarily meant for working professionals but budding (fresh) engineers who started their career in industrial plants and process plants will be equally benefitted. The book covers both hazard analysis and safety instrumentation from early stage to most advanced stage. It would be beneficial to engineers from other disciplines also in almost all industrial sectors — with special reference to process industries. For this comprehensive coverage, the book has been selected by ***IChemE*** in their series of technical books.

Acknowledgments

At the outset, the author wishes to put forward his thanks and his gratitude to ***International Electrotechnical Commission (IEC)*** and ***IChemE.***

The author is thankful to IEC for granting permission to use some of their figures from IEC 61,508 and 61,511 in the book, and would like to acknowledge as follows:

The author is extremely thankful to Institution of Chemical Engineers (***IChemE)*** for selecting the book in their series of publications. It is a great honor. The author is extremely grateful to Ms. Ann Marie of Softrel (courtesy Softrel LLC) and Professor L.J Osterweil —University of Massachusetts for granting permission for utilizing their tables and figures in the book (duly acknowledged in respective places).

The author would like to pay tribute and thanks to his professor, Dr. Shankar Sen, who could spare time in going through the book and agreed write the foreword for the book.

The author would like to thank Prof. (Dr) A.M. Ghosh, Mr. Yogendra Sasankar, Mr. Asim Sarkar, Mr. Ajay Kumar Debnath, and Ms. Rajlaxmi Basu, who could spend their valuable time in going through the write up and coming out with valuable suggestions, and sharing a lot of information with the author, to enrich the content of the book. The author is also thankful to Mr. Deb Kumar Basu for guiding and developing the cover design and with good suggestions which were extremely helpful in developing various figures in AUTOCAD. While developing the book, the author was well supported with information by Mr. Sabyasachi Guha.

The author would like to thank the entire team of Elsevier, the publisher who took all the pains in bringing up the book.

Last but not the least; the author would like to thank his children Idai (Raj) and Piku (Deb) for their continuous inspiration and support. The author would like to convey special thanks to his wife, Bani, for managing the family show with care within limited resources and encouraging the author who had to refuse all project works to dedicate their time for the book.

The author sincerely acknowledges that without all this support, it would have been impossible to publish the book.

CHAPTER

Basics of Hazard, Risk Ranking, and Safety Systems

I

1.0 INTRODUCTION

This chapter gives the overall view of the entire subject in short form. This will help in understanding the subsequent details in the following chapters. Also, a few basic issues have been detailed, such as risk ranking, and risk register. So this chapter is important for following subsequent discussions.

Assets are normally acquired against a lot of efforts, toil, and monetary cost. People always wish to protect these. Unfortunately, this is not always possible on account of hazards in various forms. None desires to lose them, so people always plan to take safety measures to protect them. Until recently, in process industry, people would incorporate necessary safety measures in the form of protections under basic process control systems (BPCSs). In the arena of industrial hazard and risk analysis, "system" is defined as a subject of risk assessment, which includes mainly process, product, facility, and environmental and logical groups. So, safety associated with it needs to be treated separately from BPCS. This is more clear from subsequent discussions. Sometimes people incorporate redundancy in the system design so that in case of failure of one, there will be others available as backup, that is, to fall back. This is not true always, as is the case with common cause failure. Later, in the 1980s people tried to interface safety systems to BPCS to have better connectivity and safety controls. Initially these were based on proprietary protocol, and later people adapted open protocol systems. Although this made a better connectivity, at the same time it made the system vulnerable to external attacks such as viruses, cyber attacks, etc. After 1995, people felt the need for integration of safety systems with BPCS, without compromising functional independence between the two, to get best secured industrial systems. Therefore, people felt it was necessary to have a standard which shall be based on safety planning and safety goals based on risk assessment. As a result of this, several international standards, IEC 61508, IEC 61511, ISA 84, etc. evolved. These standards are developed mainly for electrical, electronics, and programmable electronics (E/E/PE). The primary principle behind these standards was that there is a process which may pose risk on human, machinery, asset, and environment safety, should anything goes wrong with the process, equipment, and/or operation. Naturally, these standards are aimed at process upset, or system or equipment failure, and allow the process safety to be managed in systematic way following a risk-based management system.

Again, the global financial debacle in 2008 demonstrated how important risk management is! Risk management is an increasingly important business driver, and stakeholders are extremely concerned about it. Slowly, risk management is becoming a part of strategic decision-making, and/or has been incorporated as an embedded element in routine activity. An organizational approach towards risk management enables all concerned to consider the potential impact of all types of risks on all processes, activities, services, and stakeholders through enterprise risk management (ERM). Organizations will get benefit from what is often called the upside of risk. An ERM initiative helps all concerned to know the likelihood and severity of risk materializing as well as being better informed of strategic decisions, reduced cost of capital (otherwise due to loss), more effective financial reporting, maintaining edge over competitors, and ability to maintain brand value. This is also clear from the report of Aberdeen group research on, "operational risk management (ORM—7930 Ismail)" in November 2012, where it has been found that all good industrial houses started receiving benefits from risk management. As such, risk management is a vast area covering almost all aspects of business; in order to restrict the volume of the book, the discussion is mainly concerned with industrial plants and processes. After risk/hazard analysis, control actions need to be taken. Safety instrumented systems (SISs) play a great role in mitigating technical risks in industrial plants. Since the book is concerned with technical risk management in industrial plants, so the role of SISs cannot be overestimated. An SIS consists of a well-engineered hardware and software control system used to monitor the condition of plant within the operating limit When any risk condition arises, it triggers an alarm and will take the entire system to safe condition to mitigate all kinds of risks as far as possible. As stated earlier, SIS is a well-engineered system so it ensures that SIS is effective in mitigating risk level at moderate cost during the entire system life as shown in Fig. I/1.0-1 and I/1.0-2.

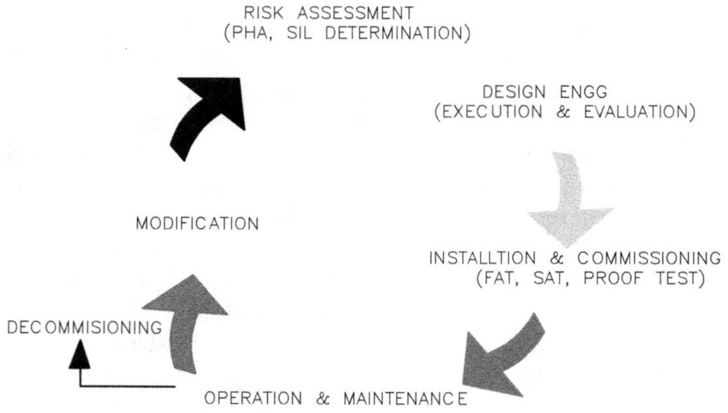

FIGURE I/1.0-1

Safety life cycle of SIS.

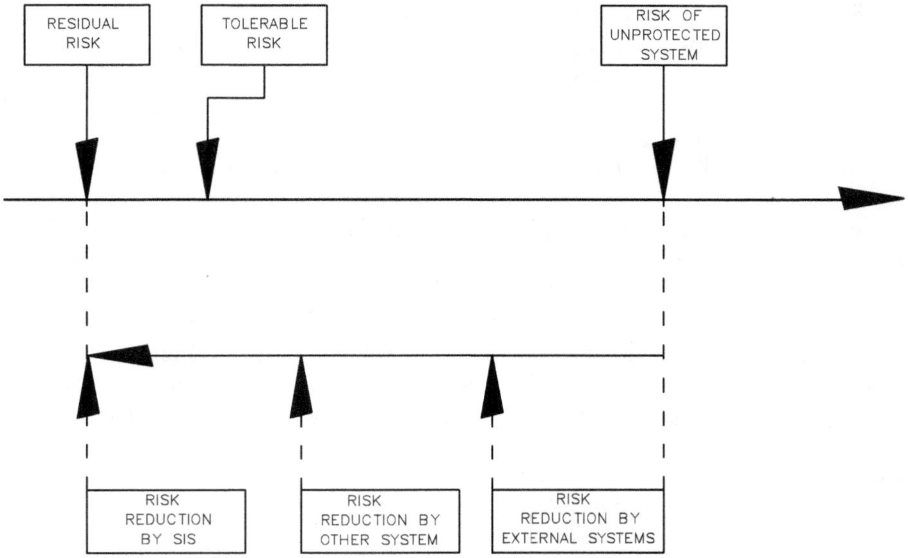

FIGURE I/1.0-2

Risk reduction by SIS (Based on standard IEC 61508 concept).

When a system is developed, it runs at a risk unless it is protected. So, *inherent risk* is the risk to an entity before any action is taken by the enterprise to change the consequence. Also every system has a tolerable risk limit. Designers aim to bring the system within, or in fact below, that risk limit by incorporating various protection measures in the system. However, even after taking action to mitigate the risk, there will be a little risk left, often referred to as *residual risk*. So, *residual risk is the remaining risk after actions taken by the enterprise*. Some of these are external (e.g., by making the control room air-conditioned so electronics are saved from harsh conditions at field). Some protections come from other technological means (e.g., rupture disk in a vessel), but major protections come through interface of BPCS with a safety system to make SIS. Readers should to keep in mind that a plant operational interlock and protection system is part of BPCS, and not really SIS. SIS interface comes from reliability analysis of the system; discussed later. From the discussions so far, it is clear that there are two parts in the whole system risk management, viz. hazard analysis and SIS. Industrial plants risk management through hazard analysis and incorporation of SIS has become a call of the day. So, in this book both these areas have been discussed at length from engineering experience (and not theoretical discussions with mathematical treatments) to give an idea to the reader about practical treatment of the issues in a comprehensive manner, as far as possible within limited pages of a single book where the reader comes across both systems at one place. This is unique when compared with other good books available in the market.

Since 1979, the author of the book has been associated with instrumentation and control systems, pertinent to different industrial and process plants in various parts of the globe. He is still in touch with modern plant design and engineering. The author wants to document his rich global experience in the form of a professional technical book so that working engineers could benefit from knowledge gathered during his tenure. In this book, effort has been made to integrate both hazard analysis methods as well as SIS. There are a number of good, valuable books available on variety of subjects related to process hazard analysis, viz. HAZOP (**haz**ard and **op**erability study), FMEA (**f**ailure **m**ode and **e**ffect **a**nalysis), etc. Also there are a number valuable and good books available on variety of subjects on SIS and SIL (**s**afety **i**ntegrity **l**evel). However, it is very difficult to get a single book or single volume of a book to cater to the requirements of plant hazard analysis (PHA) and SIS, so that practicing engineers can gain overall knowledge of the entire spectrum for day-to-day use. The book has been written in such a way that not only instrumentation and control engineers will benefit, but professionals from other disciplines such chemical, electrical, mechanical as well as safety engineering, will also benefit. In this book, all relevant topics have been covered in a logical and appropriate manner so as to gather comprehensive knowledge on the subject. From the Table I/1.0-1, the reader will come to know the scope and objective of the book, with a few details of the topics covered. However, Table I/1.0-1 has been given only to get an idea and it is not the book index. Also in order to broaden the discussions and to help the reader gain brief overview on a few topics which are relevant but not directly related with PHA and SIS, for example, enclosure classification, little JIL language, and embedded control have been included as appendices.

In Chapter I, in order to have a general idea, discussions have been generalized in nature, whereas in subsequent, relevant chapters, these have been discussed *at length* specifically for industrial plants.

2.0 HAZARD AND RISK

2.1 Discussions on Hazard and Risk

The word hazard by definition is rather confusing it stands for potential danger, or dangerous outcome! Again the word risk is related to hazard as per dictionary terms. It also means the chance of something going wrong. So these two terms are interrelated. However, in terms of functional safety, hazard stands for some event, which has the potential to cause harm to personnel, plant, and machinery and/or environment. When a big house building is dismantled, workers or people on the road have potential danger of being injured (even killed) from the falling object. Now if there is a suitable barrier placed around that location, a safe zone for people to walk around is provided. Then the people will not enter into the barrier, and would pass outside it. In this case, *risk* of people getting injured is minimal. From the preceding example, it is clear that risk is a likelihood, chance, or probability of a person getting injured. From an occupational safety and health point of view, *hazard* is a ***potential source*** of harm

Table I/1.0-1 Plant Hazard Analysis & SIS Topics

Heading	Discussion Points
I Basics of hazards, risk ranking, and safety system	Definition of hazard, risk; discussions on likelihood, consequence risk – register, matrix, & ranking. Consequence ranking, preliminary hazard analysis tolerance point—ALARP; refreshing on mathematics, fault tolerance, plant ageing, and basic functional safety fail safe operations in plants.
II Evaluation of hazard and risk analysis	Requirement PHA, identification of human/process error; existing & additional protection; factors for PHA; hazard identification and consequence; risk analysis, estimate from risk matrix; presentation of result; systematic selection of various processes.
III Qualitative hazard analysis	Discussions on methodology pros & cons of various systems like brainstorming what if; checklist, etc.
IV Guided word hazard analysis	Sequence examination, document requirement, team selection, report preparation both in manual as well as automatic methods for each of HAZOP, FMEA; HAZOP & FMEA for E/E/PE.
V Quantitative hazard analysis	Consequence spectrum, accidental event definition step-by-step procedure, probability frequency of outcomes for manual and automatic procedures for event tree analysis (ETA), fault tree analysis (FTA), and; LOPA, QRA and HRA with focus on E/E/PE.
VI Discussions on various standards for RA & SIS**; Batch process, life cycle**	Discussions on relevant standards like IEC 31010, 60812, 61025, 61508/61511, 61582, ISA 84. Life cycle (LC) analysis; SIS standard. Also batch process & ISA 88.
VII Safety instrumentation functions and systems	Functional safety of instrumentation, failure classes, reliability, safety function performance, calculations of probability of failure on demand (PFD). Special considerations for fire and gas systems. SIS for integrated system; SIS vs SIL.
VIII Safety integrity level and plant emergency	LC, RA with PFD, reliability approach, SIL calculation & assignment; automatic SIL determination; SIL for F&G SIL certification, proof test; plant emergency programs.
IX SIS requirement for field bus and components	Requirements of SIS for field bus, sensor, logic solvers, and final control elements; failure rates, specification, selection procedure. Alarm management layers of protections.
X Safety instrumentation implementation and explosion protection	Allocation of safety related functions at various stages: design and engineering, installation & commissioning operation & maintenance decommissioning—SIL validation, certification, proof testing. Hazardous area classification & elec. Safety and explosion protection. Details to include a number of standards. The discussions also include complete details of ingression protection.
XI Fault tolerance protection layers and system security	Fault tolerance concept, issues related to SIS, protection layers in SIS (to include sensors, fieldbus, logic solver and final element), security—especially cyber security for PEs, discussions on fire wall.
XII Discussions on industrial applications of SIS	Application area includes thermal power (with reference to NFPA), nuclear plants, oil & gas—offshore, petroleum refinery cement and steel plants.

or adverse health effect to the person or persons under certain condition, whereas *risk* is the ***probability*** that the person or persons will be harmed or receive adverse health effect when exposed to a hazard. *Is this not something similar to the relationship between heat and temperature?* Cigarette smokers are "n" times more likely to die from lung cancer than non-smokers. Here the probability is expressed in terms with a reference. Similarly, it can be expressed directly in terms of probability like "X" number of smokers out of, say, 10,000,000 smokers die every year at a place. Now if any smoker quits smoking, his chances or probability of survival is increased, so, he is putting a barrier.

2.2 Hazard Definition and Discussions

In this clause, efforts will be made to define hazard. As is seen from the previous clause, hazard is discussed with respect to health and occupational hazard. One thing is clear that exact definition of hazard changes from case to case, but the main theme behind it is the cause that has potential to bring out untoward situations!

2.2.1 General Definition

In general terms, one can define *hazard* as being any biological, chemical, mechanical, environmental, or physical agent has potential to cause harm or damage to humans, other organisms, plant - machinery, asset, or the environment in the absence of its control. In other words, *hazard* can be considered as a state with a set of conditions of a system, which together with other conditions in the environment, or in the environment of the system will lead to an accident. For example, an H_2 leak from a turbo generator, coming in contact with atmospheric air can cause fire or explosion near a turbine if it comes into contact with fire or locally hot surface, so smoking is prohibited near that place.

2.2.2 HSE UK Definition

In terms of industrial plants, hazard can be defined as any condition produced by or in the industry that may cause harm to property (plant — machinery, asset), human or environment. As per HSE.gov.uk, "a hazard is something (e.g., an object, a property of a substance, a phenomenon, or an activity) that can cause adverse effects."

2.2.3 CCPS Definition

As per the Center for Chemical Process Safety (CCPS), hazard is defined as, "an inherent or physical characteristics that have potential for causing damage to people, property or environment, it is the combination of a hazardous material, an operating environment, and certain unplanned event that could result an accident."

2.2.4 ISO/IEC Definition

As per ISO/IEC 51 or IEC 61508, hazard is defined as, "the potential source of harm." In IEC 61508, harm has been defined as physical injury or damage to the health of people either directly or indirectly as a result of damage to property or to the environment.

2.2.5 Major Issues and Coverage of Hazard Identification

It is always recommended to identify and record all possible hazards in the plant as well as at the work place. For such work, identification it is better to form a team comprising both experienced and fresh people. Experienced people are familiar with the job and will help to identify the hazards, whereas fresh eyes may throw light on new area of hazard. Following are the major points to be covered in hazard identifications (Canadian Center for Occupational Health and Safety [CCOHS], www.ccohs.gc.ca):

- All aspects of work
- Include all non-routine activities (e.g., maintenance, repair)
- Accident or incident/near-miss record
- Look for foreseeable unusual conditions
- Include people "off-site" (not regular)
- Include assessment groups for different levels of risk

2.2.6 Definition of Commonly Used Terms

A few terms, normally encountered in connection with plant hazard system study have been elaborated as follows:

- Accident: It is an undesired, unplanned (may not be unexpected always) event, which will result in specified level of loss (in terms of health, property, production, etc.).
- Mishap: It stands for bad luck, misfortune, etc. In terms of industry, it could be an accident, which is associated with uncontrolled release of energy and toxic material exposure.
- Near miss/incident is normally used in good sense meaning an event occurred, but it involved very minor or no loss (in terms of health, property, production, etc.).
- Safety: Freedom (or nearly freedom!) from accident or loss

2.2.7 Hazard Types

Generally hazards are classified broadly as:

- Biological: Viruses, bacteria, animal, plant
- Chemical: Physical, chemical, or toxic property of the chemical
- Ergonomic: Repetitive movement
- Physical: Radiation, fire
- Psychological: Stress
- Safety: Slipping, etc.

2.2.8 CCOHS Guidelines

When various hazards are identified, initially, it is better to study them carefully to assign risk level associated with each of these hazards, that is, to rank them according to their severity. Following are a few guidelines for such activities (based on CCOHS):

- Product information/manufacturer's data
- Past experience

- Applicable standards
- Industrial codes and practices
- Health and safety material about hazard
- Information from government or reputable organization
- Test results
- Expertise from health and safety engineer (HSE)
- Previous hazard reports (accident/near miss/incident) and hazard statistics
- Use any hazard checklist is helpful (e.g., EN1050)
- Work environment
- Skill and experience of worker
- Systems being followed
- Range of foreseeable conditions

2.2.9 *Hazard Study Issues*

It is common practice to identify the hazard control and follow up action as a part of hazard identification and preliminary hazard analysis (discussed in detail in the next chapter). In order to control hazard, one has to look for safety interfaces also. So, the following points need to form a part of initial hazard study especially for industrial or process plants, so that entire spectrum is well-covered:

- Hazardous component identification
- Possible malfunction of equipment or system including software
- Safety interface including software
- Operating condition and environment constraints if any
- Available facilities and support equipment and support system
- Operating procedure, regular test, maintenance diagnostic features
- Safety-related equipment, safe guard items
- Possible alternate approach if possible and emergency procedure

2.2.10 *Industrial Hazards*

Listed below are some of the sources of hazard encountered in industry:

- Sources and propagation of stored energy in the form of chemical, electrical, and mechanical
- Mechanical moving parts
- Nuclear radiation
- Biological hazard (bacterial growth)
- Fire/explosion
- Human error (e.g., operating error)
- Software error
- Noise of different forms
- Toxic and corrosive liquids and gas

2.2.11 *Networking Security*

Network security threat is another big hazard in modern day's practice. For example, supervisory control and data acquisition, the vital infrastructure support in major industrial set ups nowadays, faces great threat from security threats. Major security threats come on account of:

- Control networking
- Standardization and open systems
- In secured, remote connections
- Availability of public information to hackers

2.3 Hazard Analysis Approach

With this knowledge on hazards, efforts will be made to find what is hazard analysis? Hazard analysis uncovers the hazards that exist in the workplace (in this case, industrial plant) focusing on the system or project. As discussed previously, in hazard analysis, risk based decisions are taken to develop means to quantify, track, develop mitigation means, and control hazards, follow up action, verify effectiveness, and communicate. However at first it is necessary to define risk, so in the next clause same is discussed.

2.4 Risk Definition

As discussed in the previous clause, risk is the likelihood that a hazard can cause actual damage. In this case also efforts will be made to find some good definitions.

2.4.1 HSE UK Definition

As per HSE.govt.UK, risk can be defined as, "the likelihood that a hazard will actually cause its adverse effects, together with a measure of the effect. It is a two-part concept and you have to have both parts to make sense of it."

2.4.2 EPA Definition of Risk

According to EPA, "EPA considers risk to be the *chance* of harmful effects to human health or to ecological systems resulting from exposure to an environmental stressor. A stressor is any physical, chemical, or biological entity that can induce an adverse response. Stressors may adversely affect specific natural resources or entire ecosystems including plants and animals as well as the environment with which they interact."

2.4.3 ISO Definition of Risk

As per ISO guide 73 (2009), risk is the "*effect of uncertainty on object.*" With associated notes, some of them are:

- Note 1: An effect is a deviation from the expected—positive and/or negative.
- Note 2: Objectives can have different aspects such as financial, health and safety, and environmental goals, and can apply at different levels such as strategic, organization-wide project, product, and process.
- Note 3: Risk is often characterized by reference to potential events, consequences, or a combination of these and how they can affect the achievement of objectives.

- Note 4: Risk is often expressed in terms of combination of the consequences of an event (including changes in the circumstance) and associated likelihood of occurrence.
- Note 5: Uncertainty is the state, even partial, deficiency of information related to, understanding or knowledge of, an event, its consequence, or likelihood.

2.4.4 IEC Definition of Risk

There is also another definition of risk as per ISO/IEC guide 51/IEC 61508, which is rather straightforward. According to these, risk is, "*the combination of probability of occurrence of harm and the severity of that harm.*"

2.4.5 General Risk Concept

Risk can be conceived as a hazard level combined with likelihood of hazard leading to an accident and dependent on latency of exposure. The definition will be clearer when a smoker example is considered. All smokers do not suffer from lung cancer, but there is high probability of getting exposed to the likelihood. Again such chances are increased for the person smoking more cigarettes. Similarly, a person exposed to asbestos dust will have a chance for mesothelioma, and higher the exposure more will be chance of getting affected.

2.5 Risk Discussions

There are two kinds of risks; one is raw (*inherent*) risks, which represent the risk before taking any control or mitigation action. Second category risks are the residual risks, that is, the risks which could be faced after putting in place controls or mitigation actions (see Fig. I/1.0-2). It has been found that risk is a combination of likelihood of hazard, severity, exposure time, etc. In this discussion, the same has been brought in to show interference of each factor so that various terms in connection with risk assessment could be explained. The correct way of combining them is normally unknown, also parameters of each of them are not known. Through risk assessment and analysis, such issues and their combinations are determined. Risk components are therefore severity, latency (or exposure) period, and likelihood of hazard occurrence, and likelihood hazard leads to an accident. Fig. I/2.5-1 shows the four factors and their combination will generate a new risk value.

There are a few factors, which affect the risk component. Some of them are:

- Hazard complexity
- Exposure
- Energy
- Automation
- Scale
- New versus old technology replacement
- New technology and their pace of change

Under this clause, a short discussion on risk frequency, categorization, assessment, etc. shall be highlighted (based on CCOHS and British standard), whereas details for all these are available in subsequent discussions.

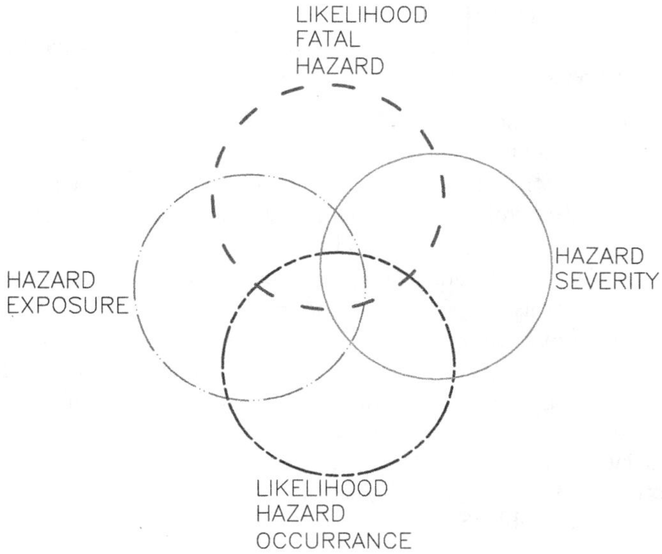

FIGURE I/2.5-1

Risk component combination.

2.5.1 Risk Frequency

This defines the likelihood of the risk, that is, it stands for the probability of risk. There are a few categories, and these are:

- Very likely: at least once in six months
- Likely: at least once a year
- Unlikely: maybe once in lifetime
- Very unlikely: May be 1%

Typical examples are shown here. Risk frequency data and release data are available in HSE (UK), OREDA, and OGP publications also.

2.5.2 Severity

Severity is loosely used to indicate the impact of risk, that is, consequence. These are slightly harmful (e.g., superficial cut, minor cut, etc.), harmful (e.g., burns, serious pains, minor fracture), and extremely harmful (e.g., major fracture, amputation). There are some other ways to categorize also. Such categorizations are listed as follows:

- Minor: Minor system damage without causing injury
- Major: Low-level exposure to personnel, activates public alarm
- Critical: Minor injury to personnel, fire or release of chemical to environment
- Catastrophic: major injury, death, big leakage (e.g., Bhopal gas leak)

2.5.3 Risk Level (Based on Action and Time)

The level of risks are often categorized based on the potential (for harm, see Clause 3.4.4 for further discussions with example). The categories are termed as follows:

- Very low: These risks are acceptable may not need any action
- Low: No control may be necessary unless these are available at low cost
- Medium: Suitable considerations shall be there to see if it risk can be lowered, where applicable, to a tolerable level within a defined time limit. However, due considerations shall be given for the additional cost for risk reduction. Whenever the risk is associated with harmful consequence, it is necessary to make sure that risk reduction controls are properly maintained.
- High: Good amount of effort is applied to reduce risk on an urgent basis within a defined time frame. It is essential to give due considerations towards the choice amongst suspending or restricting the activity or to apply interim control measure until the main risk reduction control is implemented. Whenever the risk is associated with a harmful consequence, it is necessary to make sure that risk reduction controls are properly maintained.
- Very high: Unacceptable. Substantial improvements in risk reduction control measures are necessary to reduce the risk to the acceptable level. Activities need to be halted until risk reduction control is implemented. Otherwise, work shall remain prohibited.
- Risk associated with very harmful consequences need risk assessment and analysis.
- The above categorizations are qualitative in nature. For quantitative calculations, one may need to take the help of probability and associated software, which are also available from various agencies for different applications (e.g., shell, DNV, etc., have proprietary software for their use).

2.5.4 Control Measure

In Control measure, preferred actions are to climb up the hill as shown in Fig. I/2.5.4-1 (*based on health and safety authority-www.hsa.ie/eng*).

As shown in the drawing, there are several steps listed as follows arranged in decreasing order of preference:

- Elimination of hazard: Total elimination is not always possible, but sometimes possible, for example, if unleaded petrol is used, attendants are not in the hazard of lead contamination.
- Substitute: Sometimes to eliminate hazard particular material or system, for example, a vulnerable thing can be substituted by other e.g., on account of health hazard, DEHA is substituted for hydrazine to scavenge oxygen from boiler water.
- Isolating: Isolating the hazard is achieved by restricting plant and equipment. For example, in case of hazardous area, process transmitters use flame proof enclosure or use IS circuit so that hazard does not reach electrical circuit or sufficient energy does not reach the hazardous area.

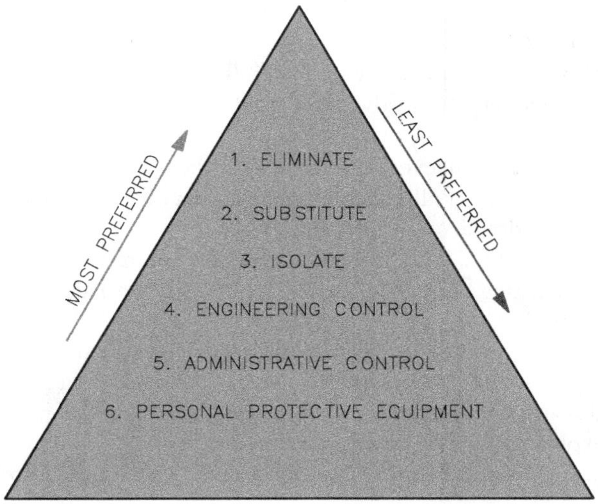

FIGURE I/2.5.4-1

Control measure for risk control.

- Engineering control: By redesigning the process and putting in barrier. Machinery guard is an example of this.
- Administrative control: Adaptation of safe control practices and procedure through appropriate training, etc.
- Personal protective equipment: Gloves, helmet, etc. are used for this.

2.5.5 Risk Target

This is a term often found in connection with plant hazard analysis. A risk target is measure that expresses the consequence of a risk in relevant terms of the project and organization concerned.

2.5.6 Risk Analysis

As per latest version of IEC/ISO 31010 (IEC 60300-3-9), risk analysis is the "systematic use of available information to identify hazard and to estimate the risk to individuals, populations, property or the environment." So essentially risk analysis finds, organizes, and categorizes sets of risks.

2.5.7 Risk Assessment

Risk assessment is one step further. This can be defined in terms of the following basic activities:

- Identification of hazard
- Analysis and evaluation of risk
- Find an appropriate the way to control and mitigate hazards. The main aim of risk assessment is to remove hazard, or reduce the risk level by adapting

necessary control measures, to move towards safety. Importance of risk assessment cannot be overestimated, because it provides a number of helpful things to prevent loss of assets (here asset stands for personnel, plant machinery, property, environment). Major advantage shall include, but is not limited to the following:

- It makes personnel at all levels knowledgeable about loss of asset due to hazard.
- It makes them aware of all possible potential hazards, their likelihood of occurrence, severity, level of risk, etc.
- Identifies risk level for each of personnel, asset, environment, etc.
- Assesses the existing control measures and gives new suggestions for change to protect asset.
- It is done in such a way that when done at early design stage (may be with little extra investment), then future damages are saved to a great amount (e.g., good save for frequent loss of production, and/or loss of property due to, say, fire, explosion, and/or environmental protection).
- It categorizes risks and helps to take appropriate control measure.

2.5.8 *Risk Assessment Procedure*

Risk assessment procedure is an elaborate process, but mainly covers the following major points: detailed discussions on them have been presented in Clause 3 in Chapter II:

- Hazard identification
- Evaluation of risk; likelihood, severity, and level of risk
- Standard operating conditions
- Emergency situation (nonstandard operation)
- Review of all associated information
- Actual and potential exposure of personnel (latency, frequency, intensity)
- Environmental impact
- Design engineering control
- Documentation

2.5.9 *Risk Management*

As per ISO guide 73, risk management stands for, "*coordinated activities to direct and control an organization with regard to risk*," defined in Clause 2.3.

3.0 RISK REGISTER, RISK MATRIX AND RISK RANKING

3.1 Preamble and a Few Term Definitions

Risk management is a buzzword now, but it is not a new idea. People used to do the same in different forms earlier also. Only in the modern days this has been looked at in a more comprehensive and systematic way. In day-to-day use, it is loosely conceived as probability of known loss. As stated earlier, that risk is an impact (mainly negative) to an asset, arising out from present action or future event and measured in terms of severity and likelihood. Importance of risk ranking in risk management, discussed in the next chapter, cannot be overestimated. Without this it is impossible to manage risk. There is every possibility that small risk may receive unwarranted priority while a large risk is unfortunately overlooked. Risk ranking is a mere number but it is to be considered in various different ways "risk" is defined, and how each stakeholder looks at it. Before starting the main discussions, it is better to define a few related terms (in the following sub-clauses) which will be necessary in understanding risk ranking.

3.1.1 Risk Appetite

The amount of risk a company or organization is willing to accept before initiating any action to address (and mitigate) the risk is referred to as risk appetite.

3.1.2 Key Risk Indicator

The key risk indicator (KRI) is an early warning to indicate that risk appetite is reached. So, this is an early warning for a potential risk, and is important in risk management.

3.2 Risk Register

Risk register is basically a record of identified risks for a project. In other terms, it is the documented response to what could happen, or occur which would stand in the way for a project to achieve its goal. The risk associated with activities and organizational strategies are identified first, then, they are graded in terms of their likelihood of occurrences and severity. The following are a few characteristics register should have:

3.2.1 Characteristics of Risk Register

- Short description of each risk along with associated consequences
- Factors influencing the likelihood and impact
- Grading of risks (discussed later in this clause) for example, low, medium, high, extreme, etc.
- Risk acceptability
- Existing and proposed actions for risk mitigation
- KRI and upward reporting factor

This is a formal document generated as a contractual document by the agencies, or it could be utilized in the organization to draw the attention of senior executives towards the emerging risk for immediate actions. There is no unique format for risk register and also it varies with organizations and basic intention for the analysis. Standard formats are available in public domain. However, as shown in Fig. I/3.2.1-1, it shall include the information discussed as follows as basic requirements. Inclusion of risk matrix, or risk level is a choice—as applicable for the project. Some use risk levels with subdivisions for yearly or quarterly periods to show how risk levels are changing after each review.

RISK REGISTER FOR < TITLE OF THE PROJECT> DOC. REF NO.: XX/NN/MMMM		SITE REF:	BY	DATE
			COMPLT'D	DD:MM:YY
		FORM NO.:	APPRVD.:	DD:MM:YY

RISK IDENTIFICATION									QUALITATIVE RISK ASSESSMENT			RISK RESPONSE ACTION		RISK MONITORING & CONTROL		
SERIAL NO.	RISK ID.	STATUS	CATEGORY	EVENT DETAILS	CAUSE DETAILS	EFFECT DETAILS	+ OR − IMPACT	PRIMARY OBJECTIVE	LIKELIHOOD	CONSEQUENCE	RISK MATRIX (L: LOW) (M: MEDIUM) (H:HIGH) (E:EXTREME)	STRATEGY	ACTION DETAILS	RESPONSIBLE PERSON	INTERVAL	STATUS REVIEW COMMENTS DATE
01	PRF 001	ACTIVE	EXTERNAL	PROJECT FUND PART LOAN NOT RELEASED IN TIME	BUDGET CONSTRAINT ALLOCATION CHANGED CURTAIL IN INTERIOR	PROJECT DELAY	(−) IMPACT	TIMING	HIGH	VERY HIGH	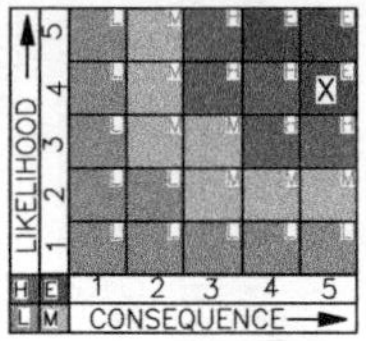	MITIGATE	PHASING IN PLANNED ALLOTMENT. HENCE WORK PHASED OUT	CHIEF PLANNING ENGINEER AND FINANCE OFFICER	MONTHLY	BANK FOLLOW UP WITH GOVT. LICENCE FINAL DEADLINE DATE DD.MM.YYYY
02	PSC 001	ACTIVE	CONSTRUCTION	UNDIDENTIFIED UTILITY TRANSPORTATION COST	ADDITIONAL CABLE COST AND LAYING COST	ADDITIONAL PROJCET COST IMPACT	(−) IMPACT	COST	MEDIUM	MEDIUM	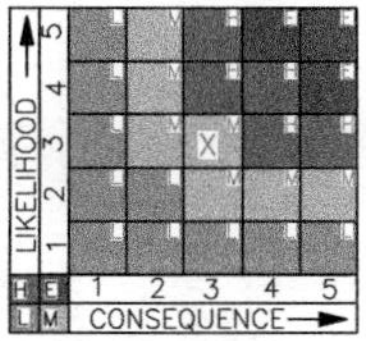	MITIGATE	PURSUE WITH UTILITY COMPANY FOR ALTERNATE ROUTE	ELECTRICAL ENGINEER	MONTHLY	LETTER TO UTILITY COMPANY DD.MM.YYYY

NOTES:

1) TYPICAL RISK REGISTER WITH ARBITRARY EXAMPLES HAVE BEEN SHOWN. STANDARD EXCEL FORMATS ARE AVAILABLE OR THIS CAN BE USED.
2) RISK REGISTER FORMAT VARIES WITH ORGANISATION AND ACTIVITIES.SUITABLE ONE FOR THE APPLICATION CAN BE CHOSEN.
3) SOME USE MONTH/YEAR–WISE RISK LEVEL IN PLACE OF RISK MATRIX. ALSO POSITIVE IMPACT E.G. EXTERNAL MARKETING EFFORT ETC. ARE INCLUDED.

FIGURE I/3.2.1-1

Typical risk register.

3.2.2 Description of Risk Register

- Risk identification: This shall include: risk ID if any, proper description of the risk involved.
- Basic cause and effect (on the project) of the risk.
- Primary objective for which the risk analysis is carried out (e.g., time, cost etc.).
- Consequence details such as likelihood of occurrence, severity, etc. to assess risk level or risk matrix (discussed in Chapter II in details).
- Mitigation and control action of the negative impacts from the risk. In case there is any positive impact, then control action may be incorporated in the register also (e.g., recruitment of marketing agent is a positive impact on marketing).
- Monitoring and control of risk response actions.

In risk matrix, there are four colored boxes, to signify four categories of risk level (discussed in next sub clause). These are derived from likelihood and impact. In the risk register shown in Fig. I/3.2.1-1, the green-colored box is for the risks with low ranking level. Similarly, yellow, magenta, and red boxes are for risks with risk ranking medium, high, and extreme, respectively. This risk level is distinctly different from consequences, which normally are divided into four to five categories and could be numbered (as shown in the figure) or named as 1, 2, 3, 4, 5, or very low, low, medium, high, and very high. Sometimes very low and low are put together and named as negligible, and other three categories may be named as marginal, critical, and catastrophic respectively (—these are only suggestive). Similarly, likelihood or probability of occurrence could be numbered as one to five, or may be termed as rare, unlikely, possible, likely, or certain. These two factors (i.e., likelihood and consequence) constitute risk level.

The impact of all mitigating actions and sources of assurance are considered before calculating the "residual" risk. Reduction of each of likelihood and consequence depends on underlying subject issue and specific control considered. When the actual control plan is not correctly implemented, then the concerned person of the risk team and manager risk team need to think of improvement for action plan implementation and/or modify the framework for control action as warranted for the project. After completion of risk register, all risk should be populated in a consolidated risk matrix (or heat map) with risk IDs. Based on this, a report summary is put before management for consideration; major points here are:

- Identify all major or key risks.
- Listing of risk controls to assess their effectiveness and implementation.
- Identify alternate way if any to address the risks.
- Improvement suggestion for risk control implementation.

There are numbers of software available in the market for risk register development and guide to ease the work of the designer. There are several agencies and universities who can undertake such job and/or provide necessary software package.

3.3 Risk Matrix

In Wikipedia, this is denoted as "risk matrix is a matrix that is used during Risk Assessment to define the various levels of risk as the product of harm probability categories and harm severity categories. This is a simple mechanism to increase the visibility of risk and assist management decision making." The definition tried to focus on basic intent of risk matrix in a simple language. However, one needs to understand that risk matrix is basically a methodology helpful during risk assignment to identify, prioritize, and manage risk level for a given project. Risk matrix may be considered as a quantitative or semi-quantitative tool for qualitative hazard analysis. It is very important to develop risk matrix design very precisely so that there will not be false sense of security after risk matrix is done. Simply if likelihood or impact of any risk is not properly defined, then as result of wrong calculation any particular risk may be considered in low risk level, but in reality it is not so. In that case one may be happy to note that it is low level hence secured—a false sense of security.

3.3.1 *Features of Risk Matrix*

A good risk matrix normally shows the following features:

- *Normally*, quantitative/semi-quantitative hazard analysis tool.
- Developed in a simple and easy to understand manner.
- Tolerable and non-tolerable ranges are clearly defined prior to developing risk matrix.
- Detailed descriptions of all consequences within the range.
- It uses orders of magnitudes and has consistent likelihood range to cover entire spectrum of potential hazards.
- Good guidance for effective hazard analysis in a qualitative manner and may not require prior knowledge for quantitative analysis. However, proper knowledge of the project for which it is done is an advantage. It shall also provide guideline additional action needed to mitigate risks with intolerable risk level, that is, to show how intolerable risk levels can be mitigated, to bring the same with tolerable range.
- It shall be designed in such a way that it has flexibility to adapt itself for various risk targets specific for the company project.
- Prior software knowledge is not essential, but it could be handled with the help of software.

3.3.2 *Risk Matrix Description*

There are several standard guidelines and published risk matrices, but at the beginning one has to decide the intent for which it is to be developed. Table I/3.3.2-1 is an example of a risk matrix available from CCPS. Tables I/3.3.2-2—I/3.3.2-4 show the explanation of risk levels, likelihood, and consequence range criteria for the risk matrix in Table I/3.1.4-1 (*based on technical review by* MS. *D. Aptikar*).

The risk matrix shown here is just an example. It varies with application area or project and organization. In the beginning, risk matrix has been defined as semi-

Table I/3.3.2-1 Risk Matrix

Frequency	Consequence			
	1	2	3	4
4	IV	II	I	I
3	IV	III	II	I
2	IV	IV	III	II
1	IV	IV	IV	III

Table I/3.3.2-2 Risk Level

Risk Level	Category	Description
I	Unacceptable	Should be mitigated by engineering and/or administrative control to risk level III or less, within a specified period (say, 6 months).
II	Undesirable	Should be mitigated by engineering and/or administrative control to risk level III or less, within a specified period (say, 12 months).
III	Acceptable with controls	Should be verified that procedures and controls are in place.
IV	Acceptable	No mitigation required.

Table I/3.3.2-3 Likelihood Ranges Based on the Levels of Protection

Likelihood Range	Quantitative Frequency Criteria (Typical)
Level 4	Initiating event or failure (e.g., Leakage/rupture)
Level 3	One level of protection (e.g., Pipe leakage, overload)
Level 2	Two levels of protection (e.g., Elec. actuator uprooting)
Level 1	Three levels of protection (e.g.,Vessel failure)

Table I/3.3.2-4 Consequence Range

Consequence Range	Quantitative Safety Consequence Criteria
4	Onsite/offsite: Potential for multiple life threatening injury or fatalities Environmental: Uncontained release with potential for major environmental impact Property (including plant): Plant damage value in excess of say, $100M units of currency.
3	Onsite/offsite: Potential for single life-threatening injury or fatalities Environmental: Uncontained release with potential for moderate environmental impact Property (including plant): Plant damage value in the range of say, $10—100M units of currency

Continued

Table I/3.3.2-4 Consequence Range—cont'd

Consequence Range	Quantitative Safety Consequence Criteria
2	Onsite/offsite: Potential for an injury requires medical attention Environmental: Uncontained release with potential for minor environmental impact Property (including plant): Plant damage value in the range of say, \$1–10M units of currency
1	Onsite: Potential for injuries requires only first aid Offsite: Noise or odor. Environmental: Contained release with local impact only Property (including plant): Plant damage value in the range of say, \$0.1–1.0M units of currency

quantitative tool. But from the observation on the matrix shown in Table I/3.3.2-5 (B.1 – IEC 61508) it is clear that it is purely qualitative. Another thing to be noted is that, in contrast to previous Table I/3.3.2-1, here severity decreases as one move from left to right.

Here, risk levels also can be defined as discussed previously, but in the reverse manner. Here, the top left part of the matrix is the highest risk area, whereas the bottom right part is lowest risk. There is no upper and lower limit. Too much increase/decrease of in the categories of risk frequency or consequence, has no practical value really. So, in practice, these are restricted to 3×3 or 5×5 matrix, which need not be a square matrix as is evident from the above (6×4). Table I/3.3.2-6 is an example of 4×4 semi-quantitative risk matrix.

Same matrices can be changed to totally quantitative when time scale has been introduced into the frequency as shown in Table 1/3.3.2-7.

The most important thing in this is that the risk target must be relevant. It shall be prepared keeping in mind specific needs for the project and/or organization. Another

Table I/3.3.2-5 Qualitative Risk Matrix

	Consequence			
Frequency	Catastrophic	Critical	Marginal	Negligible
Frequent	I	I	I	II
Probable	I	I	II	III
Occasional	I	II	III	III
Remote	II	III	III	IV
Improbable	III	III	IV	IV
Incredible	IV	IV	IV	IV

Table I/3.3.2-6 ***Semi-quantitative Risk Matrix***

Probability	Consequence			
	$1000	$10,000	$1,00,000	$10,00,000
May occur multiple times in the project	Medium	High	High	High
May occur once in the project	Low	Medium	High	High
Occurred earlier	Negligible	Low	Medium	Medium
Unlikely to occur	Negligible	Negligible	Low	Low

Table I/3.3.2-7 ***Quantitative Risk Matrix.***

Probability	Consequence			
	$1000	$10,000	$1,00,000	$10,00,000
Every month	Medium	High	High	High
Every year	Low	Medium	High	High
Once in 10 years	Negligible	Low	Medium	Medium
Once in 100 years	Negligible	Negligible	Low	Low

interesting thing is that risk targets can be combined to form common risk matrix for comparison purpose as long as the scale and range of the two axes match. When evaluating a project risk due to cost and time, and both are in terms of money, it is possible to combine them and draw a single risk matrix. However, it is better to have separate matrix for each target. In chemical plants, hazard analyses are done for harm to person, environment, or for the property, naturally the risk level in each case may be different. Hence it is recommended to use a different risk matrix.

As risk matrix application is wide it is extremely important to determine right at the initial stage, which consequence is of *interest* in the current program. These consequence lists shall include but are not limited to the following:

- Project cost
- Project time schedule
- Personal safety
- Environment safety
- Property damage/interruption of operation
- Reduced quality of project/product
- Statue or legal implications
- Company/corporate image/goodwill

Prior to developing a risk matrix, in the next step, it is necessary to do the following:

- Identification of risk
- Cause and consequence of risk (requires practical knowledge, good judgment power also quantitative calculation to help)
- Estimate likelihood of occurrence (requires practical knowledge, good judgment power, inter-discipline interaction, technical literature)
- Selection of proper range for each of consequence and likelihood with selection of unit/scale

3.3.3 *Risk Level*

From the preceding discussions, it is clear that risk level is a quantitative expression of severity of risk, and can be found by combination of likelihood and consequence, and it is embedded in the risk matrix. So it is not possible to directly measure, but can be referenced after constructing a specific risk matrix. Normally it is expressed in terms of low, medium, high, very high, and extreme. Usually more than five discrete levels of risk have no practical application. So negligible or very low are not considered. In Table I/3.3.3-1, various risk levels and associated actions has been depicted to show importance of risk level.

The use of risk matrix has been depicted in Fig. I/3.3.3-1 through a simple example of single risk target pertinent to one cement plant. In this example, it is seen that for the same cause there could be two different risk levels as shown by differences in hatch lines. In the example, the same likelihood has been considered for both the cases. Since the risk level is medium, it is accepted with suggestion of installing additional devices to reduce the risk level.

Table I/3.3.3-1 Risk Level and Reduction Control

Risk Class	Risk Level	Reduction Control
Extreme (E)	Totally unacceptable	May call for process/design change
Very high (V)	Unacceptable	Process/design change or one protective device with SIL3 (see Chapter VIII) to be incorporated (e.g., pressure safety valve Class A).
High (H)	Undesirable	Process/design change or one protective device with SIL2 (see Chapter VIII) to be incorporated (e.g., application in burner management system).
Medium (M)	Acceptable with further risk reduction	One monitoring device with SIL or of high quality (documented testing) or administrative procedure of high quality (e.g., tripping of large pump from low level of suction sump).
Low (L)	Acceptable with need for further reduction	One monitoring/protective device or administrative control (non-availability of net positive suction head for a pump low-level monitoring and open condition of suction valve).

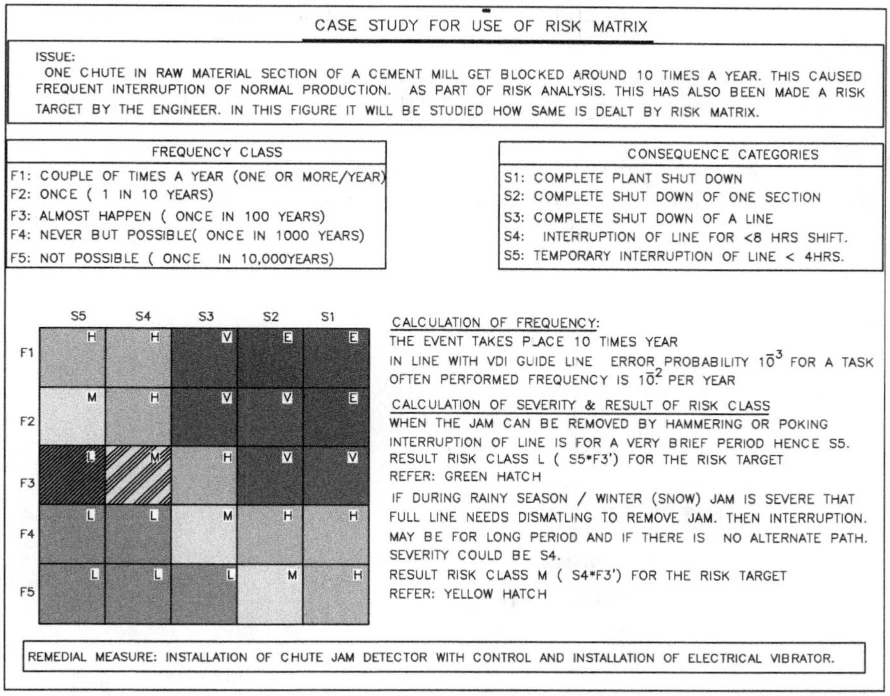

FIGURE I/3.3.3-1

Use of risk matrix.

3.4 Risk Ranking

Value is function of risk and return. Therefore, depending on the expected value of any enterprise, its risk level will vary. At any point of time, to achieve the overall goal of the organization, it will like to incur just enough risk, which matches its expected value of the organization. Naturally for this risk assessment and management (see Chapter II) is extremely important. Sound risk ranking knowledge is an essential part of risk assessment and management. If there is no proper risk ranking, then it may be that a risk of little importance gets unnecessary attention, and a vital risk gets overlooked. Now, there are real challenges for risk ranking. As is seen in the beginning that there are too many definitions of risk, depending on, from which angle it is looked at. It is more so when the risk impacts have some interrelation. Also for a consequence (arising out of a risk) has different perspective and importance amongst the stakeholders. For a project investors will be keen on return on investment whereas government authorities are more concerned about public interest. Ordering risks by their importance allows decision-makers to focus on the risk which matters most. However, it does not say anything about how to control them, and who is to do the same. Often in risk management, these are organized in terms of, say, business unit and/or as per strategic, financial, project-wise, etc. as shown in Fig. I/3.4-1, which will give an idea how risk ranking is important and utilized in risk management. In Fig. I/3.4-1A, risk has been ranked as per organization unit, whereas the same in Fig. I/3.4-1B has been organized as per risk category/type.

There are a few challenges for risk ranking. These are "too many risks," "too many definitions of risks," and "too many possible units of risks." Before main discussions on risk ranking, first the various challenges for risk ranking should be discussed.

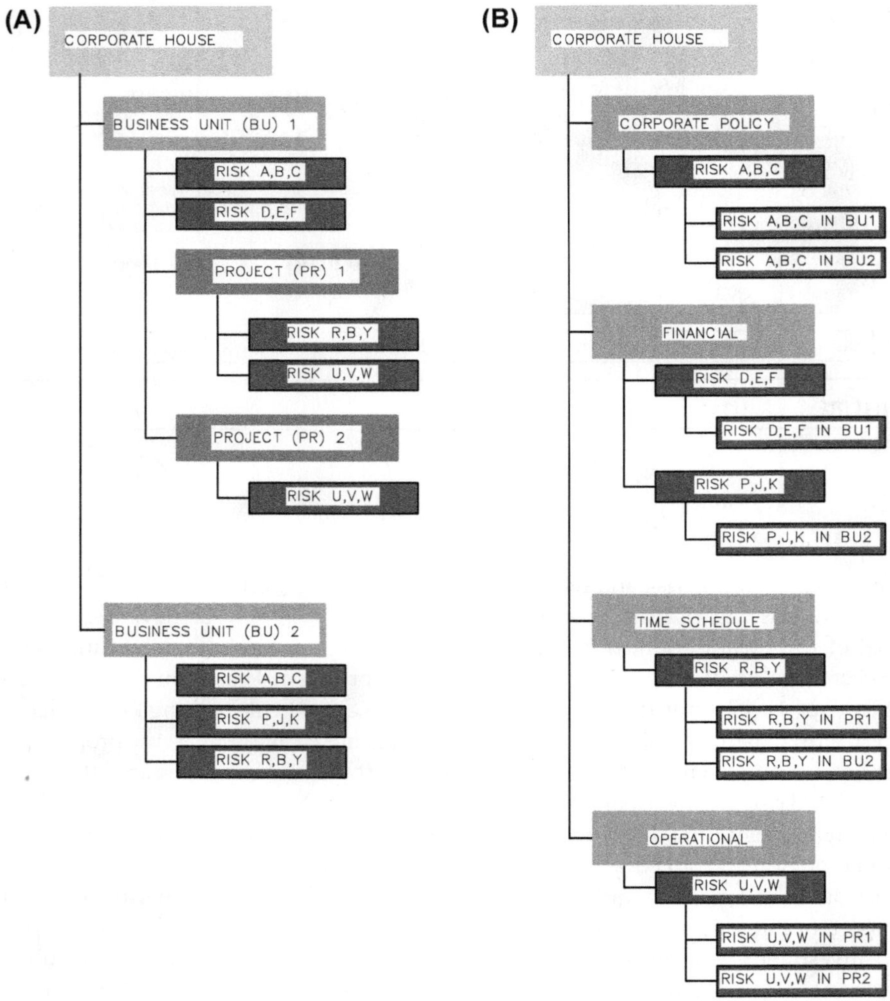

FIGURE I/3.4-1

Different types of risk ranking. (A) Risk ranking as per organisation structure, (B) risk ranking as per risk category (type).

3.4.1 Challenges for Risk Ranking

- Too many risks: The number of risk for any enterprise or agency, even private, personal life is too many. For better understanding, it is better to take a private life example. On a particular day there may be problem for child illness, reaching the office early to keep an commitment, car—making a rattling sound, and no power in the community. It's true that all will hardly come at a time but many can, so how to choose the priority. One way to do so is the sequential approach, that is, to face them as they approach. Sometimes it may yield better result but at times one may have to pay price for ignoring an important risk and paying too much attention to less important risk. Also, this moves the risks up and down in their ranking. One can easily understand if the rattling sound of the car was yesterday's problem and child illness is today's problem, one cannot handle them in sequence as car's rattling sound can wait whereas others (say, child illness or reaching office early) are more important and need immediate attention.
- Too many definitions of risk – risk unit: As discussed earlier, there are too many definitions of risks. Now when one is discussing about any industrial risks, then these could be major injury, or death of persons from an accident in the plant, or it could be catastrophic like in Bhopal India due to gas leakage. Or it could be fire hazard and property loss, or breakdown of an expensive machine on account of new technology. Since each case risk measure is different meaning that there is no *common unit* to measure risk. From these examples it is clear that, each is severe! This problem is there because there is no single measure of risk and no unit for the same. Naturally when there is no single measure/unit then how to rank them! Therefore perspective of viewing at risk is important.
- Stakeholder's perception and or too many values: Supposing risks have been characterized and ranking begins there could be difference in opinion about relative importance of them. One set of people will like to put more importance on the risks catastrophic in nature, than chronic one. Investors never want a loss of property due to fire so, they would like to put importance on the same, while statute, or environmentalists would not allow untreated wastewater to be discharged to the natural resources like river.

The problem will be clearer from Fig. I/3.4.1-1, where three risks with different probability densities have been marked as risk 1, 2, and 3. If the "mean" of the curves are considered, they can be ranked as 3-2-1. If high percentiles are used, then they will be in the order of 2-3-1 [2]. From the previous discussions, it is clear that there are too many values and based on which risk ranking can be made. So these considerations pose additional complexity on risk ranking.

3.4.2 Risk Ranking Features

Risk ranking is an important tool for risk management. It can readily offer the following usages:

- It finds its use in both qualitative as well as quantitative risk analysis.
- Risk ranking is possible for highly complex risk analysis and it is flexible enough to accommodate all kinds of risks.

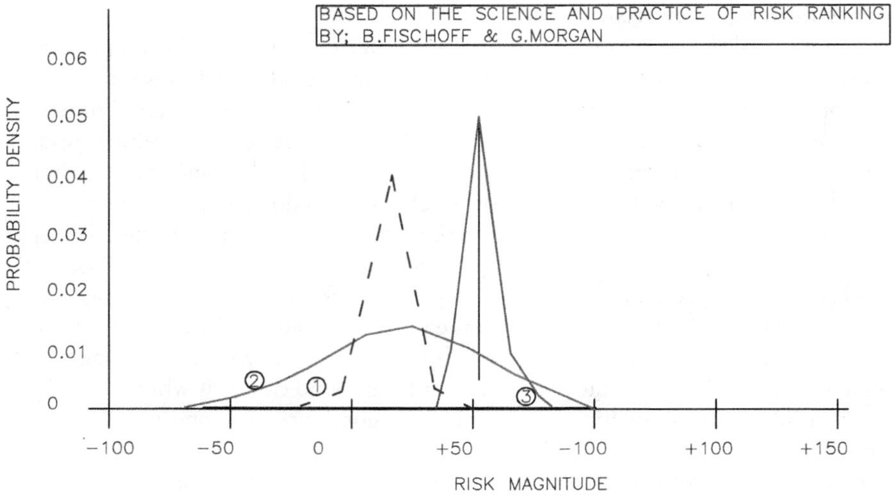

FIGURE I/3.4.1-1

Comparison of risks.

- In complex risk scenario it can be used to manage the portfolio and comparison of risks.
- There shall be uniform risk factors and evaluation criteria suitably settled before hand, prior to risk ranking.
- It provides means to prioritize risks, assign score to risk units.
- It is possible to breakdown risk into risk components, and filters them which help in risk evaluation by putting score as discussed previously.
- As seen from previous discussions it is scalable to include multiple risk factors.
- Lot of efforts are necessary to really breakdown risks into component and really evaluate them (see discussions in Clause 3.4.1).
- It is however not possible to relate risk ranking directly with absolute risks.

3.4.3 *Risk Categorization or Cataloging*

Risk management agencies or even in large organizations there could be many areas, divisions, etc. Each of them again may have several subsections. So effectively there could be thousands of risks really to be handled. Practical handling of so many risks is impossible unless they are properly cataloged. If one goes to a material store in an integrated steel plant and ask for an ammeter, the storekeeper will be at a loss! But if one specifies certain code number (may be coded according to the plant division/shop number, etc.) then it is possible for the storekeeper to issue the same. Similarly, here also risks are grouped and then to choose a set of attributes to it for evaluation. Now the problems are wide spreading; how to categorize what is the basis etc. The EPA guidebook provided some basis. In the area of risk analysis mainly similarity based categorization (concept: common properties were necessary and sufficient data to define the concept) and explanation based categorization developed in late 1980s. As told earlier, the process of categorizing huge number of risks needs value judgment. As different organization, with various stakeholders have different concerns, so, there can be no risk categorization without valued attached to it. In order to restrict our discussions to industrial scenario, some limit to this discussion is essential. Instead it is better to straight away go to the approach what and how to do risk categorizations which is important for risk ranking. For better understanding, it is better to take the help of Fig. I/3.4.3-1, where air pollution on account of industrial plants has been considered to take care of environmental effect of various industries.

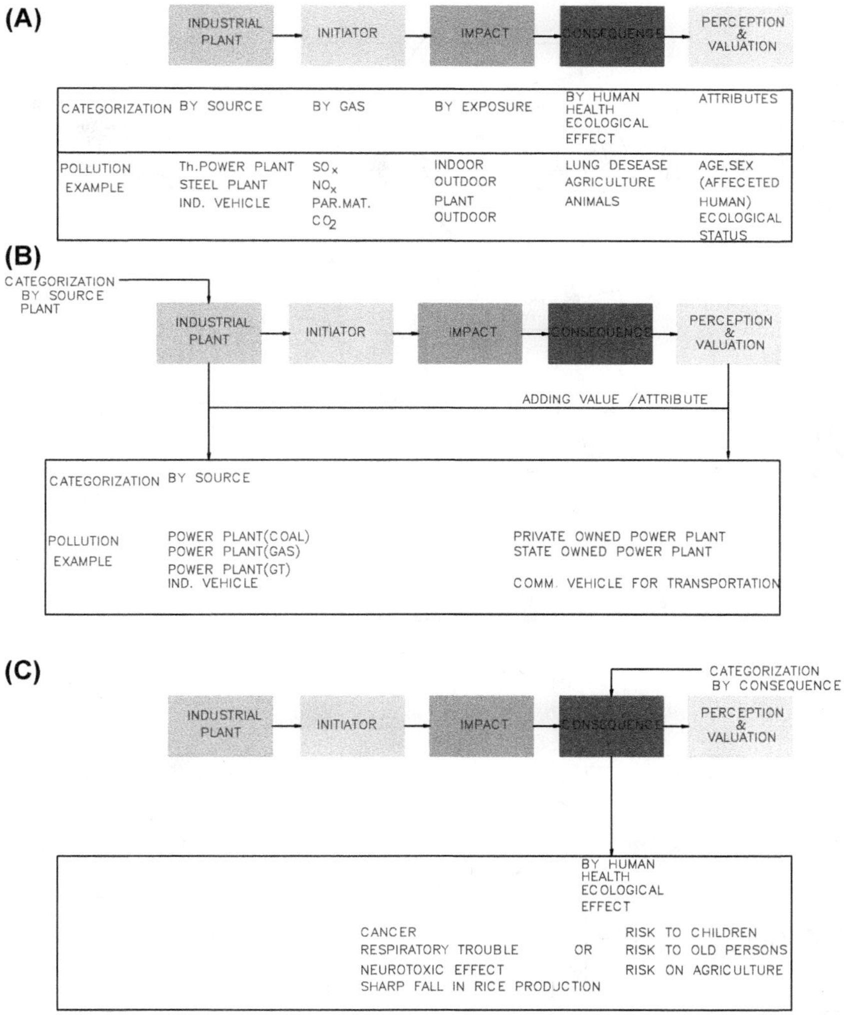

FIGURE I/3.4.3-1

Risk categorization process. (A) Standard risk categorization model, (B) change of risk categorization model (based on source), (C) change of risk categorization model (based on effects).

Fig. I/3.4.3-1 shows general risk categorization process, in which Fig. I/3.4.3-1A can be considered as general standard model, whereas Fig. I/3.4.3-1B and C show how scenario changes when categorization changes. Keeping this in mind, it is better to understand the same with the help some examples of various risk categories encountered in industrial/process plants. These have been shown in Fig. I/3.4.3-2, to

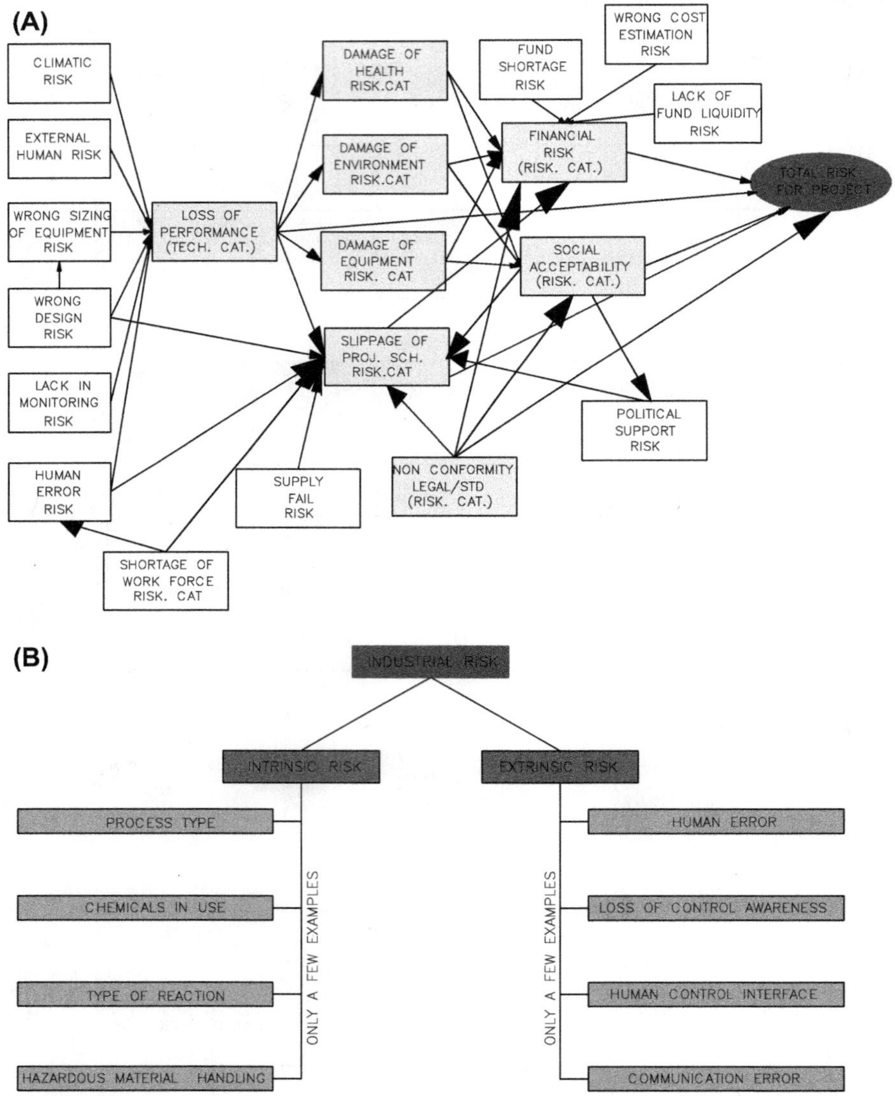

FIGURE I/3.4.3-2

Examples of risk categorization of industrial risks. (A) Interrelation of risks & categorization, (B) another way of risk categorization in a chemical plant.

get an idea about the same. Normally industrial risks can be classified into intrinsic risk and extrinsic risk. There are certain risks which are practically embedded with the process involved, for example, in thermal power plants there are chances of incomplete combustions in the furnace which can give rise to boiler explosion and for this reason every time boiler is started after purging. Similarly at some areas (say) blow out protection (offshore) area there is very high probability of fire and gas (F&G) risk so all necessary precautions, hazard area classifications, etc. are done for this. In Fig. I/3.4.3-2A, various risks one each category for a project have been shown. From the figure it is clear that with a few risks only how complex is the interrelations. In Fig. I/3.4.3-2B, another way of risk categorizations has been shown. From these, it is clear that there is no state rule by which these can be categorized, it is dependent on the project, stakeholders intention as well as company value so many things. Therefore participation of all concerned in some way or the other is essential.

3.4.4 Risk Ranking Methods

According to Grey (1995), "having identified the risk in your project, you will usually have insufficient time or resource to address them all; so the next requirement is to help you to assign realistic priority." With limited resources, it is impossible for any organization to treat all the risk with equal time and importance for all projects. Also sometimes really these are not always called for. Therefore organizations need to assess, categorize, and prioritize the risk level for the projects, so that appropriate time and effort can be utilized in managing these risks. So, when risk ranking is used in risk management and decision-making, then the same must be developed based on normally justifiable and empirically established systematic procedure and not just by some unsystematic ways by some stakeholders and experts. Risk is always perceived as a multi-attribute concept so, right after risk identification it is necessary to attach necessary attributes to each of them. Then these may be listed, which later could be utilized by the participants to rank them according to the risk. This is a risk ranking exercise. Finally, after several discussions and iterations, final risk ranking list will be developed. These generalized steps for risk ranking have been shown in Fig. I/3.4.4-1. Here one thing to be noted that this may not be possible to do in one go, instead there will be iterative process involved. Same has been shown in the figure by both side arrow head. This is based on general decision theory. There could be alternative ways also.

Another method could be by utilizing features or risk drivers for a project. Risk drivers can be conceived as the phenomenon occurrence of which may result risk consequences. Location, complexity, technological changes, etc. are driving forces. It could be clearer from the fact that similarity in project are less risky than a new one, viz. a company having a number of 500 MW units based on subcritical boiler turbo generator (BTG) will face less risk when the organization is going for a similar set of BTG. On the contrary, if it opts for (say) one 650 MW unit with supercritical technology then on account of new technology the risk will be more. This will be clearer from Table I/3.4.4-1 [3].

It is also possible to rank the risk by identifying the targets and rank them according to their likelihood of occurrence in reality. This will be clear from an example. Normally it takes at least around 3 years to build a one large thermal power

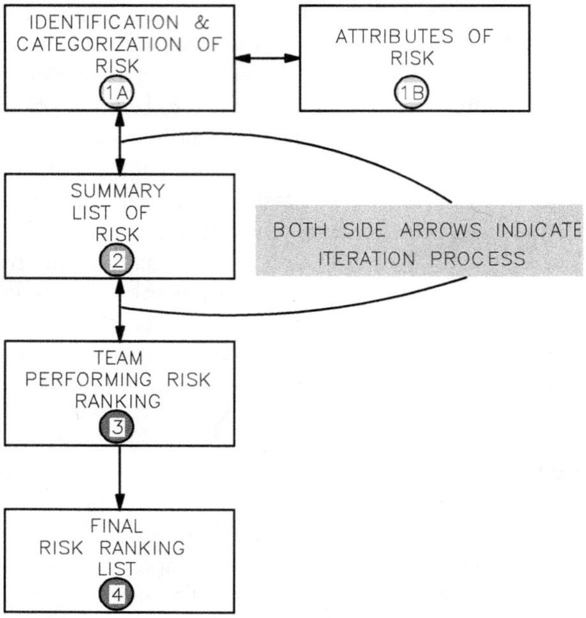

FIGURE I/3.4.4-1

Generalized risk ranking method.

plant. Now, for some election issue same has been set to be developed in one year then it is the most risky situation as time schedule is grossly insufficient. Had this been 2 years, then also risky but less than previous case as in this case schedule is likely to be inadequate (but may not be impossible if the plant is transported from elsewhere). So in this way, for various factors targets could be set and from their consequence due to not meeting the targets could be developed. On the basis of this, risk ranking manager will develop a comprehensive list of risk. It is clear from various methods of risk ranking discussed previously, it is possible to risk rate the project. What is now necessary is to put score against each of the risk drivers/risk factors (taking into account the likelihood and impact) in suitable scale (say 0—5). This will be the input to the computer then the necessary software available which will calculate the risk rating for the project. Then it is possible to display and calculate in the computer average likelihood/consequence of each factor, deriving risk rating for each factor, that is, relative levels of various risk drivers in a graphical form. Risk rating to be treated as input, not final decision. Obviously, highest risks need more attention, but managers need not spend too much money time effort for a high risk *if little* can be done for the same. Whereas middle and low risks should be given attention so that they can be further reduced.

Table I/3.4.4-1 Risk Drivers (A Few Only) [3]

	Risk Rating Lowering ⟶				
Drivers	5	4	3	2	1
Uniqueness	Prototype with new technology	Unusual (not ordinary)	Conventional	Modification of existing system	Repletion of previous ones
Complexity in deliverables	Performance based	Coordination service	Design & construction	Supply + installation	Supply
Project location	Remote in accessible	Remote but accessible	Regional access through disturbed area	Regional	Metropolitan
Hazardous material	Working with hazardous material	Possibility of handling hazardous materials	Hazardous material exists but not part of work	Unlikely to handle hazardous material	No known hazardous material
Project approval	Unidentified approval from government	Potential approval delay identified	Approval requirements known and documented	A few approval awaiting, rest obtained	No approval needed or all approval obtained

4.0 CONSEQUENCE, PHA AND ALARP

4.1 Consequence

As per the dictionary meaning, consequence stands for "result," or "relation between cause and effect," or "logical conclusion." A second meaning is closure to the sense in which consequence is used in risk management system. According to DHS Risk Lexicon, 2010 edition, consequence (or impact) is effect of an incident, event, or occurrence. The nature of consequence can vary widely and depends on the entity. Consequences include (but are not limited to) loss of life, injuries, economic impacts, psychological consequences, environmental degradation, and inability to execute essential missions, direct or indirect [see DHS Risk Lexicon, 2010 edition page (p. 10)]. From the discussions in previous clauses, it is clear that the consequence is dependent on likelihood of occurrence and severity. So consequence spectrum is the risk scenario of several events, each with associated with probabilities of occurrence. So, consequences if c_1, c_2, ... c_n are the consequences of several events, with respective probability of occurrence p_1, p_2, ... p_n then consequence spectrum will be $\Sigma c_i p_i$ where i varies from 1 to n. Through various means, qualitative or quantitative analyses consequences are analyzed. In consequence analysis, all consequences are identified not only immediate but the whole chain of event triggered by one accident or unexpected event. Consequence assessments are carried out to estimate; *the extent to which casualties or damage may occur as a consequence of an accident/unexpected event*, for example, sheering of one bearing in a turbine. This will not only make the turbine inoperative, it can make the entire unit shutdown or may cause some accident to the nearby operator. As stated earlier it also gives *conditional probability of an accident*. Not necessarily, but consequence assessments are normally carried out by expert agencies, often by a team of senior personnel chosen from the enterprise. Consequence assessment software packages are also available in the market. Mostly these software packages are little costly and proprietary in nature. "ALOHA" is a common one available in the market (developed for meeting chemical plant emergencies). Consequence, in industrial plants, refers to the extent to which risk events can affect the enterprise or corporate house. Consequence assessment criteria exclusively depend on the enterprise. For example, a hospital authority will look for medical treatment disruption or loss of reputation as major criteria. On the contrary, for an industrial plant or industrial production company, reputation is important, but disruption of in-house medical treatment may not be a major assessment criterion. On the contrary to a chemical company, leakage of gas or non-conformity to environmental statutes both could be serious and may be treated separately even though they may be related. This is because even if there is no gas leakage, production may be disrupted due non-conformance.

A short typical consequence ranking has been presented in Table I/4.1-1.

There are a few other issues which have bearing on consequence and risk management. These are vulnerability and speed onset or velocity (of the event).

Table I/4.1-1 Consequence Ranking (Impact Scale)

Consequence Category	Insignificant Rating 1	Minor Rating 2	Moderate Rating 3	Major Rating 4	Catastrophic Rating 5
Environment	No hazard material release or brief spill over but contained in the site, no environmental effect	Minor spill/hazard material release at site, no pollutant contained. Immediate clean action	Release of pollutant, medium environmental effect. Controlled by external help	Large environmental/spill (haz. mat) incident cost involvement high-no immediate life threat	Haz. mat spill not contained, immediate life threat, long-term environmental damage. Regulatory authority intervention!
Finance	0–0.1% AOB* or <50k std currency	0.1–1% AOB* or 50k to 1M std currency	1–2% AOB* or 1–10M std currency	2–5% AOB* or 10–30M std currency	>5% AOB* or >30M std currency
Human	No/minor injury (first aid enough)	Medical treatment required, may not call for hospitalization	Hospitalization, possible serious/ permanent injury	Extensive injury, threat for life even possibility for one loss of life	Long term impairment with multiple injury/one or more loss of life
Information	No impact on data (loss)/research work	Impact on data (loss) delay of work/ research	Loss of data/long delay in work	Significant loss of recoverable data/ serious delay in work	Significant loss of irrecoverable data/ significant delay in work
Management	Isolated staff dissatisfaction	General moral problem of staff	High moral problems amongst large number of staff	Senior managers quit. Not a choice of company to high position people	Good number of top managers to quit
Operating problem	No disruption of normal operation	Disruption of normal operation but manageable immediately	Disruption of normal operation no immediate solution in a week time	Disruption of normal operation without immediate solution in 1–2 week time	Major disruption of normal operation. No immediate solution <2 week time

Continued

Table I/4.1-1 ***Consequence Ranking (Impact Scale)—cont'd***

Consequence Category	Insignificant Rating 1	Minor Rating 2	Moderate Rating 3	Major Rating 4	Catastrophic Rating 5
Regulatory/legal	Not reportable to regulatory authority/ minor breach by individual	Some act infringement/ reportable without follow up	Penalty for breach of act/third party claim/ immediate corrective action implementation	Major fine number of third party claim. Requiring high cost corrective action	Significant prosecution, large fine, class action, thread for reduction in government funding and/or withdrawal of exemption if allowed earlier
Reputation	Individual/minor/ local media coverage	Local reputation damage/some media coverage/on going complaint	Nation-wide short-term negative impact/minor social/ political criticism	Nation-wide long-term negative impact/loss of market share/national investigation	Loss of business share & impact on future business nationally/ internationally badly. Continuous/ long-term criticism. Game changer

AOB, *annual operating budget.*

4.1.1 Vulnerability

The relationship of vulnerability with risk management can be easily understood from Fig. I/4.1.1-1. It is related.

The vulnerability would have at a specified point in the future to a specific hazard as a result of realizing all its current adaptive capacity through anticipatory adaption. Vulnerability refers to susceptibility of an entity to a risk event in terms of its preparedness, exposure, and adaptability. Vulnerability is very closely related with security pertinent to programmable electronic (PE) systems — the area we are interested in, so, this is dealt in detail in Chapter XI. From the figure, it is clear that higher the vulnerability, the higher the risk or consequence should the event happen. Vulnerability analysis is also carried out to assess and rate the severity and exposure. *The purpose of vulnerability analysis is to assess vulnerability against severity and exposure.* It could also determine whether current safeguards are sufficient or to augment with additional safeguards. Major criteria for vulnerability analysis are:

- Capability to anticipate
- Capability to respond
- Capability to adapt
- Capability to prevent
- Available options

Table I/4.1.1-1 gives generic idea of vulnerability ratings.

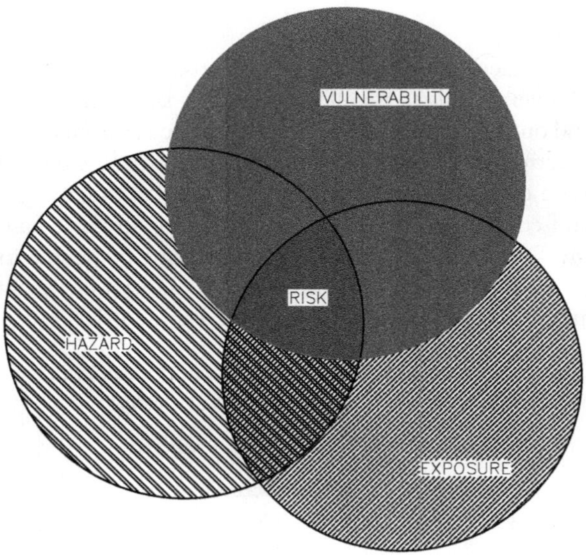

FIGURE I/4.1.1-1

Relationship between risk and vulnerability.

Table I/4.1.1-1 Vulnerability Ratings

Class (Rating)	Vulnerability for:
Very high (5)	Total lacking in planning Failure to address risk both at corporate and process level Risk response not implemented No emergency/contingency planning
High (4)	Planning for strategic risks performed only Low corporate and process level/capability to address risk Partial response implemented without achieving objective Some emergency/contingency planning in place
Moderate (3)	Sensitivity analysis for planning performed Medium corporate and process capability to address risk Response implemented part objective achieved Most emergency/contingency in place not rehearsed
Low (2)	Most options including strategic options for planning defined Medium to high corporate and process level to address risk Response implemented objective achieved, except extremes Emergency/contingency in place and some rehearsed
Very low (1)	Real options in place and strategic planning optimized High corporate and process level to address risk Redundant emergency plans in place with testing critical ones Emergency/contingency in place with rehearsal

4.1.2 *Speed of Onset*

Speed of onset is another important parameter for risk analysis. Speed of onset refers to the time a risk event tends to manifest itself, that is, the time lapse between the event to occur and one can feel the effect. This will be clear from examples: spreading of fire in large structures like hospitals, schools, etc., depending on materials present, may take little time or could be very fast. Similarly, collapsing a large structure, it is extremely fast and could happen in a few seconds. So, all these cases the speed of onset is very high. Whereas development of lung disease by people in and around a power plant with poor fly ash disposal system will be slow; a few years. So, these are also classified into various grades as shown in the Table I/4.1.2-1.

Table I/4.1.2-1 Speed of Onset (Velocity)

Class (Rating)	Normal Criteria
Very low (1)	Occurs over year(s)
Low (2)	Occurs after several months
Moderate (3)	Occurs in few moths
High (4)	Occurs in <7 days
Very high (5)	Rapid, little or no warning

4.2 Preliminary Hazard Analysis (PrHA)

The basic idea of PHA will be clarified in Fig. I/4.2-1. The main purpose of PHA is:

- To identify the extent or boundary limit of complex hazard and analyze the risks and hazards associated with all process involved.
- To assess the findings of the risks.
- Bring out the opportunity for risk reduction and control with recommendations, including owner's view on applicability of solutions (such as hazardous industry planning advisory paper HIPAP number, 6, for example).

Naturally after identification of risks, they need to be ranked according to consequence. This is an initial risk study undertaken at the early stage of the project. The result of PHA is used to focus on the important risks (e.g., to prevent an accident mainly caused by release of energy, the source of the same and to prevent the same). Also PHA provides input for detailed risk analysis.

Normally following points are considered in PHA.

- Identification of hazard elements.
- Safety-related interface with systems components including software [1].
- Environmental conditions including operating condition, and impact and interaction (i.e., impact on environment, or risk due to change in operating environment, for example, system failure to A/C failure).
- Miscellaneous tests, and emergency plan.
- System safeguard, alternative arrangement if any including the same in software.
- System component misbehavior, malfunction, etc.

4.2.1 PHA Procedure Outline

As stated earlier, the basics of PHA have been depicted in Fig. I/4.2-1.

- First stage is preparation (upper part of Fig. I/4.2-1):
    - PHA team consisting of a team leader and senior technical personnel (2-5/6) out of which normally one each shall be from operation and control and instrumentation (C&I) section, therests will be from other disciplines.
    - Scope of work and descriptions: To set the boundary values for the issue, system descriptions with plot plan, preliminary layout, P&IDs/flow diagram, block diagram.
    - List out hazardous material handling, description, detailed energy release descriptions.
    - Plant operation environment, offsite environment, including details about habitats, agriculture, etc. for assessing environmental impact.
    - Detection, monitoring, and control of risk reduction mechanism, etc.
    - Previous, similar case study for data gathering.
- The next stage is to identify the hazards or risks. In order to do the same, it is always better to break down to various subsystems and components. At this stage, all hazard, however small it may be, are listed and later ranked. It is better to provide a checklist based on EN 1050 as Table I/4.2.1-1.

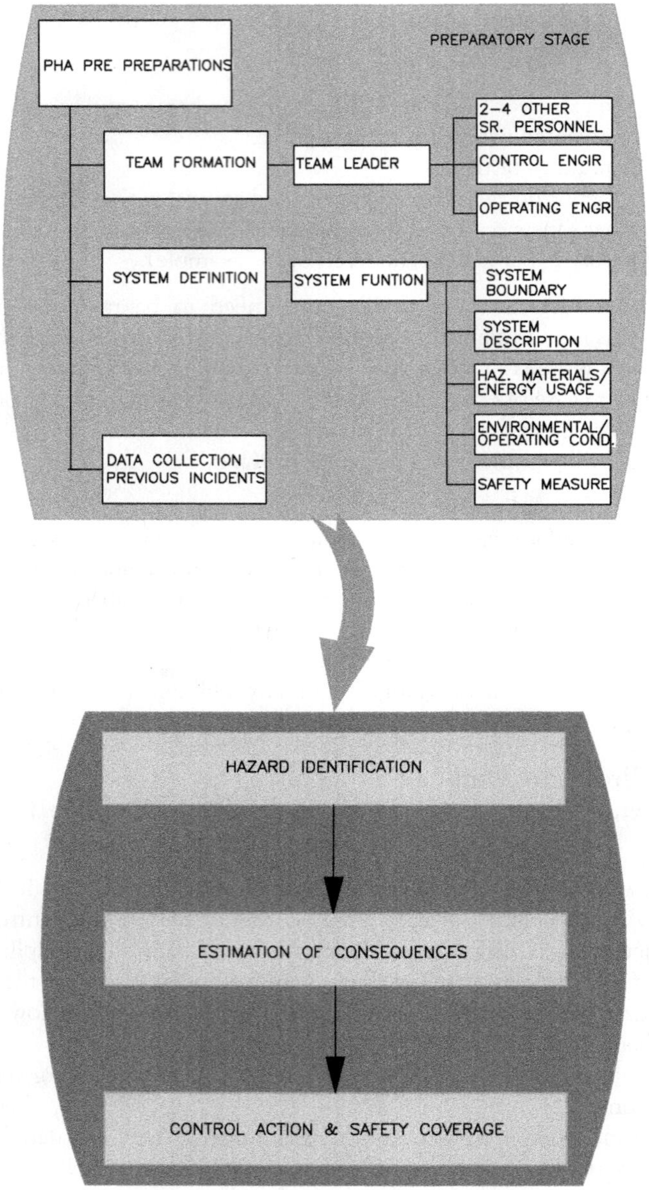

FIGURE I/4.2-1

PHA process in short.

Table I/4.2.1-1 Tentative Hazard Checklist (Based on EN 1050)

Project					**Location:**			
Main system					**Op. mode:**			
Area:	**By:**		**Checked (various disciplines)**			**Approved:**		**Date:**
Ref No.	**Hazard**	**Hazard details**	**Prob. cause**	**Action**	**Emergency Plan**	**Prob**	**Sev**	**Remarks**

Prob, *probability;* Sev, *severity checked: by all concerned.*

While carrying out PHA, the typical hazard sources shall be properly be identified. For example, source of energy is responsible for fire explosion hazards so, all sources and flow path of energy shall be identified and to check precautionary measures considered is adequate! Similarly, moving mechanical parts could be dangerous for injury due to cut, etc. Major sources of hazards could be:

- Human error (operation, movement, etc.)
- Mechanical moving parts
- Flow path of any source of energy
- Hazardous material handling
- Fire and explosion
- Gas/fuel leakage
- Nuclear/other radiations (EM)
- Op/System/Material incompatibility
- Corrosive/toxic liquid/gas handling
- Deterioration in storage
- Deterioration in machine condition
- System error including software error
- Noise/vibration
- Device collisions

It is important to collect data from previous similar systems for hazards and their analysis. Accident, near-miss report, and expert judgments in all such previous similar cases is valuable in getting ideas to identify hazards. These will provide lots of information to identify hazards in one's system also. Old data will give probable hazard, and from there it is possible to know exact analysis, lacuna, if any, emergency plans, etc. Other than these, a few materials have inherent hazardous characteristics so associated hazards could be collected from there, for example, when anybody handling radioactive material naturally associated hazard from the same is well understood. Interactions between systems, human machine, and software, need to be carefully studied. These can help in identifying the sources of error. As stated earlier, source and flow path of any form of energy must be studied very carefully to identify hazard. After these are identified these need to list in the table discussed above. After that, for each case frequency and severity estimation need to be performed and filled in. As discussed, before for each of the event one needs to find the specific (each of very high, high, etc.) severity and their frequency of occurrence to estimate consequence out of the event. [risk = severity X frequency].

Since frequency class, severity class, and risk levels have been discussed at length in previous Clauses 3.0 and 4.1 the same is not repeated. In nutshell, from all these, risks are estimated and ranked. Based on the same, one has to provide some suggestions/control action to reduce the risk level. Also emergency action plans as an alternative wherever possible and applicable should be suggested.

A brief checklist of hazards based on EN 1050:

- Mechanical hazards
  - Crushing
  - Shearing
  - Cutting
  - Friction
  - Pulling
  - Impact
  - HP ejection
  - Penetration
  - Gripping/winding
- Electrical hazard
  - Direct contact
  - Indirect contact due to fault
  - Approaching person HV device
  - Electrostatic phenomenon
  - Thermal radiation/chemicals due to short circuit
- Thermal Hazard
  - Burns and other injury
  - Damage to health and cold working environment
- Noise Hazard
  - Hearing loss
  - Verbal communication damage

- Vibration hazard; use of hand held M/C
  - Vascular/neurological disorder
  - Whole body vibration
- Radiation hazard
  - Low frequency, radio frequency/microwave
  - Infrared, visible, UV
  - Gamma ani X—ray
  - Alpha, beta, Ion beam
  - Laser
- Negative ergonomic principles
- Combined effect
- Unexpected start up, etc. Apparently it seems the task is not very difficult but in reality it needs experts view to foresee hazards and interactions are not so easily forthcoming.

After finishing the preliminary hazard analysis, it will be the turn for evaluation and going for detailed risk analysis, which will be discussed in next chapters. Prior to that, it is better to focus in how many ways the risk analysis can be performed. Normally, there are two broad headings, and these are qualitative and quantitative. Each of these can be divided into deterministic, probabilistic, and combination of the two. In fact there could be as high as 62 risk analysis methodologies for industrial plants [4]. However, a few are only mainly used in industrial hazard analysis. All these have been discussed in Chapters III—V. Out of them, HAZOP and FMEA are qualitative but have been categorized separately as guided word-based type, as entire analysis is done on guided analysis. Various hazard analysis methods under qualitative, guided-word and quantitative categories have been discussed at length in Chapters III—V. Now the concentration will be on another important principle—ALARP.

4.3 ALARP Principles

ALARP stands for "as low as reasonably practicable." The basic principle of ALARP is well understood from Fig. I/4.3-1. The triangle is a representative of increasing cumulative risk a person is exposed to. "Reasonably practicable" is a narrower range than physically possible [5]. The entire region between intolerable risk (upper tolerability limit) and negligible risk level (lower tolerability limit) is the ALARP region. At ALARP level, trouble, time,and cost of further reduction is not worthy, that is, unreasonably high. However, as long as the risk is even just above ALARP level, it is necessary to reduce the risk to bring to ALARP level.

The beauty of ALARP is easily understood from the *examples* in Table I/4.3-1. From the examples, it is clear that at various stages it has some different things to offer. However, it is necessary that periodically the ALARP shall be reviewed, because with time and technology there may be changes in ALARP level.

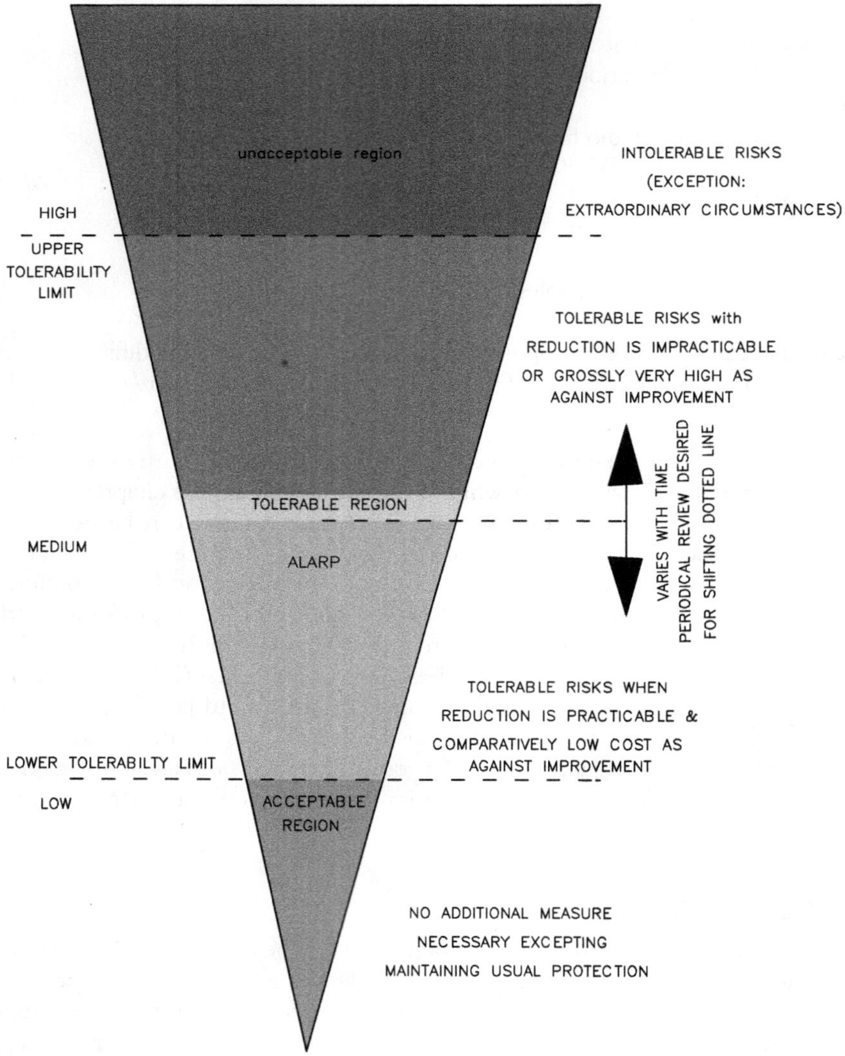

FIGURE I/4.3-1

ALARP principle.

4.3.1 ALARP Point

ALARP concept basically comes from the British health and safety system (Act 1974). It is not in true sense a quantitative method, although cost—benefit analysis is often used to get ALARP. It is a challenging subjective method, as it requires duty holders and others to exercise their judgment very carefully. In risk analysis, three factors play important role, viz. trouble, time, and cost. The breakeven point in

Table I/4.3-1 ALARP Application

Stages	Fossil Fuel Power Plant	Offshore Drilling
Conceptual	More option to choose from various alternatives such as subcritical or supercritical technology/combined cycle plant or IGCC	Choose from fixed leg platform, jack ups, subsea vessel, etc. at conceptual design stages
FEED/basic engineering	Alternative layout for accommodation of common control tower for multiple units, or say alternative ash handling system technology	Alternative layout for better weight control and piping
Detail engineering	Alternatives for safety systems incorporation, integration of control systems	Exploring possibilities for better safety, introduction of collision detect, etc.
Operation	Collecting feedback for safety improvement, frequent failure causes (e.g., introduction of drum level trips from, say, Hydrastep control etc.)	Operational improvement through integrated control system, data collections

IGCC, *integrated gasification and combined cycle plant.*

economics and accounting is the point where cost equals revenue. Similarly, ALARP is the point where risk level meets resources and effort. This is clear from Fig. I/4.3-2. In the figure, one can see that as one moves in X-axis in right hand side risk level of course decreases but cost/effort shoots up sharp line.

4.3.2 ALARP Discussions

"A risk is ALARP if the cost of any reduction in that risk is grossly disproportionate to the benefit obtained from the reduction." This popular definition of ALARP does not reflect the philosophy of the principles and the practical difficulty of the same in implementing effectively. This is clear from HSE.gov.UK, "using "reasonably practicable" allows us to set goals for duty-holders, rather than being prescriptive. This flexibility is a great advantage but it has its drawbacks, too. Deciding whether a risk is ALARP can be challenging because it requires duty-holders and us to exercise judgment."

- As discussed earlier, it largely depends duty-holders' judgment. ALARP is driven by "values." Therefore, it needs significant analysis to determine how to get into the ALARP range. Naturally, after thorough analysis, benchmarking will be helpful in doing this trade-off or balancing acts (to set ALARP point). Thus, there is no perfect ALARP. Also it varies with time and technology.

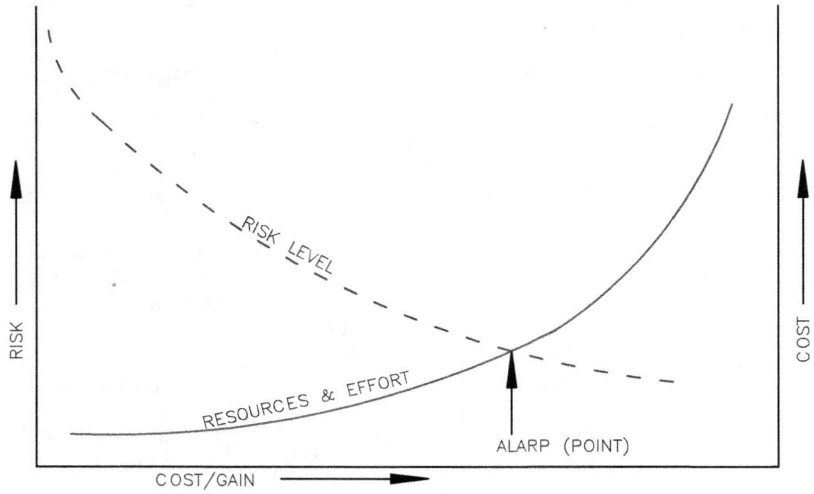

FIGURE 1/4.3-2

ALARP point.

- However, it is on the duty-holder to decide how to apply the judgment, for example, avoidance of risk by alternative solution or spending money on risk reduction, etc. The judgment is based on the following:
  - The risks are so great that it must be avoided, that is, refusal (unacceptable region in ALARP principle as in Fig. I/4.3-1).
  - Risk is so small, it can be neglected instead of spending resources or effort to reduce (i.e., acceptable zone in ALARP principle Fig. I/4.3-1).
  - Risk falling in between that stated previously for which duty-holder has to use judgment to reduce to ALARP.
- As seen from Fig. I/4.3-2, at any point duty-holders need to balance between sacrifices like trouble, time, and cost with additional risk reduction. Duty-holders need to first identify risks to be avoided, then find the cost, effort, and time to be sacrificed and compare with risk reduction involved. Such judgments are done on the basis of nature of hazard, risk involvement, safety measure, and implementation of safety measures. Therefore, it needs significant analysis to determine how to get into the ALARP range, since values are the main prime mover for this principle.

4.3.3 ALARP Process

Briefly, the entire process can be divided in to the following steps for each risk:

- To assess the damage due to risk.
- Find the methods for risk reduction (brainstorming).
- To check and balance the risk vis-a-vis sacrifice. If the risk—sacrifice is disproportionate, then the same may be curtailed; if not, then may be added in to

the list for risk reduction method. If there is no potentially feasible method it means it is ALARP.

- Now for cost benefit analysis (CBA), the cost and benefit of each of the methods listed are found out accurately.
- The next stage is the implementation and reassessment stage, when it is check to see if there is any risk in ALARP region. If no, it is ALARP; if yes then process is repeated.

4.3.4 ALARP Demonstration

Following methods are adapted for the demonstration of ALARP. This is based on Commission for Energy Regulation (CER) [5]:

- The cumulative risks the persons are exposed needs to be compared with tolerability limit as shown in Fig. I/4.3-1 as already discussed in Clause 4.3.3.
- For risks, between the upper and lower limit, demonstration is necessary provided good practices (see Clause 4.3.6) have been followed. Below lower limit, good practices will do.
- Also, it is required that to discuss and describe those measures which have not been taken into considerations, and reasons for the same. In case of changing hazard situation this is important because these may be useful afterwards to maintain ALARP in a changing scenario.
- Risk reduction methods already implemented (say as a part of control) should be assessed. Also, further risk reduction methods must be identified and implemented whenever practicable and not contradicts with good practices.

4.3.5 ALARP Assessment

With the knowledge on overall process of ALARP procedure (Clause 4.3.3), it will be easier to understand ALARP assessment:

- Competent persons having good exposure on operation, design, and having good knowledge in risk management should be included in the team for ALARP assessment.
- To identify all hazards and consider persons at risk, it is necessary to follow good practice (i.e., to follow standards, statues), otherwise there is no need to carry on. To check if there will be any major risk if no, qualitative risk analysis, and then finally to consider if further analysis is at all required.
- If there were major risk (i.e., answer was yes), then quantitative risk analysis will be carried on. If by that, risk cannot be reduced, precautionary principle needs to be applied. Otherwise, risk reduction methods are applied as discussed in Clause 4.3.3 shall be applicable.
- Comprehensive identification of hazards requires brainstorming method. It shall consider 3 Ps: person, plant, and procedure. All hazards must be specifically identified for which frequency and consequence in a risk matrix (as discussed in Clause 3.0) should be considered. For better understanding, see Fig. I/4.3.5-1.

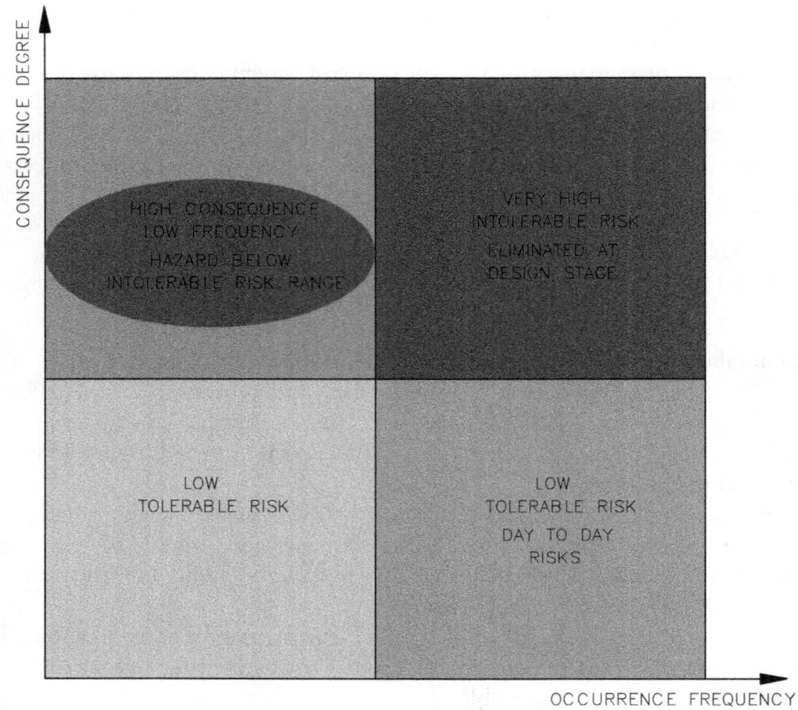

FIGURE I/4.3.5-1

ALARP assessment risk identification.

- All risk reduction methods already implemented (say as a part of protection) shall be identified. Then to add additional risk reduction methods, which are practicable and do not compromise with good practices.

4.3.6 *Good Practices*

As per CER [5]: "The recognized risk management practices and measures that are used by competent organizations to manage well understood hazards arising from their activities." Normally good practice includes, but are not limited to:

- National and international codes and standards along with their foot note—if applicable
- Guidance and recommendations from various professional bodies, for example, IEEE
- Previous case history and recommendations (with judgment)
- All such recommendations need to be followed according to hierarchy of elimination, substitution control, mitigation, and emergency plan (already discussed in Clause 1.0)

4.3.7 *Cost Benefit Analysis and Gross Disproportionate*

In the above discussions, these terms have been discussed; then what are these?

- Cost benefit analysis: This is a numerical exercise done to assess the cost of implementing a particular modification and/or design change and comparing the same against the benefit (risk reduction) achieved out of it. It is one of the tools, but cannot be the sole tool to judge whether risk reduction means is acceptable or not.
- Gross disproportionate: For each fatality, there should be one defined cost. So, when cost to avert the fatality is calculated (CBA), it is compared with the defined cost of fatality. For high-risk value, this factor could be 10, whereas for low risk value it could be 2. There is no "gross" term definition. This is done by judgment from case-to-case basis.
- A *hypothetical* example has been taken to explain both the terms: Assuming that in a plant, there has been an accident where there were fatalities, injury (numbers given in the parenthesis), etc. with compensations are as detailed out as follows:

  Fatalities \$10M (10 persons), permanently injured \$1M (50), serious injury \$ 0.1M (300), slight injury \$ 0.01M (200); rate of explosion is one in 1,00,000 years (1×10^{-5}/yr). Plant life is 25 years, then:

Fatalities:	$\$10 \times 10^7 \times 10^{-5} \times 25 = \$25{,}000$;
Permanent injury:	$\$50 \times 10^6 \times 10^{-5} \times 25 = \$12{,}500$;
Serious injury:	$\$300 \times 10^5 \times 10^{-5} \times 25 = \$7500$;
Slight injury:	$\$200 \times 10^4 \times 10^{-5} \times 25 = \$500$

Total cost to eliminate = \$45,500 since this is a major high-risk issue, DF 10 could be considered so, as against benefit of \$45,500 one needs to spend around \$455,000 is it practicable duty holders need to explain!!!

4.3.8 *ALARP Pros and Cons*

As correctly pointed out in www.hse.gov.uk (see Clause 4.3.2), it has some advantages as well as some serious difficulties. These have been highlighted as follows:

- ALARP can be applied to systems, subsystems, accident, etc.
- To confirm the risk is ALARP, one needs to demonstrate that all credible risk reduction methods are impracticable. For this, prior brainstorming sessions with experienced persons is essential, so that all risk reduction methods could be identified and defined.
- Complete application of the ALARP principle fully and accurately is very expensive and time-consuming [6].

5.0 BASICS OF MATHEMATICS

After discussions on risks, consequences, etc. it is better to take a recess and refresh memory on basics of mathematics. In risk analysis, especially in quantitative

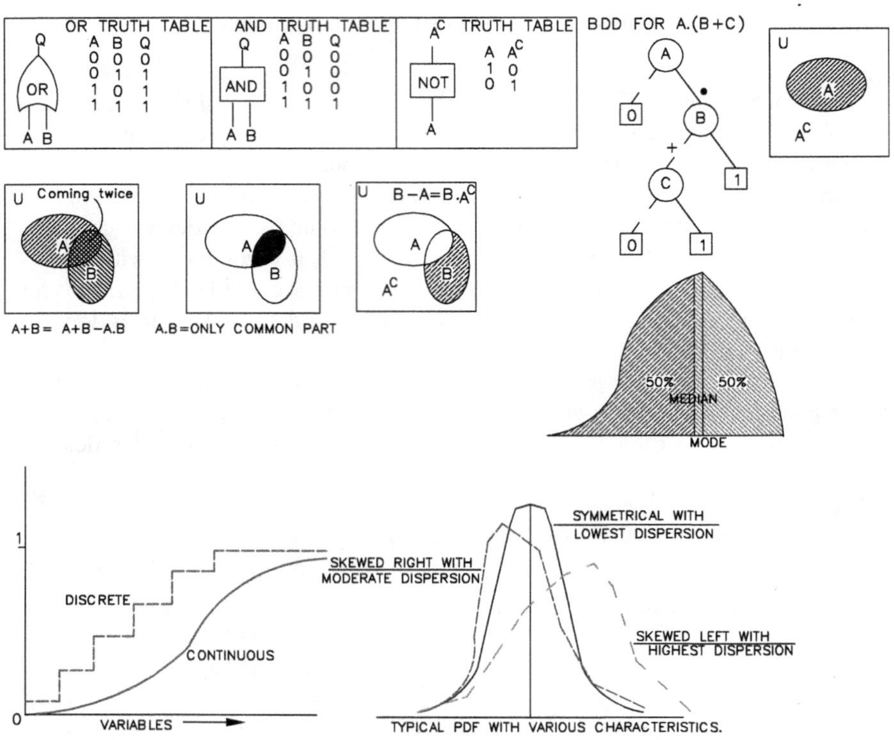

FIGURE I/5.0-1

Basic mathematical concept.

hazard analysis, for example, fault tree analysis (FTA), event tree analysis (ETA), and SIL discussions, this will be very helpful. During these *refreshing discussions relevant fundamental details about logic gates, set theory, and probability theory* will be covered at the rudimentary level only (brief to restrict the volume of the book). Fig. I/5.0-1 gives an overall view of basic mathematics discussed here, including various symbols, etc. to be used later. For further details, including detailed theoretical explanations, related textbooks and/or internet help may be sought.

5.1 Logic Gates

In Boolean algebra, there are a number of gates, out of them, OR gate, AND gate, and NOT gate functions will be mainly required so discussion on the same will be presented here. In the discussions *I/P* and *O/P* stand for input and output pertinent to gates. Generally, discussions shall be done on two inputs (except NOT gate which is basically complementary of input, e.g., x & x') but the same explanations hold good for three and higher numbers of inputs.

Table I/5.1.1-1 OR Gate Details						
	Truth Table				Symbol	
Gate	I/P A	I/P B	O/P Q	Rep. by	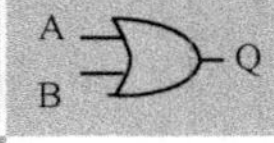	Probability
OR Gate	0 0 1 1	0 1 0 1	0 1 1 1	Q = A + B Q = A ∪ B Q = A V B	Many times OR written inside	P(Q) = P(A) + P(B) − P(A) · P(B)

5.1.1 OR Gate

OR gate is also called union. The output will be logical 1, if at least one of the input is in logical 1. Details of the same has been presented in Table I/5.1.1-1.

From the angle of probability, probability of happening of output Q will be as given in the preceding table, that is, $P(Q) = P(A) + P(B) - P(A) \cdot P(B)$. This is because of the fact that $P(A)$ will include probability of happening of $P(B)$, similarly $P(B)$ will include probability of happening of $P(A)$. So, their together happening will be twice—this calls for subtracting one together happening. This will be clear when we discuss set theory. This means that if a failure event occurs when either of event A or B fails then resultant failure will be failure probability of A or B – minus their together happening. Now following cases needs attention.

- If A and B are mutually or exclusive, then $P(A \cap B) = 0$ so, $P(Q) = P(A) + P(B)$.
- If A and B are independent, then $P(A/B) = P(A)$ or $P(B/A) = P(B)$. So $P(Q) = P(A) + P(B) - P(A) \cdot P(B)$.
- If B is dependent on A, $P(Q) = P(A) + P(B) - P(A) \cdot P(B)$.
- If $P(A)$ and $P(B)$ are of very low probability it is possible to ignore $P(A \cap B)$.
- In case of exclusive OR when both inputs are either 0 or 1, output will be 0, that is, output will be 1 *only when* 0/1 complementary input are present at A, B. then, $P(Q) = P(A) + P(B) - 2P(A \cap B)$.

5.1.2 AND Gate

It is also called intersection. The output will be logical 1 if both the inputs are logical 1. Details of the same has been presented in Table I/5.1.2-1. SO, $P(Q) = P(A) * P(B/A) = P(B) * P(A/B)$. It is better now to consider a few relevant situations as done in Clause 5.1.1.

- If A and B are mutually or exclusive then $P(A \cap B) = 0$.
- If A and B are independent then $P(A/B) = P(A)$ or $P(B/A) = P(B)$. $P(Q) = P(A) \cdot P(B)$.
- If B is dependent on A, then $P(B/A)$ will be one; $P(Q) = P(A)$.

Table I/5.1.2-1 AND Gate Details

Gate	Truth Table I/P A	Truth Table I/P B	Truth Table O/P Q	Rep. by	Symbol 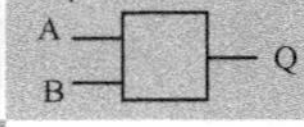	Probability
AND Gate	0 0 1 1	0 1 0 1	0 0 0 1	$Q = A \cdot B$ $Q = A \cap B$ $Q = A \wedge B$	Many times AND written inside	$P(Q) = P(A) * P(B/A) = P(B) * P(A/B)$

5.1.3 NOT Gate

It is complementary of the input, that is, for logical 1 input logical 0 will be output and for logical 0 input output will be 1. Represented by $A—A^c$ –[NOT]–.

It is useful in subtraction operation of some sets as discussed in Clause 5.2 later. It is also important in calculating the failure of component. Let there be three components for a system. Then the failure of each of the component will be represented by A^c, B^c, C^c. Since there are three components, so universal set could be $(2)^3 = 8$ that is, ABC, A^cBC, AB^cC, ABC^c, A^cB^cC, AB^cC^c, A^cBC^c, $A^cB^cC^c$. If the system failure is declared, when two-component failure occurs then last four cases stand for the same. This example is to show the importance of NOT gate.

5.1.4 Logic Gate Discussions

- With the above knowledge, it is clear that if one of the input of logic OR gate is logical 0 then output will depend on the value in the other input, for example,

$$\text{if } A = 0, \quad \text{with} \quad B = X \text{ then output will be } X \qquad \text{(I/5.1.4-1)}$$

(when $X = 0$, $Q = 0$, when $X = 1$, $Q = 1$), on the other hand, if one input (say) $A = 1$, $Q = 1$ irrespective value of:

$$B = X. \qquad \text{(I/5.1.4-2)}$$

- For AND gate, in order to have output $Q = X$, other input (say) A has to be logical 1. This is enabling input:

$$Q = 1 \cdot X = X \qquad \text{(I/5.1.4-3)}$$

(when $X = 0, 1$ then, $Q = 0, 1$). If one input $= 0$, $Q = 0$, as $Q = 0$, $X = 0$.

- Let's look at Fig. 5.1.4-1 logical conclusion.

$$Q = A \cdot (B + C) = A \cdot B + A \cdot C \qquad \text{(I/5.1.4-4)}$$

Here the lamp to glow one thing is essential, that power (I/P A) has to be ON AND SW1 (B) OR SW2(C) should be closed. This essentially means that output (lamp glowing) $= A \cdot (B + C)$. from the figure it is clear that for $A \cdot B = 1$ lamp will glow, also $A \cdot C = 1$ lamp will glow so it is possible to write $Q = A \cdot (B + C) = A \cdot B + A \cdot C$.

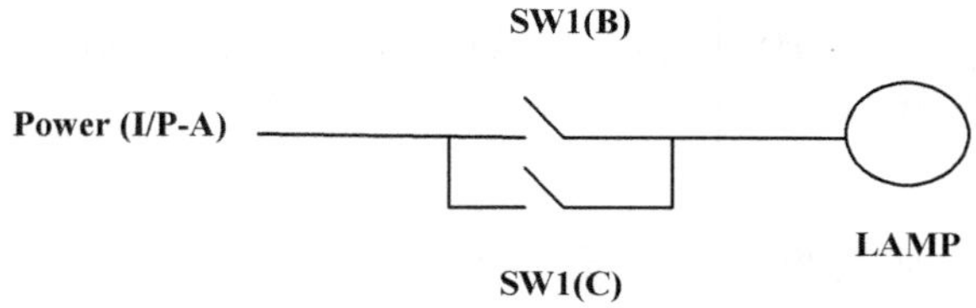

FIGURE I/5.1.4-1

Logical conclusion.

5.1.5 *Binary Decision Diagram*

In hazard analysis, especially for quantitative analysis, this recent development in representation is very helpful. So, a short discussion on the same has been presented. Binary decision diagram (BDD) is graphical representation of logical functions. So, BDD with refinements are data structure to represent Boolean or binary functions. It is binary decision trees are similar to binary trees with leaf nodes. Assuming Boolean variables A, B, and C making up input to a function. At the root node, one of the variables need to be tested and have two sub-trees one A = 0, and A = 1. Each of the two sub-trees now tested for another variable with each another two sub-trees and so on. According to fault tree handbook with aerospace applications [8], "the BDD that is used for FTA is more properly referred to as a reduced ordered BDD." Here, reduced stands for minimum and ordered means in order they appear, that is, in B + C, B will be root. It is better to note in the diagram all possibilities are shown hence not to be confused. In developing BDD, it is better to choose the following form and convention (see Fig. I/5.1.5-1 and I/5.1.5-2):

5.2 Set Theory

A collection of well-defined objects form a set. Again if within these well-defined objects, another set is formed, then the latter will be called a subset of that set from which it has been formed. This concept will be clear from the following examples:

a. Collection of natural numbers 1–6: {1, 2, 3, 4, 5, 6}
b. Collections of all rivers in the world
c. Collection of rivers in India
d. Collection of all books in a college
e. Collection of all physics book in the college
f. Satellites of the world

Here each object in the collection is called the element or member of the set. Now if examples b and c or d and e are seen, then it is clear that c is well covered in b and e is well covered in d so, "c" and "e" are the subset of "b" and "d," respectively, and symbolically represented by "⊂," that is, c⊂b and e⊂d. When an element

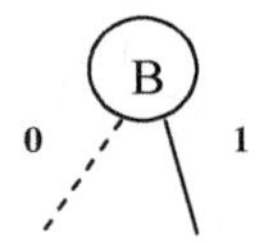

Dotted line to represent 0 & firm line for 1.

It is better to represent the equation A.(B+C)= A.B+ A.C

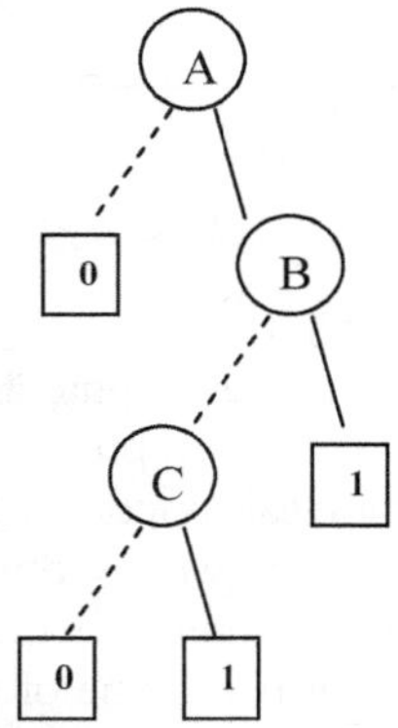

A. (B+C) will be 0 when A is zero because it is in AND 0.X=0, So, (B+C) will be in right leg. Again in B+C , B is in root and for 0 leg of B output is governed by C. (As 0+X= X). These are discussed in details below, where B+C shown separately. Also A.(B+C) has been represented by A.D assuming B+C =D. So, when D is replaced by B+C one can see BDD as shown in the left

FIGURE I/5.1.5-1

Binary decision diagram.

Lets start with OR gate B+C

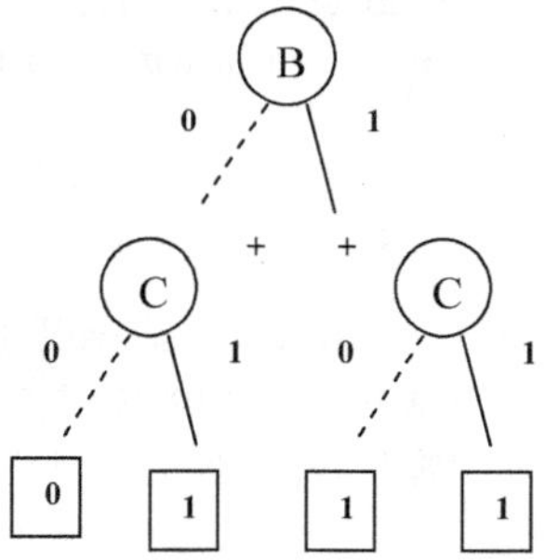

As 0+ X= X hence left leg of B has two outputs corresponding to two legs of c hence they are 0, 1. Similarly as 1+X = 1, for 1 leg of B , hence two out puts as 1s. Ref: Identity law in 5.2.1

Hence B+C will be:

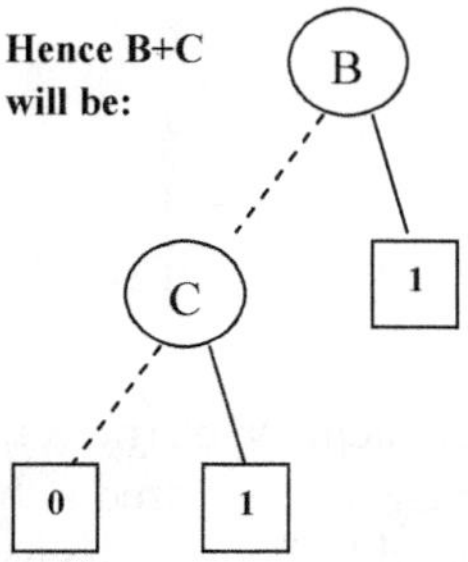

Again, Let B+C =D. As 0. X= 0 [ref: Identity law in cl.5.2.1] hence left leg of A has two outputs corresponding to two legs of D & they are 0, 0. Similarly, 1.X = X hence two out puts 0&1 corresponding to 0 & 1 values of D. Final A.(B+C) has been shown at the beginning hence not repeated.

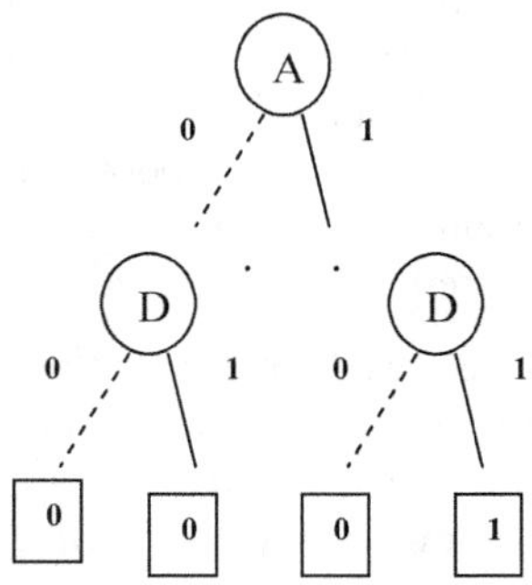

FIGURE I/5.1.5-2

Binary decision diagram explanation.

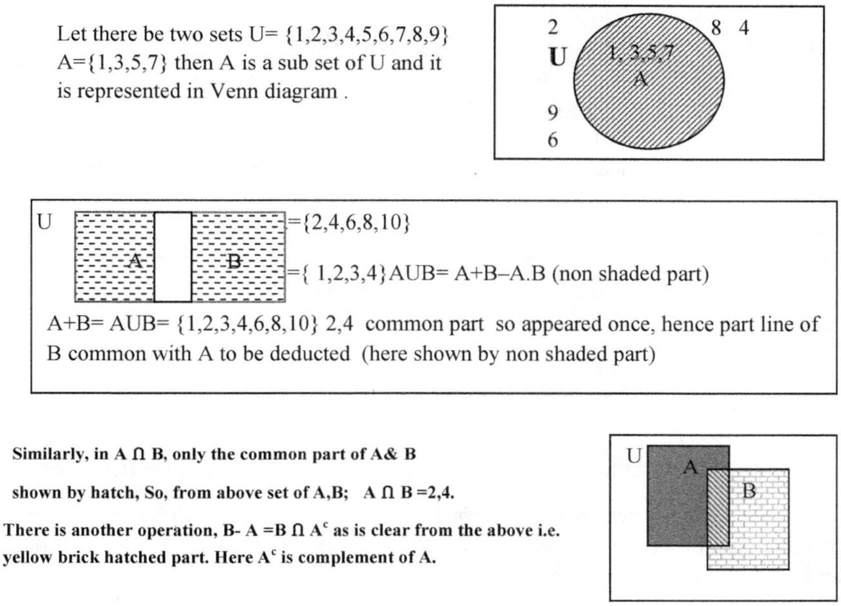

FIGURE I/5.2-1

Diagram for set theory explanation.

belongs to one set it is represented by $1 \in a$ but number 8 does not so, it is $8 \notin a$. Again set f is singleton set as it has single element. When there is no element in a set, it is void or empty set Φ (e.g., $\Phi = \{x{:}x$ is an integer and $2 < x < 3\}$). Two sets are equal when every element of a is also an element of b and every element of b is an element of a that is, a = b. The relationship between sets will be nicely understood by graphically with the help of a diagram called Venn Euler diagram or Venn diagram (see Fig. I/5.2-1).

5.2.1 *Important Laws of Algebraic Sets*

Some important laws of Algebraic Sets have been presented as follows:

- Associative law: $A \cup (B \cup C) = (A \cup B) \cup C$; $A \cap (B \cap C) = (A \cap B) \cap C$
- Commutative law: $A \cup B = B \cup A$; $A \cap B = B \cap A$
- Complement law: $U^c = \Phi$; $\Phi^c = U$; $A \cup A^c = U$; $A \cap A^c = \Phi$; $(A^c)^c = A$
- De Morgan law: $(A \cup B)^c = A^c \cap B^c$; $(A \cap B)^c = A^c \cup B^c$ [$(A \cup B)^c$ is NOR gate & $(A \cap B)^c$ is NAND gate in logical operation or in computer usage].
- Distributive law: $A \cup (B \cap C) = (A \cup B) \cap (A \cup C)$; $A \cap (B \cup C) = (A \cap B) \cup (A \cap C)$
- Idempotent law: $A \cup A = A$; $A \cap A = A$
- Identity law: $A \cup U = U$; $A \cap U = A$; $A \cup \Phi = A$; $A \cap \Phi = \Phi$ [already used in BDD]

5.2.2 Relation and Mapping

- Relation: The relation in set theory deals with the elements in a set. If A and B are two non-zero sets, then a relation R from set A to set B is defined by a Ṇ of A×B; this means R⊆A×B. If A = {1,2,3} B = {a,b,c} A×B = {(1,a), (1,b), (1,c), (2,a), (2,b), (2,c), (3,a), (3,b), (3,c)} now if R = {(1,c), (2,b), (2,c), (3,a), (3,c)} then R⊂A×B and R defines relation from A to B. There are normally different relation types; empty, identity, reflexive, symmetric, anti-symmetric, and universal relations.
- Mapping: Here f: x—y means f maps x into y. If there are two non-zero sets and if there exists a rule or correspondence f, which associates to each element of x of A, a unique element y of B then f is the mapping function of A into B [7] (see Fig. I/5.2.2-1).

5.3 Probability

It is a well-known fact that whenever a coin is tossed the chances of getting head is ½.

- Tossing a coin is one *random experiment*, in which there are different sets of possible results. Also, it is such that it is not possible to state which set of result will occur at the performance of an experiment.
- The result of random experiment is called *event.*
- These events could be *mutually exclusive*, in which two events cannot occur simultaneously. A∩B = Φ so, P(A∩B) = 0. In a throwing of an unbiased die, set A{1,3,5} and set B{2,4,6} and set C{3,6} here it is clear that set A and B cannot happen simultaneously, so, they are mutually exclusive. On the other hand, set C is not mutually exclusive with Set A and B.
- When an event is impossible to occur logically, then it is *impossible event* so P(X) = 0.
- When it is sure to happen it is *certain event* and P(X) = 1.
- If A is an event in a random experiment, then not happening of A is called *complementary event.*
- Two or more events can be said *equal probability*, if after all experiment result it is found that none has preference over other, viz. tossing of an unbiased coin.

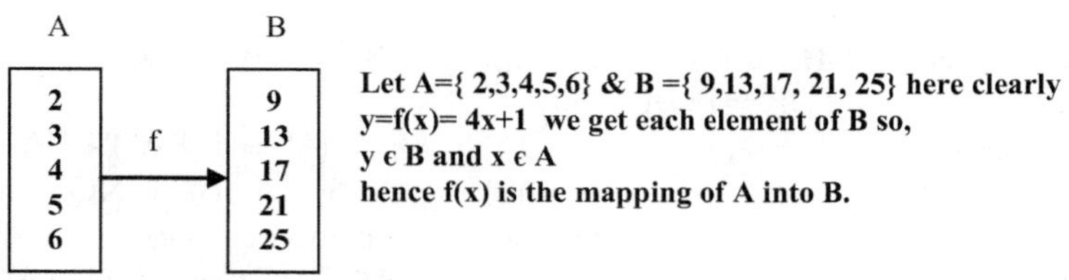

FIGURE I/5.2.2-1

Relation and mapping diagram.

Mean $\mu = \int_{-\infty}^{\infty} x.f(x)dx$ **Variance** $= \int_{-\infty}^{\infty} x^2.f(x)dx - \mu^2$

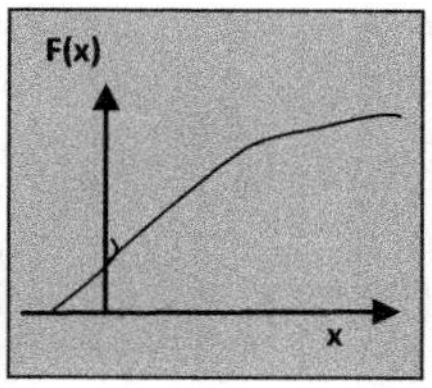

FIGURE I/5.3-1

Probability functions.

- A simple event connected with a random experiment is termed as *sample point* and set of all possible sample points is called *sample space* (Fig. I/5.3-1).

5.3.1 Probability Definition

If a random experiment E is performed n times in uniform and identical condition and if event A is connected to E in $n(A)$.

$$\lim_{n \to \infty} \frac{n(A)}{n}$$

5.3.2 Algebraic Operation of Probability

- OR operation $P(A \cup B \cup \ldots N) = P(A) + P(B) \ldots P(N)$ if these are mutually exclusive. If not, earlier it has been seen that in Clause 5.2 and the associated diagram that there will be repetition of some part (without hatch in the diagram) part so it shall be $P(A \cup B) = P(A) + P(B) - P(A \cap B)$. Similarly, for n such things will occur, in case of throwing of unbiased dice, let A = {multiple of three turns up} = {3,6} and B = {even numbers} obviously it is seen 6 is common in two places, so $P(A \cup B)$ we take occurrence of 6 will be both the places so, $P(A \cup B) = P(A) + P(B) - P(A \cap B)$.

  So, $P(A \cup B) = 1 - (1 - P(A))(1 - P(B))$. Therefore, in a generalized way for m events it will be:

$$P(A_1 \cup A_2 \cup \ldots A_m) = 1 - \prod_{j=1}^{m} (1 - A_j) \quad \text{(I/5.3.2-1)}$$

- AND Operation $P(A \cap B) = P(A) \cdot P(B)$. Therefore, in generalized ways it will be:

$$P(A_1 \cap A_2 \ldots A_m) = \prod_{j=1}^{m} A_j \quad \text{(I/5.3.2-2)}$$

- Not mutually exclusive: $P(A \cap B) = P(A) \cdot P(B/A) = P(B) \cdot P(A/B)$; when $P(A)$ and $P(B)$ are mutually exclusive then $P(A/B) = P(A)$; $P(B/A) = P(B)$ so, $P(A \cap B) = P(A) \cdot P(B)$. If not mutually exclusive, for example, an ACE is drawn from a deck of 52 cards, then probability of getting another ACE in next draw is found from this: after one draw of a card say ACE, 51 cards left when last drawn card is not replaced, so, getting non-ACE will be 48/51... and so on].

If drawing ACE is A, then will be not drawing ACE A^c then, in terms of probability it is:

$P(A_1 \cdot A_2^c) + P(A_2 \cdot A_1^c) + P(A_1 \cdot A_2) = P(A_1) \cdot P(A_2^c/A_1) + P(A_1^c) \cdot P(A_2/A_1^c) + P(A_1) \cdot P(A_2/A_1) = (4/52)(48/51) + (48/52)(4/51) + (4/52)(3/51) = 0.15$ (approx) clearly different (higher) from $P(A_1) \cdot P(A_2)$ $[(4/52) \cdot (3/51)]$

- Bayes' theorem of conditional probability can be expressed as:

$P(A/B) = \{P(B/A) \cdot P(A)\}/P(B)$, where events are A, B and $P(A)P(B)$ are independent probability and $P(A/B)$ or $P(B/A)$ are conditional probability. Refer to Appendix III also.

Let a system consists of three components A_1, A_2, A_3, contributing 0.2, 0.3, 0.5 output, respectively. If the fault contribution B by them respectively are 5%, 3%, and 1%. Then if fault occurs what is the probability that it was produced by A3?

Given the following information: $P(A_1) = 0.2$, $P(A_2) = 0.3$, $P(A_3) = 0.5$.

Fault B contribution by A_1, A_2 and A_3 we have $P(B|A_1) = 0.05$, $P(B|A_2) = 0.03$, $P(B|A_3) = 0.01$.

So, first to find $P(B)$. That may be done in the following way:

$$\overrightarrow{P}(B) = \sum_i P(B|A_i)\overleftarrow{P}(A_i) = (0.05)(0.2) + (0.03)(0.3) + (0.01)(0.5) = 0.024.$$

Now, we want to calculate the conditional probability of A_3. By Bayes' theorem:

$$P(A_3/B) = \{P(B/A_3) \cdot P(A_3)\}/P(B) = \{(0.01) \cdot (0.5)\}/0.024 = 0.005/0.024.$$

5.3.3 *Probability Mass Function*

For discrete variable X, there exists a function $f(x)$ such that $f(x) = P(X = x)$, x being the general value of X. Function of this kind satisfying the following conditions is called probability mass function of X.

- $f(x)$ 0 for all x,
- $\sum_x f(x) = 1$

$$\left[\text{mean } \mu = \sum_x x \cdot f(x); \;\; \text{variance} = \sum_x x^2 \cdot f(x) - \mu^2\right]$$

5.3.4 *Cumulative Distribution Function*

This is a useful method and represented by:

$$F(x) = P(X \le x) = \sum_{y \le x} f(y).$$

This satisfies the following conditions:

Table I/5.3.5-1 Important Probability Distributions and Their Characteristics

Name	Function	Mean (μ)	Variance (σ^2)	Remarks
Binomial	$f(x) = \binom{n}{k} p^x q^{n-x}$, for $x = 0(1)n$ $f(x) = 0$ elsewhere	np	npq	"N" = +integer, $0 < p < 1$ $q = 1 - p$
Poisson*	$f(x) = e^{-\lambda}(\lambda^x/x!)$ (for $x = 0$, 1, 2... ∞) (=0 elsewhere)	λ	λ	* for discrete random variables.
Normal	$\{1/(\sigma\sqrt{2\pi})\}e^{(-1/2\sigma^2)(x-\mu)^2}$	μ	σ^2	Where μ & σ are mean and standard deviation

- $0 \leq F(x) \leq 1$
- $F(-\infty) = 0$
- $F(\infty) = 1$
- $F(a) \leq F(b)$ for $a < b$
- $F(x)$ is right continuous

5.3.5 *Probability Density Function*

For a continuous variable X there may exist a function $f(x)$ such that, for $a \leq b$ $\int_a^b f(x)dx = P(a \leq X \ll b)$. It satisfies the conditions: (1) $f(x) \geq 0$; (2) $\int_{-\infty}^{\infty} f(x)dx = 1$.

This function is called the probability density function. It is worth noting that it indicates an area and applicable for continuous variable X at specific value (on account of definite integral), it is zero.

Some important distributions and their properties have been indicated in Table I/5.3.5-1.

6.0 FAULT TOLERANCE AND PLANT AGEING

There are a few relevant important points, which affect the hazard analysis and SIS. In this clause, short discussions on how fault tolerance and plant ageing affects safety systems in a plant are presented.

6.1 Fault Tolerance

In this part, basics of fault and fault tolerance vis-a-vis system requirements have discussed. Since fault tolerance is very much connected with PEs, so it is treated separately in Chapter XI.

6.1.1 *Fault and Fault Tolerance*

What is fault? From the system point of view, fault can be viewed some deviation from the expected behavior of the system. It is malfunction of the system. This fault may be due to human error, hardware failure, software failure, and/or design problem. Depending on timing, fault can be mainly three types:

- Intermittent fault: The fault which occurs for some time then vanishes and again reappears. These faults are mainly due to hardware. Many times the system does not function well when a part of it is hot, but if it is allowed to cool down it again starts functioning well, or system malfunctions due to one loose connection—all these are examples of intermittent fault.
- Permanent fault: It is a persistent fault say due to failure of a component/subsystem, for example, input short circuit.
- Transient fault: This type of fault occurs once then disappears.

Again depending on the nature of the fault it can be two types; silent and Byzantine.

- Silent fault: Here the output stops, that is, no output due to failure.
- Byzantine fault: Here system produces output but it gives incorrect results. Obviously it is difficult to tackle Byzantine faults.

Unless a system is fault tolerant, then a small failure of component can cause total breakdown of the system. Tolerance stands for capacity of endurance. *Fault tolerance can be conceived as capacity to endure fault in the system.* Therefore, fault tolerance is the property of the system that enables the system to continue operating properly in the event of failure of (or one or more fault within) some subsystem or component. The quality of operation of the system may continue without degradation or with decrease in performance of the system. Also if quality of performance decreases, then that decrease is proportional to the severity of fault. For fault tolerant design, the designer needs to anticipate the fault or deviation from expected behavior, and can develop necessary measures to cope with the situation. However, this is not true in all situations; because many such measures could cost heavily, and also may not be worth considering, for example, not possible to consider duplication of engine for a car or duplication of generator in a power station. A fault tolerant design enables a system to continue its intended operation, maybe at reduced level, rather than complete breakdown of the system. System fault tolerance can be resilient type (e.g., mask the process failure by replication), where the system can work uninterrupted if one or two sensor fails, for example, for 2 of 3 sensor configuration, or redundant network communication with a one link failure. Some are such that there may not be any degradation; only the operator's attention will be drawn. Such systems are transparent to the fault. This is somewhat possible in triple modular redundant (TMR) system logic or for systems with complete redundancy in sensor, logic element, and final control element (FCE). Fault tolerance is directly connected with *dependability* which

Table I/6.1.1-1 Fault Tolerance Characteristics

Name	Explanation	Example
No single point of failure/repair	System continues its operation uninterrupted for a single failure and/or during repair for that.	With uninterrupted power supply (UPS) of suitable capacity connected to a computer, computer continues its operation with main supply failure, or input supply fuse failure and repair for the same.
Fault isolation for the failing component	Ability of the system to isolate itself from the failed component and continues its operation. So, necessary fault detection and isolating devices must be provided.	It is quite common to isolate a part of the grid incase massive grid failure. PLC I/Os with galvanic/ photovoltaic isolation does the similar function.
Fault containment to prevent propagation of failure	It is possible in some cases, that on account of failure of one subsystem, the fault may propagate. So, suitable measures shall be built in to prevent such propagation.	Firewall in network in the classical example of the same. Explosion proof enclosure also does the function of containment.
Availability of recovery	A process by which failure shall be recovered.	It is possible in two ways one forward recovery when the system will be taken to a new correct state but not the last correct state. In back recovery it is brought back to the last correct state. These are often found in network communications. When computer OS fails it is possible to bring back the system by recovery at a back date when it was functioning, a simplest example from day to day experience!

consists of "*availability*," (*ready for use immediately*), "*reliability*" (*how much the system can run without failure*), "*Safety*" (*how safe the failure? What are the results, i.e., nothing serious happens for one failure*), *and* "*maintainability*" (*easiness to repair*).

Basic characteristics of fault tolerance have been given in Table I/6.1.1-1 with suitable examples for understanding.

At this point it is worth noting that the fault tolerance of a system may be implemented in both hardware as well as software pertinent to the system. Fault tolerant software is normally part of the real time operating system (OS) interface and has some special characteristics like:

- Reaction to power failure, even for a graceful shut down
- Immediate backing up for system failure
- For multiple processors working in tandem, to compare data/output for error generation and correction
- In case of backup storage immediate switchover to backup in case of failure of former one.

6.1.2 Redundancy and Replica

In most of the cases, fault tolerances are achieved through redundancy, now, it is better to look the same bit closely. It is worth noting that replica and redundancy are often used synonymously, but there is a little difference between the two.

- Replica: Normally multiple identical components/subsystems operate in parallel, and correct output is chosen via voting/quorum, for example, use of 1 of 2 or 2 of 3 transmitters (*though this is loosely termed in C&I engineering, as 2 of 3 transmitter redundancy*). In some cases, based on implementation methods of replica, these can be termed as diversity of application also.
- Redundancy: Normally multiple identical components/subsystems are kept ready for operation. In this case, one operates and others comes into operation when the former fails (failure changeover/hot standby), for example, in a programmable logic controller (PLC) with two processors, one is working and the other on standby (but tracking I/Os), so that in case of failure of one, the other will take over. There are three types of redundancies applied; these are:
  - Information: IT/network applications, viz. error correction code/parity, etc.
  - Time: Work persistently until there is success. Network transactions.
  - Physical: Hardware/software, for example, process replica discussed previously.

There will be obvious question in mind; how many replica shall be necessary to get fault tolerant system? The answer is:

- For silent fault, the redundancy shall be K + 1 so that it can withstand K faults with one survival.
- For Byzantine fault, it shall be 2k + 1so that k + 1 components will be there for voting.

In instrumentation and control, triple modular redundancy is very important for fail safe operation. Fig. I/6.1.2-1 shows the same. Here, each of the three elements are voted thrice in each stage to get the output. In network communications, especially for remote communication, there are a few other problems known as Two Army problem, Byzantine general problem, etc. The issues discussed so far basically belong to fault masking to get away with hardware fault. There is another term called dynamic recovery, in which case there shall be a special mechanism to detect hardware fault and isolate the faulty hardware and replace the same with a good one. This will be clear from an example. Say in a process control, there are two processors; one working and the other standby. If there is another processor whose main function is to act a diagnostic processor to check health of other processors, when it finds fault with

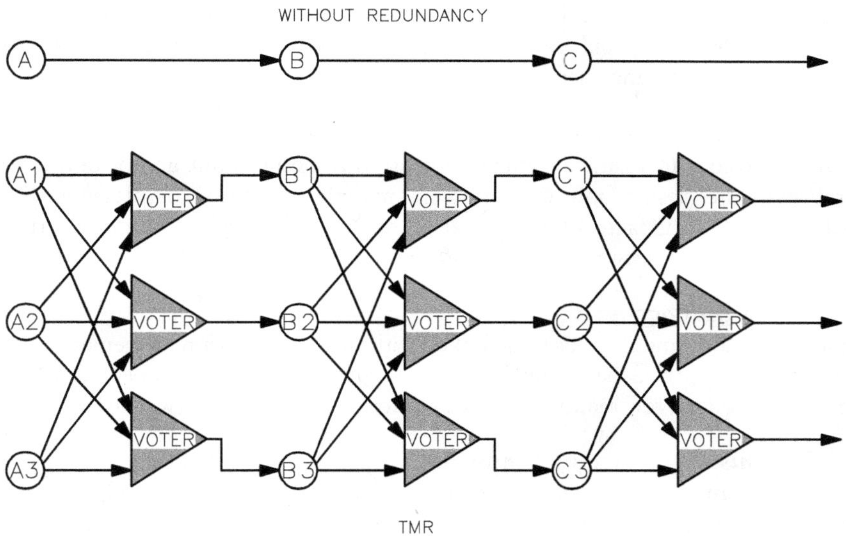

FIGURE I/6.1.2-1

Triple modular redundancy.

the working processor it switches the control to the standby processor (which also was tracking I/Os), and removes the faulty processor. Sometimes this is found to be more cost-effective than voting type. In software fault tolerance are achieved by:

- Static redundancy: In this system *n* versions of the programs are written and run for performing the same function. There will be a few checkpoints where each of the output from different versions will be voted to be selected as final output.
- In an other approach, the programs are divided into blocks which are run and tested. If any one fails the test, it will be replaced by redundant one.
- Another option is design diversity each using different set of hardware and software. All these are detailed out in Chapter XI.

6.2 Plant Ageing

Fault tolerance discussed in the previous clause are mainly related to instrumentation and control systems, etc. How to combat faulty conditions in the system? In most of these cases the situations could be defined. Similarly in mechanical systems, especially in vessels containing hazardous or corrosive materials, can withstand or tolerate up to a certain extent then they may tend to degrade. For example, on account of corrosion there could be weakening in wall thickness, or there could be structural deformity! The majority of such things happen due to plant ageing, and these could result a big accident and/or catastrophic situation. In this part, discussions will be on plant ageing. In our day-to-day experience, it is observed that many old persons are more fit compared to a younger person, both physically and mentally. Similarly, plant ageing is not solely dependent on chronological age, but also on many other factors. From petroleum safety authority, it is found that "*ageing is not about how old your equipment is; it's about its condition and how it's changing over time*." So, it is more about its degradation with time and use of the same.

6.2.1 *Contributing Factors*

Corrosion, erosion, mechanical fatigue, operation at design limit and/or beyond it, calibration failure, design fault (such as wrong material selection), obsolescence, etc. are major contributing factors for degradation of equipment/system, and/or electrical control and instrumentation (EC&I) items.

6.2.2 *Aging Management*

Plant ageing management needs to be linked with integrated asset management of the enterprise, and highly depends on how effectively site assets are inspected, tested, calibrated, and effectively maintained (both in terms of frequency as well as detailing).

6.2.3 *Various Plant Ageing Compponts*

Major equipment, structure, and other things to be managed under plant ageing system have been shown in Table I/6.2.3-1. This is an indicative list only, there can be others also, depending on plant type.

Table 1/6.2.3-1 Major Items in Plant Ageing (Category and Example)

Category	Examples	Remarks
Structural and Civil	Building structure, secondary/tertiary supporting foundation & structure for containment, structure for external impact—flooding etc. Loading/unloading point & structure.	
Process containment equipment—mechanical items (non-rotary)	Pressure/process vessels, reactors, boiler & steam system, heat exchanger, pipe, piping network, flexible hoses, utilities, column distillation system to name a few.	In majority of the cases and countries regular inspection is mandatory.
Mechanical rotating items	Pump, compressor, electric generator, fans, turbine, etc.	Some cases regular inspection is mandatory.
Safeguard items (mainly non-EC&I for BPCS)	Pressure relief valve, safety valve, associated circuit, alarm & communication system, safety instrumentation, overfill protection, flare stack, chimney, etc.	Many of these calls for regular inspection as a rule.
Electrical/C&I items	Level gauges, transmitters, and switches for BPCS & general instrumentation, power distribution, fixed hazard detection system.	In most of the cases these are subject to regular inspection.

6.2.4 *Plant Ageing Indicator*

Table I/6.2.4-1 can be used early warning for plant ageing. This is an indicative list only for looking at plant ageing in general terms.

6.2.5 *Factors*

Plant ageing factors shall include, but are not limited to the following:

- Equipment age: Time
- Obsolescence of equipment/C&I

Table I/6.2.4-1 Plant Ageing Indicator

Category	Item	Discussions
Physical condition	Damaged surface condition of equipment, poor surface painting, corrosion status, trending in leakage, trending from inspection results	In case of corrosion, erosion there will be possibility for damaged surface area. Vessel inspection, high leakage could be a result of poor maintenance, or cracking gasket damage. Result of repeated inspection can also gives indication for deteriorating condition (e.g., damage bearing—hence vibration etc.)
System availability and reliability	Frequent breakdown, loss of availability, trending in mean time between failure (MTBF), need for frequent repair, unstable BPCS	Frequent breakdown, repair and MTBF trends clearly indicate problem with ageing so to find cause for it. Instability in instrumentation could suggest either the BPCS instrumentation is obsolete or in poor condition and/or problem due to equipment ageing. This could be due to poor maintenance (discussed below) also. Needs further inspection to arrive at a conclusion.
Maintenance	Higher budget towards maintenance & repair, trends in mean time to repair (MTTR)	If more and more attention is needed either to EC&I or to equipment. Higher MTTR all indicate plant ageing.
Operational performance	Lower grade/poor product quality, high rejection, deteriorating plant operational performance	Poor efficiency, high pumping cost, unable to cope up with the requirement for product quality. Unstable operation.
Hazard potential	Action taken report from PHA, design operation reports. Incident report	When for some conditions all actions suggested could not be implemented or when implemented not improvement result noted. Also design reports, operation and incident report could be eye opener.
Energy and environmental impact	Higher energy consumed per unit production, more pollution	The efficiency of machines may be degraded hence higher energy or less pollution control efficiency, for example, ESP.

- Operation beyond designed temperature range
- Operation beyond design limits
- Bad material selection/fatigue test not carried
- Sustained operation beyond corrosion limit
- Change of services
- Extremely corrosive atmosphere
- Departure from environmental condition at design stage
- Poor maintenance of equipment/control
- Poor painting and protective materials used
- Wrong operational practice, high cycling rate of load, temperature.
- Predictable ageing

6.2.6 *Operator Action*

An experienced operator can understand the problem better so they can initiate actions such as:

- Recognition of ageing and try to identify the points where it is occurring or expected to occur.
- Advise/suggest frequent vigorous inspection with due prioritization in the inspection plan.
- Equipment de-rating and or suggest replacement.

6.2.7 *Inspection Teamwork*

A good team of inspectors from various disciplines shall be formed to take up the issue in an efficient way which shall include but are not limited to the following points for better results:

- Good coordination, leadership, and high support from management
- Maintenance of *asset register* and regular updating of the same
- Well-planned maintenance schedule routine testing of equipment for condition monitoring (e.g., vibration for rotating equipment)
- Regular inspection for corrosion data, pressure vessel, etc.
- Regular electrical equipment testing and inspection
- Regular calibration and tuning (if necessary) for C&I items, replacement of obsolete C&I equipment with compatible latest device

6.2.8 *Plant Ageing Control Discussions*

There are a few options by which the plant ageing control can be met:

- One option could be to carry on the business with periodic check and reactivate or rejuvenate the system to minimize ageing effect to a great extent and turn around. This is common practice in many chemical and process industries. This approach is not only costly, but has no foresight about the remaining plant life. In some other industries, a similar approach with more emphasis on inspection are provided to find accurate reason for ageing.

- The industries like offshore operations and nuclear plants, the issue is more critical so, they under take risk management review or ageing management review on regular interval during the design plant life, so that they always run the plant at a level much within ALARP. For risk control and management they take the help of risk matrix discussed earlier. In this method first degradation mechanisms, as discussed above, are identified. After identification, these are listed in a table. Against each of degradation mechanisms, probable consequences and safeguard/mitigation and controls, etc. are listed. Here likelihood is determined from remaining life time (see CRR 363 Risk Based Inspection [RBI], @hse.uk), for example, when remaining lifetime is say, more than 10 years, it is unlikely happening may be ignored! Severity is assessed based on guided-word later arranged in number scale. The rating very much depends on factors like impact on project, location, population, etc., hence number scale is specific for project. Corresponding to each degradation mechanism there will be one severity value and likelihood value, which will be placed in risk matrix. Several steps for this are as follows:
  - Data collection for component with degradation mechanism identification
  - Criticality selection
  - Risk assessment and ranking (risk matrix) based on remaining life
  - Risk mitigation and control asset life management

Table I/6.2.9-1 ***Progressive Ageing (Brief) [9]***

Stage 1	Post-commissioning	• Design, manufacturing, installation & commissioning issue • Early life issues (Training/trial) • Identification of potential ageing site
Stage 2	Risk-based	• Operation with design limit • Routine maintenance • Extended operating period • Upgraded risk analysis
Stage 3	Ageing	• Design limit approaching • Evidence of active deterioration • Degradation rate increasing • Repair refit modification
Stage 4	Terminal	• Accelerating and accumulating damage • Beyond design limit operating in known experience • Monitoring • Major repair, refit replacement required

6.2.9 *Progressive Ageing*

Brief discussions on plant ageing could be concluded with some ideas about progressive plant ageing from [9] in Table I/6.2.9-1.

7.0 SAFETY INSTRUMENTATION

Safety instrumentation systems form an important part of the book. In the following discussions a glimpse of these systems has been presented so that reader can get some idea about the same while going through the details in subsequent chapters. After IEC 61508/61511 came into effect, it has been found that there are two schools of thought. One group advocates for separation between BPCS and SIS. The other group does not find it wrong to combine BPCS with safety as long as it follows standards IEC 61508/61511with appropriate SIL. In reality, it is not the question of finding wrong or right but to judge the situation. This purely depends on how the stipulations in standard have been interpreted and implemented. It is quite common to have integrated systems where both systems have been combined or integrated. Also it is not that these are not permitted in the latest edition of standards. Prior to going in for complex interpretations it is better to understand the system. So, the reader's attention is drawn to Figs. I/7.0-1 and I/7.0-2. In these figures, it has been attempted to focus on the fact that apart from normal operation of the plant through BPCS, there are separate SISs so that in case of single failure of component, plant

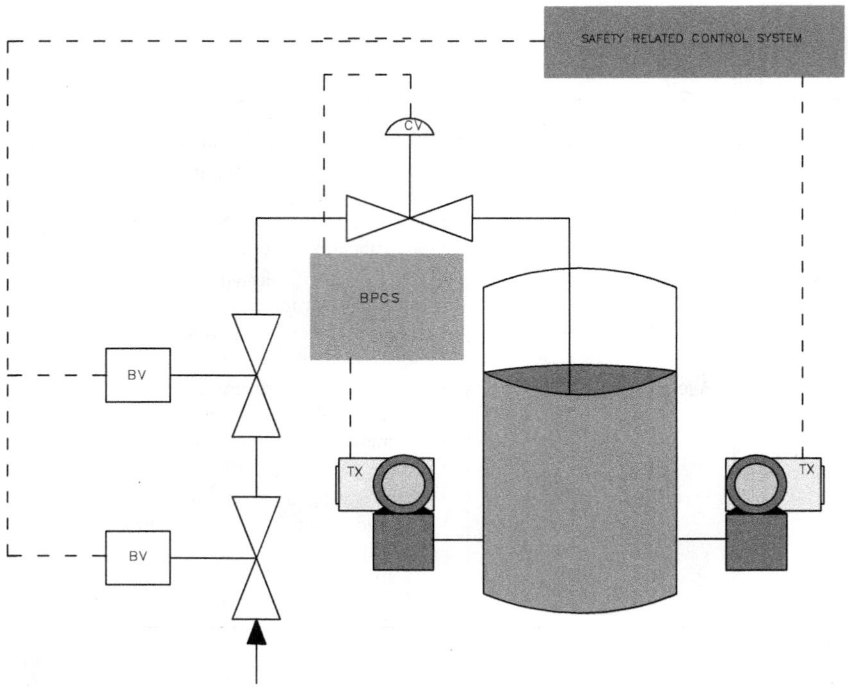

FIGURE I/7.0-1

Safety instrumentation with separate BPCS and SIS.

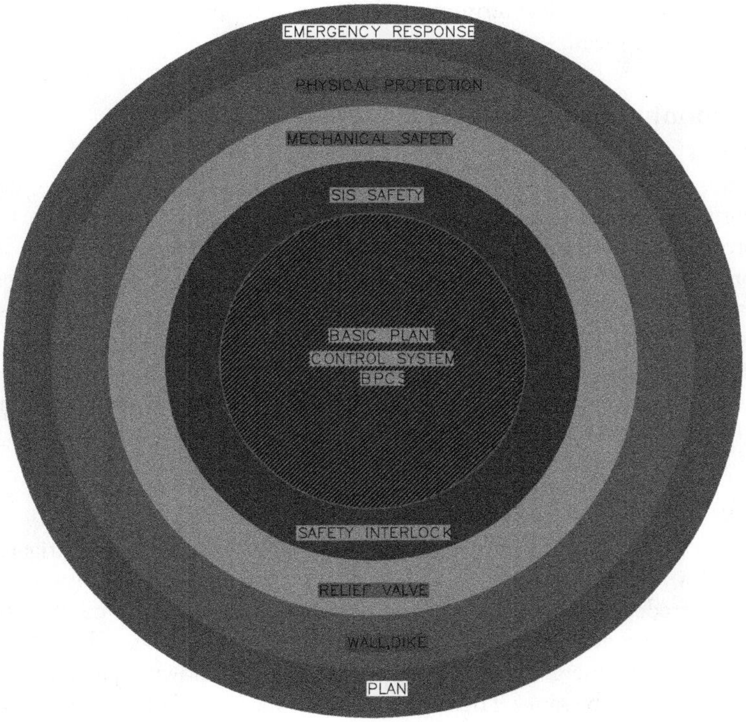

ALARM HAS BEEN CONSIDERED HERE AS A PART OF SIS (ACTUALLY TREATED SEPARATELY)

FIGURE I/7.0-2

BPCS with various protection layers.

safety is not endangered. In Fig. I/7.0-1, normal level control is carried out by BPCS and there is separate safety system. So, even if controller of BPCS fails, it is possible to take safety measure through SIS to block the feed, with the help of block valves to save the system.

Apart from these safety systems in general there are other safety measures also to be considered to combat emergency situations. Fig. I/7.0-2 shows these probable protection layers. These protection layers are:

- Alarm and safety instrumentation: In this figure alarm and SIS has been treated as single layer but *in reality they are treated separately*. Details available in subsequent chapters, independent alarm is to draw the attention of human being to take safety action. Whereas if human response fails then there will be SIS to take automatic action even may trip the system/subsystem.
- Mechanical safety such as safety/relief valve, that is, to release the material safely.
- Physical separations such as dikes, or safety walls—containment.
- Last stage—most important: Emergency plan: escape route, etc.

From the preceding, it is clear that in reality there could be stages of protection layers in the plant to meet emergencies. Before going for further details it is necessary to understand a few terms with their meanings, so that future discussions could be understood well.

7.1 Commonly Used Terms

7.1.1 *BPCS*

BPCS stands for basic process (plant) control system. This system handles the process controls and monitoring for the process. It takes the input from process sensors processes it according to control and monitoring strategy fixed at design stage to produce output for output devices/final control element, so that the process behaves according to design. Sometimes it also undertakes safety functions. According to IEC 61511, "*BPCS is a key layer of protection which responds to input signals from the process, its associated equipment, other programmable systems and/or operator and generates output signals causing the process and its associated equipment to operate in the desired manner but which does not perform any safety instrumented functions with a claimed SIL* $\geq$ *1.*"

7.1.2 *SIS*

SIS stands for safety *instrumented* system. SIS is designed to prevent or mitigate from happening of a hazardous event, by taking the process to a safe state whenever a predefined or predetermined conditions occur to the system. It is a combination of sensors, logic solvers, and final control elements. In PEs, it consists of both hardware and software. In fact, emergency shutdown system (though shown separately in Fig. I/7.0-2) will be a part of the same. There could be a number of SIF (defined next) in SIS.

7.1.3 *SIF*

SIF stands for safety *instrumented* function. SIF consists of sensors, logic solver, and final control element combination. SIF takes the system or process in to safe zone in the event of hazardous situation/event, which is determined by predefined conditions for the process (*see Clause 8.1 also, for definition as per various standards*).

7.1.4 *Functional Safety*

According to ISA, "*the ability of SIS or other means of risk reduction to carry out the actions necessary to achieve or to maintain a safe state for the process and its associated equipment.*" Also, functional safety in SIS highly depends on proper functioning of sensors, logic solver, and FCE so that reduced risk level could be achieved. In that sense, it also means proper functioning of these components also (*see Clause 8.1 also, for definition as per various standards*).

7.1.5 *SIL*

SIL stands for safety integrity level. It is a measure of performance of an SIS. It is determined by PFD for SIF (SIS). There are four SIL levels represented by number, viz. SIL 1, 2, 3, 4. Higher the SIL number, the better will be the performance and lower will be PFD value. However, with an increase in SIL number, the cost and complexity of the system increases, but risk level reduces. It is worth noting that there could be individual component PFD but not SIL. SIL is only given to a system (SIS). SIL certification can be issued by the company (self-certification allowed), or other competent authority to indicate that appropriate procedure, analysis, and calculation have been followed and compatible for use in appropriate SIL level. Proof test and full IEC certifications need further discussions presented later in the book *(see Clause 8.1 also, for definition as per various standards)*.

7.1.6 PFD

PFD is the probability that SIF/SIS fails to perform its intended safety function during a potentially dangerous condition. PFD_{avg} is normally used in calculations when regularly inspected and tested.

7.1.7 MTBF

MTBF stands for mean time between failures. It is defined as inverse of failure rate. Actually it is inverse of failure rate minus mean time to repair (MTTR). If MTTR is small, then MTBF = (1/Failure Rate).

7.1.8 SIS Implementation

As per ANSI/ISA 84 and IEC 61511, there shall be multi-disciplinary approach which follows life cycle analysis PHA before implementing SIS. So, SIS implementation requires all above activities to be performed. However, fire and gas systems automatically initiate process action for mitigation and taking the system in to safe state to natural SIS.

7.2 Control Objective and Layers of Protections

All processes, whatever the application may be, plants faces major challenges all times. Therefore, it is essential that there shall be suitable control system, which is able to keep the system running with minimum unwanted downtime. So, basically control systems are designed to meet a number of well-defined objectives.

7.2.1 Control Objectives

Control objectives shall include but not limited to the following.

- Overall safety comprising safety to personnel, equipment, environment, etc.
- Smooth and well-controlled operation of the plant with maintenance of product quality at desired productivity level with profitability.
- Monitoring, performance calculation, system diagnostics, etc.
- Response shall be limited mainly to continue operation without harm.
- Ability to manage multiple risks and face gracefully the emergency situations.

In order to meet all these requirements as per relevant standards there shall be several layers of protections in the systems so that no single failure can affect the process.

7.2.2 Layers of Protections

As shown in Fig. I/7.0-2, there are five layers of protection. In the above drawing, one layer of protection, "alarm" is missing. This is done *purposefully* to indicate that alarm may be a layer of protection but is dependent on human intervention. These protection layers are:

- BPCS: Process control (PIDs), interlocks for safety related variables, monitoring of misc. variables especially corrosion, material build up, etc., and provide safe response to instrumentation failure. This consists of sensor, logic solver, and final control element.

- Alarm system: Alarm is like wakeup call! Manual intervention is desired and not automatic action except alarm generation. Alarm storing and management could be a part of overall control system. Alarm systems may or may not use common instruments but should be independent. Alarm systems generates alarm failure for instrument and/or system failure also, for example, Tx failure or system diagnostic alarm. It keeps operator alert against hazard to people, equipment, and environment. It calls for human action. It is the first stage of protection layer to BPCS (only manual intervention necessary hence not free from possibility of human error as result of the too many alarm may make the situation more grave), just before SIS to become active, for example, pre-trip alarm.
- SIS: This is the first automatic protection layer to BPCS and second overall layer of protection. It is desired that this shall be independent of BPCS. Even if these are combined it is necessary to ensure that single failure does not take toll of safety. SIS may stop part of plant operation and/or diverts some flow safely, etc. It may have separate set of instrumentation to detect and take safety action in the event of instrument/system failure. It has to be more aggressive than BPCS for safety functions. Under SIS, there will be several interlocks and protections to save the system and in many places like off shore design, ESD is considered as last resort or ***emergency plan*** achievable through PEs.
- Diversion of flow safely: All the above systems discussed are part of electronic system requiring power. As a next layer of protection there are mechanical devices (requiring no power to meet power failure situations) to divert the flow safely. Safety and relief valves are used to in case to depressurize when SIS fails to take care (e.g., say due to control power failure). Pressure relief valve diverts the fluid for safe passage. These relief valves are spring force to close so that when pressure is below setting it is closed. Many cases rupture discs are used but in that case system needs to be closed to attain the disc. To a certain extent quartz bulb in sprinkler system does the same function.
- Containment: Many cases the hazard is not allowed to propagate hence containment functions are used. In these cases materials or hazardous events are not allowed to reach workers, etc. Dike, safety wall, etc., are examples of the same.
- Emergency plan: This is to meet the hazard events so that people are not harmed. Fire fighting, evacuations, escape routes are examples of the same. *So far as instrumentation is concerned, ESD could be considered as emergency plan achieved through PEs, hence in the book many times ESD may be referred as last resort. But reader to keep in mind that ESD belongs to SIS.*

7.3 SIS Boundary and Layout

SIS, like BPCS, consists of sensors, final control elements, and logic solver. Sensors and final control elements interface with logic solver through input and output processing section of logic solver which could be programmable electronic device (say, PLC) as shown in Fig. I/7.3-1. In this diagram, user interface and interface with

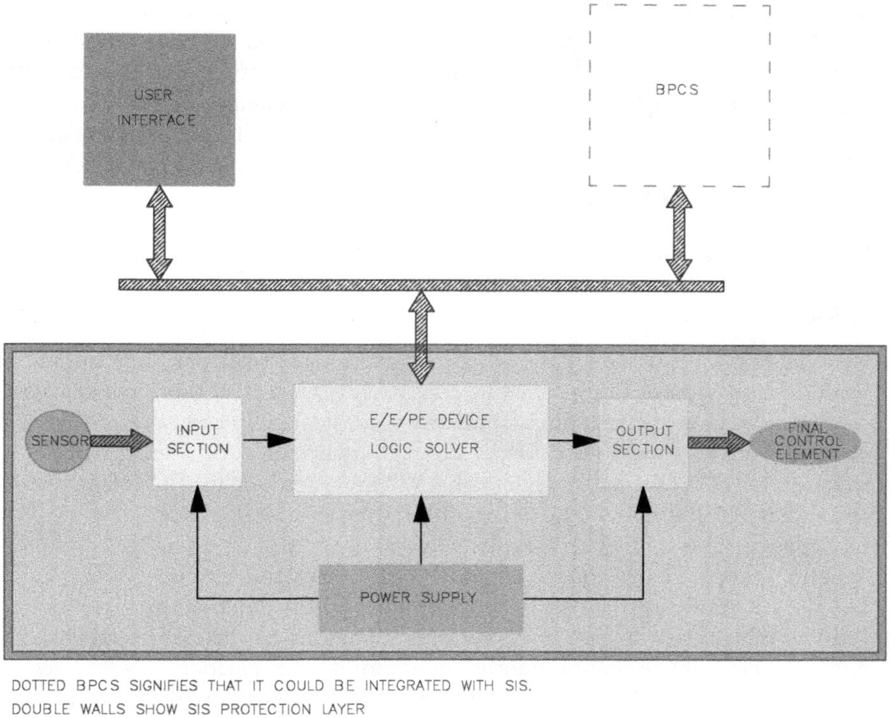

FIGURE I/7.3-1

SIS boundary and layout.

BPCS have been shown through communication bus. Here, it is interesting to note that there could be separate BPCS and SIS, but these two could be integrated as long as they meet the requirements of standards like IEC 61508/61511 or ANSI/ISA 84. In view of the same in the drawing BPCS has been shown in dotted meaning that integration is possible.

Double boxes around the SIS signifies SIS protection layer boundary. It also signifies that even when SIS and BPCS are separate but there could be communication and user interface to a certain extent.

7.4 SIS Related Issues

In this part, a short discussion shall be presented on a few major issues, which affect the SIS selection process. Before starting the discussions, it is better to understand that each plant will have unique problem/hazard and solution for the same shall be specific to the plant. This is more so, when operation criticality, financing power, etc. varies from plant to plant. A solution in one plant may not be totally acceptable for the other plant. Also in some cases there are too many options with large variations amongst them, it is very difficult to judge which one will be better for one's application. It is not purchasing one mobile phone from myriad selection possibilities, so that it can be replaced the other day. In this case of selection of SIS for a plant, not only cost but also question of lives is involved. So, it is definitely better to follow some guidelines so that a few common important issues are well covered and selection could a standard one meeting the requirements of the standards. Discussions put forward below pertains to E/E/PEs involved in SIS.

7.4.1 Hardware Issues

There are a number issues involved in selecting hardware for PEs in SIS. Some of these issues have been listed below:

- Redundancy: Only in those cases where uptime is not critical, or system can wait in those cases redundancy *may not* be considered. Other than those cases unless there is extreme cost constraint redundancy is considered. In case of sensors 1oo2 or 2oo3 redundancies are considered depending on criticality. Similar is the situation for FCE in critical applications. Whenever redundant FCE/FE is given redundant FCE/FE follows the active one so that in case of change over it immediately takes over. This is explained with the help of an example: hydraulic coupling scoop tube of a boiler feed pump (BFP) is used to control the BFP speed, i.e., the feed flow. In that case, standby BFP scoop tube follows the same for running BFP, so that in case of change over it immediately takes over and boiler does not starve. Here there is another question; if auto selection fails how to select the healthy ones. In case of processors, I/Os the selection of redundancy is more critical. Here the issue involved (mainly for logic solvers) shall include but are not limited to the following:
  - How the output from the processors will be handled; is it by hi/lo selection, or by voting? Similar will be the situation for I/Os—by diode auctioneering!
  - In case of processor how output will be inhibited and when to use it?
  - Whether these will be in dual redundant mode, one active other passive but update always so that change over can take place without bump in the system. If so, all modules associated with the redundant one will be in service, then what will happen if any module fails? Also, in case of difference between the two how to decide which one is correct and healthy!
  - Similarly, in hot standby mode how to diagnose health of two processors, better to use third system to monitor the health of other processor.
  - In going in such a way how much redundancy to be involved even with self or other diagnostic, is it possible to have 100% diagnostic feature in all? In that cases system design shall be such that there will be graceful shutdown in most (cannot be all foolproof) of the cases.
  - The issue is not too much different when one goes for 2oo3 because in one failure it is voted out but when two fail situation is similar. However, it is definitely clear from the discussions that triplicate is better options. TMR could be a better solution for critical applications as they offer better fault tolerance. See Fig. I/6.1.2-1.

- Special size considerations: This is very important for offshore applications where it is always aim to reduce the size and weight of the vessel. So in case of too much redundancy, there will appreciable increase in volume and weight so, in these applications necessary consideration must be put for it. Similarly, due consideration is necessary for leaving behind spare module, spare slot with wiring also. Flexibility: most of the systems are of variant levels of SIL. However, it is better to use same level of redundancy for all cases. So, it is guided by the redundancy demand of higher SIL.
- Environmental impact: Some of the C&I items could be placed in the field whereas for some it is not possible. When things are placed in the field wiring involvement and complexity will be less. For field mounted devices, capability to withstand harsh atmosphere need to be evaluated along with EMI effect, etc. Here, enclosure IP class play important role, so that they can be used long time in field atmosphere. In case of hazardous areas in the plant, another thing to be considered *suitable ex* proof enclosure as applicable. On the other hand, if a device is placed indoors such requirement may be less and IS or other (e.g., isolator) can be used. So cost-effectiveness is to be compared (e.g., it is always cheaper to go for an FM-approved field transmitter than to use IS circuit).
- Third-party system integration: This is basically software requirement still said here because in many cases it is necessary to share data between the systems from different vendors. If these were hardwired connections there will be lots of cabling and complex wiring which may not be acceptable to all cases (especially in case of offshore applications). So, it is important while selecting a system to have such facility. System integration in SIS is acceptable as long as it meets the requirements of the standards.
- Miscellaneous issues: There are a few issues like scan time, I/O redundancy, I/O distribution, etc. that also need attention. These will be discussed at length later in the book.

7.4.2 Software Issues

There are certain guidelines in IEC 61511/508 so, these need to be followed:

- Programming language. As IEC 61511 ladder logic, functional block diagram, instruction list, structured text, and sequence function chart are allowed so, vendor must support the same (at least one which need to be followed). Cause and effect is very good for F&G with certain advantages but may not be used for complex systems. So based on application, appropriate language is to be chosen and to see that the vendor support the same.
- Ease of programming and configuration of the system is one of the keys for system selection. Also, PLCs in SIS have some restrictions so these should be spelled out and while selecting the vendor, demonstration to show easiness and flexible configuration (will be advantageous).
- Safety and security: Since the system is meant for SIS, so the system must have enough diagnostic features to detect any probability of failure at early stage. Also the system shall have the capability to recover from such error as possible and/or

change over the system to ensure safe operation of the same. In this regard, the redundancy issue has already been discussed. Many systems offer some sense of redundancy in software also. This means in these systems it is possible to take the system to safe state in case of any program crash, etc. (this is some what like starting of computer in safe mode). There shall be sufficient built-in security to save the system from cyber attack. Also there shall be several levels of access control so that software tampering/change is not possible easily, and it would require authentication.
- Online changes registering; it shall be possible to change the program online and load it to the system on line without degradation and not bypassing any safety shutdown. This is very useful in practical use so that many systems can be added or deleted.
- Force operation. In most of vendor's systems it is possible to carryout force I/Os during commissioning and maintenance. However, this can be done only by an authorized person and this operation must be in safe condition.
- Integration of Fieldbus and other systems: As stated in connection with hardware, there suitable provisions for integrating various field bus systems, OPCs, etc. so that an integrated network is possible [for further reading see Chapter VII of Power Plant Instrumentation and Control Handbook — Elsevier].

7.4.3 *Company Issues*

This is not a purely technical issue. It is related to confidence and relationship issue. As stated earlier, there is a wide selection of vendors with different technologies to offer to achieve the same goal. How to choose one! Many times people forget that even an experienced person working in plant may not find a better solution for a problem in his plant but one vendor can! This happens because vendor normally knows his system better. Reverse is also true at times. So it is necessary both need to work in a team.

- Knowledge sharing: Frankly speaking dealing with SIS requires some good experience and exposure to good engineering practices. Vendor to be chosen in such a way that vendor should know his system as well as application of its product in the plant in the question. Reputed manufacturers have advantage here as they have good data bank for applications, which could help plant owner to get better solutions.
- New technology: It is always better to go for a new technology so that a chance of getting obsolete is less. Here there is a key question! In case a plant comprised of several sections built in "X" technology/system, and is it worth going for a new section with new technology/system! It has one advantage that people get chance for early familiarization. However there are a few problems; for a new system spare inventory need to be kept and/or there could be compatibility problem also. Also, if totally a new vendor is selected, then in addition to first two issues discussed, confidence-building will be another issue. However if the existing system discards some technology and adapt new one and the vendor is

trusted then it may be wise decision to go for it to ensure future support (as for old technology support may be withdrawn). Withdrawal of Window XP support for PC is a classical example.
- It is essential that necessary post-ordering support (technical and administrative) is absolutely essential. So for vendor selection, the same must be taken into consideration. Many times, large vendors gets the support through local representative or local companies which may not be in a position to render enough technical support. Remote operation of big vendor in such cases may finally proven to be costly.

Now after these short discussions on SIS, it will be nice to know what is functional safety and safety integrity level.

8.0 FUNCTIONAL SAFETY AND SAFETY INTEGRITY LEVEL

So far, short discussions have been presented on SISs. Naturally, *one important question comes to mind; what is safety?* According to IEC 61508, safety is the freedom from unacceptable risk (3.1.8). This is overall of physical injury or damage to health of people directly or indirectly as a result of damage of property or to the environment. Then; *what is functional safety?* Functional safety is a part of overall safety, responsible for correct operation of system or equipment in response to input changes including safety management taking into account the possibility of human error and/or system/equipment or hardware failure and environmental changes. Functional safeties usually stand for the equipment or systems that monitor safety and as and when necessary override the actual operation/application to ensure safety operation. Thus, it implements the required safety function by detecting the hazardous conditions and bringing the operation to safe state. Bearing temperature monitor in a large fan for example stops the fan from running when bearing temperature exceeds a given set value. This is an explanation, but not any definition. Now, it is better to look at it from various standards point of view.

8.1 Normally Used Terms

Normally used terms in functional safety are as follows:

8.1.1 Functional Safety

- IEC 61508: Originally functional safety was defined as part of the overall safety relating to the equipment under control (EUC, see Clause 8.1.2), and the EUC control system which depends on the correct functioning of the E/E/PE safely-related systems, "*other technology safety-related systems*" and "*external risk reduction facilities*." Later came the generalized definition of *functional safety*, which is part of the overall safety that depends on a system or equipment operating correctly in response to its inputs. Functional safety is achieved when every specified safety function is carried out and the level of performance required of each safety function is met.

- IEC 61511 (3.2.25): functional safety is a part of the overall safety relating to the process and the BPCS which depends on correct functioning of SIS and other protection layers. *This definition deviates from IEC 61508 to reflect the process sector terminology.*
- ANSI/ISA 84: Functional safety: The ability of a SIS or other means of risk reduction to carry out the actions necessary to achieve or to maintain a safe state for the process and its associated equipment.
- ISO 26262: Absence of unacceptable risk due to hazards caused by malfunctional behavior of electrical and/or electronic system.

8.1.2 *EUC*

IEC 61508 (3.2.3): The EUC is equipment, machinery, apparatus, or plant used for manufacturing, process, transportation, medical, or other activities. (EUC is different from control of EUC). To note that in IEC 61511 the term process has been used in place of EUC to reflect the difference in process sector. ISA 84 also uses the same term.

8.1.3 *E/E/PEs*

- IEC 61508 (3.2.6): E/E/PE is an abbreviation of electrical/electronic/ programmable electronic. Standard defines this as based on electrical, and/or electronic, and/or programmable electronic technology.
- ANSI/ISA 84 (3.1.56): Programmable electronics (PE): Electronic component or device forming part of a PE and based on computer technology. The term encompasses both hardware and software and input and output units. Normally these include smart sensors, PE logic solver including PLC, distributed control system (DCS), loop controller, and smart final element.

8.1.4 *Safety Integrity Level (SIL)*

- IEC 61508 SIL: An SIL is one of four levels, each corresponding to a range of target likelihood of failures of a safety function. Note that an SIL is a property of a ***safety function*** rather than of a system or any part of a system.
- ANSI/ISA 84 (3.1.77): Discrete level (one out of four) for specifying the safety integrity requirements of the SIFs to be allocated to the SILs. SIL 4 has the highest level of safety integrity; SIL 1 has the lowest. Actually, ANSI recognizes three SILs but kept four level to maintain parity with IEC.

8.1.5 *SIF*

IEC 61511: A safety instrumented function (SIF) is a safety function with a specified SIL which is necessary to achieve functional safety. IEC 61511 also includes a note to explain that this normally refers to protection systems and that if it is applied to control systems then "further detailed analysis may be required to demonstrate that the system is capable of achieving the safety requirements."

8.2 Functional Safety Discussions

There is no plant which is totally free from unacceptable risks. A few aspects have been presented below:

8.2.1 Functional Safety Concept

Basically functional safety stands on the following concept:

- All processes or manufacturing systems have inherent hazard.
- All processes or manufacturing systems have inherent quantifiable failure rate which cannot be brought to zero value.
- All processes or manufacturing systems have tolerable failure rate without causing any harm for the system. Also this failure rate is specific to the system in the question.
- For all processes or manufacturing systems, these failure rates can be categorized in terms of SIL.

8.2.2 Failure Category

Failure category in functionally safe system:

- Systematic failure: Following accounts for systematic failure and it is possible to reduce the same:
  - On account of shortcoming in system design, implementation, or manufacturing defect.
  - Absence of following any statute, standard, or even well-established engineering practices.
  - In this system failure rate can be reduced after thorough analysis and remedial measures.
- Random failure: These are basically uncontrolled:
  - Unnoticed failure, sometimes inherent with the process (e.g., secondary combustion in a furnace).
  - It cannot be reduced in systematic way, only proper attention to be given for early detection to reduce amount of loss (e.g., unprecedented grid collapse).

8.2.3 Functional Safety and SIS

Functional safety relies on active part, not on passive part. For example, a fire resistant door prevents a hazard, but is not safe instrumentation for functional safety. On the contrary, a flame scanner/switch in a utility boiler is an active system as it protects closes the fuel flow and trip master fuel relay in the event of flame failure (through logic). From the previous discussions, it is clear that there is handshaking relationship of SIS with functional safety. These cause effect relationship of functional safety and SIS can be:

- Safety function may come as a result of hazard analysis and it is to be implemented in SIS.
- Functional safety can be the result of an action taken by SIS as discussed previously for an active participant.

- Safety function may be implemented by SIS and/or by passive participants such as "*other technology safety-related systems*" and "*external risk reduction facilities*" (see Clause 8.1.1).
- Safety function allotted to SIS becomes its performance parameter in terms of timely action, effectiveness, and sustainability. This is clear from the example; an emergency pushbutton (EMPB) near say a motor is kept to stop the motor immediately (operated in case of emergency/hazard say, due to fire). This EMPB is a part of SIS and proper timely functioning and bring the motor to stop position is the performance requirement of SIS.

8.3 Safety and Safety Integrity Level

From the discussions, it is clear that safety function for any system is not responsible for the operation of the system, but if safety function fails, then said system will be vulnerable to risks. Therefore safety system is responsible to reduce the risks. How the performance of the safety function will be judged? SIL is a measure of safety system performance, or PFD for a SIF or SIS. There are four discrete integrity levels associated with SIL. The higher the SIL, the lower the PFD for the safety system, hence greater risk reduction. However, such risk reduction is at a higher cost and complexity. SIL is discrete level required for specifying the safety integrity requirement of the safety function to be allocated to *safety-related* functions of E/E/PE system. Based on the SIL of every component and architecture, SIL level of SIF is determined and based on SIL of SIF SIL for SIS is determined. So, this should be born in mind, while specifying SIL for a system. Table I/8.3-1 shows the SIL and associated PFD. In this table, both on-demand mode and continuous mode have been combined into one table, but in standard, they are defined separately. For comparing purposes, it is worth noting that risk reduction factor is inverse of PFD, that is, if PFD is 10^{-4} then risk reduction will be 10,000. What does this mean really? This means that a system will be 10,000 times more vulnerable to risk had there been no SIS to attain SIL level. So, SIL are used when implementing SIF in SIS to reduce existing risk level. The end user to ensure that implementation meets the requirement of the standards and implementation procedure to obtain necessary certification. In connection with

Table I/8.3-1 SIL Level for On-demand and Continuous Modes of Operation

Safety Integrity Level (SIL)	Probability of Dangerous Failure (per HR)—Continuous Mode	Probability of Failure on Demand—On-demand Mode
SIL 4	$\geq 10^{-9}$ to $<10^{-8}$ (RR 10^9)	$\geq 10^{-5}$ to $<10^{-4}$
SIL 3	$\geq 10^{-8}$ to $<10^{-7}$	$\geq 10^{-4}$ to $<10^{-3}$
SIL 2	$\geq 10^{-7}$ to $<10^{-6}$	$\geq 10^{-3}$ to $<10^{-2}$
SIL 1	$\geq 10^{-6}$ to $<10^{-5}$	$\geq 10^{-2}$ to $<10^{-1}$ (RR 10)

RR, *risk reduction factor (highest and lowest RR shown in bracket)*.

Common cause failures can reduce the SIF performance, and are addressed differently depending on their nature, such as systematic or random failure (see Clause 2.4.3 in Chapter IV). In standards, it is recommended that the devices, systems, or protection layers be assessed for independence, and potential for common cause failure. Common cause failure can cause multiple devices, systems, or layers to fail simultaneously. Another term is "common mode" failure (a subset of common cause failure), which describes the simultaneous failure of two devices in the same mode, for example, two redundant transmitters fail simultaneously and are disabled due to loss of signal (common mode failure) [10,11].

FIGURE I/8.3-1

Common cause failure.

discussions on standards in Chapter VI, on-demand mode and continuous mode have been detailed out.

So, with this basic knowledge on hazard, risk, and safety, it will be possible to explore next stages of detailing. Various types of failures have been discussed later in the book. Out of them common cause failure as detailed out in Fig. I/8.3-1 is interesting to instrumentation applications.

In order to get an idea, the reader may have a look on Fig. V/1.0-1.

In Chapter I, attempts have been made to give an overall view of the subject so that the reader can have some concept before studying the issues in details.

LIST OF ABBREVIATIONS

ALARP	As low as reasonably practicable
BDD	Binary decision diagram
BMS	Burner management system
BPCS	Basic process control systems
CBA	Cost benefit analysis
CCPS	Center for chemical process safety
CER	Commission for energy regulation
DEHA	Diethylhyrdoxylamine
E/E/PE	Electrical/electronics and programmable electronics
EC&I	Electrical, control & instrumentation
ETA	Event tree analysis
F©E	Final (control) element
FMEA	Failure mode and effect analysis
FTA	Fault tree Analysis
HAZOP	HAZard and OPerability study
HSE	Health safety environment
HV	High voltage
HW	Hardware

Continued

IGCC	Integrated gasification and combined cycle plant
IT	Information technology
KRI	Key risk indicator
LC	Life cycle
M/C	Machine
NFPA	National fire protection act
NPSH	Net positive suction head
OGP	Oil and gas producers
OREDA	Offshore (onshore) reliability data
PHA	Plant hazard analysis/preliminary hazard analysis (PrHA)
PSV	Pressure safety valve
RA	Risk assessment
RR	Risk reduction (factor)
SCADA	Supervisory control and data
SIF	Safety instrumentation function
SIL	Safety integrity level
SIS	Safety instrumented system
SW	Software
TMR	Triple modular redundant
UV	Ultra violet

References

[1] M. Rousand, Preliminary Hazard Analysis, NTNU/Willey, October 2005.
[2] B. Fischoff, G. Morgan, The science and practice of risk ranking, Horizon 10 (3) (2009). www.policyresearch.gc.ca.
[3] D. Baccarini, R. Archer, The risk ranking of project: a methodology, International Journal for Project Management (2001). www.elsevier.com/locate/ijproman.
[4] J. Tixier, G. Dusserre, O. Salvi, D. Gaston, Review of Sixty Two Risk Analysis Methodologies of Industrial Plants, 2002. France.
[5] ALARP Guidance a Part of Petroleum Safety Frame Work; Commission for Energy Regulation (Guidance Document CER/13/282; CER; 29 November 2013).
[6] S. Hughes, Cost Effective Application of ALARP Principle, Safety Engineering Group; ERA Technology.
[7] S.N. De, Higher Mathematics Vol. I & II (An Introduction to Calculus), Chhaya Prakshani, November 2005.
[8] Fault Tree Handbook with Aerospace Applications, NASA.
[9] Managing ageing plant through risk based inspection; http://www.chilworth.co.uk/media/219415/Hazardex-SEpt13-27-315.pdf.
[10] S. Basu, A.K. Debnath, Power Plant Instrumentation and Control Handbook, Elsevier, October 2014.
[11] A. Summer, M. Gentile, Random, Systematic and Common Cause Failure: How Do You Manage Them, SIS-Tech Solutions, LP, 2006. www.SIS-Tech.com.

Further Reading

A Structural Approach to Enterprise Risk Management (ERM) and the Requirement of ISO 31000, The Public Risk Management Association; IRM, 2010.

ALARP Guidance a Part of Petroleum Safety Frame Work; Commission for Energy Regulation (Guidance Document CER/13/282; CER; 29 November 2013).

D. Aptikar, Risk Matrix Vol. I Technical Review, 2005.

A.L. Burgenson, Risk Assessment & Management, Louza Walkersville Inc.

F. Candreva, Plant life management: ageing issue, in: Process and Asset Risk Management; DNV Conference, March 2011.

Center for Chemical Safety, American Institute of Chemical Engineers (AIChem).

Design of Microcontrollers for Safety; K. Greb and A. Seely; Texas Instruments.

EC 61508 Overview Report, Exida, January 2, 2006. Version 2.0.

H.K. Florig, M.G. Morgan, K.M. Morgan, K.E. Jenni, B. Fiscghoff, P. Fischbeck, et al., A deliberative method for risk ranking (I); overview and test bed development, Risk Analysis 21 (5) (2001).

P.K. Giri, J. Banerjee, Introduction to Statistics, Academic Publishers India, 2008.

P. Gruhn, Things to Consider When Selecting a Safety System, ICS Triplex, Houston, Texas, 2001.

Guideline Risk Register, Victoria University of Wellington.

Health and Safety Authority, Ireland; www.has.ie/eng/topics/Hazard.

HSE.GOV.UK; www.hse.gov.uk.

http://mediumrisk.com/blog/what is risk matrix.

K.W. Knight, in: 7th International Conference in Brazil, May 2012.

T. Marlin, Process Operability Class Materials, 2013.

E.M. Marszal, C.P. Well, Implementing Protective Functions in BPCS an Combined Systems, Kenexis Consulting Corporation, Colombus, OH, 2006.

M.G. Morgan, H.K. Florig, M.L. Dekay, P. Fischbeck, Categorization of risk for risk ranking, Risk Analysis 20 (1) (2000).

F. Pfenning, Principles of Imperative Computation, October 28, 2010, pp. 15—122. Lecture 19.

Plant Ageing, Management of Equipment Containing Hazardous Fluids or Pressure (rr509), TWI Ltd, ABB Engineering Services, SCS (INTL) Ltd and Allianz Cornhill Engineering, 2006. For health and safety executive.

Process Safe Book I, Rockwell Automation (AB).

Research Report 7930, November 2012. Ismail; Aberdeen Research Report.

Risk matrix as a tool of risk management in chemical industry; BASF Chemical Company Presentation.

Rocky Relationship between Safety and Security, ABB, 2013.

Safety and Functional Safetyda General Guide, ABB brochure; 1SFC001008B0201.

Tough Leadership in ERM — Risk Assessment in Practice; COSCO Enterprise Risk Management; Integrated Frame Work, 2004.

Further Reading

CHAPTER II

Evaluation of Hazard and Risk Analysis

In the previous chapter, it was established that in industry, plant hazards can cause harm to property (plant—machinery, asset), people, or the environment. So, it is important to develop some means of analyzing these and come up with a solution. Unfortunately, it is not as straightforward as it sounds. There are plenty of plant hazard analysis (PHA) techniques and each of them has certain strengths and weaknesses. Also each specific plant and associated hazard has specific requirements to be matched so that hazard analysis will be effective. In this chapter, various hazards (in generic terms) will be examined to judge their importance, conditions, quality, etc. so that out of so many techniques available for PHA it is possible to select which one is better (not the best because that needs to be done by experts specifically for the concerned plant) suited for the *type* of plant. So, discussion will be more toward evaluation of PHA techniques. Some PHA is more suited for process safety management (PSM) and is sometimes more applicable for internal fault effects [e.g., hazard and operability study (HAZOP)]. In contrast, hazard identification (HAZID) is applicable for other plants, especially for the identification of external effects and major incidents. HAZID is also covered in this chapter. As a continuation of the same discussion, it will be better to look at various aspects of risk analysis with preliminary ideas already developed in the previous chapter. In risk analysis risk assessment, control measures for safety management systems (SMSs) will be discussed to complete the topic.

1.0 PLANT HAZARD ANALYSIS PRELIMINARIES

It is better to start the discussions with the preliminaries of PHA.

1.1 Fundamentals of Plant Hazard Analysis

1.1.1 What is Hazard Analysis

Hazard analysis can be conceived as a preventive approach to identify any potential hazards, which should be realistic and relevant to the facility. Also hazards must be prevented, eliminated, or reduced. This means that in a modern plant it is necessary to detect the hazard early and there should be suitable safety measures to mitigate/control it. Now let us see how various institutions look at PHA.

Plant Hazard Analysis and Safety Instrumentation Systems.

- As per the Environmental Protection Agency: PHA is a thorough orderly and systematic approach for identifying, evaluating, and controlling the hazard of the process (involving highly hazardous chemicals). The part in parentheses signifies that such a definition is generally applicable to other plants as well.
- According to the Center for Chemical Process Safety (CCPS): PHA is an organized effort to identify and evaluate hazards associated with chemical process and operation to enable their control. This review normally involves the use of qualitative techniques to identify and assess the significance of hazards. Conclusions and recommendations are also developed. Occasionally quantitative methods are used to help prioritize risk reduction.
- From the foregoing two definitions one can conclude that PHA is a set of *well-organized and systematic* (meaning following a certain rule) rules to *identify* and *characterize* each of the *possible hazards* in the process, then to *assess* them and *prioritize* them. These must be *communicated* suitably with basic recommendations for *mitigation* and *control*. This means that in PHA, *corrective measures* are also *identified* to improve safety and this is communicated as recommendations. So, PHA is the foundation for process safety studies, and a risk management program to save the assets and control adverse effects to people and the environment.

1.1.2 *Aim of PHA*

In the world, there is no place or system that is totally free from hazard. As long as there are hazards there will be a risk of damage. People will always try to eliminate risk as best as possible. Unless one is aware of various hazards and hazardous situations, how is it possible to get rid of the same. So the basic aims of PHA are:

- Identify the potential hazard as one of the activities
- Identify equipment, instrument failure, process upset, and human error, which could manifest the potential hazard and have a detrimental effect such as loss of asset and/or negative environmental impact
- Analyze the magnitude and likelihood of hazards
- Identify and evaluate existing protection systems against risk consequences
- Document all potential hazards with recommended mitigation strategies, which when compared with the existing protection, may require additional protection
- Statutory requirements, for example, Clause 4.2 of NFPA 654 (2013) is a classic example:
  - "… Design of fire and explosion safety provision shall be based on process analysis of the facility, the process and associated fire and explosion hazards." It also says "PHA shall be documented and to be used for life of the process…"

1.1.3 *Basic Scope of PHA*

Normally, the scope of PHA shall include but not be limited to the following:

- Equipment in the process and their failure including electrical, control and instrumentation (EC&I) system [hardware (HW) and software (SW)]
- Hazards inherent to the process as well as hazards caused by reactions

- Identification of previous incidents that had likely potential catastrophic incidents
- Engineering and administrative controls pertinent to hazards and their interrelationship
- Consequences of failure of engineering and administrative controls, etc.
- Qualitative evaluation of failure of safety controls on assets, humans, and the environment
- Facility sitting (assessing explosion toxicity)
- Human failure factors
- Environmental impact on the process and hazard caused by the same
- Steps required to correct or avoid deviation

1.1.4 Major PHA Steps

The following are the major PHA steps normally followed. In this connection Fig. II/1.1.4-1 may be referred to:

- Study and preparation of PHA (some cases preceded by preliminary hazard analysis)
- Formation of team with team leader (experienced in PHA and one plant expert)
- Scope and boundary definition in accordance with system demand description. This is very important in the sense that without this the entire hazard analysis process could be a huge one and effective control will be a serious problem. Description shall include but not be limited to:
  - Specifications/detailed design requirements
  - Operational details
  - General compliance to standards and other statutory requirements
  - Applicable standard human convention factors (e.g., lamp color, etc.)
  - Accident experience and failure reports and similar plant data
- Hazard identification: Irrespective of the process chosen, this is necessary because by itemizing the hazards, analysts can imagine and develop a picture of the complexity of the plant and breadth of safety analysis required. Hazard identification gathers information about:
  - Characteristics of the hazard including inherent nature, multiplicity, etc.
  - Form and some quantification of hazard
  - Where and when in the project/facility it is present
  - How this hazard could result in an undesirable event or a chain reaction
- Collect and collate up to date *all relevant* information to support PHA
- Selection of appropriate PHA techniques out of those shown in Fig. II/1.2-1 (refer to Clause 2.0 of this chapter) is well understood
- Agreement for schedule of work also in terms of time
- Conduct regular meeting and documentation
- Depending on applicability, conduct functional analysis
- Develop preliminary hazard list, identify contributory hazard, initiator, etc.
  - Equipment failure because of hardware, software, and human factors
  - Operation and maintenance (O&M) condition

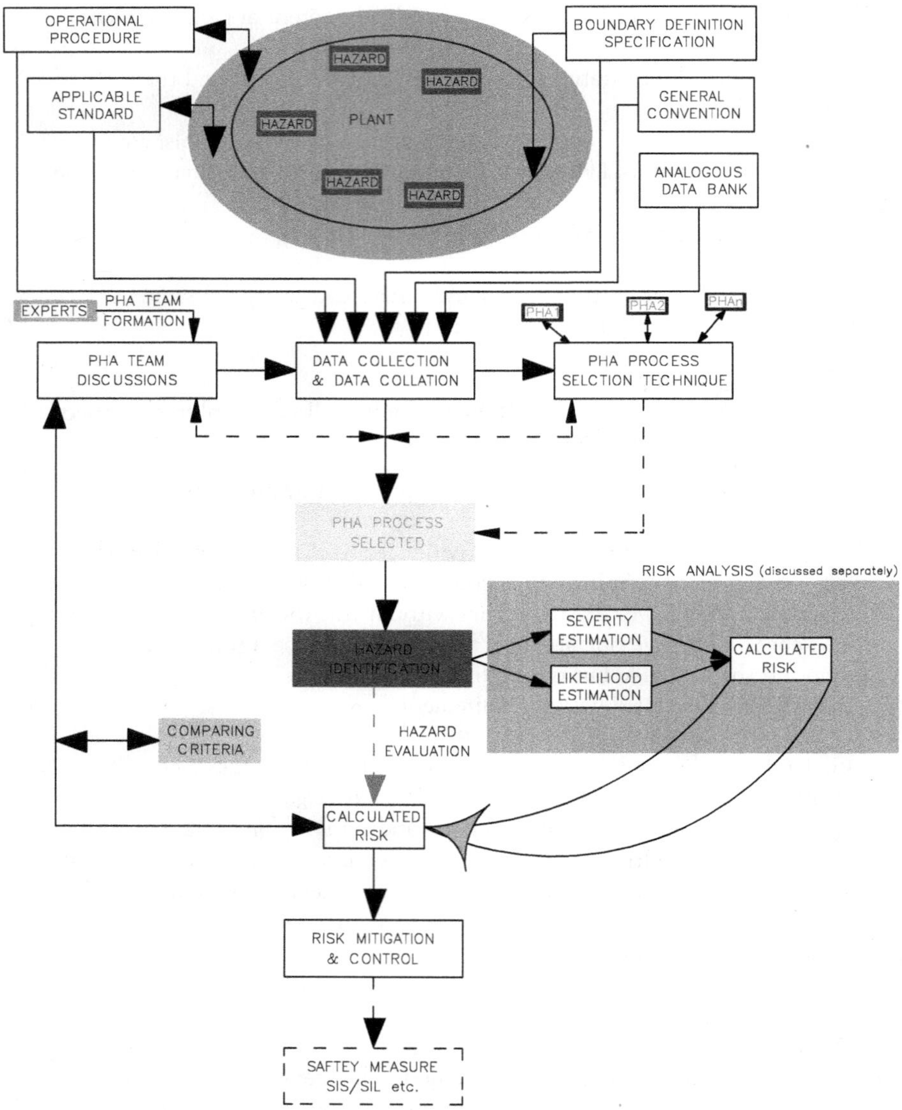

FIGURE II/1.1.4-1

Plant hazard assessment stages.

- Man–machine interface (interaction procedure)
- All other systems and procedures
- Develop safety system baseline, recommendations, and requirements for mitigating hazards

- Detailed documents of PHA result (risk assessment has been set aside as this will be treated separately in this chapter)
- Implementation [including management of change (MOC)—if called for] and follow-up

1.1.5 Management of Change (MOC)

This is another issue quite relevant to PHA especially for process plants. PHA is normally done periodically to validate that the actions suggested are still valid. In some cases periodic PHA is a statutory (e.g., Clause 4.2.3 of NFPA 654[2013]) requirement. Nevertheless, if after PHA it is revealed that some systems need to be changed to mitigate a potential hazard, as the name suggests, one needs to implement the MOC. When changes occur in any operation of a system (especially handling hazardous materials), it is essential that changes to the process must be well managed to understand and meet the challenges of these hazards and to mitigate them. This is extremely important from a safety point of view. Previously, many accidents took place on account of the nonfunctioning of the MOC. Usually, major chemical process companies have their own set procedure toward this. MOC may call for replacement in kind or change in process technology itself. Also MOC is applicable to and dependent on all PHAs.

1.1.6 Typical Output Expected of PHA

- Identification of hazards that can result in adverse consequences, improper usage, leading to accidents and even shutdown
- A quantitative assessment of likelihood and consequences discussed
- Risk ranking for major risks involved
- Development of suitable applicable criteria and specifications
- An evaluation of design hazards and the suggested corrective actions and safeguards including those in subsystems and subgroups
- A basis for program -oriented precautions, personnel protection, safety devices, emergency equipment, procedures, training, and safety requirements for facilities, equipment, and the environment
- Safety factors for tradeoff considerations
- Evidence of compliance with program safety regulations
- Develop future plans, for example, additional analyses, tests, and training

However, such results may be possible if all required information is made available at the time of PHA and there is a good team with suitable leadership to produce a result that is very lucid and easily implementable. In such cases, resource mobilization is effective and should be minimal.

1.2 Various Plant Hazard Analysis Methods

To select a PHA technique best suited for one plant it is necessary to evaluate various PHA techniques. To do this it is essential to gather basic knowledge about each of these PHA techniques. A number of such PHA techniques fall under three basic

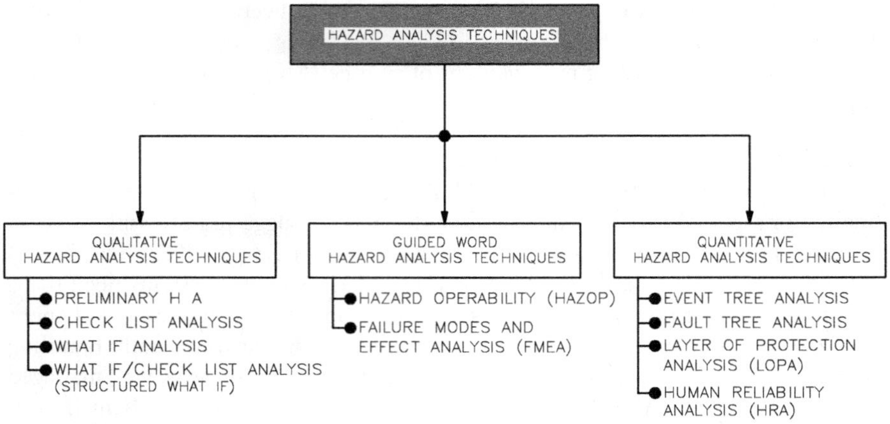

FIGURE II/1.2-1

Various plant hazard analysis (PHA) techniques.

categories, as shown in Fig. II/1.2-1. Also Table II/2.1.2, where a few quick PHA techniques and important terms associated with chemical process plants [e.g., Dow Fire and Explosion Index (FEI), etc.] have been highlighted shall be referred to.

As shown in the diagram there are three classes into which PHA techniques could be classified, namely, (1) qualitative, (2) guided words, and (3) quantitative.

1.2.1 Qualitative

Hazards may be identified through a qualitative process, either formal (part of safety assessment) or based on discussions, interviews, and brainstorming. This is normally done after formation of a multidisciplinary team. As a part of safety assessment, the main aims are: (1) to identify a potential incident/accident scenario, and (2) evaluate the scenario through detailed discussions and brainstorming to arrive at risk assessment and recommend risk mitigating procedures. In some cases when conclusive judgment cannot be made, then the system may be repeated for the final result. Some of the various methods under this category are:

- Preliminary hazard analysis: As already discussed in Chapter I (also Chapter III), this can be utilized during conceptual design, front end engineering design (FEED), or R&D stage. As stated earlier, one of the major purposes of preliminary hazard analysis is to identify the extent or boundary limits of complex hazards and analyze the risks and hazards associated with all the processes involved (see Clause 4.0 of Chapter I).
- Checklist: This is a list of attributes or check points for hazards. This list is generally prepared from risk assessments pertinent to similar projects done in the past. However, it is to be validated for applicability before it is used. Naturally, such lists need to be prepared prior to the take-up of PHA. This list is used for

acceptability by comparing the system with established norms and previous systems—hence it is a tool for assessment of hazard. Since this is a comparison, it could be used by nonexperts. Also the list may offer a wider exposure of hazards encountered in previous cases so that obvious things are not missed. However, it does not allow much imagination and as a consequence hazards that in the past were not considered or did not happen would be missed. For new projects with new technologies the checklist may not be helpful.

- "What if" (including structured "what if"): This is a brainstorming method in which a group of experienced people belonging to different disciplines participate to ask questions and/or raise concerns against possible undesired events. There is another category, which is sometimes referred to as the "what if checklist" because it is similar to "what if" but is a defined structured list prepared by the team leader. In fact it is in some respects similar to HAZOP. Here the team leader puts forward a suitable list of prompts such as "What if…" or "If some ever…," etc. It is a good alternative to HAZOP because it takes less time but can result in a good record of hazard identification. However, it is totally dependent on the experience of the team members, especially the team leader. List adaptability in any project needs to be thoroughly checked if applicable.

1.2.2 *Guided Word*

The principles of a guided word PHA are explained in Fig. II/1.2.2-1. This is a systematic method where deviation causes are initiated by guided words.

There are intended activities in a process or plant, and a list of deviations can be started immediately after the intended activity listing is complete, with the help of guided words. Next, causes and consequences are listed. Then, for each set of causes and consequences specific safety interventions are inserted and listed. Such initiating guided words could look for an event or an incorrect performance of an event that has been skipped; "none," "loss of," etc. could be the guided words. These are carried out

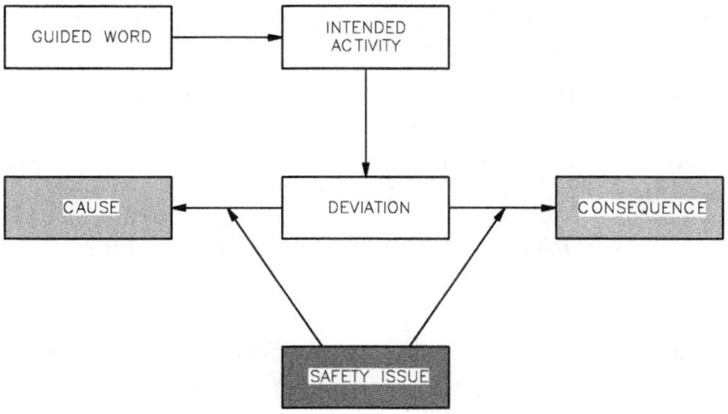

FIGURE II/1.2.2-1

Guided word plant hazard analysis (PHA) principles.

by the team and team leader who have experience in the study as well as experts from the operational group. HAZOP and FMEA (failure mode and effect analysis) are examples of PHA in this class. These are performed for startup, shutdown, and batch processes, as well as continuous processes (e.g., HAZOP in a power plant).

- HAZOP: This is a systematic guided word PHA. This is applicable to all modes of operation, process flow diagram, piping (process) and instrumentation diagram (P&ID), etc. This is quite a detailed and rigorous technique, to be developed in a step-by-step fashion. "More/less, reverse flow, or too little/too high/more/less" pressure, and other parameters are generally the guided words. As stated earlier, deviation from the intended activity is assessed for suggesting safety actions. These are well recorded during the process in established recording formats. In this method the entire system is broken down into various subsystems and units. So, the success of the study is very much dependent on the amount of information available and the skill of the team, especially the team leader. Also a good amount of preparation and time is needed for the study.
- FMEA: This is mainly used for a set of equipment or for study at a functional level to assess failure at the component level and the performance of the desired function, that is, assessment of consequences for major component failure. Since it starts from the bottom, it is often referred to as a bottom-up technique. Here also detailed system description, specifications, etc. are necessary to carry out the study. Like HAZOP this is also systematic, rigorous, and has a wide range of applications. In addition, there are established formats for record keeping. This is mainly applicable to electrical, mechanical, electronic, or computer systems (manufacturing). This is an inductive method and can handle with singlepoint failure. It is also time-consuming and expensive and depends a lot on the skill of the team leader.

1.2.3 *Quantitative*

This class is one step ahead of the other approaches discussed. Here the probability of an incident occurring is well calculated in a probabilistic manner. This is done to eliminate and/or reduce the risk to the greatest extent possible. For this, experience and a record of previous data (for similar plants) are used and hazards are calculated in terms of *number* or *rate* utilizing engineering estimates. Good use of reliability engineering is common. Since here the failure number or rates are calculated statistically it would be agreeable to have a basic knowledge of statistics and probability, as discussed in Chapter I.

- Event tree analysis (ETA): An event tree is based on the binary logic of the happening or nonhappening of an event or failure or no failure of a component. It starts with an initiating event (e.g., failure of a component or release of gas, etc.) and generates multiple scenarios. In the event tree, from each branching point, there will be only two paths (basically Y/N). Thus it starts from a single event and then consequences follow a series of possible paths, each of which is assigned a probability, and various outcome probabilities can be calculated to arrive at the final consequence probability caused by the event. A good knowledge of the system is expected of the person carrying out the analysis. There can be extensive documentation for the process. This inductive method of PHA generates specific output.

- Fault tree analysis (FTA): While the event tree is divergent in nature, the fault tree is convergent in nature, that is, it is deductive. When complementary ETA and FTA are used they are referred to as "***bow and tie***." So, here the combination of failures can lead to a specific event of interest. Here also the probability of happening will be assigned to each branch. Then, based on their combination in union/intersection (OR/AND), final event probability will be calculated. The binary decision diagram, discussed in Chapter I, could be a helpful tool in probability calculation. It is possible to address multiple failure issues in FTA. FTA is drawn from top to bottom, and the starting point is the undesired event at the top, which then goes down to have a logical tree for an immediate contributory fault, which may be caused by another fault, then another one, and so on. Theoretically, the tree could be endless but in reality it stops after a certain point. It is possible to have an extensive report of the FTA. The study can continue by having extensive training. It is moderately time-consuming.
- Layer of protection analysis (LOPA): LOPA is a systematic and structured way of quantification of risk reduction and safety integrity level (SIL) determination. Usually, it starts its work on the data developed in HAZOP analysis. For each documented undesired event with an initiating cause, it provides an independent protection layer (IPL), which will mitigate or prevent the hazard. Then, the total amount of risk can be determined. If safety instrumented function is necessary, LOPA methodology can be used to determine SIL also. From ISA 84 transaction it is found that LOPA is a simplified risk assignment tool used to evaluate the effectiveness of IPLs that are designed to reduce the likelihood or severity of an undesirable event. Quantitative PHA LOPA deals with single cause consequence pairs. Detailed documentation is possible and can be applied for continuous process.
- Human reliability analysis (HRA): With technological development and incorporation of redundancy it is possible to reduce equipment failure to a great extent. However, human behavior is not that predictable. So, there are chances that failure could occur because of human factors. This is a method by which probability is measured. It is also used in PHA. This could be quantitative as well as qualitative. Although the exact value is not certain it is estimated that error committed by a human could be as high as 60–80% (even 90%). Human performance is affected by several factors, referred to as the performance shaping factor (PSF). By this method, PSF is identified and tries to improve it. In addition to PSF, normal human error probability (HEP) is also calculated on the basis of human activity. There are so many factors that affect this analysis: accuracy, reproducibility, bias, etc. There have been several methods and each needs to be understood before application. An HRA event tree is often used. It may be informative to refer to Table V/1.0-1 (Chapter V).

2.0 EVALUATION OF PLANT HAZARD SELECTION TECHNIQUES

According to (US) Occupational Safety and Safety Administration (OSHA): "The process hazard analysis is a thorough, orderly, systematic approach for identifying, evaluating, and controlling the hazards of processes involving highly hazardous chemicals. The employer must perform an initial process hazard analysis (hazard evaluation) on all processes covered by this standard. The process hazard analysis *methodology selected must be appropriate* to the complexity of the process and must identify, evaluate, and control the hazards involved in the process." There are more or less 12 types of PHA methods. Apparently, almost all of them can serve the purpose of PHA for a particular plant. Now, the question is whether there is a better method than what has been selected for the plant. If so, will it not be a waste of money if a non-appropriate one is chosen? The answer to this question is yes! If a company is thinking of setting up a semiconductor manufacturing unit with new technology and at an early design stage it opts for the HAZOP technique, then it may not be appropriate because for semiconductor manufacturing systems FMEA, "what if," etc. may suit it better. In addition, on account of new technology at an early design stage there may not be sufficient data to carry out HAZOP. Therefore neither result will be satisfactory and it will be a waste of time and money. As a result, it is needless to argue the importance and effectiveness of the selection of an appropriate PHA for a specific application. However, the selection of an appropriate PHA method is not an easy job. There are many influencing factors that play a role in decision making. Also the experience, expertise, and leadership qualities of the person leading the PHA activity are also crucial in the selection process. There are different plant types, there are different goals for PHA, and there will be different hazard types and different levels of severity, etc. Similarly, the action called for will differ with different plant requirements. Air pollution from a thermal power plant will have a long-term effect on the environment and action to counter this should be planned for the long term, for example, installation of flue gas desulfurization, etc. On the other hand, containment failure from a chemical factory could cause havoc, for example, the gas disaster in Bhopal, India. Similarly, some systems definitely need immediate alternative action to ensure safe operation to avoid a large-scale accident, for example, aircraft control. Now, in this clause attention will be centered on discussions of selection of the PHA method and application guidelines. Here, it must be noted that in this book an endeavor has been made to cover the overall PHA for chemical process (ref 1910.119 Process Safety Management of Highly Hazardous Chemicals). Guidance on this is available in international standard ISO 31010:2009. A brief discussion on this is presented in Clause 1.1 of Chapter VI.

2.1 Preselection Consideration

A few points discussed below will be helpful in the selection procedure.

2.1.1 Basic Plant Facility and Hazard Considerations

- Plant facilities (in connection with process safety according to OSHA, facility means the buildings, containers, or equipment that contain a process):
  - Site facilities, such as, number and type of units, associated hazards, for example, for an integrated steel plant there could be a number of units such as

material preparation, sinter plant, rolling mill, captive generation units, or several units of a power plant. Note that in all cases hazard scenarios are different. In the latter case (a plant with several units of power plants), generally similar hazards are expected, whereas in the former case, different plants have different characteristics. It is worth noting that we have tried to cover plant hazards in place of process hazard alone.
  - Building or structure, types of systems and system design, etc. Here also hazards vary with the size of the building or structure, etc., for example, nature of hazard for a tall structure (say Chimney) is different from that of, say, a turbine building.
  - Process design: Chemical plants have inherent additional problems caused by reaction, etc.
  - Requirements of local or international standards, laws, etc.
  - Miscellaneous other issues
- Hazards: The following headings are generally concerned with hazards for PHA:
  - Material (OSHA guidelines given later may be referred to especially for chemical plants where 1910.119 is applicable)*
  - Energy (many plant applications with possibilities of explosion and for several manufacturing units, e.g., semiconductor manufacturing unit, say silicon gas)
  - Process
  - Synthetic external hazard (accident caused by vehicle movement)
  - Natural hazards (e.g., earthquake, tsunami, or climatic disaster)
  - *According to OSHA: "Information on the hazards of the highly hazardous chemicals in the process shall consist of at least the following:
    - Toxicity,
    - Permissible exposure limits,
    - Physical data,
    - Reactivity data,
    - Corrosivity data, and
    - Thermal and chemical stability data, and hazardous effects of inadvertent mixing of different materials."
- Checklist of general hazard components: In hazard analysis all types normally have the following checklist, given here to obtain a clearer vision of hazard. These can be classified and put against one of the categories mentioned earlier, for example, equipment failure can be classified under process, or electrical may be classified under energy.
  - Toxic chemicals
  - Exothermic/endothermic chemical reaction
  - Chemical reaction runaway
  - Very high temperature
  - Fire/explosion
  - Very high pressure (overpressure)
  - Mechanical (machine structural stability)
  - Material handling transfer
  - Drainage/spillover
  - Leakage
  - Building structure/civil

- Machine moving parts
- Vehicular movement
- Quantity of flammable materials (special hazard)
- Corrosive and erosive material handling (special hazard)
- Failure of equipment
- Failure of EC&I
- Electrical system (e.g., high-voltage isolation)
- Radiation
- Hazard caused by noise/vibration
- Environmental pollution
- Biological

- Suggestive control actions: Although this is not directly concerned with hazard analysis process selection, it may be helpful for the selection process because it is concerned with the outcome of PHA. These actions are necessary for selection of control measures to be discussed in connection with risk analysis.
  - Preventive action vis-à-vis mitigation/control action
  - Engineering prevailing over administrative action
  - Engineering design
  - Controlling at source including barrier, etc.
  - Implementable and practical control
- Phases of project/plant: With the phases of the project/plant the hazards and/or accidents also change. During construction there could be an accident, for example, the falling of a generator during the installation stage in a thermal power plant, which resulted in a few casualties as well as a long delay in job completion. On the other hand, hazards and accidents during plant operation can be different. Therefore for the PHA selection process it is necessary to consider various phases, for example:
  - New plant with new technology
  - Planning/design stage
  - Construction/installation phase
  - System qualification
  - Production/operation
  - Standby
  - Shutdown: planned or emergency
  - Maintenance
  - Decommissioning

2.1.2 Some Important Terms and Quick Methods

It is prudent that a few terms associated with PHA, mainly for PSM systems, are discussed. In Table II/2.1.2-1, a few hazard analysis methods have been highlighted such as the Dow FEI and the Mond Index, etc. for quick risk assessment in process plants.

Table II/2.1.2-1 Technical Terms Used in PHA (Especially for Chemical Process Plants)

Terms	Elaboration
Dow Fire and Explosion Index (FEI)	The Dow FEI Safety and Loss Prevention Guide was developed by Dow Chemical Company sometime in the 1980s. The Dow FEI is used as a ranking system to give a relative index of the risk pertinent to individual process units on account of potential fires and explosions. The following points are a few characteristic features of FEI: • FEI uses *material factor* (MF—discussed later), which is concerned with chemical energy and relates to reactivity and flammability, that is, the expected damage of potential fire and explosion incidents in realistic terms • It identifies the equipment that would contribute to create or escalate a flammable or explosive hazard • It indicates fire and explosion potential to the design team and concerned personnel • It is very useful for plant layout, separation of vessels, and hazard mitigation control planning • It may not be very accurate but is a useful tool and requires engineering knowledge to apply • For applying the tool a logical procedure needs to be used to study where this can be applied. It is normally applied for the following cases: ◦ Loading/unloading facility, storage tank, quench vessel, reactor and distillation column, etc. To apply the principle it is necessary that sufficient information in terms of piping (process) and instrumentation diagram (P&ID), draft layout, equipment list, and sizing is available. Normal index value 0–60 light, 61–96 moderate, 97–128 intermediate, 129–159 heavy, and >159 severe. This may be read together with a toxicity index for a chemical plant. This principle is applicable for main plants but not for auxiliaries. Also there is not much correlation with known disasters.
Material factor (MF)	MF is a measure of a material's chemical energy in terms of reactivity and flammability. It is a number between 0 and 60 and indicates energy released for explosion. For noncombustible items like water the number is 0. MF is assigned to each processing unit to take into account flammability, reactivity, and toxicity. In the Dow guide MF for commonly used materials is indicated as the starting value for FEI calculations. MF is obtained from the flammability and instability rankings according to NFPA 704 (Standard Systems for Identification of the Hazards of Material for Emergency Response).

Continued

Table II/2.1.2-1 Technical Terms Used in PHA (Especially for Chemical Process Plants) Continued

Terms	Elaboration
Mond Index	The Mond Index was developed by ICI's Mond Division as an extension of the Dow F&EI to address the toxicity hazards associated with materials in process units. This is to be used in conjunction with FEI. It is seen that when toxicity is 0–6, it is light with FEI 0–60. Toxicity >10 could be intermediate or heavy depending on associated FEI in the range of 127–158. For FEI >159 and toxicity index >10 it is severe. It relates both FEI and the toxicity Index [2]. Mond indices are classified as "Mild," "Light," "Moderate," "Moderately Heavy," "Heavy," and "Extreme" for the ranges 0–20, 20–40, 40–60, 60–75, 75–90, and >90, respectively [4].
Chemical exposure index (CEI)	The CEI is again an index used by Dow Chemical to assess products of various factors (acute toxicity, volatile portion of the material that could be released, molecular weight, distance coverage, and operating parameters like temperature, etc.) using a numeral scale.
Substance hazard index (SHI)	The SHI is an index to identify the toxic chemical substances that could be involved in a catastrophic release. The index is a simple function of vapor pressure and toxicity: The higher the vapor pressure, the more rapid will be the entrance in the atmosphere in the event of a release. The greater the toxicity, the lower the concentration required to present a hazard—so, SHI will be higher.
Hazardous material and material hazard index (MHI)	Any chemical substance that when released or misused will cause harm to the environment and/or public health. These are used in various industries. Hazardous materials may be in the form of explosives, flammable and combustible substances, poisons, and radioactive materials [5]. MHI is an index used by the State of California. It is the material vapor pressure at 25°C divided by the level of concern, which can be defined on the basis of toxicity, fire explosion, etc. [3,6].
Hazard factor (HF)	There are two kinds of hazard caused by the toxicity, fire, or explosion effect on materials, (1) General purpose, where the intensity of an accident increases on account of its presence. (2) Special purpose, where the probability of an accident increases (Clause 2.1.1).

2.2 Plant Hazard Analysis Methods Selection Criteria

In the previous Clause 2.1, some of the important issues influencing methods have been discussed. Now, we shall try to find the means by which PHA methods are normally selected taking into consideration various factors. It is advisable to refer to Fig. II/2.2-1 for an idea.

From Fig. II/2.2-1 it is seen that a team with a team leader is necessary to select the process. However, in certain cases team formation may not be mandatory. Also it is necessary that the team leader or who is leading from the front must have expertise (may be acquired through experience). There are a number of influencing factors that are responsible in the selection process. In addition, stage of the project, available

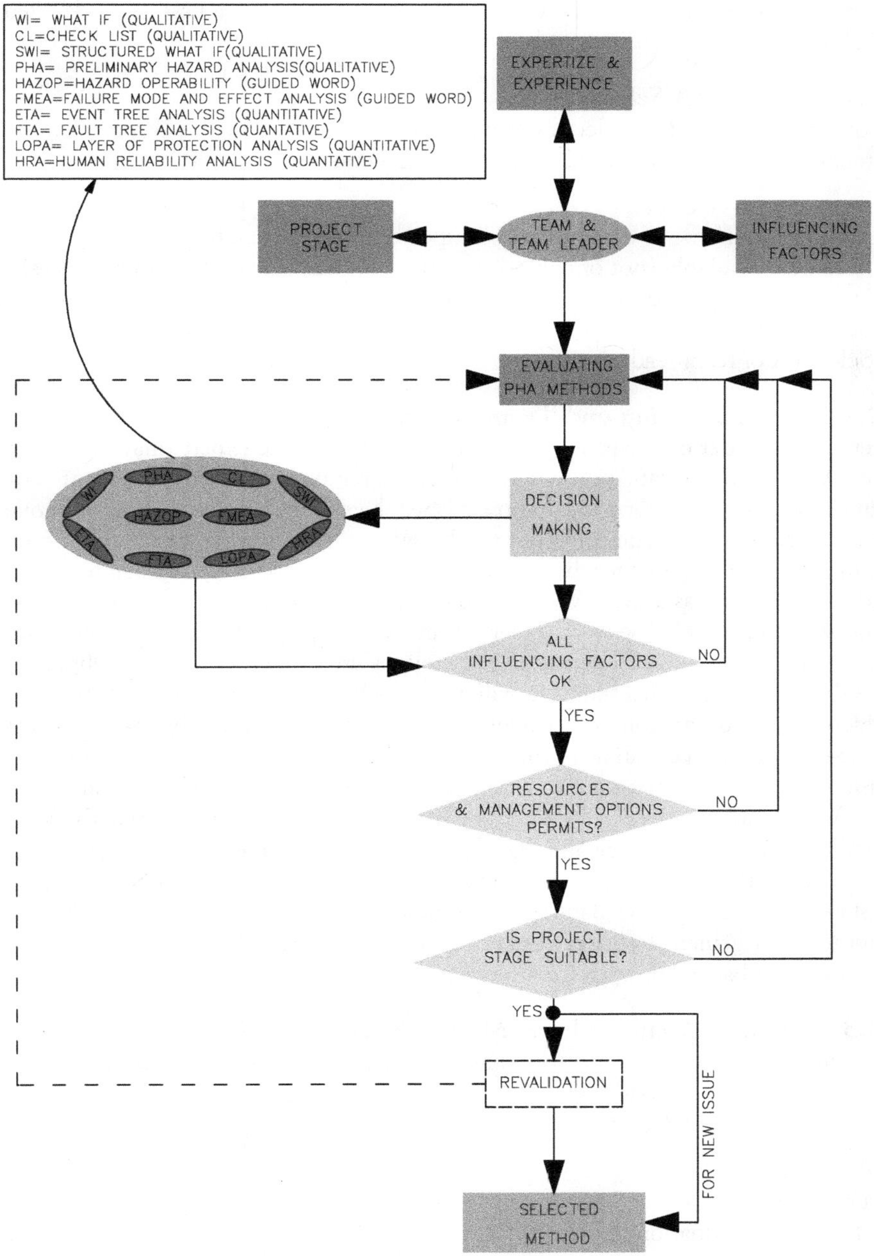

FIGURE II/2.2-1

Outline of plant hazard analysis (PHA) method selection process.

resources, and management aims influence the selection process. For detailed guidelines ISO 31010:2009 may also be referred to.

2.2.1 *Aim of PHA Selection*

At the time of the PHA selection process, it is expected that the corporate management must decide in advance and frame a general guideline for:

- Aim or purpose of PHA
- Kind of result/type of detailing to be expected from the result
- Resources available (not only funds but also information and human resources)
- Deadline of completion
- Stage of the project
- Selection of team leader/expert

2.2.2 *Team Formation and Team Leader*

After a team leader or expert is selected, it is the duty of the expert (may be outside agency) to form a team and proceed in close coordination with the management. The team leader must understand the purpose of the job very carefully and not try to fit one of the typical PHA selection frames into the system. This is important because each plant/process or company may have unique demands. Immediately, a group of people need to be selected as a team when called for (always formation of team may not be necessary). Normally in the team, experts from process/operation and instrumentation and electrical mechanical engineering disciplines are included. At the preliminary stage the team leader with the team will go through the guidelines set forth by management. It is necessary that such guidelines are modified continuously. This is ideal as long as any alternatives can meet the target in a better way. It is not uncommon that, considering the amount of information available, or based on purpose, the team leader finds that instead of process A, it is better to go for process B. It is the duty of the team leader through his experience and logic to convince management to do so. However, while doing so the team leader needs to take into consideration: (1) influencing factors, (2) stage of the project, (3) plant-specific problems, and (4) timeframe, etc. Also it is better for management to listen to the team leader as long as his/her suggestions are logical and are backed up with experience and expertise.

2.2.3 *Factors Influencing PHA Method Selections [1]*

There are a few influencing factors that directly affect the PHA selection process, as shown in Fig. II/2.2.3-1.

Major influencing factors are:

- Aim of the study
- Type of result
- Available information database
- Characteristic-matching issues
- Plant-specific problem
- Resource and management issues

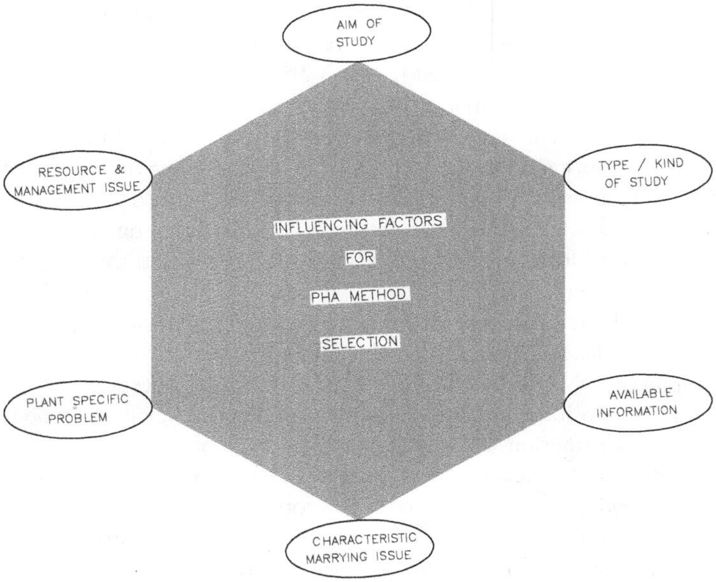

FIGURE II/2.2.3-1

Influencing factors for plant hazard analysis (PHA) method selection.

Now each of these issues will be discussed at length sequentially.

↓

- Aim of the study: As discussed earlier, it is most important that the aim of the study be well understood by the team leader otherwise it may be a complete waste of resources. There are several issues to discuss, such as the list of hazards, to meet the statute, overall risk analysis, and risk mitigation including detailed quantitative risk analysis (QRA), etc. There may be other interrelated factors such as a new plant with new technology or quantum of available information, etc. These will be discussed separately but are mentioned here because while fixing the charter they need to be considered also, for example, if for a new plant with new technology detailed quantitative risk assignment is desired, then at the initial stage this may not be possible because of lack of sufficient information. This can be done at a later stage. Therefore it is the responsibility of the team leader to select the most appropriate factor taking into account available resources.

→

- Type and kind of result: The type of result expected from PHA has been elaborated in Clause 1.1.6. All these results are not always available from all

PHA methods. Even from each of the PHA types there could be variety of results that is dependent on the outcome. These results could be just a list of hazards or accidents; they could be a detailed list with priority of action or input for a further study of safety instrumentation systems with alternatives or input for a QRA. As all methods are not capable of providing all kinds of results it will be the responsibility of team leader and the team to decide on the issue. From the discussions in Clause 1.2 almost all PHA is capable of giving a list of hazards and recommendations for mitigation. If, however, further detailing such as prioritization of risk or quantitative analysis is necessary then the team needs to opt for ETA, FTA, etc.

- Available information database: Available information plays a great role in PHA study. For example, in a process flow diagram (PFD) we have process flow and basic equipment, whereas in a P&ID we have not only process flow with equipment but also instrumentation and piping details. Again the scope of availability of information is dependent on the stage of the project. While the PFD is available at a much earlier stage, the P&ID may be available at a later stage. The following list shows the type of information and available stages. It is worth noting that as we go down the line, details of the information and stage of the project increase (i.e., time from inception increases):
  - Basic materials involved, physical/chemical property
  - Basic process or material flow
  - Experience with similar project
  - PFD
  - Equipment data
  - Plant layout
  - P&ID (this is mentioned after layout because in many cases the P&ID is slightly dependent on layout, e.g., a mud system in offshore engineering)
  - Existing details
  - Operating details/procedure/production process
  - Plant operating experience

Therefore it is clear that the quality of information and the stage when it is available also influence PHA method selection. For the same reason it is not possible to carry out HAZOP for a plant at the conceptual stage. So, if quality data are not made available, then the team and the team leader have probably only two options: they can either recommend to the authority to change the detailing desired from PHA or delay the selection and inform management. In the last example we discussed the lack of information. Similar things happen if the available data are not up to date. It is *wrong* to work with old data because such an exercise is futile. Therefore it is essential that the team leader is aware of this so that hazard evaluation performance is improved and not delayed.

- Characteristic-matching issues: Each type of plant has certain inherent characteristic issues such as continuous plant, process plant or manufacturing unit, batch processing, etc. Similarly, each of the PHA methods has some inherent characteristics for which the particular PHA method has been developed. It is not that a particular method cannot be applied to the other plant but it may not be best suited. This will be clear with some examples. Power plants and mud plants in offshore applications are all basically continuous operating process plants and HAZOP is better suited for such situation. On the other hand, FMEA is better suited for computer or electronic device production lines. Similarly, when there are questions of multiple failure, FTA could be a better choice. So, the team leader needs to see that the chosen characteristics are matched with the characteristics of the plant in question. This is somewhat like protocol matching in computer communications, where even electrical connections are made correctly; however, without proper protocol matching communications for data transfer cannot be made. For example, someone speaking in Bengali may not be understood by someone who speaks Spanish. Therefore the purpose is not met. When one speaks of the inherent characteristics of a plant this refers to:

- Type of process
- Operation style and type
- Size and complexity of the plant
- Inherent hazards, etc.
- Type of failure

- Type of process: Generally, an initial screening is possible from the type of process involved. As stated earlier, FMEA is more suitable for computer systems and HAZOP is more suitable for process systems. Naturally based on the same PHA method, selection may be possible and needs to be judged with respect to other characteristics, for example, size and complexity.
- Operational style and type: The type and style of operation also affects the characteristics of the plant and involves the team leader to check whether the PHA method is suitable. The plant may be a static or transportation type, for example, a manufacturing unit and vehicle movement in a mining operation. The involved process may be a batch process, for example, a pharmaceutical unit or a continuous process such as a power plant. In the vehicle movement case, simple systems such as "what if" would be adequate but the same is not true for a power plant—a continuous system.
- Size and complexity of the plant: Plant size and complexity play a significant role in selecting the PHA method. Naturally, it will not be a prudent decision to apply a simple PHA method, say preliminary hazard analysis, for analysis of hazards in a nuclear installation. Plant *size* mainly depends on numbers of subplants/subsystems involved, amount of equipment in each of the subsystems and/or number of steps involved. One demineralized plant involving five or six chains will be larger in size than a plant with two or three chains. Similarly, a 650-MW plant involving many auxiliaries will be larger in size than a 60-MW plant with fewer auxiliaries. With an increase in the number of auxiliaries, both

complexity and chances of failure in logic also increase. Plant *complexity* again depends on many factors, such as: operation type and condition, chemical process and/or chemistry, as well as chemical reaction and reaction rate. Complexity also depends on human interaction with the system as well as associated hazard types such as toxicity, flammability, fire, etc. Choke and kill blowout preventer (BOP) in the offshore industry are classical examples, where high-pressure flammability is involved. Similarly, complexity can be seen in ultrasupercritical boilers involving very high-pressure and high-temperature operation. Chemical process plants pose such problems because of chemical reactions.

- *Inherent hazards*: In process industries, various hazards such as toxicity, flammability, fire, explosion, etc. can be analyzed by any of the analysis processes, but if a hazard involves failure of equipment, or more than one failure, then a more detailed study may be involved and a simple system may not be suitable. In batch or manufacturing plants a similar logic applies. So, the team leader needs to adopt a different approach to address inherent hazards.
- *Type of failure*: There are several situations concerning single failure, multiple failures, etc. There may be process upset, operational fault, or hardware/software/human failure, etc. It is therefore pointless to attach importance to the fact that PHA method selections have a direct impact on these failures, for example, to handle a multiple failure situation or probability-based quantitative analysis one may have to apply FTA and/or ETA, etc. instead of FMEA and/or HAZOP.
- *Plant-specific problem*: The PHA method selection process depends on the team and team leader. So, accuracy of the study depends on the experience and skill of the team leader as well as that of the entire team. Naturally, any omission will contribute to the inaccuracy of the study and may contribute to inherent problems at the plant. So, when management decides to adopt a detailed study, they should go for a more systematic type of study suitable for their needs, for example, a utility company building a nuclear installation will invite a more detailed study than that for a thermal power plant. Some of the plant inherent problems come from operating experience, for example, when a plant is operating over a long period, then the majority of the inherent problems are well known, or could be understood from operating data. However, when a new plant comes into operation involving new technology, inherent problems may not be apparent. So, these problems will be increased and a detailed PHA method may be better. MOC is another important issue in contributing to inherent problems, for example, where a plant has been operating over a long period with few potential hazards and a number of changes have been incorporated. In this case, the chances of hazards may not be an issue and management may decline a detailed study. The author had similar experience in a cement plant. The kiln feed measurement system was old and was posing a problem but there had not been evidence of any potential hazard, just poor quality output. When the kiln feed system was later changed with an impact scale, bypassing the original one, there were not many studies because the existing kiln feed was kept in parallel so

that production was not hampered. On the contrary, for a comparatively newer plant, where there was problem of non-formation solid clinker, operational changes were incorporated after a number of studies. In summary, when there are major changes, management will adopt a systematic study for MOC (say a checklist).

- *Resource and management issues*: These are more or less financial and management issues involving:
  - Number of skilled and knowledgeable persons available
  - Financial resource of the company
  - Company culture
  - Management preference
  - Time schedule

  Quality of PHA is extremely dependent of the quality of effort of the team [1].

2.2.4 PHA at Various Stages of the Project

As discussed in Clause 2.1.1 there are several stages of the project. Naturally, at each of these stages several types of PHA may be applicable. Also from the beginning, PHA needs to be developed at various stages. In fact, the PHA is revalidated at various stages. Also at various stages there would be different appropriate PHA methods. This is because at various stages the information available varies.

- FEED stage: At the initial FEED stage, PFDs, certain equipment, etc. are fixed and it is in the process of development of a P&ID. At this stage, preliminary hazard analysis, checklists, etc. are utilized for PHA. Then, slowly instrumentation is developed based on layers of protections desired as a safeguard and interlock.
- Detail engineering: During the detailed engineering stage when plant data are more finalized, P&IDs, layout, and equipment/instrument lists are available along with respective data sheets, depending on the type of plant. HAZOP/FMEA/"what if"/structured check lists, etc. may be applied to identify the hazard or accident. If necessary, LOPA is applied to ensure safety. These are basically revalidation of PHA with more information and different PHA method. Since it is desired that most of the hazards are identified, depending on type of plant other methods may also be applied when multiple failure is taken into consideration.
- Construction stage: Normally at this stage it is not desirable to change the design; however, in reality there will be some site modification, which calls for revalidation of PHA with the changes incorporated. HAZOP could be a better solution for process industries.
- Operational life: Startup from a normal or controlled shutdown is part of the safety procedure. There may be three types of shutdown: planned, emergency, and shutdown for maintenance, as indicated in Clause 2.1.1. Each of these has certain safety issues associated with it, so, these types of plant shutdown along with startup from emergency shutdown should be checked.

- Revalidation: The PHA needs to be periodically revalidated to incorporate changes necessary for plant safety. As the plants operate there will be modifications needed. It is almost inevitable for all plants. There should be some corporate policy for MOC. MOC risk reviews are carried out as part of company policy and/or any of the PHA methods. However, this depends again on to what extent the MOC hazard study will be carried out (an example with a cement plant has already been discussed). However, normally the PHA method that was implemented earlier is also followed for MOC.
- Extended shutdown—decommissioning: Mothballing (restoration) of a plant is a step undertaken for a safety shutdown system. Not only all reactants have to be drained and neutralized but a safe state must be restored, especially for restart from an extended shutdown. Before starting, the condition of all equipment and instruments, especially those meant for safety, must be thoroughly checked, revalidated for operation, and calibrated. At this stage, a PHA review is done. Also during decommissioning when equipment is dismantled, a thorough safety check is essential to avoid any accidents. A PHA review is done prior to the decommissioning activity.

2.2.5 Decision for PHA Method Selection

As discussed, every PHA method has certain strengths and weaknesses (Fig. II/2.2.5-1) (inspired by Ref. [1]).

As shown in the figure there are a few issues regarding resources, such as number of knowledgeable persons available in the organization (if an external agency this may be less important as long as the knowledgeable staff is available within the agency; however, staff training is important). Financial support available from management is very important. In some cases, expenditure could be high, especially when an agency is deployed. Timeframe or time schedule is no way less important because if the time schedule is not met, the basic purpose may be defeated. Naturally, this may call for alternatives if possible. The management decision for selection is also critical. Here, the team leader has a key role to play. If the team leader feels that a particular method is best for the project, then the team leader needs to take the lead role in convincing management. In the end there are a few other options pertinent to PHA methods (Step 7 in the figure), which have been shown to reflect the requirements of the proposed PHA method. If there is a chance to opt for alternatives, then such selections will be helpful (e.g., some PHA methods do not need a team, for example, ETA, FTA, or a checklist that involves fewer staff, etc.—if there are alternatives then those could be selected; these are in addition to the main steps as helpful tips).

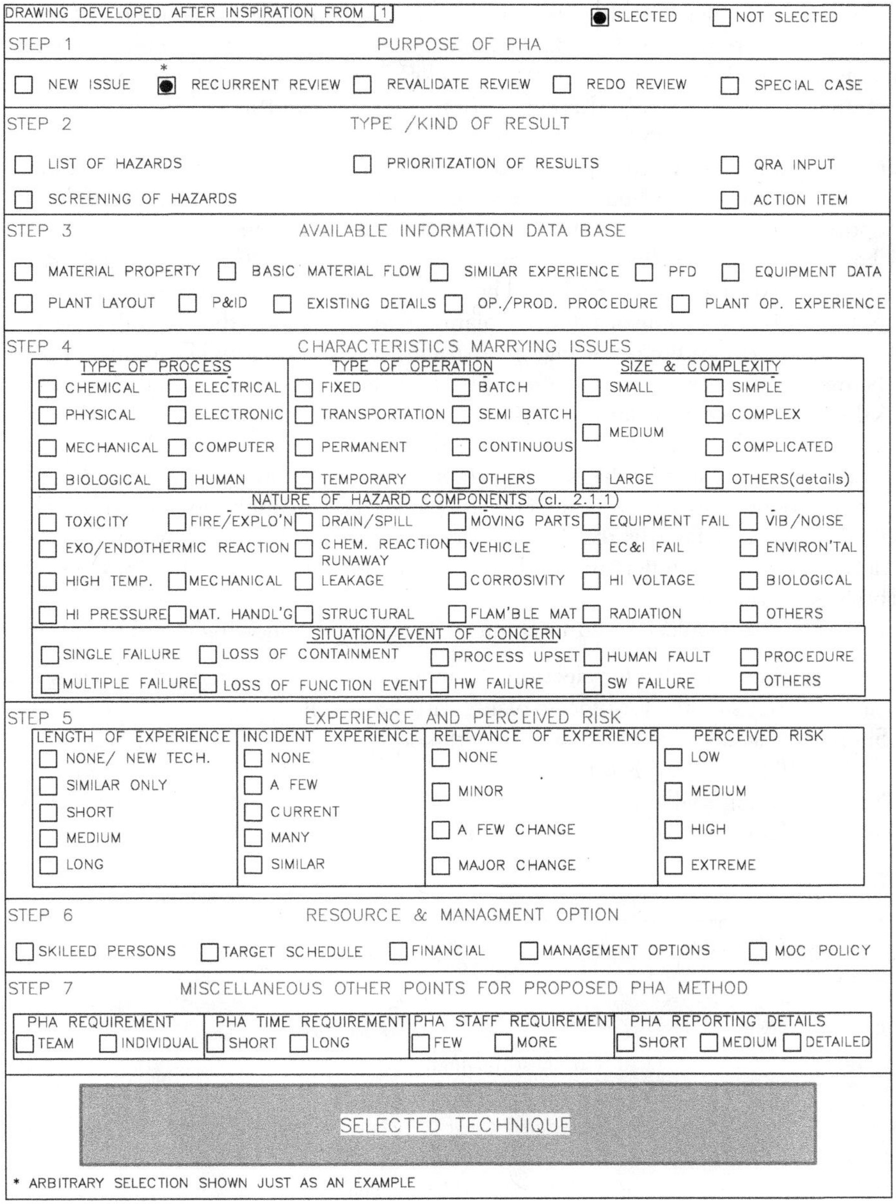

FIGURE II/2.2.5-1

Logical steps for plant hazard analysis (PHA) method selection.

2.3 Comparison of Various PHA Methods

Having discussed various features of miscellaneous PHA methods, it is better to compare their characteristics with respect to the various criteria in Table II/2.3-2 (inspired by the Process Improvement Institute Inc. Ref. [1]). This table may be referred to for various codes. In this connection it is worth noting that each of these systems has its strengths and weaknesses and is best suited for a particular type of industry or industries. The comparative studies given in the table provide ideas about their characteristics. In this connection, another important point to be noted is that it is possible to divide all plants into three categories: continuous operation, discrete operation, and batch process operation. In continuous production, materials flow continuously without interruption. Power plant operation is an example of the continuous process. There are a number of distinctive differences between this and batch process operation, for example, tank volume and time required in the batch process (refer to Chapter VI) will be greater than in the continuous process. A cement plant is an example of a discrete plant. Here, in between the raw mill section and the kiln section there is raw materials storage silo, so, when the kiln runs it may not be necessary for the raw mill to run. Similarly, because of the clinker silo, when the kiln runs, the cement mill may be idle. On the contrary, pharmaceutical production, soft drinks production, etc. are batch processes, and on the medicine label, for example, one can see the batch number. In the batch system, any mix of whatever produced in the intermediate stage must be consumed in total in batch production. Exactly for this reason, if something untoward happens in batch production, the entire batch is rejected. Based on the foregoing distinctions, PHA methods also change.

Table II/2.3-2 should be read in conjunction with the following notes:

1) HAZOP is not suitable for electronic computer/mechanical production plants
2) FMEA is not suitable for process plants with process flow
3) Simple and small plant jobs can be undertaken by PHA, FMEA, HAZOP, ETA, FTA and HRA, but this is not economical and hence is not shown
4) Normal staffing shall be two to three in a team for most cases, except HAZOP where staffing may be five to seven or more [7]

Table II/2.3-1 Codes for Various PHA Methods (Refer Table II/2.3-2 for Uses of These Codes)

Code Name Used	PHA Method
CL	Checklist (qualitative)
WI	"What if" (qualitative)
SWI	Structured "what if" (what checklist) (qualitative)
PHA	Preliminary hazard analysis (qualitative)
FMEA	Failure mode and effect analysis (guided word)
HAZOP	Hazard operability (guided word)
ETA	Event tree analysis (quantitative)
FTA	Fault tree analysis (quantitative)
HRA	Human reliability analysis (quantitative)
LOPA	Layer of protection (semiquantitative/quantitative)
QRA	Quantitative risk analysis (quantitative—different approach)

Table II/2.3-2 Comparison of Various PHA Methods: To Be Read in Conjunction With Details in Table II/2.3-1. In This Connection Annexure of ISO 31010:2009 May Be Referred to Also

Main Criteria	Subcriteria	CL	WI	SWI	PHA	FMEA	HAZOP	ETA	FTA	HRA	LOPA
Type (FTA is deductive)	Qualitative	X	X	X	X						
	Guided word					X	X				
	Quantitative							X	X	X	X
Team for all	**But individual possible***							*X	*X		
Operational mode	Continuous	X	X	X	X	X	X	X	X		X
	Batch/startup/ shutdown	X	X	X	X		X	X	X	X	X
Hazard level	High				X	X	X	X	X	X	X
	Low	X	X	X							
Simple/small plant (time in hours/days)		X	X	X							X
Complex/large plant (days/weeks)					X	X	X	X	X	X	X
Experience with system (H for high; L for low)		H/L	H/L	H/L	L	H/L	H/L	H	H	H/L	H
Information database	F: FEED, D: Detail design O: Op unit/startup	F	F/D/O	F/D/O	F	D/O	D/O	D/O	D/O	D/O	D/O
Project stages: A: All; F: FEED; E: Detail Engineering.; C: Construction/startup; O: Operation; M: Modification; D: Decommissioning		A	A	A		E/C/M	E/C/M		D/O/M		
Team leader expertise M: Minimum; MO: Moderate; E: Extensive		M	M	M	MO	MO		E	E		
Analysis quality G: Gross; S: Specific; V: Very specific		G	G	G	G	S	S	V	V		
Failure	S: Single; M: Multiple	S	S	S	S	S	S	M	M	M	
Coverage width	W: Wide coverage; P: Physiochemical	W	W	W		S	S	W	W		

X, applicable.

DO YOU KNOW??

Soveso Directive and its applications?

Major Accidents like Flixborough (UK), 1974; Seveso (IT), 1976; Bhopal (India), 1984; Basel & Mexico 1986 made people think of brining the legislature to prevent/ reduce accident in process industries dealing with hazardous material. Till now Seveso II Directive 96/82/EC is in effect. The directive was named after Seveso disaster -an industrial accident that occurred around on July 10, 1976, in a small chemical manufacturing plant near Milan in Italy.

The Seveso Directive - **Prevention of major accidents which involve dangerous substances, and the limitation of their consequences for man and the environment, with a view to ensuring high levels of protection in a consistent and effective manner. The basic goal was to protect human and environment from Industrial accident.**

Major accident: **shall mean an occurrence such as a major emission, fire, or explosion resulting from uncontrolled developments in the course of the operation of any establishment covered by this Directive, and leading to serious danger to human health and/or the environment, immediate or delayed, inside or outside the establishment, and involving one or more dangerous substances; As a result of the review process, on 4 July 2012 the new *Directive 2012/18/EU*(aka Seveso-III) was adopted which repeals the Seveso II *by 1 June 2015.* The major changes are:**

- **Updating and aligning the list of substances covered by the Directive to the EU legislation on the classification of dangerous;**
- **Strengthening citizens' rights on access to information, justice and decision-making;**
- **Improving information collection, sharing & management etc.**
- **Introducing stricter standards for inspections to ensure effective implementation**
- **Clarifying and updating of provisions, (streamlining and simplification) to reduce administrative burden.**

Who is Operator? **Operator" means any natural or legal person, including public authorities, in charge of an activity, e.g. supervising, planning to carry out or carrying out an activity;**

Hazardous Activity: **Hazardous activity" means any activity in which one or more hazardous substances are present or may be present in quantities at or in excess of the threshold quantities**

FIGURE II/2.3-1

SOVESO Directive and operator definition.

3.0 HAZARD IDENTIFICATION (HAZID) AND RISK ESTIMATE

The question may arise: because there are so many PHAs, why is there a need for HAZID? Before answering, one thing should be kept in mind: there is no conflict between PHA and HAZID. In fact many PHA may be utilized in HAZID for identification hazards. This question is best answered with the help of an example. Suppose there is a manually operated valve (maybe locked open) in a pipe line handling corrosive material, situated at a height. The valve is never/hardly operated and kept open all the time. To access the valve, there is a cat ladder. If there is a seat leakage or the valve is closed, any hazard may be analyzed using the PHAs discussed. On seeing a leakage a mechanic attends to it by climbing on a ladder but unfortunately something heavy falls on his head and he is injured (even with precautions like a helmet, etc.—such incidents are covered by human factor engineering, which deals with the application of information on physical and psychological characteristics to the design of devices and systems for human use). This accident is an external issue and not at all related to the process, and may be missed in PHA. While carrying out HAZID, the same situation may be identified and a canopy may be recommended over the area. This situation also addresses another misconception, that HAZID is used for new projects! In this case the plant was running or operating while the incident happened.

The basic concept of HAZID has been depicted in Fig. II/3.0-1.

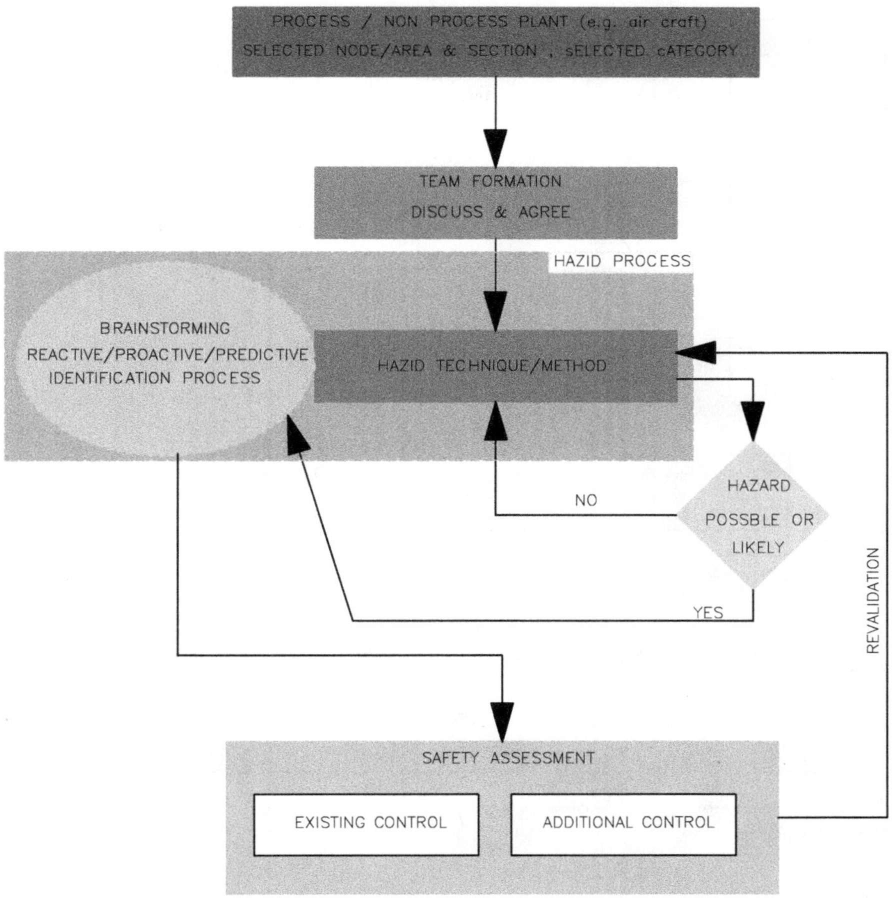

FIGURE II/3.0-1

Concept of HAZID.

HAZID is extensively used to address major accident issues, usually at a major hazard facility (MHF). In HAZID, incidents with a very low likelihood but with very high severity are also addressed so as to avoid any catastrophe at any time.

As shown in the HAZID process, hazards in process or nonprocess plants are identified usually by a group of people including workforce participation (in some

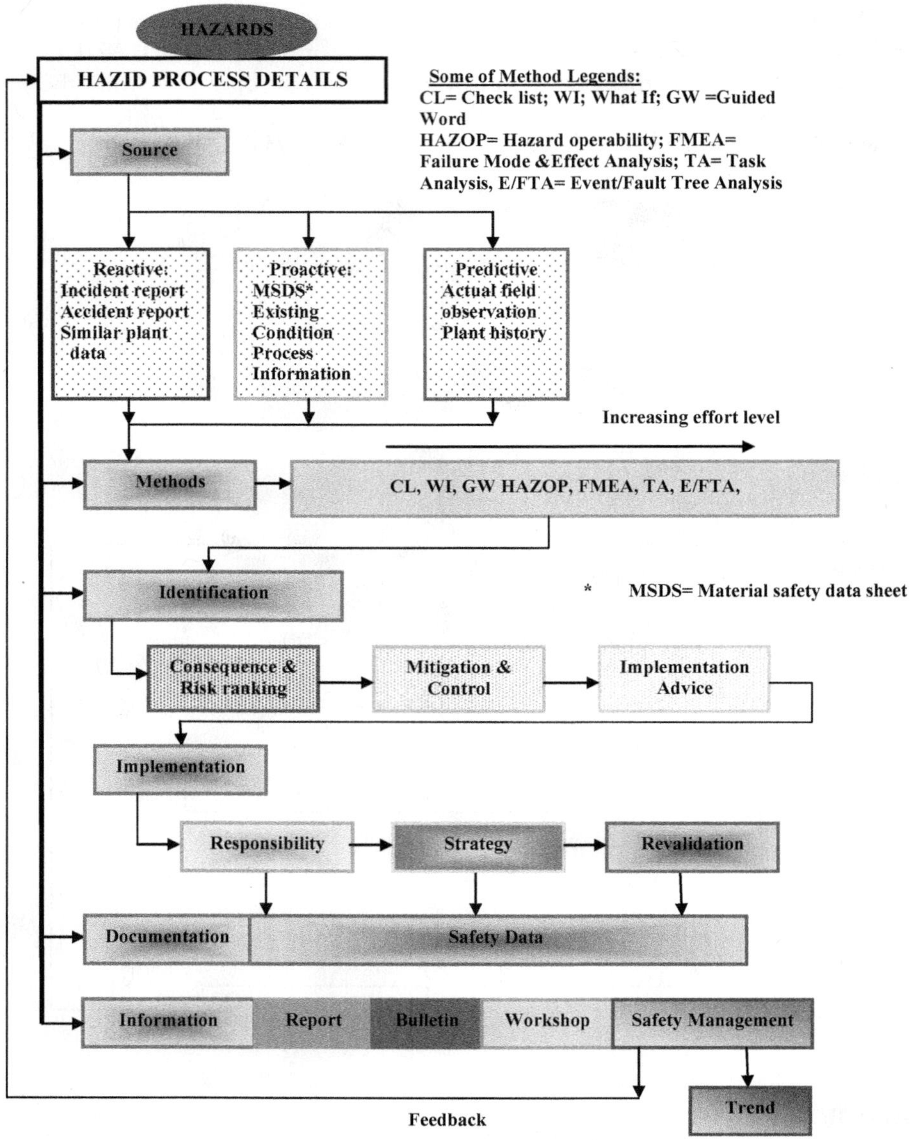

FIGURE II/3.0-2

HAZID process conceptual details.

cases this may be by an individual but workforce participation is also necessary), with the help of hazard identification techniques. When hazards are identified the risk associated is assessed and safety control systems are applied. In this clause, discussions will be restricted to HAZID, risk estimations, and task analysis methods. Further details on the HAZID concept are depicted in Fig. II/3.0-2. All PHAs are related to risk analysis but HAZID is strongly linked to risk analysis. According to CCPS [12] HAZID and risk analysis is basically pivoted around three questions: Hazard: What can go wrong?, Consequence: How bad could it be? and Likelihood: How often might it happen?

3.1 Basics of HAZID

Fundamental and relevant details regarding HAZID will now be discussed. The first point that comes in mind is: what is HAZID? HAZID basically is a study tool usually deployed at the early stage of a project. However, it can be used at other stages in the life cycle of plants also, from early project life to decommissioning. With the help of this tool it is possible to organize health and safety engineering (HSE) deliverables in the project. It identifies all hazards with a potential to cause a major accident to a facility, or to people at or near a facility and the environment. HAZID is applicable to both process- and nonprocess-based plants. It addresses not only the process hazards but also the hazards from external sources (e.g., hazard caused by the environment). Another important feature of HAZID is the involvement (direct or indirect) of workforces through proper understanding of the systems. Often people confuse HAZID with HAZOP. It is to be noted that HAZOP is mainly guided by IEC 61882 and HAZID mainly falls in the purview of the Seveso II Directive (after June 2015, Seveso II will be repealed and Seveso III will be effective) (Fig. II/2.3-1). HAZID is also a part of HSE and the Offshore Petroleum and Greenhouse Gas Storage (Safety) Regulation (Commonwealth) (OPGGS). However, HAZID is less costly, detailed, and assured when compared with detailed PHAs [12]. Some key features of HAZID are detailed in Fig. II/3.1-1.

3.1.1 HAZID Concepts and Features

- To identify all hazards, especially those that lead to a major accidental event (MEA), and to ensure that operator and workforce are well aware of the hazards at the existing facility and that new hazards are recognized before they could

To remember:
Characteristic features of HAZID: (1) Though major use of HAZID is in the early stage of the project but it can be used at all stages. (2) It is applied for both process and non process application. (3) It includes internal and external hazards also including environmental hazard. (4) Workforce understanding and participation along with operator is a key feature. (5) Success depends highly on appropriate technique selection and documentation including all assumptions and unclear issues.

FIGURE II/3.1-1

Basic HAZID features.

happen. This is to ensure workforce safety and the active participation of the workforce directly or indirectly through feedback.

- The hazard identification process should be workable and applicable for the facility concerned.
- Once the hazard is identified, the operator should ensure that proper action is taken to manage (reduce/mitigate) the issue.
- Hazard identification is necessary to instill sufficient knowledge, understanding, and awareness among all concerned, especially for those who lead an MEA, so that unwarranted situations are mitigated or prevented.
- The most suited hazard identification techniques are applied so that in-depth analysis is possible.
- A full range of hazards, modifications including health and safety-related hazards, and events are considered and output of the HAZID process is documented. Documentation should be maintained in a logical and systematic manner.
- An SMS must be employed for all hazards including health and safety hazards, not just for MEA.
- HAZID must be able to identify, select/reject, evaluate, and justify the control systems for hazard prevention/mitigation, especially for MOC.
- Identified hazards should not be ignored automatically just because a control system is in place. Also for existing systems, fresh look and inspection shall be initiated and provided. One should not ignore the existing plant as there is knowledge about it and no further knowledge is required.
- HAZID must take into account all expected modes of operation, activities, human factor, human—machine interface, system, and engineering issues.
- Many hazards occur because of a combination of failures, and so should not be ignored.
- All methods, results, assumptions, and any uncertainties must be documented in a systematic manner. Also all control systems proposed must be documented in such a way that they are clear to any third party; documentation should also demonstrate how hazards are reduced.
- HAZID is an ongoing dynamic process and applicable at all stages of the project, especially when there is any modifications and preventive control actions.

3.1.2 *Some Important Definitions and Discussions*

- Control measure: Control measures include actions that can be taken to reduce the potential of exposure to a hazard, or the control measures could remove the hazard or reduce the likelihood of the risk of exposure to that hazard being realized [8]. The National Offshore Petroleum Safety and Environmental Management Authority (NOPSEMA) defines control measures as "A control is any system procedure, process, device or other means of eliminating, preventing, reducing or mitigating the risk of any major accident events arising at or near the facility. These control measures could be process control, management decision/process, maintenance procedure, and emergency plan."

- Formal safety assessment (FSA): This is the Offshore Petroleum and Greenhouse Gas Storage (and Safety) [OPGGS(S)] regulation. It "is an assessment or series of assessments that identifies all hazards having the potential to cause a major accident event, is a detailed and systematic assessment of the risk associated with each of those hazards, including the likelihood and consequences of each potential major accident event; and identifies the technical and other control measures that are necessary to reduce that risk to a level that is as low as reasonably practicable" [9]. According to the International Maritime Organization: "FSA is a structured and systematic methodology, aimed at enhancing maritime safety, including protection of life, health, the marine environment and property, by using risk analysis and cost benefit assessment" [11]. There is another term FSA (functional safety assessment) in connection with IEC 61508 & 61511, and should not be confused with this. Functional safety assessments have been discussed in later chapters. Basically, FSA is the debating, analyzing, sharing of views, and development of consensus among the participants to develop a system. In its simplest form FSA can be divided into:
  - Identification of hazards
  - Assessment of risk
  - Risk control options
  - Cost—benefit analysis
  - Recommendations for decision making
- Major accident event (this should be read in conjunction with Fig. II/3.2.5-2): A major accident event is connected to the facility including external and natural events, which have the potential to cause fatality to persons at or near the facility including offsite (offsite is included here in light of the vastness of the major accident at Bhopal, India).
- Major incident (this is to be read in conjunction with Fig. II/3.2.5-2): A major incident is an uncontrolled incident including emission, loss of containment, fire, explosion, or high-energy release related to schedule 9/15 materials (materials are dangerous goods stored in a facility below a threshold value on account of the risk involved; for further details see Occupational Health and Safety Regulations (2007)) and which can cause a serious/immediate health and safety hazard.
- Major hazard facility: Major hazard facilities are locations where large quantities of *hazardous* materials are stored, handled, or processed, for example, an oil refinery, large chemical plant, etc.
- Safety assessment: Safety assessment process is consistent with international risk assessment standards such as ISO 31000. Safety assessment involves investigation and analysis of a major incident/accident and gives the operator a detailed understanding of all aspects of risk related to health and safety.
- Workforce: As per OPGGS(S)—an Australian legislation/standard, members of a workforce include those who are identifiable before the safety case is developed and working, or are likely to be working on the relevant facility [refer to OPGGS(S) standard]. For the development of the safety case for the facility, it is

necessary that there is consultation with and participation of members of the workforce. It may not be possible to involve all concerned but effective circulation of FSA details and feedback from the members of workforce are of utmost importance. Naturally, suitable documentation plays a major role. Sufficient knowledge, awareness, and understanding of hazard scenarios by the members of workforce cannot be overestimated. Therefore regular reviews and feedback comments are very important.

3.1.3 *Aim of HAZID*

The basic objective of HAZID is to identify hazards (all types including possible escalations) to review the effectiveness of the selected safety and control measures for achieving an as low as reasonably practicable (ALARP) risk level. This also serves as proof that the facility (process/nonprocess) is operated with all major hazards to the workforce and third parties within a tolerable risk range. HAZID involves a critical sequence of information gathering and the application of a decision-making process. These assist in discovering major hazards, which can lead to a major accident (hazard identification). It also indicates how likely it is that a major accident would occur with potential consequences. It also indicates the control measures to reduce/mitigate the hazards. So, broadly, this can be:

- Identification of hazards that can lead to a major accident event.
- To provide the operator and the workforce with awareness and a clear understanding and sufficient knowledge of all hazards including design problems. These will provide them with a means to prevent the occurrence of and/or manage a hazardous situation.
- Identify the outcomes of events and perform a risk ranking.
- Establish the clear linkage among hazards, causes, and potential events.
- Enable the identification of all hazards related to health and safety of people at or near a facility.
- Provide a means to identify, evaluate, and justify the suggested/selected control measures to mitigate or reduce the hazards based on risk ranking and acceptance criteria.
- Systematic record keeping of all hazards and their impact including those related to health and safety hazards. This systematic record must also include all uncertainty, assumptions, and hazard screening if any. These are necessary for further reviews.
- Recording of all assumptions and uncertainties for further studies.

3.2 Planning and Preparatory Stage

3.2.1 *General Considerations*

Usually, qualitative brainstorming methods are undertaken by a team with in-depth knowledge of the facility and expertise in HAZID analysis. It is quite uncommon but not impossible that an individual is doing HAZID (may be possible for ETA/FTA). As stated earlier (Clause 3.1.3) the main philosophy behind HAZID is to develop a

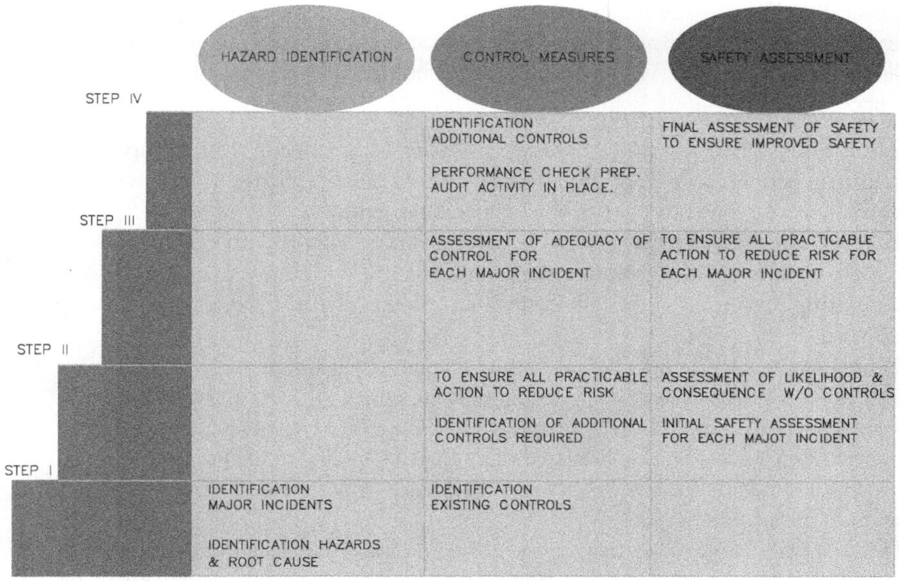

FIGURE II/3.2.1-1

Short HAZID steps.

safety assessment of the system, and safety assessment is best developed through debating, analyzing, creating, and sharing views for the development of an integrated system where involvement of the workforce is very important. Keeping parity with the diverse nature of hazards, there is diversity in HAZID analysis and in HAZID analysis techniques, so suitable system selection is valuable. Several major steps involved in HAZID have been shown in Fig. II/3.2.1-1.

In many applications it is necessary to produce a safety case outline by the operator prior to developing a safety case. (Ref MHF regulation). MHF stands for major hazard facilities (such as oil refinery, chemical plants). MHF regulation is applicable for working in these facilities in Australia, and for working safe in these areas there are some obligations to be met with. In UK similar regulation is in place in the name of control of major accident hazards (COMAH).

3.2.2 Scope

It is necessary to identify all hazards that could lead to a major incident/accident. However, this is an endless task, so the operator (who is conducting the study) is required to set the boundary for each study. If there are no possibilities, because of any activity/procedure at the facility being escalated to a major incident/accident, then these can be excluded from the study. However, this is not arbitrary! Anything that is kept outside the boundary must be properly recorded with suitable reasons for demonstrating the safety case.

3.2.3 HAZID Team Formation

Because HAZID is related to the safety case and safety assessment, team formation and number of persons in the team are very much relevant. Safety cases are the responsibility of the entire organization, so it is necessary that there is a proper mix of experts and workforce. Usually, the team consists of three to five people, one team leader, and one secretary. While forming the team the following points may be taken into account:

- The team shall be the right mix of experts and the relevant work group. For selecting people from work group it is better to have prior discussions with various work groups to choose the proper representative.
- Because HAZID takes into account external hazards, it is better to involve people from manufacturing and contractors also.
- Operating personnel with high-quality knowledge about the facility may be taken into the team.
- People experienced and knowledgeable in plant technology, control systems, maintenance, and design engineering may be taken into the team.
- As a rule of thumb, a project manager for the facility (particular section), an expert team leader, one each from design and O&M, and a mechanic/technician may be included and if necessary the team may be augmented with views from manufacturers/contractors.
- HAZID calls for good participation from workforces but it is not possible to involve all, therefore discussion points may be circulated and feedback can be obtained. This feedback could be in the form of training and control of the hazards present. In this way, the workforce can be involved. The workforce may be involved in the following cases by sharing data and receiving feedback from them:
  - Developing HAZID through interaction with workforce
  - Forming team and workshop scheduling
  - Participation in workshop
  - Review comments on the result of the workshop
  - Implementing modification and receiving feedback from shop floor workshop after analysis (involving them also)

It is desired that there is more and more workforce involvement for the awareness, knowledge, and understanding of hazards.

3.2.4 Hazard Identification Technique Selection

HAZID may not be as detailed as PHAs (detailed elsewhere) but there are a few influencing factors similar to those discussed in Clause 2.0. These are:

- Size and complexity of the facility
- Type of process and activity, for example, engineering procedure, process flow, mechanical type, electrical type, etc.
- Stages of the project in the plant life cycle
- Pilot study or safety case from a similar facility

Since hazard identification is the foundation of future courses of action, the selection of a proper identification technique is essential. The following are the features expected of hazard identification techniques:

- Systematic and structured
- Creative and lateral thinking about possible hazards

- Appropriate for the facility and stage of the job
- Possible to get maximum information
- Appropriate for people involved

The selection techniques with increasing effort level are:

- Checklist
- "What if"
- Guided word
- HAZOP
- FMEA
- Task analysis
- ETA
- FTA

These are also shown in Fig. II/3.0-2 and Table II/3.2.4-1.

Table II/3.2.4-1 Discussions on Selection Techniques

Criteria	Discussions	Remarks
Project stages in plant life cycle: Concept stage Detailed design stage Construction/startup Normal operation Decommissioning	At concept stage insufficient information so detailed systems HAZOP/FMEA are not suitable During detailed design detailed techniques will be a better choice At construction/startup and decommissioning phase better to go for task analysis Whereas in normal operation stage many are suitable[a] (see Remarks)	[a]At normal operation it will be influenced by hazard knowledge history, etc.
Complexity and size	Large system means number of operations, multitude of equipment, etc. Very simple technique may not do but complex techniques[b] may be bogged down (see Remarks)	[b]ETA/FTA
Process type	FMEA is suitable for the case to detect failure mode (say electronics/control equipment manufacturing). HAZOP is better for process systems ETA and FTA is better for multiple failure detection	All these discussed at length in Clause 2.0

For large, complex systems it is better to break down into small sections and apply suitable technique(s). Again a new system where the operator does not have any experience, or where similar plant data may be missing, then it is better to go for a detailed technique.

ETA, *event tree analysis;* FMEA, *failure mode and effect analysis;* FTA, *fault tree analysis.*

3.2.5 Documentation

Fig. II/3.2.5-1 is an important record-keeping document for HAZID analysis. This hazard register is same as the risk register discussed in Clause 3.2 of Chapter I, and details of the risk register are shown in Fig. I/3.2.1-1. Since details of the risk register are already available, it will not be repeated here. Fig. II/3.2.5-1 shows the basic structure. In the case of HAZID, hazard register is the popular term, hence it is used here. These hazard registers are available for each section of the facility. For each section, all hazards and major incidents/accidents are listed along with the probable cause. The register also contains the control measure, assumptions, etc. This will become the main document for subsequent use.

Basic requirements of the hazard register are as follows:

- All hazardous events with controls shall be numbered suitably, as shown in the hazard worksheet discussed later.
- There shall be a clear linkage between hazardous event, associated cause, and control measure, as appropriate.

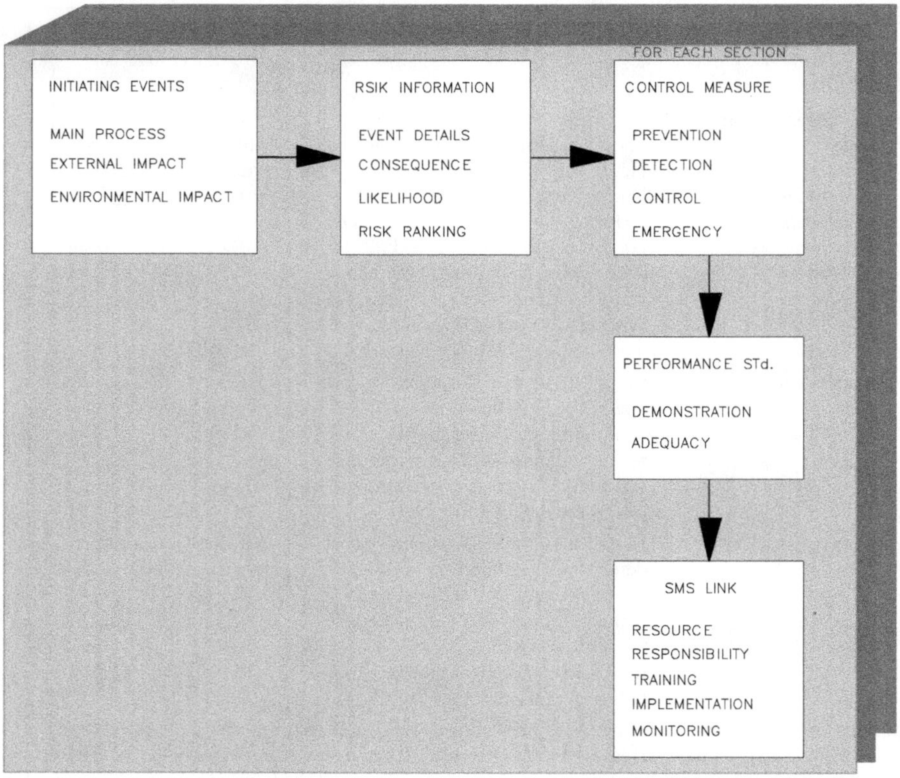

FIGURE II/3.2.5-1

Hazard register.

Table II/3.2.5-1 Section Division

Area No.	Area Details	Detailed Description	Toxic Inventory	Corrosive Inventory	P&ID Reference	Remarks
1						
2						
n						

- All documentation of the HAZID shall be maintained under the control of documentation either manually or in large cases through computer software (e.g., ProArc).
- The hazard register must contain sufficient information to support later stage use, and later time decisions can be taken on this basis.
- The document is maintainable in the sense that it is possible to revisit the document and update knowledge. So, it should accommodate any knowledge gap, assumption, uncertainty, debated issue (if any), learnings from hazards, etc.

To develop a hazard register there are hazard worksheets, which are illustrated in Table II/3.2.5-2. It is better to divide large cases into several sections; such a section division sheet may be as shown in Table II/3.2.5-1. Here the number of columns and rows are arbitrary and in reality will depend on the actual situation.

Two typical sets of HAZID worksheet are presented in Table II/3.2.5-2 Typical hazard worksheet1, in one style where the severity index and frequency index have been shown along with regulation details. In Table II/3.2.5-3 the HAZID worksheet cause and consequence details along with safeguards are shown. These are all typical examples and are presented only for understanding the typical data requirements and presentations. In the earlier discussions, major incidents and accidents were used together, but in HAZID these are normally used slightly in different contexts, which are explained in Fig. II/3.2.5-2.

3.2.6 Scheduling

As the name suggests this means fixing up a HAZID study program. Since this involves the workforce as well as an expert, proper planning and scheduling are vital. Another important issue is to allow sufficient time to complete the HAZID study with a fully defined *scope in depth* as required by the chosen HAZID *technique*. A few important guideline points are:

- Availability of expert person and workforce involved
- Need to maintain production and other activities
- Cross-shift involvement so as to adjust manning and keep activity unimpaired
- Proper mental alertness with required breaks so that fatigue does not set in
- Maintenance of continuity and consistency
- Interaction with workforce to get real feedback

Table II/3.2.5-2 **Typical HAZID Worksheet 1**

Hazard Worksheet					Accident Category: Category I Description XXXXXXXXXXXX					
ID	Hazard Description	Phase	Cause	Effect	Detection	Subcategory	Regulation	SI	FI	Comments
1.2										**Main ID may be actual/escalation issue under main ID**
1.2.1										
1.2.3										
a.b.c										

Main ID say 1.2 below actual case ID. Under main ID, etc. (just an example, or may be escalation issue).
FI, *frequency index;* SI, *severity index.*
Some typical ones shown; depending on application there may be variations.

Table II/3.2.5-3 *Typical HAZID Worksheet 2*

ID	Hazard/ Major Incident[a]	Cause	Possible Consequence	Existing Safeguard	Risk Reduction Procedure	Remarks
Main Section						
Subsection						
1.2 a.b.c						

Some typical ones shown; depending on application there may be variations.
[a]Refer Fig. II/3.2.5-2 also.

Major ACCIDENT / INCIDENT

***Major Accident*: In line with UK legislation Major Accident may be defined as**

- ***A fire, explosion or release of a dangerous substance involving death or serious personal injury to one or more persons on the installations or engaged in an activity on or in connection with it;***
- ***Any event involving major damage to the structure of the installation or plant affixed or the loss of stability of the installation or major environmental impact(s)***
- ***Any other event arising from work activity involving death or serious personal injury to five or more persons on the installation or engaged in an activity in connection with it.***

Safety assignment focuses on loss of containment, etc. so it can be slightly modified to include uncontrolled event, and accident can be caused by a sudden occurrence (an uncontrolled event) causing serious danger or harm to relevant person(s), at any risk community, property or environment immediately or at a later time. This derived from major hazardous facility (MHF) regulation. Here occurrence includes: escape, spillage, leakage, implosion, explosion, and fire. This is further to major accident defined in Clause 3.1.2.

***Major incident* is a term mainly used in connection with an MHF. As a quantitative term it shall include serious immediate risk to health and safety or extended injury. Also it includes loss of containment or the consequence of loss of containment. Here, loss of containment is discussed but in a near vicinity people may be harmed because of mechanical energy released, toxic effect, and temperature of the material even if loss of containment does not cause a fire. So, this is also major incident. MHF is mainly concerned with Schedule 9 (also 15) materials.**

Factors influencing major accidents or incidents are:

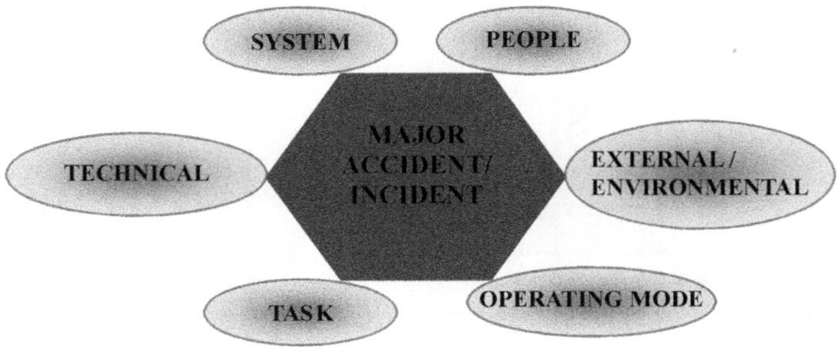

FIGURE II/3.2.5-2

Major accident/incident.

3.3 HAZID Process Description

So far the discussions have been on planning. Process description starts with collection and discussions on collected data. The basic process is depicted in Fig. II/3.3-1.

The HAZID study basically consists of the stages shown in Fig. II/3.3-1. All data and information are collected and duly reviewed by the team. They debate and analyze the data to develop an output for the hazard register discussed previously.

3.3.1 *Information Collection and Handling*

There are a few sources from where the data and information can be collected. The basis of HAZID is the comprehensive and accurate description of the facility by the operator. This includes but is not limited to the following:

- Complete process description in the form of PFD/P&ID, material safety data sheet, existing condition including any modifications carried out, etc., even before starting HAZID study.
- For an existing facility it is necessary to review all corrosion, explosion, fire incident, breakdown, equipment/control failure, and maintenance issues.
- Plant event history and/or similar plant event, major accident/incident event information or near miss issues should be reviewed. Here one point must be

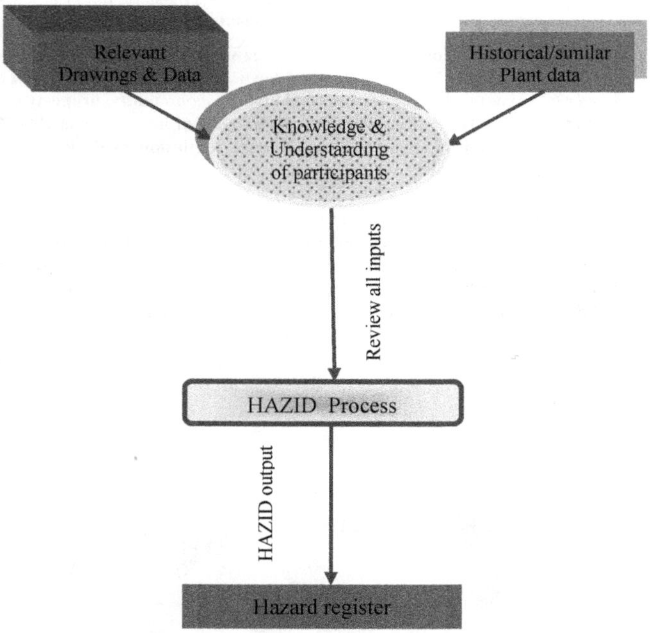

FIGURE II/3.3-1

HAZID process.

remembered: it should never be taken for granted that any incident not appearing in the history or similar plant data event *cannot* happen. So, anything that is omitted must be recorded for further review.
- Collection and compilation of all information.

Other important qualitative issues relating to the HAZID study should be properly addressed to obtain a better result. These are:

- All participants and the operator should understand the HAZID study and its purpose.
- The study being undertaken is applicable and relevant to the existing facilities.
- The study follows acceptable standard and norms.
- All knowledge gaps are identified and duly addressed.

3.3.2 General Considerations and Major Accidents/Incidents

Comprehensiveness and accuracy of HAZID is extremely important for subsequent actions such as safety assessment.

Initially, it is necessary that the facility is divided into several sections, which are manageable and logical. As shown in Fig. II/3.3.2-1, the division could be based on divisions in similar facilities. Geographical and in the case of a large complex facility logical sections may be based on similar process handling/chemicals or it could be based on similar operations and/or a combination of both, for example, in the case of an offshore mud storage facility it could be oil-based mud or water-based mud, etc. After this is done logically and in full agreement with the team and team leader, a question and answer session can be implemented. Some of these questions for each section may be as follows:

- Is there any deviation in process/activity or procedure from a design basis?
- What are the activities conducted and how could they go wrong?
- Is the presence of a hazard continuous or occasional?
- What equipment is there and how can it fail? Is there a possibility of failure because of external factors?
- What are the infrequent but abnormal activities that could be performed and how can they go wrong?
- Could there be additional hazards?
- What are the worst case events? It should be noted that not only is the design basis responsible for worst case events but also the presence of materials, extreme process conditions, failure to isolate the system, proximity, installation and commissioning issues, layout, proximity of other section/equipment, etc. could result worst case events.
- What are the major accident/incident events and how are they influenced (refer to Fig. II/3.2.5-2). A major accident/incident can be influenced by: (1) people (work culture, inadequate training, knowledge, understanding, attitude, mind condition); (2) the system (management philosophy, organizational chart, inspection, administration, communication and interaction); (3) the task (operation, construction startup); (4) the operating mode (normal operation,

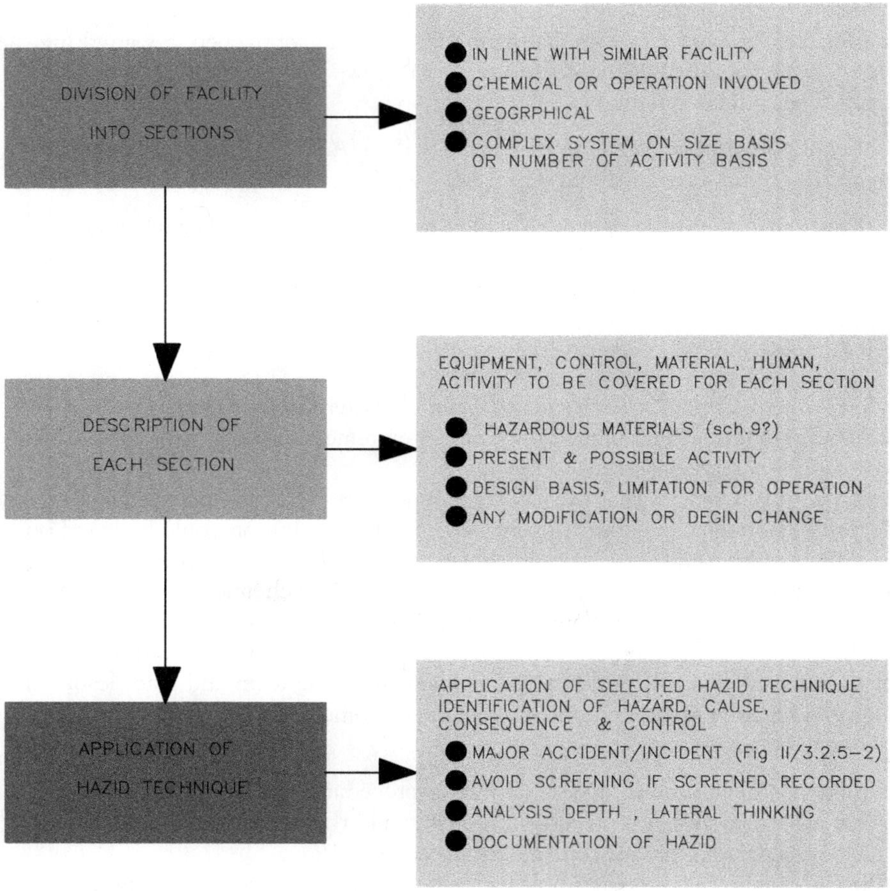

FIGURE II/3.3.2-1

HAZID process flow.

startup—when many interlocks may be bypassed—shutdown); (5) technology (equipment, controls); and (6) external/environmental conditions (suppliers, quality, political, earthquake, lightning).

- What are the possible events and their impact?
- What are the interactions with other sections and the possibility of a hazard causing aninteraction?
- How can existing controls and their impact be identified? It is helpful for a third party to understand that all identified hazards have controls and adequacy of control must be demonstrated.
- Can any additional control be introduced, and what is the impact or additional hazard introduced as a result?

- Already from Clause 3.2.5 it is known how major accidents/incidents are defined. It is now wise to consider some commonly used guided words in HAZID:
  - Unignited hydrocarbon (HC) released
  - Hydrocarbon (HC) released—fire
  - HC released—explosion
  - Toxic exposure
  - High pressure
  - High/low temperature
  - Corrosion
  - Object dropping
  - Improper access/escape
  - Radiation
  - Maintenance
  - Construction/startup
  - Explosives
  - Electrical
  - Mechanical
  - Structural
  - Effluent disposal
  - Biological and others

3.3.3 *In-depth Analysis*

The main motto of the study is that it is possible for the operator and others concerned to understand the nature of each hazard. The analysis shall be done in depth so that root cause is well identified, and appropriate remedial or control measures could be prescribed. The success of the study will not be achievable if only hazard is identified. The study shall be deep enough to indicate "When," "Where," and "Why" a hazard is present.

3.3.4 *Creative and Lateral Thinking*

In any plant there are certain obvious hazards, for example, fire hazard in shale shaker area of offshore facility or generator are hydrogen cooled generators in power plants. HAZID is not meant to identify such kinds of hazards only. There needs to be some realty and lateral thinking power while carrying out the study. Also there always exist certain chances of a hazard in a complex event, on account of a series of activities. In the study this can be revealed. These characteristics shall include but are not limited to the following:

- Challenge the existing system, such as any assumption, norm, and action, to discover any glaring weaknesses
- Thinking creatively beyond immediate experience
- Look in depth for the effect of failure of systems, controls, etc.
- Understand how a small problem can lead to a bigger problem, even to a major accident/incident, for example, a simple leakage in an airline can lead to the

closure of a final control element, which could lead to a dangerous situation if no backup or fallback is considered

- The combinational effect of various factors can lead to a catastrophe, for example, a normally running safety device is always in service but on account of a specific issue one criterion might have been bypassed during startup. Because of human error this is not reversed and because of a combination of two human errors it can lead to a dangerous situation because protective devices are bypassed. For a better understanding of combination effect hazards see Fig. II/3.3.4-1 (inspired from Guidance Note WorkSafe [10])
- Other important considerations: It is often observed that there is a general tendency to ignore certain accidental events on account of low probability of occurrence. It may be thought by some that sufficient control measures have been considered with redundancy and the likelihood of occurrence is almost

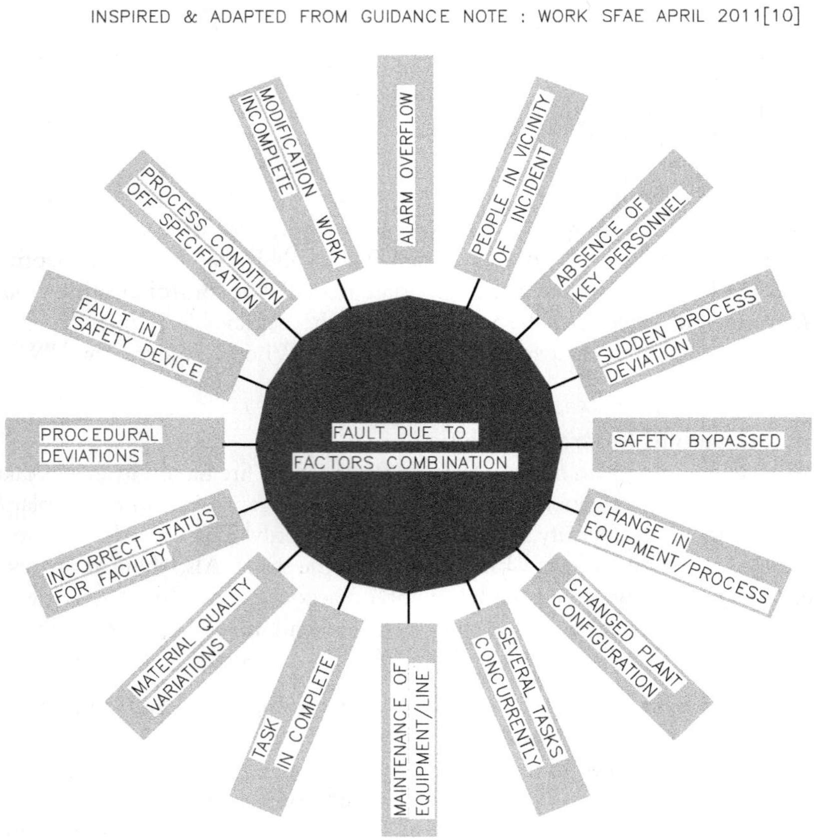

FIGURE II/3.3.4-1

Combination factors fault.

impossible. In any case, when any major accident issue is rejected on account of these considerations, in all probability such safety assignment will also be rejected by the authorities. In certain applications such as offshore, this tendency is rarely considered to reduce space and weight requirements. However, although there may be excellent control systems and protections , can anyone say for sure that it will not happen? Major contributing factors toward this logic are:

- Over time the effectiveness of controls will deteriorate.
- New control technology and action when needed may not be time tested.
- After the initial investigation, if anything is left out, and in a subsequent case it is recognized, there may be a possibility that the control system might not be adequate. Also the control system may not be properly managed on account of the absence of importance attached to it and control system requirements are properly followed.
- Potential dangers are extremely important for emergency planning.

However, initially, preliminary hazard analysis is often done for hazard screening processes to assess a major event. This will help in developing an SMS and FSA. Therefore the discussion may be concluded by saying that no potential danger situation is liable for rejection however low the likelihood of its occurrence is. Also if any assumption is considered it must be recorded for further review.

3.4 HAZID Output

The main output of HAZID is the documentation in the form of the hazard register where all hazards are listed with identification for each section, and cause and consequence and suitable linking with control measures are recorded. This is done in a logical structure and detailed in Clause 3.2.5 of this chapter, and hence is not repeated. Here, one very important point is that all assumptions, debated issues, and uncertainties must be logically recorded for further review. Another issue connected with HAZID study output is its application. The major application area of HAZID by the operator shall include but not be limited to the following:

1. An input document for safety assessment
2. Imparting knowledge with the workforce, so that hazard awareness is developed. This can be done through workshops and training. It very much depends on the operator's outlook.
3. The hazard register helps in the revising and rethinking processes, and general awareness about hazard is greatly improved.
4. Identified control measures are taken in safety assessment and assist in looking at the control system from different angles to improve its effectiveness.

3.5 Review

The operator is responsible for identification all hazards before they turn into a major event. Therefore HAZID is an ongoing and dynamic process. Whenever any new system or device is introduced and/or changed, HAZID shall be carried out to ensure that the assessed safety is not jeopardized. MOC is always an integral part of HAZID, so a facility modification or change in control needs to be carried out. If there is deficiency in control measures, HAZID needs to be augmented.

3.6 Quality Check

As in any other system, a quality check for the study is important so that it is effective and safety assessment is appropriate. The quality of HAZID may be based on input/process/output or a combination of these. The pillars on which the quality of HAZID depends are as follows:

1. Team: Qualification and knowledge, experience, training, and facility knowledge of the team carrying out the study
2. Involvement of the workforce and health and safety engineering representative, and their understanding and awareness
3. Selection of method most appropriate for the facility
4. Proper documentation and periodic review
5. Sufficient time allowed for completing the study

3.7 Task Analysis

The task analysis technique has been developed mainly to handle human factors (in HAZID and other methods). In HAZID this is helpful in addressing human factors such as human error, man—machine interface, and procedural error. Of the various hazard identification techniques, task analysis is one of the most important. The other techniques are discussed in *subsequent chapters*, but here discussions will be on task analysis. Task analysis is the study of what users need to do, in terms of actions and/or cognitive processes, to achieve a task objective. There are several factors, such as task duration, repetitive frequency, task allocation, complexity, equipment, ambience, and environmental conditions, which are required for the task and they very much influence the performance. At times, tasks are often used interchangeably with process.

- The basic purpose of task analysis is to systematically describe and analyze activities and document the procedures, processes, and resources that are used by individuals or groups to achieve the target/goal. Task analysis deals with a variety of human factors, so for HAZID and PHA the aim is to assess the human error or chances of human error and try to reduce that effect.
- Often task analysis and performance analysis are confused. They are parallel but are two different things. Performance analysis deals with future achievement while task analysis deals with the results currently being achieved.

Here concentration will be on task analysis. There are clear advantages of task analysis, but at a cost.

The advantages are:

- It gives a clear idea about the relationship between resources, process, and input with an accomplished result.
- It allows identification of what is good and bad about the current process and identifies the areas where there is scope for improvement.
- It allows a systematic review of the current task, and it will be possible to put forward recommendations for changes needed for the current task for its improvements and hazard-free operation.

The disadvantages are:

- It requires a fair amount of time and resources, which might not have been planned for.
- It is difficult and challenging to determine how the task will change because of recommendations, and what will be their impact on other parts of the system.

Broadly there are two kinds of task analysis:

- Action oriented: In this method, the observable behavior of humans along with task structure is described.
- Cognitive type: This contemporary technique is used for supervisory control and decision making. Here, the focus is on the mental process behind human observable behavior in decision making or problem solving. This also forms part of HRA.

The following important terms are defined first, for better understanding of subsequent discussions:

- Duty: The first level of breakdown of a job is the identification of duties, which are the areas of responsibilities of a person. From duty one can get an overview of the job. Normally, a job consists of five to six duties on average.
- Task: This is the first breakdown of duty, and can be further broken down into elements.
- Elements: This is the final breakdown component of a task. It gives the finer details of the task. (It is like an atom of an element.) When priorities are assigned to a duty, it permeates to applicable elements too.
- Performance influencing factor (PIF): There are a few factors, such as work environment, procedure definition, time stress, effective training, noise, long duty hours, etc., that influence the likelihood of error and human performance. It should be noted that PIFs are not automatically associated with but can influence human error. PIFs are plant specific. In applications like HRA, PIFs are rated numerically (e.g., worst 1, average 5, and best 9) to check the ratings for a procedure (written procedure, training) or working environment (such as noise, lighting, etc.). Also factors like fatigue (caused by prolonged duty hours, shift rotation, lack of rest, etc.) influence PIF, and may result in errors. The PIF scale varies from a well-defined procedural system to worst possible system without training or defined procedure, and is plant specific.

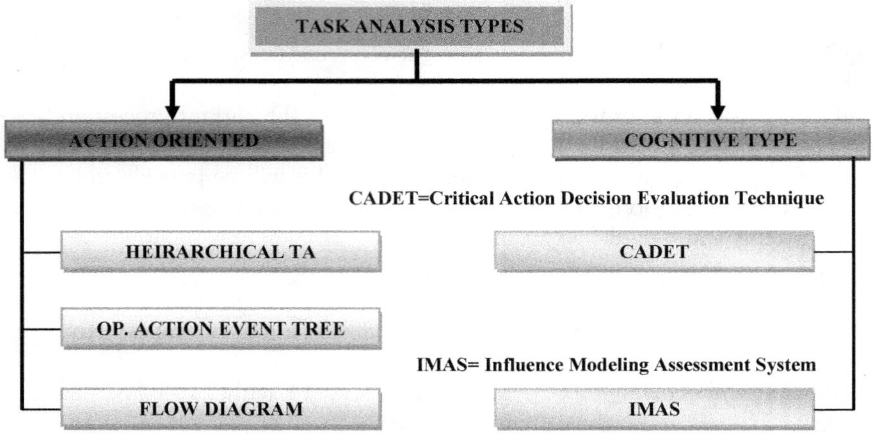

FIGURE II/3.7-1

Various task analysis types.

There may be certain preconditions for a task, and by task analysis these preconditions may be eliminated to make the system error free. Task analysis is quite often used for a new system or for modification work in a HAZID application. It can be used during detailed investigation. We will now concentrate on various systems of task analysis, as shown in Fig. II/3.7-1.

3.7.1 Hierarchical (Action-Oriented) Task Analysis

This is a top-down system. Here, there are two important terms that need to be understood:

- Operation: This refers to the action by the operator while interfacing the system or the action by the system itself.
- Plan: The condition necessary to take the action referred to as to Plan.
- Hierarchical task analysis description: Hierarchical task analysis starts its journey from the overall objective needed to be achieved. This objective is divided into a series of suboperations. The plan specifies the information source and when it is to be carried out. The question is do these suboperations need to be subdivided into finer elements? This is totally dependent on the analyst, in case the analyst feels that there is scope for error. The analyst can then further divide the suboperations into the lowest level or elements to improve the quality of task analysis. This will be clear from the example of CCPS of isolating a one-level transmitter [13], as shown in Fig. II/3.7.1-1. It should be noted that in this example, one task has been further subdivided, so that the possibility of obtaining one side of a differential pressure transmitter is avoided, following instrumentation working convention.

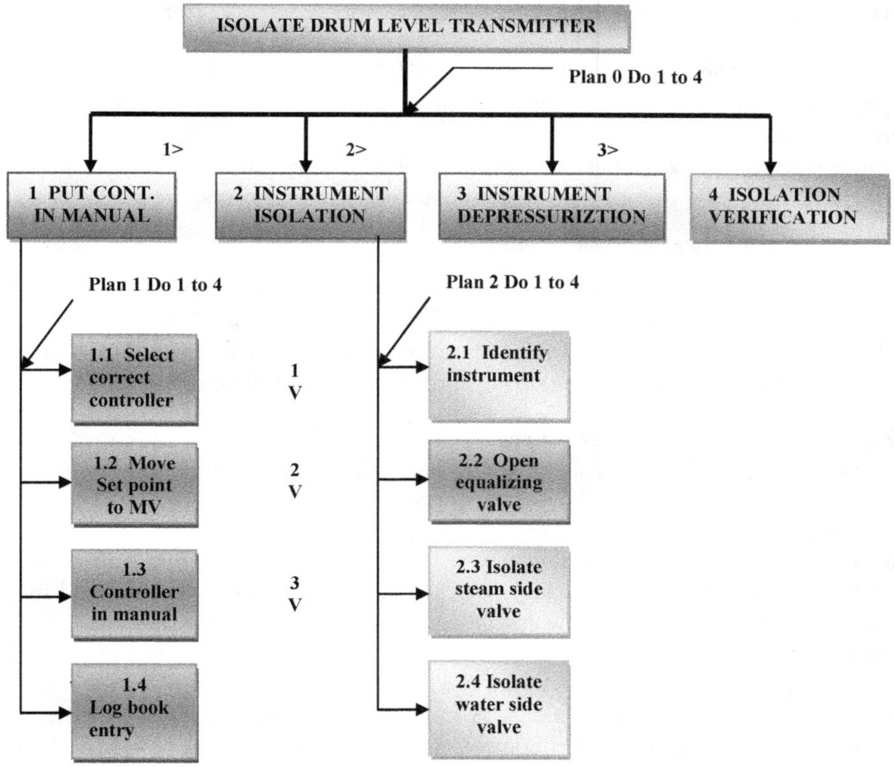

FIGURE II/3.7.1-1

Hierarchical task analysis with example for isolation of drum-level transmitter.

As shown in the figure the process starts with the main objective of isolation of a transmitter by Plan 0, which says to perform operations 1 to 4. At operation 1 it has precondition Plan 1 where it needs to do suboperations 1 to 4 (1.1—1.4) to complete operation 1. Then it goes to operation 2 and faces Plan 2, which again asks to complete suboperations 2.1 to 2.4. Note here that operation 2.2 is necessary to ensure that at no limb of the differential pressure transmitter is there a chance of much higher pressure than when carrying out suboperation 2.3 and 2.4. Also this ensures two options for transmitter drain so that the transmitter is depressurized, which is a requirement in operation 3. Finally, at operation 4 the main objective is achieved and verified. The following are the major advantages:

- It is economical in the sense that knowledge gathering and organizing can be carried out up to the point of desire.
- An analyst can concentrate on the crucial aspect of the task that is concerned with safety.

- Hierarchical task analysis can be the starting point for further error analysis methods.
- When used as input to design functional safety, hierarchical task analysis can be placed at a higher level, which helps in allocating functions to the system and people concerned.
- Since hierarchical task analysis is developed in collaboration, the analyst is well aware of the perception of people and allocated tasks and plans.

There are a few negative points also. These are:

- As hierarchical task analysis is developed with various levels of people a good amount of time is essential.
- The analyst needs a fair amount of skills training and technical knowledge about the system so that the breaking down of the tasks can be done effectively.

3.7.2 *Operator Action Event Tree*

As the name suggests Fig. II/3.7.2-1 is like an event tree diagram, which shows various decisions and actions concerned people are expected to do. This operator action tree is mainly used for post accident cases to pinpoint an operator's failure. This is very much used in accident analysis in air crashes. The level of breakdown of the main task is crucial. Too many breakdowns may pose a problem in handling. From visual inspection it is possible to identify the crucial operation that can initiate the event. The figure will give a clearer idea.

In this example, in its simplified form the actions of the operator when the drum level is high initiates the pretrip alarm. Also in this example, one set of boiler feed pumps (BFPs) and one set of feed control valves (FCVs) have been considered to

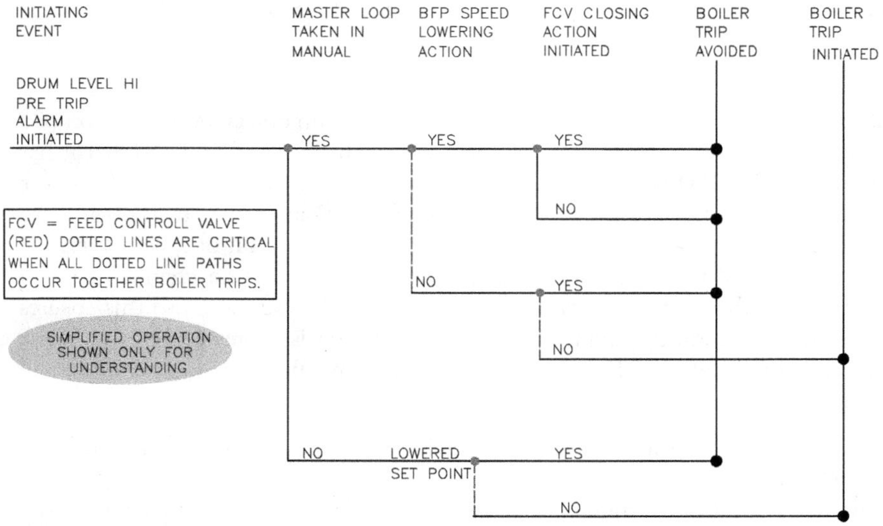

FIGURE II/3.7.2-1

Operator action event tree example.

simplify the diagram. By manual control in first action stands for master control in manual mode, whereas individual controls for BFP speed and FCVs are shown separately. This is structured information on operator actions after the event occurred and is able to identify critical action in response to an initiating event. Here it is possible to detect the error of omission but it does not throw any light on alternative actions, if any. Also if the system is complex then the event tree will be much larger and sometimes may be difficult to handle. In this example, if 3 × 50% BFPs and FCVs are considered then the event tree size will be quite large and somewhat complex because of logical selections.

3.7.3 *Flow Diagram*

This type of task analysis is applied when there are issues related to decision making, complex situation handling, time-based system interlock checking, etc. This is somewhat like a program flow chart in which action sequences are described. The system is described with the help of very short burner management logic, as shown in Fig. II/3.7.3-1.

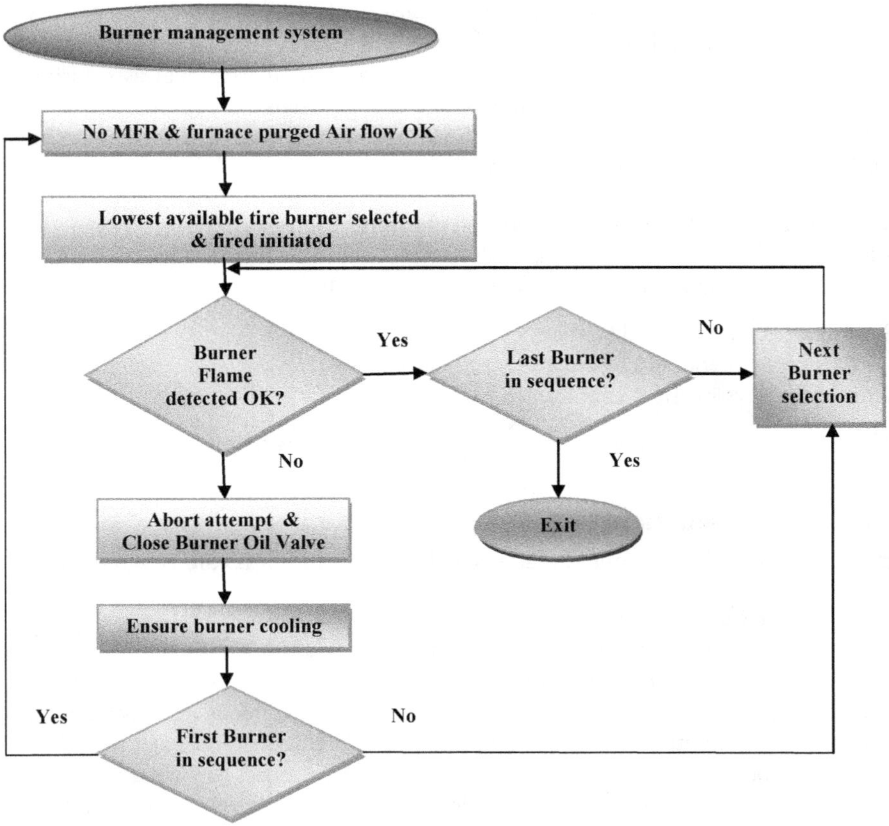

FIGURE II/3.7.3-1

Building management system flow diagram.

This is user friendly to the workers because it describes all the critical steps the operator needs to take care of to complete the process. This also helps the analyst to check whether all check points have been fully satisfied by the worker. However, if the task is too complex, unless they are broken down into smaller tasks, it is very difficult to handle.

3.7.4 *Critical Action and Decision Evaluation Technique*

The Rasmussen ladder diagram is the basis for this technique. As the name suggests it deals with critical actions and decision making. It senses the consequences. In case of failure of critical action and decision, then there may be serious consequences for safety, production, etc. Initial alertness, observation, interpretation, execution, feedback, etc. are some of the decisions and actions. For example, a pretrip alarm indicates the first stage of a problem. This is when the operator needs to take action (as shown in Fig. II/3.7.2-1). However, "absent mindedness," "distraction by any other issue," or "low alertness" are a few errors that may prevent the operator from taking necessary action in time, which may result in a boiler trip (as shown in Fig. II/3.7.2-1). Similarly, observation and interpretation of data from, say, an instrument reading are vital. During observation the operator may make assumptions (e.g., in the example of the drum level, if the assumption was that the high level was caused by swelling, which occurs at times of high demand, and actually the level rises during low demand, and the action is ignored, then the consequences may be detrimental). Also by the time operator realizes that the initial assumption was incorrect the operator needs to take alternative actions.

3.7.5 *The Influence Modeling and Assessment System [14]*

This is another kind of cognitive task analysis, and is used to elicit a subjective cause sequence model, which is a graphical representation of the operating team regarding:

- Alternative causes for a disturbance
- Various consequences as a result
- Various displays in the control room

These are used in training to acquire diagnostic solutions to problems.

3.8 HAZID Discussions

In the following subclauses, human factors and pitfalls in HAZID are covered.

3.8.1 *Human Factor*

HAZID discussions are incomplete unless human factor is addressed. Actually, human factors can be seen as interactions among people, organizations, systems, and equipment including control systems (HW&SW). Human intelligence is unquestionable, but there are also limitations in understanding, interpretation, and capacity. This varies from person to person, as well as with training and a person's particular trade, for example, when a designer is assigned a technical marketing task his performance rating may fall as that person may not be trained in technical details and coinages needed for marketing. So, from a major hazard perspective, the human role is critical and must be addressed in safety documentation. The operator should always examine foreseeable major accidents and consider the human factors involved. For example, human factors can save a plant from catastrophe through intuition and intelligence; similarly human factors can contribute to hazards at all levels. Therefore, task analysis has an essential role to play.

3.8.2 *Worst Case Scenario*

The operator needs to examine case by case the entire documentation, assumptions, etc., and the risk estimate (risk analysis) discussed later, to arrive at a worst case scenario for a facility.

3.8.3 *Common Mistakes, Pitfalls, and Suggestions*

- There is always a tendency to screen a few hazards because either their occurrence is rare or their consequence is low because of the control system. Again, the key purpose of risk assignment is to find the control measure and improve its efficiency and effectiveness. So, with such assumptions, it becomes self-defeating.
- Hazard identification should not be too generic.
- All root causes, initiating event(s), must be identified.
- It is better to conduct risk assignment/analysis, which should not be linked with HAZID.
- Large systems should be broken —down into smaller systems because the broad scope of HAZID is difficult to handle.
- Preparation, record-keeping mode, etc. must be decided beforehand to obtain the desired result. A pilot study may be helpful.
- It is better to include experts from outside along with workforce participation.

3.9 Risk Estimate (Brief)

After the hazards have been identified, it is necessary to find out the initiating event(s). Initiating events are those events that unleash the potential inherent cause of the hazard and, either directly or indirectly, cause consequential damage to the property, people, and environment. Therefore there will be a list of initiating event(s). Risk associated with these events will be estimated so that control measures can be assigned as part of risk assessment discussed later. As a preliminary step, qualitative risk estimation is a good choice. Here it is to be noted that the hazard register and HAZID worksheets discussed in Clause 3.2.5 will be helpful. From Fig. II/3.9-1, one can find the acceptable and ALARP (refer to Chapter I) events. Risk analysis is not only associated with HAZID, therefore risk analysis, assessment, and management will be discussed in Clause 4.0 in this chapter.

The risk matrix method can provide a quick understanding of the risk profile of the facility and can be based on judgment or be further investigated using more detailed information. This method has limitations also. It is not easy to incorporate the effects of risk reduction measures within the risk matrix, and it cannot be used for cumulative risks.

FREQUENCY CLASSIFICATION

FREQUENCY CLASS	QUALITATIVE ESTIMATE	DESCRIPTION DETAILS
5	OCCATIONAL	EXPECTED MANY TIMES DURING SYSTEM LIFE TIME
4	LIKELY	EXPECTED FEW TIMES DURING SYSTEM LIFE TIME
3	UNLIKELY	AT MOST ONCE DURING SYSTEM LIFE TIME
2	REMOTE	SHOULD NOT HAPPEN DURING SYSTEM LIFE TIME
1	EXTREMELY UNLIKELY	NOT EXPECTECTED DURING SYSTEM LIFE TIME

DAMAGE CLASSIFICATION

DAMAGE CLASS	QUALITATIVE ESTIMATE	DESCRIPTION DETAILS
5	CATASTROPHIC	HIGH NOS. OF DEATH ON/OFFSITE COMPLETE LOSS OF SAFETY
4	CRITICAL	ON SITE DEATHS AND/OR COMPLETE LOSS OF SAFETY
3	SEVERE	SERIOUS DAMAGE TO HUMAN COMPLETE LOSS OF SAFETY
2	MARGINAL	PARTIAL DAMAGE TO HUMAN PARTIAL LOSS OF SAFETY
1	SAFE	MINOR INJURY/NO DAMAGE NO LOSS OF SAFETY

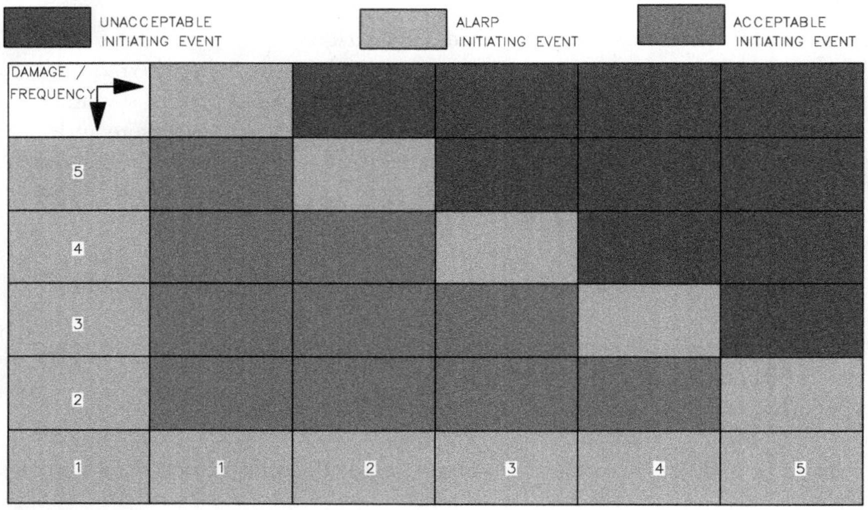

FIGURE II/3.9-1

HAZID qualitative risk estimate.

4.0 RISK ASSESSMENT AND MANAGEMENT

With the basic definition and knowledge on hazard and risk, risk matrix, risk ranking, etc., it is now possible to proceed with discussions on risk assessment and management. The first step toward risk assessment is hazard identification. This is done through knowledge of various PHA and HAZID discussed in the preceding two clauses of this chapter. Now the reader is in a position to follow discussions on risk assessment and management. In risk assessment, one part is control measures, which will also be covered in this clause. Here it is worth noting that the control measures referred to earlier do not mean process control alone but cover all physical controls for prevention and mitigation of risk (including process control). Risk analysis and management enable analysis and management of risks associated with a facility or project. No two projects or facilities will have identical risks, hence a specific task is necessary for each case to reduce/eliminate/mitigate risks. The international standard for this is ISO 31000, which is discussed in Clause 1.1 of Chapter VI.

4.1 Risk Analysis Basics

In this subclause, short discussions are presented on various definitions, meanings, etc. so that these can be used in subsequent clauses.

4.1.1 Various Terms and Definitions

It is better to refer to various standards so that the reader can get the feel of the entire scenario. For convenience, various terms are linked with the figures in the book for a better understanding.

- What is risk assessment?
  - From the Canadian Center for Occupational Health and Safety, risk assessment is a process to identify hazards, analyze or evaluate the risk associate with those hazards, and determine appropriate ways to eliminate or control hazards.
  - According to ISO 27001:2005, risk assessment combines two techniques: risk analysis and risk evaluation (defined later).
  - At HSE.gov.uk, risk assessment work has been described as identifying the harm to people in the workplace, who might be harmed and how. This is achieved by evaluation of risks and deciding on appropriate control, taking into account controls already in place, recording risk assessment, reviewing, and updates (Fig. II/4.1.1-1).
  - ISO 31000:2009 (preface) says that risk assessment attempts to answer the fundamental questions:

    What can happen and why?

    What are the consequences?

    What is the probability of their occurrence?

    Is there any factor(s) that mitigates the consequence of risk or that reduces the probability of risk?

    As per ISO Guide 73:2009—Risk Management—Vocabulary, it is the overall process of risk identification, risk analysis, and risk evaluation.
  - As per NORSOK Z013: "Overall process of performing a risk assessment including: establishment of the context, performance of the risk analysis, risk evaluation, and to assure that the communication and consultations, monitoring and review activities, performed prior to, during and after the analysis has been executed, are suitable and appropriate with respect to achieving the goals for the assessment"
- What is risk analysis?
  - According to ISO 27001; 2005: Risk analysis uses information to identify the possible sources of risks. It uses the information to identify the threats or

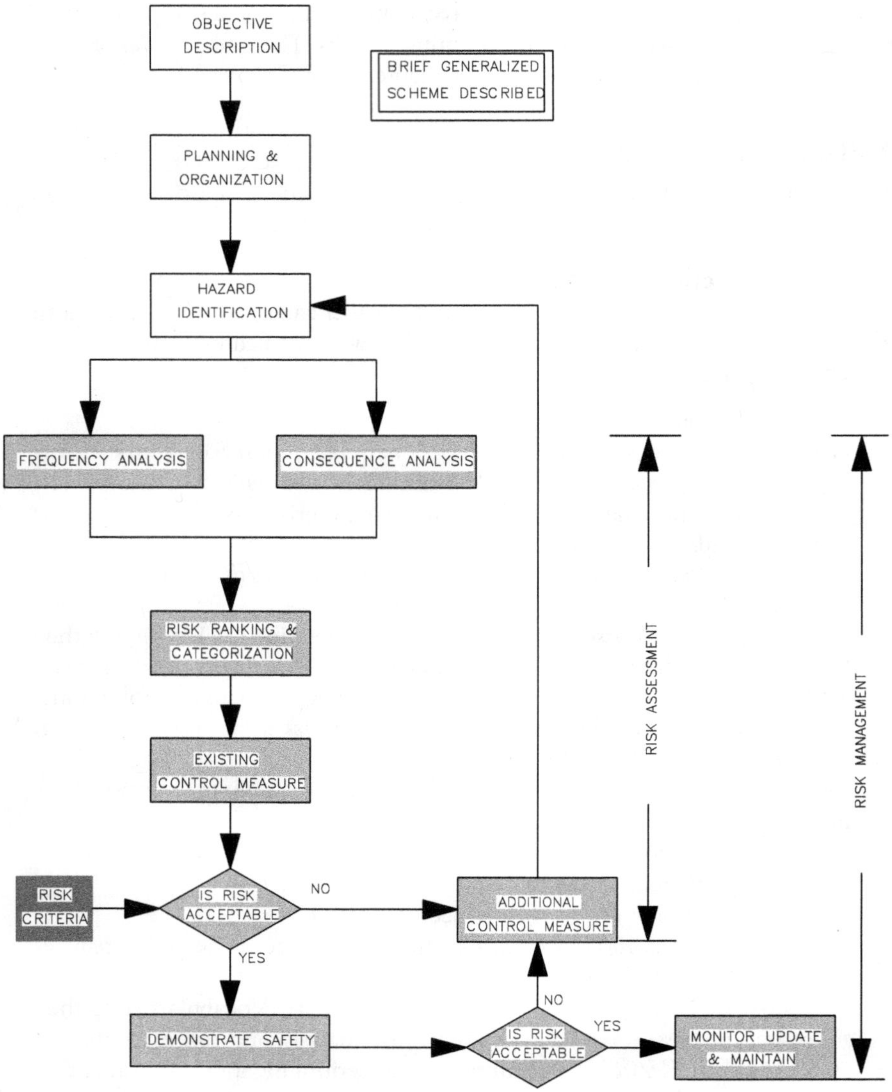

FIGURE II/4.1.1-1

Basic risk analysis process.

events that could have harmful impacts. It then estimates the risk by asking: what is the probability that this risk will actually occur in future? And what impact would it have if it actually occurred?

- As per IEC 60300-3: "Systematic use of available information to identify hazards and to estimate risk to individual or population or environment."

- As per NORSOK Z 013: "Structured use of available information to identify hazards and to describe risk." (With notes)
- What is risk evaluation?
  - According to ISO 27001:2005: A risk evaluation compares the estimated risks with risk criteria (Fig. II/4.1-1). This is done to determine how significant the risk really is. The estimated risk is established by risk analysis.
  - As per NORSOK Z013: Judgment, on the basis of risk analysis and risk acceptance criteria (Fig. II/4.1-1), of whether a risk is tolerable or not.
  - As per ISO 31000 one needs to consider project objective, whether some activity needs to be undertaken, need for risk treatment (if any), priorities of treatment, tolerability of risk to others, and decide if it is acceptable.
- What is control measure?
  - According to ISO 31000:2009 it is basically modifying the risk. (However, note that controls may be any process, policy, device, practice or other action which modify risk, and that controls may not always exert the intended or assumed modifying effect.)
  - From ISO 2700:2005 definition: Control measures are the ways to control each specific hazard. Hazard control methods are often grouped into the following categories:
    - Elimination (including substitution)
    - Engineering controls
    - Administrative controls
    - Personal protective equipment
  - Control measures in a facility may be any system, procedure, process device that will be used eliminate hazards, or prevent hazardous incidents from happening, or reduce the severity of the consequence of hazardous incidents when they occur. Control measures act like a wall between the hazard and the facility (similar to a firewall in a computer system).

4.1.2 Definitions of a Few Related Terms

- Risk assessment and safety audit: A risk assessment result can be used to predict whether the facility is safe or not. If it is not acceptable then additional control measures must be used to keep the facility safe. A safety audit is a process by which such safety claim is verified for consistency in results. Each case-by-case validation is done.
- Risk assessment and SMS: As stated earlier, with the help of risk assessment and (if necessary) with additional control, a facility will be made SFARP (so far as is reasonably practicable), that is, as far as practicable. SMS is the system to ensure that the risk levels achieved during risk assessment will be well maintained.

4.1.3 SFARP/SFAIRP

Both these abbreviations are in use. In HSE.gov.uk it is used as SFAIRP. In Clause 4.3 of Chapter I ALARP has been discussed. As per HSE.gov.uk they essentially mean the same thing, the core concept being "reasonably practicable." The term balances risk with time trouble, difficulty, and cost (together termed as sacrifice). Similar to ALARP, if sacrifice is disproportionate in relation to risk, it may be accepted but needs to be demonstrated. The major influencing factors that need to be considered are:

- Likelihood of the risk
- Severity of the consequence (harm degree)
- Availability/suitability of the ways to eliminate or reduce
- Cost for reduction/elimination
- Concerned person knowledge about the hazard and its reduction

4.1.4 Objective and Philosophy of Risk Assignment

The objectives of risk analysis and assessment are as follows:

- Develop a basis for identifying, evaluating, and ranking various risks with risk contributors.
- Define and justify control measures for the facility.
- Establish a link between the control measure and potential major incidents. This is necessary for formal safety assessment where applicable.
- Another purpose of risk assessment is to choose the appropriate tool for risk analysis and selection of control measures. For this, a preliminary study may be helpful.
- Capture knowledge about the major hazards and associated risks that can lead to a major incident. This knowledge will help in combating the challenges posed by the risks.
- Identify the concern for community, safety management issue, and emergency plan.
- Demonstrate the adequacy of control measure as well as reduce risk to ALARP.
- The result of risk assessment/analysis is used for decision making regarding additional control measures and may be required to provide specific information demanded by regulation.

The main philosophy behind risk analysis is to ensure that the entire workforce has sufficient *knowledge*, *awareness*, and *understanding* of the risks from major incidents. Understanding the risk may be accompanied by uncertainty, and by effective risk analysis, this is removed through debate, active participation, and sharing knowledge about risks and control measures. In this connection NORSOK Standard Z-013 Clause 5.2.2.3 may also be referred to. The standard also includes operation constraints and limitations, defined situations and accidents, area system and equipment classification, etc.

4.1.5 Features and Framework of Risk Analysis: The Basic Features of Risk Analysis Shall Cover

- All hazards and contributors for major incidents
- All consequences and frequency with which they happen
- All stages of the facility

In its general form, the basic framework accordingly shall include analysis of cause frequency and consequence, justification of control measure, and final evaluation, as shown in Fig. II/4.1.5-1. The entire procedure is greatly influenced by a number of factors such as people, viewing the system, appropriate selection of analysis process control measure, updating, and maintenance as shown. The operator's safety philosophy plays vital role in modifying the structure of risk analysis. If the intent is to

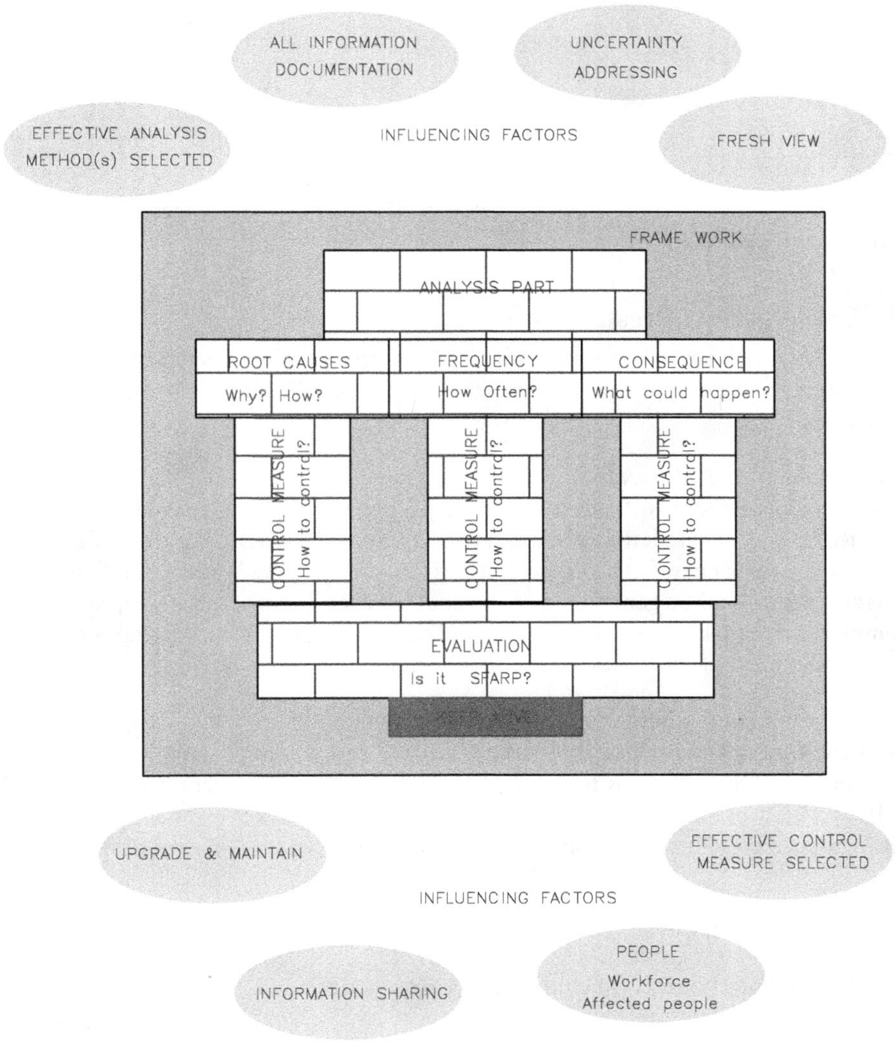

FIGURE II/4.1.5-1

Framework and influencing factors for risk analysis.

seek clearance from an authority or to meet the requirement of a standard, then it will be guided by the basis of the standard. However, it is the duty of the operator to ensure that all aspects have been covered, even if they are not covered in the standard. If the risk analysis is done for an engineering or management system, then the framework may vary.

4.1.6 *Standards*

Standards associated with risk analysis shall include but are not limited to:

- ISO/IEC 31000
- IEC 60300-3
- EN 1050
- NORSOK Z-013
- ISO 17776

The references of a few are given so that, according to need, the reader can refer to the relevant standard.

4.2 Risk Analysis Prestart Issues

For an effective risk analysis it is recommended that there shall be some planning and preparation before the actual procedure is taken up. The following basic steps may be considered.

4.2.1 *Scope and Approach*

It is very important to define the scope of the job (NORSOK Z-013 may be referred to). If this activity is undertaken for a routine/nonroutine activity of the company project, the approach will be different if it is taken as FSA. Within this limited discussion we shall take a generalized approach. However, a more rigorous approach is recommended so that operator can understand the preventive and mitigating actions, and can take well-informed decisions. Also it should help in developing SMS.

4.2.2 *Technique(s) Selection*

It is of the utmost importance that the operator selects the most appropriate technique for the facility. Various techniques are discussed in subsequent chapters (see also IEC 31010) and hence are not discussed here. For *compatibility and suitability of the techniques* for the facility, the following points should be under consideration:

- Type and complexity of the facility
- Type of likely hazards
- Selection of control measure
- Ability to rank the risks (likelihood and consequence)
- Level of details desired
- Available resources
- Technique complexity as per requirement
- Ability to handle cumulative risks

- Consistency with the facility's general approach

Depending on the requirements these could be qualitative, semiquantitative, or quantitative. As discussed next.

4.2.3 Risk Analysis Types

There are basically three types of risk analyses: qualitative, semiquantitative, and quantitative. The type suitable for a facility mainly depends on selection of the risk assessment process discussed, details desired in the assessment result, and predictive risk level.

- Qualitative: With the help of the risk matrix discussed in Clause 3.0 of Chapter I, it is possible to carry out rapid risk assessment. However, this is very project specific. Also here the numerical scale is not used. In this approach, cumulative risk assessment is very difficult. These are subjective approaches.
- Semiquantitative: Risk matrix (Clause 3.0 of Chapter I), risk nomogram, and risk graph (Fig. I/3.4.1-1): In these approaches it is possible to generate the numerical value of risks (but not absolute), to differentiate risks, and to conduct rough cumulative risk assessment. Also control measure assessment is possible.
- Quantitative: Based on calculated estimate risk value, calculation is possible. This is suitable for large complex facilities, where detailed study is required. However, it is time-consuming and expensive.
- In most cases a three tier approach is adopted, as shown in Fig. II/4.2.3-1. Initially, a qualitative or semiquantitative approach is taken to assess the risk and screen it. When risks are in a high risk zone or there is the possibility of a major accident event, then quantitative risk assessments are carried out to prescribe necessary control measures. It is quite common that in many cases a combined approach is necessary to justify consequence analysis. Mostly, when a quantitative approach is undertaken, prior preliminary analysis is done. From the diagram it is seen that whenever *all* replies to the queries shown in the diagram after qualitative analysis are "NO," then the action stops. If any reply is "YES," then the next level of analysis is carried out. A similar approach is applicable for semiquantitative and quantitative analyses (Fig. II/4.2.3-1).

In some cases detailed studies are necessary to pinpoint major incidental events, for example, aging (Chapter I). For such cases several quantitative studies will be undertaken. Some detailed study lists are presented in Table II/4.2.3-1.

4.2.4 Staffing and Workforce Involvement

In line with the requirements stated in Clause 3.2.3, there shall be total involvement of the entire workforce in the system. Only a few numbers will be directly involved in the process while others' views shall be counted through their interview feedback report. Information *sharing* is very important. The team will comprise management, supervisors, operations, maintenance groups, etc., and, wherever necessary, specialized groups. It is quite natural that each of these groups looks at things from different angles and shares different experiences. Also data is better retrieved from interviews

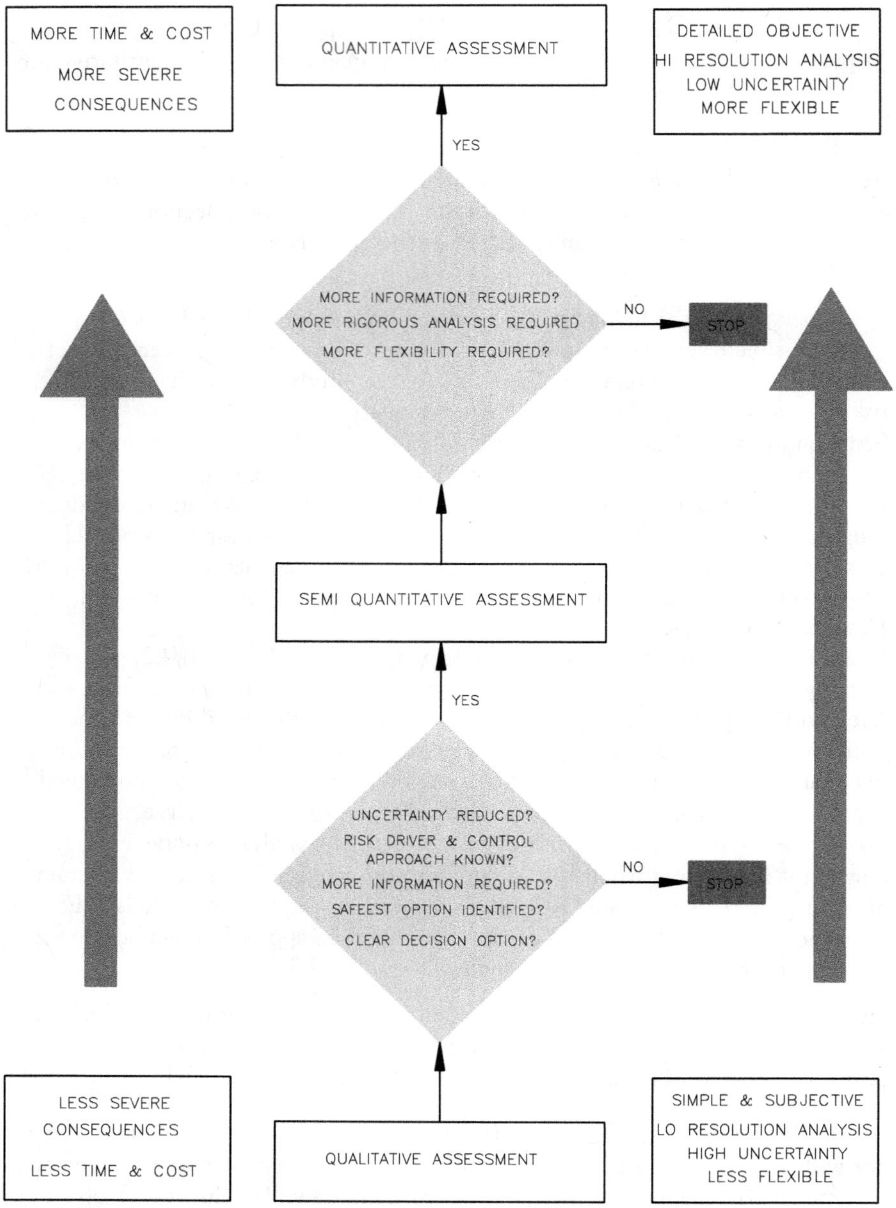

FIGURE II/4.2.3-1

Tier approach for risk assessment.

Table II/4.2.3-1 List of Probable Investigation Studies (Inspired by Ref. [16])			
Risk Driver	**Investigation**	**Risk Driver**	**Investigation**
Aging and integrity	Mechanical integrity, corrosion rate, breakdown data, reliability, inspection and maintenance issue	**Hazardous gas**	Ventilation/layout, gas/smoke ingress, wind tunnel, overpressure, gas dispersion
Process condition changes	Various HAZOP/ mechanical integrity	**Dropped/load impact**	Dropped object/ layout study/material handling
Human error	Task/HRA or procedural study	**Fire/explosion**	Electrical zone classification/ hazardous study/ equipment compliance
Control system reliability	Power supply, common mode failure		

and information from those who are not directly involved in the team but connected with the working of the facility is also helpful. It will also be better to involve contractors and suppliers because they can provide valuable information. It will be the responsibility of team members to carry out cause, likelihood, and consequence analysis. Therefore team members should possess good knowledge of the risk analysis process and be in a position to suggest control measures. So, experienced personnel should be a part of the team along with someone from the health and safety department to better organize the system.

4.2.5 *Information Flow*

After the hazards have been identified, a series of information becomes available. This information along with other relevant data gathering is very important for risk assessment. Fig. II/4.2.5-1 depicts a few sources of such information.

The collection of information will help to develop an understanding about the data within the participants so that:

- Suitable link between the control measures and the hazard causes could be developed. A part of the control measures that was not done during the hazard identification stage will be completed now.
- Available information on control measures is well understood and additional information could be developed at this stage.
- Classification of hazards can be performed so that they can be addressed in a group if possible.

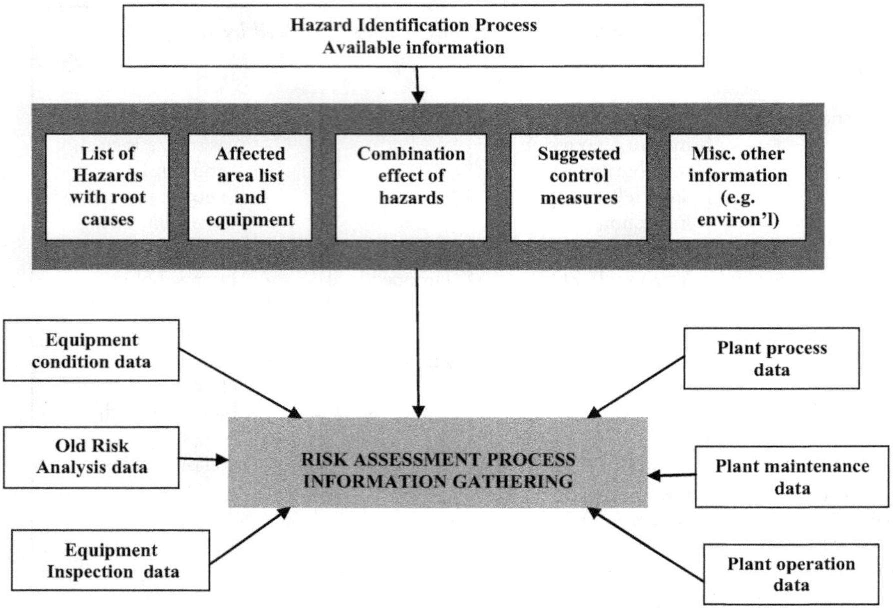

FIGURE II/4.2.5-1

Information collection for risk analysis.

4.2.6 *Combinational, Cumulative, and Individual Hazards*

Always there is no need for hazards to appear individually. In fact, an individual effect may be negligible but in combination these may cause havoc. In this connection, Fig. II/3.3.4-1 may be referred to. Another aspect on which emphasis must be placed is cumulative risk. Unless cumulative risk is assessed, the overall risk profile for the facility cannot be drawn up. This also makes it possible to evaluate key causes and control measures in the perspective of cumulative hazards. It is recommended that before starting the risk analysis process these aspects are well understood.

4.3 Outline of Risk Analysis Procedure

Basically, risk analysis means analysis of consequences, likelihood, and human factors. Control measure is also a part of risk analysis, but will be dealt separately in Clause 4.4. Also in this clause three different kinds of risk analyses shown in Fig. II/4.2.3-1 will be covered.

4.3.1 *Consequence Analysis*

As the name suggests, this judges the magnitude that deals with the scale of damage (the area/number of people affected) and the severity that deals with actual level of

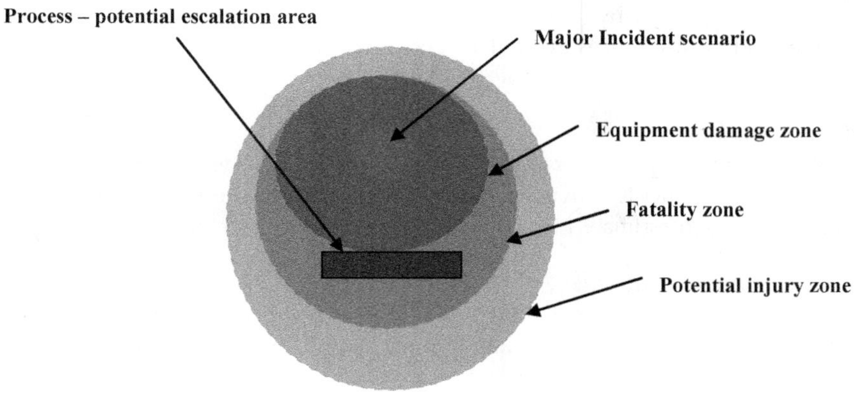

FIGURE II/4.3.1-1

Zones and consequence of incident.

damage (depth of damage). Another important factor is the escalation zone, which means intensifying the event and/or triggering another event(s). These are important areas in which to establish control measure adequacy and emergency planning.

As is seen from Fig. II/4.3.1-1, near the vicinity of the incident the severity will be more intense than further away. Also it gives an idea of the magnitude of the consequence of an event. To assure adequacy of the control measure it is necessary to realize the worst case scenario, which could be a cumulative or combination effect. Most likely the incident issue is important, hence the total range of hazards is important. Also the potential escalation zone at the interface area of the fatality zone and equipment zone is vital because severity in this zone may be wider. Qualitative estimation of consequences is done on the basis of previous experiences. In some cases, such as jet/flash fire gas dispersions, etc., detailed quantitative analysis based on computerized modeling may be called for. Various analysis types are discussed separately.

4.3.2 *Likelihood Analysis*

As already discussed in Chapter I in connection with risk matrix, in qualitative analysis, likelihood is estimated and categorized based on experience and judgment applicable for the project. Also these risk categorizations may be done on a quantitative basis as already discussed (say once in a year, etc.). In quantitative analysis the same is done based on previous records or a failure database for which quantitative PHA may be helpful. Failure occurrence data from other plants within or outside the company could be a good source of data.

4.3.3 *Nature of Injury*

While carrying out the risk analysis it is important to have data regarding the nature of injury caused by a major incident event. For this it is necessary to know what may go wrong and why (is it because of the wrong implementation of control measures or did control measures fail to act). Also it requires a whole range of outcomes, etc.

4.3.4 *Human Factor*

This is another important factor that may cause a major incident event. Unfortunately, it is rather difficult to have a quantified estimate of the likelihood of such occurrences. A rough estimate may be available from previous incidents.

4.3.5 *Screening of Hazards*

During hazard identification, screening of hazards is not recommended but based on consequences this may be done. It is important that all hazards need analysis but not at the same level. Screening will help in this regard. All screening shall be done on a real consequence basis on the assumption of effectiveness of control measures or because of likelihood of occurrence.

4.3.6 *Qualitative Risk Analysis*

From Clause 4.3.3 it is clear that the most common form of qualitative risk analysis is the risk matrix and risk ranking (Chapter I). Generally, preliminary hazard analysis (Clause 4.0 of Chapter I) is done to obtain rough knowledge about the risk profile for the facility.

4.3.7 *Semiquantitative Risk Analysis*

Semiquantitative methods are used to describe the relative risk scale. For example, risk can be classified into categories such as "low," "medium," "high," or "very high." The number of levels of risk can vary from (say) a to b. In a semiquantitative approach, different *scales* are used to characterize the likelihood of adverse events and their consequences. Analyzed probabilities and their consequences do not require accurate mathematical data. Semiquantitative assessment is useful especially if quantification of risk is difficult. At the same time, qualitative interpretation is too subjective. As already discussed, the risk graph (highly project dependent) discussed in Chapter I in conjunction with available guided risk nomogram or LOPA approach, is commonly used for this purpose. Here discussions will be mainly on the LOPA approach (LOPA will be dealt with separately later).

LOPA basically assesses single event—consequence scenarios, as shown in Fig. II/4.3.7-1. As is seen, there will be initiating events or issues that are the cause of an event, for example, electrical spark may be the initiating event, cause may be the fire, and the event may be the explosion, which may have far-reaching consequences (Fig. II/4.3.7-1).

In the figure, IPL 1—4 are independent protection layers, which are devices, systems, or actions that can prevent a scenario from spreading to an undesired consequence. All these layers are independent so that any one failure will not affect the functioning of the other layers.

As is clear from Fig. II/4.3.7-2, after the consequences are screened, consequences leading to accidents are considered. For each case, initiating event IPLs are

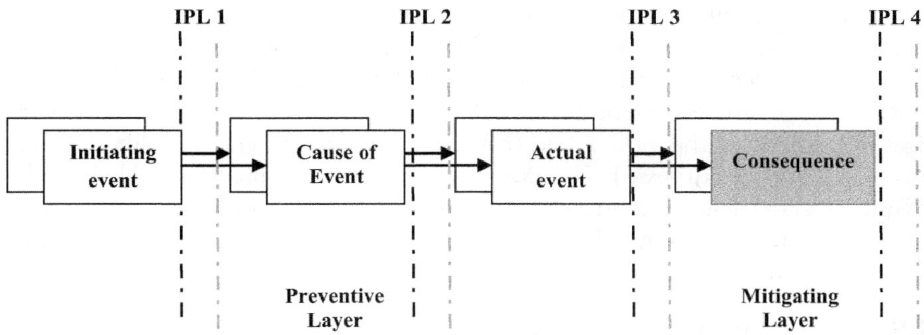

FIGURE II/4.3.7-1

Layer of protection analysis (LOPA) example for risk analysis.

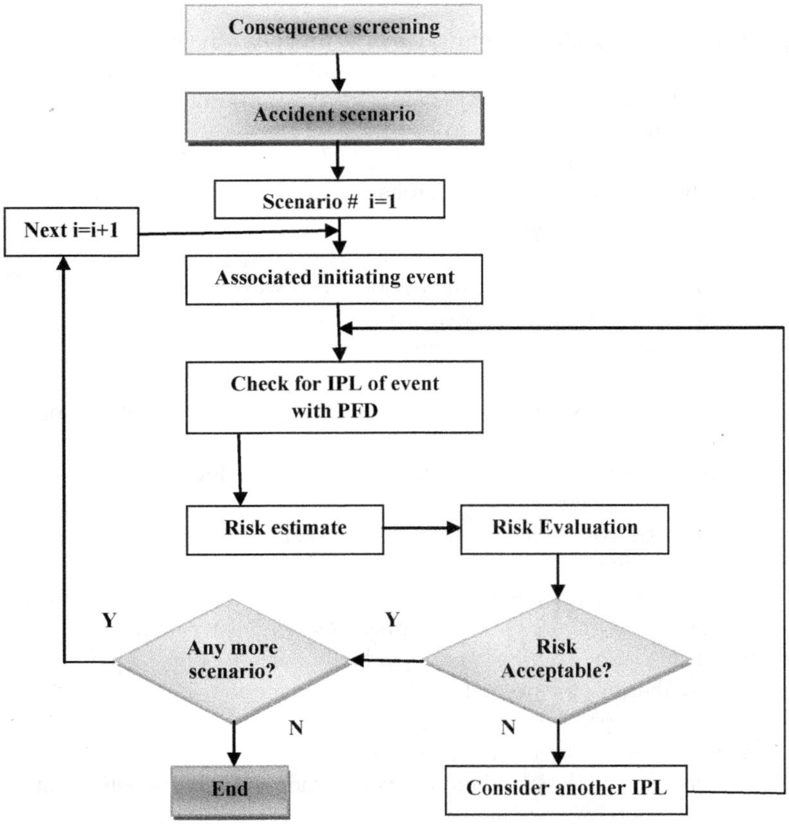

FIGURE II/4.3.7-2

Layer of protection analysis (LOPA) methodology for risk analysis.

considered. Risk estimate and risk evaluations are now done using the risk matrix, which may have quantitative scales. In the next step, associated risks are checked for acceptability. If accepted, then the next initiating event is considered. If not acceptable, then various options are considered for a new IPL, either preventing or mitigating to reduce the risk. After this exercise is over, again likelihood and consequence may be reassessed. The critical elements are: estimation of likelihood, consequence class (mainly dealing with consequence size—area or spread, fatality and cost, etc.), and tolerance limit. Total risk level can be estimated in terms of severity and probability rather in a quantitative way.

4.3.8 *Quantitative Risk Analysis (QRA)*

QRA is contemplated when further accurate studies are necessary (Fig. II/4.2.3-1). As per DNV (Clause 8.4.2 of Ref. [18]) process definition "quantitative analysis is the process for numerically analyzing the effect on overall project objectives of identified risks." Generally, QRA is performed on the basis of results obtained from qualitative analysis.

- Generally speaking, QRA is recommended in the following cases:
  - Effectiveness of various IPLs opted out, could not be established
  - Relative risk demonstration/improvement of consciousness of workforce
  - Layout issue related to escalation of hazard
  - New technology with perceived risk but lack of historical data
  - Very costly equipment/plant and probability of high risk for workforce/surrounding people because of accident
- QRA analyses numerically the effect of project risks on project objectives. A decision tree or Monte Carlo simulation is used with basic objectives:
  - Calculating probability of achieving project objective
  - Quantification of risk and then ranking for attention
  - Quantitative project outcomes with associated probability
  - This system generates quantified risk for the entire facility by cumulating from individual hazards.
  - Guidance for project management decision making, developing emergency plan because of uncertainty
  - Getting realistic cost, timeframe, and scope boundary
- Major tools and techniques are:
  - Interviewing techniques are used to quantify the probability of and impact of risks. There are normally two techniques used: one is triangular distribution where optimistic, pessimistic, and most likely cases are considered. Other is normal distribution, for normal distribution, mean and standard deviation need to be developed from collected data and placed in a normal distribution graph. All these are done on expert judgment.
  - Quantitative risk analysis and modeling techniques: One is *sensitivity analysis* and is used to determine which risk has the most potential impact. A project or facility may consist of several elements. In sensitivity analysis, effects of project objective are tested because of variation of an element when others

are in baseline values. It can highlight how a single change in one of the risk variables can make a marked difference on overall project objective. The other is *expected monetary value analysis* where each possible outcome is computed, and wherefrom the average of outcomes is determined.
   - Probabilistic and simulation: Probabilistic analysis specifies probability distribution of a single risk and then combinational distribution is considered. Generally, a Monte Carlo simulation model is used. Here a project simulation uses a model that translates uncertainty specified at a detailed level with its potential impact on the objective at the level of the total project.
   - Decision tree technique: A typical decision tree is shown in Fig. II/4.3.8-1. Here there are two major nodes: one is the decision node and the other is the probabilistic or chance node. The figure shows the decision regarding cost versus risk.
- Major utilization of QRA output: Major application areas are:
   - Probability of achieving project cost and schedule
   - Prioritizing risk according to quantified values
   - Risk trending very helpful in decision making
- Some limitations of QRA:
   - On account of statistical mistakes, it could be misleading

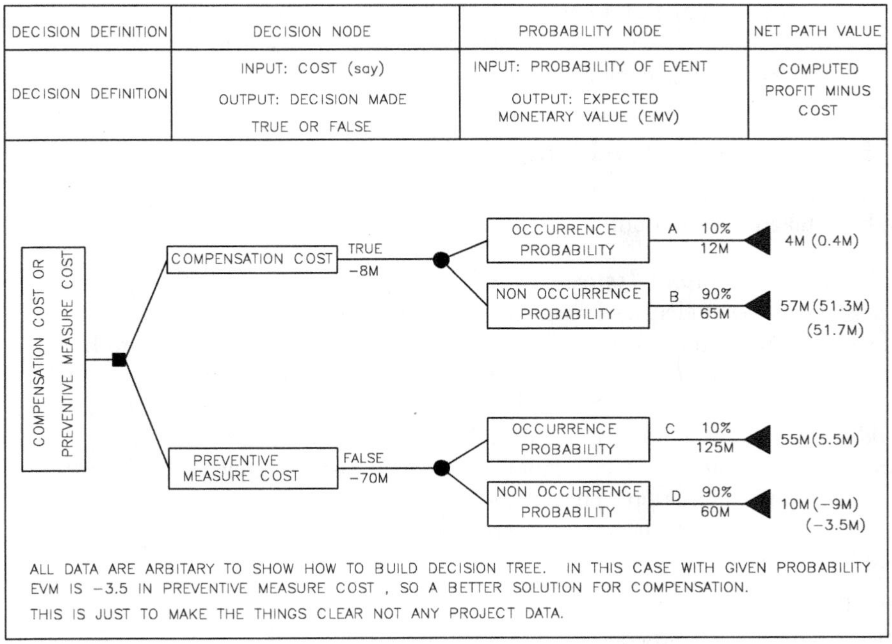

FIGURE II/4.3.8-1

Decision tree analysis.

- Possibility of workforce involvement is less
- Quality of facility management is not reflected

4.3.9 *Uncertainty Factors for Risk Analysis*

From the discussions so far it is very clear that when there is uncertainty, effectiveness of risk analysis, even the probabilistic approach, may suffer from inaccuracy. There are two kinds of uncertainties:

- Aleatory uncertainty refers to uncertainty caused by probabilistic variations in a random event.
- Epistemic uncertainty is uncertainty that comes from lack of knowledge. This lack of knowledge comes from many sources, for example, inadequate understanding of the processes, incomplete knowledge of the phenomena, imprecise evaluation of the related characteristics, etc. Epistemic uncertainties affect the values of the probabilities and frequencies of the events included in the accident scenarios, such as mechanical failure and repair rate, probability of failure on demand for a control system, or human error. There are three different cases in this regard:
  - Hardware failure: Insufficient data for quantitative analysis with expert judgment, as discussed earlier.
  - Human error: An analyst may not have sufficient refined data and characteristics.
  - Phenomenal failure or an environmental event such as gas release, earthquake etc.: Here also there may be insufficient available data regarding all processes and geographical data/experience.

4.4 Risk Assessment and Output

This is a part of risk analysis that should be done after control measures (Clause 4.5) have been taken into account.

4.4.1 *Risk Assessment Issues*

After risks have been identified based on consequences, they are screened and those need special attention are separated and QRA, etc. are performed on them. In risk assessment, the total likelihood of each incident with control measures shall be assessed. In risk assignment, the complete risk profile, highest risk incident, and the individual as well as cumulative hazards must be analyzed and evaluated. Individual hazards may be insignificant but cumulatively they may be a big issue! Therefore cumulative and combinational effects need to be considered. In the previous clause, a number of outcomes of uncertainty were discussed. There are other uncertainties caused by:

- Invalid assumption
- Incomplete hazard/consequence identification
- Improper modeling
- Old/nonupdated design document used

- Lack of process or operational knowledge or information
- Misunderstanding of the link between hazard and control measure

4.4.2 Brief Risk Assessment Goal

One of the main goals of risk assessment is to ensure that safety is brought to the system. Now it is the job of the operator to demonstrate that the risks are in SFARP.

- Risk criteria give the basis for judging tolerability for overall risk. It is better to divide overall risk criteria into three parts, as shown in Fig. II/4.4.2-1.

The top part in the figure is the unacceptable region and detailed QRA, and suitable additional control measures are necessary to reduce risk. The mid portion is the tolerable region and suitable control measures are necessary after semiquantified analysis to bring them to SFARP. Even for the acceptable zone risk analysis is necessary but qualitative will do. Adapted control measures must be shown to collectively eliminate or reduce to ALARP, with details in the safety document. Additional control measures (Clause 4.5) may be adapted for:

- Unacceptable risks
- Demonstration of risk is in ALARP
- Deficiency of existing control measure
- New technology
- Change in control measures

However, there is no prescribed methodology for such demonstration. The documents shall be detailed so that after evaluating control measures it should be possible to see that ALARP has been attained. Risk reductions can be achieved by elimination, reduction of likelihood of occurrence, or reduction in consequences.

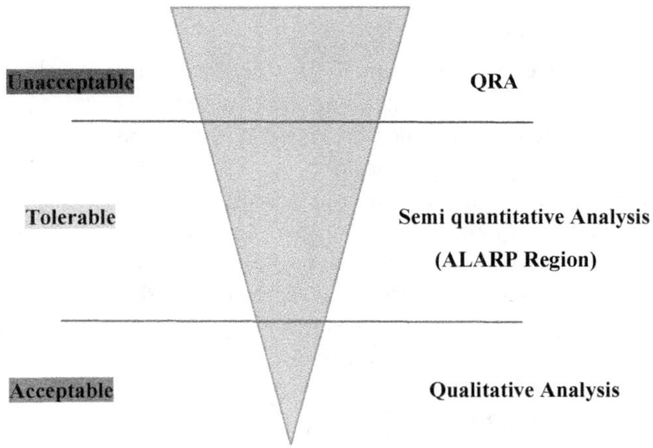

FIGURE II/4.4.2-1

Risk criteria and analysis types.

4.4.3 *Outputs and Their Uses*

Various outputs and their use from risk analysis shall be as follows:

- Identification of factors influencing hazards and control of these factors
- Establishment of defined link among hazards, major incidental events, and control measures to regulate them
- Prioritization of action to avoid major incident event
- Identification of the likelihood of major incident events
- Understanding how risk analysis can help the entire workforce to identify all the hazards and major incident events associated with the facility and how these can be combated with use of control measures
- Result of risk analysis is necessary to demonstrate that risks have been reduced to ALARP/SFARP.
- Contingency planning from risk analysis results
- Risk analysis results help in planning improvement of operating procedure/process and management. It is also helpful to tackle MOC.

4.4.4 *Risk Assessment Discussions*

Risk analysis is a live document and it should be thoroughly reviewed by the operator whenever there is any change in the system because it is the responsibility of the operator to understand all risks and control measures all the time. Also to improve the quality of risk analysis it is important to validate all hazards/major incident events, likelihood, control measures, and consequence, very rigorously. Proper communication, updating, and education/training are parts of risk management, which is an extension of risk analysis. So far helpful discussions have taken place regarding control measure, but what is control measure? The next clause provides the answer.

4.5 Control Measure

Risk assessment without control measure is like cooking without salt. Similar to reviewing risk analysis after each change, here also the reader is advised to read risk analysis after going through control measure to get a better understanding! What is control measure? First, control measure is an integral part of the facility in question. Control system may be *physical equipment* or *device*, or it could be a *system*, *process* or *procedure* meant to *eliminate/prevent/reduce/mitigate hazards* and/or their *consequences*, if they occur. It is the main tool of the risk management system or SMS. Some often refer to this as physical device! Not necessarily, it could be a procedure, for example, regular inspection to prevent corrosion is a preventive control measure. Purging a boiler is an operating procedure to avoid a hazard caused by explosion. The proper operational procedure of a kiln can help to solidify the clinker. Again an intrinsic safety barrier is a device used to prevent energy release.

4.5.1 Control Measure Characteristic Features and Associated Details

As stated previously the main function of control measure is to eliminate, prevent, reduce, and mitigate risks.

- Major characteristic features shall include but are not limited to the following:
  - Clear understanding of control measure allows the operator to gain knowledge and a better understanding of major incidents in the facility and its relationship with the control measure.
  - Identification of existing control measure and requirement for additional ones.
  - It is the basis for selection or rejection of existing controls and prescribing additional ones.
  - It is the basis for demonstrating adequacy of control measures required during demonstration of risk assessment.
  - It forms the basis for establishing a link between hazard, major incident, and control measures so that effectiveness of control measures and their impact on risk assessment/analysis can be well understood.
- In control measures there are several categories and they are placed in a hierarchical manner according to priority such as: (1) elimination, (2) prevention, (3) (risk) reduction, and (4) (risk) mitigation. Of these, the first two appear on the left-hand side of a major/top event in a bow and tie diagram (Fig. II/4.5.1-1). The first two are categorized as proactive while last two are categorized as reactive, because they work after the incident to reduce impact, hence they are shown on the right-hand side of the bow and tie diagram (Fig. II/4.5.1-1).
  - Eliminate: This enjoys the highest priority for the simple reason that it eliminates the hazard and hence there is no requirement for a control

Table II/4.5.1-1 Examples of Control Measures

Type	Engineering Control	Other Control
Elimination	Substitution with noncorrosive or nonflammable materials, safe layout to prevent escalation	Inherently safe design Plant design/operating procedure
Prevention	Barrier for dropping object, ventilation, process control, use of barriers [e.g., intrinsic safety (IS)]	Quick isolation, maintenance/operating procedure
Reduction	Process control, emergency plan, physical barriers, safety relief valves	Use of ignition suppression, procedural system
Mitigation	Firefighting, detection of fire or gas leakage	Contingency plan, emergency plan

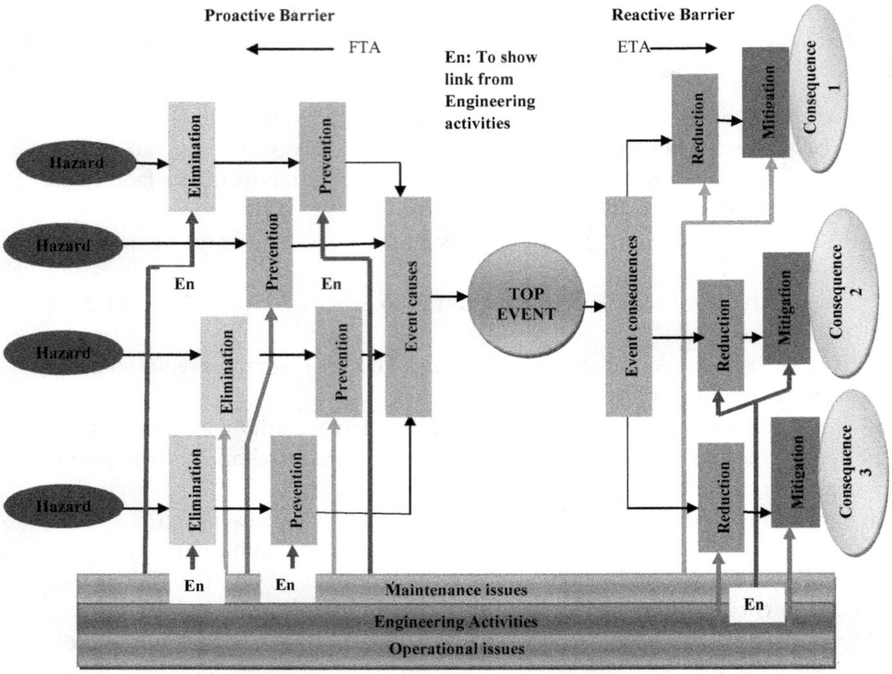

FIGURE II/4.5.1-1

Bow and tie diagram for control measure.

measure, for example, substitution with nonhazardous material if possible or plant operational procedure such as purging of boiler, use of inert gas like N_2.

- Prevention: This is the means used to prevent or remove the intended cause, or reduce the likelihood of occurrence, for example, interlocks in process control systems, use of a canopy to prevent injury caused by a dropping object.
- Risk reduction: This is used after the incident has happened to reduce the impact or severity, for example, process emergency controls and alarms.
- Mitigations are those control measures that are meant to limit the consequence after major incidents have happened, for example, passive fire protection system.

4.5.2 Various Control Measures

Generally, a single control measure is not enough so in reality layers of protection are employed through IPLs, as discussed earlier and in Chapter V. Understanding the relationship between hazard, major incident, and control measure cannot be overestimated. The following is important:

- The operator needs to understand the mechanism with which control measures act upon various hazards and manage the major incidents at various stages of the facility, be it normal operation or an emergency situation.

Also it is necessary to ensure that sufficient control measures are in place and that they are robust enough to face all the challenges. To understand the system it is better to rely on a bow and tie diagram; Fig. II/4.5.1-1. The hazard register and control have the same function. In this diagram there are a number of hazards that individually or in combination can cause a major or top event. Some of these causes can be eliminated and prevented. After the incident has happened, there could be some reactive control measure that will reduce the impact. Both proactive and reactive control measures can be from various activities such as engineering and operation and maintenance. From the figure a few points are worth noting: proactive control measures are applied before the top event and reactive barriers are applied after the top event. Also all these are applied at different stages of the project through devices, services, or procedures. Some examples are presented in the figure.

4.5.3 *Influencing Factors Related to Selections and Effectiveness of Control Measure*

There are a number of factors that influence the selection of control systems. Some them shall include but are not limited to the following:

- A simple system will obviously require a simple control measure, but if the system is complex or new, naturally a rigorous control measure assessment will be the call of the day. In any case the chosen control measure should be workable and should be best fitted for the system, and safety philosophy is well reflected.
- Existing knowledge should be properly reviewed while selecting the control measure. Even if there is previous experience, this knowledge must be reviewed because each project may have separate requirements.
- When there are too many choices, it is often not very easy to select the right one. Here it is a must to differentiate the chosen one from others to see how it best fits the project or facility. Another important factor needs consideration is that uncertainties have been reduced to acceptable limits before going for such selection. Again too many uncertainties will require further effort, otherwise the possibilities of wrong selection will be great.
- A suitable group of people through proper consultation is providing the solution, and selection team is receiving correct feedback of information from others so that proper selection is made.
- Proper documentation of all data and periodic reviews will make control measure assessment better and more effective. This is a live document, hence without periodic checks it may not be useful for future use.

4.5.4 *Selection, Rejection, and Adequacy of Control Measure With Addition Controls*

In this part the primary aim is to see that the control measures chosen are appropriate and adequate. If not, nonappropriate ones will be rejected and additional control measures will be incorporated. Several factors related to this are:

- To justify the adequacy (adequate to eliminate, prevent, or reduce risk to SFARP) of the control measure. These control measures shall be effective and viable. For these, associated influencing factors may be:

- Good experience and thorough knowledge of the usage of control measures as well as knowledge about various failure modes pertinent to the facility.
- Sound knowledge and experience of the complexity, scale, and risk profile of the facility and ability to keep pace with changes in technology over time.
- Available time and resources.

- Adherence to control hierarchy discussed, and suitable uniform distributions of control types such as engineering as well as administrative controls.
- Application of independent layers of protection matching the requirements.
- Identification of common mode failure.
- Definition of performance indicator and standards.
- Definition of critical controls.
- Layers of protection: There are many independent layers of protection provided in the control measure in addition to the basic process control system. These layers of protection make the control measures more robust. Fig. II/4.5.4-1 may be referred to for more detail. Detailed discussions are available in Chapter V.
- Common mode failure: Common mode failure refers to the failure of more than one control system on account of a common cause, which underlines the importance of independent layers of protection. However, common cause can affect both engineering and administrative controls. So, while considering the adequacy of control measures used for risk prevention/reduction/mitigation, etc. it is necessary and important to see that all such control measures are not only independent but also do not suffer from *common mode failure—discussed in later part of the book.*

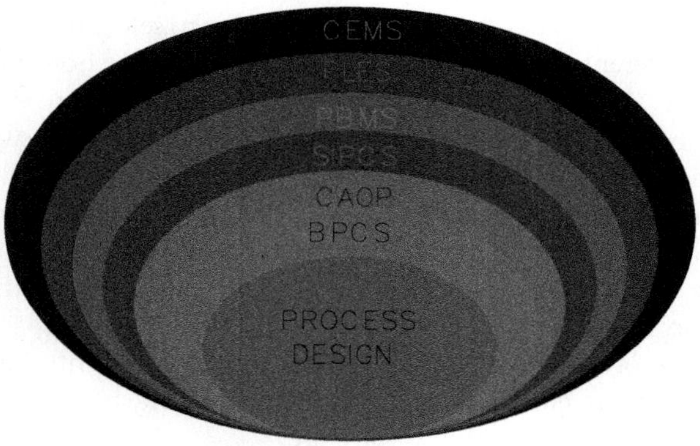

LEGNDS OF ABBREVIATIONS USED HERE

CEMS = COMMUNITY EMERGENCY SYSTEM
PLES = PLANT EMERGENCY SYSTEM
PBMS= PHYSICAL BARRIER & MITIGATION SYSTEMS
SIPC = SSAFETY INTERLOCK– PROCESS CONTROL SYSTEM
CAOP = CRITICAL ALARM AND OPERATIONAL PROCEDURE
BPCS = BASIC PROCESS CONTROL SYSTEM

FIGURE II/ 4.5.4-1

Application of LOPA in control measure.

- Control measure and life cycle: Control measures also have a life cycle. It is therefore necessary to consider that the control measure suggested is appropriate for the stage of the facility in its life cycle. In addition, various other factors such as environmental factors need to be considered.
- Critical operating parameter (COP): Many facilities try to put special focus on COP and scrutinize control measures for these. What is COP? Each piece of equipment, process, and procedure has lower and upper performance limits and the operation shall be limited within this boundary to avoid accidents. Whenever a control is designed, for example, it should meet the performance requirement (say 50—100%). Operation range below 50% and above 100% performance is not guaranteed. So, it is necessary see that these limits corresponding to each COP are not exceeded. The role of control measure in relation to COP is clear from Fig. II/4.5.4-2. The critical control parameter needs to be monitored and managed continuously for the process where immediate and continuous response from the operator is necessary.
- Critical control: All control measures have the same priority. A few are sometimes segregated as critical control measures because:
  - Control measure is relied upon for prevention/reduction/mitigation of risk with very severe consequence, or used to combat most likelihood risk.
  - A single measure to combat a major accident or a control measure to combat a number of risks.
  - Control measures with weak backup.
- Control effectiveness and viability: While selecting a control measure the following criteria are important for how well the control measure is going to perform:
  - The associated factors related to this are *functionality* and *reliability*. While functionality is associated with fitness/appropriateness of the control measure for the intended application, reliability stands for how reliable the application will be. Reliability is measured in terms of the dependence of the control measure on others, like human or other control measures. Reliability

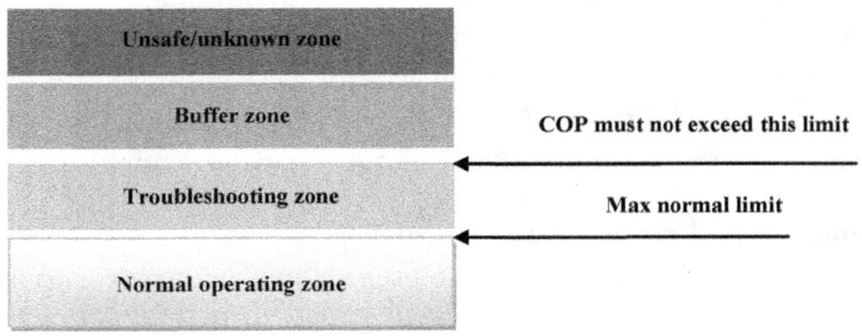

FIGURE II/4.5.4-2

Critical operating parameter.

Based on Hazard Identification, Risk Assignment and Control Measures for Major Hazard Facilities; Booklet 4, Internet Document; https://www.comcare.gov.au/.

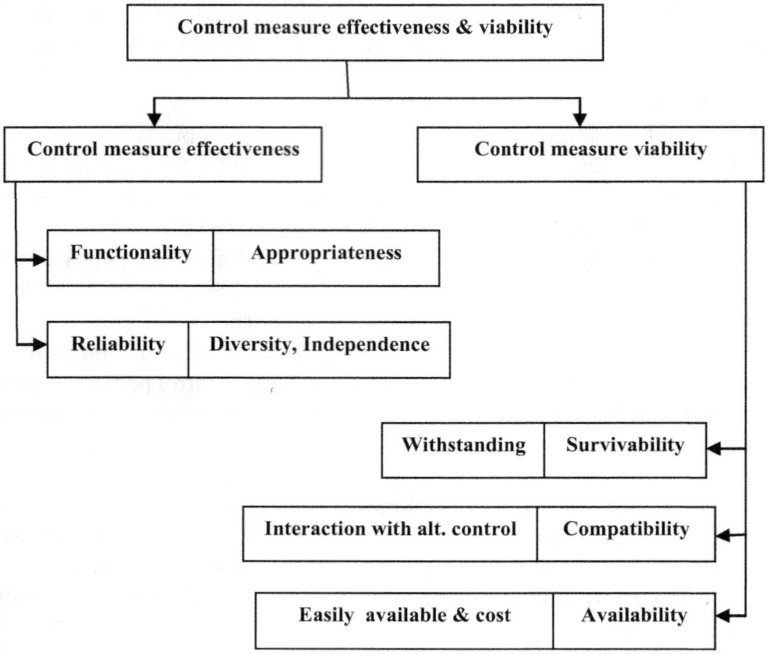

FIGURE II/4.5.4-3

Control measure effectiveness and viability.

is also measured in terms of whether it is going to fail or be affected by the failure of a single component, etc.; whenever diverse control measures are adapted then better safety is expected in one failure and may not affect the other.

- Viability assesses the practicability of the control measure. It depends on *compatibility*, *survivability*, and *availability*. Compatibility takes into account how the control measure is going to interact with alternative control. Survivability is the ability of the control measure to function as intended even during an accident for which the control measure is used for reduction or mitigation. Availability stands for how available it is and is also related to cost. For an easy understanding, Fig. II/4.5.4-3 may be referred to.

4.5.5 *Performance Indicator and Standard*

The performance of any system has a target to achieve. Here are two important things:

- Performance indicator: This is information that is used to measure the effectiveness of a control measure. This can be used to compare current with past performance. While the performance indicator is selected it should be specific,

measurable, appropriate, realistic, and timely [17]. The performance indicator is defined by the operator to:
  - Measure, monitor, or test the effectiveness of a control measure.
  - Report and suggest corrective action in case of failure.
- Performance standard: This is a target set for the performance indicator.

4.5.6 Additional Control Measure

As a part of control measure assessment it is necessary to demonstrate all control measures such as physical control (say barrier), engineering control (say process control, relief valve), and administrative control such as defined procedure, etc. During demonstration the operator needs to challenge the control measure to improvise the process. In this way, alternative controls could be taken into account, especially in those cases where all risks are not in SFARP. Helpful guidance toward both additional control measures and past disaster issues are highlighted in the following:

- Existing control measure is it fully functional
- Discarded control are really not applicable and why
- Any improvement for existing control
- Control measure's obsolescence
- Augmentation and addition of new control measure
- New control measure to modify facility

All these exercises are carried out to understand the system very closely; they could improve the suggested system and additional control measures may be included. Such reviews are important especially in cases of:

- New facility or new technology
- Loss of knowledge base for safe operation
- Obsolete system exists in control measure
- Degradation in effectiveness of control measure
- Occurrence of an incident

4.5.7 Control Measure Output

Basically control measure output consists of:

- A list of existing and additional controls with established relations among hazard, incident, and control measure
- Adequacy assessment of existing control with additional control measure list and relation
- A list of performance indicators, COP, and critical controls
- Improvement actions recommended
- A list of hazard and additional controls

These outputs can be used for:

- Risk assessment discussed earlier
- Monitoring of effectiveness of control measure
- Identification of additional control measures and their implementations

4.5.8 *Discussions*

Like risk assessment the control measure document is also a live document and should be subject to review and revision periodically or when there may be any change. Workforce involvement will be a similar involvement pattern, as discussed in Clauses 3.2.3 and 4.2.4.

4.6 Safety Management System

Discussions on risk assessment are not complete without some focus on SMS. SMS is a comprehensive and integrated system to address all aspects related to health and safety. It is to be properly documented so that this comprehensive document is accessible to all concerned. It must be compatible with the safety management objective and philosophy of the facility in question. It is mainly concerned with the control measure. It tracks errors, deviations, etc. in the control measures. It also keeps track of performance standards of the equipment and system. Therefore it is a live document and needs periodic reviews and updates.

4.6.1 *Features*

SMS manages and supports specific aspects of the facility, mainly *operational* aspects. It is concerned with control measures adopted for the facility to prevent, reduce, and mitigate risks. In a nutshell, it is an integral part of the risk management system to ensure safe operation of the facility by regulating various control measures for the facility. A major purview of SMS shall include but is not limited to the following:

- Implementation of standards and procedures at various stages (e.g., design, construction, and operation) of the facility including imparting knowledge among the workforce, supervision of tasks and procedures, resource management, and procedures to manage third parties.
- Work permit and equipment isolation procedure
- Risk assessment and MOC procedure
- O&M procedure, quality assurance (QA) of device and maintenance of critical asset
- Emergency plan and procedure
- Monitoring audit and review of SMS

4.6.2 *SMS and Control Measures*

One of the main properties in SMS is that it shall be comprehensive, so, related to control measure, SMS needs to address the following issues:

- Identification of control measures and definition of the performance standard
- Implementation of control measures. Layers of protection in control measures are checked; if one layer does not function well then another is to be brought to implement and enhance the control measure.
- Monitoring and periodic performance checks

- SMS should be in a position to identify anomalies, then report them and rectify the shortcomings. So, SMS audits the system as well as compliance against documented SMS.
- All concerned shall have easy access to and cooperation in the decision-making process.

4.6.3 *Operational Aspects in SMS*

It is a part of SMS to look into the operational aspects in the facility. A major controlling area in this part shall include but is not limited to the following:

- Hot work such as welding
- Electrical work such as high-voltage isolation, grounding
- Physical isolation
- Working in a confined place
- Issuance of permit to work
- Authorization to work or supervise

4.6.4 MOC

This is an important element in SMS. Whenever a change in the system is incorporated, then there will be possibilities that there will be changes in hazard situations as well as major incidents. Apart from these discrete changes to any system operating over time, there could be changes in the risk profile of the facility as it is a dynamic process. MOC needs to track those changes. Naturally, there will be a necessity to maintain safety and SMS. MOC may initiate a review of SMS. SMS and MOC are directly related.

4.6.5 *SMS Performance Standard*

The operator needs to develop a performance standard for SMS. This performance must be transparent, workable, and appropriate to ensure safety to people concerned and property. This performance standards needs to address a few things:

- Coverage of all aspects of SMS
- Ability to measure the effectiveness of SMS
- Suitable performance standard for critical procedure, control, and equipment
- The performance should show sufficient details.

4.7 Conclusion

The overall picture of risk analysis vis-à-vis control measure and SMS can be conceived from Fig. II/4.7-1.

Risk assessment/analysis is a vast subject and very much plant specific. It depends highly on stages in the plant life cycle, available information, and resources. Therefore it is very difficult to cover all the processes and highlight each of them. In fact, there could be as many as 62 different kinds of PHA and risk assessment

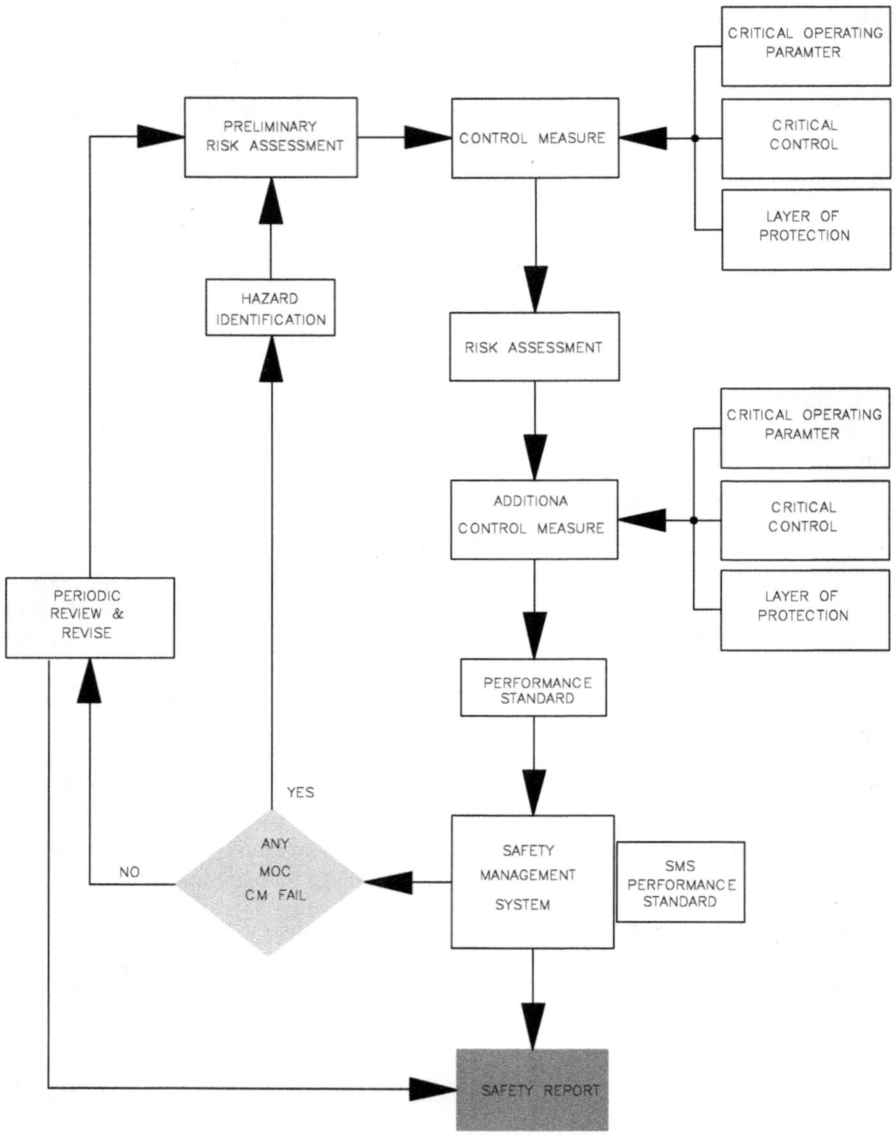

FIGURE II/4.7-1

Control measure and safety management system (SMS) in risk analysis.

methodologies available. The most popular ones, which are mainly used in industry, have been covered here. Depending on specific requirements, one may have to choose the best one at that time for the facility. Now it is time to analyze each PHA in some detail.

LIST OF ABBREVIATIONS

ALARP	As low as reasonably practicable
BFP	Boiler feed pump
CCPS	Center for chemical process safety
CEI	Chemical exposure index
COP	Critical operating parameter
DOW FEI	Dow Fire and Explosion Index
EC&I	Electrical, control, and instrumentation
ETA	Event tree analysis
FCV	Feed control valve
FEED	Front end engineering design
FMEA	Failure mode and effect analysis
FSA	Formal safety assessment
FTA	Fault tree analysis
HAZID	Hazard identification
HAZOP	Hazard and operability study
HC	Hydrocarbon
HRA	Human reliability analysis
HW	Hardware
IPLs	Independent protection layers
LOPA	Layer of protection analysis
MEA	Major accidental event
MF	Material factor
MHF	Major hazard facility
MHI	Material hazard index
MOC	Management of change
NOPSEMA	National Offshore Petroleum Safety and Environmental Management Authority
O&M	Operation and maintenance
OPGGS	Offshore Petroleum and Greenhouse Gas Storage (Safety) Regulation (Commonwealth)
OSHA	Occupational Safety and Safety Administration (USA)
P&ID	Piping (process) and instrumentation diagram
PFD	Process flow diagram
PHA	Plant hazard analysis/preliminary hazard analysis
PSF	Performance shaping factor
PSM	Process safety management
QRA	Quantitative risk analysis
SFARP/ SFAIRP	So far as is reasonably practicable
SHI	Substance hazard index
SMS	Safety management system
SW	Software

References

[1] W. Briges, Selection of Hazard Evaluation Technique, Process Imrovement Institute, Inc., February 2008.
[2] V. Asar, Hazard Assessment and Risk Management Techniques for Industries, Disaster Prevention and Management Center; GIDC India.
[3] https://www.dmv.ca.gov/portal/dmv/detail/pubs/cdl_htm/sec9.
[4] Safety Assessment Techniques and Analysis; Dow Fire and Explosion Index; http://www.cbme.ust.hk/safetycourse/download/11.05-10RiskAnalysisDowF&EIndex-Section11.5.COURSENOTES.pdf.
[5] http://gohsep.la.gov/hazardmat.aspx.
[6] Using Hazard Evaluation Techniques; ftp://ftp.feq.ufu.br/Luis/Seguran%E7a/Safety/GUIDELINES_Hazard_Evaluation_Procedures/0491X_06a.pdf.
[7] Risk Management Program (RPM); Chemical Emergency Planning, US EPA, July–August 2008.
[8] Health and Safety Authority Ireland, www.has.ie/eng/topics/Hazard.
[9] NOPSEMA Hazard Identification Guide Note; N-04300-GN107 Rev5, December 2012. https://www.nopsema.gov.au/assets/Guidance-notes/N-04300-GN0107-Hazard-Identification.pdf.
[10] Hazard Identification at a Major Hazard Facility; Guidance Note Work Safe, April 2011. www.worksafe.vic.gov.au.
[11] FSA: IMO; http://www.imo.org/OurWork/Safety/SafetyTopics/Pages/FormalSafetyAssessment.aspx.
[12] Introduction to Hazard Identification and Risk Analysis; CCPS/Topic Discussion.
[13] Guidelines for Preventing Human Error in Process Industries, Center for Chemical Process Safety; J. Wiley.
[14] D. Embrey, Task Analysis Techniques, MD Human Reliability Associates Limited.
[15] Hazard Identification, Risk Assignment and Control Measures for Major Hazard Facilities; Booklet 4, Internet Document; https://www.comcare.gov.au/.
[16] http://www.hse.gov.uk/risk/faq.htm.
[17] Guidance Note: Safety Assignment for a Major Hazard Facility; Work Safe, May 2011.
[18] http://www.dnv.com/industry/oil_gas/services_and_solutions/risk_management_advisory/safety_risk_management/quantitative_risk_assessment_qra/.

Further Reading

S. Basu, A.K. Debnath, Power Plant Instrumentation and Control Handbook, Patent, October 2014.
Center for Chemical Safety, American Institute of Chemical Engineers (AIChem).
W. Frank, Process Hazard Analysis Fundamentals, Frank Risk Solutions.
W.G. Green III, Hazard Analysis (The Process of Defining a Hazard...), in: Emergency Management Process Series, vol. 1, 2008.
Halliburton; Prevention Services; Ho11228 01/15; www.halliburton.com.
D. Hawrelak, Dow Fire and Explosion Index, 2001. Presented to ES317Y at UWO.
Hazard Identification Studies (HAZID); Safety and Risk Management Services; Germanischer Lloyd – Services/Product Description.

Hazards Analysis Guide: A Reference Manual for Analyzing Safety Hazards on Semiconductor Manufacturing Equipment, International SEMATECH Technology Transfer # 99113846A-ENG.

HSE.gov.uk; www.hse.gov.uk.

http://delcode.delaware.gov/title7/c077/index.shtml.

http://www.ccohs.ca/oshanswers/hsprograms/risk_assessment.html.

http://www.intertek.com/testing/pilot-plant/cei/.

J. Johnston, R.R. Selvage, Hazard Evaluation Techniques; LANL Safety Basis Academy; NISA.

H.R. Kavianian, J.K. Rao, G.V. Brown, Application of Hazard Evaluation Techniques to the Design of Potentially Hazardous Industrial Chemical Process, California State University Long Beach; NIOSH Instructional Module; CDC (Occupational Safety and Health) US Department of Health and Human Services.

NOPSEMA; Guidance Note on Risk Assessment; Revision 4, December 2012.

J. Perry, Project Risk Analysis and Management; a Guide by Association for Project Management, The University of Birmingham.

Process Hazard Analysis, Process Hazard Analysis, Sandia National Laboratory; Chemical Security Engagement Program.

Process Hazard Analysis: Dow Fire and Explosion Index, ChE 258 Chemical Process Safety; University of Missouri – Rolla.

Process Hazard Analysis; RPM; Chemical Emergency Prevention and Planning News Latter, July–August 2008.

L.D. Radu, Qualitative, Semi Quantitative and Quantitative Methods for Risk Assessment: Case of Financial Audit, Analele ŞTIINłIFICE ALE UNIVERSITĂłII "ALEXANDRU IOAN CUZA" DIN IAŞI Tomul LVI ŞtiinŃe Economice, 2009.

J. Ramesh Babu, Layer of Protection Analysis – an Effective Tool in PHA.

M. Rousand, Preliminary Hazard Analysis, NTNU/Willey, October 2005.

N.A. Siddiqui, A. Nanadan, N. Sharma, A. Srivastava, Risk management techniques HAZOP & HAZID study, International Journal on Occupational Health & Safety, Fire & Environment – Allied Science I (I) (July–September 2014).

M. Smith, Building Your SMS Based on Size and Complexity Hazard Identification (Try Byte Freedom to Learn).

H.J. Toups, Hazards, Accidents, Process Safety Management & Process Hazard Analysis, LSU Department of Chemical Engineering, 2003 (Significant material from SACHE Workshop).

B.K. Vaughen, J.O. Mudd, B.E. Pierce, Using the ISA 84/HAZOP/LOPA Procedure to Design a Safety Instrumented System (SIS) for a Fumed Silica Burner, GCPS, 2010.

E. Zio, N. Pedroni, Les Cahier De LASÉCURITÉ INDUSTRIELLE; Risk Analysis, July 2012.

CHAPTER III

Qualitative Hazard Analysis

To manage and control risks, it is essential that risks must be first identified. These will then be evaluated to decide their tolerable, as low as reasonably practicable (ALARP), or intolerable risks. Hazard analysis is a dynamic process. For new plant or product development, such analysis is essential and it is better to make development part of the design standard. The sooner the hazards/risks are evaluated, the less will be the expenditure toward mitigating risks. So, it is better to start this earlier in the life cycle of the project. From earlier discussions, it has been seen that there are three types of hazard analysis methods, viz. qualitative, guided word, and quantitative. Qualitative hazard analyses are done normally at earlier stages of the project and for projects that are not very large/complex. Also for large/complex projects, qualitative hazard analyses may form the basis or input for quantitative hazard analysis. Qualitative hazard analysis is normally carried out by a team (from various disciplines) to identify and evaluate major hazards with reasonable detailing and to put forward judgment on risks. Quality of analysis is very important to ensure that both quality and quantitative aspects have been given due consideration. In qualitative hazard analysis the risks are prioritized to a predefined (relative) scale. Therefore the knowledge level of the team is crucial; otherwise it will yield fruitless results, in case the team is not sure about the relative ratings. In many cases, qualitative hazard analysis is followed by quantitative hazard analysis to appreciate the depth of the issue. It has been seen in previous chapters that risks are rated, based on two things: (1) their likelihood of occurrence, that is, probability of occurrence, and (2) impact of the occurrence on the project objective. The probability scale will be within 0–1 (say) with 5 or 10 divisions, this means that 3 in the scale with 5 divisions is equal to a 60% chance of occurrence. Similarly, an impact scale may be divided into 4–5 divisions, such as catastrophic, extremely high, high medium, low, etc. From Chapter I, it is known that the risk matrix is a very useful tool. After risk ranking, recommended control measures are also suggested. In this chapter, discussions on qualitative type hazard analysis mainly cover preliminary hazard analysis [PrHA to distinguish it from plant/process hazard analysis (PHA)], "what if," and checklist. Discussions focus mainly on the characteristic features, general methodology, and their usage.

1.0 PRELIMINARY HAZARD ANALYSIS

1.1 Preliminary Hazard Analysis Definitions and Necessity

PrHA was first used in the US military. PrHA may be conceived as an organized way to identify and evaluate hazards associated with any facility. It also suggests control measures to control the risks in a qualitative manner. As per Center for Chemical Process Safety, PrHA is "an organized effort to identify and evaluate hazards associated with chemical process and operations to enable their controls. The review normally involves the use of qualitative techniques to identify and assess the significance of hazards. Conclusions and appropriate recommendations are developed. Occasionally quantitative methods are used to help prioritize risk reduction." Most of the major incidents normally do not happen because of mere failure of a system component that is responsible to maintain complex interactions within the system. Hazard analysis is a continuous iterative process throughout system development and use. It is better to begin PrHA during the conceptual development stage and the system may continue with other PHAs. Because PrHA is normally carried out at an early stage of the project to discover hazards, available information may be very limited. For an effective analysis, the team should be given a comprehensive description of the conceptual design of the plant. PrHA is done mainly to meet requirements such as:

- Standards/legislature
- Gain in confidence for a good understanding of hazards associated with the facility
- Recommendation of necessary control measure or tool to control/combat hazards
- To make sure that appropriate control measures have been prescribed for a specific hazard or set of hazards
- To ensure return on investment, that is, resources are not wasted, instead they help to reach the project objective

1.2 Preliminary Hazard: Goals and Characteristics

PrHA does not have any special technique; it uses other techniques for hazard identification.

1.2.1 *Goal of PrHA*

The objective of any hazard analysis is to identify and understand the hazards and to develop ways and means for safeguarding against them, as far as possible. The basic purpose is to determine if credible means exist (mainly through failure) that could result in an undesirable event(s) and recommend control means to combat the situation if possible. Since PrHA is evident at an early stage, it may not be possible to have a comprehensive idea of all hazards. PrHA usually addresses the following factors:

- Hazardous plant and equipment
- Environmental and external factors

- Human issues: operation, maintenance
- Facility-related issues
- Safety/control measure interface

It basically performs the following functions to identify/analyze qualitatively various hazards and recommend control measures:

- Identification of hazards
- Translation of hazards into system design constraints
- Assessment of hazards as required
- Qualitative risk evaluation and recommended control measures
- Establishment of hazard logs with recommendations

An exhaustive list of hazards is important! What is more important is a proper description of the hazard including details of sources and possible impacts. However, a list of causes should not dilute the main aim. There are various means to identify hazards (discussed later). Qualitative risk evaluation is done after analyzing the likelihood and consequence of each hazard and placing all these in a relative scale (specific for the project) to carry out risk ranking.

1.2.2 Characteristics of PrHA

- This analysis is dependent on techniques such as brainstorming.
- This technique can be applied for any activity.
- PrHA is usually carried out as a high-level analysis at the early stage of the project in its life cycle.
- Quality and availability of documentation and training of team members play major role in the success of this analysis.
- Because this is a qualitative analysis, it gives a qualitative description of hazards, and risk ranking is done after likelihood/consequence analysis in a qualitative manner.
- It is carried out by a small team whose members must have good working knowledge of the activity in question.
- PrHA depends on expert judgment for assessment and ranking of risks and their prioritization.
- PrHA is highly dependent on the knowledge of the team. Major application of PrHA is at the early stage of the project in its life cycle, when only minimal well developed information is available. Therefore its dependency on the expert's judgment is inevitable. In case real experts do not participate, because PrHA is a new technology with little past experience, then there are bound to be lot of uncertainties in the result.
- Generally, this analysis requires follow-up analysis because many of the documents are not available at the time, therefore after further information is generated, follow-up analysis is very likely. Also for large/complex systems, the output of PrHA could be the basis for further quantitative analysis or for quantitative hazard analysis.

1.3 Scope and Boundary Definition

For a facility there could be a number of hazard analyses. Naturally, it is not possible to cover all facility hazard analyses through PrHA. By scope definition and boundary, the part of the facility to be covered is defined. Operating phase stands to signify which stage of the project (i.e., in design stage, construction stage or operation stage) PrHA to be carried out. Also for a production process or system, it is very difficult to cover all types of hazard identification with the help of a single PrHA. Without scope definition, a proper understanding of the analysis and activity to be carried out is difficult. It is well established that the percentage of students graduating from a university for any course is more than students graduating from a professional institute. One of the major reasons is that university curricula are set according to a syllabus, whereas in a professional institute course coverage is fairly broad. In exactly the similar manner, when scope and boundary are well defined, the success rate of analysis will increase. Therefore it is necessary to define and specify the scope and boundary of the analysis process, otherwise the entire exercise may be futile. Scope definition will also help in concentrating on the issues properly. In doing so, associated *cumulative or combinational* risks must be taken into account. Boundary definition is important to ascertain the *interface* requirements. If necessary, follow-up analysis may have to be undertaken, for example, quantitative hazard analysis may have to be performed. The following example should make this clearer. For any industrial installation motor control centers (MCCs) are necessary for running any motor. In most cases, safety, permissive, and other interlocks pertinent to these motors are processed in a programmable logic controller (PLC)/distributed control system (DCS) i.e., logic solver meant for the particular section of the plant. Here, boundary plays a significant role. If an interface relay is placed inside the PLC cabinet then contact output from the PLC needs to be connected to a 240/110 VAC control voltage MCC, which means that a higher (more than 24 V) voltage is brought to the logic solver cabinet (also two sets of cables are necessary for relay coil and MCC status), which in many cases is not acceptable. If the relay is placed in an MCC cabinet then potential free contact with 24 VDC from the logic solver could be sent to the MCC panel and mixing of higher voltage in the PLC cabinet is avoided. In this connection Fig. III/1.3-1 may be referred to. So, the interface of instrumentation in this case is easier (e.g., Siemens/KWU turbine interface). Naturally, in this case of PrHA for instrumentation, the MCC will be out of scope but for the interface relay (though located in the MCC) it should be included as a part of the PrHA instrumentation. For an intelligent MCC, such an interface problem is reduced.

Therefore boundary selection is important. PrHA is carried out by a small team, with experts in relevant areas! Bringing in other issues will not only complicate the problem but also necessary experts' opinion may be missing. Therefore in PrHA, scope and *specific interest* for hazards/accidents are extremely important; if necessary a number of such analyses results may be combined to get an overall hazard scenario and safety control measure.

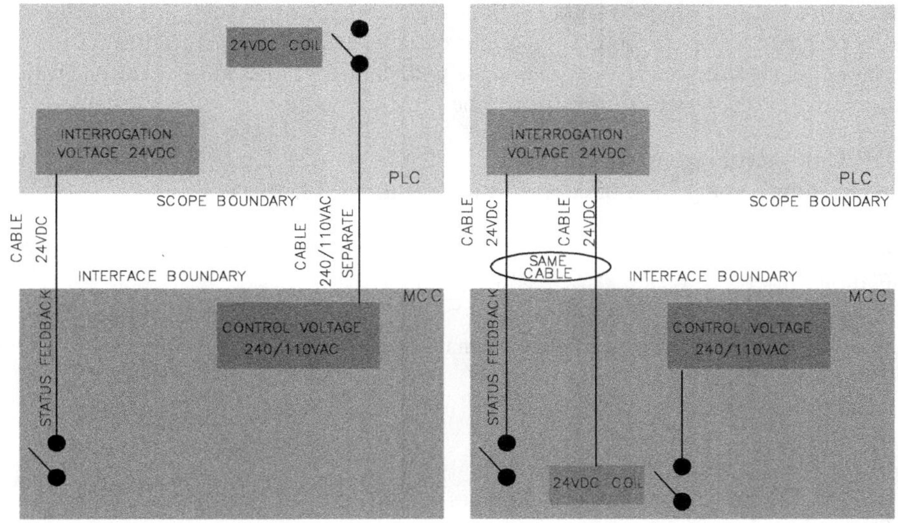

FIGURE III/1.3-1

Scope and interface boundary.

1.4 Hazard Identification

Hazard identification is the most important step in any hazard analysis method. It involves identification of all hazards, be they single or multiple (combinational or cumulative). It may not be possible to identify all hazards in one go, for example, many combinational hazards could be identified during the analysis period also (like those pertinent to detection of a failure mechanism because of a combinational effect). Within the defined boundary and scope, it has to identify all possible hazard issues. The success of the study very much depends on how comprehensive and systematic the study is. There is no hard-and-fast rule for this. It depends on the specific project; mostly it relies on scope definition related to purpose of the analysis, available information (maybe for an analogous job), and expertise of the analyzing team and their preferences. A good understanding of the process, possible hazards, how they are initiated, and most importantly how they are inflicted on the system in question is crucial. An example can make things clearer. During the morning rush hour on a weekday, nobody wants to miss their tube train to work. Now, if a person throws themself onto a train while the door is closing, there could be a hazard if the train starts without the door fully closing, because part of the body of the person may be hit mechanically or there could be an accident from high-voltage electric current from the third rail (supplying high voltage to the train motor). So, necessary design considerations are that the door area on the platform must be closed accompanied by the sound of a door closing alarm (a horn, etc.) or the train cannot start unless the door is fully closed. Here again, there may be failure of the door limit switch, so some other means must be sought to address the issue, perhaps by providing a double limit switch or guard's inspection, etc. As mentioned earlier, every facility has its own specific issue. Here, endeavor has been made to address the issues in a generalized way and readers should select the applicable one for their system.

1.4.1 Common Sources of Hazards

- Sources and propagation of energy in various forms: Electrical, mechanical, thermal, chemical
- Mechanical moving parts
- Falling objects
- Collisions
- Radiations from nuclear source/electromagnetic source
- Noise and vibration
- Material mismatch
- Reactions: Exothermic/endothermic
- Corrosive and toxic material exposure
- Loss of containment for liquids and gases
- Fire and explosion
- Biological
- Equipment/control hardware failure
- Software failure
- Human error
- Subsequent survival and escape

This is a list of typical hazards faced in the industry and production process; in addition, there could be local and environmental issues like earthquake, heavy rain, flood etc. for example, ISO 17776 for offshore.

- Hazard inventory: This part is an elaboration of the previous clause to identify a few areas where analysts need to pay attention!
- Energy source: Here an easily understandable title, type of energy, quantitative description (e.g., voltage, current/power, etc. for electrical energy), and applicable phase, etc., shall be spelt out for proper identification of hazards.
- As stated earlier, every plant has its own specific problems and issues. Naturally, based on these, each enterprise develops a hazard checklist for various plants in the same industry, or plant-wise there could be a checklist for hazards. As a starting point, a plant-specific list needs to be developed and updated to cover the required checklist for the plant. One typical generalized list of common hazards is presented in Table III/1.4.1-1.
- Operating phase and conditions: There are a number of hazards that appear in a number of phases of a project and also may be absent in many phases. On account of this, while identifying and listing the hazard it is extremely important to note which phase of the project/facility, it is. Also everyone should be aware that equipment/devices kept in a controlled environment have less chance of failure than those placed outside a facility, for example, a control device near an

Table III/1.4.1-1 List of Common Hazards

Category	Source	Effect/Impact
Energy	Chemical: oxidizer disassociation, organic compounds and explosives	Fire, explosion, exothermic reaction, production of toxic fumes, corrosion/corrosive fumes, etc.
Energy	Electrical: release of potential energy, arc, electric shock	Electrocution, personal/ equipment burns, shock, ignition of combustibles, interruption of communication, power failure and related issues, short or open circuit issues for operations
Energy	Mechanical: moving parts, sudden change in velocity, falling object, collision or impact force and energy of vehicle, etc.	Impact on personnel, equipment, disintegration of parts or components, pipes. Detonating effect, hot surface shock/damage to component, equipment, or personnel caused by friction. Human injury, equipment damage, production impedance, etc.
Energy	Thermal: change of temperature effect	Ignition, activation of chemical reaction, hazard caused by thermal expansion/contraction of solids/gases, etc.
Energy	Radiation: ionizing, electromagnetic, thermal/ nuclear, or UV	Erratic behavior of electronic equipment, safety interlock, interference, human fatality, charring of organic material, decomposition, fuel ignition, etc.
Human activity	Hazards caused by insufficient control or incorrect warning display, life support, probable action error, hazardous location	Injury or even death caused by: abrasion, cuts, bruises, burns, falls, etc. Sensory impairment or loss. Equipment/control system damage caused by maloperation, handling, or nonuse of protection
Environmental	Lightning, wind, flood, extreme temperature, gravity, humidity, etc.	Structural, equipment damage, and human injury
Degradation of materials	Corrosion, aging, embrittlement, thermal/mechanical fatigue	Change in property; corrosion, aging effect, embrittlement, structural failure, insulation breaking, fatigue failure
Chemical	Inhalation of chemical gases	Human injury or fatality, equipment damage caused by corrosion

Continued

Table III/1.4.1-1 List of Common Hazards—Cont'd

Category	Source	Effect/Impact
Vibration	Noise and vibration	Impaired hearing and other human injury, equipment damage caused by vibration, control error, communication problem caused by interference
Loss of containment	Release of energy/toxic material	Release of energy/toxic material caused by chemical energy, damage caused by contamination

acid/alkali handling area in a water treatment plant has more chance of failure than a similar device in a control room. Also there will be chances of failure of equipment operating near or beyond design limits. This means that harsh operating environments or conditions (because of high/low temperature, humidity, rain, etc.) have a serious impact on equipment, causing hazards because of failure of equipment.

- Unmitigated consequence: It is not always possible to mitigate all the hazards. There will also be some unmitigated consequences. To see if such unmitigated consequence(s) can cause a major incident, then control measures must be taken, and assessing the level of these control measures is important. This will be clear from a simple example. On loss of flame boiler is tripped to mitigate the hazard. However there may be some un-burnt fuel to cause in future. For this reason at the starting purging is carried out. So double protection and control actions are considered. Analysts encounter these judgments after PrHA is done; in this case, based on impact (likelihood and consequence), control measures need to be decided.
- Critical interlock: While identifying the hazards it is important to categorize critical interlocks (for new plants, data/experience from similar plants may be helpful) for which sufficient backup may have to be arranged as a safety measure. In many of these cases, in addition to software interlock, hardware interlocks are also provided. In some cases, even mechanical protection is provided. A critical trip interlock pertinent to a turbine is an example. For turbine overspeed, condenser vacuum, etc., in addition to electronic interlock (with sufficient hardware/software redundancy), mechanical interlocks are provided so that the equipment is saved. Choke and kill in offshore is also an example of this.
- Regulations: To meet the requirement of a regulatory board or company standards, measures are taken to enhance overall safety. Incorporation of these safety measures does not always mean that risk is adequately reduced. These measures could make the system acceptable for use, and a local regulatory board may require additional precautions. An example will elaborate the situation. Many turbine manufacturers give low forward power relays or reverse power

relays to trip the generator in case of tripping of a turbine. If the regulatory local board demands (as practiced in Germany) that the generator trips only when the reverse power relay operates, then turbine manufacturers offering low forward power relays need to incorporate reverse power relays as well (in addition to manufacturer's standards) to make the system acceptable to the local authority.

- Environmental safety and health design features: These have an impact on product design features and could become an integral part of the product to reduce the likelihood and consequence of propagation of hazards [1]. So, these are part of the product.
- Possible failure of a mechanism: During the hazard identification process it may be possible to identify some equipment and system failure modes. However, after all the data are collected analysts need to find out how a combination of events and failures can lead a hazard to an accident. All such single or multiple failures need to be identified utilizing engineering knowledge judiciously.

1.4.2 *Information Related to Hazard Identification*

The following is a list of probable documents that could be helpful:

- Plant flow diagram
- Document for operating conditions
- Plant layout
- Piping diagram (if any)
- Existing safeguard
- Design standards/norms
- Regulations
- Design sketches/schematics
- Design specification

These are useful for identifying hazards. Depending on the phase of the project and aim of PrHA, there may be variations such as:

- Characteristics/nature of hazard
- Form and quantification of hazard
- Where and when these are present
- How the event can propagate to form a major incident
- Conditions necessary for an event to become a major incident

The foregoing information in the description of each hazard makes a preliminary hazard list (PrHL), discussed next.

1.4.3 *Preliminary Hazard List*

A PrHL is generated during the initial assessment stage and is quite useful. It is updated throughout the design/technology development stages. The list is based on the sources listed earlier. Some of the PrHA output results may be as follows:

- Personal exposure to hot surface
- Loss of containment

- Nuclear radiation
- High-frequency radiation
- Injury because of falling object
- Inadvertent detonation of explosive
- Running train with door open
- Facility-specific issue

These are some of the issues that are available in different types of PrHA results and to find ways of tackling them.

1.5 Data Collection

Data collection is extremely important for hazard identification and preliminary hazard analysis. The following are some of the sources from where required data may be available. Also analysts carrying out the analysis must have a fair amount of knowledge of the functionality of the system, details of the area of concerned, system design intent, possible installation details, etc. So, an information source list is very important.

1.5.1 Specification

During the early design stage, most of the system data are not firmed up. Some analogous data may be available. However, at this stage, system specification specifying process performance, functional details, and material details will be the best source of information for the specific project. These data will be more useful during and after discussions with process engineers, hardware engineers, control engineers, and operation engineers about the specification. This will help in understanding the intent of the specification; basic materials, design criteria, and functional and operational details.

1.5.2 Design Review Interactive Sessions

The main aim of this type of session is to carry out an audit or progress assessment for:

- Compliance toward regulatory authority, design specification, clients' requirements
- Progress in achieving overall project objective maintaining safety design requirements
- Identification of hazards with their control measures adapted/prescribed
- Effect of engineering decision on safety engineering and amount of tradeoffs
- Status of current engineering documents complying with safety engineering requirements
- Identification of safety design and procedural problems that can affect overall project objective in terms of safety to personnel, property, and environment
- Progress of suppliers' documents in terms of safety engineering requirements
- Progress with respect to previous review meeting

Since this is an interactive session, the analyst with his/her team may openly interact to pose questions on various hazards with safety solutions as well as interact with any suggested modifications. Because this is done at an early stage of the project, major findings toward shortfalls in specific details can be identified. These specific shortfalls may include but are not limited to the following:

- Hazards from all energy sources
- Physical hazards
- Types of chemicals and their hazards
- Facilities for installations, accessibility
- Isolation and maintenance facility
- Critical components and controls
- Interface hazards
- Equipment failure
- Control failure/backup control
- Environmental hazards

Analysts need to use their experience and intelligence to find out specific details applicable for a particular facility!

1.5.3 *Detail Engineering Drawings and Schematics*

All applicable detail engineering drawings and documents developed by various engineering disciplines, namely, mechanical, process, electrical, control and instrumentation engineering departments, shall be routed through the safety engineering department (including analysts). It is the job of the safety engineering department to review, verify, and comment on the drawings to check that all safety-related requirements are incorporated, for example, after a fire and gas cause/effect diagram is prepared it is always vetted by all concerned especially by the safety engineering department. In fact, this is even true for fire, gas, or alarm philosophy documents. For existing products the same should be vetted by the safety engineering department. These documents can be utilized as future sources of information.

1.5.4 *Previous Analyses and Lessons Learnt*

These documents are very helpful when starting the job, especially when dealing with similar projects. Even if there are dissimilarities between them, they can be considered as a starting basis. Also lessons learnt data help in developing the current project with a more reliable database that can check near miss situations or any issue missed in earlier cases.

1.5.5 *Interview*

Interviews with personnel with extensive experience or personnel using current similar systems can throw light on many insights about the product and/or system. Such data are real-life practical data or observations and may not be available in an operation and maintenance manual. These interviews can also bring about certain operating conditions that might not have been considered even during the design stage. This will be more helpful where analysts are experienced and have good multidisciplinary knowledge so that the total impact can be understood.

1.5.6 *Miscellaneous Other Sources*

These comprise various relevant standards, customer or company standards, analogous documentation, etc.

1.6 Scenario Development and Classification

According to the Federal Aviation Administration (FAA) *System Safety Handbook*: "Seldom does a single hazard cause an accident. More often, an accident occurs as the result of a sequence of causes termed initiating and contributory hazards." To get a feeling of this statement Fig. III/1.6-1 has been reproduced and presented hereto show that there are several initiating causes that result in a hazard (undesirable event). It is not unusual that there will be a number of possible outcomes from initiating event(s) depending on several factors such as condition of equipment, people behavior, equipment location, plant layout for propagation, and even environmental conditions such as air flow direction, etc.

Another important point is to find the worst case scenario. Worst case scenario must include all foreseeable factors that could exacerbate the severity of an accident. It was mentioned earlier that there shall be a defined scope and boundary; naturally, this will focus attention on the accident of interest. Here, effort has been made to classify these hazard scenarios or risks into three basic categories, namely, people, environment, and property, based on their impact (combination of likelihood and severity). Matrix size and category divisions are purely specific for the project's and analysts' prerogative. Table III/1.6-1 gives an idea of such scenario classifications. Impacts are in decreasing order from the top in five categories.

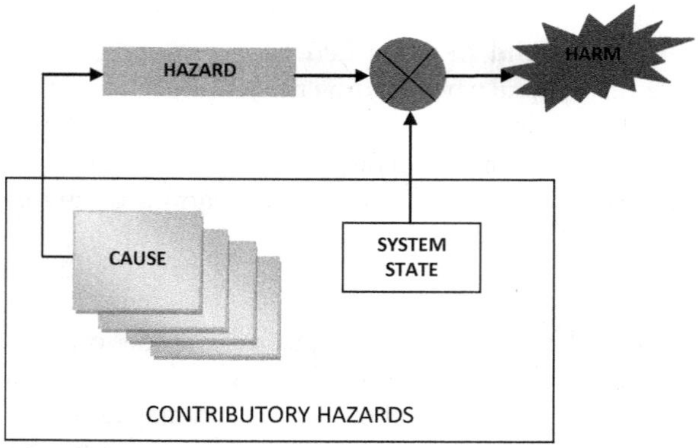

FIGURE III/1.6-1

Hazard scenario model (FAA).

Adapted from FAA *System Safety Handbook*.

Table III/1.6-1 Qualitative Estimate of Impact on People, Environment, and Asset

Asset	Environment	People
~$1.0 billion/ production loss and delay	Catastrophic impact on wide area of environment	Multiple fatalities
$100 million—1 billion/ production loss and delay	Major impact on wide area	1—2 casualties, multiple serious injuries, permanent disabilities
$10—100 million/ production loss and delay	Widespread serious impact	Major health impact and injury
$5-10 million/ production loss and delay	Substantial environmental impact	Minor health impact and injury causing lost workdays
$500k—5 million/ production loss and delay	Minor impact	Minor health impact/injury

1.7 Preliminary Hazard Analysis Methodology

Outline of PrHA procedure, has been depicted in Fig. III/1.7-1 to gain an overall view of PrHA methodology. On gathering the preliminary information with assessments, the PrHA process starts. The following are the basic steps associated with the analysis:

- Identification of nature and scale of hazards with initiating event
- Qualitative analysis of consequence and likelihood (to categorize)
- Categorization of risks and relative risk ranking
- Evaluation of existing safety/control measure
- Comparison with various risk criteria
- Suggestive additional control measures

Here, hazard identification is critical, because a single event may be benign, but cumulative or combinational effects may lead to a major incident. Consequence analysis reveals the effect of various incident outcomes. The likelihood of initiating incidents is estimated in likelihood analysis. Then, it is a question of risk interest, so it is necessary to categorize the risks as discussed in the previous clause, because one may be interested in finding human fatalities or injuries while others may be interested in loss of assets.

1.7.1 Consequence Analysis

Consequence analysis involves the analysis of the potential of hazardous incidents that could cause injuries, fatalities, and damage to assets and the environment. A sequence of failures may result in a hazard becoming a major incident, and analysts need to determine the consequences of each of these hazards. This should be based on the worst possible scenario, for example, in the case of an electrical short circuit, a

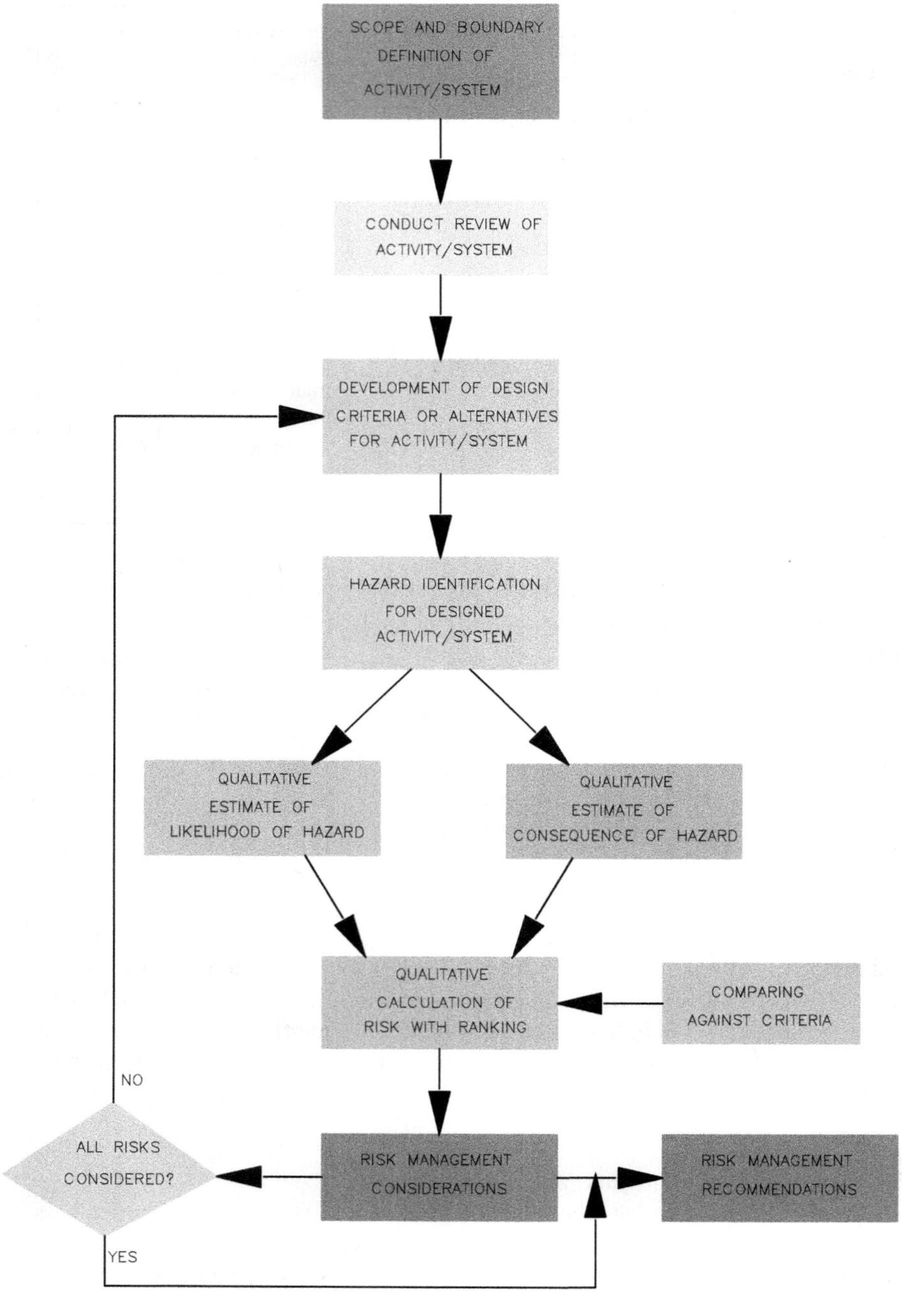

FIGURE III/1.7-1

Outline of PrHA procedure.

wire may be hot but in some cases it may lead to a fire, so fire shall be considered. In critical cases it is necessary to quantify the hazardous incidents, but in PrHA such quantification *may not* be possible, so the result of PrHA can be considered as the basis for further quantitative analysis. For each of the selected scenarios, it is required to estimate the effect of each outcome on people, the environment, and assets. On account of resource limitation, some screening of scenarios may be necessary. Therefore judicious decision making in the screening process is essential. Analysis should be accurate and realistic. For a realistic approach, an assumption may be necessary and this needs to be done very judiciously. An initiating event for one hazard is important but the sequence of events that lead to an accident may be more important. In this connection, Fig. III/1.6-1 may be helpful in understanding how several causes in a sequence develop into an incident.

1.7.2 Likelihood Analysis

Likelihood means the chance of something happening, whereas frequency means how often an actual incident happened in the past. In likelihood analysis both are used in the same way. So, for likelihood analysis some knowledge of probability is required. In most cases (unless very critical), qualitative likelihood analysis is carried out because quantitative analysis is quite rigorous and a fair amount of statistical data may be necessary. In qualitative analysis, more engineering estimates are utilized, but this is also based on probability data of the system and failure rate of equipment. Determining likelihood for an accidental or major event is important because such data provide the analysts with details regarding sequence of events and failure propagations, which will lead to a major incident/accident. There are two steps to estimate likelihood of a particular outcome of a hazardous event. The first is the likelihood estimate of an initiating event and the second is to find the probability that the initiating event will develop via event sequence (identified during hazard identification). For quantitative methods, various means like event tree analysis/fault tree analysis, etc. are used to develop a logical model. However, in qualitative analysis, generic data from a manufacturer, historical data, etc. are used. It is better to use specific plant data (if available) or historical data than generic data. Since data sources at times may not be authentic, naturally judicious engineering decision making in estimating likelihood is important. Usually, equipment failure data are expressed as failure per million hours, or failure per million of operation, etc., whereas protection failure, which is very rarely used, is expressed in terms of probability of failure on demand.

1.7.3 Categorizing and Risk Ranking

At this stage it is necessary to specify the risk category of interest. If it is health and safety, then one would go for a 5×5 risk matrix. *This is just an example.* With the help of Table III/1.7.3-1 data are derived from Table III/1.6-1.

From these data a risk matrix as discussed in Chapter I (Table I/3.3.2-5) is produced. The risks are ranked as detailed as in Chapter I. After risk ranking, safety measures as discussed in Clause 1.7.5 shall be applied.

Table III/1.7.3-1 Table for Severity in Health and Safety

Severity	Health and Safety Impact
Catastrophic	Multiple fatalities
Hazardous	1−2 casualties, multiple serious injury, permanent disability
Major	Major health impact and injury
Minor	Minor health impact and injury causing lost workdays
Negligible	Minor health impact/injury

Also data for likelihood are given in Table III/1.7.3-2.

Table III/1.7.3-2 Table for Likelihood in Health and Safety

Frequency	Occurrence With Impact on Health and Safety
Frequent	Once in every month
Probable	Once in a year
Remote	Once in 10 years
Very remote	In 10−100 years
Extremely remote	>100 years

1.7.4 Conduct Review

After the previous exercises have been completed it is time to review everything for decision making. It may be noted that it is also the job of PrHA to provide design alternatives to reduce hazards. Review of adequacy of existing control is a major activity at this stage. Also new hierarchical control measures are also done at this stage. Subsequently, at the decision-making stage, control measures may be regularized. While conducting the review it is necessary to check all necessary points. As guidance the following points may be helpful:

- Vessel: High pressure, materials contained in
- Materials: Toxicity, high reactivity, fuel corrosivity
- Environment: Rain flooding, extreme temperature, humidity, air flow, direction of air, thunderstorm
- Safety equipment and interface: Fire/explosion with detection and protection, chemical reactivity, mitigation system, enclosure class, intrinsic safety circuits, etc.
- Human factor: Error, maloperation, personal protection, accessibility
- Layout: Distance between equipment, electrical classification

1.7.5 Safety/Control Measures

After assessing the hazards and risk, the next job is to assess the existing safety/control measures or barriers in place. Obviously, this is done to evaluate the adequacy of control measures to prevent or mitigate major hazardous events. There are a number of ways and means to achieve this (see Chapter II). From experience, one way is to think of the hazards and risks and develop a picture of what could be the consequences. Once the entire picture is developed, it will be better to evaluate the existing system and come up with additional measures, if any. While developing a mental picture of the accident scenario and associated consequences, a systematic approach could be helpful. This means for each case dividing the consequences into environment, asset, and human. Then, subdivide into system response, structural, human error, structural response, equipment failure, etc. These are stated here because if the safety measure is developed along with system design as an embedded item, then its effectiveness will be much better than just adding the safety measure into the system design as an external object. As long as risk evaluation and control measures are done at the early stage of the project, then the remedial measures will be more effective and cost will be lower. On account of this, it has been found that the majority of decision making takes place at the early stage of the project.

1.7.6 *Decision Making*

A few important decisions need to be made for the following so that analysis will be effective:

- Acceptability: As discussed earlier, there are a few risks that are totally unacceptable. Also some cases are acceptable without a review, for example, low consequence. The main emphasis is on the risks that are acceptable only after review, and when there is a need for a control measure and then a review will be made for acceptability.
- As a corollary to this, one of the major activities is to identify additional control measures so that many hazards could fall under ALARP.
- Suggestions/recommendations for improvement: One such step is to explore the possibilities of applying hierarchical control measures (see Chapter II) for better safety. Another is to suggest improved engineering and administrative control including changes in equipment, operational change/improvement, or procedural changes.
- Justify additional control measures and resources deployment
- Recommendations for additional risk assessment

1.8 Output and Results

All hazards and major incidents should be properly logged. This is essential to track all hazards and suggest/modify any safety measures. These logs are specific to a particular project and there may be variations in presentations but essentially the following information is covered:

- Unit number, systems, area, activity
- Reference document number
- Scenario identification number
- Hazard description
- Initiating event/cause

- Any subsequent issue
- Possible effects
- Frequency
- Severity
- Categorization of risks
- Existing safeguard/barrier
- Recommended control measure
- Responsibility divisions
- Miscellaneous other information
- Testing/verification
- Further action
- Signature
- Target and achievement date

A typical example of a PrHA worksheet is presented in Fig. III/1.8-1.

Unit no: I	Area: Main Plant	System: Boiler	Subsystem: Flue gas
Team: A,B,C	Meeting dates 2-4/05/2009	Drawing ref.: doc 001,002,004,010	

ID	Description	Cause	Additional issues	Major effect
A				
B				
C				
D				
F				

ID	Description	Severity	Frequency	Risk Category
A				
B				
C				
D				
F				

ID	Description	Existing safety	Additional control measures	Test by date
A				
B				
C				
D				
F				

Comment	Team leader signature and date

FIGURE III/ 1.8-1

Typical preliminary hazard analysis worksheet.

Here, the details furnished are just for example. There may be other types where cause impact control recommendations are all put in one table. Now, in most cases, standard computerized formats are available and record keeping is done through the computer system.

1.9 Further Discussions

Here are a number of important and pertinent issues are discussed.

1.9.1 Timing and Software Issues

Timing of PrHA is very important. This is especially true for product development. When safety is incorporated at the early stage, then the measure will *not* only be effective and easier to implement, but it will also cost less. It is therefore recommended that the design team work in close coordination with safety staff for product development. This is true for all types of industry but especially so for *product development and manufacturing systems*. For product development, a detailed design and operating procedure needs to be clearly spelt out to minimize hazards. Also it is required to take into account possible human error and associated protection, and how statutory regulations are met shall be clearly noted.

After hazard identification, design constraints are placed and seen from the perspective of human operation, software, etc. especially for programmable electronics (PE). However, this may not completely get rid of hazards, so they are also traced at a software level. For any product safety and reliability are important issue and their intimate relationship has been explained in Fig. III/1.9.1-1. This is standard practice in product development.

Reliability and system safety: System safety analysis and control measure go hand in hand with system reliability. In fact, they complement each other. Each of them gives a lot of feed to the other. However, one cannot be substituted for the other. It is somewhat like the relation between acid and base. When they operate in close coordination with each other the result is always better in system development.

FIGURE III/1.9.1-1

Relationship between reliability and system safety.

1.9.2 Common Hazard With Effects and Controls

In industrial plants the common hazards given in Table III/1.9.2-1 are encountered.

2.0 "WHAT IF" AND "WHAT IF" CHECKLIST ANALYSIS

As the name implies, in "what if" analysis the team leader or team member raises questions in the form of "what if" to analyze various hazards associated with the system. "What if" is a creative brainstorming hazard analysis technique. It is creative

Table III/1.9.2-1 Common Hazards With Effect and Recommended Controls

Hazard Type	Potential Accidental Effects	Typical Control Measure
Energy sources	Fire explosion, electrocution/short circuiting, personal/equipment burnt, ignition, increase in toxicity	Cut off energy source quickly if possible, lockouts and interlock functions, initiate emergency handling procedure
Hazardous material (including corrosive)	Damage to human organs/systems, ignition, fire/explosion, equipment/structural damage	Substitute if possible/minimize use, use fume hoods, use of suitable containing vessel, use of material safety data sheet and label as applicable
Gaseous material handling	Overpressure, container implosion, leak, ignition, damage to human organs/systems	Use appropriate measure for secured transportation, storage and handling, minimum usage as possible, emergency procedure
Radiations and radioactive materials	Possibility of loss of controls for safety control and automation, interference, change of organic compound, human tissue damage	Emergency procedure, fume hood, suitable training, use of personal protection item, handling as per local regulation
Rapid chemical reactions	Damage to human organs/systems, ignition, fire/explosion, equipment/structural damage, propagation of hazards	Fume hood, minimum quantity use as possible, suitable automatic monitoring and alarm, quick isolation, use of personal protection item
Possibility of air pollution	Human respiratory system affected, direct environmental impact, corrosive damage to equipment	Use of personal protection item, engineering and administrative controls, use of waste trap, use of substitute material (e.g., for asbestos dust, etc.)
Noise vibration	Material failure, pressure and shock waves, valve/contact chattering	Engineering control, use of antivibration materials/isolation
High pressure/high temperature	Ignition of combustibles, initiation of other reaction, system leak, uncontrolled pressure change	Use of cooling system, emergency procedure, use of protecting device
Hazardous waste	Damage to human organs/systems, environmental impact	Use of suitable storage vessel, administrative controls, emergency procedure
Space for equipment	Escalation of hazards, fire/explosions	Use of design and layout control to prevent escalation

in the sense that skilled people need to develop from experiences how hazards can develop and pose questions in the form of "what if." In this analysis method, review is carried out on the subject process and its operational aspects after evaluating the safety of the system.

2.1 Characteristic Features and Purpose

The following general features could be attributed to "what if" techniques:

- It is a creative brainstorming approach where a group of experienced people pose questions or voice concerns over a process and operational aspects of a system hinging around safety. An experienced review team can discern major issues pertinent to the organization of a process system. It evaluates existing safety issues and comes up with additional control measures.
- Questions are usually divided into specific areas of interest and these are intended to identify the hazards/hazardous situations and major incidents for a specific area.
- The technique can be utilized to investigate almost all aspects of the design, project, or facility, for example, procedure, material, equipment, or external issues like power supply failure, or plant environmental issues.
- The "what if" process is usually done from start to end, that is, from chemical feed to the finished product, but of course within the boundary defined beforehand.
- In this analysis type, usually scope and boundary are done in such a way that the "what if" analysis, hazards for the operating system, design, maintenance, etc. are not mixed, that is, there is segregation.
- This technique is not particularly well structured. The "what if"checklist is more systematic! Thus in many cases this analysis could be followed by other techniques such as checklist.
- It is highly dependent on the knowledge and experience of the team, especially an efficient and effective team leader, for success of the analysis.
- "What if" is usually not carried out for any great length of time, because it creates fatigue in members and effective results may not be forthcoming.
- However, being simple in nature and with the possibility of carrying it out with a small group of people, it is extremely helpful when various other checklists are not available.

The purpose of the "what if" analysis technique is to identify hazards/hazardous situations, major incidents, and/or specific accidents pertinent to specific areas, which could develop into undesirable consequences. As stated earlier, this technique is very helpful when various other checklists are unavailable. This is made very clear when a small example is taken. As an example, suppose there is a heating element heating water in a heater. Now, what if the water flow stopped? The heater coil may burn out as heat will not be taken away. A precautionary measure will be to cut the power supply to the heater when there is no water flowing. Here, the situation is questioned in the form of "what if" and the answer is listed. After review a necessary precautionary measure is suggested. This is a very simple situation. However, in complex situations it is better to divide the entire scenario into smaller parts, then carry out the analysis.

2.2 Scope and Boundary

Like PrHA, here also the scope and boundary definition is important. The scope simply defines which part of a system is being analyzed. Also the presence of hazards often depends on the operating phase of the facility, so it is necessary to define the phase of the project. Therefore the boundary could be a single piece of equipment or a collection of similar equipment, etc. It has been found that the narrower the boundary, the more explicit/detailed is the analysis and control measure recommendations. However, the interfaces that cross the boundary must be examined as well. Earlier, it was stated that "what if" normally covers start to end. So, "what if" analysis can be carried out part by part as defined in each case of the scope definition (see Point 8 of Clause 2.5). In addition, it is interesting to note that often "what if" is carried out for a particular classification of consequence, for example, public safety, the environment, etc.

2.3 The Team

Like any other analysis, team formation is important as success of analysis greatly depends on team. If the team is highly experienced and has a good understanding of the process/plant, then naturally pertinent questions will be asked and as a consequence due importance will be provided. The team must understand the system and its probable failure points and inherent hazards. This will help to understand which single failure could lead to an accident. On account of too much dependence on the system by the team, it is necessary that representatives from all critical branches are well represented in the team. Also, if warranted, service personnel should be included. A safety engineering professional who is very systematic and efficient with sound experience in the analysis could be selected as team leader or facilitator, as it is normally referred to. As per Occupational Safety and Health Administration (OSHA) requirements, one member shall be an engineer with knowledge of the process and one member shall possess good knowledge of analytical systems. The team leader/facilitator should also be familiar with the team members, so there is good communication. The team leader needs to lead the team from the front and be able to coordinate all members and extract the best from them. The team leader is responsible for generation of the list and needs to screen/filter out some of the questions for the final report. The leader shall be a leader in the true sense of the word and not a dictator; the leader should give equal chances to everyone to put forward their questions freely.

2.4 Documents and Information

A few pieces of process information and documents are necessary to start the process. The following are the most important:

- Well-written process description
- General drawings such as flow diagram, layout, etc.

- Operating procedure
- Piping (process) and instrumentation diagram (P&ID)
- Schematics
- Alarm trip list
- Interlock scheme
- System/component specification
- Operation manual
- Service manual
- Past records of failure
- Similar plant history

If the facility is an existing one it is better to interview the operating and maintenance personnel to gather information. It is also a good idea to visit the site to get a first-hand information for on site as well as off site. This is especially true for environmental impact study.

2.5 Planning and Preparation Stage

Before starting the meeting with the team members, it is better for the members and especially team leaders to develop their questions. In this planning stage the following are undertaken:

1. Division of process into segments (then for each segment)
2. Development of a few questions
3. Reviewing and updating of old result for subsequent review
4. Reviewing and updating of old result for modification job
5. For each hazard, preparations of preliminary cause and effect
6. Some guidance for preparing questionnaires may come from the following:
    - Procedure: Failure to follow or followed incorrectly or changes in procedures not followed
    - Lack of work on part of operator or not trained properly
    - Procedure modification to face emergency
    - Equipment failure
    - Control system failure
    - Interlock failure without fallback
    - Utility failure or disturbances
    - Environmental effect
    - External factors
    - Human error
    - Multiple failures
7. Team leader/facilitator needs to give equal opportunity to all the members to pose questions on potential incidents/events. The answers to the questions may be discussed later (and not as they are generated) so that the thinking process is not hindered.

8. In case the process is too large, then the entire system shall be segregated before starting working on them, so that team does not have to be bogged down for any length of time.
9. There are two ways in which questions are prepared and listed. One is where all the questions are seen beforehand to ensure nothing is missed. Also this does not interrupt the thinking process. The other is where, one at a time, an issue is addressed and several questions and their safety issues are discussed. Here, there may be a problem of interruption in thinking, as discussed.
10. The boundary must be fixed before starting the meeting; also the topic in the form of a scope shall be defined. This is important if the analysis is carried out by a team with less experience, then smaller part could be manageable.

2.6 Outline of Procedure

The procedure is not a structured one and members can voice their concerns about major incidents. The following are the basics of the methodology:

1. The meeting starts with an address by the team leader/facilitator on the basic scope of the study. Team members well acquainted with the process normally briefly describe the process also. All team members need to agree to the scope and boundary defined by the team leader.
2. The team leader then begins the study usually in a beginning-to-end fashion or may follow a logical path. Each of the team members answers the questions or takes up each issue one by one. At this point, corresponding to each question, hazards are identified, consequences are evaluated, and various solutions and/or control measures are engineered. It is not necessary that all questions will be answered in one go. In some cases the issues are kept open pending additional information.
3. It is necessary to check very carefully that no potential hazard or incident is left out. Any member can voice their concern during this time, even if the same question does not appear on the list. The list may be modified depending on applicability.
4. In "what if" analysis, questions are taken from start to end within the defined boundary. As stated in point 5 of Clause 2.5, in each step "what if" questions are posed linked with procedural failure, hardware or software failure, etc. Therefore a list of questions and answers concerning the hazards and their consequences will be generated. This list shall also include recommendations regarding the safeguard/control measure. So, a tabular listing will be produced as discussed later.
5. The comprehensiveness of the analysis is dependent on the composition of the team and how exhaustive studies have been carried out. So, small interdisciplinary team members may give a better result.

2.7 Result and Documentation

As a result of any study, all findings are jotted down in the tabular/narrative form of "what if" questions and answers. After questions and answers the next task is to make a judgment regarding the likelihood and consequences. At the same time it is the job of the team to put forward recommendations regarding control measures and/or additional recommendations, if any (these may be procedural changes/material changes, e.g., asking a contractor to change valve gasket material). All these will complete the task and a report will be sent to the plant manager. One typical format for what if analysis has been presented in Fig. III/2.7-1 just as an example to gather knowledge about detailed content of "what if analysis" form.

2.8 Application Notes

"What if" is a good and simple analysis technique but is normally used for single point failure cases. On account of its simplicity it can be used for any system and process. It is one of the qualitative analysis methods for OSHA safety management regulation. It is useful for emergency planning also. It is neither systematic nor thorough nor foolproof, yet. On account of its simple approach, it is easy to operate without much training and is user friendly. However, another drawback of "what if" is its over-dependence on the team's performance. "What if" can be used for design review, equipment/control system check/evaluation, investigation of incidents, and for management of change.

2.9 Discussions

Here are a few pertinent points on the system and structure of "what if," that is, the "what if" checklist:

1. In addition to manual analysis, computerized systems are also available. There is software available for "what if" analysis: WHAT IF-PC (Primatech) [2].
2. To retain the creative nature of posing questions in meetings, it is recommended not to extend the meeting for longer than 4–5 h and day, and not more than 4 consecutive days.
3. "What if" checklist: This is a combination of both the checklist (discussed later) and "what if" analysis techniques. It combines the advantages of both. The following is the basic system description:
   - It utilizes the creative technique of "what if" and uses an experience-based checklist to give the combination a structured (systematic) creative analysis technique. The result is a comprehensive hazard analysis that is extremely useful.
   - In the checklist the experience from history, similar plants, etc. is used for analysis. If the technology or system is new and there are few analogous data available, then the checklist method may not be of much use even though it is systematic. The "what-if" part of the analysis uses the team's creativity and experience to brainstorm potential hazards and major incident situations. So,

COMPANY NAME PLATE

UNIT NO:	LOCATION	DIVISION NAME	AREA NAME

PLANT DETAILS	PROCESS/OPERATIONAL DETAILS:	
DRAWING/DOC NO.	DRAWING/DOC NO.	BRIEF SCOPE AND BOUNDARY
DRAWING/DOC NO.	DRAWING/DOC NO.	

MEETING DATE	MEMBERS	LED BY:

SL	WHAT IF	ANSWER	LIKELIHOOD	SEVERITY	CONTROL MEASURE	RECOMMENDATIONS

WHAT IF FORM	WHAT IF DOC NO:

NOTE: This form is just an example. The format may vary with company practice. However, basic information is as shown.

FIGURE III/2.7-1

"What if" analysis form.

these two are complementary to each other and fill in the blank of one system with the other.

- Usually, "what-if" brainstorming precedes the use of the checklist, although the order can be reversed. Generally, the "what if" checklist takes a little longer than "what if" when one follows the other. However, it is less than the total time taken by each of the processes.
- The selected review team is entrusted to analyze a wide range of production process plants and associated safety. The team is given basic information regarding the process and operation of the plant to be studied. The basic information includes information on hazards scenario (including information on materials and equipment), process technological details, operation procedures, design data for equipment instrumentation and control, incident experience, and previous hazard reviews (as applicable).
- A field tour of the operation is also conducted at this time.
- Usually, the team first conducts "what if" analysis. When the review team has completed listing its spontaneously generated questions, it systematically goes through a prepared checklist to augment additional questions. The answers to these additional questions are developed.
- Also the team sit together to achieve a consensus on each question and answer. After this, a list of recommendations is developed against each answer as applicable. At the same time, an additional study recommendation may be made. Like "what if," the result here is developed with all questions, answers, control measure listings, etc.

3.0 CHECKLIST ANALYSIS

In the previous clause the "what if" checklist was discussed, where it was found that checklist analysis is basically dependent on the experience of the team and previous data about the plant. Based on these a set checklist is prepared and analysis is carried out to check the deviations with respect to the checklist. This type of checklist analysis is also known as *traditional checklist analysis* and can be applied at any stage in the life cycle of the project.

3.1 Checklist Characteristic Features

1. The checklist specifies a list of hazards, common shortcomings in design, major incidents, etc. pertinent to the facility in question, for example, sudden trip of boiler feed pump (BFP), a check to see how to save a boiler from starvation, or how to provide backup for a high-pressure mud pump failure in offshore operation.
2. This list is used to evaluate the equipment, operational procedure, materials, and hazards associated with various stages of the project, for example, if the running BFP trips, the standby the BFP shall automatically cut in and it shall be designed for loaded staring (with discharge valve open).

3. Generally, a checklist is used by an industry in cases where it has some experience, for example, a utility industry having operating experience in a fossil fuel plant (say a subcritical unit) can use the experience to prepare a checklist for one of its upcoming plants. The same may even be used for *new plants or new technology*, for example, the same subcritical plant checklist may be used to develop a checklist for a supercritical unit making the necessary change.
4. A checklist is industry specific and is developed with the help of experience and plant history data. There may be generic checklists for certain types of major equipment but in industry it is unique in the sense that a checklist for supervisory control and data acquisition (SCADA) applicable for an electrical transmitting system may not be applicable for SCADA in oil and gas midstream plants. Similarly, a checklist for a nuclear plant may not be applicable for a fossil fuel power plant. Here, "may not be" means that the turbine operation part of the nuclear plant may be utilized for the fossil fuel power plant as well. However, the checklist of a refinery will be completely different.
5. A checklist is mainly used for plants covered by codes and standards/company standard practices (see the following clause), etc.
6. Usually, the replies to queries prepared in such a way are short and in the form of yes/no, not applicable/needs further investigation, etc.
7. National Fire Protection Association (NFPA) 654 may be one useful checklist.

3.2 Purpose of the Checklist

The checklist is used to check the functioning of equipment, system, and plant with respect to a set standard. Naturally, experienced personnel prepare such a list, taking into account the required codes and standards and/or company standard practices. When any deficiency is noted it is listed and analyzed for its final consequences, and suggested control measures or recommended further studies are then proposed to deal with it. From this discussion it is apparent that a checklist is more suitable for plants that are covered by international codes and standards (e.g., a propane facility—NFPA 58) and company standard practices.

3.3 Scope and Boundary

Scope and boundary definitions philosophy, discussed in Clause 2.2, may be applied here also. In the case of a complex system the entire scope may be divided into several divisions to facilitate analysis.

3.4 Prereview Preparation

Checklist analysis is a written list of items or procedures for comparing plant status with standards. Because checklists are plant specific they may vary widely in level of details covered.

1. Team formation: A team is formed by inducting expertise from different disciplines to cover the scope and boundary of analysis. An engineer having thorough knowledge and experience in the technology and operating procedures with good grasp of company standards is chosen as team leader/facilitator.
2. Information collection: To prepare the checklist it is necessary that the team has suitable access to internal standards (if any), industry guidelines, P&IDs, schematics, engineering design basis documents, operating procedures, equipment data sheets, etc.
3. Preparation of checklist: A detailed checklist for a facility is developed by a team as per the requirements of process safety management. This detailed checklist is suitably edited by the team leader. A detailed checklist should be extensive, and commensurate with the needs of the facility. However, this detailed checklist is limited by the experience of the team facilitator. Therefore it is recommended that in addition to a company-specific checklist, a generic checklist (usually applicable for the industry/equipment/system) shall also be developed so that nothing is *missed*. Generic checklists are often combined with other methods to evaluate hazardous situations. Therefore from all these, a final checklist is prepared. It is worth noting that this final detailed checklist should be treated as a *live* document and should be reviewed periodically so that any subsequent changes can be made. The final checklist verifies various aspects of process design or operation with respect to the checklist. Those aspects that do not comply with standard industrial/good engineering practices are discovered through responses to the questions in the lists. It is applied conscientiously to check how far the requirements have been met and to identify the problems that require further investigation. Once the extensive final checklist is prepared it could be used by comparatively fewer experienced people. So, preparation of the checklist is very important.

3.5 The Analysis

After the checklist is prepared, then actual analysis is carried out.

1. For an existing system the best way to start will be for team members to walk through and visually inspect the process areas to compare the process equipment and operations to the checklist items. The analysts respond to the checklist items based on observations from their visual inspections, process documentation, interviews with operating personnel, etc. Any mismatch of a process/system with respect to the specified features in the checklist needs to be noted by the analysts as deficiencies in the list.
2. A checklist analysis made prior to construction or for a new facility is usually performed during the team meeting. In this case, reviews are carried out on the process drawings, completion of the checklist, and discussion of the deficiencies.

3. After observations are noted, the team holds a meeting to list the deficiencies and discuss the issues. From the discussions the issues should be clearly understood and the team will be in a position to propose various recommendations for safety/control measures. The facilitator plays a key role at this stage. The checklist review can be done in hard copy or computerized forms could be utilized.

3.6 Common Checklist

The discussions on the checklist are concluded with a list of common checklist headings. Based on these headings, one can find a standard generalized checklist from any standards handbook and/or from the internet (Table 4.1 from the *DOE Handbook: Chemical Process Hazards Analysis* is an example) [3]. Major headings for such a checklist shall include but not be limited to the following:

1. General work environment
2. Personal protective equipment
3. Walkways
4. Floors, walls, stairways
5. Exit door
6. Portable ladder
7. Hand tools/equipment
8. Machine guarding
9. Welding, cutting
10. Compressed air system
11. Hoist and allied equipment
12. Moving vehicle
13. Environmental issue
14. Hazardous materials
15. Storage of materials
16. Material handling
17. Personal protection
18. Waste disposal
19. Major process equipment
20. Control systems
21. Maintenance issues
22. Operating procedure
23. Fire and explosion issues
24. Area classification
25. Layout issue

Against each of these headings will be specific issues that form the checklist. The hazard analysis in this chapter may not be suitable for larger and more complex facilities, hence other detailed systems may be necessary. To address this, the next chapter on guided word plant analyses will be helpful.

LIST OF ABBREVIATIONS

C&I	Control and instrumentation (engineering)
ESH	Environment safety and health
ETA	Event tree analysis
FTA	Fault tree analysis
HAZID	Hazard identification
HW	Hardware
MOC	Management of change
MSDS	Material safety data sheet
OSHA	Occupational safety and health administration (US)
PrHA	Preliminary hazard analysis (PHA)
QRA	Quantitative risk analysis
SW	Software

References

[1] Hazards Analysis Guide: A Reference Manual for Analyzing Safety Hazards on Semiconductor Manufacturing Equipment; International SEMATECH; Technology Transfer #99113846A-ENG.

[2] http://www.primatech.com/software.

[3] DOE Handbook; Chemical Process Hazards Analysis, U.S. Department of Energy, August 2004. DOE-HDBK-1100—2004.

Further Reading

V. Asar, Hazard Assessment and Risk Management Techniques for Industries, Disaster Prevention and Management Center; GIDC India.

W. Briges, Selection of Hazard Evaluation Technique, Process Imrovement Institute, Inc, February 2008.

FAA Systems Safety Handbook; Chapter 8; Safety Analysis/hazard Analysis Task/December 2000.

Hazard Assessment Checklist; https, Hazard Assessment Checklist; https://www.dir.ca.gov/dosh/etools/09-031/HazAssessCheck.pdf.

Hazardous Industry Planning Advisory; Paper No 6; Hazard Analysis, NSW Govt. Planning, January 2011.

http://www.hse.gov.uk/risk/faq.htm.

https://www.dmv.ca.gov/portal/dmv/detail/pubs/cdl_htm/sec9.

H.R. Kavianian, J.K. Rao, G.V. Brown, Application of Hazard Evaluation Techniques to the Design of Potentially Hazardous Industrial Chemical Process, California State University.

Long Beach; NIOSH Instructional Module; CDC (Occupational Safety and Health) US Dept. of Health and Human Services.

Preliminary Hazard Analysis; Appendix M; Bulliseam Operations; Illawarra Coal, BHP Billiton, May 2009.

Preliminary Hazard Analysis; Appendix N; Donaldson Coal; Tasman Extension Project; Environmental impact statement.

Preliminary Hazard Analysis; Safeware Engineering; Risk Management Series; www.safeware-engg.com/white paper/Preliminary%20hazard%20analysis.htm and www.safeware-eng.com/%20white%20paper/software%20safety%20approach.htm.

M. Rousand, Preliminary Hazard Analysis, NTNU/Willey, October 2005.

AuthorAnonymous, J.F. Shortle, Applying Qualitative Hazard Analysis to Support Quantitative Safety Analysis for Proposed Reduced Wake Separation ConOps, George Mason University, Fairfax, VA & M. Allocco, FAA, Washington, DC.

CHAPTER

Guided Word Hazard Analysis IV

Guided word hazard techniques are basically a qualitative approach but more systematic. In this method, potential design and operating hazards are identified through questions and answers with the help of a few guided words to discover what would happen if a procedure is skipped/reversed or performed incorrectly. The guided words provoke thoughts in a brainstorming session to find deviations from design intent or set operating procedure. All such deviations could lead to a potential hazard. Therefore, with the help of these guided words, hazards are identified. In this method, an experienced team leader guides the team through the entire design document or operating system of the selected part of the system falling within the boundary of the defined scope. An experienced team leader guides the team with the help of a set of words popularly known as guided words, which are applied at specified points or study nodes and are combined with system parameters (process parameters) to identify the deviation with respect to the design intent or operating procedure. After the potential hazardous situations are identified, their causes and consequences are also analyzed. A control measure or action to prevent or mitigate the hazard is prescribed. In guided word techniques, report preparation is also extremely important. Since the guided word technique involves brainstorming and is systematic, it is necessary that there shall be some background work prior to carrying out the main job. One of the basic functions is to form a team and team leader and work out suitable planning. Two types of guided word techniques will be discussed in this chapter. These are the hazard and operability (HAZOP) study and the failure mode and effect analysis (FMEA) study. In each of these cases, basic system study, system study for electrical/electronics/programmable electronics (E/E/PEs), and automation of HAZOP and FMEA will be covered.

1.0 HAZOP STUDY/ANALYSIS

1.1 HAZOP Study Feature

Dr. H.G. Lawley is very much associated with HAZOP for his contribution. HAZOP study was developed by ICI in the 1960s. Later the Chemical Industries Association supported HAZOP and produced a number of guidelines in 1977.

1.1.1 HAZOP Definitions

What is HAZOP? A HAZOP study is a structured and systematic investigation of a planned or existing plant or operation to identify and evaluate hazards that may represent risks to personnel, property, environment, or prevent efficient operation. This is a guide words-based qualitative technique carried out by a multidisciplinary team (HAZOP team) during a set of meetings. It is a versatile study technique and can be applied to a wide range of applications, both continuous systems as well as batch processes. The applicable international standard is: IEC 61882. "Hazard and operability studies (HAZOP studies)—Application guide." The basic scope of the standard is: "This International Standard provides a guide for HAZOP studies of systems utilizing the specific set of guide words defined in this document. It also gives guidance on application of the technique and on the HAZOP study procedure, including definition, preparation, examination sessions and resulting documentation and follow-up."

1.1.2 HAZOP Outline and Characteristics

The various characteristic are as follows:

- HAZOP study is a structured, systematic, and rigorous analysis of a system design and operation, by a team of experts. In this hazard analysis, system design and operations are studied stage by stage or line by line in a brainstorming, open, and creative way. A meaningful deviation is discovered by carrying out the study with the help of a set of guide words in combination with the system parameters. Here, "meaningful" is very important because all guided words (discussed later) may not be applicable to all the parameters, for example, guide word "NO" could be used for flow through a pump, but "NO" is not applicable to temperature as temperature is not physically possible!
- The team mainly concentrates on those deviations that could lead to potential hazards. Deviations are evaluated by the team, utilizing experience and judgment to pinpoint the consequences (may be risk ranked) and recommend actions for safeguards if current control measures are found inadequate or calls are made for further investigation of the problem.
- An experienced team leader develops a model for the system design or operation with pertinent information such as main design and operating procedure with details about materials, material safety data sheet (MSDS), main equipment used in process flow, and historical data about the plant hazards. The team creates the deviations for which it suggests causes, the consequences are estimated using the team's experience, and existing safeguards are taken into account. A formal record is generated.
- HAZOP is capable of identifying potential hazards and operational problems in terms of plant design, operation method/strategy, and human error. Therefore, by applying HAZOP during the detailed design stage, with special emphasis on operation and functional controls, money and time can be saved, especially during the operation stage, from add-on controls at a later date (resulting from an accident!), and there will be fewer problems in commissioning and in operations.

- In HAZOP, the entire process/design and operation is questioned with the help of guided words applied to various parameters to find the deviation. Guided words are so chosen that they focus on testing the system integrity and try to find any conceivable hazard. It generally gives total coverage for identification of potential hazards for the entire system, and necessary control measures can be applied to prevent accidents. However, sometimes it may happen that the immediate solution to a problem may not be obvious and might need further consideration (maybe by a specialist). It is more suitable for process plants but could be used for other applications including E/E/PEs.
- Choice of the right team with a team leader, details of information available, and quality of design finally determine the quality of the study report and performance of the study.
- It is a well-recommended practice by professional bodies and legislatures.
- On account of the nature of brainstorming, the study process offers a creative concept to find new deviation causes and consequences, which can be helpful in suggesting better control measures. It can be used in various plant stages/modes (only accurate sufficient information is necessary). It finds its applications for:
  - Plant design study
  - Normal plant operation
  - Reduced output operation
  - Normal startup
  - Normal shutdown operation
  - Emergency shutdown operation
  - Plant construction/commissioning
  - Special operating mode
- At present, HAZOP is extensively used by the majority of companies, especially those where engineering practice involves elevated operating parameters:
  - Oil and gas production
  - Flammable and toxic chemicals
  - Pharmaceuticals, etc. to name a few

1.1.3 Objective, Scope & Boundary and Study Stages

At the outset, discussions on the study process were presented, so that the reader can develop the concept of objective and purpose. It is needless to state that for effective analysis, scope and boundary must be defined properly.

- The basic purpose and objectives of hazards could be as follows:
  - HAZOP is used to identify the design, operating, and quality problems related to planned process. Operational problems generally come from human error, operating procedural error, or from equipment/control system reliability issues. HAZOP is well suited for such applications.
  - HAZOP is capable of identifying cause and consequence such as downtime, out of specification, monetary loss, etc.

    - It tries to find the relationship between various codes and standards with the process to check and see that no hazards remain undetected.
    - It focuses on testing the system integrity and tries to find any conceivable hazards at the design stage, normal operating mode, startup/shutdown mode, etc. It is a hazard identification technique and not a problem-solving issue (control measures are developed later through discussion—not truly a part of the technique).
- Scope and Boundary: As is clear from these discussions, the study covers the entire spectrum of the process. Therefore it is imperative to state that scope definition and boundary selection are an absolute necessity, otherwise they may be uncontrolled and will not result in a quality study. This scope covers initiation of the study, recommendation of control measure, follow-up, and implementation, that is, at various stages of the study. So, at every stage, experience and judgment of team members are essential. For large plants, this is done by dividing the entire plant into smaller independent sections and carrying out the study of each of these sections separately. This will also offer a benefit of taking the help of specialists, for example, a process engineer in a water treatment plant can better contribute to that plant. Similarly, a mechanical engineer in a coal handling plant can offer better expertise on that plant. So, it is better to divide the entire fossil fuel power plant into smaller sections as indicated. Potential hazards related to health, safety, and environmental (HSE) issues are normally addressed in HAZOP but in addition to these, the issues related to reliability of equipment/control systems, product quality, operating procedural issues, maintenance, and other related issues are often covered in the scope of this study. Also plant downtime, etc. could be calculated. The study covers normal operations and startup and shutdown conditions as well. So, proper definition of scope in all these aspects is important. It is essential that the boundary of the study must be well defined at the beginning of the study so as to achieve a better result. Boundary definition indicates which physical section(s) of the plant need to be included. Here, another important point is how will the interface be handled, that is, how will a problem outside the defined section be dealt with? In addition, if there is more than one study, then how will the interface be handled so that nothing is missed or there are no duplications? It is not easy to state here how the boundary could be defined (though some indication has been given in the example), as it is very much dependent on complexity of the plant, nature of the plant, type of process, inherent hazard interrelation among the units, standard or new or proprietary system is in the question. Also for sequential operations, manual/automatic control modes have bearing on taking decisions on boundary definition. The raw mill section and kiln section of a cement mill are apparently independent of each other with silos at interface points, so, in which section should the raw meal silo be considered? Similarly, in offshore plants, the mud section is not directly related to the drilling

or pipe-handling sections, but for the drilling operation (choke and kill), mud is essential. So, the interface part should be given proper attention when defining the boundary. Sometimes "HAZOP by difference" may be effective in some cases. Suppose one utility authority has a number of subcritical power units whose HAZOP study results are available. When the authority implements a new supercritical power unit, it will be useful and time saving to carry out "HAZOP by difference," that is, by comparing.

- Sequence of study and variations with stages: In Clause 1.1.1 the outline of HAZOP was discussed, and it is clear that the following are the basic steps:
  - Intention
  - Deviation
    - Cause
    - Consequence
  - Safeguard
  - Corrective action

  Each of these steps with an explanation is highlighted in Table IV/1.1.3-1. HAZOP studies can be undertaken at various stages in the life cycle of the project and each time the purpose is different. Fig. IV/1.1.3-1 depicts HAZOP studies at various stages.

Table IV/1.1.3-1 HAZOP Steps and Functions

Step	Explanation
Intention	Process designer to highlight plans for one section/piping and instrumentation diagram (P&ID). General scope and intention discussed. Relevant part highlighted with dotted line. Process designer to explain the part and general discussion.
Deviation	Line-by-line study commences with team leader choosing relevant guide word. Deviations with potentiality for hazards are noted. Team leader goes through all relevant guide words one by one; when all guide words are exhausted the next line is chosen and this line is highlighted firm meaning that its work is done. When all lines in this way are complete, that is, all are highlighted firm, additional words may be chosen to check the entire P&ID.
Cause	Cause for each of the deviations is identified.
Consequence	For each of these deviations, consequence (combination of likelihood of occurrence and severity) is identified through creative and brainstorming discussions. Consequences that warrant action are recorded.
Safeguard	Existing safeguard is evaluated during meeting and new control measure if any is prescribed.
Corrective action	When warranted, detailed quantitative risk analysis (QRA) or reliability analysis may be undertaken for complex systems at a later time.
NOTE	The purpose of this study is to identify hazards that require solutions and NOT the solution proper.

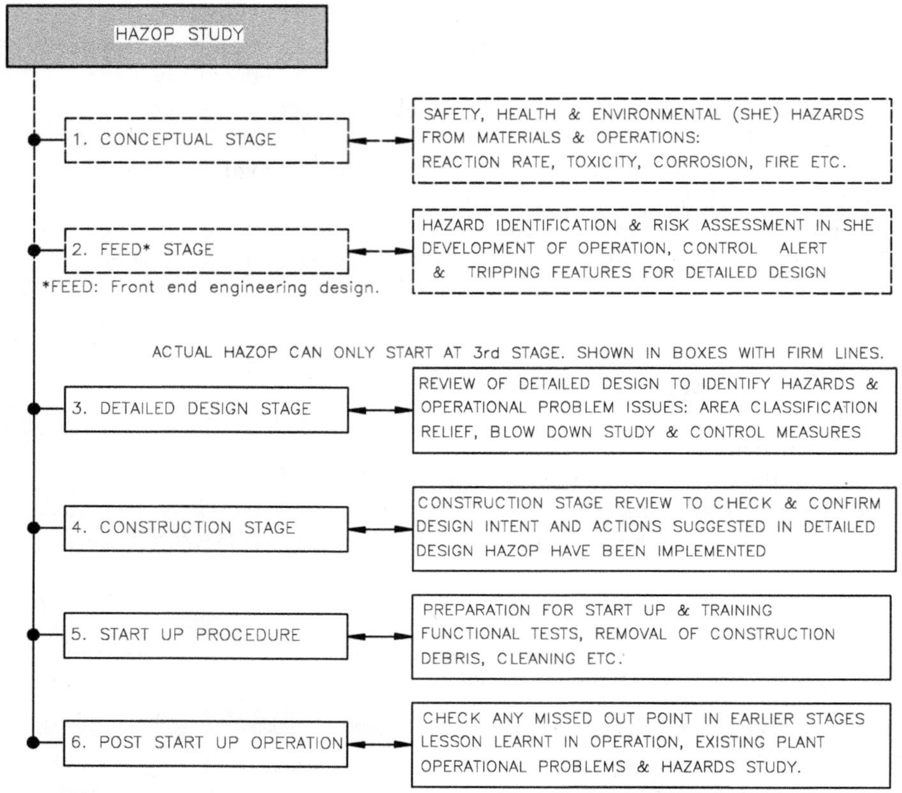

FIGURE IV/1.1.3-1

HAZOP studies at various stages.

1.1.4 Team

In this clause participants in HAZOP are discussed. The responsibility, qualification, and experience of each of the team members are discussed here. However, this is a general guideline; based on company policy and plant requirements the team formation may vary. In this connection generalized citation in Fig. IV/1.1.4-1 may be referred to.

Out of all the team members, the team leader and *scribe* (secretary) are required to possess experience in HAZOP and excellent communication skills. Generally, for plant HAZOP analysis the team is formed from people with a technical background. In most of the plants dealing with hazardous materials, and/or any other hazardous situation, a person from the HSE department is made a team member. In some countries it is mandatory to keep one person from HSE. Since in most plants, control instrumentation plays a great role in ensuring plant safety through interlock and protection, it is better to keep one person from process, operation/production

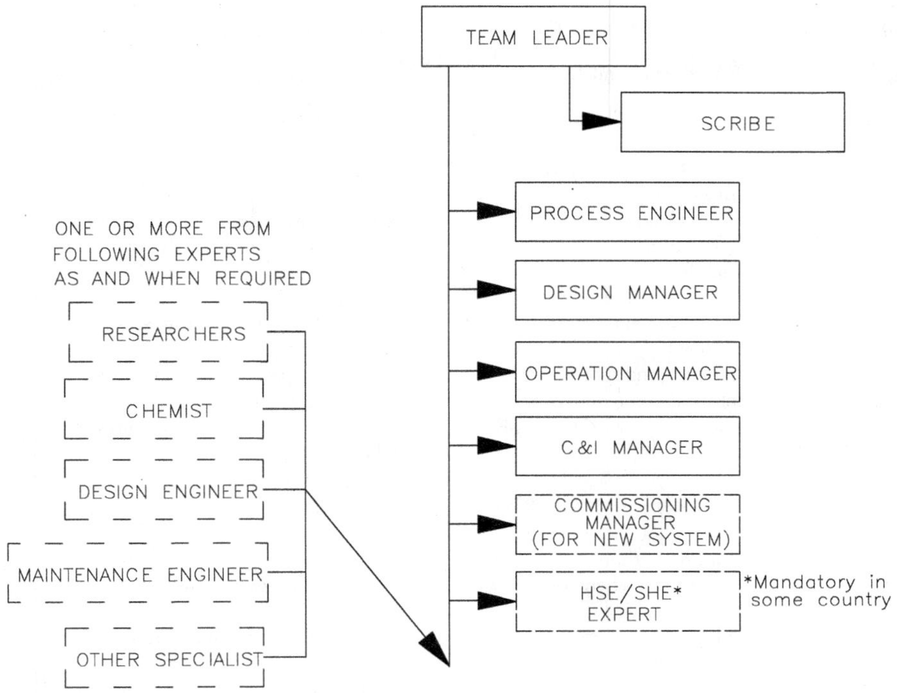

FIGURE IV/1.1.4-1

HAZOP team.

engineering, and control and instrumentation engineering. Guidelines about the qualification and responsibility of team members are given in Table IV/1.1.4-1.

1.1.5 Information Required

To take up or start HAZOP studies, information about the plant is necessary and this is detailed in the following:

- Continuous process: The following background information is helpful for the study:
  - Process and instrumentation diagram
  - Design specification
  - Other details such as process flow diagram, mass and energy balance
  - Chemical reaction details
  - Operating/control philosophy
  - Equipment/instrument specification
  - Site data and plant layout
  - Alarm/trip/relief valve set values
  - MSDS and material property

Table IV/1.1.4-1 Team Member Qualifications and Responsibilities

Member	Qualification and Experience	Responsibilities
Team leader	Meticulous and analytical skill Trained and experienced in all stages of HAZOP (acting as scribe) Quick adaptation and technical understanding capability Very good communication skills and temperament to work in a team Guidance and motivating skill, technical competence including knowledge on QRA	Developing a conceptual model, planning scheduling Proper division of the process so there will be quality production of the study report and at the same time it is not too time consuming and costly Controlling discussions at all stages with open heart Keeping focus, motivating, and helping scribe to record Judgmental power and leading the team from the front To ensure completeness (within the defined scope and boundary) and quality of the study report
Scribe	Familiarity with HAZOP Full-time job and able to start recording without depending on detection when consensus is reached To have good working relationship with team leader as helper	Taking notes of all events, recordings and documentation Establish good relations with team leader and get more time when required Production of interim report and list of recommendations Inform all details about recording Check progress and produce final report
Members	Good exposure and knowledge about HAZOP and sufficient experience in the area of his/her service Comprehensive knowledge on intent of design and/or operation Good knowledge on discipline represented	Process: Outline description of the process and intent of design/operation. Process parameters with associated design conditions Operation/control and instrumentation engineer: Depending on applicability, operation procedure, plant stability, control philosophy, details of interlock and protection, alarm lists, and other safety startup and shutdown features and requirements Design engineer: Specification details, material safety data sheet (MSDS), piping and instrumentation diagram (P&ID), layout information Maintenance engineer: Maintenance update, management of change (MOC)

  - Hazard area classification
  - Known operating problem
  - Hazard history
  - Startup/shutdown problems
- Batch process: In addition to this information, the following information is also necessary for batch process, especially when automation is associated:
  - Detailed operational method
  - Step-by-step operating procedure
  - Outline of control sequence especially for automated job
  - Time-dependent sequential operation
  - Reaction process heat and flow details
  - Physical and chemical properties of reactants
  - Design intent of each stage

1.1.6 Preparation and Planning

With the background information sought in the previous clause, there will now be a stage for preparing and planning before actual HAZOP can be undertaken. Some of these points are related to standard norms for HAZOP and some are for planning of HAZOP:

- The team shall be well informed about the intent of the study and a conceptual model shall be presented to them at the start of the meeting.
- As HAZOP study changes with the stage of the project as shown in Fig. IV/1.1.3-1, during planning this needs to be taken into consideration.
- The team leader may meet with the process engineer in advance of the study so that he/she has a good grasp of what the process is all about, and to agree a mutually acceptable basis for dividing up the P&ID into short sections suitable for individual study.
- A fresh copy of the P&ID will be put on the table at the HAZOP study as a clean drawing, and any markings made on it will be done in the presence of the HAZOP team. This second copy is then called the "HAZOP master." As discussed in Clause 1.1.3, for a HAZOP study each P&ID is divided into sections. No section should feature a process line junction (as interfacing is very difficult) and no step should cover more than one element. In a continuous process, these divisions can be done based on variation of process variables, on the basis of a junction in vessels i.e., terminal point at vessel, or at a predesired control point.
- Information on P&IDs is very important, so important information expected from one P&ID is listed here:
  - P&ID to show all piping including physical sequence and branches with valves, equipment, and instruments with final destination with associated tag numbers.
  - Miscellaneous drain and vent lines, relief valves, etc.
  - Flow directions, permanent startup/blowdown lines
  - Necessary interface points

- Another important point to be noted is that a system becomes more vulnerable during startup/shutdown, etc., so in the study special attention may be given to these phases also.
- A special feature in a *batch* process is that a single piece of physical equipment may be studied several times in HAZOP, for example, a reaction vessel, but in different perspectives, so due consideration shall be given in each situation.
- The team needs to look at the whole process and the changing conditions at each stage of the batch process. These are more important because they are all time dependent sequence, time related flow and process. A deep understanding of design intention at each stage is crucial. Without a proper understanding, a vital step or stage may be missed.
- For better results, the more divisions of the process there are, the better the quality will be. However, this may demand more time and money. The team leader needs to pay attention to this too (see the responsibility of the team leader in Table IV/1.1.3-1).
- Complexity and size of the plant within the scope and boundary of the study determine the time and cost requirement. It is the responsibility of the team leader to guide the team in such a way that the study is properly balanced. Previous HAZOP study results of similar plants could be helpful.
- A stretch of no more than 4 h in a day and no more than 4 consecutive days in a week should be deployed for one study. There shall be a number of breaks in each session so that team members can produce better results. However, there shall be no interruptions in the session. Typical timings for a HAZOP study are detailed in Table IV/1.1.6-1.

Table IV/1.1.6-1 Typical Timings for a HAZOP Study

Type of Study	Preparation	Evaluation	Documentation
Simple and small plant	8–16 working hours	1–5 working days	2–6 working days
Complex large section/ plant	2–6 working days	1–5 working weeks	2–5 working weeks

- Meetings shall be well planned and scheduled by the team leader in advance, and a comfortable conference room shall be booked in advanced. In some sessions, help from other experts may be sought.
- Some formal training on study philosophy, principles, and methodology for team members is welcome before starting a HAZOP study. There are regular courses offered by various organizations for this. Otherwise for large companies where regular HAZOP studies are carried out, new people may be included with experienced personnel to receive hands-on training. Scribes need additional training to understand *category* and *level* of recording. They need to have good computer *training* in cases where computers are used for recording purposes. The requirements of training and experience of the team leader have already been discussed.

1.1.7 *Effectiveness*

There are several influencing factors for a HAZOP study to make it effective. The majority of these influencing factors are as follows:

- One of the major issues is availability of required accurate information. If the study is to be carried out for design stage hazards or operational hazards, then accurate data shall be made available. A HAZOP study based on precooked design or on outdated data is meaningless and futile.
- For the selection of an appropriate team leader a great deal of mentoring may be necessary, so that the concerned person can perform efficiently. Without a suitable team leader, the study will be ineffective.
- The selection of the team shall be a balance of experience **and** knowledge. Team members must be from different disciplines with good insight and skills.
- How the team is able to meticulously and systematically utilize the information for identification of deviation has an impact on the study. Therefore avoidance of hazards is **directly** related to effectiveness of the study. The ability of the team for creative thinking and systematic analysis influences the quality of the report.
- Maintaining a sense of proportion in the judgment of seriousness of hazard and expenditure toward reducing likelihood are influencing factors for the quality of the report.
- Only an authoritative person (with good knowledge) should initiate a HAZOP study, so that recommended actions can be properly implemented (see last part of Fig. IV/1.2-1).
- Administrative and financial support is essential.
- Proper planning and scheduling is also an important factor to achieve better results.

1.1.8 *Pros and Cons of HAZOP Study*

The following points may be noted as advantages and disadvantages of the study:

- Advantages:
  - Helpful for hazards difficult to quantify, for example, human error
  - Brainstorming and creative
  - Systematic and comprehensive with the possibility of getting quality results
  - Possibility of quantification
  - Simpler qualitative analysis but quite accurate especially for process plants
  - Possibility of automation
- Disadvantages:
  - Requirement for a lot of information is ineffective at early stages
  - Qualitative analysis, so in many cases further quantification is necessary
  - Time-consuming and costly
  - Highly dependent on scope definition, team performance, and accurate information for comprehensiveness and quality of result
  - Inability to assess hazards caused by interaction between different parts of the system

1.1.9 Guided Word

Guided word is very meaningful in HAZOP. Guided words with meanings and comments are listed in Table IV/1.1.9-1.

- A few additional terms used for HAZOP analysis are found in Table IV/1.1.9-2.
- Some special words [1] may be:
  - Testing: Product/equipment
  - Plant equipment: Operable/maintainable
  - Instrument: Sufficient/excess/location
  - Electrical: Area/isolation/earthing
- Possible parameters with which guided words could be associated:
  - Flow
  - Pressure
  - Temperature
  - Level

Table IV/1.1.9-1 Generally Used Guided Words With Meanings

Guided Word	General Meaning	Remarks
No (not/none)	*Negation (of intent)*	No forward flow
More (higher)	*Quantitative increase*	More of any physical parameter
Less (lower)	*Quantitative decrease*	Less of any physical parameter
As well as (more than)	*Quantitative increase additional activity*	Design/operating intent achieved along with additional item
Part of	*Quantitative decrease*	Only part of intent achieved
Reverse	*Opposite of intention*	Reverse reaction/flow
Other than	*Complete substitution/ miscellaneous*	Original intention not achieved, something different happened—alternative mode of operation

Table IV/1.1.9-2 Additional Guided Words With Meanings

Guided Word	Meaning	Application
Early	Relative to clock time	Timing before intention
Late	Relative to clock time	Timing after intention
Before	Sequence order	The step (before) is effected out of sequence
After	Sequence order	The step (after) is effected out of sequence
Faster	Different (earlier) from timing intention	Faster reaction
Slower	Different (later) from timing intention	Slower reaction
Where else	Other location	Flow/transfer/source/destination

- Viscosity
- Mixing
- Stirring
- Transfer
- Reaction
- Composition
- Addition
- Separation
- Time
- Separation
- Speed
- Phase
- Particle size
- Measure
- Control
- pH
- Sequence
- Start
- Stop
- Signal
- Operate
- Maintain
- Communication
- Service

Meaningful combinations of these parameters with guided words are presented in Table IV/1.1.9-3.

Out of sequence and *missing* are additional guide words found in batch processing.

This checklist of guided words will be helpful in HAZOP analysis. In most cases, parameters and associated guided words are quite obvious! With these general ideas and HAZOP study philosophy in mind, it is better to concentrate on the actual procedure.

1.2 HAZOP Methodology Description

HAZOP study methodology is discussed in this clause with Fig. IV/1.2-1.

1.2.1 Intent of the Study and Study Process

It is necessary that the team begins the study with full intention in mind. The team must know and understand the exact scope and boundary of the study and the stage of the project for which the study is intended. With available information or having a good knowledge of the system in question, it will be possible to develop a conceptual model of the system for which the study is intended. As discussed earlier, the team leader actually leads and imparts the detailed idea to the team members through a *conceptual model* and clears up any doubts about the entire system. Then a *full*

Table IV/1.1.9-3 ***Meaningful Combination of Parameters and Guided Words***

Guided Words/ Parameter	More	Less	None	Reverse	As well As	Part of	Other Than
Agitation (mixing)	Fast	Slow	No		As well	Part	Other
Communication	More	Less	No		As well	Part	Other
Composition					As well	Part	
Distributed control system (DCS) failure			Fail				
Drain/vent	Long	Short	No				
Flow	Hi	Lo	No	Reverse	Deviation	Conta-mination	Deviating materials
Level	Hi	Lo	Empty		Different level		
Maintenance			None				
Phase				Reverse	As well		Other
Pressure	Hi	Lo	Vacant		ΔP		Explosion
Reaction	fast	Slow	No				
Startup/shutdown	Too fast	Too slow			Action missed		
Time	Long/ late	Short/ early	Sequence skip	Reverse	Action missed	Extra action	Wrong timing
Utility fail			Fail				
Vibration	Hi Hi	Lo Lo	None				

Hi, *High*; Lo, *low*; Δp, *differential pressure*.

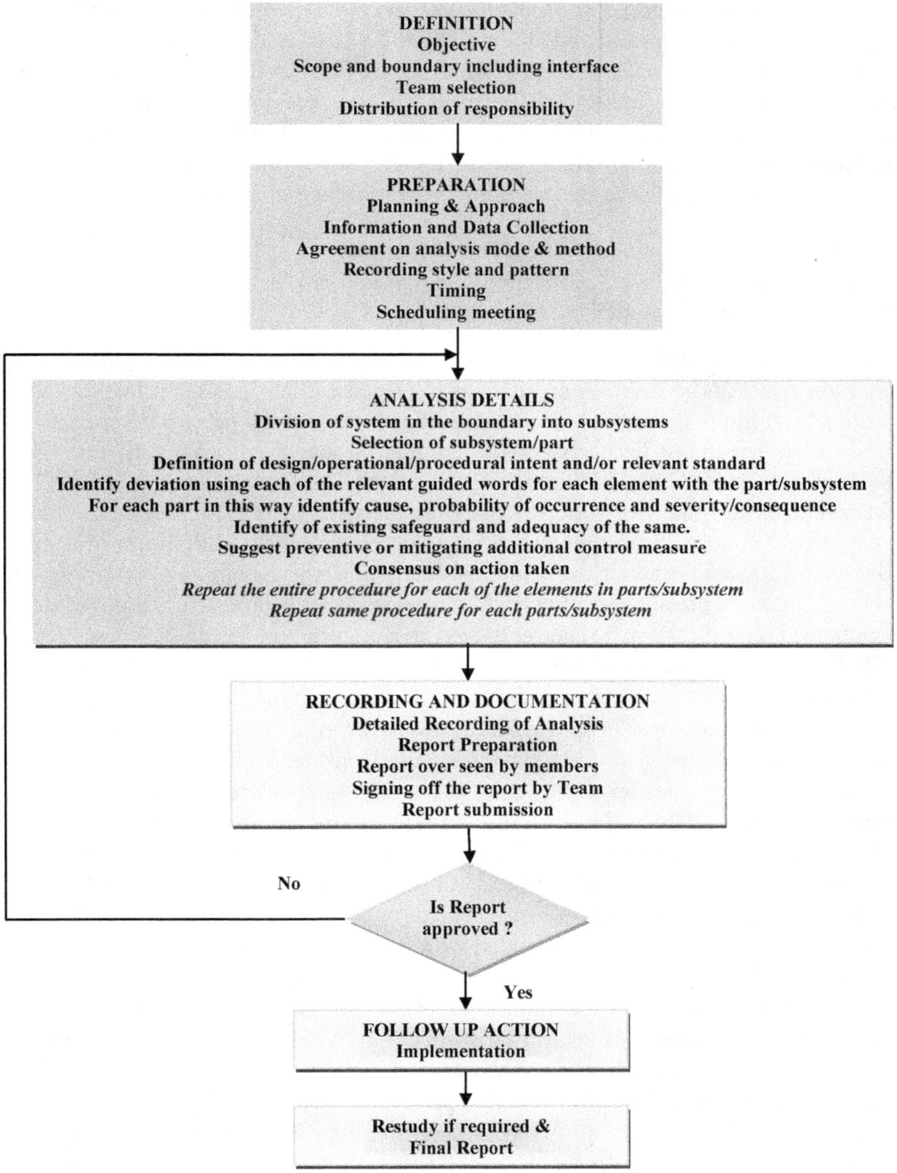

FIGURE IV/1.2-1

Overall view of HAZOP study.

description can be presented by the process person with all key parameters and HAZOP study *intention* formulated and recorded. To get a quality study it is better to have a detailed and comprehensive study intention. The intention of the study may be focused on equipment, material, conditions, controls, changes, etc. All these shall be spelt out in detail. During a detailed description, the scope and boundary definition along with interface points, which are extremely important, must be elaborated clearly, so that nothing is missed and unnecessary wastage of time is avoided.

1.2.2 HAZOP Methodology and Risk Assessment

There are several steps in HAZOP methodology and risk assessment, which starts with finding deviations to identify hazards.

- Deviation and hazard identification: The team leader is responsible for starting and terminating the meeting on time, as discussed earlier. The team leader begins by calling on the process engineer to give the overall process description and the design intent for each P&ID. The team leader then defines the first section to be studied in detail and may put a dotted yellow line on the HAZOP Master to ensure that everyone is looking at the same section of the plant. The team leader then calls on the process engineer to give the design intent and detailed information about the contents of the line, the design and operating temperature and pressure, etc. For each of the sections, deviations are discovered. For hazard identification, it is better to refer to the third box "Analysis Details" in Fig. IV/1.2-1. The team leader writes onto the HAZOP Master the number of the action in *red*, next to the item to be changed. If the change is agreed by the meeting, he/she also marks up the HAZOP Master in *red* with the change to be made. When all the guide words have been used and the study of a subsection is complete, the team leader should *color yellow* all the completed sections. The team leader then defines the second section to be studied in detail and calls on the process engineer to give a detailed process description. The extent to which the hazards will be evaluated, ranked, and solved varies with company policy, objective, etc. defined at the planning stage.
- Combination: In the drawing a guided word is *selected and combined* with a parameter of the element/line. This means that a *meaningful* deviation is to be created with the help of the combination of a guide word and one parameter. Naturally, these combinations can be done either by selecting a parameter first, then combining one guide word as shown in Fig. IV/1.2.2-2A, or a guide word may be chosen first and combined with a parameter as shown in Fig. IV/1.2.2-2B. For a complete flow diagram of HAZOP methodology Fig. IV/1.2.2-2 should be viewed in conjunction with Fig. IV/1.2.2-1.

  Thus from the discussions, it is clear that guide words and parameters are combined to get a meaningful deviation. Here, the word "*meaningful*" is very important in the sense that if the combination of guide word and parameter gives a deviation that is unrealistic or incredible, then it shall be discarded, for example, "NO" with "TEMPERATURE." It is important to select *a set so that none will be missed*. For this, various standard combinations discussed earlier or additional

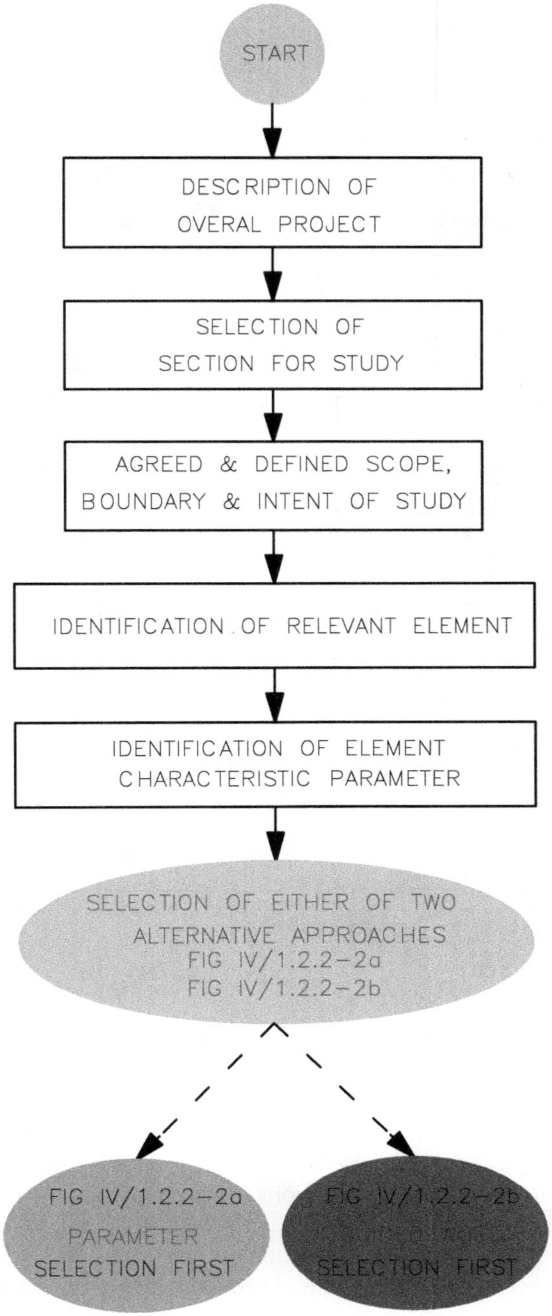

FIGURE IV/1.2.2-1

HAZOP methodology.

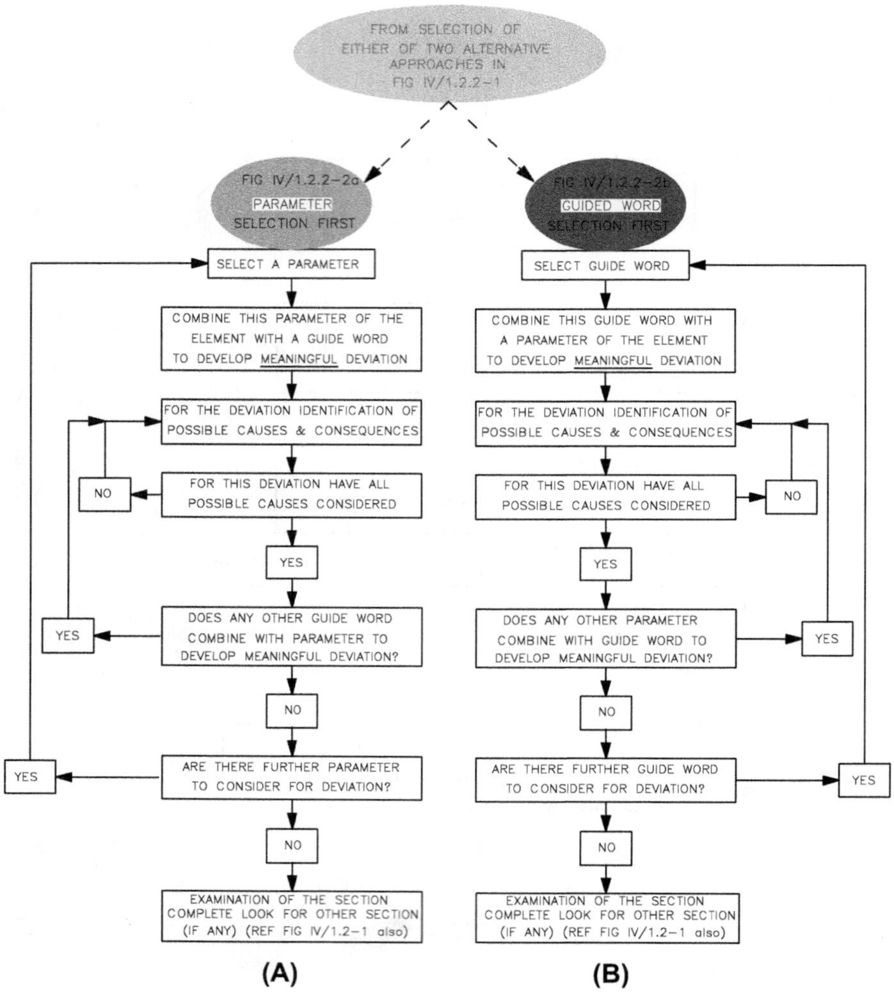

FIGURE IV/1.2.2-2

HAZOP methodology alternative approaches. (A) Parameter selection first, (B) guide word selection first.

alternative combinations may be tried. HAZOP is a creative process, so new combinations to find meaningful deviations are necessary for the process.

- After a credible/meaningful deviation is identified, then it will be necessary to find the causes.
  - If the consequences of the deviation are trivial, then it may be meaningless and a waste of time.

- For any cause that is not like what has been discussed previously, there shall be a brainstorming session to find all probable causes. There shall be a positive attitude from all the members even if for some reason a team member may be responsible, for example, if the cause is "lack of maintenance," then the member from the maintenance group needs to have a positive attitude.
- Human error, equipment/control system failure, or procedural mistakes should be dealt with properly.
- Available data bank/old history standards, etc. will be helpful. However, since HAZOP is a creative process, a data bank/old history should *not* hinder the brainstorming session/creativity. Data from these sources are helpful to ensure no cause is overlooked.

- After hazard identification, risk assessment is performed. Acceptable risks are arrived at by balancing frequency of occurrence and severity, as discussed in Chapter II. It is quite possible that such assessment is done outside these meetings. Here, a realistic hazard is important and it is very much related to frequency. So, the term realistic hazard varies greatly with industry, company/company policy (to an extent), legislature, and country. In any case, the cause needs to be recorded even if it is very low frequency. All major causes and consequence are noted. In a HAZOP study the cause and consequence discussions are very closely coupled. For better results, these may have to be covered, in conjunction with safeguards, in a separate meeting—after the main study of all the lines is over.
- In a HAZOP study, after all causes of the deviations pertinent to a parameter/guide word are covered, another parameter and guide word combination is sought. If a parameter (/guide word) is chosen first, then the next guide word (/parameter) is combined with it to obtain a meaningful deviation. If all such combinations are complete, then next parameter (/guide word) is chosen and the process continues until all parameters (/guide words) are complete, as shown in Figs. IV/1.2.2-1 and IV/1.2.2-2.
- As shown in Figs. IV/1.2.2-1 or IV/1.2.2-2 the next line/subsection is chosen and the process is continued in this way until the study of all lines and/or subsections is complete (i.e., the entire procedure is repeated for each element in the part/subsection and for all subsystems within the scope and boundary of the study).
- The need to receive and incorporate action responses may be decided in a special coordination meeting.

1.2.3 Consequence Analysis

As discussed earlier, cause and consequence are closely coupled. However, in some approaches safeguards are not considered initially and only consequences are considered pertinent to all causes. Finally, safeguards are applied to obtain the final recommended action. Normally, it is the prerogative of team leaders to decide when to carry out the consequence analysis. In most cases it is seen that such exercises are done after each subsystem study is complete, so that all causes and consequences can be carefully studied. It is necessary to see if the system goes beyond the intended operating range or into the danger zone. The consequence may be immediate or long term; it may be within the plant and outside the plant (especially environmental impacts). Another important part is the development of consequences and how operating personnel are informed of these through use of a pretrip alarm, interlock, and trip sequence. A sequence of event recording is done to check the system. This will be helpful in detecting human error. In consequence analysis, various safeguards play a major role and shall be seen together.

1.2.4 Safeguards and Recommended Actions

There are several ways to look into the problem:

- To ignore the existing system, identify the cause and consequence of the hazard, and then apply the safeguard such as an alarm, interlock trip, etc. In this method there is very little chance that anything is missed.
- To consider all safeguards and based on the situation come up with an additional recommended action.
- In any case, risk is assessed qualitatively by utilizing the experience and judgment of the team using risk matrices, as discussed earlier. However, if there is a necessity to assess a *risk* whose frequency is very low but severity may be high, this will be recommended for further analysis by QRA.
- All recordings, namely, causes, consequences, and recommended actions, must be recorded clearly and unambiguously, so that at a later date a third party may understand. Assumptions if any must be recorded also. When a safeguard is adequate, causes and consequences are recorded. If a safeguard is inadequate, then additional recommendations may be needed. There are various ways these recommendations are given in and outside the meeting, as shown in Fig. IV/1.2.4-1.
- As seen in the drawing, in some cases the recommended actions are left for decision by a respective field expert. In some cases where the recommended actions are warranted by a standard and/or when there is consensus, then actions are taken in the meeting balanced are left for outside meeting decision. In some cases a team leader may be confident that all decisions may be taken in the meeting. Specific recommended actions are always preferable; however, for certain cases where there are repetitions, generic recommended actions may be called for.

1.2.5 Influencing Factors During HAZOP Study

There a number of factors that affect the study intensely. Some of these points were discussed earlier. In Table IV/1.2.5-1 these are put forward in a consolidated manner.

1.2.6 Generic Example

As discussed earlier, it is difficult to consider a section of a plant within the scope of this book. For this reason a simple generic example has been chosen. A shell and tube-type heat exchanger is used in most plants. In the heat exchanging process, as

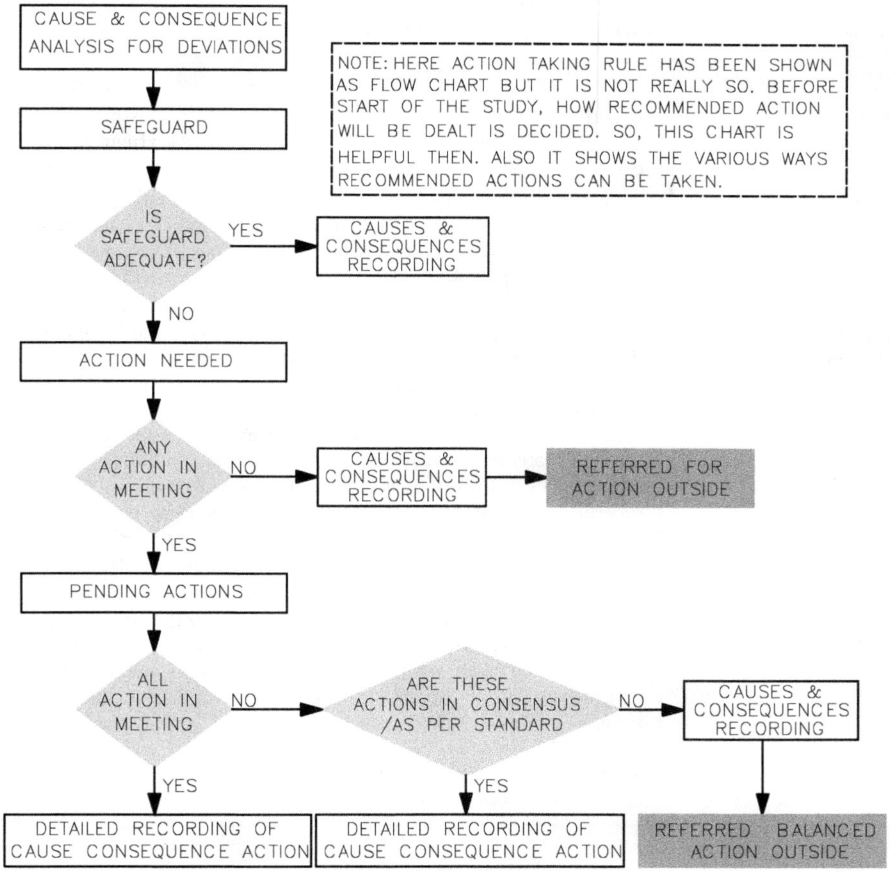

FIGURE IV/1.2.4-1

Recommended actions in HAZOP study.

shown in Fig. IV/1.2.6-1, fluid passes through the tube and cooling water passes through the shell to cool the process fluid to the desired temperature with suitable control (not shown). Here, a few points should be noted. In the particular example, only the HAZOP worksheet has been shown for cooling water. In this case, three of the most relevant guide words—"None," "More," "Less"—have been associated with parameter flow. Here, flow parameter is chosen first, then the relevant guide word has been associated with it to obtain the deviation.

So, with a little background about the HAZOP worksheet it is better to concentrate on how HAZOP reports are produced and presented.

1.3 Reporting and Follow-Up

Like all other hazard analysis methods, reporting in HAZOP is also very important; it shows the actual work carried out by the team. The purpose of HAZOP documentation is to record and represent the discussions by the team during the study meeting.

Table IV/1.2.5-1 Influencing Factors During HAZOP Study

Factor	Discussions
Motivation	Experienced team leader must be able to motivate to extract the best out of the team members for a quality result. Also high-quality communication and positive openness of team members are critical.
Time	Time affects the system in two ways: sufficient time shall be allowed to perform the work. Also since this is a creative exercise it should not be continued for a long period. The study should always be carried out with fresh minds. Also team members should be expected to work out of hours to achieve results.
Scope and boundary definition	For a new project or for a modification work, scope and boundary definitions are extremely important, especially for a connected system. There shall be a clear description of the system, intent of study, and study envelope. This is because any modification in one subsystem may affect other subsystem, for example, a modification in a feed water system may directly affect a condensate system. The same applies if this is modification work, for example, if a superheater temperature control is modified, then one has to keep in mind that it may not only affect reheat temperature control but also heat distribution in a once-through supercritical boiler.
Action	Details regarding recommended action were discussed in the previous clause. It is recommended that all actions during the study should have consensus. The team leader needs to decide what actions will be taken within the study limit and how much shall be allocated to outside experts. Also whatever action is recommended it must be relevant and unambiguously defined, so that at a later date a third party may understand. Also while recommending any action, all aspects like hardware/software failure/procedural mistake, and/or human error need to be considered.

One thing that must be kept in mind is that the report shall be very clearly and systematically recorded so that it can be used by others in the future. Major content must include what, why, how, when, and by whom. Regarding reporting format, it is important to note that the report may be used for a number issues, such as:

- Action implementation
- Link for future HAZOP
- Contractual document to meet standards
- Safety database for the plant
- Future training
- Troubleshooting

Therefore recording and report preparation can never be overestimated. It is worth noting that all the background information based on which study has been carried out should be part of the dossier. This will be helpful for further reference as well as to mark the changes in the future. Therefore the amount of information in the HAZOP final report will vary from case to case. Each section shall be covered fully with a selection of guided words and a heading.

(A)

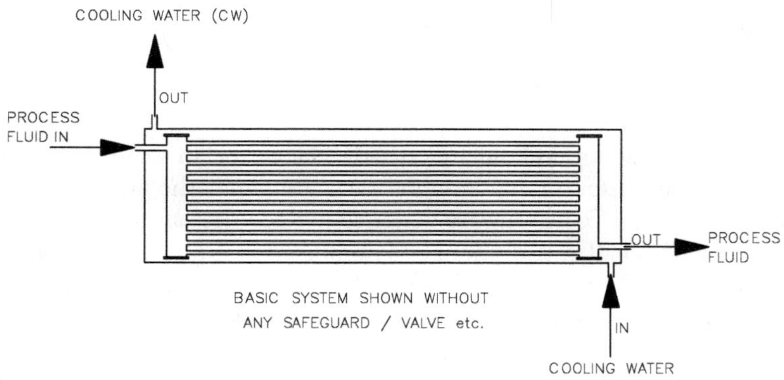

(B)

HAZOP WORK SHEET

GUIDE WORD	DEVIATION	CAUSES	CONSEQUENCES	ACTION
NONE	NO COOLING WATER FLOW	INLET VALVE NOT OPEN NO FLOW IN COOLING WATER HADER COOLING WATER PUMP FAIL	PROCESS FLUID TEMPERATURE HIGH PRODUCT QUALITY UPSET	INSTALL C.WATER FLOW METER INSTALL TEMPERATURE MONITOR AT INLET & OUTLET FOR ALARM & TRIP PROCESS FLUID FLOW AT HI HI TEMP.
MORE	MORE COOLING WATER FLOW	FAILURE OF I/L CONTROL VALVE FAILURE OF CONTROL SYSTEM	PROCESS FLUID TEMPERATURE LO LO PRODUCT QUALITY UPSET	INSTALL C.WATER FLOW ALARM HI INSTALL TEMPERATURE MONITOR AT INLET & OUTLET FOR ALARM & TRIP PROCESS FLUID FLOW AT LO LO TEMP.
LESS	LESS COOLING WATER FLOW	FAILURE OF I/L CONTROL VALVE FAILURE OF CONTROL SYSTEM	PROCESS FLUID TEMPERATURE HI	INSTALL C.WATER FLOW ALARM LO INSTALL TEMPERATURE MONITOR AT INLET & OUTLET FOR ALARM

FIGURE IV/1.2.6-1

Generic example of HAZOP. (A) Shell & tube heat exchanger, (B) HAZOP work sheet for shell & tube heat exchanger (shown only cooling water flow part).

1.3.1 Recording Format

Discussions are normally recorded in tabular format, as shown in the generic example in Fig. IV/1.2.6-1. There may be variations in recording format based on the scope; however, the following information is generally included:

- **Reference number**
- Guide word
- Parameter
- **Deviations**

- **Causes**
- **Consequences**
- Safeguard
- **Action required**
- Action allocated

The column headings in bold are essential. However, in some cases, either a guide word or parameter is put as a subheading just above the table, as shown in Fig. IV/1.3.1-1A and B. It is always better to include the safeguard column in tabular format. Also all entries should be numbered for proper understanding as well as referencing. To facilitate risk matrix formation for risk ranking, frequency of occurrence, severity, etc., other columns may be added.

(A) **Plant No: XXXX** **Sub section YYYYY**

Area :ZZZZ **STUDY REF NO: LMLMLMLML111111**

Element: Heat exchanger ABACAD **Guide word: NONE**

No	Parameter	Deviation	Causes	Consequences	Safeguard	Action

(B) **Plant No: XXXX** **Sub section YYYYY**

Area :ZZZZ **STUDY REF NO: LMLMLMLML111111**

Element: Heat exchanger ABACAD **Parameter: FLOW**

No	Guideword	Deviation	Causes	Consequences	Safeguard	Action

FIGURE IV/1.3.1-1

Typical recording format for HAZOP Study. (A) Recording format of HAZOP with guide word. (B) Recording format of HAZOP with parameter.

Recording formats shown in the figure will form the part of the report after these are finalized. Before finalizing, draft recordings should not be made a part of the dossier. Follow-up is another important aspect of the HAZOP study. Typical formal HAZOP follow-up action close-out is presented in Fig. IV/1.3.1-2.

Plant No: XXXX **Sub section YYYYY**

Area :ZZZZ **STUDY REF NO: LMLMLMLML111111**

HAZOP Study Follow up report **Report by:**

No	GW	Deviation	Action required	By	Action taken	Checked	Date

FIGURE IV/1.3.1-2

HAZOP follow-up action close-out format.

1.3.2 *Detailing in Record*

The amount of detailing to be included in the record is highly dependent on purpose. Several modes of HAZOP tables are:

- Deviation by deviation HAZOP table
- Cause by cause HAZOP table
- Exception only HAZOP table
- Action only HAZOP table

  Another way HAZOP tables (generally in use) can be categorized is as follows:
  - Full record: A full record system is very useful for subsequent usage. Generally, full records are done to keep parity with company norms or when it is required to meet the requirements of legislature. In fact, from an auditing perspective it is necessary to ensure that HAZOP is documented as fully as possible. Each deviation created by combining a guide word with the parameter provided gives meaningful deviation. In full recording, any deviation observed by the team is recorded, even if it is insignificant. Also all guide words are listed, even if not used for deviation finding. In detailed recording, safeguards are more likely to be maintained, as the function and purpose are recorded in detail in the HAZOP. Naturally, this type is time-consuming and costly.
  - Record by exception: This is the other extreme of recording. Here, these will be listed only when the team recommends an action. These are done very quickly but have little value for general or subsequent use because they are the result of immediate and shorter meetings. However, reporting is simpler and less costly. Usually, this type of report is done for short study completion time.
  - Intermediate: In this there will be sufficient discussions within the team and incidents are noted even if no action is recommended. If there is any action it is noted, but here deviations are also noted that are realistic but well guarded by the safeguard system in service (no action). This can be used at a later stage for general and audit purposes.
  - One thing that should be borne in mind is that all the entries shall be very clear and unambiguous, so that there is no difficulty with future interpretations. Short or brief entries may lead to ambiguity at a later date.

1.3.3 *Computerized Recording*

The use of computers for HAZOP is quite common. In addition to automated HAZOP (discussed later), computerized recording is also available. The recording programs are basically special spreadsheets. This transition has resulted in thorough and quick recording of HAZOP studies. There are a number of features of use of computers for recording:

- The computer can be connected to a projector, thus it is possible to display what is being recorded to all the team members.
- Guide word and parameter combinations can be selected very quickly.

- During discussions it is possible to alternate between design intention, boundary, interface issues, earlier study details, etc. Also a huge databank can be made available within a few keystrokes. These are extremely helpful for the members conducting the study. HAZOP software makes it possible to review every deviation and refer back to common scenarios, and where common scenarios are met use set pieces of text to save time.
- Recommendations are easily retrievable and can be converted to actual documents.
- The scribe should be familiarized with such computer programming.
- HAZOP Manager V 7.0 and HAZ1508 from Rowan House Limited in the UK are just two examples of HAZOP software. These programs provide a framework for the study and perform the recording functions.

1.3.4 Follow-Up Action

Generally, follow-up action is performed by authorized personnel from line management. However, the team leader of the HAZOP team at times is given responsibility to pursue line management personnel for action implementation. For keeping track of action follow-up, Fig. IV/1.3.1-2 may be followed.

If any action is rejected or could not be implemented, it must be suitably logged in the record and duly signed by a responsible person. There shall be a proper check for management of change (MOC) or a new system before this action is commissioned.

1.4 HAZOP Discussions

1.4.1 HAZOP Timing (When to conduct)

The objectives of the study and benefit from it determine the timing of a HAZOP study. When the process design is complete, a full study procedure may be applied. Operating procedures may be examined to ensure that all eventualities have been considered. Modifications generally benefit from a rigorous study. Often an apparently simple, uncomplicated modification can give rise to a greater problem than it was intended to solve. Existing plant and new equipment are other examples of topics that may benefit from the study. Therefore a project may be studied several times in its lifetime with different objective and purpose.

1.4.2 HAZOP in Different Applications

Here, HAZOP in different plant applications has been outlined. HAZOP applications in E/E/PEs are discussed separately later.

- Existing unit periodic study: Periodic HAZOP studies of existing plants are not uncommon; rather they are legally binding to ensure that the plant runs safely throughout its lifetime. This is important, specifically when there is a change in operation procedure, material feed, product type, and/or major changes in the system. In the following cases, to ensure safety, such periodic studies are often done:
    - Major incidents
    - Major or many modifications

  - Design deficiencies
  - Inadequate previous HAZOP study
  - Plant not running smoothly

  HAZOP is time-consuming but can give a more comprehensive output. In any case, time period/completion time for such a periodic HAZOP study is specified. Also the next periodic study dates need to be finalized.
- Similar or repeat design: It is quite possible that different units are set up by a company with a similar (or even the same) design. One company may have two or three different types of (say) 500 MW plants at different locations, or in the same location one is a drum-type boiler and the other is a once-through boiler. In all such cases HAZOP by difference, that is, detailed comparison of the two systems may be helpful. However, the team must be well aware of variations between the two systems.
- MOC: Management of change is another area where HAZOP is done. If the modifications are done in the plant (operating procedure, materials, catalysts, sequence, software, etc.), then the safety of the plant because of such modifications is often checked with the help of HAZOP. In such cases the modifications are considered as a new design and a HAZOP study needs to be carried out. Here, one thing needs to be kept in the mind that if such changes are minor/small, then the study can be done easily, but if the modification is large, then extent of the HAZOP study may be greater. If the plant had a HAZOP study done earlier, then such information may be very useful. Another factor is the scope and boundary definition. In many cases, some modification may have made wide changes in the system. In these cases, it is possible to consider a HAZOP study up to a certain distance (extend) from the point of modification, but that must be agreed upon by all members of the team. Otherwise HAZOP for the entire affected area needs to be carried out—it depends on what has been changed (MOC in case I: from 3 × 50% boiler feed pump (BFP) to 2 × 100% will be different from MOC in case II: from 2 × 60% induced draft (ID) fans to 3 × 50% ID fans just if we look at the complications of electrical bus transfer first case complication is less than the second one).
- HAZOP study for procedure: This is normally a detailed method and should be properly defined. This study is usually carried out by a well-balanced team from different disciplines. The information required includes but is not limited to procedure definitions, up-to-date P&IDs, and other relevant drawings and documents. Also the aim of the study (depth of study and major area of interest such as HSE, accident, etc.) needs to be well defined before starting a step-by-step procedure. Each stage/step and action is examined with the guide word to identify meaningful deviation for analysis for hazard identification. In the case of a HAZOP study for procedure, human error is an important factor.
- Drain vent and interconnections: It is quite common that drain vent relief from different equipment is interconnected through common piping. Occasionally, proper interconnections are not shown distinctly in any single P&ID; instead

they are spread over a number of P&IDs. Also it is likely that there may be mismatch in material (compatibility), pressure, and mode of operation. So HAZOP studies in these cases are complex. This is because one needs to see interface management between P&IDs, fluid mismatch, and simultaneous operation of two pieces of equipment. It is therefore necessary that there shall be one P&ID where all these, for example, material release, fluid incompatibility, pressure mismatch, simultaneous operation, etc., shall be clearly marked or in each P&ID fluid material, pressure, release mode, and interface point shall be well defined. For HAZOP, study parameters (with their guide words in parentheses) are shown: Pressure (More), Flow (No, Reverse, More), Phase (Change), Temperature (Low, High). The HAZOP study needs special attention for these cases and so are discussed separately.

- Major issues related to HAZOP in commissioning are dealt with in the main HAZOP study:
  - Removal of construction debris
  - Purging and cleaning
  - Testing of major equipment

    In a test run different fluids may be used, so during HAZOP connected with commissioning the following guide words may be useful: Density: Higher/Lower; Noise: Higher; Debris: More/Some; Contamination: Oxygen/Inert—Source and Disposal; Cleanliness: More/Less; Stress: More, etc.

1.5 Computer HAZOP

Conventional HAZOP are useful for process and equipment failure but are not really suitable for control systems utilizing computers. This is, not suitable for E/E/PEs deployed for plant control systems. In view of the current system architecture of control systems, it is better to refer to such control systems as *E/E/PEs* rather than computer control. Conventional HAZOP does not take into account the potential hazard caused by failure of system/system components, and does not consider control logic failure. It would be dangerous to go ahead with conventional HAZOP considering E/E/PEs as a black box, because many HAZOP findings are concerned with control and safety applications. Therefore by considering E/E/PEs as a black box many such safety critical potential hazards will be missed. E/E/PEs are much more flexible than conventional hardware control systems and many high-end sophisticated controls are possible by control systems based on E/E/PEs. E/E/PEs provide many benefits to the controls and at the same time introduce additional hazards also. In 1994 Nimmo developed a computer HAZOP (CHAZOP) system meant specifically for computers, that is, E/E/PEs. A framework similar to conventional HAZOP is used for CHAZOP with **different interpretation and different guide words**. In conventional HAZOP the guided word "NO" when associated with parameter "Flow" indicates no flow in the pipe. While in CHAZOP "NO" is associated with control signal/data flow to signify *different* meanings. There are *two* types of CHAZOP: Preliminary and Full CHAZOP. The obvious question is what, why and when should this CHAZOP be developed and implemented? The CHAZOP concept is shown in Fig. IV/1.5-1.

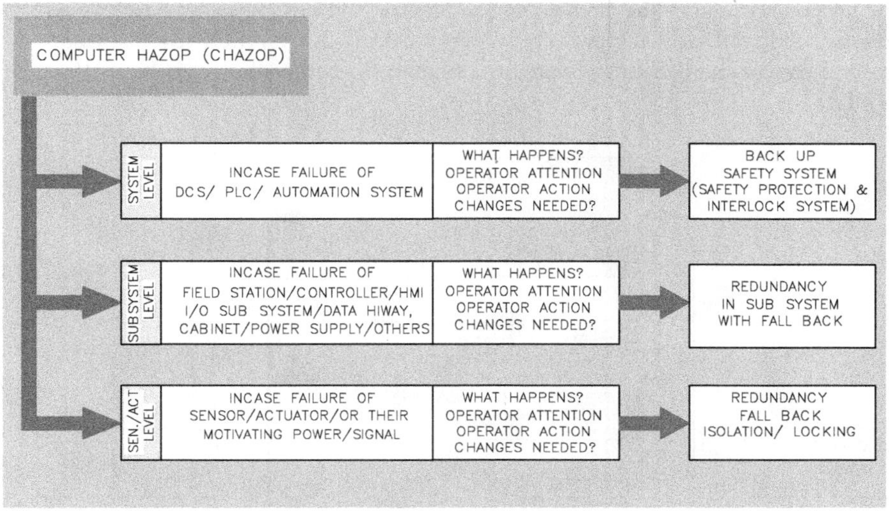

FIGURE IV/1.5-1

CHAZOP concept.

1.5.1 CHAZOP Conceptual Details (What, Why, and When)

Basically, HAZOP may be considered as an initial investment at the beginning to save money at a later date.

- What—the purpose: CHAZOP has a framework similar to HAZOP. There are several forms of CHAZOP for E/E/PEs but most of them are based on guided words driven by the HAZOP process. During earlier discussions it was seen that there are several protection layers for a process/plant. Out of these, CHAZOP is carried out on basic plant control systems (**BPCS**s). The CHAZOP method has been developed to generate *confidence*, that the instrumentation and control system has been designed, installed, operated, and maintained using good engineering *practices* and is capable of performing at the higher end of the allowable *range* of operation. Like HAZOP it is a team-based study and the review can be done at different stages of the project life cycle. At the early stage it is carried out to improve design; at the postdesign stage it is carried out for design verification. During the operational stage it is carried out to understand the impact of modifications/upgrades, etc. Therefore CHAZOP can be considered as a structured review process (based on HAZOP framework) intended to examine the potential threat to the claimed performance of the BPCS, with the help of an experienced team utilizing previous experiences. Thus CHAZOP encompasses general threats, hardware, software, and human factors, as shown in Fig. IV/1.5.1-1.

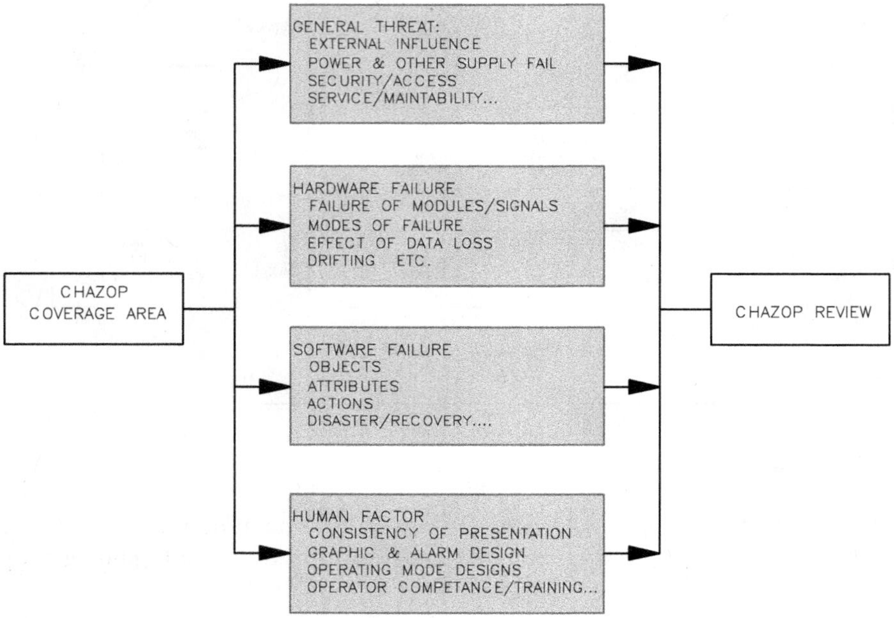

FIGURE IV/1.5.1-1

CHAZOP areas.

Various points considered in CHAZOP are different from HAZOP. These are:

- External factors: Temperature, humidity, ducting, smoke, etc. directly influence the performance of the control systems. Also if there is power, air, or hydraulic system failure, then control systems cannot work. Even if E/E/PEs are not secured or there is unauthorized access, then control systems will fail to work. All these are *not* covered in conventional HAZOP, but they are very much *applicable* to **CHAZOP**.
- Hardware failure: Detection of hardware failure/modes of failure of modules [e.g., input/output (I/O) module], etc. will have a direct impact on the performance of control systems. Naturally, in CHAZOP, such major failures, redundancy criteria, nature of fallback, etc. need to be considered.
- Software failure: There may be errors in the selection of objects, attributes in files, etc. and such software failure is extremely important in control systems and CHAZOP. This will be clear from a simple example: in HAZOP no flow

means no flow in pipelines, but in the case of CHAZOP it may mean *no data flow* and this could lead to a catastrophe.
- Human factors: The display/log presentations have lot of bearing on operator action. Alarm presentation is extremely important so that many human errors can be avoided. Recommendations by the Engineering Equipment and Materials Users' Associations (EEMUAs) are very useful here. A typical recommendation of EEMUA in connection with alarm system has been depicted in Fig. IV/1.5.1-2.

Engineering Equipment Materials Users' Association (EEMUA): Alarm system objective:
Alarm shall
"be relevant to the user's time indicate clearly what response is required be present at a rate the user can deal with"

FIGURE IV/1.5.1-2

Alarm philosophy as per the Engineering Equipment Materials Users' Association (EEMUA).

- Why—the benefits: The major benefits of HAZOP discussed earlier lie with the wider set of views about the hazards and associated recommended actions by the team. In addition to these, CHAZOP provides additional benefits, which make it more useful:
  - The CHAZOP team brings to notice a number of points that otherwise would be missed but could be a reason for a potential hazard. In HAZOP a P&ID is analyzed, and now associated control and safety are connected with the E/E/PEs. In HAZOP, no flow in pipe is considered to be a potential hazard in one section, but what happens if there *is no data flow in E/E/PEs? The complete system or the even complete plant may fail!*
  - CHAZOP may help in developing an easier and flexible control or a requirement of redundancy in the system.
  - CHAZOP is able to provide more flexible operational features during normal operation and special operations. Also in some cases E/E/PEs allow mistakes to be tolerated.

  All these focus on the advantages and opportunities of E/E/PEs, which are also associated with a number of hazards such as common mode failure, snapping of data link, etc. Another important issue is that the spectrum of failure of E/E/PEs is quite different from process equipment failure or standard hardware failure. As a result there is a need for in-depth knowledge of E/E/PE systems for carrying out CHAZOP.
- When—the timing: Usually HAZOP and CHAZOP are done separately. However, some effort to integrate both or to coordinate the two activities closely is beneficial. Like HAZOP, CHAZOP can be carried out at all stages in the life cycle of the plant. However, greater benefit is achievable when this is carried out

at an early stage of the plant. At an early stage one may proceed with an idea that a distributed control system (DCS) or programmable control system (PLC) will be used as BPCSs possibly with a separate protection and safety system, which may be hardwired or E/E/PEs. Now if CHAZOP is applied at this stage it is possible to identify the criticality of functions and to develop a suitable system architecture, which will be beneficial to the proposed control system. At this stage it may be possible to integrate a protection system through network integration using different techniques and/or redundant data highways, etc. Because of the network integration capability and flexibility of modern E/E/PEs, they can offer a wide variety of system architectures. In view of this, it is thought by some that system architecture can be developed at a later date. Sometimes this may not be very fruitful. In any case, a detailed CHAZOP study at various stages is necessary. At an early stage, when flow diagrams only are developed, then all details are not available, so the overall perspective of E/E/PEs can be developed. At this stage, CHAZOP may be termed preliminary CHAZOP. When a P&ID is developed, then certain design details are available, various control loops are identified, system architecture can be developed, and many options can be looked into. However, it has been found in most cases of E/E/PEs that people rely more on the opinion of E/E/PE designers who rely on the help from HAZOP study results and their own experience. After completion of coding, the entire system design is firmed up. In modern practice, people typically apply CHAZOP at this stage. Also at this stage, quality assurance (QA) techniques/acceptance tests are done for formal acceptance of the system. The vendor should also be involved. It is necessary to *note* that at this stage big changes may be costly too. From experience it may be recommended that application of CHAZOP at this final stage may not be wise; instead, earlier CHAZOPS can be finalized. From the discussion, it is clear that at every stage, CHAZOP for E/E/PEs can be applied and each has some pros and cons. The best time for CHAZOP to be carried out depends on the following:

- Project type, that is, green field or retrofit
- Project authority/team
- Technical: Flexibility and network integration capability of the E/E/PE system
- Implementation in house/vendor

In the following clauses preliminary CHAZOP and full CHAZOP will be discussed along with other related details such as team formation, preparation, etc.

1.5.2 Team Formation

In many places HAZOP and CHAZOP are integrated. In this case, there should be one knowledgeable person for E/E/PEs. Again, for small or preliminary CHAZOP, the requirement for a knowledgeable person for E/E/PEs may not be mandatory; instead an efficient HAZOP team can to some extent cope with these requirements. However, for large systems, integration of HAZOP and CHAZOP may not be possible at all. Where attempts are made to integrate, then the team may be too large and it may not be possible for all the members to be present at the same time. In addition, when CHAZOP is discussed, that is, detailing about E/E/PEs, those not having sufficient knowledge may lose interest and fatigue may set in. So, for large projects, HAZOP and CHAZOP are done separately. An intermediate approach could be to include one person knowledgeable in E/E/PE as a representative in HAZOP, who may also be present in CHAZOP, so that some uniformity is maintained. When in a large project several parallel teams are working together such action is not feasible. Normally, a team for CHAZOP consists of one team leader, one scribe, and one member each from the design/implementation team. Inclusion of one member from the vendor at a later design stage for CHAZOP may be *beneficial*. Usually, in the team an independent observer is also included for verification and validation. A team approach carries more weight in decisions on complex issues. As in the case of HAZOP discussed earlier, all members must be allowed to express themselves freely, especially the designer who should not be defensive in obtaining a better result from CHAZOP.

1.5.3 Preparation

Before starting CHAZOP some preparations, as discussed in connection with HAZOP, are necessary. The basic information to be collected may be as follows:

- For preliminary CHAZOP the following documentations are necessary. Note that in most cases these are similar to what has been discussed for HAZOP.
  - Basic flow sheet
  - Basic equipment details
  - MSDS and associated chemistry
  - Proposed site layout with tanks
  - Utilities and their location details
  - Basic control *philosophy*
  - Overall system *architecture*
  - Alarm and trip schedule
  - *Safety and protection* details
  - Functional specification
  - Linking with DCS/PLC
  - Tentative I/Os
  - *Interface* points with types
  - Environmental conditions

  Here, it is noted that out of these documents a few are also needed for normal HAZOP, so the same documents may be shared.
- Full CHAZOP: Almost all information required for HAZOP as found in Clause 1.1.5 will be necessary here. In addition, the following documentation is necessary for detailed CHAZOP for both new projects and retrofitting projects. For retrofitting projects, where new E/E/PEs are to be implemented, it may be very difficult to get all the necessary information, especially the latest HAZOP data and/or up-to-date P&IDs. Therefore in those cases, the old control philosophy needs to be *updated* to marry with current system requirements. For

these retrofitting cases, proper and detailed specifications and a clear control philosophy definition are very important. Also those control schemes and protective systems that were added after HAZOP must be clearly defined and elaborated.

- Details of control *subsystems*
- Details of *interface*
- Details of cabinets
- Data highway details
- I/O channel details
- I/O and remote I/O card details
- Controller/processors
- Filing system
- Memory system/storage media
- Backplane connection details
- Power supply and communication card details
- Communication link speed, type
- Field bus details
- Server details
- Remote link details
- Cabling and data highway plan
- Alarm and trip schedule
- Utility supplies and distribution
- Peripherals and human/machine interface (HMI)
- Displays and logs
- Redundancy and fallback facilities
- Failure monitor and diagnostics
- Security system
- Software and associated details
- Environmental protection
- Control philosophy document
- Documents necessary for HAZOP (Clause 1.1.5)
- Miscellaneous other

This list covers four parts of CHAZOP, as shown in Fig. IV/1.5.1-1. In some cases, depending on requirements a few other necessary documents may be required.

1.5.4 Guide Word and Questions

HAZOP for process is characterized by the procedure of combining guide words with process parameters. It is usual to use a similar framework for both HAZOP and CHAZOP. There are a few drawbacks in using HAZOP, namely, time, cost, and safety gap. Out of these three, the first two can be minimized by automating the process. It is difficult to remove safety gap because there is no assurance that hazards will not be missed. These are more prominent in the case of E/E/PEs, mainly for ambiguity in interpretation. In HAZOP, "NO" is a very common guide word. In process HAZOP, "no flow" means that there is no flow in the pipeline but in the case of CHAZOP it may mean "no measurable flow" or "no display for flow" or *actually* "no flow" or there is "no flow in the ini-file." In the case of CHAZOP one needs to consider various words such as "objects" (e.g., pump/heat exchange), "attribute" (e.g., flow, velocity temperature), and actions (e.g., operator action). Therefore in CHAZOP similar guided words are used but with different meanings and/or implications. Also to get better results it is necessary to add more guided words. All expressions shall be less ambiguous and complete so that hazards are not missed. Later a few words were added such as "early"/"late" to indicate an event or action relative to time and "before"/"after" to indicate the ordering of an action or event (Table IV/1.1.9-2). A few guide words used in various CHAZOP studies are listed in Table IV/1.5.4-1.

Table IV/1.5.4-1 CHAZOP Guide Word for Hardware/Logic System and Human Factor

Guide Word	Deviation for Hardware/Logic System	Deviation for Human Factor (Action)
No	No signal or no action	No information or no action
More	More signal or more action	More information or more action
Less	Less signal or less action	Less information or less action
Wrong	Wrong signal or wrong action	Wrong information or wrong action

Inspired by S. Yanag, W.H. Chung, Hazard analysis and support tool for computer controlled processes, Loyuborough University, Journal of Loss Prevention in the Process Industries, 1998; Elesevier.

There shall be two sets of CHAZOP study: preliminary CHAZOP to be done at an early design stage and full CHAZOP to be done at a subsequent stage. In full CHAZOP the following divisions are considered:

- Computer system/environment
- I/O system
- Complex control system

Accordingly, there will be different sets of questions for each. These are presented in Table IV/1.5.4-2.

For I/O signals a separate set of questions will be applicable, as shown in Table IV/1.5.4-3.

A questionnaire for a complex system is enumerated in Table IV/1.5.4-4.

Table IV/1.5.4-2 ***Questionnaire for Computer and Environment System***

Failure	System	Question	Remarks
Gross	Whole machine/ system*	What should happen? Will operator know? What should operator do? Will the failure propagate to other machines/systems or to safety system? Any change needed?	*In an offshore control network programmable logic control corresponding to vertical pipe handling fails
Random	Cabinet/crate Controller or control subsystem I/O card Operator—console Watchdog timer Power supply Other utilities	What should happen? Will operator know? What should operator do? Will the failure propagate to other machines/systems? Any change needed?	All questions are applicable for each of the items listed

Inspired by S. Yanag, W.H. Chung, Hazard analysis and support tool for computer controlled processes, Loyuborough University, Journal of Loss Prevention in the Process Industries, 1998; Elesevier.

Table IV/1.5.4-3 ***Questionnaire for I/O Signals***

Signal/Actuation	Deviation	Question
Signal	Low	Does it matter? Will the operator know? Action required by operator or other system?
Signal	High	***All above questions applicable here also***
Signal	Drifting	***All above questions applicable here also***
Signal	Invariant	***All above questions applicable here also***
Signal	Bad	***All above questions applicable here also***
Actuator	Driven failure high	***All above questions applicable here also***
Actuator	Driven failure low	***All above questions applicable here also***
Actuator	Drive stuck	***All above questions applicable here also***
Actuator	Drive drifting	***All above questions applicable here also***

Inspired by S. Yanag, W.H. Chung, Hazard analysis and support tool for computer controlled processes, Loyuborough University, Journal of Loss Prevention in the Process Industries, 1998; Elesevier.

Table IV/1.5.4-4 Questionnaire for Complex Control Scheme

Complex Scheme	Points to be Considered
Purpose and method of operation, e.g., sequence of operation interlock, etc.	Safety-related function
Point of operator access	Set point/cascade make or break
Limit application	Careful use of limit for good safeguard and/or early warning
Other scheme interaction	Startup/shutdown/normal operation, timing issue, synchronization, required/expected operation
Controller tuning	Initialization/winding up
Relationship with trip and alarm—action in the event of major plant upset	Loss of utility, spurious/correct operation of emergency shutdown
Unauthorized access protection and others	Spreading over large system (>1 controller subsystem)

Inspired by S. Yanag, W.H. Chung, Hazard analysis and support tool for computer controlled processes, Loyuborough University, Journal of Loss Prevention in the Process Industries, 1998; Elesevier.

1.5.5 CHAZOP Procedure (Outline)

General brief outlines are discussed here. It is recommended that both preliminary as well as full CHAZOP are carried, so both the systems have been included in the discussion.

- Preliminary CHAZOP: This is undertaken at the early stage when not enough documentation is available. The following issues are addressed:
  - Overall proposed system architecture including functionality of the system is discussed. In this the following points are taken into account:
    - Machine/main system
    - Basic functions of all subsystems
    - Redundancy criteria
    - Diversity factor
    - Protection issues
    - Both PEs as well as non-PEs
  - Total review of the system including protection and interlock, alarm, and safety aspect. In this connection Fig. IV/1.5.1-2 may be referred to.
  - Study of the system performance in the event of failure of main system and/or machine and/or failure of power supply and/or other utilities.
- Full CHAZOP: The main procedure is to study the system with the help of guide words and questionnaire, discussed in Clause 1.5.4 in detail. The designer presents the required information to the team in the form of various drawings and documents so that the study can be carried out. As before, such a complete

failure of the machine and/or its components, etc. is studied systematically. In this connection it is to be noted that in complex control schemes it is necessary to study and monitor system performances also. For CHAZOP study purposes there are some differences in considerations between continuous process, sequential process, and batch process. The technical aspects of these differences are discussed and highlighted here:

- Continuous process: The system measurements are the continuous type, hence I/Os are continuous signals so I/O signals are checked for "bad" (e.g., signal out of range, say beyond 4—20 mADC, "transmitter accuracy" and condition, etc.). In addition to these the redundancy of transmitters is different to that in the case of pure binary type signals (switches). For controllers, tuning parameters are checked for stability and correctness as well as against changes in process conditions. Overall control philosophy, interaction with other systems and safety aspects such as safety interlock, loop performance, etc. are monitored.
- In the case of sequential controls, startup, running, and shutdown modules are checked for operator action. All critical timings (including waiting time and monitoring time) and major equipment interactions, etc. are monitored. For each sequential step module, similar considerations are applicable also. In addition to this the overall sequence activation/deactivation sequence and communications are also considered during the study.
- For batch process (see Clause 3.4 of Chapter VI), in each step (depending on applicability), the foregoing considerations need to be checked. Additionally, those discussed in Table IV/1.5.4-4 need to be considered. Checking is time-consuming but *necessary* as the system needs to cope up with various situations in the batch process.

  For the basic concept of CHAZOP Fig. IV/1.5-1 may be referred to see that the safety system has been kept separate from BPCS, to give backup in case of BPCS failure. At lower levels, redundancies with fallback facilities have been considered at controller as well as at field level. There are a few issues such as ambiguity, incompleteness, nonsensicality, and redundancy that are very much present in conventional HAZOP and should be avoided *prior* to applying data from HAZOP directly to CHAZOP. On a case-by-case basis, there may be additional factors to be considered for CHAZOP. It will be prudent to consider only those combinations that will give meaningful deviation, and repetitive combinations should be avoided to limit the time and fatigue in discussions.

1.5.6 CHAZOP Discussions

From the foregoing discussions it has been seen how safety aspects are dealt with in various layers and redundancies. Here, failures are taken into account, but control logic is not handled explicitly. There are no means to verify the logic requirements. For hazard analysis it is necessary that process is ***well*** represented. In the process and related controls, safety subsystems must be identified and suitable logic built for the safety system. It should be possible to *verify* the safety logic. In safety logic, all safety critical events shall be identified and then a questionnaire for the safety critical events shall be applied over the entire life cycle, as shown in Fig. IV/1.5.6-1.

P&IDs are the starting point in most cases but they can at best give the structure of control systems. Control logic and plant behavior are not clear from P&IDs. Some use a signed directed graph (SDG). An SDG consists of nodes to represent process variables, and a signed directed branch symbolizing the relationship of variables. However, this can only give the qualitative relationship of variables in the process. There is another method, known as event time diagram (ETD), that gives components of computer system hazards, that is, I/O levels, communication, and the computer with which control logic components are related. The process control event diagram

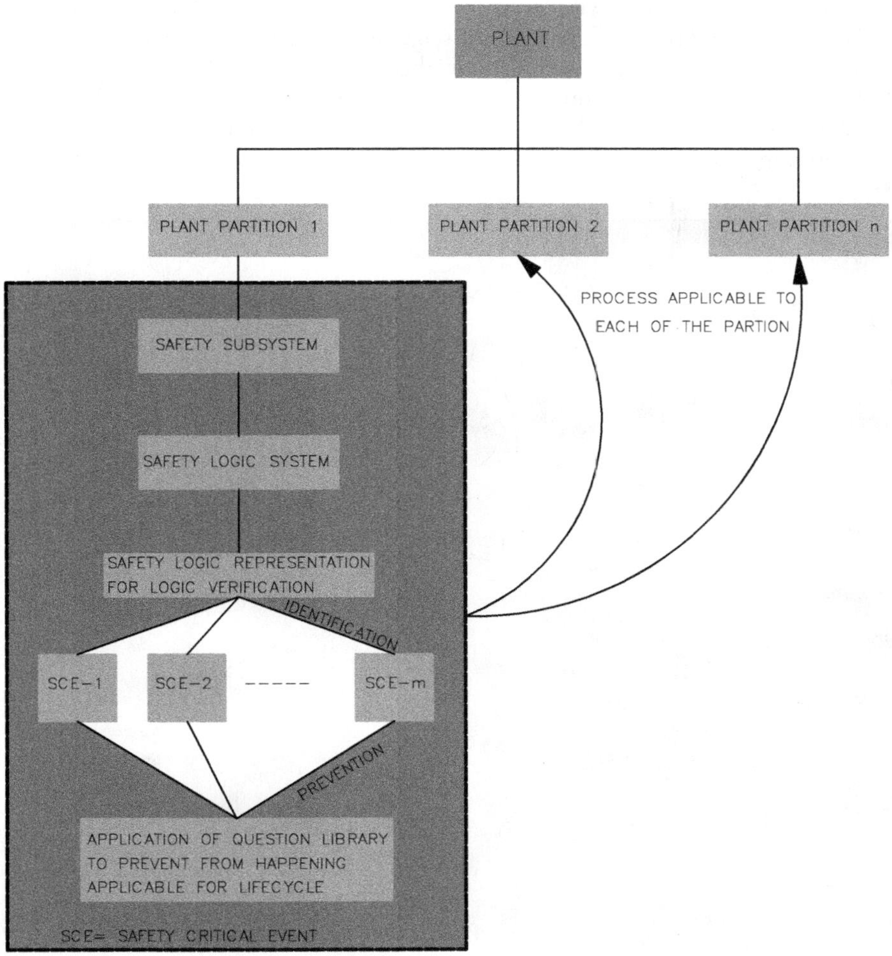

FIGURE IV/1.5.6-1

CHAZOP hazard analysis framework.

(PCED) by Chung and Yang takes the advantages of SDG and ETD to complement P&ID for CHAZOP. In this diagram, intervention level by operator, HMI display, communication, main processor (/controller/computer), sensor actuator, and the various process divisions are shown at various levels. All control logic components are related to these with the help of arrow heads to represent actual control logic, which can be verified in the diagram. After identification of control logic, it is necessary to discover what can go wrong and its effect. Following HAZOP principles, deviations are found from each control action in the PCED [3]. If control logic under certain deviations does not satisfy safety requirements, then these are safety critical events for which necessary questionnaires are used to identify deviations and then necessary steps are taken to control action. These questionnaires will be separately filled in for each of the generic objects such as HMI, communication, sensor, actuator, process, etc. For each of these objects there will be different stages like design, implementation, operation and maintenance, modification, etc. In this connection Fig. IV/1.5.6-2 may be referred to. All these questionnaires will help to pinpoint probable events and accordingly control actions may be taken. Details can be found in Ref. [3]. This process is in theory very easy but in reality it is not so, especially for complex systems. Also for

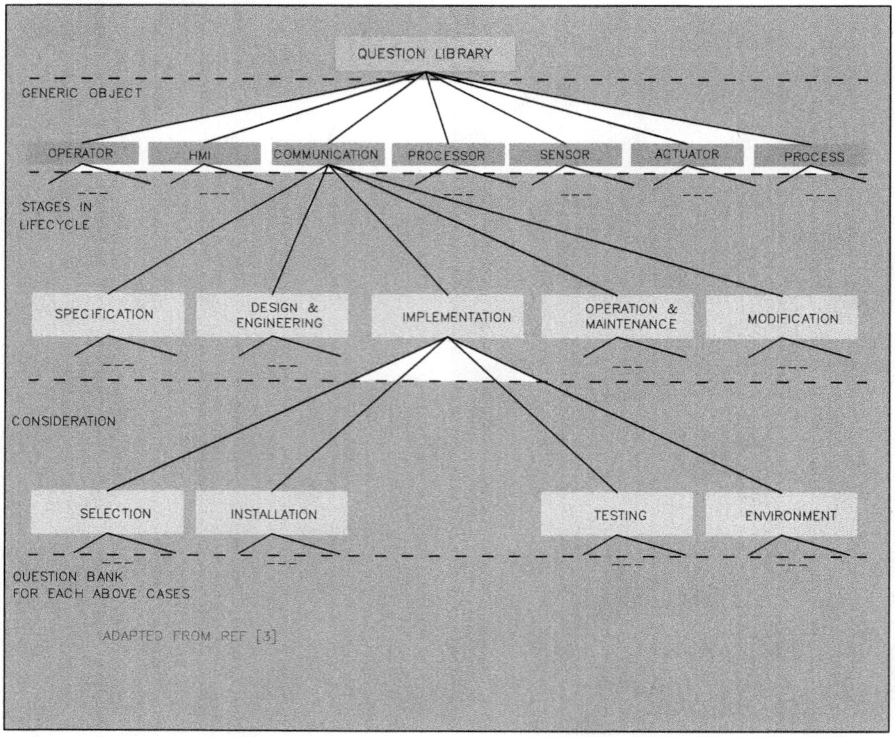

FIGURE IV/1.5.6-2

Question library.

large plants it is time-consuming and costly. So, efforts have been made to automate the system. Now, we will look at the intelligent HAZOP system.

1.6 Intelligent HAZOP (HAZOP Automation—The Concepts)

As discussed earlier, the time and cost for HAZOP analysis is quite large, especially for complex/large plants. Automation in HAZOP is capable of reducing both time and costs. Also it can provide a detailed and consistent report free from human error. Since much time and effort are saved, the team members can concentrate more on complex plant issues. Also it is possible to integrate various other technical programs, for example, computer aided design (CAD). There are a number of issues to be addressed before automating HAZOP. Some of these issues shall include:

- Generic/general approach for plant hazards
- Different approach toward control/computer system (see Clause 1.5)
- Recognition of the different approaches for continuous plant
- Approach difference for batch process
- Inclusion of plant-specific data such as:
  - P&ID
  - Material property
- Combining general system with plant-specific data to:
  - Find abnormal causes
  - Find adverse consequences
  - Assess hazard propagation

It is quite some time since various methods were developed for automating the HAZOP process. Some methods included computer-based systems. We shall now see how these systems have been developed.

1.6.1 Short Discussions on Developmental Process

Back in 1987, Parmer and Less developed a rule-based approach for automation of HAZOP. They developed knowledge representation regarding the propagation of faults/failures from each of the process units, such as pumps, heat exchanger, etc. They tried to include control systems such as sensors, control valves, controllers, etc. In this method they partitioned P&ID into several sections. In one section, a line was chosen for hazard analysis. They tried to find cause from an initial event in the process unit and consequence in the terminal unit. So, hazard analysis was restricted to the limit of the line under study. This is in contrast to the HAZARD analysis principle, which continues to the end of the process. Hence it could show immediate effects/consequences. They used FORTAN language for computer programming. In 1989 Water and Ponton studied automation of HAZOP through their simulation approach. For practical application, on account of its complexity, it did not become so popular. The rule-based approach was also introduced by Heino and Suokas in 1990. They developed rule-based logic to search for the cause and devised it in "If deviation type" and condition is like that… then probable cause will be… Naturally, in industrial applications, where there are so many possibilities, there will be too many rules to search for a cause. As a result this approach was difficult to handle. Also here, cause has more importance than consequence, which is the opposite in the normal HAZOP approach. A modeling language for chemical process and reactions was developed by Henning and Leone in 1990. This was mainly applicable to chemical processes involving reactions. Nagel in 1991 developed inductive and deductive type rules for hazard identification in chemical reactions in chemical plants. The rules although ideal for chemical processes involving complex reactions were not widely used for general industrial applications. In 1994–96 Venkatasubramanian and Vaidhyanathan developed an object-oriented model based on HAZOP expert systems with different approaches for continuous and batch processes. In 1997 Dimitradis, Shah, and Pentelides developed a quantitative approach toward HAZOP automation. In this approach it was necessary to identify the disturbance profile leading to hazards. In many cases, especially with nonlinearity, it was difficult to solve the problem. In the same year Faisal and Abbasi produced TOPHAZOP, a knowledge-based two-compartment general and specific software tool for projects. Objects, attributes, causes, and consequences were included in the model. While objects are developed in a frame structure with their attributes, causes and consequences are generated through rule-based systems by combining the rule with the framework. In 1998 Srinivasan, Shah, Dimitradis, and Venkatasubramanian developed a hybrid knowledge-based mathematical model, and in 1999 Turk developed a discrete model to capture hazards in continuous and sequential chemical process. In addition, there have been several other approaches for automating HAZOP: McCoy (1999), Bartolozzi (2000), Kanga (2003), and Zhao (2005). Of these the HAZOP expert-based system is applicable in a generalized way and is more acceptable on account of its quick process. However, this system requires a huge database and it is not very easy to integrate with CAD. In subsequent clauses, expert system, P&ID integration, and model-based systems will be touched upon. However, wide use of automated HAZOP in industrial scenarios has a long way to go! Some of the developments in this area will be highlighted in later clauses, enabling the reader to grasp some knowledge of automated HAZOP. Digraph shown Fig. IV/1.6.1-1 is often used in automated HAZOP.

A directed graph in short is known as a digraph. It is quite popular in event/fault tree analysis, yet it is used in automated HAZOP analysis also. The majority of the system displays in automated HAZOP digraph are used. In fact digraph is a technique for using the matrix representation of symbolic logic models to analyze functional system representation. There are several steps involved: first, determination of a combination of systems/subsystems, and construction of a failure propagation path. Then, construction of a digraph model in the success domain to convert into a digraph model in the failure domain. Digraph models are made independently to generate a single/double cut set. Lastly, from these cut sets a minimal cut set is determined.

FIGURE IV/1.6.1-1

Digraph techniques.

1.6.2 *HAZOP Expert*

HAZOP expert is an object-oriented model-based intelligent technique used to automate the HAZOP process. It is well known that no two plants are identical, hence the problems will be different. At the same time it is also true that the majority of time and effort is spent on routine work in a HAZOP study and there is commonality in this routine work. So, by automating this routine part (which is a major part), time, effort, and cost can be saved. But how can this routine work be matched with plant-specific issues? Another important issue to be noted is that automation is not meant to replace the HAZOP team but to supplement it, so that experts instead of being bogged down with routine work can concentrate on how to address the complex issues pertinent to the plant. To understand HAZOP expert it is better to refer to Fig. IV/1.6.2-1.

- Description: HAZOP expert is a HAZOP digraph model-based system. The digraph represents the plant process with the help of a directed graphical model in which ways and means, that is, knowledge to identify abnormal causes and adverse consequences, have been built in. The HAZOP digraph is built in a context-independent manner to cover a wide range of applications. The following discussions shall be read in conjunction with Fig. IV/1.6.2-1. The key part of the model is the graphic user interface (GUI), which consists of three subparts, namely, material and P&ID editor, HAZOP digraph model developer, and reporting section. Details of the subsystems of the GUI can be found in Table IV/1.6.2-1.
  Once P&ID and material data are inputted, then the associated HAZOP model for the particular plant is generated with P&ID (inputted) connected automatically and ready for HAZOP analysis. So, a lot of knowledge acquisition time and effort are saved. The user can initiate HAZOP analysis in any line in the P&ID. The next part is Knowledgebase, which could be an Microsoft (MS) Access database consisting of information derived from old HAZOP analyses, related to process operation, associated process units, operating issues, failure modes and conditions, etc. Knowledgebase consists of two parts: generic and project specific. Knowledgebase is capable of suggesting probable solutions. Knowledgebase directly communicates with the GUI to help the user to carry on HAZOP analysis. The third part is the inference engine, which is like a search engine. It has a fault propagation algorithm, which in conjunction with Knowledgebase helps the user to retrieve the generic and specific abnormal causes and associated adverse consequences. This deviation propagates from the starting process unit to the end equipment, as defined by the user through the GUI. The user accesses the system through I/O units, as shown in Fig. IV/1.6.2-1.
- Discussions: In manual HAZOP it is very difficult, if not impossible, to consider process variable deviation in each pipe valve, etc. Normally, it is grouped and put in a node to study. Naturally, in HAZOP expert, there will be many more deviations generated when compared with the manual counterpart. So there is less/no chance of missing an issue because of human error. These are all positive

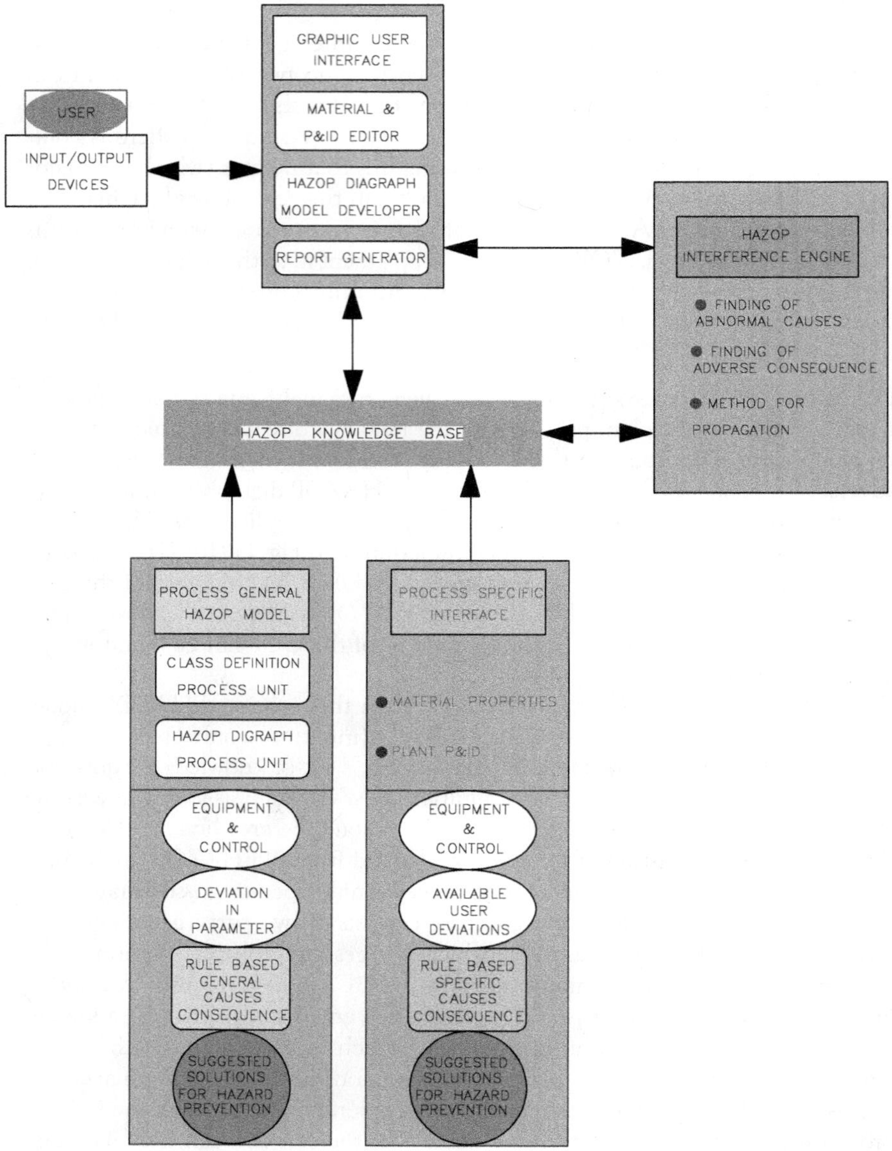

FIGURE IV/1.6.2-1

Basic structure of HAZOP expert.

Table IV/1.6.2-1 Graphic User Interface (GUI) Subsystems

GUI Subsystem	Function	Remarks
Material and P&ID editor	The piping and instrumentation diagram (P&ID) editor is meant to develop P&IDs* specific for the plant with the help of the editor. The material editor is used for inputting material property data (such as flammability, toxicity, corrosion, etc). In case these data are available in any database then they can be imported.	*Direct import of P&ID in computer aided design (CAD) is possible.
HAZOP digraph model developer	The model library has generic models for various process units and control systems such as process pump, heat exchanger, tanks, vessels, pipes, valves, controller actuator, etc.	
Report generator	The basic function of this is to generate reports in a desired format.	

sides, but there is a negative side also. In manual HAZOP, many of the useless causes and consequences are filtered out at the beginning, on account of specification, operating condition data, etc. In this system there will be so many unnecessary causes and consequences. As a result people are now thinking of utilizing a semiqualitative type so that unnecessary causes and consequences generated because of pure qualitative methods can be avoided.

- Batch process (see Clause 3.4 of Chapter VI): Batch process is quite different from continuous process, for example, in continuous process there is hardly any direct role of the operator in the process steps. In continuous process the operator controls, monitors, and diagnoses faults in the process. On the contrary, the operator has a great role to play in processing steps. There may be the requirement of a subtask initiation by the operator. Any omission or wrong operation such as wrong sequence selection, wrong duration selection, even wrong material selection can lead to process hazards in the batch process. Unlike continuous plants, batch process is both discrete and characterized by a temporal nature. Plant status changes with time in a defined sequence. P&ID cannot totally describe and get a full idea of the plant. So, it is necessary to have an *instruction* sheet and *sequence* chart both of which form the recipe. HAZOP in batch process mainly concentrates on finding plant maloperation and process variable deviation. A model-based expert system for automating HAZOP is referred to as HAZOP expert. Batch HAZOP expert (BHE) architecture is similar to what has been discussed for continuous process and includes:

- Knowledge of the organizational framework (expert system)
- Inference mechanism
- Graphical user interface comprising:
  - P&ID graphical editor
  - Product *recipe* graphical editor
  - Process chemistry editor
  - Subtask *digraph* model editor

Two terms are important here: **Recipe Petri net**, which describes the sequence of various tasks, and **Task Petri net**, which defines the sequence of subtasks within a task. For each subtask there is one digraph model built to get a cause and effect relationship between the associated variables. These are shown in Fig. IV/1.6.2-2. For Petri net definition Fig. IV/1.6.2-3 may be referred to. Fig. IV/1.6.2-2 shows Petri net knowledge base representation of a recipe with a number of tasks and subtasks. Each subtask is then a digraph model for HAZOP analysis. Plant maloperation can be generated by using this figure.

Semiquantitative reasoning with required information like operating parameter, hazard critical data for equipment, and materials is used to filter out

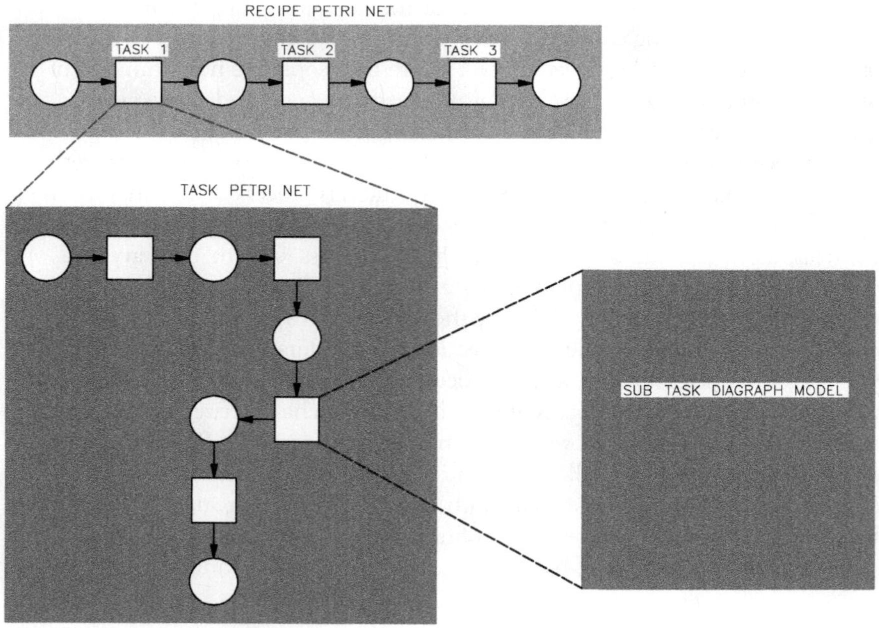

FIGURE IV/1.6.2-2

Petri net-based batch HAZOP expert (BHE) hierarchy.

Petri net is a tool to model concurrent systems and their reasoning. Petri net has its application in different types of modeling, qualitative and quantitative analysis, supervisory and coordinate controls, planning and scheduling, and hybrid system design.

FIGURE IV/1.6.2-3

Petri net.

less possible consequences. For a Petri net task with too many subtasks, it is always better to go for HAZOP expert for filtering and ranking the consequences.

1.6.3 LDG HAZOP Automation

HAZOP expert discussed earlier has wide application, but for success it depends too much on a database. Also actual integration of CAD P&ID into the system is not so easy with normal P&ID, which depends too much on drafting, and manual input of process-specific data invites human error. It is now possible for various software packages to have a communication link to other external software packages. So, Intergraph's Smart Plant P&ID (SPPID), which is an asset-centric rule-driven CAD P&ID, is very helpful in integrating P&ID into the computer system. It depends more on plant data than drafting. This is helpful in exchanging specific plant data into the system. From this, an initial model based on signed directed graphs (SDGs) was developed. It had a few shortcomings. Subsequently, a layered directed graph (LDG) model-based HAZOP expert system (LDG HAZOP) was developed to *overcome* the shortcomings of SDG. The LDG model qualitatively highlights cause and effect relationships between process deviations with the help of HAZOP guide words. Here also a user interface with the system with the help of an I/O system at the user interface module is directly linked with document (DOC) and LDG modules, as shown in Fig. IV/1.6.3-1.

LDG qualitatively tries to find cause and effect relationships between process deviations generated by HAZOP guide words. LDG HAZOP is a web-based multiclient expert system for HAZOP developed in Java language [4]. The main subunits connected to the user interface are the DOC and LDG modules. The DOC module is mainly used for the word processing unit. DOC modules consist of (1) result/report generation subsystems, (2) a document management subsystem, and (3) a hint generator.

Various submodules of the LDG module are as follows:

- LDG library: LDG modules of various process units it is somewhat like general knowledge base in HAZOP expert discussed earlier.
- Process description (PD): PD of the process in the question is inputted to generate a *process-specific* database. (This can be compared with project-specific interface in HAZOP expert.) Based on data defined by PD, an algorithm is developed to match a model in the library. Equipment interconnections are done based on process-specific data in this module.

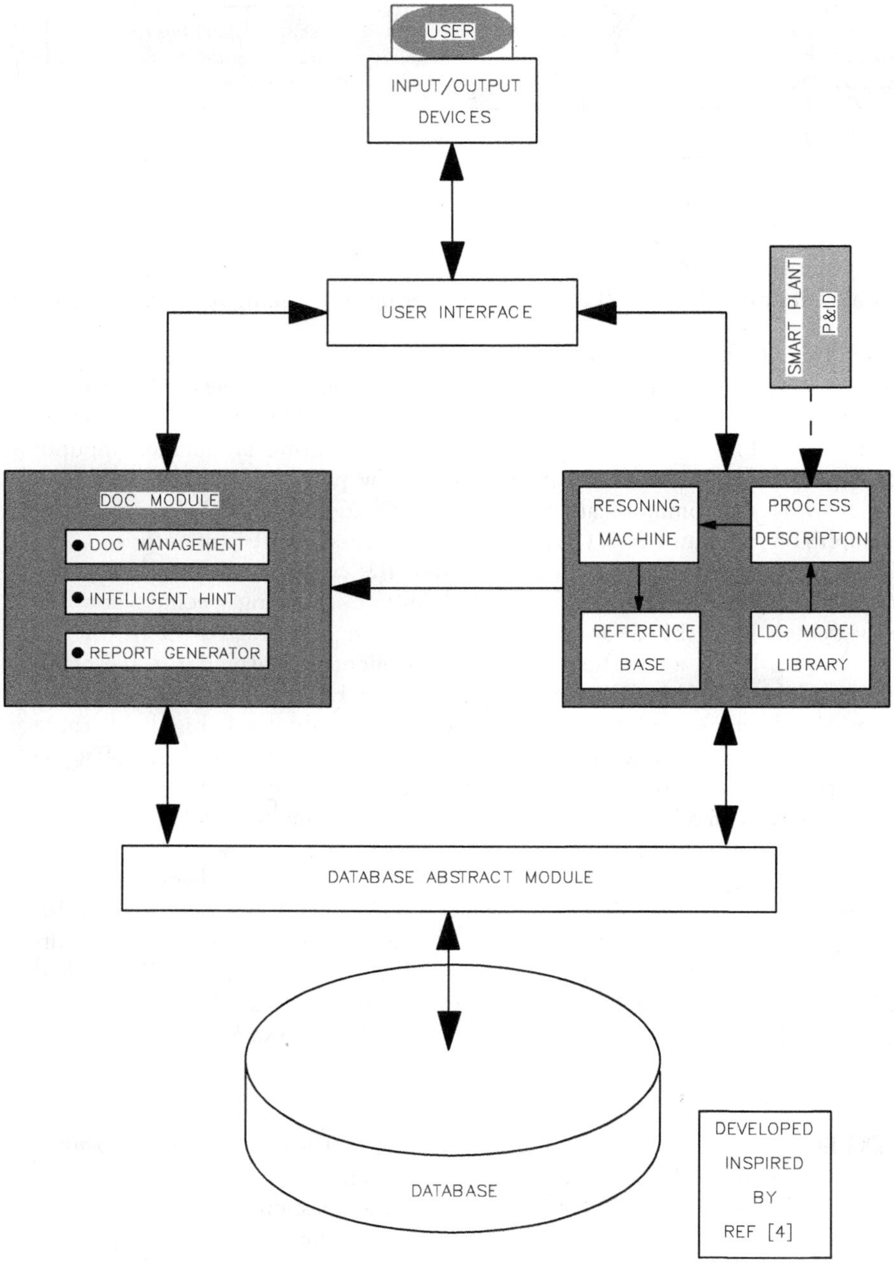

FIGURE IV/1.6.3-1

LDG HAZOP model.

- Reasoning machine: This is used to start an automatic HAZOP study, after LDG models are properly linked.
- In the next stage, human validation and modification is done.
- Reference base (REF base): A DOC module can generate a report by directly importing REF base when desired by a human expert. It also generates intelligent hints. A *similar* function was also present in HAZOP expert.
  SPPID interfaces LDG module through PD, as shown in Fig. IV/1.6.3-1. Thus it is possible to directly import project-specific data from SPPID.

1.6.4 Multilevel Flow Modeling-Based System

Another way HAZOP can be automated is by multilevel flow modeling (MFM). MFM is described in Fig. IV/1.6.4-1 and it is widely used in risk analysis.

> *Multilevel flow modeling (MFM): This method can be used in various hazard analysis methods, for example, fault tree analysis, HAZOP, etc. The multilevel flow model consists of goals and functions: a goal describes the purpose of the systems and subsystems; function represents the ability of the system/subsystems in terms of flow of mass, energy, and information. MFM establishes the relationship between goals and functions so that goals can be achieved. MFM also gives the relationship between functions and subgoals, which describes the conditions. Mass and energy flow are used to model the function of the plant/subplant and their activity. Flows of information are used to model control system and operator actions. There are standard symbols to develop models for the process. Nowadays MFM is widely used in risk analysis. One of the uses of MFM is in automatic HAZOP analysis.*

FIGURE IV/1.6.4-1

Multilevel flow modeling.

MFM is capable of representing functional knowledge with an easier understanding of the process. It could provide an efficient paradigm for facilitating HAZOP studies and for enabling reasoning to identify potential hazards in safety critical operations. MFM is a qualitative reasoning model that could be used to assist the HAZOP team by ensuring coverage and consistency. MFM divides the system into subsystems according to the functions in terms of goals, relations, and process units. MFM provides a set of reasoning rules that can be used to perform an automatic HAZOP study and reveal the potential hazards and casual paths of a hazard in a visual way.

A typical automated HAZOP study principle-based method is shown in Fig. IV/1.6.4-2. MFM is a functional modeling method, used to represent goals and functions of process plants that involve interactions between flows of material, energy, and information. Various representative symbols typically used in building an MFM model are shown in Fig. IV/1.6.4-2B. As shown in Fig. IV/1.6.4-2, qualitative reasoning in an MFM model is based on representation of process knowledge. The rule-based interference engine is used to find causes and consequences pertinent to hazards. With slight modification the same model can be used for similar plants. Another unique feature of MFM is that it can be applied to control systems to identify the hazards.

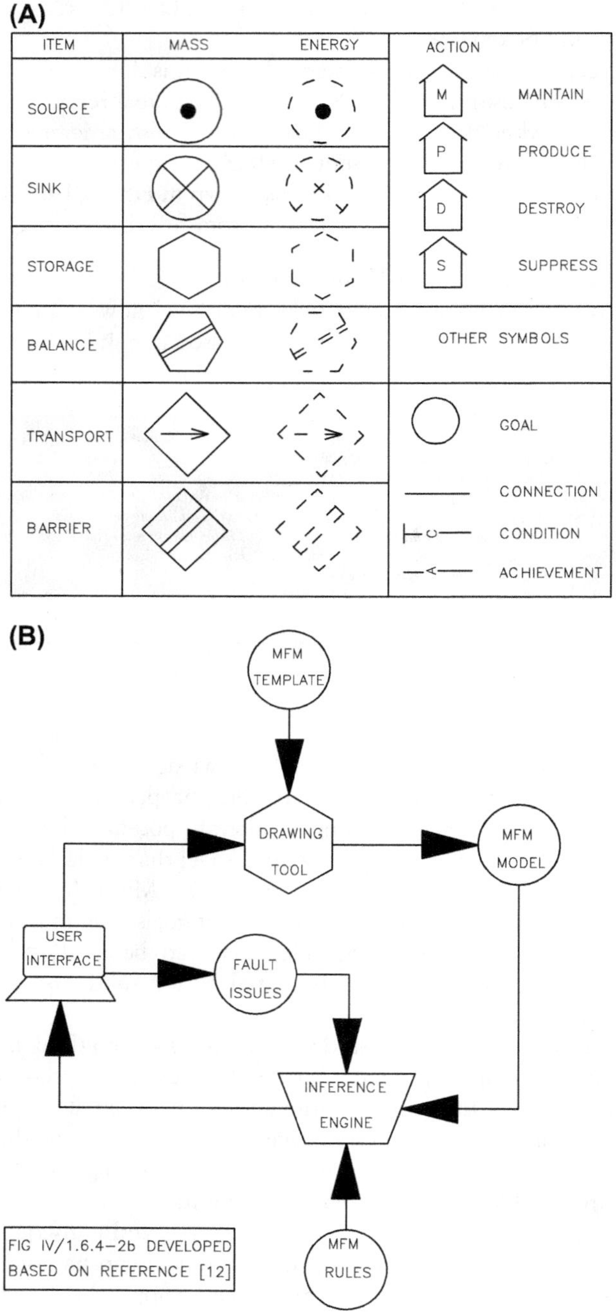

FIGURE IV/1.6.4-2

Multilevel flow modeling (MFM) method in HAZOP. (A) MFM symbols (general), (B) MFM blockdiagram for HAZOP.

Now, efforts have been made to develop suitable software for various methods of plant (process) hazard analysis (PHA). There are a number of papers available where through suitable software it is possible not only to automate one PHA method but to integrate several automated PHA methods such as event tree/fault tree (Chapter V) with HAZOP or HAZOP with FMEA, which will be discussed in the next clause in this chapter.

2.0 FAILURE MODE AND EFFECT ANALYSIS

FMEA is an analytical method used to identify potential problems in the product and in its process of development. It is an inductive method used for identification of hazards of a system with single point failure. When criticality analysis is added with FMEA it is known as failure mode effect and criticality analysis (FMECA). It was used as early as 1950 in reliability engineering. FMEA/FMECA is mainly used for manufacturing, product development, etc.

2.0.1 *Primary Objective and Purposes*

The basic objective of FMEA/FMECA is early identification of catastrophic, critical and other potential failures so that these can be eliminated or mitigated through design and manufacturing process at the earliest possible time. There are three kinds of FMEA/FMECA, namely, functional FMEA/FMECA, design FMEA/FMECA (DFMEA/DFMECA), and process FMEA/FMECA (PFMEA/PFMECA) (see Clause 2.0.4). The aim of functional FMEA/FMECA is to improve the design of the system. Similarly, DFMEA/DFMECA always looks to see that there is improvement in the design of subsystems/components. The objective of PFMEA/PFMECA is to improve the manufacturing process. Thus it transpires that the major objective of FMEA/FMECA shall include:

- Improvement in production planning
- Improvement in production and process controls
- Loss of production downtime
- Loss of performance of production line
- Loss of quality of product
- Production performance degradations
- Improvement in testing and verification
- Improvement of maintainability of production machinery
- Online diagnostic feature

Successful FMEA/FMECA is helpful in identifying the potential failure mode from experience with a similar product or production process or on common physical failure logic. FMEA/FMECA is an iterative process and supports maintainability, safety, and logistic analysis. Also it is carried out at various stages of manufacturing of the product, so it is important to coordinate and ensure that at no time or stage effort is duplicated in the same program.

2.0.2 *History of Development*

The US military and aerospace engineering used FMECA way back in the 1950s to tackle critical issues and prevent hazards (Military Procedure MIL-P-1629 "Procedures for performing a failure mode, effects and criticality analysis," dated November 9, 1949). Also the National Aeronautics and Space Administration used both FMEA and FMECA in various forms. The 1960s and 1970s witnessed the use of FMEA in other industries as a potential tool in reliability engineering. The automotive industry, especially Ford, used this technique extensively during this period. In 1988 when ISO 9000 pushed formalized quality control and management, FMEA gained importance. QS 9000 compliant components were bought by various automotive industries and since then it became a part of quality planning. In 1993 automotive industries published FMEA standards. The latest version of IEC 601812 is the relevant standard followed internationally for FMEA.

2.0.3 *FMEA Definition*

What is FMEA? FMEA is an analytical, inductive reasoning, single point of analysis and a part of reliability engineering. FMEA is an analytical method designed to identify, fully understand, and analyze:

- All potential failure modes of the various parts of a system or end use for a given product or process
- The effect of these failures on the system
- The risks associated with these identified failures, their causes, and effects on the system
- The issues and prioritize them for corrective actions
- How to avoid the failure and/or mitigate the effects of the failure on the system (according to priority)

  When criticality analysis is also done it is referred to as FMECA. In fact, prioritization is done mainly based on criticality analysis so that the most critical issue is addressed first. FMEA reduction may be in terms of reducing severity or lowering the occurrence, or both. When the root cause is addressed, then it will be almost impossible to occur. So, in a nutshell, FMEA could be used as a guiding tool to the complete set of actions in product development process so that risks are either avoided or mitigated in the systems, subsystems, or components (to an acceptable limit).

  There a few other definitions such as:

> FMECA is a technique used to identify, prioritize, and eliminate potential failures from the system, design or process before they reach the customer.
>
> **Omdahl (1988) [5]**

> FMECA is a technique to "resolve potential problems in a system before they occur."
>
> **SEMATECH (1992) [5]**

So, it is seen that FMEA/FMECA needs to address the following issues/information:

- Items
- Functions
- Failure (all)
- Effect of failure
- Causes of failure
- Detection (*criticality*)
- Risk priority number (*RPN*)
- *Criticality* analysis (FMECA)
- Current controls
- Recommended actions
- Implementations
- Calculation of new RPN
- Other related issues

In this connection one may note that ISO 9001:2000 para 8.5.3 requires a documented procedure for preventive action. It also demands that preventive actions must be in proportion to the effect of the failure. The FMEA/FMECA approach satisfies ISO 9001 para 8.5.3, so for ISO 9001:2000 FMEA/FMECA may be applied, but it is *not a specific* ISO 9001 requirement, since the basic purpose of FMEA or FMECA is the same and the procedures are similar. However, one should note that there are some differences between them, which are listed in Table IV/2.0.3-1.

There are different types of FMEA/FMECA, discussed next.

2.0.4 Types of FMEA/FMECA

There are three types of FMEA/FMECA: functional FMEA/FMECA, DFMEA/DFMECA, and PFMEA/PFMECA. Apart from these there are two other types of FMEA: service FMEA and SWFMEA. In service FMEA the focus is on service issues. SWFMEA (discussed separately in later clauses) focuses on software issues. Functional FMEA/FMECA is also known as concept FMEA/FMECA or system FMEA/FMECA. Some literature shows two types of FMEA/FMECA in the sense that one is functional and other is hardware FMEA/FMECA (where both PFMEA/PFMECA and DFMEA/DFMECA are considered under hardware FMEA/FMECA). These different types are a result of changes in analysis pattern and assessment, but the basic concepts/approaches are the same.

- Functional (/system/concept) FMEA/FMECA: This actually determines the correct failure step. It is done at the conceptual stage when design is not frozen to fix hardware. Naturally, at this stage there will be functional analysis. The following points may be noted about this type of FMEA/FMECA:
  - This is done on a function/conceptual basis at an early stage, when hardware details are not defined.
  - It determines the potential failure associated with functions of the conceptual proposal.
  - It focuses on the functional interface between various systems and subsystems, that is, interaction among multiple elements (even component level interfacing with other systems) and systems including human interface.

Table IV/2.0.3-1 Differences Between FMEA and FMECA

Point of Difference	FMEA	FMECA
Approach	It is the primary step for FMECA	It is more apt for hazard control
Determination	It is determined by the risk priority number (RPN) (see Clause 2.0.8)	Qualitative: Comparing failure mode in criticality matrix (discussed later) Quantitative: Item criticality = Σmode criticality (MC) where: MC = Unreliability × probability of loss × Mode ratio of unreliability Criticality matrix is necessary
Criticality	No criticality	Criticality addressed and ranks them to help to take action based on priority
Criticality matrix	Here criticality matrix is not required	Criticality matrix is necessary
Type	It is mainly used for manufacturing processes than to implicate design processes. More concerned with process and design	It is used for the system, hence requires system operation information. More concerned with operator safety
Failure	It can account for multiple failures	It analyses individual failure one by one
Human factor	Limited human factors addressed	Human factors not addressed
Nature	It emphasizes more on prevention	Emphasis on detection and control
Level	Improves production quality, production process	Helps to develop planning such as logistic support, maintenance and test planning
Value addition	Reduces nonvalue added operation look for customer satisfaction	Reduces nonvalue added operation. Internal customer satisfied
Time	Less time	More time
Information	Cross-functional team is necessary	Information from cross-functional team is required but machinery information is more important

- It is meant to ensure that both physical and irreversible damage is not propagated on account of a single failure.
- It ensures safety and system integration.
- It looks into the relationship between various units/subassemblies, etc. to ensure there is no conflict or damage caused by the propagation of any break/failure. It also looks into various services in the production process and human interface, as shown in Fig. IV/2.0.4-1A.

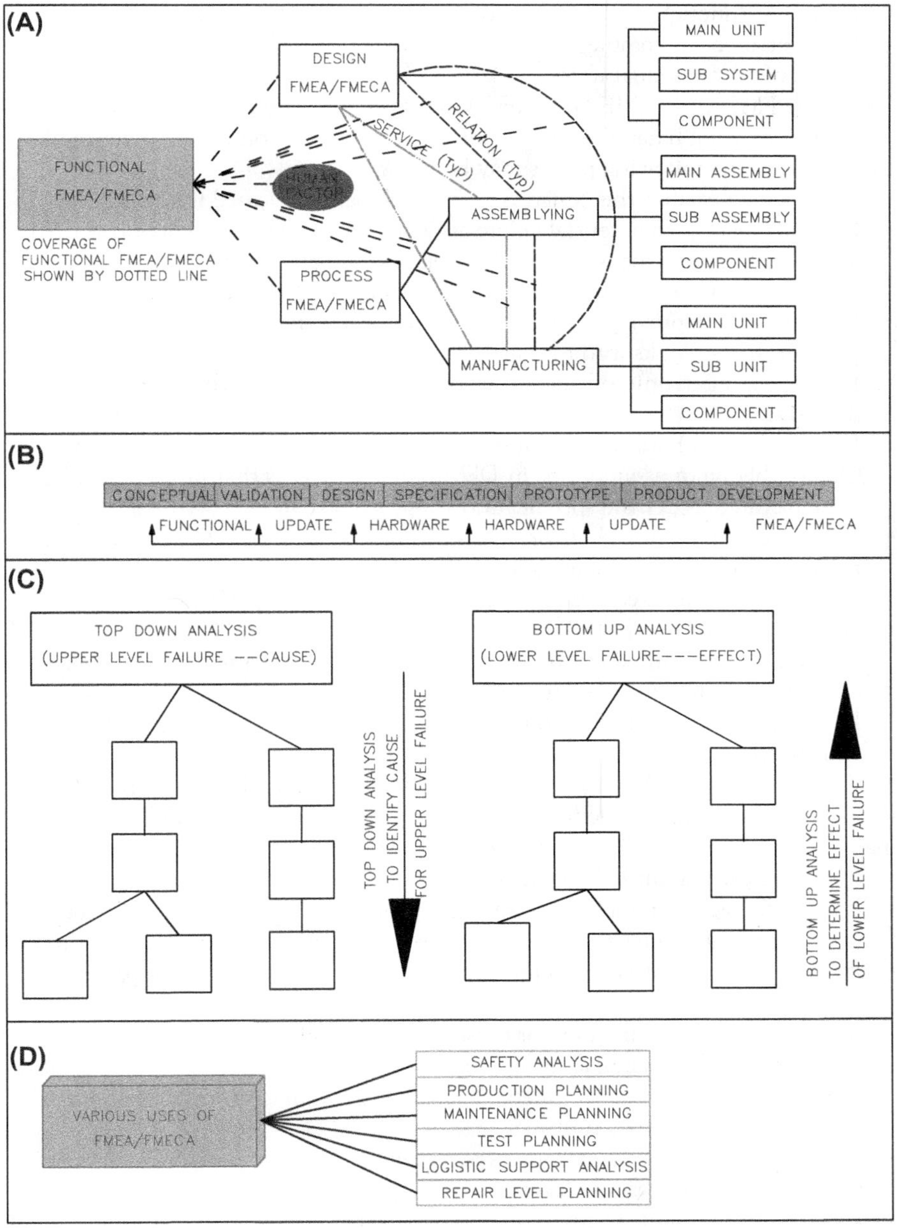

FIGURE IV/2.0.4-1

Types of FMEA/FMECA. (A) Relation amongst FMEA/FMECA types, (B) design FMEA/FMECA concept, (C) process FMEA/FMECA concept, and (D) uses of FMEA/FMECA.

   - Normally, functional analysis is "top down" (Fig. IV/2.0.4-1C).
   - It can be used for complex system analysis when it is *not* possible to identify failure of all individual items.
- DFMEA/DFMECA: DFMEA/DFMECA is used to identify the design risks including potential hazards, possible failures, and degradation of performance of the product. The following points may be noted:
   - It is carried out after the design is completed, before released for production.
   - It mainly focuses on potential failure associated with proposed function of the application. It also focuses on design deficiencies, possible failures, and degradation of performance of the product.
   - It sees how to improve product design, system safety, reliability, and life of the product. It also looks into interface issues.
   - It is done at main unit, subsystem, and component levels, as shown in Fig. IV/2.0.4-1A.
   - It analyses both hardware and function, as well as a combination of both.
   - One of the major assumptions in DFMEA/DFMECA is that product will be manufactured as per the specification.
   - It is carried out with the help of a cross-functional team. Also there is prework to do so that the time taken is less.
   - It should be completed concurrently with design process. Various check points pertinent to DFMEA/DFMECA are depicted in Fig. IV/2.0.4-1B. All such checks are carried out during the preliminary design review meeting, critical design review meeting, etc.
   - It can be a top-down and bottom-up analysis (as shown in Fig. IV/2.0.4-1C), whichever is suitable.
- PFMEA/PFMECA: PFMEA/PFMECA is used at the assembly line or in a manufacturing process to identify deficiencies related to assembly and/or manufacturing process. The following points may be noted:
   - It may be carried out before the design is finally released for production.
   - It is conducted when DFMEA/DFMECA provides special characteristics for a new technology, a new tool in use, or for planning process control for a plant.
   - It is used to look into certain processes involving customer care or patient care in medical equipment.
   - It focuses on potential failure or possible failure in the process of the assembling and/or manufacturing process.
   - It gives special attention to safety issues in any manufacturing process step.
   - It improves the manufacturing process.
   - It ensures manufacturing is done as per design requirements in a safe manner with minimal down time and rework.
   - It takes into account incoming materials, transporting, and storage tool maintenance.
   - It assumes that the design is sound.
   - It also addresses operator errors in the assembling/manufacturing process.

2.0.5 Timing Considerations

Initially, FMEA is carried out at the conceptual or design stage. Throughout the entire product development cycle, changes and updates are inevitable for the product and process. These changes may introduce new failure modes. Therefore it is important to review and update the FMEA in special cases, such as when a new product or process is being initiated or changes are made to the operating conditions the product (or process) is expected to function in. The product and process are interrelated, which means when there is a change in the product design, the process is impacted and vice versa. So, when a change is made to either the product or process design, FMEA needs to be carried out. The FMEA is a living document. Therefore it is important to review and/or update the FMEA when:

- A new product or process is being initiated (at the beginning of the cycle).
- Changes are made to the operating conditions the product or process is expected to function in.
- A change is made to either the product or process design.
- New regulations are instituted.
- Customer feedback indicates problems in the product or process.
  But in this case the basic questions should be answered by the FMEA. These questions are:
  - How can each part fail?
  - What are the mechanisms for these modes of failure?
  - What are the effects of these failures?
  - Is the failure safe or unsafe?
  - How is the failure detected?
  - How is the failure compensated for?
  - What are the preventive actions?
  - What are the actions taken to mitigate failure?

As discussed, at different stages of the production cycle there will be an FMEA exercise. Naturally, the details with which FMEA/FMECA at different stages will be carried out will vary and accordingly the results will also vary.

2.0.6 Application, Benefits, and Limitations

Where could FMEA/FMECA be applied and what is it for? These are very pertinent questions, which are answered when one goes through the application area of FMEA/FMECA. There are a few distinct advantages or benefits from FMEA/FMECA. However, there are limitations too. All these are discussed next to get a good appreciation for the system:

- Application of FMEA/FMECA: FMEA/FMECA contributes to improve product and production process to achieve better quality, better reliability, enlarged efficiency with increased safety, and enhanced customer satisfaction at comparatively lower cost. Therefore FMEA/FMECA is a tool that has been adapted in various ways for different applications and purposes, some of which are listed here:

- Evaluate the requirements obtained from the customer or other participants in product development and manufacturing process to attain quality product without potential failure
- Develop corporate memory and the success of future products through tracking and managing potential risks
- Comply with the requirements of ISO 9001 discussed earlier
- Optimize maintenance and testing planning for repair and logistics management
- Address various quality control issues and as a total quality management (TQM) tool
- Help with design choices (tradeoff analysis)
- Develop and evaluate diagnostic systems
- Other industry uses: Automobile manufacturing industry, aerospace, medical, E/E/PE production, nuclear, and other manufacturing industries

- Benefit of FMEA/FMECA: The following are some of the immediate benefits obtainable:
  - The final "safe" product is the target! FMEA helps to identify and eliminate or control dangerous failure modes, minimizing damage to the system and its users.
  - Puts more emphasis on failure prevention and prevents it (or similar) happening in the future.
  - An increasingly accurate estimate of probability of failure will be developed, with the help of FMEA/FMECA.
  - Collection of information not only reduces future failure but also increases the knowledge base.
  - Quality, reliability, customer satisfaction, and safety of the product will be improved.
  - Time and cost will be reduced because of timely identification and correction of potential failures: Decreases development time and redesign costs, reduces warranty cost, and decreases waste and nonvalue -added operations.
  - Contributes to the development of control plans, testing requirements, optimum maintenance plans, reliability growth analysis, and related activities. Financial benefits are also a result of the benefits discussed.
  - Team work and idea exchange will be developed.
  - Improved production yield results better profitability.
  - Improves company image and provides an edge to the company for competitiveness.
- Limitations: There are also a few limitations of FMEA/FMECA:
  - When used in top-down mode it can only identify major faults/failures but not all—especially when compared with fault tree analysis (discussed in the next chapter). When used in bottom-up mode it can augment fault tree analysis.
  - It reveals so many failure scenarios, which otherwise may not be significant, so some wastage of effort!

- It does not take care of operator fault or *human* factor.
- As long as a process is operating in the normal way it cannot sense any danger.
- Critical failure modes and causes and effects not *recognized* by the designer will not be addressed
- It is not able to discover complex failure modes involving *multiple* failures within a subsystem.
- It does not address multiple failures and effect individually, hence it does not combine failure and its effect will not be addressed.
- Since in FMEA/FMECA corrective action and risk ranking are done on the basis of RPN, which is a product of severity, occurrence, and detection, it is possible that risk with less severity may get more priority (through RPN) than risk with more severity.
- It is unable to report expected failure *intervals* of particular failure modes up to the upper level subsystem or system.
- It is a cross-disciplinary team effort, but when it is dominated or turns out to be a one man show, then the result may not be good.
- Customer satisfaction is a benefit of FMEA/FMECA, but if such satisfaction does not cause improvement in the product, then FMEA/FMECA is not fruitful.
- Timing is very important in FMEA/FMECA; if it is too late then the purpose will be defeated.
- It is a long process and time may not yield good results.

2.0.7 Ground Rules

One of the criteria in developing an FMEA/FMECA procedure is to establish ground rules. The analysis team leader must clearly document ground rules and assumptions, and get them approved prior to starting the FMEA/FMECA analysis. Normally, ground rules are applicable to system/equipment, its environment, and objective and analysis procedure. When a piece of equipment is excluded, then the reason must be documented for future use. Ground rules are an agreement for level/*depth* of analysis, any specific criteria for system/equipment, equipment condition, and the mission of the project, as discussed earlier. These are project specific, but in generic terms they are:

- All inputs including software are at their normal value.
- Normal utilities are available.
- Consumables are present in sufficient quantity.
- Only *one* failure mode exists at a time.
- Objective or mission of the item under analysis (e.g., printed circuit assembling part)
- The phase of the mission, the analysis will consider (suppose a product goes from shop 1, 2...) then for which phase the analysis to be carried out)
- Operating time in the mission
- Severity category for classification
- Failure mode distribution

- Fault detection concept and methodology
- Vendor data for source of part failure rate

2.0.8 Standards and Standard Terms

In this clause, standards/guidelines for the analysis of and general terms in use in connection with FMEA/FMECA shall be discussed.

- Standards and guidelines: There are several applicable standards and guidelines available for FMEA/FMECA analysis, some of them are listed here. One may note that these are not standards for FMEA/FMECA alone. Also many of them are for reliability engineering and some may give guidelines for FMEA/FMECA:
  - IEC 60812 "Procedures for failure mode and effect analysis (FMEA)"
  - MIL-STD 1629 "Procedures for performing a failure mode and effect analysis"
  - BS 5760-5 "Guide to failure modes, effects and criticality analysis (FMEA and FMECA)"
  - SAE ARP 5580 "Recommended failure modes and effects analysis (FMEA) practices for non-automobile applications"
  - SAE J1739 "Potential failure mode and effects analysis in design (design FMEA) and potential failure mode and effects analysis in manufacturing and assembly processes (process FMEA) and effects analysis for machinery (machinery FMEA)"
  - MIL-STD -785 "Reliability program for systems and equipment development and production"
  - SEMATECH (1992) "Failure modes and effects analysis (FMEA): a guide for continuous improvement for the semiconductor equipment industry"
  - Other industry-specific guidance may be available from: MIL-STD 1543, IEEE std 352:1975, ANSI N411.4:1976, and SAE G-11
- General terms in FMEA/FMECA: The following terms are commonly used in FMEA/FMECA; short discussions on these are also presented to understand the meaning associated with them.
  - Corrective action: A documented design, process, or procedure change used to eliminate the cause of a failure or design deficiency. It could also mean the action to mitigate the risk of hazard caused by failure (effective reduction of severity or probability of occurrence).
  - Criticality: A relative measure of the consequences of a failure mode and the frequency of its occurrence.
  - Criticality analysis: A procedure by which each potential failure mode is ranked according to the combined *influence* of severity and probability of occurrence.
  - Detection: The detection of failure mode by a built-in system, maintainer, and/or operator including its dormancy period.
  - Detection method: The method by which a failure can be discovered by the system operator under normal system operation or by a maintenance crew carrying out a specific diagnostic action.

- End effect: The consequence a failure mode has upon the operation, function, or status at the highest indenture level.
- Failure cause: The physical or chemical processes, design defects, quality defects, part misapplication, or other processes that are the basic reason for failure or that can initiate the physical process by which deterioration proceeds to failure.
- Failure effect: The consequence a failure mode has upon the operation, function, or status of a system or equipment or the adverse effect the customer may face.
- Failure mode: The way in which a failure is observed. It describes the way the failure occurs, that is, the manner in which the product/process fails to perform the intended function. It is the result of failure mechanisms.
- Indenture levels: An identifier for system level. Levels identify or describe the relative complexity of an assembly or function. Complexity increases as it is closure one to failure point.
- Local effect: The consequence a failure mode has on the operation, function, or status of the specific item being analyzed.
- Mission phase operational mode: The statement of the mission phase and mode of operation of the system or equipment in which the failure occurs.
- Next higher level effect: The consequence a failure mode has on the operation, function, or status of the items in the next higher indenture level above the specific item being analyzed.
- Occurrence: This is related to probability of the failure mode and cause. It is expressed in terms of probability of occurrence.
- RPN: This is expressed in FMEA/FMECA in terms of product of severity, occurrence, and detection ranking in the scale of 0–1000.
- Severity: The measure of seriousness of the effect of failure mode (consequence of a failure classified by the degree of injury, property damage, system damage, and mission loss that could occur); it is described in FMEA/FMECA in the scale of 1–10.
- Single point failure: The failure of an item that can result in the failure of the system and is not compensated for by redundancy or alternative operational procedure.

2.0.9 General Discussions

- TQM compatibility: FMEA is also often considered in terms of quality control. FMEA can accommodate various elements of total quality (control) management (TQM):
  - Customer requirement
  - Engineering specification
  - Component specification
  - Suppliers' requirements and control
  - Development of DFMEA with improvement
  - Development of PFMEA with improvement

- Identification of potential failure
- Control measures and reduction of failure

As discussed at the beginning, FMEA is considered as a part of reliability, safety, and quality engineering and it started its journey a long time ago.

2.1 FMEA/FMECA Methodology

At the outset it is important to fix the target for which the study is undertaken. Normally, these targets are: environment, equipment, personnel, product, and product development. For each target, the approach may be different. These are necessary for project definitions, as shown in Fig. IV/2.1-1. Also it is to be noted that FMEA/FMECA is best suited for subsystems rather than an entire generic system. It is almost impossible to carry out FMEA/FMECA on an entire semiconductor manufacturing process but better to carry it out on a semiconductor product in a manufacturing unit and achieve greater customer satisfaction. Therefore it is very important to fix the target as well as study boundary. Requirements for a complete definition of the system shall include but not be limited to the following:

- Identification of internal and interface functions for:
  - Each mission
  - Mission phase
  - Mission times
  - Operational modes
  - Primary and secondary mission objectives
- System performance at each indenture level
- System restraints
- Failure definitions

These descriptions of the system should describe:

- Service use profile
- Equipment utilization
- Expected mission time
- Function and output of each item

Conditions that constitute part and full system failure should also be determined. The system indenture levels must be identified to complete the FMECA.

Like HAZOP, here also a team is entrusted to carry out the study. Another important point to be remembered is that this is usually a *live* document so that any changes, etc. that can affect safety/risks can be *reassessed* with the help of the analysis. So it is not a case of once it is done the responsibility is over; it needs to be updated. FMEA/FMECA methodology is presented in Figs. IV/2.1-1 and IV/2.1-2. It is worth noting that these two drawings should be **viewed** in conjunction, as one is related to the other. Fig. IV/2.1-3 is presented to depict PFMEA and its relation with DFMEA.

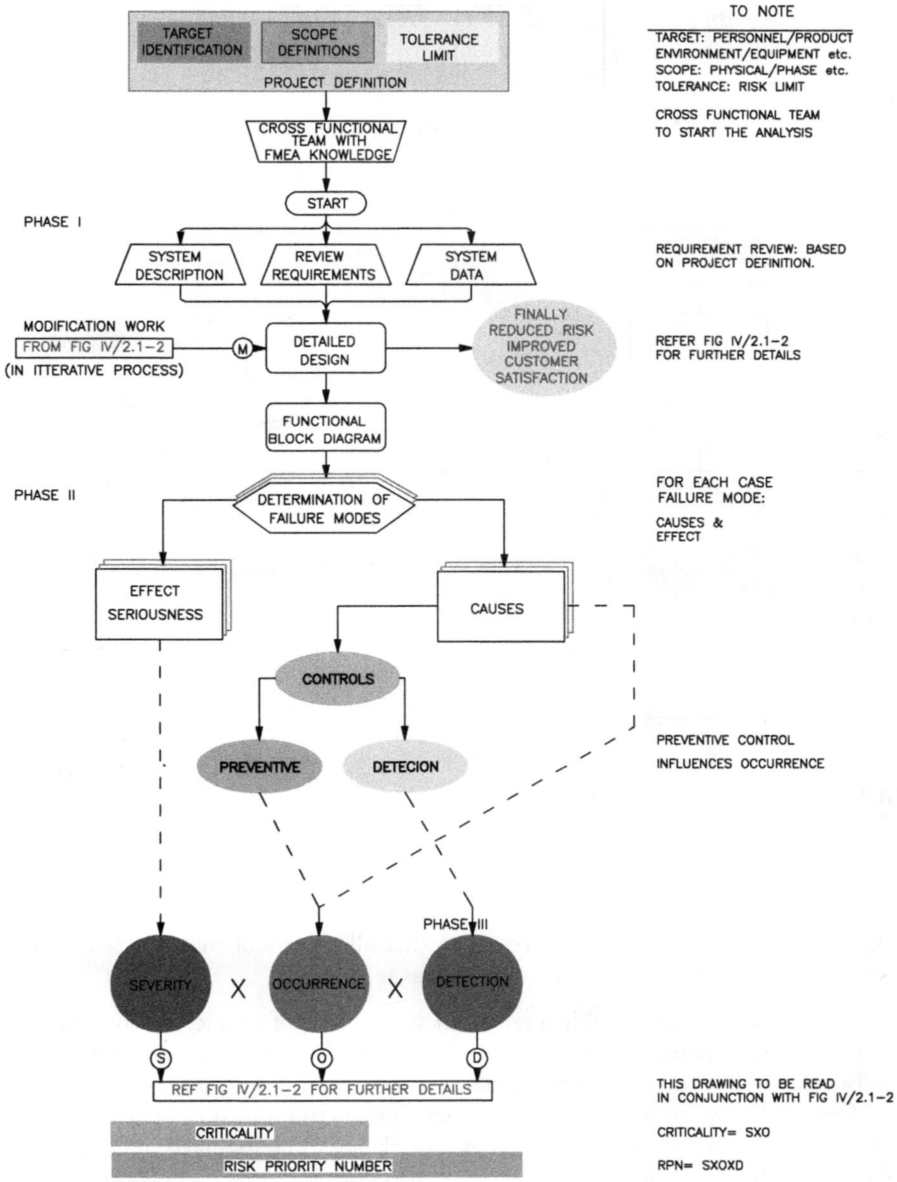

FIGURE IV/2.1-1

FMEA/FMECA process flow 1.

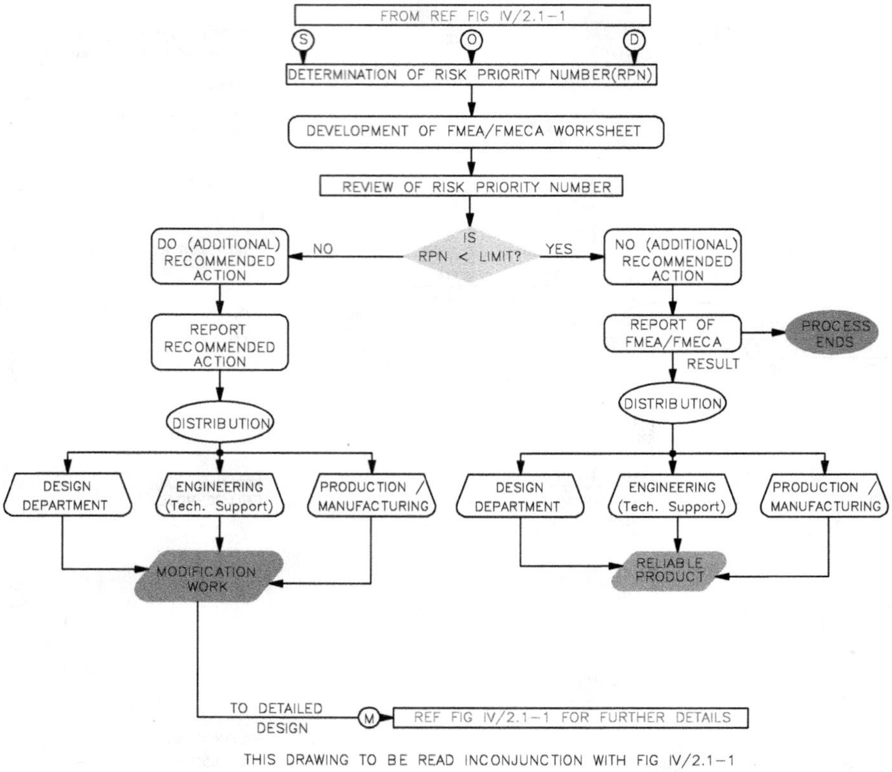

FIGURE IV/2.1-2

FMEA/FMECA process flow 2.

2.1.1 *Team*

Here, a cross-functional team is necessary so that all aspects of the process can be addressed. The major points in this regard are as follows:

- It is recommended to assemble a cross-functional team of people with diverse knowledge on different aspects such as product, product process, safety, service, software, and marketing pattern including customer requirements and preference. Everyone involved in the process/production needs to be involved. However, it does not mean that *all* these people are to be involved in *all* meetings, which will cause wastage of time and effort. However, people involved in any step of the process need to be present in the discussions of that step, so that advice and feedback from the persons concerned are available.
- The team should cover the following functional aspects:
  - Design and engineering
  - Purchase

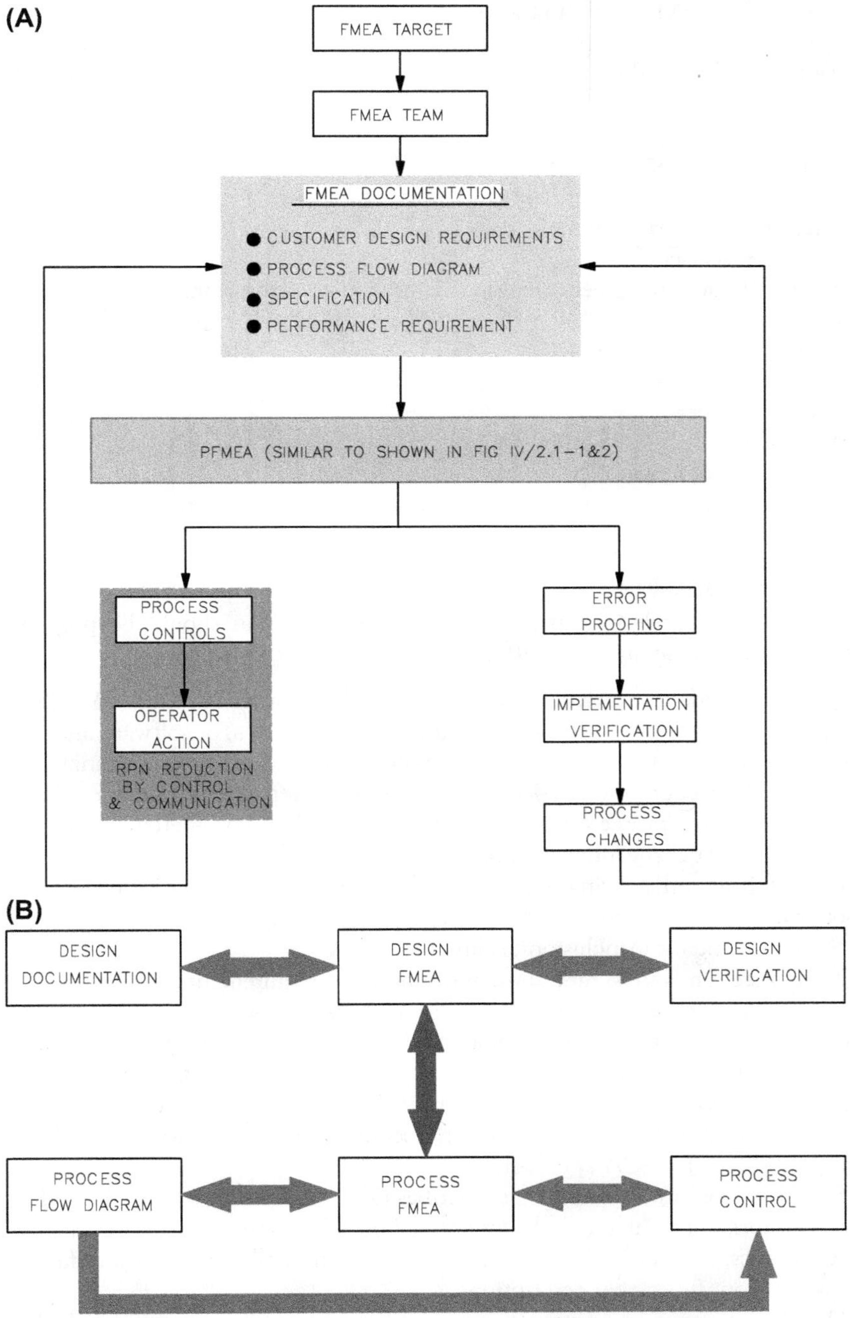

FIGURE IV/2.1-3

PFMEA and relationship with DFMEA. (A) Process FMEA, (B) PFMEA relationship with design FMEA.

- Manufacturing/production process
- Quality control
- Testing and reliability
- Software (as applicable)
- Maintenance
- Sales
- Marketing
- Technical support
- Customer services

- Generally, the following personnel may be included in the team:
  - Design engineer
  - Quality control engineer
  - Production engineer
  - Safety engineer
  - Necessary people from other departments
  - Others (as applicable)

  The listed people are only indicative and there may be changes as per requirements.

2.1.2 Prerequisites

Before starting the analysis it is necessary that a few issues should be properly addressed, so that the analysis is fruitful and not a waste of time and effort.

- Scope and boundary: For defining the scope and boundaries of FMEA/FMECA the major questions are: Is it for conceptual, design, process, or software and services? Also the purpose of the study shall be questioned. The scope of analysis shall take into account the physical boundaries, operating phases (operational or startup/shutdown phase, etc.), and any other assumptions considered in the referendum. In brief the following points shall constitute the scope and boundary of analysis. It is worth noting that all **interface** points should be included in the scope even if these are beyond the physical boundaries defined.
  - System boundaries (inclusion/exclusions)
  - Main system missions, functions, and functional requirements
  - Operational and environmental conditions to be considered
- Information/documentation requirements: For a successful FMEA/FMECA analysis a number of documents are necessary. The variety of information/documentations necessary for DFMEA/DFMECA and PFMEA/PFMECA will be different. Also there will be variations in documentations for functional and hardware FMEA/FMECA approaches:
  - System-related information that describes the system needs to be collected. This information shall include but not be limited to: drawings, specifications, schematics, layout drawings, component lists, terminal points and interface information, functional descriptions, etc.
  - Detailed information on previous/similar designs from various sources, including failure reporting and analysis corrective action system data,

interviews, etc. Sometimes similar risk analysis projects by other methods may be helpful.

- Warranty and recall information
- Plant problems, specific technical problems of production including safety-related issues, case reports on scrap and rework including bad quality report
- Depending on applicability, software problems and related issues in the plant or in similar plants
- Service problem report as applicable
- Other documentations shall include but not be limited to those listed in Table IV/2.1.2-1.

***Table IV/2.1.2-1 Documentation for Hardware and Functional FMEA/FMECA* (Typical Indicative Only)**

Hardware FMEA	Functional FMEA
Complete process knowledge	System definition document
Reliability and functional block diagram	Reliability and functional block diagram
Boundary block diagram	Boundary block diagram
Schematics and wiring	Ground rules and assumptions
Process flow diagram (as applicable)	Software specification
Total parts and component lists	Functional details
Process control description/ philosophy/lists	Operator actions
Interface points with short discussions	Interface point relations

2.1.3 FMEA/FMECA Approaches

As discussed earlier in Clause 2.0.4, there are two different approaches for FMEA/ FMECA with different purposes. One approach is top down, while the other is bottom up. Complexity of design, development stages, and amount of data availability actually *determine* which approach is better suited. In some complex large systems it is not uncommon to use both in tandem. These approaches can be applied at any level. Functional analysis may be considered as an input for determining failure mode in both approaches. The top-down approach is better suited for functional analysis.

- Functional approach: In this approach the functions of each unit are considered for analysis after classifying them each with numbers of failure modes. When exact hardwires in the systems are not properly known/identified, or for complex systems, this approach is utilized. Also it can be *tailor* made to apply to those cases where design is immature or part designed subassemblies are considered. However, for each of such cases the type of approach and level of analysis need to be properly tailored. Development of a functional block diagram (FBD) is helpful in this approach. From the FBD, analysts define and identify the functions and associated failure modes. Analysis is completed by determining potential failure modes, their causes and effects through brainstorming methods. For criticality analysis it is necessary to determine potential failure *probability* in terms of % of time and *failure rate* in terms of numbers per million hours. These two will determine occurrence probability through multiplication. Effect or severity is determined by *propagating* the potential failure of each functional unit through each of their higher-level indentures. This again is a brainstorming method. Further analysis can be done by utilizing the critical matrix and/or quantitative method. Now, failure isolation, compensation and detection, etc. are recorded for determination of RPN. After each functional block is analyzed in any of the indenture levels, output is recorded. Output of the analysis will be in the form of a list containing hazards (to be eliminated or reduced), undetectable features, critical single point failures, etc.

- Hardware approach: In the previous approach, immature design was referred to. In this connection it is worth noting that if in the immature design some subassemblies are fully developed, then for these subassemblies the hardware approach will be a better choice. In the hardware approach, hardwires are identified in the associated drawings. The hardware approach is normally used in a bottom-up manner, that is, analysis starts at the lowest level and then slowly moves upward according to indenture levels. Here, all the hardwires with their unique identifier are listed. This list shall also include for each item their potential failure modes. This approach is only *possible* after designs and documentations are made available. In this connection, Table IV/2.1.2-1 may be referred to. Analysts complete the analysis by identifying potential failure modes and causes of failure of each hardware item in the system. As discussed, effect or severity is determined by propagating the potential failure of each functional unit through each of their higher-level indentures in a brainstorming manner. Further analysis can be done utilizing critical matrix and/or quantitative methods. Now, failure isolation, compensation and detection, etc. are recorded for determination of RPN. Output of the analysis will be in the form of a list containing hazards (to be eliminated or reduced), undetectable features, critical single point failures, etc.

2.1.4 *Failure Mode and Effect Analysis—Process and Procedure*

To understand the system it is better to refer to Figs. IV/2.1-1 and IV/2.1-2 (to be viewed in conjunction as they are related to each other). The major steps involved are listed in Table IV/2.1.4-1, which may also assist in reading the drawings.

- Structuring of the system: Depending on the objective of the analysis it is better to divide the system within the boundary (including interfaces) into subsystems. At times it is better to divide the systems in terms of an FBD (especially for a functional approach). Such subdivisions are shown in Fig. IV/2.1.4-1.

Table IV/2.1.4-1 Major FMEA Steps (Figs. IV/2.1-1 and IV/2.1-2)

Phase	Step	Description	Remarks
–	–	Project definition and team formation	With scope and boundary
I	1	Prerequisites: Information collection and FBD development	FBD = functional block diagram
II	2	For each functional unit/hardwire identify how the failure can happen	Potential failure mode determination
II	3	For each failure mode identify the root cause and consequences in the system, customer, product production process, software (as applicable), personnel safety, environment, regulation, and services. Effect is determined in terms of severity (S) and occurrence probability (O) for each cause	Here, effect is determined by severity rating in terms of (say) 1–10. For occurrence probability a 1–10 scale may be used
II	4	There will some existing process controls to prevent/reduce some of the failures. To see how it affects the occurrence probability and how it detects the cause of or the failure mode, The detection rate (*D*) is determined	Here also a scale of 1–10 is used
III	5	Determination of risk priority number (RPN) = S × O × *D**	*Criticality = S × O
IV	6	Check RPN with limit set for the project.	Fig. IV/2.1-2
V and VI	7 and 8	Now there are two options when RPN is within the limit (for each case) Recommended actions are not necessary, otherwise there will be some recommended action. Based on the recommended action/control, necessary measures are taken and implemented. New RPN is calculated with recommended action to see RPN is within the limit	After the same again the above process is reiterated

For each of these blocks the analysis will be carried out and results will be produced as a part of the worksheet. These are the examples of the block diagram mentioned in Fig. IV/2.1-1. Potential failure modes are determined by scrutinizing the output or functional output identified in the block diagram thus constructed.

- Failure mode identification: What is failure mode? A failure mode is the manner in which the item or system or their operations potentially fail to meet or deliver the intended task or function and/or associated requirements. Failure mode identification of each of the items along with interface is necessary. As stated,

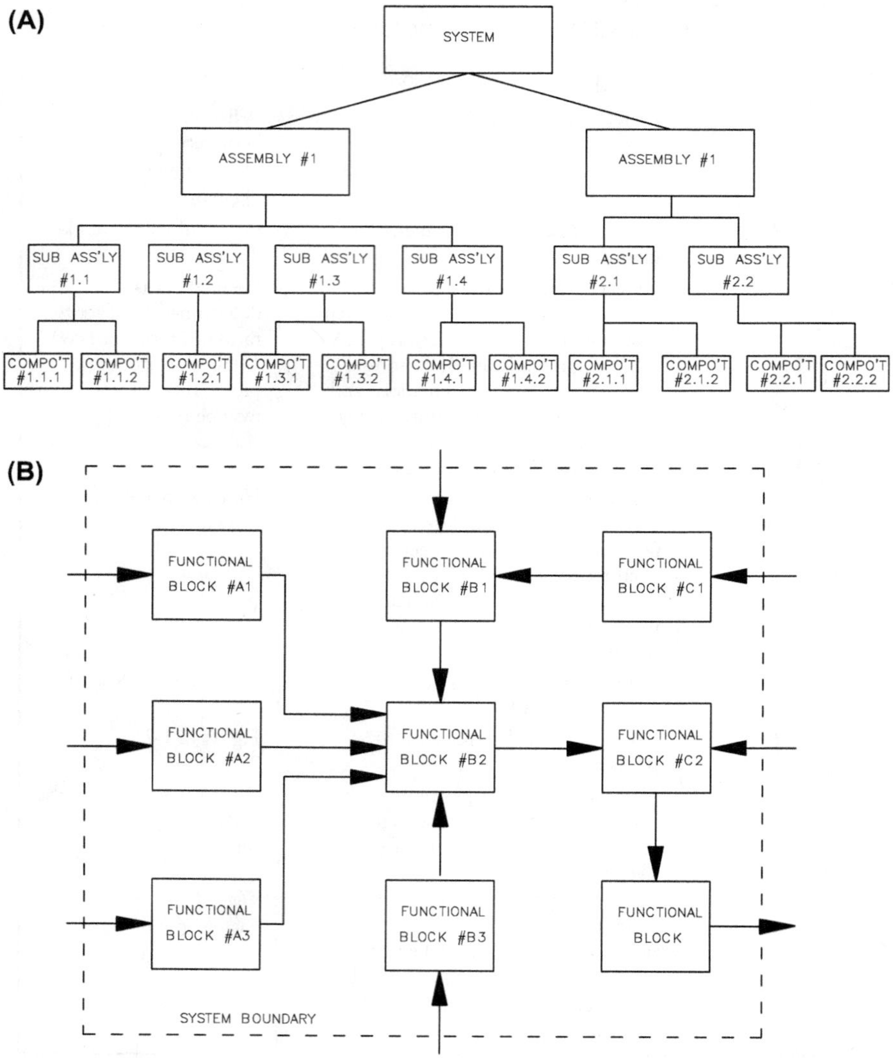

FIGURE IV/2.1.4-1

System structure for FMEA/FMECA. (A) Division of systems for FMEA/FMECA, (B) functional block diagram for FMEA/FMECA.

this is done by scrutinizing the output of each item. It is important that all probable failure modes of each item are identified. Not only identification of all probable failure modes of each item including interface is necessary, but *also* their effect on immediate function, overall system, mission, etc. needs to be assessed.

The following generic check points are important. However, for a specific project there may be variations in the checklist.

- Premature operation
- Performance degradation
- Failed to perform within given limit
- Performed undesired function
- Failed to operate in time
- Wrong timing in sequential operation
- Intermittent operation
- Failed to start or stop in proper time
- Loss of output
- Degraded output
- Lower operational capability

This list is necessary for critical analysis to determine failure mode probability, as discussed earlier.

- Cause: Cause is the specific reason for the failure and can be found by asking "*why?*" In DFMEA, design deficiency and lack of specification are the major causes of various failures. Naturally, manufacturing process/subassembly process deficiencies and lack of safety measures could be the cause of failure. One needs to explore the issue until the root cause is identified.
- Effect/consequence analysis: Effect is the consequence (already discussed in Chapter I) of failure on the system, end user, and environment. Each failure mode can have multiple effects on the system or its parts. There can be a single-level effect and/or three-level effect, such as on local, next higher level, and end effect (Table IV/2.1.4-2). As discussed earlier, the consequence or effect of failure of each item *must* be studied in the next higher levels. For PFMEA/PFMECA, effect needs to be considered at manufacturing, subassembly level as

Table IV/2.1.4-2 Effect/Consequence Levels and Categories

Item	Description
Level	
Local	Effect from the failure mode affects the level or indenture under consideration
Next higher level	Consequence on the next higher level to the level of indenture under consideration
End	Effect of failure on operation and function/status of the whole system
Category	
System failure	Complete catastrophic failure of the operation of the system as a whole
Degraded operation	Although system objective or mission is not affected it has an effect on system operation
System status fail	Operator is unaware of actual system status
No immediate effect	No effect on system operation

well as for the system. The failure may affect several indenture levels as well as the mission. Even personnel safety, maintenance issues may be affected. Analysts need to apply all their efforts to find these effects. There are several categories in which the consequences are to be judged. These are elaborated in Table IV/2.1.4-2. It is sometimes possible that the effect may not fall into any of the categories discussed in Table IV/2.1.4-2; in these cases the specific project may have to be tailored.

2.1.5 Criticality Analysis

As seen in Chapter I, criticality is determined by combining severity with occurrence. Criticality analysis starts as an integral part of the early design and is updated as the design progresses. Criticality analysis throws light on a relative measure of significance of the effect a failure mode has on the successful operation and safety of the system. To perform a criticality analysis, it is necessary to complete consequence analysis to get the effect of failure mode on various indenture levels. In fact, to perform criticality analysis it is necessary that FMEA is complete and information on the system such as system mission, definition of failures, severity categories, part failure rate information, etc. is available. Criticality analysis can be completed using either a qualitative or quantitative approach. The level of availability of part configuration and failure rate data will determine the analysis approach to be used. The qualitative approach is used when specific part or item failure rates are not available. The quantitative approach is used when there is sufficient failure rate data available to calculate item criticality numbers.

2.1.6 Criticality Matrix (Qualitative Approach)

Occurrence classification/severity classification—from the discussions in Chapter I, it is seen that occurrence is a ranking number associated with the likelihood that the failure mode and/or associated causes of failure will occur. DFMEA/DFMECA looks for occurrence during the design phase of the product, whereas in PFMEA/PFMECA, the same is applicable during the production process. FMEA identifies failure modes in terms of probability of occurrence. These are actually relative values rather than absolute values, because they are measured in a defined scale. The probability of occurrence of each failure is grouped into discrete levels:

- Level A—Frequent
- Level B—Reasonably probable
- Level C—Occasional
- Level D—Remote
- Level E—Extremely unlikely

Severity is the (relative) ranking of seriousness of effect for each failure mode. This will be determined independently of occurrence classification. So, in severity analysis, severity is categorized as follows:

- Catastrophic I Death or system loss.
- Critical II Severe injury, severe occupational illness, or major system damage.
- Marginal III Minor injury, minor occupational illness, or minor system damage.
- Minor IV Less than minor injury, occupational illness, or minor system damage.

From these two data categories a criticality matrix as detailed in Chapter I (in the name of risk matrix) is formed. In many cases, both probability of occurrences and severity categories are numbered in the scale of 1—10 and a criticality matrix can be formed for the same purpose. As these are already covered in Chapter I they are not repeated here. Analysts need to understand that these are for reference only. Analysts use their judgment of failure mode frequency for each specific application. The analyst should tailor the analysis to focus on significant components or subassemblies where failures will result in undesirable system-level effects. Based on these judgments, analysts develop risk categories for the specific application.

2.1.7 Quantitative Approach for Criticality Analysis

In the quantitative approach, alpha and beta values representing failure mode ratio and failure effect probability, respectively, are necessary. Alpha represents the probability, expressed as a decimal fraction, that the given part or item will fail in the identified mode. Beta represents the conditional probability that the failure effect will result in the identified criticality classification, given that the failure mode occurs. In the quantitative approach the value of each failure mode criticality number is defined as:

$$C_m = \alpha * \beta * \lambda_p * t \; [6]$$

where C_m, failure mode criticality; α, failure mode ratio; β, the conditional probability of mission loss; λ_p, part failure rate; and t, duration of applicable mission phase expressed in hours or number of operating cycles. Part failure rates for major items are available in the market or obtainable from the supplier, with the help of the equation C_m to be calculated.

2.2 FMEA/FMECA Detection and Risk Priority Number

2.2.1 Detection

Failure modes with a higher RPN value are more important than the failure modes with lower RPN value. In Figs. IV/2.1-1 and IV/2.1-2 it is seen that RPN is constituted on three factors, namely, severity, occurrence, and detection. Out of these three factors the first two constitute "criticality," discussed earlier. When criticality is multiplied by "detection" (rate) it is RPN. So, what are detection and detection rate?

Detection is related to causes of failure and controls, as shown in Fig. IV/2.1-1. Thus there are two ways to look at it: preventive and detection control. In prevention, with the help of existing controls, failure modes are prevented, whereas the other way detects the failure and takes corrective action before it reaches the customer (see Fig. IV/2.2.1-1). FMEA/FMECA identifies the method by which occurrence of failures/failure modes is detected by the operating personnel. Audio

Customer: End users, OEM plants, supplier plants, and government agencies are major customers who can contribute to develop functional requirements. Precise definition of functions, requirements, and specifications.

FIGURE IV/2.2.1-1

Customers.

visual alarm annunciation is an example of this. If there is a high level in a drum, then feed water flow to the boiler needs to be regulated to avoid the possibility of water ingress to the turbine. A drum level high alarm warns the operator for action. This is just an example of detection. *However, it is more to do with the alarm system than FMEA itself*. So, warning devices, automatic sensing devices, and tripping devices are examples of detection systems. Any other devices that could warn the operator about failure modes are also detection methods e.g., declaration of bad input in MooN selection device is also an example of fault detection. It is not always possible to get such indications. From MIL STD 1629 it is seen that FMEA/FMECA needs to determine if the failure can jeopardize the mission/objective or personnel safety, if there is no indication available to the operator. Also if, in the case of failure without indication, system operation is not affected, then FMEA/FMECA needs to explore a second possibility (if it exists), which can provide some indication to personnel! There are three types of indications:

- Normal: To indicate that system operation is safe and normal
- Abnormal: To indicate any malfunction or failure
- Incorrect: An indication that malfunctions have occurred but there is no fault. It could also be an indication that the failure has occurred but there is a *fault* in the system.

2.2.2 *Detectability*

This is the probability of a failure being detected before the impact of the failure on the system or process. Detectability is rated according to the ability to detect the effect of the failure mode or the ability to detect the failure mode itself. In other words, it is detection rating.

2.2.3 *Risk Priority Number (RPN)*

In addition to the other risk assessment tools discussed in Chapter I, an organization may choose to develop risk ranking tables based on RPN to assist the decision-making process. The RPN approach is an alternative to the risk matrix, also found in FMEA/FMECA. In FMEA/FMECA, the analyzing team assigns each failure mode numeric values that quantify likelihood of occurrence, likelihood of detection, and severity of impact. So, each failure mode has a numeric score to quantify (1) likelihood of failure occurrence, (2) likelihood of failure undetected, and (3) severity of harm or damage the failure mode may cause. All ranks are given on a scale from 1 to 10 (or 1—5). The specific rating descriptions and criteria for the ranking of occurrence (O), severity (S), and detection (D) are defined by the organization, FMECA standard, and/or the analysis team to fit the products or processes that are being analyzed. Fig. IV/2.2.3-1 shows typical S,O and D in scale of 10, for RPN calculation facility. Here, one thing worth noting is that in the case of occurrence likelihood, generally, component failures in E/E/PE, etc. are expressed in terms of once in number of years, whereas in other cases it is expressed in terms of number of items failed per (say) 1000 items.

When using this risk assessment technique, it is important to remember that RPN ratings are relative to a *particular* analysis. Therefore RPN in one analysis is

RANK *	SEVERITY [SCALE1(LEAST) TO 10 (MOST) *ZERO SCALE NOT DEFINED
1	MINOR FAULT, NOT ANY DISCERNABLE EFFECT ON PRODUCT/SYSTEM PERFORMANCE, AND CUSTOMER MAY NOT NOTICE AT ALL!
2, 3	FAULT CAUSES LITTLE DETERIORATION OF SYSTEM /PRODUCT PERFORMANCE & MAY BE NOTICED BY DISCRIMINATING CUSTOMERS (~25%) WITH LITTLE ANNOYANCE AND DISSATISFACTION. ALSO MINOR REWORK MAY BE CALLED FOR.
4, 5, 6	FAULT CAUSES SOME DETERIORATION OF SYSTEM /PRODUCT PERFORMANCE, MAY BE NOTICED BY MOST CUSTOMERS (~75%) WITH ANNOYANCE AND DISCOMFORT. SOME PART MAY CALL FOR REWORK, SOME PART (<100%) SCRAP, DAMAGE OF EQUIPMENT. UNSCHEDULED REWORK AND SOME WASTAGE.
7, 8	FAULT CAUSES GREAT DETERIORATION OF SYSTEM /PRODUCT PERFORMANCE, INCLUDING NON FUNCTIONING/ INOPERABLE. HIGH CUSTOMER DISSATISFACTION. EVEN MAY CALL FOR TOTAL REWORK, UP TO 100% SCRAP, DAMAGE OF EQUIPMENT. SERIOUS DISRUPTION OF WORK HOWEVER NO VIOLATION OF REGULATION/ GOVT. NORM.
9, 10	OPERATOR/ EQUIPMENT SAFETY ENDANGERED WITH (9) /WITHOUT (10) WARNING. GOVT./ REGULATION VIOLATED

RANK *	OCCURRENCE LIKELIHOOD [SCALE1(LEAST) TO 10 (MOST) *ZERO SCALE NOT DEFINED
1	EXTREMELY UNLIKELY AND FAILURE HIGHLY UNLIKELY. 1 IN 1.5M
2	RARE LIKELIHOOD OF FAILURE. 1 IN 150K
3	VERY LOW LIKELIHOOD OF FAILURES I IN 15–30K
4	LOW LIKELIHOOD OF FAILURE 1 IN 2–5K
5	OCCASIONAL FAILURE LIKELY 1 IN 400–800
6	MEDIUM NUMBERS OF FAILURES LIKELY 1 IN 80–150
7	MODERATELY HIGH NUMBER OF FAILURES LIKELY 1 IN 20–50
8	HIGH NUMBER OF FAILURES LIKELY 1 IN 8–10
9	VERY HIGH NUMBERS OF FAILURE LIKELY 1 IN 3 TO 6
10	FAILURE ALMOST CERTAIN. (1 IN 3 OR LESS)

OCCURRENCE MAY BE EXPRESSED AS FAILURE PER UNIT OF TIME e.g.1IN1000 YEARS OR 1PER MONTH.

RANK *	DETECTION LIKELIHOOD [SCALE1(LEAST) TO 10 (MOST) *ZERO SCALE NOT DEFINED
1, 2	VERY HIGH PROBABILITY THAT WITH KNOWN CONTROL/VERIFICATION EXISTING DEFECT/DEFICIENCY OR FAILURE MODE CAN BE DETECTED. ALMOST CERTAIN TO DETECT DEFICIENCY.
3, 4	HIGH PROBABILITY THAT WITH KNOWN CONTROL/VERIFICATION EXISTING DEFECT/DEFICIENCY OR FAILURE MODE CAN BE DETECTED. GOOD CHANCE OF DETECTION/ VERIFICATION DEFICIENCY.
5, 6, 7	MODERATE PROBABILITY THAT WITH KNOWN CONTROL/VERIFICATION EXISTING DEFECT/DEFICIENCY OR FAILURE MODE CAN BE DETECTED. MODERATE POSSIBILITY OF DETECTION/ VERIFICATION EXISTS
8, 9	LOW PROBABILITY THAT WITH KNOWN CONTROL/VERIFICATION EXISTING DEFECT/DEFICIENCY OR FAILURE MODE CAN BE DETECTED
10	VERY LOW PROBABILITY THAT WITH KNOWN CONTROL/VERIFICATION EXISTING DEFECT/DEFICIENCY OR FAILURE MODE CAN BE DETECTED (ALMOST ZERO PROBABILITY OF DETECTION)

FIGURE IV/2.2.3-1

Ranking of severity, occurrence, and detection for RPN.

comparable to RPNs in the same analysis but it may *not* be comparable to RPNs in another analysis. So, it is not possible to share these numbers with other applications.

- The RPN $= S \times O \times D$. Higher RPN is worse than lower RPN, that is, higher RPN numbers signify more risks. Here, one needs to note that RPN (simple product of three parameters) is not a perfect representation of risks because these number assignments are subjective and not continuous. Another interesting factor worth noting is that RPN (in its simplest product form, i.e., RPN $= S \times O \times D$) cannot assume any number, for example, the RPN value cannot be 13 because this is a prime number >10 (since the scale is 10, and 13 can have only factors 13 and 1). Another interesting fact is that with same RPN value, risks are not same, for example, $(S)(O)(D) = 2 \times 6 \times 10 = 120$ and $8 \times 3 \times 5 = 120$; in one case severity is at the lowest end, whereas in other case it is at the higher end of the scale. Naturally, in the second case, severity is more significant than in the first, though both have same RPN.
- An organization may consider issues with high severity and/or high occurrence ratings to represent a higher risk than issues with high detection ratings. Therefore basing decisions solely on the RPN (considered in isolation) may result in inefficiency and/or increased risk. In view of this, for better assessments, instead of using a simple product form, many companies use other calculation methods suitable for the application, that is, in some cases it is quite possible to use a weighted sum so that specific weightings could be more transparent and the result is more accurate and may be free from the limitations discussed previously. A particular analysis team may choose to supplement or replace the basic RPN methodology with other related techniques, such as revised RPNs, the occurrence/severity matrix, ranking lists, risk ranking tables, and/or higher-level RPNs. All of these techniques rely heavily on engineering judgment and must be customized to fit the product or process that is being analyzed and the particular needs/priorities of the organization.
- FMEA worksheets will typically identify whether corrective action is required based on a combination of severity, occurrence, detection, and/or RPN values. After RPN assessment, recommended actions are suggested. After implementation of the recommended action, the entire issue is reassessed to get an indication the effectiveness of the corrective action. Naturally, with a revised set of severity, occurrence and detection ratings a new RPN is calculated. From these two values it is possible to get % reduction in RPN:

  $\%\text{reduction in RPN} = 100 * \{RPN_{initial} - RPN_{revised}\}/RPN_{initial}$

  Let initial S, O, and D values be 7, 8, and 6, so RPN = 336 and revised S, O, and D values be 7, 5, and 4, so RPN = 140.

  Therefore %reduction in RPN = 100*(336 − 140)/336 = 58.3%. From here it can be concluded that RPN is a method to assess the *relative risk for a particular analysis* and is a helpful tool. Also there can be several revised methods or techniques to calculate this and apply it for the analysis best suited. Another important term, explained in Fig. IV/2.2.3-2 is "error proofing." Readers are advised to take a note of this as the same will be referred to in subsequent chapters.

Error proofing refers to the implementation of a failsafe mechanism to prevent a process from producing defects. The philosophy behind error proofing is that it is not acceptable to make even a very small number of defects, and the only way to achieve this goal is to prevent it from happening in the first place.

FIGURE IV/2.2.3-2

Error proofing.

2.3 Controls and Recommended Actions

2.3.1 Controls

Controls refers to the existing controls in place for elimination or mitigation of risks. There can be two types of controls, such as preventive control, responsible for preventing risks that occur through interlocks. In this case there could be automatic sensors to detect failure during (or after) the process, and mitigate the risk, before any catastrophe could occur e.g., surge protector in electrical circuits. There can be a number of controls and/or interrelated controls associated with various failures and failure modes. Control actions can be broadly categorized as follows:

- Prevention-type design controls are meant mainly to prevent cause, failure mode, or effect in the product design with the help of current controls or planned actions. They are intended to reduce the likelihood of occurrence, so they have an effect on *occurrence* ranking. System interlock is an example.
- Detection-type design controls are meant to detect failure mode or cause in the product design with the help of existing controls and/or automatic sensing or planned actions. These are done before the product design is released for production. Detection controls are intended to increase the likelihood that the problem will be detected before reaching the customer. So, detection-type design controls are used as input to the *detection* ranking.
- Prevention-type process controls are meant mainly to prevent cause, failure mode or effect in the manufacturing or assembly with the help of current process controls or planned actions. They are intended to reduce the likelihood of occurrence, so, has effect on occurrence ranking. System interlock could be an example.
- Detection-type process controls are meant to detect failure mode or cause in the manufacturing or assembly process with the help of existing controls and/or automatic sensing or planned actions. These are done, before shipment from the plant. So, these controls are used in the detection ranking.

2.3.2 Review Work

After conducting FMEA/FMECA and preparing a worksheet it is necessary to review the entire system so that necessary recommended actions can be placed. For review work, normally the *designer* takes the lead. At this juncture, personnel from most of the key sectors are involved. For guidance the following is a list of personnel:

- System design engineer
- Hardware design engineer
- Software design engineer (as applicable)
- Service engineer
- Testing engineer
- Quality control engineer
- Commissioning engineer
- Maintenance engineer
- Safety engineer
- Production/manufacturing engineer
- Reliability engineer
- Project manager

  In addition, if necessary, personnel from a corporate management group shall be involved for necessary support. The main purpose of the review is to study the worksheet (discussed later) with special reference to risk matrix or RPN with the main objective of taking decisions on the following issues:
- To decide on acceptance of the system
- To recommend necessary actions so that RPN can be reduced by:
  - Reducing likelihood of occurrence
  - Increasing chance of failure detection
  - Reducing the effect of potential failure

After the review a number of recommended actions are discussed and suggested. Based on the discussions the recommended actions are implemented. When improvements are decided, the FMECA worksheets have to be revised and the RPN should be updated. Now, one needs to look forward to receiving details about the recommended actions.

2.3.3 *Recommended Actions*

In an FMEA, "recommended actions" are the tasks prescribed by the FMEA team that can be used to reduce or eliminate the risk associated with a potential cause of failure. Recommended actions take into account the existing controls, the relative importance (prioritization) of the issue, and the cost and effectiveness of the corrective action. There can be multiple recommended actions for each cause. The main aim of recommended actions is to reduce the value of RPN. The type of recommended action will greatly depend on the *specific* project.

2.3.4 *Action Taken*

This stands for the specific action implemented to reduce the risk at an acceptable level. These specific actions must bear proper correlation with actions recommended and should be assessed to see how RPN and what % of RPN has been reduced.

2.3.5 FMEA/FMECA Worksheet

With this knowledge, it is possible to explore the development of the FMEA/FMECA worksheet, whose various columns were discussed earlier. In Clause 2.4 the FMEA/FMECA worksheet is described and it shall be read in conjunction with Fig. IV/2.4-1.

2.3.6 Fault/Failure Isolation

Once the failure is detected it needs to be isolated. In this approach the operator is allowed to isolate the system when failure is detected. In products such possibilities are mentioned in troubleshooting process flow. Failure isolation may require action or series of actions to isolate the system. At times there can be a builtin automatic failure detection and isolation system. As an example, when observing grid failure, it is possible to isolate one local system/grid, or in the case of failure of a transmitter it can be isolated from the system. These are examples of automatic isolation systems. Thus in both cases, when sensing the failure, action for isolation is necessary either by the operator or by an automatic device. In the same way some products (especially PEs) can be isolated by operator or automatic.

2.3.7 Compensation

This is a provision at times kept in the design and the operator can intervene to reduce the effect of failure. Compensating provision in the system can halt the fault or negate the effect of malfunction/failure. Such compensating provisions can be in the form of *backup* or *redundancy* and/or *safety monitoring* and *alarm systems*. It is a common experience that in the case of failure developing in a normal PC, at times the device goes into/starts in safe mode; this is an example of a compensating provision. Similarly, automatic selection of a redundant transmitter/logic solving processor in (say) a PLC with an alarm may be an example of *redundancy and safety monitoring* (discussed earlier) in compensating provisions.

2.4 FMEA/FMECA Worksheet

Preparation of a worksheet for FMEA/FMECA is extremely important. A successful analysis must have a well-defined worksheet. One typical composite worksheet (based on Quality—One document guidelines) is shown in Fig. IV/2.4-1.

The worksheet shown in the figure is a generic worksheet with possible details and explanations. When carrying out FMEA/FMECA it is important to decide on the worksheet type. Often, organizations or customers have their own say. For analysis the following information is necessary:

- Description of the unit:
    - Reference number
    - Function
    - Operational mode
- Description of failure:
    - Failure mode
    - Failure cause/mechanism
    - Detection of failure

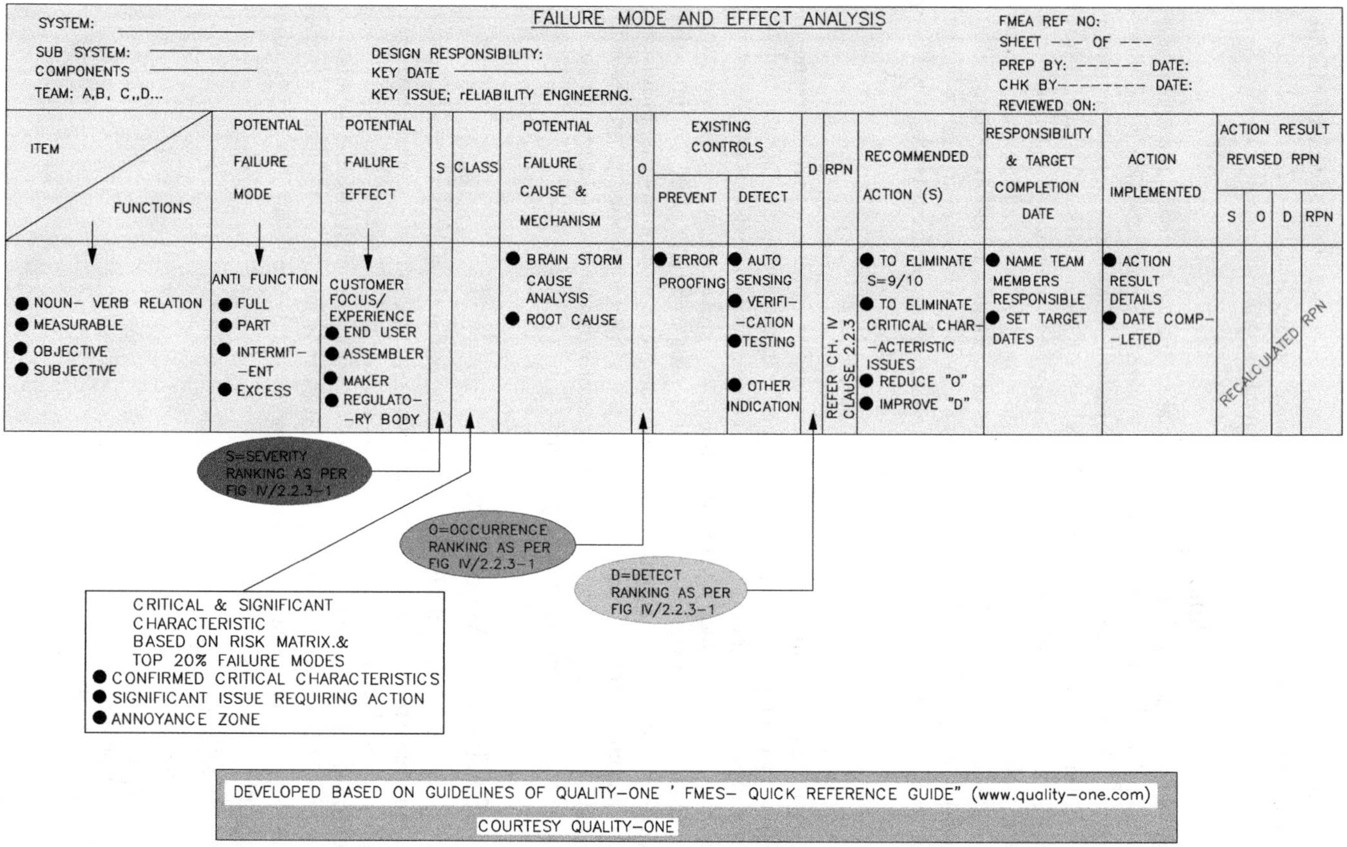

FIGURE IV/2.4-1

A typical FMEA/FMECA worksheet.

- Effect of failure on:
  - System
  - Subsystem
- Severity ranking
- Failure occurrence
- Risk reducing measure
- Comments

First, one needs to identify the unit with the help of an ID so that it can be pinpointed and later can be referred. The functions of the element are listed. It is important to list all functions. Then, based on operational mode, various failure modes are listed, as discussed in earlier clauses. For each case, cause and effect analysis is done to discover severity, occurrence, likelihood, etc. Since all these have been discussed in earlier clauses, and explained in the figure, they are not repeated. One important thing to note here is the "Class" column. It sometimes shows YC or C or N, etc. YC is for criticality and significance. Normally, criticality matrix is chosen and critical characteristics are confirmed. In DFMEA/DFMECA when criticality is present, then PFMEA/PFMECA needs to be carried out for verification. If in PFMEA/PFMECA the same is present, then a suitable control plan shall be developed to eliminate the criticality as far as possible. Similarly, in some cases "N" is placed in class column to indicate noise, meaning that it is a risk that should be mitigated, but is largely out of the control of the development team, whereas "C" stands for controllable.

In Tables IV/2.4-1 and IV/2.4-2 a number of worksheet examples are presented.

2.5 FMEA for E/E/PEs

One of the key issues in software development is maintaining quality assurance of developed software systems. With the introduction of embedded systems, electronics frankly speaking has turned from hardwired to software-based systems. In embedded systems, with minor changes in the lone codes, one can meet the challenging demands of the customer. Such devices now include small handheld devices (smartphone), household equipment, smart instrumentation, and even safety critical control systems of hazardous plants and thermal/nuclear power plants. So, the advancement and proliferation of computer and information technology has made it possible for specified functions of systems including safety critical systems to be software driven. Naturally, these safety and performance critical systems demand *exact* limits and accurate measurements. Average and statistical data may not suffice. It is logical to invest more in the failure analysis of safety critical systems, but in general detailed failure analysis from the design stage to completion will reduce overall manufacturing cost when cost of changes are considered at design, implementation, and postimplementation stages. Failure analysis right from the early design stage will help to identify weaknesses in the design, and mitigating actions could be undertaken. Another important issue here is that, in most cases, hardware failure modes are normally well understood on account of familiarity with their failures and associated

Table IV/2.4-1 Typical Example of Worksheet of FMEA in Line With IEC 60812

Indenture level: Sheet no: Operating mode:			Designed by: Item: Revision:					FMEA Prepared by: Approved by:				
Item ref.:	Item description/ function	Failure entry code	Failure mode	Possible failure causes	Symptom detected by	Local effect	Effect on unit output	Compensating provision against failure	Severity class	Failure rate F/Mh	Data source	Recommendations and action taken

Table IV/2.4-2 Typical Example of SWFMEA					
Product ref. System ref.: Subsystem ref.:				Prepared by: FMEA Approved by: Date	
Reference no./ ID	Component details	Fault	Causes	Failure effect	Safety measure

consequences. The failure pattern of software is in many cases are *not known as well as people are less accustomed/familiar*. One of the peculiar difficulties with software systems is the **complex relationship** between faults and effects. A minor fault may completely crash a software system; on the contrary, one very complex and almost invisible fault may have subtle and long-lasting side effects. Also software failure may occur because of systematic faults (Chapter VII), in which case it may be difficult to apply statistical analysis. For SWFMEA there is no predefined, linear flow that can be copied from a work instruction. It is not meant to cover 100% of design but to focus on occurrence and severity in consequences (especially for safety critical applications). SWFMEA is *a systematic tool to anticipate problematic issues and improve validation and remove design defects* and *not a substitute* for software *reliability methods*. It is better to understand and address a few fundamental issues related to software systems and automation systems.

2.5.1 Related Terms and Definitions

In connection with software, normally a few terms such as fault, error, failure, etc. are used very loosely, which create anomalies. Also in SWFMEA a few terms will be encountered. It is better that all these are clarified at the beginning:

- Coding: The objective of coding is to translate or transform system design into codes in any higher-level language to test the codes. Programmers in any organization adhere to a well-defined and standard style of coding called coding standard.
- Error: A measure of the estimated difference between the observed or calculated value of a quantity and its true value, that is, deviation from true value. In hazard analysis an *error* is a deviation from the required function of a system.
- Failure: Failure is a lack of success to achieve a desired goal. In this case a *failure* occurs when a system fails to perform its required function.
- Fault: An unattractive or unsatisfactory feature present in a piece of work. Or a *fault* may be seen as a weakness or defect in the system.
- Software development: Software development is characterized by:
  - Time: Time required to perform the task
  - Size: Volume/size of the work produced
  - Defect/fault: Number, type, time to remove, point of removal

- Software life cycle: Fig. IV/2.5.1-1B (both hardware and software life cycles are compared). In the hardware case it is basically a bathtub curve (comprising random fault, early life manufacturing, or design defect and wearout during the end of usual life— see Chapter VII). Software does not really *wear out* (like a hardware system) but it *deteriorates*, though *not* as a function of time but as a

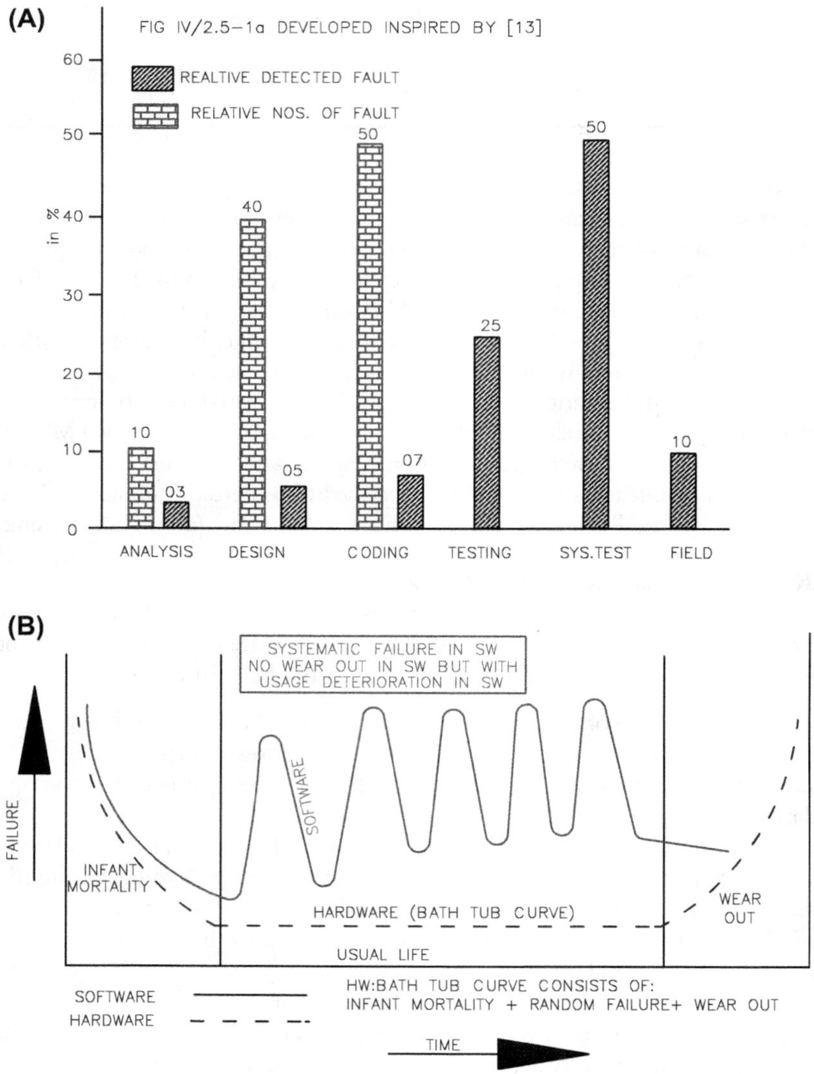

FIGURE IV/2.5.1-1

Software fault occurrence detection and life cycle. (A) Relative software fault occurrence & detection, (B) software life trend compared with hardware.

function of ***usage***. Deterioration of software is affected by latent defects and adjustment of the software with new requirements and maintenance. Of course, PEs/software may have systematic (mostly) and random faults, as discussed later.

- Source and object code: Source code is any collection of computer instructions written using human readable language. It is written in one or more programming languages. Source code is a program written to facilitate the work of the programmer. It is transformed by the compiler into low-level machine language and is object coded.

2.5.2 Cost Impact

Timely performed SWFMEA is much better and more cost-effective than crisis management at a later date. In the early phases of software development, changes cost less. It makes the system preventive in the sense of identifying the fault before it turns into failure. The cost comparison in terms of percentage is shown in Fig. IV/2.5.1-1A. The figure reveals that the number of faults is more in the design and coding phase, whereas detection is revealed more at a later stage. So, in the case of FMEA done at an early stage, many of these faults will be identified early when the cost of fault correction is much less. This is because at a later stage the cost of corrections increases sharply (not shown in the figure).

2.5.3 Software Failures and SWFMEA

Failures normally encountered in software are:

- Systematic failure (for more details see Chapter VII): Systematic failure in systems comes from faults in design and/or manufacturing. Often best practices are not followed; instead, to cut costs short cuts are taken. Such practices often lead to systematic failure. From IEC 61508 it is found that systematic failure is failure related in a deterministic way to a certain cause, which can only be eliminated by a modification of the design or of the manufacturing process, operational procedures, documentation, or other relevant factors. Human errors are the contributing factor for systematic failure during development or manufacturing process/operation. Unless removed, systematic faults reappear whenever matching circumstances are available. Because it is difficult to predict when matching circumstances will appear, it is very difficult to *predict* the occurrence of systematic faults and their effect on the safety of the system. Software faults are systematic, so neither prediction nor statistical calculation is possible. Each device has many known *opportunities* for systematic error. For new technology, there is the potential for many unknown (or as yet unidentified) failures. When issues associated with *interconnectivity, communication, and support* systems are added, there are generally a large number of potential systematic failures. Demonstrating the safety of software relies upon assessing the likelihood of this type of fault. To meet the demand of the customer, especially for safety critical applications, the software within safety critical (especially control and automation) applications is increasing in size and complexity. Naturally,

systematic faults are more and more *prevalent*. The statistical prediction of systematic faults, for software risk quantification is not possible. So currently it is argued that the acceptability of software is based on the *suitability* of the development processes followed! *Therefore the role of FMEA cannot be overestimated. (Unfortunately not many standards/norms for SWFMEA are available.)* Systematic faults are mainly:
  - Manufacturing defects (hardware/software error/fault)
  - Incorrect/improper specification
  - Wrong implementation
  - Operation and maintenance
- Random failure (see Chapter VII): Random failures are project specific in the sense that they depend on the process and its use. From IEC 61508 it is found that a failure occurs at a random time, which results from one or more degradation mechanisms. Random failures are mainly caused by physical damage/changes such as wearout, thermal stress erosion/corrosion, etc. These are applicable for hardwires of E/E/PEs in automation systems. The rate of failure of random failures normally cannot be reduced; instead for random failures focus should be on their *detection* and *handling*. Statistical data handling and treatments can be applied to random failures, hence risks associated with random failure can be calculated. This is not possible in the case of software with systematic faults.
- Common cause failure (see Chapter VII): This is a kind of fault that causes multiple devices/systems to fail simultaneously. Common cause failure may be random or systematic. This is discussed in Fig. I/8.3-1 in Chapter I, Chapter VII, and in Ref. [9].
- Failure modes: Hardware failure modes have been discussed at length in this chapter. Software failure modes may be "data" and "event" failure modes and these may be repetitive in nature, because they may be caused by systematic failure. At times, redundancy may not really serve the purpose of avoiding such failures.
- Failure rate: Determination of failure rates in software is challenging. For hardware failure rate, IEC 62380 may be consulted. For software, there is hardly any set of faults or fault models that could be applied industry wide.
- SWFMEA from different points of view: Table IV/2.5.3-1 shows how SWFMEA can be conducted from different points of view.

2.5.4 *Timing for Effective SWFMEA*

- When: It is an established fact that most benefit of SWFMEA can be achieved in the early phases of design, (ideally) before coding but after major requirements are frozen. From a cost point of view this can be seen in Fig. IV/2.5.1-1A. When performed early, FMEA can reveal weakness in design and solutions can be sought in both hardware and software to avoid costly design changes in later phases. Also this will offer flexibility in finding reliable, most cost-effective, and appropriate solutions. Additionally, because this is done at an early stage it is possible to have better test planning including additional tests, if any, for the

Table IV/2.5.3-1 SWFMEA From Different Points of View

FMEA Viewpoint	Product Level Viewpoint	Identifies Failure Related to	Life Cycle Timing
Functional	Requirements	Timing sequence, faulty data, erroneous error messages for a component	SRS completion
Interface	Interface between two components	Timing sequence, faulty data, erroneous error messages between two components	Interface design specification completion
Detailed	At class or module level	All the above plus memory management, algorithm, I/O database issues	Detailed design code is complete
Production	Process-related failure during product development	Problems with many defects and/or ability to meet a schedule, execution and tools	Any time
Maintenance	Changes to the software	Problems when software is modified, installed, updated	During maintenance
Usage	User friendliness and consistency, documentation	Software/documentation is too difficult or inconsistent to be used properly	As early as possible, as all these issues will influence design

SRS, *software requirement specification.*
Courtesy of SoftRel LLC. Copyright SoftRel, LLC 2010. This material may not be reprinted in part or in whole without written permission from Ann Marie Neufelder (permission: E-mail from Ms. Ann. Marie dated 24th June 2015).

new solution. Also more energy can be channeled to reduce RPN. SWFMEA also finds its use in later stages of the project. In the following cases SWFMEA with some amount of detailing is carried out on:

- New systems or reuse in different systems
- Safety critical systems
- Performance critical systems
- Large and high cost systems

- Effective way: The following points are helpful for effective SWFMEA:
  - Establish all requirements for software, analysis, coding, etc. (as applicable) when it is undertaken
  - Identify failure modes caused by software

- Software quality: Walking through SWFMEA it is possible to improve quality requirements, design, and coding.
- Coding testing and inspection quality: SWFMEA allows software engineers to anticipate software problems based on input/output. This can improve codification. It is possible to take SWFMEA failure modes as a test case. For inspection a checklist is used. SWFMEA can augment the checklist and/or an inspection checklist can help in finding failure modes.
- Identify missing points in design and requirements defined
- Undefined/unwritten assumptions
- Interface analysis (software)

2.5.5 *Difference Between HW and SW in FMEA Approaches*

As discussed earlier there are some differences between hardware and software FMEA. The major points are enumerated in Table IV/2.5.5-1.

Table IV/2.5.5-1 *FEMA Comparison Between Hardware and Software*

Point of Comparison	Hardware	Software
Level	At functional, interface, part level, even at application specific integrated circuit (ASIC) level	Only possible at functional level
System	Applicable for system free from failed components	System with software fault that can lead to potential failure
Checks	Failure modes mainly as per wearout, aging, thermal stress, corrosion, etc.	Functional failure/systematic software faults
Consequence	Consequence at entire system	Consequence at entire system
Action	Criticality and measures to mitigate consequences	Criticality and measures to mitigate consequences

2.5.6 *SWFMEA Preparatory Work*

Like normal FMEA, this also needs some preparation in the sense that a team formation and information collection phase is necessary.

- Team formation: Generally, the following members are included in the SWFMEA team:
  - Team leader or facilitator: Good knowledge and experience with SWFMEA
  - System development/management personnel: Responsible for the system project
  - Software engineering team members: Members consisting of representation from developer, testing, services, and domain expert. They need to have good knowledge of the subject matter, and be familiar with design, coding, testing,

software requirement specification (SRS), interface, etc. The domain expert should be knowledgeable about the use of the system in question.

- Meeting: Normally, the facilitator/leader initiates the meeting and others need to check that no part is missed out. The meeting shall be maintained at a high level. It is better for the leader or facilitator and a few other members to prepare an FBD. Since all members will not be present for the FBD, it is better that this is displayed. During discussions, modifications may be suggested; to carry out such modifications software may be sought. When a group of engineers start to discuss issues, questions are bound to be asked. Some of the issues will pass quickly while some may take little time. Usually 30–60 min will be spent for each SRS statement at a functional level, while each interface variable/detailing in a module may take a little more time.
- Information required: As in the case of a normal FMEA, a fair amount of information is required, which shall be freely available. In line with guidelines from IEC 60812 the following is the basic documentation necessary for a successful SWFMEA:
  - System boundary limit
  - System structure
  - Representation structure
  - System operation. Maintenance controls
  - Modeling
  - Environment
  - Functional structure of the system
  - FBD
  - Failure significance
  - Compensation provisions
  - Other details as required

2.5.7 SWFMEA Procedure

Generally, there is not much difference in procedure between SWFMEA and FMEA, so this is discussed in short only. Challenges are discussed at length and the major points are as follows:

- The software in question is divided into components/modules/functions with descriptions.
- To understand software structure in terms of system software and application software, Fig. IV/2.5.7-1B may be referred to for an appreciation of the idea, which will be different for different cases/applications.
- Now for each component/module as applicable to define Failure Modes – What can go wrong here? And corresponding to each possible failure (brainstorming) and faults assigned to each and every function to find out the effect asking what will happen then?
- The hardware failure mode issue is rather straightforward. The most difficult part is discovering the failure mode for software. One of the first articles regarding

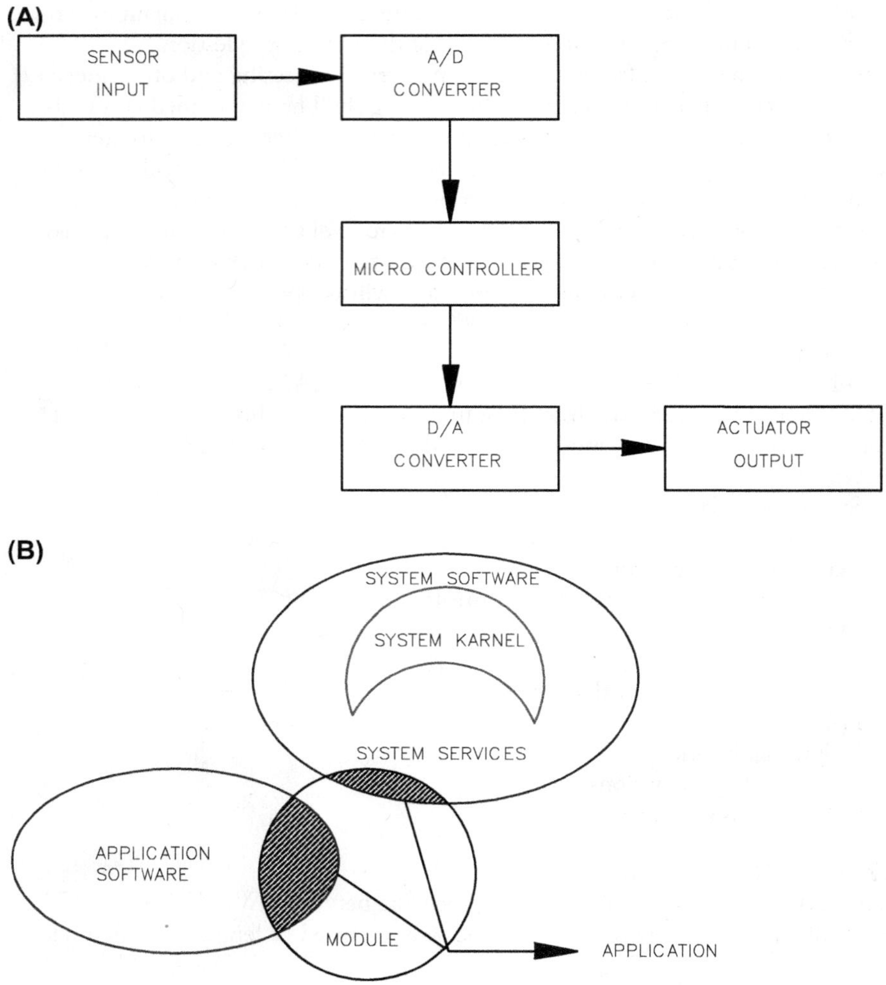

FIGURE IV/2.5.7-1

Microcontroller and software basics. (A) Basic micro controller configuration, (B) software fundamentals.

software failure mode and effects analysis (SWFMEA) was given in Reifer (1979) [8] and from that concept it could be on the basis of:

- Computational
- Logic
- Data I/O
- Data handling
- Data definitions

- Database
- Interface
- Others

  Or on the basis of Risord (2001) [8]:
- The operating system stops.
- The operating system stops with a clear message.
- The operating system stops without a clear message.
- Programs run with the wrong result.
- Program runs apparently correctly but gives the wrong result.

These details are examples of a few methods but there are a few others also:

- In the next phase the root cause is discovered: Why will that happen?
- For criticality analysis and RPN severity of failure effect, likelihood of occurrence detection probabilities are discovered.
- Challenges: The major challenges in SWFMEA are:
  - Classifying software failures into appropriate failure mode
  - Difficulty in assessing likelihood of occurrence
  - Estimating of detection probability

The manifestation of an inherent software fault depends on the *operational* profile of the system as well as on the software *itself*. Because the failures are systematic in nature and cannot be treated statistically it is very difficult to predict the frequency of occurrence of triggering a fault into a failure, that is, when and how many times the fault will trigger into failure. (*This frequency is usually not known. Luke (1995) proposed that a proxy such as McCabe's complexity value or Halstead's complexity measure be substituted for occurrence. Luke argued that there is really no way to know a software failure rate at any given point in time because the defects have not yet been discovered. He stated that design complexity is positively linearly correlated to defect rate. Therefore Luke suggested using McCabe's complexity value or Halstead's complexity measure to estimate the occurrence of software defects* [8].) From this one may note that it is very difficult to predict the probability of detection, except a part that could be found by self-diagnostic methods in control systems. Empirically, such variations are shown in Fig. IV/2.5.7-2 adapted from "Software FMEA—Opportunities and benefits of FMEA in development process of software-intensive technical systems; O. Mackel; Siemens AG [7]." In this they adapted two new parameters "A" & "E" in place of likelihood and detection probability. Variation of these two parameters with complexity is shown in the figure in the reference. For further details [7] may also be referred to.

2.5.8 Discussions

Software is part and parcel of digital systems. There are now various kinds of software available in the market and each with different characteristic features. Software failure in a microcontroller will not be the same as/similar to system software in a conventional microprocessor based PLC/DCS. So, in this part a brief outline about such variations from an application point of view are presented.

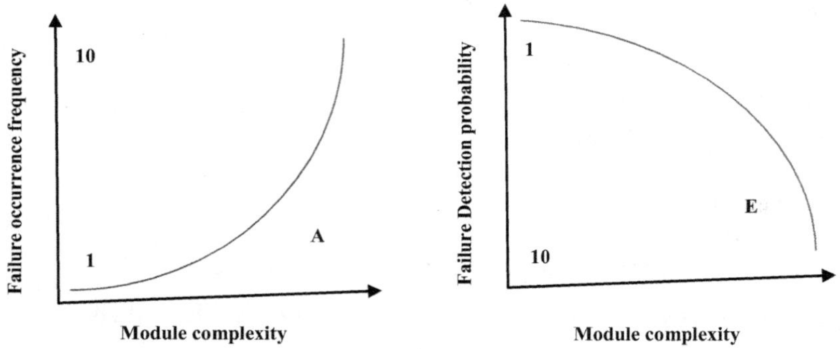

FIGURE IV/2.5.7-2

Thumb sketch of variation of likelihood and detection probability with complexity of module.

Drawn inspired by document referenced O. Mackel, Software FMEA—Opportunities and Benefits of FMEA in Development Process of Software–Intensive Technical Systems, Siemens, AG.

- Embedded systems and microcontroller: Embedded system is simply an advanced dedicated digital system to provide a fixed specific task. These embedded systems are found in handheld cell phones and safety critical control systems, for example, dedicated control/PLC. Now manufacturers have come up with these embedded systems, because system components residing within the control system make it possible to have greater flexibility and functionality (e.g., ABB 800xA) [9]. System components in automation terms and safety terms are classified as E/E/PEs. A microcontroller is a digital PE component and it also belongs to embedded system. Currently, these embedded microcontrollers are used in almost all PE devices, which pleases ever-demanding customers because it means very few changes are made to the software. A generic model of a microcontroller is shown in Fig. IV/2.5.7-1A. The safety concepts start at the system level, then flow to subsystems, and finally to the microcontroller unit consisting of hardware and software. Software requirements may be allocated as a system safety function, with hardware and software integrity checks.
- Architectural rendering and analysis of the system: Software architecture refers to a structured conceptual representation of a software system. Software architectural rendering of a system serves as a framework from which more detailed design activities can be developed. Software architecture first defines top hierarchical or modular components of the system that are sufficient to represent the system. The details of each modular component could then be addressed in further design. Failure analysis could start at the top. After analysis at the first level, modifications are suggested and implemented. After this level, work is complete, and then one can proceed to the next level down; in this way the entire system can be analyzed. With this, discussions on FMEA for E/E/PE is concluded to explore the possibility of automating FMEA.

2.6 Automated FMEA (The Concepts)

In this dynamic global market there is tremendous all-round competition. To keep pace with this sustained growth, customer loyalty is extremely important. To gain customer loyalty, producers need to ensure high-quality and reliable products at competitive prices. FMEA is a classic system safety analysis technique employed to improve productivity and achieve quality products. New product design and development is a highly iterative process. Also this demands substantial heuristic *knowledge* of components in the areas of customer requirements, product design specifications, production, etc. So, there will be high dependence on high standards of specific knowledge and experience of designers. Conventional FMEA (even with software applications to replace some clerical jobs such as report preparation, etc.) has many shortcomings. Two of these shortcomings are the enormous time taken to complete the analysis and associated cost. Because of this it is hardly possible to use FMEA more than once (usually done only once during the detailed design phase) because of cost and schedule constraints. The majority of these shortcomings can be circumvented with the introduction of automation in the FMEA process. Before going to further, it is better to look at the major advantages of automated FMEA as detailed next.

2.6.1 *Major Benefits of Automated FMEA*

- Multiple iterations possible from concept stage to detailed design stage
- Enables early identification of potential problems
- Single point failure
- Better assessment of effects
- Better assessment of criticality and RPN (fuzzy approach)
- Less dependence on human expertise
- Less probability of missing failure mode (large database)

Some efforts have been made to compare the two kinds of FMEA: manual and automated. Since in automated FMEA there are several methods, many of the comparison points mentioned in Table IV/2.6.1-1 may differ and there could be a few more additional points too.

2.6.2 *Reasons for Automated FMEA*

Why Automated FMEA? Further to what was discussed earlier, the traditional approach has a few other serious drawbacks, for example, it is not suitable for reuse and information retrieval is extremely difficult for reuse especially when traditional FMEA grows. Also it is difficult for traditional FMEA to meet the demands of frequent changes in design. At the same time, FMEA is not much use when it is conducted at an early stage and information availability is scant. So, traditional FMEA created after engineering is done may be of no value if it does not cater for the requirements of changes in design as stated earlier. Therefore traditional FMEA may be a nonvalue-added activity created just *to satisfy the contractual requirements of customers*!

Therefore the requirements of automated FMEA are now being appreciated and work toward this approach is increasing. Major work in this area has been found in

Table IV/2.6.1-1 Comparison of Manual and Automated FMEA (Generic)

Data Item	Generated Manually	Generated Using Automatic Tool
Identification	ID or number	
Item	Unfailed state item	Unfailed and multiple degraded states items
Failure mode	Description of how model failed, mostly direct	Name and description of error model. Both direct as well as combinational possible
Immediate effect	Manual description of immediate effect	Name and properties of transitioned state
Subsequent effect	Only two levels: immediate level and end level	Not limit; subsequent states and transitions
Detection	Description of any detection methods	Better probability of detection
Compensating provision	Description of any mitigation method	Compensating provisions (i.e., recovery) described in the context of subsequent transitions
Severity level	Assigned at any time in the analysis	Contained in state property, and/or through set rule (fuzzy membership)
Expert/database	Depends highly on experts	For initial rule setting expert advice is necessary, then depends mainly library and stored database

automotive, aerospace, and medical disciplines. Until now, to the best of the author's knowledge, there has been no real established method whereby one can get guidelines for automated FMEA for use across industries. However, when reviewing the literature a few different types of study have been noticed. On account of the limitations in the scope and size of this book, a few will be discussed very briefly. Readers are advised to go through various literature referenced at the end of the chapter for further reading. In the following subclauses a few automated systems are discussed to gain a conceptual idea about the procedure.

2.6.3 *Semiautomatic Approach*

Papadopoulos [10] proposed this semiautomatic FMEA generic model (which includes P&IDs, FBDs), which has been developed and stored in the library for reuse. It is possible to generate the model at various stages of the design life cycle. The models that provide the basis for the analysis should identify the topology of the system in its hierarchical structure. First it is best to establish the local failure behavior of components in the model as a set of failure expressions to show how deviations of component outputs can be caused by internal malfunctions and deviations of component inputs. After inserting local failure analyses, a model is used to automatically determine how local failures propagate through connections in the model and cause functional failures at the outputs of the system. The fault trees (which are automatically constructed by traversing the model of the system moving backward from the final elements of the design, i.e., the actuators, toward system inputs) show how functional failures or malfunctions at the outputs of the system are caused by logical combinations of component failures, as shown in Fig. IV/2.6.3-1. With reference to Fig. IV/2.6.3-1 "the top events of these fault trees represent system failures. Leaf nodes represent component failure modes while the body of intermediate events (and intervening logic) records the propagation of failure in the system and the progressive transformation of component malfunctions to system failures" [10].

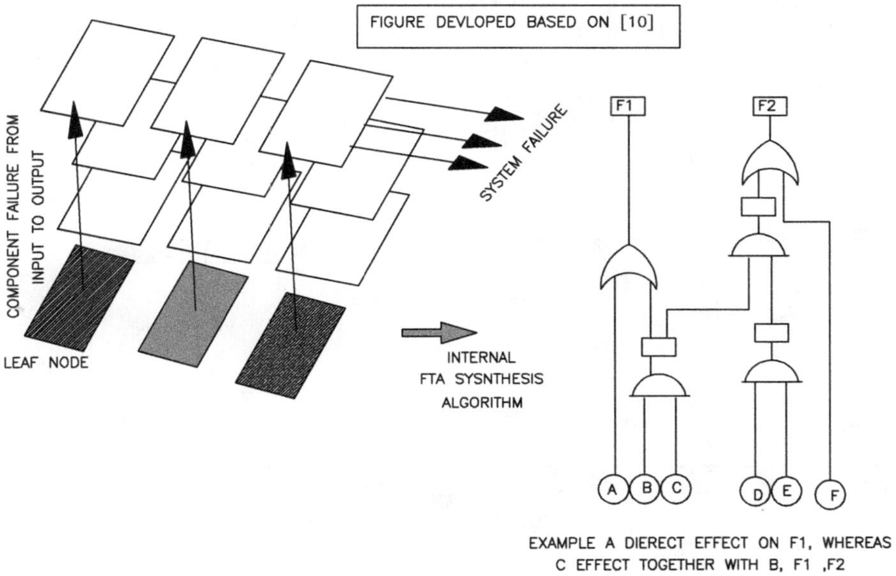

FIGURE IV/2.6.3-1

Semiautomatic approach for FMEA.

For large and complex systems, fault trees tend to grow very large and it is difficult to inspect and interpret them meaningfully. In the final step, a body of fault propagation logic is removed from the analysis by an automated algorithm, which translates the network of interconnected fault trees into a simple table of direct relationships between component and system failures with the help of MATLAB Simulink (Fig. IV/2.6.3-2). It is possible to synthesize more than one interconnected

Simulink developed by MathWorks is a block diagram environment for multi domain simulation and Model-Based Design. It supports simulation, automatic code generation, and continuous test and verification of embedded systems. It provides a graphical editor, customizable block libraries, and solvers for modeling and simulating dynamic systems and is integrated with MATLAB, enabling you to incorporate MATLAB algorithms into models and export simulation results to MATLAB for further analysis

FIGURE IV/2.6.3-2

MATLAB Simulink.

system fault tree simultaneously. The second significant extension made was the addition of a minimal cut-set calculation algorithm.

2.6.4 *Fuzzy Approach*

The application of fuzzy logic introduced by Zadeh (1965) in various fields is well established. In control systems the application of fuzzy logic is well known. In our daily lives we experience linguistic variables, for example, in control systems an experienced operator will often say that speed/temperature was increased minimally to resolve an issue. The operator rarely says that the parameter was increased by *x* units. So it is often very difficult to express linguistic variables objectively. The fuzzy approach is very convenient in this situation. (For an understanding of the fuzzy approach, Fig. V/4.3-1 and discussions in Clause 4.3.2 of Chapter V may be referred to.) A typical fuzzy approach for FMEA is shown in Fig. IV/2.6.4-1.

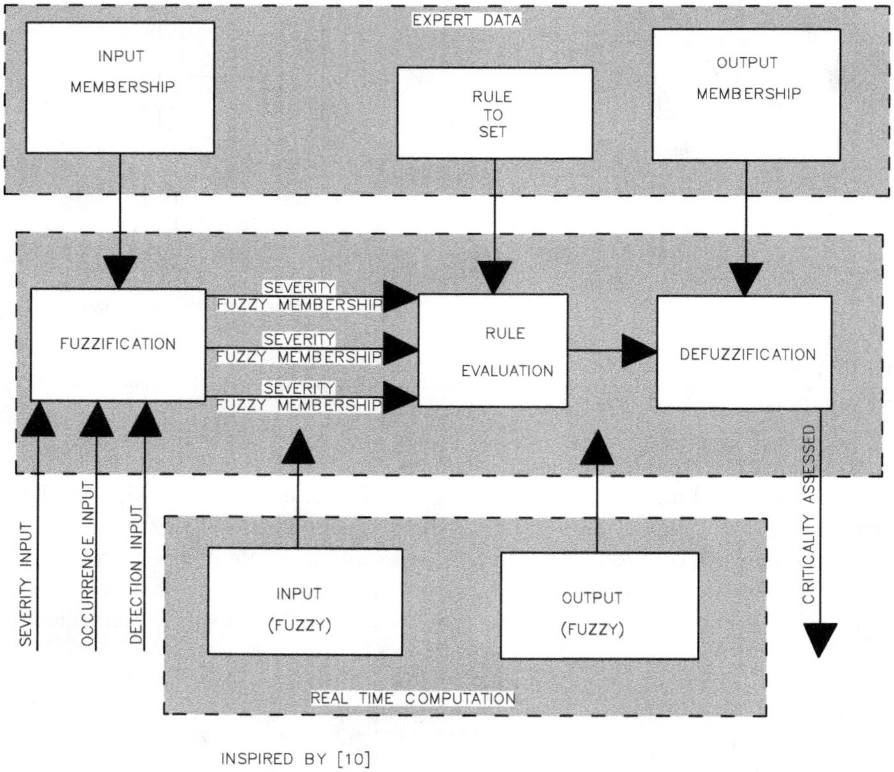

FIGURE IV/2.6.4-1

Fuzzy approach for FMEA automation.

To get a feel of the fuzzy definition set, detection likelihood, as shown in Fig. IV/2.2.3-1, has been transformed into a fuzzy definition. A typical fuzzy membership is shown in Fig. IV/2.6.4-2. Actual fuzzy values are derived based on the fuzzy rule set. Fuzzy inputs are evaluated using a rule-based set, so that criticality and RPN calculations can be made. In the fuzzification process, with help of crisp ranking, set $S * O * D$ is converted into fuzzy representation so that these can be matched with the rule base. Here, the "if then" rule has two parts: an antecedent (which is compared to input) and consequence (which is the result). On the other hand, in the defuzzification process, the reverse takes place. It is possible to automate FMEA using fuzzy logic and rule-based systems. The rule allows quantitative data such as occurrence to be easily combined with judgmental and quantitative data (such as severity and detectability) very easily and uniformly. The rule based on the linguistic variables is more expressive and useful (for further reading see Ref. [11]).

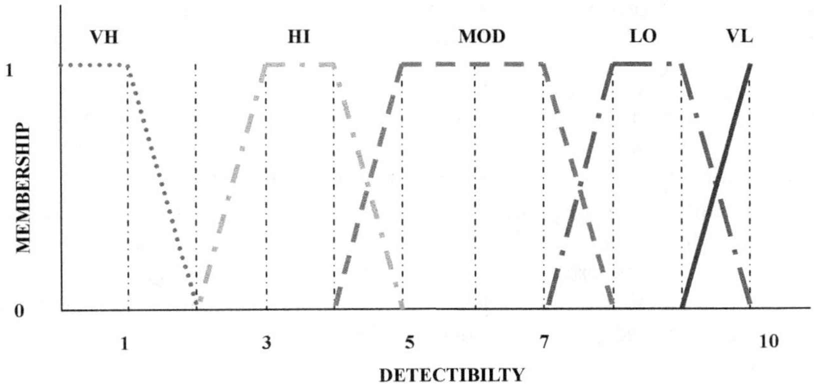

FIGURE IV/2.6.4-2

Fuzzy set definition (for detection likelihood).

2.6.5 *Automated FMEA and Little-JIL Process Definition Language*

In this method, automated FMEA is done with the help of process modeling in sufficient detail using Little-JIL process definition language (see Appendix IV). Little-JIL process definition is a hierarchy of steps, representing a single unit of work. Every step specifies all artifacts and resources it uses in its interface. A step without any substeps is called a leaf step. Each nonleaf step has a sequencing badge that indicates the execution order of its substeps. Artifacts are objects passed between different steps via four parameter types, IN, OUT, IN/OUT, and Locals parameter. In Little-JIL, resources are special kinds of artifacts for which there is contention for access [11].

In this automated FMEA three basic steps are applied:

- Identify failure modes: First, appropriate failure modes are identified by Little-JIL process definitions. At each step, the basic elements of a Little-JIL process definition have an interface, which is to specify the artifacts it uses as parameters along with each type. Thus the failure modes related to artifacts for each step are created.
- Identify potential effects for each failure mode: The effects derivation algorithm consists of two phases:
  - Phase 1. Here it is necessary to construct the artifact flow graph (AFG) from the Little-JIL process. The AFG can be easily constructed by traversing the process tree with an algorithm. The AFG is used to determine whether an artifact is data dependent on another artifact.
  - Phase 2. Derive FMEA information using the AFG: After generating the AFG for a Little-JIL process definition, the effects for each failure mode can be identified. It is straightforward to determine incrementally the artifacts to watch at the steps that could be contaminated by traversing the AFG. Therefore the fault of these artifacts is defined as the effects of the given failure mode.

Depending on the requirements the formats of FMEA information may vary. The top level of an effect tree lists the step name of each process definition step. The second level lists the failure modes of each step. The third level lists the effects of each failure mode. The next and all subsequent levels list the effects resulting from the failure of their parent. This expansion continues until there is no subsequent effect that can be propagated, according to the AFG. Inspecting all effects of each failure mode should help identify effects that could result in significant damage. For further reading the document referenced as [11] may be read, from which this description has been derived.

For Little-JIL process definition language refer to Appendix IV.

2.6.6 *Automatic FMEA Model*

Software often has very nonuniform qualities in terms of the effects of potential failures. Efforts have been made to develop an automatic model for software FMEA. The major steps involved are:

- Automated model construction: Very difficult if not impossible in certain software
- Injection and propagation of faults: Software model dependent
- Identification of system level effects

There are a few other methods that have also been tried to automate SWFMEA—work in in this area is in progress. MATLAB simulink elaborated in Fig. IV/2.6.3-2 are very useful in automating hazard analysis process and researchers are taking the help of the same

LIST OF ABBREVIATIONS

AFG	Artifact flow diagram
BHE	Batch HAZOP expert
BPCS	Basic plant control system
CAD	Computer aided design
CHAZOP	Control hazard and operability
DCS	Distributed control system
DFMEA/DFMECA	design FMEA/FMECA
DOC	Document
EEMUA	Engineering Equipment and Materials Users' Association
E/E/PEs	Electrical/electronics/programmable electronics
ETD	Event time diagram
FBD	Functional block diagram
FMEA	Failure mode and effect analysis
FMECA	Failure mode effect and criticality analysis
FRACAS	Failure reporting and analysis corrective action system
FTA	Fault tree analysis
GUI	Graphic user interface
HAZOP	Hazard and operability study
HMI	Human/machine interface
HSE	Health, safety and environmental
IEC	International Electrotechnical Commission
I/O	Input/output
LDG	Layered directed graph/
MFM	Multilevel flow modeling
MOC	Management of change
MSDS	Material safety data sheet
P&ID	Piping and instrumentation diagram
PCED	Process control event diagram
PFMEA/PFMECA	process FMEA/FMECA
PD	Process description
PDF	Process flow diagram
PHA	Plant (Process) hazard analysis
PLC	Programmable logic control
QRA	Quantitative risk analysis
RPN	Risk priority number
SDG	Sign directed graph
SWFMEA	Software FMEA
SPPID	Smart plant process & instrumentation diagram
SRS	Software requirement specification
TQM	Total quality management

References

[1] Hazop Guide to Best Practice; Guidelines to Best Practice for the Process and Chemical Industries, EPSC; IChemE; Chemical Industries Association; F. Crawley University of Strathclyde and WSAtkins, M. Preston Eutech and ICI Technology & B. Tyler S&T Consultants; Institution of Chemical Engineers, 2000.

[2] S. Yanag, W.H. Chung, Hazard analysis and support tool for computer controlled processes, Loyuborough University, Journal of Loss Prevention in the Process Industries (1998). Elesevier.

[3] P. Lucas, J. Walkington, T. Atkinson, Taking the control system for granted — ensuring the integrity of sub-SIL instrumented functions, in: IChemE Symposium Series No. 155, ABB Engineering Services, 2009.

[4] S. Rahaman, F. Khan, B. Veitch, P.R. Amyotte, ExpHAZOP + Knowledge Based Expert System to Conduct Automated HAZOP for Oil & Gas Operations, Dalhousie University.

[5] Quality-One International, N2 document; http://quality-one.com/fmea/.

[6] FMEA-FMECA, Your guide for FMEA resources; http://www.fmea-fmeca.com/types-of-fmea.html.

[7] O. Mackel, Software FMEA—Opportunities and Benefits of FMEA in Development Process of Software—Intensive Technical Systems, Siemens, AG.

[8] H. Pentti, H. Atte, Failure Mode and Effect Analysis of Software-Based Automation Systems, VTT Industrial Systems, August 2002. STUK-YTO-TR-190. https://www.julkari.fi/bitstream/handle/10024/124480/stuk-yto-tr190.pdf?sequence=1.

[9] S. Basu, A.K. Debnath, Power Plant Instrumentation and Control Handbook, Elsevier, October 2014.

[10] Y. Papadopoulos, D. Parker, A method and tool support for model-based semi-automated failure modes and effect analysis of engineering design, in: Brisbrane Conference in Research and Practices in Information Technology, vol. 47, University of Hull & Christian Grante; Volvo Car Corporation; Australian Computer Society Inc., 2004.

[11] A Failure Mode and Effect Analysis Technique for Process Defined in the Little JL Process Definition Language, D. Wang, J. Pan; Nanjing University, China; G.S. Avrunin, L.A. Clark; University Of Masschsetts, USA.

[12] R. Srinivasan, V. Venkatasubramanian, Automating HAZOP analysis of batch chemical plants. Part II. Algorithm and application, Computers & Chemical Engineering 22 (9) (1998). Elsevier.

[13] K. Strong, Using FMEA to improve software reliability, in: PNSQC 213 Proceedings, August 2013.

Further Reading

Dr. AA, Hazard and Operability (HAZOP) Study; http://www.cheme.utm.my/staff/arshad/images/lecture/Safety/hazop_method.ppt.

P. Andew, HSE Contract Research Report No 26/1991; Guidance for HAZOP Procedures for Computer Controlled Plants, KBS Process Technology Ltd., 1991.

N. Bidokhti, How to Close Gap Between Hardware and Software Using FMEA, Cisco Systems, 2007.

J. Cohen, R. Ferguson, W. Hayes, White Paper: A Defect Prioritization Method Based on the Risk Priority Number, Carnegie Mellon University, 2013.

L. Cui, J. Zhao, R. Zhang, The integration of HAZOP expert system and piping and instrumentation diagram, IChemE, Process Safety and Environment Protection (2010). Elsevier.

Development of a Fuzzy FMEA Based Product Design; K.S. Chin, A. Chang; City University of Hong Kong; J.B. Yang; University of Machester; The International Journal of Advanced Manufacturing Technology, 2008.

DOE Handbook; Chemical Process Hazards Analysis, U.S. Department of Energy, August 2004. DOE-HDBK-1100—2004.

Failure Mode and Effect Analysis; K Crow; DRM Associates; Failure Modes and Effects Analysis Guide, PQRI Manufacturing Technology Committee e Risk Management Working Group, May 2008.

Failure Mode and Effect Analysis; Understanding the Fundamental Definitions and Concept of FMEAS.

Failure Mode Effect and Criticality Analysis; AD-278—508; CRTA-FMECA; Concurrent Engineering Series; Reliability Analysis Center — DOD Information Center, 1993.

FMEA Info Center, Handbook on FMEA; http://www.fmeainfocentre.com/.

K. Greb, A. Seely, Design of Microcontroller for Safety Critical Operation, Microcontroller-Texas Instruments; ARM Techcon, Internet document.

Hazardous Industry Planning Advisory; Paper 8; HIPAP 8; HAZOP Guidelines, January 2011.

Hazards Analysis Guide: A Reference Manual for Analyzing Safety Hazards on Semiconductor Manufacturing Equipment, International SEMATECH, Technology Transfer # 99113846A-ENG.

R. Homkes, D. Evanecky, H. Kraebber, Applying FMEA to software, in: Proceedings of American Society for Engineering Education Annual Conference & Exposition, Purdue University College of Technology, 2005.

http://en.wikipedia.org/wiki/Hazard_and_operability_study.

http://www.effectivefmeas.com/uploads/Failure_Mode_and_Effects_Analysis__FMEA__for_publication.pdf.

N. Huges, E. Chou, C. Price, M. Lee, Automating Mechanical FMEA Using Functional Models, University of Wales, UK; American Association for Artificial Intelligence, 1998.

B. Jenkins, Automating FMEA: Next Generation Failure Analysis from NANEVA, Ora Research LLC (2013).

T. Kletz, HAZOP and JAZN: Identifying and Assessing Process Industry Hazards, third ed., Institution of Chemical Engineers, 1992.

T.A. Kletz, P. Chung, E. Broomfield, C. Shen-Orr, Computer Control and Human Error, IChemE.

A. Lam, C. Vogl, C. Dimpfl, Automated Generation of Failure Modes and Effects Analyses from AADL Models of Failure Modes and Effects Analysis from AADL Models, Aerospace Corporation, April 2012.

R.N. Liin, L. Morten, J. Niels, J.S. Bay, A goal based methodology for HAZOP analysis, Nuclear Safety and Simulation 1 (2) (June 2010).

L.S. Lipol, J. Haq, Risk analysis method: FMEA/FMECA in the organizations, International Journal of Basic and Applied Sciences IJBAS-IJENS II (5) (2011).

J.A. McDermid, M. Nicholson, D.J. Pumfrey, P. Fenelon, Experience with the Application of HAZOP to Computer—Based System, Internet document.

G. Mckay, Process Safety Management & Risk Analysis; http://www.cbme.ust.hk/safetycourse/download/09.pdf.

G. Mckay, Process Safety Management and Risk Hazard Analysis, HAZOP study analysis course; Internet document.

M. Mraz, FMEA-FMECA, University of Ljubljana; Management Center Innsbruck, Ljubljana, Spring 2005.

A.M. Neufelder, Software Failure Mode and Effect Analysis Overview, Softrel LLC, 2010. www.softrel.com.

S.H. Nggada, Software failure analysis at architecture level using FMEA, University of Hull, UK, International Journal of Software Engineering and Its Applications 6 (1) (January 2012).

R.H. Perry, Preliminary Hazard Analysis and Risk Assessment for Wood Centre Development, Southwood Resources, 1998.

Preliminary hazard analysis; Safeware Engineering; Risk management series; www.safeware-engg.com/white paper/Preliminary%20hazard%20analysis.htm and www.safeware-eng.com/%20white%20paper/software%20safety%20approach.htm.

C. Price, N. Snooke, An automated software FMEA, in: International System Safety Regional Conference, Aberystwyth University, Singapore, April 2008.

Process Control and Safety Group,UTM; Kutch Chemical Industries Limited, India.

Dr. T.C. Rajan, Software FMEA and Software FTA – an Effective Tool for Embedded Software Quality Assurance, Mahindra Satyam, White paper; Internet document.

M. Rausand, HAZOP; Hazard Operability Study, Norwegian University of Science and Technology, October 2005.

M. Rausand, System analysis e failure mode effect and critical analysis, in: System Reliability Theory, Wiley, October 2005.

Reliability Engineering Resources; Reliability Hotwire, December 2004 weibull.com, issue 46.

M. Rodriguez, J.L. de la Mata, Functional modeling for risk analysis, in: 17th Annual Symposium on Computer Aided Process Engineering, Elsevier, 2007.

J.F. Shortle, Applying Qualitative Hazard Analysis to Support Quantitative Safety Analysis for Proposed Reduced Wake Separation ConOps, George Mason University, Fairfax, VA & M. Allocco, FAA, Washington, DC.

Software Failure Mode and Effect Analysis; from the Reliable Expert, The Omnicon Group Inc., Internet document; www.OmniconGroup.com.

V. Venkatasubramanian, J. Zhao, S. Viswanathan, Intelligent systems for HAZOP analysis of complex process plants, Computers & Chemical Engineering (2000). Elsevier.

Wikipedia, FMEA, Internet document.

S. Yang, P.W.H. Chung, Hazard analysis and support tool for computer controlled processes, Journal of Loss Prevention in the Process Industries, Elsevier, 1998.

S. Yang, P.W.H. Chung, Hazard analysis and supporting tool for computer controlled process, Journal of Loss Prevention in the Process Industries 11 (1998). Elsevier.

CHAPTER

V

Quantitative Hazard Analysis

1.0 INTRODUCTION

A system is a part of the universe within a certain domain in space and time. What is an environment? Outside the frontier of the system is the environment [1]. Here, system shall have an identity, that is, deterministic. There shall be an external boundary to the system. An external boundary is determined by what aspect of system performance is of concern. This is stated here because for quantitative hazard analysis, boundary definition is extremely important. Also, the interface part needs to be considered (See Fig. V/3.0-1). The process definition for qualitative risk analysis is: "Qualitative Risk Analysis assesses the priority of identified risks using their probability of occurring, the corresponding impact [...] as well as other factors such as the time frame and risk tolerance [...]". On the contrary, quantitative risk analysis (QRA) as per DNV is: "Typically, a QRA can be defined as the formal and systematic approach of identifying potentially hazardous events, estimating the likelihood and consequences of those events, and expressing the results as risk to people, the environment or the business."

Quantitative analysis is a step further from qualitative analysis. QRA is used when greater accuracy is desired and/or when qualitative hazard analysis is not quite suitable. QRA is done when more details about the risk is known. Probabilistic risk assessment (PRA) is also used. The basic objective of QRA is to achieve maximum safety by eliminating or mitigating hazards or accidents with the help of preventive and corrective controls. Precise probability of occurrence of accidents is calculated, and the risk probability is expressed using a number or rate. Hazards and accidents are first identified, then estimated with the help of engineering estimations, and experience and/or documented history of similar projects or equipment. Probabilistic safety assessment (PSA) is a classic methodology, which describes accident sequences with the help of fault tree and event tree analyses. Event tree analysis (ETA) and fault tree analysis (FTA) are widely used in industrial plant modeling, identification of accidental incidents, and evaluating reliability of safety systems. In a true sense, layer of protection analysis (LOPA) is not a quantitative but semi-quantitative method used to determine the risk of individual hazard scenarios. In this chapter, quantitative analysis systems shall be taken up. All these analyses can be arranged according to their utilities. Based on AIChE guidelines, in Fig. V/1.0-1, disposition of various

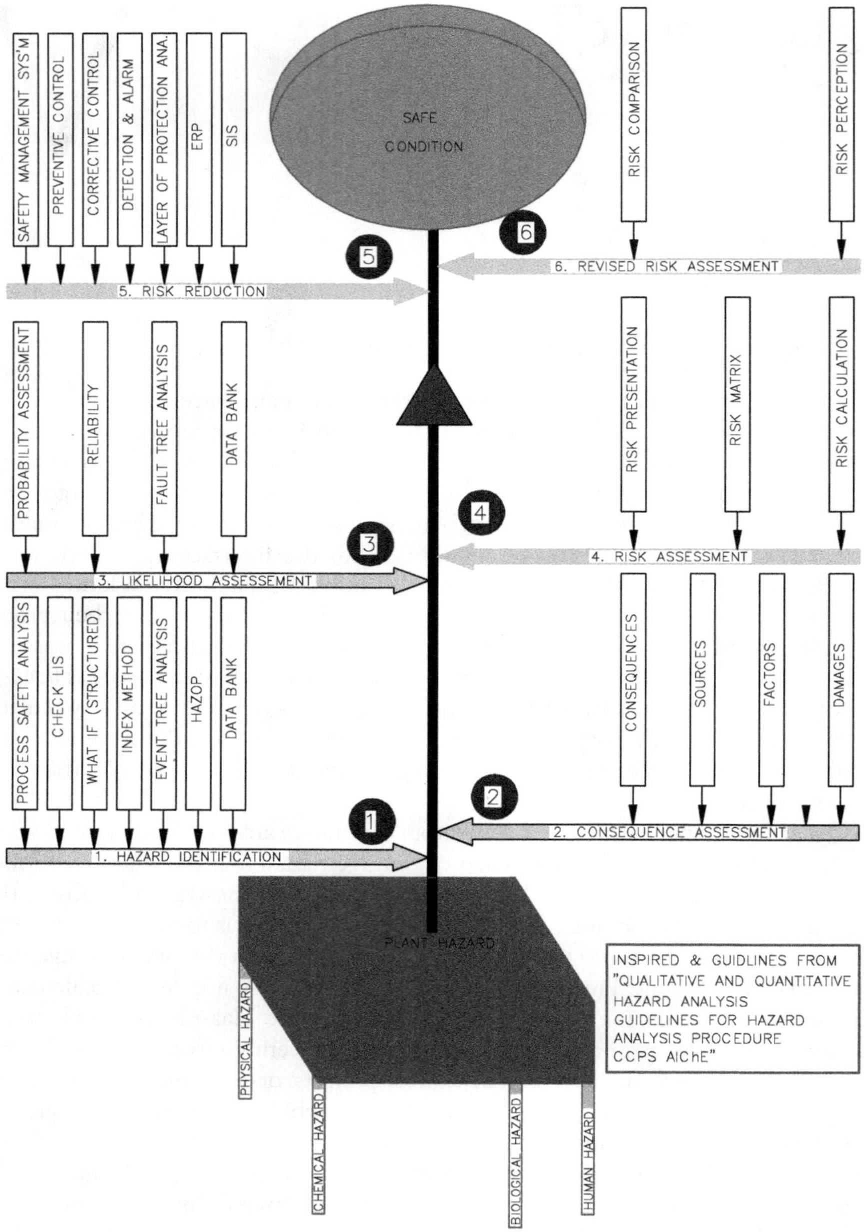

FIGURE V/1.0-1

Dispositions of various hazard analysis systems.

Table V/1.0-1 *Hazard Analysis Utility Table*

Analysis Type	Conceptual/Early Design Stage	Design Stage	Operational Stage	MOC
PHA (PrHA)	G	NR	NR	G
What if	G	NR	F	G
Check list	F	F	G	F
HAZOP	NR	G	F	G
FMEA	NR	G	G	F
ETA	NR	G	G	F
FTA	NR	G	G	F
HRA	NR	G	G	F

F, *fair;* G, *good;* NR, *not recommended.*

analysis systems have been depicted. This figure also shows the various stages of how plant hazards are handled and a safety system is arrived upon.

As stated earlier, each type of hazard analysis is suitable for a range of use. Table V/1.0-1 further illustrates this point:

The discussions on QRA starts with ETA.

2.0 EVENT TREE ANALYSIS

2.1 Event Tree Analysis Method

ETA is a very valuable system to analyze consequences as an outcome of a failure, undesired event, or an accidental event outlined in Chapter II. This is a binary-based, logical system. The binary decision diagram (BDD), discussed in Clause 5 of Chapter I, can be used to obtain outcome details. To get a general idea about ETA, refer to Fig. V/2.1-1.

2.1.1 What is an Event Tree Analysis?

ETA is an inductive, forward logic approach in reliability engineering. It is a graphical representation of a logical model used to identify and quantify the possible outcome resulting from an initiating event, which has happened (or has not yet happened) on account of failure of a component, an undesired event, or an accidental event. ETA, based on binary logic, is used as a quantitative method of fault propagation modeling. It starts with an initiating event, and defines possible outcomes for each sequence of events. Barriers or safeguards are placed into the event tree to prevent or mitigate unwanted outcomes. With the help of ETA, it is possible to check the effectiveness of the barrier in preventing or mitigating the consequence. There are some other issues involved, which also influence the propagation of fault. This would be clear from an example; in the case of a fire, if the surroundings have explosive chemical mixtures, a flammable product, and/or wind flow, then the fire will propagate very quickly.

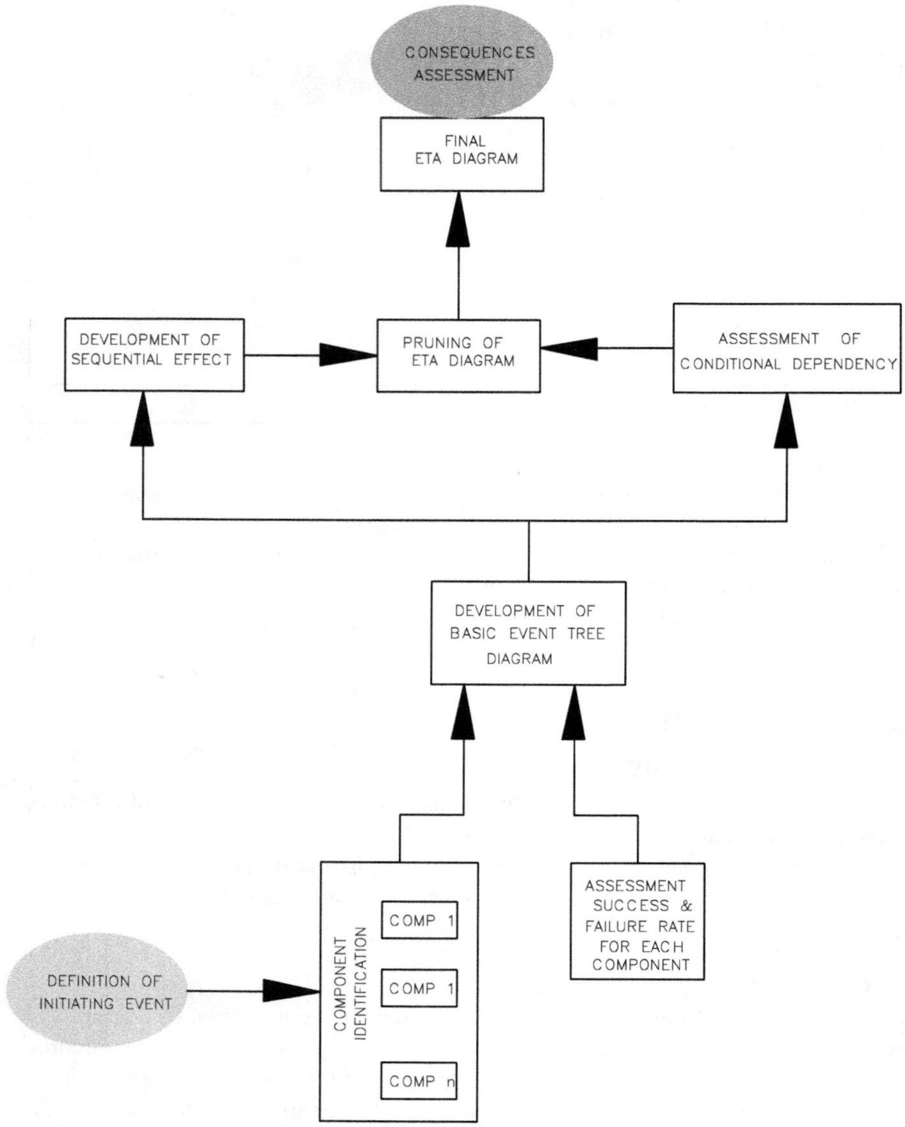

FIGURE V/2.1-1

Block diagrammatic representation of event tree analysis.

On the other hand, a fire may not propagate much if the wind flow is less, and or it is a reducing atmosphere. These are examples of external factors that influence fault propagation. Sprinklers are a *barrier*, whereas wind flow is a *factor*. So, in an ETA diagram, the barriers and safeguards and their functioning and factors are taken into

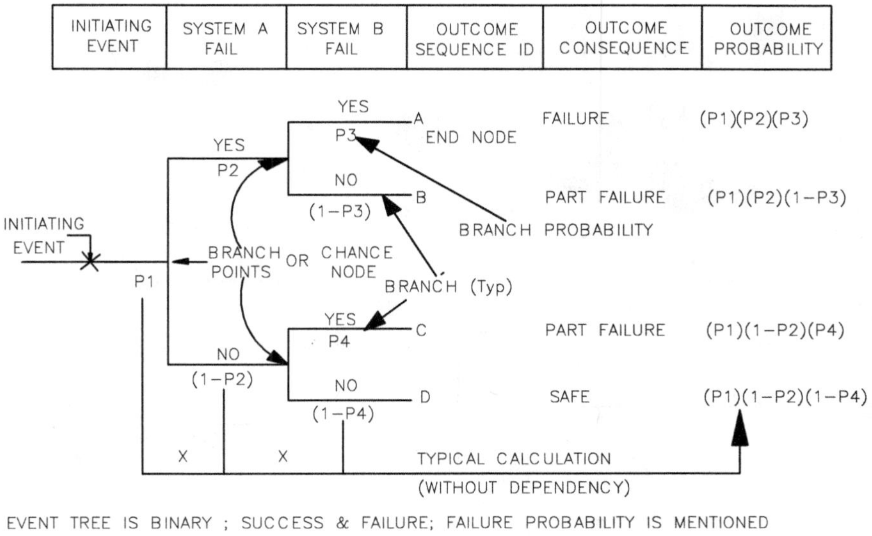

FIGURE V/2.1.1-1

Basic event tree diagram.

account. With the help of an ETA diagram, it is possible to identify potential accident scenarios, and time sequences of events in fault propagation caused on account of an initiating event. It is possible to calculate the probabilities of all outcomes so that necessary precautionary measures can be planned. In this connection, Figs. V/2.1.1-1 and V/2.1.1-2 may be referred to. In some cases, it is possible to find the weakness in design or weakness in procedure etc.

2.1.2 Terms and Definitions

During discussions in this chapter, a number of terms will be referenced, so it is better to clarify these terms at the beginning. Definition and clarification of a number of terms in connection with ETA have been enumerated as follows:

- Accidental event: An accidental event is the first major deviation from a normal situation. An accidental event may lead to a number of consequences including death or severe damage to property or the environment. In process industries, accidental events are mainly caused by the release of energy and chemical materials, for example, gas leakage or fire breakout (or it may be caused by physical means such as a falling object). What, where, and when are three "Ws," which are very important to be noted in case of accidental events. "What" stands for, "what is the accidental event; is it fire breaking out or gas leakage?" In each case, the factors and barriers to be considered would be different. "Where" refers to the location in which the accidental event took place; "is it in the control room, or in a hazardous zone? Or is it a highly populated area?" etc. (the Bhopal Gas disaster is an example). For plant operations, it is important to note when

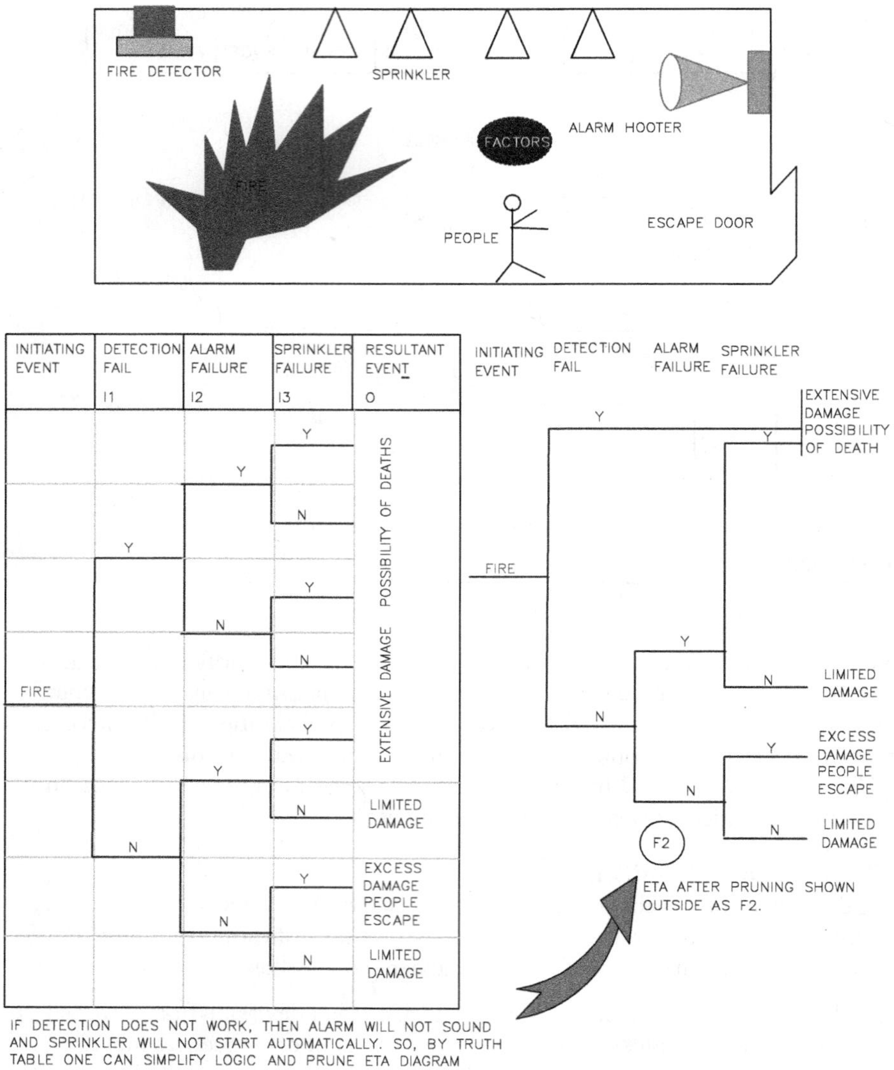

FIGURE V/2.1.1-2

Event tree analysis diagram development with example.

the accident took place; "is it during normal running, during start up/shutdown, or during offline maintenance?" How the accident happened is extremely important to note, so that necessary safeguards can be planned. "How" stands for whether an accident has happened due to human error (what about safety interlock?), equipment failure, or operation failure/process upset, etc. A few other questions are also related to these accidental events, such as, "what are consequences, and how does it propagate?" and whether there is dependency on other system(s).

- Barriers/safeguards: Generally, all properly designed systems have several safeguards/barriers as preventive or control measure to reduce consequences of potential accidental or undesired events. Depending on functionality of barriers/safeguards, the propagation of accidental events or faults may vary, and accordingly outcomes will be different. In ETA, barriers are described in a negative way, that is, barrier does not function, say, "sprinkler does not work!" Generally barriers can be classified into the following categories with examples:
  - Automatic detection: Fire detectors
  - Safety system: Sprinklers
  - Preventive system: Fire resistant low smoke (FRLS)/low smoke zero halogen (LSZH) cable
  - Alarm system: Annunciation
  - Preventive/procedural system: plant safety interlock
  - Operator action: Manual intervention
  - Mitigating control: Corrective control

The items listed here are only a few examples, and there may be some functional overlapping, for example, an alarm may help people to escape in case of fire; at the same time, an alarm may also call for manual intervention.

- Influencing factors: As discussed earlier, there are a few other issues that also influence a sequence of events. These are called influencing factors, such as presence of an explosive mixture, wind direction, gas ignition, or people present in the area and their numbers. It is therefore necessary to list and place these factors in the event tree diagram in the sequence in which they appear. If there is a fire, then to see if the fire spread quickly or not. If the fire did not spread quickly, the scenario may be different than if it had spread quickly. Here, influencing factor(s) in worst case are also described, that is, the wind was blowing in the direction of fire spread.
- Pruning: From the discussions in the previous items, it is clear that for each of the factors or devices (to be referred to as components), there will two states; success or failure. So, if there is "n" number of components, then there will be 2^n numbers of branches, as shown in Fig. V/2.1.1-2, for three components there are eight branches. Now if the numbers of components grows, then it will be difficult to handle the tree. An event tree can be pruned by eliminating all the branches that have no conditional probability. It is recommended pruning should be done when the event tree has been quantified, so that it is possible to estimate whether pruning the tree will have a significant effect. When pruning is done at the time of construction of the tree, then it poses difficulties to the analyst who cannot know in advance the effect of pruning will have on end results.
- Branch point/chance node: Referring to Fig. V/2.1.1-1, it is can be noted that there is a branching point in the event tree. This is usually designated by a circle (not shown) at the end of a branch indicating the occurrence of an unknown event. This is also called chance node.

- Branch: A possible event is represented by a line segment, preceded by a branch point or chance node, that is designated as a branch. It is a subset of the sample pace for all possible outcomes associated with a random variable. These are represented by *thick lines* in Fig. V/2.1.1-1 (and Fig. V/2.1.1-2).
- Pathway: A pathway is a unique sequence of events representing a possible set of events.
- End node: An end node is a possible end state in the sequence of an ETA diagram.
- Branch probability: The branch probability is the probability of the event represented by the branch, due to the occurrence of the events to its left in the event tree. See Fig. V/2.1.1-1.

2.1.3 Aim of Event Tree Analysis

Event trees are mainly associated with accidental incidents. They are applied to both pre- and post-accidental incidences. In case of pre-accidents, they try to predict the outcomes and helps to design barrier systems. For post-accidental incidents, they help to find the range of likelihood and potential outcomes. From the above discussions it is clear that ETA helps for decision-making on:

- Identification of all possible accidental incidents including risk importance analysis.
- Comprehensive risk quantification and calculation.
- Acceptability of the systems.
- Identification of better barriers to be provided to improve system performance in accidental incidents. Recommendations for barriers, and against influencing factors.

2.1.4 Application Area

Major application areas of ETA are as follows:

- Major accidental hazards in plant application, and health and safety application.
- Technological risk analysis.
- Identification of improvements in barrier and safety functions.
- Optimization of resource deployment for improvement of control measures, as it can identify traceability of failure progression path for single failure.

2.1.5 Advantages and Limitations of Event Tree Analysis

There are both advantages and limitation of event tree analysis:

- Advantages:
  - It enables the assessment of multiple co-existent faults and dependent events.
  - It can be used to determine the likelihood of multiple consequences.
  - It is applicable for almost all technical systems.
  - It helps in identification of single point potential failure.
  - It helps to optimize deployment of resources for improvement of controls.

- It provides a visual model that can be used to qualitatively understand the sequence of events.
- It prioritizes risks to identify the major sources of risk.
- It provides sensitivity analysis, the possibility of the automatic variation of event failure, and repair data between specified limits.
- It estimates a frequency of occurrence for key accident scenarios and can produce a comprehensive risk calculation.
- The effect of recommended actions can be immediately incorporated into the model to evaluate their effectiveness.

- Limitations:
  - There is no standard for graphical representation for event tree analysis (e.g., for a fire example, instead of detection, sprinkler failure can be taken up first, then the tree will be different, though the end result will be the **same**, but analyzing the pattern may be **different**).
  - ETA can only be applied after the initiating event is identified, that is, it cannot identify an initiating event. Also, operating conditions are foreseen by the analyst.
  - Only *one* initiating event can be analyzed in one analysis.
  - It identifies multiple pathways, but level of loss in any pathway can be identified after further analysis, and does *not* automatically come out of ETA.
  - ETA may not be suitable for *common cause failure* cases.
  - For dependent events, it is likely that subtle dependent events may be missed.
  - ETA cannot identify omissions.

2.1.6 *Initiating Event*

An initiating event is an event that creates disturbances in the system, and based on performance of various barriers and other mitigating systems can lead to potential damage to the system, directly or under the influence of other factors. Fire and gas leakage are examples of initiating events.

2.1.7 *Event Tree Analysis Discussions (Dependence)*

- Dependence: To understand the following discussions, it is better to recapitulate the basic mathematical discussions (probability) given in Clause 5.3 in Chapter I. Referring to Fig. V/2.1.1-1, it is found that outcome probabilities are multiplied by path-independent probabilities (as shown). These are very simple as long as they are *independent*. Also, it is easier if the dependence is strong, for example, $P(A/B) = 1$ or $P(A/B) = 0$. The procedure is complicated when there is weak dependence. For such cases, fault tree is helpful. Referring to Fig. V/2.1.7-1 some simplification is done as shown; System A (let the failure and success be represented by suffix f & s respectively): $P_2 = P_{2f}$, $(1 - p_2) = p_{2s}$, and System B: $P_3 = P_{3f}$, $(1 - p_3) = p_{3s}$.

When there is weak dependence, then there will be conditional probabilities. So, outputs are: Outcome 1 = $P(3f/2f) \cdot P(2f) \cdot P_1$; Outcome 2 = $P(3s/2f) \cdot P(2f) \cdot P_1$; Outcome 3 = $P(2f/3s) \cdot P(3s) \cdot P_1$; and Outcome 4 = $P(2s/3f) \cdot P(3f) \cdot P_1$.

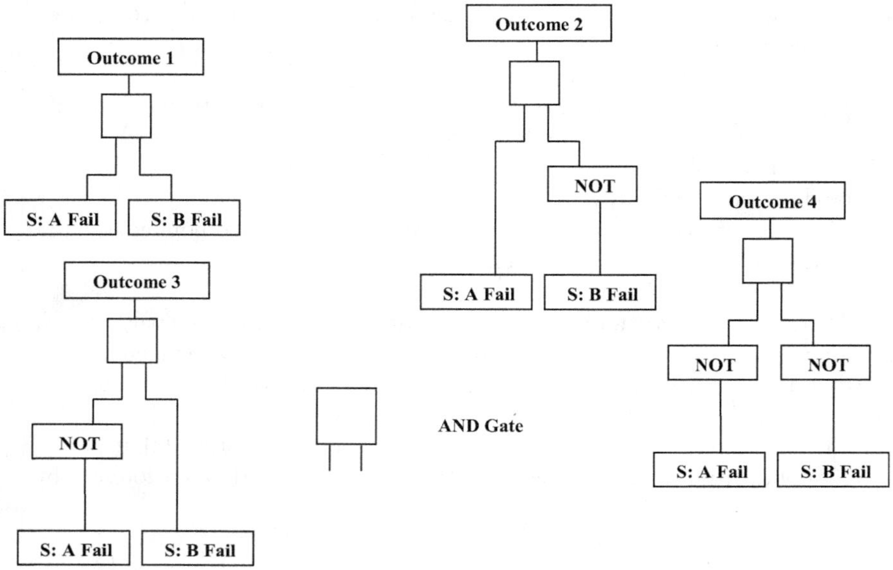

FIGURE V/2.1.7-1

Fault tree analysis representing event tree analysis.

Many analysts avoid, as much as possible, operator NOT for constructing fault trees (FTs). The existence of negated events in an FT makes its logic "non-coherent," a coined term. This is by no means a misnomer but at times are misleading. The use of a negated event in the FT logic implies that a cutset containing the original event, cannot occur if another cutset containing the negated event occurs as well. This not only creates mutually exclusive sequences of events, but its subtlety encourages the analyst to believe that non-coherent FT reduction, solution and interpretation may be carried out following the usual guidelines of coherent diagrams... but entails flawed assumptions that become evident only after the FT logic is scrutinized in detail [14].

FIGURE V/2.1.7-2

Non-coherent fault tree.

Also, from here, it is clear that there is close relation between ETA and FTA. This is detailed out to show how to estimate branch probability especially when there is dependencies. So, on account of the weak dependence, quantification of ETA depends on the accuracy of non-coherent FTA (refer Fig. V/2.7.1-2), and detailed scrutiny without misleading assumptions is necessary. From International Electrotechnical Commission (IEC) 60300-3-1 one gets that particular care is necessary when dealing with dependability. The core elements of dependability are: *availability*, *maintainability* and *reliability* of the item in the question. Based on characteristics, event tree can be:

- Functional event tree: Mitigating functions.
- System event tree: System intervention.
- Phenomenological event tree: On physical phenomena.

2.2 Event Tree Construction

An event tree is a graphical representation of event sequences with probabilities, following an initiating event. An event tree provides the systematic model of the time sequence of event propagation through a series of barriers, normal plant actions, and/or operator intervention and incidental consequences. Now, short discussions have been put forward to see how an event tree is constructed with the help of a number of steps; there are basically seven steps involved, and these are defined in Table V/2.2-1.

Event tree construction is well depicted in Fig. V/2.2-1, where various barriers have also been shown. The figure also depicts how barriers influence the outcome probabilities.

Steps involved in Event Tree construction:

- Step I: There are various hazard identification techniques, discussed earlier in the chapter, that can be applied to identify the initiating event. Suppose there is a gas leakage resulting in a fire. In this case, gas leakage is the main initiating event,

Table V/2.2-1 Steps Involved in Event Tree Construction

Construction Steps	Basic Functions
Identification of initiating event	Various hazard identification techniques
Identification of barriers and influencing factors	Identification of various safety functions of the system, better known as barriers. Also, influential factors, if any, need to be identified, for example, wind flow and direction for fire hazards
Details of construction	Chronological development and progression of incidents following an initiating event
Outcomes and their classification	Develop all possible outcomes from ETA and their grouping/classification
Probability estimation	Finding **conditional** probabilities of each branch path with respect to preceding path, and then finally arrive at probability in each path
Quantification of outcomes	Final outcome probability is estimated by multiplying path probabilities
Testing of outcomes	Testing of the result with experience and commonsense, especially when any non-coherent FTA is involved

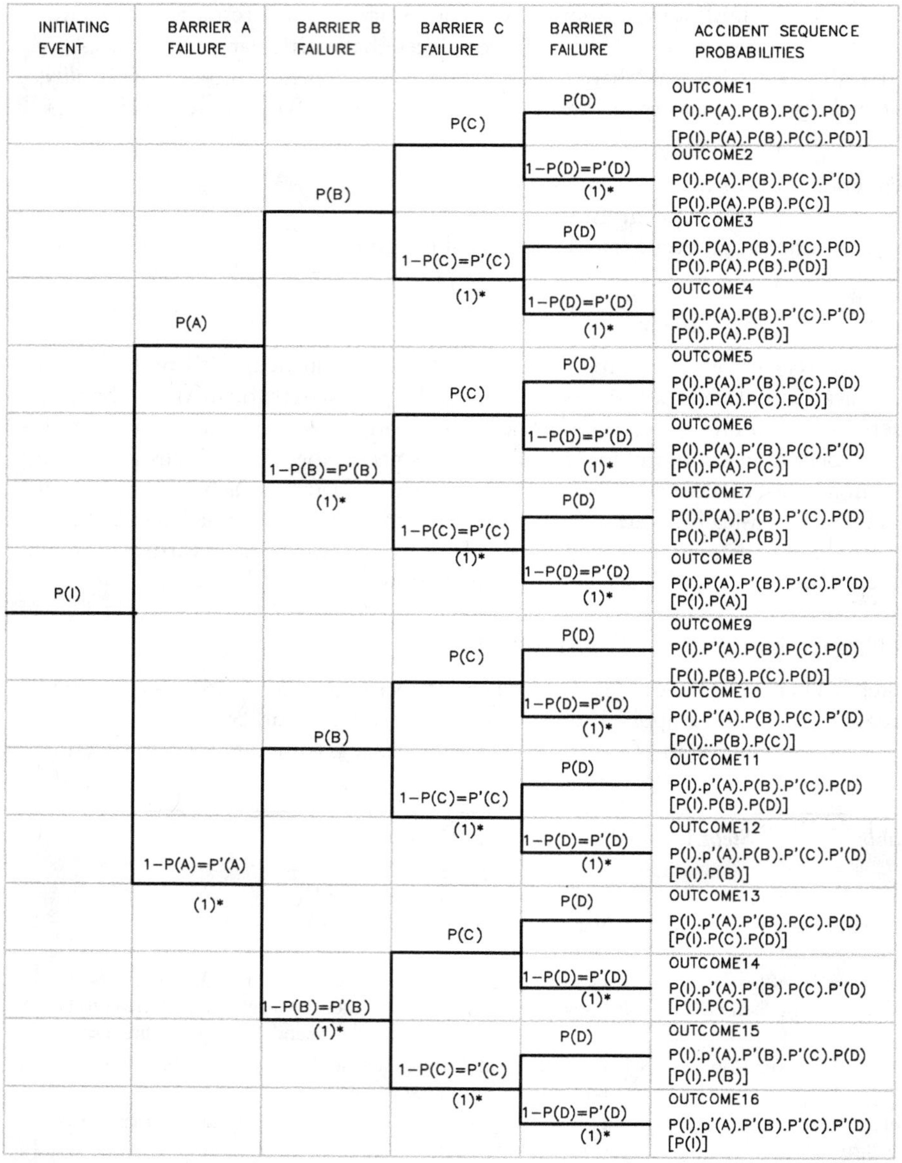

*IF PROBABILITY OF FAILURE IS TOO LOW THEN, 1-P(A/B/C/D) WILL BE ~1. SO, WITH THIS APPROXIMATIONS, NEW SIMPLIFIED PROBABILITIES ARE SHOWN IN PARENTHESIS, AGAINST EACH VALUE IN SEQUENCE PROBABILITY.

FIGURE V/2.2-1

Construction of an event tree.

and fire is a sequential event. So, in this case, ETA needs to be initiated from gas leakage, and not from fire starting, because necessary barriers to prevent gas leakage are required.

- Step II: Identification of barriers and influencing factors. For each system, there are a number of safety functions or barriers built in to the systems. So, even if a hazard has happened, these barriers interrupt further progress of the hazard. A few barriers include automatic safety systems, alarms, operator action, detection systems, containment, etc. A sprinkler is a specific example as well. In plant controls, there are a number of such safety features built in as shown in Fig. V/2.2-2 (to explain barrier and functions and their sequence of appearance). In this figure, a separator of offshore drilling is shown where liquid (water and oil) is separated from gas, and taken out through different outlets as shown. The lower part is actually oil and water that will be taken out, while gas will be taken out from top. A mixture of gas, oil, and water comes out from which are sent to the separator. It is connected with a choke and kill. So there is a possibility that the separator may be highly pressurized. Also, there is a possibility of explosion of the separator due to the high pressure. There are three stages of barrier here; in the first stage, if the pressure inside separator is high, then the first inlet valve will be closed. If pressurization is not arrested, then pressure safety valves will operate. In the last stage, a rupture disk will operate for safety. So while developing an event tree, it is necessary to identify these barriers. In addition to the barriers, one needs to give attention to influencing factors, for example, blowing wind in the direction of fire may increase chances of fire hazard in the case of gas leakage. On the contrary, wind blowing away, in the same example, may help to dilute gas in case of, say, LPG leakage!

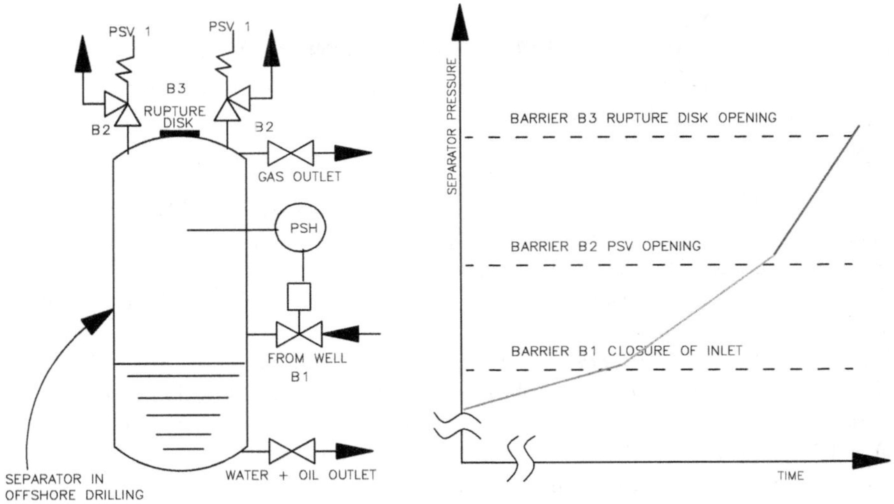

FIGURE V/2.2-2

Various barriers in plant applications.

- Step III: Details of construction. An event tree is a graphical representation of event sequences with probabilities, after an initiating event. Commencing with the initiating event, the event tree is constructed from left to right. First, potential resulting accident sequences are listed out. Then, one needs to list out barriers and influential factors. For post-accident analysis, final outcomes actually correspond to a range of incident outcomes. All stages' (intermediate/final) headings are placed at the top of the analysis list, as shown in Fig. V/2.2-1. At each node, several alternatives need to be considered for arriving at an outcome for the node. As this is a binary system, at each node corresponding to each component successes and failures are considered. Barriers and incidents are named as A, B, C, and D from left to right. After construction of the entire event tree, analysts eliminate a number of obvious issues such as an impossible issue to prune event tree as shown in Fig. V/2.2-3. Pruning has been explained with examples in the next clause.
- Step IV: Outcomes and their classification. It is not compulsory, but at times outputs are categorized, and similar outcomes are placed in one group.
- Step V: Probability estimation. Other than an initiating event, for each heading, conditional probability is estimated (in similar fashion discussed in Clause 2.1.7 above). Probability of each heading is 1.0, so, probability of success (P_s) for one heading will be $1 - P_f$ when probability of failure is P_f. Each barrier or factor (worst case) is described by a *negative statement* so, P_f is defined for each heading.
- Step VI: Quantification of outcomes. Deriving the Boolean expression corresponding to each accident incident, assuming an AND logical gate. Each outcome is determined by multiplying initiating event frequency by *conditional* probabilities of each pathway to the outcome.
- Step VII: Testing of Outcomes. To get better results, it is recommended to test the result with common sense, expert opinion, and history.

2.3 Event Tree Analysis (Quantitative Analysis): Evaluation

With the basics about event trees covered, now it is necessary to understand the probability of arriving at various outcomes.

2.3.1 *Basic Calculations*

If $P(\mathrm{A})$ is probability of failure of component A, then its success probability is $1 - P(\mathrm{A})$ as probability of failure + success of any component is 1.

So, for example, in Fig. V/2.2-1, Outcome $2 = P(\mathrm{I}) \cdot P(\mathrm{A}) \cdot P(\mathrm{B}) \cdot P(\mathrm{C}) \cdot P'(\mathrm{D}) = P(\mathrm{I}) \cdot P(\mathrm{A}) \cdot P(\mathrm{B}) \cdot P(\mathrm{C}) \cdot \{1 - P(\mathrm{D})\}$ by *ANDing* of all probabilities in the pathway.

Similarly, when gone by possibilities, Outcome 2 will be: $\mathrm{Min}\{\prod(\mathrm{A}), \prod(\mathrm{B}), \prod(\mathrm{C}), [1 - \prod(\mathrm{D})]\}$ whenever probabilities or possibility of failure is too low, for example, in the order of $<10^{-4}$, then probability of success will be 1, that is, if $P(\mathrm{A}) < 10^{-4}$ then $1 - P(\mathrm{A}) \sim 1$, so, the outcome will be $P(\mathrm{I}) \cdot P(\mathrm{A}) \cdot P(\mathrm{B}) \cdot P(\mathrm{C})$. Hence, there will be a simplified event tree diagram as shown with simplified outcomes within parenthesis in Fig. V/2.2-1.

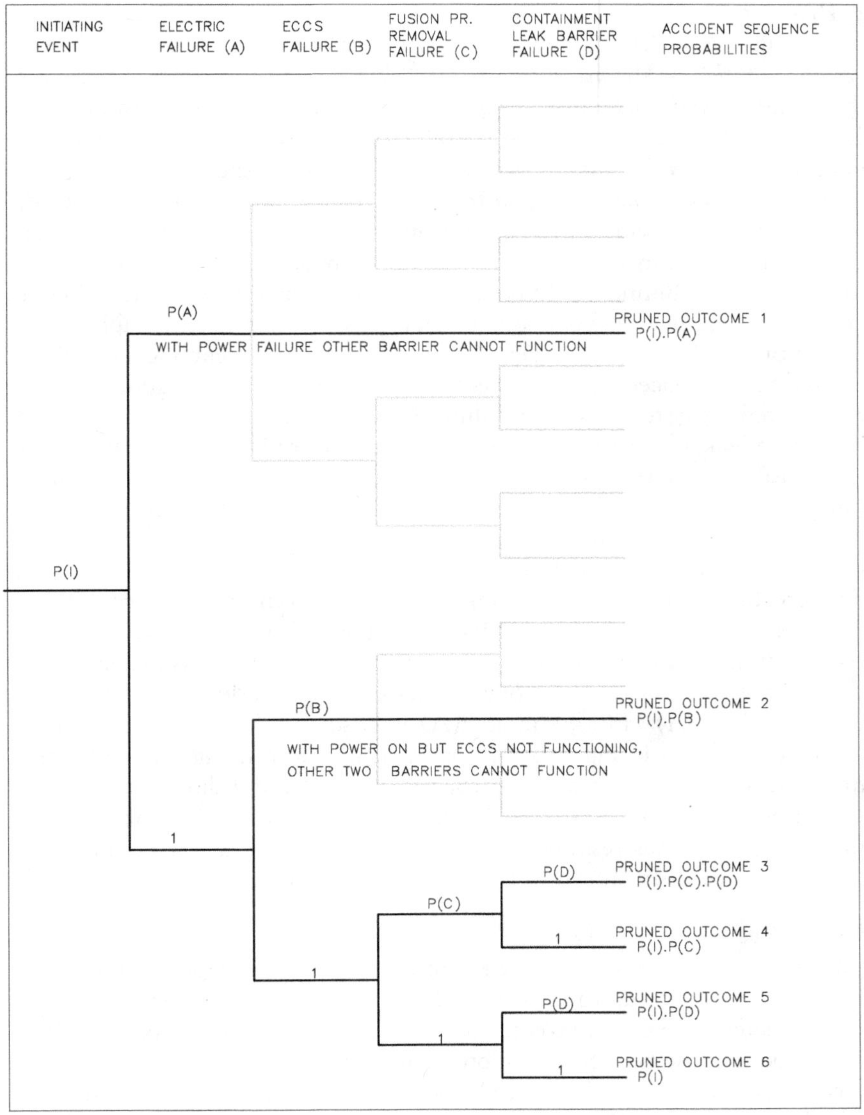

PRUNING OF EVENT TREE WITH LOW PROBABILITY OF BARRIER FAILURE. IN ETA FAILURE PROBABILITY USUALLY PUT FIRST & SUCCESS PROBABILITY LATER. FOR COMPARISON BACK GROUND LINES ARE KEPT & OUTCOMES ARE KEPT TO SHOW HOW BY PRUNING TREE HAS BEEN MADE SLIM & REDUCED.

FIGURE V/2.2-3

Construction of an event tree with pruning.

2.3.2 *Pruning*

As discussed in Clause 2.1.2, for "*n*" number of components, then there will be 2^n number of branches, as shown in Fig. V/2.1.1-2, for three components there are eight branches. Now if the number of components grows, then it will be difficult to handle the tree. An event tree can be pruned by eliminating all the branches that have *no* conditional probability. It is recommended *pruning* should be done when an event tree has been quantified, so that it is possible to estimate whether by pruning the tree there will any significant effect. There are certain obvious inferences the analysts can find so that a part of probabilities, that is, impossible occurrences, can be eliminated. This can be better understood with the help of a very common example of loss of coolant in a typical nuclear plant (simplified for clear understanding), as shown in Fig. V/2.2-3. In the figure, components A, B, C, and D have been replaced by components electric failure (power failure), emergency core cooling system (ECCS) failure, fusion product removal failure, and contamination leakage, respectively. In a typical nuclear plant, major basic safety issues are: to control reactivity, removal of decay and stored heat, protect contaminant integrity, maintain primary coolant integrity, and scrub reactive material from containment atmosphere. In order to simplify the issue, four major issues are addressed as indicated above, that is, electric failure (power failure), ECCS failure, fusion product removal failure, and contamination leakage. If there is total power failure, then all other components cannot function, so probability of Outcome 1 (pruned) happening is $P(\text{I})$. Similarly, when power is available but ECCS fails, then the other two components cannot function, so Outcome 2 (pruned) is $P(\text{I}) \cdot P(\text{A}) \cdot P(\text{B})$, and as $P(\text{A})$ is 1, so it is $P(\text{I}) \cdot P(\text{B})$. All other outcomes are as shown in Fig. V/2.2-3. From this, it is seen than for four components there should have been $2^4 = 16$, but as is seen and discussed, obvious impossibilities have been discarded, so after pruning, a simplified event tree with six numbers of outcomes resulted. In order to have better understanding, the original $2^4 = 16$ has been shown in light color in the back ground.

2.3.3 *End Events*

There are various outcomes, and for each of the frequencies or probabilities, and consequences are estimated. Table V/2.3.3-1 shows the end events and associated consequences after the event tree is constructed, pruned, etc. Frequency of each of the episodes is placed in the necessary location. Now analysts need to find the impact of the outcome in terms if loss of life, production, material, equipment, and environment. As discussed earlier, these consequences may be classified as nil, low, medium, and high-impact against each of the headings such as loss of lives, loss of materials, etc. Based on the same, decisions are made regarding further actions such as control actions, or additional barriers, etc.

2.3.4 *Decisions*

At this stage, it is important to make decisions regarding:

- Judging the acceptability of the system. Here, the acceptability is judged after exploring the possibilities to reduce accidental events.

Table V/2.3.3-1 *End Events and Associated Consequences*

Event Tree Diagram	Outcome ID	Frequency	Loss of Lives				Loss of Material				Loss for Environment			
			Nil	Lo	Med	Hi	Nil	Lo	Med	Hi	Nil	Lo	Med	Hi
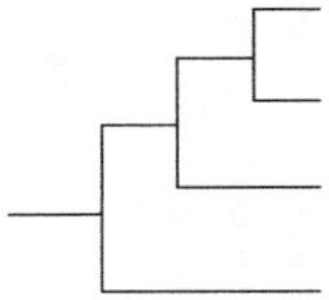	1 2 3 4	$P(1)$ $P(2)$ $P(3)$ $P(4)$												

- Introduction of additional barriers to improve the situation. At this juncture, several alternative arrangements such as preventive, detection, and/or mitigating controls are explored, for example, use of intrinsic safety circuit, increasing redundancies, etc. are considered so that failure probabilities can be reduced or eliminated. These control measures could also reduce impact on life, material, or environment.
- Recommendations: Analysts can put forward their recommendations at this stage, as discussed previously.
- Justification: Whenever any additional resource deployments are done, they need to be well-justified so that corporate management people are well-aware of the situation, and also convinced about the recommendations, etc.

As discussed earlier, there is hand-to-hand relationship of ETA with FTA, so now discussions will be on FTA, which is another QRA system.

3.0 FAULT TREE ANALYSIS

FTA was originally developed in 1962 at Bell Laboratories, under a U.S. Air Force Ballistics Systems Division contract to evaluate the Minuteman I Intercontinental Ballistic Missile (ICBM) Launch Control System [2]. This method is widely used in the area of safety engineering, to quantitatively determine the probability of safety in an undesired or accidental event referred to as a top event. With reference to Fig. V/3.0-1A, at the beginning of the chapter the importance of boundaries is discussed. In the figure, there is some gap between the external boundary and actual problem part (actual problem boundary). This is the interface part, and is quite important in any risk analysis. In the case of FTA, it is also important to define limits of resolution. A system is usually divided into subsystems, and in some cases a part may be subdivided into smaller parts. Now, the smallest part determines the limit of resolution and internal boundary. If, in one case, fault level is to be determined up to card level, and in another case it is necessary to determine the fault up to component level, naturally in the latter case, resolution will be higher. Limit of resolution has been shown in Fig. V/3.0-1A.

FTA is a backward logic, deduction-type hazard analysis method. In the case of ETA (see Clause 2.0), forward, inductive-based logic has been discussed. Both inductive and deductive types have been illustrated in Fig. V/3.0-1B, with explanations of why they are called forward or backward logic. In hazard analysis, a cause creates accidental events, which results in consequences. They can be looked at from different perspectives also. An inductive method is from specific (event) to general (consequences). On the contrary, in a deductive method, it is from general (event/consequence) to specific (set of causes). In the case of an inductive method, analysis is carried out from event to consequence in forward direction. In a deductive method, analysis is carried out to find the causes from event in the reverse order.

FTA is a deductive technique, which focuses on a particular event or consequence. The purpose is to identify initiating events that lead to this consequence. A completed fault tree shows many scenarios, all with the same consequence. In plants, there could be a number of such consequences, so, analysts need to look for all them separately;

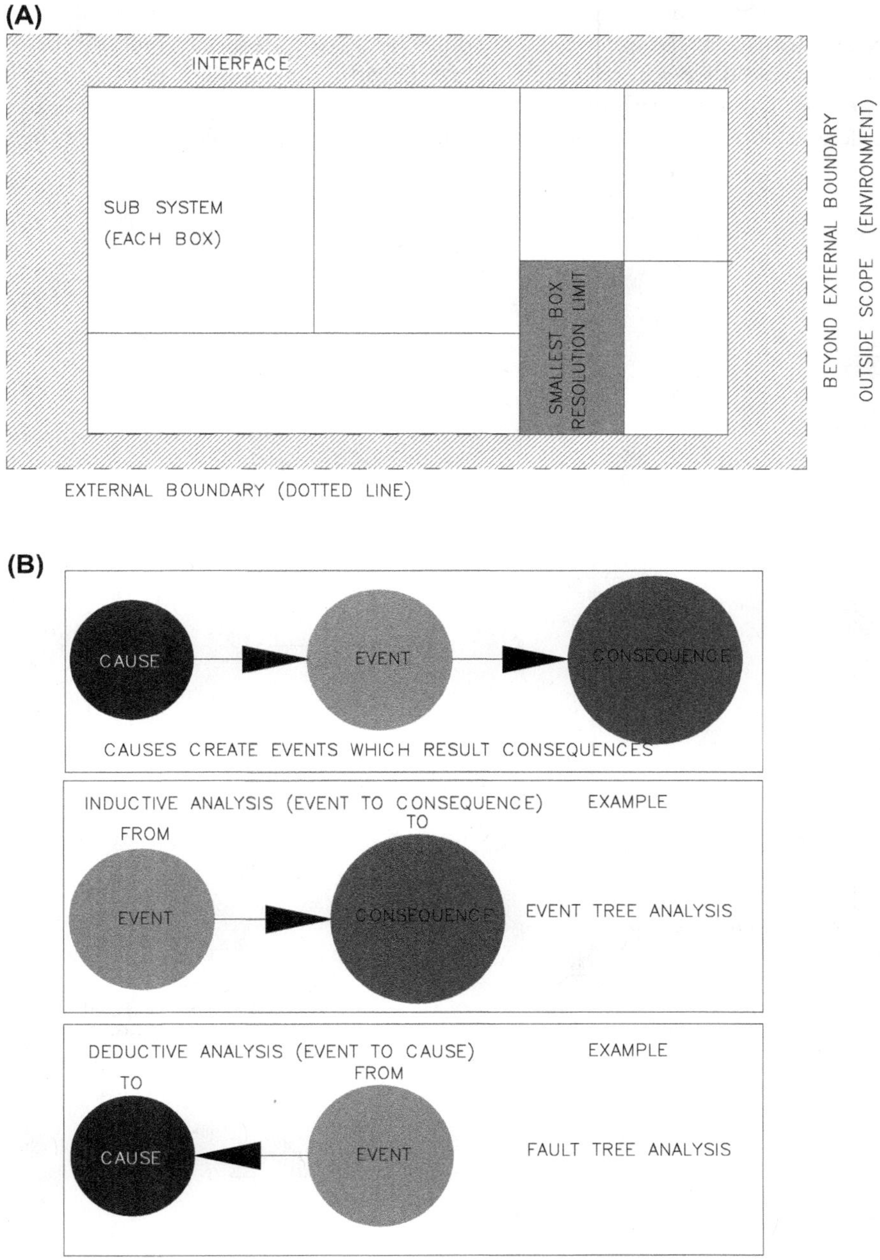

FIGURE V/3.0-1

An approach to fault tree analysis. (A) External boundary & resolution limit, (B) deductive analysis—FTA.

this is an inherent disadvantage of FTA. In contrast, with this analysis it is possible to arrive exhaustively at the exact cause of failure. This is a distinct advantage. In this deductive method, logical approach is taken, with some specific requirements. All these are illustrated with the help of a logic diagram, for which it is necessary to have detailed look at the symbols and legends shown in Fig. V/3.0-2. These are the

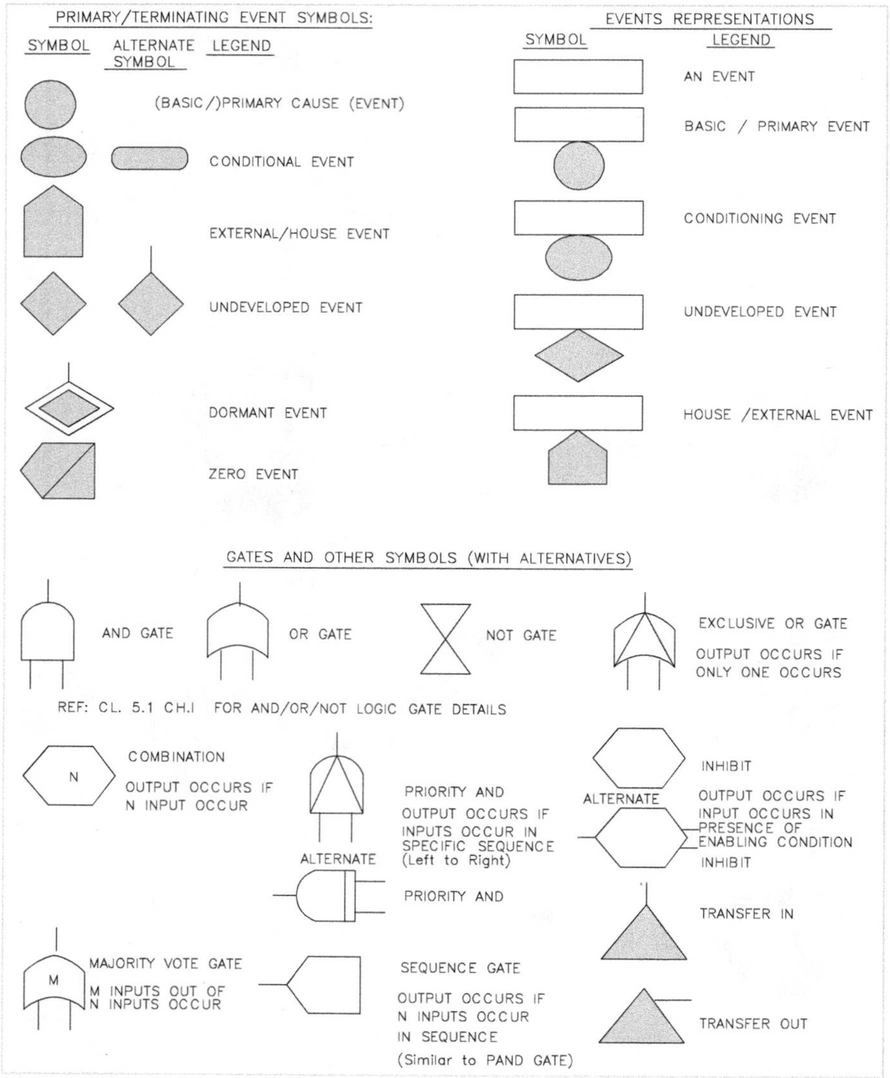

FIGURE V/3.0-2

Symbols and legends for fault tree analysis.

symbols normally used in FTA, may be little different from mathematical symbols shown in Chapter I. Some typical representation examples have been shown in Fig. V/3.1.4-2 also.

FTA is somewhat similar to fish-bone analysis another hazard analysis method (Fig. V/3.0-4). When compared with fishbone analysis shown and discussed in Fig. V/3.0-3 and Fig. V/3.0-4 respectively, FTA is more formal and specific, that is, it resolves basic causes for the accidental event or consequences.

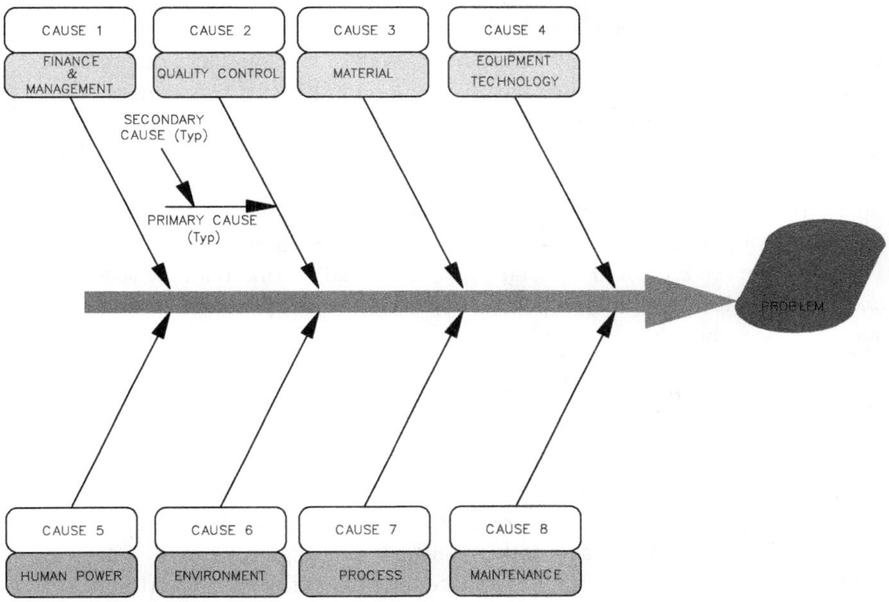

SHADED BOXES ARE FOR A TYPICAL MANUFACTURING PROCESS, WHILE OTHERS ARE GENERAL CAUSES (1–8)

FIGURE V/3.0-3

Fishbone diagram example.

For a serious problem, it is important to explore the causes for it before really thinking of a solution for the same. In 1968, Kaoru Ishikawa developed a diagram based on brainstorming and a type of mind map for considering all possible causes of the problem. This is known as fishbone or cause–effect analysis diagram. Fishbone diagrams provide a structured means for discussion around the potential causes of the problem. The fishbone technique is easy to implement and creates an easy to understand visual representation of the causes, categories of causes, and need. The simplicity of a fishbone diagram is also its weakness—as it may be difficult to represent the truly interrelated nature of problems, and causes in some very complex situations. Commonly it is used in product design and quality defect prevention. Equipment, process, people, machines, materials, measurements, and environment are some of the categories of causes commonly used.

FIGURE V/3.0-4

Fishbone analysis diagram discussions.

Now in the following subclauses a few important basic issues of FTA will be addressed:

1. Major utilities: FTA is a clear graphical representation of the system to understand the logic leading to the top event or undesired state. Analysis is carried out in a top—down fashion with interactions to individual components. FTA not only is capable of resolving the problems into specific causes, it also can quantify failure probabilities. FTA can optimize the resources to ensure safety and upgrade the product and process, that is, the system as a whole. The FTA can be used as a design tool that helps to create requirements as well as assess design deficiencies. Thus, FTA also has role as a diagnostic tool to identify and correct causes of the top events, and can be used to develop diagnostic manuals and processes. FTA can be used to model system failures in risk management. An event consists of an initiating event and subsequent system failures, so with the help of FTA logical representation it is possible to develop a PRA model so that probability of accident, that is, top event, can be determined. Based on operating procedure, FTA can be used to get human interactions with the system, and is capable of predicting human error probability.
2. Failure space: For ETA and especially people normally work in failure space. For the following reasons, analysts prefer to work in failure space rather than in success space, though one complements the other:
   - Success depends on a number of factors; for example, throughput, efficiency, etc., so finding the threshold is difficult.
   - Failure threshold is easily determined.
   - Failure can be resolved into more discrete items.
   - Quantification of failure thus is easier.
   - It is easier to come to a concurrence regarding failure than success.
3. Fault and failure: For FTA, a few other issues are also important, such as defining what is a fault and what is failure. It is better to explain the issue with the help of an example. A pump is delivering water to the system from a sump whose level is monitored by a level switch. Case 1: If the pump fails to operate due to damage in bearing or impeller, the result is that there will be no water supply. Case 2: If the pump stops, or trips, from a level switch signal, this will also result in no supply of water. The two cases are not same. In the first case, the pump could not deliver water on account of the failure of the pump. In the second case, the pump really does not have a failure. There may be for some reason, a low sump level, and the level switch operated and stopped the pump. In the second case, it is a system fault for which undesired event took place. So, it could be stated that *all failures are faults, but all faults are not failures.*
4. Failure modes and failure mechanism: A failure mode is the failure state of the system or component. Lack of water supply is a result of a failure state of the system; if there is no water supply due to failure of the impeller, then it is the initiator hence the failure mode (impeller failure). Now if impeller of the pump fails due to corrosion, then it is the failure mechanism. So, failure mechanism stands for the mechanism by which the failure occurs.
5. Active and passive components: There are two kinds of components encountered in a process; active and passive. Active components, such as a

pump or valves are active because they can originate or modify the system; for example, opening and closing of valves modifies the flow in the pipeline, hence the system. On the other hand, passive components such as pipeline, electrical wiring, and structural members are used as for transmission of signal of material, etc. Passive component operations have direct impact on the system.

6. State of system versus state of component faults: A "state of component" fault is associated with a component. A "state of system" fault is associated with more than one component of the system.

With this knowledge, detailed discussions can occur.

3.1 Fault Tree Analysis Basic Details

In this clause, advantages, limitations of FTA, cut set, minimal cut set elaborations, and basic methodology will be covered so that readers can get an idea about the entire process. From the discussions, it is clear that FTA depicts a diagrammatic, logical view of the multiple casualties, failures, and interactions with individual components in a top—down fashion. Some pros and cons of FTA are detailed below.

3.1.1 *Advantages of Fault Tree Analysis*

- Graphical representation of FTA, helpful in understanding the logic.
- Starting from top event, users can select the point of interest and find the root cause for the same.
- Monitoring and control of the safety performance of a complex system is possible, and it can optimize resources.
- FTA shows compliance with system safety and reliability requirements.
- FTA provides framework for detailed qualitative and quantitative analyses, which give complementary information.
- Cut set (Clause 3.1.3) method can be deployed to identify components which provides the highest probability contribution to the unwanted top event.
- Results can directly prioritize the contributors leading to the top event; creating the critical equipment. parts, and events lists for different importance measures.
- By mathematical modeling of FTA, a number of modes could be arrived at; repairable, non-repairable, and stand-by modes.
- There could be consideration of sensitivity cases for modifications to system components, architecture, and component testing intervals.
- Easy conversion of a system to other systems is possible, for example, reliability block diagram, and system model for evaluation of nuisance trip rates.
- Provides detailed reports and records of the systematic analysis and basic causes for the undesired or top event.
- Easy availability of software tools for performing FTA in computer. Also, on account of its operation following Boolean logics, it can be integrated with many other systems, for example, power flow integration in nuclear power plant applications.

3.1.2 *Limitations of Fault Tree Analysis*

- Large amount of effort and time is necessary for a complex system.
- Possibility of *omissions* failure event probability.
- Partial failure is nonexistent.
- For calculating error probability of top event, probability of all basic events are necessary, which are not always possible to obtain.
- There are chances of error in calculation, especially when performed manually.
- Difficult to handle *large* systems on account of time and effort.
- Different nature of FTA will result when same problem is looked over by different safety personnel.

3.1.3 *Cut Set and Minimum Cut Set*

A cut set is a listing taken directly from the fault tree of the events, all of which must occur to cause the top event to happen. Cut sets are the unique combinations of failures that can cause a top event to happen. A minimum cut set (MCS) is the smallest combination of primary or basic events that can result in the top event or undesired incident. This means that when any basic event is removed from an MCS, then the remaining events collectively are no longer a cut set. Thus:

- Each MCS is a combination of primary events.
- The complete set of MCS provides complete set of causes of the top event.
- It links the top events with primary events.
- Single-element MCS identifies a single failure cause of top event (e.g., valve failure to open—event E, in Fig. V/3.1.5-1).

3.1.4 *Basic Methodology*

A basic block diagram depicting several steps involved has been depicted in Fig. V/3.1.4-1, which is a block cum flow diagram for FTA methodology and is explained below. Before starting the analysis, it is important to first understand the process and try to identify the top event for which FTA is to be carried out. The important issues here are:

- Set a basic target objective.
- Definition of scope and boundary.
- Internal boundary (resolution).
- Understanding of the process and identification of top event.
- Associated interface parts to be covered.

The starting point is the identification of a top event. The next step is to resolve the FTA by defining the causal events. Also, it is necessary to define initial state of the system (see Clause 3.2.7). The next important step is to set up the ground rules (see Clauses 3.2.1—3.2.6) as listed as follows:

- The immediate, necessary, and sufficient rule.
- Clear statement rule.
- No miracle rule.

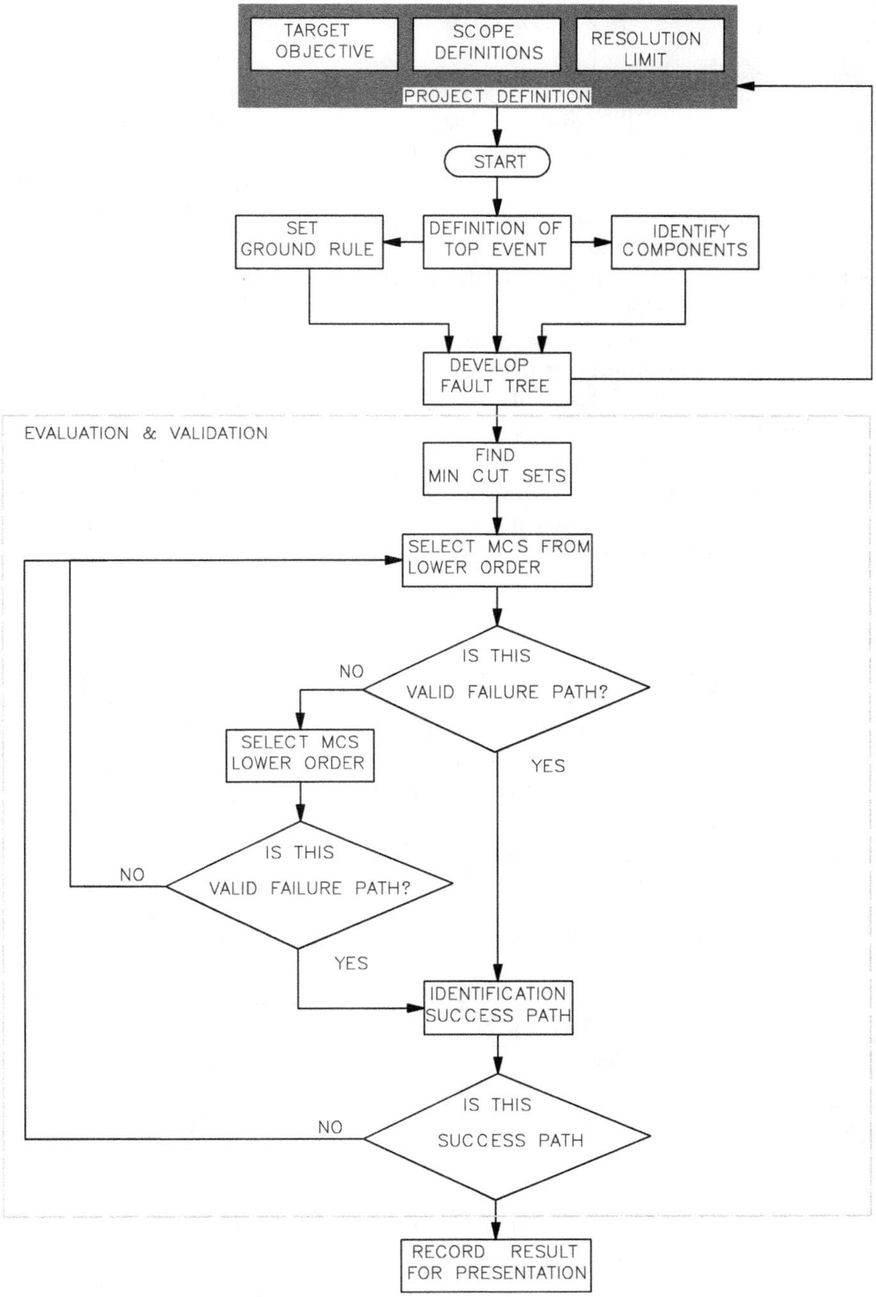

FIGURE V/3.1.4-1

Block cum flow diagram of fault tree analysis methodology.

- Complete the gate rule.
- No gate-to-gate rule.
- Component/system fault rule.

All the ground rules will be discussed *later* with examples. At this point, only the above diagram is explained. In the next step after developing FTA, one needs to find the MCS. To find the MCS, one needs to start with the lower-order MCS and check whether it is a valid path to reach the top event, that is, valid failure path. In this way, various cut sets are found and the MCS is identified. After these are done, the entire system is recorded for presentation. The fault tree is terminated either at basic event or at undeveloped event (see Clause 3.2.7). The flow process shown in Fig. V/3.1.4-1 is for basic understanding of the system only. In reality, the fault tree is evaluated by transforming it into an equivalent set of logic equations. The equation is expanded and simplified to be a sum of products (SOPs), which are further simplified using various laws applicable in Boolean algebra, so that final equations provide a basis for qualitative and quantitative evaluations. Based on symbols and legends used in FTA (Fig. V/3.0-2), some of the additional representation examples have been shown in Fig. V/3.1.4-2 so that readers can get idea about typical representations in FTA.

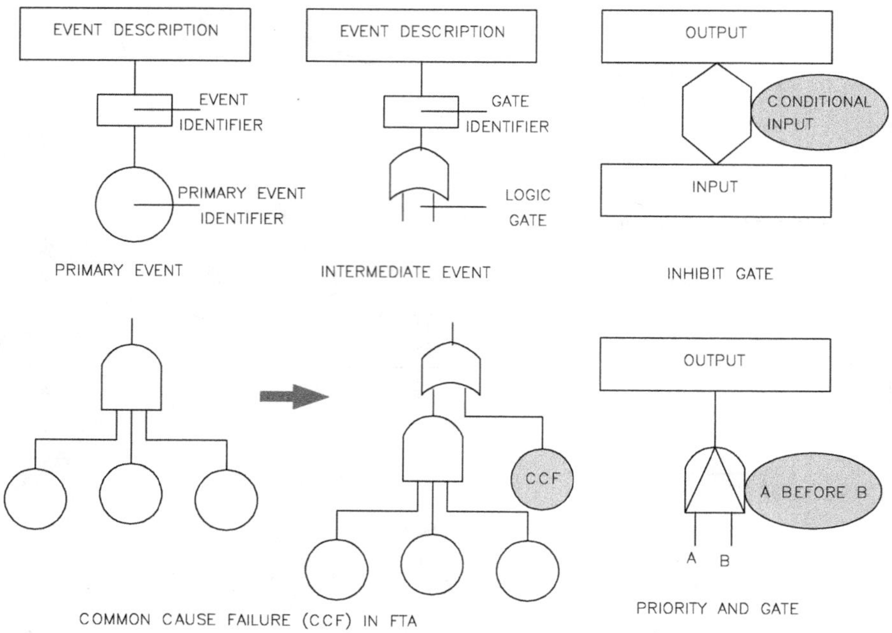

FIGURE V/3.1.4-2

Some fault tree analysis representation examples.

3.1.5 *Explanation With Examples*

Basic methodology discussed previously shall be explained with some examples as shown in Fig. V/3.1.5-1. In this figure, two examples have been tried. In the first case, a typical fire takes place.

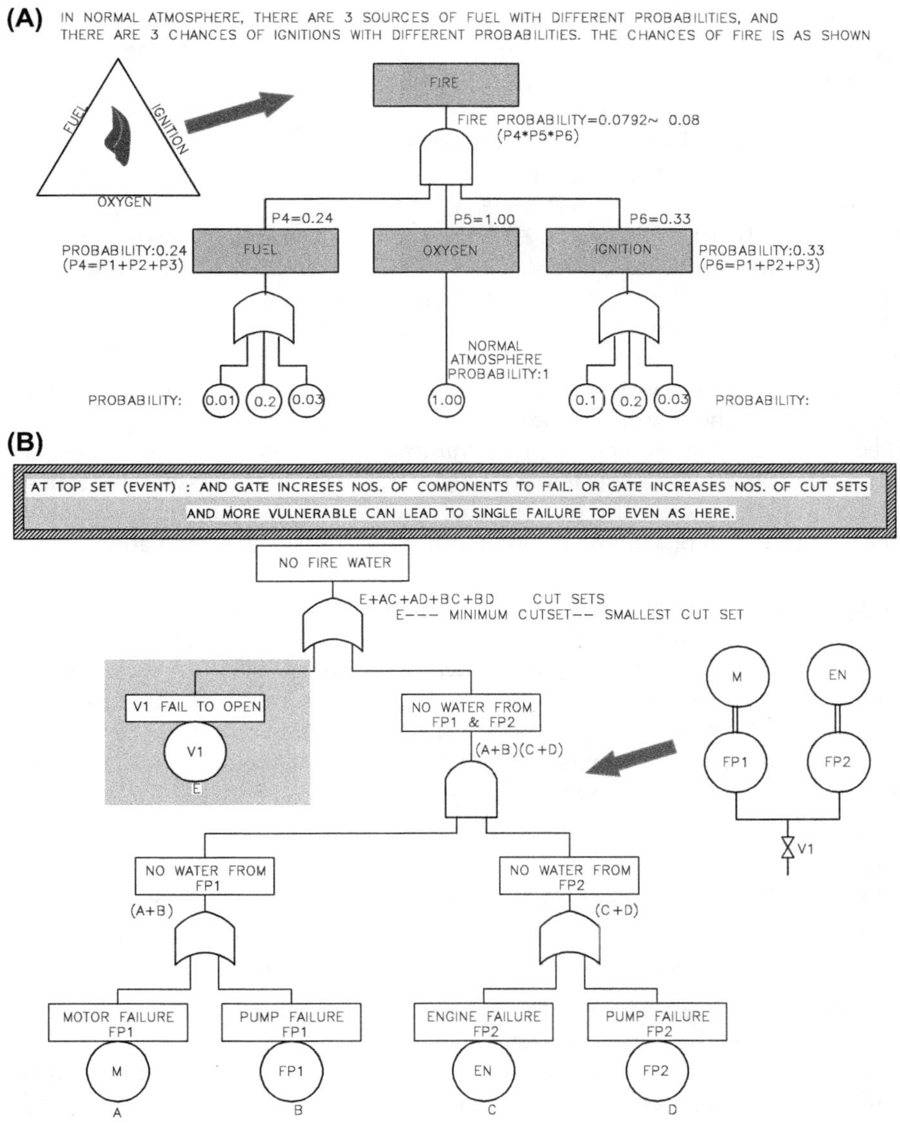

FIGURE V/3.1.5-1

Fault tree analysis methodology with examples. (A) FTA methodology example for fire, (B) FTA methodology example for fire water with cut sets.

It is well known that there shall be combinations of three things, viz. fuel, oxygen, and ignition possibilities, to create fire. Assume there are three sources of fuel (may be flammable substance also), with probabilities as shown (0.01, 0.2, 0.03). Similarly, assume there are three possibilities of ignition (such as high temperature) with probabilities as shown (0.1, 0.2, 0.03). From there, the probability of fuel source is 0.24 ($0.01 + 0.2 + 0.03$); similarly, probability of ignition is 0.33 as shown. As it is in normal atmosphere as stated in the figure so, oxygen probability is 1.00. Therefore the probability of fire is $0.33 * 0.24 = 0.0792 \sim 0.08$.

Now, as oxygen is 1, various combinations of fuel and ignition possibilities will lead to fire. If fuel sources are designated as A_1, A_2, and A_3, and ignition possibilities as B_1, B_2, and B_3; then fire can occur in a combination of $A_1B_1 + A_1B_2 + A_1B_3 + A_2B_1 + A_2B_2 + A_2B_3 + A_3B_1 + A_3B_2 + A_3B_3$. These are cut sets. Here it is seen as there is an *AND* gate at top set, so *at least* two components are necessary and in this case, component-wise all cut sets are the same but quantitatively, the A_2B_2 combination has the highest probability of fire.

In the next case, there are two fire—water pumps. One is motor driven, and the other is engine driven. Both are connected in a common pipeline where there is a common discharge valve as shown in Fig. V/3.1.5-1. Now, before going for any logical details, looking at the flow diagram it is clear that if the valve fails to open, then there will be no supply of fire—water. The most vulnerable component is the valve, as per the logic (*to be read in conjunction with the Fig. V/3.1.5-1*). For expressing things in a simpler way, each basic event is represented by A, B (motor and pump, respectively):

FP1 to fail: A + B; but water supply can be there if FP2 combination work. Similarly;

FP2 to fail: C + D water supply can be there if FP1 combination work.

But if both fail, then there will not be any water. Also, if the valve fails, there will be no water. Thus, from the pump side, water to fail is a combination of $(A + B)(C + D) = AC + AD + BC + BD$. Therefore, a top event can happen if any one of AC/AD/BC/BD or E is present. Hence more vulnerable. So, AC/AD/BC/BD—all these are cut sets, and E is the MCS as it has the least component (E). When this is compared with first example, it can be noted that:

- When a top event has an *AND* gate, then there will be more components in the cut sets.
- When an *OR* gate is in the top event, then there will be more cut sets, and in some cases this may lead to single failure (vulnerable), as is seen from the example if a valve fails to open it will result in a top event.

Another interesting feature can be noted. If there is another valve kept parallel to the current valve in the aforementioned example, then pump failure logic will remain as it is, but water supply shall stop when both valves fail. If the events for two valve failures are E_1 and E_2, then in this situation a top event will be any one of AC/AD/BC/BD or E_1E_2. Thus, in this situation, all cut sets will have two components. From here it is established that redundancy also changes failure probability and cut sets.

Referring to Fig. V/3.1.5-1, one may note that there is validation for a correct failure path. In the aforementioned example, it is seen that AB is not really a valid failure path in the sense that if motor and pump fail, then a top event will not occur. As mentioned earlier, the fault tree is evaluated by transforming it into an equivalent set of logic equations where such an invalid path will not occur. In order to understand how a fault tree is transformed into an equivalent set of logic equations, and cut sets are determinted, refer to Fig. V/3.1.5-2.

From the diagram, associated logic is:

$$(C+D)(AC+DB)(AB) = C\cdot AC\cdot AB + C\cdot DB\cdot AB + D\cdot AC\cdot AB + D\cdot DB\cdot AB; \quad [\text{since } A\cdot A = A]$$

The equation will be simplified to ABC + ABCD + ABCD + ABD since ABCD is covered by ABC + ABD.

So, final cut sets are:

$$\mathbf{ABC + ABD}. \qquad (V/3.1.5\text{-}1)$$

Now it is time to go into details about FTA postulates.

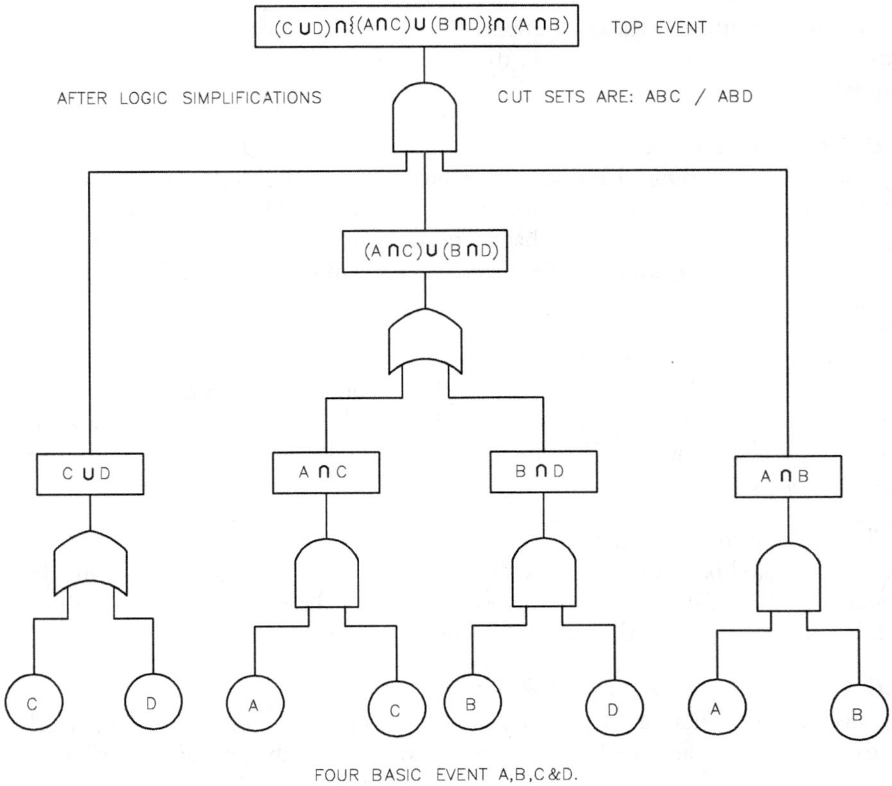

FIGURE V/3.1.5-2

Fault tree analysis and cut set.

3.2 Fault Tree Analysis Postulates

Major assumptions and ground rules have been discussed thus far. It is necessary to follow these for development of an FTA diagram.

3.2.1 *Immediate, Necessary and Sufficient Causes*

- Immediate stands for the event closest in space, time, and derivation of the event above. These are not basic causes of the event, but some intermediate stages or mechanism for the event above it; that is, while constructing the tree, not to skip past event.
- Necessary cause refers to no redundancy in the statement or gate linkage, and the event above could not result from a subset of causal, that is, for construction, to include what is necessary.
- Sufficient: The events will, in all cases, at all times, cause the event above; that is, not to include anything more than minimum necessary.

3.2.2 *Clear Statement Rule*

The purpose of this rule is to define an event box statement very clearly and precisely. According to this rule, fault event must be stated as fault with description and timing of the fault condition. (1) What is the state of fault system/component? This describes the operating condition of the component, and (2) when is the system/component in fault state? This describes condition of the system with respect to the component in the question.

3.2.3 *No Miracle Rule*

If the normal functioning of a component propagates a fault sequence, then it is assumed that the component functions normally. The main philosophy behind such a rule is to prevent one from thinking that there could be some abnormal situation that will prevent fault propagation, or there will be some failure in other components to prevent fault propagation.

3.2.4 *Complete the Gate Rule*

All inputs to a particular gate should be completely defined before further analysis of any one of them is undertaken. If not, then either logic will be incomplete or some condition may be omitted. FTA insists on level-by-level analysis so only after completing one level, the next lower level should be undertaken.

3.2.5 *No Gate-to-Gate Rule*

Gate inputs should be properly defined fault events, and gates should not be directly connected to other gates. As such, there may not be a direct effect on the analysis result, but for large systems this may be a cause of confusion.

3.2.6 *Component/System Fault Rule*

If the answer to the question, "can this fault consist of a component failure?" is, "yes," then the event is classified as a "state of component fault." If the answer is, "no," then the event is classified as a "state of system fault." As a general rule, when energy originates from a point outside the event, it may be classified as a "state of system" [3].

3.2.7 Important Points for Fault Tree Analysis

The points below are not postulates but are important issues for FTA. A few pertinent points for FTA shall be discussed here:

- Initial state: The initial state for the FTA refers to the initial states of components (e.g., switch open), initial conditions, and initial inputs assumed (e.g., power available). Initial state also involves what components are assumed to be initially in operation, initial commands, if any, environmental conditions (e.g., normal atmosphere assumed in fire example), and operational conditions or existing failure conditions, if any. It may also include historical details about the component.
- Failure category: Failure can be classified into three categories such as primary, secondary, and command. Primary faults are those which occur within the design category of the component, for example, when an electronic component is designed to operate up to a temperature T°C; and if it fails at a temperature t°C < T°C, then it is a primary failure. But if it is operated at T_1°C (T_1°C > T°C), then the failure is secondary failure, and usually it is not analyzed unless abnormal conditions are not included in the modeling. Command failure refers to proper operation of the component but at the wrong time or place.
- FTA termination: Fault tree is terminated either at basic event, which represents the lowest level events, or it is terminated at undeveloped event (which is not developed for further event).
- Human errors: Usually there are two basic types of human errors, viz. "an error of omission" is not doing a correct action; and "an error of commission" is doing an incorrect action. In FTA, human errors are treated like component failure. When there is a human interface with the component or system, then the failure modes need to be expanded to include failure induced by the human.
- Common cause failure: As the name implies, it is the common deficiency for which multiple components suffer. These can be four types:
    - Common design or material deficiency, which results when multiple components fail to perform a function or fail to meet design data.
    - Common installation flaw causes multiple components to be misaligned or functionally inoperable.
    - Common maintenance error may result in multiple components misaligned or malfunctioning after maintenance.
    - Common environment issues such as vibration, radiation, moisture, or contamination that causes multiple components to fail.

Common cause failure (CCF) has been shown in Fig. V/3.1.4-2. CCF is quantified with the help of a factor β = "beta factor," which is the probability that a failure cause results in multiple failures. β values range from 0.3 to 0.01 when CCF susceptibilities exist. If there are three redundant components each with failure 10^{-5}, then failure rate would have been $10^{-5} \times 10^{-5} \times 10^{-5} = 10^{-15}$ but with a CCF, factor it will be less, that is, individual failure rate is multiplied by β so, if $\beta = 0.01$ then CCF would be $10^{-5} \times 10^{-2} = 10^{-7}$. As seen above, there are several categories into which CCFs can be grouped. Such categories are:

- Atmospheric/climatic susceptibility (Humidity, seismic, flood, storm, sea corrosion)
- Contamination
- Design/material deficiency
- Energy source
- High temperature
- Human error (e.g., mal-operation)
- Installation degradation
- Location
- Maintenance degradation
- Manufacturer (manufacturing)
- Mechanical disturbance (e.g., Vibration, wear)
- Radiation
- Test degradation

3.3 Fault Tree Analysis Evaluation

There are various purposes of FTA, and there are several ways to evaluate the same. Refer to Fig. V/3.3-1 for further explanation.

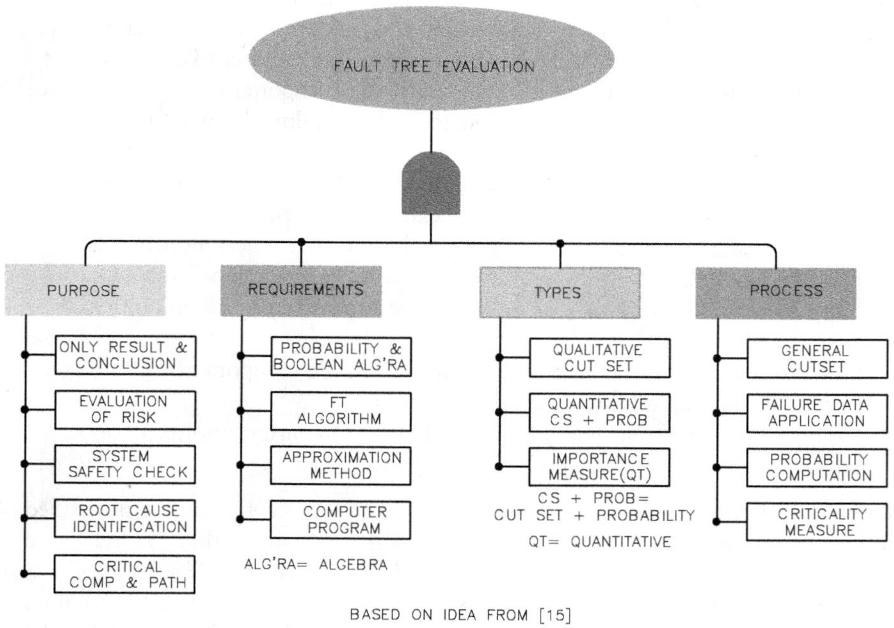

FIGURE V/3.3-1

Fault tree evaluation.

After the construction of the failure logic, basic understanding of the system failure logic is more or less clear to the analyst. It gives not only overall view about the failure logic, but also helps to obtain the critical failure and its path. An account about the purpose of FTA logic has been clearly shown in Fig. V/3.3-1. For the purpose of evaluation, a fault tree is transformed into an equivalent set of logic. By successive substitution, each gate event is expressed in terms of the basic event as shown in Fig. V/3.1.5-1 and Fig. V/3.1.5-2. Finally, entire logic is simplified (utilizing Boolean algebraic law), and expressed as a SOP as shown in Eq. (V/3.1.5-1) in Clause 3.1.5. This could be considered as the starting point for the qualitative and quantitative evaluation process.

3.3.1 Qualitative Evaluation

Qualitatitive evaluation is characterized by the following:

- More subjective than quantitative values or numerals.
- Listing of minimal cut sets.
- Qualitatively analyze cut sets for specific issue.
- Cut sets are arranged according to order; lower cut set has *more* importance.
- Assign importance to components based on how many times it appears in cut set.
- Cut sets are analyzed for:
  - Weak points in design.
  - Bypass safety feature.
  - Finding root cause combination.
  - CCF issues.

The cut sets are arranged in order. Depending on types, they are arranged as follows (typical):

- Human error (1)
- Failure of active elements (2)
- Failure of passive elements (3)

When in combinations they can be arranged as shown in Table V/3.3.1-1.

Here one needs to note that the single component "failure of an active unit" has been given the lower rank than human error (in item 2 in Table V/3.3.1-1), as that cut set contains two components. So from here, no of component failure gets first preference in ordering a cut set than the order discussed in Table V/3.3.1-1. Computer codification is used to complete the cut set ordering. Therefore in many cases, only cut sets with single, double, or at the best triple components are listed (computer time increases with number of components very quickly). Also, as the number of components increases, probability decreases.

- Common cause effect: While assessing MCS, it is assumed the components are independent, but it may not be so! If there is a calibration error in an instrument, then a number of instruments may be affected, and it is not unlikely that some of these instruments may appear as a primary cause in more than one MCS. In order to get rid of such problems for large systems, people

Table V/3.3.1-1 Event Ranking According to Type of Basic Events

Rank	Base Event 1	Base Event 2	Base Event 3
1	Failure of an active unit		
2	Human error	Failure of Active unit	
3	Human error	Human error	Failure of Active unit
4	Human error	Human error	Failure of Passive unit
5	Human error	Failure of Active unit	Failure of Active unit
6	Human error	Failure of Active unit	Failure of Passive unit
7	Failure of Active unit	Failure of Active unit	Failure of Active unit
8	Failure of Active unit	Failure of Active unit	Failure of Passive unit
9	Failure of Active unit	Failure of Passive unit	Failure of Passive unit
10	Failure of Passive unit	Failure of Passive unit	Failure of Passive unit

divide the components according to category (discussed in Clause 3.2.7 above), then codify by computer.

3.3.2 *Quantitative Evaluation*

In quantitative evaluation, after MCSs are obtained, its probability is estimated for failure probability of a top event. Thus, it is done in a sequential manner as follows:

- Determination of component failure probabilities.
- Determination of minimal cut sets probabilities.
- Calculation of top event probabilities.

As stated earlier, that logic will be expressed as SOP, so each MCS will be constituted by product of basic events, and all MCSs will be as a sum mode to form top event. Thus by *ANDing* basic events, an MCS will be available, and by *ORing* MCS, top event probability will be available. In mathematical terms:

$P(\mathrm{T}) = \sum P(\mathrm{M}_k)$ = sum of minimal cut set probabilities for "k" nos. minimal cut sets.

Any MCS, say "k", is $P(\mathrm{M}_k)$, then $P(\mathrm{M}_k) = P(\mathrm{E}_1), P(\mathrm{E}_2)\ldots. P(\mathrm{E}_i)$ for "i" number. of basic events in any MCS (M_k). Here E_i stands for ith basic event. Again for the mathematical formula relating to probability of "$P(\mathrm{T}) = \Sigma P(\mathrm{M}_k)$" and "$P(\mathrm{M}_k) = P(\mathrm{E}_1) \cdot P(\mathrm{E}_2)\ldots \cdot P(\mathrm{E}_i)$," Eqs. (I/5.3.2-1 and I/5.3.2-2) of Chapter I may be reviewed. Here, it has been assumed that the events are independent. In this case, it is assumed that failure rates are constant and they can be expressed as constant value such as "a" per hour or per cycle. In case time dependency (e.g., wearing out/ageing) in failure, same needs to be considered suitably. In some cases then more sophisticated calculations such as Weibull, gamma distribution, etc. may be referred to.

3.3.3 *Event Types For Evaluation (Tips)*

There are normally five events that take place, and these are:

- Non-repairable unit.
- Repairable unit.

- Frequency of event.
- On-demand probability.
- Hidden failure unit (periodical testing).

Each of these types with its formula is discussed briefly as follows:

- Non-repairable unit: If for unit "*i*" which is non repairable, and failure rate is λ_i, then probability of event will be: $P(E_i) = 1 - e^{-\lambda_i T} = \lambda_i T$, where T, exposure time.
- Repairable unit: If unit "*i*" is repaired when failure occurs, and failure rate is λ_i mean time to repair MTTR $= \tau_i$, then:

$$\text{Failure probability } P(E_i) = \lambda_i * \tau_i/(1 + \lambda_i * \tau_i) \simeq \lambda_i * \tau_i'$$

- Frequency: If an event happens occasionally without a specific duration, then frequency is f_i. However, if the duration is known, then it is like a repairable unit.
- On-demand probability: The unit "*i*" not active during normal operation, but may be subject to demands then, P (unit "*i*" fails on request).
- Hidden failure: If failure rate hidden is λ_i and test interval is τ_i, then:

$$\text{Basic event probability} = \lambda_i * \tau_i$$

With this evaluation of FTA concluded, an FTA in programmable electronics (PE) shall be discussed.

3.4 Fault Tree Analysis for E/E/PE Concepts and Application Notes

Use of FTA for physical failure in hardware is well proven. In this section, discussions will be mainly on use of FTA in PEs; to be specific on FTA for software. In today's industrial automation, almost all control systems deploy PE systems. Therefore, software is very critical for control systems. They are increasingly used to handle safety-critical system functions also. As a natural consequence, a high number of hazards in such systems are known to be caused by software which controls the systems. In this section, the focus will be on PEs, for example, programmable logic controller (PLC) faults. PE system faults can be categorized in three classes depending on whether they are caused by:

- Hardware failures of the control system.
- Hardware failure of input/output.
- Design flaws in the logic, or software, of the controller; either a result of coding errors or misinterpretation of control requirements.

In this connection, refer to Fig. V/3.4-1. In the example, one PLC automation has been considered.

The first two classes of fault are currently considered in a classical FTA already covered, so not repeated here. Another important issue here is that research on *software* FTA is not as mature as it is for hardware or classical fault tree, so discussions

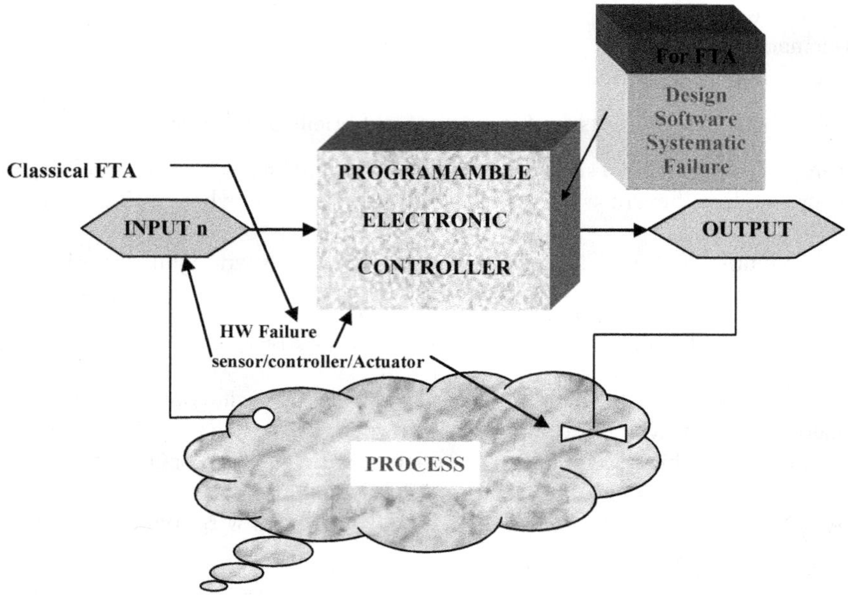

FIGURE V/3.4-1

Automated system: Fault tree analysis in programmable electronics.

needs to be based on some research work done in the area. Only a few suggestive causes and possible measures are discussed here. Lot more work may have to be done in future!

3.4.1 *Software Faults*

Software related failure is highly dependent on the systems chosen. (If one looks very carefully, software is internal to the system and alone cannot do any harm. But its *interface and interaction* with hardware and the outside world give rise to all the hazards). Software faults are usually analyzed and dealt with through external hardware safety systems or with software self-checking and defensive coding techniques derived out of analysis. Exhaustive categorization and analysis of software fault may be *impractical*, if not impossible. A good software design is **critical** to success in narrowing this scope. So it is necessary to select issues, for example, to consider the timing of the failure in relation to when it becomes a hazard. Also, in most cases a detect-and-correct approach for mitigation is applied in place of preventive action. This means that up to a certain time, fault is allowed to exist. So, it is necessary to ensure that this length of time is acceptable for the failure to exist before becoming a hazard. However, if the failure becomes an immediate hazard, then in the system architecture it is necessary to keep some *redundancy* or some *layer* of software protection. In the following points, application area of FTA in software will be discussed.

- Level of failure: In the majority of cases, systems are analyzed for a single level of failure. However, some industrial safety systems and standards demand multiple levels of failure pertinent to critical safety categories, leading to a redundant design. Even for a single level of failure, consideration should be given to how long a backup system keeps the device operational, that is, the length of time for repair once the backup system has taken over. Many systems and mitigations are designed to fail safely, that is, safe shutdown. However, if the shutdown is a hazard itself? In that case, it is important to *identify* any requirement that must be met even in the presence of failures!
- Memory corruption: One very common failure is memory corruption. As corruption is random and produces many unpredictable results, it is difficult to handle this exhaustively in FTA. So a better approach is to capture where *specific* memory corruption causes a hazard and defend against this using detection of corruption and correction. *Data* memory may be corrupted, prompting *invalid* values or decisions. Corruption in *pointers* initiates cascading effect for further corruption. Stack corruption can affect the program counter, leading to out of order execution. A processor with memory management unit (MMU) support is a great first layer of protection. It can isolate areas of concern so that the risk is of corrupted memory causing a safety hazard is lower, and allowing for more focused review and testing. Cyclic redundancy check (CRC) checked, and refreshing the output data out of a timer that is checked by the watchdog.
- Corrupted code: This can lead to out of order or incorrect execution, or exceptions from invalid op codes, etc. A very common mitigation is to perform a **CRC** of the code when powering up or periodically during runtime [4].
- Out of order execution: Conditions are checked and met before issuance of an order for changing the state of an output or multiple outputs, in specific order. A software failure may cause out of order execution. One of the possibilities could be the things which set programmable counter. It is better to detect the problem, in a generalized way and focusing on detecting the problem around the code path, which if executed out of order could cause a hazard.
- Timing: It is one of the most important things in software. FTA may be focused to detect what can go wrong to affect timing in design flaws, for example, corrupted timers, bad code execution, or unexpected flood of other events. A watch dog timer could be a good solution for the this.
- Corrupt peripherals: Peripherals such as analog to digital converter or input/output (I/O), where I/O may be interchanged or wrongly connected. This is somewhat similar with memory corruption, so CPU with MMU can be helpful.

3.4.2 *Design Flaws*

With the idea about software flaws discussed above, it is time to marry these with FTA. Assuming the correct inputs, there can be erroneous output at the controller that may be due to a flaw in designing the software. All these are design flaws; a flaw in software logic is a systematic fault, which can be reproduced every time the conditions that trigger the error in the control logic are present. These conditions are

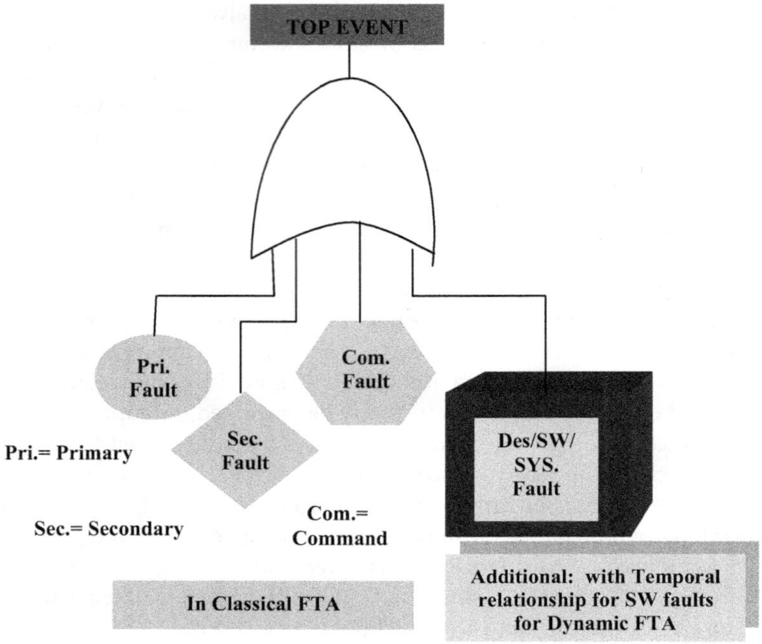

FIGURE V/3.4.2-1

Requirements for fault tree analysis in programmable electronics.

typically sets of correct inputs (assumed), which by triggering the error in the embedded software result in a fault manifested as an omission or commission of controller outputs and/or a deviation of output from correct timing or value. In order to identify such design or software faults, analysts need to assume that even with correct input information the controller fails, delivering erroneous outputs. One needs to integrate these flaws into the fault tree structure for further investigation and eventual elimination of these flaws in relevant parts of the control logic. It is important to note that out of large numbers of errors in a program, only a small portion of those errors that will trigger faults that can contribute to the hazard investigated as a top event in a particular fault tree. A simple classical FTA is represented by primary, secondary, and command faults, whereas FTA with design flaw needs *another* head as shown in Fig. V/3.4.2-1. The additional head is for design flaw, systematic, or software fault with a temporal relationship. What is temporal relationship? For including such controller faults in FTA, it is necessary to extend the vocabulary of FTA. In classical FTA, it is possible to represent a proper sequence with the help of priority *AND*, but in case there is a requirement such that output will be logical *one* when an input is *now* logical one and input *was* one for past *n* instants. So it is necessary to include vocabulary for notion of time and event ordering.

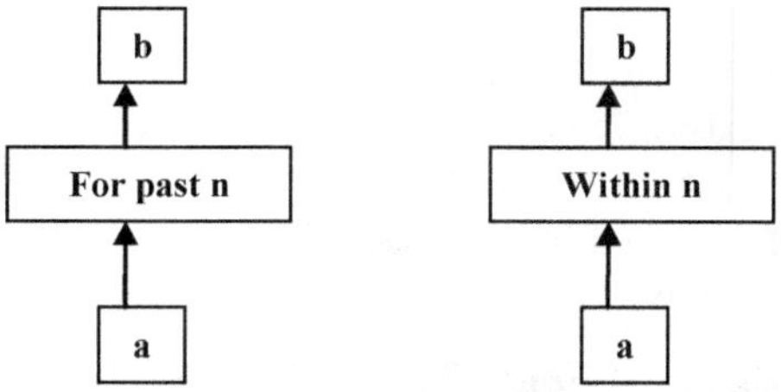

b will be 1 if a is now 1 and was 1 in last n instants or at some instant within last n instants.

FIGURE V/3.4.2-2

Temporal (relationship) gate in software fault tree analysis.

As discussed earlier, a controller may produce erroneous output in response to a set of correct input, on account of an error in the embedded software logic, or may also be due to software fault discussed in Clause 3.4.1.

In classical FTA, it is irrelevant in which order the basic failures occur, but in software FTA it is important. This is normally taken care of by temporal relationship among the events, especially because in PEs it is a sequence of flow of information as typically seen in a ladder diagram or functional block diagram (FBD) in PEs. If there is an error in timing or sequence, it will result error in the controller output, that is, if there is incorrect timing, it is an error of the system. Typical temporal gates have been shown in Fig. V/3.4.2-2. Since these kinds of FTAs show dynamic information about the events, these are referred to as dynamic fault trees. From the figure it is clear that for same input based on timing there could be different output. In classical FTA this is not taken care but through temporal relation ship this can be established. In software this is additional requirement and is very important.

3.4.3 Software Fault Tree Analysis in PEs: Example

As per IEC 61131-3, structured text, ladder diagram, FBD instruction list, etc. are programming language for PLCs. FTA is generally performed before codes are generated, and it is done in automatic/semiautomatic mode. There are two kinds of views in looking at software (SW) FTA. These are:

- Fault-oriented view: The fault-oriented view shows feasible faults in FBD design, for example, input and outputs are incorrectly connected.
- Cause/effect (CE)-oriented view: The CE-oriented view illustrates how incorrect values may be propagated through an FBD network.

3.4.4 Software Faults (Typical)

Typical SW faults could be as listed below:

- Input/output faults:
  - Incorrect I/O Address, variable or value.
  - Incorrect ordering of switching inputs.

    - Incorrect timing.
    - Range maximum/minimum as input.
- Incorrect operation/comparison (e.g., arithmetic operation).
- Omission/misplacement (inclusion/exclusion).
- Routine error: Interactions among failure modes (wrong execution order).
- Other errors (input type error/data range error).

3.4.5 *Model-Based Synthesis of Fault Tree Analysis*

For PEs, another way to look into the problem is to fuse the idea of computer hazard and operability study (HAZOP) into FTA. This is done by considering the system model, then analyzing it for safety and from there, develop an FTA model. This could be done in *semiautomatic* or *automatic* mode, as shown in Fig. V/3.5.1-1. The idea is to develop a method which will assess an evolving programmable design that starts at the early stages of the design lifecycle. However, for this good support tool both in software as well as from the system is necessary. The left-hand side model provides the basis for the analysis of the system. This model describes the hierarchical decomposition of the system into subsystems, and basic elements for material energy or data flows. The model also depicts the functional decompositions into lower sub-functions. With architectural information about the allocation of functions to hardware and composite elements, the model starts to represent PEs. To analyze this model at a particular stage of the design, the failure behavior of each component in the model has to be determined. This could be accomplished by Matlab Simulink Ref. [5,6]. From this model, it is possible to automate FTA, which is discussed in next clause.

3.5 Automated Fault Tree Analysis Concepts

In the case of automated FTA, for that matter automated hazard analysis, there is a distinct advantage over its manual counterpart when there is a large system, and when there is complexity especially with PEs (involving software). A few points in favor of automated systems has been listed in Table V/3.5-1. The problem is, there are not many established automated hazard analysis methods for different hazard analysis techniques.

Although SW tools are available for computerization of FTA, literature searches reveal that it is quite difficult to construct fault trees when the steps are complex, and fault propagation is not very clear. There are a number of different approaches have been chosen, which are discussed in brief in the following subclauses.

3.5.1 *Model-Based System*

As shown in Fig. V/3.5.1-1, these are actually somewhat similar to computer HAZOP, and mainly used for PEs as discussed in Clause 3.4.5 previously.

Referring to Fig. V/3.5.1-1, one may note that it starts with the model. It depicts the hierarchical decomposition of the system into sub-systems and basic elements meant for communication of material energy or data flows, and is the basis of analysis. When architectural information is embedded in this model, it would represent

Table V/3.5-1 Comparison of Automated and Manual Analysis		
Issues	**Automated Analysis**	**Manual Analysis**
Understanding of specification	Uniform understanding of specification	Possibility of different understanding and interpretation
Introduction of error	Normally lower chance of error introduction	Possibilities of additional error introduction during analysis
Implementation	Uniform implementation	Inconsistent implementation
Time taken	Long time usually necessary	Less time consuming
Evaluation	Improved error free evaluation	Prone to error in evaluation
Reliability	Improved & reliable safety assessment	Comparatively less reliable safety assessment

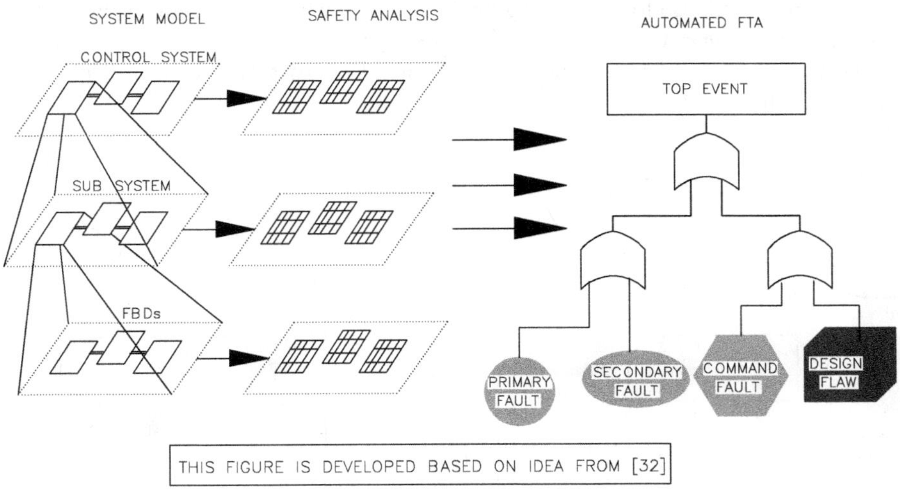

FIGURE V/3.5.1-1

Modeling technique for fault tree analysis automation.

programmable entities. So, if it is controller then in the model, interdependencies of various tasks and how to handle input data are shown in the model. To analyze this model at a particular stage of the design, the failure behavior of each component in the model is determined. In the next stage, hardware and task analyses are shown. During the application of this technique on a component, each output of the component needs to be examined for potential deviations from the intended normal behavior. The specific failure modes of each output are determined as the behavior of the output is scrutinized for potential deviations. Thus, from these, an automated model for FTA is developed. The analysis effectively assists the systematic improvement of the failure detection and recovery mechanisms in the system. Standard software package such Matlab Simulink has been used. For further reading refer Refs. [7,8].

3.5.2 *Code-Based FTA*

Code-based FTA is typically applicable for SW, that is, for PEs. Mainly on account of cost-effectiveness, software safety is applied in the early design stage. However, there is a flip side of the same. During that stage, it is not possible to pinpoint precisely what computation is for. This is possible after coding. Also after coding, it is possible to decompose hazards more precisely. As shown in Fig. V/3.5.2-1, a fault tree is generated from the code. In the next step, fault is injected in to the non-root nodes to cause hazards. It is backward approach. So, catastrophic outputs or events are determined. One needs to work backward to show either the output cannot occur, or probability is very low. In this method, with the help of source codes and user defined hazards fault tree is built. It is then fed into the mitigator, where user-defined operational details are fed also for estimating frequency with which the event in the root will occur. The mitigator uses an extended propagation analysis (EPA) algorithm, which uses the injection of artificial fault (SW + HW) to test software tolerance towards the untoward event.

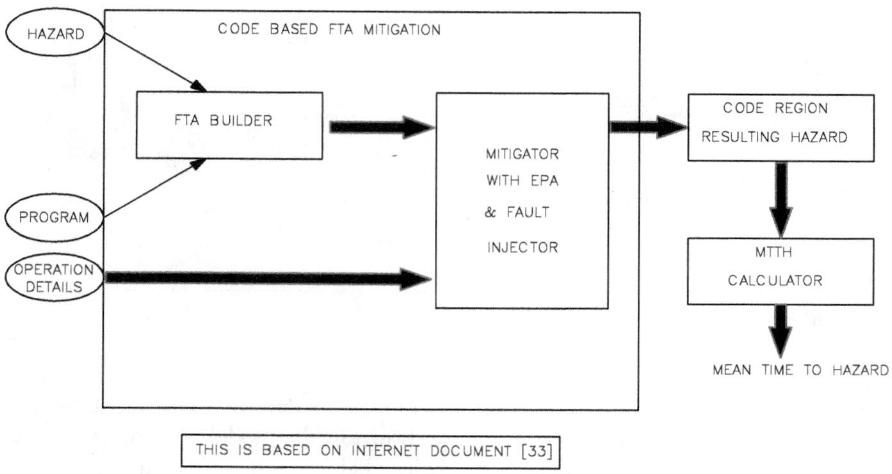

FIGURE V/3.5.2-1

Code-based mitigation in fault tree analysis.

3.5.3 *Architecture Analysis Design Language and Use in System Modeling*

For automation of hazard analyses such as FMEA/FTA, use of architecture analysis design language (AADL) is quite common. Modeling language AADL is standardized by SAE (2008 version). Major uses of AADL includes, but not limited to, the following:

- Design of hardware and software.
- Criteria analysis, for example, performance, safety, and security.
- Evaluation purposes.
- Description of fault events such as propagation, error event.
- Safety analysis and evaluation.
- Integration of models.

Application areas of AADL are shown in Fig. V/3.5.3-1.

There is extension of AADL for fault description, such as error event and propagation. As discussed earlier, errors may be different types such as value error, out of range, and inconsistency. These errors may propagate from sensor to controller and then to the actuator. Error behavior may be classified as error related transition, use of error type and propagation rules. Major application area of AADL may be:

- Preliminary system safety assessment: Functional hazard analysis, FTA.
- System safety assessment: FMEA, FTA with quantitative rule.
- Evaluation of safety analysis process:
  - FTA: Use of composite error behavior at FTA nodes, component error behavior for error event, etc. Help to generate complete FTA with focus on AADL sub-components

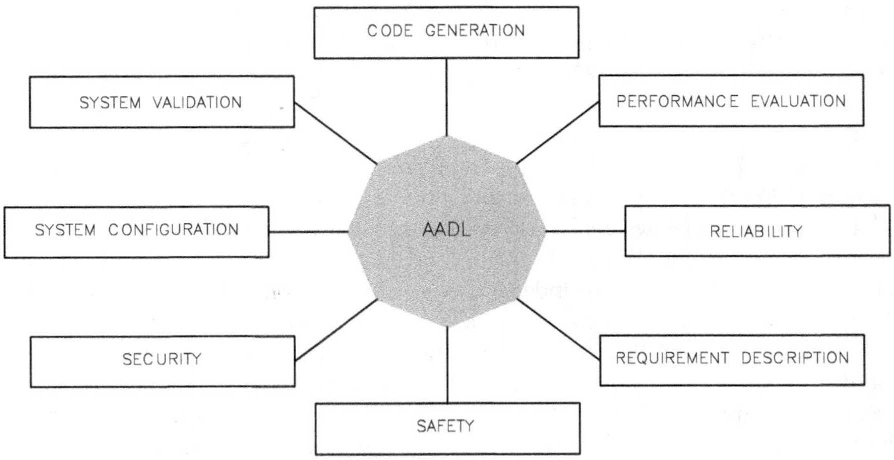

FIGURE V/3.5.3-1

Architecture analysis design language in safety automation.

- FMEA: Use of component behavior, internal event, error propagation rule. Traverse all error paths.

Major supporting tool for AADL is open source AADL tool environment (OSATE).

Based on this language model-based FTA, automation has been developed. OSATE tool may be used to generate fault tree. The generation tool is designed to be flexible and can be re-targeted to more than one fault tree analysis tool. The portion of the tool that extracts the system instance error model can be reused to generate different types of safety artifacts, such as Markov Chains.

3.5.4 *Miscellaneous Other Systems*

From technical papers, there are several other types of FTA automation are also found. Probabilistic fault tree (PROFAT) is one approach where simulation algorithm ASII has been utilized. For the complex cases where there is unavailability of reliability data for the number of specific equipment, generic probabilistic data are used. Also, probabilistic data for human error are used. However, fuzzy approach in this regard is also found (especially for human error). Instead of specific values for human error hybrid approach with fuzzy logic is quite effective. In fuzzy logic, set failure rates are defined in linguistic way, which is more realistic. For human robot, offshore applications FTA with fuzzy approach produce better result in automating the process.

There is another well-proven technique for hazard handling, which is layer of protection. It shall be discussed in next clause.

4.0 LAYER OF PROTECTION ANALYSIS

LOPA is based on the assessment of a single event, and the associated consequence scenarios. LOPA is a rather simplified form of quantitative hazard analysis technique normally applied for the cases where the system is too complex or the consequence too severe to be handled by HAZOP. As a matter of fact, it is a quantitative analysis technique, but lying almost at the lower edge of quantity analysis technological scale. It is in between HAZOP and QRA. This is shown in Fig. V/4.0-1B. At times, it takes input from HAZOP and its output could be used as input to QRA.

Conceptually, it analyzes the hazard events, then offers a number of independent protection layers (see Fig. II/4.3.7-1) so that risk frequency is reduced and system is acceptable. In LOPA, each of the independent protection layers (IPLs) are tested and analyzed individually to see how it provides prevention/mitigation of hazards for which they are used. Thus IPLs could be preventive as well as mitigating types. LOPA is mainly concerned with mitigating type protection layers. In a later part of this clause, it will be discussed that, in LOPA each of the protection layers are analyzed for their probability of failure on demand (PFD), which is very relevant in the context of safety integrity level (SIL) and associated international standards like International Society of Automation (ISA) 84 and IEC 61508/61511. In fact, it was initially

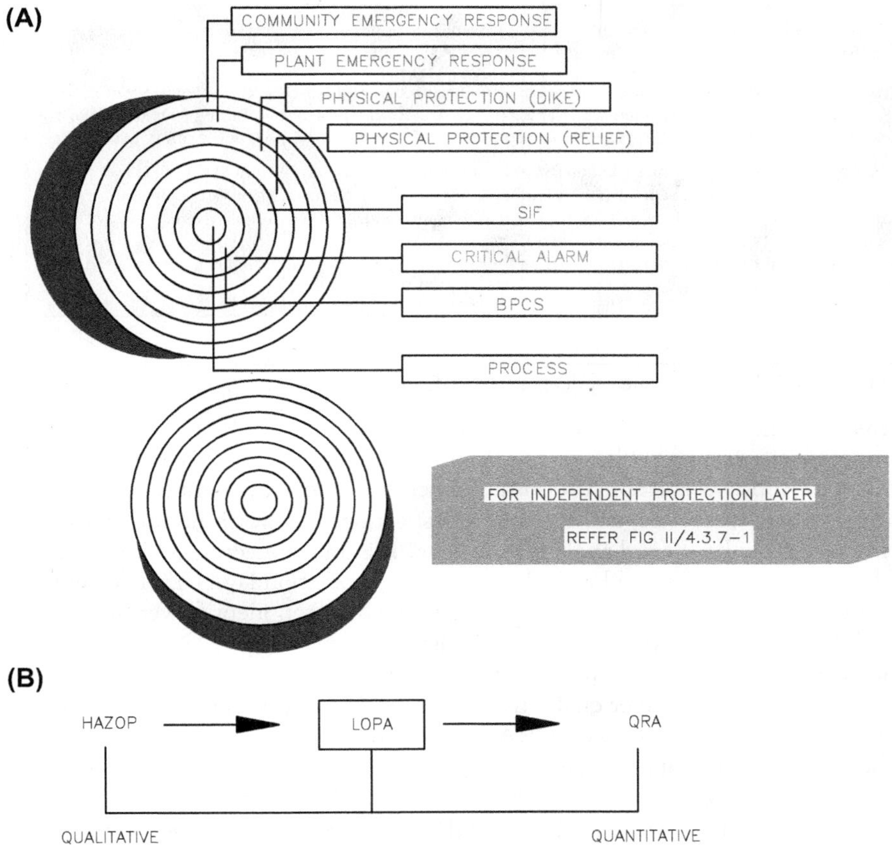

FIGURE V/4.0-1

Layer of protection analysis concept. (A) LOPA onion, (B) position of LOPA between qualitative & quantitative hazard analysis.

developed for SIL calculations for electrical, electronic, and programmable electronic systems (E/E/PEs). Reference to this analysis was first seen in the CCPS publication, "Guidelines for safe automation systems for chemical processes" in 1993. In order to complete the discussion in this clause a few terms have been defined in Fig. V/4.0-2. Detailed discussions on these have been included in later chapters.

With these preliminary ideas, it is better to look into the system more closely.

4.0.1 LOPA Conceptual Discussions (see Fig. V/4.0-1)

- *What is LOPA?* LOPA is a simplified semi-quantitative risk assessment technique and a hazard analysis tool, mainly deployed to assess the adequacy of IPLs (a requirement set forth in CCPS document). As LOPA is a semi-quantitative method, it uses numerical quantities to estimate the parameters needed to

Reliability: **Reliability is the probability that a system can perform a defined function under stated conditions for a given period of time.**

Probability of failure on demand**: Probability of failure on demand (PFD) is the probability of a system failing in response to a demand. The average probability of a system failing in response to a demand in a specified time interval is referred to as PFD_{avg}. Therefore, PFD = 1 − safety availability.**

Safe failure fraction**: Safe failure fraction (SFF) is a term from the IEC 61508 and IEC 61511. As per IEC, "safe failure fraction is the ratio of the (total safe failure rate of a subsystem plus the dangerous detected failure rate of the subsystem) to the total failure rate of the subsystem." There are four types of random hardware failures: Safe undetected (SU); safe detected (SD); dangerous detected (DU); and dangerous undetected (DD). Detailed discussions are available in Chapters VI–VIII.**

FIGURE V/4.0-2

Reliability, probability of failure on demand and safe failure fraction (IEC).

calculate the necessary risk reduction (PFDs), as well as taking the benefits of simplifications, expert judgment, and tables, etc. This flexible tool can be used in different contexts and applications also, such as improved planning, management of change, and incident investigation. It is worth noting that it is not really another hazard analysis technique! It is an engineering tool to ensure that the risks are mitigated with the help of the protection layers. LOPA can be conceived as a cost-effective engineering tool which can identify, verify a number of IPLs, which can lower the frequency and consequences of hazardous incidents. LOPA provides specific criteria and restrictions for evaluation of IPLs, and not in a subjective or qualitative manner. This process is much more cost-effective when compared with QRA. LOPA also supports the use of mitigating factors or conditional modifiers, which may not be strictly protection layers, but are effective in reducing the risks.

- The conceptual foundation of LOPA is that a system needs to be well-protected by different layers of protections as shown in LOPA onion diagram in Fig. V/4.0-1A. The system consists of protection layers including a basic process control system (BPCS), critical alarms for human intervention, safety instrument function (SIF), physical protection such as relief and dike, and emergency response of the plant and community. BPCS is the control system used for normal operation such as PLC or DCS. An input sensor senses the changes in the process, and sends the signal to the control systems, which generates output to operate the actuator. There could be manual intervention to operate the plant also. Thus, in this way the normal operation of the plant goes on. If the process goes out of control, BPCS (operator) may initiate actions to stabilize the situation. Alarms monitoring is considered another protection layer. In case of deviation of parameters or equipment operating beyond range, alarms are generated. The alarms could be two types. One is a pre-trip alarm, which alerts the operator who may intervene to stop the hazardous development e.g., Drum level hi is a pre-trip alarm so that operator can save the boiler from tripping from Drum level hi hi. Another is a trip alarm, which also brings to the notice of the operator that some equipment has tripped and situation may get out of hand, so that operator can take necessary measure. For example, when a pulverizer is tripped, a trip alarm is generated. Now, the operator needs to see that boiler control system has taken action to reduce the load; if not, operator needs to intervene. Note that the alarm system has to be in *another* loop than the BPCS in order to be independent (CCPS: 2001; IEC 61511: 2003). SIF is one safety instrumented system (SIS) comprised of sensors, logic solver(s), and actuating (final) elements. SIS can be looked upon as an independent protection shell for machinery or equipment. The inherent idea is that protection of equipment and/or system needs to be independent, for example, separate trip logic (PLC) independent of main DCS so that in LOPA, SIFs are considered as protection layers. Physical protection (other than E/E/PEs) includes equipment like pressure relief devices, a rupture disc, etc. Other physical protections includes dikes, blast walls, etc. These have their function after the release or explosion has occurred. All these are physical protection, and are considered protection layers in LOPA. In all plants, especially in process or chemical plants, there are emergency response plans to combat the situation. These include, but are not limited to; combat procedures, evacuation plans, equipment and medical treatment to help the exposed personnel to escape, or to mitigate damage or injury. Similar plans, when offered by the community, are community emergency plans. The last two emergency plans are the final layers of LOPA as shown in Fig. V/4.0-1A. LOPA incorporates the reliability of the existing barriers to determine the reliability of the needed SIF. LOPA, however, **never** decides the types and numbers of protection layers to be incorporated. It totally depends on the situation at hand (CCPS).

- Use of LOPA: Similar to HAZOP, LOPA is most effective when it is used during front-end engineering after a process flow diagram and preliminary P&IDs are prepared. In case of existing projects, LOPA are used during or after the HAZOP review revalidation. As LOPA lies between HAZOP and QRA, it is "typically applied after a qualitative hazards analysis (e.g., HAZOP) has been completed, which provides the LOPA team with list of hazard scenarios, with idea about their frequencies of occurrence and associated consequence descriptions." However, "typically" means just that; LOPA can also be used to analyze scenarios that originate from any source, and it can be applied at any point of time in the lifecycle of a project, including design option analysis and incident investigations. LOPA can also be applied when a hazard evaluation team (or other entity) believes either of the following:
  - A Scenario is too complex for the team to make a reasonable risk judgment using purely qualitative judgment.
  - The consequences are too severe to rely solely on qualitative risk judgment.

"Scenario as too complex" may be considered when there is dearth in understanding of initiating event, or sequence of events, independent of safeguard (to be considered as IPL). LOPA is also be used as a screening tool prior to a more rigorous QRA (e.g., chemical process quantitative risk analysis) method. When used as a screening tool, each scenario will first go through LOPA, and then certain scenarios will be targeted for a higher level of risk assessment. This means that decision for QRA will be determined in LOPA. A well-documented procedure, which sets the criteria and team requirements, is helpful for LOPA in a successful manner. This procedure must incorporate criteria for initiating event frequency evaluations and required PFD for IPLs to be chosen. This procedure development may be little time-consuming. Referring to Fig. V/4.0.1-1, it is seen how with the help of various IPLs (discussed later) the risks are reduced. It is worth noting that IPLs can reduce

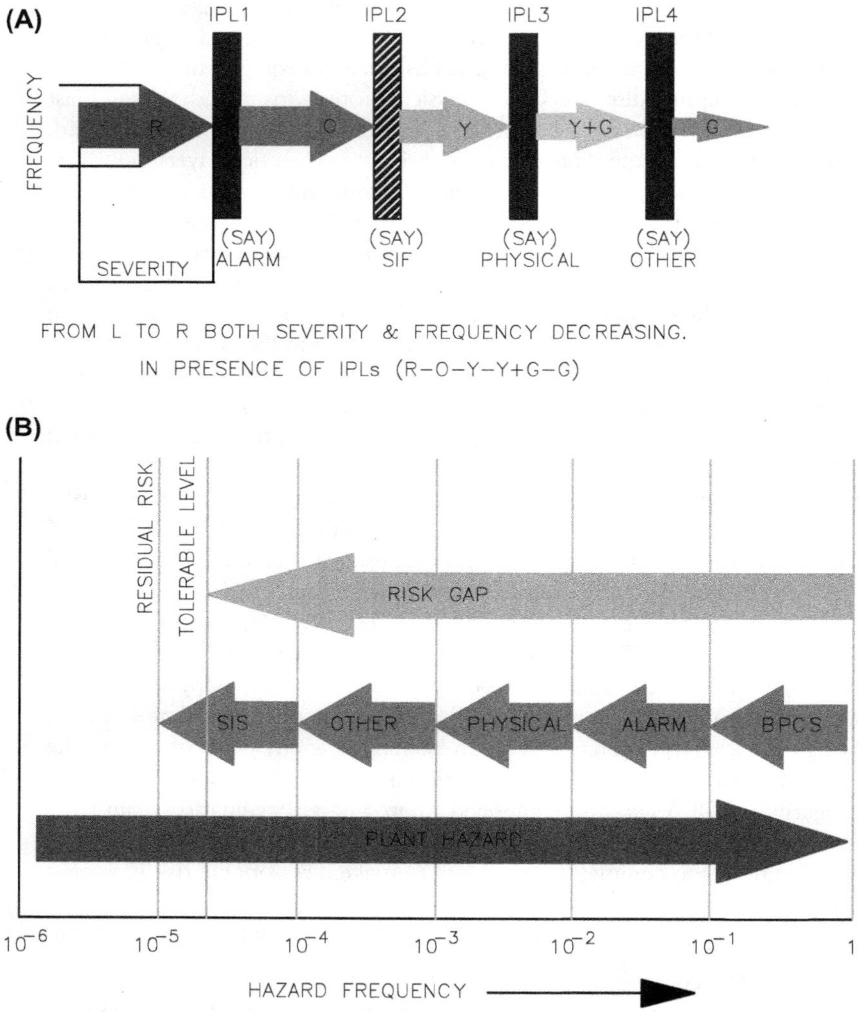

FIGURE V/4.0.1-1

Risk reduction by independent protection layers in layer of protection analysis. (A) Risk reduction for IPLs, (B) risk gap removal by IPLs.

both frequency as well as consequences indicated by reduction of thickness (dikes and blast walls are examples for the same "scenario as too complex;" also see Fig. VIII/1.0.2-2), and the length of arrow respectively. Fig. V/4.0.1-1A may be referred to for the same.

Fig. V/4.0.1-1B shows how one after another, IPL reduces the chances of risk and the system is brought back to ALARP region. A typical actual process example has been presented in Clause 1.1.2 (Fig. VII/1.1-3) of Chapter VII.

4.0.2 Commonly Used Terms

- Process deviation: According to Norsok Z-013 (2001), an accidental event is defined as an "event or chain of events that may cause loss of life, or damage to health, the environment or assets." So, one may interpret process deviation consists of events or chain of events for which there will be deviation from accepted performance and may cause loss of life property or environment. From task management point of view process deviation could be interpreted as deviation from agreed upon requirement or course of action. It also provides an idea about quantum of such deviation so that performance could be judged [13].
- Scenario: According to CCPS (2001), a scenario describes a single cause—consequence pair from the HAZOP. In LOPA terms, a scenario comprises a single initiating event—consequence pair. In other words, it is the process deviation to the impact event (discussed later) inclusive of associated causes for the same.
- Initiating event: In hazard analysis, an event could be the occurrence of a deviation which may lead to an accident. So, the initiating events are the causes for which there is the process deviation. The initiating events may be or may not be the most basic underlying root-causes, but are the results of the root causes. According to CCPS there are three types of initiating events or causes:
  - External events, for example, earthquakes, hurricanes, flooding, lightning, fire, and explosion in an adjacent area, vehicle impact, third party interference, etc.
  - Equipment failure, for example, mechanical failure, BPCS component failure, software failure, mechanical failure, vessel, piping failure, performance failure, utility failure, etc.
  - Human failure: Omission or commission; failure to complete the task properly.
- Enabling event: Enabling events or conditions in which are neither failures nor protective layers are necessary to be present in order for the scenario to proceed or unleash one scenario. These factors or conditions do not directly cause the scenario. This will be clear from an example: In the case of leakage of inflammable gas, then there is a chance of fire only when there is a source of ignition. Here, the initiating event is the leakage of gas, and source of ignition is the enabling cause.
- Impact event: As the name implies, it refers to the consequence due to an event, that is, impact of an event. According to CCPS (2001), it is: "the ultimate

potential result of a hazardous event. Impact may be expressed in numbers of injuries or fatalities, environmental or property damage, or business interruption." As per IEC 61511, an impact event is equivalent to the consequence in the HAZOP study. From these two, one may infer that impact event is the unwanted consequence of accidental incident. This impact event could be intermediate or ultimate consequence. So, one can argue that impact event starts from first harm to personnel, property, or environment, and goes up to ultimate consequence. Immediate impact event in Bhopal incident towards a human may be loss of life, but still the impact event is passing through generations!

- Intermediate likelihood: Intermediate event is the occurrence of the end consequence with protection layers but without the SIF under consideration. The intermediate event likelihood is the frequency per year of the occurrence of this event.
- Mitigated event likelihood: Mitigated event is the occurrence of the end consequence with all protection layers, including SIF (proposed), in place. The mitigated event likelihood is the frequency per year of the occurrence of the event.
- Independent protection layer: In addition to conditional modifier and mitigating factor (influencing the likelihood of event), IPLs, which reflect the reliability of the protection layer, are responsible for performance of LOPA. In LOPA, the concept of the IPL is important, and should be well-defined and understood. CCPS definition of IPL is: "An IPL is a device, system or action which is capable of preventing a scenario from proceeding to its undesired consequence-independent of the initiating event or the action of any other layer of protection associated with the scenario. The effectiveness and independence of an IPL must be auditable."

  Effective in preventing the consequence when it functions as designed. An IPL (device/system/action) needs to satisfy the following criteria:
  - Independent: Independent of the initiating event and the components of any other IPL already claimed for the same scenario, and no common cause failure.
  - Specificity: The IPL is capable of detecting, preventing, or mitigating the consequences of specified hazardous event, that is, one protection layer designed to prevent or mitigate the one sequences of one potential hazardous event.
  - Dependability: IPL reduces identified risk by a known amount that is, it must act as per intended design.
  - Auditability: IPL must be designed to permit validation of function and PFD (including drills for human error), at a regular periodic manner (i.e., IPL is auditable).

  From the discussions, it is clear that all IPLs are safeguards, but not all safeguards would qualify as IPLs. CCPS also provides detailed guidance on how to determine whether a safeguard constitutes an IPL for a given scenario. Training and certification, normal testing and inspection, maintenance, procedure, communication, fire protection, etc. may be safeguards, but are not considered IPLs. On the other hand, examples of IPL are: Critical alarm with specified operator response, SIF/SIS, relief device, blast wall, dike, etc. Preventive IPLs working satisfactorily may stop a scenario from developing any further. Mitigation IPLs working satisfactorily reduce magnitude of consequence but cannot stop the event altogether.

According to IEC 61511, an IPL must have the same inherent characteristics. As per IEC 61511 part 1, BPCS can be considered as one protection layer with a risk reduction factor <10. Also as per IEC 61511, IPL must provide at least 100-fold of risk reduction (10 for a protection layer) and have functional availability of at least 0.9 (IEC 61511, 2003). According to IEC 61511, an IPL is just a protection layer with stricter requirements to availability and degree of risk reduction. When this is considered with respect to previous discussions isn't it little confusing? Protection layers may work in concert with other layers, but for IPL this is missing!

4.0.3 *Pros and Cons of LOPA*

- Advantages of LOPA shall include, but not are not limited to, the following:
  - It is possible to have a rough estimation of the likelihood of an event and estimate of consequences due to the event, so that it is possible to determine tolerability.
  - For risk-based decisions, it is a useful tool at various stages such as design, safety operating procedure, management of change incident investigation, and emergency response planning.
  - As LOPA is focused on scenario-related plant risks, it is capable of identifying safety issues otherwise not possible in other qualitative plant hazard analysis. It identifies safety requirements for operations, and processes and with the help of IPLs. So it focuses on the most critical safety systems.
  - It gives more precise cause—consequence pairs compared to risk matrix and the risk graph.
  - LOPA gives clear ideas about SIL so, LOPA is used as a primary means for SIL selection necessary for SIS.
  - LOPA is capable of determining an objective for risk with the help of set of rules.
  - LOPA helps in resolving disagreements of other qualitative hazard analyses.
  - LOPA allows determination of IPL effectiveness, and often identifies acceptable alternatives to the SIS, with the help of other IPLs such as mechanical/physical IPL (e.g., relief devices). It allows cost/benefit analysis, for selecting alternatives for risk reduction.
- Some limitations of LOPA
  - Risk comparison is valid only if same LOPA method and comparison are based on same risk criteria.
  - Numbers generated in LOPA are not precise in quantitative terms.

- Criteria for risk tolerance must be established before LOPA process starts. Therefore there may be inconsistency when criteria changes.
- On account of a simplified approach, may not be applicable for all scenarios, especially for very large, complex systems.
- It is not intended to be a hazard identification tool, and depends on methods used to identify the hazardous events.
- May be less than FTA, but it is also a time-consuming method of analysis.

4.0.4 *LOPA Discussions*

While on the discussions on LOPA and IPL, it is important and better to have some understanding about how BPCS and other protection layers should be treated in accordance with various standards and guidelines. A few pertinent points are reproduced here, and the reader is advised to refer to concerned standards/guidelines for details:

- As per HSE.UK guidance (technical measure): the following points are important for control, alarm, and protection systems:
  - Control system:
    - The logic elements may be distributed, linked by communications, or marshaled together and may be in the form of distributed control systems (DCS), supervisory control and data acquisition (SCADA), computers (including PCs), or PLCs.
    - The control system should be sufficiently independent of the safety systems.
    - Consider all expected normal and upset modes of operation, including start up and shut down.
    - The dangerous failure modes of the control system should be determined. This should include the analysis of I/O, power, and communications failure.
    - Consideration should be given to change control and backup systems.
    - Consideration of survivability during hazardous events and emergency responses.
  - Alarm systems:
    - The alarm system should be designed in accordance with IEC 61508 to SIL 1 or 2, with the designated reliability.
    - The alarm system should be independent from the process control system and other alarms unless it has also been designated safety related.
    - The operator should have a clear written alarm response procedure for each alarm.
    - The alarms should be presented in an obvious manner, distinguishable from other alarms, and have the highest priority.
    - Human interface: The human interface should be suitable. Alarms may be presented either on annunciation panel, individual indicators, visual display unit screen, or programmable display device.
  - Protection systems (trips and interlocks): Protective tripping systems provide a defense against excursions beyond the safe operating limits. Trips should not be self-resetting unless adequate justification has been made. Protective

interlocks prevent those control actions which might initiate a hazard from being undertaken by an operator or process control system, and are by nature self-resetting.

- Independence: Protective systems should be sufficiently independent of the control system or other protective systems (electrical/electronic or programmable). Where there is an interface, then the method of achieving independence should be defined, and common cause failures adequately considered.
- From IEC 61511 Part 2, the following points are important:
  - For BPCS: Any claim for risk reduction should be justified by the consideration of the integrity of the BPCS.
  - BPCS: The procedures used for configuration, modification, operation and maintenance; it is important to ensure the access security and change management are provided.
  - Human factors must be carefully considered.
  - Claims for risk reduction should be supported by a documented description of the operator response, that there is sufficient time for the operator to take corrective action, and assurance that the operator will be trained to take the protective actions.
- Common mistakes:
  - Sharing sensors or valves in both systems (viz. BPCS & SIS)
  - Operator action twice; that is, operator action at hi is same as hi hi so should not be considered twice; an independent protection layer.
  - Inappropriate operator credit.

4.1 Layer of Protection Analysis Process

One of the objectives of LOPA is to develop functional safety of the system and assess SIL. It is better, therefore, to have some concept about functional safety from IEC 61511-2 standards (see Clause 5.2.6.1.1):

"The use of Functional Safety Assessment (FSA) is fundamental in demonstrating that a Safety Instrumented System (SIS) fulfils its requirements regarding safety instrumented function(s) and Safety Integrity Level (SIL). …The following attributes are considered good practice for Functional Safety Assessment:"

- A plan should be generated for each FSA, identifying the scope of the assessment, the assessors, the competencies of the assessors, and the information to complete the system.
- The FSA should take into account other standards and practices, which may be contained within external or internal corporate standards, guidelines, procedures, or codes of practice. The FSA plan should define what is to be assessed for the particular assessment, system, or application area.
- The frequency of FSAs may vary across different system developments but as a minimum should always take place before the potential hazards being presented to the system."

From the standard, it is also noted that FSA frequency and rigor shall be variable with respect to complexity, safety significance, etc. The standard also requires proper definition of overall objective scope and a team to carry out the exercise. In this section, attempts will be made to cover all of these. LOPA procedure can be developed from *safety life cycle* of IEC 61508, from "Concept Development" through "Safety Allocation."

It is worth noting that there have been several different approaches towards the LOPA process, but overall coverage shall include the following in all these approaches, as major objectives are the same:

- Documented hazard analysis.
- Identifying and listing of scenario and impact event.
- Identifying initiating cost.
- Development of IPL.
- Determination of frequency, PFDs, etc.
- Evaluation and SIL assessment.

4.1.1 Scope, Boundary Definition and Risk Assessment Criteria

The scope definition section requires the definition of the boundary of the process and equipment under control being assessed, together with its control system. Since in the majority of cases the input to LOPA is taken from preliminary hazard analysis or HAZOP, the scope is more or less otherwise developed from previous analysis. It is necessary to ensure that equipment under control and its environment are sufficiently understood along with scope and boundary including interface before detailed assessment commences.

Risk acceptance criteria are very important to start the system. It also sets the scope and boundary to a great extent. The acceptance criteria are decided based on the requirements from the company, authorities, and customers. Acceptance criteria should be established for all types of consequences as safety, environmental, property, and other economical factors.

4.1.2 Team and Meeting

As in HAZOP, LOPA is also performed by a multi-disciplinary team, conducting several meeting sessions. The team at least should consist of:

- Operating and production management personnel.
- Process/mechanical engineer.
- I&C (design) engineer.
- Maintenance engineer.
- Health safety and environment (HSE) risk analyst/specialist.

It is needless to mention that the chosen team must have good knowledge about the entire process, and some knowledge about plant instrumentation. As LOPA is specialized, subject and allocations of IPLs need some good knowledge about LOPA. It is necessary that one of the members should be LOPA expert. Out of the team members, one should be chosen as the leader and one as secretary. Persons with other expertise may take part in the analysis at different points in the analysis when needed. As stated, in the case of HAZOP, the meetings are usually run in several sessions and it is less time-consuming (may be less time than FTA). The team needs to prepare a spreadsheet as per standards (IEC 61508/61511). Like HAZOP, meetings shall be well-planned, and a proper schedule to be maintained. Care should be taken to see that team is not under fatigue and can take the decisions freely. In this connection, see Clause 5.2.2.2 of IEC 61511-2: "The skills and knowledge required to implement any of the activities of the safety life cycle relating to the safety instrumented systems should be identified; and for each skill, the required competency levels should be defined. Resources should be assessed against each skill for competency and also the number of people per skill required."

4.1.3 Documentation

Depending on applicability, the information requirements under shall be under following heading:

- Process definition
- Basic design
- P&I diagram
- Physical and chemical property data
- Control philosophy
- Control system details
- Alarm and trip details
- Maintenance plan
- Operating conditions
- Plant layout details
- Plant location
- Applicable standards
- Legislative
- Risk details
- Historical data on plant
- Hazard analysis report (if any)

LOPA is used to evaluate scenarios that have been identified in a prior hazard identification, viz. HAZOP. It is important to note that if necessary team needs to investigate whether there is any further need to explore the possibilities of hazard identification. In that case, additional documentation or missing applicable documentation, as listed previously, need to be collected afresh. A first step in the LOPA study is commonly to screen these scenarios, (usually) on the basis of consequences. This needs to be examined in the light of risk acceptance criteria.

4.1.4 Detailed Procedure

A detailed, step-by-step procedure for LOPA shall be discussed in the following paragraphs. It is better to have a detailed scrutiny of Fig. V/4.1.4-1.

- Step 0: In Fig. V/4.1.4-1, it has been shown that LOPA starts from plant hazard analysis (PHA)/HAZOP results. This is the normal case, but it could be started afresh also. In both the cases, it is necessary to fix up the scope and boundary as

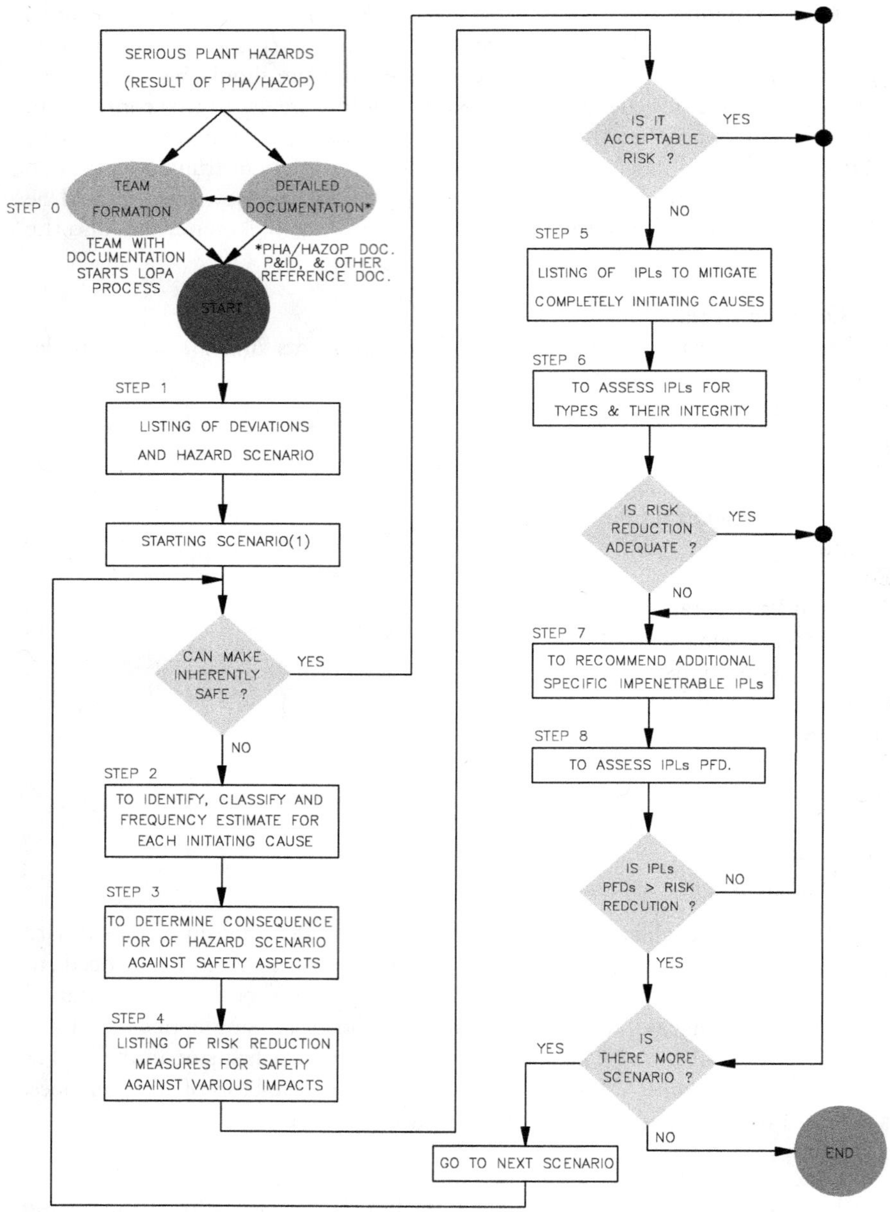

FIGURE V/4.1.4-1

Brief layer of protection analysis process.

discussed above. In the figure, one stage, Step 0, has been shown when the team formation and documentation, etc. are done. Therefore, the starting point really is to develop and fix up of risk acceptance criteria, and then to go through the collected documents to check the adequacy. In case information is inadequate, then it is better to collect necessary documents before starting the process discussed next.

- Step 1: Identify the consequence to screen the scenarios by transforming all technical data for development of scenario. At this step, it is necessary to list out deviations. Since LOPA typically evaluates scenarios that have been developed in a prior study, a first step by the LOPA analyst(s) is to screen these scenarios. Based on consequences, screening is normally done. The consequence is typically identified during a qualitative hazard review (e.g., HAZOP study). Next the consequence (including the impact) is evaluated and its magnitude is estimated by the analysts. It is better to explicitly estimate the impact to people, property, and environment with detailed estimate for likelihood for harm. Depending on the policy, some companies stop for each scenario to look at the magnitude of a release of material or energy, etc. Then an accident scenario is selected as first case. LOPA is applied to the scenario. One scenario is taken at a time. The scenario describes a single cause—consequence pair.
- Step 2: In this step, a hazard is checked whether it can be made inherently safe, then the issue is resolved. If not, then at this stage it is necessary to find details about the incident by identifying the initiating event of the scenario by classifying and determining the initiating event frequency. Here, one important issue is to assess the consequence, available from perhaps a, previous HAZOP. The initiating event will lead to the consequence (given failure of all of the safeguards). For frequency, an estimation needs to take into account the background aspects of the scenario, such as the frequency of the mode of operation for which the scenario is valid. In the majority of cases companies have some guidance on frequency estimation.
- Step 3: One pair of initiating cause and impact events should be evaluated at the time. At this stage, determination of the consequence of the hazard scenario is done. This is an evaluation in terms of safety, environmental, and economic losses. Safety and environmental impacts are guided by specific standards of the country, but economic loss prevention is a company decision. The economic risks are assessed to meet loss prevention goal. Cost—benefit analysis can be used to determine what the actual design should be for an IPL. After having clear understanding of frequency and consequence risk, a matrix or graph may be used to determine whether the risk is acceptable, or whether IPLs are required for further risk reduction. Quantitative targets can also be used for the assessment. Here one may *note* that when in quantitative target, if it is found that the consequence is such that there is the possibility of single and/or multiple on-site fatality, and likelihood is after some mitigation in the order of 10^{-5}, then QRA may be necessary. In such cases, LOPA may not be carried out, and such scenarios may be eliminated by screening and recommending it for QRA.

Acceptable risk criteria play a good role here. This situation is not shown in the referred figure.

- Steps 4 to 8: At this stage, listing of IPLs meeting the independence, specificity, dependability, and auditability requirements are done. An example will make it clear; common sensing elements should not be used for both BPCS and alarm, when BPCS failure is one of the causes for an impact. List the IPLs that can completely mitigate all listed initiating causes first. Now the PFD is determined. When IPL criteria are satisfied, the PFDs are added in the LOPA worksheet. Some estimates of PFDs can be found in standard tables that are available, for example, CCPS - 2001 (PFD data for BPCS failure, lightning, regulator failure, and more data are available). Also, plant-specific data can be used. The PFD is a measure of the risk reduction that can be obtained using the IPL. Since SILs are measured in terms of PFD, it could be considered as equivalent so far as SIS is concerned. In this connection, ANSI/ISA 84.01-1996 (2), and IEC 61511 (3) may be referenced. Pre-approved PFDs help in saving time. The inherent process design and reduction factors need to be evaluated and given credit. However, assigning credit for such issues is difficult at times. The following modifiers shall be taken into account:
  - Occupancy
  - Ignition probability
  - Time at risk (not continuously in operation)

The occupancy factor is calculated for the risk graph (IEC 61511, 2003). For flammable hazards, ignition probability shall be considered. In all cases of unsure issues or conservative values, *one* is chosen in work sheet.

- As shown in Fig. V/4.1.4-1, the process is in a loop. When there are more than single impact event-initiating causes, then after finishing one pair, the next pair is chosen for repeating the same process discussed previously. In this process, the LOPA team provides specific implementable recommendations which are optional for the company.
- One point is important that for better, consistent results, there should be a risk estimate *mathematically* combining the consequence, initiating event, and IPL data. Other factors may be included during the calculation, depending on the definition of the consequence. With these discussions on the procedure, it is concluded and next the work sheet based on IEC 61511(2003) can be discussed.

4.1.5 Worksheet (for LOPAIEC 61511)

The most common worksheet for LOPA is as per IEC 61511. It has been presented as follows in Table V/4.1.5-1. A short description of each of the columns have been also presented for explanation and understanding.

- Column 1: Impact event description: Normally, this comes from previous PHA/ HAZOP for various consequences.
- Column 2: Severity level: The severity of the impact event is entered and is classified; Minor (M), Serious (S), or Extensive (E) by the analysts

Table V/4.1.5-1 Layers of Protection Analysis Worksheet (IEC 61511)

Protection Layers												
1	2	3	4	5			6	7	8	9	10	11
Impact event description	Severity level	Initiating cause	Initiation likelihood	General process design	BPCS	Alarm etc.	Additional mitigation (restricted access)	High integrity additional mitigation (dike, pressure relief)	Intermediate event likelihood	SIF integrity level	Mitigated event likelihood	Notes

Headings are only shown in rows in which necessary data to be completed.

Table V/4.1.5-2 Severity Level

Severity Level Category	Consequences
Minor (M)	Restricted to local area potential for extending to broader area if no corrective action initiated
Serious (S)	Any serious injury or fatality on/off sites
Extensive (E)	Five or more than in severity than serious (above)

(see Table V/4.1.5-2). These classification norms are the same as discussed in connection with the risk matrix, hence not repeated. These can be done as per risk graph approach also.

- Column 3: Initiating cause: Direct initiating causes of the impact event are listed.
- Column 4: Likelihood: Initiating cause likelihood; occurrences per year are entered here (see Table V/4.1.5-3).
- Column 5: Independent protection layers (see Table V/4.1.5-4): This column consists of three sub-columns. As discussed previously, when protection layers satisfy the IPL criteria (and reduces risk from initiating cause), credits are given in terms of PFD value in the worksheet. In all cases of unsure issues or conservative values, *one* is chosen in work sheet; against PFD value. Process design, BPCS, and alarms are placed in column 5 from left to right in sub-columns. Whenever any one of these is capable of reducing or mitigating the

Table V/4.1.5-3 Initiating Cause Likelihood

Initiating Likelihood Category	Typical Frequency of Occurrence (per year)
Low	$f < 10^{-4}$
Medium	$10^{-4} < f < 10^{-2}$
High	$10^{-2} < f$

Table V/4.1.5-4 Typical Independent Protection Layer Probability of Failure on Demand

Independent IPL Type	PFD
Control loop	10^{-1}
Relief valve	10^{-2}
Trained human performance	10^{-2}
Operator response to alarm	10^{-1}
Vessel pressure rating above external & internal maximum challenge when regular inspection on	10^{-4}

risk from initiating events, due PFD credit is entered in the corresponding sub-column.

- Column 6: Additional mitigation: Restricted access, with their PFDs when known, are entered here.
- Column 7: High integrity additional mitigations: High integrity mitigation IPLs, such as dike pressure relief, are entered here.
- Column 8: Intermediate event likelihood: The occurrence of the end-consequence with the existing or planned IPL without SIF. It is the frequency per year of the occurrence of intermediate event. It is calculated value, which is multiplication of values from Column 4 and values from Columns 5, 6, and 7. The calculated number is in events per year. This is used to compare with the criteria. When intermediate event likelihood is greater than the corporate criteria, additional mitigation is needed. Inherently safer designs are considered prior to introducing new SIF.
- Column 9: Safety integrity level: When a new SIF is needed, the SIL is calculated by dividing the corporate criteria for this severity level by the intermediate event likelihood (column 8) and result is entered here.
- Column 10: Mitigated event likelihood: The occurrence of the end-consequence with all protection layers in place, including the proposed SIF. The mitigated event likelihood is calculated by multiplying columns 8 and 9 and entering the result in column 10 as per year of the occurrence of this event.
- Column 11: Notes: Special notes (if any) are kept in this column against each impact event.
- In this way, the worksheet is filled in for each impact event. Last would be the total risk.

4.1.6 *Evaluation of Layers of Protection Analysis Worksheet and Worksheet Data*

After the LOPA process and LOPA worksheet, one obtains the following result from the analysis:

- Detailed identification of initiating cause causing the impact event.
- Estimation of initiating cause frequency, severity of consequences, and PFD for them. Analysts get a clear idea about the order of magnitude of harm or end scenario from various cause impact pair. This helps in developing safeguards in the form of preventive and mitigating IPLs.
- Formal designation of an IPL based on independence, effectiveness, and auditability.
- Assessment of IPL adequacy and recommending additional safeguards/IPL based on risk based decision.

A few typical data associated with a worksheet has been presented in Tables V/4.1.5-2—V/4.1.5-4 (inspired by data from Ref. [9]) to get an idea about the issues.

4.1.7 Layers of Protection Analysis Discussions

- LOPA—HAZOP integration: LOPA can be effectively used for determination of SIL. Also, LOPA and HAZOP require a similar database. Therefore, if the two are integrated, it will give cost and time benefits. Now the question is if some software means is deployed to integrate them, the nearly 25—30% savings in terms of monetary value as well as time is achievable. Such software tools are available in the market. If HAZOP analysis is recalled, it will be seen that in HAZOP also some safeguards are sought. These can be protection layer (PL) for LOPA. Inclusion of any IPL in LOPA worksheet column requires evaluation to check the IPL characteristics. In LOPA, the impact event likelihood and consequence severity are classified. Here they can be obtained from HAZOP as well. From the discussions, it transpires that by using software tools, many of the stages of LOPA and HAZOP can be done automatically. Integration offers following:
  - Time savings.
  - Cost savings.
  - Common database.
  - Faster data transformation.
  - Easier and better quality checks.

There is also lot of commonness between the two worksheets as shown in Table V/4.1.7-1. Only some adjustments are necessary, and this can be done suitably in software.

***Table V/4.1.7-1** Commonness in Worksheet Between LOPA and HAZOP*

LOPA	HAZOP
Impact event	Consequence
Initiating cause	Possible cause
Frequency of initiating cause	Consequence likelihood
Severity level	Consequence severity

- Risk judgment/categorization: CCPS publication gives four basic categories of criteria:
  - Criteria that places risk characterizations per scenario in matrices, with parameters of frequency and consequence as guides.
  - Criteria that specify a maximum allowable risk (e.g., risk of fatality or financial loss) per scenario.
  - Criteria that specify a minimum number of IPLs (or IPL credits) for any specific scenario.
  - Criteria that specify a maximum cumulative risk for a process or geographical area. From these judgment is taken to see if any further action is needed.

4.2 Layers of Protection Analysis Safety Integrity Level and ISA

Discussions put forward in the above clauses mainly based on IEC 61508/61511. Similarly, ISA standards also have direct reference to LOPA. Here, a short discussion on this is provided. ISA 84 implementation begins with an early PHA. The PHA team then assesses the adequacy of existing safeguard. If it is inadequate, then action items are developed to guide the engineering team to an appropriate solution. The most useful way is to go for "layers of protection analysis." The main requirement is that each protection layer must be designed to function independently from the other protection layers so as to ensure protection even after failure of one or more PL. Like any other method, cost-effectiveness of risk reduction is achieved when each IPL is maximized for risk-reduction at minimum cost. The challenge is to select protection layers that yield the best cost-to-benefit ratio, while achieving the ultimate goal of reducing process risk to a tolerable level. For SIS for risk reduction, the requirements of ISA 84 must be implemented. The SIL is assigned by the owner/operator to the SIS. ISA 84 has three discrete SIL performance ranges, as shown in Table V/4.2-1.

The SIL is related to the average probability of the SIS failing on process demand (PFD_{avg}). From here, one gets that SIL 1 must achieve a minimum PFD_{avg} of 0.1, which means that the SIS has a probability of failing one in every 10 times that it is needed. SIL 1 represents the lowest acceptable performance. SIL 3 represents the highest recognized performance as per ISA. The SIL is affected by the following:

- Device integrity (i.e., failure rate).
- Redundancy and voting.
- Functional testing frequency.
- Diagnostic coverage.
- Other common causes, for example, systematic factors, human error.

These five factors represent the major design decisions. In a sense, these parameters can be considered "*degrees of freedom*" in the design of the SIS, while the SIL is the design constraint established by ISA 84 which is a performance based standard and has a few requirements listed below in a simpler manner (similar to other standards):

- Process control and SIS components and functions must be separated.
- Minimum hardware redundancy.
- Controlled access to SIS components and functions.

Table V/4.2-1 Safety Integrity Level and Average Probability of Failure on Demand

SIL	$PFD_{(avg)}$
1	10^{-1} to 10^{-2}
2	10^{-2} to 10^{-3}
3	10^{-3} to 10^{-4}

- The use of digital communication is prohibited!
- Restricted information (transfer for operation) between process control and SIS.

It is important to have quantitative verification of the SIL at each step to achieve PFD_{avg}.

ANSI/ISA 84.01-1996 also establishes specific requirements for the design, installation, operation, and maintenance of SISs. Well-documented technical information are necessary to develop the cases for mitigating risks and to demonstrate that the SIS was designed to meet those requirements, and to show that the operation and maintenance practices are established to maintain the SIS. Detailed discussion on these will appear in Chapter VI.

4.2.1 *Critical Alarm in Layer of Protection Analysis*

Following points are important in connection with critical alarm in LOPA. In this connection, ISA guideline is also important. These are discussed here. Clause 4.2 of Chapter VII also deals with the same in greater detail.

- Alarm management system plays key role toward the maintenance of plant safety. ANSI/ISA-18.2 standard: "*Management of Alarm Systems for the Process Industries*" (ISA-18.2) could be considered as a foundation for successful design, implementation, operation, and maintenance of alarm systems in a process plant. Operator's response to an alarm is considered as IPL in LOPA. This is important link of ISA with LOPA as the performance of critical alarms can impact the design of the SIS also. If its performance is poor, then result from LOPA could deteriorate.
- The critical alarm management system is very important to operator action. Alarm load, nuisance alarms, and alarm priorities have tremendous effect on an operator's corrective action. For an operator response to be considered an IPL, it needs to meet the following requirements:
  - Specific: The alarm must be specifically designed to prevent the consequences under consideration.
  - Auditable: It should be proof-tested and maintained.
  - Independent: The alarm shall be independent of all other protection layers; no common sharing.
  - Dependable: The ability to prevent occurrences of consequence dependably.
- The PFD for the operator's response to an alarm is dependent on:
  - The probability of alarm system failure.
  - The probabilities of operator to successfully detect, diagnose, and respond to the alarm.

The first issue is related to hardware and software of an alarm management system. Once the alarm is annunciated, a series of steps must be performed by the operator.

The following issues are responsible for human action, and associated error (if any):

- Important single or number of decisions in a short time under emergency conditions.
- Physical or mental capability.
- Clarity and transparency of data and display.
- Job training.

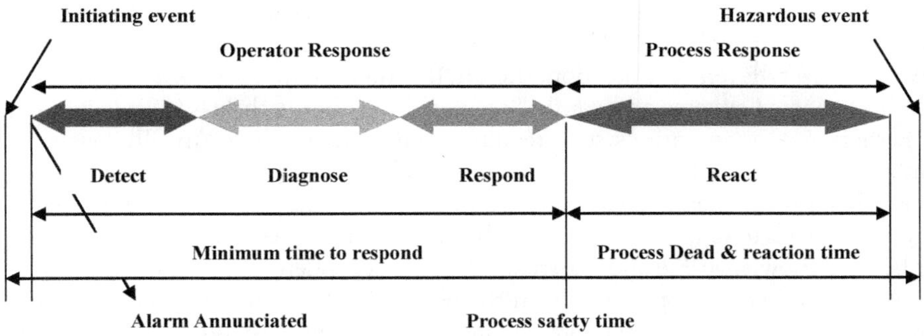

FIGURE V/4.2.1-1

Critical alarm and operator response.

This figure has been developed as inspired by Fig. 6 from T. Stauffer, Making the Most of Alarms as a Layer of Protection, Director Alarm Management Services, exida LLC.

Timing is another issue. It is the operator's response time and process safety time that are discussed for clarification in the following and in Fig. V/4.2.1-1 [10].

- Timing: It is also important to define a minimum operator response time, which is the minimum amount of time that must be provided to the operator for him or her to respond to any alarm in the system.

From Fig. V/4.2.1-1, it is clear that in order for the operator to have a clear response, he or she has to perform a number of functions such as detect, diagnose, and respond. Similarly, process safety time requires additional timing for process dead and reaction time, thus, a key requirement for a safety IPL alarm to be valid is:

$$T_{\text{Detect, Diagnose, Respond}} + T_{\text{Deadtime}} < T_{\text{Process Safety Time.}}$$

and

$$T_{\text{Process Safety Time}} > T_{\text{Minimum Operator Response Time}}$$

From this, another aspect is clear that operator action is important for any critical alarm, which needs to be perfect, and follow ISA guidelines mentioned previously. A few recommendations of ISA have been listed below:

- Maintenance and testing to ensure alarm integrity.
- Classification helps manage critical alarms.
- Eliminate "nuisance" alarms.
- Practice makes perfect.
- To make alarm response procedure available online
- Minimize alarm load on operator by monitoring alarm system performance.

4.2.2 *Final Discussions*

OSHA process safety management regulations require process hazard analysis.

- How system is?
- What and how many protections are needed?
- Which categories of protection are needed?

LOPA can provide answers to these questions in a risk-based approach. LOPA used in compliance with IEC 61511(/ISA 84) standards, requires the allocation of safety functions to layers of protection, the SIL determination. Thus, LOPA and SIL are well-connected systems and can be integrated together with HAZOP. Is it the last? Perhaps not, as an entire system requires good understanding. An SIL concept including use of LOPA is narrowed down to SIL classification based on PFD calculations. This is easier than to go for inherent safety with a complete change of process design! One needs to remember that for such calculations, it is essential that the team must be very knowledgeable and experienced. Also, dependency on an operator may be a major issue in catastrophic results in the past. Actual proper implementation of the result is very important. Therefore, the main idea is to understand the system and try to incorporate inherent safety designs as far as possible.

4.3 Intelligent Layer of Protection Analysis: The Concept

The automation approach for LOPA is similar to that in HAZOP, and many tries to integrate the two. In order to standardize an analysis system, it is necessary for large complex systems to go for intelligent LOPA systems based on knowledge-based system. Here only conceptual part will be discussed to limit the volume of the book. Generally, standard software systems are available to for the intelligent LOPA analysis method. However, in most of the cases this software also integrates other analysis methods, for example, HAZOP. Major capability enhancement includes, but is not limited to the following:

- Improved focus on process safety.
- Quick, complete risk assessment accurately.
- Complete analysis including vulnerability tests.
- Possibility to share historical knowledge database in future cases.
- Consistent terminology, lesson learned knowledge.
- Consistency over entire risk management process
- Centralized knowledge-based system, knowledge-based initiatives, planning and compliance and can handle management of change (MOC).
- Enterprise—wise shared framework.

It is needless to argue that the intelligent system will allow uninterrupted business progress with regular assessment of risk locally as well as globally for the entire business area. Various risk assessments can be linked, and possible implementation of tracking of all mitigation techniques and their results. There are two major approaches. These are the model approach and the fuzzy approach, and sometimes a combination of both when there is question of integration of various analyses are concerned.

4.3.1 Model Approach

Out of many intelligent methods, modeling is one of the best options. The model approach makes it possible to offer better visibility for better structuring the in-depth lines of defense by establishing barriers appropriate to each elementary situation of the accidental process. The model of accident scenarios can be done in a generic manner for analysis. Identification of accident scenarios will be based on the development of accidental processes based on the occurrence of exposure event, initiator event and the top event. These events have the ability to cause the change in position of an entity between, initial situation (IS), exposure situation (ES), hazardous situation (HS) and accident situation (AS). Evaluation of the scenario offered by several techniques and mathematical models of risk prediction have been developed for these [11].

4.3.2 *Fuzzy Approach*

Another approach is fuzzy approach for LOPA. In LOPA, which can quantify the risk quickly utilizing multiple types of logic, fuzzy logic is used extensively, for example, the Sugeno model. With the help of membership functions of fuzzy logic, knowledge base can be quantified in an accurate manner and expert systems can be built based on fuzzy logic to achieve better decision-making, and getting accurate results out of analysis so that it could be used as a screening tool for QRA. As shown in Fig. V/4.3.2-1, the approach is similar to what has been discussed in connection with FMEA. Here also, input is taken into the system for "fuzzification" based on fuzzy rule and membership. Knowledge base uses expert systems, based on the same inference engine or decision-making unit, to produce output which is taken for "de-fuzzification" before actual output is produced. It is better to understand what membership and rule based system are:

Unlike the Boolean set of 1 or 0, fuzzy logic represents a continuous spectrum with the help of fuzzy sets. In line with human conclusion and decisions, fuzzy logic (developed by Zadeh) utilizes inexact information, not on a pure mathmatical approach. In fuzzy approach, a set is defined in terms of its membership. Fuzzy automation systems keeps fuzzy rules in its knowledge base, and apply the same on the process inputs to give output based on a fuzzy reasoning process. Let there be a typical fuzzy set "A" with various elements (a, b, c, d), having membership functions 0.2, 0.3, 0.6, and 1. Naturally, in the set, the element with a membership function 1 is a full member, and the others are partial members. Out of partial members "a" has the least weightage, and d has the highest weightage. To limit the discussion, and make the system generalized, let the elements be large negative (LN) to large positive (LP), as shown in Fig. V/4.3.2-1B (moderate, slight negative or positive represented by MN, SP etc.), in seven classes with zero as middle. Referring to Fig. V/4.3.2-1B, an element in 40% will have 0.4 SP and 0.6 MP, and similarly when rate of change is 15% will have 0.4 ZE and 0.6 SP as shown in Fig. V/4.3.2-1B. Fuzzy control takes heuristic approach, and on account of its closeness due to human thinking, it has wide acceptance. For further details on this, the standard book on fuzzy system may be referred to. As per Sugeno, model output Z corresponding to the two inputs, x, y is given by $Z = ax + by + c$, where a, b, and c are constant values. To get a weighted value, inputs need to pass through membership functions discussed previously. This final output is the weighted (W) average of inputs based on their membership. So, in fuzzy method final output is $Z = \sum_{i=1}^{i=n} Wi * X_i \Big/ \sum_{i=1}^{i=n} Wi$.

With this, discussions on LOPA are coming to an end to have a glance through other analysis methods.

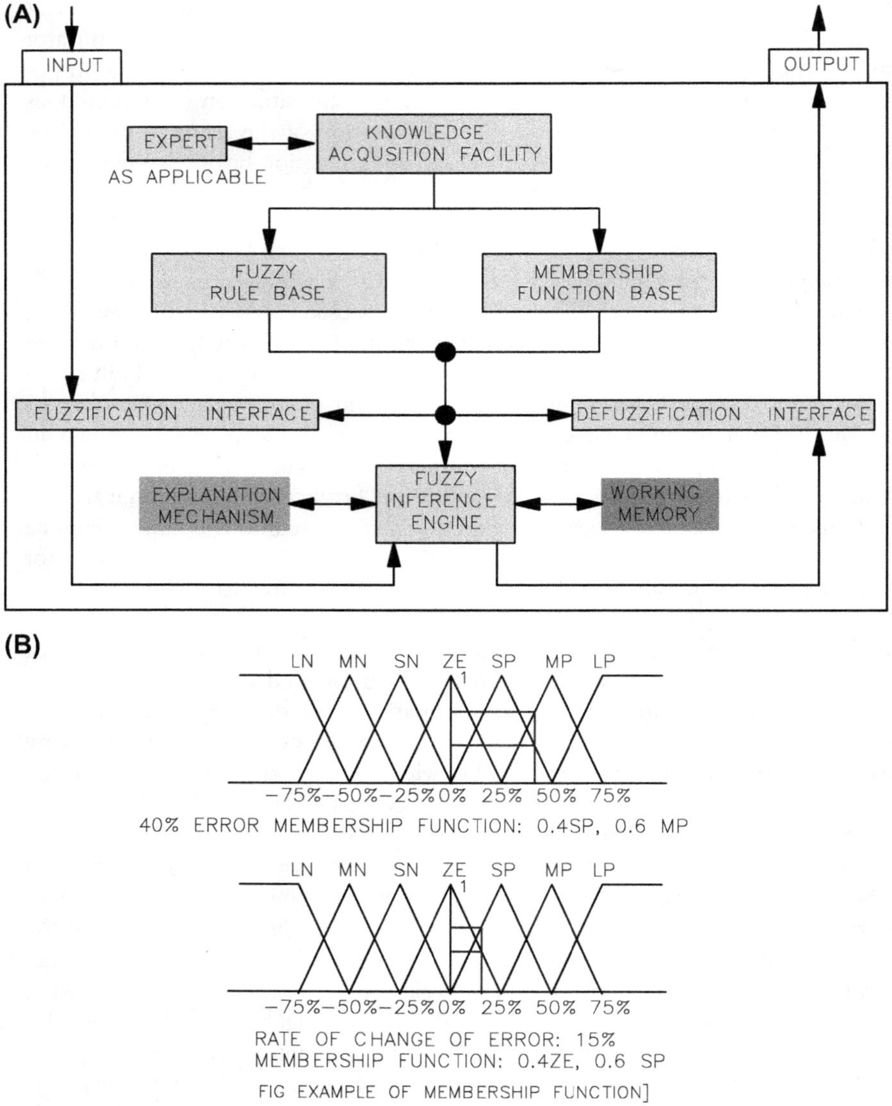

FIGURE V/4.3.2-1

Fuzzy approach for layer of protection analysis. (A) Typical architecture of fuzzy approach, (B) understanding of fuzzy approach.

5.0 QUANTITATIVE RISK ANALYSIS IN BRIEF

From DNV, one defines quantitative analysis as follows; "a QRA can be defined as the formal and systematic approach of identifying potentially hazardous events, estimating the likelihood and consequences of those events, and expressing the results as risk to people, the environment or the business." In a way, it can be conceived as a quantification of effects of risks on project objectives.

5.1 Requirements for Quantitative Risk Analysis

Usually these are completed after qualitative hazard analysis is over to address the following issues:

- Determination of probability of achieving the project objective.
- Risks prioritization.
- Improve the identification of opportunities and threats (ISO31000:2009).
- Identification of realistic cost, schedule, and/or scope and cost—benefit analysis.
- Comply with relevant legal and regulatory requirements and international norms (ISO31000:2009).
- Project outcome quantification with associated probability.
- Can be modified to fit specific need of a company.
- Improve governance (ISO31000:2009).
- Effectively allocate and use resources for risk treatment (ISO31000:2009).
- Good selling tool to management, and have less chance of disagreement.
- Determination of contingency and contingency planning.
- Improve organizational learning and resilience (ISO31000:2009).
- Guidance for project management decisions under various uncertainties.

5.2 Method Outline

There could be several methods for QRA based on the objective of the analysis. In process plants, this can be carried out, with any or all of the objectives listed below in line with DNV.

- An analysis of the severity/consequence of accident scenarios.
- Predicted number of fatalities/casualties for each scenario.
- Individual, group, or societal risk.
- Potential loss of life.
- Location-specific risk.
- Further analysis of accident scenarios that are ALARP (As Low As Reasonably Practicable).
- Preventative/mitigation measures.
- Sensitivity of results to uncertainties and assumptions.

In order to carry these out in most of cases, QRA and modeling techniques, sensitivity analysis, monetary analysis, decision tree, and simulation are used. Major coverage in these methods shall include probability modeling, and consequence modeling, cost benefit analysis and risk mitigation techniques. Typical QRA has been depicted in Fig. V/5.2-1. As there will be a requirement of very large database (e.g., tangible—intangible asset of the company, historical attitude of the company, to name a few), there will be a need for a suitable software tool to be utilized by an

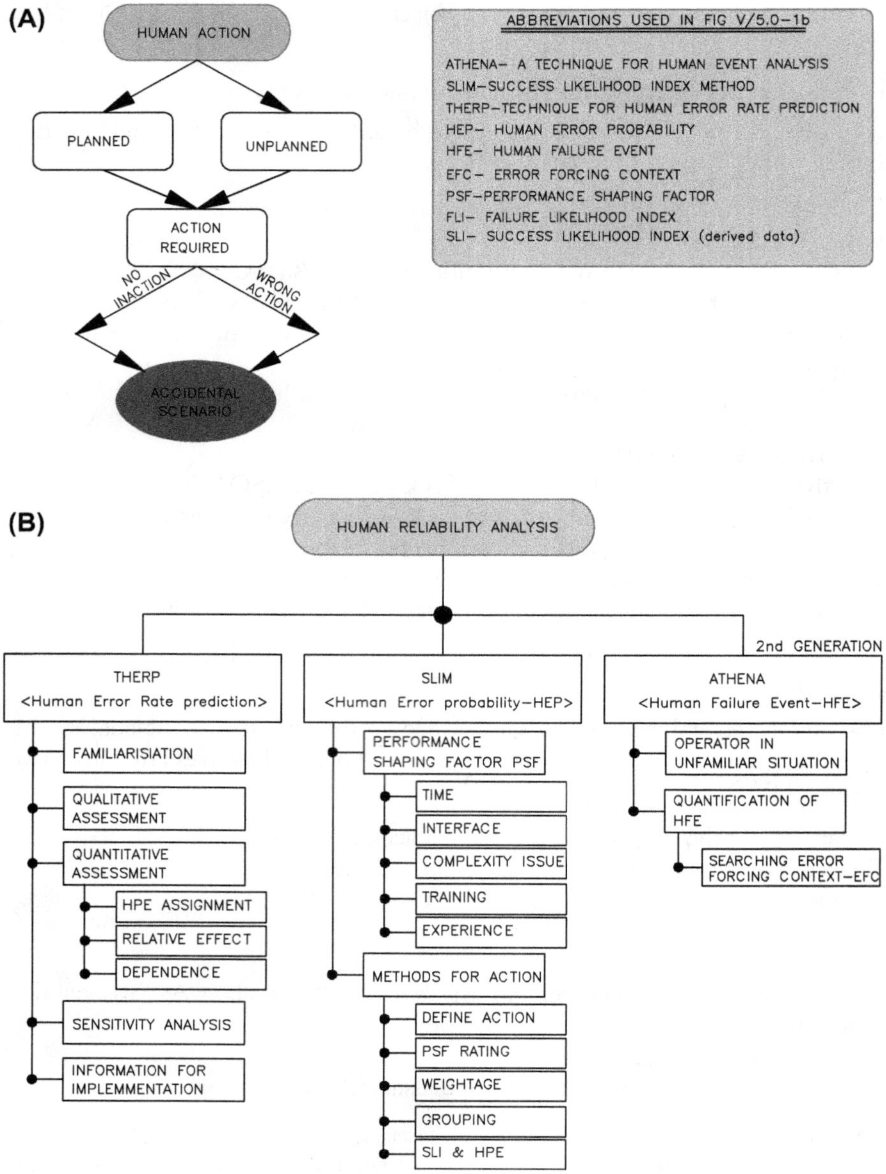

FIGURE V/5.2-1

Quantitative risk analysis methodology.

experienced person capable of handling the same. QRA is common in the places where there is threat from explosion and toxicity. In this connection, it is better to note that major bottleneck in QRA could be:

- Inherent lack of data, and lacking in analyzing capability.
- Proper quantification of risks.
- Compromise monetary expected value.

5.3 Documentation and Information Collection

A number of documentations are necessary to carry out QRA:

- Organizational and environmental process assets.
- Project scope statement.
- Risk management plan.
- Project management plan (cost and schedule).

Like various other methods, information is gathered through interview (to get the idea of probability and impact of risk on project objectives from concerned persons), collection of detailed documentation, and expert judgment.

5.4 Major Output From Plant Quantitative Risk Analysis

Major output for plant or process technical QRA shall include, but not be limited to, the following issues:

- Risk register updates (to indicate risk result of QRA).
- Probabilistic analysis of the project (including time and cost).
- Forecasts of potential project schedule and cost results.
    - Project duration and completion time.
    - Cost estimate.
    - Contingency requirement.
    - Contingency planning.
- Prioritized list of quantified risks.
- Trends in quantitative risk analysis results (after repetition).

A brief QRA process has been depicted in Fig. V/5.4-1. Probabilistic QRAs are also used. In the probabilistic approach, first a deterministic model is developed, then numerical values are replaced by probability structure and distribution patterns to get the output.

6.0 HUMAN RELIABILITY ANALYSIS IN BRIEF

From various internationally published estimates, it is found that in the world, out of the total number of plant accidental scenarios, around 60–90% are on account of human failure in different forms, and the rest are on account of technical deficiencies of equipment and control systems, or on account of other issues. Naturally, human reliability assessment or analysis claims is a major focus. In Fig. V/6.0-1A, a short

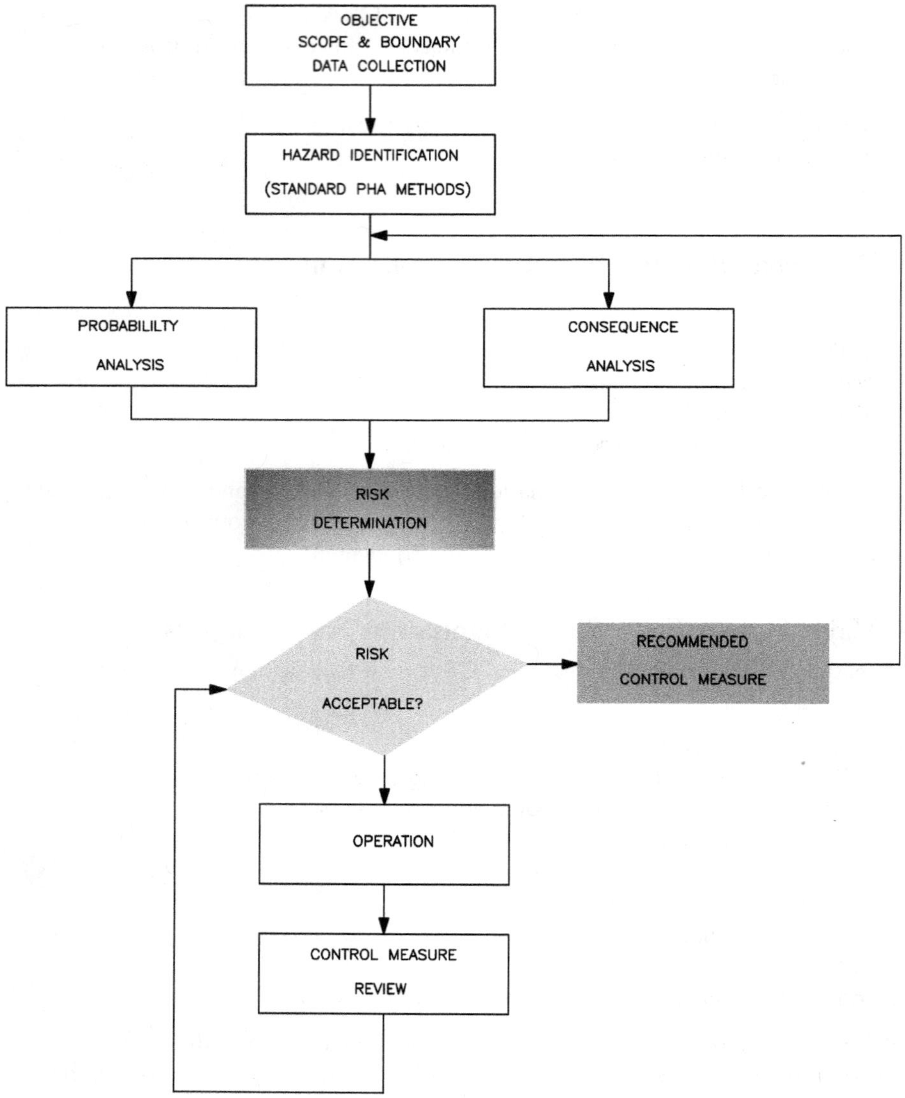

FIGURE V/5.4-1

Quantitative risk analysis process.

depiction (Ref: Clause 6.4 also) has been presented to demonstrate understanding of how does it happens.

In almost all the PHA methods discussed so far, human error is considered as part of it, for example, human error and other systemic errors during operations, maintenance, testing, and restorations should be considered in assigning the PFD for IPL. For the manufacturing process, human errors have direct impact on the product, and

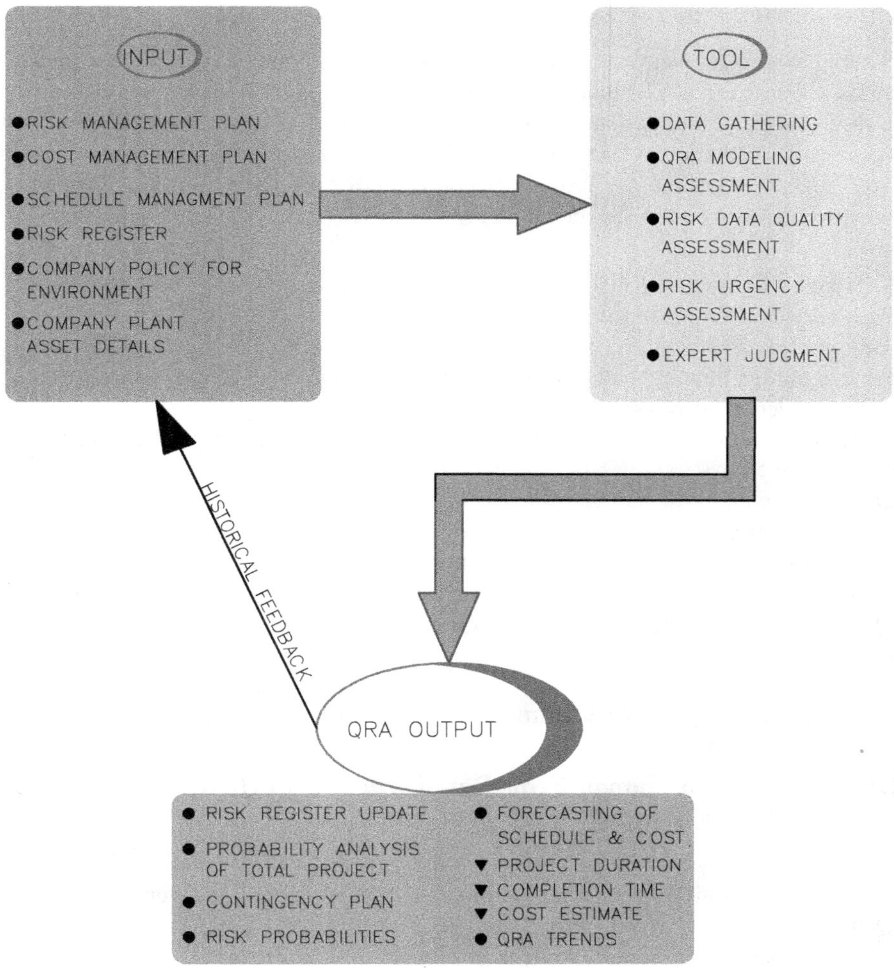

FIGURE V/6.0-1

Human reliability analysis fundamentals. (A) Human action and accidental scenario, (B) HRA method.

this may increase rejection rate and reduce productivity, sales, and company reputation. The basic objective of an HRA is to evaluate the reliability of an operator's action, and find the contribution of it in total reliability of the system. In doing so, HRA also evaluates the following:

- Predict human error rates.
- The degradation to human–machine interface systems due to human errors.
- The degradation of equipment functioning on account of, operational procedures and practices.

- Other human characteristics influencing the system behavior.

There have been several methods used to assess human reliability. Out of various methods, the technique for human error rate prediction (THERP) is in use since the beginning and still quite popular. Many of the HRA methods have been developed for specifically for various plants, for example, nuclear action reliability assessment (NARA). Short-working methods of important HRA methods, used as general purpose, in majority plants are shown in Fig. V/6.0-1B.

6.1 Short Human Reliability Analysis Steps

- Problem definition: Define the issue in the question.
- Task analysis: Description of task and understanding of the system by analyst.
- Error detailing: Listing of all probable errors and classification of various kinds of errors to identify various failure criteria.
- Development: For analysis preparation; development of risk tree and/or models.
- Qualitative analysis.
- Quantitative analysis:
  - Impact assessment: Impact of error on the system.
  - Total risk contribution from each action.
- Error reduction methods, for example, redesign, training.
- Quality assurance using suitable techniques.
- Documentation.

6.2 Brief Description of Commonly Used Human Reliability Analysis Methods

6.2.1 *Technique for Human Error Rate Prediction (THERP)*

The following are major steps to be followed:

- Familiarization: System information analysis.
- Qualitative analysis by task analysis, and development of human event tree.
- Quantitative Analysis: Sub-steps:
  - Assignment of nominal human error probability (HEP).
  - Impact/relative effect on performance shaping factor.
  - Assessment of dependence.
  - Assessment of success/failure probabilities.
  - Determination of recovery factor.
- Sensitivity analysis.
- Documentation.

Note that there are standard guidelines available for HEP (such HSE UK or Norwegian Oil and Industry Association (OLF) provide such guidance for offshore/onshore explorations) and a dependence actor.

6.2.2 *Success Likelihood Index Method*

The purpose of this section is to assess HEP, and following steps may be performed:

- Defining the "actions."
- Performance shaping factor (PSF) rating (scale of 0-10) and weightage (lo, med, hi): PSF is influenced by work force deployment, safety culture, behavioral safety,

work design (shift change, work permit etc.), training and experience, procedural guidance, time adequacy, human—machine interface, task complexity, and stress factor.
- Classifying the actions.
- Calibrating success likelihood index (SLI) as per defined formula.
- Transforming SLI into HEP as per standard formula in logarithmic scale.

6.2.3 *ATHENA*

Following are the steps followed for analysis by ANTHENA:

- Definition of scope, then definition and understanding of the issue.
- Describe probabilistic risk assessment accident scenario.
- Definition of associated human failure event (HFE).
- Assessment of human performance with associated information.
- Identification of deviations of the PRA scenario.
- To identify potential recovery factor.
- Estimation of HEPs for HFEs and apply to PRA.

6.3 Variations in Human Reliability Analysis

There have been a variety of human reliability analyses/assessments. All of them have been evolved based on their usage and applications in various types of plants. A few of them have been developed specifically for a particular plant, while a few are in use as a general way.

- HRA category: A list of major HRA methods is available in HSE.UK-rr679 (2009). Out of so many HRA methods, few of them have been listed in the following clause. There are mainly three types of HRA methods, under which all of them can be categorized.
  - 1st generation: HRA under this category are mainly concerned with the behavioral aspect of human, for example, THERP.
  - 2nd generation: HRA under this category mainly deals with the conceptual side or on cognitive aspect of human, for example, ANTHENA.
  - Expert judgment: There is another category of HRA which utilizes expert judgment, for example, SLIM-MAUD.

6.4 Human Reliability Analysis Types, Principles and Usages

A few important HRA types are listed in the following section.

6.4.1 *1st Generation: Behavioral Type*

- ASEP: Accident sequence evaluation program. It is almost THERP (discussed in later), but requires fewer resources. It is mainly developed in the United States for nuclear program. It can be used by non-HRA specialists. Both pre- and post-accident quantification of HFEs are done in ASEP. Usually, it gives conservative output. It is used for nuclear plants.
- HEART: Human error assessment and reduction technique. Relatively quick to apply, and understood by engineers for quantification of human errors. It is a generic one and has general use.

- SPAR-H: Standardized plant analysis risk human reliability analysis: It is useful for the cases where detailed assessments are not called for, as it does not identify or model HFEs. It is possible to quantify HEPs for pre-initiator and post-initiator HFEs. It was initially used for nuclear application, but now has wider applications [12].
- SLIM: Success likelihood index method. It is an HRA quantification technique by which HEPs are quantified. For taking actions, this may be utilized in conjunction with multi-attribute utility decomposition (MAUD), discussed later. Here, SLI (Ref: Clause 6.2.2) is calibrated. It should actually be considered under expert judgment type. It has wider application as it is somewhat generic.
- HCR: Human cognitive reliability (operator reliability experiments, ORE). These are actually developed to quantify post-initiator human actions (e.g., actions performed by control room crews associated with emergency and abnormal operating procedures). These were meant for nuclear applications.

6.4.2 2nd Generation: Cognitive Aspect

- ATHENA: A technique for human error analysis. ATHENA is one HRA method which has been developed to improve the state-of-the-art in HRA, especially with respect to how realistically HRA can represent the kinds of human behaviors seen in accidents and near-miss events. It requires a good amount of resources. ATHENA approach incorporates the current understanding of why errors occur, based on the work of earlier pioneers, and substantiated by reviews of a number of significant accidents. It has been developed for mainly for nuclear application, but now it is used in generic manner.
- CREAM: Cognitive reliability and error analysis method. In CREAM, the operator model is more significant and less simplistic than that of first generation approaches. It can be used both for performance prediction as well as accident analysis. CREAM is used for evaluation of the probability of a human error for completion of a specific task. There is good application of fuzzy logic in this method. It was again started for nuclear application but has wider applications, too.

6.4.3 Expert Judgment

SLIM-MAUD: SLIM (as well as FLIM) method requires expert judgment and when they are used with an interactive computer program called multi-attribute utility decomposition (MAUD). It is called SLIM-MAUD.

6.4.4 Current Generation

There have been further developments in NARA meant for nuclear applications. This type removes the limitation of 2nd generation HRAs. Now, an advanced version of HEART is used for the nuclear field. Also, there are a few other types such as justified human error data information, cognitive environmental simulation, etc. but these are not yet publicly available [12].

LIST OF ABBREVIATIONS

Abbreviations presented here to be read in conjunction with Fig. V/6.0-1 for human reliability analysis.

ADC	Analog to digital converter
BDD	Binary decision diagram
BPCS	Basic plant control system
C&I	Control and instrumentation (engineering)
CCF	Common cause failure
CCPS	Center for chemical process safety
CRC	Cyclic redundancy check
DCS	Distributed control system
DNV	Det Norske Veritas (Norway)
E&I	Electrical and Instrumentation
ECCS	Emergency core cooling system
EEMUA	Engineering Equipment and Materials Users' Association
E/E/PE	Electrical/Electronics/Programmable Electronics
EPA	Extended propagation analysis
ETA	Event tree analysis
FRLS	Fire resistant low smoke (cable)
FTA	Fault tree analysis
HAZOP	Hazard and operability study
HEP	Human error probability
HFE	Human failure event
HRA	Human reliability analysis/assessment
HSE	Health safety and environment
HW	Hardware
I&C	Instrumentation and control
IEC	International Electrotechnical Commission
IPL	Independent protection layer
ISA	International Society of Automation
LOPA	Layer of protection analysis
LSZH	Low smoke zero halogen (cable)
MCS	Minimum cut set
MMU	Memory management unit
MOC	Management of change
MTTR	Mean time to repair
NASA	National aeronautics and space administration
P&ID	Piping and instrumentation diagram
PFD	Probability of failure on demand
PHA	Plant (process) hazard analysis/preliminary hazard analysis (PrHA)
PRA	Probability risk analysis

Continued

PROFAT	Probabilistic fault tree
PSA	Probabilistic safety assessment
PLC	Programmable logic control
PRA	Probabilistic risk assessment
PSF	Performance shaping factor
PSV	Pressure safety value
QRA	Quantitative risk analysis
SDG	Sign directed graph
SFF	Safety failure fraction (IEC)
SIF	Safety instrument functions
SIL	Safety integrity level
SOP	Sum of products
SW	Software
VDU	Visual display unit

References

[1] S. Basu, A.K. Debnath, Power Plant Instrumentation and Control Handbook, Elsevier, October 2014.

[2] R.B. Patil, L.Y. Waghmode, P.B. Chikali, T.S. Mulla, An overview of fault tree analysis (FTA) method for reliability analysis & life cycle cost (LCC) management, IOSR Journal of Mechanical & Civil Engineering (IOSR-JMCE).

[3] W.E. Vesely, F.F. Goldberg, N.H. Roberts, D.F. Haasl, Fault Tree Handbook; US Nuclear Regulatory Commission, 1981.

[4] B. Schmidt, Software Fault Tree Analysis, 2007.

[5] H. Bezecny, D. Inverso, V. Maggioli, G. Rabe, A. Weinert (Eds.), Guidelines for the Use of Programmable Controllers in Safety- Related Systems, vol. 13, October 1997. White paper 6009.

[6] I.B. Santiago, J.M. Faure, Y. Papadopoulos, Including Systematic Fault into Fault Tree Analysis, December 2008. HAL Id: hal-00348072, https://hal.archives-ouvertes.fr/hal-00348072.

[7] Y. Papadopoulos, M. Maruhn, Model-based Synthesis of Fault Trees from Matlab - Simulink Models.

[8] Y. Papadopoulos, J. McDermid, A. Mavrides, C. Scheidler, M. Maruhn, Model-based semiautomatic safety analysis of programmable systems in automotive applications, in: Proceedings of ADAS 2001, the International Conference on Advanced Driver Assistance Systems, IEEE Publications, Birmingham, UK, September 2001. CFP # 483.

[9] A.A. Fredericson, The Layer of Protection Analysis (LOPA) Method, Independent Consultant - Member of Safety User Group Network, April 2002.

[10] T. Stauffer, Making the Most of Alarms as a Layer of Protection, Director Alarm Management Services, exida LLC.

[11] D. Djamal, B. Lylia, B. Abdelkarim, Towards a better approach for mastering industrial risks from modeling accidental process to integrating safety analysis techniques supporting the identification of intelligent safety decision, International Journal of u- and e- Service, Science and Technology 8 (3) (2015).

[12] HSE.UK; rr 679; Human Reliability Analysis.

[13] Task management guide. http://www.taskmanagementguide.com/glossary/what-is-process-deviation.php.

[14] S. Oliva, Non-coherent fault trees can be misleading, Journal of System Safety 42 (May—June 2006). Redondo Beach, California.

[15] C.A. Erison II, Fault Tree Analysis, September 2000. Internet document, http://www.thecourse-pm.com/Library/FaultTreeAnalysis2.pdf.

Further Reading

T.R. Albuquerque, A.C.F. Guimarães, M.L. Moreira, Event and fault tree model for reliability analysis of the Greek research reacator, in: 013 International Nuclear Atlantic Conference – INAC 2013, Cidade Universitária, Recife, PE, Brazil, November 2013.

J.D. Andrews, S.J. Dunnett, Event Tree Analysis Using Binary Decision Diagrams, Loughborough University, UK.

A.A. Baig, R. Ruzil, A.B. Buang, Reliability analysis using fault tree analysis: a review, International Journal of Chemical Engineering and Applications 4 (3) (June 2013).

S. Basu, A.K. Debnath, Power Plant Instrumentation and Control Handbook, Elsevier, October 2014.

H. Bezecny, D. Inverso, V. Maggioli, G. Rabe, A. Weinert (Eds.), Guidelines for the Use of Programmable Controllers in Safety-Related Systems 13, October 1997. White paper 6009.

K. Bingham, P. Goteti, Integrity of HAZOP and LOPA/SIL Analysis: Best Practice Recommendations, The Instrumentation, Systems and Automation Society, October 2009.

E.D. van Breukelen, R.J. Hamann, E.G. Overbosch, Qualitative Fault Tree Analysis Applied as a Design Tool in a Low Cost Satellite Design: Method and Lessons Learned, May 2006.

CCPS, Layer of Protection Analysis: Simplified Process Risk Assessment, Wiley Publications.

C.S. 'Chip' Howat, An Introduction to Fault Tree Analysis in Risk Assessment, University of Kansas, 2008.

B. Danher, Using Fault Tree and Event Tree to Manage Risks.

K.A. Dejmek, Fault Tree Analysis as a Tool for Safety Instrumented System (SIS) Performance Evaluation, Wilfred Baker Engineering, Inc.

D. Djamal, B. Lylia, B. Abdelkarim, Towards a better approach for mastering industrial risks from modeling accidental process to integrating safety analysis techniques supporting the identification of intelligent safety decision, International Journal of u- and e- Service, Science and Technology 8 (3) (2015).

Evaluation of Human Reliability Analysis Methods Against Good Practices, NUREG-1842, Final Report, U.S. Nuclear Regulatory Commission Office of Nuclear Regulatory Research Washington.

J. Faber, Fault Tree Analysis with Moby/FT, University of Oldenburg, 2005.

Fault Tree Handbook with Aerospace Applications, Prepared for NASA Office of Safety and Mission Assurance NASA Headquarters, Washington, DC, August 2002.

Fishbone Diagram; Needs Assessment Knowledge, Internet document; http://siteresources.worldbank.org/WBI/Resources/213798-1194538727144/9Final-Fishbone.pdf.

A. Franks, Lines of Defense/Layers of Protection Analysis in the COMAH Context, Amey VECTRA Limited for the Health and Safety Executive; http://www.hse.gov.uk/research/misc/vectra300-2017-r02.pdf.

A.A. Fredericson, The Layer of Protection Analysis (LOPA) Method, Independent Consultant – Member of Safety User Group Network, April 2002.

P. Fuchs, J. Kamenicky, T. Saska, D. Valis, J. Zajicek, Some Risk Assessment Methods and Examples of Their Application, Technical University of Liberec; http://risk.rss.tul.cz/.

D.N.D. Hartford, G.B. Baecher, Risk and Uncertainty in Dam Safety, Technology and Engineering; Google Book.

R. Hill, Cause and Effect Analysis, Mind Tools; http://www.mindtools.com/.

K.D. Hornbacher, When is quantitative project risk analysis necessary? How effective leaders deploy@risk for excel and project, in: Palisade Risk Conference, University of Pennsylvania, Las Vegas, Nevada, November 2012 (Internet document).

HSE.UK; rr 679; Human Reliability Analysis

A. Joshi, P. Binns, S. Vestal, Automatic Generation of Fault Trees from AADL Models.

G. Kennedy, Human Reliability Assessment, October 2006, 06ELD061/06ELP461.

A. Ketsakorn, W. Methom, Event tree analysis for analyzing probable fire scenario in the plastic manufacturing extra building. International Journal for Current Research and Academic Review, October 2014. ISSN:2347–3215.

C.A. Lassen, Layer of Protection Analysis (LOPA) for Determination of Safety Integrity Level (SIL); Stud. Techn., Aker Solutions; NTNU, June 2008.

P. Lucas, J. Walington, T. Atkinson, Taking the control system for granted – ensuring the integrity o sub SIL instrumented functions, in: IChemE Symposium Series No 55, ABB Engineering Services, 2009.

J. Marshall, An Introduction to Fault Tree Analysis (FTA), Product Excellence Using 6 Sigma; Module, The University of Warwick, 2012.

Minimum Cutset, Reliability Hotwire; wiebull.com (63) (May 2006).

Y. Oh, J. Yoo, S. Cha, H.S. Son, Software safety analysis of function block diagrams using fault trees, Reliability Engineering and System Safety; Elsevier (2005). www.sciencedirect.com.

R. Ouache, A.A.J. Adham, N.A.B. Azizan, Hybrid layer of protection analysis and bow tie analysis with fuzzy approach for quantitative risk assessment, International Journal of Advanced Research in Engineering and Technology (IJARET) 5 (10) (October 2014).

Y. Papadopoulos, J. McDermid, A. Mavrides, C. Scheidler, M. Maruhn, Model-based semi-automatic safety analysis of programmable systems in automotive applications, in: Proceedings of ADAS 2001, the International Conference on Advanced Driver Assistance Systems, IEEE Publications, Birmingham, UK, September 2001. CFP # 483.

Y. Papadopoulos, M. Maruhn, Model-based Synthesis of Fault Trees from Matlab – Simulink Models.

R.B. Patil, L.Y. Waghmode, P.B. Chikali, T.S. Mulla, An overview of fault tree analysis (FTA) method for reliability analysis & life cycle cost (LCC) management, IOSR Journal of Mechanical & Civil Engineering (IOSR-JMCE).

J.T. Perez, A. Madewell, Critical P&ID reviews using HAZOP/LOPA methodology – overkill or underrated? in: Prepared for Presentation at American Institute of Chemical Engineers 2013 Spring Meeting, April 2013.

C. Pietersen, The risk of LOPA and SIL in the process industry, Safety Solutions Consultants; Mary K ay O'Connor Process Safety Center, in: International Symposium; Beyond Regulatory Compliance Safety Second Nature, October 2008.

Quantified Risk Assessment Technique Part 2 Event Tree Analysis ETA, The Institute of Engineering and Technology; Health and Safety Briefing No 26b, August 2012.

M. Ragheb, Event Tree Analysis, October 2013. Internet document, http://mragheb.com/NPRE%20457%20CSE%20462%20Safety%20Analysis%20of%20Nuclear%20Reactor%20Systems/Event%20Tree%20Analysis.pdf.

J. Ramesh Babu, Layer of Protection Analysis — an Effective Tool in PHA, Internet document: http://citeseerx.ist.psu.edu/viewdoc/summary?doi=10.1.1.522.1335.

M. Rausand, A. Hoyland, System Reliability Theory Models, Statistical Methods, and Applications, second ed., Wiley, 2004.

M. Rausand, System analysis event tree analysis, in: NTNU; System Reliability Theory, Wiley, 2005.

A. Reusch Berg, Applicability of Layer of Protection Analysis to Determine Safety Integrity Levels in the Process Industry, NTNU, 2007.

I.B. Santiago, J.M. Faure, Y. Papadopoulos, Including Systematic Fault into Fault Tree Analysis, December 2008. HAL Id: hal-00348072, https://hal.archives-ouvertes.fr/hal-00348072.

B. Schmidt, Software Fault Tree Analysis, 2007.

E.O. Schweitzer III, B. Fleming, T.J. Lee, P.M. Anderson, Reliability Analysis of Transmission Protection Using Fault Tree Methods.

C.G. Soares, Safety Reliability of Industrial Products, Systems and Structures, CRC Google Book; (Internet document).

T. Stauffer, Making the Most of Alarms as a Layer of Protection, Director Alarm Management Services, exida LLC.

A.E. Summers, Introduction to layer of protection analysis, SIS-TECH Solutions; Mary Kay O'Conner Process Safety Center Symposium; Texas A&M University, Journal of Hazardous Materials (October 2002).

B.K. Vaughen, J.O. Mudd, B.E. Pierce, Using the ISA 84/HAZOP/LOPA procedure to design a safety instrumented system (SIS) for a Fumed Silica Burner, in: Prepared for Presentation at American Institute of Chemical Engineers, 2010 (Spring Meeting).

W.E. Vesely, F.F. Goldberg, N.H. Roberts, D.F. Haasl, Fault Tree Handbook, US Nuclear Regulatory Commission, 1981.

J.M. Voas, K.W. Miller, An Automated Code-based Fault Tree Mitigation Technique, Internet document; https://www.cigital.com/papers/download/safecomp95.pdf.

M. Wass, Practical guidelines for practical guidelines and procedure for SIL ranking under IEC 61508/61511, AK EHS & Risk, Aker Kvaerner Engineering Services Ltd, in: Symposium Series No 150, IChemE, 2004.

CHAPTER

VI Discussions on Standards for Risk Assessment and Safety Instrumented Systems

From Chapter I, discussions about basic knowledge of risk assessment (RA) can be gathered. After discussions on various process hazards analyses (PHAs), now it is time to move on to safety instrumented systems (SISs). To understand SIS, it is necessary to know why there are requirements for SIS and how it is done. There are risks in every sphere of life, and people globally are trying hard to avoid risks at times. Since it is not possible to completely eliminate risks, research has been ongoing to develop some systematic way to reduce risk levels. Industrial plants are not out of it! Therefore, a number of industrial standards have been established to study various technical problems in a systematic way and suggest some solutions. Each of these standards sets forth certain regulations, which need to be followed to get the best results. The present form of SIS is in fact a result of such standards. Various standards, for example, IEC 61508/61511, set forth the regulations for functional safety, which is the prime requirement of SIS. Before starting the discussions on SIS, it is necessary to understand various requirements set forth by these standards. In this chapter, an overview of various standards shall be discussed. This chapter is not a substitute for any of the standards, but will be helpful in understanding and interpretation of the corresponding standard and associating the standard with RA, PHA or SIS in a better way. There may be a number of standards on the same topic by different authorities, and apparently they are different, and an endeavor has been made in this book to marry such differences, for example, IEC 61511 and ISA 84.00.01. In fact, this chapter will somewhat act as bridge between discussions made in previous chapters and various chapters to follow. As various requirements for SIS have been derived from these standards, so it precedes any discussions on SIS so that subsequent discussions are meaningful to the reader. The chapter has been arranged sequentially as follows:

- Standard requirements for risk assessments;
- Standards for PHAs;
- ISA 88: Batch process;
- Functional safety IEC 61508 and functional safety IEC 61511
- ISA 84

1.0 STANDARD REQUIREMENTS FOR RISK ASSESSMENT

It is not that there was no risk assessment earlier. The importance of risk assessment has increased many times mainly on account of the reasons described in Table VI/1.0-1.

Table VI/1.0-1 Factors for Risk Assessment

Category	Issues Involved
External context	Financial and cost, legal, regulatory standards, international/national/local influence, competition, stakeholder requirements
Internal context	Time, organizational objective, project process, activity and associate objective, schedule control, company standards or policy, contractual obligation, stakeholder influence, product reputation, and liability
Risk management	Objectives, scope, responsibilities, methods, defining risk criteria or measures, tolerance levels, scheduled control, views or requirements of stakeholders, time

- Responsibility: It is the responsibility of the employer to ensure the health and safety of all employees in the workplace. Similarly, employees also have certain responsibilities.
  - Responsibilities should be clearly identified. People responsible for particular HSE, especially where expertise is called for.
  - Clear direction and demonstration to be understood by others.
  - To ensure that understanding of respective responsibilities by workers, supervisors, and team leaders are provided with necessary resources and time to carry out the same.
  - To ensure that well-defined lines of communication are laid down amongst people with different responsibility levels.
  - To set objectives and define accountability.
- Planning: Risks have direct or indirect impact to an organization in the long, medium, and short terms, which are actually associated with strategy, tactics, and operations, respectively. Strategic planning is long-term planning, and is done usually for a minimum of 3—5 years. The method by which an organization achieves its objectives is known as tactics, and is covered by medium-term planning. Projects, product development, etc. are associated with tactics of the organization. Day-to-day operations of the organization are covered by short-term planning. There could be strategic long term planning also by the company, say for maintaining work culture or goodwill.
- Now, what is risk management? Detailed discussions on risk management have been presented (in Clause 4 of Chapter II); in brief, risk management consists of the following main headings as also shown in Fig. VI/1.0-1.
  - Hazard identification
  - Risk assessment

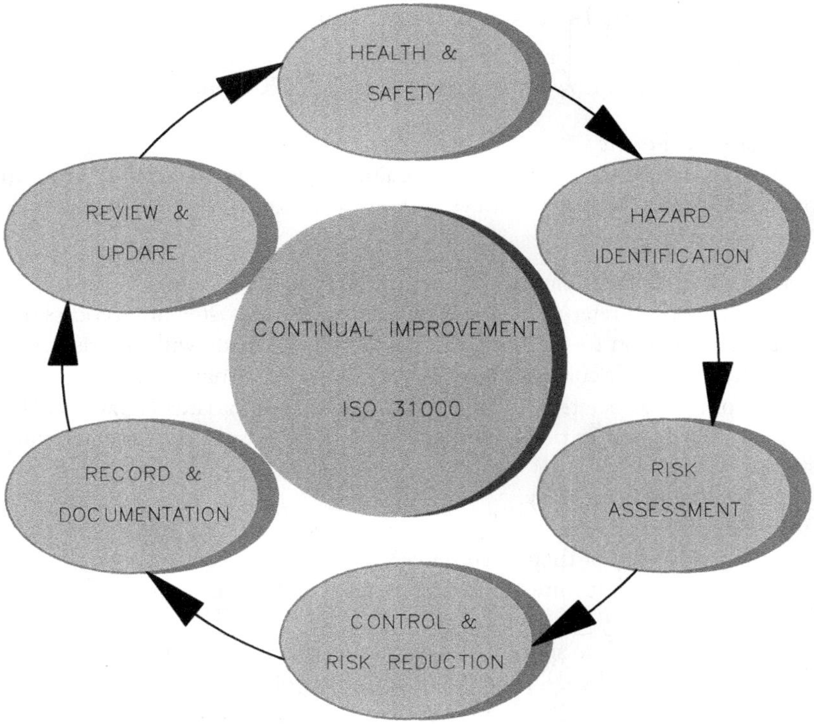

FIGURE VI/1.0-1

Brief risk management process.

- Control and risk reduction
- Recording and documentation
- Review and update

As discussed earlier (Chapter I, and Clause 4 of Chapter II; reader may read this part of discussion in conjunction with Clause 4.0 of Chapter II), one part in risk assessment is the risk evaluation or risk comparison. In risk evaluation, risk classification is very important, as it helps to recognize accumulations of similar risks. Risk classifications can be done on the basis of issues such as financial, operational efficiencies, and market planning. With the help of risk classification, it is possible to pinpoint the vulnerable elements in strategic, tactical, and operational planning. As all these have been covered in detail in earlier chapters, they will not repeated again. "Continual improvement" is an important aspect, and it will "enhance key elements of the risk management framework to either improve current processes and/or progress towards a more mature risk management framework" [2]. In the following discussions, such improvement policy of the standards will be covered. Basic risk management process has been described in Fig. VI/1.0-1.

1.1 Risk Assessment Issues and Approach to ISO 31000

There are a few requirements for risk assessments. All of these are discussed as follows. The discussions also cover how an approach to ISO 31000 could cater to them.

1.1.1 *Important Terms*

However prior to discussing risk assessment requirements, it is necessary to bring the attention of the readers to a few terms that may be encountered in during the discussions. These are:

- Risk acceptance: The decision to accept risk (ISO Guide 73).
- Risk assessment (also refer to Clause 4 of Chapter II): A systematic process of organizing information to support a risk decision to be made within a risk management process. It consists of the identification of the hazards, analysis, and evaluation of risks associated with the exposure to these hazards (ICH O9) [1]. ISO 31000:2009 says that risk assessment attempts to *answer* the fundamental questions:
  - What can happen and why?
  - What are the consequences?
  - What is probability of their occurrence?
  - Are there factors that mitigate the consequence of risk or that reduce the probability of risk?
- Risk communication: The sharing of information about a risk and/or risk management between the decision-maker and the stakeholder for performance improvement, based on well-informed consensus with respect to risk.
- Risk evaluation: The comparison of the estimated risk to given risk criteria. For this, a quantitative or qualitative means and scale can be used to understand and determine the significance of the risk and associated impact.
- Stakeholder: Any individual, group, or organization that can affect, be affected by, or perceive itself to be affected by a risk. Decision-makers might also be stakeholders [1].
- Risk owner: The person, organization, or entity accountable and with authority to manage a risk.
- Miscellaneous others terms: There are a few other important terms such risk, severity, and consequence risk management. Since these are already discussed at length in previous chapters, they are not repeated here.

1.1.2 *Risk Response*

There will be risks in all spheres of life, now the question is how to face them. The options are popularly known as the 4 Ts. These options are:

- Mitigate (Treat): To take control measures and/or corrective action to reduce the severity and/or likelihood of the risk element; in the majority of the cases this is done.
- Avoid (Terminate): Completely discard the action so that risk is eliminated.
- Transfer: Shift the impact on other individuals or other entity.

- Accept (Tolerate): Accept the risk without any corrective action. This is also seen for as low as reasonably practicable (ALARP).

1.1.3 Requirement of Standards

For fruitful results, it is necessary that risk management will be a part of organization culture to mandate leadership and commitment from the board. This requires well-coordinated strategic planning taking in to consideration the tactical and operational objectives. This will obviate assignment of risk management responsibilities throughout the organization. Also it should be related to accountability, performance measurement, and rewards. This will also help in enhancing efficiency at all levels. This will give rise to a good and appropriate risk architecture, strategy, and protocols. In order to successfully implement all of these, a structure is required. So, there have been a number of standards developed. Some of the standards are:

- ISO 31000 2009: Risk Management Principles and Guidelines
- ISO/IEC 31010:2009: Risk Management; Risk Assessment Techniques
- IRM/Alarm/AIRMIC 2002: A Risk Management Standard
- COSO 2004: Enterprise Risk Management; Integrated Framework
- OCEG "Red Book" 2.0: 2009: A Governance, Risk, and Compliance Capability Model

Out of these standards, the first two will be discussed, as they are an internationally agreed-upon standard for the implementation of risk management principles. An important note is that ISO 31000 is the basis of most of the standards applicable for risk assessment, for example, NORSOK Z013: 2010 indicates some specific guidelines, but is based on ISO 31000. When one refers to Clause 4.1 of NORSOK Z013:2010, one can see "*Although risk management in general is a subject beyond the scope of this NORSOK standard, the same structure, principles and model as the one used in ISO/IEC 31000 have been applied for the processes of performing a risk and emergency preparedness assessment covered by this NORSOK standard. The main difference is that the element "risk treatment" is not covered.*"

1.1.4 ISO 31000 Family and Scope

Currently, the ISO 31000 family consists of the following standard:

- ISO 31000:2009: Principles and Guidelines on Implementation
- ISO/IEC 31010:2009: Risk Management; Risk Assessment Techniques
- ISO Guide 73:2009: Risk Management; Vocabulary

The following are some excerpts from the ISO 31000:2009 scope so that intent can be well understood.

> This international standard can be applied to any type of risk, whatever its nature, whether having positive or negative consequences.
>
> Although this international standard provides generic guidelines, it is not intended to promote uniformity of risk management across organizations. The design and

implementation of risk management plans and frameworks will need to take into account the varying needs of a specific organization, its particular objectives, context, structure, operations, processes.

This international standard is not intended for the purpose of certification.

1.1.5 *Advantages of ISO 31000*

"When implemented and maintained in accordance with this Standard…" *[introduction: ISO 31000:2009]*. The following advantages or main themes are achievable by the organization:

- Higher likelihood of objective achievement
- Proactive management
- Awareness for identification and treatment of risks
- Improvement: Identification of opportunities and threats
- Compatible risk management practices
- Compliance of regulation and international norm
- Improvement in controls, finances, governance, and stakeholders' confidence
- Reliable decision-making and planning
- Effective resources management for risk treatment
- Improved operational effectiveness and efficiency
- Enhance health and safety environmental protection performance
- Improvement in loss prevention
- Improved organizational learning; resilience

Let a brief over view of the ISO 31000:2009 standard start with Fig. VI/1.1.5-1 (which is also Fig. 1 of the standard):

Fig. VI/1.1.5-1 has been developed based on Figure 1 of the standard. As shown in Fig. VI/1.1.5-1, there are three basic parts of the standard, viz. principles, framework, and process. Keeping parity with the preceding, further discussions shall be covered in three parts.

1.2 Risk Assessment Standard ISO 31000

Short discussions on ISO 31000 principles have been elaborated in the following sub-clauses.

1.2.1 *Principles of ISO 31000*

There are 11 points presented in the left-most box of Fig. VI/1.1.5-1 (points a through k). The heading of each of the following sub-clauses indicates the key sense *(not the exact point)* of each of the points described in the standard under principles.

- Create value: Risk management stands to achieve the objective with protection for process systems and people.
- Integral part: Risk management is an integral part of the organization process and not an additional task with which to be compliant.

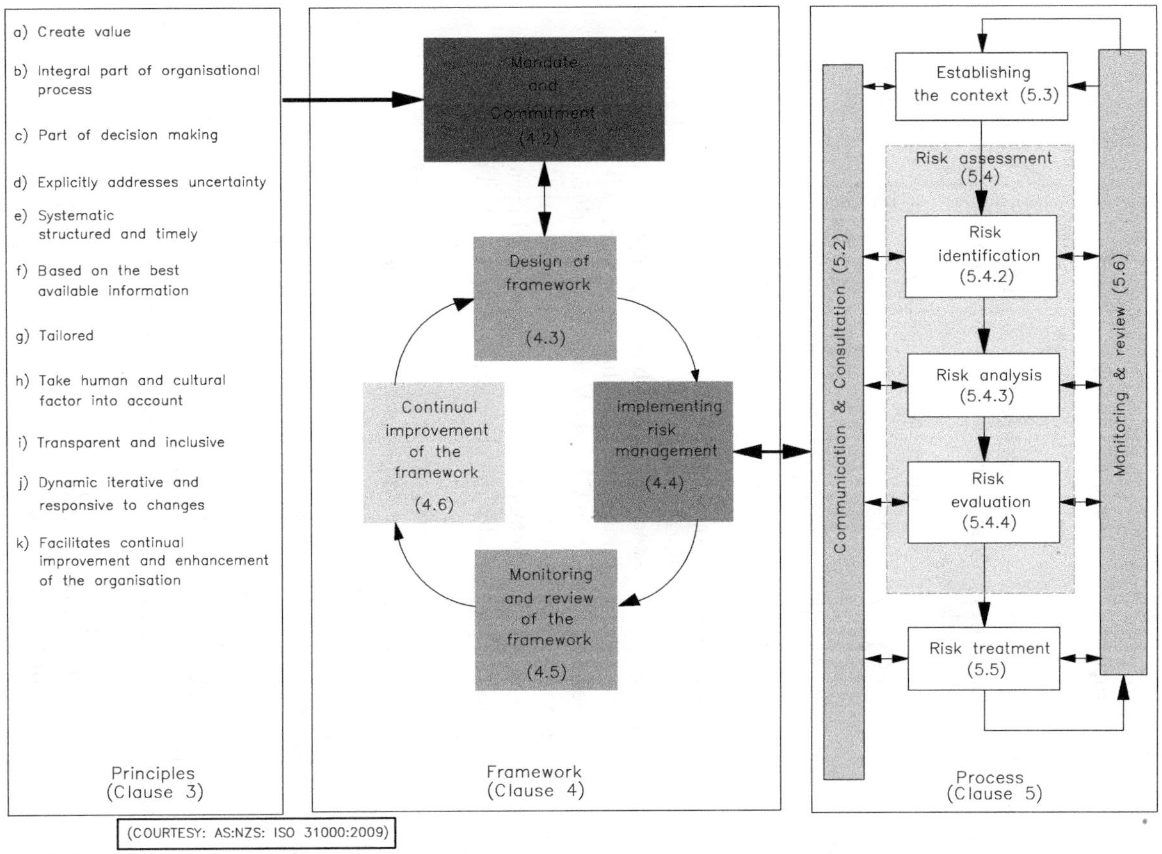

FIGURE VI/1.1.5-1

Principles, framework, and process of ISO 31000.

- Decision-making: Risk management, being an integral part of organization, is a part of decision-making to help decision-makers with informed choices of alternatives to set priorities and allocate limited resources.
- Address uncertainty: No matter what be the level or nature of uncertainty, it will be addressed by risk management explicitly.
- Systematic and structured: Systematic and structured approach of risk management produces consistent, comparable, effective, efficient, and reliable results.
- Best available information: Risks are managed based on information from various sources. First to start with easily available information, then increase the same with the increase in risk level.
- Tailored: Risk management shall be tailor-made to suit the external and internal context discussed earlier (see Table VI/1.0-1), and should corroborate with the company risk profile and risk appetite (see Chapter I), that is, risk management shall be tailored to the environment.
- Human factor: Risk management recognizes the capability, perception, and intention of human force in the organization. Human resources of an organization makes it different from others.
- Transparent and inclusive: Appropriate and timely involvement of stakeholders at different levels of the organization makes risk management successful and up to date. Risk management must set out clear job profile, appraisal, and employment contract. In this connection guidelines from HSE, UK and OGL discussed with reference to HRA in Chapter V may be referred to.
- Dynamic and iterative: Regular review of risk register, framework, internal auditing etc. to keep risk management accurate, relevant, and up to date at all times. For this incorporation of all time changes, external/internal events, and review is necessary.
- Continual improvement: Risk management must support and encourage continual improvements. The organization shall encourage development and implementation strategies, which will help to improve the maturity of management in all aspects including risk management. Continual improvement is one of the key features of principles.

1.2.2 ISO 31000 Framework

This is a cyclical process in the form of, "plan, do, check, act" (PDCA) method. As described in Clause 1.1.4, it is not the intention of the standard to detail all risk management processes, but to help the organization to integrate risk management within the overall management system following the stipulated framework.

- It starts with mandate and commitment (4.2); major issues are policy issue and risk management plan standards. This goes hand in hand with design framework (4.3).
- Design framework (4.3): Design framework starts with the understanding of the organization and its context.

- Understanding of organization: As stated earlier, the context can be external, which includes but is not limited to the stakeholders' requirements, key drivers, legal, sociopolitical, and environmental issues. There is internal context also, such as organizational objectives, project process, activity and associated objective, and organizational culture. The next item is risk management policy establishment.
  - The policy shall be simple, understandable, achievable, and auditable. The policy shall be well-coordinated and matched with the requirements of all stakeholders.
  - It has documentation requirements also, and this includes but is not limited to the following:
    - Accountability
    - Responsibility
    - Policy linking management of conflicts
    - Risk management performance measurement
    - Reporting process
    - Review process
- Accountability: Here risk owners are identified. The basic issues are:
  - Identification of risk owners at all levels
  - Allocation of authority and resource to manage to the risk owners
  - Performance measurement
  - Reporting
- Integration into organizational process: Strategic planning, policy development, organization-wide risk management plan.
- Resources: This shall include funding, people, time, process, tools, and information.
- Establishing external as well as internal communication and reporting mechanism.

- Implementation of risk management: Major points to be considered here shall include but are not limited to the following:
  - Appropriate timing
  - Parity with organizational strategy and tactics
  - Standards and regulations
  - Training
  - Implementation
  - Established monitoring system

Implementation is largely determined by the nature, scale, and complexity of the organization.

- Monitoring and review: This is an integral part of the whole system, and all the issues must be reviewed. Today's small issue may become a disaster tomorrow! When ignoring environmental impact, one emphasizes development, then disasters come. In the world there are so many examples of the same! Also, as part

of the risk management process, risks and controls should be monitored and reviewed on a regular basis to verify that:
  - Assumptions about risks remain valid.
  - Assumptions on which the risk assessment is based, including the external and internal context, remain valid.
  - Expected results are being achieved.
  - Results of risk assessment are in line with actual experience.
  - Risk assessment techniques are being properly applied.
  - Risk treatments are effective.
- Continual improvement: There has been pronounced focus on continuous improvement in risk management through all aspects of it.

1.2.3 ISO 31000 Process

Risk management is an integral part of corporate management, and should be an embedded part of the organizational culture. The process consists of:

- Establishing the context
- Risk assessment, comprising:
  - Risk identification
  - Risk analysis
  - Risk evaluation
  - Risk treatment
- Communication and consultation
- Monitoring and review
- Establishment of context: This shall include but is not limited to the following:
  - Objective
  - Environment
  - Identification of stakeholder
  - Standards, government regulation
  - Planning and policy
  - Consequence criteria
  - Management structure
- Risk assessment: As risk assessment has already been discussed at length, it does not require repetition here.
- Communication is not really crisis communication. Main issue here is to avoid all conflicts. The sharing of information about risk and risk management, amongst the stakeholder, to improve performance based on well-informed mutual decisions with respect to risk.
- Monitoring and review: Already discussed in connection with framework, and does not require repetition here.

With these in mind, the focus now will be on standard ISO 31010.

1.3 Standard: ISO 31010:2009

At the outset it is better to tell the reader that they should consult latest version of the standard available at the time of reading. This standard deals mainly with selection of various risk assessment techniques. The ISO 31010 standard is a supporting standard for ISO 31000, and provides guidance on selection and application of systematic techniques for risk assessment.

> "This standard is not intended for certification, regulatory or contractual use.
>
> ...This standard does not provide specific criteria for identifying the need for risk analysis, nor does it specify the type of risk analysis method that is required for a particular application."

1.3.1 *Advantages of Risk Assessment as per* ISO 31010:2009

- Understanding the risk in order to select risk treatment, and to assess risk impact on project objectives.
- Information for decision-makers.
- Identification of major risks contributors and weak links in systems.
- Comparison of risks in alternative systems, technologies, or approaches.
- Communicating risks and uncertainties.
- Establishment of priorities.
- Assistance in incident prevention from post-incident investigation.
- Selection of different risk treatments.
- Compliance to regulatory requirements.
- Risk evaluation for acceptance by when comparing with pre-defined criteria.
- Assessing risks for end-of-life disposal.

1.3.2 *Risk Assessment*

Discussions on risk assessments have been presented in Clause 4 of Chapter II. Various processes involved in risk assessments in accordance with ISO 31010:2009 have been presented as follows:

- Risk identification
- Risk analysis
  - Controls assessment
  - Consequence analysis
  - Likelihood analysis and probability estimation
  - Preliminary analysis
  - Uncertainties and sensitivities
- Risk evaluation
- Documentation
- Monitoring and reviewing risk assessment
- Application of risk assessment during life cycle phases

Based on the above criteria, the selection of various PHA methods (see Clause 2 of Chapter II). One thing to be noted that control system to be reassessed after the analyses are over.

1.3.3 *Selection of Assessment Techniques*

Here, some of the major issues dealt with in the standard have been listed. In this connection, *Table A2 of the standard* in conjunction with Clause 2 of Chapter II of this text may be referred to, for further details. In general terms, suitable techniques should exhibit the following characteristics:

- Justifiable and appropriate to the issue in the question.
- Result of the technique to enhance risk understanding and risk treatment.
- Technique should be traceable, repeatable, and verifiable.

The relevance and suitability of the technique selected should be well-defined. For integration of different studies, it is necessary that the results the techniques and outputs of various studies be compatible and comparable. After objectives and scope of risk assessments are complete, as per ISO 31010, one needs to take into consideration the following factors:

- The aim and objectives
- Need-based selection of decision-makers
- Risk: Type and range
- Consequences and their magnitudes
- Resources: Human, expertise, and other as required
- Risk assessment: Modification/updates
- Regulatory and contractual requirements

Other factors or constraints influencing selection of risk assessment techniques are:

- Available resources: Skills and experience, team capability, time, and available funds
- The nature and degree of uncertainty: Quality, quantity, and integrity of information
- The complexity of the application: Dependent on understand the consequential impacts and risk dependencies
- Application in life cycle phases: Life cycle phases have different needs and require different techniques, so at what stage the same is applicable is important. If it is in design and development phase, it contributes to:
  - Ensuring risks are tolerable
  - The design refinement
  - Cost-effectiveness
  - Impact upon subsequent life-cycle phases.

Type of risk assessment technique is also discussed in Clause 2 of Chapter II. Now, after gaining some knowledge about technique selection standards, it is better to focus attention on the various standards applicable for different PHAs.

2.0 SHORT DISCUSSION ON STANDARDS FOR PROCESS HAZARDS ANALYSES

In this clause, a short discussion on relevant international standards for some of the various PHAs shall be covered. Also, one needs to note that in most cases, the reference is taken from ISO 31000, so there will be commonality among them.

2.1 Dependability and Associated Standard

By definition, *dependability* is a measure of system's availability, reliability, and maintainability. In fact, dependability encompasses availability, reliability, safety, security, and maintainability. As a general follow up of this, one can understand that in order to increase dependability, a system needs either to avoid mistakes or failure, or in case of failure, it must be quickly detected and rectified so as to limit the damage caused by the failure. From this discussion, it is clear that there is direct correlation of the same with PHAs. A list of dependability techniques and their characteristics based on international standards (IEC-60300-3-1:2003) has been presented in Table VI/2.1-1, so that subsequent discussions can be developed. There are a few items of note for going through the following table. Firstly, this table is a combination of two tables in the standard. Only IEC standards have been listed and covered, although there are other standards also. Only a few PHAs have been included. Here, **BU/TD** stands for bottom up/top down, and **ql/qn** stands for qualitative and quantitative analysis, already discussed in previous chapters. Here, those standards have been included for which a short overview is presented here. Additionally, for hazard identification (HAZID), NORSOK standard Z-013 has also been covered. Dependability standards prepared by IEC TC56 subcommittee have organized IEC 60300, entitled dependability management, in three parts for management of dependability.

IEC 60300 consists of a few parts as detailed as follows with respect to current status, and their revisions in progress for newer editions

2.1.1 IEC 60300-1

Dependability management. Part 1: Dependability program management. Initially, mainly performance issues including availability, reliability, and maintainability were covered. Major topic headings include: Dependability management system, management responsibility, resource management, product realization, measurement analysis, improvement, and appendices. Currently, IEC 60300-2 is *withdrawn* and included in Part 1. IEC 60300-1:2014 establishes a framework of dependability management. It now includes products, systems, processes, and services involving

Table VI/2.1-1 Dependability Techniques and Characteristics (IEC 60300-3-1:2003)

Technique	Standard	Life Cycle	Aim	BU/TD*	Cause Effect Relationship Exploration	Dependency Modeling
HAZOP	IEC-61882	Early	ql	BU	Exploratory	No
FMEA	IEC-60812	Across	ql	BU	Inductive	No
FMECA	IEC-60812	Across	qn	BU	Inductive	No
ETA	IEC-62502	Across	ql/qn	BU	Inductive	Yes
FTA	IEC-61025	Across	ql/qn	TD	Deductive	No

hardware software and human factors also. This standard provides guidelines to management and their personnel for optimization of dependability.

2.1.2 IEC 60300-2

Dependability management. Part 2: Dependability program elements and tasks. Originally dependability management of product design development, evaluation, and life cycle model to describe product development and project phases. Now this is withdrawn, [http://tc56.iec.ch/about/standards0_1.htm], and subsumed within IEC 60300-1.

2.1.3 IEC 60300-3

Dependability management. Part 3: Application guide: There are several sections of the standard, each providing guidelines for different aspects, for example, Section 1 is for dependability guide on methodology, an overview of common dependability with advantages and disadvantages. Similarly, Sections 2 and 3 are for data collection, life cycle, etc. Section 6 deals with the software aspect of dependability. Section 9 is on risk analysis; replaced by ISO 31000. Sections 10 and 12 are for maintainability and integrated logistic support, and they are revisions for newer editions.

2.1.4 Vocbulary

An essential aspect of all IEC standardization is vocabulary. IEC 60050(191), is now replaced by IEC 60050-192:2015, which gives the general terminology used in the field of dependability. The terms are generic and applicable to all fields of dependability methodology, including electro—technical applications. The document is not an exhaustive vocabulary for all IEC standards.

Now, in the following clauses, various standards pertinent to different PHAs shall be discussed in brief. Mostly IEC standards will be covered. *Each of these standards contains* scope, n*ormative references, terms, definitions, and abbreviations. These are not individually mentioned in the discussions while narrating the major headings of a particular standard.* For these, the reader is advised to refer to the corresponding standard.

2.2 NORSOK Z013: Hazard Identification

HAZID method is a qualitative risk assessment method. It is a common and frequently used technique within the petroleum industry and in offshore processes. NORSOK Z013:2010 (mainly based on *ISO/IEC 31000*) is quite relevant in this sector of industry. Therefore, here relevant parts of the standard related to HAZID have been included in brief. The requirements to the hazard identification are as follows:

- Identification of hazards should included irrespective their control under organization.
- Tools and techniques suitable for identification of relevant hazards and the established context.

- Basis for a HAZID may be:
  - Use of check lists and accident statistics.
  - Experience from previous analyses/assessments, safety inspections, and audits.
  - Internal/external incident reports.
  - Step-by-step methodologies such as HAZOP/failure mode and effects analysis (FMEA).
- The system basis for HAZID shall be established. Also associated personnel should know and understand system.
- For all relevant hazards identification, requirements for discipline-wise participation should be established.
- Scope of HAZID and documentation: Already discussed at length in Clause 3 of Chapter II, hence are not repeated here.

2.3 IEC 61882: Hazard and Operability Study

Hazard and operability study (HAZOP) is a guided word (word or phrase which expresses and defines a specific type of deviation from an element's design intent) technique of PHA. To understand the process, salient features of the standards can be divided in to various parts as listed as follows (basic intent of the standard):

- To facilitate the examination, a system is divided into parts so that design intent of each part is adequately defined. The size of each chosen part is governed by system complexity and hazard severity, viz. a complex system or a system with higher hazard severity will have a smaller size of each part, and a simpler system will have larger part. Guide words and associated elements are related through a matrix so that combinations of the two are available in each cell of the matrix as already discussed in Chapter IV. To achieve comprehensive hazard identification, the elements and their associated characteristics need to cover all relevant aspects of the design intent, and guide words cover all deviations to cover the entire system under study.
- Design representation: "Design intent" is the main baseline, and should be accurate and complete in all respects. Prerequisites for examination that are an accurate and complete design representation, which provides a descriptive model of the system adequately including descriptions of the system under study, its parts and elements, and identifying their characteristics. This representation may be of the physical design or the logical design. The representation of the system consists of two basic parts:
  - The system requirements
  - A physical and/or logical description of the design.
- Application: Originally HAZOP technique was developed for systems involving material flow in the process industries. However, its area of application has steadily widened in recent years to include even software

application systems involving transport systems such as road and rail. It has also expanded to examining different operating sequences and procedures assessing administrative procedures in different industries including medical devices.

- Limitations: Some of the limitations of HAZOP as per the standards include the following:
  - When there are serious hazards involving interactions between a number of parts of the system, there is no *guarantee* that all hazards or operability problems will be identified in a HAZOP study. So, a complex system should not depend *entirely* upon HAZOP studies during different system life cycle phases.
  - In cases of systems with high inter-linked systems, adequate local mitigating action may not cover the real cause. Therefore, accidents are possible, and in such cases the problem can be circumvented by carrying forward the implications of deviations from one part to another. In practice, such *carry-forward* may not be possible or easy to achieve.
  - HAZOP study depends on the ability and experience of the study leader and the knowledge, experience, and interaction between team members. HAZOP only considers parts that appear in the design representation. Activities and operations which do not appear in the representation are *not* considered.

In the standard guidelines regarding the use of HAZOP in various phases of life cycle, the preparatory phase and procedure, along with various guide words and their associations with various elements, have been detailed. Flow charts are presented as examples for guidance. Use of HAZOP in various industries has been detailed in the appendix of the standard. HAZOP procedure in line with the standard is presented in Fig. VI/2.3-1 (see also Fig. 1 of standard).

2.4 IEC 60812: Failure Mode and Effects Analysis and Failure Mode, Effects, and Criticality Analysis

IEC 60812:2006, "analysis techniques for system reliability — Procedure for failure mode and effects analysis (FMEA)," describes FMEA and failure mode, effects, and criticality analysis (FMECA), and provides guidance as to how these techniques may be applied to achieve various reliability program objectives. Major issues include:

- The basic principles.
- Identification of appropriate terms, assumptions, failure modes, and criticality measures.
- Procedural steps necessary to perform an analysis.
- Examples of the typical forms used.

Failure is the loss of the ability of an item to provide its required function. FMEA is a logical process for identification of failure modes of the elements of a system with focus on causes of failures and the failure effects. FMECA is an extension of FMEA, where quantitative estimations of the likelihood and the severity of each failure mode

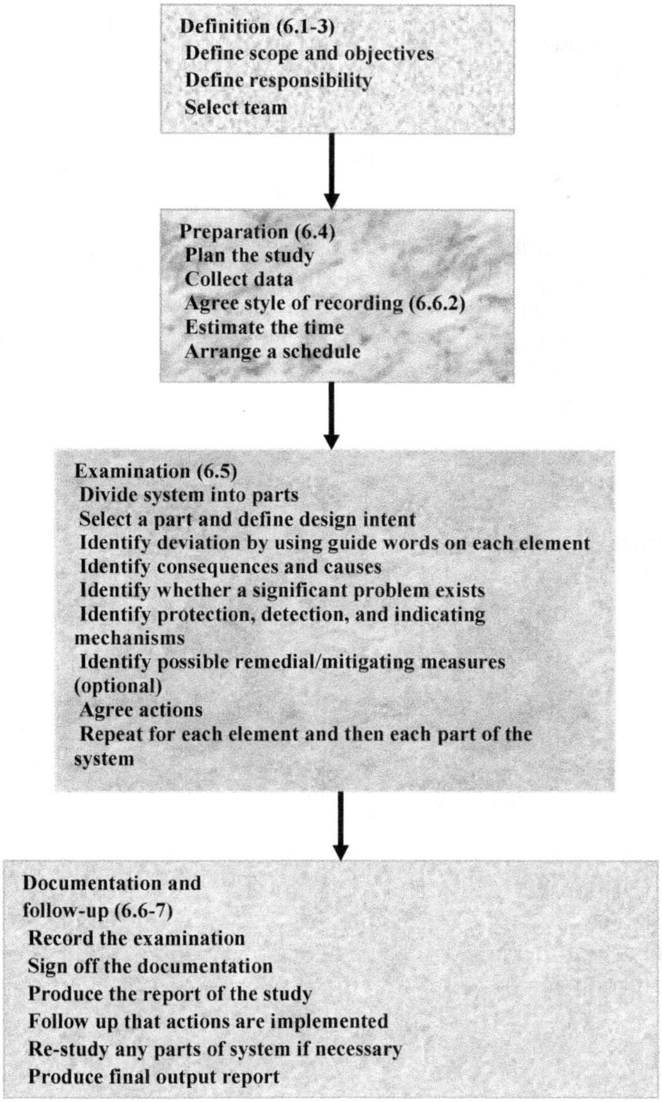

FIGURE VI/2.3-1

Hazard and operability study process.

are carried out by combining these factors. It is possible to get a measure of its criticality, so that control measures could be identified and prioritized. These analysis techniques are very versatile, and are applicable to product development, manufacturing processes, and services also. They are most cost-effective when performed at the early stage in the life cycle, and from time-to-time updated. Major section headings include:

2.4.1 *Overview*

Purpose and objectives of the analysis. This analysis can be initiated as soon as the systems is defined enough to be presented. Here timing is very important. It also defines the purpose and scope for carrying out analysis.

2.4.2 *Failure Mode and Effect Analysis*

General considerations, preliminary tasks, FMECA, and report analysis. Major issues involved shall include but not limited to: Basic background rule, FMEA with relevant worksheet, reporting and updating. It also provides detailed flow diagram for FMEA. Critical matrix and FMECA are also included.

2.4.3 *Other Considerations*

Common-cause failures, human factors, and software errors and consequences of system failure.

2.4.4 *Application*

Benefits, uses, limitations

2.5 IEC 62502: Event Tree Analysis

IEC 62502 defines the basic principles and procedures for the dependability technique known as event tree analysis (ETA). ETA is a bottom up technique.

2.5.1 *ETA Standard Headings*

Event tree analysis has IEC 62502 standard, which mainly consists of the following headings as in fault tree analysis (FTA) IEC 62502 standard:

- General description.
- Benefits and limitations of ETA.
- Relationship with other analysis techniques: Combination of ETA and FTA, layer of protection analysis (LOPA), etc.
- Development of event trees: General, various steps in ETA.
- Evaluation: Preliminary remarks, qualitative analysis; managing dependencies; quantitative analysis.
- Documentation.

2.5.2 *IEC 62502 Coverage*

The following topics are covered in the standard for ETA:

- Definition of terms, symbols, and methods of graphical representation.
- Specification of the procedural steps for construction of the event tree.
- Details about assumptions, limitations, and benefits of the analysis.
- Identification of relationships with other dependability and risk-related techniques.
- Guidelines for the qualitative and quantitative aspects of the evaluation.
- Practical examples.

This standard is applicable across industries where the dependability and risk-related measures for the consequences of an initiating event must be assessed.

2.6 IEC 61025: Fault Tree Analysis

A fault tree is an organized representation of conditions and other factors causing or contributing to developing a defined outcome known as "top event." A fault tree is a form for clearly understanding and analyzing as necessary to facilitate identification of factors affecting investigation of a top event:

- To identify the factors affecting reliability and performance, when used in reliability analysis such as design deficiency.
- To identify the events affecting more than one functional component which could destroy the benefit of specific redundancies, or affects more than two parts which otherwise could have been independent.

2.6.1 IEC 61025 Headings

For FTA, standard IEC 61025 mainly consists of the following headings as in FTA IEC 61025 standard:

- General: Fault tree description and structure, objectives, applications, and combinations with other reliability analysis techniques, for example, FTA-FMEA, FTA-ETA, etc.
- Development and evaluation: General considerations, required system information, and fault tree graphical description and structure.
- Fault tree development and evaluation: General, system familiarization, fault tree developments, fault tree construction, and failure rates in FTA.
- Identification and labeling in a fault tree.
- Reporting.
- Appendices for symbols and detailed procedure.

2.6.2 Root Cause Analysis

From the standard it is clear that the major scope of this analysis includes root cause analysis and reliability analysis.

2.6.3 FTA as a Tool

FTA can be used as a tool such as:

- Determination of logical combination of events leading to a top event.
- Investigation of a system under development and anticipate, prevent, and mitigate potential causes of the top event.
- System analysis for reliability.
- Probabilistic risk assessment.

With this, short discussions on various standards pertinent to PHA have been concluded. Batch process is different from other continuous systems. Therefore, it is necessary to understand the system first before applying safety assessment for the same in the following clause.

3.0 BATCH CONTROLS AND ASSOCIATED STANDARDS

When discussions are to be developed for safety instrumentation functions, associated controls (logic solver) cannot be kept aside. Now, for discussions on control systems, special attention is necessary for batch process and controls. Plant processes and productions can be categorized as discrete process, continuous process, and batch process. Out of these three types, batch process is more complex in the sense that the problems are logistically more challenging. In order to meet this challenge, complex issues are divided into simpler steps, then solving them, and all the solutions are integrated suitably to arrive at a solution for the complex issue. In this way, batch controls follow a special pattern and established standards. Definitions of various processes, with examples, are discussed as follows:

- *Discrete process*: In a discrete process, finite quantities of discrete parts or products are manufactured or produced. Manufacturing of a smart phone, car, or airplane are examples of a discrete manufacturing process.
- *Continuous process*: A continuous process is a streamlined process that involves ongoing production of end products or services. So, a continuous process produces a continuous flow of free-flowing product. Thermal power generation, raw meal to clinker production in cement plant are examples of continuous processes. However, for continuous process in chemical, oil, and gas industries, there is technical report, TR106.00.01-2013, which is being developed for risk reduction also (see Clause 4.0 of Chapter XII).
- *Batch process*: Batch process involves production of a *finite quantity* of free-flowing product. In a chemical process plants, this involves production of discontinuous "charge-wise" production of chemical product. Pharmaceuticals and food products are typical examples of the batch process products. In such cases, usually a batch number, date, etc. are mentioned in the products. The major standard for batch process is ISA 88. According to ISA 88, batch process is "a *process that leads to the production of finite quantities of material by subjecting quantities of input materials to an ordered set of processing activities over a finite period of time using one or more pieces of equipment.*"

In view of this, it is necessary to include pertinent, short discussions on batch controls. The discussions presented here are mainly based on ISA 88.00.01-03. At the end of the discussion, brief coverage of HAZOP in batch controls will be covered.

3.1 Batch Process Details

A batch process can be conceived as an event-driven process with a finite beginning, middle, and end. Like any other processes, there will be modulating controls, sequential controls, and interlocks to meet the control needs. However, in batch controls, there are a few special and salient aspects normally encountered, such as time-based process conditions, transition phenomena, storage, and reactions. Prior to discussing further detail, it is better to address and explain a few important issues and their relevance.

3.1.1 *Commonly Used Terms in Batch Process and Controls*

Commonly used terms in this process have been discussed as follows. It is worth noting that majority of the terms discussed here are ***based on*** ANSI/ISA 88.

- Batch: A finite quantity of material that is being produced, or that has been produced by a single execution of a batch process. An entity that represents the production of a material at any point in the process may be intermediate or final.
- Batch control: Control activities and functions that provide a means to process finite quantities of input materials by subjecting them to an ordered set of processing activities over a finite period of time using one or more pieces of equipment.
- Control module: The lowest level grouping of equipment in the physical model that can carry out basic control, for example, valve, pump, or controller.
- Control recipe: It is the execution of the recipe (defined later) which will define single batch of a specific product (see Fig. VI/3.1.3-1).
- Equipment module: The equipment-specific functionality that provides the actual control capability for equipment including procedural, basic, and coordination control, and is *not* part of the recipe. An equipment module may involve one or more control modules around a piece of process equipment. In dosing issue, a set of valves or pump flow measuring instrument is an equipment module.
- Exception handling: Exception handling is the procedures and/or functions that deal with plant or process contingencies and other events which occur outside the normal or desired behavior of batch control.
- Formula: A category of recipe information that includes process inputs, process parameters, and process outputs (see Fig. VI/3.1.4-1).
- General recipe: A type of recipe that expresses equipment and site-independent processing requirements (see Fig. VI/3.1.3-1).
- Header: Information about the purpose, source, and version of the recipe such as recipe and product identification, creator, and issue date.
- Lot: A unique amount of material having a set of common traits (character and quality).
- Master recipe: A recipe for producing a batch product utilizing same set of equipment (see Fig. VI/3.1.3-1).
- Operation: A procedural element defining an independent processing activity consisting of the algorithm necessary for the initiation, organization, and control of phases.
- Phase: The lowest level of procedural element in the procedural control model, for example, charge, mix, heat, and reaction.
- Process cell: A logical grouping of equipment that includes the equipment required for production of one or more batches. It defines the span of logical control of one set of process equipment within an area.
- Recipe: The necessary set of information that uniquely defines the production requirements for a specific product. This means that it dictates the batch controls how to make the product, and there will be a recipe for each final product. Recipe composition is shown in Fig. VI/3.1.4-1. There are four types of

recipes defined in this standard: general, site, master, and control (detailed out in Clause 3.1.3 below).
- Site recipe: A site-specific recipe (see Fig. VI/3.1.3-1).
- Unit: Unit is collection of associated control modules and/or equipment modules and other process equipment, in which one or more major processing activities can be conducted. Units are presumed to operate on only one batch at a time. Units operate relatively independently of one another. Unit is applicable for both the physical equipment and the equipment entity. Examples of unit activities include react, crystallize, and make a solution.

3.1.2 *Batch Process Models*

Model concept in a batch process helps in understanding its hierarchical structure, and is specified in Part I of ISA 88. This structure will assist in dividing the systems into smaller parts. In line with the standard, Fig. VI/3.1.2-1 shows the various model types. Fig. VI/3.1.2-1A (or Fig. 1 of ISA 88) shows how the entire process has been divided into smaller parts for analysis of the system. As shown, the process is subdivided into process stage, process action, etc. This is a process model of batch process. The entire process can be divided in terms of models in three ways, viz. physical model, procedural model, and control activity model, as shown in Fig. VI/3.1.2-1B. Therefore, it is better to start the discussion on the three types one by one.

- Procedure model (Fig. VI/3.1.2-1B(i)): As the name implies, it gives a procedure hierarchy. It consists of an ordered sequence of unit procedures to specify the strategy for batch making. A unit procedure specifies the ordered sequence of the functions, or operations, within a unit for a batch of production. Operation consists of ordered sequence phases, to initiate significant physical or chemical changes in a batch. Phase, which is the smallest element of procedural control, can initiate one or more actions. Master or control recipe is *responsible* for the data and procedure for batch production. Master recipe, which may be derived from the general or site recipe, is stored in the control system for developing the control recipe. One of the elements in a recipe is formula, which consists of process input, process parameter, and process output as shown in Fig. VI/3.1.4-1. Another important issue in recipe is header, which contains information like product and grade identifiers, originator, and date of issue. Detailed composition of recipe shall be discussed later.
- Physical model (Fig. VI/3.1.2-1B(ii)): As per ISA 88, "physical model that can be used to describe the physical assets of an enterprise in terms of enterprises, sites, areas, process cells, units, equipment modules, and control modules."
  - Enterprise: An enterprise, which is a collection of one or more sites, may have process cells, units, equipment modules, and control modules. Enterprise determines what products will be manufactured, and at which sites.
  - Site: A site is a physical, geographical, or logical grouping determined by the enterprise, with boundaries determined mainly by organizational or business criteria.

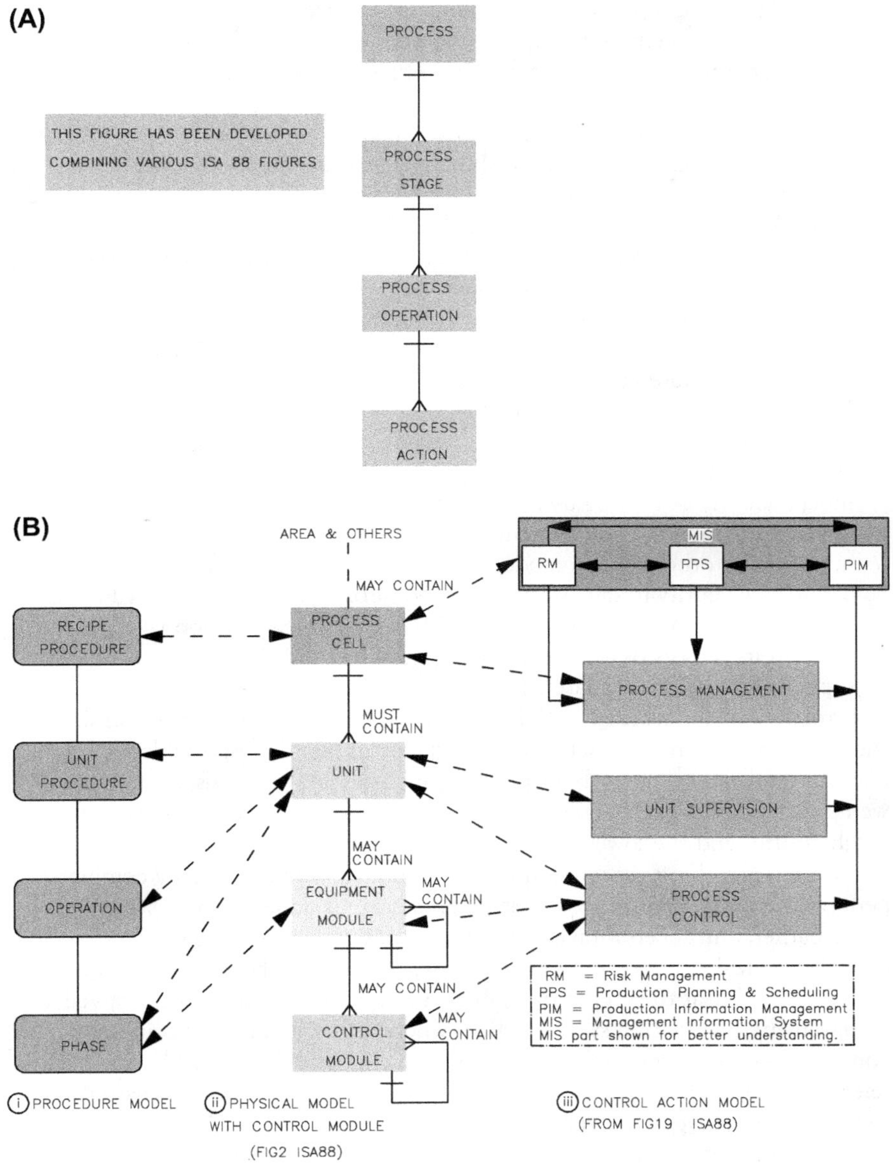

FIGURE VI/3.1.2-1

Model types of batch controls. (A) Process model (ISA 88 Fig. 1); (B) various batch control models.

- Area: An area is a physical, geographical, or logical grouping at a site, and contains process cells, units, equipment modules, and control modules. In this hierarchical model, *commands* flow from higher to lower levels, whereas *information* flows from lower to higher levels.
- Process cell: A process cell consists of a number of units, for example, storage tanks, mixing tanks, which as a whole need to execute a single recipe from start to finish batch.
- Unit: A unit usually consists of a number of process equipment or equipment modules, for example, pumps, valves. A unit is normally centered around a major piece of equipment which is:
  - Frequently operated/contains entire batch, or
  - May operate on or contain only a portion of the complete batch, or
  - Cannot operate but contains more than one batch at a time.
- Equipment module: An equipment module is a collection of control modules or other equipment modules. It contains necessary processing equipment to carry out definite numbers of minor phases. An equipment module may exist independently or may be used by multiple units.
- Control module: A control module usually consists of collection of sensors, actuators, or other control modules, and associated processing equipment to act as single entity from control system point of view. Though procedures do not run on control modules yet, it establishes direct connection with process through its sensors or actuators. The control of a unit is carried out by unit supervision as discussed later.

- Control activity model (Fig. VI/3.1.2-1B(iii)): The control activity model shows the hierarchy of control functions. The top part is mainly responsible for batch planning and management. Basic planning includes raw materials, equipment, as well as manpower. The middle part actually regulates production steps, hence batch control and the lowest part in the hierarchy is the process control and monitoring part. Unit supervision and process management actually combined perform the batch control. Basic recipe processing is done at this tier only. As stated earlier, unit supervision controls the unit. Details regarding the functional and hierarchical structure of control activity model have been elaborated in Table VI/3.1.2-1 at the end of this sub-clause. The top tier can be conceived as management information system (MIS), where as the middle tier could be conceived as supervisory information system (SIS). For details of MIS and SIS, refer to Chapter VII of [4]. After model discussions, it is better to direct attention to recipes, recipe types, and recipe composition. In the figure, it is seen that there are arrowheads among the various blocks in the control activity model. These arrowheads show signal and information exchange in the various modules in the model.

The dotted arrowheads show the functional interactions among the models shown in Fig. VI/3.1.2-1 (depicted for better understanding of the models).

Table VI/3.1.2-1 Control Activity Module Hierarchy and Functions [3]

Tier	Level Name	Module (Fig.)	Function	Activity
Top	Planning stage	Recipe management	Production management	Recipe management, general recipe management
Top	Planning stage	Production planning and scheduling	Process/product management	Production and inventory planning production scheduling
Top	Planning stage	Production information management	Process/product management	Batch history management, information collection and management
Middle	Batch control	Process management	Batch management	Recipe generation/ selection Batch execution supervision unit coordination report generation
Middle	Batch control	Unit supervision	Unit supervision	Unit allocation and coordination
Lower	Monitoring & control	Process control	Process control	General process control and monitoring functions, for example, modulating control, on–off control, sequential control, & safety interlock

3.1.3 *Recipes*

There are four standard recipes. These are general recipe, site recipe, master recipe, and control recipe. The definitions of all these have already discussed in Clause 3.1.1. In this connection, also refer to Fig. VI/3.1.3-1 (see also ISA 88 Fig. 8).

The recipes can also be divided according to the following:

- Single procedure single formula
- Single procedure multiple formula

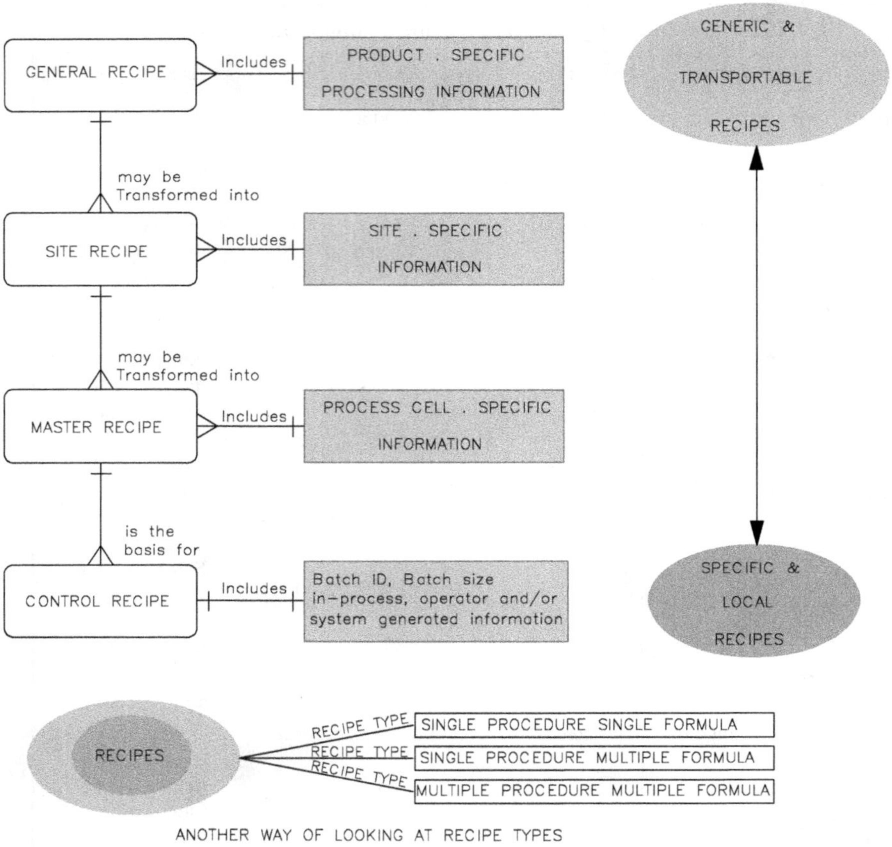

FIGURE VI/3.1.3-1

Types of recipes and inter-relations.

- Multiple procedure multiple formula

  Now, a short description of each type of recipe and their utilization:

- General recipe: General recipe, which is helpful for production planning and costing, is an "enterprise level recipe," and the basis for "lower-level recipes". It does not have equipment-specific knowledge. Raw materials, their relative quantities, and required processing are included in a general recipe, but does not contain specific information about site or the equipment available. In a general recipe, the quantities and equipment requirements are specified as normalized values and in terms of the attributes (e.g., process requirement, material of construction) respectively.
- Site recipe: As site recipe is tailored for each target location, it is derived from a general recipe to marry the conditions of a particular manufacturing location

with detailing for site-level, long-term production planning and scheduling. However, it may also be created directly without general recipe. Site-specific things such as the language or local raw material differences, regulation, or unit of measure are accommodated as site-specific variances. An example will make it clear; supposing a product is manufactured at site A (say in South Korea with *Kimchi*), and an intermediate product from A will be sent to site B (say in India with Basmati rice) for final processing, then each site will extract a portion each from general recipe. A site recipe may include only that part of a general recipe which is actually implemented on the site, for example, header, formula, procedure, equipment requirements, and other information, for master and control recipes.

- Master recipe: The master recipe, which is targeted to a process cell or a subset of the process cell equipment, can be derived from a general or site recipe. With the necessary process and product knowledge, it can be developed as stand-alone recipe also. Based on ISA 88 standard, some characteristics of master recipes include the following:
  - Multiple master recipes are possible when each is derived from a site recipe, to cover part of the recipe implementable in a process cell at the site.
  - Sufficient adaptation of the properties of the process cell equipment to ensure the correct processing of the batch, by marrying the master recipe information with the functionality of the specific set of process cell equipment.
  - The formula data may be specified as *normalized*, *calculated*, or *fixed values*.
  - The master recipe may contain product-specific information, for example, process input information or equipment requirements.
  - The master recipe is essential, as without it, control recipes cannot be created, hence no batch processing.
  - The master recipe exists either as a set of written instructions or as an electronic entity, irrespective of manual or automatic operation of batch manufacturing equipment.
- Control recipe: The control recipe, which contains product-specific process information with necessary detailing (e.g., actual material equipment information—quantity, sizing etc.), is a *specific* version copy of a master recipe, modified to meet the requirements for scheduling and operational information to be specific to a single batch. Modifications of a control recipe can be made over a period of time based on scheduling, equipment, and operator information, for example, adding, adjusting parameter. The procedural relationship amongst the various recipes has been shown in Fig. VI/3.1.3-2.

As shown in Fig. VI/3.1.3-1, master and control recipes work on procedure while general or site recipe works on the process itself. In a similar fashion, general or site recipes work on various process stages, operation, and actions. Master and control recipes work mainly with unit operation and phases (functions). In the figure, a one-to-one relationship has been shown between the stages. There can be one or multiple relationships or interactions also. When considered for a particular product manufacturing, the distribution of recipes can look like what has been shown in Fig. VI/3.1.3-3.

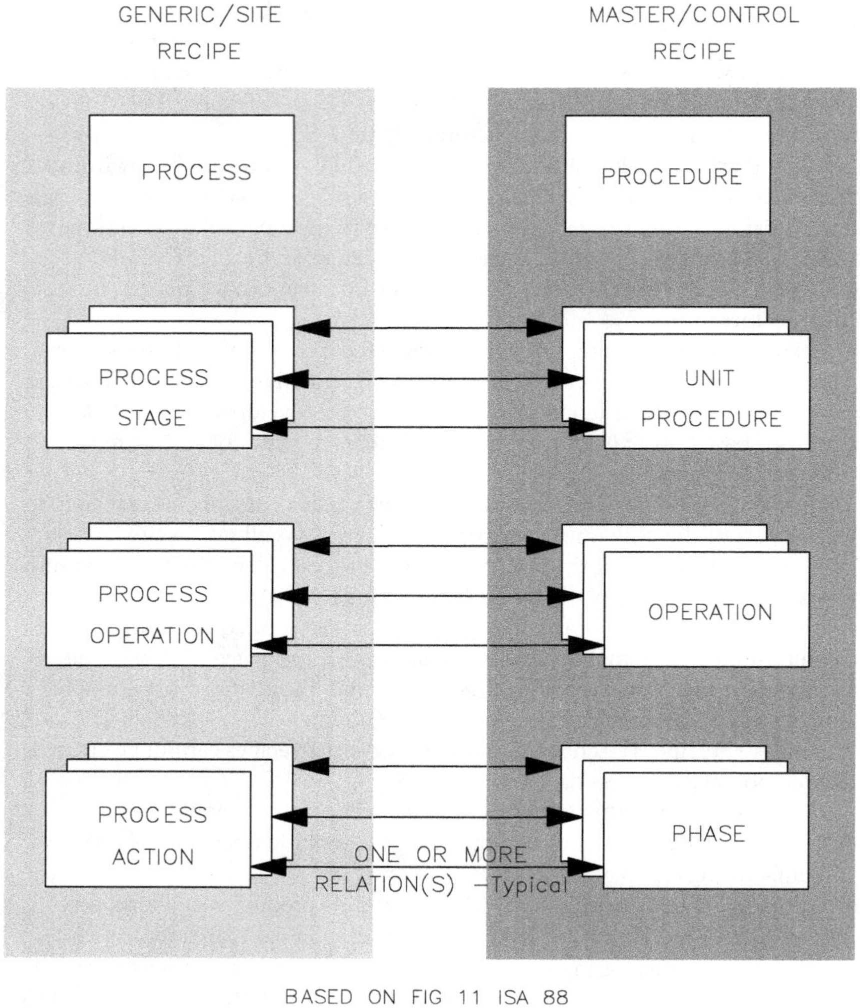

FIGURE VI/3.1.3-2

Procedural relationship between recipes.

As shown in Fig. VI/3.1.3-3, the same product may be manufactured at two places with some variations in product types. Here, production at two places has been shown; it could be made at several places also, e.g., same generic medicines may be manufactured at different countries (may be little change in composition to meet

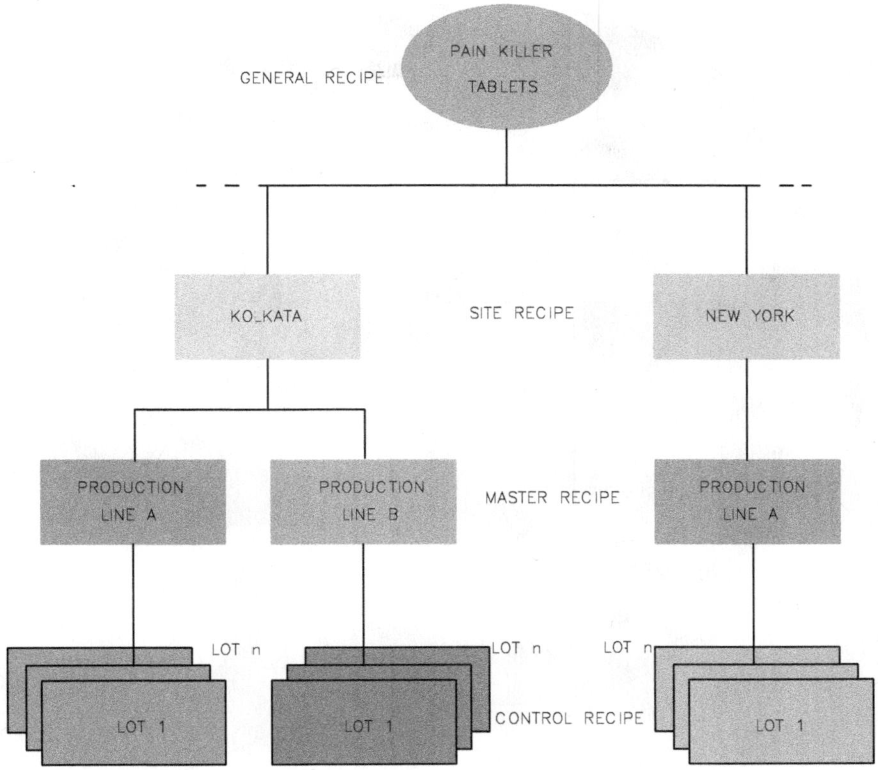

FIGURE VI/3.1.3-3

Application of recipes in production process.

country regulation) and may be available in different names. Thus, dotted lines have been shown at the top. Variations in types of products have been shown by different shades.

3.1.4 Recipe Composition

Recipe composition has been shown in Fig. VI/3.1.4-1. Basically, it consists of the following:

- Header: Header stands for the administrative information in the recipe:
  - Administrative information
  - Recipe identification
  - Product identification
  - Version identity
  - Originator

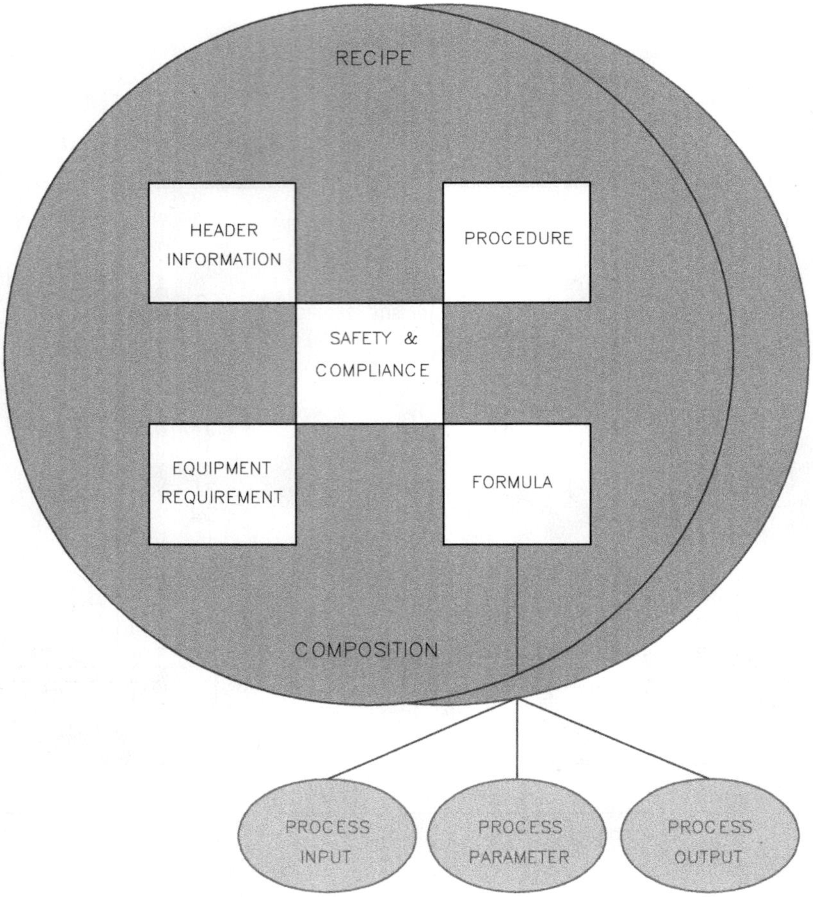

FIGURE VI/3.1.4-1

Recipe composition.

  - Issue date
  - Approval status
  - Process summary
  - Other information
- Procedure: The recipe procedure provides the control definition for a process. General and site recipe procedures as discussed earlier, are non-equipment specific. The master and control recipe procedures are developed around procedural elements, with direct relationship to equipment. The combinations of these procedural elements are used to define a procedure. Determination of these procedural elements to become a part of the procedure is application specific and is influenced by a number of factors.

- Safety and compliance: This stands for the safety requirements of the batch process and procedural compliance. Safety is the process and equipment safety suitably addressed and discussed in connection with control systems to be discussed later. Process compliance mainly stands for compliance with respect to information, equipment, etc. discussed in procedure. Here, it is important that procedure gives the orderly set of unit procedures, which in turn provides an orderly set of operations. In compliance, such sequence maintenance is also important.
- Equipment requirement: Equipment requirements identify the equipment to be used to implement a specific part of the procedure. At the master recipe level, the equipment requirements are specified, as allowable equipment in process cells. Based on guidance from the general or site recipe, it is possible for the master recipe to identify "the equipment of the train rather than the full range of equipment in the process cell," [ISA 88]. At the control recipe level, the equipment requirements are the same as, or a subset of, the allowable equipment in the master recipe.
- Formula: Formula mainly consists of process input, process parameter, and process output. A process input provides the identification and quantity of a raw materials as well as energy and other resources for the product. A process parameter specifies various parameters such as temperature, pressure, time, set points, comparison values, or in conditional logic pertinent to the product, "but does not fall into the classification of input or output," [ISA88]. A process output is the identification and quantity of a material and/or energy expected to result from execution of the recipe. The other pertinent data may include but not limited to: environmental impact and intended outputs in terms of quantity or yield.

3.1.5 Batch Process Structural Variations

Structurally, the batch process may be conceived as:

- Single series stream
- Parallel stream
- Multiple pathways

All these types of streams have been shown in Fig. VI/3.1.5-1A—C, respectively. As shown in Fig. VI/3.1.2-1A, in single stream a number of units are placed in series. Therefore, the batch process moves from one unit to another serially, following programmed instructions. Parallel stream systems can be conceived as a number of isolated serial single stream system discussed previously and shown in Fig. VI/3.1.5-1B. In a multiple—pathway system shown in Fig. VI/3.1.5-1C, there is no fixed movement of batch along any fixed path; it is based on the availability of a unit of the type required. There could be usage of common resources, that is, a piece of equipment or service that is used by more than one. Common discharge header for a utility such as a steam header, for example.

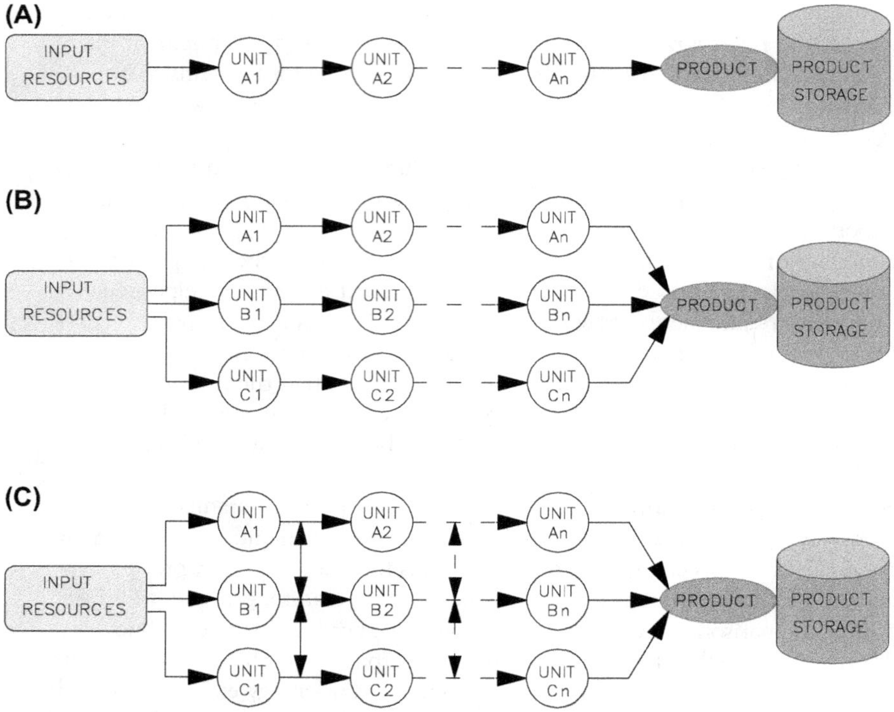

FIGURE VI/3.1.5-1

Batch process structural variations. (A) Single stream batch process; (B) parallel streams batch process; (C) multiple pathways batch process.

3.2 Batch Process Automation

Like other systems, there are broadly three distinct types instrumentation and controls. These are:

- Controls
- Interlock and protection
- Monitoring alarm

Controls are again three types, viz. modulating/regulatory, discrete, and sequential.

3.2.1 *Control Systems*

As stated previously, three type of controls will be addressed:

- Modulating/regulatory controls: As the name suggests, in modulating/regulating controls, based on controller output, final elements or actuators modulate their openings to keep process variables as close as possible to the set points at all conditions, including process and load disturbance. These types of controls are mainly used for regulating process variables, viz. maintaining steady flow during charging, or maintenance of agitator speed during mixing. PID and other advanced control algorithms for improved performance are used in modulating/regulatory controls.

- Discrete controls: Discrete controls, often called on/off controls, are used for controlling process equipment, such as valves, that have only a limited number of stable states. Often these are operated with necessary interlocks. In a batch control environment, discrete control functions can be used as directed by steps in phase logic. For example: say during *n*th step, a valve may be opened to discharge a fixed amount of charge, and after discharging the valve is closed from a time signal!
- Sequential controls: Sequential control functions perform real-time control of equipment to carry on a batch process to move a process through a succession of distinct states. An example of sequential control is opening a valve and running a pump for material transfer, material at any step, then after completing the step (may time controlled or quantity controlled) moves on to other step. Sequential controls in a batch process could be for major processes such as mixing, heating, and reaction. A phase may consist of a number of sequential control steps to manipulate equipment within a unit boundary. At time, it may be necessary that multiple units work in a synchronized fashion, for example, the transfer of material from one unit to another, with each unit having its own phase. Here, suitable communication between units is essential for synchronized operation.

3.2.2 *Interlock and Protection*

Like any other process, in a batch process there are also interlock and protection. Though functionally these are more or less same as discrete control, interlock and protection mentioned here is mainly used for safety purposes. For this reason, they are discussed separately. Interlock functions could be a part of discrete controls and/or sequential controls (say to open a valve in a step of sequential logic or for path selection depending on some other condition), or may come into operation when warranted, for example, an interlock may be used for selection of a pump when a running pump trips (may be for safety purpose of the batch). Protection functions are implemented for plant and personnel-related safety, and may be independent of the product or the state of the batch under manufacture. Therefore, these functions are always active at all times after they are set. When plant configuration or personnel safety considerations are changed, then these can be changed and set again. These functions generally override other interlocks that may be active only during certain process phases or conditions.

3.2.3 *Monitoring*

As the name suggests, monitoring the normal operation of the plant and apprising operating personnel regarding the condition of the plant equipment and process is active at all times. The function that checks the plant and process conditions on a regular basis is called monitor function. Alarm is a part of monitoring function. In batch process, the normal logic is monitored and when there is deviation it is exception logic. Exception logic, as the name implies, specifies control functions that are required to take care of failure conditions. Exception logic can be simple or elaborate. *Annunciating for the operator regarding abnormal logic and waiting until the device is back to normal (fixed) is an example of simple exception logic.*

3.2.4 *Batch Process Automation Discussions*

In the previous discussions, the functional part of process instrumentation and controls has been covered. Now the question is, how all these are implemented! In modern days, complete hardware implementation of entire thing is obsolete, even though in some cases, especially for protection issues, hardware protections are used as backup. Mostly these are implemented in any of the followings:

- Programmable logic control
- Distributed control system
- Personal computers

Frankly speaking, in the present era of modern controls and integration, it is rather difficult to really distinguish between them. So depending on control functions, intelligent control type may be decided. For further details, refer to Chapter VII [5].

3.3 Batch Process Discussions

With this basic idea, now it is time to discuss a few salient features pertinent to batch process.

3.3.1 *Comparison of Batch Process and Continuous Process*

The distinguishing factors of the two systems have tabulated in Table VI/3.3.1-1.

Table VI/3.3.1-1 Distinguishing Points Between Batch and Continuous Processes

Issues	Batch Process	Continuous Process
Materials	Both free flowing, as well as non-free flowing materials	Usually free flowing materials
Feed	Raw materials fed prior to reaction, and in stages	Constant feeding of material in entire process
Sizes	Usually large investment and relatively larger installation	Could be with lower investment and smaller installation also
Final product	After finish of stage(s)	At all times during reactions
Controls	Relatively simpler	May be complex at times
Product development	Could be used for new product, on experimental way	Normally done after looking at economical feasibility
Final product category	Variations in product category possible	More or less permanent type
Trouble shooting	Batch repair is relatively simpler and may not affect entire process	On account of interconnection it is rather difficult and fault repair may have to be carried out after stoppage of the system

3.3.2 *Reliability*

The reliability of a batch control system is more critical than its counterpart of continuous control system. This is because, in case of failure, the fallback system needs to know the *exact state of a batch* to continue the production or to bring to a safe condition. Standard reliability is defined as follows:

$$R(t) = e^{-\lambda t}$$

where $R(t)$, reliability; λ, failure rate; t, time.

This definition, however, does not take into account that a piece of equipment can be repaired and put back to service. The definition of availability takes care of that:

$$A = MTBF/(MTBF + MTTR).$$

where A, availability; MTBF, mean time between failures; MTTR, mean time to repair. The degree reliability needed for a batch control system is a function of criticality of the process. Safety interlock, device control, and sequence controls are very critical from a fault tolerance point of view. Similarly, inter-processor communication and continuous controls are also moderately critical from fault tolerance point of view.

3.3.3 *Batch Process Standards*

For batch processing, ANSI ISA 88.00.01 (,02,03) and IEC 61512-01,02,03 are common. There can be equivalence between the standards in the sense that ISA 88.00.01 is more or less equivalent in coverage to IEC 61512-01. ISA 88 has four parts:

- ISA 88.00.01 Batch control Part 1: Models and terminology.
- ISA 88.00.02 Batch control Part 2: Data structures and guidelines for languages.
- ISA 88.00.03 Batch control Part 3: General and site recipe models and representation.
- ISA-88.00.04 Batch control Part 4: Batch production records.

It is needless to elaborate that to produce a product under batch process, e.g., ice-cream, one needs to know how to make the product (recipe), what are the tools needed (equipment), and how to run equipment (process). These are well guided through the standards. The terms discussed above and models are based on the ISA 88-01. Similarly, ISA 88 Part 2 discussed the formal representation of entities specified in Part 1, e.g., recipes, equipment, etc. The standard also specifies the information exchange requirements between recipes, process, and equipment. Later, the standard deals with a graphical representation of procedure. Part 3 deals with recipe types. The last part of the standard is about the record. Since discussions of this main clause are based on this standard, these are not repeated again.

3.4 Hazard Analysis for Batch Process

Batch process relies highly on procedures to be followed, and in batch process, state changes are much more in normal operation than a continuous plant. In batch process, deviations can develop both from process variable deviation as well as from procedure deviation. In its simpler term, a (chemical) batch process can be conceived of three major operations, viz. charge, reaction, and discharge operations. Therefore, a HAZOP can be carried out thrice in terms of charging operation and discharging, as shown in Fig. VI/3.4-1.

3.4.1 *HAZOP Approach*

The standard HAZOP approach of scrutinizing the system line by line, may not be suitable ideal for batch processes. Main distinguishing characteristics of batch process, as stated earlier, are: operating procedures, operator actions, and discrete event characteristics of a batch process. There can be additional guide words to create deviation in case of charge operation, for example, "*don't charge, more charge, less charge*, or *charge other than*," etc. (associated with *one* charging material). Another important issue in batch process is the sequence of operations, and associating suitable guide words to be applied to each of the steps such as mixing, heating, reacting, and settling. As discussed in connection with batch process, time and variations in rate of change pertinent to each step play an important role in creating deviations in HAZOP. Since the majority of HAZOPS in batch process are currently intelligent, it is extremely important to take care of deviations from software issues. Following the guidelines discussed previously, a typical HAZOP approach for batch process units has been shown in Fig. VI/3.4-1. Also, in this connection, discussions in Clauses 1.5.5 and 1.6.2 of Chapter IV may be referenced.

3.4.2 *Ruled Based Automated HAZOP*

A rule-based system for automated batch HAZOP is another approach for batch process. In this system, state-based simulation units and CHECKOP rule-based system can be used. In order to limit the size of this book, this is not detailed out here, but an interested reader may refer to Ref. [7].

3.4.3 *Modeling Approach*

Another option is modeling approach, where a qualitative model is developed to include batch process. One of the objectives of such models is to concentrate on the understanding of how hazardous scenarios develop in batch processing plants. For this, different types of information can be integrated within a model to allow reliable prediction of the behavior of a system. Software tools are used for modeling of batch plant systems viz. HAZOP expert—LDGHAZOP. Based on the

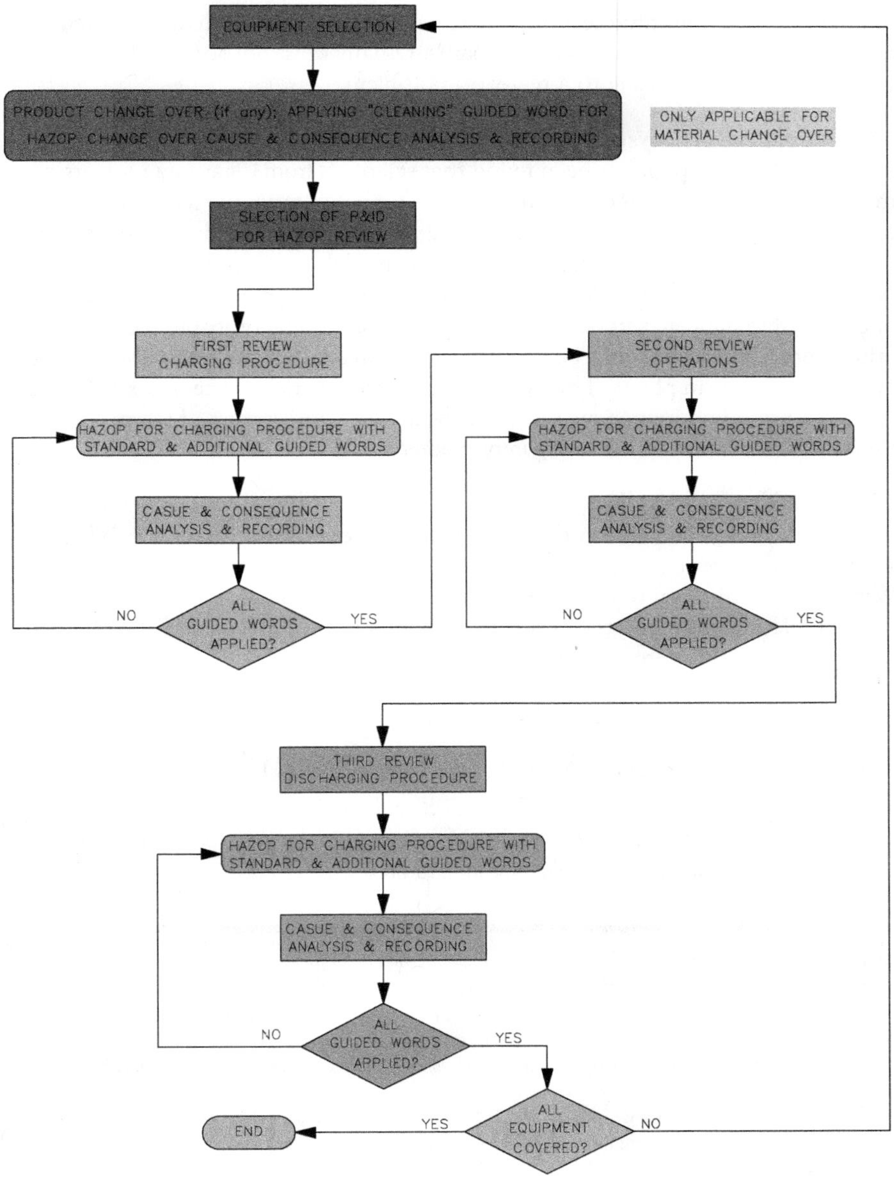

FIGURE VI/3.4-1

HAZOP for batch process.

model, model-based reasoning techniques are used to acquire knowledge required to capture the operation of batch processing plants. Usually these are done on a test-case basis before applying the same to a real-world problem.

3.4.4 *Fault Tree Analysis Approach*

From a study of literature, it has been found that scientists from Okayama University, Japan developed a dynamic model to utilize FTA for batch process. According to the paper [6], batch processes are divided into several safety objects linked to operation level. Dynamic simulation and fault propagation models are generated to evaluate safety objects. With the help of this model, FTA method using house event-time, an event expected to happen, can be developed for calculating the probability of failures. The time-dependent failures can be considered as unavailability of safety objects that can lead to accidents in plants. The rank of safety object performance index (PI) can be estimated using importance measures. PI shows the prioritization of safety objects that should be investigated for safety improvement in the plants.

3.4.5 *Petri Net Approach*

Petri net (see Fig. IV/1.6.2-3) on the other hand, is well-known for its capability in modeling discrete event systems, so the same can be applied.

With this, the discussions on batch process and hazard analysis is concluded, and next on to safety instrumentation systems for which discussions on IEC 61508 and IEC 61511 cannot be overestimated.

4.0 FUNCTIONAL SAFETY STANDARDS: IEC 61508

At the beginning of the discussion on IEC 61508 and 61511, the author wishes to put forward the following acknowledgement to IEC for granting permission to use the figures of their various standards mentioned in the box for this book.

The gap between calculated risk in a process or plant and tolerable risk is normally met by:

- Changes in process design
- Other risk mitigation methods
- Safety instrumentation systems

For instrumentation engineering, a major concern is with safety instrumentation systems. Safety instrumentation systems offer functional safety to the associated plant systems. Using "functional safety" to describe the action of a protection system is a relatively recent innovation, after the introduction of safety standards. With the publication of safety standards for manufacturing and process industries, the users become more knowledgeable about safety issues and are focusing SISs to satisfy their needs with more cost-effective ways through integration with control systems, scalable architectures, and less frequent proof testing. A protective system needs to address overall health of safety loops in an integrated safety solution from sensor to actuator including logic solvers. The best way to reduce risk in a manufacturing plant is to design inherently safe processes. However, risks prevail wherever there are hazardous or toxic materials stored, processed, or handled. Having accepted that no system is completely immune to failures, one needs to take necessary measures to ensure that even in case of failure, it should provide a safe condition. *During working tenure I found many do not like to put much importance to this issue, in design of may systems and/plants (e.g., fossil fuel power plant designed even in 2007), because the concerned plant does not handle with explosives or toxic materials. This is* **not** *a correct approach. Important issue here is that all enterprises especially industrial ones* **must** *adapt safety life cycle in their system. This will provide ultimate benefit to the system in the long run*. In this connection, it is important to note that there are mainly kinds of failures encountered in electrical/electronic/programmable electronic (E/E/PE) systems. These are random failure and systematic failure (discussed at length in Chapter VII).

Random failure usually refers to permanent failure due to non-functioning of system components, and these are more related to *hardware* failure. Probabilistic performance-based approach could be one way to address this. For E/E/PE systems, safety integrity level (*SIL*) is considered for such purposes.

Systematic failure normally occurs on account of *design* failure, including incorrect specifications, using a component not fit for the operation, and or due to error in software. *Safety life cycle* is adapted for systematic faults. So safety standards meant for E/E/PEs take care of both. SISs (Ref. Chapter VII) are developed to prevent or mitigate hazardous events to protect people or the environment, or prevent damage to process equipment. In this connection another important issue is SIL (Chapter VIII), which is a discrete level for specifying the safety integrity requirements of safety functions, but is *not a measure* of risk. SIL provides means for risk reduction to a tolerable level. The fundamental question, in case of functionally safe instrumentation, is how *frequently* failures of function will lead to accidents. The answers can be:

- For functions with a low demand rate, the accident rate is a combination of the frequency of demands, and the probability the function fails on demand (PFD).
- For functions that have a high demand rate or operate continuously, the accident rate is the failure rate.

Safety standards give guidance on best practice and offer recommendations, without absolving users' responsibility for safety. These standards deal with technical issues as well as the planning, documentation, and assessment of all activities required to manage safety throughout the entire life of a system. The safety life cycle an important issue in safety instrumentation, and can be categorized into three broad areas. The first is the analysis phase. In this phase, identification of hazards and hazardous events, the likelihood these hazardous events and potential consequences are handled. Also in this phase, the availability of a layer of protection, as well as the need for any SISs and the allocated SIL are covered. The second phase is realization, focusing on design and fabrication of the SIS. The final phase is operation, which covers startup, operation, maintenance, modification, and eventual decommissioning of the SIS. These phases encompass the entire life-cycle process of the safety system from concept through decommissioning. In the following clauses, concentration will be on various standards on functional safety and safety life cycle. The entire discussions have been arranged as follows:

- IEC 61508 (Clause 4 of Chapter VI)
- IEC 61511 (Clause 5 of Chapter VI)
- ISA 84 and interrelation with IEC (Clause 6 of Chapter VI)

In this clause, the focus will be on IEC 61508 international standard for "functional safety of electrical/electronic/programmable electronic safety-related systems" (developed in the year 1998). This standard covers ***"those aspects to be considered when electrical/electronic/programmable electronic systems (E/E/PESs) are used to carry out safety functions."*** It is important to note that IEC 61508 safety standard is applicable to a wide range of industries, and is focused mainly for the *supplier* community. The standard has been developed in such a way that it can be used as standalone standard or industry/company specific standard (e.g., IEC 62061) can be developed from it. The overall framework of the standard is shown in Fig. 1 of IEC 61508 Part 1. Various parts of the standard have been shown in Table VI/4.0-1.

There are a few differences of this standard with IEC 61511, which is also meant for similar application but specifically for *process* industries. The difference between the two has been illustrated in Table VI/4.0-2.

This standard is mainly concerned with the E/E/PE safety-related systems where failure of that systems would have *serious impact* on the safety of *persons and/or the environment*, or in those cases where consequence could create *serious economic implications*, and in those cases standard could used for the *protection of properties including equipment or product*. Another important issue here is that this standard IEC 61508 does not cover safety issues, viz. electric shock, hazardous falls, long-term exposure to a toxic substance, or E/E/PE systems where, ***"a single E/E/PE system is capable of providing the necessary risk reduction, and the required safety integrity of the E/E/PE system is less than that specified for safety integrity level 1 (the lowest safety integrity level in this standard)."*** Currently, functional safety has been an integral part of the project execution for safe and successful applications. Now, it is time to concentrate on the philosophy behind IEC 61508.

Table VI/4.0-1 Various Parts of IEC 61508

Part	Details of Coverage	TR	OR	CR	SI
1	General requirements: Development of overall safety requirements; concepts, scope definitions (7.1–7.5) (clause numbers—Typ.) allocation of safety related requirements to E/E/PEs (7.6); specification of safety requirements (7.10) installation commissioning, safety validation, operation and maintenance, modification and retrofit, decommissioning or disposal of E/E/PEs (7.13 and 7.14). Operation, maintenance modification (7.15–7.17)	Yes	Yes[a]	Yes	
2	Realization (requirements) phase for E/E/PEs (HW& others!), safety related system	Yes		Yes	
3	Realization (requirements) phase for E/E/PEs safety related software	Yes		Yes	
4	Definitions and abbreviations		Yes		Yes
5	Risk based approaches to the development of safety integrity requirements	Yes			Yes
6	Guidelines of application (Parts 2 and 3)	Yes			Yes
7	Overview of techniques and measures	Yes			Yes

CR, *compliance required;* HW, *hardware;* OR, *other requirements;* SI, *supporting information;* SW, *software;* TR, *technical requirements.*
[a]*For Documentation (5), Management of functional safety (6), and functional safety assessment (8) in Part 1 (OR).*

Table VI/4.0-2 Differences Between IEC 61508 and IEC 61511

Issues	IEC 61508	IEC 61511
Nature	Generic safety standard	Specific safety standard
Industry	Wide range of industries	Process industries only
Divisions	Technical and other requirements	Mainly analysis, realization, and operations
Coverage	Safety related systems including external risk reduction systems	Mainly on safety instrumentation systems
Focus	Major focus on supplier community (i.e., manufacturing)	Primary focus on system designer, integrator, user, and end user

4.0.1 *IEC 61508 Philosophy*

The following issues are the main pillars on which this safety standard is standing:

- System life cycle
- System subdivisions
- Functional safety
- Risk reduction
- SIL
- ALARP

4.0.2 *System Life Cycle*

The standard provides a good life cycle model with required details. This will be discussed separately, later. However, there is no specific guideline to fit the same to a typical system. For this, one needs to develop requirements, design specification, coding (as applicable), implementation, integration, maintenance, modification, and decommissioning in accordance with the model prescribed.

Fig. VI/4.0.2-1 shows the safety life cycle phases as per IEC 61508.

At the outset, it is better to keep in mind that although shown here, Box 11 pertinent to safety related systems and other risk reduction facilities, it is out of scope of the standard so it is shown by the *dotted line* in Fig. VI/4.0.2-1.

The main background ideas for safety life cycle are to develop and document a safety plan, execute of the plan, and document execution details, continuation of follow-up of safety plan from concept through to decommissioning with appropriate documentation throughout the life of the system. Any change along the way also needs to follow the pattern of planning, execution, validation, and documentation.

4.0.3 *System Subdivision*

Four system subdivisions as per the standard are:

- Equipment under control (EUC)
- EUC control system (EUCCS)
- Safety-related system (SRS)
- None of the above

4.0.4 *Functional Safety*

The following are the basic points under this heading:

- Safety treatment to ensure safety for functions.
- Safety functions enact to ensure a safe state for the EUC, against hazardous events.
- Safety function enact to try to inhibit the hazardous events.

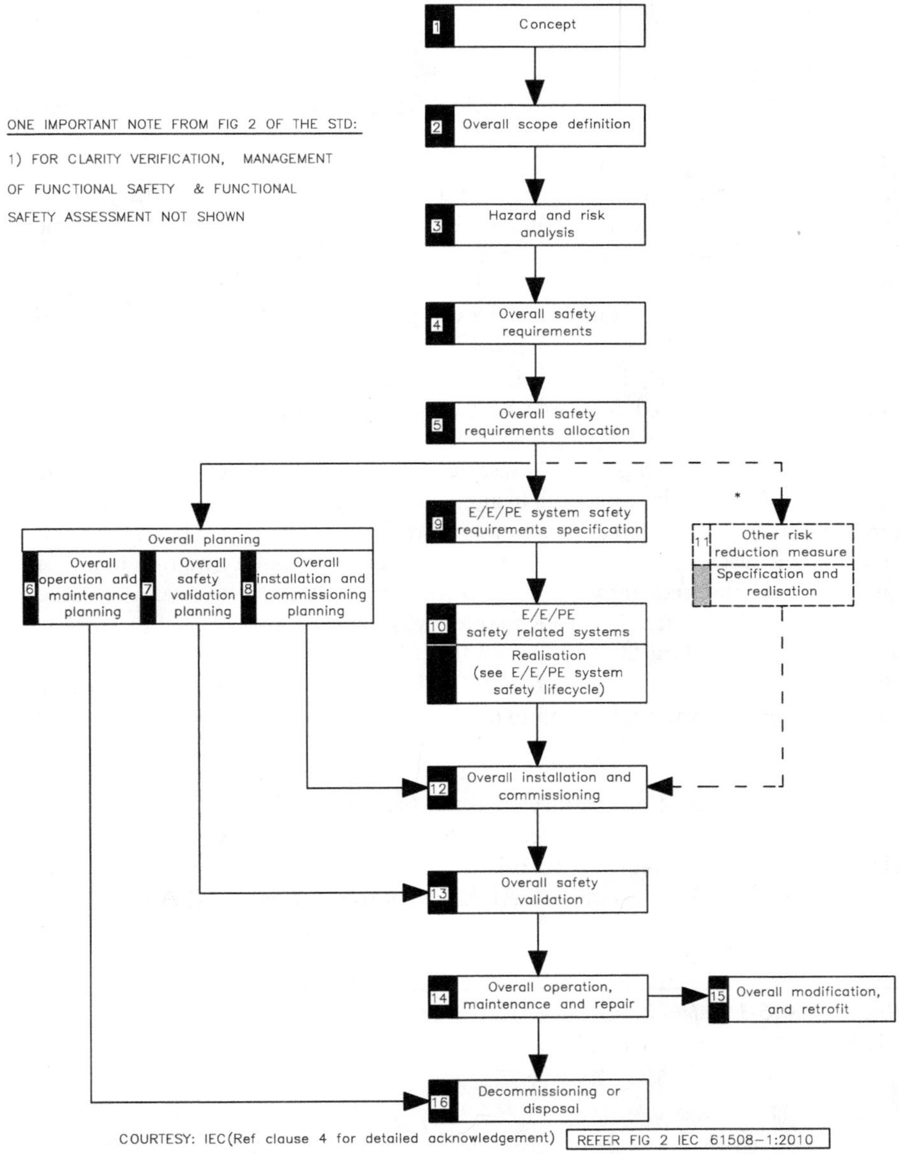

FIGURE VI/4.0.2-1

Safety life cycle phases IEC 61508.

4.0.5 *Risk Reduction*

There is no system with zero risk. Safety functions are concerned with risk reduction. Developers need to assess EUC risk, tolerable risk, and residual risk. Residual risk shall be brought down to ALARP.

4.0.6 *Safety Integrity Level*

Each SRS is assigned an SIL representing the probability of fulfilling safety functions, that is, it is the reliability of its safety functions. The SIL is done based on required risk reduction from EUC risk at least to the tolerable risk. There is quantitative difference between continuous-operation (high-demand) functions and low-demand functions. Development of SRS with a designated SIL requires a certain development process. There are four levels of increasing reliability (SIL 1—SIL 4); frankly SIL 0 specifies nothing! SIL and associated risk reductions are defined in Table VI/4.0.6-1. Before looking at the table, it is important to note the meaning of low-demand and high-demand. One of the ways it can be explained is in terms of proof test frequency, which refers to how often the safety system is completely tested and insured to be fully operational. In case of low-demand, the frequency of demands for operation made on a safety-related system is no greater than *twice* the proof test frequency, whereas in case of high-demand or continuous mode the frequency of demands for operation made on a safety-related system is *greater than twice* the proof frequency. However, this way of expressing modes has been revised in IEC 61508—4:2010 (Clause 3.5.16). Current definitions are explained in Clause 1.1.3 of Chapter VIII. In Part 4 of the standard, while defining low-demand and high-demand modes, it takes the help of an arbitrary one year to distinguish between high- and low-demand.

4.0.7 *The as Low as Reasonably Practicable Principle*

The ALARP region lies between acceptable and intolerable, in which the system developer is required to reduce the risk to be "as low as reasonably practicable (ALARP)." IEC 61508 directs towards ALARP. For the concept of ALARP, refer to Chapter I.

Table VI/4.0.6-1 *Safety Integrity Level Assignment Table (Also Refer to Table VIII/1.0-1 for Availability)*

SIL Number	Low-Demand Mode[a] (PFD_{avg})	Continuous Mode Probability of Failure/ Hour (PFT)	Risk Reduction Factor[b] (For PFD_{avg})
4	$\geq 10^{-5}$ to $<10^{-4}$	$\geq 10^{-9}$ to $<10^{-8}$	10^5 to 10^4
3	$\geq 10^{-4}$ to $<10^{-3}$	$\geq 10^{-8}$ to $<10^{-7}$	10^4 to 10^3
2	$\geq 10^{-3}$ to $<10^{-2}$	$\geq 10^{-7}$ to $<10^{-6}$	10^3 to 10^2
1	$\geq 10^{-2}$ to $<10^{-1}$	$\geq 10^{-6}$ to $<10^{-5}$	100 to 10

[a] PFD_{avg}, *probability of failure on demand (average).*
[b] *Similar for PFT also.*

4.0.8 *Contents of IEC 61508 Parts*

In this sub-clause, major content shall be elaborated:

- Part 1: From Table VI/4.0-1, it is needless to say this is the main part of the standard. It deals with general compliance, management of functional safety, and functional safety assessment. What this means to a company is shown in Fig. VI/4.0.8-1, and to have a feel about the standard.
- Part 2: This part is realization part. This part mainly deals with the hardware safety part of E/E/PEs. It covers a detailed safety life cycle for hardware and aspects of assessing functional safety for the hardware. Part 2, Annex A, includes "control of failures during operation," and "techniques and measures" respectively. Annex C covers the calculation of diagnostic coverage factor (what fraction of failures are identified by the hardware) and safe failure fraction. All these annexure are required for compliance.
- Part 3: This is a realization part for the software aspects of safety-related systems. Detailed software safety life cycle has been covered. Annexure A covers "techniques and measures." Design and coding details are provided in tabular form in Annexure B. All the annexure are required for compliance.
- Part 4: It mainly covers various definitions and abbreviations.
- Part 5: This part provides informative Annexes A through E which contain discussion and example methods for risk, safety integrity, tolerable risk, and SIL selection. It presents several techniques of SIL selection by both quantitative and qualitative methods. This part also provides qualitative risk graph and severity matrices, etc.
- Part 6: Part 6 provides guidelines on the application of Parts 2 and 3 via informative Annexes A to annexure E. Annex A gives a brief overview of Parts 2 and 3, and example flowcharts of detailed procedures for implementation. Annex B provides example techniques for calculating probabilities of failure for the safety-related system.
- Part 7: Part 7 contains important information for those doing product development work on equipment to be certified per IEC 61508. Annex A addresses control of random hardware failures. Annex B covers the avoidance of systematic failures through the different phases of the safety life cycle. Annex C provides a detailed overview of techniques for achieve high software integrity.

Functional safety management as per IEC 61508 group of standards for a company means!

Clear safety policy including measurements and systems
Clear organizational structure with clear safety responsibilities
Appropriate techniques and measures in place for conformance to the standard
Identification of non conformance issues and corrective actions are under taken.

FIGURE VI/4.0.8-1

Functional safety management for company.

Short details and elaboration on different parts have been presented in the following sub-clauses.

4.1 IEC 61508-1

Part 1 is for general requirements and safety life cycle. Major points are discussed in the following sub-clauses:

4.1.1 *Conformance (Clause 4)*

This standard details about conformance requirements such as "to conform to this standard it shall be demonstrated that all the relevant requirements have been satisfied to the required criteria specified (for example safety integrity level) and therefore, for each clause or sub clause, all the objectives have been met."

4.1.2 *Documentation (Clause 5)*

This clause specifies the necessary information to be documented for safety-related systems for all phases of E/E/PE systems. This clause also details out the necessary information to be documented in order that the management of functional safety, verification, and the functional safety assessment activities can be effectively performed. All these documents shall be *accurate* and *concise*, *easy to understand*, *suitable* for the purpose, *accessible* and *maintainable*.

4.1.3 *Management of Functional Safety (Clause 6)*

This clause specifies requirements for the management and technical activities during the overall, E/E/PEs, and software safety lifecycle phases. It is also to specify the responsibilities of the persons, departments, and organizations responsible for each overall, E/E/PE system and software safety lifecycle phase or for activities within each phase. Basic requirements/issues shall include but are not limited to those listed below. It is advisable to refer associated clause (6) of 61508-1: Latest revision (2010) to note the changes with respect to previous revisions.

- Overall strategy, evaluation methods, and communication within organization.
- Identification of persons, departments, and organizations responsible for E/E/PEs and software safety life cycle.
- The overall E/E/PEs or software safety lifecycle phases to be used.
- Documentation.
- Measures and techniques used.
- Functional safety assessment activities.
- Procedure for follow-up and resolution for recommendations after analysis.
- Competent personnel (training, etc.).
- To ensure hazards and near-miss situations are analyzed, and suitable recommendations are made to prevent repetitive occurrence.
- The procedures for analyzing operations and maintenance performance.
- Periodic functional audit.

- Procedure for modification and authorization, that is, management of change (MOC).
- The procedures for maintaining accurate information on potential hazards and safety-related systems.

4.1.4 Safety Life Cycle Requirements (Clause 7)

Safety life cycle, according to this standard, can be considered as a cyclic process or closed loop comprising in cyclic fashion of identify—analyze—design—verify (***comparable with "plan, do, check, act" ISO 31000***). Referring to Fig. VI/4.0.2-1, it is seen that:

As a measure of risk reduction, the overall safety life cycle includes:

- E/E/PE safety-related systems
- Other technology safety-related systems
- External risk reduction facilities

The parts related to E/E/PE safety-related systems have been shown in Fig. VI/4.1.4-1 and same for software has been shown in Fig. VI/4.1.4-2. Fig. VI/4.1.4-3 is to establish relation between the two systems. Table 1 *in* the standard elaborates the objective scope, reference sub-clause, input/output, etc. for each of the phases.

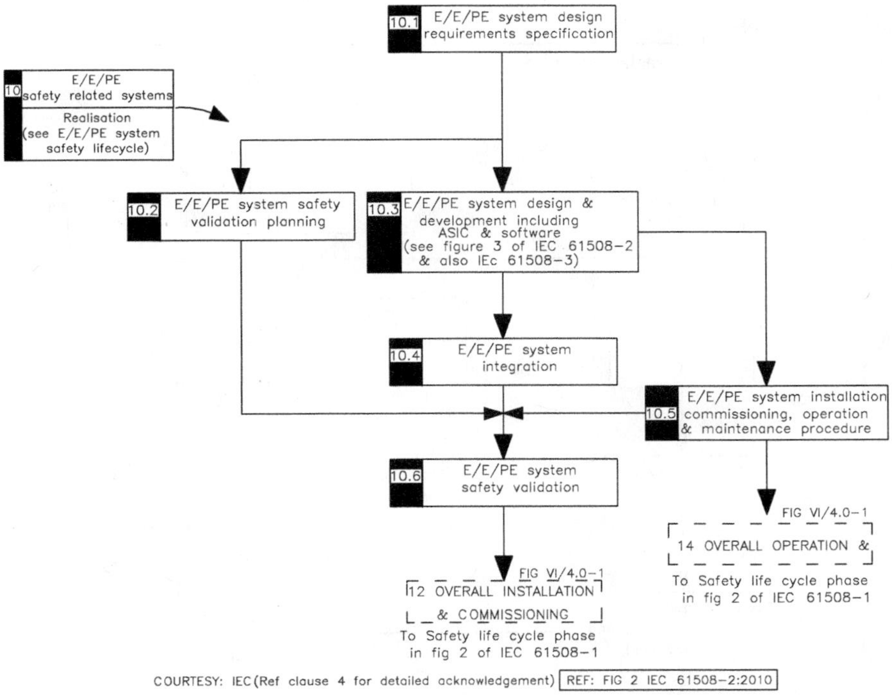

FIGURE VI/4.1.4-1

E/E/PE system safety life cycle realization phase.

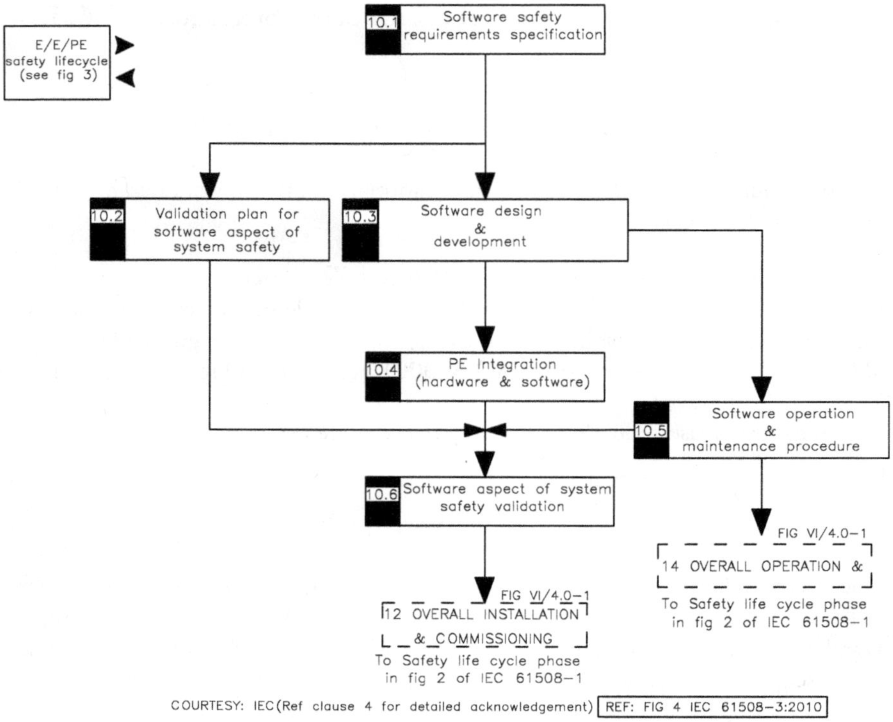

FIGURE VI/4.1.4-2

Software safety lifecycle in realization phase.

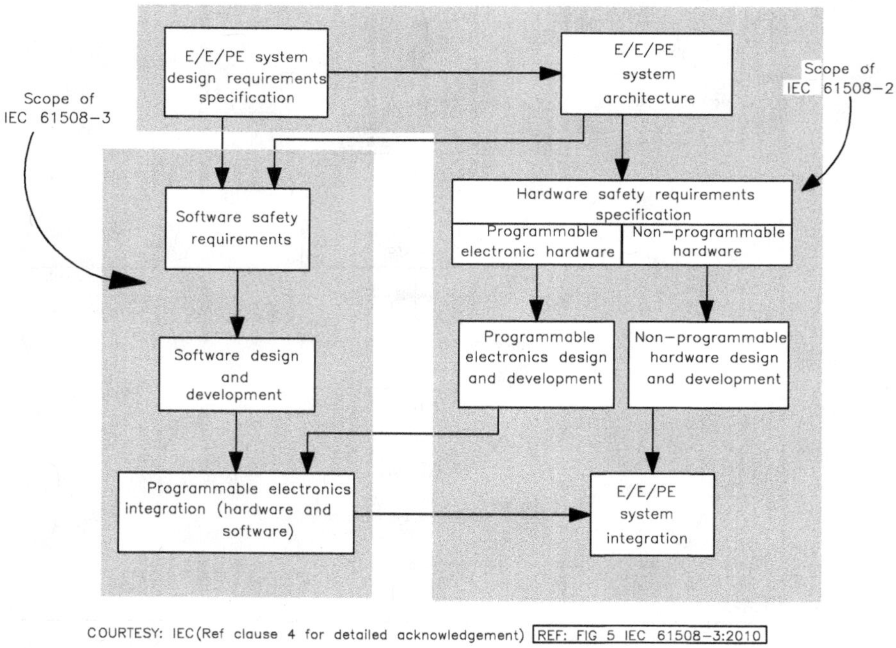

FIGURE VI/4.1.4-3

Relationship between and scope IEC 61508-2 and 3.

The overall life cycle discussions in the standard mainly covered in this main Clause 7, having 17 major sub-clauses. Now coming back to main life cycle phases in Fig. VI/4.0.2-1, it is seen that the first part of the safety life cycle is basically the analysis part comprising concept, scope for the system/EUC, hazard/risk analysis, creation of overall safety requirements, and identification of specific safety functions to prevent the identified hazards safety requirements allocation. The middle part is realization activities (Clause 7.10) as detailed in Figs. VI/4.1.4-1 and VI/4.1.4-2, are dealt with in Parts 2 and 3 discussions. The next part of the life cycle is related to installation and commissioning (Clause 7.13). Then comes the validation (Clause 7.14), operation and maintenance (Clause 7.15), modification, retrofit (Clause 7.16), and finally, decommissioning (Clause 7.17).

4.1.5 Functional Safety Assessment (Clause 8)

The objective of this assessment is to investigate and conclude regarding the level of safety achieved by the E/E/PE safety-related systems. The process requires that one or more competent persons be appointed to carry out a functional safety assessment. These individuals must be suitably independent of those responsible for the functional safety being assessed, *depending* on the SIL and consequences involved. The standard includes the requirements for each of the consequences as well as qualifying requirements of the personnel in Clause 8.2.14 and Tables 4 and 5 therein. The functional safety assessment shall include all phases and life cycle activities to get the outputs. The plan requires specifying the personnel along with their level of independence, and the competency required. The assessment may be done in parts after each activity or group of activities. The functional safety assessment activities shall be planned and consistent. The following major points are to be covered:

- All work done since last assessment.
- The plans for implementation needing further functional safety assessments.
- The previous recommendations and verification that changes have been made.

4.1.6 IEC 61508 Annex A (Informative)

An example of documentation structure; this consists of general, safety life cycle document structure (with three tables for various documentation requirements), physical document structure, and list of documents.

4.1.7 IEC 61508 Annes B (Informative)

(Informative): Competence of persons; this comprises objective and general considerations to cover:

- Engineering knowledge in the application, technology, and safety engineering.
- Knowledge of the legal and safety regulatory framework.
- Idea about the consequences of safety-related system failure.
- Safety integrity levels of safety functions in a project.
- Previous experience and its relevance to the job.

4.1.8 Annex C (Informative)

This part deals with the details of Bibliography.

4.2 IEC 61508-2

IEC 61508-2 stands for IEC 61508 Part 2, which deals with the requirements for safety-related hardware. Keeping parity with the standard safety life cycle will be basis for compliance. Box 10 of the safety life cycle shown in Fig. VI/4.0.2-1 (Fig. 2 of IEC 61508-1), and is expanded in Fig. VI/4.1.4-1 (Figs. 3 and 4 of IEC 61508-1) to detail out for E/E/PEs which consists of hardware and software; hence there are two similar figures. In this part, the hardware safety life cycle is focused for design of the control hardware for safety systems. For the overall safety life cycle, there are requirements for functional safety management plan, safety requirements specification, and verification and assessment activities. So, one needs to concentrate on Fig. VI/4.1.4-1. Short descriptions of each of the boxes have been presented in Table VI/4.2-1 (for details refer to Table 1 of IEC 61508-2:2010) with direct reference to *Fig. VI/4.1.4-1*.

***Table VI/4.2-1** Realization Phase of E/E/PEs*

Box Number	Title	Short Objective	Clause Ref.
10.1	Safety requirement specification	To specify the requirements for each E/E/PE safety-related system, in terms of safety functions and safety integrity, to achieve the required functional safety	7.2.2
10.2	Safety validation planning	To plan the validation of the safety of the E/E/PE safety-related systems	7.3.2
10.3	Design development	Design of E/E/PE safety-related systems for safety functions and safety integrity	7.4.2 to 7.4.9
10.4	Integration	Integration and testing of E/E/PE safety-related systems	7.5.2
10.5	Installation, commissioning, operation, and maintenance procedures	Procedure development for the functional safety of the E/E/PE safety-related systems is maintained during operation and maintenance	7.6.2
10.6		Validation of E/E/PE safety-related systems meet, the requirements for safety in terms of the required safety functions and the required safety integrity	7.7.2
–	System modification	To make correction enhancement (MOC)	7.8.2
–	System verification	Testing and evaluation of outputs	7.9.2
–	Functional safety assessment	Judgment of achievement of functional safety for E/E/PE system	8

4.2.1 *Specification (Clause 7.2)*

The safety specification (Clause 7.2) shall include details on both the safety function and the safety integrity level. E/E/PEs safety requirements need to be expressed in clear, precise, unambiguous, verifiable, testable, maintainable, and feasible way so that these are comprehensible. Major issues shall include:

- Procedures to achieve or maintain a safe state for the EUC
- Throughput and response time performance
- E/E/PE safety-related system, operator, and other system interfaces
- All information for functional safety, at all relevant modes of operation of EUC
- Behavioral mode and hardware/software interactions for E/E/PE safety-related systems.

The E/E/PES safety integrity requirements specification shall contain:

- The safety integrity level for each safety function
- The low/high-demand operational mode for each safety function
- The proof testing of the E/E/PE hardware; requirements, constraints, functions, and facilities
- The electromagnetic immunity limits and the extremes of all environmental conditions.

4.2.2 *Safety Validation Planning (Clause 7.3)*

Applicable for safety-related E/E/PE:

- Requirements defined in associated specification.
- Procedures implementation for validation of each safety function, the pass/fail criteria, and safety integrity correctly.
- The procedures to be applied to validate that each safety function for tests.
- The required environment, and all necessary tools and equipment.
- Test evaluation procedures (with justifications).
- Test procedures and performance criteria for electromagnetic immunity limits tests.
- Policies for resolving validation failure.

4.2.3 *Design and Development*

The design of the E/E/PE safety-related system including the overall hardware and software as detailed in Figs. VI/4.1.4-1 and VI/4.1.4-2 are to be followed. Major issues here are:

- Hardware safety integrity; the architectural constraints (dealt separately later) and probability of random hardware failure.
- Systematic safety integrity issues including avoidance of failures, control of systematic faults, system behavior on fault detection, and proven equipment issue.
- Achievement of a safety integrity level (hardware and software).
- SIL determination with further detailing (Clause numbers 7.4.3.1—2, 7.4.3.6—12 of the standard).

- Architectural constraints is described in Clause 7.4.3.1 of the standard and major issues are:
  - Hardware fault tolerance of N means that $N + 1$ fault could cause a loss of the safety function.
  - A single fault directly leads to the occurrence of one or more subsequent faults.
  - For hardware fault tolerance, faults with very low likelihood, in relation to the safety integrity requirements, may be excluded (to be justified and documented).
  - The definition of safe failure fraction of a subsystem.
- Definition of types A and B of subsystems (Clause 7.4.3.1.2) are detailed in Table VI/4.2.3-1.

In Table VI/4.2.3-2, hardware safety integrity: architectural constraints, has been presented. This table has been developed combining Tables 2 and 3 of IEC 61508-2.

Table VI/4.2.3-1 *Definition/Requirements of Type A and B Subsystems*

Conditions	Type A (All Necessary)	Type B (Any One)
Failure mode	Well-defined for constituent components	For at least one constituent component not well-defined
Faulty sub-system	Behavior completely determinable	Cannot be completely determined
Dependable failure data	Sufficient failure data available from field	Insufficient failure data available from field

See Clause 1.3.1 of Chapter VIII also.

Table VI/4.2.3-2 *Hardware Safety Integrity: Architectural Constraints*

	Type A (*as per Table VI/4.2.3-1*)			Type B (*as per Table VI/4.2.3-1*)		
	Hardware Fault Tolerance[a]			Hardware Fault Tolerance[a]		
Safe Failure Fraction	0	1	2	0	1	2
<60%	SIL 1	SIL 2	SIL 3	Not allowed	SIL 1	SIL 2
60% to <90%	SIL 2	SIL 3	SIL 4	SIL 1	SIL 2	SIL 3
90% to <99%	SIL 3	SIL 4	SIL 4	SIL 2	SIL 3	SIL 4
>99%	SIL 3	SIL 4	SIL 4	SIL 3	SIL 4	SIL 4

[a]*A hardware fault tolerance of* N *means that* N + 1 *faults could cause a loss of the safety function* (See Clause 1.3.1 of Chapter VIII also).

4.2.4 Requirements for Avoidance of Failure (Clause 7.4.2)

Major issues in line with SIL for the chosen design method shall include but are not limited to the following:

- Transparency, modularity, and features to control complexity.
- Clear expressions for functionality, subsystem interfaces, sequencing, etc.
- Clear documentation and communication.
- Verification and validation.

4.2.5 Integration (Clause 7.5)

The major issues here are:

- The integration of safety-related software into the PES (next part)
- Appropriate documentation of the integration testing
- Modifications or change shall be subject to an impact analysis

4.2.6 IEC 61508 Part 2 Annex

Part 2 of the standard also has annexure such as control of failures during operation (Annex A), avoidance of systematic failures during different phases of the life cycle (Annex B), and diagnostic coverage and safe failure fraction (Annex C).

4.3 IEC 61508-3

IEC 61508-3 stands for IEC 61508 Part 3, which deals with the requirements for safety-related software issues. Software life cycle has been shown in Fig. VI/4.0-4 (Fig. 4 of IEC 61508-3) with interrelation shown in Fig. VI/4.0-5 (Fig. 5 of IEC 61508-3). This standard, IEC 61508-3, also depicts software safety integrity and the development life cycle (the V-model); Ref: Fig. 6 Part 3 of the standard.

4.3.1 Software Quality Management System (Clause 6.0)

In this clause, basics of software quality management have been covered in the standard.

- The functional safety strategy planning, procurement, development, integration, verification, validation, and modification in line with the requirement of intended SIL.
- Software configuration management system must include:
  - Software changes for specified requirements for software safety.
  - Demonstration of achievement of required software integrity.
  - Maintenance of all documentation and source code, including the safety analysis, software specification, design documents, and software source code modules, as well as change or modification requests.
  - Prevention of unauthorized modification.
  - Approve or reject proposed modification.
  - Analysis of impact for proposed modification.
  - Documentation of release of safety-related software.

4.3.2 *Software Lifecycle*

A software safety life cycle overview has been presented in Table 1 in the standard 61508-3. This part of the standard addresses the safety-related issues and life cycle for software systems. Like the previous part, Box 10 of main life cycle has been expanded to discuss the issues. Some major issues out of that have been illustrated in Table VI/4.3.2-1.

4.3.3 *Software Specification (Clause 7.2)*

Validation and Planning (Clause 7.3): There are two clauses here, which are discussed separately.

- Software specification (Clause 7.2): As seen from the previous table, the first part in realization phase of software starts with the requirements of software specification. Specification, including terminologies, shall be clear, precise, verifiable, testable, maintainable, and feasible, and detailed enough to allow

Table VI/4.3.2-1 ***Software Life Cycle Overview (Part) IEC 61508-3***

Box No	Title	Short Objective	Clause Ref.
10.1	Software safety the requirements specification	To specify the requirements for safety-related software system safety functions and integrity. Also to specify the requirements for software safety functions and systematic capability for each E/E/PE safety-related system to implement and to achieve SIL specified	7.2.2
10.2	Software safety validation and planning	Plan to validate safety related software	7.3.2
10.3	Software design and development	Architecture: Fulfillment of SIL requirements and review and evaluate requirements of software from hardware architecture Support tools and programming language selection Detailed design development for software and fulfillment of requirements including same for individual software module Implementation of codes Module testing Software integration testing *(see standard)*	7.4.3 through 7.4.8

Table VI/4.3.2-1 Software Life Cycle Overview (Part) IEC 61508-3 Continued

Box No	Title	Short Objective	Clause Ref.
10.4	PE integrations (hardware and software)	Integration of software in to PEs combination of hardware and software compatibility and to meet the requirements of the intended SIL	7.5.2
10.5	Software operation and modification procedure	Information and procedure to ensure functional safety of the E/E/PE safety-related system is maintained during this phases	7.6.2
10.6	Software safety validation	To ensure that integrated system complies with the intended SIL	7.7.2
–	Software modification	Correction and enhancement (MOC)	7.8.2
–	Software verification	Testing and evaluation	7.9.2
–	Software functional safety assessment	Judgment of achievement of functional safety for the software aspect of E/E/PE system	8

design. All modes of operation, and relevant constraints between the hardware and the software must be listed. The requirements of specification needs to detail out all software self-monitoring, any diagnostic tests, etc. to be performed on the hardware, and testing of critical functions, including online testing of safety functions.

- Software validation and planning (Clause 7.3): Detailed plans for demonstration of software satisfying the safety requirements should be specified. A combination of analysis and testing techniques must be specified in the plan. This shall include assessments to be done. The major issues here are:
  - Equipment
  - Timing of validation
  - Who will do the validation
  - Modes of operations to be validated
  - Abnormal conditions foreseen
  - Safety-related software to be validated
  - Software specification references
  - Expected results; pass/fail criteria

4.3.4 *Software Design and Development (Clause 7.4) and PE Integration (Clause 7.5)*

Discussions start with software design and development then, PE integration will be discussed.

- Software design and development (Clause 7.4): The design method shall allow and support modularly clear and unambiguous expression of functionality, sequencing, timing constraints, data structures, design assumptions, and their dependencies as per good engineering practices. The architectural design and description must include:
  - Component interconnections
  - Fulfillment of requirements for intended software SIL of sub-systems/ components
  - Review and evaluation of software/hardware interactions
  - Design feature for maintenance of the safety integrity of data
  - Integration tests to ensure requirements for software

Support tools such as set of integrated tools, configuration management tools, testing tools, and programming languages including compilers must meet the safety integrity needs of the software. Design development needs to take care of overall complexity, testability, and safe modifications capability. For different SILs, the overall design is only valid for the *least* stringent SIL. The design features include software proof tests and all online diagnostic tests as specified. All changes in requirements must be documented and approved. Standards codes, which need to be readable, understandable, and testable meeting requirements specified (also tested during design), shall be deployed. For certain specific issues mentioned in the standard, readers are advised to go through the same.

- Programmable electronic integration (Clause 7.5): Major issues are integration of software into hardware of PEs, and hardware and software in combination meet the requirements of intended SIL and associated testing such as test data, test criteria, corrective actions on test failure case, tools, etc.

4.3.5 *Software Operation and Modification Procedure (Clauses 7.6 and 7.8)*

Authorization is necessary for software modification under the procedures specified during planning. Major issues involved in authorization shall include but are not limited to: Hazard to be affected, and proposed change with necessary reason (duly documented). It is necessary to ensure that the required SIL is maintained. In this connection for detailing, clause number 7.8 of the standard may be referenced. The modification process involves an analysis on the impact of the proposed on functional safety and how much of the safety life cycle must be repeated.

4.3.6 *Software Safety Validation (Clause 7.7)*

Software validation, which involves overall checks for software safety requirements with appropriate documentation, may be done separately or with system validation. Approved quality validation plan shall be in place and all must be documented properly. For separate validation, major issues are:

- Record of the validation activities
- Software safety validation plan version
- Test environment: tools and equipment
- Planed test reference

- The results with discrepancies
- Change request for discrepancies

4.3.7 *Software Verification (Clause 7.9)*

This is for process tests and evaluation of the results of the software safety life cycle. Major verification issues shall mainly include the following:

- Software safety requirements
- Architecture design
- System design
- Software module design
- Source code
- Data
- Software module and integration testing
- Hardware integration testing
- Software validation

4.3.8 *Functional Safety Assessment (Clause 8)*

"The objective and requirements.of Clause 8 of IEC 61508-1 apply to the assessment of safety-related software."

4.3.9 *IEC 61508 Part3 Annex*

There are two annexure (normative): Annexure A: guide to the selection of techniques and measures (many data in tabular form); and Annexure B: detailed tables.

Before concluding the discussions on software development in line with the standard, it is worth mentioning that software is developed in stages, which follow one another giving the so-called name, "waterfall model of development," comprising: requirement specification and analysis, design specification and analysis, program coding and verification, hardware/software system integration, and maintenance and modification.

4.4 IEC 61508-4

IEC 61508-4 stands for IEC 61508 Part 4, which deals with abbreviations and definitions (self-explanatory).

4.5 IEC 61508-5

IEC 61508-5 stands for IEC 61508 Part 5, which deals with examples of methods for the determination of SILs (informative). Part 5 is comprised of five annexes, viz. A through E. These annexes describe some key concepts and various methods for SIL calculations. A list of the these has been presented below:

- Annex A: Risk and safety integrity: General concept
- Annex B: ALARP and tolerable risk concepts
- Annex C: Determination of SILs: A quantitative method
- Annex D: Determination of SILs: A qualitative method: Risk graph
- Annex E: Hazardous event severity matrix

4.6 IEC 61508-6

IEC 61508-6 stands for IEC 61508 Part 6 which deals with guidelines in the application of Parts 2 and 3 (informative). Part 6 provides more detailed explanations and examples on how to comply with Parts 2 and 3. This part is also comprised of mainly annexes:

- Annex A: Application of Parts 2 and 3
- Annex B: Example technique for evaluating of probabilities of failure (with calculations for PFD and PFH; Appendix B2)
- Annex C: Calculation of diagnostic coverage: Worked example
- Annex D: A Methodology for quantifying the effect of hardware-related common cause failures in multichannel PE systems
- Annex E: Example application of software integrity tables of Part 3

4.7 IEC 61508-7

IEC 61508-7 stands for IEC 61508 Part 7, which deals with overview of techniques and measures (informative) with annexes:

- Annex A: Overview of techniques and measures for E/E/PEs: Control of random hardware failures
- Annex B: Overview of techniques and measures for E/E/PEs: Avoidance of systematic failures
- Annex C: Overview of techniques and measures for achieving software integrity
- Annex D: Probabilistic approach to determining software safety integrity for pre-developed software. This basically deals with probabilistic approach for SIL of proven software

There have been some changes of this standard in its revision in 2010, so it is better to consult the latest revision of the standard to note the changes. It is not possible to compare the two revisions, yet some of the major issues are highlighted here:

4.7.1 *Systematic Capabilities*

Systematic capabilities is a newly developed concept and is a measure (on a scale of 1–4) that the systematic safety integrity of an element fulfills the given safety function, considering the instructions stated in the product safety manual.

4.7.2 *Competence*

Now for this, the following criteria need to be met:

- Organizations involved on safety system projects or activities shall appoint one or more persons with responsibility for one or more phases of the safety life cycle (per IEC 61511).
- All persons, departments, or organizations shall be identified, the responsibilities clearly defined and communicated.
- Activities related to management of functional safety shall be applied at phases.

- All persons undertaking specific activities shall have the appropriate competence which is documented.

4.7.3 *Security*

In case the hazard analysis, identifies unauthorized action, a security threat (reasonably foreseeable), then a security threats analysis should be carried out.

Further discussions on this standard have been presented in Clause 2.1 of Chapter VIII.

IEC 61511 provides similar standards, specifically for process industry and this will be discussed in the following discussions in Clause 5 as follows.

5.0 FUNCTIONAL SAFETY PROCESS STANDARDS: IEC 61511

At the beginning of the discussions on IEC 61508 & 61511 author wishes to put forward the following acknowledgement to IEC for granting permission to use the figures of their various standards mentioned in the box for the book.

Functional safety standard for process IEC 61511 has a good number of requirements taken from IEC 61508. In this connection, a number of figures referred to in ANSI/ISA 84.00.01 are also comparable i.e., *Figs. VI/5.0.3-1 and VI/6.0-1* actually belong to IEC 61511, but shown there to indicate that these two standards are same, and only deviations are shown (in the standard) by striking of the deviation and addition by underlines (these are explained in Clause 6.0) in the standard. The standard can be divided in to mainly two parts; technical requirements and support parts. To get a good feel of the system, it is better to refer to Fig. VI/5.0-1 (Fig. 8 of the standard:2003). From this figure, it is clear that there are 11 phases, of which phases 9—11 (covered in Clauses 5—7); are mainly supportive. Phases 1—8 (covering Clauses 8—18), are part of main technical requirements. Also, entire technical requirements can be divided into three sections, viz. analysis, realization and operation. Similarly, there are different stakeholders for each of these phases. All these have been indicated in Table VI/5.0-1.

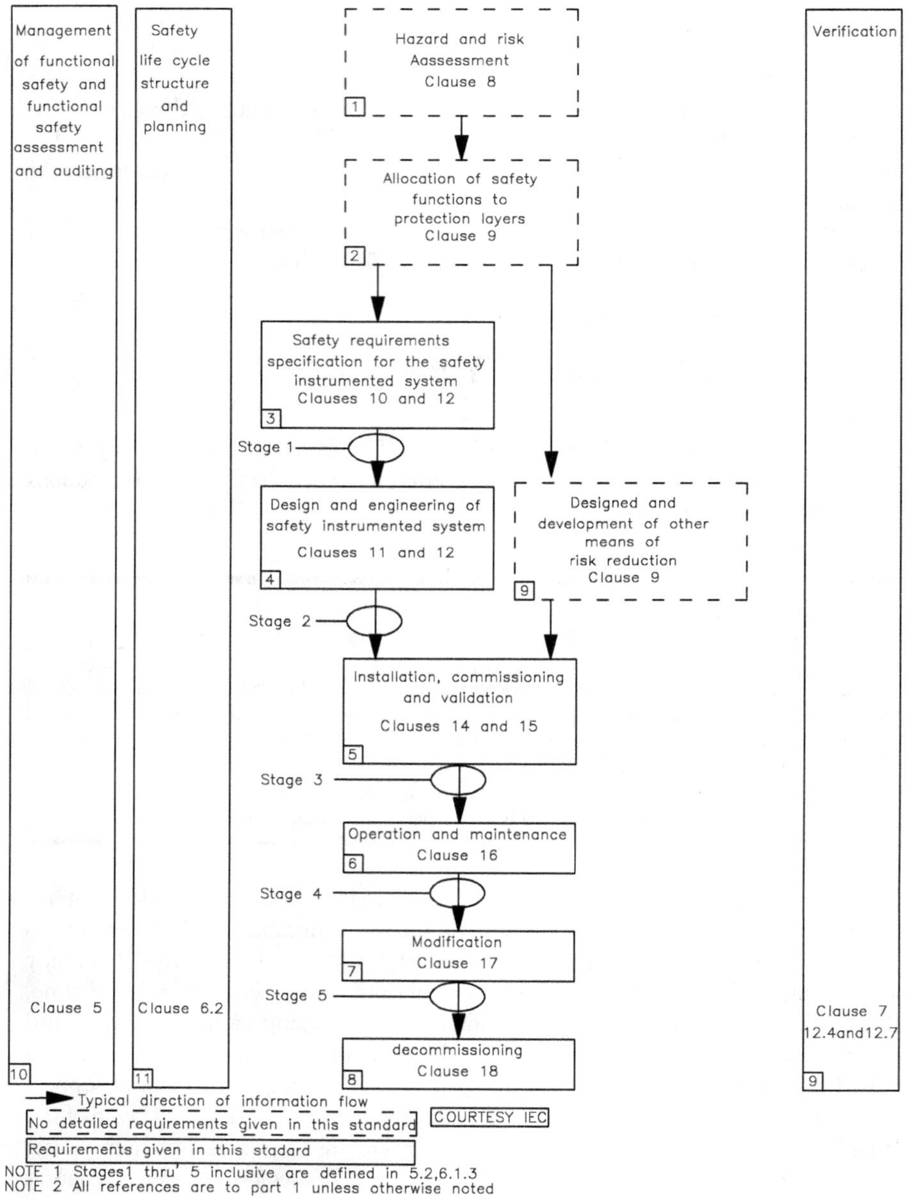

FIGURE VI/5.0-1

Safety life cycle phases of IEC 61511.

Table VI/5.0-1 *Elaboration of Framework of IEC 61511 (Based on Fig. 1 of Standard)*

Clause	Ph	Heading	Activity	TR/SP	PC	Responsibility
2		Reference		SP		
3		Definition abbreviation		SP		
4		Conformance		SP		
5	10	Management of functional safety		SP		ALL
6	11	Safety life cycle requirements: structure and planning		SP		ALL
7	9	Verification		SP		ALL
8	1	Hazard and risk assessment (concept scope)	Hazard identification and requirement specification	TR	A	End users and operators
9	2	Allocation of safety functions to protection layers		TR	A	
10 12	3	Safety requirement for SIS (see also **Clause 12**)	Configuration, meeting requirements	TR	A	Engineering and equipment supplier
11 12	4	Design and engineering SIS (see also **Clause 12**)		TR	R	
13 14 15	5	Installation, commissioning & validation (see also **Clauses 13 and 15**)		TR	R	
16	6	Operation and maintenance	Operate maintain modify		O	End users and operators
17	7	Modification			O	
18	8	Decommissioning			O	

A, *analysis*; Cl, *clause*; O, *operation*; PC, *phase category*; Ph, *phase*; R, *realization*; SP, *support part*; TR, *technical requirement*.

Here it is noted that each phase number is given in a small box in left bottom most corner of each box in Fig. VI/5.0-1. Also, applicable clause numbers are indicated in each of the main boxes (with the exception of Clause 1 being separately discussed). There are three parts of IEC 61511. Some detailing of Part 1 has been elaborated in Table VI/5.0-1. Parts 2 and 3 of the standard stand for, "guidelines for the application of Part 1" and "guidelines for determination of required SIL," respectively, hence these two parts are not really relevant for the table.

Clause 19, belonging to Part 1, is for information requirements, and Annex A of Part 1 is for differences with IEC 61508.

5.0.1 *Basic Scope Discussions*

This clause has been included only to highlight some of the key features in the scope of the standard, but not to detail out the entire scope for which readers need to refer Clause 1 of the standard. Major features are:

- It elaborates on the requirements for achieving functional safety, but not responsibility.
- The standard provides relationship between IEC 61508 and 61511 (see *Figs. 2 and 3* of the standard (***Fig. VI/6.0-2A and B***)).
- It also gives outline of relationship between safety instrumented functions and other functions; see *Fig. 4* of the standard (***Fig. VI/5.0.3-1A***).
- The standard applicable for application software, has limited variability or fixed programmed but *not* for manufacturers, safety instrumented systems designers, integrators, and users that develop embedded software.
- It specifies system architecture, hardware configuration, application software (user and integrator of SIS), and system integration, requirements for safety instrument functions (SIFs) including human factor, and safety life cycle. It also specifies the techniques and measures for SIL.
- It also establishes numerical targets for average PFD and frequency of dangerous failures per hour for SIL.
- Standard also may be applied in non-safety applications such as asset protection.

5.0.2 *Relation of IEC 61508 and 61511*

As stated earlier, this standard addresses the same functions and utilities as discussed in connection with IEC 61508. The major difference is that, this is specifically meant for process industries in contrast to generalized nature of IEC 61508. Also it mainly addresses the *end users and integrators*, and not manufacturers. Differences between IEC 61508 & IEC 61511 have been detailed out in Table VI/4.0-2. Other than those, there is a lot of commonality between the two, which is clearly depicted in Figs. 2 and 3 of the standard. *Since ISA 84.00.01 and IEC 61511 are almost identical, such similarities have been shown in Fig. VI/6.0-1 (mainly for ISA 84 applicable for IEC 61511, with **necessary minor changes**)*. For the process industry sector, *for* SIS standards, IEC 61508 should be referred to, whereas safety instrumented designers, integrators, and users need to refer IEC 61511. Similarly, in this sector for hardware, standards are developed as per IEC 61508-2; for proven hardware, IEC 61511 is to be followed. In

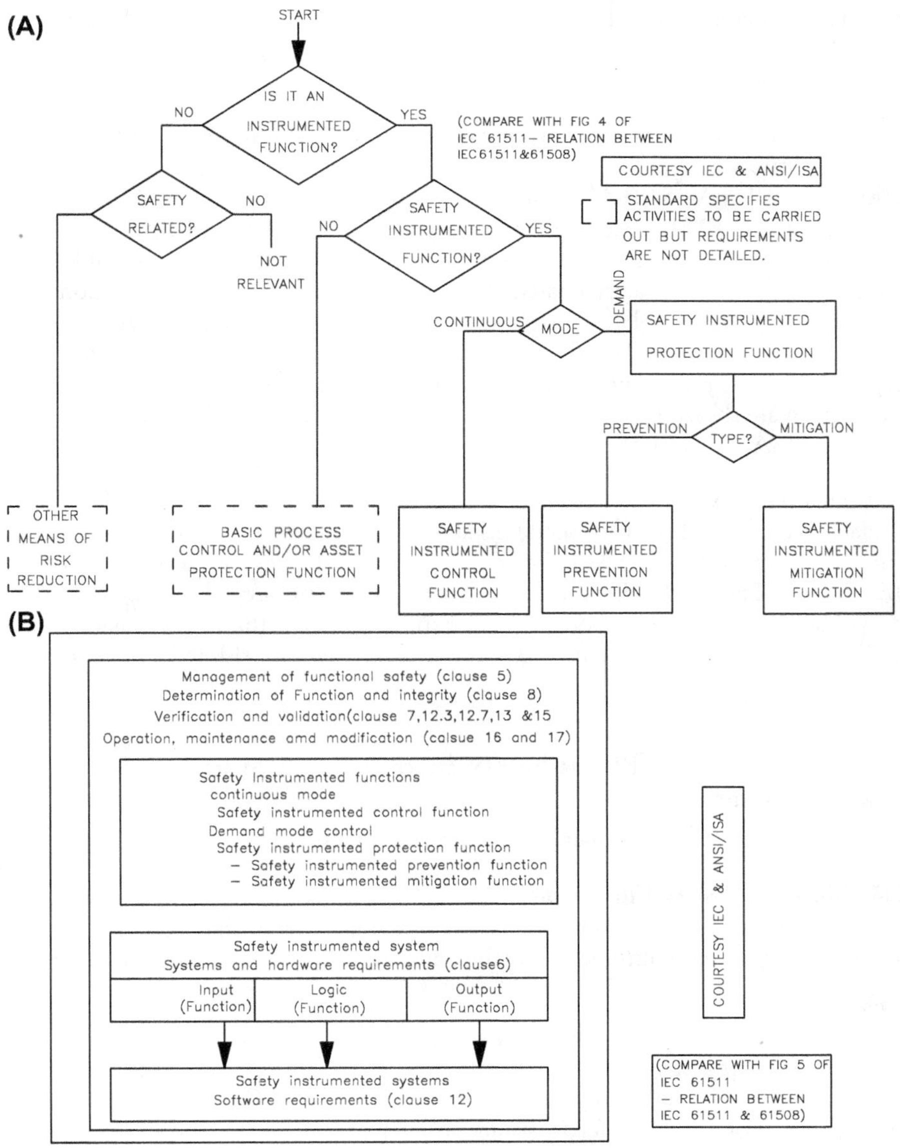

FIGURE VI/5.0.3-1

Understanding of ISA 84.00.01. (A) Relationship between safety instrumented function and other functions (Fig. 4 of ISA 84.00.01); (B) relationship between system, hardware and software of ANSI/ISA-84.00.01 (Fig. 5).

the similar manner for software, also for embedded software, application software with full variability language IEC 61508-3 is to be followed otherwise this standard to be followed as shown.

5.0.3 *Relationship Between Safety Instrumented Functions and Other Functions*

For this, refer to Fig. VI/5.0.3-1A. This is shown in connection with ISA 84, but it is applicable for IEC 61511 (Fig. 4). At the beginning, it is tested whether it is for instrumentation. If not, then for safety, other means of risk reduction is used. If it is for instrumentation but not safety-related, then basic controls for asset protection is chosen. If it is for safety related instrumentation, then, for continuous mode safety instrumented control function is the choice. On the contrary, for demand mode, safety instrumented preventive or mitigating functions are chosen. Other means of risk reduction are shown by the *dotted box*. These explanations (Clauses 5.0.2 and 3) are valid for ISA 84.00.01 also.

5.0.4 *Discussions*

The clause number 1 deals with scope discussed above. Clause numbers 2—4 cover definitions abbreviations etc. As indicated Table VI/5.0-1 Elaboration of framework of IEC 61511. These are self-explanatory, in the following sub-clauses IEC 61511 parts will be discussed briefly. Another important issue here is that this standard also includes LOPA, which may be used to address no SIF cases for risk reductions (e.g., Phase 2).

5.1 IEC 61511-1 (Without Safety Instrumented Systems and Software)

Part 1 of this standard covers mainly the following issues:

5.1.1 *Management of Functional Safety (Clause 5)*

Identification of management activities and ensuring meeting of the requirements of functional safety, are the main objective of the clause. Major issues in this connection are:

- General:
  - The policy and strategy for identification of safety
  - Evaluation of achievement
  - Communication in the organization
  - Assurance of usage of safety instruments
- Organization and resources:
  - Persons, department organizations
  - Competence to carry out the job
- Risk evaluation and management
- Implementing and monitoring
  - Hazard analysis and risk assessment
  - Assessment and auditing

- Verification
- Validation
- Post-incident/accident activities
- Procedure for quality management
- Evaluation of SIS performance

- Assessment auditing and revision
- SIS configuration management

5.1.2 *Safety Life Cycle Requirements (Clause 6)*

Phase definition and establishment of requirements for safety life cycle activities is one of the major objectives which also includes planning and technical activities to ensure SIS meet the safety requirements. Major issues are listed in the Table VI/5.1.2-1 (see Fig. VI/5.0-1 or Fig. 8 of standard).

Table VI/5.1.2-1 ***Safety Instrumented System Life Cycle Overview, Part (IEC 61511-1)***

Box	Title	Objective	Clause Ref.
1	Hazard and risk assessment	Determination of sequence of event leading to hazard, hazard events with risk involvement with the same, risk reduction and safety function for risk reduction.	8
2	Allocation of safety functions to protection layers	Allocation of safety function to protection layers and each SIF, and associated SIL.	9
3	SIS safety requirement specification	To specify requirements to each SIS in term of required SIF with associated safety integrity to achieve functional safety.	10
4	SIS design engineering	SIS design to meet SIF and associated safety integrity.	11,12.4
5	SIS installation, commissioning and validation	SIS integration and testing. Validation of SIS in all respects of requirements in terms of SIF and associated safety integrity.	12.3,14,15
6	SIS operation and maintenance	Maintenance of functional safety during operation and maintenance.	16
7	SIS modifications	Correcting, enhancing and adaptation of SIS to ensure that require SIL is achieved and maintained.	17
8	Decommissioning	Reviewing and ensuring that appropriate SIF is maintained.	18
9	SIS verification	Testing and evaluation of output at a given phase with respect to products and standards for the given input.	7,12.7
10	SIS functional safety assessment	Investigation and arrive at judgment that functional safety is achieved by SIS.	5

5.1.3 Verification (Clause 7)

This involves demonstration by reviewing, analysis, and testing that output requirements for each phase is achieved. Major issues here are:

- Procedure and techniques
- When to carry out the activity
- Responsibility and independence of agencies involved
- *Items and basis* for verification
- Non-conformance *handling*
- Analysis and supporting tools
- Verification planning
- Verification results

5.1.4 Process Hazard and Risk Assessment (Clause 8)

The main objective of this clause includes: event sequence leading to the hazards, the hazardous events of the EUC and associated control equipment; EUC risks associated with the hazards, risk reduction requirements, safety functions, and SIF. Major issues are:

- Hazard and risk assessment to include:
  - Hazard identification
  - Factors for hazards
  - Likelihood and consequences
  - Operating conditions
  - Risk reduction measures
  - Assumptions
  - Safety function allocations
  - Identification of SIF
- Dangerous failure rate of basic plant control system (BPCS)
- Hazard and risk assessment results

5.1.5 Allotment of Safety Function to Protection Layers (Clause 9)

The main objective is to allocate safety functions to protection layers, to determine SIF, and for each SIF determine the associated SIL. Major coverage of this clause shall mainly include:

- Allocation will result in specific safety preventive/mitigating function to specific layer and risk reduction target for each SIF
- Allocation of SIL and risk reduction factor for each SIF with details as mentioned in Table VI/4.0.6-1 SIL assignment table
- Additional requirements for SIL 4

5.1.6 *Factory Acceptance Test: FAT (Clause 13)*

SIS and software *separately* in Clauses 5.2 and 5.3, respectively. So, in order to complete the discussions on IEC 61511-1, the next clause factory acceptance is discussed here. The basic objective of a factory acceptance test (FAT) is to carry out tests of the logic solver and associated software together to ensure it meets the safety requirement specification. The major issues here are:

- Need for FAT to be specified during design stage
- Planning of FAT to include:
  - Type of test including functional test
  - Test case, description, and data
  - Other system dependence and interface
  - Test environment and tool used
  - Logic solver configuration
  - Tests criteria
  - Test fail corrective action procedure
  - Competence of test personnel
  - Physical location
- Defined version of logic solver
- Actual test major points
  - Version of logic solver
  - SIF performance characteristics
  - Test procedure description
  - Detailed record of test activities
  - Tools and equipment used
- Test results
- Any modification done, then analysis for impact

5.1.7 *Information and Documentation Requirements (Clause 19)*

This clause deals with availability and documentation of information so that all the phases of safety life cycle, validation, verification, and safety assessment are well-carried out. Major issues are availability of all required documentation, which shall be:

- Well-described equipment, systems, and installation details
- Accurate
- Easy to understand
- Suitable for the intended purpose
- In an accessible and maintainable form

In next clause, detailed requirements about SIS shall be discussed and that also is a part of Part 1 of IEC 61511.

5.2 IEC 61511-1 (Safety Instrumented System Part Only)

SIS and SW need special attention so these are discussed separately. In this clause, IEC 61511 Part 1, SIS-related clauses will be addressed (i.e., Standard Clause nos. 10,

11, 14—18 shall be discussed here, Clause 12 related to software shall be discussed in Section 5.3). Major headings under this clause shall include:

- Clause 10: SIS safety requirement specification: According to the standard, here the requirements for SIF shall be specified in required details.
- Clause 11: SIS design and engineering: Standard discusses designing one or multiple SISs to provide SIF and meet SIL requirements.
- Clause 14: SIS installation and commissioning: In the standard, this clause deals with installation and commissioning of SIS as per specification, so that they are ready for validation.
- Clause 15: SIS safety validation: Validation through inspection and testing of installed and commissioned SIS, and associated SIF achieved the safety requirements specified.
- Clause 16: SIS operation and maintenance: Main aim is to ensure that required SIL of each SIF is maintained during operation and maintenance so that functional safety is maintained.
- Clause 17: SIS modification: Modification of any SIS should be planned, reviewed, and approved prior to changes so as to ensure required safety integrity is maintained even any changed made to SIS.
- Clause 18: SIS decommissioning: Main issue here is to ensure that before decommissioning, any SIS proper review is carried out and authorization is obtained so SIF is operational during decommissioning.

In sub-clauses that follow, basic issues pertinent to all these will be touched upon. For details standard needs to be consulted.

5.2.1 *SIS Safety Requirement Specification (Clause 10)*

Allocation of SIF will be driving force for safety requirement specification. Major issues here are as follows:

- Description of all the safety instrumented functions.
- Identification account of *common* cause failures.
- Definition of the *safe-state* of the process for each SIF.
- Definition of *individual* safe process states (for concurrency).
- Source of *demand* and demand rate for SIF.
- Proof-test *intervals*.
- *Response* time requirements for the SIS.
- SIL and mode of operation (demand/continuous).
- Description of SIS measurements and trip points.
- Description of SIS output actions and criteria.
- Functional relationship: Process inputs and outputs, and logic.
- Math functions and permissive.
- Manual shutdown, to energize or de-energize to trip requirements.
- Post-shutdown SIS reset requirement.
- Maximum number of allowable spurious trip rate.
- Failure modes and response of the SIS.
- Starting up and restarting the SIS procedure.
- Interfaces between SIS and other systems.

- Requirements for overrides/inhibitors/bypasses, and their clearance.
- Specification of *action* necessary to achieve or maintain a safe state in the event of faults detected, including human factor.
- MTTR for the SIS.
- Location, spares holding, and environmental constraints.
- Identification of the dangerous combinations of output states of the SIS.
- Extremes of all environmental conditions for SIS, including electro magnetic interference (EMI)/radio frequency interference (RFI), etc.
- Identification of normal and abnormal modes, and SIF needed to support requirements for SIF necessary to survive.

5.2.2 SIS Design and Engineering (Clause 11)

Some of the major issues are listed below:

- For both safety and non-safety instrumented functions, all hardware and software affecting negatively in any condition is to be treated as a part of SIS.
- If SIS with SIFs of different SIL, then the shared or common hardware and software shall conform to the highest SIL unless otherwise proven.
- If BPCS is not part of the standard, then it is to be separate and independent to the extent that the functional integrity of the SIS is not compromised.
- Requirements for operability (including human factor), maintainability, and testability to be addressed.
- The human capabilities and limitations to be considered.
- SIS design shall be such that process will be in safe state unless SIS is reset.
- Unless directed, manual means shall be there to face emergency.
- Design of SIS considers dependence and independence of SIS with BPCS/other systems.
- Device is SIS should not be shared for BPCS *normally*.
- Requirements for system behavior on detection of a fault as per standard guidelines.
- Requirements for hardware fault tolerance as per guides given in this standard (see and compare with Table VI/4.2.3-2 meant for IEC 61508), but SIL 4 as per IEC 61508.
- Requirements for selection of components and subsystems, field devices, and interface refer guide in the standard.

5.2.3 SIS Installation and Commissioning (Clause 14)

Here the major activities shall include:

- Installation and commissioning planning:
    - Activity details
    - Procedure, techniques, and measures
    - When to carry out
    - Activity responsibilities of concerned agencies
- Safety instrument installation as per norm

- SIS commissioning in accordance with planning to take care of:
  - Proper earthing/grounding
  - Well-connected operational energy source
  - Removal of packing material and no physical damage
  - Well-calibrated instruments
  - Field devices operational
  - Logic solver I/O operational
  - Interface and peripherals operational
- Appropriate recording
- Design information updated "as built" (as applicable)

5.2.4 SIS Safety Validation (Clause 15)

Validation activity is comprised of the following:

- Validation planning:
  - Various modes of operation: Preparation, start up, shut down, maintenance, and abnormal condition
  - Procedure, measures, and techniques
  - When to carry out
  - Activity responsibilities of concerned agencies
  - Basis documentation
- Validation planning for application software
  - Software needing validation before operation
  - Technical strategy, for example, auto/manual, static/dynamic, statistical
  - Validation environment
  - Pass/fail criteria
  - Policy and procedure, etc.
- Instrument used for high-accuracy validations
- Validation of SIS and associated SIF
  - Device performance as per specification
  - SIS documentation
  - SIF performing as per specification
  - Correct computation included in SIS
  - SIS provides proper alarm including diagnostic alarm and display
  - Proper shutdown sequence activated
  - Bypass, start up override, and manual shutdown operational
  - Well-documented proof test procedure
  - Correct operation of SIS in utility failure, EMI, etc.
- All software safety requirements correctly performed
- All information and results properly documented
- After software validation, prior to identified hazard present following shall be performed
  - Process bypass functional activity
  - Process isolation valves in start up condition
  - All test materials removed
  - All forces and enabling condition removed

5.2.5 SIS Operation and Maintenance (Clause 16)

Major issues to be covered shall include:

- Operation and maintenance (O&M) planning for SIS:
  - Normal and abnormal operation modes
  - Proof testing and other maintenance activities
  - Procedure, measure, and techniques
  - When to carry out
  - Activity responsibilities of concerned agencies
  - Verification: Meeting O&M procedure
- O&M procedure based on safety planning to take care of:
  - Actions or constraints to prevent unsafe conditions
  - Demand rate and system failure information
  - Procedure information during fault/failure condition/tests/audit
  - Maintenance procedure when fault occurs to SIS
  - Maintenance/calibration of test instrument during O&M
- Operator and maintenance personnel training
- Analysis of discrepancies between expected and actual behavior of SIS
- Revision of O&M procedure as applicable
- Procedure for proof testing
- Proof testing and inspection

5.2.6 SIS Modification (Clause 17)

Planned modification can be done on SIS, provided prior authorization procedure and controlling procedure are in place. Major other issues are:

- Clear well defined work and hazards must in place.
- Established analysis for impact on functional safety in place.
- Proper authorization and proper information (reason of change, hazards, history of configuration, etc.) must be maintained.
- Well-trained, qualified persons should be deployed.

5.2.7 SIS Decommissioning (Clause 18)

Major issues here shall include:

- Prior authorization procedure and controlling procedure.
- Clear, well-defined work and hazards must in place.
- Established analysis for impact on functional safety in place.
- Result of analysis to be used for safety planning.

With this discussion on SIS for IEC 61511, it is concluded and the start discussion of software in IEC 61511 is in the next clause.

5.3 IEC 61511-1 (Software Part Only)

Let the discussions begin with Fig. VI/5.3-1, software safety cycle in process industries (courtesy of IEC). The figure consists of two parts, one is software safety cycle (Fig. VI/5.3-1A to compare with Fig. VI/4.1.4-1B), and relation between software safety life cycle and SIS safety life cycle. Both Fig. VI/5.3-1A and B have been developed on the basis of Figs. 11 and 10 of IEC 61511, respectively.

5.3.1 *IEC 61508 Software Objective*

Basic objectives of application software safety requirements, as well as the relationship between SIS and application software safety life cycle, shall be discussed. Objectives have been enumerated in Table VI/5.3.1-1 (see Table 7 of IEC 61511-1, courtesy of IEC).

5.3.2 *Application Software Safety Specification (Clause 12.2)*

Application software safety specification (Clause 12.2): Typical SIS sub-system architecture in line with the requirements of IEC 61511 has been shown in Fig. VI/5.3.2-1.

Major requirements for application software safety specification may be as follows:

- Development of application requirements
- Input these requirements for each SIS:
  - Specified safety requirements of SIF
  - Requirements from specific architecture
  - Requirements from planning
- Detailed application software mainly to include:
  - Supporting functions
  - Capacity and response time
  - All modes of operations
  - Proof and diagnostic tests
  - Software self-testing
  - Other devices with SIS monitoring
  - Periodic testing, etc.
- Structuring and expressions of software safety requirements
- Proper equipment selection so that following are possible:
  - Achieving and maintaining safe state
  - Annunciation and fault management
  - Capacity and response time requirements
  - Off/online *periodic* testing of SIF
  - SIS modification
  - Interface to non-safety functions
  - Maintenance of SIL for all above functions

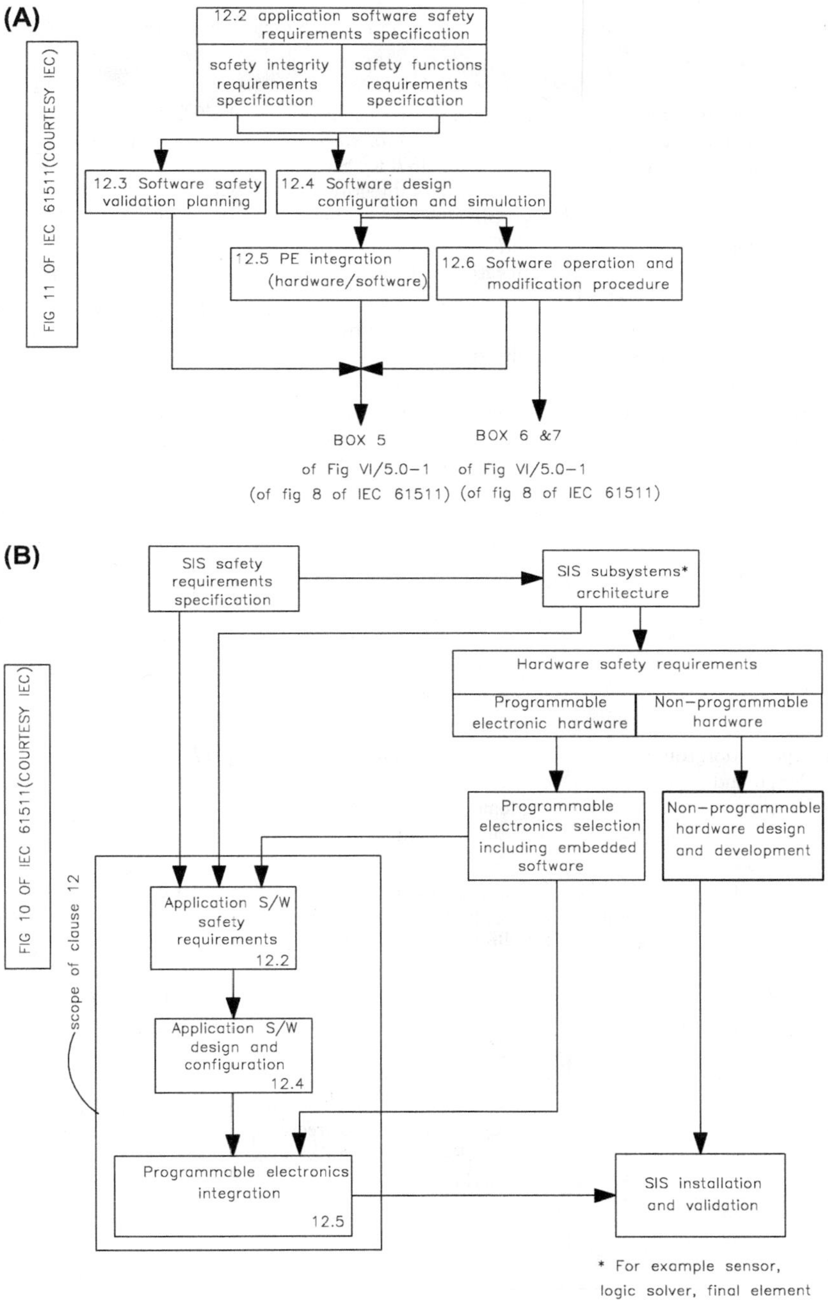

FIGURE VI/5.3-1

Software safety life cycle in process industries. (A) Software in realization phase; (B) safety life cycle relation between software and safety instrumented system.

Table VI/5.3.1-1 A Objectives of Application Software Safety Life Cycle

Box	Title	Objective	Ref. Clause
12.2	Application software safety requirement specification	Requirements of specification for software SIFs for each SIS function required for SIFs. Software safety integrity for each SIF allocated to that SIS.	12.2.2
12.3	SIS safety validation	Validation of SIS including application software meet the safety requirements.	12.3
12.3	Application software safety validation planning	Planning for validating application software.	12.3.2
12.4	Application software design and development	To create software architecture, this fulfils the software safety requirements as well as the requirements placed by hardware architecture.	12.4.3
12.4	Application software design and development	Support tool and programming language: Configuration, library, and management simulation and test tool necessary for software safety life cycle. Also procedure for program development of application software.	12.4.4
12.4	Application software design and development	Application module development and implementation of application software to fulfill application safety.	12.4.5
12.4	Application program development using full variability language	Program development and test FVL only: Implement FVL fulfilling software safety requirements specified.	12.4.6 and 12.4.7
12.4	Application program development using full variability language	Application software testing to verify **1.** Achievement of software safety. **2.** To show that all application programs interact correctly to perform the intended function.	12.4.6, 12.4.7 and 12.7
12.5	PE integration (hardware and software)	Integration of software to the target PE.	12.5.2

Box numbers in the table refer to box numbers shown in Fig. VI/5.3-1A). Ref. Clause: Requirements.

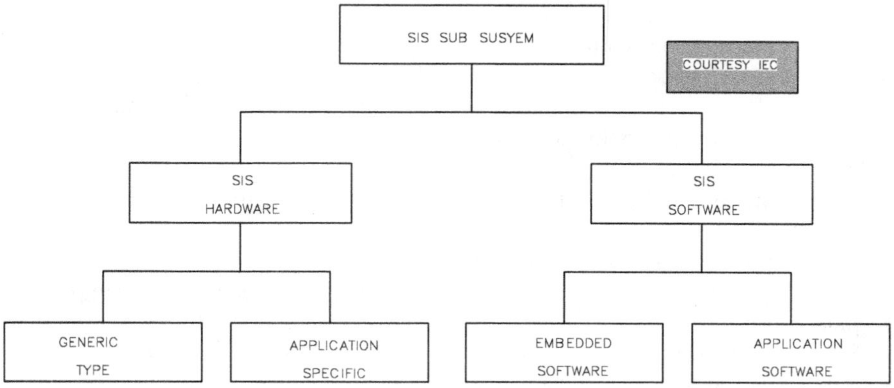

FIGURE VI/5.3.2-1

Safety instrumented system sub-system architecture (Based on IEC 61511 Fig. 13).

5.3.3 *Application Software Safety Validation (Clause 12.3)*

As shown in Table VI/5.3-1 software validation has roles to play for both SIS as well as application software. These are discussed *in Clause 15 of the standard* (which is discussed in Clause 5.2.4 previously).

5.3.4 *Application Software Design and Development (Clause 12.4)*

This is an important aspect of the standards to include a number of issues listed below:

- General requirements with testing verification and validation as per IEC 61508-3
  - Consistent with restrictions and development tools.
  - Design method and application languages mainly include limited variability language (LVL) or fixed program language (FPL) with coverage for modularity, functionality, logic expression, sequencing, information flow, verification, and validation.
  - Design achievements.
  - Compliance to IEC 61508 (FVL) or IEC 61511 (FPL, LVL), depending on language or successful implementation in similar applications.
  - Information details such as legal entity, logic convention, description, and traceability.
- Requirements for application software architecture to be developed based on SIS specifications, subject to consistent operation under constraints posed by SIS architecture, design tools, safety manual, and sub-system requirements. Additionally, it shall provide:
  - Detailed descriptions of internal structures and components identified with associated connections, processing order, software module in SIS but not used in SIF and non-SIF function (and to make sure these do not hamper SIF operation).

  - Set of measures and techniques for application software.
  - Maintenance of safety integrity.
- Requirements for support tools, user manual, and application languages: Suitable tool, application programming language configuration managements simulation, and provision for auto testing are major issues here to cover, as well as the following:
  - Availability of suitable tool to provide relevant services throughout the life of SIS.
  - Application programming language: Well-defined compiler/translator matching with application characteristics, supporting good programming practices and design methods, and should proscribe generic software features.
  - Set procedure for use of tool with due consideration for constraints.
  - Safety manual to address: Diagnostic features, watch dog, certified safety library support SIL. This shall also include detailing of requirements and limitations of supporting tools and programming language.
  - Suitability of the tools could be verified.
- Requirements for application software development: Detailed information such as detailed requirements, application software and architectural requirement including fault tolerant functionality and application logic, etc. to be provided. These will also include:
  - Structural development including: Modularity and testability of functionality, traceability application functions, and safe modification capability.
  - Robustness of each module capability to full I/O interface, and system configuration check.
  - Structural test for each application software module.
  - Application software shall be readable, and understandable with capability to meet requirements and design principles.
  - Application software shall be reviewed to check conformance to design principles and requirements.
- Requirements for application software module testing: This shall be done through review, simulation, and testing such as configuration (input through logic to output), and result shall be readily available.
- Requirements for application software integration testing: This needs to be carried out to check proper interaction of module, components sub-systems, and results of test must indicate that all requirements specified are satisfied. Any *modification* done during integration *testing* shall be subject to test of safety impact.

5.3.5 Integration of the Application Software with SIS subsystem (Clause 12.5)

Integration of application software with SIS test is to be carried out as early as possible in the safety life cycle. Major issues are:

- Testing details: The following points need to be addressed:

  - Division of application software in to manageable sets
  - Test criteria and test types
  - Test environment, tools, program, and configuration
  - Procedure for corrective action
- Any software modification done shall be subject to software safety impact analysis.
- Test information shall include:
  - Configuration items
  - Supporting tools
  - Involved personnel details
  - Test results
  - Judgment that objectives and requirements are met
  - For failure result, corrective actions
  - Detailed recording
  - Re-test/verification results.

5.3.6 *FPL and LVL Software Modification procedure*

It is mainly guided by assessment auditing and revision, and SIS configuration management described in Clause 5.1.1 earlier, that is, Clauses 5.2.6.2 and 5.2.7 in IEC 61511. As per standard some additional requirements are:

- Prior to modification, analysis of impact on safety of process and software design
- Safety planning and re-verification
- Modification and re-verification in line with planning
- Updating of all affected documents
- Details of SIS modification made available

5.3.7 *Application Software Verification (Clause 12.7)*

This is mainly guided by Clause 7 of IEC 61511 (Clause 5.1.3 above). Some features include:

- Results of each phase shall be verified for:
  - Adequacy of output for particular life cycle phase
  - Adequacy of review/inspection testing for coverage of output
  - Compatibility of output at different phases
  - Correctness of data
- Verification to address testability, readability, and traceability
- Integration of non-safety function and process interface

5.4 IEC 61511 Discussions (Details of Parts of the Standard)

In this clause, the remaining portion of Part 1 along with other parts of IEC 61511 will be discussed.

5.4.1 IEC 61511 Part 1

The discussions made so far on IEC 61511 belong to Part 1 of the standard. Apart from that, there is one annexure, Annex A (informative) which discusses difference with IEC 61508.

5.4.2 IEC 61511 Part 2

Based on Part 1 the standard provides guidelines for "the specification, design, installation, operation and maintenance of Safety Instrumented Functions and related safety instrumented system as defined in IEC 61511-1. This standard has been organized so that each clause and sub clause number therein addresses the same clause number in IEC 61511-1."

5.4.3 IEC 61511 Part 3

This part of the standard provides the guidance for determination of required SIL.

With this, the discussion on IEC 61511 is concluded, and next to have a look at ANSI/ISA 84.00.01, and find how it is related to this standard.

6.0 FUNCTIONAL SAFETY STANDARD: ANSI/ISA-84.00.01

This standard is the same as IEC 61511 only. All of the text of IEC 61511-1 (2003) has been retained, but deviations are shown by strikeout and added parts are indicated by putting underline below the added text in the standard. One of the major additions into this standard with respect IEC 61511, is Clause 1, "*for existing SIS designed and constructed in accordance with codes, standards, or practices prior to the issue of this standard (e.g., ANSI/ISA-84.01-1996), the owner/operator shall determine that the equipment is designed, maintained, inspected, tested, and operating in a safe manner*." This clause is similar to one contained in ISA S84.10-1996. This popularly referred to as **Grandfather** clause (in the previous paragraph, a portion of the write-up is shown in italics as is found in the original standard to indicate some addition from IEC 61511). This will be clear from the following figures which actually are from *IEC 61511*, but when used in ISA there are some changes in Fig. VI/6.0-1B, where IEC 61511-1 has been struck off (not shown here), and ANSI/ISA reference has been put in. In order to bring these to reader's notice, this figure has been placed under this standard.

The situation with Fig. VI/6.0-1 is similar. Here also, IEC 61511is struck off and replaced by ISA/ANSI, but within bracket reference of IEC 61511 and IEC 61511 (mod).

As discussed earlier, these standards are identical, so it is meaningless to repeat the discussions. Discussions in Clause 5.0 are also valid here. There were some differences between ISA 84: 1996 and IEC 61511 [8], but with the introduction of ISA 84.00-0.01:2004 such differences have been removed.

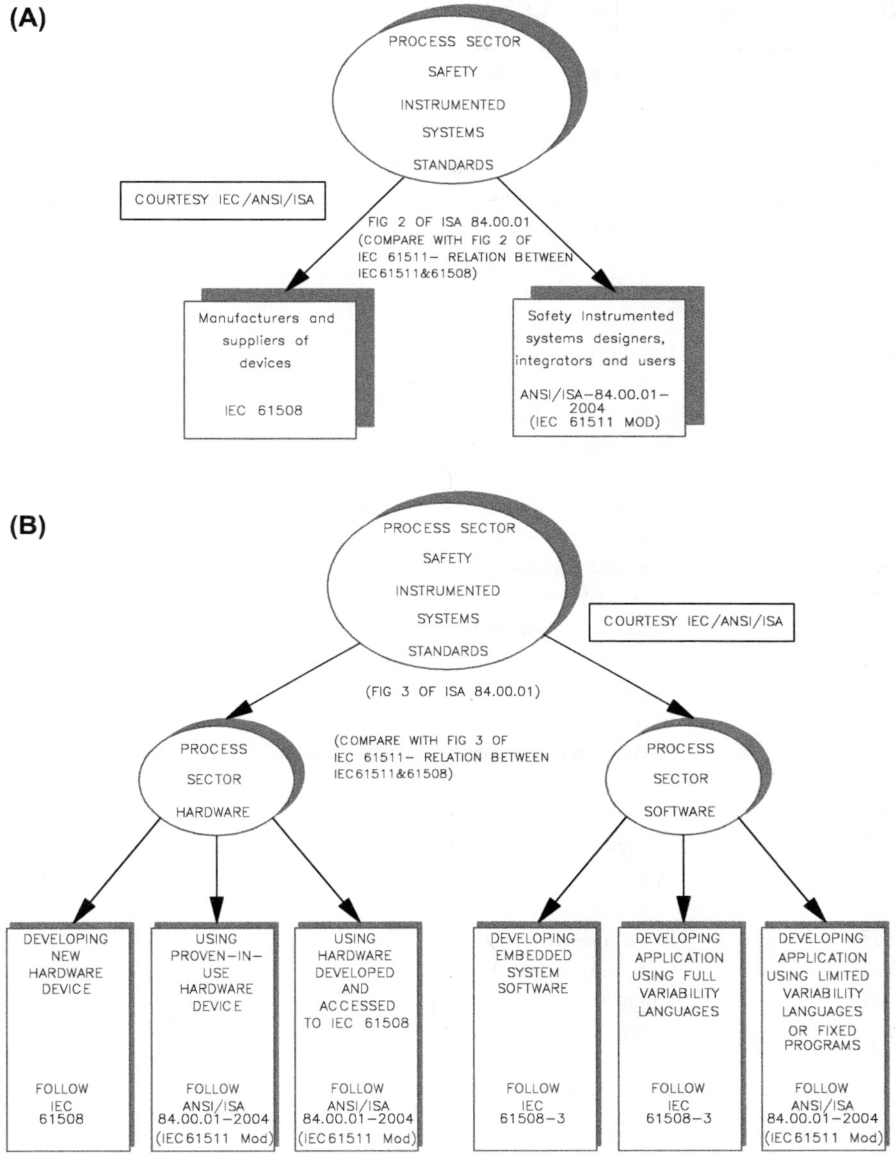

FIGURE VI/6.0-1

Relation between ANSI/ISA 84.00.01 (2004) with IEC 61508 (/61511). (A) Relation between IEC 61508 and ISA 84.00.01 (2004)-1; (B) relation between IEC 61508 and ISA 84.00.01 (2004)-2 (IEC 61511 modified).

LIST OF ABBREVIATIONS

ALARP	As low as reasonably practicable
BPCS	Basic plant control system
BU/TD	Bottom up/top down
C&I	Control and instrumentation (engineering)
COSO	Committee of sponsoring organizations of the trade way commission
CPU	Central processing unit
DCS	Distributed control system
DNV	Det Norske Veritas (Norway)
E&I	Electrical and instrumentation
E/E/PE	Electrical/electronics/programmable electronics
EMI	Electro magnetic interference
ETA	Event tree analysis
EUC	Equipment under control
EUCCS	EUC control system
FME(C)A	Failure mode and effect (criticality) analysis
FPL	Fixed program language
FTA	Fault tree analysis
FVL	Full variability language
HAZOP	Hazard and operability study
HET	House event-time
HW	Hardware
I&C	Instrumentation and control
IEC	International Electrotechnical Commission
I/O	Input/Output
LOPA	Layer of protection analysis
LVL	Limited variability language
MIS	Management information system
MOC	Management of change
MTTR	Mean time to repair
O&M	Operation and maintenance
P&ID	Piping and instrumentation diagram
PFD	Probability of failure on demand
PFH	Probability of failure per hour
PSA	Probabilistic safety assessment
PHA	Plant (process) hazard analysis
PI	Performance index
RA	Risk assessment/analysis
RFI	Radio frequency interference
SFF	Safety failure fraction (IEC)
SIF	Safety instrument functions
SIL	Safety integrity level

SIS	Safety instrumentation system OR supervisory information system (DCS)
SRS	Safety-related system
SW	Software
WRT	With respect to

References

[1] A.L. Burgenson, Risk Assessment and Management, Lonza Walkersville Inc.

[2] Risk Management Update ISO 31000 Overview and Implications for Managers, Inconsult, Internet document; http://www.inconsult.com.au/wp-content/uploads/ISO-31000-Overview.pdf.

[3] B.G. Liptak, Process Control and Optimization vol. II; Chapter 8.3; B.A. Jensen (1995, 2004).

[4] S. Basu, A.K. Debnath, Power Plant Instrumentation and Control Handbook, Elsevier, 2014.

[5] S. Basu, A.K. Debnath, Power Plant Instrumentation and Control Handbook, Elsevier, November 2014.

[6] D. Rizal, S. Tani, K. Nishiyama, K. Suzuki, Dynamic simulator for evaluation of safety objects in batch process. Okayama University; European Symposium on Computer Arded Process Engineering — 15, Elsevier Science B.V., 2005.

[7] Y.F. Wang, J.Y. Wu, C.T. Chang, Automatic Hazard Analysis in Batch Operation with Perti Nets, Elsevier, Tiwan, ROC, December 2001.

[8] Difference between IEC 61511 AND ISA 84.01-1996; A.E. Summers; President, SIS-TECH Solutions, LP.

Further Reading

A structured approach to Enterprise Risk Management (ERM) and the requirements of ISO 31000, The public Risk Management association; IRM internet document.

ANSI/ISA S88.01 Batch Standard, D. Moffatt-Invensys Foxboro and G. Skovira-EZSoft Inc., ANSI/ISA S88.01 Batch Standard - ISA, Philadelphia, September 2014.

Application Functional Safety IEC 61511: ABB paper; Internet document.

S. Bernardi, J. Merseguer and D.C. Petriu, Dependability Analysis Techniques; Model- Driven Dependability Assessment of Software Systems, Springer; http://www.springer.com/978-3-642-39511-6.

D. Brandl, Batch Control Using the ANSI/ISA 88 Standard, BR&L Consulting, 2004.

Controlling Risks: Safety Systems, USPAS, January 2012 (Internet document).

A.G. Foord, W.G. Gulland, C.R. Howard, T. Kellacher, W.H. Smith, Applying the latest standard for Functional Safety - IEC 61511; 4-sight Consulting.

Functional Safety: The next edition of IEC 61511; M. Generowicz Engineering Manager I&E Systems Pty Ltd.; HazAus2015/1507.

Guidelines for Risk Assessment and Safety Statements, Health and Safety Authority, Dublin, January 2006.

HAZOP in a batch process, Internet document; http://www.processoperations.com/HazIDRA/HZ_Chp04b.htm.

Exida; Version 2, IEC 61508 Overview Report; a Summary of the IEC 61508 Standard for Functional Safety of Electrical/Electronic/Programmable Electronic Safety-related Systems, January 2006.

IEC 61511 Implementation e The Execution Challenge; T. Shephard and D. Hansen e Mustang; Control magazine e May 2010.

A. Kianmehr, N. Nabhani, F. Jaderi, Qualitative risk assessment approach based on HAZID method in Sirri Island gas compression andexport facility, Technical Journal of Engineering and Applied Sciences, 2013. TJEAS Journal-2013-3-23/3405-3409.

L. Kunimatsu, Risk Management Basics e ISO 31000 Standard, CRISC; Ford Motor Company.

P.B. Ladkin, An Overview of IEC 61508 on E/E/PE Functional Safety, Causalis and University of Bielefield, 2008.

B.G. Liptak, Process Control and Optimization vol. II; Chapter 8.4; A. Ghosh (1995, 2005).

S. McCoy, D. Zhou, Risk Assessment of Batch Processing Plants, Loughborough University, UK; Gradient projects; Computer science; Internet documents; http://gradients.lboro.ac.uk/projects/Batch/index.php.

070 – Norwegian Oil and Gas Application of IEC 61508 and IEC 61511 in the Norwegian Petroleum Industry; Norskolje&gass No.: 070 Established: 01.02.01 Revision no.: 02 Date revised: 29.10.2004.

Overview of ISO 31000 2009 Risk Management Standard, Praxiom Research Group Limited, 2010. Internet document, http://www.praxiom.com/iso-31000-sum.htm.

C. Palmer, P.W.H. Chung, J. Madden, A Ruled Based System for Automated Batch HAZOP Studies, Symposium series 154, IChemE, 2008.

F. Redmill, Redmill Consultancy, An introduction to the safety standard IEC 61508, Journal of the System Safety Society 35 (1) (1999).

Reduce Risk with a State-of-the-Art Safety Instrumented System, ARC White Paper; ARC Advisory Group, September 2004.

Safety Integrity Level IEC 61508/61511 Catalog; PepperlþFuchs.

The Concepts of IEC 61508; an Overview and Analysis, RVS; University of Bielefield, 2005.

CHAPTER

Safety Instrumentation Functions and System (Including Fire and Gas System)

VII

1.0 SAFETY INSTRUMENTED FUNCTION, SAFETY INSTRUMENTED SYSTEM, AND SAFETY INTEGRITY LEVEL DISCUSSIONS

It is interesting to note the heading of the chapter! It is named like this to indicate that *safety instrumented functions (SIFs—in plural) constitute a safety instrumented system (SIS)*. From the discussions in previous chapters, readers have come across varieties of definitions of accidents, hazards, and risks from different perspectives. Also there were a few things common such as the accidents are not always negative, and there are always aim to avoid accidents so that there is not loss to the system, personnel, and environment. At the back of the mind, we always try to develop and incorporate some things to reduce risks and to maintain work safety. Therefore, safety work involves some activities, measures, and techniques, which can contribute or help to reductions in losses in different forms, and human injury or fatality. There are quite a good number of elements involved in work safety; these include, but are not limited to the following:

- Types of consequences to include:
  - Personnel: Injury, fatality, mental effect, reduced health
  - Environmental damage: Contamination, exposure
  - Function: Total/partial loss, production loss, machinery loss, risky operation
  - Economic loss: Direct, indirect loss of costly machine, property
  - Others: Loss of reputation, goodwill, opportunity
- Responsibility: Mainly includes:
  - Persons: Design, operation, services, victims, and others
  - Community suffering impact
  - Organization: Directly, policy, planning, etc.
  - Governing bodies: Standardization bodies, regulators
  - Interested bodies, for example, labor union, consumer forum
  - Other units: Researching bodies, statistical bodies, etc.

Plant Hazard Analysis and Safety Instrumentation Systems.

- Arena: Staying and performing activities:
  - Workplace
  - Caretaking
  - Education
- Sources of risk: Refer to Chapter I
- Action taken:
  - Political
  - Social
  - Legal/legislature
  - System analysis
  - Risk management (main interest area)
- Others

1.0.1 *Risk Reduction*

At this point, the main interest is to go about to see how, with help of risk management, the safety work is ensured.

Fig. VII/1.0.1-1 shows various methods of risk reduction in a common figure to include all risk reduction methods. Here, SIS is of main concern to us, so it is shown separately (in dark box). SISs are one of the most commonly used, engineered safeguard systems offering good flexibility to the designers. On account of safety life cycle requirements of IEC 61508/61511, and for better SIS design, experts need to analyze the risk associated with process under control at the beginning. SISs are risk-based systems. When in the subject, it is better to address the first barrier, then to SIFs. Barrier functions are planned for prevention, regulation, and mitigation of undesired events. In safety barriers, such barrier functions are used to combat undesired events. A safety function could be a technical or organizational function, human action, or a combination of them, used to reduce risks. Therefore, safety functions are a type of barrier.

1.0.2 *Safety Barriers for Risks*

These barriers could be conceived simply for a fire situation:

Several safety barriers for fire include:

- Smoke detection system
- Heat/thermal detection system
- Exit signal/exit gate

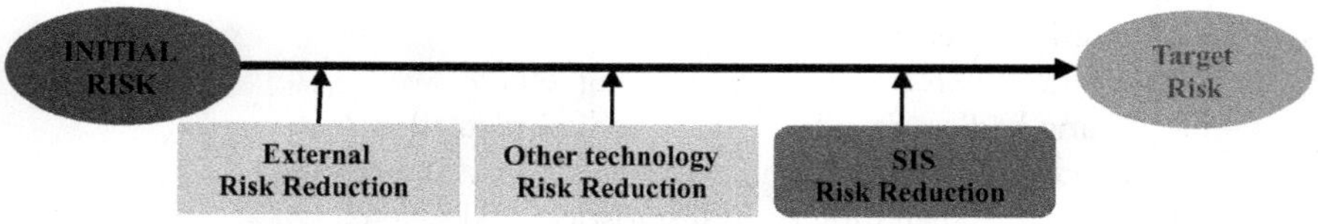

FIGURE VII/1.0.1-1

Typical risk reduction methods.

- Manual call/alarm
- Smoke control—air handling
- Fire door/fire wall
- Sprinkler system
- Evacuation/emergency plan

Out of these, the first four are detection and alerting safety barriers, whereas last four are protection type safety barriers. So, from here it could be argued that these safety barriers can be classified into various forms such as:

- Active/passive barrier
- Proactive/reactive
- High-demand/low-demand
- Technical, human, or organizational action

All of these can also be seen from the examples of fire (e.g., detection parts are proactive whereas protection parts are reactive). In fact, the safety barriers can be categorized according to their influence of safety. Safety barriers need to be evaluated before selection. These safety barriers actually are provided by layers of protection. From the discussions on layers of protection analysis (LOPA), it is known that safety barriers or independent protection layer (IPL) can be credited with risk reduction if they are:

- Effective in preventing the consequence
- Independent of the initiating event
- Independent of other credited independent protection layers (IPLs) for a given scenario
- Auditable

1.0.3 *Safety Instrumented System Discussions*

SIS is one of such safety barriers, which may be applied in combination with other technology or external risk reduction. The discussions so far were generalized in nature. Now it is time to shift the focus towards the standards which deal with electrical/electronic/programmable electronics (E/E/PEs). In both the standards IEC 61508 and 61511, safety functions or functional safety related to E/E/PEs are not always seen in isolation but in conjunction with other technology and external risk reductions. According to Clause 3.2.68 of IEC 61511-1:2003, safety function is defined as, "function to be implemented by an SIS, other technology safety related system or external risk, reduction facilities, which is intended to achieve or maintain a safe state for the process, with respect to a specific hazardous event." The reader might have noted from the discussions on IEC 61511 in Chapter VI that in the standard SIS and SIF have been used interchangeably. One more definition from IEC 61511-1:2003 is also of interest for coming to the discussions on SIF. According to IEC 61511-1:2003, Clause 3.2.71, "safety instrumented function (SIF): safety function with a specified safety integrity level which is necessary to achieve functional safety and which can be either a safety instrumented protection function or a safety

According to IEC 61511, *safety instrumented control function* stands for safety instrumented function with a specified SIL operating in continuous mode which is necessary to prevent a hazardous condition from arising and/or to mitigate its consequences. *Safety instrumented control system* is instrumented system used to implement one or more safety instrumented control functions. Also *safety instrumented system (SIS)* is instrumented system used to implement one or more safety instrumented functions (SIF). An SIS is composed of any combination of sensor(s), logic solvers, and final element(s).

FIGURE VII/1.0.3-1

Safety instrumented system details as per IEC 61511-1:2003.

instrumented control function." So, there are basically two types of SIFs, viz. the first one is: *safety instrumented protection function*, — operating in the demand mode, and other one is: *safety instrumented control function* (see Fig. VII/1.0.3-1), — operating in the continuous mode.

From the preceding discussions, it transpires that there are two kinds of safety functions; on-demand and continuous controls. Let's the examine a boiler/furnace fire rate that is controlled as per demand continuously to make sure that the furnace environment is always air-enriched, and the air—fuel ratio is maintained. This is safety instrumented control function, and could be a part of basic plant control system (BPCS) also. On the other, hand flame out (total) in a boiler or furnace could cause release of toxic and inflammable gas like CO. This is specific hazard, and there is demand on safety instrumented function (part of a system) through which the automatic protective interlock will trip or close the main fuel valve to stop any inflow of fuel to the furnace, and brings the system to a safe state. This action will initiate master fuel relay (MFR) trip. Another important issue here is that corresponding to each hazard, there shall be one specific SIF, for example, closure of main fuel in the event of loss of flame is one SIF; now, if there is very low feed water flow, or very low drum water level, then through automatic protection interlock, MFR will be tripped. From here, one can conclude that SIS consists of a number of SIFs which are included for each set of hazards. So far as MFR is concerned, all separate SIFs have the same result (tripping of MFR), which in turn operates different actuators to bring the system to safe state. By definition, each SIF needs to have one SIL and depends on amount of risk reduction by the SIF for the particular hazard. The SIL is selected based on the risk posed by the hazard. From previous chapters, it is well known that risk is constituted by the consequence and frequency of occurrence of the hazard. Therefore there will be different SIL for different SIFs. Obviously highest SIL will result safer SIS! Now there could be some situations when an SIF may have multiple initiating cause scenarios. In such cases, the highest SIL corresponding to scenarios will be used. Refer to Fig. VII/1.0.3-2 for details about the interrelation amongst them.

In Fig. VII/1.0.3-2, instead of safe state, safer state has been indicated to mean that by SIF or SIL, etc., one may reach to an acceptable hazard state. Now it is time to look a little deeper into all of these terms and their interrelations in the subsequent clauses with special reference to applicable international standards *IEC 61508/IEC 61511* (ISA 84 included).

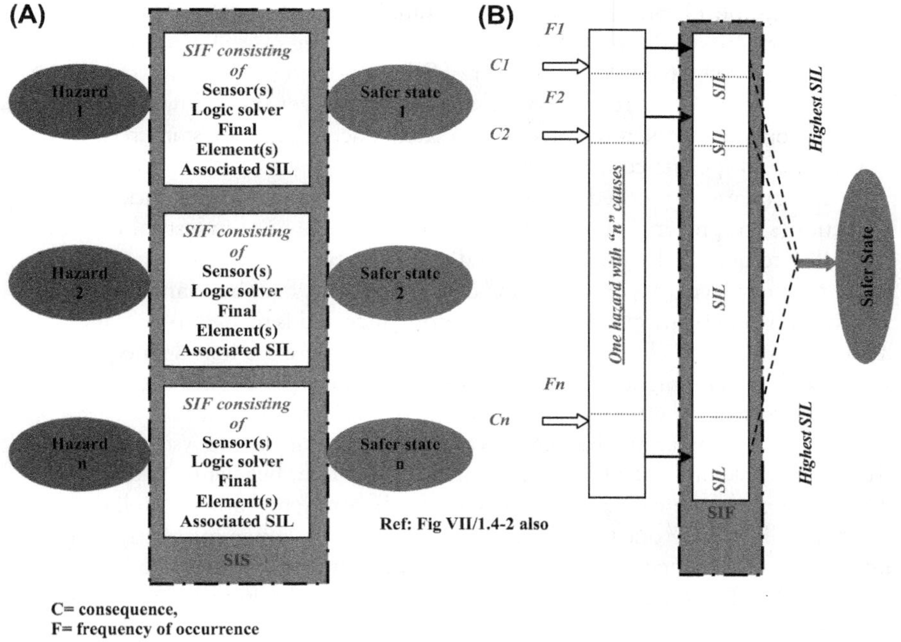

FIGURE VII/1.0.3-2

Relationship of SIS, SIF, and SIL. (A) SIS structure, (B) SIL Selection.

1.1 Safety Instrumented Function and Safety Instrumented System Functional Aspects – Discussions

When first published, many automation professionals felt confused as to what the terms SIF and SIS really meant. Definitions as per IEC have already been elaborated in previous clauses. However, even the standard uses the term SIS and SIF somewhat interchangeably in places. Therefore it will be prudent to look into these terms a little in detail.

1.1.1 Various Terms with Application Notes

- Safety functions: Safety functions are derived from the hazard analysis and provide actions taken by the safety system specific for each hazardous event. Safety functions are implemented through the combination of SIS or SIF, other technology, and external reduction facilities. Another similar term, "functional safety," when associated with SIF, relates to E/E/PEs and is to describe the safety system dependent on the correct functioning of constituent elements of SIF, viz. the logic solver, sensors, and final elements to attain the desired risk reduction level. Successful execution of SIF discussed later will result achievement of functional safety.
- Safety instrumented function: SIF stands for safety instrumented function. A set of control equipment intended to reduce the risk due to a *specific hazard*. SIF is designed to control, prevent, or mitigate a specific hazard and takes the process to

a tolerable risk level. SIF is composed of a combination of sensors, logic solvers, and final elements with an assigned SIL level depending on the amount of risk that needs to be reduced. A number of SIFs constitutes SIS as shown in Fig. VII/1.0.3-2. Some issues related to SIF shall mainly cover the following:

- SIF automatically takes the process to a safer or acceptable risk state in case of violation of specific condition(s).
- Permit process to proceed in a safe manner under permissive conditions.
- Action(s) to prevent/mitigate the consequences related to industrial hazard.
- Ability to detect, decide, and act as dictated by SIL of the function.

• Safety integrity level (SIL): As per IEC 61511, each SIF shall have an associated SIL, which is a measure of safety system performance and is related to probability of failure on demand (PFD) for the associated SIF. There are four defined SILs: SIL -1, 2, 3, 4. The higher the SIL number, the lower the PFD for the safety system, indicating a better system performance. Also, it has been found that higher the SIL number, the higher the cost and complexity of the system will be. SIL is applicable and calculated for to entire SIF system, but *not* on individual products or components. As told earlier, each SIF is assigned an SIL. Reliability and availability of SIF due to SIL are achieved by design, design installation, and testing. SIL is also dependent on architectural constraints—detailed out in Chapter VIII. The following [2] points on SIL are worth noting:

- SIL 0/none: lowest risk
- SIL 1: 95% of the SIFs
- SIL 2: 5% of SIFs
- SIL 3: <1% (off-shore platforms/nuclear)
- SIL 4: highest risk (nuclear industry)

• Safety instrumented system (SIS): SIS is meant to prevent, control, or mitigate hazardous events and take the process to a safe state when predetermined conditions are violated. An SIS can be one or more SIFs, which is composed of a combination of sensors, logic solvers, and final elements. Other common terms for SISs are safety interlock systems, emergency shutdown (ESD) systems, and safety shutdown systems (SSDs). So, SIS is used as a protection layer between the hazards of the process and the public. SIS or SIF is extremely important when there is no other non-instrumented way of adequately eliminating or mitigating process risks. As per recommendations of standards IEC 61511:2003 (or ANSI/ISA-84.00.01-2004), a multi-disciplinary team approach following the safety life cycle, conducts hazard analysis, develops layers of protections, and implements an SIS when hazardous events cannot be controlled, prevented, or mitigated adequately by non-instrumented means.

• Proof testing interval: To achieve an SIL during the design phase calls for many design *assumptions* to predict the SIF performance. Operating conditions will affect these assumptions and hence SIF/SIS integrity level. SIS-specific monitoring and proof tests must be conducted as per managed procedures (see HSE.UK for guidelines) that ensure the tests are done correctly, consistently, and safely to verify all design assumptions at regular intervals referred to as *proof test interval*. Periodic proof testing of the entire system, including sensors, logic solvers, and final elements, is necessary to *ensure* the continued and demonstrable integrity of each SIF. The testing frequency is variable for each SIS depending on the technology, system architecture, and target SIL. The proof-test interval is an important component of the PFD calculation for the system. Failed devices must be properly bypassed, repaired, and then returned to a fully operational state and proof testing. It is more likely that higher the SIL, the more frequent will be proof testing.

- Competency structure for functional safety: During the discussions on IEC standards, it was seen that there should be a responsible authority to implement the safety principles. For organizations, a company safety authority (CSA) is charged with this responsibility. Core principles endorsed by senior management are collectively referred to as "strategic competency principles," which are based on a multi-tiered approach to demonstrating functional safety capability. A typical competency pyramid is shown in Fig. VI/1.1-1. There are four strategic competency principles:
  - Benchmark current practice: Assessment against IEC 61508 and IEC 61511 to establish the scope of the task.
  - Implement safety standards: Specification and implementation of program of work to achieve accredited certifications for each of the organization for functional safety management systems.

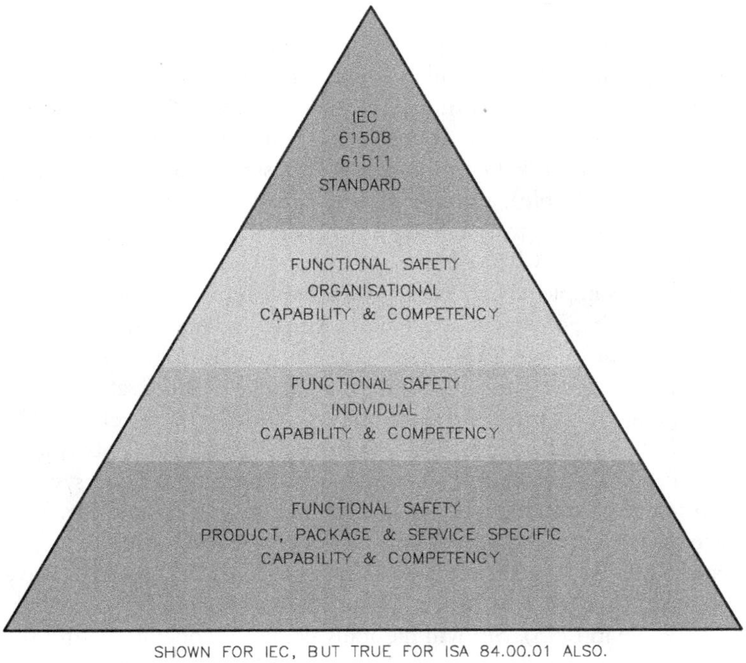

FIGURE VII/1.1.1-1

Competency structure for functional safety.

Figure has been developed inspired by ABB doc (Functional Safety Handbook, ABB Limited, 2008). Courtesy ABB.

- Establish individual competency: Safety engineers of the organization to get certified functional safety engineer status through competent authority following their safety program.
- Manage third party integrators (if any)

  The key points/issues of these strategic competency principles [3] are:
  - To use certified products
  - To employ competent (certified) persons
  - To implement safety systems through the certified organization

    These are stated here to understand by the manufacturers and system integrators how to get the accredited certificates.

1.1.2 *Basic Principle Discussions*

The SIS is a part of instrumentation and control system which when detects that the process is out of control and there could be possibilities of hazards, it automatically returns the process to a safe state. It could be the last line - or near last line - of defense against a process hazard. Normally it is better to keep it out of BPCS and treat separately. However, there could be cases where these may be integrated with BPCS also (let the discussions start with separate SIS). The last line or near last line of defense is what differentiates an SIS from the BPCS, which is used for normal regulatory process control. Fig. VII/1.1.2-1 shows the relationship between BPCS and SIS with respect to the process.

From the preceding example, it is clear that each of the BPCS and SIS will consist of sensors, logic solvers, and final elements. Both need to work independently. Usually BPCS regulates the process to keep the parameter of interest to the desired set point. If for some reason, the system is going out of control range, then SIS will act upon to bring the process back to the safe state. Here, one thing to be kept in mind that safe state could be an ALARP (as low as reasonably practicable), that is, in safe operating state, or in some cases, system shutdown. So, safe state is defined in that sense. This is explained with the help of an example in Fig. VII/1.1.2-2.

In the preceding example, flammable fluid is taken out of 10 m high tank, which is being fed by a line having one control valve to regulate and maintain the level in the tank, say, at 5 m. In case of high levels, there is an alarm at the 6 m level. Now there is the possibility of explosion if the tank is full. So, there is pressure safety valve AND also there is a separate level loop with set point at a point higher level than alarm set point — say at 7 m — to close the shut off valve in supply line. This is a classical example of LOPA with different IPLs, which are used to push the systems from process risk to tolerable risks, as typically shown in Fig. V/4.0-3 and Fig. VII/1.0-1. Here as shown, there are two different controls one is a modulating controls under BPCS to regulate the flow control valve (2nd in the inflow line, i.e., RHS valve). On the other hand, SIS will shutdown the inflow by closing the shut off valve (1st in the inflow line, i.e., valve in LHS). SIS will naturally come into operation after the level is above BPCS set point as well as the alarm set point. So, here SIS will act as next-to-last defense, which is the operation of plant shutdown valve (PSV). Here, IPLs are BPCS, alarm, SIS, and PSV. Independent level control shutdown function is implemented in an SIS. Here it is worth noting that the components, which include the sensor components, the logic solvers, as well as the final control elements, used in

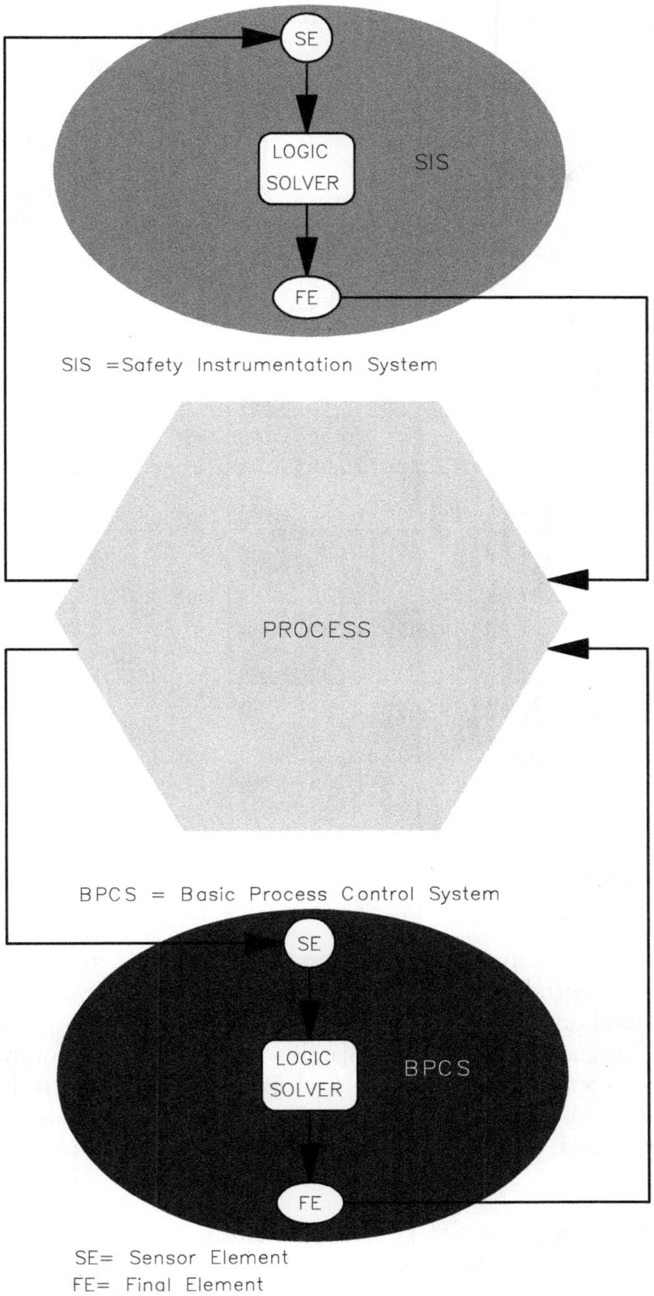

FIGURE VII/1.1.2-1

BPCS and SIS for process.

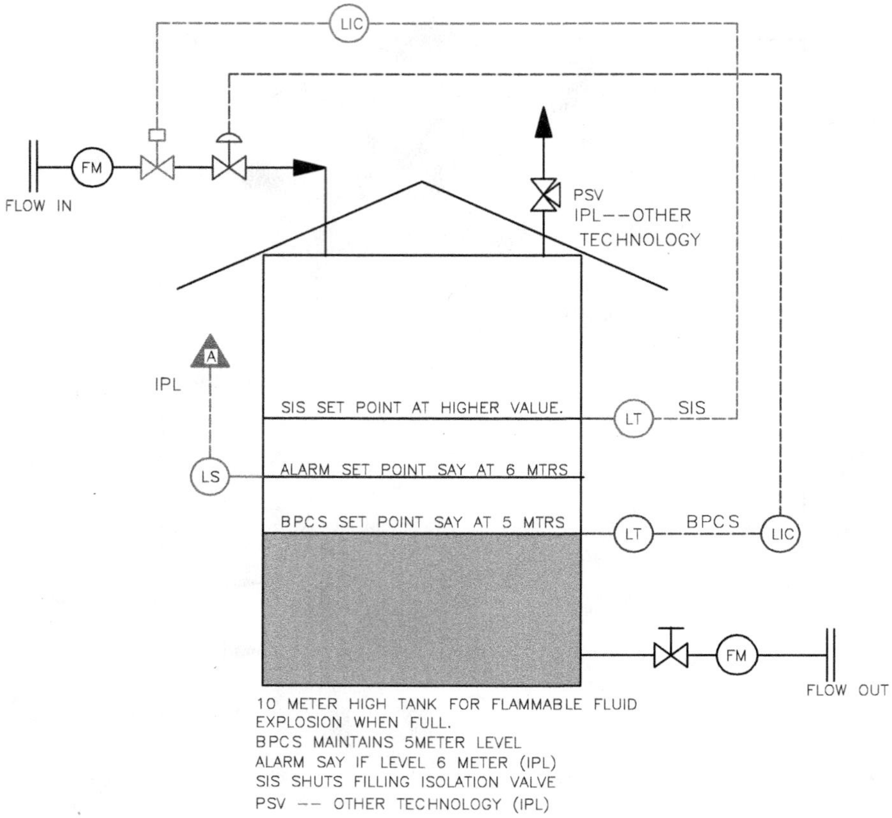

FIGURE VII/1.1.2-2

Example of BPCS and SIS in process.

the BPCS and the SIS are separate, physically and functionally — hence independent. SISs are *generally* independent (both physically and functionally) from the BPCS to ensure that any condition, which might result in an out-of-control process parameter in the BPCS, is safeguarded by the SIS, regardless of the functioning of the BPCS, that is, even if BPCS fails SIS can provide a safeguard. Another issue is that SIS is acting on demand. In the given example even if operator misses alarm there is safeguard against hazard. SIS is not necessarily programmable only; it could be non-programmable also.

1.1.3 Scope, Boundary and Safety Requirement Specification

In various plant analyses, it has been seen that prior to start of the same, it is necessary to clearly define the scope and boundary for the study. Similarly, the gap between the organization's functional management system and the requirements as per the standard needs to be assessed. Only after the gap assessment, proper implementation of standard is possible. In this connection, Clause 6 of IEC 61508-1:2010 and Clause 5 of IEC 61511-1:2003 may be referenced. It is clearly stated that the two basic objectives are to specify the responsibilities and activities. Clause 6.2.17 of IEC 61508-1:2010 calls for quality management also. So, some of the pre-requisition towards such quality checks shall include, but is not limited to, the following:

- Hazard analysis and risk assessment previously done to ensure systematic development of requirements of specifications which is a key deliverable.
- The hardware for SIS (sensors and final elements also applicable) consisting of logic solver, I/O termination cabinet, power supplies, and communication module.
- The subsystem used including logic solver and accessories, needs to be third-party certified as per standard.
- Safety integrity data (PFD-hardware fault tolerance), reliability data for all devices with clear definition of the same for logic solver in the manual.
- Overall hardware architecture as per specification and software is limited variability language type (Ref: Chapter VI) with suitable library as certified functional blocks which are protected against unauthorized modification.
- Coding standard: as per IEC 61131-3 standard.
- Development tool support and facility for development application software with functional blocks.
- Development environment to facilitate creation of version and management facilities.
- Facility to integrate such as downloading and data configuration as per requirements.

SIS safety requirements necessary at early bid proposal stage, are well-defined in Clause 10.3 of 61511 part 1. These requirements offer significant benefits to both the parties (e.g., party with responsibility to develop SRS as well as the party for integration process) needing SRS. Salient requirements from 10.3.1 of 61511 part 1 are listed as follows:

- Description of SIF
- Addressing common cause issues
- Safe state for each SIF
- Source of demand with demand rate for each SIF
- Proof test interval
- Response time requirements
- SIL with mode of operation
- SIS measurement and trip point and output action
- Functional relation between process I/O and logic solver
- Manual shutdown requirements
- Energize/de-energize to trip requirements
- Resetting after SIS shutdown
- Allowable spurious trip
- Failure mode and desired SIS response
- Interface between SIS and other systems

IEC 61508 does not make a direct *correlation* [3] with the required level of rigor and competence. However, it has been found that more rigorous competences are necessary in the event of any one.

- Greater consequence
- Higher SIL
- Novelty of design procedure/new design

Therefore, it is better to go for specification of competence level for a well-organized company. "Knowledge," "experience," "training," and "qualification" are major attributes for competence level to be considered. The reader may refer Clause 8.2.9, Table 4, and 5 of Clause 8.2.18 IEC 61508-1:2010 for further details in this regard. From the discussions so far, it has been found that various types of failure play a great role in assessing SIF or SIS. So, better to address various kinds of failures and how to address them!

1.2 Failure and Failure Classes

One of the major reasons for the need to develop standards IEC 61508/IEC 61511 (ANSI/ISA 84.00.01-2004) was to devise some means to minimize the **propagation** of device failure through design, operating, inspection, and maintenance practices into system. Naturally this calls for reader to understand various types of failures systems normally need to face with special reference to E/E/PEs.

1.2.1 Introductory Discussions

All plant/process controls have a range within which the control systems would operate. However, there may be situations or emergencies when systems cross such ranges (e.g., when there is a catastrophe in a grid, it is almost impossible to keep the governor control of any connected sets within given range—one similarity) then, hazardous conditions may occur. Obviously in such situations, there may be failure in one part and there will be high possibility that this propagates and may give rise to a hazardous situation. Therefore it is needless to say that SIS must address all such situations right from the beginning, so that there will lesser consequences to human life, environment, and assets. From IEC 61508/61511 discussions, it is clear that after identification of SIF, the risk reduction allocation is related to its SIL so that these two factors can establish a benchmark for system design, production or operation, and other management functions to be used throughout the SIF life cycle. In industrial application basically there are two major types of failure, viz. random failure and systematic failure. The said benchmark defines the (1) minimum performance against random failures, and (2) the rigor of the protective management to reduce the potential for systematic errors. A few common causes for random and systematic failures are listed as follows, and for further details ISA may be referenced [10]:

- Random fault
  - Stress (over)
  - Ageing

- Systematic fault
  - SIF deign error
  - Hardware error in design or implementation
  - Software error
  - Error in management of change (MOC)
  - Installation error
  - Human interaction error

There is another kind of failure, popularly known as *common cause failure* (CCF), which could be either random or systematic, and is responsible for failure of multiple devices, systems, etc. People often use redundant transmitters (say two of three) for SIF protection purposes. As long two of three transmitters are working, then it is okay and the system can work to take care of SIF; but if two transmitters fail then the situation is different and bad. Such kinds of failures are often found when the device specifications are violated, in multiple devices simultaneous failures can occur (and redundant instruments may fail). This is known as a CCF. In this competitive market, plant designers are very keen in designing the system optimizing system design, which calls for simplification and cost-effectiveness. Such optimizations often make the systems with lesser safety margin. "As less and less safety margin is built into the design, greater importance is placed on the precision of the predictive failure calculation" [4]. So, device failure is no longer restricted to benchmark for performance, but is a **critical** parameter for process design.

Characteristics of each SIF device, the interactions among its components, and the properties of the system itself are responsible for SIF performance. The following issues are responsible for performance and failure probability of SIF:

- Random failure rate of devices
- Design parameter
- Systematic failure
- Common cause failure (Random failure of multiple devices)

Independent failures of devices are analyzed to identify roots, and it is possible to reduce reoccurrence of failure. Random failures are physical failure while mostly systematic failures (design failure) occur due to human error, mainly hence a local issue. CCFs are application and/or installation specific. In IEC 61511, adapts different strategies to handle various failure types to make SIF devices are useful in the life cycle as long as these are installed and managed as per design basis and operating procedure. Now, the focus shall be on failure characteristics to conceive how these affect SIF performances.

1.2.2 Bathtub Curve Discussions

The lifetime of an entire population of products often is graphically represented by a set of curves collectively called the *bathtub curve*. Bathtub curve has been depicted in Fig. VII/1.2.2-1. The bathtub curve consists of three periods. First is an infant mortality or burnt-in period with a decreasing failure rate showing early-life failure. These

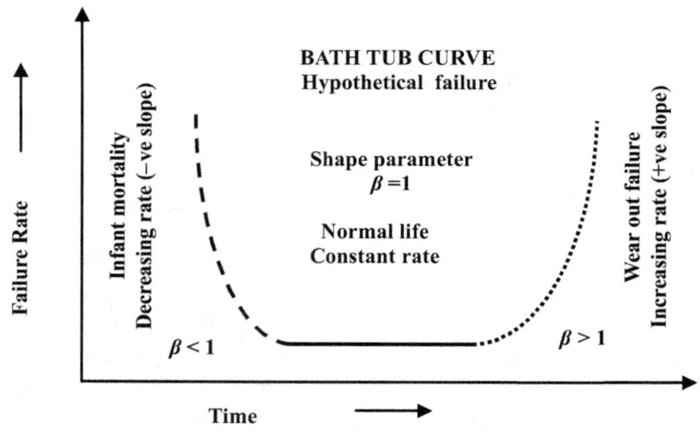

FIGURE VII/1.2.2-1

Bathtub curve for failure rate.

early life failures are known as burnt-in or infant mortality failures occur *on account of defects such as design defect, material defect, or error in assembly/manufacturing flaws in the components*. These could be detected during burn-in and function testing prior to release from production. *Device handling during shipment, unpacking, storage, transportation to work site, and installation may be responsible for infant mortality failure*. The second part is a relatively constant failure rate referred to as normal life period or "useful life." Normal or useful life failures mainly occur on account of random events with an increase of the stress on the device beyond its physical limits. Here it is worth noting that in many cases there are failures during normal life due to infant mortality failures. This is a generalized curve, not calibrated for any particular product life, with the relative failure rate of population of products over time.

Individual units may have different failure rates; some may fail during early stage while some may survive until normal life or even during wear out stage. Failure categories of different individual units have been shown in Fig. VII/1.2.2-2 (to show variations).

The third part is the wear-out period that exhibits an increasing failure rate. During the wear-out failure stage, the failure rate is positive, meaning that it is dominant over the useful life failure rate, showing the ending normal life of the device. Wear-out may be due to fatigue or depletion of materials (such as lubrication depletion in bearings) in mechanical devices. Electromechanical devices may have a gradual increase in failure rate, and depends on how well their mechanical components are inspected and maintained. In cases of electronic/programmable devices, these rate of changes may be very sharp due to the large number of aging components. A product's useful life is limited by its *shortest-lived* component. It is the manufacturer's responsibility to ensure proper functioning of the device under specified conditions, through the intended product life. The infant mortality period is crucial

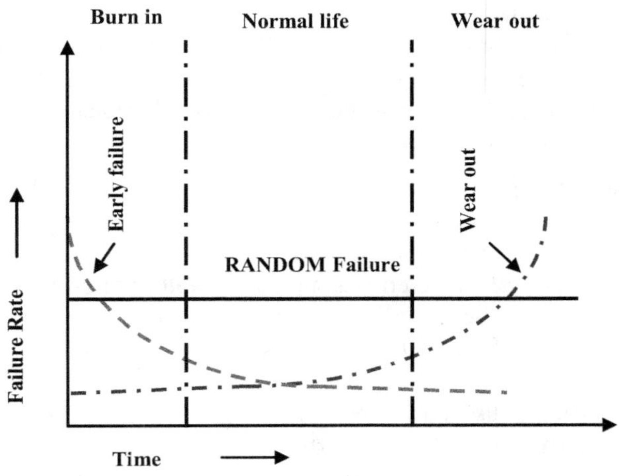

FIGURE VII/1.2.2-2

Typical component failure curves.

to the manufacturer/marketing company, on account of customer dissatisfaction (number of failures in short time), and warranty expense. In view of these and constant failure rate in normal life, it is desirable that the bottom of the bathtub to be as low as possible. Major issues related to the bathtub curve have been listed as follows:

- Infant mortality
  - Defects elimination
  - Appropriate and adequate specifications
  - Sufficient design tolerance
  - Sufficient component de-rating
  - Early stress testing — design evaluation

Some typical testing methods are, for example: highly accelerated life test (HALT), highly accelerated stress test (HAST), and highly accelerated stress audit (HASA) adapted in different industries. In electronic industries; high temperature test and temperature cycling tests are very common for checking acceleration. Also, electronic items are tested for higher than operating voltage to find the accelerating factors.

- Wear-out:
  - Electronic items wear out after a reasonable period of operation.
  - In mechanical assembly, it is often seen that a particular component fails much sooner than expected life, but when replaced assembly can a last longer period.
    A Weibull distribution can be used to assess actual situations so, in the following clause a brief discussion on the same has been presented.

1.2.3 Weibull Distribution

One of the most popular statistical methods is Weibull distribution for failure analysis. It is very useful for failure analysis. Waloddi Weibull invented the Weibull distribution in 1937 and claimed that his distribution applied to a wide range of problems. Salient issues on the same are as follows:

- Outline: Life of product population is predicted with the help of "Weibull" analysis in which a statistical distribution is attempted to fit into life data from a representative sample of units. Then same data set can be used for estimation of important life parameters/characteristics such as reliability or probability of failure at a specific time, the mean life, and the failure rate. For Weibull data analysis, the following information are required:
  - Product life data.
  - Selection of fitting a distribution pattern based on applicable parameters.
  - Estimate of the parameters for fitting the distribution to the data.
  - Results and plots generation for estimation of product life characteristics.
- Weibull analysis scope and data types handling include but are not limited to the following:
  - Plotting the data
  - Interpreting the plot
  - Failure forecasting
  - Spare parts forecasting
  - Corrective action planning
  - Maintenance planning
  - Replacement strategies
  - Warranty analysis
  - Controlling production processes
  - Supports CAD\CAM
  - Censored or suspended data
  - Mixtures of failure modes
  - Extremely small samples (as small as one failure)
  - Early data missing
- Failure types of Weibull analysis include but are not limited to the following areas:
  - Development activity
  - Production/service
  - Mechanical/electronic devices
  - Materials and human failures
  - Lightning strikes
  - Human error
  - Quality control
  - Design deficiencies
  - Defective material
  - Warranty claims

- Weibull distribution discussions: The Weibull model can be single, double, triple, and/or mixed type. Various types of mathematical modeling can be used. The typical, basic formula has been presented as follows:

$$f(t) = \frac{\beta}{\eta}\left(\frac{t-\gamma}{\eta}\right)^{(\beta-1)} e^{-\left(\frac{t-\gamma}{\eta}\right)\beta} \qquad \text{(VII/1.2.3-1)}$$

Where $f(t)$ stands for probability density function, $f(t) > 0$. From this it is clear that it is a case of three-parameter model. Shape and locations can be varied in accordance with selection of range of parameter for various variables viz. η (>0), defines the bulk of the distribution parameter (scale), β (>0), determines shape (also often called Weibull slope) and γ, defines the location of the distribution in time and has wide variation.

A Weibull distribution is a generalized distribution, as each type of product population provides different types of information about the life of the product, and different life data analysis method may vary. It is quite normal that life will be in unit of time but it may not be the case always, for example, life of lubrication in automobile is dictated by kilometers running of the vehicle and not in time units. Time is a common measure of life, so often referred to as "times-to-failure." There could be different (e.g., three) types of life data, each type provides different information about the life of the product; "with 'complete data,' the exact time-to-failure for the unit is known (e.g., the unit failed at 100 h of operation). With 'suspended' or 'right censored' data, the unit operated successfully for a known period of time and then continued (or could have continued) to operate for an additional unknown period of time (e.g., the unit was still operating at 100 h of operation).With 'interval' and 'left censored' data, the exact time-to-failure is unknown but it falls within a known time range [5]". Based on statistical methods, characteristic parameters are calculated to fit a life distribution to a particular data set. For further details on this generalized statistical distribution, a standard book on statistics may be referred to. After the same is done one can use to get the following results:

  - Mean life
  - Failure rate
  - Reliable life (reliability parameter)
  - Reliability given time
  - Probability plot
  - Failure rate versus time
  - PFD plot
- The following are major advantages of Weibull distribution:
  - It can provide reasonably accurate failure analysis and failure forecasts with extremely small samples.
  - It provides a simple and useful graphical plot of the failure data.
  - Weibull analysis may be useful even with inadequacies in the data.
  - Identification of mixtures of failures modes.
  - Investigation of alternate aging parameters,
  - Construction of a Weibull curve when no failures have occurred.

1.2.4 Failure Types and Random Failure

As stated earlier, there are two types of failure; random failure and systematic failure. There is another kind of failure called common cause failure. Common cause failure may be random or systematic failure. However, in ISA -TR84.00.02-2002 - part 1, this has been categorized under physical failure. However, in the note CCF of systematic failure has been mentioned. In this sub-clause, discussions shall be restricted to random failure only. The other two types will be separately in the subsequent sub-clauses. Random failures are physical failures. Random failure is characterized by unpredictable failure of device or component such as electronic card failure. As per ISA-TR84.00.02-2002 - part 1, "a failure is classified as physical when some physical stress (or combination of stressors) exceeds the capability of the installed components." Random failures normally exhibits following characteristic features:

- Physical stress due to:
  - Unusual/abnormal process conditions
  - Corrosion or material deterioration
  - Adverse environmental conditions
- Occurrence any time during life
- May not be detectable during accepting stage
- No particular pattern for occurrence

More and more information needs to be collected to identify the trend which can be used to minimize random and systematic failures. Random failures can be detected by diagnostics internal to the devices, externally-configured diagnostics, and proof tests. Usually PEs includes a number of such *diagnostic* features to combat random failures. Device redundancy is most common form to back each other up if a failure in one device occurs (provided there is no CCF discussed later). Redundancy often is used with voting schemes and diagnostics to help verify which between two or among several devices is operating correctly in the event of a fault. Redundancy and fault tolerant subsystems are used to reduce failure in devices to affect SIF. Diagnostics help in identifying the presence of current or imminent faults communicates the information to O&M personnel to avoid deterioration of SIS performance. Diagnostic programs are often run in a given interval, which is determined as per manufacturer's recommendations, good engineering practice, and system needs. It has been found that observed random failure rate often higher than actual random failure on account of interference from other systematic failures also.

1.2.5 Systematic Failure

Systematic faults occur due a combination of conditions resulting in a reproducible failure of the system, and are most often attributable to software issues in programmable safety systems. This failure may be a result of some error in design, operation or production process, installation and/or maintenance. Improper implementation of MOC at any stage could be responsible for systematic failure also. Device manufacturing errors can be addressed by diversity; this increases the SIF complexity. Diversity can be applied to sensor, I/O technologies, control and software platforms, and even product development teams. Incorrect specification, implementation, production or operational process, or installation and/or maintenance errors, which could be the cause of systematic failure, cannot be addressed by diversity discussed above. When complexities increase in SIS, probabilities of occurrence of such failure are more. Also, as the complexity increases, chances of error detection decreases. Therefore systematic failures bear some direct relationship with SIF *complexity*. In case of new technology, obviously there will be possibilities for many unknown failures, such as interconnectivity, communication software compatibility, etc. So, it is needless to tell that the chances of systematic failure will be high. Systematic failures include many types of errors, such as:

- Specification: Inadequate or incorrect design basis or inaccurate software specification.
- Manufacturing: Defects, hardware or software errors contributed during manufacturing or system operation.
- Implementation: Incorrect installation, programming error, interface issues, and/or not following standard.
- O&M: Inspection/testing flaws.

The preceding cases, there are some errors in any stage and are detectable. However, it may so happen due to complex combined operation that it is sometimes difficult to follow the failure pattern by testing. Hence, it is unpredictable, and systematic failure will again occur when matching conditions are forthcoming. In such cases these become intrinsic to the device hence associated SIF and hence associated SIS.

1.2.6 Common Cause Failure

This needs special attention in the case of E/E/PEs and for associated SIS. According to IEC 61511, CCF is defined as, "*a failure which is the result of one or more events, causing failures of two or more separate channels in a multiple channel system, leading to a system failure*." Also, as per IEC 61511(1), life cycle recommendations for the devices, systems, or protection layers need to be *assessed* for independence and the *potential* for CCF. So, understanding of independence and common cause are very important as at times they are interrelated. Any lack of independence will result potential for a CCF. Similarly, when a common cause is identified, it means there is lack of independence. CCFs mainly occur due to external influences, and could cause more than one system component or layer of protection to fail. So, CCFs could develop serious threats to reliability of SIS, and could be responsible for simultaneous failures of redundant components and safety barriers. Common cause failure is represented by β ($\beta = \lambda_c/\lambda$; where λ_c stands for CCF failure and λ stands for total failure meaning β is factor for CCF. Often, β factor model is used to determine CCF. On account of CCFs, SIS may fail to perform functional duties demanded by a process. All CCFs have the potential to reduce the SIF—SIS performance; but they are treated in different ways, and also strategies for coping up with the situation may be different depending on the nature of the failure (e.g., systematic or random). Diversity is most often cited as a means to counter common cause and systematic failures. Here it is interesting to note from an example; in any application, two pressure transmitters are used in redundant mode each with different technology, say, one is working in capacitance principle and the other is working in piezoelectric principles, but both failed, that is, loss of signals due to diaphragm failure. Here there is diversity in transmitters- but they fail due to common mode failure which is a sub set of CCF, and in this case there may be error in design or selection of proper transmitter pressure range selections *(in this case diaphragm failed, even two technologies could not save as the pressure* ***range*** *selection was wrong!)*. From here, it is clear that CCFs must be identified during the design process and the potential impact on the SIS functionality must be understood. Functional testing and inspection are also important activities, especially when operating in low-demand mode for revealing SIS failure. These functional testing and inspections greatly affect CCFs, as CCFs can be identified and corrected by efficient testing and inspection. Inadequate procedures and human errors could result simultaneous failure of components in SIF/SIS. In IEC 61508/61511standards, one does not find any mention of the critical time aspect in their definition of CCF, but in nuclear power plant critical time between two consecutive inspections have been included and is important for CCF. There are two types of CCF to be addressed:

- Single points of failure; one malfunctioning device causes an SIF failure.
- Single events leading multiple failures in a redundant subsystem.

From experience it could be argued that CCF may be introduced in:

- Design stage:
  - Improper understanding of failure mechanism and response
  - Improper selection of hardware
- O&M stage due to:
  - Human error
  - Inadequate testing
- Environmental effect beyond design basis

CCFs have variations among the industries and there lot of disagreement among the experts regarding the actual definition and causes of CCFs. Some of the issues towards these related to instrumentation shall include but are not limited to the following:

- Incorrect calibration of sensor
- Stoppage of common point in process taps
- Improper selection
- Improper maintenance
- Environmental stress
- Serious process condition variations

Major factors attributable to CCFs shall include:

- Affected components are unable to perform as required
- Multiple failures within redundant configuration
- First in line failure
- Failure to occur within critical time interval
- Failure due to single underlining defect/phenomenon
- The effect of failure to leads to major disability to perform as required

There are mainly two major factors behind CCFs, and these are actual root cause and coupling issues. From international common cause failure data exchange (ICDE) classification of common causes (Nuclear Energy Agency—NEA, 2004/2011), one gets that the following are classes of root cause [6]:

- State of other components
- Inadequacy in design, manufacture, or construction
- Human actions
- Maintenance
- Internal to component
- Procedure inadequacy
- Abnormal environmental stress
- Others

Similarly, the classes of coupling factors shall include [6]:

- Same or similar hardware
  - Hardware design
  - System design
  - Hardware quality deficiency
- Same or similar operational conditions
  - Testing and maintenance schedule
  - Testing and maintenance procedure
  - Testing and maintenance staff
  - Operation procedure
  - Operational staff
- Same or similar environmental conditions:
  - Internal
  - External
- Others

Based on the root cause and classification of coupling, there are three separate sets of checklists for CCF avoidance, viz. preparation, execution, and restoration. There are cause-defense matrices available, which shall include CCF, root cause, coupling factor, defense alternatives impact, and cost, etc. For further details [1,6,7] may be referenced. With this discussion on failure types concluded, we move on to look at reliability issues.

1.3 Reliability Theory and Safety Instrumented System Issues (Including Reliability Block Diagram, Risk Graph)

Functionally, reliability engineering is responsible for the development of reliability requirements for the system and design of the system or product to meet the reliability requirements. Reliability engineering also establishes reliability program, and carry out analysis to monitor the actual reliability of the system or product throughout life cycle of the same. Reliability engineering is closely related to safety engineering and to system safety. It deals with the estimation and management of high levels of "lifetime" engineering uncertainty and risks of failure. Also, major areas covered in reliability engineering are [8]:

- Reliability strategies
- System or product design
- Failure modes and effects analysis
- Reliability modeling and estimation
- Reliability testing
- Quality assurance strategies
- Work management and execution
- Continuous improvement

No system is totally immune to failure; with SIS in operation one can expect that it will always provide safe conditions when a failure occurs. Reliability with respect to SISs can be conceived as its ability to command an output to a safe state on a process demand within desired response time without causing a spurious action such as nuisance process trip. With this basic idea, in the following clauses include brief discussions to present how both the issues could be addressed to get the best out of the two. At the beginning, it is better to understand meaning of various commonly used terms in connection with reliability.

1.3.1 Common Terms in Reliability and SIS with Explanations

A few selected commonly used terms frequently encountered in reliability engineering and SIS as per standards are:

- Failure rate: The total number of failures within an item population, divided by the total number of life units expended by that population, during a particular measurement interval under stated conditions. Failure rate can be considered as the frequency with which an engineered system or component fails, expressed in failures per unit of time. It is time-dependent. In this connection, for the definition of the same, from probability point of view, may be referred to in IEC 61508-4:2010.
- Dependability: A measure of the degree to which an item is operable and capable of performing its required function at any (random) time during a specified mission profile, given the availability at the start of the mission. In other words, it is a measure of RAM of a system, where RAM stands for reliability, availability, and maintainability. Each of these three terms has been defined separately in this clause.
- Availability: A measure of the degree to which an item is in the operable and committable state at the start of the mission, when the mission is called for at an unknown state. Availability can be conceived as the probability that the system is operating properly when it is requested for use. In terms of failure, availability is the *probability* that a system has *not* failed or is undergoing a *repair* action when it needs to be used. From this, one can infer that if a system has a *high availability* then it should also have a *high reliability*. So these two terms go hand in hand. The relationship between availability, maintainability, and reliability has been shown in Fig. VII/1.3.1-1.

RELIABILITY	MAINTAINABILITY	AVAILABILITY
DECREASING	CONSTANT	DECREASING
CONSTANT	CONSTANT	NO CHANGE
CONSTANT	INCREASING	INCREASING
CONSTANT	DECREASING	DECREASING
INCREASING	CONSTANT	INCREASING

FIGURE VII/1.3.1-1

Relationship among RAM members.

- Maintainability: The measure of the ability of an item to be retained or restored to specified condition when maintenance is performed by qualified personnel in a prescribed procedures and resources, at each prescribed level of maintenance and repair. Also, in simpler terms, it can be seen as a measure of ease and rapidity with which the system, equipment, or subsystem can be restored to operational condition following a failure. It depends on the design characteristic of the equipment or system provides other issues discussed at the beginning are made available. Relationship of availability, maintainability, and reliability has been shown in Fig. VII/1.3.1-1.
- Mean time between failure (MTBF): It is intended to measure only the time a system is available and operating. It is the mean number of live units (during which all parts of the item perform within their specified limits), during a particular measurement interval under stated conditions. It is average time excludes the time spent waiting for repair, being repaired, being re-qualified, and other downing events such as inspections and preventive maintenance, and so on [9]. Calculation basis of MTBF/MTTF is the same. **MTBF:** $\boldsymbol{\theta = T/F}$, where: θ, mean time between failure; T, total running time/cycles/miles/etc. during an investigation period for both failed and non-failed items; F, the total number of failures occurring during the investigation period.

- Mean time to failure (MTTF): This refers to reliability for non-repairable items: It is the expected time to failure for a *non-repairable system*.
- Mean time to repair: The average time required to troubleshoot and repair the equipment or system and return to normal operating condition. It is the sum of corrective maintenance times at any specified level of repair, divided by the total number of failures within an item repaired at that level, during a particular interval under stated conditions.
- RAM: Here RAM stands for *reliability*, *availability* and *maintainability*. These three factors collectively referred to as RAM are major attributes towards system design with great impact on sustainment, life cycle costs as well as system's ability to perform the intended mission and affect overall mission success.

  Relationship among the members of RAM has been shown in Fig. VII/1.3.1-1.
- Reliability: Reliability is the probability of zero failures of an item over a defined time interval (or mission). From the definition one can infer:
  - The probability of failure-free performance under stated conditions.
  - The probability that a system, equipment, or device can perform its intended function for a specified interval under stated conditions.

  Reliability represents the probability of components, parts, and systems to perform their required functions for a desired *period of time* without failure in *specified environments* with a *desired confidence*. Reliability accounts for the time in which operating system or device under specified conditions, may fail. The relationship between availability, maintainability, and reliability has been shown in Fig. VII/1.3.1-1, which is self-explanatory.
- Safe failure fraction and diagnostic coverage: Both of these terms are related to SIS and reliability, and both of them have been derived from IEC 61508-4:2010. Safe failure fraction is the *ratio* of average safe failure plus average dangerous failure detected *and* safe failure and dangerous failure. Diagnostic coverage stands for the *fraction* of dangerous failure detected by on line automatic tests. In order to understand, it is better to take the help of Fig. VII/1.3.1-2.

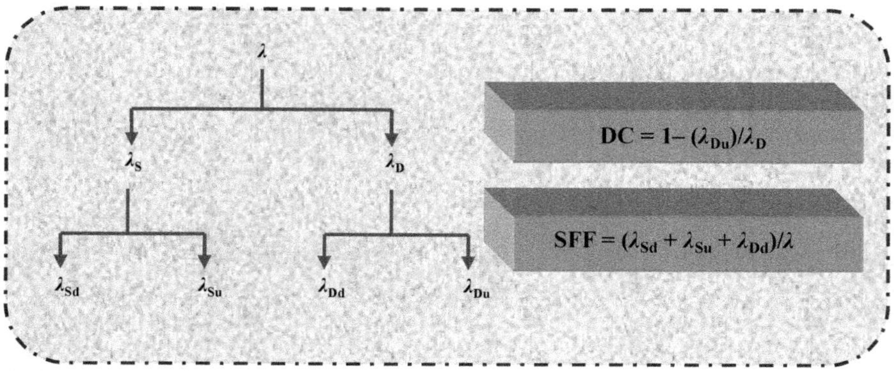

FIGURE VII/1.3.1-2

Failure rate safe failure fraction and diagnostic coverage with their relationship. For symbol legend refer to Table VII/1.3.1-1. Standard followed: IEC 61508:2010.

Table VII/1.3.1-1 ***Lists of Symbols for Failure Rate and Related Terms***

Various Terms	Symbols Used to Represent		
	Detected	Undetected	Others
Safe failure (λ_S)	λ_{Sd}	λ_{Su}	NA
Dangerous failure (λ_D)	λ_{Dd}	λ_{Du}	NA
Safe failure fraction	NA	NA	SFF
Diagnostic coverage	NA	NA	DC

These symbols have been used in Fig. VII/1.3.1-2.

Failure rate is represented by λ. There are two categories, viz. dangerous failure, where probability of safety action decreases and EUC (Appendix I) may not in safe state (λ_D), and safe failure, where EUC will be in safe state represented by (λ_S). Again each of these failure rate types has two subdivisions, viz. detected (d) and undetected (u). Various symbols used have been detailed out in Table VII/1.3.1-1.

- Reliability—utility: Uncertainty may have a very adverse impact. It is therefore necessary to have studied the reliability of product and services. Poor quality product or service may have serious impact to include but not limited to:
  - Safety
  - Property, environment, and personnel
  - Profit margin
  - Cost of repair and maintenance
  - Delay in deliverables
  - Customer confidence
  - Competitiveness
  - Reputation

The following are the major take off from reliability studies, which is very pertinent and important for SIS studies as both are coupled with each other.

- Reliability is a study and measure of uncertainty. So, one needs to take the help of probability theory and some knowledge on the same will be helpful.
- Reliability is a type of quality over time and necessity to be attached to products and services for better results and arriving at safe situation.
- Reliability helps to identify the cause of failure and during design stage; the same needs to be addressed so that EUC is in safe state at different failure conditions.
- Cost of unreliable design has serious impact on business as discussed previously.

1.3.2 Reliability Theory and Measurements

In Chapter I, basics of probability density function or PDF, cumulative distribution function (CDF), etc. have been discussed; with this knowledge it is better to explore some mathematical treatment. So these are not repeated, if necessary the reader may recapitulate.

- Short mathematical treatment: Since reliability is a time base function, let cumulative distribution be represented in terms of time hence, $\boldsymbol{F(t)}$, where:

$$F(t) = \int_{-\infty}^{t} f(t)dt. \quad \text{(VII/1.3.2-1)}$$

  whereas by definition survival function:

$$R(t) = 1 - F(t). \quad \text{(VII/1.3.2-2)}$$

  Hazard function $h(t)$ is other important function in reliability; it is the conditional probability of failure in the *interval t to* $(t + dt)$, given that no failure has occurred in between 0,t:

$$h(t) = \frac{f(t)}{R(t)} \quad \text{(VII/1.3.2-3)}$$

  There are a number of distributions such as normal, exponential, etc. that could be applied. Several types have been listed in brief in Table VII/1.3.2-1. There could be other distributions also. However, in order to limit the size of the book discussions shall be restricted to only one type, so exponential distribution is chosen (in the table) for discussions, and also Weibull modeling can be done by exponential distribution.
- Reliability block diagram and Laws of Reliability: Another important issue is reliability block diagram. *What is reliability block diagram (RBD)?* An RBD is a graphical representation of the components or subsystems of the system and their relation from a reliability point of view. The diagram represents the functioning state (i.e., success (S) or failure (F)) of the system in terms of the functioning states of its components/subsystems. RBDs are mostly used for analysis. Necessary software packages are available to handle these. Here, both independent modes

Table VII/1.3.2-1 Major Distribution in Reliability

Distribution	Formula	Remarks
Exponential	$f(t) = \lambda e^{-\lambda t}$ $F(t) = 1 - e^{-\lambda t}$ $R(t) = e^{-\lambda t}$ $h(t) = \frac{\lambda e^{-\lambda t}}{e^{-\lambda t}} = \lambda$	Failures occur in random intervals and the expected number of failures is the same for long periods. $h(t)$ is constant λ, $1/\lambda$ is MTBF (θ). So, 63.2% of items will have failed by time $t = \theta$.
Weibull	The survival function: $R(t) = e^{-\left(\frac{t}{\eta}\right)^{\beta}}$	To compare with Eq. (VII/1.2.3-1) in Clause 1.2.3, where influencing factors have been elaborated ($\gamma = 0$). Also refer to Fig. VII/1.2-1, bathtub curve, for the following: When $\beta = 1$, the hazard function is constant can be modeled by $\eta = 1/\lambda$. When $\beta < 1$, decreasing hazard function when $\beta > 1$, increasing hazard function.

and dependent modes can be analyzed in the similar fashion. In practice of instrumentation and control engineering, there are several components and the measuring system is complex, and there may be serial or parallel components. With the following example, these shall be examined.

- Series system: In series system, several subsystems may be in series, viz. sensor, conversion unit, signal processing unit, etc. The reliability of components in series needs to take into account the probability of individual failures in a given period as shown in Fig. VII/1.3.2-1.

  From the figure, one gets that if:

$$\mathrm{Rs}(t) = \mathrm{R1}(t) * \mathrm{R2}(t) \tag{VII/1.3.2-4}$$

  when generalized:

$$\mathrm{Rs}(t) = \prod_{i=1}^{n} \mathrm{Ri}(t) \tag{VII/1.3.2-5}$$

  where Rs(t) is the reliability of the system and Rn(t) is the reliability of nth subsystem at a given time t. Let for a system, with three subsystems having reliabilities as 0.9, each then $\mathrm{Rs}(t) = 0.9 * 0.9 * 0.9 = 0.729 = 0.73$. Whereas if they are parallel, then:

- Parallel system: In a parallel system, subsystems are in parallel as shown in Fig. VII/1.3.2-2. There would have been serious improvement in reliability as shown in the figure.

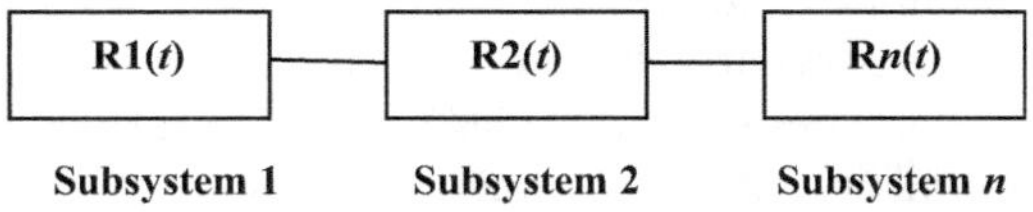

FIGURE VII/1.3.2-1

Series components in reliability block diagram (RBD).

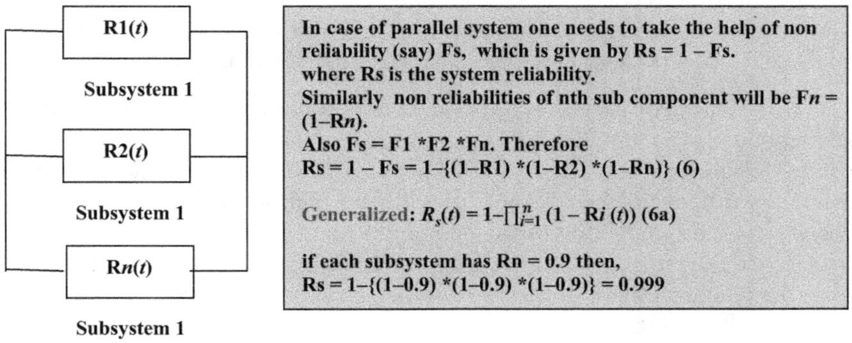

FIGURE VII/1.3.2-2

Parallel components in RBD.

- MooN system: In this system, there are N devices from which M numbers of devices need to be used. In thermal power plants of say 500 MW, there are eight pulverizers and six numbers to be made available for full load operation.

$$R = 1 - \sum_{i=0}^{m-1} \binom{n}{i} R^i (1 - R)^{n-i} \qquad \text{(VII/1.3.2-6)}$$

  Further details will be available in SIL discussions (Chapter VIII) for calculation for PFDs.

- General guide for reliability design shall include but is not limited to the following:
  - Component selection.
  - Load—strength relationship with safety margin.
  - Least possible complexity.
  - Design optimization.
  - Analysis of failures.
  - Single-point failure treatment.
  - Lessons learned.
  - Avoidance of systematic fault.
  - Diversity (CCF issue).
  - I&C manufacturing quality, etc.

1.3.3 *Reliability of Safety Instrumentation Systems*

It is important to understand the relationship between the failure types or causes, and failure rates or effect of failures. Failure types discussed in Clauses 1.2.4—1.2.6 are failure causes for a hazardous event, which would create two types of failures, viz. safe failure and dangerous failures shown in Fig. VII/1.3.1-2. Dangerous failures call for SIS so that system will be safe state. This relationship has been depicted in Fig. VII/1.3.3-1.

From the preceding, it is clear that SIS and reliability go hand-in-hand. A functional safety assessor (FSA) assesses the safety assessment/analysis report (SAR) after SIS is developed to check the system reliability and SIL. In order to achieve a tolerable risk level, the end-user, who needs to install SISs in the plant, initiates SIS projects. Detailed process outline for the same has been elaborated as follows:

- Based on management decisions, a risk analysis is performed to determine the risk level of the plant (or EUC).
- Decisions are undertaken to install SIS as required, based on risk analysis report.
- It is important to decide how reliable the SIS and each subsystem should be.
- Assessments of the reliability of the design are carried out to ensure that the reliability requirements are achieved. It is neither expected nor possible to finalize the whole thing in one go; naturally it may be made final by iterative method. At this point, suppliers play great roles for supplying the failure data, or suggesting optimum solution from suppliers' experience.

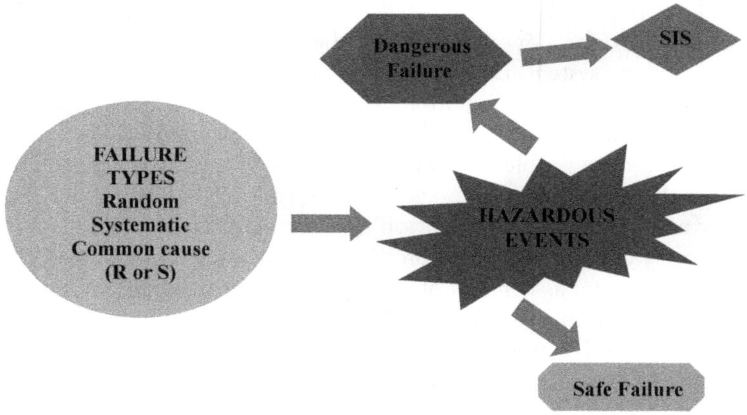

FIGURE VII/1.3.3-1

Relation between failure causes and effects.

- SIL allocation comes next and is usually carried out jointly by end-user, consultants (if any), and analysts.
- Subsequent to SIL allocations, an SRS is prepared. The SRS serves as the governing document in the SIS development, and specifies the detailed requirements for the SIS reliability.
- Based on the SRS, an SIS design is developed which is the basis subsequent steps.
- Based on finalized SIS design, SRSs are developed and floated for suppliers to respond.
- When as per SRS materials are supplied, these are installed.
- In order to demonstrate that the reliability requirements in the SRS are fulfilled, **SAR** with all the evidence and arguments including reliability calculation are prepared.
- For final reliability demonstration, the end-user, system integrator, component supplier, and FSA will be involved to assess the SAR independently and verify the arguments.
- FSA decides to approve or reject the SAR based on joint assessment.
- When a SAR is rejected, SIS revisit and modification is necessary for a better system.

Often SIL is considered as the overall measure for SIS reliability (safety integrity). From the standard, it is clear that there are three important issues in system integrity, and these are *hardware*, *software*, and *systematic* integrity. SIS reliability *quantification* is influenced by many factors, for example, selection of RBD, etc. On account of dependence on environmental conditions and human interaction, it is not always easy to keep independence between SIS and rest of the system. Naturally, it is important to be aware of some side effects introduced by

SIS, that is, additional problems introduced by SIS. In modern times for floating oil installations people go for a dynamic positioning (DP) system, in place of conventional mooring system. DP systems are independent systems, but needs to interact with integrated control systems for exchange of many data. Therefore, it is needless to tell assessment of reliability of the system is extremely important. Now, if there is some interface problem between the two systems, then additional complications will arise are they not? Also at times, spurious actions from SIS make operators very much disturbed. In addition to the possible safety hazards, the SIS may also spuriously operate to disturb production and lead to losses and costs. Therefore, importance for assessment of reliability of SIS cannot be overestimated. There are several methods to assess reliability of SIS such as RBD, FTA, LOPA, and risk graph. RBD, FTA, and LOPA have already been discussed, so let's concentrate on *risk graph* now.

1.3.4 Risk Graph

In Chapter I, risk matrix has been discussed. Risk graph is similarly used to categorize risks; additionally it can be used to determine SIL in process industries (especially when there many hazards to analyze). The discussions are presented naturally in a generalized way, however for each application one needs to calibrate the same. In this approach, there are four sets of parameters, viz. "C," "F," "P," and "W" (see Fig. VII/1.3.4-1), each arranged as a column. All of these parameters "describe the nature of the hazardous situation when safety instrumented systems fail or are not available. One parameter is chosen from each of four sets and the selected parameters are then combined to decide the safety integrity level allocated to the safety instrumented functions" (Annex *D1of IS/IEC 61511-3:2003*). Let an example with exposure to radioactive radiation case be analyzed.

Based on *consequence*, the first the hazard is divided; these are C_a through C_d. Such categorization is based on injury /fatality as shown in the figure. Then exposure time is under consideration, that is, how much time a person is exposed to the hazard (e.g., radiation) represented by F as shown. These are rare exposure or frequent exposure. The hazard event is tested for probability (represented by P as shown) of avoidance. Both the exposure rate and probability of avoidance are combined with all four consequences as shown. Now based on frequency (represented by W as shown) of occurrence, SIL are decided. Compare the same with risk matrix, where consequence and frequency of occurrence were main considerations only, but SIL were not really assigned from there. The reader's attention is drawn to top right side SIL box to the bottom left SIL box. The bottom left box is "b," meaning not recommended operation, that is, no single SIF is sufficient (should be avoided). On the other hand, the top right side says no safety requirement. The preceding description is based on a generalized case, and for specific case one needs to calibrate the same. This is a semi-quantitative approach. The reader may refer to Annex D of (IS)/IEC 61511-3:2003 for further details. In this connection, D3 of

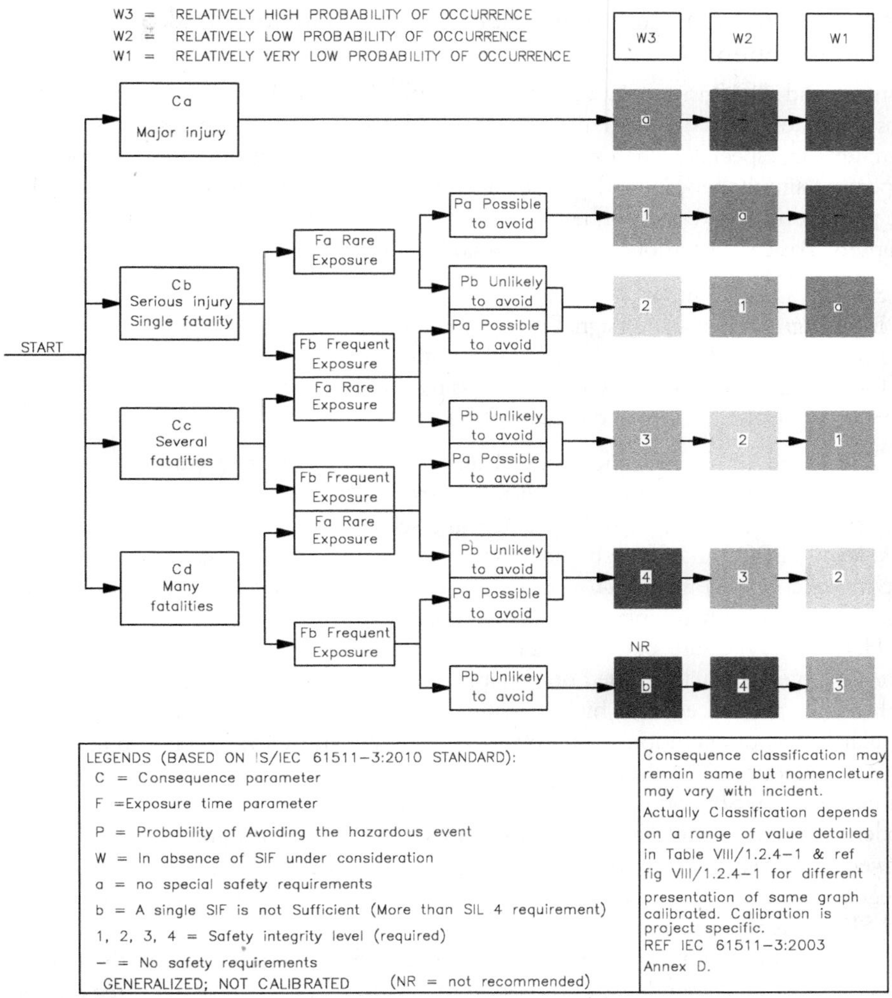

FIGURE VII/1.3.4-1

Risk graph (generalized not calibrated).

IEC 61511-3:2003 the stated standard may be referenced for calibration. Also, Tables D1 and D2 of the stated standard may be referenced for explanation of C, F, P, etc. SIL determination by risk graph has been detailed in Clause 1.2.4 in Chapter VIII also. With these discussions on reliability concluded, now it is time to move on to see how instrument components need to be considered for SIS requirements.

1.4 Component Functional Safety (Safety Instrumented System Components)

Control and managing of industrial plants with the right I & C components and subsystems for optimal operational efficiency and safety is not an easy task. It is a quite complex job, especially when system availability and avoidance of spurious tripping is important simultaneously with plant safety. Conventionally, there are two parts in the plant control and safety. These are BPCS and SIS as already discussed in previous chapters. Basically each of these systems consists of:

- Sensors as signal input
- Input signal interfacing (signal processing and interface)
- Logic solver and associated communications
- Output signal interfacing (signal processing and interface)
- Final control elements as recipient of output command

The scope of an SIS encompasses all instrumentation and controls that are responsible for bringing a process to a safe state in the event of an unacceptable deviation or failure. For better understanding see Fig. VII/1.4-1. In this figure, *only* (two layers) BPCS and SIS components have been taken into consideration (and *not mechanical device like safety valve, etc*). This is the same previous example (Fig. VII/1.1.2-2), so no further explanation is necessary.

Here, high pressure petroleum product is transferred to a tank, and incoming control valve is regulated based on the level in the tank. If there is a high level, the tank will be pressurized so at high pressure, the shutoff valve is closed. Usually there are alarms before operation of shutoff valve (not shown here), which also acts as a layer of protection. One thing interesting here to note, is that there is a *communication channel* between SIS and BPCS. This has been kept intentionally to show that in the modern era it is possible to integrate the two different systems and/or sharing data between the two systems which otherwise are kept separate for safety and safety integrity. From Fig. VII/1.0.3-2, it is seen that there is interrelation among SIF, SIS, and SIL. This is made clearer in Fig. VII/1.4-2.

Another important aspect of SIS is to determine SIL so as to make sure that safety is available when demanded. According to IEC 61508-4:2010, there are three modes of demands. Empirically it can be considered as: frequency of demand; <1/year — low demand, >1/year — high demand, and where the safety function retains the EUC in safe state as a part of normal operation. So, SIL is a statistical representation of the safety availability of an SIS at the time of process demand and is a major constituent of SIS with a number of factors to influence with:

- Device design feature
- Device integrity
- Diagnostics feature
- Systematic failure
- CCFs
- Testing

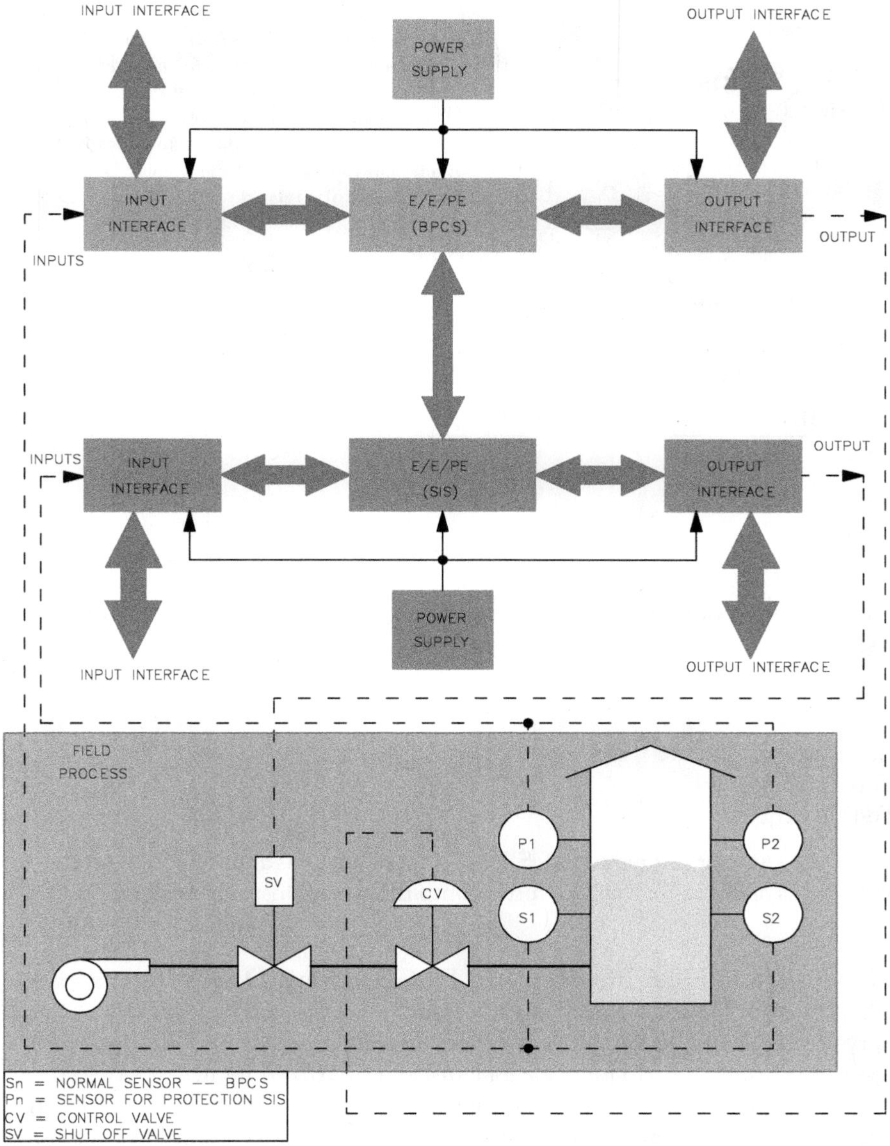

FIGURE VII/1.4-1

BPCS and SIS components in process application.

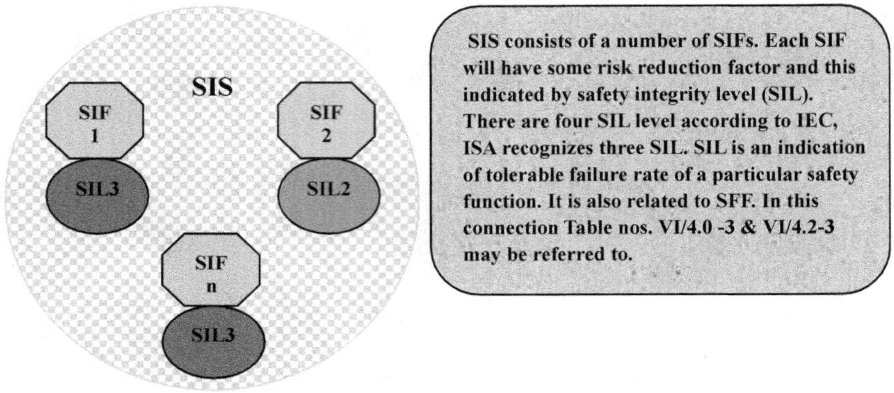

FIGURE VII/1.4-2

SIS, SIF, and SIL relations (2).

- Installation
- Operation
- Maintenance

The proportion of time that the system is operational depends on component or subsystem:

- Failure rates
- Failure modes
- Hardware architecture
- Voting scheme
- Testing frequency

Brief procedure for risk reduction, SIL assignment, etc. has already been discussed. Now the focus of discussions will be turned onto the type of instruments better suited for SIS.

1.4.1 *Sensor and Final Element (non PE) Electronic Device Details*

Initially, IEC 61508 was developed as a safety and life cycle standard, which was not industry-specific. Later, there have been three separate divisions, viz. IEC 61511 for process industries, IEC 61513 for nuclear plants, and IEC 62061 for machinery sector. IEC 61511 has clearly put forward some guidelines for hardware to be used in SIS. In this standard, there are clearly two divisions, for example, PEs and non-PEs (in loose sense to distinguish from logic solver, because nowadays sensors and final elements also have programming features) like sensors and final control elements. In this part, sensors and final control elements are considered, while PEs will be discussed separately in next clause.

On account of advancement in technology, people go on adding feature after feature to the systems. It has been found many times these cause additional problems to the system, viz. in the form of spurious trips. It will not be prudent to consider that the existing standards are obsolete, instead people need to introspect whether the reliability of SIS has been properly evaluated! A well-designed system would increase the availability of plant safety by reducing spurious trips which occur on account of failure of SIS in correctly judge the safety situation. This depends largely on hardware architecture considered. Plant data suggest that majority of SIS problems are due to field devices such as sensors and final elements yet people put more attention on logic solver than field devices. This is also another issue. Modern smart field devices provide lot of information to the operator as well as to logic solver to address these issues. In modern generation, transmitter intelligence is built-in with sensors as well, in addition to the same in electronics of the transmitter. This enables characterization of the sensor, and it is possible for early detection and prediction of fault in the sensor. In this connection, Fig. VII/1.4.1-1 may be referenced. This is just an example to show that now it is possible to diagnose low supply pressure to final element, degraded voting of transmitter, impulse line plugging, etc. quickly. On account of this, field diagnostics and use of field bus systems make it possible for system integration of SIS with BPCS also.

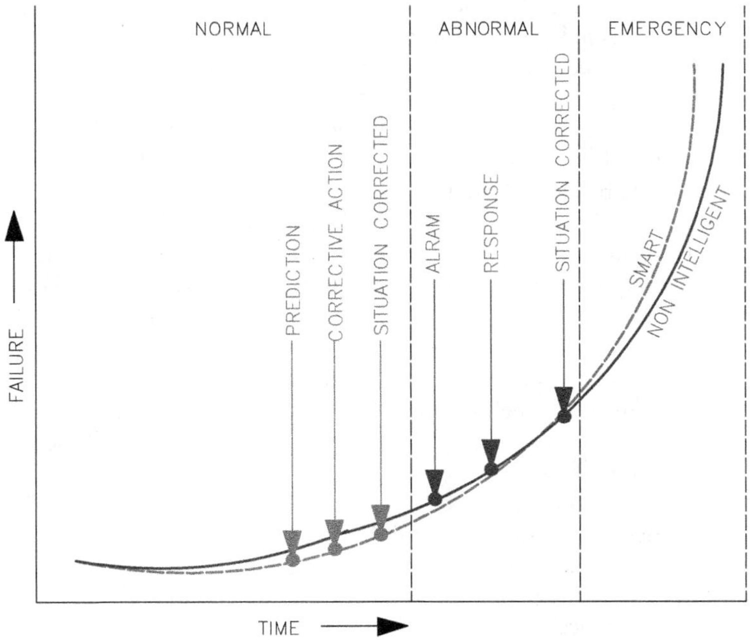

FIGURE VII/1.4.1-1

Smart instruments with diagnostics.

Inspired by and courtesy. D. Walker, Using Smart Safety Instrumentation System (SIS) to Make Better Operational Decisions, Emerson Process Management, March 2009.

- Non-PE general requirements: Now coming back to the specifics, one needs to note that when a device is designed and developed as per IEC 61508, for safety purposes, it is the **responsibility** of the manufacturer to specify *safety level, capability, and limitation* of the device including wetted parts. When these devices interface with process, it is the **responsibility** of the end-user to ensure that the *interface is proper*; there are no untoward incidents coming from the process such hammering, corrosions, H_2 embrittlement, etc. The end-user also needs to prove that there is no *undetectable failure in the process*.
- Sensors: Third party validated supplier data reflects laboratory results, and not real-world failure causes. Even failure rate data specifying "*x* years" does not specify failure causes. After finalization of requirements of SIL, subsystem selection process is undertaken. During that time, it is better to look into the following as a thumb rule [12]:
  - Replace discrete switches with transmitters
  - Improve common cause strength
  - Use diversity
  - Use diagnostics
  - Add redundancy

  As there has been modern improved technology already in use for transmitters with a few improved effect discussed previously, naturally, much greater improved performance is expected of a transmitter compared to a switch. Not all common causes can be addressed, yet some common causes like high temperature failure can be addressed by selecting a robust transmitter, transmitters with built in ambient compensation capability, high temperature failure prediction capability (diagnostics), installing transmitter away from hot spot, etc. These are a few things which could be looked into to avoid common causes, and the same shall be considered during selection process. Depending on applicability and cost consideration, redundancy could be afforded. While selecting redundancy, due consideration needs to be put towards architecture for hardware redundancy and diversity in technology (e.g., 1oo2, etc. and say, for level measurement DP transmitter and capacitance probe — just for example to elaborate architecture and diversity points in redundancy). For further details on sensor redundancy, refer to [13]. The main idea behind all of these is to improve return on investment in terms of safety. Major issues involved for sensor performance shall include but are not limited to the following:
  - Overall transmitter performance
  - Process interface (discussed previously)
  - Sensor (robustness and characterization)
  - Transmitter electronic performance
  - Quality control process
  - Selection criteria and process
  - Installation practices
  - Maintenance

Computing power in modern microprocessors make diagnostic features more and more powerful so that they go beyond detecting component failures to evaluate the performance of the complete measurement system, including common process interface problems. Some examples of these diagnostic features shall include but are not limited to:

- Pressure: Plugged impulse line detection and prediction
- Pressure-level: Leaking diaphragm seal detection
- Temperature: Fast, predictable detection of failed resistance temperature detector (RTD) or thermocouple (T/C); sensor drift
- Vortex flow: Detection of application changes (viscosity, density, etc.)
- Magnetic flow: Detection of faulty ground or electrodes or high process noise
- Coriolis flow: Detection of slug flow or tube coating
- pH: Detection and prediction of faulty electrodes

- Final elements: From the Offshore Reliability Equipment Database Association (OREDA), one can get that SIS failure due to sensor and logic solvers are 42% and 8%, respectively. So, a balance of about 50% of SIS failure occurs on account of final control elements making it to top the list (followed by sensor issues). Therefore, final elements generally are least reliable in the chain. Hazardous conditions of a plant are sensed by process sensors, and are reported to the logic solver, which, based on its wisdom set by the control logic, triggers (say) shutdown response. It is the duty of the final control elements to carry out the command through decisive and swift action. Normally, chopper-type final elements are chosen for such shutdown actions. For example, main fuel flow control for steam generators is done by a control valve, and there is separate shut off valve for trip action, that is, control and trip actions have separate valves. Similarly, in case of turbine control, there is a separate emergency shutoff valve (ESV) and control valve. However, there are situations in process plants where the same valve is deployed for both purposes. It is possible to take care of both the actions through single valve as shown in Fig. VII/1.4.1-2A.

In this connection, only one possibility (air to open fail close) has been shown for all the valves shown in Fig. VII/1.4.1-2. Here, valve types chosen are air to open/fail close valve and air connection is from the top. However, similar things are possible for air to close/fail open type valves or stay put type valves also. Interested reader may go through Chapter VI of "Power Plant Instrumentation and Control Handbook" (Elsevier) for further details. During system design, all elements of each of the SIF loops need to be addressed. Referring to Fig. VII/1.4.1-2A, it may be noted that based on control demand, the control valve is throttled and pneumatic signal is sent to the control valve via solenoid valve, which when is energized gives the path "E" for pneumatic signal to reach control valve (positioner). In case of requirement for shutdown, the electrical signal to the solenoid valve may be withdrawn (zero signal), which will cause path shown by "DE" to be opened so that air pressure from the control valve is vented and control valve closes. So, single valve is used for both purposes.

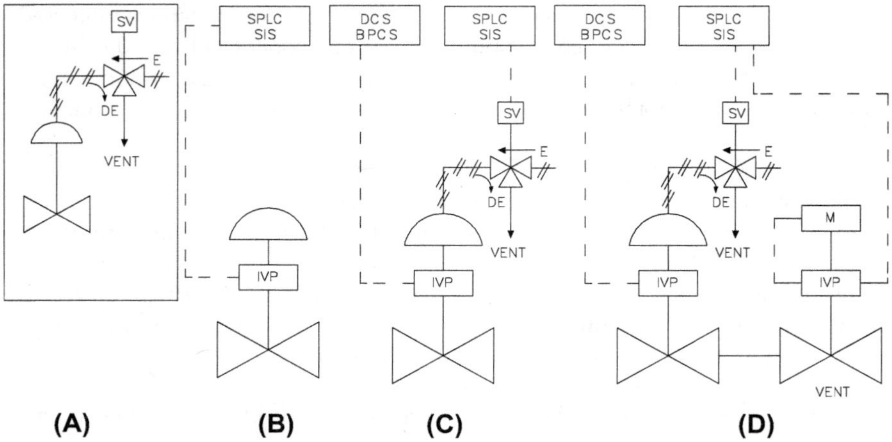

NOTE: VALVES SHOWN HERE FOR TOP AIR SUPPLY AIR TO OPEN (FC) VALVE. OTHER CONFIGURATIONS WILL BE SIMILAR TO THIS. FOR FURTHER DETAILS CHAPTER VI OF POWER PLANT INSTRUMENTATION AND CONTROL HANDBOOK (ELSEVIER MAY BE REFERRED TO)

FIGURE VII/1.4.1-2

Final elements in SIS.

Since final element is the weakest link, so proper selection of the same is extremely important. The following three configurations are very popular. In all cases, control valves are equipped with intelligent valve positioner (IVP). These are:

- Single valve used as safety function: Referring to Fig. VII/1.4.1-2B, one may note that the control valve is in ON/OFF duty, and in case of emergency, shutdown command will be sent to IVP from safe PLC (SPLC).
- Single valve used for both control and safety purposes: Refer to Fig. VII/1.4.1-2C. In this case, the control valve is required to do its normal duty based on signal from DCS, where from signal is sent to the control valve via solenoid valve similar to what has been described for Fig. VII/1.4.1-2A. So based on command from SPLC, control valve is closed.
- Control valve as redundant SIS final element with one ESV: Here, DCS sends signal to the control valve for normal operation. In case of emergency, the control valve solenoid valve as well as ESV gets the signal for closure. So, based on SPLC command, both the control valve as well as ESV will close as shown in Fig. VII/1.4.1-2D.

1.4.2 Programmable Electronic Device Details

"Safe PLC," "fail safe controller;" there are so many names by which PEs in SIS are known. Frankly, these are various names given by the vendor to bring distinction to their products. However, in these controllers or PLCs, there will be choices for redundancies, high-end designs (taking care of safety requirements), and advanced diagnostic features. Mostly these are third party certified logic solvers for use in SIS. The fail safe PLC/controllers are used for the functional safety of machines or plants. They try to ensure fail safe operation and comply with the relevant standards: EN 954-1 up to category 4, IEC 62061 up to SIL 3 (fail safe and fault tolerant), and EN ISO 13849-1 up to PL e (discussed in subsequent sub-clause). Usually there would be two independent systems for control and safety functions in line with the standard. Simatic S7 of Siemens, Pluto or Independent Hi safety system of ABB, Guard Logix, AAdvance/ Plant Pax of Rockwell Automation, and Quantum Safe PLC of Schneider Electric are examples of safe PLCs. Now systems such as Simatic S7 of Siemens are available where it is possible to simultaneously process both the standard and the safety programs using single set of CPU. Major features include but are not limited to the following:

- High productivity with less injury
- Machine architecture reduces system footprint
- Machine safety functions with speed and analog monitoring
- Advanced programming possibilities
- Better system integration
- Reduction of components: Coexistence of standard and safety programs in one controller
- Single PROFIBUS or PROFINET standard
- Safety communication Conform to SIL 3/Category 4
- Fast system configuration and commissioning
- Flexible, modifiable, and implementable software-based safety functions
- Faster reproducibility

Application areas may include both processes as well as machinery automation. In machinery area, application shall include mainly:

- Automobile industry
- General mechanical engineering
- Building of special machines
- Series mechanical engineering, OEM
- Plastics processing
- Packaging industry
- Food and beverage industry

A short discussion on safe PLC has been presented in Clause 2.4 of Chapter VIII also.

1.4.3 *Machinery Automation (IEC 62061 and EN ISO13849)*

IEC 62061 is a generalized standard for machinery applications. In case of factory/machinery automation in place of SIL, other symbols are more pertinent, discussed here, under major issues related to machinery safety, shall include but are not limited to:

- Protection of operator of machine and people around it.

- To avoid people injury, property, and business loss for unintended energy of the machine.

IEC covers the safety of any assembly where there is at least one moving component for safety-related electrical control systems. New EU Standards for Machine and Device Manufacturers covers functional safety with through IEC 62061 and EN ISO 13849, with both having some root with IEC 61508. European standards (viz EN/ISO 12100, IEC 62061 each have one or the other type A/B/C) for safety of machinery have typically has following structure: type A standards for basic safety; type B specific standard comprising: B1 for particular safety such as distance, temperature; B2 for safeguarding such as interlock, two hand control; type C machine safety. This directive also specifies when a product can use "CE" markings. Risk assessment and risk reduction for machines are available in EN/ISO 12100. IEC 62061:2005 is derived from IEC 61508 for safety of machines— functional safety of safety-related E/E/PEs. On the other hand, EN ISO 13849—1:2006 is also for machine safety — safety-related parts of PEs, control systems, and all types of machinery, regardless of the technology or energy type employed (electric, hydraulic, pneumatic, mechanical, etc.). One of the major distinction points between IEC 62061 and ISO 13849 is that reliability measure. In the case of IEC, it is SIL, and in case of ISO 13849, standard is represented by performance level (PL). In PL, with suffix "r," it indicates the required performance level (PL_r). *What is PL?* PL is a measure of the reliability of a safety function, having distinct five levels (a—e). PL e gives the best reliability required at the highest level of risk. PL is a technology-neutral concept applicable for electrical, mechanical, pneumatic, and hydraulic safety solutions. The structure for the components in the block is assessed to determine category B, 1 to 4 for diagnostic coverage (DC). The typical relationship between *diagnostic coverage* and *categories* are as listed as follows:

- For Category B DC: none
- For Category 1 DC: none
- For Category 2 DC: Low, medium
- For Category 3 DC: Low, medium
- For Category 4 DC: High

For calculation of PL of the system, details on the system's structure, that is, category, $MTTF_d$ (suffix "d" stands for dangerous failure), and DC, are necessary. (In case of ISO 13489 PL level is assessed mainly based on principles similar to risk graph and in ISO 62061 risk matrix style are normally followed.)

There is possibility of confusion regarding use of two standards viz ISO 13489 & IEC 62061. Normally when medium other than electrical system, ISO 13489 may be more appropriate. Whereas for customer demand for demonstrating safety IEC 62061 may appropriate. For safety-related control systems, standard components can be used as it is allowed as per standard also. However, safety components offer the advantage of reducing workload *as the* ***safety-oriented assessment, and analysis of the components*** *used, is carried out by the producer of the safety components.* For functional safety, the systematic integrity of components is taken into account, in addition to *the use of a suitable category, the implementation of necessary fault detection* and *the* *consideration of failure rates/probabilities*. When machinery manufacturers use standard components for safety functions, it is definite duty to *assess* compliance with safety-related requirements as per standard. This will call for considerable effort, if not practically impossible, especially when detailed producer's information is not always available. For systematic failures, it is important that the components to be used operate correctly under all the operating and environmental conditions (temperature, humidity, vibrations, electromagnetic interference, etc.) including operating limit of the components (that can be expected). There are no general rules for using standard components in safety related applications. With standard components, it may be difficult to ascertain DC. SPLCs often offer the advantage of supporting necessary tools for safety-oriented programs. Finally, while using standard components and complying standards one should never ignore the influences of systematic approach and environmental effects.

1.4.4 Systems for Nuclear Power

The potential hazards to the public from nuclear power plants (NPPs) — especially in terms of the risk of having to evacuate significant areas of land for many years — are in general greater than those for any other potential industrial hazards (Exception! Bhopal disaster where impact is still going on through generations). Major safety cases are embedded with C&I architecture and it leads to space requirements and system separation requirements. Therefore C&I architecture must be addressed at early stage of the design. The typical order for C&I systems shall mainly cover:

- Definition of the overall C&I architecture, including space and separation requirements
- Functional specifications for C&I systems
- Detailed technical specifications for C&I systems
- Procurement action
- Detail design and manufacture
- Works testing
- Site acceptance test
- Installation and commissioning.

Some details regarding safety classification in NPP has been described in Chapter XII. There are two major systems of NPPs related to safety aspects of the plants; these are:

- Reactor protection system (RPS)
- Engineering safety feature actuation systems (ESFAS)

Normally, a dedicated sensor, logic solver, final elements, and human—machine interface (HMI) are assigned for checking the abnormality of the system. There shall be separate control and protection systems, and such separation shall include both space, physical separations, as well as electrical isolation. For RPS, in addition to separation discussed there shall be diversity applied to take care of CCFs. Also, separate route diverse logic solvers are dedicated. Even different shutdown systems are planned, for

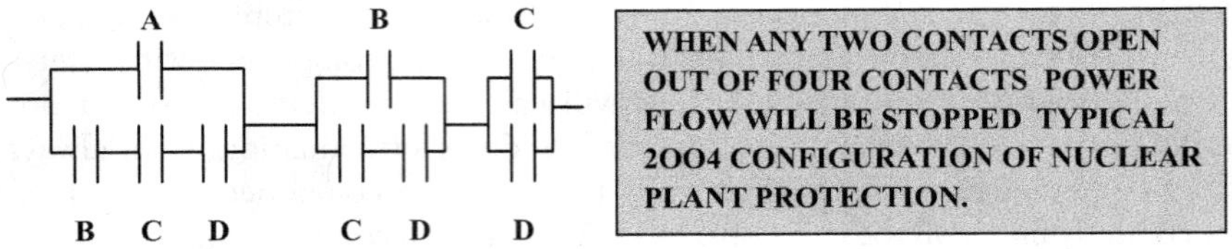

FIGURE VII/1.4.4-1

Typical 2oo4 in nuclear power plant protection.

example, boron versus controlling rod. There shall be both physical as well as electrical separation between different systems indicated as follows. Typical SIL ratings are as follows (with redundancy for RPS):

- Control systems: SIL 2/3
- Primary RPS: SIL 3 or 4 redundancy 2oo4
- Primary RPS: SIL 3 or 4 redundancy 2oo4
- ESFAS: SIL 3 or 4 redundancy 2oo4

Typical 2oo4 protection in nuclear power plant has been shown in Fig. VII/1.4.4-1. Now concentration will be put on redundancy architecture and associated PFDs.

1.4.5 Hardware Redundancy Architectures

It is better to start the discussions with Fig. VII/1.4.5-1 from where it is clear that just by adding devices in redundant mode, the safety availability does not always increase, as there are pros and cons in each of the applications as shown; in the first case, chances are more for spurious trip, in the second case, it is the other way round. So, a detailed study of various redundancy modes will be necessary. Dual configuration shown appears to be safe, but could also be the root cause of spurious trips due to random failures. Now the aim is to find advantages and disadvantages of implementation of various hardware redundancy architectures (industry specific and international standards). It is very useful to understand a few terms such as *redundancy*,

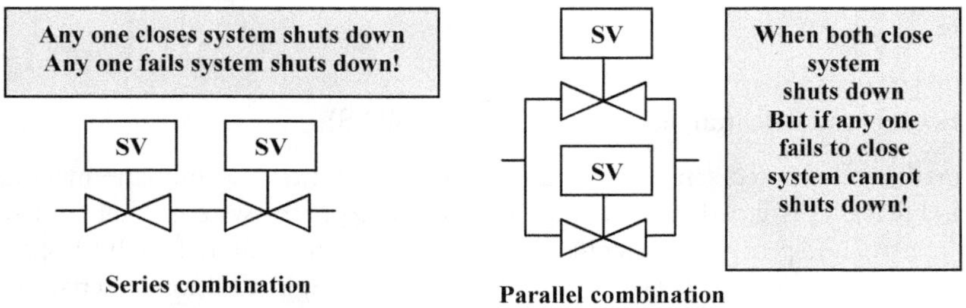

FIGURE VII/1.4.5-1

Pros and cons of redundant devices for understanding.

multiplicity, and diversity. From United States Particle Accelerator School (USPAS) one can get:

- Redundancy: Calls for use of *identical* SIFs to achieve higher safety reliability.
- Multiplicity: Use of *multiple shutdown* paths or protection devices.
- Diversity: *Different types of devices* used to reduce the probability that multiple or redundant devices can be affected by common failure modes.

Also, the architecture needs to take care of:

- Implementation of system at high level
- Fault tolerance
- Physical environment
- Physical constraints
- Redundancy, multiplicity, and diversity
- Final elements

In the case of *de-energize* type logic set of contacts, *opens* to trip or shutdown the system (in case of *energize to trip* logic, *contact closure* will be considered for trip; also series, parallel configuration types will be reversed). When ***trip*** is associated with proper cause it is safe failure. When this is NOT, then it is spurious/nuisance trip or shutdown. As per IEC 61508: *Demand or failing element commands output to a safe state.*

Now a few configurations will be discussed. Note that only empirical formulas are given here detailed calculation basis as in line with IEC 61508-6 AnnexB:2010 has been presented in Clause 1.3.7 of Chapter VIII.

Following symbols and their meanings will be helpful in studying various configurations discussed in Figs. VII/1.4.5-2 to 1.4.5-4.

λ for failure rate with suffices DU/DD, etc. for dangerous failure detectable or undetectable as discussed earlier. MRT, mean repair time, TI, test interval, PFD_{Avg} with usual meaning already discussed. Also "S", Sensor, I/P, input LS, Logic solver, O/P, output, AC, Actuator, P, Power, etc.

- 1oo1: This stands for one out of one configuration. This is simplex device. Analysis details are available in the accompanying Fig. VII/1.4.5-2.
- 1oo2: IEC 61508 requirement: *One demand or one failing element commands output to a safe state.* In this configuration de energize trip circuit has been considered.

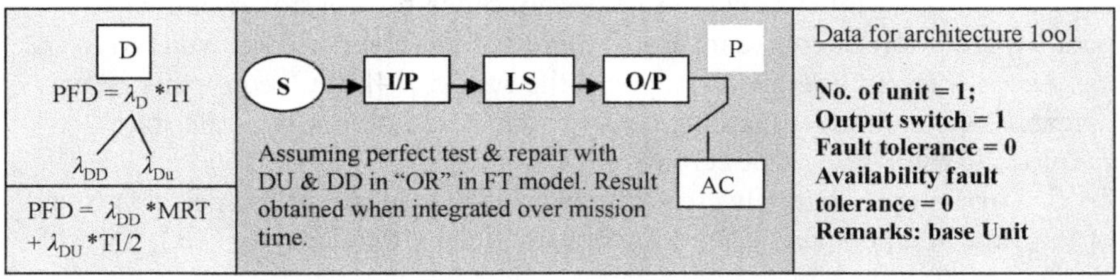

FIGURE VII/1.4.5-2

Configuration details of 1oo1 hardware architecture.

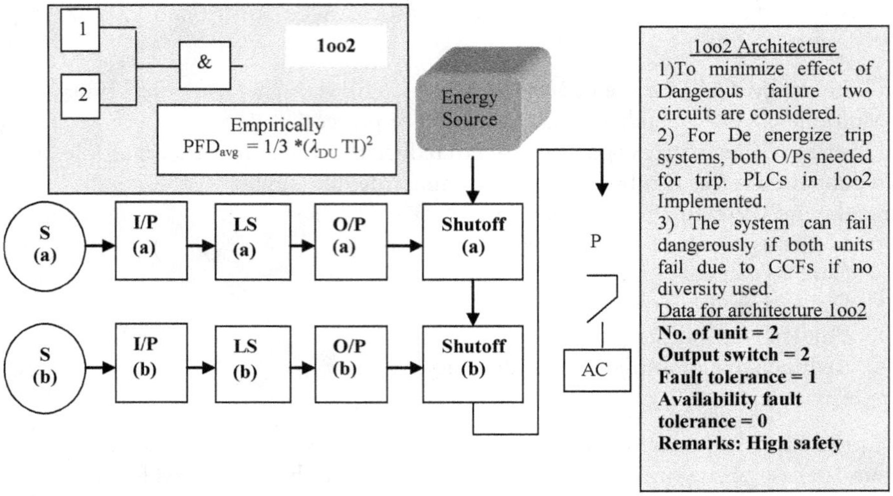

FIGURE VII/1.4.5-3

Configuration 1oo2 block diagram.

- Basic block diagram of 1oo2 has also been shown in Fig. VII/1.4.5-3 with an explanation given in the RHS box. Basic symbolic representation of 1oo2 has been shown in the (top) box (in Fig. VII/1.4.5-3). Empirical formula of PFD_{avg} is also shown. In the preceding figure, perfect test is considered. CCF factor β and proof test coverage, etc. have not been considered in arriving at the PFD value. There is another version of the same, that is, 1oo2 with diagnostic represented by 1oo2D. From data available through IEC, it shows that the highest diagnostics 1oo2D systems can achieve is 90%.
- 2oo2: This dual controller configuration is used in energize-to-trip protection systems. The outputs of two controllers are wired in parallel. If one controller fails with its output de-energized, the other is still capable of energizing the load. A disadvantage of the configuration is in failures of controller where the output is *energized* (discussed at the beginning), when controller fails (short circuiting). This configuration is not suitable for de-energize to trip protection systems unless each unit is of an inherently fail safe design. Since the controllers are wired in parallel, any short circuit (dangerous) failure of the components results in a dangerous (outputs energized) failure of the system [14]. IEC requirements: two demands or simultaneous failing elements command output to a safe state. Empirically the PFD_{avg} will be: $\boldsymbol{\lambda_{DU} * TI}$.
- 2oo3: This is very commonly used configuration. There is another special version of the same and popularly known as triple modular redundancy (TMR). Typical configuration of both of them has been depicted in Fig. VII/1.4.5-4.

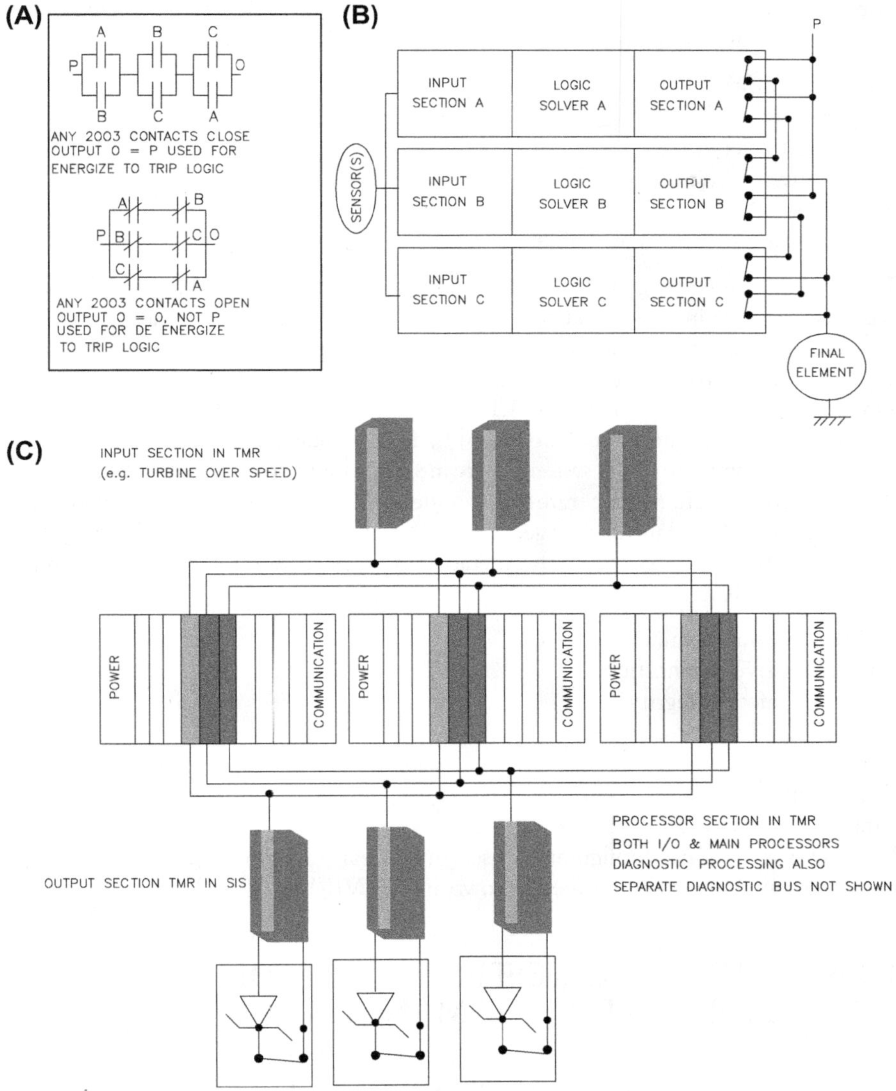

FIGURE VII/1.4.5-4

2oo3 and triple modular redundancy configurations. (A) Contact configuration of contacts, (B) typical 2oo3 architectural block diagram, and (C) triple modular redundancy (Typical).

In Fig. VII/1.4.5-4A, configuration of contacts both NO type and NC type have been shown. In the former case, a combination of two of them are in parallel with three such probable combinations are in series. In case of NC contacts, de-energize to trip the configuration is just reverse. A typical block diagram of 2oo3 architecture has been shown in Fig. VII/1.4.5-4. Here the architectural data are as follows:

- Number of units: 3
- Output switches: 6
- Safety fault tolerant: 1
- Empirically PFD_{avg}: $(\lambda_{DU} * TI)^2$
- Available fault tolerant: 1
- Remarks: Safe as well as available

TMR is a specialized version of 2oo3 with much more stringency implemented. TMR logic is quite popular in power and other industrial plants (e.g., TMR logic is used in many governor controls of turbine, GE gas turbine, or overpressure protection of KWU turbine). In case of sensors, both 2oo3 and TMR are also applicable. TMR voting logic principles have been shown in Fig XI/1.1-2. For details, Chapter 3 of "Power Plant Instrumentation and Control Handbook" [13] by Elsevier may be referenced.

In case of complete TMR system, there are three main processors (MPs) as well as I/O processors. Each MP operates in parallel with the other two. I/O control processors manage the data exchange. Triple I/O bus systems connect the trident systems. The MPs use I/O data in the memory for the voting process. Feature and advantages of TMR are as follows:

- No single point of failure
- Comprehensive diagnostics
- Very high safety integrity
- High availability
- Low maintenance cost
- Online repair
- Simple programming
- 2oo4: This is a typical configuration used in nuclear power plants. A typical configuration of the same has been shown in Fig. VII/1.4.4-1.

2.0 FAILURE MODES, EFFECTS, AND DIAGNOSTIC ANALYSIS AND SYSTEM INTEGRATION ISSUES

2.1 Failure Modes, Effects, and Diagnostic Analysis

Failure Modes, Effects and Diagnostic Analysis (FMEDA) is an extension of FMEA (Chapter IV) to identify online diagnostics techniques and the failure modes relevant to SIS design. Exida developed FMEDA in the 1990s. Using the component database that has failure rates, failure modes, and useful life, the FMEDA product generates a product failure rate as well as product failure modes, diagnostic coverage, and even useful life. This is a useful design verification tool and means to provide more precise input to reliability and safety modeling. Also this is extensively used by the user and in projects for safe instruments so that safety life cycle is well maintained by the product. It is necessary to know failure rate as function of failure mode. In other words, FMEDA can be described as a systematic technique for problem identification and related systems diagnostic coverage. Component level FMEDA can provide will determine failure mode of the module. So, module level FMEDA can predict the failure mode of device. FMEDA is used to calculate various failure rates and SFF. So, FMEDA results are depicted in SIL certification documents such as functional safety assessment (FSA) document. FMEDA is a procedure for the detailed determination of error causes and their impact on the system and can be very efficiently used in the early stages of systems development to identify weaknesses. From the FMEDA, failure rates are determined and consequently the SFF is calculated for the device. This is where FMEDA is used with its mathematical models and calculation methods, aimed at estimating the residual error probabilities and residual error rates resulting from malfunctions. For functional safety standard, IEC 61508, probabilistic analysis is necessary and for this huge database on failure data about the product is necessary. Unfortunately, such a database for all products is not always available. OREDA provides a good amount of data. Engineers using a good component database, these numbers can be determined and predicted more accurately than with only manufacturers' field data on previous products [15]. FMEDA is best when reviewed and certified by a third party, although self-declarations can be done by the manufacturer.

2.2 System Integration Issues

The topic is very sensitive. It has been found that there two schools of thought towards system integration. One group is of the opinion that *when BPCS is designed as per IEC 61508 with appropriate SIL, combined BPCS and SIS should be acceptable and would consider as a good engineering practice*. *The other group argues that an absolute separation of functionality between the BPCS and SIS is necessary*. *Also there shall be independence and diversity in protection layers*. *The integrity requirements for combined systems are extremely high because there are failure modes that will simultaneously generate a hazard and also disable multiple protection layers*. However, some communications may be allowed, as long as no common mode failures can occur. In this connection guidelines from www.61508.org is important and may be referred to.

2.2.1 Center for Chemical Process Safety Recommendations

According to the Center for Chemical Process Safety (CCPS) recommendation SIS separation from the BPCS is required to ensure that safety and environmental aspects of the SIS are consistent with user, manufacturer, local, national, and international standards [16] as:

- Minimize the effects of human error on the SIS from normal BPCS activities.
- Protect safety system software from unintentional changes by isolating the PEs-based SIS from process control induced programming changes in the BPCS.
- Provide access security.
- Ensure that SISs are maintained safely and correctly.

- Facilitate stand-alone testing and maintenance of the SIS and BPCS.
- Ensure security and integrity, allowing the PEs-based SIS to achieve a level of security and integrity equal to or better than a direct-wired SIS.
- Minimize CCFs (both hardware and software).

2.2.2 Requirements of IEC and Other Related Standards

With modern digital technology, it is very easy to integrate BPCS and safety system into one common platform or as an integrated system. Some of the major potential benefits include [17]:

- Seamless integration
- Time synchronization
- Elimination of data mapping duplication
- Common HMI
- Compatible configuration tools
- Minimized set of spare parts
- Single operator and maintenance training requirements

However, in such cases there is every possibility that there may be some *compromise* with safety system/SIS. These integrated solutions may offer some cost benefit but may be at a *cost* protection sacrifice against common cause and systematic design and application errors.

- From IEC 61511-1 Clauses 9.5.2 and 11.2.4, one can infer that BPCS shall be designed to be separate and independent without sacrificing safety integrity and avoidance of systematic and common fault. Also various other standards elaborate in the similar line.
- NFPA 85 also calls for "the logic system performing the safety function for burner management shall not be combined with any other logic."
- API 14C – "Two levels of protection should be independent of and in addition to the control devices used in normal operation."
- API Recommended Practice 554 (1995) – Process Instrumentation and Control, "the shutdown and control functions should be in separate and independent hardware."

An integrated control/safety system with same operating system may have systematic design and common cause errors. Another important point here is that, third-party certificate of compliance of the equipment is not really sufficient, *especially* for end-user who has the ultimate *responsibility* with the plant operating company's management, not the vendor. International standards call for compliance of SIS logic solvers to IEC 61508. Such third party certification does not cover:

- Application-specific issues as it is certified in isolation.
- Spurious trip vulnerability—specific for application.
- Issues with avoidance of CCFs for SIS and BPCS if common platform is used.
- The systematic errors inherent in using the same platform.

Also, long-term management of an embedded SIS is more about maintaining the legality. MOC will be really complex and complicated. Cyber attack vulnerability is also an important issue. Remote process monitoring, diagnostics, and maintenance and asset management through corporate WAN (web connectivity) have become an efficient tool causing highly increased vulnerability from hackers, worms, etc. for the "integrated" control system and SIS. Therefore, catastrophic failure may not be unlikely. In this connection, Clause 5.0.2 of Chapter IX may be referenced also.

2.2.3 About Existing Systems

There are a large amount of installations which have been in place for so many years much before performance-based standards like IEC 61508/61511 came in to effect. According to ISA-TR84.00.04-2005 Part 1, Guidelines for the Implementation of ANSI/ISA-84.00.01-2004 (IEC 61511 Mod), "for existing SIS designed and constructed in accordance with codes, standards, or practices prior to the issuance of this standard (e.g., ANSI/ISA-84.01-1996), the owner/operator shall determine that the equipment is designed, maintained, inspected, tested, and operating in a safe manner." The Technical Report focuses on:

- Confirmation: Risk analysis has been done to determine qualitatively or quantitatively the level of risk reduction needed for each SIF in the SIS.
- Confirmation: Assessment of the existing SIF, to determine required risk reduction delivery.

As per ISA-TR84.00.04-2005 Annex A.2 when any of the following conditions is met, detailed review is necessary:

- Modifications to the process unit that impact process risk managed by the SIS.
- Modifications to the control system that impact protection layers used to achieve safe operation.
- When an incident or near miss investigation has identified an SIS deficiency.
- When the review of another process unit designed according to similar practice has identified an SIS deficiency.

2.2.4 Discussions

Having said so many words about separation between BPCS and SIS, in reality it is found that the systems *may* have to be integrated at some level, so as to put forward some interface point for the plant personnel. Why so? BPCS and SIS using separate diverse hardware and programming for their unique operation of I/O, but need to be integrated at some level to exchange data and information so that plant operating and maintenance personnel can manage the issues easily. Such integration obviously calls for some amount of wiring and data mapping via serial communication or OPC, but much less wiring than separate systems which at times calls for more maintenance efforts also. Again this could also be vulnerable to error. So, a well-suited SIS needs to be developed. Also, in support of system integration, modern safety solution providers try to take advantage of fieldbus systems having "built in"

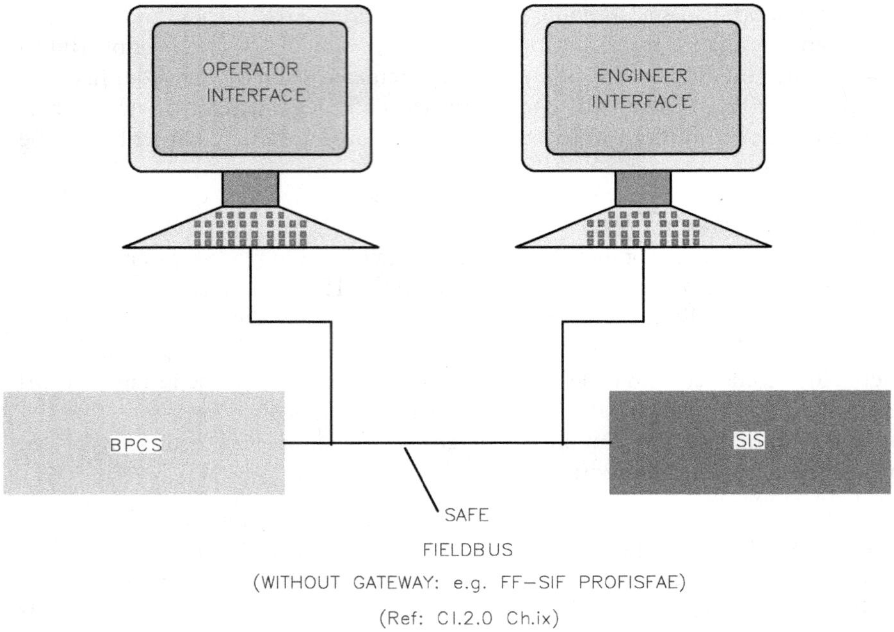

FIGURE VII/2.2.4-1

Integrated BPCS and SIS via Fieldbus.

safety to reduce wiring costs, improve diagnostics, and better manage assets. See Fig. VII/2.2.4-1.

These features of safe fieldbus systems (discussed in Chapter IX) attracts manufacturers towards system integrations mainly:

- For the reason that integrated safety systems can improve equipment availability, increased overall equipment efficiency.
- Improved return on investment.
- Harmonization of international safety standards, so that manufacturers and end-users to develop globally acceptable safety solutions.
- For limiting liability exposure and improvement of public image.

For further details on safety fieldbus systems, which would play important role in both BPCS and SIS, Clause 2.0 of Chapter IX may be referenced. From a security point of view, a short discussion about separation and integration of BPCS and SIS has been covered in Chapter XI also. Now it is time to move to some specifics: say of fire and gas system.

3.0 FIRE AND GAS SYSTEM IN SAFETY INSTRUMENTED SYSTEMS

Fire and gas (F&G) detection and prevention systems are critical in maintaining overall safety. Fire and gas system (FGS) are very important tools for safeguarding all plants and production facilities, especially those handle flammable and toxic materials. Facilities with inherent fire risk require the installation of FGS to mitigate these hazards. Offshore/onshore oil and gas facilities, refineries, chemical plants, tank farms and terminals, pipelines, and power plants are examples of a few facilities where FGS are extremely important (it in no case means that elsewhere these are not important). An FGS stands to continuously monitor a fire, or combustible or toxic gas release, etc. FGS is a mitigation safety layer, to reduce the consequences of hazardous events. Another major role of FGS is to initiate and automate ESD. FGS safety system comprises detection, logic control, and alarm and mitigation functions. Logic solver is the central control unit of the overall FGS detection and control system. The controller receives signals from monitoring devices/manual call point (MCP) deployed for detection. The controller processes these signals and takes the required actions to initiate alarms and mitigate the hazard as typically shown in Fig. VII/3.0-1.

In the present day, people take the help of new digital technology and computing systems to reduce the number of spurious alarms and to improve detection rates. In the modern day, FGS detection systems are generally PEs, with much higher safety availability, mitigation effectiveness, and safety integrity. It is possible to integrate FGS with the overall process safety strategy, ESDs, etc. As discussed earlier, there are two schools of thought regarding integration of these systems; many people, especially end-users, prefer to separate independent FGS system and ESD systems described in Fig. VII/3.1-2. However, bus communication to ESD may be accepted. Plants need to comply with strict standards and codes such as NFPA, API, and SEVESO II. There have been efforts towards ***appropriate*** integration with other systems keeping in mind the safety as well as efficiency. The FGS shall have *communications* with the DCS for making F&G graphics and alarms to the operator. This is in addition to independent displays for central monitoring of F&G as well as for emergency when DCS display may not be available. System architecture for such system discussed previously, has been depicted in Fig. VII/3.0-2.

The plant FGS, with addressable detection systems has been integrated with fire logic solvers. On the other hand, plant evacuation and system/site security center have been connected to FGS over bus for communication purpose only, without disturbing the independence of the same — as in the case of ESD. Here it is worth noting that the main focus is on FGS, so it has been shown in a different network connected with other via *gateway*. The reader needs to keep in mind that the network shown on the right-hand side of FGS is other network, where *ESD, SIS, and DCS have been shown, but in* ***no*** *case these are common system*. Each of them,

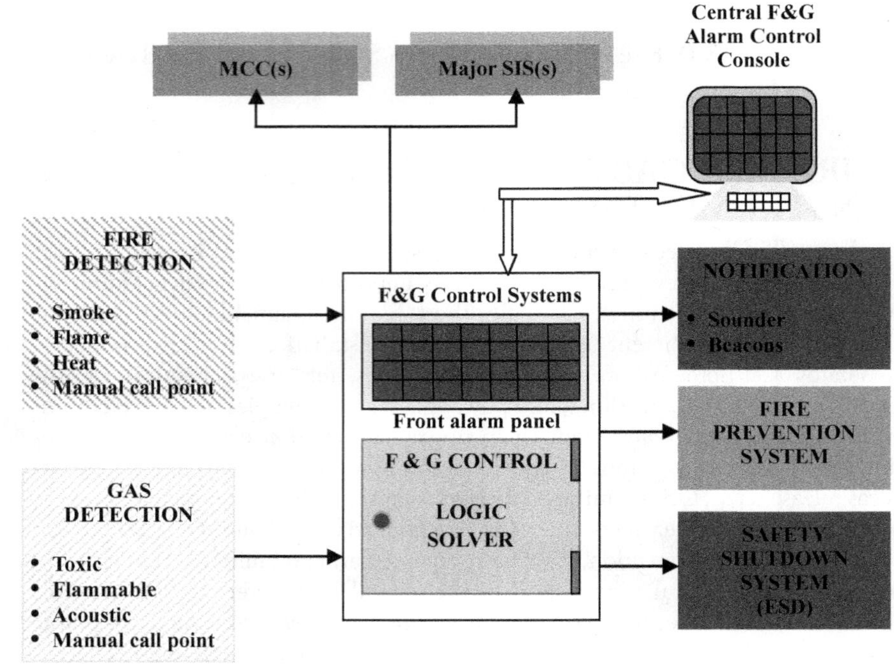

FIGURE VII/3.0-1

Basic block diagram for integrated fire and gas systems.

Inspired by Integrated Fire and Gas Solution - Improves Plant Safety and Business Performance, Honeywell Process Solutions; White Paper, April 2009; https://www.honeywellprocess.com/library/marketing/whitepapers/FireGasSystem_Whitepaper_April09.1.pdf. Courtesy Honeywell process solutions.

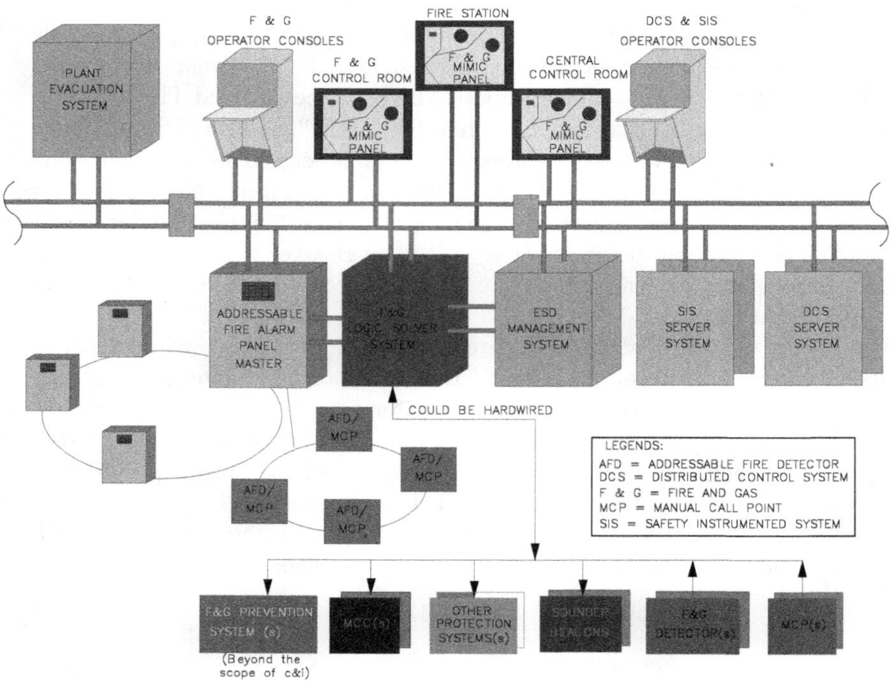

FIGURE VII/3.0-2

System architecture of typical integrated FGS.

as per standards, needs to be independent only for simplification common network has been shown with gateway. However, in this connection, integration in network should not be confused. Serial communications with ESD and addressable systems have also been shown in case bus integration for communication is *not* possible. For implementation of the IEC 61511 and S84.01 standards is becoming increasingly prevalent for F&G detection systems. An ISA technical report TR84.00.07 provides guidance on the evaluation of FGS effectiveness. A good FGS normally has an **SIL 3**-certification fire and gas logic solver. FGSs should be capable of early detection warnings of explosive and health hazards, including gas releases, thermal radiation, minute traces of smoke in sensitive, etc. and would provide audible and visual alarm to alert plant personnel about the gravity of situation. Since addressable detector can develop an envelope of the abnormal situation, chances of false alarm will be less when addressable detectors are deployed. The FGSs have interfaces with other systems, so that necessary protection signals are sent to various systems such as MCC(s), various control/protection systems viz. BMS. F&G automatically initiates executive actions through preventive systems to minimize escalation of hazardous situations. Integration at higher level can provides plant-wide SIS point data, diagnostics, etc., and could curtail the cost. "Safety system platforms based on quadruple modular redundant (QMR) diagnostic technology execute automated safety functions and provide the interfaces and input functions for standard connection of a wide range of fire and gas detection devices" [18]. FGS requires quire good amount of engineering for better results. Some of the major issues in engineering those need attention, include but are not limited to the following:

- FGS philosophy documents
- Performance criteria selection
- Assessment of detector technology
- Detector placement study
- Detector coverage assessment
- Cause—effect diagram
- Assessment of FGS functional safety
- FGS specifications
- Testing and maintenance supports

Why are these stated at the beginning? This is because of the fact that it has been found that people are more worried about SIL rating of F&G detector. It is an important issue but there are a few other critical issues as well. One needs to keep in mind that it is very critical in an FGS that sensors are placed optimally, otherwise hazardous gases and flames are not adequately detected and in that case the SIF/SIS will not be effective. Therefore, it is absolute necessity that proper engineering exercise has been carried out in selecting the detector and their placement. FGS is a subject by itself it is very difficult really to cover all aspects of the same. Here endeavored has been under taken to cover salient features of FGS in reference to SIS. The discussions will be mainly on detection and processing part which are related to instrumentation and not much on fire fighting system or prevention system, that is, mechanical engineering area.

3.1 Fire and Gas Systems: From Safety Instrumented System Perspective

Prior to the start of discussions on FGS, it is better to refer Fig. VII/3.1-1 for familiarization of commonly used terms in connection with FGS. Normally people require from a unified systems combining FGS and ESD so that there will be cost savings as well as uniform operator interface. However, many standards require that ESD (Ref Fig. VII/3.1-2) to be treated separately as an independent system with some communication maybe at the bus level, with due necessary security for access for each of these systems. For that reason in ESD and FGS have been shown in two different networks, with serial communications shown between them. However, common engineering tools may be used to reduce operational risks. Now discussions on some of the requirements pertaining to various standards will be taken up to see how FGS can be treated as SIS.

Addressable Device:. **A fire alarm system component with discrete identification that can have its status individually identified or that is used to individually control other functions.**

Flame Detector: **A radiant energy–sensing fire detector that detects the radiant energy emitted by a flame.**

Fire–Gas Detector: **A device that detects the presence of gases in an area, often as part of a safety system.**

Gas Detector: **A device that detects the presence of a specified gas concentration. Gas detectors can be either spot-type or line-type detectors.**

Heat Detector: **A fire detector that detects either abnormally high temperature or rate of temperature rise, or both detects gases produced by a fire.**

Fixed-Temperature Detector: **A device that responds when its operating element becomes heated to a predetermined level.**

Smoke Detector: **A device that detects visible or invisible particles of combustion.**

Flammable gas: **Gas or vapor which, when mixed with air in a certain proportion, will form an explosive atmosphere.**

Lower flammable limit: ***V*olume fraction of flammable gas or vapor in air below, which an explosive gas atmosphere does not form, expressed as a percentage. This is popularly known as lower explosive limit (*LEL*).**

Sensor: **Assembly in which the sensing element is housed and that may also contain associated circuit components. Sensor that is not integral to the main body of the apparatus is remote sensor.**

Upper flammable limit: **Volume fraction of flammable gas or vapor in air above, which an explosive gas atmosphere does not form, expressed as a percentage ---upper explosive limit (UEL).**

FIGURE VII/3.1-1

Selected commonly used FGS terms from standards.

Emergency shutdown system (ESD): **ESD consists of sensors, logic solvers, and final elements for taking the process, or specific equipment in the process, to a safe state when predetermined conditions are violated. ESD takes the process to safe state by isolating, de-energizing, shutting down, or de-pressurizing the process unit or process equipment. ESD is a preventive risk reduction process and is normally an independent system, i.e, Independent set of sensors, logic solvers, and final elements. However it can have communication with other systems through serial communication/ system bus.**

FIGURE VII/3.1-2

Emergency shutdown system.

3.1.1 *Fire and Gas System as Independent System (NORSOK)*

In the previous discussions, it has been stated that the two systems, viz. ESD and FGS shall be independent. In support of the same, in Clause 12.4.9 of NORSOK S001: 2008, one can get that "the F&G detection system shall operate as an independent system. F&G detection safety instrumented functions shall be functionally and physically segregated from other systems or functions but can be *integrated as a node* within the SAS:

- F&G units (logic solver) shall not be used for other than F&G related safety functions.
- F&G field devices (arrangement of sensor and final element) shall be separate and independent from other devices, and used for the safety function only including activation of equipment such as fire dampers."

However, common HMI for safety system operator is permitted. These are elaborated to indicate that FGS is treated as another safety system — SIS. Also, Clause 10.4.7 of NORSOK S001:2008 demands that "ESD system shall operate as an independent system. ESD functions shall be functionally and physically segregated from other systems/functions but the ESD logic solver can be an *integrated node* in the overall SAS:

- ESD units (logic solver) shall not be used for other than ESD functions
- If applied for ESD, sensor accessories (e.g., process tapping, impulse lines, air supply branch-off and power fuses) shall be separate from other functions such as process control devices and process shut down devices." Therefore, while taking up the designs of FGS or ESD, similar requirements need to be kept in mind. Another important issue in this regard is the interpretation of the standard, for example, here common HMI stations are allowed. However if such HMIs are used then some amount of data mapping is necessary which may be treated as a sort of integration! In this connection, Clause 2.2.4 above may be referenced.

3.1.2 *NFPA 72 for Fire Detector Systems*

From the developmental work of the ISA S84 WG 6 Committee, it is found that attempts have been made to treat mitigating system FGS as SISs, and the system requires an appropriate SIL rating (as per IEC 61511 for performance criteria) be achieved. The NFPA 72:2013 (incoming 2016 version) deals with application, installation, location, performance, and maintenance of fire alarm systems and their components. Earlier standards were mainly perspective, but now same standard has focused on performance system to support IEC 61508/61511. However, the latest edition NFPA 72 Standard, Annex B, provides a "performance basis for location and spacing of fire detectors" so that "fire growth rate (B2.3.2.3) and fire signature" can be developed with respect to the "component characteristics," and "damageability" of the target space including occupants, equipment, structures, etc. This will be necessary for determination of "location of specific types of detector to meet the objective established for the system." So that standard focuses on:

- Location and spacing of types of detectors to define performance goals and objective
- System designing and evaluation

For various goals and objectives, tables Table B.2.2.4.1(a, b, c) of Annex B of the same standard may be referenced. The standard also focuses on various types of detectors (e.g., Clause 3.3.266 on smoke detectors), executive software and site-specific software, etc. From "Application of IEC 61508 and 61511in the Norwegian industries—OLF 70," one gets that in Norwegian oil industries, fire detectors have been SIL(2) rated (see Clause 7.6 and Appendix A8 of OLF 70). From the preceding, one gets some idea about performance-based fire detection systems, but for gas detection system the reader may look into other standards discussed in the following sections.

3.1.3 *Gas Detection System: IEC 60079-29/EN 50402 and Norwegian PDS (OLF) Handbook*

As discussed earlier, if "detection of release of gas" is considered as a safety function, then it could be included as a part of IEC 61508/61511 also, and a necessary SIL rating could be established. Standard IEC 60079-29 and standard EN 50402 deal with gas detection systems (Clause 3.4 of Chapter X, gas detection systems have been discussed). In these standards, the focus is mainly on detector hardware and efforts have been made to *come out* with some ratings. However, this is not an easy task on account of some other issues associated with the same discussed here. It is known that for SIL2, minimum risk reduction is >100. Now there are two issues here. One is *detector* failure, and the other is *detection* failure. From various data, it has been found that detection coverage in many cases are <90% (e.g., typical detection coverage is 60%), and mitigation effectiveness is <90%. So, this would pose problems in SIL ratings discussed in Clause 3.2. *Why not higher coverage?* There are several factors which may affect the detection, even if the detector is working perfectly but cannot see the gas. Such factors may include: *wind (direction)* can dilute the gas, *physical obstruction*, *bad location* of the detector, etc. Therefore, it is necessary to take all of these into consideration. In this connection, *BS EN 60079-29-2* may be helpful for location of detectors which provides some guiding points to focus on:

- Sensor and sampling points shall be located in such a way that accumulation of gas could be detected before it can create a significant hazard.
- Location shall be done in consultation with people having knowledge on gas dispersion system and knowledge on plants and equipment.
- Combined effect of source and propagation needs to be considered.

- To find where is the source; inside or outside structure.
- Local environmental conditions.
- Access points such as window or door.
- Access for maintenance.
- Occupation health and safety requirements.
- Sensors to be located near potential source of release of gas, and those places where there is probability of accumulation of gas.

There are some guidelines and data available from PDS method handbook and OLF standards from Norwegian oil industries (Application of IEC 61508 and 61511in the Norwegian industries—*OLF 70** including appendix A8 and A9 of the same document); some selected guiding points are:

- Development of SIL requirements: Sub-function gas detector need for SIL (Section 7.6) (Table 7.1 continued minimum SIL requirements – global safety functions*)
- Systematic fault data PDS handbook provides generic fault data for gas detectors
- Also voting pattern (Moo*N*) for gas detectors

3.1.4 *Major Steps for Performance-Based Fire and Gas System Analysis*

The following points may be noted for performance-based FGS analysis (in line with ISA TR 84.00.07 guidelines):

- Screen to determine if FGS function is necessary
- Identification of risk scenario
- Analysis of consequences
- Analysis of hazard frequency
- Unmitigated risk assessment
- Identification of device for risk reduction
- Initial FGS design
- Assessment of detector coverage
- Assessment of FGS safety availability
- Mitigated risk assessment
- FGS design modification

Detector types and coverage: For gas detection to be effective, it is needless to tell that proper detector type is extremely important. Selection of appropriate detector technology and deployment of the right number of detectors at the correct/appropriate location, so that the system can respond properly on demand is extremely important. There are two kinds of detector for coverage:

- Geographical coverage assessment: This is the fraction of the geometric area (in a particular elevation) of a defined monitored process area which if a release were to occur in a given geographical area [21]. Geographic coverage is homogeneous, but not sensitive to the frequency of hazard in any specific area.
- Scenario coverage assessment: It is defined as the fraction of the released scenario that would occur as a result of the loss of containment in a defined monitored process area that can be detected by detectors considering the frequency and magnitude of release scenarios and the defined voting arrangement [21].

3.1.5 Perspective and Performance-Based Systems

For prescriptive standards for FGS, there was a set of comprehensive rules for system design. Mainly fire systems and often some gas detection systems were designed according to these rules. Though these types of designs are good for occupied buildings, they have a number of shortcomings for process industries mainly in *open areas*, for example, fuel storage tanks where really there are maximum chances of flammable gases. It is really challenging for gas detection systems, when one tries trying to use risk analysis techniques that are compliant with ISA84/IEC 61511 standards for SISs. There are various sources of leakage in fuel storage tank farm. Many leaks may arise from:

- Holes due to corrosion
- Faulty flange or valve due to high pressure
- Fugitive emission from tank

Naturally, it is difficult to characterize them, and also LOPA does not include initiating events due corrosion, etc. *Detection coverage factor, related to detection effectiveness, the number of detectors, and the placement of detectors* at a given elevation are very important for performance-based systems. For gas detectors, this also depends on the type of gas, gas density relative to ambient air, etc. Detection response time is very critical as the FGS must operate within its mitigation limits. Also, ESDs demand for a definite specified response time. Performance approach for FGS has two parts:

- Fire: Fire part is taken care of by NFPA 72:2013, as discussed earlier. From the standard, one gets that performance-based design is a design of a system for achieving specific level of performance and not "one-size-fits-all," as in perspective design. This is flexible type of design requires the designer to carry out the risk assessment for determination of level of performance desired then start designing the system according to the standard to achieve the required level of performance.
- Gas detection: From ISA TR 84.00.07, one can get that gas detection systems can be designed similar to an SIS, when detector coverage is considered as an additional performance metric. In this way one needs to assign targets for safety availability (equivalent to SIL), targets for detector coverage. Also, performance-based standards calls for other factors (normally encountered in SIS) like diagnostics, redundancies and voting, and verification. All these enable the validation of detector coverage of the gas detection system design. ISA TR 84.00.07 suggests that coverage by the initial design detector layout to be calculated and *verified* to ascertain that it can meet the target — if not, to modify the same. Also the safety *availability* (~SIL) for each function needs to be calculated and verified in line (similar) with the IEC 61511 standard.

The FGS is inevitable for the purpose of maintenance of operational integrity of the overall facility. The FGS functions to mitigate the effects of a hazard to address mainly:

- Human safety
- Loss of property/business

- Environmental Impact

As discussed earlier, there are two failures; detector failure and detection failure. Detection failure is of more importance in case of performance-based system because even if detector is perfectly sensing, it may be imperfect on account of many issues already discussed.

3.1.6 *Description*

For performance-based systems, one needs to carry out risk analysis, and risk is calculated as the product of frequency and consequences. The consequence reduction can be achieved by installing detection systems for fires, explosions, toxicity, etc. Large leaks can be detected by pressure sensors. Medium and larger leaks are easier to detect due to the size of the gas cloud than smaller leaks. The most dangerous situations arise from small to *small—medium* leaks for which importance of gas detection cannot be overestimated. However, in addition to mitigating FGS, preventive ESD (Fig. VII/3.1-2) and blow down systems are also activated to minimize ignition probability. The best way to locate a gas detector is where the frequency or chances are more. However, determination of the number and location of gas detectors is not a simple task as it is influenced by a number of factors like: gas type, location, direction, wind direction, wind speed, and leak rate cost, coverage, leak detection time, etc. Depending on applicability, the same logic is true for fire detectors also. Latest version NFPA 72 is available for as good guide. ESD mainly reduces the frequency, and mitigation FGS reduces the consequences. System response time is also very critical. For risk scenario assessment, risk matrix and LOPA are effective and good tools for the same. Then comparing with the company's risk guidelines, it will quickly indicate the acceptability of the system. When mitigated risks satisfies the company's risk criteria, the design for this hazard is complete, otherwise improvement by increasing detector coverage, or improving FGS safety availability will be necessary. Geographic detector coverage, discussed earlier, is selected on the basis of design criteria of fraction of total area coverage (e.g., gas cloud of less than X meters at YY% LEL (lower explosive limit) and voting logic for the detectors; i.e., 1ooX or 2ooX, etc.). Based on the total area, the fraction that is covered by 1ooX, 2ooX, etc. determines the overall geographic coverage [20]. Scenario detector is voted (i.e., 1ooX or 2ooX) to detect the hazard. Each identified hazard is addressed in this manner, and the overall coverage is determined by the ratio of detectable release frequency to total release frequency (for all hazards) [20].

Availability $= 1 - PFD_{avg}$, where PFD_{avg} is the average value of the probability of dangerous failure over the proof test interval. Like other SIS, for FGS, the PFD_{avg} for each safety function is the summation of the sensor PFD_{avg}, the logic solver PFD_{avg}, and the PFD_{avg} of the mitigation devices. Although it is not mandatory to obtain SIL for FGS detector yet, some standards demand for the same.

3.2 Fire and Gas System and Safety Integrity Level Issues With Short Planning Discussions

The controversy of associating SIL with mitigating layer FGS is still on. People in the industry are divided into two schools of thought — one supporting the issue, and the other is not in a mood to accept. However, one thing is for sure, that it is most important for end-users to ensure the selection of proper sensors and their placement within FGS. Incorrect placement of FGS detectors will result inadequate detection of hazardous gases and flames, then the SIF/SIS issues will have no significance, whatever SIL level for detectors be chosen.

Instead of going into the debate, it is better to make a case study with very optimistic scenario with detection coverage more than 90% and detector safety availability 98% with complete mitigation effectiveness (overall system response probability to prevent/***mitigate*** hazardous situation). From Fig. VII/3.2-1 and calculations, it is found that even with these assumptions SIL of FGS here is SIL1. There may be some *improvements* with application of redundancy and voting circuits. Some of these are discussed here. When there is a single sensor, when it senses, action is taken. When there are two sensors, action will be taken when both senses the leak or fire. This of course decreases the probability of nuisance alarm, but at the same time reduces the chances of detection (see Fig. VII/1.4.5-1 for similarity). Similar redundancy approach may be applied for LEL of gas, that is, if any one of (say) three detectors senses >*X*% LEL then it will be taken. FGS equipment manufacturers always like to favor attaching SIL rating to their product so that they can have some edge over others in this steep competitive market. Also, a few standards discussed earlier demand SIL rating! So, FGS detection manufacturers are enhancing the safety of their products and improving their processes by designing to comply with IEC 61508 standards. There remains enough room for which legitimate arguments could be made both for and against classifying FGS as an SIS and attaching SIL numbering to them. The opposition group feels that there are inherent limitations of FGS detection

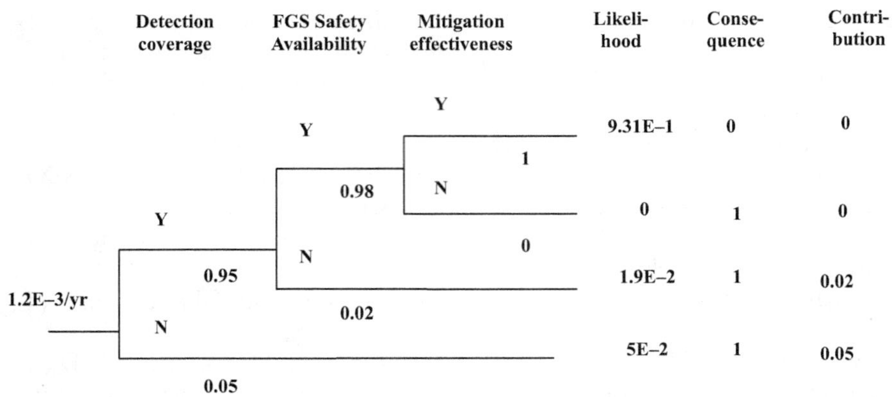

So, total weighted average is approximately 0.019 + 0.05 = 0.069 = 0.07. Therefore risk will be 1.2E–3 * 0.07 = 8.4E–5/yr. Likelihood of mitigated risk is 93% and its frequency will be 0.93 * 1.2E–3 (~1.116E–3). Pfd = 1 – availability = 1 – 0.93 = 0.07. Therefore risk reduction factor is 1/0.07 = 14.28 (hence <100) so SIL 1.

FIGURE VII/3.2-1

Event tree for FGS model for SIL estimation.

equipment; FGS could be only augmentation to a plant safety system and not as safety critical as SIS. They feel that on account of too much importance placed on the hardware of FGS, unnecessarily instills a false sense of security. In that case, *is it necessary to have a separate standard to address new performance-based systems specifically for FGS to adequately address the issue of detector technology, detector placement, coverage, and related issues!* Those who advocate for FGS as SIS, feel that FGS detectors are an effective mitigation layer and should fall within the scope of IEC 61508/61511. It can be conceived as "a facility chooses to implement an FGS as a layer of protection while understanding the inherent limitations of the technology, then the system should conform to the guidelines put forth in both 61508 and 61511" [22]. From the discussions in Fig. VII/3.2-1, one thing is clear, that merely buying SIL based sensors, logic solvers, etc. alone won't give much improvement in results. In this connection, refer to Clause 3.1 of Chapter VIII also.

3.2.1 *Gas detector guiding points*

General guiding points may be as follows:

- Gas type: Noxious, explosive, or both?
- Source analysis: Potential sources of leakage determined by process hazards analysis (PHA), no small, unmonitored source left.
- Positioning strategy: Point detectors only or hybrid approach of point and line of sight infrared
- Number of sensors: Duplication/triplications taking into account failure/maintenance.
- Voting strategy
- Measurement technique
- Requirement of aspirators
- Alert and shutdown levels requirements (%LEL)

3.2.2 *Gas Detector Positioning Guidelines*

Some positioning guidelines may be as follows [23]:

- Coverage Density: For greater numbers of potential sources of leakage spacing may be ~5 meters gap between detectors.
- Toxic gases, which are heavier than air, monitor them around head breathing zone.
- Positioning close to ground for vapors from combustible liquids from density point of view.
- Normally H_2, NH_3, and CH_4 gases are lighter than air, so detectors shall be near ceiling for gas accumulation.
- For forced ventilation, the gas detectors should be sited at the inlet and in the ducting.

NFPA 72 provides good guidelines for fire detectors. In order to achieve real functional safety, the safety life cycle analysis, which will be discussed next, could be helpful.

3.3 Performance-Based Fire and Gas System — Discussions

As the name implies, performance-based design is a type of design meant to achieve specified level of performance. It is different from "one-size-fits-all" basis of perspective type of approach. A typical performance-based FGS has been shown in Fig. VII/3.3-1. It is not that performance-based systems prescribe any specific solutions, instead it is meant to demonstrate that the proposed design meets the design objective.

From Fig. VII/3.3-1, it is possible to comprehend that in this approach it is possible to translate design objective into quantifiable parameters necessary to set limits parameters and also to develop means to estimate the performances of deliverables in the proposed design. These estimated design performance parameters are compared with the required performance parameters set forth for the design. On the contrary, perspective type approach mainly looks for the following area:

- Codes: To specify functional objectives and performance requirements to reflect expectations.
- Guidelines: Separate documents adapted for reference of accepted methodology.
- Tools: They provide accepted methods for development, review, and design verification as per good engineering practice and guidelines.

Both the systems have been shown conceptually in Fig. VII/3.3-2 to have a comparative look. The left side approach is perspective approach, whereas the right side approach is a *performance-based approach*. For basic design, there are two ways to look into the problem; one is preventive, when necessary precautions are taken in design to reduce the probability of leakage and ignition. The second way is to control and mitigate the consequence of hazardous situation. This is the main concern of FGS. FGS is used to reduce release inventory, reduce consequence due to un-ignited release (cryogenic leaks), and ignited releases.

Typical comparison between the two systems has been presented in Table VII/3.3-1.

Having discussed so far various aspects of both types of design, it is now time to concentrate on performance-based systems. Short discussions on system design basis and discussions on various stages described in Fig. VII/3.3-1 shall be taken up. For this, a detailed study is necessary for development of design basis pertinent to FGS. This will help in proper definition of design objective, performance criteria in terms of detector layout, and logic functionality which needs to be defined in such a way to achieve the target risk level in line with company standards [19].

3.3.1 *Design Philosophy Document*

Probably FGS philosophy document is the first step for development of actual design basis before developing the specification for the same. For FGS design philosophy, good guidance is available from DEP 32.30.20.11-Gen (Shell). FGS philosophy is developed for basically meaningful appropriate definitions of various tools and techniques. Through this document, various safety policies of the company, desired procedural details are properly expressed in line with applicable standards (e.g., IEC

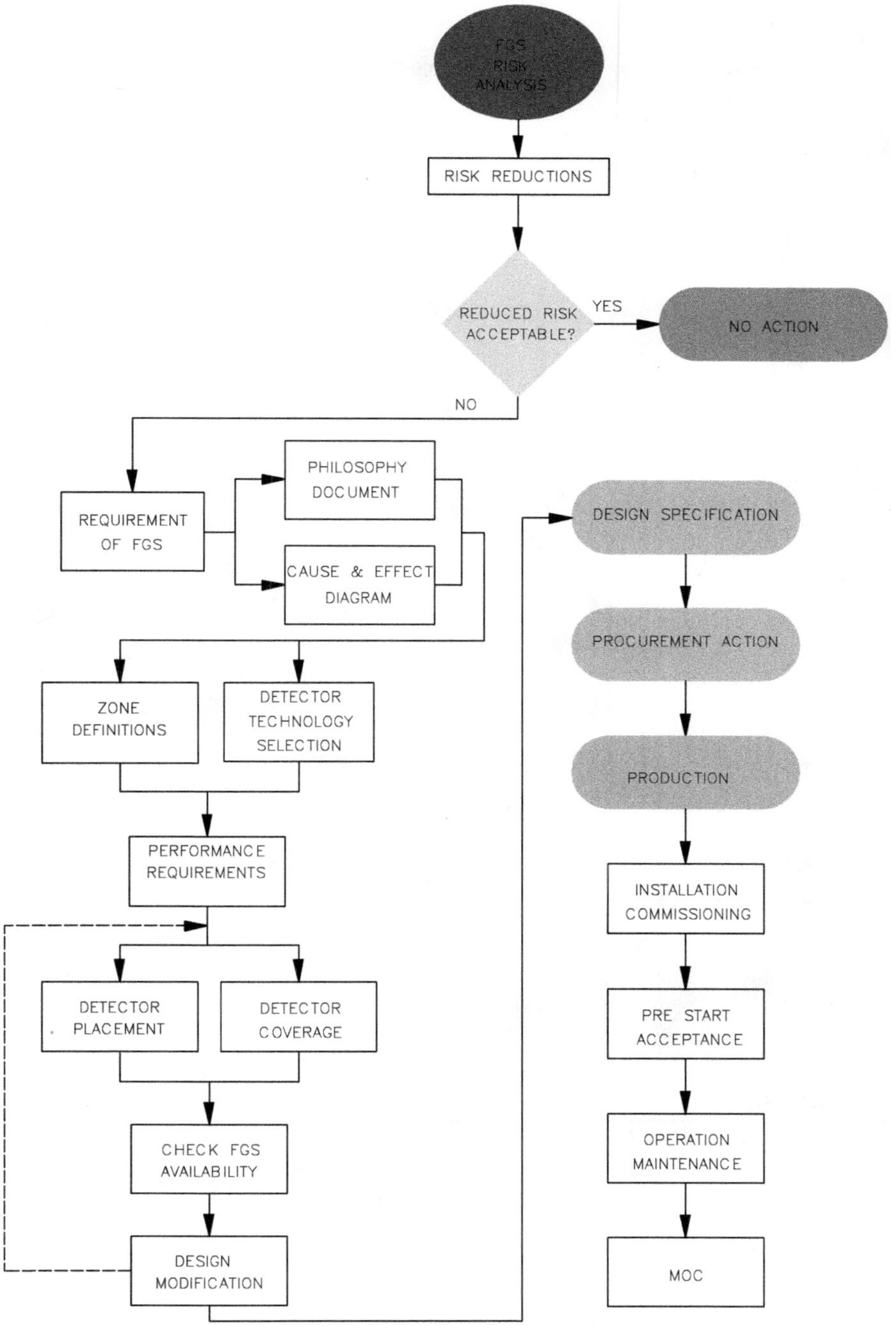

FIGURE VII/3.3-1

Performance-based FGS.

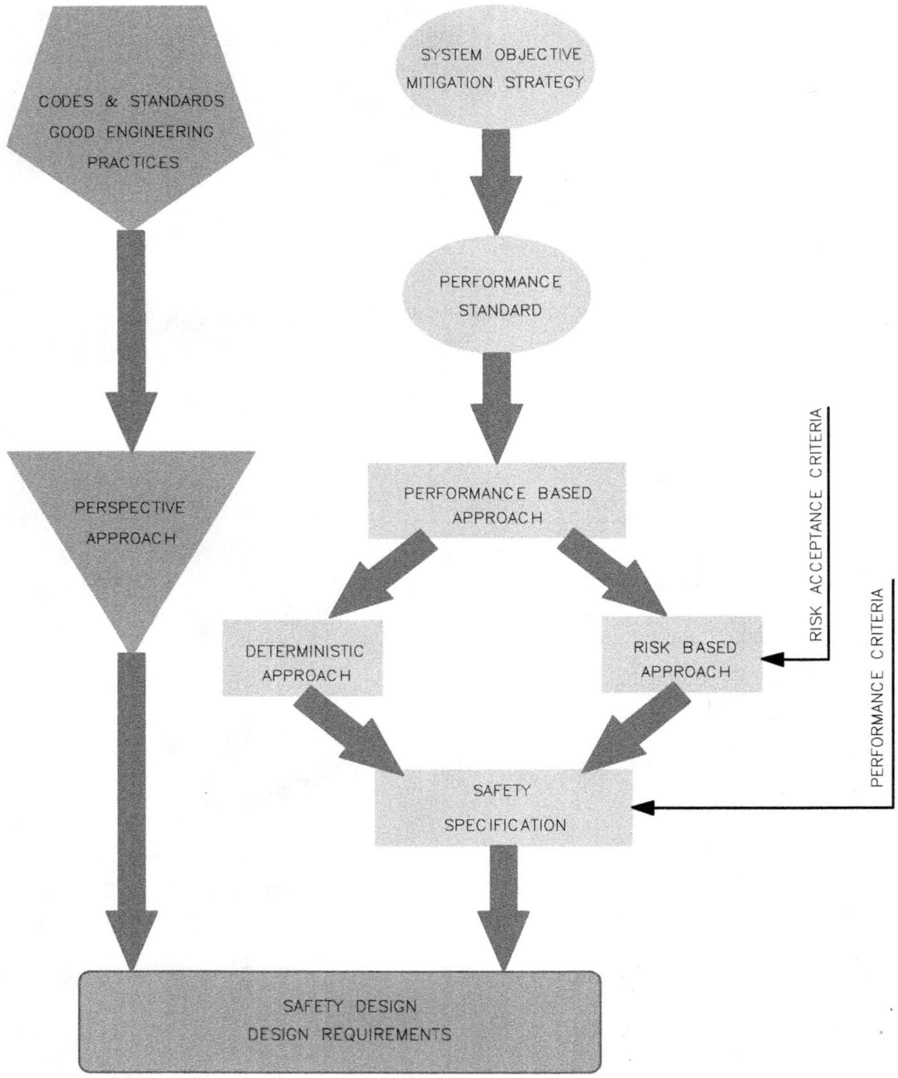

FIGURE VII/3.3-2

Different approaches for safety design.

Developed inspired by J. Hocquet, Y. Le Gourrierec, L. Paris, Challenges in using risk- and performance-based design methods for FLNG safety engineering, Technip; www.gastechnology.org/Training/Documents/LNG17-proceedings/12_07-Jerome-Hocquet-Presentation.pdf.

Table VII/3.3-1 Comparison Between Perspective and Performance-based Approaches

Issues	Perspective Design	Performance-based Design
Philosophy	Focused on means and method	Focused on objective and performance
Application	Conventional, suitable for within building application but limitation in open space	Flexible and can cope up with requirements including open space
Type	Straightforward, code based, well-known and controlled method	Definition of objective and associate performance requirements is essential
Strength	Compliance and endorsement is to achieve. Approach is generalized	Optimization, cost reduction possible. Specific approach as it is not one-size-fits-all
Weakness	Implicit objective difficult to handle any special case and longer procedure for accepting any deviation	Definition of acceptance criteria is difficult hence acceptance is difficult. Requires more resources. Time consuming during engineering phase
Safety life cycle	Not applicable	Safety management system required during entire life cycle

61511 compliant). Therefore once this is developed can be used as main guideline documents for all projects with specific changes required for specific facilities. The basic purpose of FGS philosophy shall include but is not limited to the following:

- Method/procedure standardization.
- Hazards identification criteria.
- Process conditions and constraints.
- Quantify requirements of sensors.
- Sensor types and usage criteria.
- Redundancy and voting guideline.
- Procedures and methods guidelines.
- Zone definition and grading criteria.

Therefore, good emphasis is required during its development. General coverage of the document shall include, but is not limited to, the following:

- Detection philosophy general (and *specific* for facilities — as applicable)
- Quantity for fire detectors and flammable and toxic gas detectors for the facility
- Detector voting philosophy during normal operation as well as the start-up and shutdown periods
- Types of detectors
- Selection philosophy for detector types and technology
- Short, qualitative specification of detectors

- FGS system architecture
- Compliance with IPS philosophy and relevant standards

FGS functionally performs a number of functions such as monitoring of fire, smoke, and/or sensing accumulation of gas. On detection generates alarms and initiates mitigating actions. It consists of sensors, logic solvers, and final elements to initiate mitigating actions. Additionally, it also interfaces with other systems in IPS and SIS. Within the system, it diagnoses the fault of sensors, etc., and at times inhibits the same or supports manual inhibition with alarm in each case. As stated earlier, philosophy documents lists out quantity of various types of FGS sensors at various locations in a tabular form with various facility areas as rows. The columns indicate various types of detectors such as:

- Flammable gas detector — line of sight, point
- Toxic gas detectors
- Optical fire detector
- Smoke detector

Each cell indicates the quantity and type of detectors in the area. Selection of each type of FGS detector will be done on the basis of detector technology assessment study (strength, weakness, and applicability) from the list of all available detector technologies for fire detection and gas detection. A comparison of the performance affecting attributes of each zone (discussed later) against the relative strengths and weaknesses of each detector type shall be prepared to select detector technology for the zones. The design philosophy document may only *list of out types* based on broad category of detectors and *technology selection* may be done at later part in design basis documentation also. Short, general guidelines for sensor selections have been enumerated as follows:

- Flammable detector: In open/process area the hydrocarbon leakages occurs and care needs to be taken to ensure that they are not very likely to escalate into fires or explosions within the confines of the plant. This objective is attended by layout optimization during the design stage, so that this leakage and air mix is not ignited by electrical or any other source of ignition. Also, it is important to promptly detect the leakage so that leakage is arrested immediately, without further escalations. For further details, refer to Clause 3.4 of Chapter X.
- Toxic gas and other releases: Release of toxic gases into the atmosphere will generally create a health hazard. The same is the case for un-ignited release such as cryogenic leaks in floating liquefied natural gas (FLNG). However, with H_2S and CO, situations for fire or explosion may arise also. There are two categories such as category which creates immediate health hazard other long-term health hazards. So, all these call for suitable sensor selections.
- Fire detectors: Depending on application and property of the detector fire detectors are selected based on matching. Major types are IR fire detectors, smoke detectors, heat detectors, etc. to name a few.

- Fire and gas mapping for coverage verification: Fire and gas mapping is necessary to verify that the coverage targets that were selected for each fire and gas zone are achieved by the proposed design. Design philosophy documents provide some guidance for the same. A few points are listed as follows:
  - Layout and orientation to check sufficiency of coverage, unimpeded view of detectors.
  - Sufficient range selected for process and environmental condition.
  - Geographic coverage verification for FGS detector array.
  - Use of computer aided tools for mapping for analysis and modeling, etc.

  Some guidelines for accurate modeling may be considered here. Actually these may be done at later stage under design basis documents.
- Detection coverage criteria: From DEP 32.30.20.11-Gen September 2014, one may get some indication regarding coverage:
  - 90% for a single detector alarm based on *N* detectors in the zone.
  - 85% for two or more detectors alarming based on *N* detectors in the zone, where N ($\geq$3) is the number of detectors in the zone.
- Alarm levels, detector settings: Depending applicability of standards and company standards various limits for alarm level 1, 2, etc. are set. There are three terms important here; these are ceiling, short-term exposure limit (STEL), and time weighted average (TWA). Refer to Fig. VII/3.3.1-2 for definitions and explanation of these terms. Some cases LELs, are set below STEL level and some refer to TWA. Here, one needs to note that presence of too much oxygen is also not permitted. According to OSHA, the presence of O_2 (in %) should be restricted to little above 23%. The alarm settings are done and indicated in philosophy documents. Some of ***typical indicative*** alarm setting has been indicated in Table VII/3.3.1-1.
- Detector voting: In the early part of the discussions, it has been indicated that there are pros and cons of adapting Moo*N* redundancy. Major issues were that no case should be missed, especially in low and medium leakage. At the same time, spurious trips shall be avoided as far as possible. These two are conflicting demands for detector voting logic. As a part of normal operation, associated always tries to diagnose fault of the detectors, either directly or via zonal address modules, etc. Whenever a detector is found faulty, it shall be kept out of activation or may be inhibited. Also, manual inhibition is possible. For better understanding, see Fig. VII/3.3.1-1. Each of the faulty or inhibited detectors are also put in an OR logic for taking care during voting. To reduce the number of point detectors, a strategy of "2oo*N*" shall be applied where normally two detectors are required for confirmed action. All detectors shall be voted discriminative. That is, where one detector develops a fault, or the signal is lost, then that loop is automatically reverted to "high" or "alarm," so that where only two detectors are provided in one fire zone, then the system defaults to "1oo1" arrangement. So, in *N* numbers of detectors, action will be taken when there is sensing in $\geq$2 detector OR when sensing is in one detection AND $\geq$ 1 detectors are in fault or inhibition state (e.g., if $N = 4$ and one detector is in fault and one

Table VII/3.3.1-1 Typical Alarm Settings

Gas	Ceiling[a]	TWA	STEL
CO	35 ppm	100 ppm	100 ppm
H_2S	10 ppm	10 ppm	15 ppm
SO_2	2 ppm	2 ppm	5 ppm
O_2	Low 19.5% high 23.5%		
LEL	10% LEL		

[a]Ceiling: Maximum gas concentration to which a worker can be exposed (see Fig. VII/3.3.1-2 also).
Courtesy ***Biosystems*** *- L. Boettern, Where to set your confined space gas detector alarms; Biosystems Application note no: AN20010828, August 2001.*

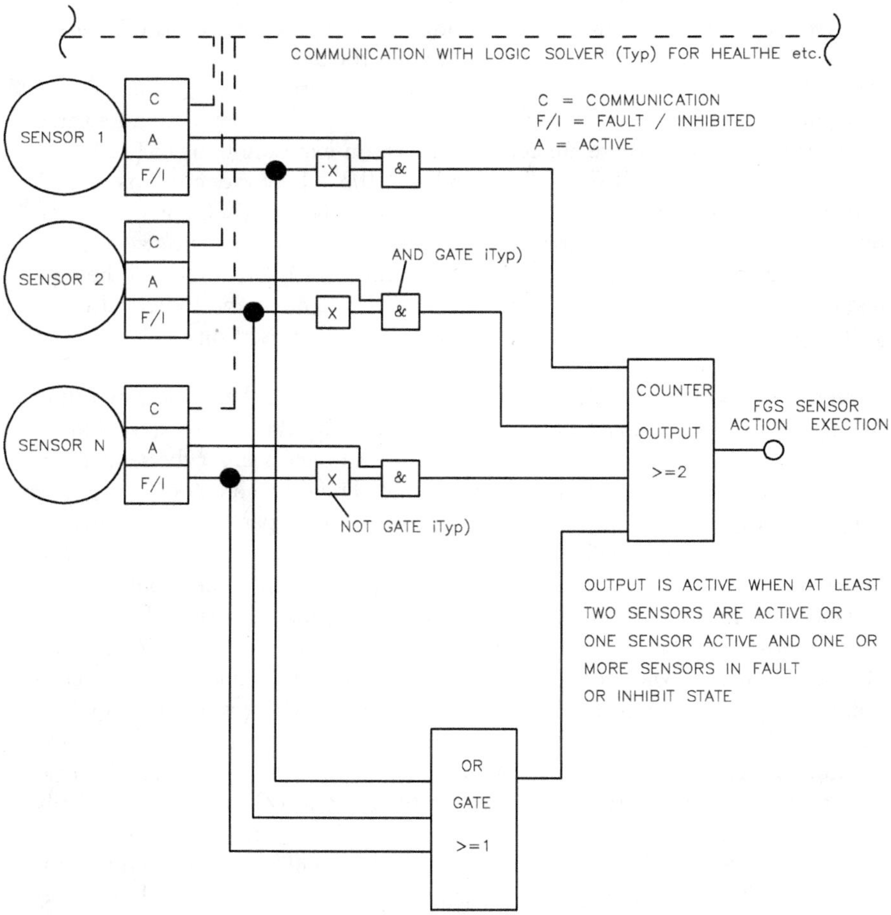

FIGURE VII/3.3.1-1

Typical safe voting logic for FGS sensors.

Threshold Limit Value (TLV): **Refers to the airborne concentration of substances and represents conditions under which it is believed that nearly all workers may be repeatedly exposed day after day without adverse health effects.**

TLV – TWA (Time Weighted Average): **This is the average amount of gas that a worker can be repeatedly exposed to in a normal 8-hour day, in a 40-hour week, without adverse health effects.**

TLV - STEL (Short Term Exposure Limit): **The gas concentration that most workers can be continuously exposed to for a 15-minute time period without suffering adverse health affects that would impair selfrescue or worker safety. This limit should not be repeated more than 4 times per day and there should be at least 60 minutes between individual STEL exposure periods.**

TLV – Ceiling: **The highest gas concentration to which workers may be exposed. Ceiling TLVs should never be exceeded and they take precedence over all TWAs and STELs.**

FIGURE VII/3.3.1-2

Threshold value definitions and units.

detector senses). The same logic has been shown in Fig. VII/3.3-1 which is self explanatory.

- Cause—effect diagram: Cause and effect diagrams are normally included as basic documents in philosophy documents. FGSs are typically used to initiate typically the following functions automatically or manually:
  - Audible and visual alarms/messages at appropriate places
  - Messages of emergency situations to local community services
  - Starting of firefighting equipment, for example, fire water pumps
  - Activation of safeguarding and depressurization (blow down systems)
  - Automatic closure of air ventilation inlets
  - Personnel evacuation systems (barriers, howlers, etc.)

  All of these are very well *visualized* in "cause and effect diagram" which is presented in a tabular matrix form. Normally, various causes such as manual call point, heat (say) rate of rise, flame detection (say) infrared, etc., whereas effect gives the details like audible alarm, audio visual alarm in DCS, etc. Generally these documents need approval (from competent authority).
- HMI, IPS, and operational philosophy: It is quite clear from the heading that details of HMI needs to be described. These descriptions mainly cover various control center and specific functionality associated with each of them. It also details out in each case the major displays and controls with necessary hierarchy. Obviously, for such philosophy development overall plant operational philosophy play key roles and the same must be very well-described. Also, HMI is greatly related with system architecture. Therefore brief design details about IPS system architecture will be necessary and its interface with other systems such SIS, DCS, etc. are also necessary.
- Standards and interface: Details of standards based on which the philosophy document has been developed *must* be spelled out very clearly. It shall also state at which points compliances are necessary at various stages (discussed in subsequent clauses) such as specification stage, design stage, installation stage, etc.

It is now clear that philosophy document is essentially the main stepping-stone for further design.

3.3.2 *Zone Categorization and Setting Up Performance Requirements*

After FGS philosophy documents are prepared, it is necessary to categorize the zone of the facilities. Risk analysis is also necessary to set the performance target of each FGS function. To get proper idea about the extent of consequence of typical scenario from risks, consequence modeling is very important tool.

- For zone classification process flow diagram (PFD), P&IDs, cause and effect diagrams, area classification of plants for electrical apparatus (ref. Chapter X) and plot plans are required documents for zone classification. In process plants, different areas have different leakage or release hazards, for example, hydrocarbon storage area will have flammable gas release whereas sulfur recovery units pose a H_2S toxic gas hazard. So, it is imperative to define and segregate the zones from each other.
- Similar equipment grouping with similar process hazards
- FGS system actions – confirmed fire or confirmed gas releases
- Deck-wise differentiation
- Segregation of hazards
- Classified electrical equipment (see Chapter X)
- Based on occupancies, for example, turbine house/living quarters
- Performance requirements: Risk analysis and consequence modeling: Based on philosophy document, zone categorization details actual risk analysis that needs to be carried out. This is necessary to set the performance target for each of FGS function. The risk analysis documentation are developed based on company policy procedure with philosophy document in mind. Based on risk of safety and/ or commercial impacts of fire and gas releases in the zone, each FGS function could be assigned with SIL target if applicable! As a part of the risk analysis for determination of detailed and extents of the typical scenarios for the facility, consequence modeling is very important. Proper understanding of the potential consequences of a hazard here is essential, so that an appropriate degree of integrity can be specified. Typical outcomes of these modeling shall include, but are not limited to, the following:
  - Pool fire
  - Jet fire
  - Combustible gas cloud
  - Toxic gas cloud
  - Boiling liquid expanding vapor explosion

3.3.3 *Safety Availability Safety Integrity Level*

As **applicable**, detailed SIL calculations are performed and this needs to be scrutinized and verified to ensure that the proposed design and associated testing and maintenance philosophy achieve the selected safety availability (SIL) targets. For these calculations equipment database shall be utilized and help from software may be taken. The personnel carrying out such calculations and verifications should be qualified and competent to take up the job. Also, they should be well aware of various requirements of IEC 61511, etc. in the next step specification to be developed.

3.3.4 *Functional Specification*

Entire study details discussed so far will be translated in to a detailed technical specification where all the requirements shall be appropriately stated. From experience, it has been found that a good specification will contain both functional requirements and integrity requirements. If too much of hardware and software details are specified, then not only the procurements will be problematic, and manufacturer would have very limited scope to work with and contribute. Instead, if the functional and integrity requirements are clearly and unambiguously stated, the manufacturer will get enough space to suggest some good and economical solutions, keeping functional requirements unchanged. This shall be developed in general accordance with Clause 10 of the IEC 61511standard. In addition to the general requirements, the following documents also form part of the specification:

- FGS logic cause and effect diagrams
- FGS detector lists with types, locations, orientations, and set points
- FGS detector layout diagrams
- FGS verification and validation

3.3.5 *Verification and Validation*

The objective of the verification and validation activity is to ensure that the installed FGS system is in compliance with standards, and with the FGS specification having various performance parameters. There are two distinct phases. The first phase shall be a verification of installed equipment against the FGS requirements of specifications. During this phase, all detectors will be viewed at the facility in order to verify that they are the correct type, and installed in the proper orientation and location as per specification. The second phase of this task will involve validation testing. The validation testing shall be accomplished by physically activating each detector by either simulating a fire with a fire simulator, and gas detectors with calibrated gas of the supplier. In each case, the audio-visual alarm, etc. is to be verified also.

4.0 ROLE OF ALARM SYSTEM IN SAFETY INSTRUMENTED SYSTEMS

From ISA technical reports, one gets *an alarm is an audible and/or visible means of* ***indication*** *- to the* ***operator*** *about equipment malfunction, process deviation, or* ***abnormal condition****, requiring a* ***response***. As per Annex F of IEC 61511-3:2003, an alarm is an independent protection layer. In this connection, F14.6 of IEC 61511-3:2003 may be referred to; "the alarm can be counted as a protection layer since it is located in a different BPCS controller than the temperature control loop." Therefore, it is needless to tell an alarm system has great role to play in SIS, so any SIS discussions is incomplete without the discussions on alarm system. Safety controls, alarms, and interlocks are process safety safeguards incorporated as part of IPS to arrive at and maintain a safe state for process. Unlike safety interlocks, where actions are automatic, in case of alarm the action part lies with the operator and is manual action. Alarm requires operator's action in accordance with a response procedure to stop the hazardous event propagation. When an operator's response to an alarm is included in a layer of protection analysis, operator reliability in responding to the alarm should also be considered in making claims for safety alarms. Operator process knowledge and capability to take correct timely action are part of the behavior of the alarm system as a whole, and need to be validated using tests, drills, or simulated environments. Here the design, management, and performance greatly influence operator action as well as the level of actual risk reduction. For example, if the operator is flooded with alarms and out of them many are spurious alarms, the operator is highly disturbed and thereby runs high probability that even a very knowledgeable operator might miss some important issues. The result of poor alarm management will affect SIL, and as a consequence SIF. In this regard, the standard ANSI/ISA-18.2, "Management of Alarm Systems for the Process Industries" (ISA-18.2) may be referenced. The basic intent of this standard is to improve safety of the process through alarm system. Ineffective alarm systems are often found as contributing factors for accidents. Like FGS for alarm design, also a philosophy document (to describe objective, guidelines and working of alarm system), is also developed as a primary documents in many projects. One of the major issues is alarm rationalization. Rationalization involves comparing potential alarms with various criteria set in the philosophy document and generate alarm database. All alarm points are not rationalized, so, for this it is necessary to identify the potential alarms for rationalization process. Alarm systems related to process does not *vary* even with variations in process. This is because is that alarm response is really not a function of the specific process, but man—machine interface. In order to treat alarm system as SIF consisting of sensor, logic solver, and final control element one needs to consider alarm system, operator, and action together. Therefore basic parts of alarm systems are:

- Physical sensor
- Logic solver to compare process value with set point
- HMI for audiovisual signal
- Operator to detect, diagnose, and respond
- Logic solver to receive the response via HMI
- Communication of operator response through HMI command
- Logic solver to interpret the same, and obey the command

Details about alarm system with life cycle have been discussed in Chapter IX under SIS components, keeping in mind various guidelines, for example, EEMUA 191 and standards like ISA 18.2:2009 and IEC 62682:2014.

LIST OF ABBREVIATIONS

ALARP	As low as reasonably practicable
BPCS	Basic plant control system
C&I	Control and instrumentation (engineering)
CCF	Common cause failure
CCPS	The Center for Chemical Process Safety
DC (in SIS)	Diagnostic coverage (in SIS)
DCS	Distributed control system
DEP	Design and engineering practice
DNV	Det Norske Veritas (Norway)
E&I	Electrical and instrumentation
E/E/PE	Electrical/electronics/programmable electronics
ESD	Emergency shutdown system
ESFAS	Engineering safety feature actuation systems
ETA	Event tree analysis
EUC	Equipment under control
FGS	Fire and gas system
FLNG	Floating liquefied natural gas
FMEDA	Failure modes, effects and diagnostic analysis
FSA	Functional safety assessment/assessor
FTA	Fault tree analysis
HALT	Highly accelerated life test
HASA	Highly accelerated stress audit
HAST	Highly accelerated stress test
HMI	Human machine interface
HW	Hardware
I&C	Instrumentation and control
IEC	International Electrotechnical Commission
I/O	Input/output
I/P or O/P	Input or output
IPL	Independent protection layer
IPS	Instrumentation protection (function) system
IVP	Intelligent valve positioner
LEL	Lower explosive limit
LHS	Left hand side
LOPA	Layer of protection analysis
LS	Logic solver
LVL	Limited variability language
MFR	Master fuel relay (for boiler/furnace systems)
MOC	Management of change
MTBF	Mean time between failure
MTTF	Mean time to failure

Continued

MTTR	Mean time to repair
NO	Normally open (contact)
NC	Normally close (contact)
NPP	Nuclear power plant
OEE	Overall equipment efficiency
OLE	Object link & embedded
OPC	OLE process control
O&M	Operation and maintenance
OREDA	Offshore Reliability Equipment Database Association
PFD	Probability of failure on demand or doc/process flow diagram
P&ID	Piping and instrumentation diagram
PL	Performance level
PLC	Programmable logic controller
RA	Risk assessment/analysis
RBD	Reliability block diagram
RHS	Right hand side
ROI	Return on investment
RPS	Reactor protection system
SAR	Safety assessment/analysis report
SAS	Safety automation system
SFF	Safety failure fraction (IEC)
SIF	Safety instrument functions
SIL	Safety integrity level
SIS	Safety instrumentation system Supervisory information system (in case of DCS) [13]
SPLC	Safe PLC (programmable logic controller)
SRECS	Safety related electrical control systems
SRS	Safety-related system/safety-related specification
STEL	Short-term exposure limit
SW	Software
TMR	Triple modular redundancy
TWA	Time weighted average
USPAS	United States Particle Accelerator School
WRT	With respect to

References

[1] A. Summers, M. Gentile, Random, Systematic and Common Cause Failure: How Do You Manage Them, SIS-Tech Solutions, LP, Process Safety Progress, December 2006.

[2] J.L. Bergstrom, Safety Instrumented Systems (SIS) and Safety Life Cycle, Process Engineering Associates, LLC, September 2009.

[3] Functional Safety Handbook, ABB Limited, 2008.

[4] D.J. Wilkins, The Bathtub Curve and Product Failure Behavior Part One - the Bathtub Curve, Infant Mortality and Burn-in, ReliaSoft Reliability Field Consultant, November 2002. Reliability hotwire e magazine issue 21.
[5] Life Data Analysis (Weibull Analysis); An Overview of Basic Concepts, Reliability Engineering Resources; http://www.weibull.com/basics/lifedata.htm.
[6] M. Lundteigen, M. Rausand, Common cause failures in safety instrumented systems on oil and gas installations: Implementing defense measures through function testing, Journal of Loss Prevention in the Process Industries 20 (2007). Elsevier.
[7] A.E. Summers, Estimation and Evaluation of Common Cause Failures in SIS, November 1999. SIS-tech solutions; "safeguard safety instrumented systems," chemical engineering progress.
[8] Reliability Engineering; Exponent — Engineering and Scientific Consulting; www.exponent.com.
[9] Reliability Basics, December 2008. Reliability hotwire e magazine; issue 94, http://www.weibull.com/hotwire/issue94/relbasics94.htm.
[10] ISA-TR84.00.02-2002-Part 1; Safety Instrumented Functions (SIF) — Safety Integrity Level (SIL) Evaluation Techniques Part 1: Introduction, ISA, June 2002.
[11] D. Walker, Using Smart Safety Instrumentation System (SIS) to Make Better Operational Decisions, Emerson Process Management, March 2009.
[12] S.R. Brown, M. Menezes, Selecting Transmitters for Safety Instrumented Systems, ISA, Houston, 2001.
[13] S. Basu, A.K. Debnath, Power Plant Instrumentation and Control Handbook, Elsevier, November 2014. http://store.elsevier.com/Power-Plant-Instrumentation-and-Control-Handbook/Swapan-Basu/isbn-9780128011737/.
[14] H. Cheddie, Safety Instrumented Systems Verification: Practical Probabilistic Calculations, IHS.
[15] W. Goble, The Exida FMEDA Process - Accurate Failure Data for the Process Industries, Exida, January 2015; http://www.exida.com/Blog/the-exida-fmeda-process-accurate-failure-data-for-the-process-industries#sthash.IFhMeQez.dpuf and http://www.exida.com/Blog/the-exida-fmeda-process-accurate-failure-data-for-the-process-industries.
[16] Implementing Protective Functions in BPCS in an Combined Systems; Edward M. Marszal Christopher P. Weil Kenexis Consulting Corporation.
[17] Integrating Control and Safety: Where to draw the line; Robin McCrea-Steele, TÜV FSExpert Invensys-Premier Consulting Services; http://www.iceweb.com.au/sis/IntegControl&Safety.pdf.
[18] Integrated Fire and Gas Solution - Improves Plant Safety and Business Performance, Honeywell Process Solutions; White Paper, April 2009, in: https://www.honeywellprocess.com/library/marketing/whitepapers/FireGasSystem_Whitepaper_April09.1.pdf.
[19] Model RFQ for Performance Based Fire and Gas System Design Basis; Development and Validation; Revision 0, Kenexis, June 2011.
[20] Fire and Gas Detection and Mitigation Systems; L. Beckman; TÜV FSExp; SafePlex Systems Inc.
[21] P. Walsh, A. Kelsey, A review of leak detection for fuel storage sites, Health and Safety Laboratory, UK; ECM/2008/08.
[22] Can a Fire and Gas System be Considered a Safety Instrumented System? General Monitors; http://www.gmigasandflame.com/sil_info_102.html.

[23] Planning and Designing Gas Detection -Technical Engineering References for Instrument and Fire & Gas Design Engineers; Planning and designing of gas detection systems – for instrument and fire & gas engineers; ICEweb; http://www.iceweb.com.au/F&g/GasDect/Gas%20Detector%20Planning%20and%20Design.htm.

[24] J. Hocquet, Y. Le Gourrierec, L. Paris, Challenges in using risk- and performance-based design methods for FLNG safety engineering, Technip; www.gastechnology.org/Training/Documents/LNG17-proceedings/12_07-Jerome-Hocquet-Presentation.pdf.

[25] D. Walker, Using Smart Safety Instrumentation System (SIS) to Make Better Operational Decisions, Emerson Process Management, March 2009.

Further Reading

ABB Pluto Safe PLC Literature, ABB.

A New Approach to Machine Safety: EN ISO 13849-1:2006 e Safety-related Parts of Control Systems; Schemersal; Wuppertal/Wettenberg in March 2006.

B. Adamski, White Paper: Is the Risk Worth It? Beware of the New Safety Instrumented Suppliers, Invensys Premier Consulting Services.

H. Bezecny1, D. Inverso, V. Maggioli, G. Rabe, A. Weinert, Guidelines for the use of programmable logic controllers in safety-related systems, in: European Workshop on Industrial Computer Systems: Technical Committee 7 (Safety, Reliability and Security); Working Paper 6009; Version 13, October 1997.

V. Beck, Performance-based Fire Engineering Design and its Application in Australia; invited lecture; Fire safety science. Proceedings of the Fifth International Symposium; http://www.iafss.org/publications/fss/5/23/view.

C. Cassiolato, Reliability of Measuring Systems and Safety Instrumented Systems, SMAR Industrial Automation, 2011. Technical article SMAR.

Controlling Risks Selecting a Safety Integrity Level; Controlling Risks: Safety Systems, USPAS, January 2012.

Controlling Risks: Safety Systems, USPAS, January 2012.

A. Coleman, Safety Instrumented Systems: Control Valves as Final Elements, Emerson Process Management; Chemical Engineering, January 2011.

Fire and Gas Systems; ARMEH Catalog; http://customer.honeywell.com/techlit/pdf/62-0000s/65-0312.pdf.

Gas Detection Handbook, fifth ed., MSA.

S.N. Ganesan, Performance-based Gas Detection System Design for Hydrocarbon Storage Tank Systems, Kenexis DMCC; Oil & Gas Industry Conclave e IOCL, Delhi, India; http://www.isssource.com/wp-content/uploads/2012/06/062312KenexisPaper1.pdf.

S. Gillespie, Safety Instrumented Systems, Shell Global Solutions, UK, Measurement, Instrumentation and Automation Business Group.

W. Goble, J.C. Grebe, White Paper: Hybrid Fault Tolerant Architecture New Levels of Performance, Availability and Safety Integrity, August 27, 2008. Exida; RTP 3000 System; Report No.: RTP 06/10e16 R002 Version V1, Revision R2.

P. Gruh, SIL rating for fire and gas system hardware - are we breaking up the wrong tree? ICS Triplex.

P. Gruhn, SIL Ratings for Fire & Gas System Hardware – an Introduction to ISA TR84.00.07, Rockwell Automation, Houston, TX.

Guide to Safety Analysis for Accident Prevention, 2013. Lars Harms-Ringdahl IRS Riskhantering; AB.

Honeywell Process; Marketing white paper; Honeywell; https://www.honeywellprocess.com/library/marketing/whitepapers/FireGasSystem_Whitepaper_April09.1.pdf.

B. Hoekstra, Reliability with Respect to Safety Instrumented Systems, Yokogawa Global Safety Solutions Center, The Netherlands, October 2004.

IAEA Safety Standards Series no SSG-30, Safety Classifications of Structures, Systems & Components in Nuclear Power Plants, International Atomic Energy Agency, 2014.

Integrated Fire and Gas Brochure; Fire and Gas Solutions - One Stop Shop for World Class Solutions Honeywell; www.honeywellautomationindia.com.

Is your current safety system compliant to today's safety standard?; L.M. Duran; ABB Houston, TX; White Paper ABB; https://library.e.abb.com/public/10a4ad7d4b51a371c1257a060069bf3b/3BUS095609_en_Whitepaper_-_Is_your_SIS_compliant.pdf.

H. Jin, A Contribution to Reliability Assessment of Safety-instrumented Systems, NTNU, Trondheim, September 2013.

R. Kamphaus, Comparison of Integrated versus Dedicated Turbine Safety System, Woodward Inc., USA; http://pennwell.websds.net/2013/vienna/pge/papers/t5s5o4-paper.pdf.

B. Mostia Jr., The Safety Instrumented Function: An S-World Worth Knowing; Understand the SIF to Control Confusion, Complexity, and Cost of Safety Instrumented Systems, Control for the Process Industries, August 2002.

Overview of Safety Standards: New Machinery Directive 2006/42/EG, Panasonic Electric Works Europe AG, 2010.

D. Perry, Selecting " Sensors" for Safety Instrumented Systems as per IEC 61511 (ISA 84.00.01-2004), Emerson Process ManagementdRosemount Division; (internet document); http://www2.emersonprocess.com/siteadmincenter/PM%20Rosemount%20Documents/SelectingSISSensors.pdf.

Process Safe Book 1; Functional Safety in the Process Industry Principles, Standards and Implementation, Rockwell Automation.

Profisafe and Profibus for High Availability and Safety in the Process Industries, ARC White Paper; ARC Advisory Group, April 2009.

M. Rausand, Reliability of Safety - Critical Systems: Theory and Applications, J. Wiley.

Reliasoft, Blocksim; http://www.reliasoft.com/BlockSim/rbd_analysis.htm.

Safety Instrumented System Engineering Handbook, Kenexis Consulting Corporation, Columbus, OH, 2010.

Safety Instrumented System: Can they be integrated but separate; M. Kuchle, T. Macdaugall; Spartan controls.

Safety Instrumented System Fire & Gas; Emerson Process; write up; http://www2.emersonprocess.com/en-us/brands/deltav/sis/applications/pages/fgs.aspx.

Safety in control systems according to EN ISO 13849-1 Machine Safety - Jokab Safety Products, ABB.

Siemens S7 Technical Brochure. Siemens.

T. Stauffer, P. Clarke, Using alarms as a layer of protection, in: 8th Global Congress on Process Safety, Exida, 2012.

The ISA-18.2 Alarm Management Lifecycle, Exida, September 2011.

J. Thomson, Nuclear Power Station Control and Instrumentation Safety Systems Architecture e an Overview, March 2012. Safety in Engineering, www.safetyinengineering.com.

Understand the Function of Safety Instrumented Systems SIF, SIS, SIL, SRS, SLC? Whatdoes all these mean.

Using Reliability Block Diagrams to Analyze Dependent and Independent Failure Modes,Reliability hotwire; http://www.weibull.com/hotwire/issue27/hottopics27.htm.

D.J. Wilkins, The Bathtub Curve and Product Failure Behavior Part Two - the Bathtub Curve, Infant Mortality and Burn-in, ReliaSoft Reliability Field Consultant, December 2002. Reliability hotwire e magazine issue 22.

R.I. Williams, Hardware Configuration of Redundant Safety Integrated Systems, Control Engineering (June 2013).

CHAPTER

SIL, PE, FGS, and Plant Emergencies VIII

1.0 SIL METHODS

It is clear from previous discussions that hazard analysis and risk assessment study determine the need for a safety instrumentation system (SIS). SISs are developed and used for risk reduction, through prevention and mitigation, in the event of hazardous situations to protect people or the environment or prevent property loss. The international safety standard IEC 61508 specifies the safety life cycle—a systematic approach for the development of an SIS. Risk assessment is a part of this safety life cycle. *To what extent can a device or process be expected to perform safely, and, in the event of a failure, to what extent can the process be expected to go to the safe state?* These issues are addressed by the term safety integrity level (SIL). Therefore SIL gives a measure of safety risk or risk reduction to a tolerable limit for a given process. The IEC 61508 standard also specifies the measures, such as "fault avoidance" (systematic faults) and "fault control" (systematic and random faults) [2], to be taken in the design of safety functions consisting of sensor, logic solver, and final element. Referring to IEC 61508 (see Chapter VI) one knows that SIL depends highly on two major factors, namely, *hardware failure tolerance (HFT)* and *safe failure fraction (SFF)*.

- HFT: This is the ability of hardware to continue to perform a specified safety function in the presence of faults or errors. HFT of N means, that $N + 1$ faults will cause loss of safety function for the unit.
- SFF: This is the ratio of the average failure rates, *safe plus dangerous detected failure*, and *safe plus dangerous failure* (IEC 61508)—already discussed in the previous chapter.

Details related to the definitions of various fault types referred to earlier were discussed in the previous chapter. Probability of failure on demand (PFD_{avg}) provides a measure of SIL, and it has variations with low demand and continuous modes. For this, Table VI/4.0-3 may be referred to. The interrelationships between SIL, SFF, and HFT were elaborated in Table VI/4.2-3. Here, one important point is that redundancy could help in attending higher SIL but there may be additional requirements of attending systematic safety including software integrity. These will be discussed later but it is important to know what PFD_{avg} is:

- PFD_{avg}: This is the *probability* of a functional unit or system *failing* to respond to a *demand* for action arising out of a potentially hazardous condition, that is, a device will fail to perform its specified safety function *when it is asked* to do so. PFD_{avg} is used in calculations for system reliability. When there is probability of failure, then there has to be a question about availability. What is *availability*?
- Availability (see also Chapter VII): This is defined as the probability that equipment will perform its task. The interrelationship between SIL, PFD_{avg}, and availability is presented in Table VIII/1.0-1, which should be read in conjunction with Table VI/4.0-3.

The relation between SIL and certain factors is depicted in Fig. VIII/1.0-1; it is like a fire triangle!

Another important issue is that because sensors and final elements are further exposed (in the field) to physical and chemical loading and parameter variations, as already discussed in the previous chapter, failure rates are high for sensors and final elements. Hence a larger share, as shown in the pie chart in Fig. VIII/1.0-1B, is accountable for failure of sensors and final elements. The relationship between SIL and the safety life cycle will be focused on in the following clauses.

1.0.1 *Safety Life Cycle for SIL (an Idea)*

Any discussions on SIL will not be fruitful unless the concept of safety life cycle is well understood. Life cycle is a closed loop engineering process designed to optimize the design of the SIS and to increase safety. The concept was introduced and incorporated by international standards IEC 61508 and 61511(see Chapter VI). A safer plant will have low systematic errors, with decreased cost of engineering and increased process uptime. Safety life cycle defined in international standards through multiple steps takes functional safety from concept to decommissioning for functional safety management. In this approach, the key performance level parameter is SIL, which is selected for each SIF within SIS. Therefore SIL is a *fundamental* parameter in the safety life cycle approach. Each SIF is assigned its own SIL required to prevent (or mitigate) the specific hazard associated with the function. The integrity level requires

***Table VIII/1.0-1** Interrelationship Between SIL, PFD_{avg}, Availability, and Consequence*

SIL	PFD_{avg} [c]	Availability (%)	Consequence (Fatality)
SIL1[a]	10^{-2}–10^{-1}	90–< 99	Minor on-site injury
SIL2	10^{-3}–10^{-2}	99–<99.9	Major on-site injury or may be fatality
SIL3	10^{-4}–10^{-3}	99.9	Multiple on-site fatality
SIL4[b]	10^{-5}–10^{-4}	>99.99	Fatality in community

See also Table VI/4.0-3.
[a]*Lowest SIL.*
[b]*Highest defined SIL.*
[c]*For high/continuous mode (probability of failure per hour).*

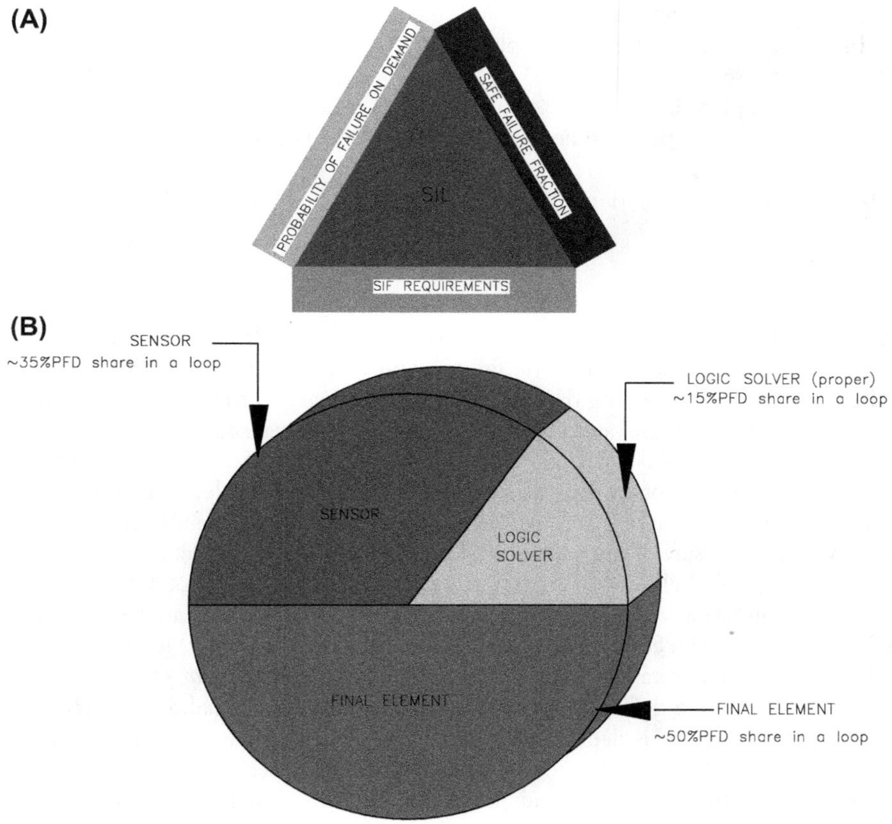

FIGURE VIII/1.0-1

SIL relations. (A) Major factors for SIL, (B) failure rate in loop (SIL requirement).

or defines the performance for that function in terms of its ability to achieve a tolerable risk target. Safety life cycle tasks are continuously performed during the entire life of the process to obtain lower operation and maintenance costs by selecting the right architecture and technology. In the safety life cycle process, periodic testing intervals need to be defined. SIL requires periodic testing intervals. Interrelations between SIF and SIL were shown in Fig. VII/1.0-3. At this stage it is better to focus on life cycle. With reference to Chapter VI, it is known that safety life cycle can be divided into three phases:

- Analysis phase:
  - Identification and estimation of potential risks
  - Evaluation for tolerable limit (standard/regulation)
  - Checking for layers of protections
  - Assignment of SIL for residual risks
  - Development of safety requirements specifications (SRSs)
- Realization phase:
  - Development of conceptual design

- Development of detailed design
- Installation and commissioning planning
- Operation phase:
  - Validation planning
- Startup review and operation and maintenance (O&M) planning:
  - SIS startup
  - Operation and maintenance
  - Periodic tests
  - Modification
  - Decommissioning

As stated earlier, safety life cycle verification concepts result in safer, more cost-effective designs with fewer systematic failures and more uptime for processes. For SIL, the specification containing the safety integrity requirements for safety functions needs to be performed by the safety-related systems. The safety integrity assessment in the safety life cycle is related to SIL. So SIL is tied in with the safety life cycle right from the very beginning, is relevant throughout the entire safety life cycle, and both go in hand to hand. In the case of a safety-related system there should be a lower rate of unsafe failures. A measure of the rate of unsafe (dangerous) failures is defined in Part 4 of IEC 61508 as "the likelihood of a safety-related system satisfactorily performing the required safety functions under all the stated conditions, within a stated period of time." In the standard an SIL is defined as "a discrete level (one of four) for specifying the safety integrity requirements of safety functions." Thus an SIL is a *target* probability of dangerous failure of a defined safety function. The standard demands that whenever a safety requirement is defined it should have two components: its functional component and its safety integrity component. Therefore SIL sets the intended target for the developers. SILs also guide and define the *rigor* with which the development process needs to proceed. After risk analysis the requirements for risk reductions are established. This is then translated into SIL for further development of the process. SIL sets the target safety requirements and achieves safety, but the beauty of the life cycle process is that it does not stop here. It also needs to be demonstrated, which is definitely a step toward confidence building. Various product suppliers come up with various SIL certified products. End users/integrators complete loop comprising sensor(s), logic solver, and final element(s). Therefore it is important to them to know the SIL of each of these components—because all of them together determine the SIL for the particular SIF.

1.0.2 *Outline of SIL Selection Fundamentals*

During risk analysis, the required SIF functions are itemized in the SIF list. Once this is done, the SIL selection process can be started as discussed next:

- Identification of factors: At this stage, the SIFs are thoroughly checked one by one to identify the factors/parameters affecting the risk for these hazards. Then, one makes the selection of an appropriate SIL (which defines the performance criteria for the SIF) to achieve a desired risk tolerance threshold. As per IEC

61508/61511, SILs are order of magnitude bands of PFD_{avg}, which also reflects the amount of risk reduction of a preventive safety instrumented function.

- Non-SIS: Mainly two parameters, namely, consequence and likelihood, which affect risk, are considered. The consequence is the potential severity of the hazard. The likelihood is the frequency of occurrence. Risk graphs/risk matrices are used for these purposes. The inherent risk can be reduced by non-SIS risk reduction. To assess the risk, one is required to know and evaluate the effectiveness of all non-SIS risk reduction measures to ensure that the risk is reduced to as low as possible before application of any SIS. In other words, it is required to assess whether an SIS is necessary to further reduce the risk. Non-SIS risk reduction methods could be consequence reductions such as a dike, whereas blast walls or blast-resistant control buildings could reduce likelihood.
- When one looks at the defined SILs given in Table VIII/1.0-1 or Table VI/4.0-3 one notices that SIL1 has a PFD_{avg} of 10^{-2}–10^{-1}, whereas SIL2 has a PFD_{avg} 10^{-3}–10^{-2}. This signifies that there is a change in the *order of magnitude* of risk reduction level—see the relevant associated table. SIL can be arrived at by direct calculation methods also, discussed later. Here, only various means are discussed.
- Risk matrix: One of the ways to determine SIL is with the help of a risk matrix, discussed many times previously. A typical calibrated risk matrix is shown in Fig. VIII/1.0.2-1.

THIS IS A CALIBRATED MATRIX, OBVIOUSLY 1ST COL & BOTTOM ROW WILL BE ZERO AS THEY ARE EITHER RELATED WITH EITHER "0" FREQUENCY OR CONSEQUENCE. NEXT COLUMNS ARE FILLED IN FROM TOP ROW WITH 3,2,1 ORDER. SUBSEQUENT COLUMNS ARE FILLED IN SAME ORDER. EACH ELEMENT IS THE ORDER OR REDUCTION NECESSARY. IN LOW DEMAND MODE MAX RISK REDUCTION IS OF THE ORDER 5 SO, 6&7 IN THE MATRIX IS SHOWN WITH HATCH

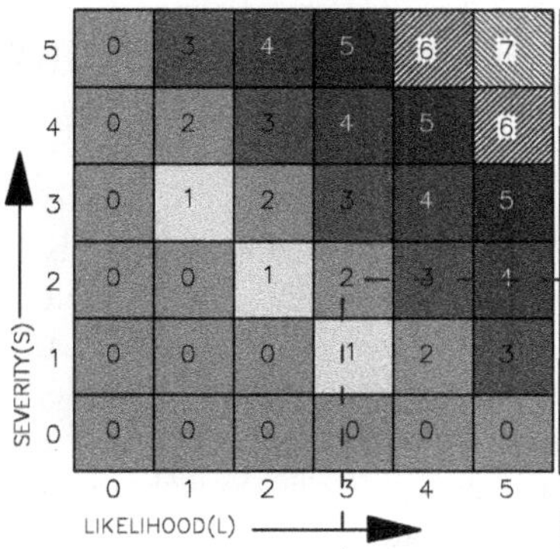

SEVERITY TABLE IN THE EXAMPLE

S	CATEGORY	HEALTH	ENVIRONMENT	FINANCIAL	TMEL
5	VERY HIGH	MULTIPLE FATALITIES	LARGE RELEASE V.H. CLEANING PERM. DAMAGE	ABOVE 500 MILLION	1E-06
4	HIGH	SINGLE FATALITY	LARGE RELEASE H. CLEANING SEVERE DAMAGE	ABOVE 50 MILLION	1E-05
3	MODERATE	SEVERE INJURY	LARGE RELEASE GOOD CLEANING LESS OFFSITE	ABOVE 5 MILLION	1E-04
2	LOW	MODERATE INJURY	MOD. RELEASE MOD. CLEANING ON SITE DAM.	ABOVE 0.5 MILLION	1E-03
1	VERY LOW	MINOR INJURY	SMALL RELEASE MIN. CLEANING	ABOVE 0.05 MILLION	1E-02
0	NONE	NOT SIGNIFIC-ANT	NONE	NONE	NA

LIKELIHOOD TABLE IN THE EXAMPLE

LIKELIHOOD	0	1	2	3	4	5
DESCRIPTION	NONE	VERY LIKELY	UN LIKELY	OCCAT-IONAL	FREQU-ENT	VERY FREQUENT
FREQUENCY	NA	1000Y	100Y	10Y	1Y	0.1Y

INITIATING FREQUENCY & CONSEQUENCE SELECTED

INTERSECTION IN THE MATRIX INDICATES THE ORDER OF RISK REDUCTION REQUIRED.

HERE METHOD USING RISK MATRIX IS SHOWN ONLY. OTHER METHODS ALSO DISCUSSED IN WRITE UP

FIGURE VIII/1.0.2-1

Risk matrix and SIL.

From previous knowledge of the risk matrix it is known that there are two parts. On the *x*-axis is placed the likelihood of occurrence of each of the risk issues associated with SIF, whereas the consequence pertaining to the same risk is entered on the other axis. Both likelihood and consequence are categorized into five divisions in this calibrated risk matrix. Thus from the list the consequence and likelihood are selected and the intersection of these two will determine the amount of risk reduction necessary, that is, the risk matrix will indicate the risk reduction factor (RRF) needed.

- Risk matrix discussions: Readers may refer to the consequence table in Fig. VIII/1.0.2-1 and note that there is a column (shaded) that indicates tolerable maximum event likelihood (TMEL). TMEL denotes the frequency at which a consequence of that magnitude is tolerable, that is, the *higher* the consequence category, the *lower* the TMEL value—if it is 1E-02 for a very low category, then it will be 1E-06 for the highest category of consequence. So, utilizing the TMEL approach one needs to determine only the consequence category.
- Risk graph: This is another approach to arrive at the SIL. Fig. VII/1.3-6 may be referred to see how SIL can be arrived at from the graph. In the risk graph, time of exposure to the hazard and possibility of avoidance are considered. Here also intersection of probability of occurrence and consequence will directly give the SIL in contrast to the risk matrix where usually one achieves the risk reduction order.

1.0.3 Layer of Protection Analysis (LOPA)

Layer of protection analysis (LOPA) is a well-established method that could be used to determine the SIL. In each independent protection layer (IPL) there is *a* risk reduction. Some of the standard risk reduction methods may be physical means etc. SIF/SIS is one of the ways protection layers could be posed for electrical/electronics/programmable electronics (E/E/PE). In this connection, Clause 4.2 of Chapter V and Fig. V/4.0-3 may be referred to. Typical IPLs for LOPA are shown in Fig. VIII/1.0.3-1. In this connection Fig. V/4.0-3 may also be referred to.

Here, each IPL is capable of reducing the risk by a factor. LOPA can be used in two ways, namely, implicit LOPA and explicit LOPA. Implicit LOPA is used in the risk matrix approach; through intersection of probability and consequence one obtains the order of magnitude as discussed. Readers may refer to the consequence table in Fig. VIII/1.0.2-1 and note that there is a column (shaded that indicates TMEL). TMEL denotes the frequency at which a consequence of that magnitude is tolerable, that is, the higher the consequence category as discussed above. So, utilizing the TMEL approach one needs to determine only the consequence category. Explicit LOPA is used in the TMEL approach, where risk figures are actually calculated. Details are discussed in Clause 1.2.3.

1.1 Risk Analysis and SIL

In this part an effort will be made to establish the relationship between risk analysis and SIL assessment in the light of the safety life cycle—the requirement of IEC 61508/61511 or ISA 84.00.01.

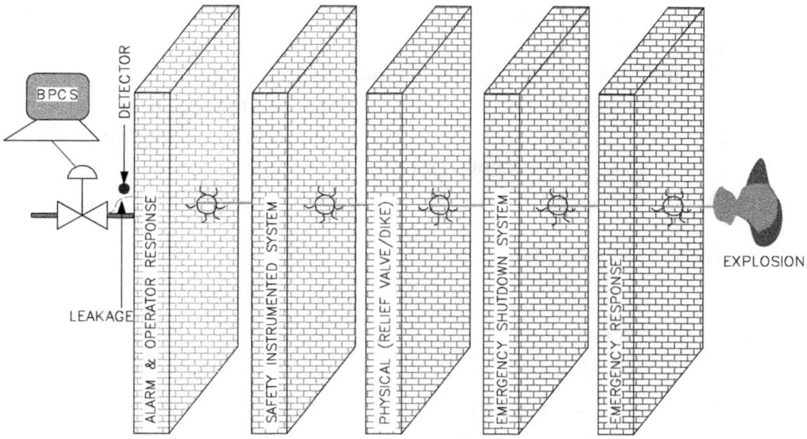

FIGURE VIII/1.0.3-1

Examples for layers of protection. *BPCS*, basic plant control system.

1.1.1 *Hazard Analysis and Risk Assessment*

Conceptual process design is the starting point of the safety life cycle. Based on the conceptual design document process, safety information documentations such as PFD, piping and instrumentation diagram, block flow diagram, safety set points, etc. will be developed. Process safety information will be used as the main input to the plant hazard analysis (PHA)/risk assessment. PHA has already been described in previous chapters as a qualitative or semiqualitative analysis of plant hazards by team engineers from various disciplines and with varied experiences. PHA finds the deviations from design intent and determines if they could result in a potential hazard for which the consequences of those hazards are identified and documented. The preventive and mitigating safeguards are considered to prevent those potential hazards. These qualitative risk assessments are done utilizing guidelines such as risk matrix, etc. for risk ranking and categorization. When considered safeguards are inadequate, the team recommends additional safeguards or means (at times even design modification) for reducing or eliminating risks. The entire PHA is documented suitably by a secretary. The PHA report will be the basis of the next step in the SIS engineering life cycle. With respect to the SIS safety life cycle, it is important for the *selection of an SIL* target to be *consistent with PHA* studies performed for the plant process. Also the information in these studies can provide valuable assistance and insight during the SIL selection effort. All these details are available in previous chapters.

1.1.2 *SIL in Safety Life Cycle*

Safety standards, such as IEC-61508 & 61511, specify an overall approach to the task of identification of potential hazards and applying safety within a facility or process plant. This approach calls for the concept of the safety life cycle (recapitulated in Clause 1.0.1), which directs the users to take into account all of the required phases of the life cycle. When any claim for compliance with the standard is made, it is necessary that all issues are taken into *account* and fully *documented* for assessment. In fact, the standards provide the required framework and directives for the application of the safety life cycle to cover concept, commissioning, operation, maintenance, and decommissioning.

1.1.3 Risk Reduction and SIL

From discussions on various PHAs, the standards that risk can be conceived as are:

- Probability of occurrence of a hazard
- The degree of severity of harm

So, Risk = Consequence * Frequency.

When consequences are not implicit within the accident description, then they are made explicit with the intention to reduce the risk. The overall mechanisms for risk reduction are then broken down into specific safety functions to achieve a defined task. One such division is SIS, as is seen in Fig. VIII/1.0.3-1, which is mainly applicable for E/E/PE with which the standard is concerned. Depending on the risk reduction strategies implemented, some control functions are dedicated for safety purposes—such control functions are designated as safety functions. These safety functions constitute SIFs in SIS. In parallel with assignment of the overall safety requirements to specific safety functions, a measure for the *dependability or integrity* of those safety functions is necessary as per the standards. This measure stands for safety integrity level or SIL. From IEC 61508, one understands that safety integrity (SIL) is the probability of an E/E/PE, a safety-related system satisfactorily performing the specified safety function under all the stated conditions within a stated period of time. Therefore SIL is a measure of the *reliability* of the safety function performing to specifications and the *time* for that response to take place. It is easier to identify and quantify possible conditions and causes leading to failure of a safety function than to guarantee the desired action of a safety function [3]. So, SIL often is measured in *reverse* of the definition stated earlier. There are three modes for SIL. *Low* demand mode, *high* demand mode, and *continuous* mode. In high demand mode and continuous mode, failure rates are measured in terms of probability of failure per hour (PFH), as shown in Table VI/4.0-3 (see Clause 3.5.16 of IEC 61508-4:2010). The three demand modes in detail are:

- Low demand mode: Where safety function is only performed on demand to transfer equipment under control (EUC) into a specified safe state and the frequency of demand is less than one per year. Fig. VII/1.4-1 is an example of demand mode of operation. This is because normally the level in the tank will be maintained by sensors S_n and control valve C_v, but in case of a high level to avoid a hazardous situation, separate protection sensors P_n will come into play and close the shutdown valve S_v. So safety comes on demand.

- High demand mode: Where safety function is only performed on demand to transfer EUC into a specified safe state and the frequency of demand is greater than one per year.
- Continuous mode: Where the safety function retains EUC in a safe state as part of normal operation in the same system. Fig. VIII/1.1.3-1 is an example.

In continuous mode, as in the burner management system (BMS) shown in Fig. VIII/1.1.3-1 (self-explanatory with the associated note), normal boiler operations such as the burner cutting in and out are done with the help of the BMS. Also whenever there is flameout, that is, not sensing of flame by a number of detectors, from a safety point of view the boiler should trip by stopping the fuel supply. Such functions are done by the BMS as shown. Similarly, at the time of startup, purging is done by starting and running the FD fan, which is also done by BMS. So, in this example, the BMS is meant to carry out both *normal* as well as *safety* functions/operations. For further details on the BMS, the author's book [1] may be referred to. Now, the question is how to associate the relevant SIL with the safety function that is being applied to balance a particular risk. For this there are ways and means that have been outlined Clause 1.0 and also further explained with details in Clause 1.2.

1.1.4 IEC and SIL Assignment

International standards such as IEC 61511-3:2003 put forward various ways and means for SIL assessments, as listed in Table VIII/1.1.4-1.

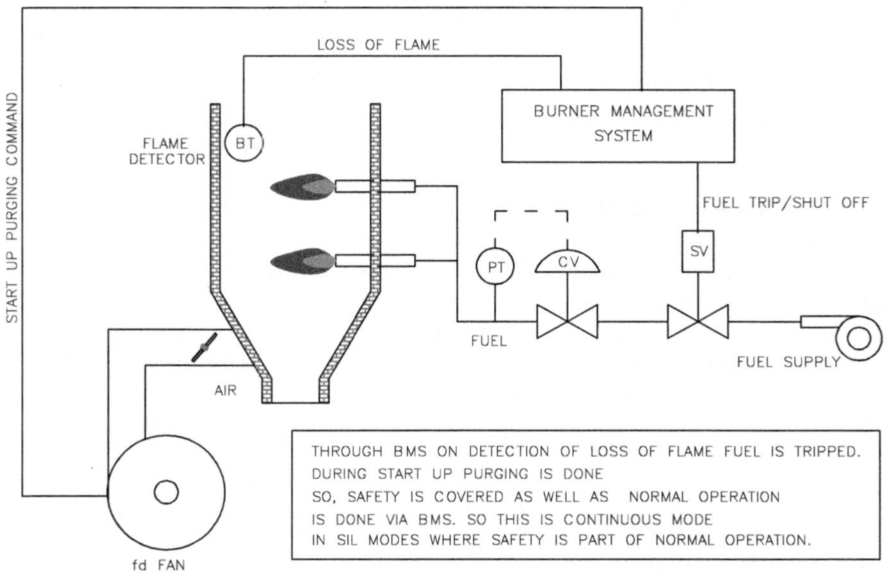

FIGURE VIII/1.1.3-1

Burner management system (BMS) continuous mode. C_v, control valve; P_n, protection sensors; S_n, sensors; S_v, shutdown valve.

Table VIII/1.1.4-1 SIL Determination Methods as per International Standards

IEC Annex	IEC Reference	Heading	Method
Annex B	IEC 61511-3:2003	Semiquantitative method	Fault tree
Annex C		Safety layer matrix method	Safety layer matrix
Annex D		Determination of the required SIL—semiqualitative method: calibrated risk graph	Risk graph
Annex E		Determination of the required SIL—a qualitative method: risk graph	Risk graph
Annex F		Layer of protection analysis	LOPA

LOPA, *layer of protection*; SIL, *safety integrity level.*

1.2 SIL Determination Techniques

As discussed, there are many ways SIL assessments can be done. Generally, the following are the major steps involved to determine the SIL for SIF:

- To carry out the PHA of the process by selecting any suitable method of PHA discussed in earlier chapters (see Chapters II through V)
- Reveal the study result with cause, consequence, recommended safety, etc.
- Identification of hazards and existing safeguards
- Identification and initiating cause and development of hazardous scenario
- Quantification efforts for hazard frequency and safeguard reliability

With these recapitulated points in mind, in the following subclauses various methods will be discussed with examples. During the discussions, low demand mode and continuous mode will be touched upon at least in one of the methods discussed.

1.2.1 Direct by Calculation

Based on the failure rate, the risk tolerable frequency can be estimated. Such things are possible in simpler cases, as given in the example later. Referring to the storage tank example as shown in Fig. VII/1.4-1 (or Fig. VIII/1.2.2-1), let this hydrocarbon storage area be manned 2 h per shift, that is, 6 h a day. On account of overpressure there may be leak that could cause an explosion and (say) out of 15 such leaks there will be one explosion causing a fatality. Analysis indicates that every 6 years there is an explosion. Let the tolerable frequency of explosion be 10^{-5}. The required PFD can be found by direct calculation:

Therefore fatality rate (chance) $\lambda = 1/6$ (pa) * 1/15 * 6/24 = 0.00277 = 2.77×10^{-3}
Tolerable frequency = 10^{-5}; so PFD = $10^{-5}/2.77 \times 10^{-3} = 0.36 \times 10^{-2} = 3.6 \times 10^{-3}$. Hence SIL2.

Similarly, in continuous mode for the case of a boiler, if on account of loss of flame there is accumulation of CO and there could be an explosion, 1 in 500 cases leads to a fatality. If the boiler tolerable failure rate is 10^{-5} pa, and if failure rate because of loss of flame is λ, then $10^{-5} \geq \lambda * 1/500$. So $\lambda = 500 * 10^{-5} = 5 \times 10^{-3}$, that is, SIL2.

1.2.2 *SIL Determination by Fault Tree (See Annex B—IEC 61511-3:2003)*

From the PHA result, one after another the hazard event will be taken up and a fault tree will be tried to establish how the top event or hazardous event is reached. Then an event tree is developed with safeguard reliability values to check whether the existing system will meet the target level; if not, additional SISs with necessary SILs (SIS is mentioned in place of the safeguard system to show our interest in the E/E/PE system and SIFs) will be added to recheck whether it will meet the target with other existing safeguards. *The question may arise as to how to select an SIL and why, when the purpose is to know the method of such selection!* The answer is simple: looking at the order of risk reduction, one needs to estimate (with *minimum* risk reduction possible in the SIL) the SIL, then to apply and check whether it is satisfactory, which may be an iterative way of doing this. As an example, help from same field process described in Fig. VII/1.4-1 is taken. It is redrawn and presented as Fig. VIII/1.2.2-1 in suitable way but the process is the same and hence is not described again.

Here, a relief valve has been added to divert the fluid to flare and it is assumed that it works perfectly without failure. As SIS is our concern, for simplicity

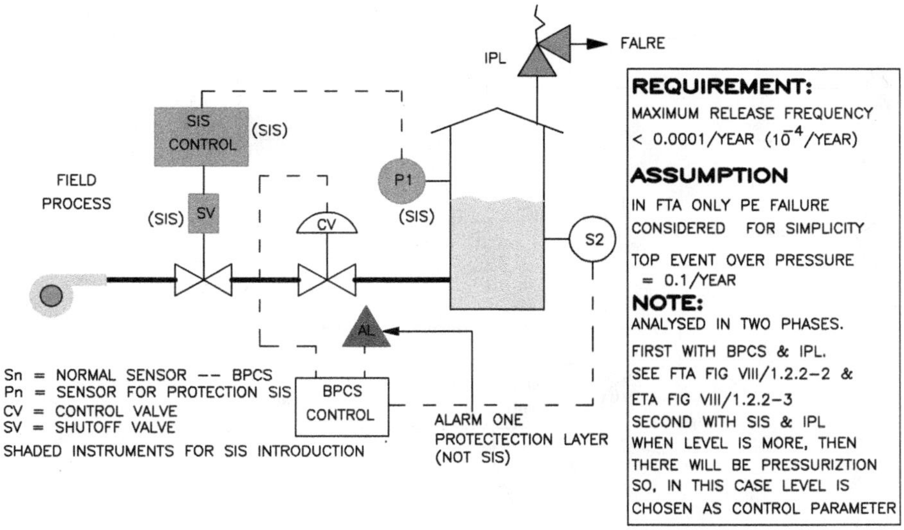

FIGURE VIII/1.2.2-1

Safety integrity level (SIL) by fault tree analysis (FTA).

such an assumption has been made. Also during initial analysis and in fault tree analysis (FTA) it is presumed that system is operating with a basic plant control system (BPCS) to control vessel level, *without SIS* (duly marked). In the BPCS a high alarm has been added to draw the attention of the operator. This alarm is a safeguard (one layer of protection). If there is any problem, there will be a high-level alarm for the operator to respond to. When the level is too high there will be very high (over) pressure, and the relief valve (protection layer) should divert the fluid to flare. In this process, *maximum failure frequency is assumed to be* 1×10^{-4}*/year*. Failure associated with the BPCS has been depicted in a fault tree drawing, as detailed in Fig. VIII/1.2.2-2. So, overpressure can be caused by failure of BPCS components and an external factor—fire. Initially, there is no SIS, so an event tree corresponding to BPCS and IPL for the process has been shown in Fig. VIII/1.2.2-3. Both the figures are self-explanatory with assumptions and hence are not described again. From Fig. VIII/1.2.2-3 it is found that maximum release frequency to the environment could be as high as 19×10^{-4}/year (more than 1×10^{-4}/year — sum of events 3 & 5). This is not acceptable. So, SIS with SIL2 has been chosen. SIS is shown by the shaded part in Fig. VIII/1.2.2-1.

Since the minimum risk reduction in SIL1 is 10 (10—100), to get the target figure is insufficient, so SIL2 has been chosen with minimum risk reduction 100

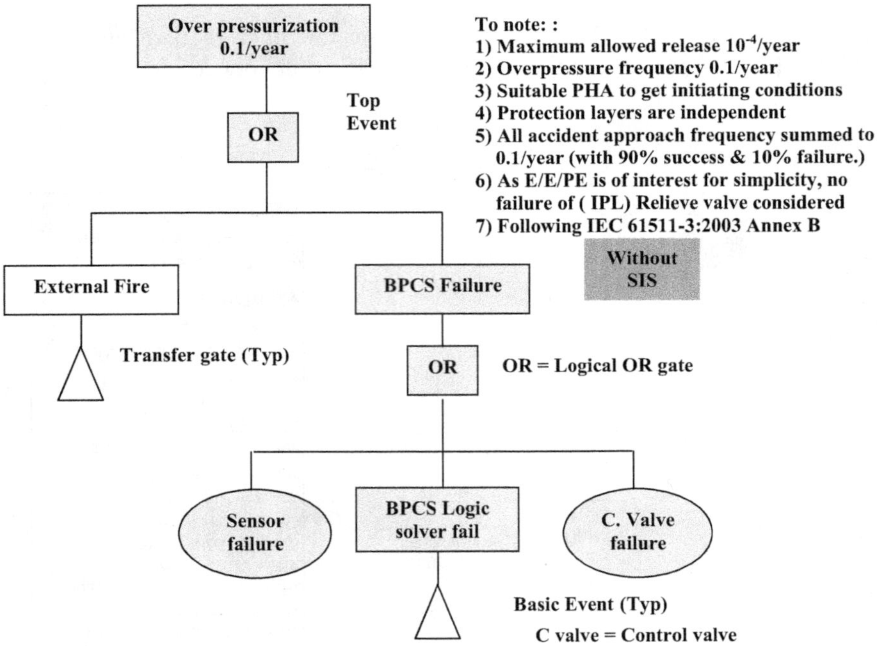

FIGURE VIII/1.2.2-2

Fault tree for overpressure example (Fig. VII/1.2.2-1). *BPCS*, basic plant control system; *C Valve*, control valve; *E/E/PE*, electrical/electronics/programmable electronics; *IPL*, independent protection layer; *PHA*, plant hazard analysis; *SIS*, safety instrumentation system.

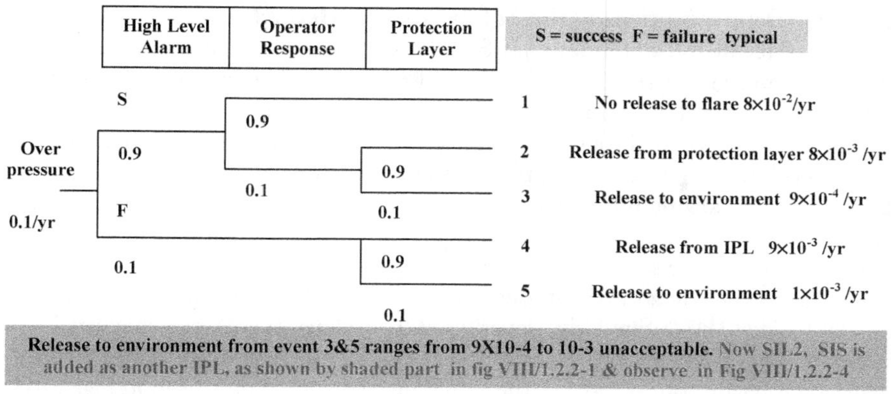

FIGURE VIII/1.2.2-3

Event tree for field process (Fig. VIII/1.2.2-1) with independent protection layer (IPL) but without safety instrumentation system (SIS).

(100—1000). Now a new event tree diagram, Fig. VIII/1.2.2-4, with both IPL and SIS in place is considered. Here, it is seen that the final risk frequency is 0.19×10^{-4}/year. Therefore the chosen SIL2 for SIS is acceptable. SIS comprising sensor, logic solver, and final element will be used.

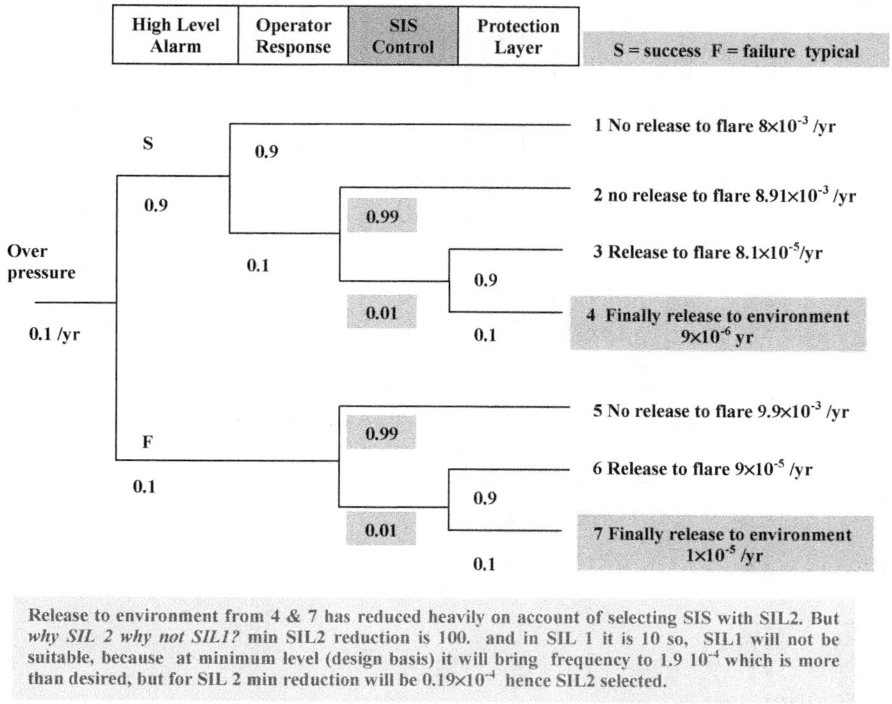

FIGURE VIII/1.2.2-4

Event tree for field process (Fig. VIII/1.2.2-1) with independent protection layer (IPL) and safety instrumentation system (SIS) for frequency reduction.

1.2.3 SIL Determination by the Safety Matrix Method

Readers may recall the discussions on SIL determination from the risk matrix in Clause 1.0.2 and Fig. VIII/1.0.2-1 where a matrix comprising various orders of risk reduction is shown and the intersection of consequence and likelihood gives the order of risk reduction needed through SIL. As shown in the example, the intersection called for a risk reduction of order 2, that is, minimum SIL2 will be necessary. Little is elaborated here following IS/IEC 61511-3:2003 Annex C. As indicated earlier, after risk analysis, various hazardous events are listed. Then, likelihood and severity impact are categorized as follows.

Likelihood is categorized without any protection layer as "Low (L)", "Medium (M)," and "High(H)." Similarly, severity of hazardous events is categorized as "Extensive (E)," "Severe (S)," and "Minor (Mi)." After these are done, various protection layers are considered and the actual matrix is the form shown in Fig. VIII/1.2.3-1. This is a safety risk matrix, hence protection layers including SIS have been

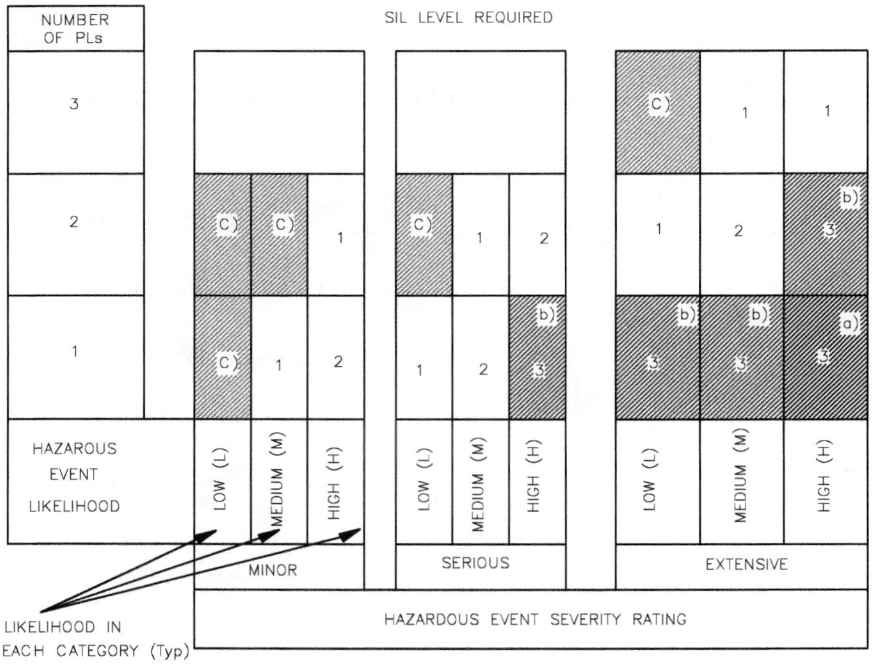

a) ONE LEVEL SIL3 SIF IS NOT SUFFICIENT. ADDITIONAL MODIFICATIONS ARE NECESSARY (SEE d)

b) ONE LEVEL SIL3 SIF MAY NOT PROVIDE SUFFICIENT RISK REDUCTION ADDITIONAL REVIEW REQUIRED

c) SIS INDEPENDENT PL IS NOT NECESSARY

d) NOT SUITABLE FOR SIL 4

DEVELOPED IN LINE WITH ANNEX C IS/IEC 61511–3:2003

FIGURE VIII/1.2.3-1

Safety risk matrix for safety integrity layer (SIL) determination. *PLs*, protection layers; *SIF*, safety instrument functions.

considered. This may not be suitable for SIL4. Readers are advised to *note* the order of SIL category placements (numbers in the box refer to IPL). Depending on the category of SIL requirements in the figure, different hatches have been provided. Category c does not require any special attention whereas categories b and a need special attention. In fact, category a modification is called for. Readers can compare this figure with Fig. VIII/1.0.2-1 and may notice that these use the same approach, only the presentation is different as IPLs were not shown in earlier figure.

For this implicit LOPA (similar approach has been discussed in Clause 1.0.3) the calculation may be as follows:

$$\text{SIL} = \text{Risk Reduction (Severity-Likelihood matrix)} - \sum \text{IPL credit.}$$

where the credit point of IPL is related to PFD, for example, credit 1 for PFD = 0.1 and credit 2 for PFD of IPL = 0.01. Since 1/PFD is risk reduction, each IPL credit will change the order of risk reduction required, therefore they are taken out (subtracted). SIL rating e.g., risk reduction, will be done on the remaining order of risk. Similarly, in the case of TMEL, which signifies the frequency at which the event is tolerable, explicit LOPA is used. The SIL value can be arrived at by:

$$\text{SIL (RR)} = F/\text{TMEL};$$

where F stands for frequency of event occurring without an SIS to protect.

$$F = \Sigma_i \left(\text{IE}_i * \prod \text{IPL PFD}_j \right)$$

where IE stands initiating event frequency of *i*th numbers and $\prod \text{IPL PFD}_j$ = continuous product of the PFDs *j*th number of protecting layers (as applicable) for each initiating event. To understand the situation it is better to take an example. Say for the top event in Fig. VIII/1.2.2-2 there are three issues related to the top event. Pertinent to the BPCS there were three initiating causes; let there be four arbitrary safeguards for them. Then for each of the initiating causes one needs to check the change of frequency of occurrence after safeguards are in place, as shown in Table VIII/1.2.3-1.

Table VIII/1.2.3-1 PFD Calculation (Example)

Initiating Cause	Frequency (n/year)	Arbitrary IPL				Modified Frequency
		1	2	3	N	(Product of 1 to N) * Frequency
A	0.1	1	0.01	0.1	1	10^{-4}
B	0.1	0.1	0.01	1	0.1	10^{-5}
C	0.4	0.1	0.01	0.1	1	4×10^{-5}
Subtotal event modified frequency						15×10^{-5}
Target						3×10^{-5}
Maximum PFD_{avg} for SIF = modified frequency/target						5
So, SIL (SIL1 provides risk reduction = 10 min)						SIL1

PFD_{avg}, *probability of failure on demand;* SIF, *safety instrument functions;* SIL, *safety integrity level.*

1.2.4 *SIL Determination by Risk Graph*

In Chapter VII, knowledge of the uncalibrated risk graph was gathered. Now it is necessary to calibrate this graph to utilize it for generalized use. It is extremely important to note that actual calibration is case specific. This means that general guidelines may have to be suited for a specific use. This calibration may be done according to IEC 61508-3:2003 guidelines. One of the major parameters is consequence (C) which can be classified as C_A, C_B, C_C, and C_D. Each of the issues has a specified range. Such range calculations are done by multiplying the number of people present in the exposed area with vulnerability factor *V* specified based on the nature of hazards:

$V = 0.01$ small release of flammable/toxic material
$V = 0.1$ large release of flammable/toxic material
$V = 0.5$ large release of flammable/toxic material
with probability for catching fire
$V = 1$ rupture or explosion

Other parameters are: occupancy (*F*), which stands for proportion of length of time, and area exposed to hazard occupied during normal working hours (exposure of personnel). The probability of avoiding a hazardous event is represented by *P*. Demand rate *W* is basically to factor for frequency estimation. Guidelines as per IEC 61511-3:2003 are presented in Table VIII/1.2.4-1.

Table VIII/1.2.4-1 *Guidelines for Calibration Parameter (Basis: IEC61511-3:2003)*

Risk Parameter	Symbol	Classification	Comments
Consequence (C)	C_A	Minor injury	Injury and death, normal healing time
Consequence (C)	C_B	0.01–0.1	Injury and death, normal healing time
Consequence (C)	C_C	0.1–1.0	Injury and death, normal healing time
Consequence (C)	C_D	>1.0	Injury and death, normal healing time
Occupancy (*F*)	F_A	<0.1	Rare occupancy
Occupancy (*F*)	F_B	>0.1	Frequent and permanent occupancy
Probability (*P*)	P_A	>90%	Good probability of avoidance
Probability (*P*)	P_B	<10%	Poor probability of avoidance
Demand (*W*)	W_1	0.1 day/year	Demand rate
Demand (*W*)	W_2	0.1–1 day/year	Demand rate
Demand (*W*)	W_3	1–10 days/year	$W > 10$ days/year—higher integrity

Readers may refer to Fig. VIII/1.2.4-1 to look into the order of SIL placement.

With reference to Fig. VIII/1.2.2-1, let in this filling station, a person visits for maintenance and stays there for say about 1 h/day. Let the chance of overfill is about say maximum a day in 10 years. When the person is present, there is no chance that they could avoid an explosion if it happens. What is inferred is (1) it is a case in W_1 category as maximum frequency is 0.1 day/year, (2) as person stays

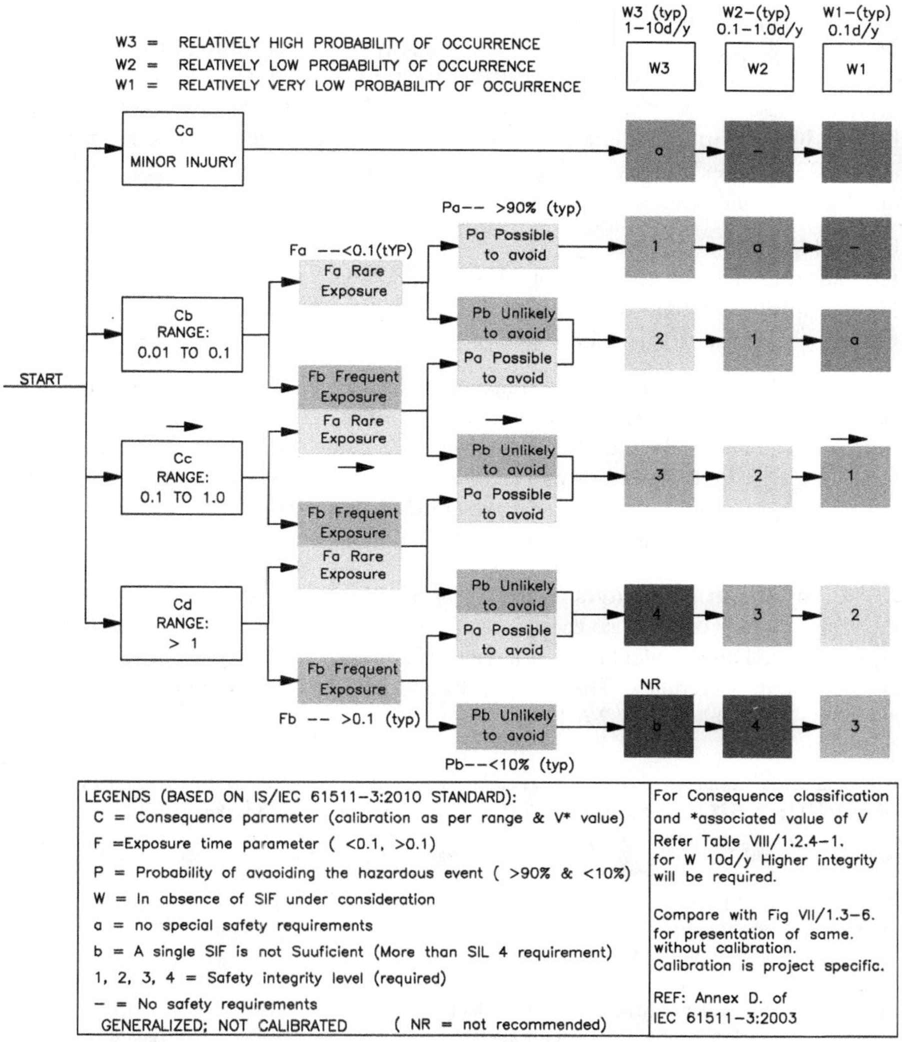

FIGURE VIII/1.2.4-1

Calibrated risk graph for safety integrity level (SIL) assignment.

1 h/day meaning it 1/24, that is, <0.1 so occupancy is F_A, and (3) probability of avoidance is <10%. Therefore the route to arrive at SIL value is by following Fig. VIII/1.2.4-1 will be C_C–F_A–P_B–W_1, hence the necessary SIL for SIS will be SIL1. There is another way of synthesizing the risk graph qualitatively as detailed in Annex E of IEC 61511-3:2003.

1.2.5 LOPA for SIL Assessment

IEC 61511-3:2003 Annex F brings about other ways for determining SIL for SIF as protection layer. For this a detailed LOPA is to be performed. LOPA was discussed in Chapter V and a LOPA report was presented in Table V/4.1.5-1; these will be the basis of finding the intermediate likelihood of an initiating event. Table F4 of IEC 61511-3:2003 Annex F gives some guidelines regarding the PFD of several items as detailed in the following:

- Control loop: PFD 10^{-1}
- Human interface (trained and not stressed) PFD: 10^{-4}–10^{-2}
- Human interface (stressed) PFD: 0.5–1.0
- Operator response to alarm PFD: 10^{-1}

After applying all IPLs, if it is seen that a new SIF is necessary, this is applied and its SIL is calculated by *dividing* the corporate criteria for this severity level of an event by the intermediate event likelihood. The previous IEC guideline figures may be used for various IPLs. Finally, it is necessary to *add up* all the mitigating event likelihoods for serious and extensive impact events that present the same hazard. In view of the initial discussions of this chapter and discussions in Clause 4 of Chapter V, it is not repeated.

1.2.6 Comparison of Various Methods of SIL Determinations

During the previous discussions there have been several means and methods of SIL determinations. Direct calculation, for example, is not always possible especially when the system is complex. The advantages and disadvantages of various methods are presented in Table VIII/1.2.6-1.

***Table VIII/1.2.6-1** Comparison of Various SIL Determination Methods*

Methods	Advantage	Disadvantage
Safety matrix	Not pure numerical, SIL and PHA in same session Easy to conceive	Approach is too simplistic Not integrated with safety life cycle MOC difficult to handle
Risk graph	Semiquantitative, no special calculation Person with a good feel for application can use this Quick assessment and not much detailed study necessary	Subjective Credit for low safeguard may not be well taken care of Not very well-defined SIL calculation basis
LOPA	Semiquantitative, but use quantitative logic for analysis Explicitly accounts for risk mitigation Numerical approach possible Clear idea on safeguard and initiating event Results rigorous next to FTA	More time and effort Numerical tools necessary
FTA	Object oriented and powerful Easy to understand With mathematical base Rigorous results	Knowledge of probabilistic application is necessary

FTA, *fault tree analysis;* LOPA, *layer of protection analysis;* MOC, *management of change;* PHA, *plant hazard analysis;* SIL, *safety integrity level.*

SIL assessments are required for various purposes and at various levels, so, from the point of view of *usage*, comparison among the various methods discussed is presented in Table VIII/1.2.6-2 [4].

Table VIII/1.2.6-2 Uses of SIL Determination Methods

Purpose	Safety Matrix	Risk Graph	LOPA	FTA
Initial analysis	X	X	X	NR
Detailed analysis	NR	NR	X	X
Multiple causes and protections	NR	NR	X	X
Dependencies	NR	NR	NR	X
SIL/PFD_{avg}	SIL	SIL	PFD_{avg}	PFD_{avg}
Human factor	NR	NR	X	X
SIL	1	1	1&2	>1

FTA, *fault tree analysis;* LOPA, *layer of protection analysis;* NR, *not recommended;* PFD_{avg}, *probability of failure on demand;* SIL, *safety integrity level;* X, *acceptable.*

We will now concentrate on the determination of SIL calculations pertinent to systems.

1.2.7 *SIL Verification*

After SIL selection it is necessary that this be verified. As Fig. VIII/1.2.7-1 shows there are five major parameters that affect the SIL achieved:

- Component selection: This determines the suitability of components for application, especially in safety.
- Fault tolerance: This is mainly guided by redundancy and voting circuit.
- Functional testing: Functional testing decreases the probability of failure as latent failure is identified.
- Common cause failure is highly attributable to failure so diversity may be chosen.
- Diagnostic coverage: This is a proof test and decreases the probability of failure.

1.3 SIL Calculation for Safety Instrument Loop

It is quite a common feeling in people's minds that there are special requirements for SIL instruments. Alternatively, some put a lot of stress on selecting SIL for a logic solver as it controls the loop. What is a safety loop? To answer these issues, it is important to understand how the safety of a loop can be assessed. In this clause the emphasis will be on calculation of SIL for the safety loop and what this actually means.

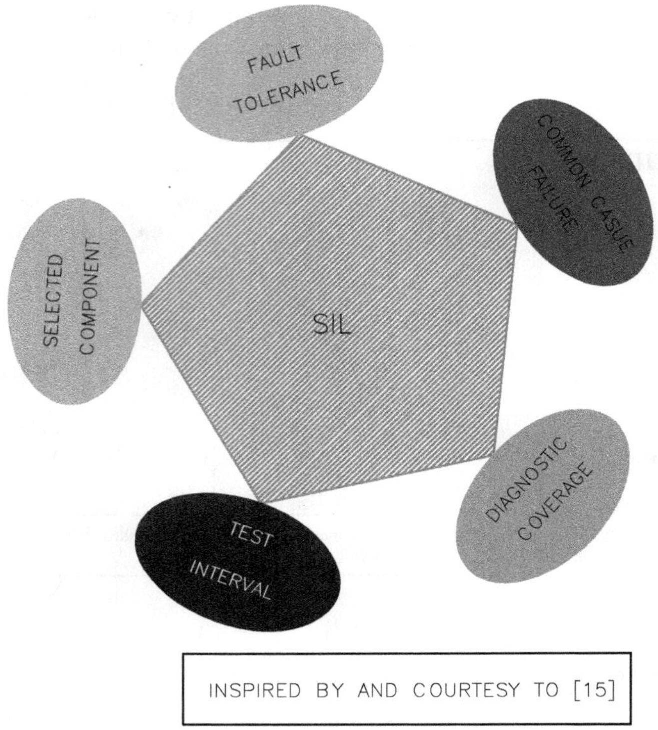

FIGURE VIII/1.2.7-1

Safety integrity level (SIL) affecting parameters.

From Fig. VIII/1.3-1 it is seen that a typical safety loop consists of a sensor (say PFD_1), input signal processing section of the programmable logic controller (PLC) (PFD_2), (main) logic solver (PFD_3), output signal processing section of the PLC (PFD_4), and final element (PFD_5). Therefore PFD of the safety loop will be the sum of

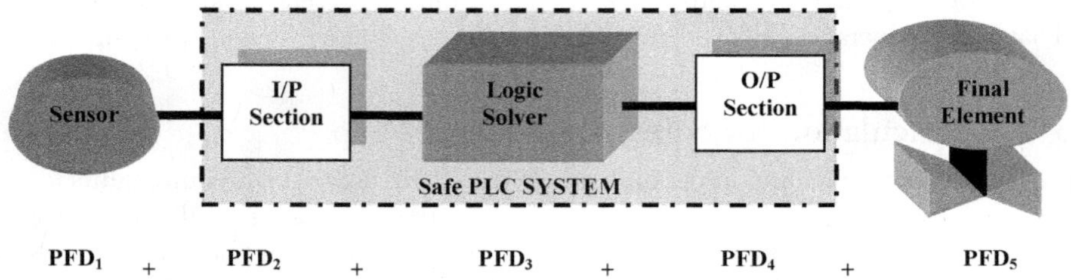

FIGURE VIII/1.3-1

Probability of failure on demand/safety integrity level (PFD/SIL) for a safety loop (typical). *I/P*, input; *O/P*, output. For share of SIL requirements see Fig. VIII/1.0-1B.

all these PFDs. What does this really mean? As shown in the figure, the PFD_{avg} of the loop shall be the sum of all PFDs. This statement means nothing. Suppose that the sensor section has maximum SIL2, the logic solver SIL3, and the final element is SIL3. Then what could be the SIL of the loop? Since the sensor section has SIL2, without any hesitation one can understand that the loop maximum SIL will be guided by the minimum SIL of all devices. Hence it will be maximum SIL2. Again by changing the sensor configuration this can be changed, for example, by selecting redundancy in the sensor the SIL may be upgraded. This means that safety loop architecture is also a factor. This will be clear from Fig. VIII/1.3-2.

From Fig. VIII/1.3-2 it is clear that the architecture constraint is one of the issues that need to be considered while assessing the SIL of a safety loop. The SIL limit of a product is affected by a number of parameters such as: architectural constraints, HFT, SFF, PFD_{avg} (for low demand mode)/PFH (for continuous mode), and maintenance interval. These will be discussed in the following subclauses.

1.3.1 Architectural Constraints

In this connection Tables VI/4.2-2 and VI/4.2-3 may also be referred to. Clause 7.4.4 of IEC 61508-2:2010 discusses the architectural constraints in SIL assessment. From (Clause 7.4.4.1.2 of) the standard one gets A subsystem can be regarded as *Type* A if, for the components required to achieve the safety function.

- The failure modes of all constituent components are well defined; *and*
- The behavior of the subsystem under fault conditions can be completely determined; *and*
- There is sufficient dependable failure data from field experience to show that the claimed rates of failure for detected and undetected dangerous failures are met.

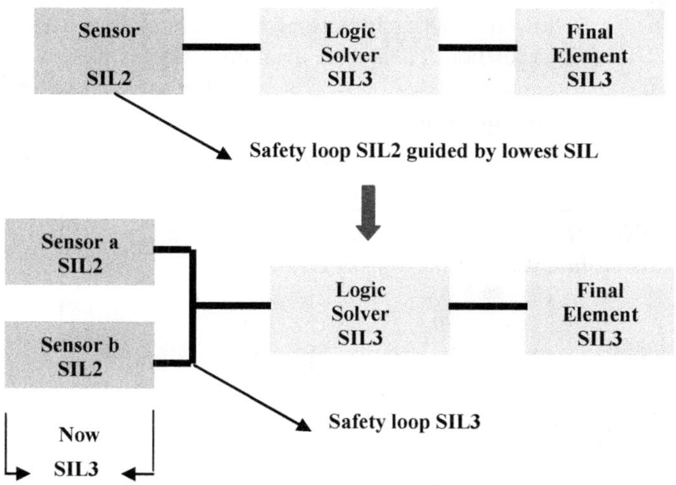

FIGURE VIII/1.3-2

Safety loop architecture.

Table VIII/1.3.1-1 Safety Integrity of Hardware Architectural Constraints Type A Elements

SFF of an Element	HFT 0	HFT 1	HFT 2
<60%	SIL1	SIL2	SIL3
60% to ≤90%	SIL2	SIL3	SIL4
90% to ≤99%	SIL3	SIL4	SIL4
>99%	SIL3	SIL4	SIL4

HFT, *hardware fault tolerance;* SFF, *safety failure fraction.*

The associated table for fault tolerance for Type A (Table 2 of the standard) is reproduced in Table VIII/1.3.1-1 (see also Table VI/4.2-2).

From (Clause 7.4.4.1.3 of) the standard one gets B subsystem referred to as *Type B* to be applied in case of:

- The failure mode of at least one constituent component is not well defined; or
- The behavior of the subsystem under fault conditions cannot be completely determined; or
- There is insufficient dependable failure data from field experience to support claims for rates of failure for detected and undetected dangerous failures.

The associated table for fault tolerance for Type B (Table 3 of the standard) is reproduced in Table VIII/1.3.1-2 (see also Table VI/4.2-2).

1.3.2 *HFT*

Why are all these discussed here? What are the outcomes of the discussions? These are discussed here to show how systems architectural constraints change the SIL (as shown in Fig. VIII/1.3-2). (Further details with diagrammatic examples are cited in the standard.) The outcomes from these discussions show that there are several factors discussed above, influence the determination of the SIL of a loop, results from failure

Table VIII/1.3.1-2 Safety Integrity of Hardware Architectural Constraints Type B Elements

SFF of an Element	HFT 0	HFT 1	HFT 2
<60%	Not allowed	SIL1	SIL2
60% to ≤90%	SIL1	SIL2	SIL3
90% to ≤99%	SIL2	SIL3	SIL4
>99%	SIL3	SIL4	SIL4

HFT, *hardware fault tolerance;* SFF, *safety failure fraction.*

modes, effects and diagnostic analysis (FMEDA), programmable electronics (PE) or highly integrated systems (can be classified as Type B).

What is the significance of HFT? HFT is an indicator safety function quality. HFT = 0 single-channel use and HFT = 1 redundant version; a minimum of two hardware faults need to occur to cause a safety loss, etc. These were discussed right at the beginning of the chapter.

1.3.3 SFF

From Tables VIII/1.3.1-1 and VIII/1.3.1-2 it is clear that SFF has a direct relationship with HFT. SFF signifies the fraction of safe device failures. SFF of 80% means that 80 out of 100 device failures do not affect the safety function of the device. So, SFF and HFT together determine the safety level pertinent to the device. Referring to Table VII/1.3-1 (for symbols), in line with IEC 61508, one can say that:

$$\text{SFF} = (\lambda_{\text{Sd}} + \lambda_{\text{Su}} + \lambda_{\text{Dd}})/(\lambda_{\text{Sd}} + \lambda_{\text{Su}} + \lambda_{\text{Dd}} + \lambda_{\text{Du}})$$

or:

$$\text{SFF} = 1 - (\lambda_{\text{Du}})/(\lambda_{\text{total}}) \qquad \text{(VIII/1.3.3-1)}$$

From Tables VIII/1.3.1-1 and VIII/1.3.1-2 it is seen that if SFF exceeds 60%, 90%, or 99%, SIL1, SIL2, or SIL3 is obtained, respectively, for $HFT = 0$. Self-declaration for SIL is permitted by IEC 61508 for SIL1 but requires the third-party certification for SIL2 and higher. So, if there are a number of components in a safety loop, then the SFF for the safety loop can be discovered:

$$\text{SFF of loop} = 1 - \Sigma\lambda_{\text{Du}}/(\Sigma\lambda_{\text{S}} + \Sigma\lambda_{\text{D}}).$$

If for a transmitter the given data are $\lambda_{\text{Du}} = 195$ FIT, $\lambda_{\text{Total}} = 809$ FIT, and an isolator $\lambda_{\text{Du}} = 21$ FIT, the $\lambda_{\text{Total}} = 207$ FIT where 1 FIT $= 10^{-9}$.

Therefore:

$$\Sigma\lambda_{\text{Du}} = 216\ \text{FIT} \quad \text{and} \quad \Sigma\lambda_{\text{S}} + \Sigma\lambda_{\text{D}} = \Sigma\lambda_{\text{Total}} = 1016\ \text{FIT}$$

hence:

$$\text{SFF of loop} = 1 - (216/1016) = 0.79 \rightarrow 78\% \qquad \text{(VIII/1.3.3-1a)}$$

1.3.4 PFD_{avg} and PTI

As discussed many times earlier, PFD_{avg} is an important parameter for SIF and it is dependent on proof test interval (PTI). This means that if a device has a PFD_{avg} of 4×10^{-4} for T_{proof} 1 year the failure probability of the safety function for the device is 0.0004 within 1 year. A similar calculation is also applicable for PFH.

1.3.5 Time Dependency and MTTF

From IEC 61508-2/6:2010-Annex B one sees that there will be variations of PFD based on PTI. Assuming a probabilistic approach, one may get: PFD: $Q(t) = 1 - e^{-\lambda dt}$; here time ($t$) stands for the PTI. The shorter the time interval, the better will be the result. It also depends on failure rate λ. The basis for development of the

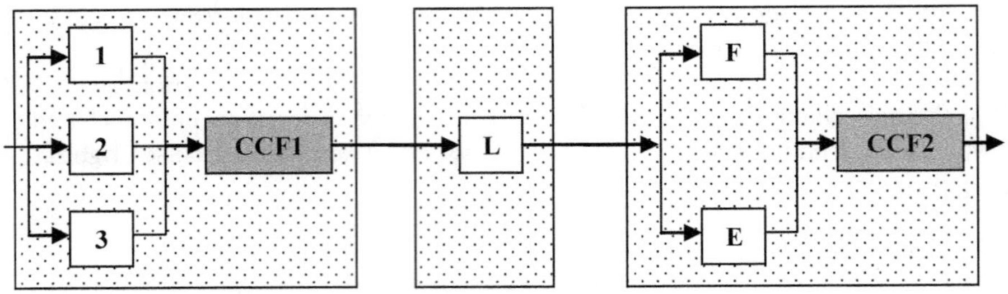

Based on IEC 61508 -2/6 Annex B : 2010

FIGURE VIII/1.3.5-1

Reliability block diagram for safety loop. See text for meaning of abbreviations.

time-dependent value of PFD is the reliability block diagram (RBD) of a complete safety loop shown in Fig. VIII/1.3.5-1 in line with IEC.

PFD_{avg} is derived from the ratio of $MDT(T)/T$ where $MDT(T)$ stands for the mean downtime over a time interval of $0-T$. So, mean downtime of the system is the sum of mean downtime of each bigger block (shown in *dotted hatches*).

1, 2, and 3 are triple failure, F and E are double failure, and common cause failure (CCF)1, L, and CCF2 are single failure. For low demand failure:

$$PFD_{avg} = PFD^{123}_{avg} + PFD^{L}_{avg} + PFD^{FE}_{avg}. \quad \text{(VIII/1.3.5-1)}$$

This is true as long as they are not parallel. In the case of a parallel combination both need to fail. Therefore if mean downtime within the block is (say) 1, 2, 3 or F, E, one needs the help of the following equation as an example:

$$MDT_{(FE)} = \int PFD_F(t) * PFD_E(t)dt \quad \text{(VIII/1.3.5-2)}$$

Then put the value in Eq. VIII/1.3.5-1 to obtain PFD_{avg}. Also from the standard one gets that for continuous/high demand mode:

$$PFH = 1/T\int_o^T w(t)dt \quad \text{(VIII/1.3.5-3)}$$

$$\text{for } PFH \ll 1, \text{ one gets } PFH \simeq 1/MTTF \quad \text{(VIII/1.3.5-4)}$$

One may refer to ISA-TR84.00.02-2002—Part 2 (Clause 5.1) for details of the calculation of a safety loop in various combinations. Returning to the original issue of PFD_{avg} of a safety loop one needs to know the test procedure implemented and test intervals so that one can arrive at the maximum SIL for the device or subsystem. Table VIII/1.3.5-1 shows the data from two components: one pressure transmitter and one isolator in series. Let us now check the SIL for the subsystem comprising these two components (pressure transmitter and isolator); for checking purposes the test interval as per FMEDA has been taken as 1/2/5 years (data). The selection with reasoning is elaborated in the table itself. The data shown in the table are taken from a reputable manufacturer's data in SIL certifications.

Table VIII/1.3.5-1 SIL for Safety Loop of Two Components

PTI T_{proof}	Pressure Transmitter (1)		Isolator (2)		Subsystems (1 and 2)	
	PFD_{avg}	SFF	PFD_{avg}	SFF	PFD_{avg}	SIL
1	2×10^{-4}	95%	9.21×10^{-5}	>89%	2.92×10^{-4}	SIL2 as SFF <90%. Also for PTI = 5 years PFD_{avg} is even higher
2	4×10^{-4}	95%	1.8×10^{-4}	>89%	5.8×10^{-4}	
5	8×10^{-4}	95%	4.6×10^{-4}	>89%	1.26×10^{-3}	

PFD_{avg}, *probability of failure on demand*; PTI, *proof test interval*; SFF, *safety failure fraction*; SIL, *safety integrity level*.

PTI is usually decided by the end user who needs to check the manufacturer's FMEDA result/certificates carefully to fit into the end user's system.

(Another way to look at PFD_{avg} is as follows: if F_{np} = frequency of hazardous event with no protection and F_t = tolerable frequency of hazardous event, then $PFD_{avg} = F_t/F_{np} = 1/RRF$).

For certification of the products, tools like FMEDA and Markov models are used. An FMEDA extends the FMEA techniques to include online diagnostic techniques and identify failure modes relevant to safety-instrumented system design. These PFD calculations may be carried out by commercial software tools (e.g., "SILence" by HIMA). In all these cases, bottom-up approaches were undertaken. Also it has been found that there has been a serious lack of reliable failure rate data and their variations among lab specifications, lab/field and field specifications. There are wide variations in the data obtained from these different sources. Therefore it is suggestive to use a *band of data* instead of discrete PFD and other failure rate data [5].

1.3.6 PFD Calculation Method

In Clause 1.4.5 of Chapter VII, PFDs for various configurations were discussed. In this clause, short calculation methods are discussed. These discussions are based on ISA-TR84.00.02-2002—Part 2 [8]. The following issues need to be looked into:

- Identification of the hazardous event for which the SIS is acting as a layer of protection.
- Identification of SIL for each SIF required for each hazardous event.
- Listing of components having an impact on each SIF such as sensor, logic solver, and final elements. SIL for SIF is assigned by the PHA team.
- Using the SIS architecture, the PFD_{avg} for each SIF is calculated with the combination of sensors, logic solver, final elements, power supply, and any other components.
- Determine if the PFD_{avg} meets the SRS for each SIF.
- If required, modify SIS and recalculate to meet the requirements specified in the SRSs as per the standard for each SIF.

- If SIS reliability impacts the consequence of concern, determine the expected spurious trip rate for system components and combine to obtain mean time to failure (MTTF) for the SIS.

As per IEC 61508-6 Annex B:2010, PFDs for various architectures shall be presented.

1.3.7 *Architecture for Low Demand Operation and Associated PFD (in line with IEC)*

In this clause, short calculation methods based on IEC 61508-6 (Annex B): 2010 have been depicted. For detailed calculation, standards should be referred to.

- 1oo1 (Fig. VIII/1.3.7-1)

$$\lambda_D = \lambda_{Dd} + \lambda_{Du} = \lambda/2 \quad \text{(VIII/1.3.7-1)}$$

$$t_{CE} = (\lambda_{Du}/\lambda_{Dd})(T_1/2 + \text{MRT}) + (\lambda_{Dd}/\lambda_D)\text{MTTR} \quad \text{(VIII/1.3.7-2)}$$

$$\lambda_{Du} = \lambda_D(1 - \text{DC}) \quad \text{(VIII/1.3.7-3)}$$

$$\text{PFD} = 1 - e^{(-\lambda_D * t_{CE})} = \lambda_D * t_{CE} \ \text{as}\ \lambda_D * t_{CE} \ll 1 \quad \text{(VIII/1.3.7-4)}$$

for 1oo1;

$$\text{PFD}_G = \lambda_D * t_{CE} \quad \text{(VIII/1.3.7-5)}$$

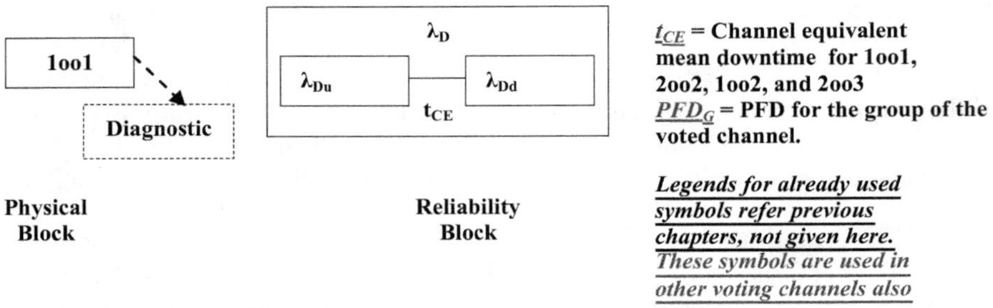

FIGURE VIII/1.3.7-1

1oo1 low demand mode voting circuit.

- 1oo2 (Fig. VIII/1.3.7-2)

$$t_{CE} = (\lambda_{Du}/\lambda_{Dd})(T_1/3 + \text{MRT}) + (\lambda_{Dd}/\lambda_D)\text{MTTR} \quad \text{(VIII/1.3.7-6)}$$

$$\text{PFD}_G = 2((1 - \beta_D)\lambda_{Dd} + (1 - \beta)\lambda_{Du})^2 t_{GE} * t_{CE} + \beta_D\lambda_{Dd} * \text{MTTR} + \beta * \lambda_{Du}(T_1 + \text{MRT}) \quad \text{(VIII/1.3.7-7)}$$

where β = the fraction of undetected failures that have a common cause; β_D = those failures that are detected by the diagnostic tests, the fraction that have common cause, and t_{GE} = voted group equivalent mean downtime (hour) for 1oo2 and 2oo3 architectures.

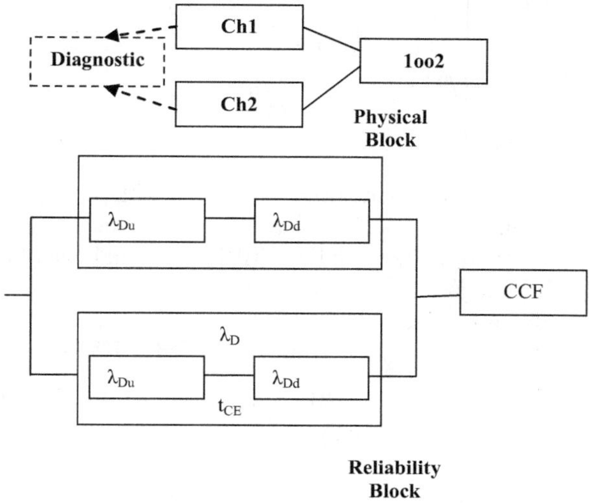

FIGURE VIII/1.3.7.-2

1oo2 low demand mode voting circuit. *CCF*, common cause failure.

- 2oo2: In an RBD it is just a series of two reliability blocks each comprising 1oo1 RBD. Therefore:

$$PFD_G = 2\lambda_D * t_{CE} \quad \text{(VIII/1.3.7-8)}$$

- 1oo2D: Two parallel channels both need to demand a safety function before it can take place under normal conditions. If a diagnostic test detects that one channel is faulty, output voting is adapted so that the overall output follows that given by the other channel. Channel comparison and switching may not be 100% decided by the *K* factor, standing for efficiency of comparison λ_{Sd} (Fig. VIII/1.3.7-3).

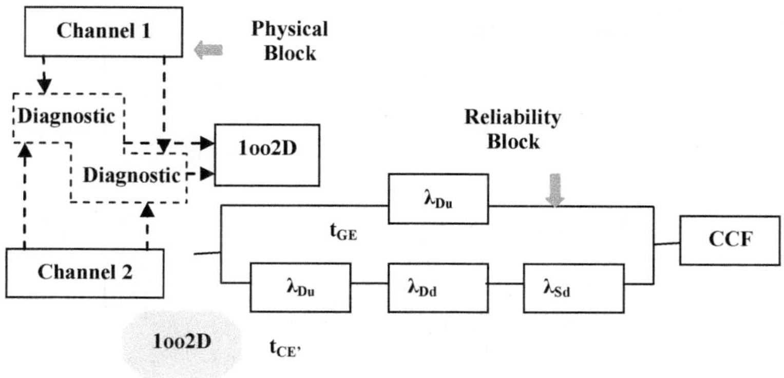

FIGURE VIII/1.3.7-3

1oo2D low demand mode voting circuit. *CCF*, common cause failure.

$$t_{CE'} = \{\lambda_{Du}(T_1/2 + MRT) + (\lambda_{Dd} + \lambda_{Sd})MTTR\}/\{\lambda_{Du} + (\lambda_{Dd} + \lambda_{Sd})\} \quad (VIII/1.3.7\text{-}9)$$

$$t_{GE'} = T_1/3 + MRT \quad (VIII/1.3.7\text{-}10)$$

$$PFD_G = 2(1-\beta)\lambda_{Du} * ((1-\beta)\lambda_{Du} + (1-\beta_D)\lambda_{Dd} + \lambda_{Sd})t_{CE'}t_{GE'} + 2(1-K)\lambda_{Dd}t_{CE'} + \beta\lambda_{Du}(T\,21 + MRT) \quad (VIII/1.3.7\text{-}11)$$

where $t_{CE'}$ = channel equivalent downtime for 1oo2D, $t_{GE'}$ = voting group equivalent downtime for 1oo2D, and K = factor of success of auto test in 1oo2D. Similarly for 2oo3 shown in Fig. VIII/1.3.7-4 will have PFD_G as shown in Eq. VIII/1.3.7-12. With the discussions on SIL calculation on SIS is concluded. Readers attention is drawn to the check list for CCF and systematic failure (based on ISA TR 84), shown in Fig. VIII/1.3.7-5.

$$PFD_G = 6((1-\beta_D)\lambda_{Dd} + (1-\beta)\lambda_{Du})^2 * t_{CE} * t_{GE} + \beta_D * \lambda_{Dd} * MTTR + \beta * \lambda_{Du}(T_1/2 + MRT) \quad (VIII/1.3.7\text{-}12)$$

1.4 SIL Certifications and Standards

Starting from the safety loop, let us examine how for a particular instrument/device SIL certifications are given and what is the minimum that needs to be seen. During

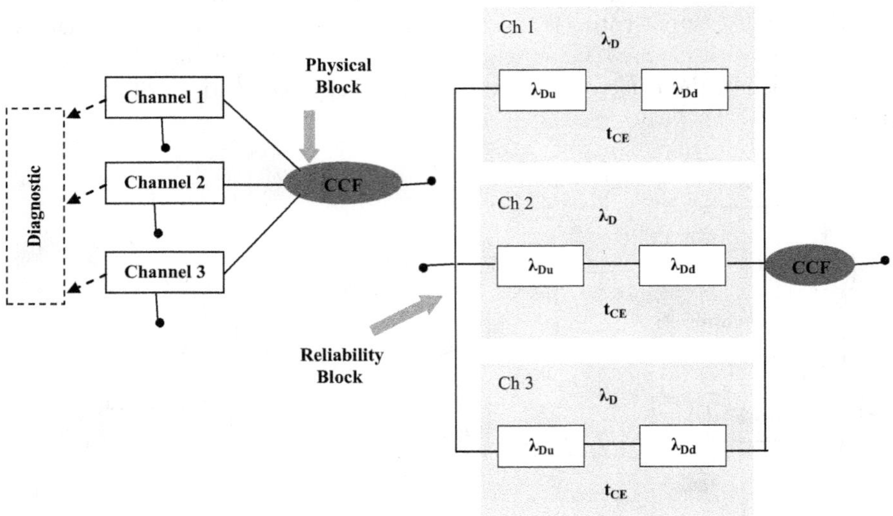

FIGURE VIII/1.3.7-4

2oo3 low demand voting circuit.

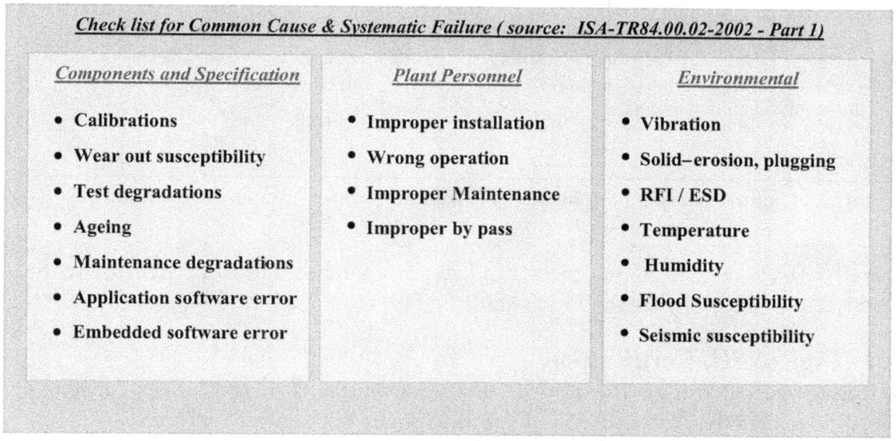

FIGURE VIII/1.3.7-5

Checklist for common cause and systematic failures. *ESD*, emergency shutdown; *RFI*, radiofrequency interference.

ISA-TR84.00.02-2002—Part 1.

discussions a sensor has been chosen, because it is responsible for an appreciable share (see Fig. VIII/1.0-1) of PFD_{avg} for the SIF.

Referring to Fig. VIII/1.4-1 it is seen that the safety loop comprises sensor SIL2, logic solver SIL3, and final element SIL2. (For better understanding of a safety loop Fig. VIII/1.4-2 may be referred to.) Here, focus is on the sensor to check that the certificate conforms to the requirements. As discussed earlier and shown in

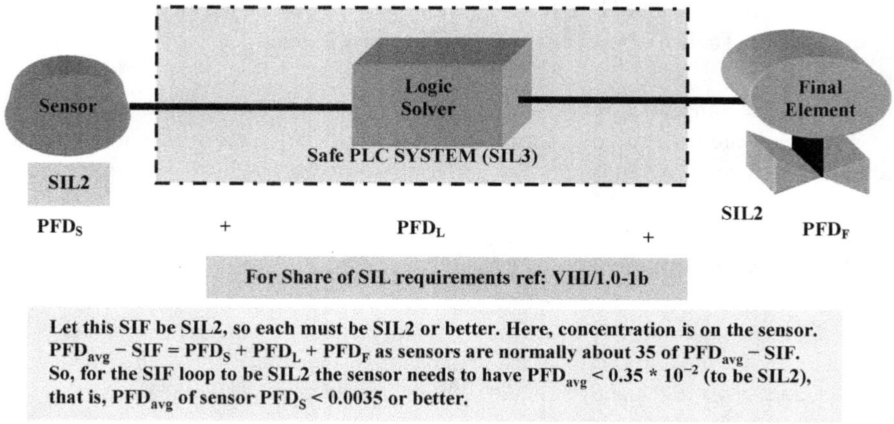

FIGURE VIII/1.4-1

Conformity of component safety integrity level (SIL) and safety loop (SIF). *PLC*, programmable logic controller; *PFD*, probability of failure on demand.

Functional block: **The smallest element of an SIF whose failure can result in failure of the safety instrumented function. A functional block may consist of a number of functional block elements. In general, in the description of the SIF/SIS the sensor, logic solver, and final elements are considered as one functional block.**

FIGURE VIII/1.4-2

Function block for safety instrument functions (SIFs). *SIS*, safety instrumentation system.

Fig. VIII/1.0-1B, nearly 35–40% of PFD requirements come from sensors, so to meet SIL2 for the loop, sensor PFDS shall be less than $0.35 * 10^{-2}$.

1.4.1 Typical SIL Certification

The following SIL certification is for a smart transmitter. The third-party certifying authority declares the transmitter as Type B (a smart transmitter with software—see Table VI/4.2-2 for definitions). SIL2 with HFT 0 (single connection use) has an SFF of **75%**. This example was chosen to show how IEC *proven* use can be used. Readers should note that this certification is given by a world renowned certifying authority to a manufacturer who is also world renowned. It has PROVEN IN USE, which is allowed by **IEC 61511**. The certification declaration states:

> A generally accepted distribution of PFD_{avg} values of a SIF over sensor part, logic solver part and final element part assumes that 35% of the total SIF PFD_{avg} value is caused by sensor part…
>
> The pressure transmitter XXX with 4–20 A output is considered to be a type B component with HFT = 0.
>
> Type B components with SFF 60 to <90% must have a hardware fault tolerance of 1 according to Table 3 of IEC 61508-2 for SIL (sub-) systems.
>
> As the pressure transmitter XXX is supposed to be a proven in use device an assessment of hardware with additional prior use demonstration for the device and its software was *carried* out. The prior use investigation was based on *field return data* collected and analyzed by … manufacturer. This data cannot cover *process connection*. Prior-use justification for the process connection **still needs to be done by end user** (to make a note by reader)…
>
> *According to the requirements of IEC 61511-1 first edition 2003-01 Section 11.4.4 and the assessment described in Section 5.1 the Type B pressure transmitter XXX with a hardware fault tolerance of 0 and SFF 60 to <90% is considered suitable for use in SIL2 safety function*. The **decision** on the usage of prior-use device, however, is **always with the end user** (to make a note by reader).

SIL certificate of conformity: (reader to note the data furnished)

Manufacturer: NNNNN
Product name Pressure transmitter XXXXX
Functional safety according to IEC 61508/61511
declaration…
SIL2 Type B HFT 0 (one channel application)

Transmitter Type	Meas. Range	SFF	PFD_{avg}	$\lambda_{Dd} + \lambda_S$	λ_{Du}
XXX	10 mBar	75%	$8.54 * 10^{-4}$	614 FIT	195 FIT

Now to see how FMEDA/proof test results are declared!
Summary of version "*n*" results:

T_{Proof} = 1 year	T_{Proof} = 5 years	T_{Proof} = 10 years
PFD_{avg} = 8.54E-04	PFD_{avg} = 4.26E-03	PFD_{avg} = 8.5E-03

The common requirements are [5]:

- CE Declaration of Conformity
- Safety manual
- Product documentation
- FMEDA test

In the certificates, the certifying authority declares how the *failure rates* are assigned in the previous example "field failure date using only operational hours from the warranty period of operation." Also how confidence interval calculations are done is declared. How the failure data are *compared* with industrial databases is explained by the certifying authority. Assuming the connected logic solver *can* detect fail high or fail low, then they can be *considered* as safe failure or dangerous but detected failure [6].

1.4.2 Understanding the Certificate

SIL is a dependability measure of the overall safety function performed by the safety system in question. So, it is necessary that all *critical* attributes of a device are stated in the certificate to indicate suitability of the device in SIL-rated safety functions. SIL is not just the probabilistic failure data. Also note that the SIL number mentioned in a certificate is normally arrived at after a *number* of assumptions on overall safety and conditions required for SIL allocation. In fact, IEC 61508 does not directly mention requirements for certificates; instead it puts more importance on functional safety assessment (FSA).

1.4.3 FSA Discussions

As per Clause 8 of IEC 61508-1:2010 the objective of an FSA is to specify the activities necessary to investigate and arrive at a judgment on the **adequacy** of functional safety achieved by E/E/PE. It is not a mere one-page certificate but a report to cover all necessary information for the user of the product. The FSA report is auditable to ensure that all relevant clauses of IEC 61508 are *traceable*. As per the standard, an FSA is to be carried out by appointed person(s) and applied in all phases throughout the E/E/PE system and software safety life cycle including documentation, verification, and management of functional safety. An FSA at least needs to look into the following issues:

- Work done since previous assessment
- Plan or strategy to be implemented for a further FSA of the overall E/E/PE system and software safety life cycle
- Recommendations as per previous assessment and to check to what extent the changes have been made to meet these recommendations
  The standard uses another term similar to SIL, "*systematic capability*" of an element. The systematic capability number refers to the rigor of the documentation and quality process used throughout the product's development to avoid *systematic* failures.

1.4.4 HFT, SIL, and SFF

HFT, SIL, and SFF are common terms found in SIL certificates. Each has its own meaning as already discussed. These meanings are further discussed here to clear up any confusion. One important issue is that suppose a certificate says a device SIL @ HFT1. Generally, HFT1 means that it can withstand a single failure signifying that there will be *redundant* devices needed to achieve SIL3, but it does not say that with redundant devices SIL3 will be achieved, or what SIL will be achieved when redundancy is used. An interesting issue arises in the case of a final element. Suppose a final element has λ_{open} and λ_{close} 60 and 600 FITs. Then SFF could be 60/660 = 9% or 600/660 = 91%! So, the terms "safe failure," "dangerous failure," and hence "safe failure fraction" for an instrument are only *relevant* when there is *knowledge* of the target application [7].

1.4.5 Data and Data Sources

Referring to the underlined part of Clause 1.4.1, it is seen that the certifying authority mentions the sources of data. Sources of component failure data are **extremely important** especially when they are qualified through "**proven instruments**." This is because these data at times are industry specific and have *variations* with industrial use. The sources of data must be checked to confirm that the database is appropriate for the current intended location and application and environmental condition for intended use of the instrument. It has already been stated that probability of failure PFD_{avg} is dependent on the PTI; therefore end user needs must be taken care of with respect to FMEDA results. Every compliant instrument must have a "safety manual" duly referred to in the certificate. It is critical to use the device only in accordance with the safety manual (the certified failure data is usually invalid otherwise) [7].

2.0 SIL POLICY AND PROGRAMMABLE SYSTEMS

Safety in process plants, production plants, and in machinery industries is mainly governed by the following:

- Moral obligation: Avoidance of any harm to personnel and environment
- Laws: Local and international legislations and regulations
- Sound economic reasons

These safety needs are to be kept in mind from conceptual design stage through all stages of the safety life cycle already discussed. As mentioned at the beginning of the chapter, the logic solver has nearly 15% of the total PFD demands, yet these need special attention because these are *Type B* devices where both hardware and software need proper consideration. When people talk of programmable systems, naturally logic solvers come to mind! However, smart sensors and final elements also fall under this category.

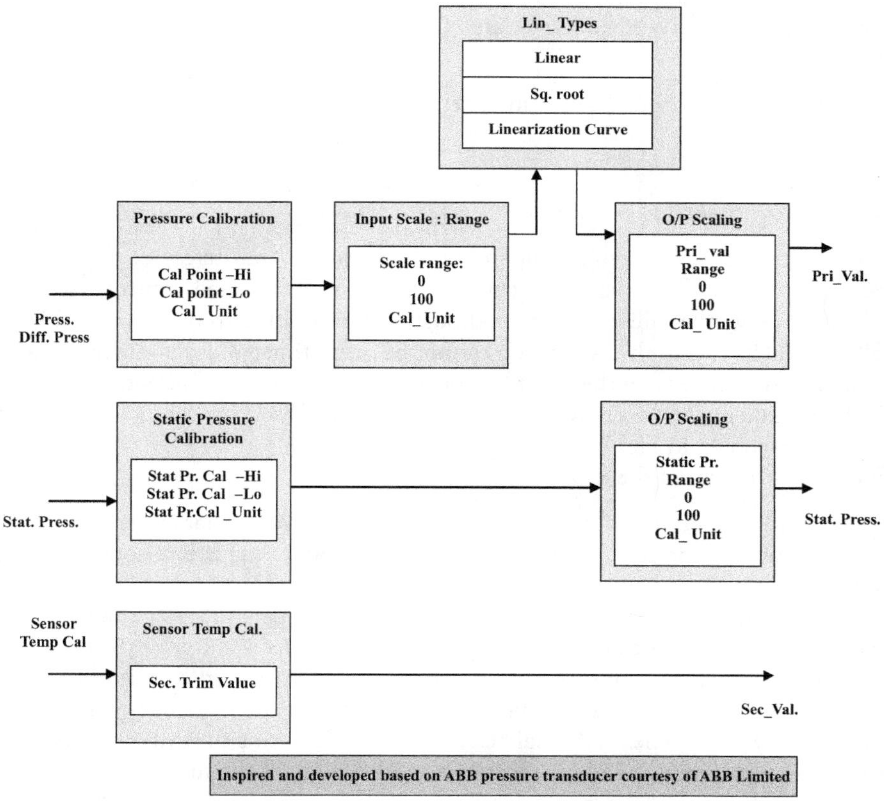

FIGURE VIII/2.0-1

Typical intelligent pressure transducer block diagram.

A typical functional block diagram (refer Fig. VIII/1.4-2 for explanation of functional block) for a simple pressure transmitter (courtesy ABB) is presented in Fig. VIII/2.0-1. Here, it is shown how even in a simple pressure transducer the characteristics are attached to each block and these are connected to the output via the software system. This simple example has been chosen to show the role of hardware and software and their interrelationship in PE.

The main issue here is that for any programmable system, both hardware and software aspects are equally important. International standards for PE take into consideration both hardware and software. In contrast to previous discussion, now the discussions both hardware and software issues related to safety system and SIL will be focused on to show both their importance and their interactions. Another important issue is when any PE safety-related system needs to implement both safety and nonsafety functions, then hardware and software shall be developed and treated as *safety related* unless there is enough independence between the safety and nonsafety functions, and it can demonstrate that:

- These are functionally independent.
- Implementations are separate and independent.
- Nonfunctioning of any nonsafety-related function will not affect any safety-related function.

Also in PE systems there shall be integrity in terms of:

- Hardware integrity
- Software integrity
- System integrity such as operation

The logic solver also plays an important role in the overall safety of the loop discussed previously. The logic solver could be a PLC or distributed control system (DCS) and/or any other allied system. Although the logic solver is only responsible for 15% of PFD its role in the safety loop cannot be overestimated. It mainly handles safety functions after sensing the hazard through a sensor, until final action is taken via the final element. Therefore it is better to look into the basic building blocks of a logic solver, as shown in Fig. VIII/2.0-2.

The basic structure of the logic solver consists of:

- Process behavior and interface: All the aspects of the process that are to be controlled and monitored during normal operation and during failure condition to be considered.
- Human—machine interface (HMI): This relates to the operator's action and communication with the process. Correct operational aspects and the operator's failure to react must be considered in this subsystem. In addition to this, the subsystem needs to interface for engineering and programming changes.
- Functionality of PE: The functional behavior of the PE during normal operation as well as operation with failures shall be considered under the functional processing subsystem. All possible conditions must be considered so that PE can take the system into a safe state under failure conditions. Also here, all plant-

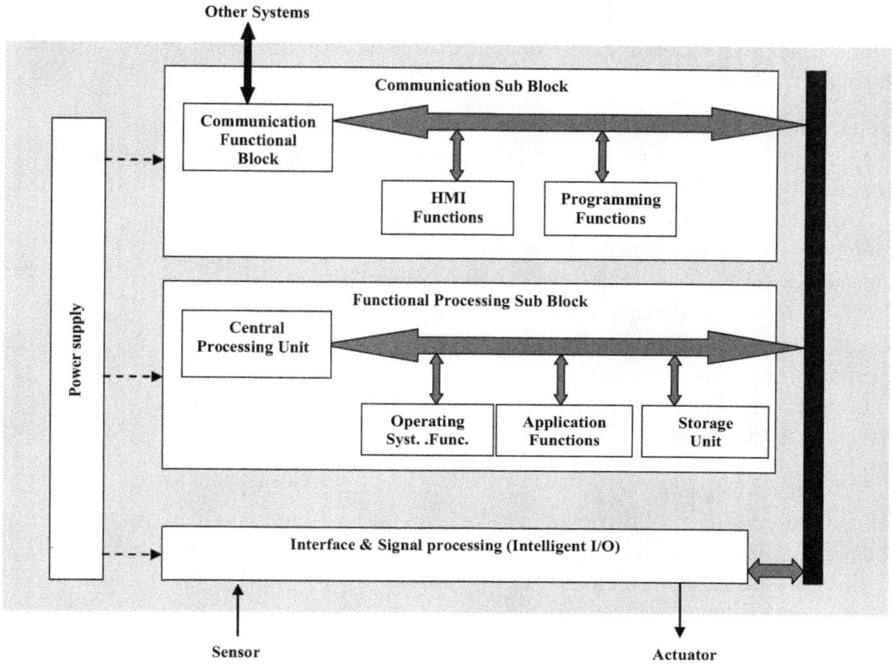

FIGURE VIII/2.0-2

Logic solver block and functional diagram. *HMI*, human–machine interface; *I/O*, input/output.

related parameters and other data storage systems play an important role in maintaining system functioning. Systems operating software and application software are part of the system so that the PE can function in a desired manner.

- Communication with other systems: In many cases the system is connected and integrated with other systems, therefore proper communication with other systems is important.
- Internal data bus: All these subsystems are connected and data are shared via an internal data bus, so that it is possible to exchange data throughout the system.

Another important issue here is that a PLC can be used for safety function after proper PHA only. The basic technical requirements of specifications for a PLC or logic solver are beyond the scope of this book. Interested readers may refer to Chapter VII of [1], where details are discussed. Here only safety aspects of a PLC/logic solver are discussed based on IEC 61508/61511. The requirements of safety specifications shall cover both function and safety properties to be considered during the design phase for achievement of functional safety for PE systems.

2.0.1 Functional Safety Requirements

The functional SRS should include the following as a minimum to achieve functional safety of a PE system:

- Required safety functions to achieve functional safety
- Application of safety functions as protection (low demand) and/or continuous control
- All applicable modes of operation of EUC—normal and under failure
- Failure behavior and expected response
- Hardware and software interface and constraints (if any)
- Environmental conditions related with safety functions (if any)
- Operator interface and communication
- Response time and throughput
- Interface with non-PE systems for achievement of safety function
- Periodic testing interval

2.0.2 Safety Integrity

As already discussed, if a PE system is to be used for both safety as well as nonsafety applications it has to be designed for a safety system unless it can be demonstrated that there is adequate independence between the two. Also for PE systems,

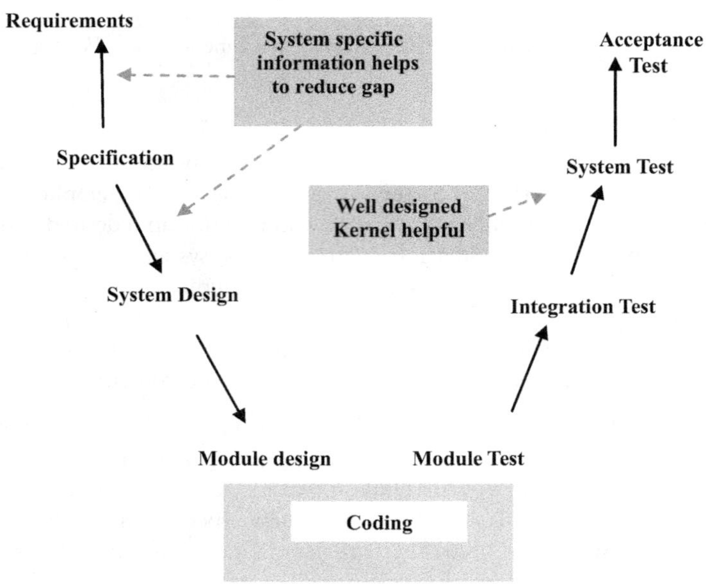

FIGURE VIII/2.0.3-1

Software design process—basic concept for a V model.

Idea from Quantum Safety PLC Safety Reference Manual, Schneider Electric, November 2007.

hardware, software, and system integrity need to be demonstrated, as already discussed (see IEC61508-2:2010 Clause 7.4.2.3).

2.0.3 *Software Characteristics*

Based on the basic structure of a PLC/logic solver shown in Fig. VIII/2.0-2, the software design process shall be as shown in Fig. VIII/2.0.3-1. From the figure it is clear that first the system requirements are to be developed to the main specification. Based on the specification the main system design including module design is developed so that coding can be done. Again based on the coding module, testing is carried out. Afterward, successful module testing work for integration is undertaken. System test is the last step based on which system is accepted. It is worth noting that the V model of IEC 61508 also provides a similar model and is shown in Fig. VIII/2.0.3-2. Software for the PLC in safety applications is characterized by its *size*, application-specific *language*, limited *variability*, and use of *kernel*. Since kernels are fixed programs this could be made rigorous, and because of its many uses it could also be made economical.

In Fig. VIII/2.0.3-2 the V model approach for system software as per IEC 61508 is shown to focus on the software development process in safety applications. The international standard makes the generic approach for all safety life cycle activities for

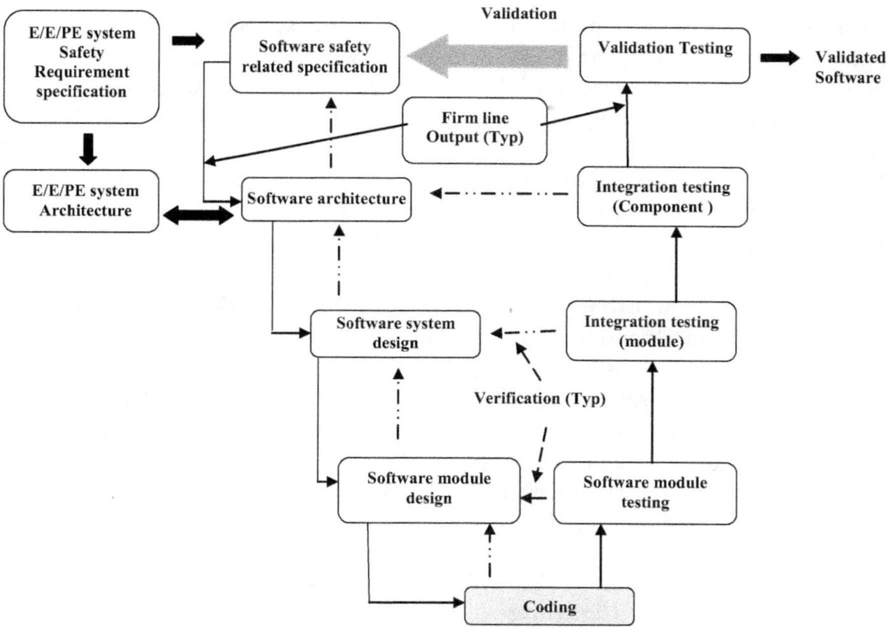

FIGURE VIII/2.0.3-2

Software systematic capability and development. *E/E/PE*, electrical/electronics/programmable electronics.

E/E/PE systems. The framework of the approach is very robust and can handle all activities for rapidly changing technologies from concept through installation and operation up to decommissioning. The following is a short discussion on various requirements of the standard to see how in a generic manner the approach toward PE systems needs to be undertaken. By comparing Figs. VIII/2.0.3-1 and VIII/2.0.3-2 it is clear that structurally they are similar only if V model system architecture is taken into account.

2.1 Discussions on IEC Requirements for PE Systems and SIL Policies

IEC 61508:2010 is the main international standard considered for PE systems in the safety life cycle. *All these clauses may be read in conjunction with Clause 4.0 of Chapter VI. In the following subclauses general requirements for PE systems from IEC 61508-1: 2010 are briefly discussed.*

2.1.1 *Overall Safety Life Cycle Requirement Issue*

Here, the requirements for the overall life cycle are discussed in brief highlighting major issues (see the standard).

- For each identified hazard event, necessary risk reduction, integrity requirement, and safety functions to ensure the required functional safety shall be specified. These shall finally constitute the overall safety for the system.
- All possible common cause failures need to be considered during allocation of safety functions. For this various E/E/PE and other technology safety-related systems shall be based on functionally diverse technologies, if these are to be treated as an IPL.
- For PE safety-related systems with different SIF of different SIL, then if independent implementations cannot be demonstrated, the *requirement* of the highest SIL will be applicable.
- An architecture with a single E/E/PE safety-related system of SIL4 is permitted if:
    - There has been an explicit demonstration, by a combination of appropriate analytical methods and testing, of the target SIL failure measure *OR*.
    - There has been extensive operating experience of the components used as part of the E/E/PE safety-related system in a similar environment and, as a minimum, used in a system of comparable complexity level *AND* there is sufficient hardware failure data, etc. to allow sufficient confidence in the hardware safety integrity target failure measure that is to be claimed.
- The SIL for E/E/PE systems shall be strictly as per Tables 2 and 3 of IEC 61508-1: 2010.

2.1.2 *Safety Validation Planning*

Safety validation planning: This is necessary to optimize the validation phase and to guarantee the feasibility of the validation. A plan shall be developed that shall include the following:

- When the validation shall take place
- Who shall carry out the validation
- Specification of the relevant modes of the EUC operation and relationship with E/E/PE safety-related system, including the following as applicable:
  - Preparation for use, including setting and adjustment
  - Startup
  - Communicate
  - Automatic
  - Manual
  - Semiautomatic
  - Steady state of operation
  - Resetting
  - Shutdown
  - Maintenance
  - Reasonably foreseeable abnormal conditions
- Specification of the E/E/PE safety-related systems that need to be validated for each mode of EUC operation before commissioning commences.
- Technical strategy, measure, and technique for validation, etc. for confirmation that allocated safety functions are carried out correctly.
- Required environment for validation activity
- Pass/fail criteria
- Policy and procedure for evaluating results
- Proper documentation

2.1.3 Overall Installation and Commissioning Planning

A detailed plan for installation and commissioning shall be developed to specify the following:

- Installation planning:
  - Installation schedule
  - Responsibility for installation of different parts
  - Installation procedure
  - Installation sequence
  - Criteria for declaring installation activities complete
  - Procedures for the resolution of failures and incompatibilities
- Commissioning planning:
  - Commissioning schedule
  - Responsibilities for commissioning of different parts
  - Commissioning procedure
  - Relationships to the different steps in the installation
  - Relationships to the validation

2.1.4 *Safety Requirements Specification*

Major issues related to the SRS shall include but not be limited to the following points (in this connection Fig. VIII/2.0.3-2 may be referred to):

- Main requirements derived from allocation of safety requirements (Clause 7.6 of the standard).
- Functional safety requirements and SILs.
- Safety specification shall be made *available* to the *developer* of the E/E/PE system.
- Specification shall be structured and developed in a clear, precise but comprehensive manner written in formal or natural language.
- The functional safety details shall include:
  - Comprehensive requirement sufficient for design
  - Detailed description
  - Response time performance
  - Operator interface and interface with other systems (if any)
  - All related information that can influence functional safety
  - Relevant modes of operation of EUC
  - Expected modes of behavior of the E/E/PE system
- The functional safety details shall include:
  - SIL for each SIF and target failure value if possible
  - Low demand/high demand modes of operation
  - Required duty cycle and lifetime
  - Requirements, constraints, functions, and facilities
  - Environmental constraints including electromagnetic interference, etc.
  - Limitations and constraints for realization (if any).

2.1.5 *Modification and Retrofit*

The modification and retrofit phase shall be initiated only by the issue of an authorized request under the procedures for the management of functional safety (Clause 6 of the standard). Various issues related to these requests are:

- The affected identified hazard
- The proposed change (hardware and software)
- The reasons for the change such as:
  - Functional safety below specification
  - Systematic fault experience
  - Change in safety legislation
  - Modifications to the EUC and use
  - Modification to the overall safety requirements
  - Analysis of operations and maintenance; performance below target
  - Routine functional safety audits
- Impact analysis to assess impact of modification with documented result
- Authorization for management of change

2.1.6 *Functional Safety Assessment*

For functional safety assessment the issues involved shall include but not be limited to the following:

- A team to carry out a functional safety assessment to ensure the functional safety achievement.
- All persons involved to have access to functional safety assessment information equipment.
- Functional safety assessment to be applied to all phases of the life cycle.
- If tools are used, then they shall be subject to functional safety assessment.
- Work done since last assessment and strategy for next assessment.
- Plan of functional safety assessment to specify:
  - Those to undertake the functional safety assessment
  - The outputs from each functional safety assessment
  - The scope of the functional safety assessment.
- Prior approval for safety assessment.

2.2 Realization Phase (IEC 61508-2:2010)—Hardware

Hardware: This part is intended to be used only after thorough understanding of Part 1 of the standard. This part specifies all the requirements pertinent to the safety life cycle for E/E/PE systems, especially in terms of realization for E/E/PE systems. It also provides necessary information for operation and maintenance.

2.2.1 *Conceptual Approach*

This develops an understanding of EUC and associated hazards.

- Acquiring familiarity about EUC
- Identification of likely sources of hazard
- Collecting information about identified hazard including hazard caused by interaction with other equipment
- Safety regulations

2.2.2 *General*

Design Requirement Specification: As told earlier the specification shall be precise but comprehensive and written in clear, natural/formal language. In this connection Fig. VIII/2.0.3-2 may be referred to for software (SW) issues.

- Details of hardware and software necessary to implement safety functions including the following:
  - Requirement of hardware and software for subsystems and their integration
  - Throughput performance and response time criteria
  - *Accuracy* and *stability* of measurement and controls
  - E/E/PE systems and operator *interface*
  - All modes of behavior, including failure behavior of the E/E/PE system
  - Hardware and software *interactions* and their *constraints* (if any)

  - Architecture of the *subsystem* to meet the architectural constraints
  - Any specific requirement for startup of the E/E/PE system
  - Modeling parameters
  - Action taken during dangerous failure
  - Equipment capability to meet extremely harsh environmental conditions
- Progressive design specification with caution to avoid mistakes
- Implication imposed on architecture on account of design requirements

2.2.3 Validation Planning

Planning steps necessary to demonstrate requirements are met and the major issues are as follows:

- Safety and design requirements specification
- Procedure to demonstrate/pass fail criteria
- Required environment for testing
- Test and performance evaluation procedure
- Policy for resolution of validation failure

2.2.4 Design and Development

This promotes the basic requirements for design and development of E/E/PE systems including hardware and software architecture, sensor, actuators, and PE including embedded software, application specific integrated circuit (ASIC), etc. Major issues related to this shall include but not be limited to the following:

- All of the requirements for the design and development of safety-related systems:
  - Hardware safety integrity: Architectural constraints and *quantification* of random failure
  - *Special* architecture for integrated circuits with on-chip redundancy
  - Systematic safety integrity
  - System behavior on fault detection
  - Data communication process
- Highest SIL of SIF will determine the requirement for hardware and software unless independent implementation can be demonstrated. Also see Clause 2.0.2.
- Independence between safety functions, design documentation for the methods of implementation, and associated justification are required.
- Developer needs to review the following to ensure *adequacy* in the specification:
  - Safety function
  - SIL
  - Equipment and operator interface
- Interaction between hardware and software shall be identified, justified, and documented.
- All techniques and measures should be justified and documented as applicable.
- After initial design, an analysis shall be undertaken.

- When ASICs are used in a safety function the safety life cycle of the ASICs shall be studied.
- Synthesis of elements to achieve required systematic capability (**edition 2**):
  - Systematic safety integrity
  - Systematic capability (SC) of element SC N.
  - Maximum capability after combination: $N + 1$.
- Hardware safety integrity architectural constraints:
  - A hardware fault tolerance N means that $N + 1$ faults can cause loss of safety function.
  - To consider a single fault when one may cause subsequent faults
  - Definition of Type A and Type B elements (*Table VI/4.2-2*)
  - When estimating SFF, diagnostic test interval/ratio of test rate and demand rate are to be considered for low demand and high demand cases, respectively.
- Method/procedure for claiming maximum SIL (see Clause 7.4.4.2 of Part 2: 2010)
- Method/procedure for minimum hardware fault tolerance (see Clause 7.4.4.3 of Part 2:2010)
- Requirements for quantifying the effects of random hardware failure—for each safety function, achieved safety integrity related to system because of random hardware failure and random failure of communication are to be estimated. It shall be less than the target failure measure (see Clause 7.4.5.1 of Part 2:2010).
- Estimate of achieved failure measure (see Clause 7.4.5.2 of Part 2:2010)
- Diagnostic test interval for any subsystem (see Clause 7.4.5.4 of Part 2:2010)
- Various other requirements are (with clause number of IEC 61508-2:2010 in parentheses):
  - Avoidance of systematic faults (Clause 7.4.6)
  - Control of systematic faults (Clause 7.4.7)
  - System behavior on detection of fault (Clause 7.4.8)
  - System implementation (Clause 7.4.9)
  - Proven in use elements (Clause 7.4.10 and Clause 1.4.1)
  - Data communication (Clause 7.4.11)

2.2.5 E/E/PE System Integration

This clause is meant to integrate as per the design and test accordingly the related issues as follows:

- Testing of modules to ensure correct interaction of modules for desired safety function
- Integration of safety software
- Appropriate group of techniques and measures for avoidance of error during integration
- Any change during integration is subject to impact analysis.
- Appropriate documentation during integration for the following:
  - Version test specification
  - Acceptance criteria

- Version of the E/E/PE system under test
- Result of each test
- List of tools and equipment with calibration data
- Discrepancy between expected and actual result
- Analysis and decision on tests

2.2.6 *Operation and Maintenance Procedure*

This includes criteria to maintain functional safety during the O&M period. The major concerns are:

- O&M procedure to include major issues:
  - Routine action and replacement of elements
  - Action and constraints to prevent unsafe state
  - Maintenance of documentation for system failure, results of audit, etc.
  - Maintenance procedure to be followed during fault diagnosis and repair, revalidation or revalidation when original element not available, etc.
  - Tools necessary for maintenance and revalidation
- Assessment of O&M procedure impact on EUC
- Continuous upgradation from audit and test results
- Routine maintenance procedure to be evaluated by systematic method

2.2.7 *Safety Validation*

This is to validate that all requirements of safety functions and integrity are maintained. Major issues to address are:

- Prepared plans
- Calibration of measuring equipment as per standard
- Adequate implementation of safety requirement as per design
- Appropriate documentation covering:
  - Version of the E/E/PE system
  - Safety function under test
  - List of tools and equipment with calibration data
  - Result of each test
  - Discrepancy between expected and actual value
- In case of discrepancy in validation, analysis and decision taken to be documented
- Availability of validation test to developer of EUC
- Techniques and measures for avoidance of discrepancy during validation tests

2.2.8 *Modification*

There shall be proper documentation for all such modifications. Some of the major issues shall include but not be limited to the following:

- Detailed specification for modification
- Impact analysis
- Approval of changes

- Progress of changes
- Test cases—revalidations
- Changes to procedure and documentation
- Manufacturers to maintain a system for changes following a defect
- Revalidation after changes

2.2.9 *Verification*

The main objective of verification is testing and evaluation of a given phase. Major issues involved shall include but not be limited to the following:

- Concurrent planning with development of the phase with criteria, techniques, and tools
- Activities to be performed to ensure correctness and consistency of the product
- In each design and development phase to show that functional and integrity requirements are met and design requirements are verified
- Documentation of the result of each verification
- Determination of completeness and consistency of E/E/PE safety-related systems with design requirements

2.3 Realization Phase (IEC 61508-3:2010)—Software

Software: This is applicable to safety-related software of E/E/PE systems to be studied in conjunction with Parts 1 and 2 of the standard. Under this are the requirements of safety-related software in terms of software safety functions for each PE (say a subsystem of a PLC or different PLCs), which will be used to implement the required safety function. It is also necessary to check the software systematic capability for each PE system. In this connection Table VI/4.3-1 and Figs. VI/4.0-4, VI/4.0-5, and VIII/2.0.3-2 may be referred to for understanding the various steps toward software development.

2.3.1 *Software Safety Requirements Specification*

This clause mainly deals with specification of software safety requirements and systematic capability. The major issues here are:

- Specification for safety-related software shall be derived from specified safety requirements of E/E/PE systems.
- Specification shall be sufficiently detailed enough for design and development to include but not be limited to the *following*:
  - Safety function
  - Architecture of the system
  - Hardware safety integrity
  - Software systematic capability
  - Capacity and response time
  - Equipment and operator interface
- To address independence CCF *analysis* to be carried out
- All relevant modes of operation of EUC of E/E/PE systems

- Safety and relevant constraints between software and hardware as applicable
- Software specification shall also include but not be limited to the following:
  - Software self-monitoring
  - Monitoring of PE sensor and actuator
  - Periodic testing of safety function
  - Software function to execute proof test
- Software safety specification to specify safety property of product (not project) such as:
  - Functions enabling EUC *to* achieve and maintain safe state
  - Function to detect and annunciate management of fault
  - Function for management of software itself
  - Online testing of safety function
  - Periodic offline testing of safety function
  - Functions allowed for safe modifications
  - Capacity and response time
  - Interface of software with PE and nonsafety functions
  - Safety-related communication
- Where software safety is expressed or implemented by configuration data (set point), then data shall be consistent with requirement and within permitted range and defined in a manner compatible with underlying software.
- Operational parameters shall be protected against corruption, unauthorized changes, etc.

2.3.2 *Validation Plan for Software Aspect of System Safety*

This is to develop a plan for validation of the safety-related software aspect of the system. Depending on applicability, various issues discussed in Clause 2.1.2 are valid. Additional points are as follows:

- Technical strategy for validation of safety-related software shall include the choices of:
  - Automatic/manual technique
  - Static/dynamic technique
  - Analytical/statistical technique
- As a part of the procedure for safety-related software validation, scope and content shall be done on the basis of mutual agreement.

2.3.3 *Software Design and Development*

This is an important phase for safety-related software, as it has a number of purposes such as:

- Software architecture to fulfill specification requirements for safety-related software
- Evaluation of requirements by hardware architecture on safety-related software
- Selection of suitable tools for required SIL
- Design and implement requirements of safety-related software
- Verification that requirements of safety-related software are achieved.

- Configuration of PE system fulfills specified requirements for software systematic capability.

Depending on applicability, various requirements discussed in Clauses 2.1.4 and 2.2.4 will be valid here. The foregoing bullet points specify the objective, whereas those listed next are applicable for various requirements for software design and development:

- Generally, the conformance to software design and development lies with the supplier. Major technical design features shall include but not be limited to the following:
  - Abstraction/modularity—for complexity control
  - Functionality
  - Information flow between elements
  - Sequence and timing information
  - Timing constraints
  - Concurrency and synchronization for shared resource
  - Data structure and properties
  - Design assumption and properties
  - Exception handling
  - Behavioral and structural view
  - Comprehension by all concerned
  - Verification and validation
  - Testability and capacity for safe modification (modularity/encapsulation features)
- Design representations shall be based on unambiguously defined notations.
- As far as possible safety-related software shall be *simple*.
- Commensurate with safety-related software there shall be self-monitoring in software (data flow).
- For a mix of safety and nonsafety functions, all are to be treated as *safety* unless there is adequate independence as discussed earlier.
- System capability shall be equal or above SIL, if not possible by one element then elements may be in combination.
- For pre-exsiting software reuse, Clauses 7.4.2.12 and 7.4.2.13 of IEC 61508-3: 2010 may be referred to.
- Software architecture design: This defines the major elements and subsystems and their interconnections. Major issues here are:
  - Depending on development process, conformance to supporting tools including programming language (PL) lies with multiple parties.
  - Software architecture shall be established by the supplier/developer in detail to include various requirements such as an integrated set of techniques and measures, partitioning of subsystems, etc., as per Clause 7.4.3.2 of IEC 615008-3:2010.

- Requirement for supporting tools including PL: For details Annexes A, B, and C of Part 3 of the standard and Annex F of Part 7 of the standard may be referred to. A few major issues related to this may be as follows:
  - Software online support tool is an element of safety-related software.
  - Software offline tool is a coherent part of safety-related software.
  - Software offline tool for safety-related software shall be justified.
  - An assessment should be carried out for offline support tool.
  - The results of tool validation shall be well documented.
  - PL for development of safety-related software shall be based on a PL coding standard that specifies good programming practices (for PL features see later).
  - Suitability of coding and automatic translation shall be assessed.
  - Configuration management.
- PL: As per the requirement of SIL, for software a suitable PL should be chosen to include major features like:
  - Well-assessed and fitted standard translator suitable for the purpose
  - Use only defined language features
  - Features to detect syntax/design error, etc.
  - Well matched with design method
- Software system design: This enables one to achieve the partitioning of major elements in architecture into a system software module. The nature of detailed design and development will vary according to software development activities.
  - Depending on the development process, conformance to software system design lies with *multiple parties*.
  - Requirement specification, architecture design, and validation planning shall be carried out *prior* to the start of software system design.
  - The software developed shall have modularity, testability, and safe modification.
  - Software elements/subsystems in software architecture design shall be refined based on partitioning into software modules.
  - Software integration tests shall be specified.
- Code implementation: Source code shall be readable and understandable and shall follow the properties defined in Clauses 7.4.4 and 7.4.5 of the standard (IEC 61508-3:2010).
- Software module testing: This shall be carried out as per specification to show whether each module is performing the intended function. All test results shall be documented and corrective action shall be initiated.
- Software integration testing: This is shall be specified during the design and development phase. Major issues here are as follows:
  - Division into a manageable integration set
  - Test cases and test data
  - Types of tests
  - Test environment, tools, and programs
  - Test criteria
  - Procedure for corrective action

- Documentation of test results
- Any modification subject to impact analysis

2.3.4 *Programmable Electronics Integration (Hardware and Software)*

The basic purpose of this is to integrate software with the PE system and combine software and hardware in the PE system. Some of the major issues related to this are as follows (see Clause 2.3.3):

- Software and PE integration test specification to include but not be limited to:
  - System splitting to integration level
  - Test case and test data
  - Test types
  - Test environment, tools, and support
  - Test criteria for judgment
- Distinction points in test specification:
  - Activities in developer's premises and at site
  - Merging of software in PE
  - Addition of interfaces for sensors/actuators
  - Application of safety-related E/E/PE to EUC
- Any modification during integration testing shall be subject to impact analysis.
- Proper documentation of test cases and results
- Failures must be documented and any modification shall be subject to impact analysis.

2.3.5 *Software Operation and Maintenance Procedure*

See Clause 2.2.6.

2.3.6 *Software Aspect of System Safety Validation*

This is done to ensure compliance with system safety specification to take into account:

- Completeness of validation
- Correctness of validation
- Repeatability
- Precise definition of validation configuration

  Because validation is already established as per Clause 2.2, it need not be repeated. Otherwise, based on applicability, see Clause 2.2.7 for the software aspect of system safety.

2.3.7 *Software Modification*

This is intended for correction, enhancement, and adaption in validated software. Major issues involved shall include but not be limited to:

- Selection of proper techniques and measures to ensure:
  - Completeness with respect to requirement
  - Correctness with respect to requirement

  - Freedom from intrinsic design faults
  - Unwanted behavior avoidance
  - Verifiable and testable design
  - Regression and verification coverage
- Prior availability of modification procedure
- Modification is initiated by authorized request at the safety planning stage for:
  - Hazards that may be affected
  - Proposed modification
  - Reason for modification
  - Reasons discussed in Clause 2.1.5
- As per the plan, modification is to be carried out and documented to include details like:
  - Modification request
  - Software configuration history
  - Result of impact analysis
  - Deviation from normal operation/condition
  - Documentation affected by modification
- Assessment of modification based on impact analysis result

2.3.8 *Software Verification*

This is required for SIL to test the correctness and consistency of output with respect to input of the phase. Major issues to address are:

- Software verification planning, concurrent with development
- Affecting factors that shall include:
  - Project size
  - Degree of complexity
  - Degree of novelty of design
  - Degree of novelty of technology
- Sound documentation
- All previous phase information for next phase verification
- Major issues for verification:
  - Software safety requirement
  - Software architecture
  - Software system design
  - Software module design
  - Code
  - Data
  - Timing performance
  - Software module testing
  - Software integration testing
  - PE integration testing
  - Software aspect—system validation
  - Data structure

2.3.9 *Functional Safety Assessment*

See Clause 2.1.6.

2.4 Short Discussions on the Safe PLC System

From Chapter VII readers may form an idea about what a safe PLC is and why it is necessary. In this clause short discussions will be presented on the internals of a safe PLC. It is needless to mention that there may be quite different configurations of safe PLCs from different manufacturers. Naturally, it is impossible to cover all these configurations within the limited space. Therefore discussions will be based on one internationally popular PLC in a generalized manner to understand the *fundamentals only (so that further discussions on logic solvers in Chapter IX will be easier).* For details of PLCs and intelligent control systems readers may refer to [1].

From IEC 61508 it is clear that a safety-related logic solver/system is a system that is required to carry out at least one specific function to be kept at an acceptable level. A system becomes safe if random, systematic, and common cause failures do not lead to malfunctioning of the system. Safety functions are executed to keep the system in a safe state even during failure. Another issue related to safe PLC is its SIL certification from an authorized third party. Other agencies like FM/ANSI USA, TUV Germany and HSE UK puts stringent requirements for safe PLCs. Generally, such certifications are done for some specific use and for certain SIL numbers. For safety functions, safety modules, discussed later, are used. If it is necessary to combine nonsafety functions in the same logic server, it is necessary to use so-called noninterfering modules for the nonsafety functions.

The name noninterfering module is mainly associated with the Modicon PLC but similar modules are also available with other PLCs such as the Independent Hi from ABB. There are a number of differences between safe PLCs and normal PLCs. To start the discussion it is better to look at Fig. VIII/2.4-1 (inspired by the Quantum safe PLC of Schneider Electric). In this figure it is seen that there are two sets of processors, not in primary and hot standby mode but in 1oo2 mode. There is diversity in the technology between the two processing units each with its own memory for executing logic functions. When both execution results are equal, then only output will be sent to the actuators, otherwise it will stop. On account of diversity, two executable codes are generated independently with diversity of compilers, so that systematic error in code generation may be avoided. Also these two diverse central processing unit (CPU) configurations could help to detect both systematic errors in code execution and random errors in the PLC. Additionally, on account of two random access memories (RAMs), it is possible to detect the random error in RAM. Summarily safety systems developed with standard controllers requires additional engineering time, I/O hardware, and wiring to support the safety portion of the application. This is addition to HW/SW needed to run the application.Therefore safety PLCs used to have redundant processors, Flash and RAM continuously monitored by a watchdog and synchronous detection circuit. Usually, there are two modes of operation, namely, safe mode and maintenance mode. In the former mode, safety functions are executed and during this time any modification, etc., is not allowed. In maintenance mode, program changes, etc. could be done. In non-SIL-rated applications, all five languages mentioned in IEC 61131-3 are available for use in normal PLCs. In SIL applications, in a safe PLC usually there is a restriction in the use of PLs. Similarly, in the case of a safe PLC usually a dedicated power supply is

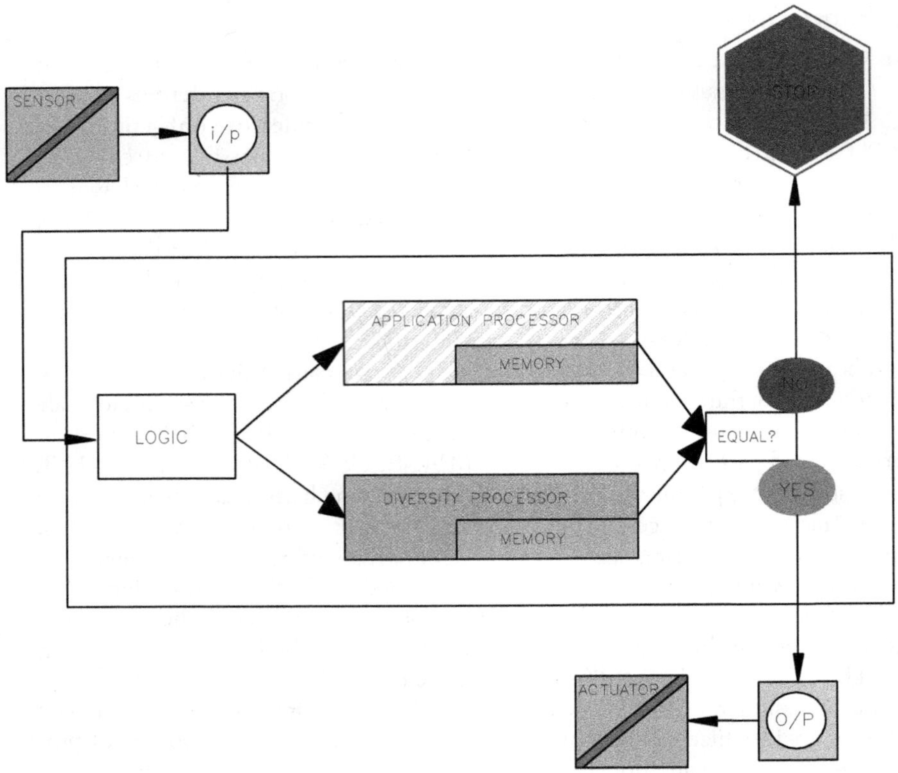

INSPIRED BY COURTESY: SCHNEIDER ELECTRIC

FIGURE VIII/2.4-1

Safe programmable logic controller (PLC) internal architecture. *I/P*, input; *O/P*, output.

used. For a better understanding, the distinctions between standard PLC and safe PLC are found in Table VIII/2.4-1.

2.4.1 Redundancy I/O Modules

In safe PLCs, safety inputs/outputs (I/Os) are distributed in a highly modularized and flexible I/O system to ease installation modules and cabling. These modules and their termination units can be mounted and combined in many different configurations to fit any space requirements or meet the needs of any safety application. Also to enhance reliability and availability safe PLCs utilize an internal "output" circuit with each input for driving inputs both high and low for very short cycles during run time to verify their functionality testing e.g., PLC from Rockwell automation. The digital output module provides both normally energized [emergency shutdown (ESD)] and normally deenergized [fire and gas (F&G)] outputs. Also for outputs in safe PLCs there may be a test point (watch dog) after redundant safety switches located behind

Table VIII/2.4-1 Comparsion of Standard and safe Programmable Logic Controllers (PLCs)

Feature	Standard PLC	Safe PLC
Processor	May be single or Dual	Dual cross wired for better safety
Input	Single or redundant	Usually redundant. Internally each input may be tested with short period during runtime
Output	Usually single if double, automatic safe mode selection may be absent	Usually with redundant safe switches with monitoring facility. In case of failure, automatic transfer to predefined safe state is possible
Program execution	Program is executed in one type of processor	Executed separately in two diverse processors each with memory
Firm wire	Normal firm wire	Safety firm wire
Programming language	As per IEC 61131-3 all five languages	Usually two to three types such as choice of Ladder diagram, functional block diagram or structured etc.
Warm start	Allowed	Warm start is normally not done

the output driver. Safety switches are controlled by a unique microprocessor. When a failure is detected PLC will automatically acknowledge system failure and safe PLC will default to a known state on its own e.g., safety PLC of Rockwell automation. Analog inputs may support a Highway Addressable Remote Transducer (HART) for easy calibration checking and diagnosis with configurable access. Digital inputs support local time stamping of signal changes for high accuracy sequence of events (SOEs) logging. Hot-swap and redundancy of I/O will increase the system availability and allow maintenance activities without process interruption. To increase availability, redundant I/O modules can be used. Fig. VIII/2.4-1 shows the redundancy of I/Os for safe PLCs.

Fig. VIII/2.4.1-1A shows analog inputs. Two sensors are used for higher availability through two sets of analog input. To get better results two inputs may belong to two different modules for better availability. Similarly in the case of digital inputs, two sets of sensing actuators are channeled through two different inputs belonging to two different modules. In cases where a common sensor is used, the same sensor may be channeled to two different digital input channels with a common

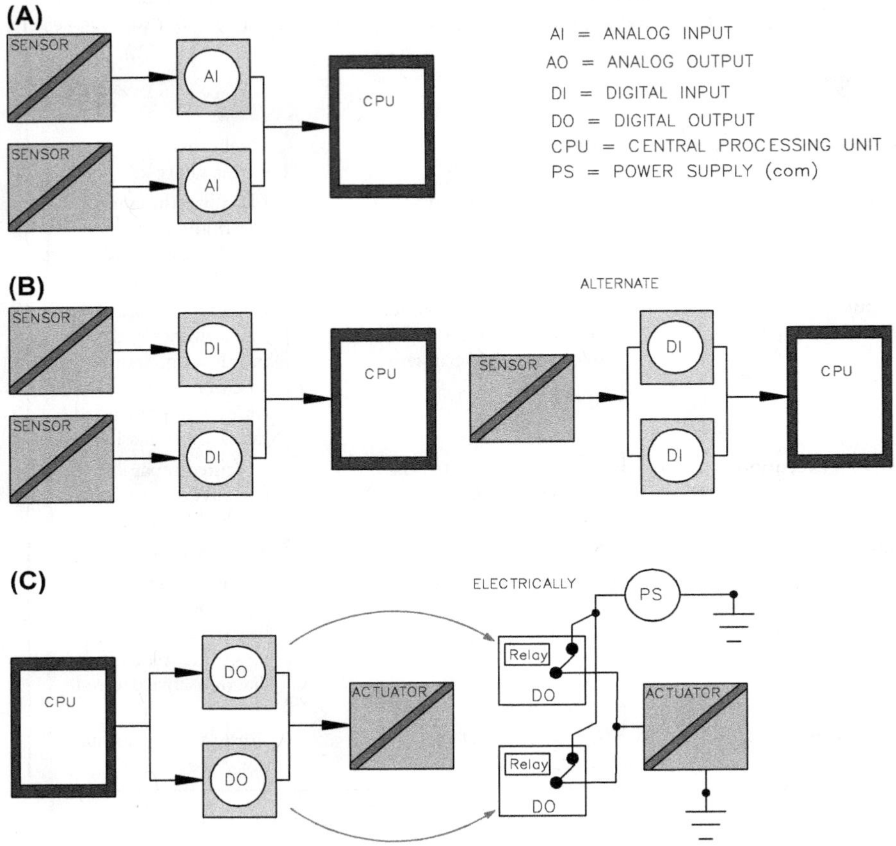

FIGURE VIII/2.4.1-1

Input/output (I/O) redundancy for a safe programmable logic controller (PLC). (A) AI redundancy for safe PLC (typ), (B) DI redundancies for safe PLC (typ), (C) DO redundancy for safe PLC (typ).

power supply. In the case of dual sensors it is possible to use different power sources. In the case of dual output the same signal from the CPU is used by both the output modules or it could be from unique processor discussed above. Here, the power supply may be connected to the actuator through two different paths as shown.

2.4.2 *Safety I/O Modules*

The typical architecture of a safety input module is shown in Fig. VIII/2.4.2-1. Here, each input is checked in two separate systems each having its own microprocessor and memory unit and driving an analog-to-digital converter, etc. Inputs are selected after they are voted and sent to the communication media (say) back plane. Diagnostic features are important.

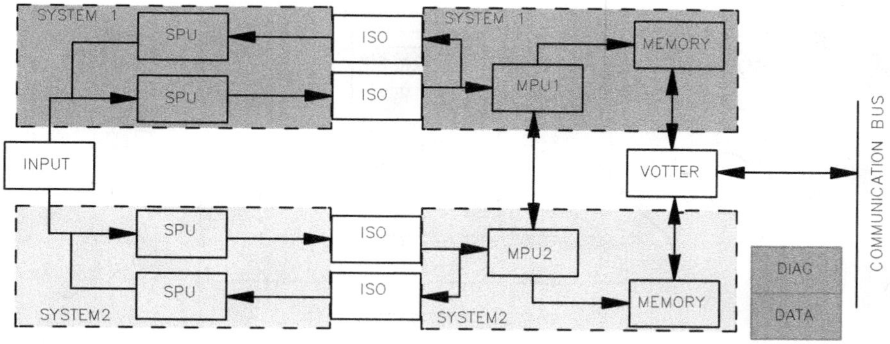

FIGURE VIII/2.4.2-1

Safe input module (typical). *SPU*, signal processing unit; *MPU*, microprocessor unit; *ISO*, isolator; *Diag*, diagnostic.

Depending on use, diagnostics shall include:

- Out of range
- Broken wire
- Overload
- Field power
- Cyclic redundancy check for CPU to I/O communication
  At power-up and during run time, I/O modules execute self-check and pulse run for outputs. When power-up self-check is not passed it is not considered fit and will not be in operation.

2.4.3 CPU Configuration

There are many possible configurations for PLCs such as standalone CPU, redundant CPU in hot standby mode, etc. Normally safe PLCs have dual (cross wired) processors in Fig. VIII/2.4.3-1 redundant CPUs, that is, primary CPU and standby CPU, are shown. Also in I/Os there can be redundancy, as shown in the figure. Fig. VIII/2.4-1 is an example of a nonredundant CPU and I/O.

Of two safety CPUs in a hot standby configuration, one CPU is the primary CPU and the other is the standby CPU. In hot standby mode there is a fiber optic link between the CPUs to exchange data. There are two modes of operation, safe mode and maintenance mode, and in each mode the running of the CPUs is as follows [9]:

- Run primary CPU: The primary CPU executes the user logic to control the process.
- Run standby CPU: The standby CPU is working fine and ready to take over the process if the primary one stops. When the standby CPU becomes the primary one, it automatically enters into safety mode.
- Run offline: This is a temporary mode activated manually or the CPU detects itself. In run offline state, the PLC executes the complete project but the I/Os are inactive.

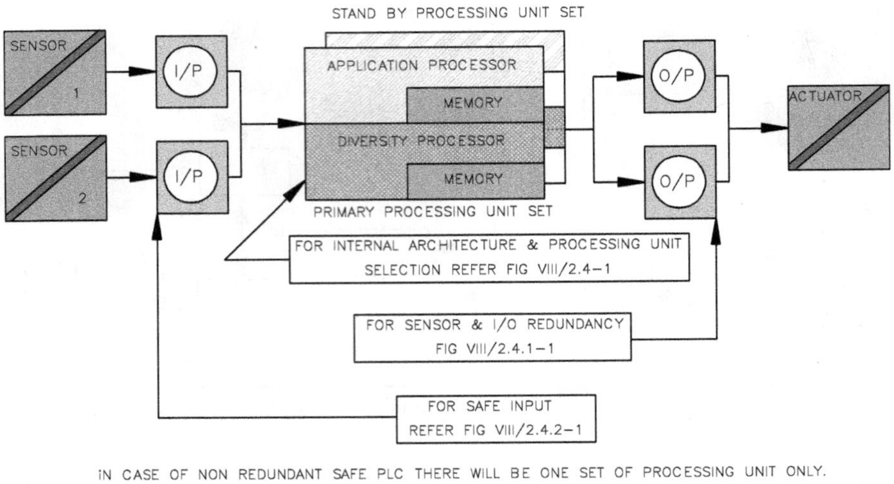

FIGURE VIII/2.4.3-1

Functional details of a safe programmable logic controller (PLC) configuration. *I/P*, input; *O/P*, output.

- Stop offline: In this mode, no PLC execution is performed. This can happen to both CPUs simultaneously.
- In case of a detected error, the PLC enters into a halt state when running in maintenance mode and an error state when running in safety mode [9]. Also it is possible to bring the Logic solver in know safe state automatically as discussed earlier.

2.4.4 Some Features of a Safe PLC

The following are some of the features from internationally reputed safe PLC manufacturers:

- Available redundancy with high availability
- Very high diagnostic coverage without hardware fault tolerance factors or voting
- Communication modules for safety peer-to-peer interfacing with third party equipment
- Certified libraries for streamlined engineering ensure sustainable solutions over the application life cycle
- Controlled builders for safety engineering
- Security and embedded firewall
- Time stamping for SOE

2.4.5 Triple Modular Redundant Logic Solver

The most popular configuration in safety applications is the triple modular redundant logic solver/PLC. Triconix-Tricon (Invensys) uses this configuration. Its major application is in BMS, ESD, F&G, and turbine controls (e.g., GE frame VI). The following are various features of this logic solver:

- Fault tolerance is achieved through triple modular redundancy (TMR) technology. TMR employs three isolated, parallel control systems and extensive diagnostics integrated into one system. The system uses 2oo3 voting to provide high-integrity, error-free, uninterrupted process operation with no single point of failure. The extensive diagnostics are inherent and transparent to the programmer. This operation is based on the principle of safety and high plant availability. This control system offers a comprehensive suite of software solutions with a set of robust and user-friendly tools to help manage and maintain their safety systems. Having the right information at the right time can help clients mitigate risks, avoid costly process downtime, comply with regulations, and help achieve production targets [10].

  Major features shall include but are not limited to the following:
  - Increases reliability and availability, performance and profitability
  - Meets safety targets
  - Mitigates environmental impact
  - Assists in avoiding penalties
  - Certifies up to SIL3 applications as per IEC 61508
  - Easily integrates with DCS
  - Online module replacement and hot swapping of modules
  - Comprehensive family of I/O modules
  - Supports HART protocol
  - Fast program cycle time

3.0 FIRE AND GAS

FGS and mitigation systems are key components in the overall safety and operation of any production facility and its on-site personnel. Unfortunately, it is not understood properly with the overall perspective. However, now performance-based approaches are seen in standards like Annex B of NFPA72. Here, performance-based detectors are considered. Still it is often found that people are satisfied when they achieve (say) SIL3 certification for SIS and are not much bothered about FGS. Is this enough? Actually, consideration must be given to overall safety including FGS maintenance of instruments, etc. to get a proper result. In the industry as such there is controversy over considering FGS under SIL certification. As discussed in Clause 3.2 of Chapter VII, FGS is a mitigating method, while SIL certification, etc. discussed in IEC 61508 is mainly for preventive layers. However, IEC 61511 includes all safety layers in the standard. However, incorrect *detection technology*, *wrong detector placement*, and/or *environmental influences* (see Clause 3.0 of Chapter VII) can prevent the product from detecting a hazardous event, even if it is working properly. When a hazard is undetected, the appropriate safety action cannot be taken. As a result, many argue against inclusion of FGS under SIL certification as it would be meaningless. Therefore some feel that a new performance standard needs to be developed and adapted for FGS to cover adequately the issues of detector placement, coverage, etc. Those in favor of inclusion of FGS under the standard say that FGS detectors are an effective mitigation layer and should fall within the scope of IEC 61508/61511. It is now debatable that any facility wanting to consider this as an SIS after giving due considerations to the inherent limitation of the detection system discussed, should argue that the selected FGS should conform to the guidelines put forth in both in the international standards discussed. The performance-based approach for FGS is noticed from the work of the ISA S84 WG 6 F&G committee toward guidelines for end users to apply the performance-based concepts of the S84.00.01-2004/IEC 61511 Standard to Fire and Gas System implementation and verification [11]. Other standards toward these shall include but not be limited to the following:

- DIN EN 54-2, 1997: Fire Detection and Fire Alarm Systems
- BS EN 50402; Rev. 5, 2005 (Gas Detection)
- IEC Draft 60079-29 (Gas Detection)
- OLF Recommended Guidelines for the Application of IEC 61508 and IEC 61511 in the Norwegian Petroleum Industry; Rev. 2, 2004 (Appendices A.8 and A.9)

The last three standards also address SIL for FGS [11]. The basic approaching rule toward this for FGS could be:

- Definition and identification of the hazard with frequency
- Hazard instrumentation with proper coverage
- Risk reduction calculation
- Assessment of residual risk
- Risk modification to cater for risk criteria

It goes without saying that the number of detectors and their placement for detection coverage is extremely important because incorrect placement of the detectors may miss detection of gases and flames, hence the SIF/SIS will not be effective, regardless of the SIL of the system. Detailed assessment of detection coverage calls for detailed analysis of potential sources of fire or gas leakage. Various approaches of detection coverage have been discussed in Chapter VII and are not repeated. Another important aspect in FGS is that the *F&G logic solver operates in the deenergized state* (as opposed to the energized state for ESD applications) and is required to *energize its outputs to implement the safety function*. This needs to be kept in mind while discussing FGS. As such, using SIL certified devices (in accordance with the requirements of IEC 61508) has some definite advantages. So, FGS detector manufacturers may improve the safety of their products by designing to comply with IEC 61508 standards. End users could design safer FGS by conforming to IEC 61511 standards and keeping in mind detection coverage and environmental influence. With this approach, the heart of functional safety is preventing and mitigating risk to an acceptable level via the safety life cycle [12] and it is possible to carry out safety performance verification for FGS. During initial discussions it was said that selection of proper technology for detectors, proper detector placements, and environmental influences are more important than arguing for SIL for FGS. In the subsequent discussions how detection coverage influence SIL and gas detection system will be discussed.

3.1 FGS Model and Detection Coverage

The basic model consists of an event tree with three basic components: detection coverage, hardware response, and mitigation effectiveness, as shown in Fig. VIII/3.1-1.

Like other SIS, for FGS the PFD_{avg} for each safety function is the summation of the sensor PFD_{avg}, the logic solver PFD_{avg}, and the PFD_{avg} of the mitigating devices. Each of these contributions will be determined using its respective dangerous undetected failure rate, voting architecture, and PTI. It is better to examine the issue.

From earlier discussions it is clear that $PFD_{avg} = PFD_{Sensor} + PFD_{LS} + PFD_{FE}$.

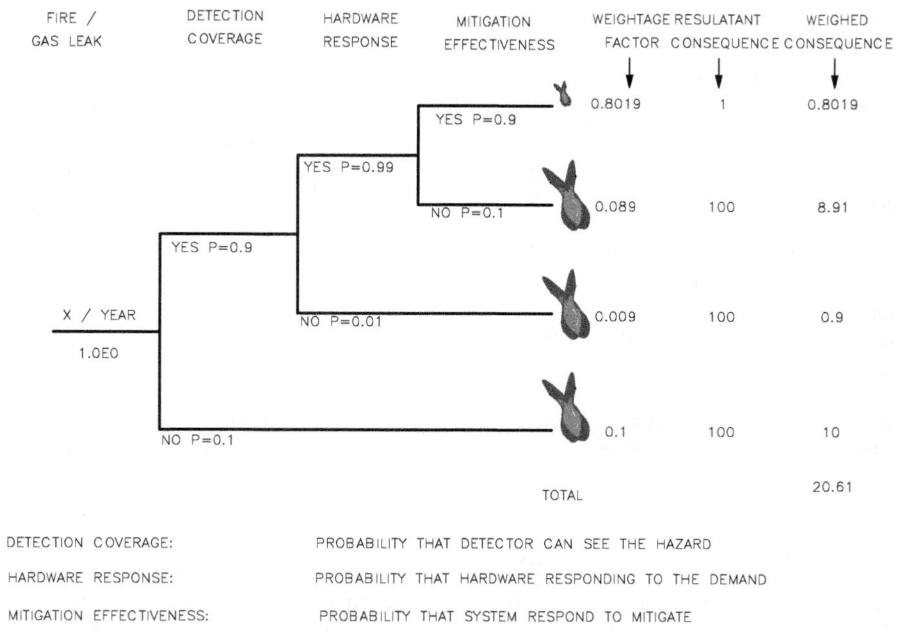

FIGURE VIII/3.1-1

Fire and gas system (FGS) performance analysis model.

So FGS availability can be calculated by Eq. VIII/3.1-1. However now focus will be on how to arrive at SIL for FGS from the discussions that follows:

$$\text{FGS availability} = 1 - PFD_{avg} = 1 - (PFD_{Sensor} + PFD_{LS} + PFD_{FE}) \qquad \text{(VIII/3.1-1)}$$

The three factors are as follows:

- Detector coverage: The probability of the device actually seeing the hazard.
- Hardware response: The probability of the hardware responding properly to the demand, that is, availability.
- Mitigation effectiveness: The probability of the overall system's response for mitigation (/prevention) of the hazardous event.

The figure depicts details of the method of calculation. In this case, a minimum detection coverage of 90% has been considered with a hardware response of 99% and mitigation effectiveness of 90%. From the calculation it is clear that if the detection coverage is less than 90% and the mitigation effectiveness is less than 90%, then overall risk reduction will never be greater than 10, and in that case it is meaningless to debate on the issue because the minimum RRF is 10 ($0.9 * 0.9 = 0.81$; so $1 - 0.81 = 0.19 = \text{PFD}$, hence $\text{RRF} = 5.26 < 10$). It can be shown that with a detection coverage of 100% only there is effectiveness of SIL1, 2, 3, etc. and it could be shown that RRF could be 1010, 101, and 10 for SIL3, SIL2, and SIL1, respectively. Even when detection coverage is 99%, SIL divisions have no significance and it reduces to at best SIL1. So, one may conclude that detection coverage even after environmental influences would be of utmost importance as far as SIL rating is concerned. From a study report of the Health and Safety Executive in the United Kingdom it has been found that out of the total gas leakage in offshore facilities about 60% was detected. Therefore 40% remains undetected. Because the report does not cover this it may be caused by other influences. There are different methods of estimating detection coverage, and many variables to consider, such as:

- Size of the area to be monitored
- Number of detectors
- Density of gas
- Wind speed
- Number of leak sources

It has been found that detection coverage is very high for single sensors and a catastrophic release. Detection coverage is very low for multiple sensors detecting medium or small releases.

3.2 Gas Detection Discussions

From the above discussions it is clear that detection coverage is extremely important for FGS. Gas detection is also a part of FGS. So, short discussions on this are presented here. There are several categories under which gas monitoring is done. During the discussions this will be highlighted. In Clause 3.1.3 of Chapter VII placement of gas detectors at various locations was touched upon. The following discussions will supplement this topic.

3.2.1 *Gas Detection Category*

In an industrial scenario, mainly three kinds of atmosphere and gas types exist and these are:

- Combustible atmosphere—Combustible/flammable gas
- Toxic atmosphere—Toxic gas
- Oxygen deficient/enriched atmosphere—Oxygen (deficiency)
  Major issues related to these will be discussed in the following subclauses.

3.2.2 *Definitions of Terms*

It is better to note the following related terms, which will be used in subsequent discussions. A few other terms have already been defined in Chapter VII.

- *Confined space*: An area large enough for an employee to bodily enter to work but has restricted areas of entry/exit, and is not designed for *continuous* human occupancy.
- *Explosion*: Rapid uncontrolled combustion process that generates high temperature, volume of gas, and a pressure or shock wave.
- *Explosive range*: The range between the substance's lower explosive limit (LEL) and upper explosive limit (UEL), and is capable of sustaining combustion (see also Clause 3.4 of Chapter X).
- *Flash point*: The minimum temperature at which a liquid gives off enough vapor to form an ignitable mixture with air (reaching 100% LEL) (see Clause 3.4 of Chapter X).
- *Oxygen-deficient atmosphere*: Atmosphere containing less than 19.5% oxygen by volume.
- *Oxygen-enriched atmosphere*: Atmosphere containing more than 20.8% oxygen by volume.
- *Permissible exposure limit* (in ppm): An airborne concentration of contaminant that most workers can be exposed to repeatedly in a normal 8-h day, in a 40-h week, without adverse health effects.

3.2.3 *Major Gas Hazards*

Indicated in Table VIII/3.2.3-1 are the major gas hazards in plants.

3.3 Combustible Gases

Combustible gases are capable of causing explosion. The following are the major issues (further details can be found in Clause 3.4 of Chapter X).

Table VIII/3.2.3-1 Major Hazards From Gas Leakages

Hazards	Sources	Monitoring	Purpose
Explosive gases	Presence of combustible gases/vapor, leakages, maloperation of process	Explosion	Worker facility and community safety
Toxic gases	Leakage, fugitive emission, and process defect	Human protection	Worker and community safety
Process malfunctioning	Possible fault in process or operation	Industrial process	Process control and human hazards
Environmental degradation	Oil leaks to sewer and other water bodies, etc., acid/gas emissions	Environmental	Environmental safety (e.g., particulate or SO_2 monitoring at stack of thermal power plant)

3.3.1 *Major Issues Related to Combustible Gases*

The major issues are as follows:

- Explosive hazard
- Avoidance of explosion; atmospheric levels should be maintained below LEL for each gas/O_2.
- Detectors used are of two categories: (1) infrared detector (2) electrocatalytic detector.
- Combustible gas monitors should annunciate an alarm way ahead in time before actual hazardous condition.

3.3.2 *Combustible Atmosphere and Characteristics*

For a flame to exist, the following conditions must be met:

- Fuel source (any flammable gas/vapor)
- Oxygen supply (>10%) to oxidize or burn the fuel
- Source of ignition (heat) to start the process

  In industrial plants the major sources of ignition may be from open flames (e.g., lighters, burners, welding torches, etc.) or radiations from hot surfaces, sunlight, or a spark from electrical appliances. Other pertinent issues are as follows:
  - The presence of gas or vapor: It is worth noting that vapor is applicable to those cases when a substance is in a liquid/solid state at room temperature.
  - Vapor pressure and boiling point: These two substance properties are also determining factors for flame. The lower the vapor pressure, the lower will be the hazard. In this case, a highly sensitive instrument will be required because lesser molecules may not be detected. The lesser the boiling, the more chances for evaporation, hence the more chance of a hazard.

- Flash point: Gas and vapor cannot ignite automatically until they reach flash point. Below flash point the substance will release fewer vapors to ignite. When it is above flash point, a spark is sufficient to cause ignition. So, the lower the flash point, the higher is the chance for flame or ignition.
- Autoignition: When heated beyond a certain point a flammable fluid can ignite automatically with its own heat energy without external ignition. The temperature at which the fluid ignites spontaneously is the autoignition point.
- Explosive limits: To cause an ignition, a sufficient amount of gas/vapor is necessary. On the contrary, if the quantity is too great to displace air, then ignition may be prevented. So, ignition mostly takes place between the lower and upper explosion levels.
- Vapor density: This is an important issue for instrument placement.

3.4 Toxic Gases

Toxic gases are harmful to workers. So, from a human health point of view these are very important to measure.

3.4.1 *Toxic Gas Measurement Issues*

Characteristic toxic gas-related issues are as follows:

- Human health hazard: Worker exposure must be monitored.
- Measured in the parts per million (ppm) range.
- Toxic gas monitors are required to alert workers before the gas reaches a harmful concentration.
- Many toxic gas monitors can calculate the average exposure over time.
- Monitoring instrument types could be infrared, electrochemical, solid state, and photoionization types.

3.4.2 *Toxic Gas Atmosphere*

A toxic gas can cause damage to living tissue, the central nervous system, severe illness, or even death. Some toxic gases are not visible, cannot be smelled, and/or they may not have an immediate effect but can be the cause of death. So, without sensing using instruments, it may not be wise to rely on human sensing alone. Life threats are a time-based exposure, so their measuring ranges are time-weighted average, threshold limit value, etc., as defined in Chapter VII and not repeated here.

3.5 Oxygen Deficiency/Enrichment

Ambient air contains an oxygen concentration of 20.8% by volume. When the oxygen level falls much below this it causes an oxygen-deficient atmosphere, whereas when it is much greater than atmospheric O_2 it is an oxygen-rich atmosphere. Neither of the two is ideal.

3.5.1 *Oxygen Deficiency*

When the oxygen level falls below 19.5% of the total atmosphere, the area is considered oxygen deficient. Oxygen deficiency may also be caused by:

- Rust
- Corrosion
- Fermentation
- Other forms of oxidation

As materials decompose, oxygen is drawn from the atmosphere. This deficiency may have a gradual or sudden impact depending on the condition. Oxygen-deficient atmospheres can be dangerous or fatal when inhaled. Oxygen deficiency may cause increased breathing/heart rate, poor muscular movement, nausea, or even death.

3.5.2 *Oxygen Enrichment*

When the oxygen concentration rises above 20.8% by volume, then it is oxygen enriched. On account of the higher oxygen level, the likelihood and severity of a flash fire or chances of explosion is significantly increased.

3.5.3 *Instruments Used*

Electrochemical, electromagnetic instruments can be used for oxygen measurement.

3.6 Gas Detector Placements

Major issues related to the detector placement process shall include but not be limited to the following (for combustible gases, Clause 3.4.3 of Chapter X may also be referred to).

3.6.1 *Basic Procedure*

The following are basic issues related to this:

- Development of philosophy document (see Chapter VII)
- Analysis of the potential gas hazard of the facility
- Identification of potential discharge sources and contact areas
- Consideration of air flow and gas concentration pockets

3.6.2 *General Guidelines*

- Place sensors where the highest gas concentration or gas buildup is likely to be. For sample collection, entrances or fresh air vents are to be avoided.
- Sensor to be close to the source.
- Combustible gas sensors to be located between the potential leak and the ignition source.
- Toxic oxygen deficiency sensors to be placed between the source and populated areas.
- Ease of access for maintenance requirements.

- Locate away from high temperature, high vibration, and radio frequency interference (RFI).
- Avoid contamination/dust, etc.
- Storage and installation and wiring practice to follow manufacturer's instructions.
- Air intakes are a good location for sensors.
- Factor in the vapor density of the monitored gases.

3.6.3 *Thumb Rules*

- Gas density greater than air (e.g., CO_2, heavy hydrocarbon), place sensor close to ground.
- Gas density same as air (e.g., N_2), place sensor as per air current or near breathing level.
- Gas density less than air (e.g., H_2, CH_4), place sensor near ceiling.

We next move on to energy to activate ESD/plant emergencies!

4.0 PLANT EMERGENCIES

According to the Occupational Safety and Health Administration [13]: "A workplace emergency is an unforeseen situation that threatens your employees, customers, or the public; disrupts or shuts down your operations; or causes physical or environmental damage." There could be a number of reasons why this is so, such as:

- Natural calamity, for example, flood, earthquake, heavy storm
- Fires
- Toxic gas releases
- Chemical spills
- Accidents caused by radiation
- Explosions
- Sociological/political disturbances

Another important issue here is that there shall be suitable means to alert all concerned about the emergency situation. For this, alarm systems, public address systems, and emergency communication systems are usually deployed. To combat the situation it is always better to plan and be prepared. An emergency action plan is a well-designated plan prepared by employers and employees to ensure safety to all personnel concerned, the facility, and the surroundings (as far as possible). This relies on team work, where personnel from various departments and management representatives need to participate. When developing an emergency action plan, it is always better to look at all possible potential emergencies that could occur in the facility. An emergency action plan shall be prepared separately for each site and each facility. When preparing an emergency action plan, detailed information from various sources is required. These shall include but not be limited to the following:

- All potential sources of emergencies
- Hazard assessment
- Results and documentation of PHA
- Detailed list of potential sources of hazards

Plant emergency is quite a large topic having variations with respect to specific plants. It is hardly possible to cover all these in the limited space of this book. As per the IPL concept, basically, plant emergency is mainly concerned with evacuation. In that sense, ESD is a predictive safety system. From an instrumentation point of view an ESD system is the main system deployed during an emergency and this will be discussed here. Here, only ESD philosophy and general requirements are discussed because specific details of ESD depend on the plant type/facility, for example, the ESD for offshore facilities will be different to that of other plants.

4.1 ESD Fundamentals

Although this may not be really a plant emergency like evacuation, it is discussed here to show the importance of the system. The basic purpose and requirements of the ESD system are discussed. ESD is mainly used to save personnel, the facility, and the environment.

4.1.1 Objective of ESD

The objective of the ESD system is to prevent escalation of a number of abnormal conditions into a major hazardous event and to limit the extent and duration of the abnormal conditions. The ESD system needs to react rapidly in the event of such abnormal conditions for the following reasons: to stop the operation of the applicable process and isolate it from incoming or outgoing connections; to prevent fuel or flammable gas flow or isolation and control the ignition source; to shut down equipment and systems to reduce the likelihood of an unwanted event from occurring, continuing, or escalating; and to bring the systems to a predefined safe state, shown in Fig. VIII/4.1.1-1. Depending on applicability, the ESD will arrange to remove the pressure and hydrocarbon from equipment through venting and flaring.

The safe state is the state at which a system can be put out of its current operational state. The safe state has a system-specific lower hazard potential than the operational state. At the safe state there will be the lowest amount of energy involved. However, it is not always possible to obtain the safe state without involving any danger, just by switching the device off (e.g., an airplane).

FIGURE VIII/4.1.1-1

Safe state.

4.1.2 *Reasons for Selection of Appropriate Technology for ESD*

When selecting or replacing an ESD system, consider the following [14]:

- Can the current control system protect personnel, equipment, and the environment in the event of an emergency?
- Is it more cost-effective to run and maintain than the current system?
- An ESD system needs to comply with current government regulations and industry standards.
- Shutdown of the operation of the facility because of selection of poor technology for ESD system can be costly.
- Cost of litigation for an accident because of the use of inadequate safety technology can be enormous?

4.1.3 *Functional Safety*

From the foregoing it is clear that the ESD system will take all associated systems to the safe state, but what will happen if the ESD system fails? The interface of the ESD and the system itself shall be designed in such a way that all connected systems will be pushed to a safe state for the facility upon failure of the ESD system to achieve a *fail safe* condition. Such ESD system failures could be caused by an open/short circuit of the ESD system I/Os, loss of power, or loss of emergency communication with associated systems [e.g. Clause 2.1.3 Section 4 and Table 4-2 (safest conditions) of DNV-OS-A101:2013].

4.1.4 *Operation Under Normal Conditions*

As discussed earlier the ESD system normally operates in a "normally energized" condition. Therefore initiating P_B, limit switch contacts, etc. shall be a normally open (NO) contact, but a few contacts such as contact from the motor control center (MCC), overspeed, reset/stop push button, etc. are normally closed (NC)-type contacts (refer Fig. VIII/4.1.4-1).

4.1.5 *ESD Interface and Utility Requirements*

The ESD system is an independent system but it has interfaces with the following safety systems/functions:

- ESD initiation system (manual/auto)
- Process safety system (SIS)

Normally open (NO) contact is a contact that is open or in a nonconductive state when it or the device operating it is in a deenergized or relaxed state. How is this recognized in a changeover contact device? An open contact (nonconductive between two assigned terminals) *at no energy and no flow condition.* It is NO contact. Similarly, just the reverse condition holds for normally closed (NC) contacts (conductive state when it or the device operating it is in a deenergized or relaxed state), that is, closed contact at *no energy no flow condition.*

FIGURE VIII/4.1.4-1

Normally open/normally closed (NO/NC) contact types.

- FGS
- Control system (DCS/PLC)
- Blow down (BD) and flare/vent system (as applicable)
- Public address system
- Alarm and emergency communication system
- Other systems such as heating, ventilation, and air conditioning
  From a utility point of view the ESD system is dependent on the following utilities:
  - Electric power supply system (uninterrupted power supply)
  - Hydraulic power
  - Instrument air quality

4.2 General Requirements

General requirements for ESD shall include the following:

4.2.1 Major Components of ESD

The following are major components of ESD system:

- Manual input devices (push buttons)
- Field devices (e.g., level, pressure sensors)
- Interfaces to various systems (discussed earlier)
- Central logic solver to interact with inputs and interfaces for creating output signals. The ESD logic solver provides a visual indication of initiated inputs, activated outputs, and alarms.
- Output actuators (e.g., relays, valves, dampers, etc.)
- Power supply
- Adaption of suitable redundancy to be certified as functional safety from fault tolerance
- Suitable developmental tools and user library
- Provisions for supporting remote I/Os preferably with fiber optics
- Well-organized diagnostic system to increase system availability and reliability

4.2.2 Available Systems at ESD

During ESD there are some facilities such as emergency communication, etc. that should not be shut. This is industry specific and is normally decided based on regulations/standards followed for the facility, for example, in North Sea explorations, DNV *guidelines* are followed in offshore facilities.

4.2.3 Activation Mechanism

As discussed earlier the following are the major initiating mechanisms (auto/manual):

- Manual push buttons located at strategic locations including central point
- Automatic initiation through various interface systems
- Plant safety systems

4.3 Performance Requirements

4.3.1 *Performance Requirement Issues*

The following points are the basic performance requirements usually considered (depending on the application there could be other requirements):

- ESD shall be fault tolerant (single fault—no effect) with suitable redundancy certified for functional safety
- Minimization and localization of the effect of equipment failure
- Continuous operation
- High availability and lower probability of failure.

4.3.2 *Implementation Issues*

The following points may be considered for designing the system:

- Use of equipment and device with a proven track record
- Application of required redundancy and voting logic to improve availability
- Fail safe mode of operation in logic
- Regular testing of the logic (online and offline as applicable)
- Ability to operate as an independent system with suitable fail safe interface to other systems discussed earlier.

4.3.3 *ESD Levels*

The entire ESD system may be divided as EDS 1, 2, 3, etc. as seen in offshore applications. These ESD levels shall have a hierarchy. The levels are governed by the associated standards.

4.4 RELIABILITY ISSUES

The economic impact of a spurious or nuisance trip of an ESD system can be disastrous. An ESD system is an important layer of protection to prevent and prevent hazardous situations from occurring. So, it is needless to mention that the ESD system must be extremely reliable and function on demand. During an emergency, it must put the process in a safe state in orderly fashion. Also ESD systems design shall be based on a fail safe independent system, that is, ESD systems are such that during a failure of a component the process reverts to a condition considered "safe" and not a vulnerable serious hazardous event. Reliability and availability are major parameters for ESD system performance. Reliability is a function of system failure rate (its reciprocal) and mean time between failures. Spurious trip conditions may initiate a so-called fail safe incident that may result in accidental shutdown of equipment or processes. However, undetected process design errors or operations may initiate dangerous incidents that may disable the safety interlock and may even cause accidental process fluid release/equipment damage/fire/explosions. Reliable and fail safe design of the ESD system:

- Reduces accidental incidents to low levels or as low as reasonably practical
- Maximizes system availability through proper testing and maintenance
- Reduces hazard rates to an acceptable level

The specification of fail safe for final elements can be designed as fail close, fail open, or stay put/fail lock, that is, the last position before failure depending on application. Usually, ESD valves are the fail close type for isolations (in some cases it may be fail open). In this connection Fig. IX/6.0-1 may be referred to. Also, just as an example, Clause 10.4.2 of NORSOK standard S-001:2008 may be referred to (normally used in offshore facilities). Some of the characteristics of ESD valves especially for offshore facilities (NORSOK standard) are:

- ESD valves position indication shall be made available both remotely and locally.
- ESD valves shall have either a spring return or local accumulators for fail safe function.
- ESD valves shall have defined criteria for leakage rates based on safety criticality.

4.5 PROTECTION ISSUES

ESD system components are located in areas where there are direct fire exposures, so there should be provisions for fire protection measures to ensure integrity during ESD operation and the duration of the major efforts to control the emergency. In this connection, API standard 607 may be referred to for fire protection pertinent to valve systems.

4.6 ESD CONFIGURATION AND HMI

Typical configuration for an ESD system is shown in Fig. VIII/4.6-1. All field inputs from process variables and outputs to various actuators' MCC shall be connected to the ESD system with the help of an I/O bus. Depending on applicability there could be redundancies in various systems and the communication bus, as shown in Fig. VIII/4.6-1. Redundancies are shown with shadows. In the case of other systems where there are multiple systems, these are also shown with the help of shadows. Some systems (say an alarm system) may have an RS 485/422 link. In this configuration an attempt is made to highlight that the system is an independent system but can have an interface with other systems as well. At the top layer, Ethernet has been considered for HMIs, data server (if any), SOE, and other open systems on the same IEEE 802.3 Ethernet. There could be object linking and embedding process control (OPC) clients and these could be serviced by an ESD logic server, which will have dual or TMR redundancy built in.

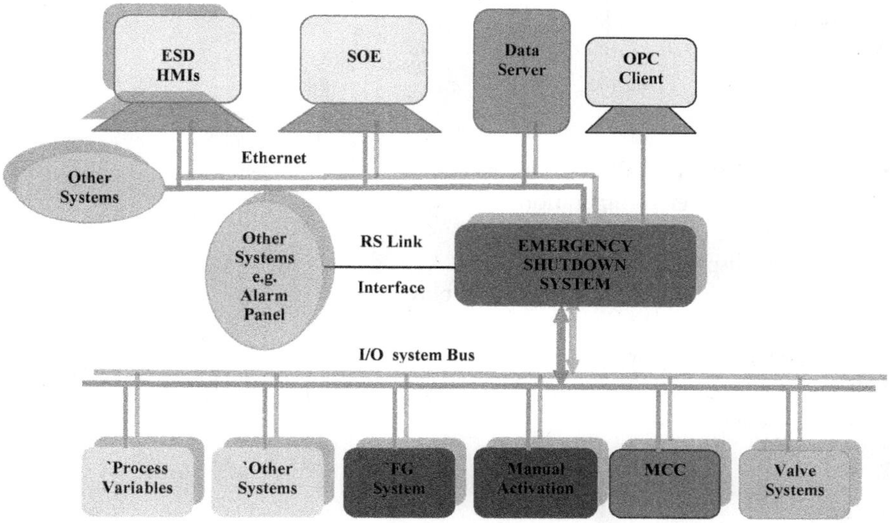

FIGURE VIII/4.6-1

Typical emergency shutdown (ESD) configuration with interface. *ESD*, emergency shutdown; *FGS*, fire and gas system; *HMIs*, human-machine interfaces; *I/O*, input/output; *MCC*, motor control center; *OPC*, object linking and embedding process control; *SOE*, sequence of events.

Developed and inspired by Emergency Shutdown Systems Solutions, Superior protection of personnel, equipment and the environment; Invensys; Triconex catalog.

The main operator interface to the ESD system is via an operator workstation located in the main control room and/or in an ancillary control room. Modern intelligent workstations are deployed for this service. The operator's functions and information exchange and changes through workstations shall include but not be limited to the following:

- Display and control of process variables and actuators
- Alarm acknowledgment and priority
- Print initiations
- Online generation of database
- Online generation of displays
- Generation of diagnostics and displays
- Authenticated configuration changes (as applicable)
- Tag identification and description
- Type points
- Point address
- Scan frequency and priority
- Signal processing

- Engineering unit
- Alarm limits
- Constants
- Fail safe position
- Various displays:
  - Overview display of all shutdowns
  - System parameter configuration
  - Master menu
  - Summary display
  - Communication and system status display
  - Report summary
- Various reports and logs:
  - Event log
  - Operator action report
  - Alarm log
  - Diagnostic report
  - Tag report
  - Historical report
  - Database cross-reference

The system normally have redundant communication channels and well-designed diagnostic systems to report. ESD is very important for process plants especially where there are chances of fire and explosion hazards such as in offshore plants.

The brief discussions on SIL, PE systems, and plant emergency systems are now concluded. With these concepts on SIS and SIL discussed in Chapters VII and VIII, respectively, it is time to look at various safety loop components with SIL ratings in SIS applications. These are covered in the next chapter.

LIST OF ABBREVIATIONS

AI/O	Analog input/output
ALARP	As low as reasonably practicable
BPCS	Basic plant control system
C&I	Control and instrumentation (engineering)
CCF	Common cause failure
CRC	Cyclic redundancy check
DC (in SIS)	Diagnostic coverage (in SIS)
DCS	Distributed control system
DDT	Derived data type
DEP	Design and engineering practice
DI/O	Digital input/output
EDT	Elementary data type
E/E/PE	Electrical/electronics/programmable electronics

E&I	Electrical and instrumentation
EMI	Electromagnetic interference
ESD	Emergency shutdown system or electro static disturbance
ET(A)	Event tree (analysis)
EUC	Equipment under control
FBD	Functional block diagram
FGS	Fire and gas system
FMEDA	Failure modes, effects and diagnostic analysis
FSA	Functional safety assessment/assessor
HC	Hydrocarbon
HFT	Hardware fault tolerance
HMI	Human machine interface
HVAC	Heating ventilation and air conditioning
HW	Hardware
I&C	Instrumentation and control
IEC	International Electrotechnical Commission
I/P or O/P	Input or output
IPL	Independent protection layer
LD	Ladder diagram
LEL	Lower explosive limit
LHS	Left hand side
LOPA	Layer of protection analysis
LS	Logic solver
MFR	Master fuel relay (for boiler/furnace systems)
MOC	Management of change
MRT	Mean repair time
MTTR	Mean time to repair
NO/NC	Normally open/normally close (contact)
OPC	OLE process control
O&M	Operation and maintenance
OREDA	Offshore reliability equipment database association
PEL	Permissible exposure limit
PFD	Probability of failure on demand or doc: process flow diagram
PFH	Probability of failure per hour
P&ID	Piping and instrumentation diagram
PLC	Programmable logic controller
PSI	Process safety information
PTI	Proof test interval
RA	Risk assessment/analysis
RBD	Reliability block diagram
RFI	Radio frequency disturbance
RHS	Right-hand side

Continued

RRF	Risk reduction factor
SAR	Safety analysis report
SFF	Safety failure fraction (IEC)
SIF	Safety instrument functions
SIL	Safety integrity level
SOE	Sequence of event
SIS	Safety instrumentation system/supervisory information system (in case of DCS)
SPLC	Safe PLC (programmable logic controller)
SRS	Safety-related System/safety-related specification
STEL	Short-term exposure limit
SW	Software
TMEL	Tolerable maximum event likelihood
TMR	Triple modular redundancy
TWA	Time weighted average
UPS	Uninterrupted power supply
WRT	With respect to

References

[1] S. Basu, A.K. Debnath, Power Plant Instrumentation and Control Handbook, Elsevier, November 2014. http://store.elsevier.com/Power-Plant-Instrumentation-and-Control-Handbook/Swapan-Basu/isbn-9780128011737/.

[2] IEC 61508/IEC 61511 Safety Integrity Level, Functional safety in the process industry—risk reduction with safety instrumented systems, Endress + Hauser catalog.

[3] A. Ingrey, P. Lerévérend, A. Hildebrandt, Process Automation Manual; Safety Integrity Level; SIL 61508/61511, Pepperl+Fuchs.

[4] P. Lucas, Safety Integrity Level, ABB Engineering Services; Astra Zeneca, Sweden, March 2007.

[5] D. Düpont, L. Litz, P. Netter, Sources of Mistake in PFD Calculations for Safety—Related Loop Typical, July 2007. Research gate conference paper, http://www.researchgate.net/publication/4282858.

[6] SIL – Safety Classification; Inor-member of Krohne group; http://inor.com/pdf/SIL%20-%20Safety%20Classification.pdf.

[7] Understanding 'SIL' Certificates; SIRA Certification, Advertisement feature cover story; Connectingindustry.com/industrialcompliance.

[8] ISA-TR84.00.02-2002-Part 2; Safety Instrumented Functions (SIF)-Safety Integrity Level (SIL) Evaluation Techniques Part 2: Determining the SIL of a SIF via Simplified Equations: Introduction, ISA, June 2002.

[9] Quantum Safety PLC Safety Reference Manual; Schneider Electric, 11/2007.

[10] Invensys hardware data sheet; Tricon Safety Instrumented System (SIL3).

[11] L. Beckman, Fire and gas detection and mitigation systems, TÜV FSExp SafePlex Systems, Inc., in: Presented at 54th International Instrumentation Symposium, ISA, 2008.

[12] SIL 102: Can a fire and gas system be considered a safety instrumented system? General Monitors; www.generalmonitors.com.

[13] How to Plan for Workplace Emergencies and Evacuations; Occupational Safety and Health Administration; https://www.osha.gov/Publications/osha3088.html.
[14] Emergency Shutdown Systems Solutions, Superior protection of personnel, equipment and the environment; Invensys; Triconex catalog.
[15] Safety Instrumented System Engineering Handbook, Kenexis Consulting Corporation, Columbus, OH, 2010. http://www.kenexis.com.

Further Reading

An Introduction to Functional Safety and IEC 61508; Application Note; AN9025; MTL.
H. Bezecny, D. Inverso, V. Maggioli, G. Rabe, A. Weinert, European Workshop on Industrial Computer Systems; Technical Committee 7 (Safety, Reliability and Security); Guidelines for the Use of Programmable Logic Controllers in Safety-Related Systems, October 1997. Working Paper 6009.
Controlling Risks Selecting a Safety Integrity Level, Controlling Risks: Safety Systems, USPAS, January 2012.
Fundamentals of combustible gas detection, A Guide to the Characteristics of Combustible Gases and Applicable Detection Technologies; White Paper; General Monitors, July 2006. http://s7d9.scene7.com/is/content/minesafetyappliances/Combustible%20Gas%20Detection%20White%20Paper.
Gas detection handbook, fifth ed.; Mine Safety Appliances (MSA) e The Safety Company.
W.G. Gulland, Methods of Determining Safety Integrity Level (SIL) Requirements - Pros and Cons, 4-Sight Consulting, April 2004.
How to implement a safety lifecycle; R, 2007 How to implement a safety lifecycle; R. Ali; Fisher-Emerson process management, Valve Magazine (Summer 2007).
ISA-TR84.00.02-2002-Part 1, Safety Instrumented Functions (SIF) e Safety Integrity Level (SIL) Evaluation Techniques Part 1: Introduction, ISA, June 2002.
Is your current safety system compliant to today's safety standard?; L.M. Duran TUV FS Eng; ABB Houston; https://library.e.abb.com/public/10a4ad7d4b51a371c1257a060069bf3b/3BUS095609_en_Whitepaper_-_Is_your_SIS_compliant.pdf.
Mitsubishi Electric, Melsec WS. Series, Safety Controllers User's Manual, 2014.
Process Safe Book 1; Functional Safety in the Process Industry Principles, Standards and Implementation, Rockwell Automation.
PLC Open Safety, PLC Open - Technical Committee 5eSafety Software Technical Specification; Part 1: Concepts and Function Blocks, 2006. Version 1.0 e Official Release.
Quantum Safety PLC Safety Reference Manual, Schneider Electric, November 2007.
M. Rausand, Reliability of SafetyeCritical Systems: Theory and Applications, J. Wiley.
SIL-Safety Instructions SM/265/7/9/SIL-EN Rev. 03; Pressure-Transmitter 2600T.SIL2 IEC 61508; ABB; Exida; management summary; FMEDA and prior use assessment.
SIL Determination techniques reports; White paper; ACM facility safety-a division of ACM Automation INC.
SIL 104: Impact of Gas Detection Coverage on SIF SIL Rating; E. Choo, CFSP; General Monitors Systems Asia, Singapore.
Technical Report; ISA-TR84.00.02-2002-Part 3; Safety Instrumented Functions (SIF)- Safety Integrity Level (SIL) Evaluation Techniques Part 3: Determining the SIL of a SIF via Fault Tree Analysis, June 2002.
Technical Review; Control System Design; A condensed guide to automation control system specification, design and installation. Part 1: system identification; T. Evalsky.

Using Control Logix in SIL2 Applications; Safety Reference Manual, Rockwell Automation, April 2004.

Understanding of Safety Integrity level, Magnetrol; special application series.

Your Safety.Our Future, TEESSIDE Section, Institute of measurement and control; http://www.61508.org/wp-content/uploads/2013/03/IMC-Teesside-2012-SILCalculations-Easy-or-Difficult.pdf.

CHAPTER

Safety Instrumented System Requirements for Fieldbus and Control Components IX

1.0 GENERAL

On account of the presence of dangerous materials like flammable gases and corrosive chemicals, many industrial plants inherently are involved with varying amount of risk. In some plants, like (say) boilers where there may not be much involvement of hazardous materials yet due to mal or wrong operation, there are possibilities of hazard, for example, hazard due to CO formation in furnace due to incorrect operation in oxygen deficient atmosphere. Safety instrumented systems (SISs) are specifically designed to protect personnel, property, and the environment by reducing the likelihood or severity of impact in a hazardous event. Also, SIS is supposed to take the system to a safe state to avoid an accident. Each year, fires and explosions cause loss of huge property in various plants. Also such hazardous situations cost human life, loss of property, degradation of environment, and huge expenditure as litigation charges and compensation. So, corporate managements recognize the financial rewards of utilizing a properly designed process system that optimizes reliability and safety. The increased capability of modern instrumentation coupled with huge costs as result of industrial accidents compelled SIS to a new level of breadth of application especially in light of IEC standards for safety (61508 and 61511). Major challenges could be the balance between attending the maximum safety and making the same economically viable.

It is one of the functions of SIS to see that such flammable/toxic materials are isolated in during hazardous events. SIS consists of a number of safety instrument functions (SIFs), which call for safety integrity as per the international safety standards. SIL achievement in accordance with the requirements of IEC 61508 is required to be demonstration for each SIF. From previous discussions in Chapter VIII, it is clear SIL is dependent on:

- Architectural constraint: Associated factors: Safety failure fraction (SFF) and hardware fault tolerance (HFT)
- Target failure measure, expressed in PFD (probability of failure on demand)/PFH (probability of failure per hour)
- Systematic capability: Each element or component carrying safety function and methodology.

For any safety loop comprising several components, the safety integrity level (SIL) achievement is a joint responsibility of end-user and supplier, as will be clear from Table IX/1.0-1. *Why discussing this here?* These are discussed here to show that equipment manufacturer/system integrator or end-user is not only responsible for the same in isolation. In a safety life cycle, there are several phases involving several activities. So, at various stages there will be involvement of either end-user or supplier/manufacturers. The same issue has been elaborated in Clause 1.0.1; refer to Fig. IX/1.0-1 also.

Table IX/1.0-1 Phase-wise Responsibility of End-user and Supplier

Phase	Responsibility	Activities
Pre-design Phases 1–5 & 9	End-user	Hazard identification, specify requirements, & setting up target SIL
Design & installation Phases 6–8, 10–13	Manufacturer/ supplier	Development as per requirements and get target SIL
O&M Phases 14–16	End-user	Operation, maintenance, modification, & SIL maintenance

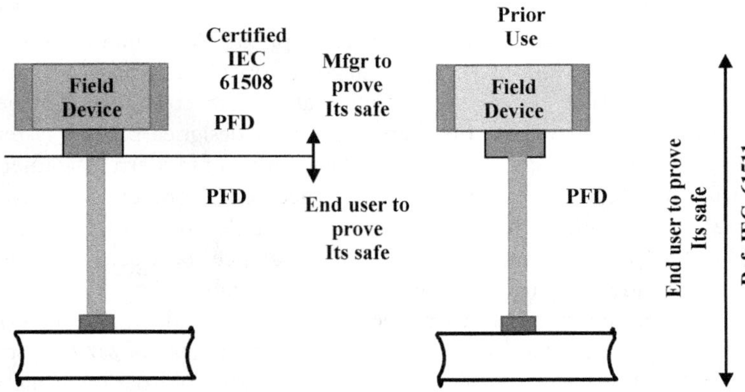

FIGURE IX/1.0-1

Options for field device selection and responsibilities.

1.0.1 *Process/Plant Safety Requirement Specification Details*

When one concentrates on IEC 61511-1:2003 (Clause 10.3), one would find that the standard specifies safety requirements, and the following are some of these safety requirements for SIF/SIS:

- All SIFs required for the required functional safety
- Criteria for successful operation at all conditions

- The extreme environmental conditions
- Major interferences such radio frequency interference, electrostatic disturbance, etc.
- Electrical area classifications
- Identification of process safe state for each of these SIFs
- Demand and demand rate for each SIF
- SIL and associated mode (continuous/low demand) of operation for each SIF
- To check the interaction of concurrent occurrence of safe states (e.g., multiple relief to flare)
- Requirements towards common cause failure (CCF)
- Requirement for proof-test intervals
- Response time requirements related to process safe state
- SIS process measurements and control with set point
- Functional and logical relationship between process input and output
- Various process operating/permissive conditions
- SIS process output actions
- Shutdown requirements (manual and auto protection)
- Resetting requirements after shutdown
- Energize or de-energize to trip logic
- Allowable spurious trip rate (maximum)
- Desired response from SIS on failure modes
- Interfaces between the SIS and other systems
- Various modes of plant operation and associated instrument function for each SIF
- The application software safety requirements IEC 61511-1:2003
- Requirements for overrides/inhibits/bypasses related to safe state maintenance
- MTTR for SIS taking all issues
- Avoidance of dangerous combinations of output for SIS
- Operational procedures
- Safety instrumented function necessary for survival in major accident
- Extreme environmental condition

1.0.2 Component Related Issues

As the title of the chapter signifies, in this part discussions will be on various components. According to the OREDA-97 report — well recognized in the world, 92% of the time these failures were due to the non-performance of their *safety function* upon *demand* by the sensors and final elements. Even if one looks from a conservative perspective as per Fig. VIII/1.0-1, at least 85% of faults are due to failed device. In modern automation systems, most of these field devices sensors and final elements are smart devices, and they are connected to the logic solvers with the help of fieldbus systems for some obvious reasons (discussed later). Naturally, these fieldbus systems need to be safe fieldbus systems. Therefore, in addition to the discussions on sensors, logic solver, and final elements, it is also necessary to address safe fieldbus systems to complete the discussions. According to IEC 61511, there are two groups considered.

One is programmable electronics (PE) (mainly logic solver), the other is non-PE devices or equipment. Here there is a point to be noted; currently most of the instruments are smart, so they fall into type B instruments category (already discussed Clause 1.4 of Chapter VIII). There are two options for end-users to select field devices (sensors and final elements):

- Components selected in SIL application in accordance with IEC 61508-1 and 2.
- To cater to the requirements of IEC 61511 (11.4 and 11.5.3 to 11.5.6).

The difference between the two selections is based on who is to bear the responsibilities for the burden of proof. This is clear from Fig. IX/1.0.2-1.

From the preceding figure, it is clear that in case of prior use, referenced in IEC 61511, it is the user who is responsible.

However, here details presented mainly concentrate on sensors and final element in their generic terms.

1.0.3 Safe Fieldbus (FB)

Fieldbus is more than just a network and a digital replacement for 4—20 mA technology. The same fieldbus system, with additional technological developments, is used in safety application under the name of *safety fieldbus* system. For this reason, Erich Janoschek of TÜV Rhineland, describes a safety fieldbus system as “a safety fieldbus network as a specialized mail messaging process.” On account of clear advantages of fieldbus system, the entire industry connected to modern instrumentation systems is highly impressed and interested towards use of fieldbus systems in safety applications also. Major advantages of fieldbus and safety fieldbus systems are [2]:

- Easy to use, and easy future expansion and modification.
- More extensive data exchange to and from the field devices at a much faster rate.
- Improved diagnostics.
- Reliable, space-saving, cost-effective, and deterministic.
- Greater flexibility in system design, space savings, and layout.
- Openness and interface capability makes it possible to integrate multiple products from different vendors in a system.
- Increased uptime and plant utilization resulting from improved asset management.
- Improved maintenance and test data for reporting.

Nearly 90% of the causes for failure are due to field devices. Naturally, checking the health of the I/O, field devices, and valves are obvious. The system should also incorporate capabilities, such as sensor validation and environmental condition monitoring for conditions that can cause sensor degradation, and monitoring of impulse line blockage. All these are possible with the help of a safe fieldbus system. Safe fieldbus technology is a unified infrastructure to manage data, communication, plant assets, and plant events while providing highly distributed control functionality and interoperability between devices and subsystems. For safety protocol transmission, it is important how it is sent and received in a corruption-free way rather than the medium itself. So what is more important is the safe field bus *technology*.

System integrity ensures that the running safely, with maximized reliability and a proactive maintenance strategy. In an emergency hazardous situation, the system shuts down in a logical and controlled manner with avoidance of the creation of waste, loss to the facility, and damage to personnel and the environment. At the same time, it is important to avoid spurious/nuisance trips, which has negative effect on production. To manage the show, it is at times important to have global access to the process data. At the same time, well-*secured* data should be presented in a manner that is easy to understand and act upon. Open, scalable integration addresses the need to drive out custom integration costs and the requirement for best of breed applications to work seamlessly together in an open environment. As an open automation infrastructure, safe fieldbus technology addresses all of these requirements in several different ways [4].

1.0.4 Sensor

Field sensors are used to collect information necessary to determine if an emergency situation exists. The purpose of these sensors is to measure process parameters (e.g., temperature, pressure, flow, etc.) or monitor the process condition/status, used to determine if the equipment or process is in a safe state. Sensor types range from simple pneumatic or electrical switches to smart transmitters with on-board diagnostics. These sensors in some applications are dedicated to SIS. Manufacturers' approach, in line with IEC 61508 towards the development of sensor, shall include but is not limited to the following:

- Development of safety requirements based on end-user specification
- Instrument hardware and architecture design as per standard-part 2 rule
- Software control verification validation in line with standard-part 3
- Diagnostics and verification of the same
- Manufacturing controls to ensure to the desired safety requirements
- Control of management of change (MOC) to meet the requirements
- Failure modes, effects and diagnostic analysis (FMEDA) to determine SFF, PHF
- Details of proof test requirements
- Get third-party certification as per functional safety assessor (FSA) discussed in Chapter VIII
- Development and issuance of safety manual

Normally, manufacturers notify the certifying authority regarding the issuance of the certificates. The certifying authority checks that all the requirements of the standard are met (see Clause 1.4 of Chapter VIII), and issues the certificates. As per IEC 61511, users can develop other criteria for certifying SIS loop components including requirements for the selection of components and subsystems based on prior use (Clause 11.5.3 of the standard). The main issue here is the "appropriate evidence" for subsystems and components that these are suitable for safety instrumentation. The standard also specifies the required documentation pertinent to the following issues mainly (see Clause 1.4 of Chapter VIII):

- Manufacturers
- Quality
- Management

- Configuration management
- Adequate specification and identification of components
- Hardware/software identification
- Awareness evaluation by end-user for when modification is done by manufacturer
- Demonstration of performance of subsystem/components (proof)
- Similar operating profile
- Physical environment
- Volume of operating experience
- Update and monitoring
- Addition with proper experience
- Removal when non-satisfactory performance is reported

  However, there is significant value to the end-user for the product "designed as per IEC 61508."

1.0.5 Logic Solver

The purpose of a logic solver, as a component portion, I/O signal processing unit, main sets of processors pertinent to SIS, determine the actions to be taken based on the information gathered. Highly reliable logic solvers can provide both fail-safe and fault-tolerant operations. A controller reads signals from the sensors and executes pre-programmed actions to produce output to final elements for preventing a hazard event from happening. Logic solver subsystems include but are not limited to the following:

- Processor/memory units
- Communication processors (as applicable)
- I/O signal processing boards
- Peripherals:
  - Human machine interface (display)
  - Printers/loggers
- Link interface
- Gateways/network support switches
- Termination units
- Racks, cabinets
- Power supplies system
- Gateways/network support switches
- System software:
  - Software libraries
  - Programming tools
  - Communication protocols

Suitable safety manuals shall be there to help utilizing the facilities. When there is any non-certified equipment, then the PFD calculations are necessarily performed to show the contribution of such logic solver is within acceptable limit. Whereas for certified item pre-calculated and verified data are made available. For software applications, V model of the standard needs to be followed; for this Fig. VIII/2.0.3-2 may be referenced. In case of non-certified systems, necessary precautions shall be taken to ensure safe communication, which is well-documented and verified in case of certified systems. For SILs above SIL1, certified relays, safety solid state, and safety programmable logic controllers (PLCs) would be necessary. Really to achieve SIL3, redundancy in safety PLC is necessary. For existing systems, it is not possible to replace the entire systems with a new certified logic solver. In that case, a few issues need to be addressed:

- Hazard analysis to check whether existing safety functions are covered; if not, to find the reasons and get the ways.
- Study isolation between safety functions and basic plant control system (BPCS) except for low-risk cases where this may not be suitable.
- To check and document proof test procedure and set proof test interval. Suitable measure towards this is necessary.
- Record of failure rates from maintenance, etc.

Many industries used to maintain separation between the process controls and safety systems; however the latest edition of the standard does not specify the same in black and white. So, some integrations and interconnections between the systems are seen. For such integration/interconnections, special measures shall be implemented to avoid adverse effects between SIS and non-SIS systems, or between various SIS and applications. If suitable special measures are implemented, a limited degree of interconnection can be allowed.

1.0.6 Final Elements

Not only because final elements contribute 50% of PFD share, but also final control elements are the key components of any control loop in any system, be it BPCS or SIS. Therefore selection of final elements needs special attention. Final element implements the action determined by the logic system. This final control element of interest to SIS is typically a pneumatically/hydraulically actuated on-off valve operated by solenoid valve(s). But it could be other types also. It is important to keep in mind the applicability SIL assignment to final elements. When needed, assignment criteria may be applied. Using perspective of IEC 61508 and 61511 same assignment criteria could be fixed.

From the discussions, it transpires that each element of safety loop contributes to PFD for the safety loop. Naturally, PFD from the valve or final elements needs to be considered for SIL. This is very easy simple statement. One needs to look into the issue slightly in depth. From use point of view there can be mainly three pattern of usage of final elements, such as:

- Single element used for process control under safety system (on/off)
- Single element in normal conditions modulates under BPCS and for safety controlled by safety system through safety interlock
- Two sets of valve one for modulating control under BPCS another for safety under SIS

In the first two cases, failure of the final element places demand on SIF so it is needless to tell this calls for SIL rating. In the third case, final element under BPCS provides additional hardware fault tolerance for *higher* SIL application. To the end-user, here mean time to fail (MTTF) data is important to determine SIL level. Thus it is clear why SIL is important for final elements, which in fact is not restricted to valve damper, but could be anything such as circuit breaker, etc. also. So, for each application, suitability of the same is important for consideration. Major issues related to selection of final element selection shall include but are not limited to the following:

- Type: As suitably applicable for the application
- Architecture in line with IEC 61511-1:203 clause number 11
- Control design: Locking arrangement to avoid unauthorized operation
- Partial stroke test as functional test

1.0.7 Typical Failure Data

The data presented here are typical values only. Actual data needs to be collected. The data given here in Table IX/1.0.7-1 is for the reader to get an idea about the issue. Major sources of these data are from Refs. [1] and [3], and balance data are from the plant data collected.

Table IX/1.0.7-1 ***Typical Failure Rate for Some Common Field Devices***

Item Details	Failure Rate (Hr)	Safe Failure (%)	MTTF (Year)	Remarks
Pressure switch	4.0E-06	40	20–30	
Temperature switch	5.0E-06	40	20–30	
Flow switch	7.0E-06	60	20–30	
Level switch	5.0E-06	55	20–30	Float/ displacer
Pressure transmitter	1.5E-06	>50	55	
DP transmitter	1.5E-06	50	40–60	
Flow meter	3.5E-06	25	50	Coriolis meter
Level instrument	7.0E-06	40–50	40–60	Displacer
RTD	5.0E-08	81.6	60–80	
Thermocouple	1.5E-06	95	60–80	
Temperature transmitter	5.0E-06	30	70	
I/P converter	4.0E-06	40		
Solenoid valve	2.0E-06	60	25–30	
Ball/Butterfly valve	3.0E-06	55	30	
Gate valve	2.0E-06	45	50	
Globe valve	2.5E-06	55	50	

Now with these basic ideas in mind, it is time to go for little details into each system components and subsystems. Starting with safe fieldbus system, discussions will be presented on field instruments, alarm systems, logic solvers, and final elements also in sequence.

2.0 SAFE FIELD BUS SYSTEM [2]

What is fieldbus? Why is it so important? Fieldbus may be conceived as a digital communication system for various field devices such as sensors, actuators, and field control systems. Fieldbus has to address a few needs, such as device addressing, two-way communications to exchange data related to the parameter as well as status of the field device, communication speed, etc. Obviously fieldbus cables are not "just wires." In modern instrumentation and control system, it is like one main artery in the human body.

2.0.1 Fieldbus Basics

The following are basic transmission features noted from this standard combining with the field intrinsically safe concept (FISCO):

- Each segment should only have one power supply.
- No power fed to the bus when a station is sending.
- In steady state, all field devices consume constant current.
- Field devices are passive current sinks.
- Passive line termination at both ends of the main bus line.
- Linear, tree, and star topology allowed.
- To increase reliability, redundant bus segments can be designed.

In order to have some idea about the requirements of the standards IEC 1158-2 - Table IX/2.0.1-1 may be referenced:

For further details, explanation given in "*Power plant instrumentation and control handbook; Elsevier*;" may be referenced.

Table IX/2.0.1-1 IEC 1158-2 Features

Features	Requirements	Features	Requirements
Data transmission	Digital, bit-synchronous, Manchester coding	Data security	Preamble, error-free start, and end delimiter
Speed	31.25 kbps voltage	Cable	Two wire-shielded twisted pair
Explosion protection	IS EExia and ib EEx d/m/p/q	Number of stations	32/Segment (maximum could be 126 addresses available)
Topology	Line/tree or combination	Repeater	4

2.0.2 Safe *Fieldbus Basics*

From the preceding discussions, it is clear that there are certain advantages of using fieldbus in modern instrumentation system now; it is time to look into how same fieldbus could be utilized in *safe* instrumentation application. The advantages discussed above (Clause 1.0.3) is well *applicable* for safe fieldbus system also, *additionally* there are some other advantages which shall include but are not limited to the following [4]:

- Safety aspects:
  - Self — diagnostics features of devices
  - Real time identification of dangerous failures
  - Valve partial/full stroke testing possible
  - Reduced manual proof testing burden
- Operability:
  - Trip on demand only
  - Reduction in nuisance trips
  - Device self diagnostics
  - Multi-drop architecture
  - Installation flexibility
  - Asset management tools
- Cost reduction
  - Multi-drop architecture
  - Installation flexibility
  - Asset management tools

For safety instrumentation functions, fieldbus systems pass beta testing. (Prototype designed product tested at customer site for evaluation of the product.) Also, safe fieldbus systems are required to *meet* the IEC 61508 standard for functional safety systems for specified SIL so that suppliers can build their systems on the same. Also, protocols are subject to *third*-party approval for safety applications. There are a number of fieldbus systems recognized in the world. Also, there are other system integration tools and protocols such OPC and HART, respectively. HART is a protocol, while OPC is a standard described in brief in Fig. IX/2.0.2-1. Now, initially each of them has their own ways and means to communications in intelligent way. There are some similarities in their network protocol but there was no common platform. These made it rather difficult to integrate various control systems, which is of tremendous demand not only from technological point of view but also from economical point of view. In 2003, major players in fieldbus systems joined hands and

OPC is an independent platform and treated as interoperability standard for secure, reliable data exchange and seamless flow of information amongst various devices from multiple vendors in industrial automation field.

FIGURE IX/2.0.2-1

OPC standard.

thereby there was formation of field device integration (FDI) (discussed later). In order to understand the same, it is better to explore and see how these were dealt and how two standards were married.

2.0.3 *Electronic Device Description Language and Field Device Tool*

Electronic device description language (EDDL – IEC 61804) is supported by the Fieldbus Foundation, HART Communications Foundation, and PROFIBUS & PROFINET International, and then by OPC Foundation. These are supported by almost all devices in process controls using fieldbus. Electronic device description (EDD) was in use for the support device diagnostics and calibration. Now, EDDL interpreter recognizes EDD and helps to interpret the same as a sort of XML (eXtensible mark up language) for intelligent field devices [5]. EDDL is a programming language used to create device descriptions (DDs) to describe the attributes of field devices in an operating system (OS) and human–machine interface (HMI)-neutral environment. EDDL is independent of OS and does not require any machine level execution. EDDL also supports automated procedures or wizards with built in "method calls." There have been cross-functional teams that have tried to leverage the power of EDDL across different protocols. Field device tool (**FDT**) is accepted as IEC 62453 to support greater than 16 protocols including Foundation Fieldbus, HART, and PROFIBUS indifferent applications including process and factory automation. FDT is capable of supporting graphical representation of information, and can work with complex devices, for example, radar-level gauges. The FDT Group provides a common environment for accessing some features of intelligent field devices in a *generic* manner (regardless of supplier and communication protocol). The essential parts of FDT technology are the frame application (FDT frame), device type managers (device DTM—as driver) and CommDTM (for communication), available for field devices and communication equipment. The strength of FDT/DTM technology can support a high level of diagnostic information and its independence from any communication protocol, and the software environment of the host system. Both these two technologies are complementary to each other.

2.0.4 *Field Device Integration*

In September 2011, FDI Cooperation, LLC, was formed with representations from the Fieldbus foundation, FDT Group, HART Communications Foundation, OPC Foundation, and PROFIBUS & PROFINET International. Also, major process automation suppliers like ABB, Emerson Process Management, Endress + Hauser, Honeywell, Invensys, Siemens, and Yokogawa lent their support.

The mission of the organization is [5]:

- Standardization of FDI under IEC
- FDI specification
- FDI tool kits for manufacturers
- Develop high quality technology for FDI support, independent of and common to various communication existing protocols

- Providing state-of-the-art technology with full backward compatibility
- Ensure stability, interoperability, and compatibility of FDI-based products

FDI ***integrates*** various technologies such as EDDL, FDT/DTM, and the structured OPC UA (Unified Architecture) standard IEC 62541 to address the following issues:

- Independence on:
  - Platform
  - Operating system
  - Host system
  - Protocol
- Compatibility with technologies like:
  - EDDL
  - FDT/DTM
  - OPC UA
- An open system specification
- An international standard
- Full access to field device capabilities
- Integrated Development Environment (IDE) to include:
  - Host interpreters
  - Package conformance tools
  - Host conformance tools
  - EDDL profile harmonization and extensions
  - Common EDDL file format

FDI technology has been conceived to provide *scalable solutions* users can deploy in different applications and for tasks associated with *all phases of their lifecycle* from configuration, commissioning, and diagnostics to calibration.

2.0.5 The Device Package

The main focus of FDI architecture is the device package shown in Fig. IX/2.0.5-1. Each field device is represented by a single FDI device package containing up to four different elements. Major issues here are:

- Electronic device description (EDD) — an information model
- Optional user interface plug-ins (optional)
- Attachments (optional)
- Miscellaneous other major issues:
  - EDDL harmonization
  - FDT interoperability
  - EDDL interoperability
  - Registration and conformance testing
  - Integrated development environment

  In order to restrict the volume of the works it is rather difficult to elaborate this discussion. Interested readers may look into foundation fieldbus site for further details. Now short discussions will be presented on different safe fieldbus systems.

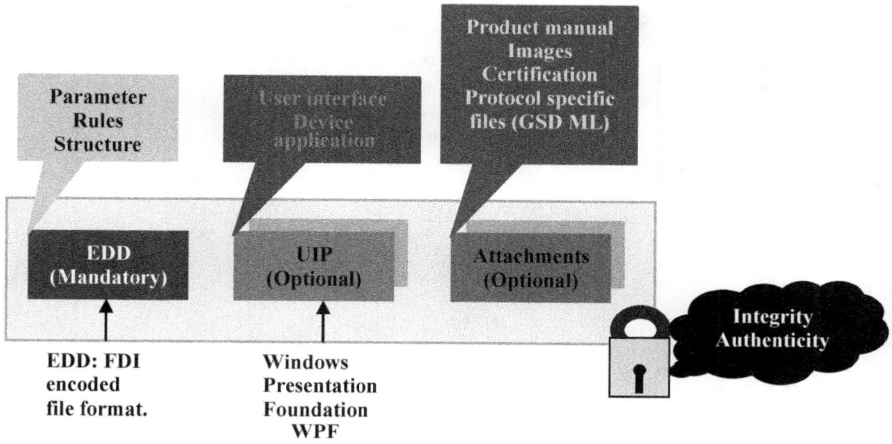

FIGURE IX/2.0.5-1

Device package.

Developed based on white paper from Field Device Integration (FDI) Making Device Management Easy; fdi_white_paper.pdf; Foundation fieldbus; http://www.fieldbus.org/images/stories/enduserresources/technicalreferences/documents/fdi_white_paper.pdf. Courtesy: Foundation fieldbus.

2.1 Foundation Fieldbus

Foundation fieldbus consists of H1, which interconnects field equipment and high-speed ethernet (HSE 100 Mbps). Link device is an HSE device to connect one or more H1. It has been incorporated in international standard IEC 61158-2. So, two types of foundation fieldbus systems are:

- H1 bus based on IEC 61158-2 standard uses field devices like sensors, transmitters, and I/Os.
- HSE bus is an HSE bus that provides integration of high-speed controllers, subsystems via link device, data servers, and workstations [2].

2.1.1 Foundation Fieldbus Basics (General)

Two bus systems, H1 bus and HSE bus, are discussed as follows.

If necessary, the reader may go through the standard book on networking, to recapitulate the fundamentals on the same with fieldbus.

With respect to ISO, OSI model foundation fieldbus H1 consists of three layers: (1) physical, (2) data link, and (3) user application, as shown in Fig. IX/2.1.1-1. Layers 3 to 6 are missing [2].

The user application is made up of a function block and device description, and it is directly based on the communication "stack" shown.

The function of physical layer is mainly to translate message to physical signal and vice-versa. The second layer is responsible for combining several technologies that together control transmission of data on the fieldbus. The last layer is for the user to

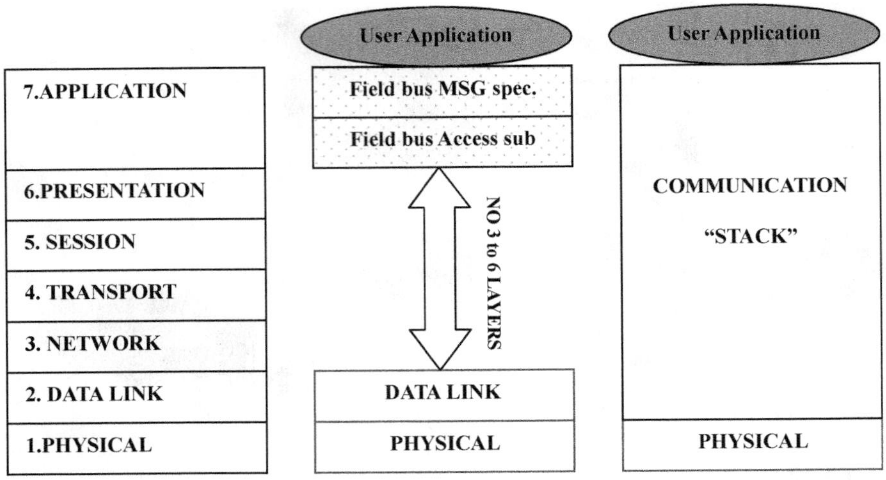

FIGURE IX/2.1.1-1

Foundation fieldbus model [2] - by author.

interact with other layers and go through application issues. The user layer contains resource blocks, transducer blocks, and function blocks which are responsible for device capabilities. A few key features of H1 bus and HSE bus are as follows:

- For the common power and IS (where applicable) for the H1 bus, minimum voltage for communication is 9 V. H1 bus communication speed is *31.25* kbps with *Manchester-encoded* data; standard is IEC 61158-2. Token passing is the communication mode deployed in an H1 bus.
- There are two types of devices in the H1 bus: basic unit and link master. Max *32* devices per H1 bus in line, tree, and star topology.
- Maximum length of H1 bus (sum of trunk and spur) is *1900* m without a repeater with the maximum possible spur length of 120 m

HSE is based on standard Ethernet topology with a high speed of 100 Mbps for communication some of the key features are as follows:

- Asset management functions like diagnostics, calibration, identification, etc.
- Since the system is an open system, it is interoperable enabling user to mix various subsystems to form an integrated system.
- Peer-to-peer communication enables communication between two devices without a central computer system.
- Since it is standard Ethernet, there is no need for any special tools and cables.

2.1.2 Foundation Fieldbus—Safety Instrument Function Background

It started in 2006, when FF got type approval for protocol from TUV. Fieldbus foundation demonstrated foundation for SIF on May 2008 in the Netherlands. This project initiated good response from end-users in process industries as they were interested in CAPEX/OPEX benefits along with open and interoperability of fieldbus technology in SIF. SIF in foundation fieldbus (FF) meets the IEC 61508 standard for functional safety systems with SIL level up to **SIL 3** (to the best of my knowledge is at the time of this book). So, systems can be built for process instrumentation as per IEC 61511. There has been third-party approval of the protocol from TÜV (approval) with suppliers offering for approval of their products. In December 2012, foundation declared foundation *SIF ITK 1.2* toolkit for troubleshooting and debugging fieldbus instruments, and provides all hardware and software required to ensure a manufacturer's complete device interoperability as specified by the Fieldbus Foundation's official registration testing procedure. Foundation fieldbus safety instrumented functions (FF-SIF) is for the application of foundation fieldbus (FF) technology in safety control systems. General guidelines in line with ISA TR84.00.06 have been presented in Clause 2.2.4 as follows.

2.1.3 *Role as Process/System Integration*

Process integrity addresses the requirements such as maximizing reliability and preventive maintenance to keep the plant running at its best. In case of an abnormal situation, when there is need for plant shut down, then it is to be done in controlled manner so that there is no damage to people, environment and loss of property including avoidance towards creation of wastes. Also it is important to avoid spurious nuisance trips, which negatively affect productivity but do not enhance safety. In modern instrumentation systems, there is need for global data access in a presentable and well-understood manner. This system helps in developing open scalable, seamless integration at optimum costs without dispensing off the system safety. FF-SIF is an open automation infrastructure to address all these requirements.

2.1.4 *FF-SIF in Safety Functionality and Integration*

The value of integrated safety functionality: FF-SIF provides a common platform for SIS and process automation with common HMI, engineering, and asset management tools, ***but*** separate logic execution. Benefits extracted from the system include but are not limited to the following:

- Reduction of installation cost
- No cost for custom integration
- Unified view of control
- Low cost and effort for asset management
- Low life cycle cost and inventory
- Common support towards:
  - Training
  - Maintenance
  - Services

In this approach there will not be separate costs for interface, maintenance and synchronization. In this common platform and in FDI effort, it is possible to

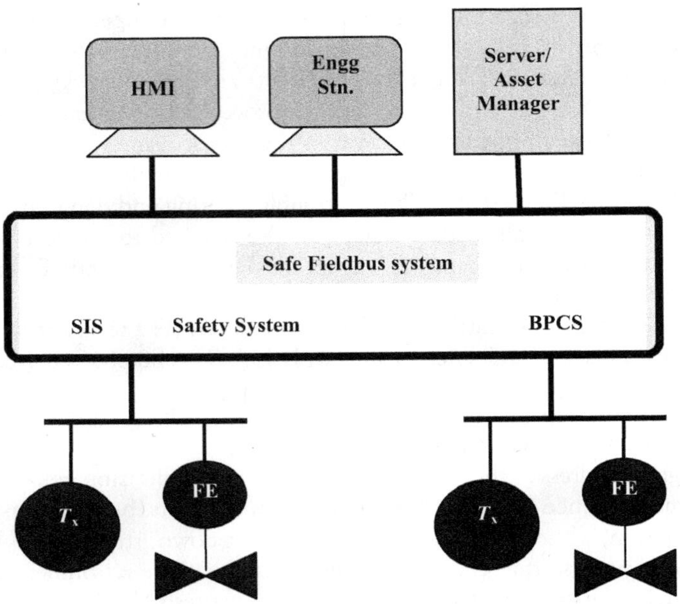

FIGURE IX/2.1.4-1

Integration of BPCS and SIS through safe fieldbus system.

use completely different systems for control and safety systems. See Fig. IX/2.1.4-1. This also offers advantages as follows from CCF point of view:

- Special software-independent failures
- Security – no effect of change in BPCS on SIS, and vice versa.
- Special requirements SIS:
  - Extended diagnostics error-checking
  - Data storage protection
  - Fault tolerance

FF-SIF provides better diagnostics to minimize nuisance trips and offers potential to integrate directly into standard control architectures with improved asset and event management and lesser chances of failure.

2.1.5 Communication Channel

The safety layer is essentially built into the network and it exists between the communication stack and the application according to the IEC 62280-1 standard. As the protection layers are built in to the network, even non-safety devices can be added to the safety network. SIF devices are required to register with foundation. For registration SIF device must undergo a rigorous test process.

2.2 PROFIBUS BUS/PROFINET

PROFIBUS family is another proven and popular fieldbus system in use in industrial automation system. The PROFIBUS system has three compatible versions that operate all together seamlessly.

- PROFI FMS (Field Message Specification): This is a high-end application level communication. It provides object-oriented transmission with structured data, loading, and control programs. It is the extension of other family members such as PROFIBUS DP.
- PROFIBUS DP (Decentralized Periphery): Low-cost field and controller level communication. Majority of PROFIBUS devices belong to this category.
- PROFIBUS PA (Process Automation): This is a cost-effective, two-wire connection used for field devices to communicate.

PROFIBUS is based on **RS 485** and speed is restricted to maximum *12* Mbps. For IT applications, PROFINET, (comparable to Ethernet) with speed up to *100* Mbps is used. For safety critical applications, PROFIBUS/PROFINET has PROFISAFE.

2.2.1 Communication Channel [2]

Structural details of PROFIBUS have been presented in Fig. IX/2.2.1-1 [2]. Like foundation fieldbus, here also there are different bus systems, viz. FMS, DP, and PA.

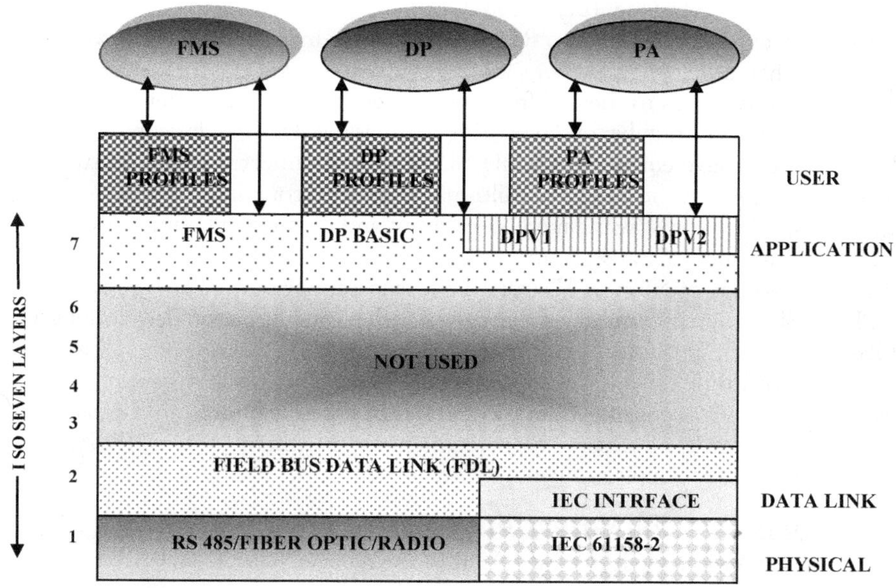

FIGURE IX/2.2.1-1

PROFIBUS layer structure [2].

The discussions presented as follows are basic details for structure, operation etc. For further details, standard books or [2] may be referenced.

- Structural details: From the preceding figure, it is clear that the two layers, physical and Fieldbus data link layer (FDL), in all three types of PROFIBUS are nearly identical. When compared with the ISO/OSI model, it is clear that layers 3 to 6 are missing in PROFIBUS, where FDL handles transmission protocol, data security, and error detection. Original DP has been extended to include these functions, referred to as DPVs, found in the application layer. Of 0 to 128 device numbering, 126 is reserved for generic devices and 127 is used for broadcast messages. In line with limits from IEC 61158-2 and RS 485, there will be 32 devices per segment.
- Topology: Topology can be linear in RS 485 in daisy chain, and can be made a tree as per IEC 61158-2. The circuits can include repeaters, couplers, and optical link modules (for transforming optical signals into electrical signals). Installed between PROFIBUS DP and PA, is a *signal coupler*, which is necessary as both systems have different techniques for: using IS, transmission, and powering scheme. Features of signal coupler are as follows:
  - Electrical isolation between safe and IS bus
  - It is slave to PROFI DP
  - Baud rate adaption 31 kbps (PA) to 1200 kbps
  - PA bus powering
  - Transmission technique: IEC 61158-2
- PROFIBUS operational aspects: PROFIBUS operates in a master slave mode with more than one master. Master stations control the network communication and the slave responds to the call from the master. Each of the masters can communicate with a number of slaves. The system is highly democratic, meaning that all masters have equal rights while all the slaves connected to any master have equal rights. In the case of a multi-master system, communication is regulated by the token passing method — cyclic and acyclic communication: network configuration and start-up.
- General station description (GSD) files: To ease the configuration, all supporters of PROFIBUS are standardized, and these are called general station descriptions (GSD). These GSD files are a tremendous help during interoperability [2]. During the planning stage, GSD files are loaded to the master. If necessary, the reader may consult documents from PROFIBUS International for further details. Now it is better to explore some basic requirements pertinent to PROFIBUS/PROFINET in safety applications.

2.2.2 PROFISAFE: PROFIBUS/PROFINET

It is better to gather knowledge about PROFIBUS and PROFINET.

- PROFIBUS: As already discussed, Profibus has three parts, viz. FMS, DP, and PA and run on RS 485 (*twisted pair or four wires with power, optical is also possible for FMS and DP*) at 5 V level and depending on PA/DP type, the speed is 31.25 and 9.6 kbps—12 Mbps, respectively.
- PROFINET: It is an open standard of Ethernet adapted for real time DCS/PLC application. So, it runs on Ethernet technology. It has speed up to 100 Mbps.

- PROFISAFE: PROFISAFE is an application layer that describes the communications between fail-safe devices. Supports safe communication over open standard buses PROFIBUS (DP, PA) and PROFINET—TUV Certified to IEC 61508 SIL 3/EN 954-1 Cat 4.

2.2.3 PROFISAFE Network

As discussed earlier, in PROFISAFE, safety protocol is provided as a part of the communication protocols for both PROFIBUS/PROFINET. Therefore, end-users do not need separate networks BPCS and safety system; instead they can integrate both the systems into a single network as shown in Fig. IX/2.2.3-1.

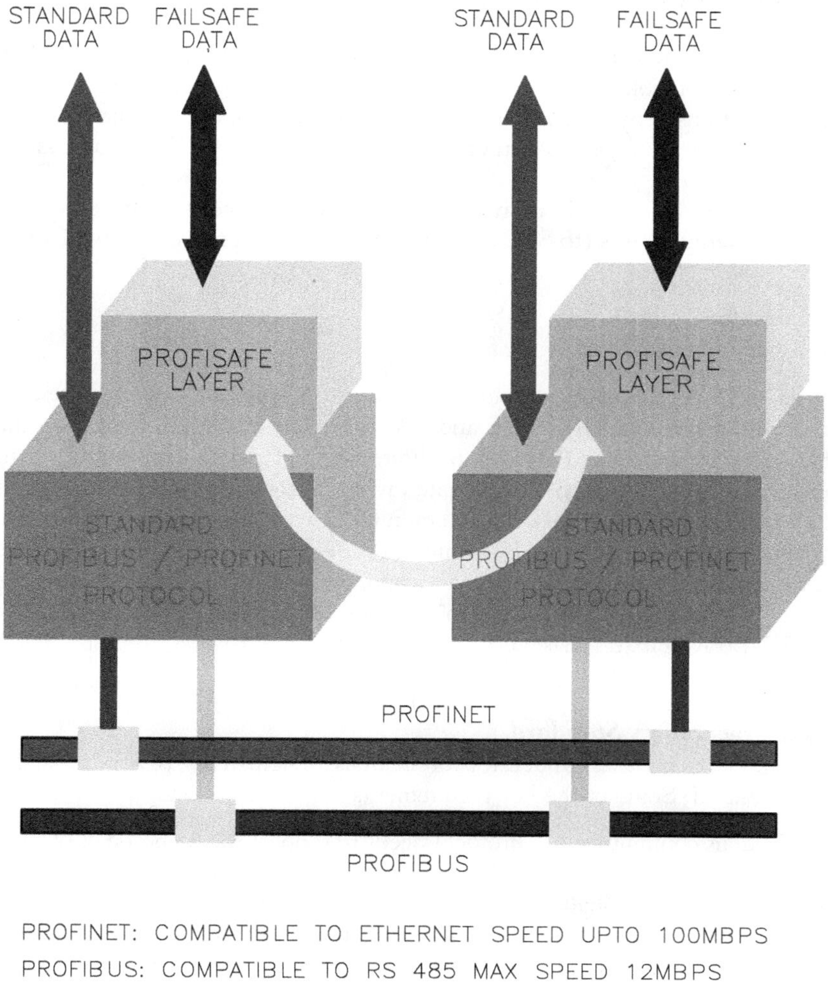

FIGURE IX//2.2.3-1

PROFISAFE communication pattern.

Developed based on PROFISAFE document. Courtesy: PI south east AIS; PI PROFIBUS PROFINET; Profisafe; South East Asia; http://www.profibus.com/nc; Internet document.

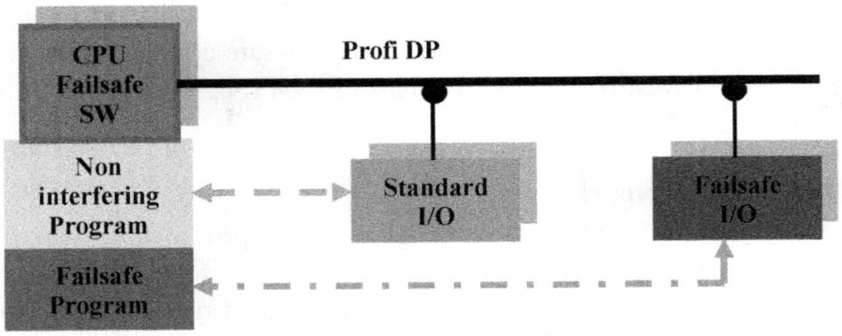

FIGURE IX/2.2.3-2

PROFISAFE network (typical).

PROFISAFE is an extended version of the standard PROFIBUS/PROFINET communication protocol. PROFISAFE is a application layer (profile) that describes the communications between fail-safe devices and addresses special requirements in application layer, for safety-related information necessary to conform to strict safety standards.

PROFISAFE in its application layer has the following *elements* to meet the reliability requirements of SILs (to SIL3) prescribed by international safety standards:

- Message numbering
- Data consistency checks
- Avoidance of network messaging faults

As PROFISAFE communication protocol has the built in safety features, all connected devices in PROFIBUS DP and PA, as well as PROFINET, can use the same for safety application. This also facilitates uses of devices in two buses to communicate without the need for any gateway and or bridge. A similar thing has been shown in Fig. VII/2.2.4-1 also. In common CPU with failsafe software, standard and fail-safe I/Os can be connected by common PROFIBUS DP. In application, there shall be two parts; one fail-safe which shall communicate with fail-safe I/Os, whereas so-called non-interfering application can be there for standard I/Os as shown in Fig. IX/2.2.3-2. PROFISAFE has TUV Certification to IEC 61,508 SIL 3/EN 954-1 Cat 4.

2.2.4 Fieldbus Safety Standard

ISA TR 84.00.06 provides guidelines for fieldbus design. This is applicable for ***fieldbus in general***. Some of the key requirements may be as follows:

- Safety fieldbus communication protocol needs to support the highest SIL of the SIS when it is certified. So far, SIL3 certification is available.
- Programmable devices shall be interoperable and *non-proprietary*.
- Devices may exist in a network in safety- and non-safety-related applications, provided non-safety-related devices are non-interfering (as shown in Fig. IX/2.2.3-2).

- Fault tolerance is optional.
- Enough security shall be there for inadvertent changes.
- Diagnostics shall be "transparent" to the user and with possible reporting.
- Online replacement of devices.
- System shall be "testable."

2.2.5 *High Availability for PROFIBUS Process Automation*

For higher availability applications, PROFIBUS PA can be configured in ring architecture to ensure communication, even if part of the network cable is disabled due to (say) physical damage. "Active field distributors (AFD)" are used to integrate field devices via four short-circuit-proof spur line connections in a PROFIBUS PA ring with automatic bus termination [6]. The PA bus may be connected to ProfiDP (single or redundant) with the help of DP/PA link to complete the communication. In this kind of architecture, in the case of a ring break the end of the remaining segment is terminating so that network communication can continue.

2.2.6 *Smart Diagnostic Features*

Intelligent diagnostics of field devices can provide lots of information about the device through predictive maintenance. Two important issues are information extraction and evaluation of the extracted information. Process device manager — a software tool available, for configuring, commissioning, diagnosis, and maintaining intelligent process field devices networked with PROFIBUS. Process device manager is helpful in getting the information. Plant asset management helps to predict and assess the health of plant assets by monitoring asset condition periodically and it can identify potential problems before affecting the process or leading to a catastrophic failure. With this discussion on fieldbus system is concluded to concentrate on field devices connected to fieldbus. So in the next clause, discussions will be on field sensors.

3.0 FIELD SENSORS FOR SAFETY INSTRUMENTED SYSTEMS

An SIS is a collection of SIFs, each of which is one or more components designed to execute specific safety-related tasks designed to bring an industrial processes to a safe state in the event of hazardous conditions. One major issue of SIS is to balance and optimize two different parameters, viz. maximization of facility safety (against), maximization of economic benefit. A safety system needs to provide reliability for the intended protective function, at the same time ensuring higher operational availability of the process itself.

Another important issue one should keep in mind that the safety data provided by manufacturers for their devices, with due validation by a third certifying authority are laboratory results. Laboratory safety data, provided by manufacturers may be useful predictor safety data pertinent to control system but real world installed safety is always much worse. "Proven-in-use" data includes real-world failure causes; however it tends to be *conservative*, as it must cover the whole range of the category, from 20-year-old pneumatics to the latest smart technology [7].

3.0.1 Choices for Sensor Types

Instruments/sensors for SIS are separate and distinct from those used for BPCS, to ensure an independent layer of protection safety. In modern SIFs, continuous process transmitters in conjunction with comparator circuits are used for hazard detections, in places of process switches. The ability to easily change set point, better diagnostic capability, and continuous health checks make process transmitters better suitable and first choice for users. A "healthy" transmitter show changing output signal; in contrast to that, process switches provide no indication of "healthy" operation. So, when a process switch needs to change states at set point, it might also be failed and incapable of registering a dangerous situation! It indicates that process switches suffer from *covert* (hidden) failure. Basic purpose of diagnostics is to detect covert failures as *overt* (detected) failure. Since smart transmitters can be provided with diagnostics so also it is *preferred* in SIS applications.

3.0.2 Redundancy and Voting of Sensors

In SIS applications, importance of reliability and availability cannot be over-estimated. In such cases, redundant transmitters may be deployed in 1oo2, 1oo3, 2oo3 or 2oo4 (typical for nuclear plants). Out of these, 2oo3 is a very good and common choice but voting of sensors in 2oo3 architecture demands special attention. High, low, average, and median are typical selection criteria for "voting" modules. The following may be considered as a guideline for selection of voting circuits for different kinds of applications in dangerous/hazardous conditions or events (from safety point of view):

- High selection: Larger measured value in the hazardous condition.
- Low selection: Smaller measured value in the hazardous condition.

Thus with above approach, 2oo3 reduces to 1oo3. Now, *average selection* is rather problematic, especially when one of the redundant transmitters happens to fail in the "safe" direction, making skewing (e.g., if 2, 3, 4 units are three values, average is 3 units now dangerous situation is say >4.5, then if the third transmitter fails then average would be 2.5 so going away from dangerous value). The *median select* criterion is very *useful* in safety systems because it effectively ignores any measurements deviating substantially from the others and is effective 2oo3 selection. Median value selection is also possible for discrete switches and/or for O/P of comparator circuit of transmitter.

3.0.3 Field Device Diagnostics and Its Impact

The concept of device diagnostics to diagnose overt and covert component failures has undergone sea changes in the present days. Apart from monitoring own performances, smart instruments are capable of predicting a great deal of information about what's going on with a process, well beyond the specific variable measurement or control. Some smart instruments are able to detect deterioration in a key component or realize *sensor drift*. Others look at the process and might spot *changes*, for example, differential pressure sensors can measure process noise and determine the normal level. Today's smart instrument can send a warning when they begin to notice such changes. Powerful diagnostic information is now available to optimize overall operation and maintenance. A similar model has been shown in Fig. IX/3.0.3-1.

(1) The lowest level with individual field devices diagnostic information. (2) Groups of individual loops working together to provide basic functions. (3) It is the operational level. (4) Based on above information, one can optimize overall operation and maintenance effectively at level 4.

From a safety point of view, these are very effective in the sense that smart devices now have sophisticated diagnostics, and are able to predict when a problem was developing, such as an impending component failure or sensor drift. Some diagnostic functions monitor and record the amount of force necessary to close a valve or the amount of time spent in any given position or voltage problems, etc. So, it is possible for the operating personnel to get insights into the process in their dashboard (beyond the basic process variable). Now with developed communication and networking with fieldbus, it is possible to send information to the control or asset management systems. These are achievable with the help of less expensive programmable electronics.

Some of the specific issues related to field devices are as given in the Table IX/3.0.3-1.

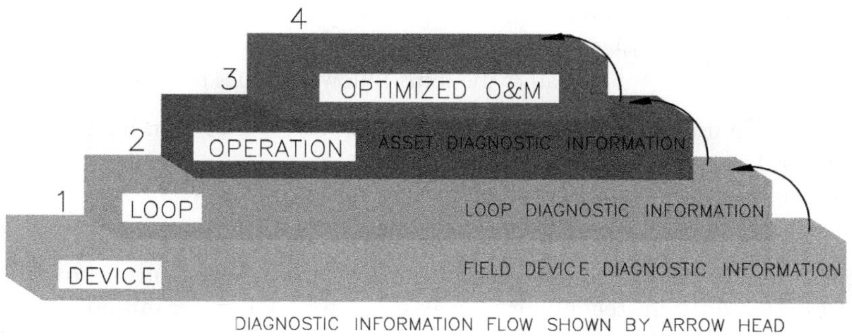

FIGURE IX/3.0.3-1

Smart device diagnostic information for overall performance.

Developed inspired by A. Ajmeri, Smart Instruments and Device Diagnostics: How Well Is Your Plant Using Information, Yokogawa; Plant Services, March 2015; http://www.plantservices.com/articles/2015/smart-instruments-and-device-diagnostics/.

Table IX/3.0.3-1 Diagnostic Prediction of Measurement Problem

Measurement	Diagnostic Issue	Measurement	Diagnostic Issue
Pressure	Plugging of impulse line	Vortex flow	Change in medium
Pressure/Level	Diaphragm leakage	Coriolis flow	Slug flow, tube coating
Temperature sensor	Coating in thermo well, drift, sensor open/short	Magnetic flow	Faulty electrode/ ground, process noise, tube coating
Differential pressure	Process noise, zero drift	pH	Faulty electrode, cleaning

3.1 Technical Issues for Sensors

There are a few important technical issues need attention of designers. Some of the pertinent issues have been addressed in the following clauses.

3.1.1 Factors Influencing Selecting Sensors

It is needless to argue that measurements affect the safety of a measurement loop. Therefore, it is important to look into the pertinent issues, listed as follows, which influence selection of sensors for various measurements:

- Use of transmitters, not process switches (Clause: 3.0.1)
- Common cause strength
- Diversity
- Built-in diagnostics
- Redundancy and availability

The reason for selection of transmitters in place of process switch and issue related to redundancy already discussed in Clause 3.0.1 and 2. High ambient temperature may be a common cause; to combat against the same following issues may be considered:

- Robustness: To use robust transmitters (to protect electronics from high temperature)
- Installation: Install away from hot point
- Diagnostics: Suitable software to predict failure due to high temperature.

Diversity is an important point to eliminate CCF. Redundancy with instruments and devices using different technology is very useful for CCF or for minimizing the risk that a single failure mode will affect both. Of course, the diverse, backup technology should be strongly resistant to the failure modes that affect the primary technology. So, diversity in technology may be used, for example, for level measurement differential pressure transmitter and displacer type may be deployed in redundant mode so that higher availability as well as due to diversity, CCF will be less. This is true, provided the difference in sensitivity and response time of two types of instruments are suitably taken care of. The key issue always is safety return on investment (ROI), hence balancing between safety and life cycle cost.

3.1.2 *Factors Influencing Performance*

The following issues are a few examples of factors that influence performance of sensors, therefore need proper attention:

- Transmitter performance: Mere overall accuracy and other details may not be sufficient. Designers need to take care of many other data, for example, there may be loss of performance due to orientation change — positioning in tilted condition. A few other issues affect the transmitter performance shall include but not limited to: Ambient temperature variations, high static pressure, drifts, and stability. Drifts may be due to voltage variations, noise pick up, etc.
- Process interfaces: When there is plugging in the line, correct pressure is not reaching the transmitter, so even if transmitter is okay, the actual reading may be erroneous. Similarly if there is coating, thermowell or magnetic flow meter could not sense, and again produces erroneous result. In modern smart transmitters, separation of the sensor and electronics is possible, so the sensor may be directly connected to the process, with the electronics located at suitable place, this could sense such changes to reduce the effect.
- Sensor robustness quality control: Corrosion can cause overt failure of a process transmitter. Sudden overpressure in differential pressure transmitter can cause zero drift. Similarly, instrumentation practice is important, for example, if technician ignores the important role of equalizing valve, then sudden pressure may come on either side of a differential pressure transmitter causing damage (when without opening equalizing valve, either side valve is closed or opened suddenly). The last issue is related to maintenance practices. To circumvent these issues, robust sensors may be deployed.

3.1.3 *Installation and Maintenance Issues*

In order to have easy access earlier, transmitters are installed with long impulse line. This is no longer valid as modern transmitters are less prone to drift or fail (actual critical failure of hardware is around 15% and software 34%) and these can be accessed from control room via fieldbus. As more problem may be with impulse line so old concept needs change.

The user needs to quantify the calibration frequency based on its application for getting better results. The user can minimize their risk of sub-optimal selection, installation, and maintenance by working with suppliers who can bring to bear substantial measurement expertise and experience, and strong local technical support [7].

3.2 Safety Instruments Documentation

Standard specification sheets detailing measuring range, performance parameter, material specification, etc. are available in standard books and or from manufacturers'/ engineers' documentation. Since the purpose of the book is safety applications, such standard documentations are not discussed here. If necessary, interested readers may go through [2]. Here, discussions shall be on various safety parameters for process transmitters. Naturally, during the discussions references will be made to standard process transmitter suppliers for data. Most of the parameters are extracted from renowned manufacturers. From the discussions it will be clear that in case of safety application, only authorized persons are allowed to handle the instruments, otherwise safety may be endangered. In this part, discussions will cover various symbols used, basic contents of the manuals, and various safety related parameters in brief.

3.2.1 *Symbol Descriptions and General Instructions*

There are a number of symbols (related to safety) used, and each of them has specific meaning or significance. Such commonly used symbols (related to safety) are listed in Table IX/3.2.1-1. Some of such indications may be as follows:

> **Warning**: "I read this manual before working with the product. For personal and system safety, and for optimum product performance, make sure you thoroughly understand the contents before installing, using, or maintaining this product." Or
> **Caution**: "Product described in this document is not for use in nuclear application."

Apart from the above symbols, at various places there will be following notifications in the manual. These are collected and reproduced here in brief; actual one may be more elaborative. Therefore, detailed instruction manual needs to be read and understood before handling the instruments even by authorized personnel.

- Authorized personnel: Only persons trained and having knowledge about the regulations are authorized to use and handle the instrument.
- Appropriate use: Each instrument is made for specific purpose and it should be used for the same. Operational reliability is ensured only if the instrument is properly used according to the specifications in the operating instructions manual as well as possible supplementary instructions.
- Incorrect use: There may be application specific hazard if used incorrectly.

Table IX/3.2.1-1 Symbols and Instructions for Instruments

Symbol	Significance
i	Helpful additional information: Tips, note:
	**Caution**: If ignored, there will be fault or malfunction. **Warning**: If ignored, injury to persons and/or serious damage to the instrument. **Danger**: If this warning is ignored, serious injury to persons and/or destruction of the instrument can result.
	Explosion applications.

- Safety Label: Safety markings and tips must be followed.
- General safety instruction: The high tech device demands for strict following of instruction and guidelines mentioned.
- Safety instruction for oxygen use.
- CE conformity.
- NAMUR recommendations.
- Detailed product description:
  - Configurations and commissioning tips
  - Enabling and disabling operations
  - Principles of operation
  - Standard application details
  - Standard installation details
  - Mounting details
  - Standard parts details
  - Transportation and storage
  - Connection details
  - Wiring details
  - Precautionary measures
  - Setup and adjustments
  - Set of program
  - Saving of adjusted parameters

3.2.2 *Functional Safety Details*

Under this clause, various functional safeties and their interpretations will be discussed. Details given here are *extracted* from various safety instrument manuals of process instrument manufacturers of international repute, so that readers get the feel of actual data in use in various plant instrumentation.

- Manuals usually mention the certification details in the front such as "additional instructions for IEC 61508-certified device.
- SIL details in the transmitter body. There are various ways and means to describe SIL number in the transmitter. Some mention the same as last letter and digits in the model number of the transmitters in the name plate, for example, IA transmitter: "*IGP10-T20D1F-M1**S2***;" last code "S2" indicates SIL option. Whereas in ABB transmitter identification of 266 (2600T series), pressure transmitter IEC 61508-certified are identified as "....The "*Output*" characteristic as per product datasheet is to be codified with digits *8 or T* ... Product Code with digit "8" (additional options required) and Product Code with digit "*T*" (no additional options)". Also, the significance of SIL number is mentioned in the manual, for example, *pressure transmitters certified according to IEC 61508 are suitable to be used in SIL2 applications as single channel, and in SIL3 applications with architecture* **1oo2**. It shows that there will be change in SIL application with changes in architecture, which was already discussed in previous chapters. This is true because for a single component, it is not correct to define a SIL level. SIL refers to the complete safety loop, and single device is designed to achieve the

desired SIL level in the entire safety loop. Naturally with variations *in architecture of the loop SIL requirements will* ***change***.

- Safety function: The instrument could be used in safety-critical applications to measure process pressure. If the process value is invalid due to an internal failure of instrument, the system is to go into safe/alarm state (NAMUR NE43 — taken from ABB 2600T) and the malfunction *must* be shown as warning message on the display (if applicable) and as a variation in the output.
- Miscellaneous other functional safety information in manual: The following are a few pertinent functional safety information found in instrument manual meant for safety:
  - Management of functional safety including safety planning focusing on:
    - Policies and strategies for safety achievement
    - Safety life cycle activities, with responsible personnel
    - Relevant procedures to life cycle phases
    - Audits and follow-up procedures
  - Safety information from plant owner:
    - Overall safety life cycle information
    - Applicable laws and standards
    - System safety requirements
    - I/O system response time
    - System structure
    - Safety requirement allocation
    - Safety routine

The total *system response time* is determined by sensor detection time, logic solver time, and actuator response time, and this total time must be **less** than *process safety time* to get the safety.

- Design verification details: Third party certification and inspection report for compliance with IEC 61508:2010.

The safety parameters listed in the safety manual, the SIF designer SIL achieved using the PFD_{avg} considering the architecture, proof test interval, proof test coverage, automatic internal diagnostic, repair time, and failure rates of the entire equipment are included in the SIF. The HFT must be checked and taken into consideration by the SIF designer to ensure that each subsystem within the SIF is in compliance with the *minimum* HFT requirements.

There are a few other functional safety details provided in the manual; these are:

- Installation:
  - Environmental limits
  - Mechanical installation and system completion
  - Application limits
- System wiring
- Commissioning overall system
- Write protection (external/internal)

- Factory settings
- Output current limits (*NAMUR 43 Standard—ABB 2600T*)
- Overload condition (*adjustable*)
  - Lower limit $\leq$3.8 mA
  - Upper limit $\geq$20.5 mA
- Alarm current (*adjustable*):
  - Lower set point <3.6 mA
  - Upper limit >21 mA
- Faults outside the functional safety
- Operation details:
- System operating discipline
- Maintenance
  - Preventive
  - Routine
  - Unit replacement
- Modification request
- MOC
  - Process
  - Procedure
  - Documentation
- Training requirements
- Proof testing
- Error messages

With these functional safety requirements in mind, now discussions shall be on various safety and reliability related parameters important for SIS applications.

3.2.3 *Safety and Reliability Data*

Some safety and reliability data from internationally reputed manufacturers have been presented to note the type of data and the range of data normally associated with safety sensors.

- Safety data: The following are some of the safety data:
  - Safety accuracy: 1–2%
  - Response Time: 750 ms (maximum)
  - Damping set to 0 s
  - Updates: Every 50 ms

  Notes associated with these data may be presented as warning with suitable symbol mentioned above.

  Note: **WARNING** to ensure safe fault monitoring, the DCS should be able to detect *fail high* and *fail low* conditions as malfunction indicators.
- Reliability data: Failure data in line with IEC 61508 has been presented in Tables IX/3.2.3-1–IX/3.2.3-3. Table VII/1.3-1 may be referred to for legends of failure symbols.

Table IX/3.2.3-1 Typical Failure Data-1 as per IEC 61508

Device	λ	λ_S	λ_{Dd}	λ_{Du}	SFF
Transmitter: XX ...	366 FIT	2150 FIT	1160 FIT	3680 FIT	90.06%

Table IX/3.2.3-2 Typical Failure Data-2 as per IEC 61508

Transmitter Type	Measurement Range	SFF	PFD_{avg}	$\lambda_{Dd} + \lambda_S$	λ_{Du}
XXX	10 mBar	75%	$8.54 * 10^{-4}$	614 FIT	195 FIT

Table IX/3.2.3-3 Typical Proof Test Results

$T_{[Proof]} = 1$ year	$T_{[Proof]} = 5$ years	$T_{[Proof]} = 10$ years
$PFD_{avg} = 8.54E\text{-}04$	$PFD_{avg} = 4.26E\text{-}03$	$PFD_{avg} = 8.5E\text{-}03$

There could be another type/way as shown below Table IX/3.2.3-2 (see Chapter VIII also).

PFD_{avg} may be specified in the manual, for example, 1.6E-3; proof test interval: 1 year (The PFH for single transmitter is 1.87E-7.) or PFD_{avg} may be presented in the certificate with associated proof test interval (see Chapter VII) as shown in Table IX/3.2.3-3.

Manufacturers also specify useful lifetime of components contributing to λ_{Du}; (viz. in IA transmitter of Invensys). Also suggested preventive maintenance details such as periodic proof testing procedure are included as a part of the manual.

Like sensor alarm annunciation also play a major role in safety instrumented system as protection layer. Now focus will be on alarm systems.

4.0 ALARM SYSTEMS IN SAFETY INSTRUMENTED SYSTEMS

In this clause, brief discussions on alarm and alarm management systems will be covered. Alarm system is an essential part of the operator interfaces to all types of industrial facilities. Alarm provides vital support to the operators by early warning about plant abnormalities and/or situations which need their attention so as to prevent, control, and mitigate many effects of abnormal situations. So as far as facility operations are concerned, it is personally felt that pre-trip alarm is of more significant for operator's action than the actual trip alarm. This is because of the fact that after pre-trip alarm, the operator can take certain measures to prevent tripping—hence downtime. However, in case of trip alarm, the operator is to take care of post-trip actions only, and downtime may not be saved. This will be clear from one BMS action: When there is high furnace pressure pre-trip alarm, the operator can regulate induced draft (ID) fan control or regulate the air/fuel flow to prevent boiler from tripping, whereas in case of very high furnace pressure boiler will trip, then operator needs to see post-trip actions are suitably taken but downtime could not be prevented. This suggests that for operating personnel pre-trip conditions need more attention even if the priority assignment may be lower.

What really is alarm management? Alarm management typically encompasses:

- Development of alarm philosophy documentation
- Definitions of nuisance alarms and standing alarms
- Definitions for alarm priorities based on impact and reaction time
- Define procedures for management of changes to the alarm system
- Alarm rationalization and implementation
- Creation of alarm response procedure
- Design of advanced alarm system, viz. dynamic alarming
- Establishment of benchmarking for alarm system performance
- Alarm system performance monitoring and assessment
- Auditing alarm system practices
- Training procedure for technical personnel
- Creation of reporting procedure and record keeping

In the preceding points, alarm life cycle is not mentioned but with introduction of various standards on alarm management, alarm life cycle has become part and parcel of it. Naturally, the same will be dealt in the latter part of the discussions. Before that, let us find out *what benefit does alarm management offer?*

Major benefits of alarm managements include, but are not limited to the following:

- Reduction in unplanned downtime
- Prevention of damage to equipment and system
- Reduction in avoidable of maintenance of equipment
- Increased productivity
- Production optimization and increased throughput
- Savings on SIL requirements
- Energy savings
- Reduction in penalty due to regulatory incidents

EEMUA 191, first published in 1999, has become the first globally accepted and leading guide to good practice for all aspects of alarm systems. To date, the latest edition of EEMAUA is 2013 edition which includes guidance for HMI management techniques, alarm configuration, and also include appendices covering alarm suppression; geographically distributed processes; intelligent fault detection; requirements for alarms; and supporting checklists. Starting from 2003, ISA also worked towards this to bring out a standard ISA 18.2:2009. Unlike other ISA standards, it is not really about hardware or software. ISA-18.2 is about the *work processes* of people. ISA-18.2 is a comprehensive standard developed per stringent methods based on openness, balancing of interests, due process, and consensus [8].

From ISA-18.2 one gets alarm definition as "*an audible and/or visible means of* ***indicating*** *to the* ***operator*** *an equipment malfunction, process deviation, or* ***abnormal condition requiring a response***." In the preceding definition, indication, operator abnormal condition, and operator's response need special attention. The same definition has been graphically represented in Fig. IX/4.0-1 for better understanding.

According to layer of protection analysis (LOPA), alarm is one of the independent protection layer (IPL) used. So in SIS importance of alarm system is quite obvious. In order to have brief but comprehensive discussions on alarm system entire discussions have been sub-divided into:

- Alarm objectives feature and performance indicator
- Alarm systems and LOPA implication
- Alarm life cycle
- Alarm philosophy
- General discussions on alarms

Typical alarm with high—low dead bans have been presented in Fig. IX/4.0-2. There are limits for alarm pertinent to a process parameter when the same is exceeded in both directions (for high and low) alarms are generated. However, there will be a dead band in each side when it is to be considered as normal.

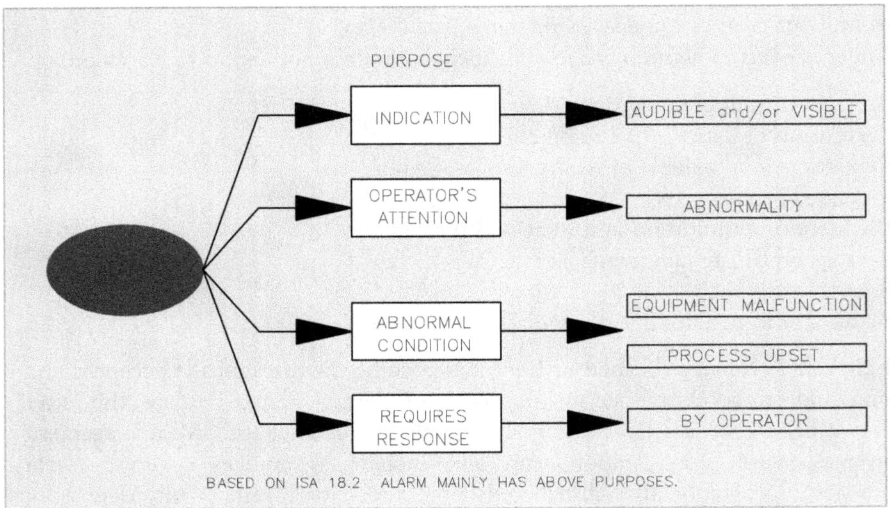

FIGURE IX/4.0-1

Alarm definition.

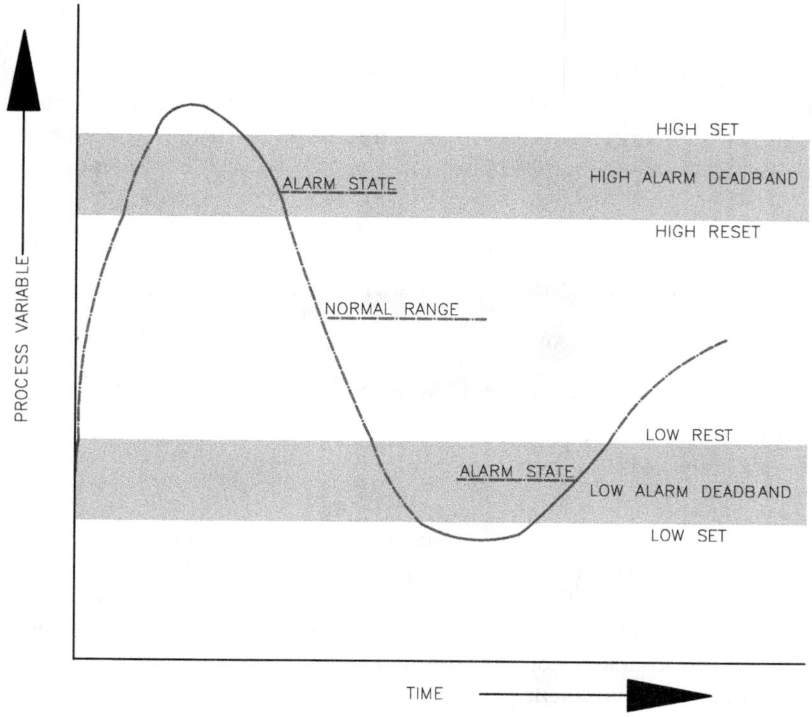

FIGURE IX/4.0-2

Typical process alarm with dead band.

4.1 Alarm Objectives Features and Performance Indicator

One understands the requirement and importance of an alarm but in real life applications very little attention given to operator response to alarms, or associated justification for the same. Also, to be on a safe design platform, many alarms are *created* with practically little or no engineering efforts. As a result of the same there is high risk for alarm *floods* and nuisance alarms. Even if a good operator may not be able to handle the situation! This obviously has consequential adverse effects on product quality, process efficiency, equipment protection, environmental incident, and personnel safety. A survey by *Automation World* revealed that 68% of respondents indicated that alarm overload negatively affects their ability to properly operate the process (*Taken from Alarm Management — 31st Annual European AIChE Seminar - April 23rd 2015*). In this connection, the definition of alarm discussed above as per ISA 18.2 is very important. It is quite often that a few alarms are used to bring to the attention of operator with some message, for example, "standby pump started" to keep operator informed that there is no system abnormality but as a standby pump has started operator needs to investigate and keep stand by utility. So, this is a for information message.

On the other hand, bearing temperature in pump "X" high is really an alarm as it demands the operator's attention and action in the form of (say) either increased lubrication or reduced load of the concerned pump. In the latter case, prompt action by the operator is necessary as this is an alarm. In this connection, Fig. IX/4.1-1 may be referenced.

ISSUES	OPERATOR MUST ACT	INFORMATION (operator)
ABNORMAL CONDITION	**ALARM**	**ALERT**
EXPECTED ACTION	**PROMPT**	MESSAGE
EXAMPLE	***Main Pump Bearing temp hi***	*Auto starting of standby pump*

FIGURE IX/4.1-1

Issues related to alarm and alert.

EEMUA 191 is only guidelines, meaning a set of guidance and recommendations whereas ISA 18.2 is a standard and later in 2014 November IEC developed an international standard IEC 62682 which has adapted majority part of ISA 18.2 including alarm life cycle (defining work process should be in place). Naturally regulatory bodies and insurance companies adapted this standard as "good engineering practices" for alarm management.

4.1.1 Objectives

For alarm objective development, one needs to carry out objective analysis to ascertain:

- Causes — expected/unexpected
- Consequence — direct and ultimate
- Corrective action
- Acknowledgement

When there is no ultimate consequence and does not require any corrective action then it is *not* an alarm which also conforms to the definition given above. However, when it is an alarm the there shall be some *objective* associated with the same. Basic objective of alarm system normally shall include safety, usability, performance monitoring, and return on investments.

- Safety aspect: Clear identification of safety to the people, plant equipment, and property and environment.
- Usability: It is an important issue in the sense that the alarm needs to indicate that it is:
  - Relevant for the user
  - Defined response for the alarm
  - Rate of appearance must commensurate with ability of the operator to handle

- Performance to be evaluated during design stage and commissioning period to ensure that the alarm is effective all operational conditions and stages of the facilities.
- Alarm engineering should be done in such high standard to achieve all these goals with full risk assessments to get proper return on investment (ROI).

As a corollary to the above following terms are very important for alarm design:

- Defined response
- Adequate time to respond
- Relevant, unique, and useful
- Well-defined and prioritized, and if applicable, grouped.

These objectives and various standards to be discussed is *valid* for process industries irrespective of whether facility is continuous or batch process because of the fact that alarm is a function of HMI and not of the process. Now it is better to look into the key features of various guidelines (e.g., EEMUA) and standards (e.g., ISA 18.2 and IEC 62682). Since almost 90% of ISA 18.2 has been adapted in IEC 62682, only one is discussed.

4.1.2 *Features and Performance Indicator for EEMUA*

Major issues here are as given in the following points, and could complement those in ISA 18.2:

- In textbook format, with worked examples
- Alarm philosophy and design principles
- Details about alarm design and various risk assessment approaches
- Alarm system specification
- Design implementation issues
- Design details about field sensor and displays
- Performance assessment and alarm management with scope for improvements
- Performance matrices and questionnaires
- Separate section for alerts
- Alarm suppression hazard study
- Risk assessment and priority determination
- Separate section for batch process

In case of EEMUA performance indicator comprises of the following factors:

- Average alarm rate
- Maximum alarm rate (following an upset)
- Percentage of time alarm rates outside acceptable target rate
- EEMUA also recommends that for a control system with >1000 configured alarms should have standing alarm <10 and shelved alarm (other than maintenance shelved alarms) <30.

What is standing alarm? And what are shelved alarms?

- *Standing Alarm*: An alarm which is in an active alarm state for a significant period of time, for example, 4–8 h [10] is referred to as standing alarm.
- *Stale alarm* (also see Table IX/4.1.2-1): When an alarm goes in to alarm state but never return to normal state within 24 h time. This is similar to standing alarm.
- *Shelved alarm* is an alarm resulted on account of alarm suppression technique/mechanism initiated by operator to temporarily suppress a set of alarms until an underlying problem can be corrected. Shelved alarms must be presented to the operator. (Such shelving should be done with due *authorization* for with detailed reasoning noted and displayed. A *review* of the same after change of shift *should* be undertaken). When equipment is under maintenance alarms associated with it may be shelved.

  Tables IX/4.1.2-2 and IX/4.1.2-3 show the long term average alarm rate and alarm rate following an emergency/upset (in line with IEEMU191 Guidelines).

Table IX/4.1.2-1 ISA 18.2 Performance Metrics

Metric	Target Value	
Annunciated alarms per time	Target value likely to be acceptable	Target value max. manageable
Per 10 min per operating position	~1	~2
% of hours containing >30 alarms	<1%	
% of 10 min containing >5 alarms	<1%	
Max no. of alarms in 10 min period	10 or less	
% of time alarm system in flood condition[a]	<1%	
% contribution on the top 10 most frequent alarms on alarm load	1–5% max with action plan to address	
Quantity of chattering[b] and fleeting[c] alarms	0 action plan if it occurs	
Stale alarm (see Clause 4.1.2)	<5% on any day action plans to address	
Priority distribution	3 priorities: ~80% lo, ~15% med, ~5% Hi or 4 priorities ~80% lo, ~15% medium, ~5% Hi <1% highest	
Unauthorized alarm suppression	0 outside controlled o approved methodology	
Improper alarm attribute change	0 outside approved methodology or MOC	

In chattering and fleeting alarms suitable time delay at input will give better results.
[a]*Alarm flooding condition: is that condition when 10 or more alarms are received by an operator within 10 min.*
[b]*Chattering alarm is on then off and on again in a short period of time (e.g., <1 min).*
[c]*Fleeting alarms turn on and off very quickly, but do not necessarily repeat-mainly occur due to contact chattering.*

Table IX/4.1.2-2 Average Alarm Rate Steady State Operation (IEEMU191 Guidelines)

Long Term Average Alarm Rate Steady State Operation	Alarm per Operator		Acceptability
	No. per Hour	No. per 10 min	
More than 1 per min	>60	>10	Very likely to be unacceptable
1 in 2 min	30	5	Likely to be over demanding
1 in 5 min	12	2	Manageable
Less than 1 per 10 min	<6	<1	Very likely to be acceptable

Table IX/4.1.2-3 Average Alarm Rate Following an Upset (IEEMU191 Guidelines)

No. of Alarm Displayed in 10 min Following a Major Plant Upset	Acceptability
More than 100	Definitely excessive and very likely abandonment of the system
20—100	Hard to cope with
Less than 10	Should be manageable, but if each requires complex action by operator then it is difficult

4.1.3 *Features and Performance Indicator of Standard (ISA 18.2)*

Major issues here are as given in the following points which could complement those in EEMUA 191:

- Major focus on an alarm system life cycle
- Similarity in life cycle development with IEC 61511
- IEC 62682 developed based on this standard
- Performance indicators are very clearly defined
- Section on compliance
- Detailed requirements for alarm philosophy document
- Alarm system requirements specification
- Alarm identification and rationalization
- Alarm and HMI design details with advanced methods
- Implementation, operation, and maintenance
- Monitoring and assessment
- Management of change handling and auditing
- Standard hence with lesser examples

  For performance metrics, Table IX/4.1.2-1 may be referenced.

4.2 Alarm Systems and Layer of Protection Analysis Implications

With introduction of IEC 62682, an international standard on alarm management, there will be focus on the role of alarm system as a key layer of protection to assure safe and effective asset operations. When an alarm is included in a LOPA, it is important to evaluate its risk reduction, in terms of the behavior of the alarm system as a whole for its effectiveness in guiding the operator to take the correct action. As discussed earlier in Clause 4.0 of Chapter VII, when design, management, and performance of the alarm system are not satisfactory, then the level of actual risk reduction will be less than expected. Alarm flooding/nuisance alarms are also very disturbing as for this many critical alarms may be missed. SIF and alarms systems when use as IPL has similar goal of risk reduction, they do have another similarity for example, priority of alarms could be considered *similar* to SIL of SIF! Like any SIF alarm systems also comprise sensor, logic solver, and final elements.

4.2.1 Operator Response to Alarm

Alarm system is more related with work process. In case of alarm there are two sets of logic solving *issues*, viz. electronic system which getting signal from sensor decides the alarm and alarm type. The other one is the human brain, which actually initiates action. So, here human intelligence is also acts as logic solver to initiate the action. In this connection. Fig. IX/4.2.1-1 may be referenced. When there an untoward even

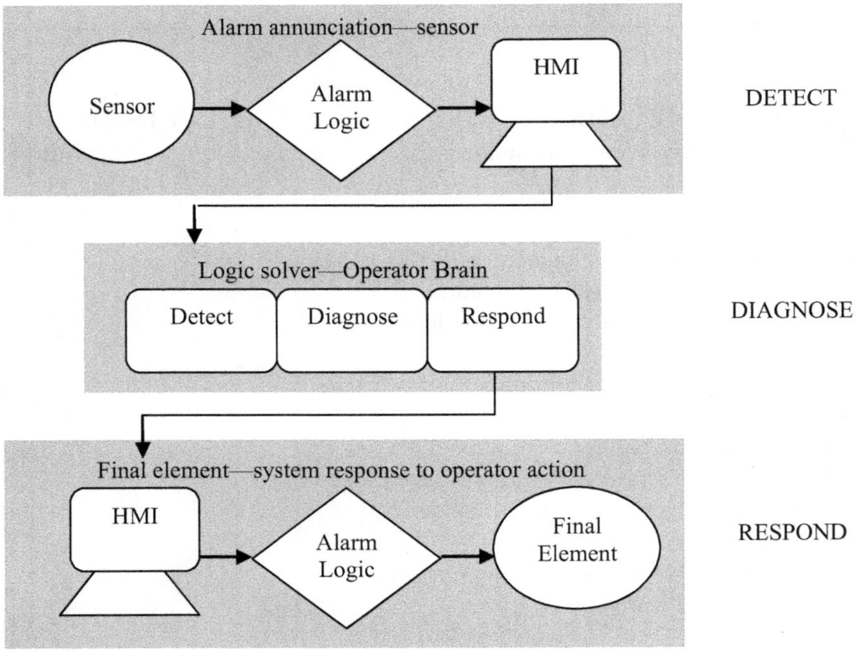

FIGURE IX/4.2.1-1

Model of operator response to alarm.

occurs and sensor senses the same, it is presented to the logic solver which draws the attention of the operator through display and sound in HMI present. As shown in the figure these three, viz. sensor, logic solver, and HMI actually detects the fault and present it to operator. All *three* in combination is responsible for alarm detection because:

- If any of sensor, alarm logic, or HMI fails alarm could not be generated
- If alarm is disabled, or wrong attributes present, alarm would not be generated
- Shelving, suppression of alarm by operator will also cause non-appearance of alarm

Operator detects the alarm at HMI whose display design and alarm management design are vital in getting the operator's detection. Now "logic solver – human intelligence" comes in to action, as operator is correctly to understand the issue and use operator's expertise or experience to quickly determine the consequence and initiate corrective action. This is the diagnosis of the problem.

Now the operator needs to react in terms of (say) opening/closing of a valve and/or starting stopping of pump, etc. However, the operator's response will be through HMI physical logic solver (say PLC/DCS) receives the operators command via HMI and takes necessary action as per logic via final element. All these are shown in Fig. IX/4.2.1-1 with the help of arrowheads. One thing from the preceding discussions, it is clear that alarm system is highly dependent on human factor which is extremely vulnerable under stress, etc. Major reasons for such stresses could be on account of:

- *Quick* decision-making under emergency
- Taking multiple decisions in *short* time
- Suffering from lack of/over *confidence*
- *Boredom* from stereotype of job
- Not well-defined or conflicting data or condition (*confusion*)
- Lack of *training* (especially in new technology).

4.2.2 Process Safe Time

Now with the help of Fig. IX/4.2.2-1, an attempt will be made to see how process safe time is determined. When an untoward incident occurs after a small gap of time, alarm appears with the help of sensor, alarm logic, and HMI. The small gap told here is sensor detection time and reaction time of alarm logic and HMI. After appearance of alarm in HMI there will be some time for operator's detection, diagnosis, and response as shown in Fig. IX/4.2.2-1. Now after operator responds actual thing to happen there will be reaction time, which comprises of process dead band and system reaction time. Sum total of operator action time and reaction time will be minimum necessary for any corrective action to happen. So, if hazardous event occurs after these time from $T = 0$, then system will be safe. Thus, process safe time will be as shown in Fig. IX/4.2.2-1.

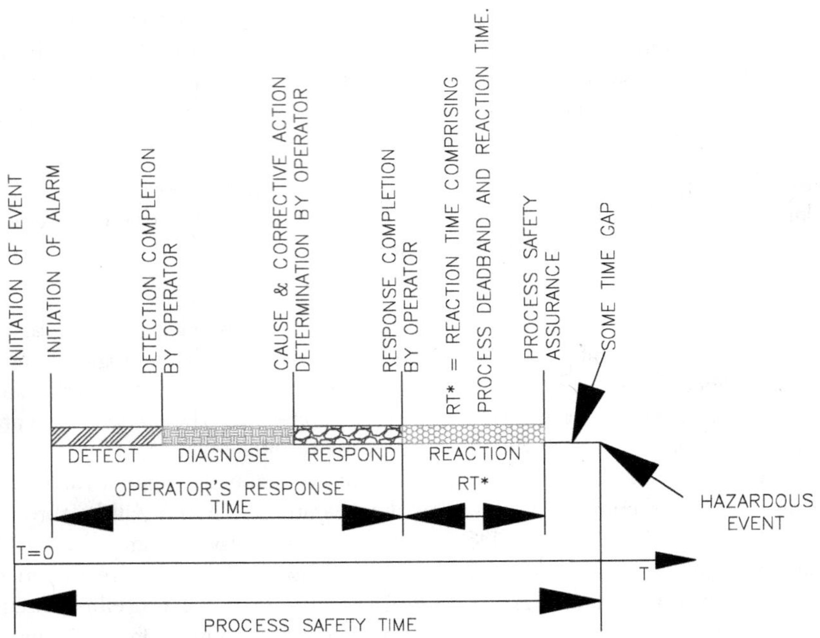

FIGURE IX/4.2.2-1

Process safe time determination.

4.2.3 Factors for Alarm Systems as Independent Protection Layer

When alarm system is considered as an IPL, the following check list needs to be fulfilled:

- The alarm must be a *valid* one.
- *Rationalization* of alarm system.
- Performance of alarm system is measured and proven to be *adequate*. "Specific," "dependable," "independent," and "auditable" are the main criteria for IPL and discussed in Chapter V. Now, the same needs to be applied for alarm system also.
- **Specific**: As in alarm system, is more of a work process of human so it should contain one *set* of cause and consequence associated with each alarm to avoid any confusion with operator.
- **Dependable**: For alarm system to be dependable, alarm response procedure shall be *well-defined*, alarms should be *rationalized* (Clause 4.3), needs to have good HMI design, MOC handling, and performance alarm systems must be *measureable* and checks are made to ensure this is *adequate*.
- Independent: To become an independent systems many use *separate* set of hardware/software for alarm system with *separate* HMI which may fall back in

case normal HMI fails. Use of *diagnostic* in the system to warn early, in case of any failure in alarm system could be considered also.

- Auditable: Testing, classification, and prioritization of alarm and *performance measurements* are major issues related to this point. Each alarm is checked against its becoming stale, standing shelved, etc. needs to be considered.

4.3 Alarm Life Cycle and Rationalization

Alarm management system discussed at the beginning of this main clause, encompasses a huge area in alarm engineering. Out of them, from day-to-day activities in the control room people frequently come across the following terms:

- Alarm overload: More alarm than manageable
- Alarm floods
- Nuisance alarms
- Chattering alarms
- Standing/stale alarms
- Redundant alarms
- Alarms which have no response
- Alarms with the wrong priority

A suitably designed alarm management system needs to handle all these successfully so as to yield better results. Actually alarm management concept mainly comes from standard ISA 18.2:2009 — now also as international standard IEC 62682: 2014. One of the milestone activities of the standards is alarm life cycle, which has been presented in Fig. IX/4.3-1.

4.3.1 *Alarm Life Cycle Discussions*

As shown in Fig. IX/4.3-1 there are several parts of alarm life cycle. This life cycle is similar like that in IEC 61511 also. With reference to Fig. IX/4.3-1, basically there are three major sections based on their usage. These are (1) Alarms comprising A, B, C, and I mainly for master alarm database and MOC; (2) D, E, F, and G for alarm suppression and dead band, etc. (3) H for alarm analytic and metrics.

Each of the blocks shown in the figure has been discussed below:

- Philosophy: Alarm life cycle starts with the development of an alarm philosophy. It is facility specific document (to follow specific guidelines of ISA 18.2) provides guidance for alarm management practices for the *facility*, in all life cycle stages. It includes definition of an alarm, the alarm handling capabilities of the control system, and other site specific details. The philosophy document establishes the processes for planning and documentation pertinent to other life cycle stages.
- Identification: Though not defined in ISA 18.2, yet this is stage involves review of documents (e.g., P&ID reviews, PHA report, etc.) to ensure that the results are useful for *rationalization*. Also this database is helpful for documenting cause, consequence, and the time to respond for each identified alarm. For existing system existing alarms will be *included*.

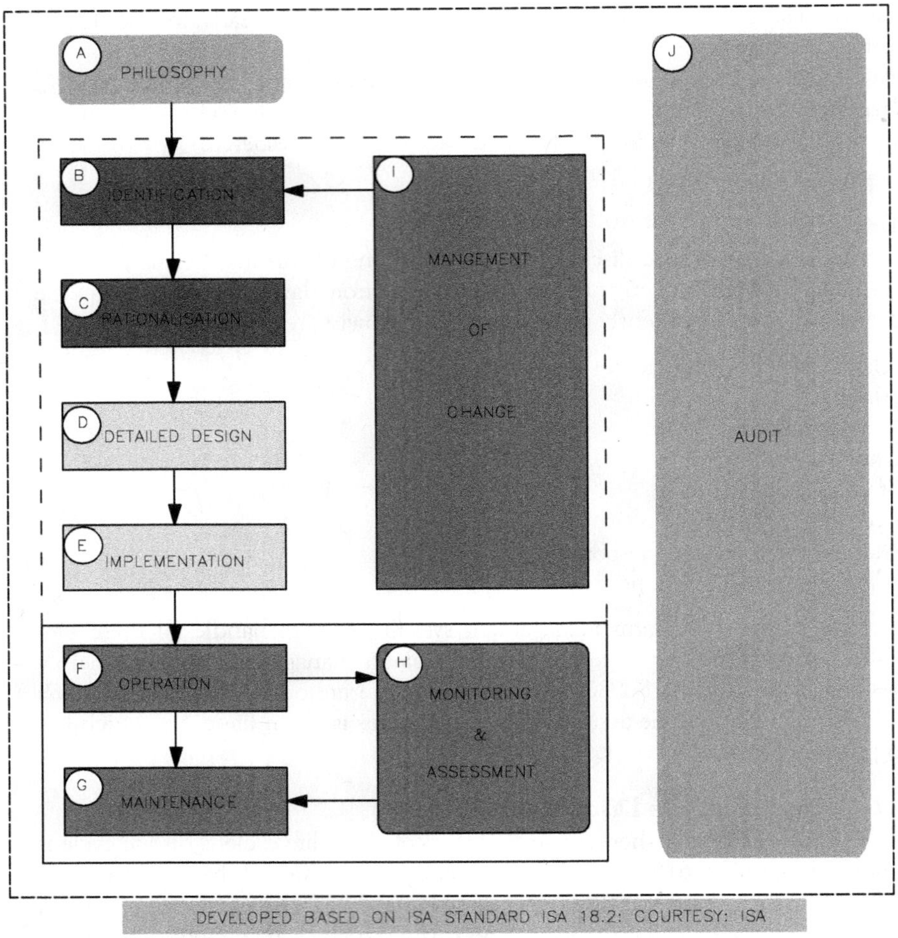

FIGURE IX/4.3-1

Alarm management life cycle.

- Rationalization stage: Each potential alarm is tested against the criteria in alarm philosophy to justify that it meets the *requirements* of being an alarm. The consequence, response time, and operator action are documented. Alarms are analyzed to define their attributes, for example, set point, priority (based on severity and time to respond), type, and classification. A typical rationalization scheme has been shown in Fig. IX/4.3.1-1. Other associated issues are training, testing, documentation, or data retention. The results of the rationalization are documented in a master alarm database which can be used as auditable demonstration of good engineering practices, alarm consistency, etc. [9].

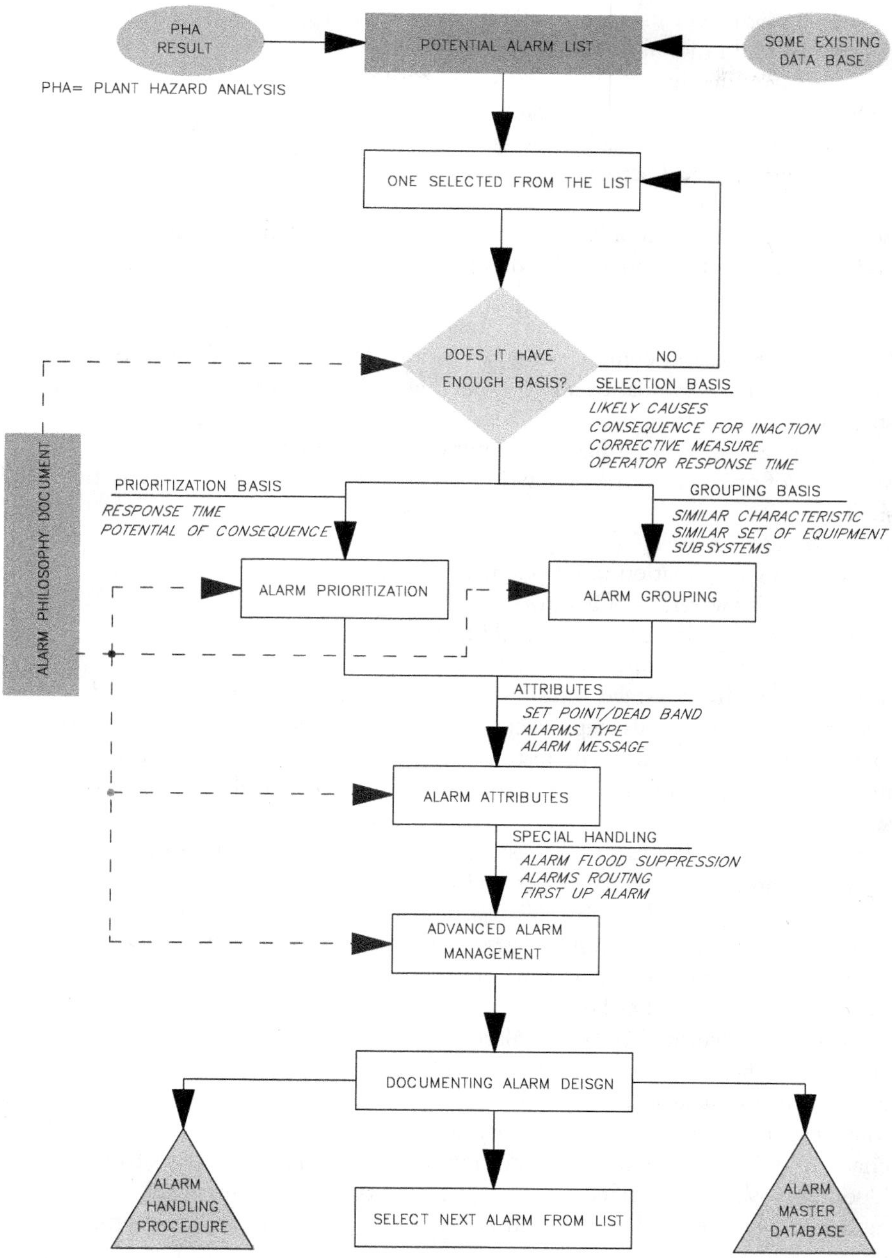

FIGURE IX/4.3.1-1

Alarm rationalization.

- Detailed design: At this stage an alarm needs to cater to the requirements in philosophy and rationalization documents for:
    - Basic alarm design
    - Setting up of attributes (set point, dead band, etc.)
    - Conducive operator graphics design
    - Elimination of bad actors
    - Advanced alarm design

  So, this stage is for accurate translation of requirements in to an alarm. Poor design or incomplete design will end up with problematic issues for alarm management.
- Implementation: In this stage, the alarms are put into operation with operator training, testing, and commissioning. Testing and training will be applicable for new systems, new alarms addition, and any change in process design. This stage also includes activation of alarm designs in the running system.
- Operation: At operation stage an alarm performs its function of drawing attention of operator to the presence of an abnormal situation. Key activities shall include:
    - Alarm management by system tools
    - Accessing information of rationalization stage
    - Operator's awareness of alarm
    - Avoidance of alarm before it could happen
    - Ensuring expected operator action
- Maintenance: In this stage the alarm does not perform its functions needing operator's action. In this stage, recommended procedures to remove an alarm from service and return an alarm to service, etc. are done. A major issue here may be ensuring visibility accountability for restored critical alarm to active service, etc.
- Monitoring and assessment: This stage is meant to gather data from the operation and maintenance stages and comparing alarm system performance with performance *goals* in the *philosophy* document. In order to respond effectively, an operator should be presented with no more than one to two alarms every 10 minutes. So, the major key issue is the rate alarms are presented to the operator. Another key activity at this stage is to identify "nuisance," "chattering," "fleeting," or "stale" alarms. So at this stage suitable analytics and facility specific metrics are tailored to analyze raw alarms. This is used for key performance indicator and reporting.
- Management of change: Major activity of the stage involves authorization for all changes to the alarm system, viz. addition/deletion of alarms, changes to alarms. Upon approval the modified alarm is processed through the stages of rationalization, detailed design, and implementation again and documentation updating is necessary.
- Audit: The audit stage is mainly meant for periodic review of the work processes and performance of the alarm system with a goal is to maintain the integrity of the alarm system throughout its lifecycle and to identify areas of improvement. Audit process may call for updating philosophy document to reflect any changes carried out during audit.

4.3.2 *Alarm Rationalization*

From the discussions so far it is clear that in order to an actual alarm there is need for alarm rationalization. Detailed process is shown in Fig. IX/4.3.1-1.

- The process starts with review of potential alarm against criteria set in the philosophy document to make sure that these are necessary and appears in master alarm data base (MADB). For this review of various documents and PHA reports are helpful.
- From the data base, one alarm is selected and tested for validity and basis test along with objective analysis. There is difference between alert with alarm (see Fig. IX/4.1-1), according to the standard definition of alarm (ISA 18.2). So conditions are analyzed such as abnormal condition, requiring operator action, etc. Other major factors here are consequence, corrective action, confirmation, and operator response time.
- Based on the response time and potential consequence the selected alarm is prioritized.
- Based on facility, alarms may be grouped as per location, functionality etc.
- In the next stages the attributes are assigned and if there is any special handling same is assigned. Until this stage all activities are performed as per guideline available in alarm philosophy document developed for facility specific.
- Entire alarm design is then documented including alarm response procedure, MADB.
- Next another alarm is chosen and same procedure is repeated.

4.4 Alarm Philosophy

According to ISA 18.2: 2009 (as well as for IEC 62682:2014), alarm philosophy document is one of the most important documentation for alarm design for any facility. Alarm philosophy document is facility specific document. From ISA 18.2 one gets the basic content/topic to be covered in philosophy document. Table IX/4.4-1 accounts for such requirements as per ISA 18.2.

From the table, it is quite clear how standard itself guides philosophy documents because it is very important for further system design on alarm systems.

4.4.1 *Basic Principles and Purposes*

Alarm philosophy encompasses basic definitions, key work practices, and various processes as well as the alarm system performance goals and details about the, roles, and responsibilities. This facility *specific* document provides guidance for a consistent approach to alarm management and defined activities of the alarm management life cycle. Major positive issues (see Fig. IX/4.4.2-1 and Table IX/4.4-1), addressed by alarm philosophy shall include but are not limited to the following:

- Optimally designed alarm system to:
  - Cater to user needs
  - Meet ergonomic requirement
  - Define relevant user's role
  - Understand easily

Table IX/4.4-1 ***Requirements of Alarm Philosophy Contents (ISA 18.2/IEC62682)***

Content	Required	Recommended
Purpose of alarm system	6.2.2	
Definitions	6.2.3	
References	6.2.4	
Roles and responsibilities for alarm management	6.2.5	
Alarm design principles	6.2.6	
Rationalization	6.2.7	
Alarm class definition	6.2.8	
Highly managed alarms (or site equivalent)		6.2.9
HMI design guidance	6.2.10	
Prioritization method	6.2.11	
Alarm set point determination		6.2.12
Alarm system performance monitoring	6.2.13	
Alarm system maintenance	6.2.14	
Testing of alarms	6.2.15	
Approved advanced alarm management techniques		6.2.16
Alarm documentation	6.2.17	
Implementation guidance	6.2.18	
Management of change	6.2.19	
Training	6.2.20	
Alarm history preservation	6.2.21	
Related site procedures		6.2.22
Special alarm design considerations		6.2.23
Alarm system audit	6.2.24	

Associated ***sub-clause number*** *has been given in applicable* ***required/recommended*** *col^{n}.*

- Alarm design consistency
- Risk management goals/objectives
- Good engineering practices
- Effective alarm rationalization
- Efficient operator response to alarms
- Prioritization of alarms
- Identification, justification, and classification of alarms
- Alarm conditions and attribute specification
- Design of alarms
- Implementation and testing of alarms
- Commissioning of alarms
- Training, maintenance, testing
- Operation and use of alarms

- Maintenance and testing of alarms during service
- MOC for alarm systems
- Performance monitoring and assessment of alarm system
- Auditing of alarm system performance

4.4.2 Characteristics of Good Alarm System

A good alarm system is characterized by the details shown in Fig. IX/4.4.2-1.

4.4.3 Operator's Role

An operator has a range of different activities from plant operation, fault identification, coordination of maintenance, etc. to name a few. Again such tasks depend on plant *state*, such as normal condition, start up, upset/emergency shutdown, or controlled shutdown. When there will be disturbance in near vicinity of target control, it is control system which acts automatically to mitigate disturbances to keep the plant close to target operating conditions. However, in case significant disturbances, plant may go beyond operating range of the control system and at this time alarm system has great role to play in bringing attention of the operator for operator's intervention. This happens usually before complete upset of the plant so that operator has some time to respond. However when timely action is not taken finally disturbance may call for ESD.

4.4.4 Alarm Design Principles

The basic objective of alarm system is to direct the operator's attention to abnormality in plant conditions requiring timely assessment and guide operator in taking action.

Relevant	**Not spurious or of low operational value**
Unique	**No duplication**
Timely	**Not too long before response needed, not too late to do any thing**
Prioritize	**Indicates the importance the operator to deal with the problem**
Understandable	**Clear and easy to understand message**
Diagnostic	**Identifying the problem that has happened**
Advisory	**Indicative of the action to be taken**
Focusing	**Drawing attention to the most important issues**

FIGURE IX 4.4.2-1

Characteristics of good alarm system.

Keeping issues discussed in Clause 4.4.2 in mind, the following points could be helpful design [11] tips to achieve the objectives:

- Context sensitive
- Functional aspects of alarm:
  - Alert
  - Apprise
  - Action guiding
  - Relevant
  - Useful
- Clear consequential effect of abnormality
- Adequate time to respond
- Rationalized alarm
- Alarm developed as per guideline
- Attention towards CCF especially for sensors actuators
- Proper operator training to interpret
- Periodic performance monitoring measures
- Defined and authorized MOC
- Issues related to redundant equipment
- Alert for individual component failure
- More attention for nos. of equipment necessary

4.4.5 *Alarm Human Machine Interface*

The major issues on HMI for Alarm Systems, which is mainly derived from ISA 101 (standard on HMI) shall include but are not limited to the following:

- Alarm states, priorities, and types
- Acknowledgement and silencing
- Alarm sound system
- Alarm information and message
- Alarm suppression; shelving, out of service
- Summary display functionality
- Miscellaneous alarm-related functionality

Many functionality items are mandatory and some are recommended. There could be varieties of displays suitable for specific plant applications. Some of the typical alarm displays could include but not limited to the following:

- Alarm tree view: The process alarms in a hierarchical manner for operator to browse the alarm status configuration etc.
- Active alarm list: Triggered but not cleared alarms in tabular form
- Acknowledged alarm list:
- Message history

4.5 General Discussions on Alarm Systems

In this part, discussions will be on miscellaneous issues related to alarm systems. Here, one thing to be noted that various types and descriptions of alarm systems are

Table IX /4.5.1-1 Dead-band and Delay Timing for Alarms (ISA 18.2)

Parameter	Dead Band (% of Parameter Range)	Delay (On/Off) (in Seconds)
Pressure	2	15
Temperature	1	60
Flow	5	15
Level	5	60

out of scope of this book. For this any standard book (e.g., Chapter VII of [2]) or ISA standard may be referred to. A few pertinent issues are as listed in the following.

4.5.1 *Dead Band and Delay*

To avoid nuisance alarms etc. it is customary to use some dead-band and On/Off delays. Process experts judgments are used to fix these values however some guideline for four commonly encountered parameters could be as in Table IX/ 4.5.1-1.

4.5.2 *Alarm State Indications*

Table IX/4.5.2-1 shows typical alarm state indications in line with ISA 18.2.

4.5.3 *Alarm Suppression*

At times it is necessary to suppress some alarms. There are three categories of them; they are:

- Shelving: This alarm suppression technique/mechanism used by an operator to manually suppress an alarm temporarily during operation process for a definite

Table IX/4.5.2-1 Alarm State Indication (ISA 18.2)

Alarm State	Audible Indication	Visual Indication		
		Color	Symbol	Blinking
Normal	No	No	No	No
Unacknowledged (new) alarm	Yes	Yes	Yes	Yes
Acknowledged alarm	No	Yes	Yes	No
Return to normal state Ind.	No	Optional	Optional	Optional
Unacknowledged latched alarm	Yes	Yes	Yes	Yes
Acknowledged latched alarm	No	Yes	Yes	No
Shelved alarm	No	Optional	Optional	No
Designed suppression alarm	No	Optional	Optional	No
Out of service alarm	No	Optional	Optional	No

period of time for operational facility (e.g., low flow alarm when associated pump is not in operation). Shelving of alarm can be done for short-term maintenance purpose also. Visibility and accountability of such alarms is very important for safe operation of the plant.

- Suppression by design: Any mechanism within alarm system preventing an alarm from an operator's visibility based on plant state or other condition. This is a part of advanced design.
- Out of service: The state of alarm during which the indication is suppressed manually one reason could be long-term maintenance/shut down of associated.

4.5.4 *Some Other Alarm Types and Techniques*

There are a few other alarm types which are used as a part of alarm management. Some of them are listed as follows:

- Dynamic alarming: Dynamic alarming techniques are used for elimination of alarm floods by automatic suppression of redundant and consequential alarms resulting from anticipated equipment malfunction or process abnormality.
- First-out alarm: First-out alarming permits the first occurring alarm in a related group of alarms to annunciate and *may* suppress the rest. Here the word "may" has been intentionally used, because in certain plants rest alarm may stay but blinking is stopped after acknowledgement, for example, High furnace pressure caused boiler trip is a first-out alarm, then after acknowledgement all alarms are displayed as steady display only first issue blinks until first up acknowledgement push button is pressed.
- Conditional alarming techniques help in eliminating, for example, low pressure alarm will not be active until an upstream pump has been on for sufficient time to generate the expected pressure and to prevent the low pressure alarm from activating each time the pump stops.

  With this, the discussion on alarm system is concluded, and next to concentrate on logic solver issues from hardware and software points of view.

5.0 SAFETY INSTRUMENTED SYSTEM LOGIC SOLVER

It is a well-known fact that in order to achieve plant safety, it is necessary that in SIS loop all the three basic components, viz. sensor, logic solver, and final element must be reliable. Although from the reports it has been found that 8–10% of failures are due to logic solver, yet proper care is taken to see that logic solver is all the time available. Even though failure rate of logic solver is lower, yet it is important because in SIS loop logic solver actually takes the corrective actions. In order to achieve the combination of safety and fault tolerance, a reliable SIS is required, which can bring the plant to a safe state when necessary. Until recently, the thought of a safety system brings in mind the images of TMR systems, which obviously would demand heavy capital expenditures. Now, manufacturers offer wide range of safety certified devices that can be integrated into very cost-effective solutions. Comprehensive range of safety instrumented plant automation products and services provides a comprehensive range of products and services for fail-safe and fault tolerant applications in the process industry and in factory automation. Good logic solver in safety application may be characterized by following factors:

- Engineering
- Communication
- Diagnostic
- Safety
- Security
- Robustness

5.0.1 Factors for Selection of Safe Logic Solver

When thinking of logic solvers in safety applications, an obvious choice becomes use of a safe PLC. However, for smaller and simpler cases, there could be separate loop for each SIF and safety and process controls are separate. In bigger and/or more complex facilities, for example, petrochemical plants or large power plants, such an approach may not really be feasible. In those cases, obviously people may need to go for safe PLC/DCS in safety applications. (PLC and DCS are two separate controls, meant for different applications, yet in the modern days, the difference between them has been highly narrowed down. *Chapter VII* of *[2]* deals with the issue with reference to industrial plant application in great detail.) For very large and complex facilities integrated approach could be a better solution. In many cases, it has been found that not only these are integrated but they are integrated with IT systems also. However, in such cases, there are always security threats from malware or unauthorized access which can give rise to many issues such as espionage and stealing of password, sabotage, etc. These have been discussed separately in Chapter XI.

- Single loop approach: At times, in smaller and less complex systems, a discrete logic device for each loop could be a better solution, as it avoids the complications and expense of a programmable solution. One of the major necessities for functional safety is to engineer the protection layers with minimum complexity (due to safety-related functionality). The first step toward this is to design the overall concept for the minimum number of safety instrumented loops, avoiding the unnecessary use of more complex technology and reducing interdependency between loops and keeping safety and non-safety functionality separate. Such design avoids higher burden on the architectural design. It is always better to consider straightforward approach avoids the development cost of application programming (plus software maintenance, upgrades, configuration management, and back-ups) and the need for specialist competence in operation and maintenance of the programmable platform. Installation, validation, and commissioning of complex programmable systems also require specific competence and procedures, which can make the functional safety management (FSM) system more onerous to set up and maintain [13]. However, this solution may not be valid in cases of complex plants.

- Safe PLC approach: Safe PLC brings the benefit of conventional PLC systems to complex safety application, replacing earlier hardwired relay based systems to bring automated process to safe state. This safe PLC may or may not reside in a single controller chassis. However, such safe PLCs provide flexibility in programming as well as familiar easy to use environment for the programming personnel. Summarily, any safety designed around standard PLC (or DCS) will cost more time, engineering efforts, I/O HW, wiring support etc. Safe PLCs have redundant processors (may be dual cross wired), flash memory, RAM etc., with continuous monitoring watch dog and detection systems. Inputs may be checked with high and low signal for short duration for functional test verification. Also outputs may be connected through safety switches with test facility. In this approach, BPCS and SIS use different hardware and separate engineering tools. The two systems are connected by a gateway for data exchange as shown in Fig. IX/5.0.1-1.
- Integrated approach: At a plant location, an owner needs to meet the regulations to ensure safety to personnel, the environment, as well as to protect the facility. Also owner needs to meet the life cycle activities and thoroughly document the same. In addition to these, maximum availability of the system to be ensured. If for the timing, controversial issue of BPCS—SIS integration is kept aside, then naturally to meet all these needs, a separate or non-integrated approach for safety system alone may not be cost-effective not only for two separate sets but also for additional cost due to safe PLC as discussed above. Also fieldbus systems made system integration much easier, so for complex systems integrated approach is taken and it found out to be cost-effective also. Through judicious selection of the system hardware, software, integration policy, and proven track record, it is possible to develop a system without compromising system safety. As owner/end-user will be finally responsible for facility operation, it is up to the judgment of the end-user to finally take the decision on the same. In the following sub-clause, a few points for making such judgment have been presented.

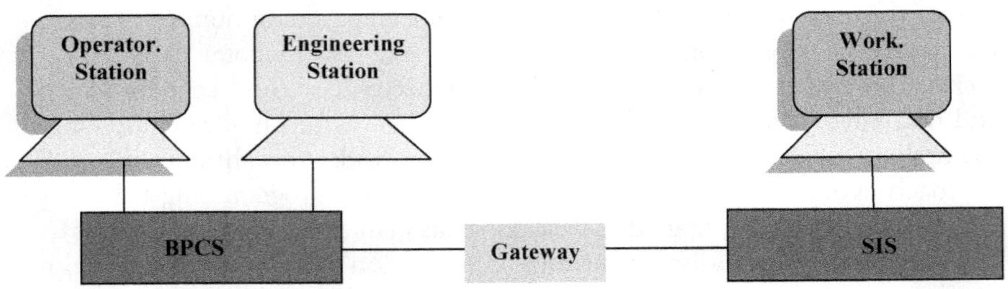

FIGURE IX/5.0.1-1

BPCS-SIS interface approach.

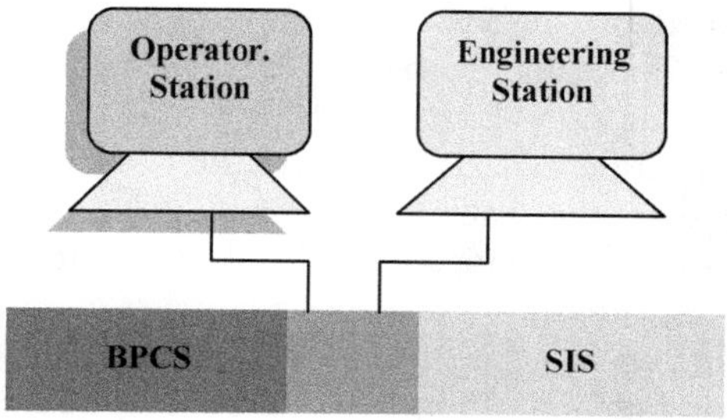

FIGURE IX/5.0.1-2

Common BPCS—SIS.

Here there are two approaches; integrated and common.

- Integrated: BPCS—SIS are based on different sets of hardware, but have a uniform communication system and use a common engineering tool as shown in *Fig. VII/2.2.4-1*.
- Common: BPCS and SIS are combined in the process control system, using common hardware (controller, fieldbus, I/O). However, standard and safety-related programs are executed in parallel and ***independent*** of each other. A typical configuration for this has been depicted in Fig. IX/5.0.1-2.

5.0.2 *Separation or Integration of BPCS With SIS—Judgment*

As part of natural discussions, a question (already discussed in Clause 2.2 of Chapter VII) of whether the two systems, viz. BPCS and SIS are separate independent or if there are possibilities for integrating them. Frankly speaking, there is no clear advice/directive available from latest versions of IEC 61508/61511 or ISA 84. In the first edition of ISA 84, clear separation was mentioned, but in later revision, such directive is not clear. Clearly there are two schools of thought, and manufacturers may advocate for integration but it is better for end-user to utilize own judgment in deciding the same. This is because end-user is ultimately responsible.

- Advantages of separate SIS and BPCS: Major positive points in favor of separate SIS and BPCS shall include, but are not limited to the following:
  - Independent failure: Less chance for simultaneous failure.
  - Security: Prevents changes in BPCS to cause changes in SIS – here in modern integrated systems they are kept separate even though same platform is used.
  - Different requirements for controllers: SIS is designed to fail in a safe way, whereas BPCS controllers the main focus on system availability.
  - SIS has extensive diagnostics, special software error checking, protected data storage, and fault tolerance. In BPCS, the same may not be required with that extensiveness.

- Advantages of integrated system: Some advantages of integrated system have been discussed already. These mainly shall include, but are not limited to the following:
  - One common engineering tools and approach for BPCS and SIS.
  - Lesser engineering effort for common systems.
  - Lesser chance of error when compared with two different systems.
  - Common controller platform.
  - Common data mapping.
  - Direct and seamless communication between SIS and BPCS.
  - Automatic integration of safety-related alarm and messages with time stamping.
  - Possible flexible modular redundancy approach.
  - Redundancy in I/O is independent of that in CPUs.
  - On account of cost-effective design higher redundancy and fault tolerant design can be utilized.
  - Effective higher security and access protection possible for common platform.
  - Based on SIF requirement, redundancy is selected.
  - Safety is not a function of redundancy.
  - Utilization of safe fieldbus system.
  - Reduction in space, hardware and wiring, assembly, installation engineering, and overheads hence significant cost savings for the complete life cycle of the plant.
  - Possibility for increased flexibility and fault tolerance.
  - Cost-effective and financial benefit.

5.0.3 *Logic Solver Type Variations With Safety Integrity Level*

The standard also includes fault tolerant tables to show what configurations of logic solvers and field devices are suitable for different integrity levels. See Table IX/5.0.3-1.

Table IX/5.0.3-1 Technology and Configurations to Meet Safety Integrity Level [15]

SIL Level	Sensors	Logic Solver	Final Element
SIL 1	Switches, transmitters	Relay, safety solid state, PLCs, safety PLCs	Standard valves
SIL 2	Fault tolerant sensors, transmitters with comparison, safe transmitters	Relay, safety solid state, and safety PLCs	Fault tolerant valves, simplex valves with partial stroking
SIL 3	Fault tolerant transmitters	Relay, safety solid state, and redundant safety PLCs	Fault tolerant valves with partial stroking

5.1 Technical Details of Logic Solvers

Studies have shown (e.g., studies by Aberdeen Group) that safety and productivity go hand in hand. Three key elements found in best-in-class plants shall include mainly:

- Safety Culture
- Appropriate risk management strategy
- Improved technologies for safety and productivity

Keeping these points in mind, it is the time to look into the details of logic solvers. In view of extensive use of PLC as logic solver in majority of the plants, our discussion will be focused on the same but may be applicable for DCS configurations **also**. During discussions whenever controllers are referred to it means PLC controller and associated I/O modules. *Discussion presented here is based on real-life practical systems from internationally renowned manufacturers. In order to limit the volume of the book, short discussions will be put forward, so if necessary the reader may refer standard books on PLC for intricate details. Book [2] includes details on PLC.*

5.1.1 Controller Requirements

Safety-related controllers in conjunction with safety or fail safe I/O modules are used for critical and hazardous applications where an incident can result in danger to persons, and/or damage to plant and environment. These safety-related controllers can work with the safety-related distributed I/O system (may be with internal verification for input or output via safety switches as described in Clause 5.0.1—safe PLC approach), or directly with fail-safe transmitters connected via the fieldbus. These controllers are supposed to detect faults both in the process and their own internal (self-diagnosis) to the system. It is the duty of the same to automatically set the plant to a safe state in the event of a fault. These controllers need to work in multitasking environment – may be in a mix of standard BPCS or safety-related applications, if integrated operation is permitted by the end-user. The programs of BPCS and SIS must be functionally separate, so that faults in BPCS applications have no effect on safety-related applications and vice versa. Special tasks with very short response times can also be implemented [14]. For safety applications controllers and I/O modules need to individually certified by third party and to comply SIL 2/SIL 3 (as the case may be – SIL 4 only for nuclear application) as per IEC 61508. For safety-related applications a few restrictions are followed such as:

- Mutual influencing is prevented
- BPCS program ***must*** be separated
- Data exchange is done by *special conversion* blocks

  Also, in certain systems, the safety functions are executed twice in different processor sections of one CPU through redundant, multi-channel command processing. Potential errors are detected by the system during the subsequent comparison of results [14]. When safety programs are carried out in different controllers of a plant, they can also carry out safety-related communication with each other over the plant communication bus – industrial Ethernet as discussed in Chapter VII.

5.1.2 *Redundancy*

In order to increase availability redundant subsystems may be deployed. The redundant controllers may work in 1-out-of-2 principle (or dual cross wired) comprise two subsystems of identical design. They could be kept electrically isolated from one another, and are synchronized over fiber-optic cables. In the event of a fault, there should be a bumpless transfer from the active subsystem to the backup subsystem. Though it is possible to keep two subsystems but in the same rack, or spatially it is better to keep them apart so that in case of fire both systems may not be lost.

5.1.3 *Integrated BPCS and SIS Configurations*

As discussed many times earlier, there two kinds of configurations possible; one integrated system and separate system. In this sub-clause, integrated systems have been discussed. Within integrated type, there could be redundant and simplex systems in terms of controllers. In case of simplex system, there would single controller (normally avoided in safe PLC). Redundant configurations of controllers and I/Os have been shown in Fig. IX/5.1.3-1.

In this configuration, two controllers are in 1oo2 mode. Also there are three categories of I/Os; one simplex, viz. direct connections for LS, and SV and CV via fieldbus. Dual connections of inputs for LTs are shown via two different fieldbus systems Whereas triple redundancy of PTs are via two fieldbus systems and one direct connection. Thus in the I/O subsystems of both the controllers receive these inputs for processing. These two controllers are also in redundant mode. I/O redundancy can be achieved via direct connections or through different fieldbus systems residing at two different subsystems. Other important issues here to note are in I/O sub systems there is a mix of safety as well as standard functions. However, two controllers are isolated as discussed earlier. Referring to Fig. IX/5.1.3-2, one may note that in case of SIL3 application, there is need to use isolation module for isolating bus module having various types of devices in different usages. Also, in case of mixing standard modules with safety modules, isolation modules are used (this is based on *Siemens ET 200M* guide line for information only).

5.1.4 *Separate and Independent SIS*

Dedicated SIS separate from BPCS have been presented in Fig. IX/5.1.4-1. In this approach, even several other functions such as FGS, ESD, and other SIS can be kept independent from each other. Independent SIS functions are (for example marked as set "a" and "b"). These SIS functions can exchange data with the help of Ethernet switch, Network bridge, and group Ethernet switches as shown by dashed double arrow heads (i.e., data exchange between SIS a and SIS b). Here (it is assumed that) BPCS is connected to SIS via Ethernet. The data exchange with the industrial Ethernet (hence BPCS) takes place with the help of Ethernet switch and corresponding coupling controller as marked by black dotted double arrowheads.

There are a few benefits of this type of system, listed as follows:

- Dedicated safety system separate from BPCS
- Data exchange possible with BPCS

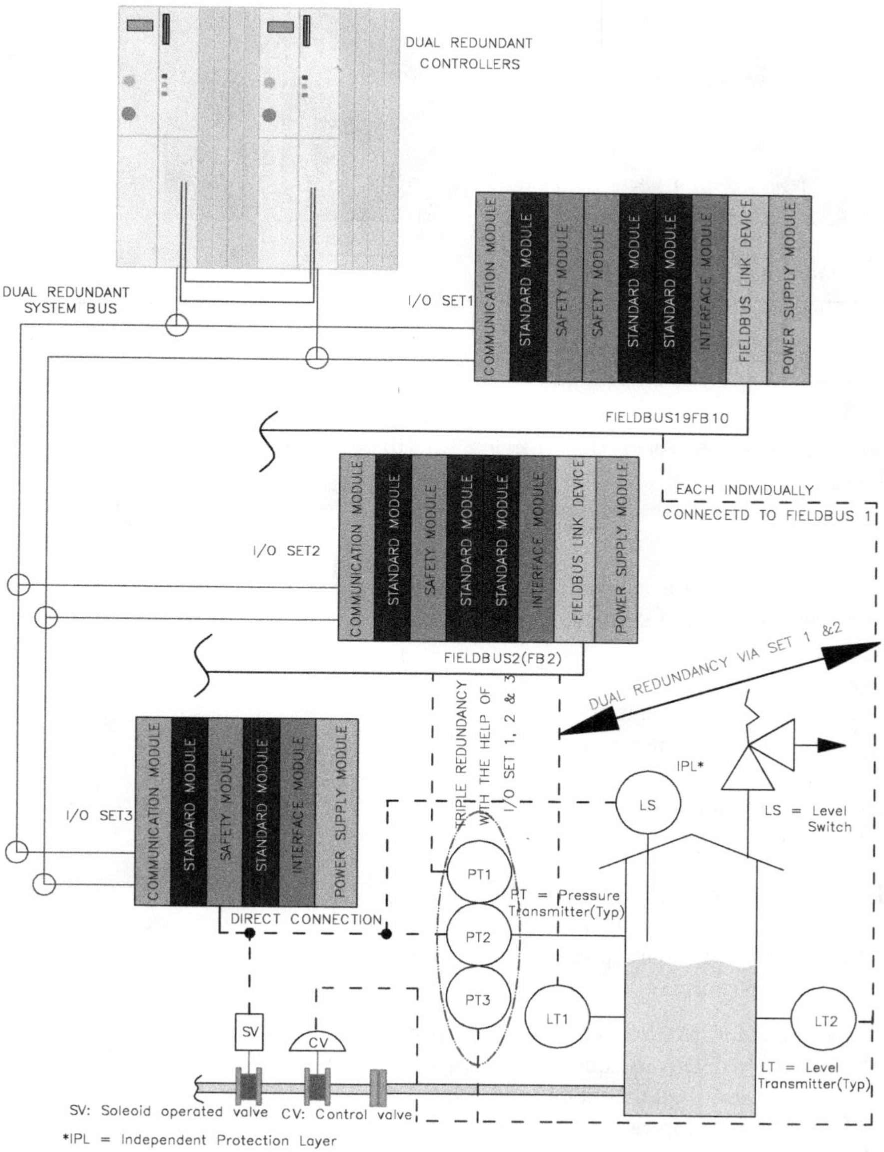

FIGURE IX/5.1.3-1

Safety controller configurations.

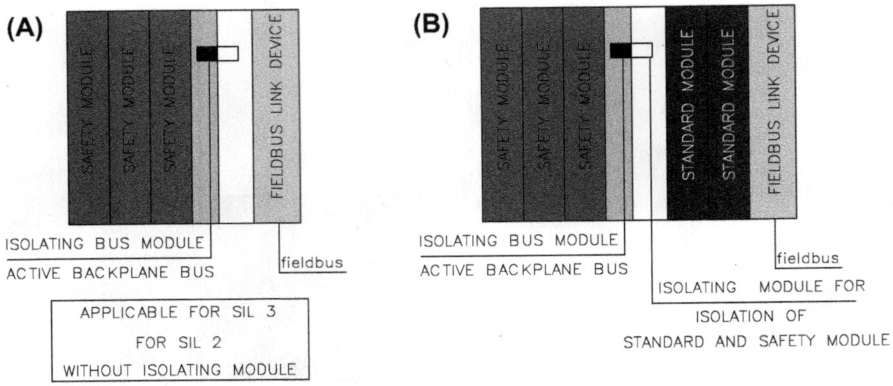

FIGURE IX/5.1.3-2

Module dispositions. (A) For safety modules in SIL 3, (B) isolation between standard & safety (fail safe) module.

Developed based on idea adapted from Siemens ET 200M modules. Courtesy Siemens.

- Each safety system is isolated from the other
- Immune to failure of control network
- CCF chances are less
- Quicker update
- Predictable communication traffic
- Redundant communication
- Extensive diagnostics
- Scalable in smaller increments
- SIL 3 applications

5.1.5 PLC Selection Fundamentals

Selection of best-suited safe PLC for a facility may have certain specific factors. However, following generic points are normally considered for selection of safe PLCs in different applications.

- Type of sensor/actuator signal (e.g., self-testing facility of I/O)
- Power supply details including back-up
- Grounding and shielding details
- Housing and enclosure details types
- Fault tolerance
- Configuration flexibility
- Redundancy: Quality (e.g., error latency) and levels
- Addition and on line module replacement
- Uniform remote I/O
- Line monitoring
- Safety communication
- Connectivity

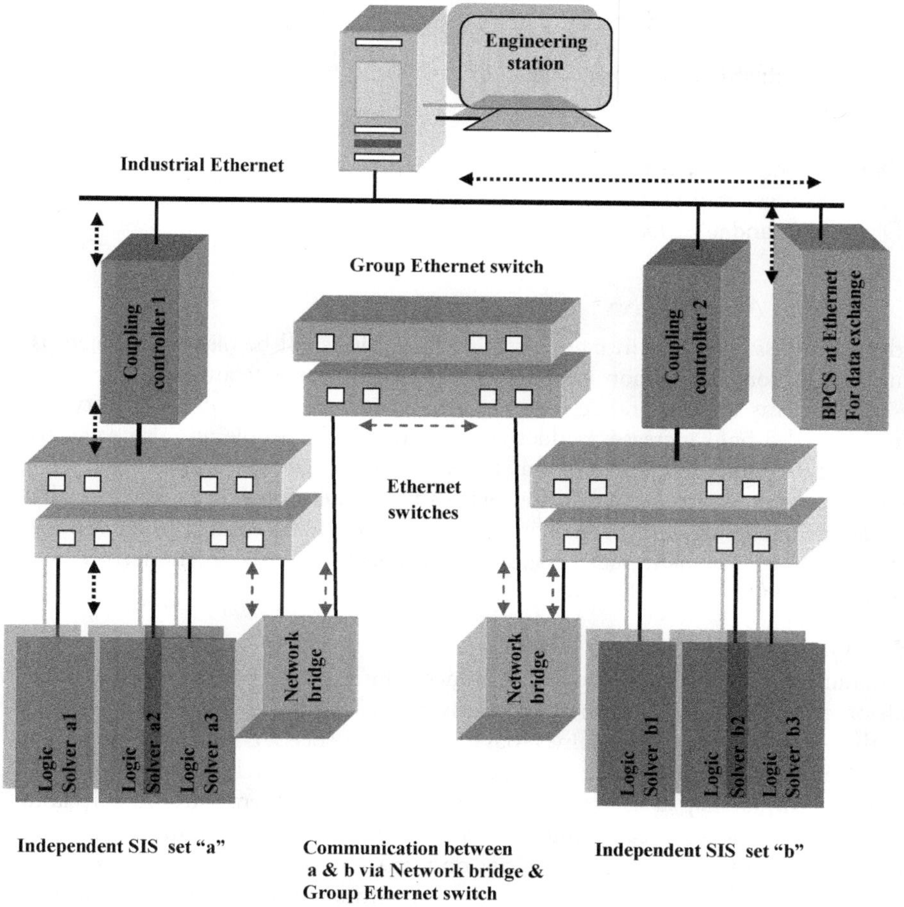

FIGURE IX/5.1.4-1

Separate and independent SIS functions.

Developed based on idea from Delta V SIS: Courtesy Emerson process management.

- Buses and their compatibility
- Time synchronization
- Environmental aspect
- Security aspect
- Third party certification
- Programming language
  - Fully-defined language
  - Subset of standard language
  - Limited variable application program
  - Control/data flow: Software safety
  - Developmental tool support

- Software design and life cycle
  - Software design specification
  - Software module design specification
  - Coding
  - Testing
- Online changes of software
- Simulation capability
- Diagnostics and asset management

5.2 Factory Automation

In this clause, safety PLCs in manufacturing industries shall be discussed. Safety is required in factory automation in manufacturing industries as requirements to meet the requirements of laws and regulations. People now realize that such safety requirements also help increase productivity and profit. So people are also eager to implement the same. In earlier days safety requirements necessary to meet the regulations were treated separately, but now, with advancement of technology especially with developments in fieldbus and other safety automation systems, people go for safety integrated systems in factory automation. Keeping pace with the trends, discussion will be focused on the same.

5.2.1 *General Discussion*

Minimum level of safety is necessary to meet the regulations, but with a little addition to this can increase their productivity and profit. This is because of the fact that modern production lines have very high-speed throughput and any downtime especially on account of safety could be very costly. Safety system is crucial for the facility as it could pinpoint the exact problem very quickly and efficiently. Safety systems can bring the machine or facility into safe state in case of failure/dangerous failure. This means not only the costly production line is saved but also associated personnel are saved coupled with large reduction of downtime. Although separate safety components could be used for smaller systems with two to three functions, as they have limited diagnostics may not be suitable for larger systems. Integrated and embedded safety solutions are well engineered and have more intelligence to offer much better solutions. In current trend, people go for safety networking for the entire production lines and process.

5.2.2 *Basic Safety Requirements*

A major objective for safety requirements are in compliance with all relevant safety standards to adequately reduce of risk and to secure suitable shield against liability due to losses in various forms. Functional safety involves the part of the safety of a machine or plant that depends on the correct function of its control or protection equipment. The standards available to user are the latest versions of:

- EN 62061
- IEC 61508
- ISO 13849-1

A solution towards this could be a safety-integrated system to embed and integrate safety into the standard functions. Table IX/5.2.2-1 gives an overall view of the entire process.

An example of the same could be SIMATIC of Siemens. The core of the same comprises a wide range of standardized products and systems [12]. Fail-safe controllers and associated system have the following functions:

- Comprehensive self-tests and self-diagnostics.
- Simultaneous execution of standard and safety programs in a single controller **with** programming isolation.
- Fail safe I/O modules with dual channel internal design.
- *Integrated processors* monitor each other and automatically test the I/O wiring.
- In the event of a fault, F-module places system in a safe state.
- Failsafe power modules supply load voltage to the potential group.
- Safety-related communication is possible by means of the safe fieldbus system.
- One fieldbus may handle both standard and safety-related data.
- IEC approved FBD, LD, etc. programming languages are used.

5.2.3 Features of Safety Integrated Systems

The following are the general features in safety integrated systems. The features given here is extracted parts from a practical factory automation system from a manufacturer of international repute:

Cost savings: Safety integrated system is implementation of safety in integrated automation to provide an edge to the machine manufacturers in a competitive environment. In such systems, on account of advanced engineering following advantages are achievable for machines and plants:

- Implementation of new technological requirements in easier and faster way
- Easier up-gradation, retrofitting, and modernization

Table IX/5.2.2-1 Various Phases for Safety System of Factory Automation

Phase	Requirement
Design and engineering	Risk analysis and reference to relevant standards to identify the risk of the system and countermeasure for the same.
Installation and start up	From risk analysis to identify components needed for making the hazardous zones safe. Safety verification for CE marking.
Operation and maintenance	Operator to follow occupational safety guidelines and the documents for following the machine directives by manufacturer.
Maintenance	Required maintenance performed quickly to reduce downtimes and standstills as short as possible. Fault must be eliminated quickly to ensure production up and running continuously with minimal downtime.
Upgrading	To keep the plant up-to-date through inspections, documentation, consulting, and defining remedial measures.

- Modular expansion
- Safer machine
- Higher productivity
- Higher profit

  Such advantages could be visible throughout the complete life cycle.

5.2.4 *Safety Evaluation Tool*

The safety evaluation tool is a *free* online calculation tool for the IEC 62061 and ISO 13849-1 standards, and offers valuable support for the implementation of safety functions on machines. The third party approved online tool guides are available for step-by-step support from definition of the safety system structure through selection of the components, all the way to determination of the achieved safety integrity level (SIL/PL) [12]. The following are the major benefits of it:

- Safety functions evaluation in less time
- Fast access to product data
- Calculation as per standards
- User-friendly
- Comprehensive, predefined libraries

5.2.5 *Modern Trends*

Modern factory and industrial automation is more flexible and open and meant towards higher productivity. These systems replace conventional relay or solid state control systems, with extensive use of open programmable electronics in all types of scalable and demanding solutions. Modern safety integrated safety-related controllers of various designs with a variety of hardware and software architectures, including PC-based solutions, cover all sectors and types of factory automations. Associated standards for factory automations are:

- IEC 61508: SIL
- IEC 61511
- EN 60204-1
- EN 62061: SIL
- NFPA 79, NFPA 85
- ISO 13849-1: PL

  The subject of safety PLC is not only very interesting, but very much pertinent to SISs. So, detailed study on the same cannot be overestimated. However, in order to limit the volume of the book, it is necessary to restrict the discussions on safety PLC here and focus on various requirements pertinent to final elements.

6.0 FINAL CONTROL ELEMENTS FOR SAFETY INSTRUMENTED SYSTEMS

During SIS design, all constituents for each SIF must be addressed especially final elements. According to the published reports, final elements account for nearly 50% of SIS failure. Final control elements consist of control or shutdown valves/damper, actuator, and associated accessories such as solenoid valve, positioner, etc. As final control elements are the weakest link in SIF, hence SIS, therefore, final elements and associated devices must be selected with proper care so as to improve the reliability, availability, and reduce risks of the associated SIF.

A dangerous condition is sensed by process transmitters to initiate the logic solver. Based on a response from the logic solver, the final control elements need to move with decisive and swift action. Such positive response may be regulated by standard control valves and/or by an emergency isolation or shutoff valves. In modern instrumentation, a digital positioner, associated with final element, plays an important role. With intelligent actuators with positioners, majority of earlier problematic issues associated with solenoid valves, pressure monitoring, partial stroke testing, etc. are well-taken care of. Now it is better to have a closer look on these.

6.0.1 *Use of Valves as Final Control Elements*

In the majority of the cases, valves are used as final elements. In SIS designs, valves are found in following ways in applications as final elements:

- Single control valve for as control and on/off valve
- Single valve (may be control type) as on/off valve
- Control valve as redundant final control element

  For some smaller applications, common valve is used for both the purposes of regulation and shutdown, that is, using the valve to regulate fluid flow during normal operation and fully stroke (either open or closed depending on the application) during a shutdown condition. Also, in cases where there is separate shutoff valve still control valves may have shutoff interlock for additional safety, third options mentioned above, is an example of the same.

6.0.2 *Safe Position of Final Control Elements*

Solenoid valves are used to play important role in cases of emergency shutdown and are used for actuations. In this connection, Fig. VII/1.1-3 or IX/5.1.3-1 may be referenced. Now for all such cases, there will be fail safe condition of these valves, viz. fail open or fail close and fail lock (last position—mainly for modulating valves). In fail lock condition, regardless of the valve's "natural" failsafe state, the system makes the valve to lock positioner air inside the valve actuator as shown. However, on account of leakage, this position cannot be held permanently over very long period. Various safe failure conditions of the control valves have been depicted in Fig. IX/6.0.2-1. In the figure, control valves have been shown, however this equally applicable for all other valves/dampers, etc. also. The figure is self-explanatory; for further details Chapter VI of [2] may be referenced.

As final element is the most vulnerable component in a safety loop so, redundancy is another important issue here. Use of redundant valve in series and parallel mode is a related to process concerned but in any case it is normally a costly affair. However, proof testing at regular intervals could be an economic alternative to ensure reliability and availability of final element in the safety loop. This proof testing for final elements is popularly known as partial stroke testing discussed later in this chapter.

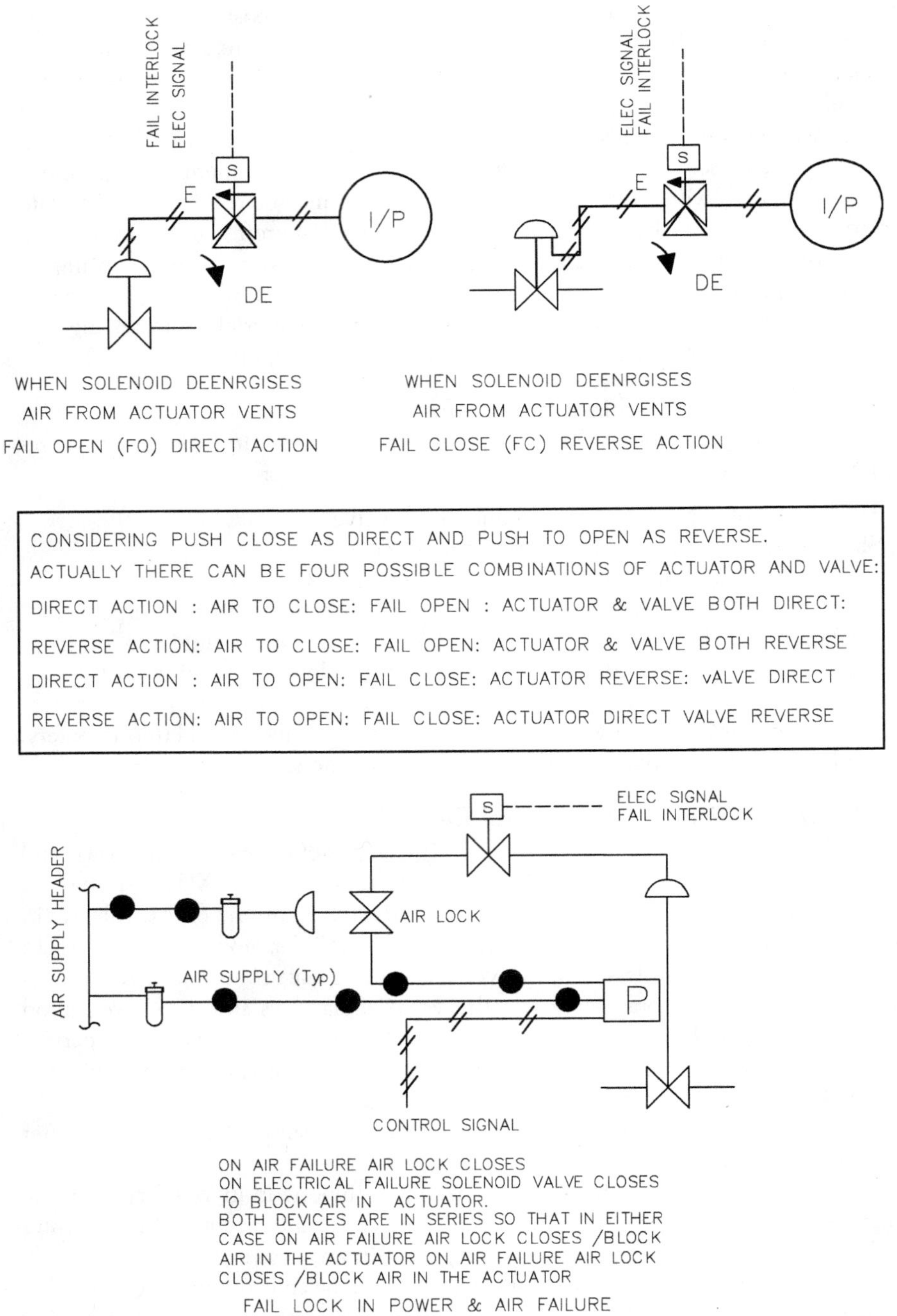

FIGURE IX/6.0.2-1

Safe failure of valves.

6.0.3 *Role of Digital Positioner and Control Valve in Safety Loop*

The dangerous condition in process is sensed by the sensors, and reported to the logic solver for its response. Under normal conditions, the control valve regulates the flow; however, in case of an emergency shutdown signal is sent to both shutdown valve as well as to control valve (via solenoid as a safety interlock). As such for shutdown valve, solenoid valve is the main actuator. In modern instrumentations, the digital positioner not only replaces I/P converter, but also plays a great role in achieving better safe shutdown conditions. Fig. IX/6.0.3-1 shows the role of valves as final elements in a safety loop. In this figure, the control valve with positioner and solenoid valve for interlock have been shown, whereas in case of shutoff valve, solenoid valve has been replaced by digital positioner. This is shown intentionally to show that control valve shutdown signal has been used as safety interlock meaning that control valve can be used as a redundant shutdown valve to ensure better availability and to have two types of devices to minimize CCF. In shutdown valve digital positioner has been used in latter options to show that digital positioner can be used with integral solenoid valve. It is worth noting that in case of control valve, solenoid valves could be *replaced* by digital positioner with integrated solenoid valve also.

6.1 Various Possible Valve Configurations in Safety Instrumented Systems

As discussed in Clause 6.0.1, control valves can be used as final control elements in SIS in various forms. The same has been detailed in Fig. IX/6.1-1. In the first case, mainly used in smaller applications, a control valve has been used both for control purposes as well as it receives the safety interlock from separate SIS. As shown in Fig. VII/1.4-4, usually in such cases control valve is also provided with a solenoid valve through which safety interlock is performed by full closing or opening of the control with the help of solenoid valve by venting air from the actuator of the control valve. In this case also, the same is done only instead of separate solenoid valve, this is achieved by a ***digital positioner*** with integral solenoid valve. In the second case, BPCS uses separate control valve for normal control and the other valve (which may be a control valve even) is used as on/off valve and is operated with the help of digital positioner. Normally, shutdown valves have high shutoff capability (bubble tight leakage). In the third case, the control valve is normally used by BPCS for regulating purpose. There is a separate SIS shut off/ shutdown valve with on/off actuator and integrated digital positioner. For shutdown purposes command from SIS goes to this on/off shutoff valve. As a matter of redundancy same command is also sent to the control valve, which with the help of integrated digital positioner carries out the safety interlock by full opening or closing of the control valve.

Since the two valve types are different, CCF chances will be minimized. In both cases of II and III, there are two valves; the only difference is that in case III the control valve additionally shares the burden of SIS interlock as redundancy to SIS.

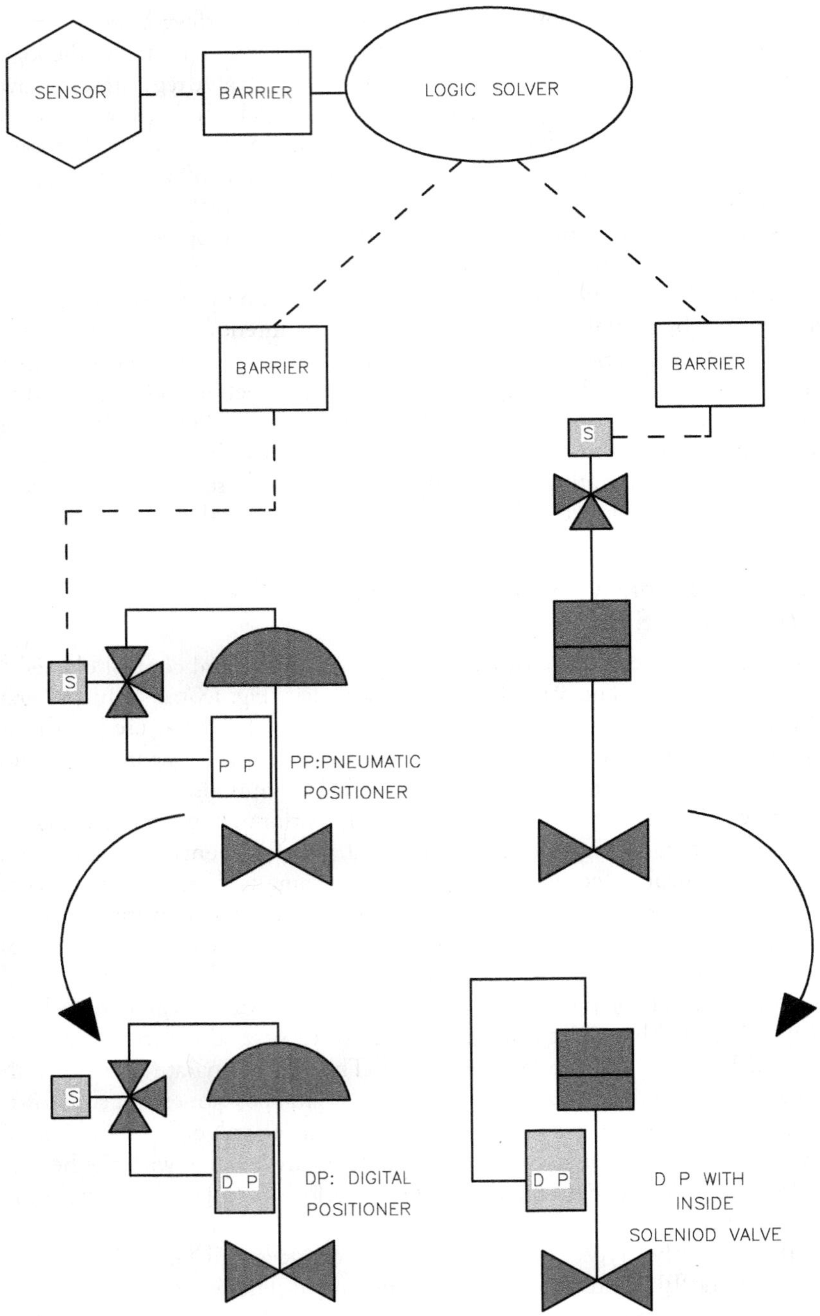

FIGURE IX/6.0.3-1

Role of final elements in safety loop.

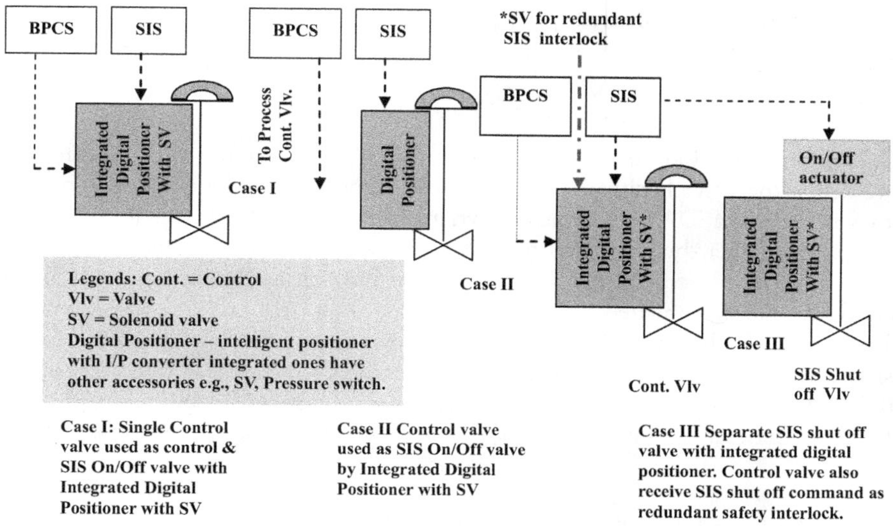

FIGURE IX/6.1-1

Various valve configurations in SIS.

6.2 Testing of Movement for Shutdown Valve

Whenever a device is not operated for long time, it is quite likely that after a long time when someone wishes to run the same it may, or may not, meet the demand. This is especially true for mechanical or moving items. Final control elements **in** SIS under normal circumstances are not moving. They remain in one position without mechanical movement and even never are called into action. Sticking in last position for non-use for quite some is typical failure of shutoff valves, so, may fail on demand. As a result of the same, in a dangerous condition leading to an explosion, fire, and/or a leak of lethal chemicals and gases to the environment. As discussed earlier, proof testing frequently at regular interval is necessary to improve reliability and availability of safety shutdown valves. Traditional methods of testing could only highlight on the movement but does not provide any internal valve diagnostic data such as valve friction, process build up, pressure status, valve torque, etc.

Typically SIS valves are subject to following conditions:

- Increased packing friction
- Seizing of shaft in bearing
- Bearing corrosion
- Fluid buildup on shaft
- Fracture valve shaft/stem
- Spring issue: Broken or permanent setting
- Linkage breakaway friction
- Air exhaust slow or blocked

- Spring return actuator dented
- Increased valve breakaway friction
- Bending of shaft/stem
- Increased friction of closure element in seal

When the issue is looked at closely, then it may be noted that it is very difficult to test to detect the valve health. Full 100% travel, it can only be done during annual shutdown. From FMEDA tests results, shown in clause number 1.4.1 of Chapter VIII, it is seen that higher the proof test interval higher will be PFD value, meaning that when such testing interval is high then it will have direct nearly linear impact on SIL value which is directly related to PFD value.

In a running facility or plant, if testing is to be performed on solenoid operated on-off shut down valves to assess the health of valve, and then only way is to carry out partial stroke testing of the valve. This may give rise to chances for spurious tripping of the plant. In the following sub-clauses, partial stroke and full stroke testing shall be covered.

6.2.1 *Partial Stroke Test*

With a partial stroke test (PST), it is possible to improve the safety, as failures are detected at an earlier stage than when only proof testing is conducted. It is possible to maintain the same SIL rating even with longer proof test intervals. Partial stroke testing can be two types; one is manual PST, and the other is automatic computerized PST *with* digital positioners. The modern trend is to opt for the second option. People have realized the need of digital positioner so for detailed analysis automatic computer programmed PSTs are preferred. Partial stroke testing can be done while the plant is running. Usually the valves are so designed that it has nearly 10% cushion so, usually if on/off valve is stroked, (say) 10% from its open position in closing direction without influencing the process or at least without shutting down the process. For better diagnostic coverage, testing procedure is carried out in an elaborate way. This 10% change in valve position does not affect the process, yet provides evidence that the valve can still be moved. The valve status and response to mechanical movement can be monitored during the test. However, the chances of unplanned shutdown cannot be totally ruled out. So, precise control and termination of the partial stroke tests are necessary. In this connection, the following points are worth noting:

- Time for actuator pressure to settle to a defined value before partial stroke
- Ramp time to limit the speed of movement
- Maximum time for test
- Maximum permissible valve position
- Overshooting above position test is terminated
- Minimum permissible air pressure in the actuator [17]

The stroking procedure can be specified in detail by entering the corresponding parameters such as initialization, start point, end point, PST travel, speed of travel, dead time, friction Hi/Lo limits and type of stroking — ramp or step response. Normally conditions, viz. test duration, overshoot and minimum actuator pressure, are defined for the test. There are a few other parameters included in the testing procedure such as schedule PST, PST interval, new PST schedule, etc.

The PST can supply a number of parameters which are very helpful in analysis so that health conditions of the shutdown valve can be inferred; see Fig. IX/6.2.1-1.

When PST is performed at regular intervals, it can be considered as diagnostic test in place of proof test as it helps in improving undetected failure rate, that is, diagnostic coverage (DC) and SFF.

Data gathered from PST can be listed as follows:

- Valve in closed position
- Valve transit timing
- Dead time
- Valve position
- Deviation from set point
- Actuator pressure
- Overshoot
- Response time

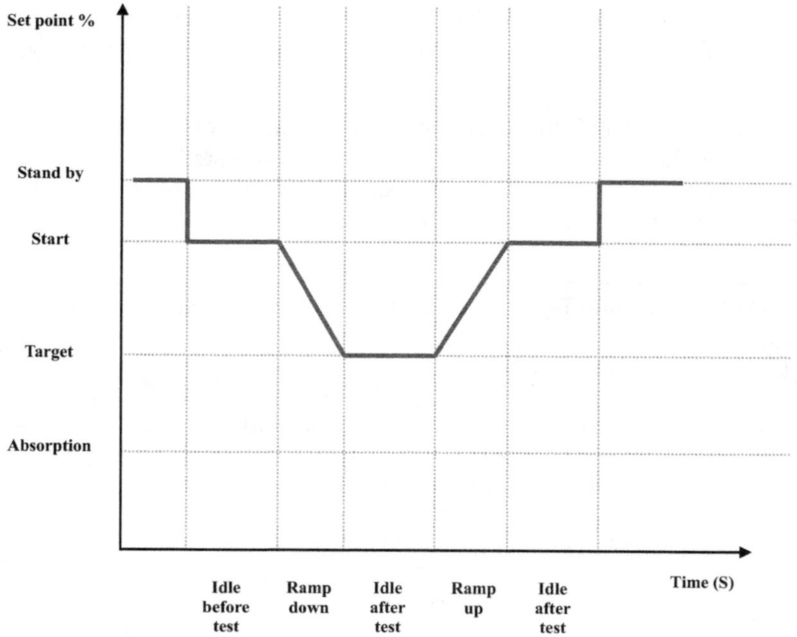

FIGURE IX/6.2.1-1

Partial stroke testing methods.

Developed based on the idea from Partial stroke testing for final elements; T. Karte, I.J. Kiesbauer; Samson AG; Conference Proceedings of "Petroleum and Chemical Industry Conference (PCIC) Europe, 2005, courtesy of Samson AG.

- Spring range
- Break away torque
- Friction
- Droop
- Operating hours
- Permissible temperature

  All these data are available from step response and ramp response tests.

6.2.2 *Full Stroke Test*

SIFs in SIS are tested at regular intervals to prove their full functioning. A full stroke test (FST) involves the valve being closed under supervision and then the test result being documented. As stated earlier, normally proof test interval is one year in most plants. However, longer test intervals are also common in some plants (e.g., petro chemical industry). In modern plants, these tests are now accomplished automatically with the help of digital positioner to provide detailed information such as dead time, closing time, end position, and uniform valve stroking over the entire travel range, etc.

6.2.3 *Discussions on Partial Stroke Test and Full Stroke Test*

Table IX/6.2.3-1 provides various data available from PST and FST. The table elaborated as follows is based on data from the Instrument Engineers' handbook by B.G. Liptak [18].

- Monitoring spurious trips: In modern plants, the automatic online data analysis is done with the help of DCS/PLC system and associated data logging methods. Naturally all the tested results are well-documented and can be retrieved for future analysis. During PST after the test is done, it is well-recorded in the

Table IX/6.2.3-1 Failure Types Tested at Partial Stroke Test and Full Stroke Test [18]

Failure Type	Failure Mode	PST	FST
Valve packing is seized partial- or full-stroke	Valve fails to close (or open)	X	X
Valve packing is tight partial- or full-stroke	Valve is slow to move to closed or open position	X	X
Air line to actuator crimped	Valve is slow to move to closed or open position	X	X
Air line to actuator blocked	Valve fails to move to closed or open position	X	X
Valve stem sticks	Valve fails to close (or open)	X	X
Valve seat is scarred	Valve fails to seal off		X
Valve seat contains debris	Valve fails to seal off		X
Valve seat plugged due to deposition or polymerization	Valve fails to seal off		X

system, supposing the test results show there is high friction then same shall be annunciated as well as in preventive maintenance report same will be listed. Similarly in case of an unplanned shutdown, the same is also documented.

6.2.4 *Diagnostic Features in Partial Stroke Testing*

Additionally, there a few features involved which make digital positioner better when compared with the same with solenoid valves these are:

- When actuator is activated by solenoid valve, extra air pressure is applied to overcome force in return spring. However, this can give rise to dead time during closing and could be critical level in some cases.
- In case of digital positioner, the same may not be needed and it is capable of maintaining opening. This shortens the closing time.
- Also digital positioners have diagnostic programs which are capable of small changes around the set point to detect the friction force changes.
- These are elaborated in Table IX/6.2.4-1, and reference [16] is very helpful.

From the preceding discussions, it is clear that smart/intelligent digital positioners have important roles in valve testing. So, in next clause focus will be on smart/intelligent digital positioner.

6.3 Smart/Intelligent Digital Positioner

Smart digital positioners can provide necessary data for predictive diagnostics to improve system availability. Smart digital positioners/controllers have capability to reduce spurious trips during testing and also maintains safety availability, should a "safety demand" arise, during testing.

A typical modern smart digital positioner has been depicted in Fig. IX/6.3-1. When a test is in progress and demand arises, the safety shut down valve will go to its fail-safe position. In order to address similar such issues as well as various diagnostic features discussed previously, smart digital positioners have been introduced in the

Table IX/6.2.4-1 Fault Detection by Diagnostics

Fault	Diagnostic Means
Valve friction	Step response and hysteresis tests
Valve leakage	Packing chamber by pressure switch. Also change in friction
Valve shut off change	Zero shifts
Set point deviation	Raw data
Packing degradation	Travel counter
Actuator spring	Breakage by travel or pressure characteristics
Unstable performance	Travel sensing

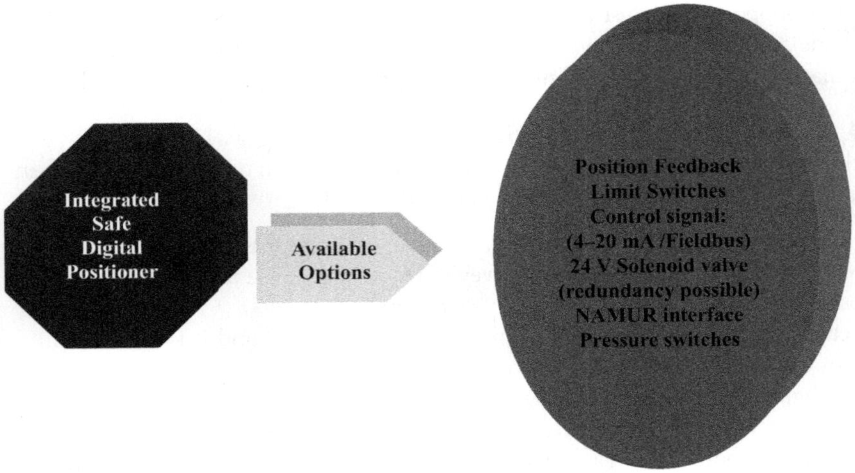

FIGURE IX/6.3-1

Integrated safe digital positioner components.

SIS loop to find health of final element. These smart digital positioners are communicating, microprocessor-based current-to-pneumatic instruments with internal logic capability. These are also capable of accommodating (rather integrating) a number of other accessories like solenoid valves, pressure switches, etc. On account communication capability, smart digital positioners exchange information in the loop, with the help of digital communication means such as HART protocol. In smart digital positioners, there are in-built sensors for valve travel, pressure (supply and output pneumatic pressures), and pneumatic relay motion. Microprocessor of smart digital positioners process these data in a desired way and give the power of not only internal diagnosis but also diagnosis pertinent to valve and actuators. These types of positioners that they can provide advance information/warning to maintenance people for preventive maintenance. These can also predict a dangerous situation before it happens. These in conjunction with PST can develop various means to improve DC and SFF, discussed earlier. These types of positioners allow the test intervals to be ***preprogrammed*** *and* ***automatic*** to meet the target SIL. It also makes the system tests as diagnostic test rather than proof tests. On account of automatic testing the chances of spurious trips are much less, however it is suggestive to initiate the test manually. While performing partial stroke test by smart positioner, it collects travel and pressure data during the test periods. It is possible to monitor valve performance trends. So, after each PST, these can be automatically analyzed in the system in conjunction with HMI. As discussed earlier, data collected during the test will provide inferential information about inertia force, required pressure to actual movement of valve travel, sticking in shaft/bearing area, linkage dead motion, etc. It is possible to save the results for printout or later use. In nutshell, a smart digital positioner can provide and address the following important issues:

- Complete diagnostic health information on the final element and positioner
- An immediate alert notification when necessary
- Documentation of all testing
- Complete documentation of any emergency event

Smart digital positioners may have integrated solenoid valves pressure switches as detailed in Fig. IX/6.3-1. With this, the discussion on smart digital positioner is concluded with the idea that these are extremely helpful for DC and SFF improvement and meeting higher SIL requirement. As a matter of fact, this is an interesting subject by itself and with this, discussions on SIS component are concluded. Now concentration will be on implementation part of IEC safety standard.

LIST OF ABBREVIATIONS

AI/O	Analog input/output
ALARP	As low as reasonably practicable
BPCS	Basic plant control system
C&I	Control and instrumentation (engineering)
CCF	Common cause failure
CRC	Cyclic redundancy check
DC	Diagnostic coverage
DCS	Distributed control system
DI/O	Digital input/output
DP	Diff. pressure/decentralized periphery—PROFIBUS
EDDL	Electronic device description language
E/E/PE	Electrical/electronics/programmable electronics
EEMUA	Engineering Equipment and Materials Users' Association
E&I	Electrical and instrumentation
EMI	Electromagnetic interference
ESD	Emergency shutdown system *or* electro static disturbance
EUC	Equipment under control
FBD	Functional block diagram
FDT	Field device tool
FF (SIF)	Foundation fieldbus (SIF)
FMEDA	Failure modes, effects and diagnostic analysis
FMS	Field message specification (PROFIBUS)
FST	Full stroke testing
HC	Hydrocarbon

Continued

HFT	Hardware fault tolerance
HMI	Human machine interface
HVAC	Heating ventilation and air conditioning
HW	Hardware
I&C	Instrumentation and control
IEC	International Electrotechnical Commission
I/O	Input/output
I/P or O/P	Input or output
IPL	Independent protection layer
IT	Information technology
LD	Ladder diagram
LEL	Lower explosive limit
LHS	Left hand side
LOPA	Layer of protection analysis
LS	Logic solver
MADB	Master alarm data base
MOC	Management of change
MTTF	Mean time to fail
MTTR	Mean time to repair
NO	Normally open (contact)
NC	Normally close (contact)
OPC	OLE process control
O&M	Operation and maintenance
OREDA	Offshore reliability equipment database association
PA	Process automation—PROFIBUS
PFD	Probability of failure on demand or doc: Process flow diagram
PFH	Probability of failure per hour
P&ID	Piping and instrumentation diagram
PLC	Programmable logic controller
PST	Partial stroke testing
RFI	Radio frequency disturbance
RHS	Right hand side
RRF	Risk reduction factor
SFF	Safety failure fraction (IEC)
SIF	Safety instrument functions
SIL	Safety integrity level
SIS	Safety instrumentation system/supervisory information system (in case of DCS) [9]
SOE	Sequence of event
SPLC	Safe PLC (programmable logic controller)
SW	Software
TMR	Triple modular redundancy
UIP	User interface plug ins
UPS	Uninterrupted power supply
WRT	With respect to

References

[1] Safety Instrumented System Engineering Handbook, Kenexis Consulting Corporation, Columbus, OH, 2010. http://www.kenexis.com.

[2] S. Basu, A.K. Debnath, Power Plant Instrumentation and Control Handbook, Elsevier, November 2014. http://store.elsevier.com/Power-Plant-Instrumentation-and-Control-Handbook/Swapan-Basu/isbn-9780128011737/.

[3] ISA-TR84.00.02-2002-Part 1, Safety Instrumented Functions (SIF) – Safety Integrity Level (SIL) Evaluation Techniques Part 1: Introduction, ISA, June 2002.

[4] ARC White Paper, Foundation Fieldbus Safety Instrumented Functions Forge the Future of Process Safety, ARC Advisory Group, September 2008.

[5] Field Device Integration (FDI) Making Device Management Easy; fdi_white_paper.pdf; Foundation fieldbus; http://www.fieldbus.org/images/stories/enduserresources/technicalreferences/documents/fdi_white_paper.pdf.

[6] ARC White Paper, Profisafe and PROFIBUS for High Availability and Safety in the Process Industries, ARC Advisory Group, April 2009.

[7] Selecting Transmitters for Safety Instrumented Systems, S.R. Brown Du Pont & M. Menezes Rosemount: ISA Huston, 2001.

[8] Understanding and Applying the ANSI/ISA 18.2 Alarm Management Standard; PAS understanding of ISA 18.2; https://www.isa.org/standards.

[9] Alarm Management, DeltaV Distributed Control System, White Paper, Emerson Process Management, August 2015.

[10] Alarm Management; ABB; Meet the experts, September 2011.

[11] Philosophy of ITER Alarm System Management; Iter; 3WCD7T, February 01, 2013. http://static.iter.org/codac/pcdh7/Folder%201/11-philosophy_of_ITER_Alarm_System_Manageme_3WCD7T_v2_1.pdf.

[12] SIMATIC Safety Integrated for Factory Automation, Standard and Safety Technology in One System, Brochure, Siemens, April 2011.

[13] Safety Instrumented Systems, The "Logic" of Single Loop Logic Solvers; Moor Industries Worldwide, Brochure, Moore Industries-International, Inc., January 2014.

[14] Technical Brochure, Safety Integrated for Process Automation: Reliable, Flexible, Easy, Siemens AG, April 2008, www.siemens.com/process-safety.

[15] P. Gruhn, Is Your Existing Safety Instrumented System Up to Current Standards? Rockwell Automation. Technical literature; http://literature.rockwellautomation.com/idc/groups/literature/documents/wp/proces-wp012_-en-p.pdf.

[16] T. Karte, J. Kiesbauer, Smart Valve Positioners and Their Use in Safety Instrumented Systems, 2009. Samson W023040en.pdf; Industrial valve.

[17] Partial stroke testing for final elements; T. Karte, I.J. Kiesbauer; Samson AG; Conference Proceedings of "Petroleum and Chemical Industry Conference (PCIC) Europe, 2005.

[18] B.G. Liptak, Instrument Engineers' Handbook: Process Control, Chapter 6.10; Emergency partial-stroke testing of block valves; A.S. Summers.

Further Reading

2600T Series Pressure Transmitters Safety Manual for 266 Pressure Transmitters Certified According to IEC 61508; Safety Manual SM/266HART-SIL-EN Rev. L.

A methodology for the achievement of Target SIL, ABB Limited; https://library.e.abb.com/public/c96a3bd3a6797850c1257bd4002e623a/GBABB_PAOG_Brochure_Safety_SIL%20Methodology.pdf.

A. Ajmeri, Smart Instruments and Device Diagnostics: How Well Is Your Plant Using Information, Yokogawa; Plant Services, March 2015. http://www.plantservices.com/articles/2015/smart-instruments-and-device-diagnostics/.

Alarm Management Standards e are you taking them seriously? Advanced SolutionWhite Paper; Honeywell; https://www.honeywellprocess.com/library/marketing/whitepapers/honeywell-alarm-manager-alarm-management-standards-taken-seriously-wp817.pdf.

Alarm Philosophy Document Template, Exida Consulting LLC; http://www.exida.com/images/uploads/exida_Alarm_Philosophy_Sample_(Oct_2011).pdf.

M.M.S. Anand, Electronic Instruments and Instrumentation Technology, PHI learning Pvt. Ltd, January 2004.

Application of IEC 61508 and IEC 61511 in the Norwegian Petroleum Industry, The Norwegian Oil Industry Association, October 2004. No. 070.

ARC White Paper, Profinet: An All-encompassing Industrial Ethernet Solution, D.W. Humphrey, November 2005.

ARC White Paper, Reduce Risk with a State-of-the-Art Safety Instrumented System, ARC Advisory Group, September 2004.

H. Bezecny, D. Inverso, V. Maggioli, G. Rabe, A. Weinert, European Workshop on Industrial Computer Systems; Technical Committee 7; (Safety, Reliability and Security); Guidelines for the Use of Programmable Logic Controllers in Safety-related Systems, October 1997. Working Paper 6009; Version 13.

K. Bingham, Partial Stroke Testing with Positioners and/or Logic Solvers, April 2014. http://www.acm.ca/resource-centre/articles/partial-stroke-testing-positioners-andor-logicsolvers.

J. Cahill, Credit for Partial Stroke Tests in Verification Process, Emerson Process Expert, 2011, in: http://www.emersonprocessxperts.com/2011/01/credit_for_part/.

A. Coleman, Safety Instrumented Systems: Control Valves as Final Elements, Emerson Process Management; Chemical Engineering, January 2011.

CT534: Alarm Management How to Create an Effective Alarm Management Program, T. Stauffer (Exida), March 2015. Rockwell Document, https://www.rockwellautomation.com/resources/downloads/rockwellautomation/pdf/events/process-summit/CT534_How_to_Create_Effective_Alarm_Management_Program.pdf.

DeltaV SIS Logic Solver; Product Data Sheet, Emerson Process Management, June 2014.

DeltaV SIS Safety Network Components, DeltaV SIS Process Safety System, Emerson Process Management, January 2015. Product data sheet.

P. Gruhn, Is Your Existing Safety Instrumented System Up to Current Standards? Rockwell Automation; write up.

P. Gruhn, Things to consider when selecting a safety system, in: ISA Safety Division Symposium, 2009.

I/A Series_ Pressure Transmitter Functional Safety Manual, Invensys, June 2009. MI 020—357, http://resource.invensys.com/instrumentation/documentation/eib/mi/mi_020-357.pdf.

IEC 62682 and ISA 18.2 Alarm Standard, The Expected Impact of the IEC 62682 Standard on Alarm Management for the Process Industries, ARC Industry Forum, March 2015.

Logic Solvers for Overpressure Protection; Moor Industries Worldwide; Brochure; Moore Industries-International, Inc.; www.miinet.com.

J. Love, Process Automation Handbook: A Guide to Theory and Practice, Springer.

T. Moor, Machine Safety in Factory Automation, IHS, HIS Engineering 360, March 2015.

PI PROFIBUS PROFINET; Profisafe; Southeast Asia; http://www.profibus.com/nc; Internet document.

Predicting health of final control element of safety instrumented system by digital valve controller, in: Presented at ISA Automation West, R. Ali – Fisher Controls Int'l., LLC, 2004. www.isa.org.

Process Safe Book 1; Functional safety in the process industry principles, standards and implementation, Rockwell Automation.

M.Rausand, Reliability of Safety-Critical Systems: Theory and Applications, J. Wiley.

Reliability Data and the Use of Control Valves in the Process Industry in Accordance With IEC 61508/61511; T. Karte, E. Nebel (SAMSON AG) and M. Dietz, H. Essig (Infraserv Höchst); Samson; Special print from atp e Automatisierungstechnische Praxis, vol. 47(2) (2005).

A. Rentcome, Process Safety & Critical Control: Rockwell Automation Process Safety Platforms, Rockwell Automation, 2012.

Safety Instrumented Functions and Systems, IAM Search, September 2015. http://iamechatronics.com/notes/general-engineering/482-safety-instrumented-functions-andsystems.

Selecting "sensor" for safety instrumented system as per IEC 61511 (ISA 84.00.01–2004); D. Perry; Emerson process managementdRosemount division.

T. Stauffer, P. Clarke, Using alarms as a layer of protection, in: Prepared for Presentation at 8th Global Congress on Process Safety, Exida Asia, April 2012.

Technical Report; ISA-TR84.00.02-2002-Part 3; Safety instrumented functions (SIF)-safety integrity level (SIL) evaluation techniques. Part 3: determining the SIL of a SIF via fault tree analysis, June 2002.

The Layer of Protection Analysis (LOPA) Method, Look for Best Practices and Guidelines on How to Use the LOPA Method as an Alternative to Mitigate Risks, A. A. Frederickson-Independent Consultant e Member of Safety Users Group Network, April 2002.

Updated Alarm Management Guidelines; Control Engineering Europe, December 2013. http://www.controlengeurope.com/article/65594/Updated-alarm-managementguidelines.aspx.

User Specification; Rowan House Ltd. Technical Paper; http://www.rowanhouse.co.uk/SpecificationIEEGuidelines.html.

What You Need To Know About SIS and Partial Stroke Testing of ESD Valves, Metso Automation; write up; www.metsoautomation.com.

CHAPTER

X

Instrumentation Safety Implementation and Explosion Protection

EXPLANATION

With the introduction of safety standards IEC 61508 and 61511 (for process industries), there is a defined need for proper implementation of safety systems embedded into the main system. The safety life cycle has various phases. Phases 1 and 2 have been discussed at length in previous chapters (Chapter VI and Chapter VII and to a certain extent in Chapter IX). In this part, detailed discussions have been presented to include Phases 3–7, that is, from safety-related systems (SRSs) to modifications. This has been done purposefully so that prior to looking at the detailed implementation part of the standard, readers need to have some knowledge of the safety instrumented system (SIS), safety integrity level (SIL), and their implementation in various instrumentation components. So, this part of the discussions in conjunction with previous chapters will complete the topic of IEC 61508/61511. Safety instrumentation cannot be complete without discussions on explosion protection. With reference to IEC 60079-(0, 10, 14, 15, 17, etc.) and NEC (497, 499, 70, etc.), electrical area, classification of plant, explosion protection, etc. also have been included as part of this chapter to make the system complete in all respects. In view of this, these are presented in two sections. Section 1 for IEC implementation and Section 2 for explosion protection.

SECTION 1: IEC 61508/61511 IMPLEMENTATION

Introduction: People all over the world are working very hard to obtain optimum solutions toward implementation of design, installation, commissioning, and operation and maintenance (O&M). Here, the word "optimum" is very important. Unless proper planning is done right from the beginning there is every possibility that the system could be under- or overdesigned. In the former case, basic safety might be compromised, whereas in the latter case it will be costlier. Another issue is that many people believe that by increasing redundancy, greater safety will be achievable. It is true that the greater the safety requirements, the greater the availability should be. However, the reverse may not be true. This will be clear from the following simple example.

In some cases, even with 2oo3 sensor redundancy, without certification (keeping aside proven in use as per IEC 61511), a system will only achieve SIL1. Hence if this architecture is chosen for higher SIL applications, then in reality it is below target SIL2 or 3.

IEC 61511 is the basis of the design and selection of safety devices and this standard is mainly developed for *end* users. End users have no hand in design and internal testing, but they need to assure themselves that a device as per IEC 61508 (or the IEC 61511 standard) has been selected. So, end users may have to depend on third party assessments. One of the main objectives of the IEC 61508 standard is to design a "safe" system. The system shall be designed in such a way that it either can tolerate internal failures, and carry out the safety function, or if it cannot it can at least notify an operator via an alarm. So, it is essential to study how a safe system can fail. According to IEC 61508, equipment can fail because of three types of failures (see Chapter VII):

- Random: Random failures are two types, one permanent (after repairing it goes) and one conditional, that is, when a condition (say high temperature) exists, it exists, but when a condition disappears, then the fault disappears. So the latter condition is difficult to handle.

  As per IEC 61508, random failure can be handled in two ways, namely:
  - SIL measures to be implemented to control failures that might occur in hardware (IEC 61508 Part 2).
  - A qualitative and quantitative failure analysis of the hardware—failure mode and effects analysis with documentation. It is necessary to carry out a probabilistic reliability calculation to determine the average probability of failure on demand of the safety function.
- Common cause: Diversity means that the safety function is to be carried out in a redundant manner with different hardware/design principles/even completely different technologies (e.g., one a mechanical system, the other an electronic system so that in the former system the EMC effect will not exist). In practice a real common cause is difficult to find because the failures of a multichannel system must by definition of a common cause occur at exactly the *same* time. The same hardware will always have different strengths and thus fail at a slightly *different* time. A well-designed safety system can take advantage of this gap in time and detect one failure before the other failure occurs.
- Systematic failure: This is related in a deterministic way to ascertain the cause that can be eliminated by modification of the design/manufacturing

process/operational procedures and other relevant factors. A systematic failure can exist in hardware and software. Systematic failures are found during testing, development, or proof testing. The IEC 61508 standard addresses systematic failures in the appendix of Parts 2 and 3. The philosophy behind this is that when the required measures to avoid failures are implemented correctly, then there is hardly any possibility for systematic failures. Further, it is better to look in detail at how safety requirement specification is developed. The discussions presented in this chapter may seem to overlap with discussions in Chapter VI (especially with Clause 5.2). In reality this is not so, because the discussions in the previous chapter are purely on the standard points, whereas the *discussions here are from an implementation point of view.*

1.0 SAFETY REQUIREMENT SPECIFICATION

From a health and safety study report in the United Kingdom, it has been found that, of the total number of accidents involving control systems, 44% resulted from specification-related issues, 15% from design and implementation issues, 6% from installation and commissioning, 15% from O&M issues, and 20% from changes after commissioning (indicated anticlockwise in the figure).

From the percentage failure contribution diagram in Fig. X/1.0-1 it is clear that the development activities for specification need special attention. First, we shall look at how a system specification is developed from system requirement definitions, then we shall look at the requirements for a safety system, that is, SRS.

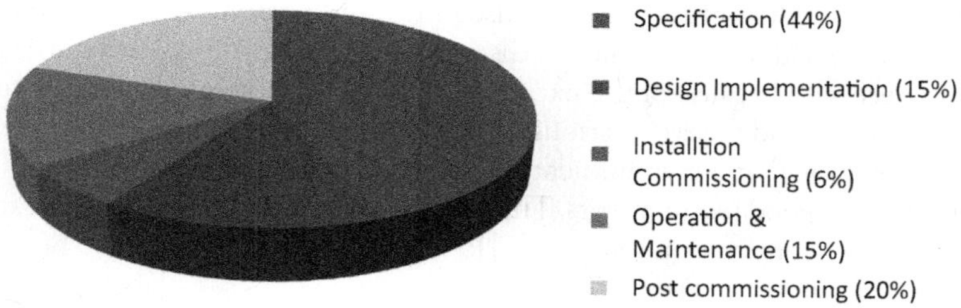

FIGURE X/1.0-1

Failure contributions.

1.0.1 System Specification Development

Defining requirements to establish specifications is the first step. It is necessary that requirements are established in a systematic way to ensure their accuracy and completeness. Requirements and specifications are very important components. There is a difference between "requirements" and "specifications."

- Requirements: A requirement is a condition needed by a user to solve a problem or achieve an objective. After a thorough study, user's requirements should be clarified and documented to generate the corresponding specifications. For establishing the requirements, the following brief discussions are helpful and important.

  Discussing requirements with the customer is very important. Without agreement between the customer and the designer, requirements cannot be established. Ambiguous requirements can be caused by missing requirements, ambiguous words, introduced elements etc. It is better to remove any ambiguity at the earliest opportunity. One of the most important things is to *ask questions*. Context-free questions are high-level questions (e.g., what environment is this product likely to encounter and what is the tradeoff between time and value?). These questions are posed early in a project to obtain information about global properties of the design problem and potential solutions. Errors developed during the requirements or specifications stage may lead to errors in the design stage, so, when an error is detected, the engineers must review the issue to fix the problem as soon as possible. This may cost additional time but could indicate errors in assumption or other requirements.
- Specifications: A specification is a document that specifies, in a precise, verifiable manner, the requirements, design, behavior, or other characteristics of a system, and often the procedures for determining whether these provisions have been satisfied.

  For establishing requirements and developing specifications, the appropriate people must be involved. A meeting should be arranged where a facilitator can help to resolve issues like conflicts, personality clashes, intergroup prejudices, etc. On account of the highly complex issues involved, the facilitator should possess knowledge from different engineering disciplines and possess experience in the area. So, people with high technical skills are necessary. Also good communication skills are necessary because of the exchange of information between the customer and the designer, and to settle various issues among participants. Therefore people with both technical and communication skills would be necessary for the establishment of good requirements. The following are the major steps necessary for establishing the correct requirements [11]:
  - Ambiguity poll
  - Technical review
  - User satisfaction test
  - Existing products

  For a definition of system requirements the following points will be useful:
- Definition of mission: Identification and establishment of the primary operating mission along with alternative and secondary missions
- Parameters and performance: Defining operational characteristics and functional requirements
- Use requirements: Marrying system with user requirements
- System details: Quantification of equipment, personnel, area requirements, etc. and location details
- Operational life cycle

Costing, failure rates, mean time between maintenance, maintenance downtime, the environment (environmental condition), etc. are major factors that affect the development of requirements of specifications. Fig. X/1.0.1-1 shows typical stages involved in developing specifications from the definition of requirements.

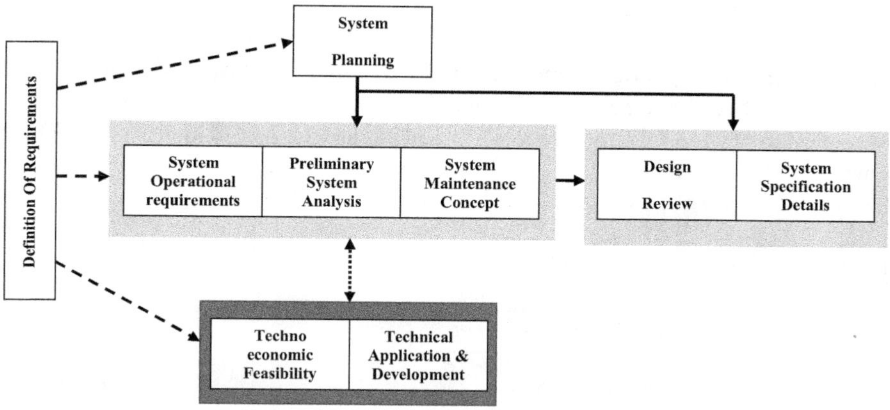

FIGURE X/1.0.1-1

Stages of development of specifications.

1.0.2 Functional Safety Challenges

In spite of the fact that almost all requirements for SIS in process and other industries are stated in the international standards, it is still challenging for engineers to justify and estimate investments in SIS during initial phases [1]. So, the frontend engineering and design team starts with a predefined SIS design when the process conceptual design is not well defined. A predefined SIS (architecture, technology, testing philosophy) could lead to a nonoptimal SIS design, that is, the SIS could be overdesigned or underdesigned. Thus this could be a starting point, but subsequently this needs to be corrected through a systematic approach toward developing a specification. As seen in Fig. X/1.0-1, a poor quality specification is responsible for major accidents. The two major challenges for SIS design are [9]:

- Reduction in random hardware failures
- Reduction in systematic faults

The former could be addressed by appropriate selection of equipment, while to address the latter a good quality specification needs to be developed. It has been found that 60% of SIS failures are embedded before commissioning and 85% of such faults are related to engineering. Therefore importance and requirements for a systematic approach in developing a specification cannot be overestimated.

1.0.3 Certified Functional Safety Experts

As per IEC 61511 one sees that "Persons, departments, organizations involved in safety life cycle activities shall be competent to carry out the activities for which they are accountable." Therefore involvement of certified functional safety experts at an early stage is helpful in appropriate equipment selection and proper framing of specifications. With proper planning, companies can manage inherent operational risks and avoid production outages, property damage, environmental impact, and damage to personnel. Experts help to obtain certified hardware, software, field devices, and engineering services [such as development of SRS, safety instrument function (SIF) designs, and SIL verification calculations] needed to meet the requirements of the safety system life cycle.

1.0.4 *Functional Approach Toward Safety Requirement Specification*

In line with IEC 61511 the requirements of SIF, hence SIS, can only be determined after completion of the conceptual process design and process hazard analysis (PHA), as shown in Fig. X/1.0.4-1.

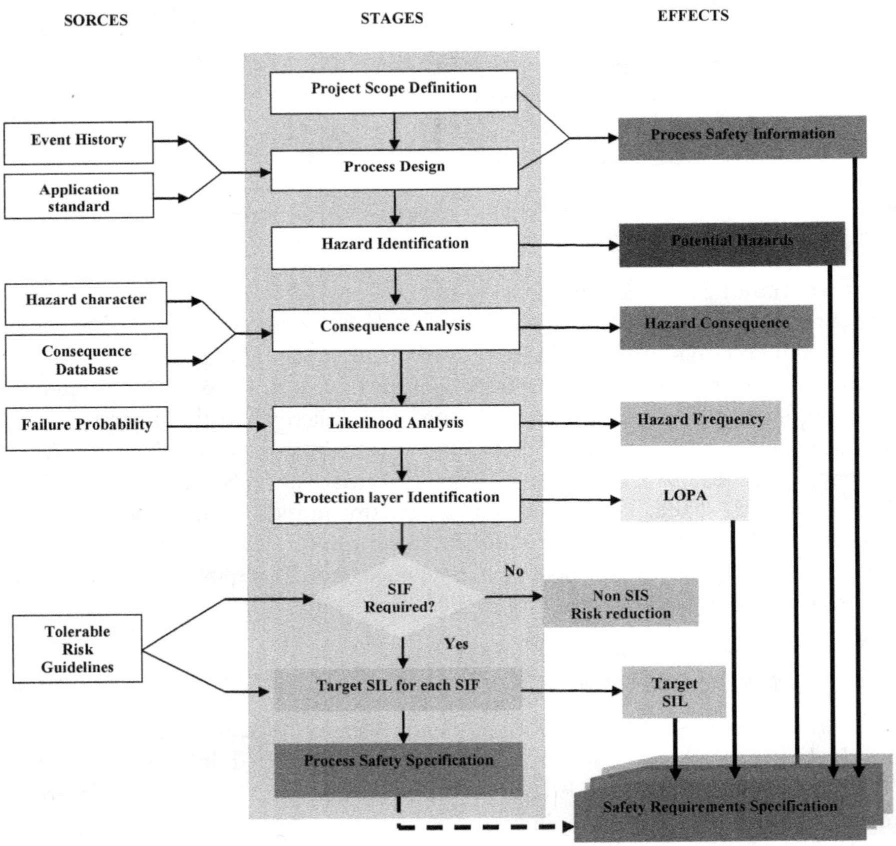

FIGURE X/1.0.4-1

Approach to safety requirement specification (SRS) as per IEC 61511. *LOPA*, layer of protection analysis; *SIF*, safety instrument functions; *SIL*, safety integrity level; *SIS*, safety instrumentation system.

The system starts with project/process definitions and process conceptual design. The event history and application standards are applied to the conceptual design to obtain the process safety information. Now, the entire system is subject to PHA to obtain details about various consequences. After applying the available protection layer the situations are reassessed for SIF. As discussed earlier, target SIL assessments are also done. So, finally, all the following components form the SRS for SIS:

- Process safety information
- Potential hazard
- Consequence of hazard
- Available protection layer
- Target SIL

  Tolerable risk guidelines actually help in deciding the SIF requirements and associated target SIL.

1.1 Safety Requirement Specification

According to IEC 615111:2003 (Clause 3.2.78) SRS is defined as "specification that contains all the requirements of the safety instrumented functions to be performed by the safety instrumented system" and the basic objective of the SRS is to specify all the requirements of the SIS to be utilized during detailed design and safety information systems. In this clause coverage shall be on the design engineering part, that is, requirements to design SIS to provide the necessary SIFs and meet SIL. SRS forms the basis for further planning for the design of SIS. Also because SRS includes the associated tests and test criteria, it also forms the basis for verification and validation. The implementation phase starts with the design and detailed planning of the SRS (and other measures) for reducing risk. Design and planning are followed by installation, commissioning, and validation of the plant.

Since different people work as per SRS during different phases of the safety life cycle, it is necessary that SRS is developed in an easily understandable manner. SRS includes the *functional description* of the safety functions as well as all the conditions that cause them to be triggered. In addition, determination of the SIL is part of the detailed consideration of each safety function.

The cause and effect method (see API RP 14C guideline) is one of the easier, proven, and extremely effective ways for the description of safety functions and shutdown conditions. No special programming knowledge is required for this. So, process specialists can also directly define their requirements. The causes are defined in the rows and probable effects are shown in columns. The linking of several causes and the definition of the relationship between causes and effects is carried out at the intersections of the rows and columns.

Fig. X/1.1-1 shows another way, where various system requirements and SIF requirements in conjunction with SRS inputs form SRS to get an SRS output. Now we shall look deeper into SRS and discuss how various inputs and requirements form SRS to establish details in the form of SRS output.

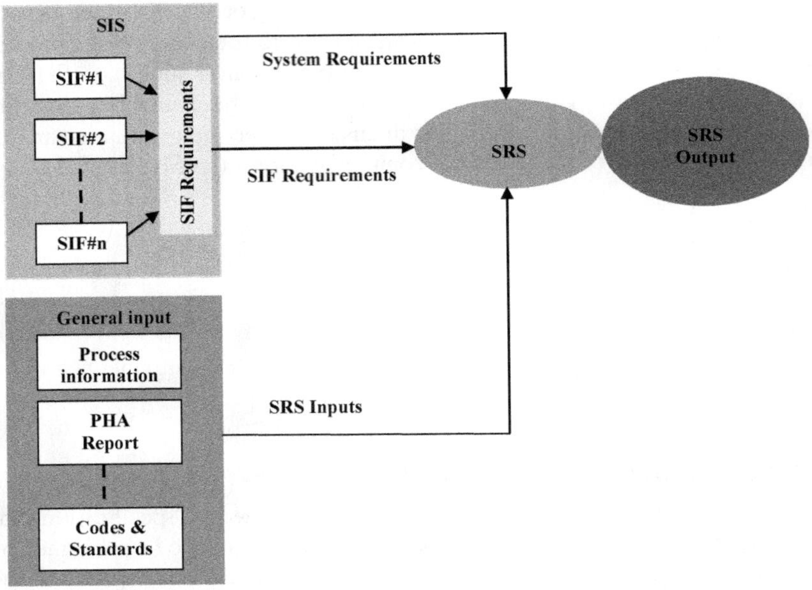

FIGURE X/1.1-1

SRS blocks. *PHA*, process hazard analysis; *SIF*, safety instrument functions; *SIS*, safety instrumentation system; *SRS*, safety-related system.

1.1.1 SRS Development Personnel

Who is responsible for development of SRS documentation? As safety is very much related with process, some think that this should be developed by a process contractor. It is better if it is developed by a team of engineers. As this end user is finally responsible for it so, it is better if it is initiated by the end user's instrumentation engineer. There will be a great deal of information exchange and interaction on the issue, so the end user's instrumentation engineer needs to be involved to coordinate with in-house process engineering and the safety engineering department personnel. In addition, the SIS supplier engineering contractor and process supplier engineers should be involved in developing the SRS.

1.1.2 SRS Documentation

A number of documentations are generated for SRS development. However, the quantum of such documentations depends highly on the complexity of the project. So it is not possible to be precise about such documentations. However, generally, documentations under the following headings are usually generated in connection with SRS developmental activities [5–7]:

- Design details
- Architectural details
- Reliability study
- Target SIL/availability

- Hardware specification
- Software development details
- HMI details
- Installation and testing
- Operating procedure
- Maintenance requirements

  All such documentations are developed for *each* SIF. There are three components or parts in SRS requirements and associated documentation, such as:
  - General
  - Functional
  - SIL

    So the basic format for SIS consisting of *n* SIFs can be conceived as shown in Fig. X/1.1.2-1.

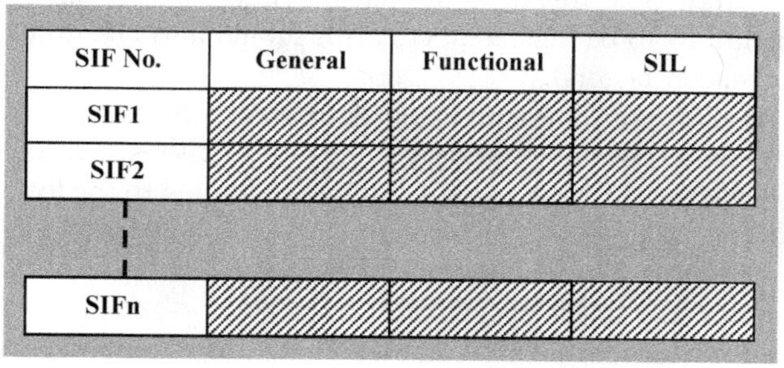

FIGURE X/1.1.2-1

Typical safety-related system (SRS) format for *n* safety instrument functions (SIFs) in a safety instrumentation system (SIS).

1.1.3 SRS Input

These are general requirements or input necessary to develop SRS. These shall include but are not limited to the following:

- Process information details: Process description with well-defined scope and boundary and supported by relevant drawings is necessary to acquire the process parameters. Special/specific conditions for the process also need to be described.
- Regulations: Local/national regulations and international codes and standards under whose purview the facility falls must be included as it is necessary for compliance.
- Environmental condition: The environmental condition expected for the facility especially for extreme conditions should be mentioned.
- PHA report: The reports of PHA conducted shall also be treated as an important input because they highlight various hazards, occurrence frequency, and consequences.

- Operational modes: Various modes of operation of the plant/facility including normal and abnormal operations as well as startup/shutdown mode shall be included.

1.1.4 SIF Requirements

Based on clause 10.3.1 of IEC 61511-1:2003, the information listed here is mainly applicable for each SIF:

- List of all SIFs in SIS, along with description of SIF
- Unique name or ID for SIF
- Process description of SIF
- The required target SIL for SIF
- Demand mode and demand rate for SIF
- Safe state of the process for SIF: As per IEC 61511-1:2003 "state of the process when safety is achieved" (the description should address in detail the starting, stopping, or running of a stream, sequential shutdown, which valves to be operated, which flow to be regulated, etc.; this should also take into consideration human factors).
- Proof test interval for SIF: This is an important issue. Proof tests are meant to test the function as far as possible (could be dangerous in a shutdown condition). Major issues related to this shall include but not be limited to the following [12]:
    - Description of the proof test procedure
    - Precaution and safety to be under taken for proof test
    - Required proof tests in life cycle
    - Human aspect: Consequence because of wrong proof test procedure
    - Documentation of proof test activity
- Process safety response time consists of three components:
    - Process time constant and dead time in process response
    - Electrical timing: Scan time of controller, or control system delay time
    - Mechanical inertia
- Reset requirements for SIF
- Definition and clearance for bypass/inhibit/override for SIF
- Alarm and trip [3] details for SIF including:
    - Architecture
    - Number and type of input
    - Measurement range
    - Trip level
- Fail safe and fault tolerance for SIF
- Additional SIF requirements to describe how it works:
    - Process parameter normal range
    - Process output and action
    - Operator interface and action
    - Energize/deenergize to trip
    - Action on loss of power
    - Response on overt fault

- Requirement of diagnostic coverage (DC) to achieve target SIL
- Requirement of maintenance and testing

1.1.5 SIS Requirements

SIS requirements shall include but not be limited to the following. Also many of these will be determined from corresponding data of various SIFs under SIS:

- Dangerous combination of outputs
- SIS and other system *interface*
- Feasible mean time to repair (MTTR) for SIS
- Defined failure modes and response from SIS
- Sources of common cause failure (CCF)
- Response time of SIS for safe state
- Manual shutdown requirements
- Maximum allowable spurious trips
- Action to bring to safe state in case of fault detection

1.1.6 SRS Information

Based on the inputs discussed in Clause 1.1.3—1.1.5 an SRS shall be developed, which shall include at least the following information in line with IEC 61511-1:2003:

- Description of all the SIFs necessary to achieve the required functional safety
- Requirements to identify and take into account CCFs
- Definition of safe state for each SIF
- Definition of individual safe process state, which when it occurs concurrently may create a separate hazard (e.g., multiple relief to flare system).
- Assumed sources of demand and demand rate
- Requirements of proof test intervals
- Response time of SIS to bring the system to a safe state.
- SIL and mode of operation (on demand/continuous) for each SIF
- Description of process measurements and trip points
- Description of SIS process output action and criteria for successful operation
- Functional relationship among process input and output including logic, mathematical functions, and permission
- Requirements for manual shutdown
- Requirements for energize/deenergize to trip
- Requirements for resetting after trip [e.g., for purging in burner management system (BMS)] [5]
- Failure modes and desired response of SIS (e.g., first out alarm)
- Maximum allowable spurious trips
- Procedure for starting and resetting SIS (if any)
- SIS interface with other systems [e.g., with basic plant control system (BPCS), fire and gas system (FGS), etc.]
- Description of the modes of operation of the plant and identification of SIFs for each mode

- Requirement, definition, and clearance for bypass/inhibit/override; to specify any action needed to achieve and maintain safe state for such actions human factors should be taken into account
- Application of software safety function as per IEC 61511-1: 2003 (Clause 12.2.2)
- Feasible MTTR taking into account location, spares, etc.
- Dangerous combination of outputs to be avoided
- Extreme environmental conditions (to take into account temperature, humidity, electromagnetic interference, radiofrequency interference, etc.) [4]
- Identification of normal and abnormal modes of operation and requirements to survive a major event
- Definition of SIF necessary to survive a major accident [8]
- Identification of security measure during hazard and risk analysis to prevent malevolent and unauthorized actions [8]

The discussions on SRS will not be completed until specific reference is made to software development. For software development, the IEC recognizes (Clause 12.2.2 of IEC 61511-1:2003) three types of software and three software languages as depicted in Fig. X/1.1.6-1.

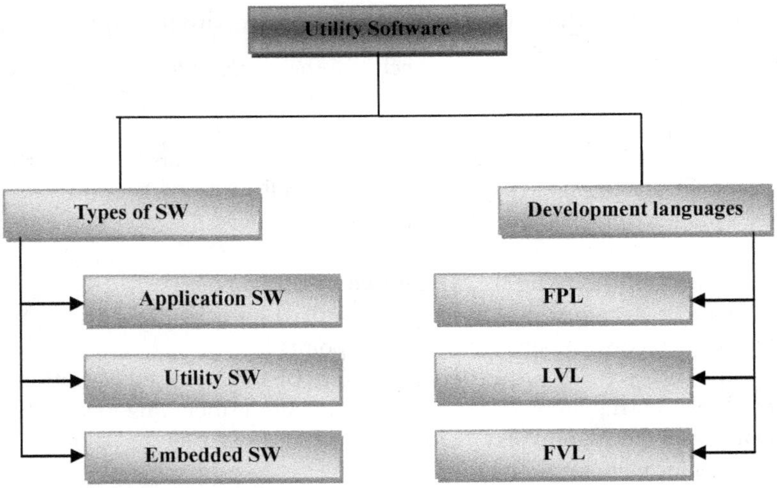

FIGURE X/1.1.6-1

Software types and languages.

1.1.7 *Alarm Annunciator*

The basic characteristics of alarm systems have already been covered at length in Chapter IX. Since an alarm is one of the most fundamental issues in a control system, it will be discussed here again in this context. In this clause, focus will be on independent alarm systems that can be applied to critical applications independently of a programmable logic controller/distributed control system (PLC/DCS). In spite of so much technological developments and modern instrumentation and control, because of the potential consequences to plant and personnel, "hard" alarms are still not only accepted but also considered as primary alarming functions and a backup to DCS and PLC strategies in critical emergency shutdown (ESD) and SRS (independent of PLC/DCS) in many industrial sectors/applications. One of the probable reasons could be that "soft" alarms can be susceptible to common-mode failures (in contrast to hard alarms). This is shown in Fig. X/1.1.7-1 where an alarm (and a trip signal) is generated from a transmitter with the help of a limit value monitor and is independent of PLC/DCS.

In view of the CCF effect, many feel that "soft" alarms may be inadequate in providing the degree of protection demanded for some critical applications. "Soft" alarms are based on scanning input whereas "hard" alarms provide continuous supervision of an individual process signal. In some fast-changing applications, the computer's scanning speed or network throughput time may be inadequate. Also "hard" alarms could be set very easily compared to soft alarms in which there is a need to understand programming and the programming errors that may occur. Alarms have a dead band (the difference between the set and reset point) and this can be adjusted easily in hardwired alarms. With a simple "cause and effect" action, limit alarms can be economically used in a wide variety of basic and complex applications [3]:

- Warn of a hazard with "hard" alarm when the signal goes beyond a set limit
- Create an independent trip system to avert central power failure or DCS shutdown
- Back up "soft" alarms from DCS or PLC

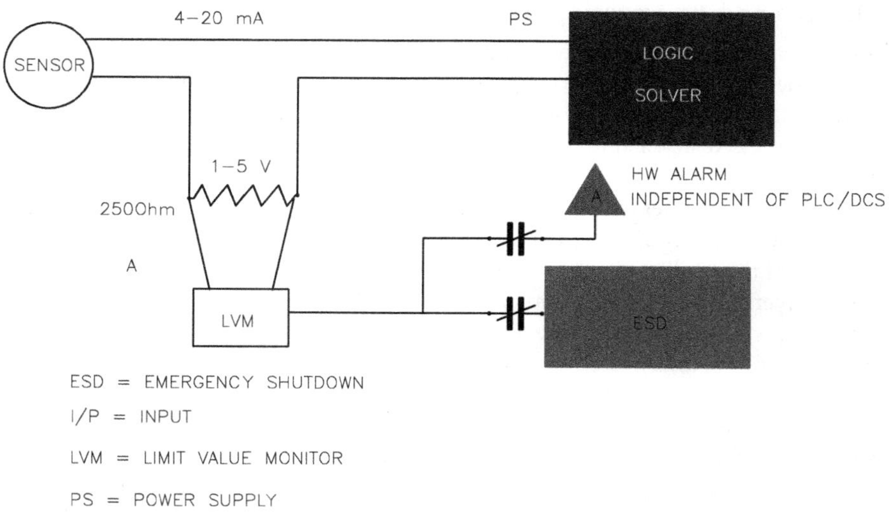

FIGURE X/1.1.7-1

Hardwired alarm for safety. *DCS*, distributed control system; *ESD*, emergency shutdown; *LVM*, limit value monitor; *PLC*, programmable logic controller; *PS*, power supply.

- Reliable without CCF and cost-effective
- Monitor parameter annunciates when parameter rate of change exceeds a selected rate.

There are various types of hardwired alarms in industrial use:

- Input fault alarm
- Rate of change alarm
- Self-diagnostic alarm
- Window alarm
- Average/differential alarm
- Fail safe/nonfail safe alarm

Most of these are standard alarms and details are available in any standard instrumentation book. When the average of two or three input signals exceeds a set point, an average alarm is generated. A differential alarm is generated when the difference between two input signals, such as two resistance temperature detectors (RTDs), exceeds a specific value. Fail safe and nonfail safe alarms [3] are clearly depicted in Fig. X/1.1.7-2. In the case of a fail alarm sensed by a normally closed contact as shown, the relay deenergizes. However, this also happens when power fails. So in a fail safe alarm it is also annunciated because power failure is not a normal condition. For this reason this is fail safe.

In the case of a nonfail safe alarm if power fails and during that period a fault occurs, the relay cannot operate to annunciate. So this is nonfail safe. These alarms are standard issue. With this, discussion on SRS is concluded and we can now move on to discussions on design and engineering.

1.2 Design and Engineering

Having completed SRS it is time to implement it. This implementation naturally starts from the design and engineering phase. Before tackling the technical details it is better to understand this phase with one hypothetical example of a tank-filling process, as discussed in Chapter VII (Fig. VII/1.1-3). To understand design implementation, Fig. X/1.2-1 may be referred to.

As stated in the banner of the figure, when the pressure inside the vessel, sensed by two pressure sensors (SIL2) in redundant mode, exceeds the set point, it will initiate the stopping of the feeding pump and close the supply line block valve with the help of a logic solver. Here, in the example, sensors, final elements, and the safety functions to be performed are all identified i.e., the logic solver. Minimum of five requirements are necessary discussed subsequently. Until now only three have been specified. Let the process be SIL3 safety and the process safe time be 6 s. It is common practice to consider *half of process safe time as time of response* for the safety system. Therefore the safety response time necessary is 3 s. So, all the minimum five requirements, namely, sensor, logic solver, final element, SIL requirement, and process response time, are specified. Here, sensors are redundant architecture (see Fig. VIII/1.3-2 for two SIL2 sensors in redundant mode equivalent to SIL3) and can be considered as SIL3. Fig. X/1.2-1 shows how SRS requirements have been translated into design engineering in implementing the requirements. The drawing also shows

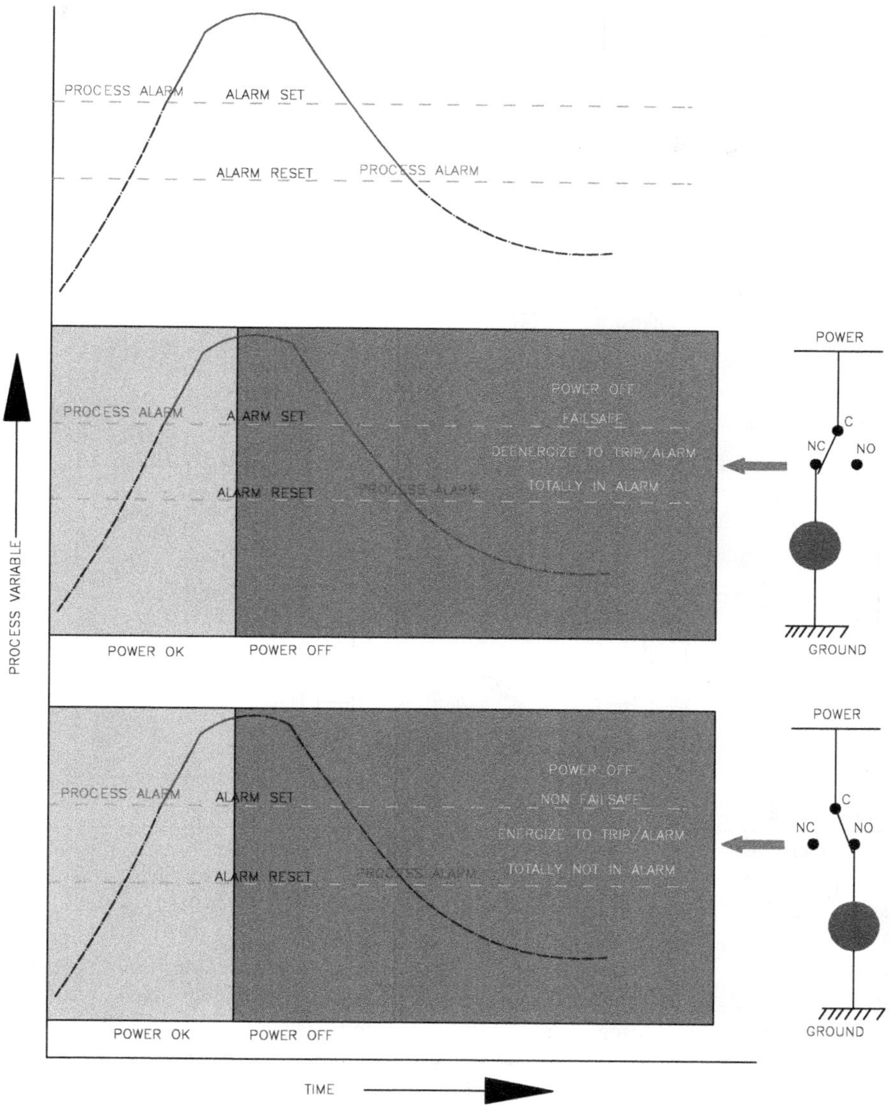

FIGURE X/1.1.7-2

Fail safe and nonfail safe alarms. *C*, contact; *NC*, normally closed; *NO*, normally open.

how the wiring has been done and how commands are issued. In the case of field bus application there will not be individual physical wiring exactly as shown in the drawing; instead in those cases, software implementation, that is, software communication, needs to be taken into account. In this example, details regarding the logic

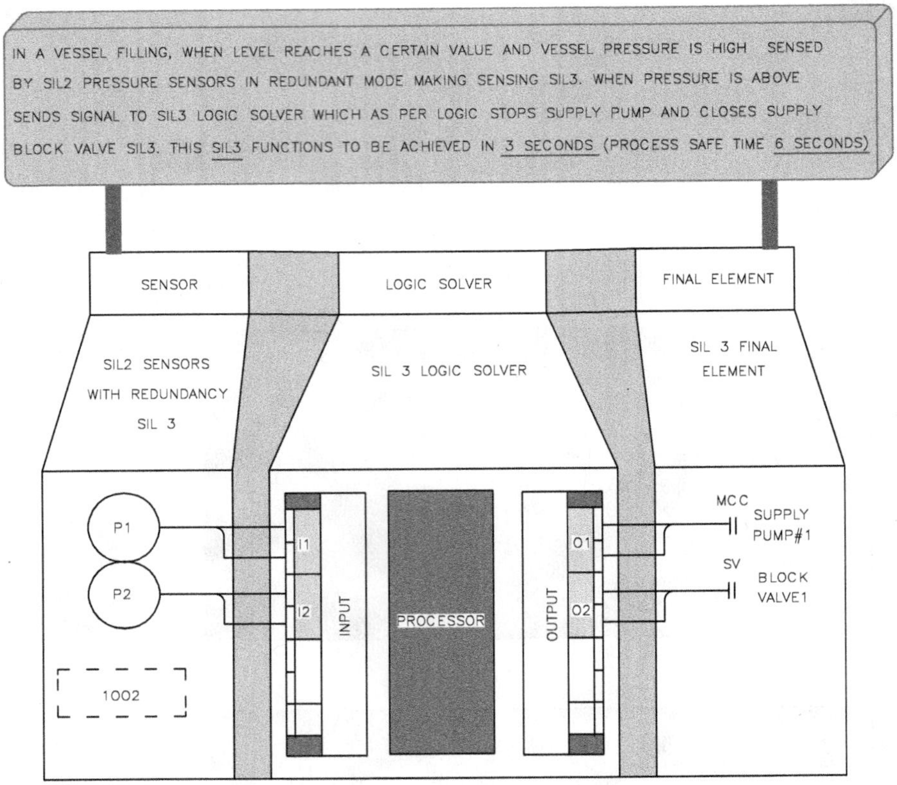

FIGURE X/1.2-1

Design implementation. MCC, Motor control center; SV, solenoid/shut down valve etc. *Based on T.V. Capelle, M.J.M. Houtermans, Functional Safety: A Practical Approach for End-Users and System Integrators, HIMA Paul Hildebrandt GmbH Co KG, June 2006. White Paper.*

solver are missing. To look deeper into the logic solver and other components let us visit Fig. X/1.2-2 which shows much detailing about logic solver issues such as hardware/software/other integrating details. As stated earlier, in cases of SIS application it is better to deploy safe PLC, which is now offered by all major manufacturers as indicated in Fig. X/1.2-3. In the case of sensors and final elements, in addition to wiring, software communication details have been shown to take care of the field bus system, if any. Thus from the two examples an attempt has been made to establish how various requirements in SRS could be implemented in design and engineering. The two examples are meant to give the idea that at the design and engineering stage proper planning shall be developed to validate all SRS requirements and testing at every subphase. Now, discussions will be focused on various issues related to design and engineering systems—the first step toward implementation.

1.2.1 Resources

Because this is the design phase, various vendors will be involved such as engineering consultants, process vendors, control systems, field equipment suppliers, etc. As per

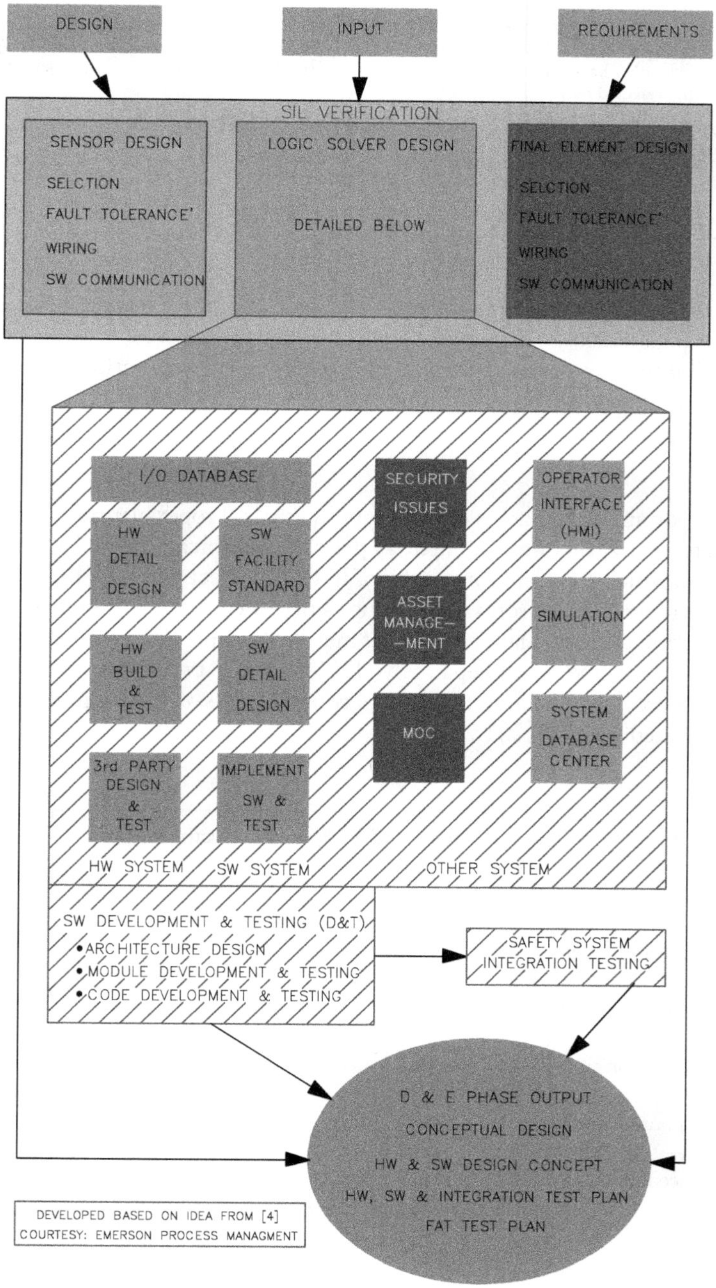

FIGURE X/1.2-2

Outline of design and engineering process. *D and E*, design and engineering; *FAT*, factory acceptance test; *HMI*, human–machine interface; MOC, Management of change; *SIL*, safety integrity level.

Safety system: With the introduction of international standards on safety systems for electrical/electronics /programmable electronics (E/E/PE) systems, safety systems have gained so much importance that all major automation suppliers either have a separate department or a whole subsidiary company to meet the challenges. A few examples are presented here:

ABB, *Safety Systems*; Emerson, *Smart SIS*; GE Fanuc *SafetyNet*; Honeywell *Safety Manager*

Invensys, *Triconex*; Rockwell, *ICS Triplex*; Siemens, *SIS*; Yokogawa, *ProSafeRS*

FIGURE X/1.2-3

Safety system companies.

IEC, there shall be competent key personnel with defined lines of responsibilities from all concerned organizations. Those who are significantly involved must understand their responsibility in analysis, assessment, validation, or verification thoroughly; otherwise the results may be disappointing. In this connection, Part 3 of IEC 61508 may be referred to.

1.2.2 *Planning*

Planning is important in all phases of the safety life cycle. Proper planning is required to demonstrate that SIS is implemented as per SRS. Verification shall run parallel with validation, which starts after integration. In this connection Fig. VIII/2.0.3-1 (V model) may be referred to for software development. At each phase, results should be verified. Validation and verification activities may be documented directly as a part of SIS planning or may be listed separately. IEC endorses two types of document, namely, the SIS progress plan and the test procedure. Details of documentation in this regard are indicated in Table X/1.2.2-1.

***Table X/1.2.2-1** Validation and Verification Documentation*

SIS Progress Plan	Test Procedure
SIS validation and verification activities	Description of test procedure setup
Timing activities takes place	Environmental requirements
Procedure for verification	Test strategies
Person/organization responsible	Responsible person for test and assessor
Cross-reference with test procedure	Pass/fail criteria
X	Steps to verify all safety requirements are listed
X	Steps to verify correct operation under all modes of operation and abnormal conditions

SIS, *safety instrumentation system*; X, *not applicable*.

1.2.3 *Design and Engineering Inputs and Requirements*

In this clause, various inputs and requirements for this phase shall be discussed:

- Input: At this phase, inputs come from SRS discussed previously; and these shall include but not be limited to the following:
    - Accurate specification for functional details of SIF
    - Specific SIL for each SIF
    - Process detailing
    - Environmental requirements
    - Responsibility chart for person/organization
- General requirements: Listed are the major general requirements for the design phase:
    - Nonqualified BPCS shall be independent of SIS.
    - Operability, maintainability, and testability to facilitate human factor requirement.
    - HMI design to take care of human limitation and capability.
    - Human limitation and capability considerations for specific task
    - Unless prohibited in SRS, final element can be operated manually.
    - Non-SIS affecting SIS negatively needs to meet highest SIL.
    - Any device in SIF whose failure can cause loss of process control *cannot* be used in BPCS.
    - Minimum test interval requirement
    - Placing process in a safe state until it is reset.
    - Loss of power (see Clause 1.2.4)
- Subsystems and component requirements: The following are the basic requirements:
    - To comply with IEC 61508-2 and 3:2010
    - All components of each SIF shall have a minimum tolerance level.
    - Suitability to be demonstrated for selected subsystem/components.
- Software design requirements:
    - Defined interface and communication (see Clause 1.2.4)
    - Verification with respect to SRS:
        - Input/output (I/O) point types and service
        - Installed spare capacity
        - All SIF listed in SRS in tabular form
        - All items with SIL description
    - Verify area allocation consistent with BPCS
    - Consistent alarm philosophy and priority with BPCS
    - Alarm has sufficient details for operator response.
    - Verify if any HOLD has been considered.
    - Ensure operator interface details have been considered.

1.2.4 *Important Considerations for Design and Engineering Implementations*

There are a few important points to note while designing a system:

- Interface: The interface between sensor and logic solver and the interface between logic solver and final elements must be properly defined especially when there is external equipment like an ex barrier, isolator, etc., whether or not an ex barrier is part of the sensor or logic solver. In offshore practices, in line with guidelines from [13], it is the point where filed devices terminate. Naturally, an ex barrier, etc. will be part of the logic solver. In any case this must be well defined.

- Field devices: A few interesting points are listed:
  - Redundancy: In the case of *specially certified* sensors, redundancy may be reduced. Redundancy is more a part of availability than safety, as already discussed.
  - Mounting consideration: This is very important to avoid common-mode failure because of environmental conditions, attaining true parameters, etc.
  - Tubing/other lines must be properly sized and saved from mechanical damage. Pneumatic tubes must have optimum design so that they have sufficient capacity to open/close valves and not be oversized so as to cause response delay.
  - Alarm: Any SIL device when bypassed, inhibited, and overridden must be annunciated. Also such activities shall be protected against unauthorized access.
- Energize/deenergize to trip: There are some cases where energize to trip may be necessary and a power supply monitoring facility is necessary, whereas in some cases deenergized to trip is used. In some BMSs, for example, tangential tilt burner of combustion engineering (CE) make, a separate power supply for unit (DC) and corner (AC) is used to avoid this situation. In a BMS/furnace safeguard and supervisory system, power supplies are driven from two sources as stated. In the case of AC failure when corner logic final elements are not functional, by monitoring the AC supply and using DC supplies the boiler could be tripped. Similarly, in case of DC failure when unit logic final elements are not functional, the DC supply is monitored and tripping is done via the AC logic system. Further details are available in [2].
- Utility: Due consideration shall be given to redundancy in a power supply where it can improve system availability. However, in such cases due consideration should also be given to proper protection, fault level, and changeover.

1.2.5 *Software Development and Testing*

This forms an important role in the design and engineering phase. Based on software design requirements, proper architecture is developed for further development and testing of software, as shown in Fig. X/1.2-2. The major issues are:

- Module development
- Module testing
- Code development
- Code testing

1.2.6 *Design and Engineering Phase Output*

The major design issues finalized at the end of this phase are shown in Fig. X/1.2-2.

From Fig. X/1.2-2 it is seen that (based on design, inputs from SRS, and various requirements) at the end of the design and engineering phase the following are the major design details used in the later part of the life cycle:

- Conceptual safety design of the facility
- Hardware conceptual design
- Software conceptual design
- Hardware and software testing plan
- Integration testing plan
- Factory acceptance test (FAT) planning
- FAT testing details

1.2.7 *Documentations*

Major documentation in this phase shall include but not be limited to the following:

- Safety analysis report
- System description
- System block diagram
- Control system architecture (topology)
- System operational writeup including:
  - Modes of operation
  - Abnormal conditions
- Component failure rate
- Voting
- Diagnostic coverage
- MTTR
- CCF
- Testing description
- Other project-specific documents

1.2.8 *Factory Acceptance Test*

This is a part of the design phase. IEC 61511-1:2003 Clause 13.1.1 defines the objective of FAT.

- Objective: "The objective of a factory acceptance test (FAT) is to test the logic solver and associated software together to ensure it satisfies the requirements defined in the safety requirement specification. By testing the logic solver and associated software prior to installing in a plant, errors can be readily identified and corrected."
- Recommendations: FAT is to be specified during the design and engineering phase and should take care of the following with suitable planning:
  - Development of integration test *joint* effort from designer and supplier
  - Activities following the design stage but prior to the installation stage
  - Applicability to subsystem of SIS irrespective of programming facility
  - Testing to take place in factory environment

- Planning for FAT takes care of the following aspects:
  - Type of test
  - Test case
  - Test description
  - Test data
  - Dependence on others
  - System/interface
  - Test environment
  - Test tools
  - Logic solver configuration
  - Criteria for test completion
  - Test location
  - Corrective action procedure for test failure
  - Competence of test personnel
  - Defined version of logic solver
  - Each SIF and performance characteristics
  - Each test: Test procedure and description
  - Each test's chronological description of activities
  - Test tool, environment, and interface details for each test

All tests should be properly documented as per IEC guidelines. In this connection Clause 5.1.6 of Chapter VI may be referred to.

The position of FAT is between the design stage and the installation stage. So, with discussions on design phase now complete, we next move on to the installation and commissioning in the life cycle realization phase.

1.3 Erection Commissioning and Validation: Safety Life Cycle

This phase comes after the design phase. In this phase there is another subactivity, mechanical completion. On account of special attributes, mechanical completion is treated separately. In reality, mechanical completion is an interface between construction and commissioning, as is elaborated in Fig. X/1.3-1.

1.3.1 *Objective of the Phase (E&C)*

This phase ensures that SIS has been installed as per SRS and commissioned for final validation. Various issues related to the objectivity of this phase shall include:

- Installed as per specification and associated drawings

Mechanical completion: ***This is a part of management for verification of approved designed installation. It basically represents the interface between construction and commissioning and should include all interface checks. It is the transition from installation phase to precommissioning phase, but in many cases it is seen as a part of commissioning. On account of some special attributes of mechanical completion, it is treated and planned separately.***

FIGURE X/1.3-1

Mechanical completion.

- Mechanical completion (Fig. X/1.3-1) finished and system ready for final validation
- Validation through inspection and testing of installed SIS to ensure that the system meets SRS requirements

1.3.2 Personnel

Well-identified personnel/departments/organizations need to carry out the activities and review the work. Also they must be well aware of the responsibilities bestowed on them. All necessary authorities (e.g., Det Norske Veritas) should be informed and involved as necessary.

1.3.3 Planning

There are two planning stages involved here, namely:

- Installation of mechanical completion and commissioning:
  - Completion of all activities related to the foregoing
  - Procedures, measures, and techniques for installation and commissioning
  - Timing for activities to take place
  - Responsibilities as discussed in Clause 1.3.2
- Validation planning to define activities in this subphase (IEC 61511-1:2003):
  - Validation activities including validation of SIS with respect to SRS and implementation and recommendations
  - All modes of operation of the process and equipment including:
    - Setting and adjustment
    - Startup, automatic, manual, semiautomatic, steady state of operation
    - Resetting, shutdown, maintenance
    - Reasonably foreseeable abnormal conditions, and those identified in risk analysis
  - Procedures, measures, and techniques
  - Timing for activities to take place
  - Responsibilities as discussed in Clause 1.3.2
  - Reference information for validation
  - Additional validation activities:
    - Identification of software for validation
    - Information on technical strategies, measures, and procedures
    - Environmental requirements in which validation should take place
    - Pass/fail criteria
    - Evaluation policy and procedure

      Detailed discussions of these are available in Clause 5.2.4 of Chapter VI.

1.3.4 Installation and Commissioning Activities

Major installation and commissioning activities in the implementation phase shall include but not be limited to the following:

- All installation activities completed as per plan
- Proper connections to earth (grounding)

- Proper connection and operational state of energy sources
- Transportation activity stopped and packing materials removed
- No physical damage present
- Proper *calibration* of instruments
- All field devices, I/Os, and logic solvers *operational*
- *Interface* to other systems operational
- Proper recording of *commissioning* activities
- Evaluation of *nonconformity* including installations by competent person
- If the foregoing nonconformity has no impact, then design documents need to be updated; in the case of an impact the documents need modification.

1.3.5 Validation Activities

Major activities related to validation shall include but not be limited to the following:

- SIS performs under normal/abnormal conditions
- No adverse interaction of BPCS with other systems to impact SIS
- Proper established communication between BPCS and other systems
- All SIF components (hence SIS) including redundant channels perform as per SRS
- SIS documentation consistent with installed system
- SIFs perform as specified with invalid process variables
- Proper shutdown sequence activated
- Proper alarm and display by SIS
- Correct computations included in SIS
- Bypass and manual shutdown function correctly
- Reset of SIS functions as per SRS
- Proof test interval included in maintenance procedure
- Diagnostic alarm performs as required
- SIS performs as required in case of loss of power
- SIS goes to desired state on power resumption
- SIS EMC immunity as per SRS

1.3.6 Outcome

There should be an appropriate record that SIS has been properly installed (as per design and installation plan), commissioned, and validated. For validation the following points are also very important:

- Validation plan version
- Tested hardware and software version
- Test specification version
- Each SIF function number referenced to SRS
- Tools and equipment used
- Calibration data for each of the foregoing
- Result of each test
- Discrepancy between expected and actual result
- Analysis and evaluation of discrepancies

The major responsibility of this phase lies with manufacturers in coordination with engineering consultants, owners, and suppliers. Next, O&M issues will be discussed to complete the discussion on the implementation part of the life cycle.

1.4 SIS Operation and Maintenance

1.4.1 Objective

The basic objective of this phase is to ensure that:

- SIS is operated in accordance with requirements of SRS throughout SIS life, and
- The required SIL of each SIF function is well maintained (i.e., not deteriorate below specified integrity level) with maintenance of functional safety of SIS.

1.4.2 Requirements

There are a number of requirements for the O&M phase. Also there shall be a number of considerations for operation of the system and considerations for maintenance of the system. All these requirements along with separate considerations for O&M are detailed in Fig. X/1.4.2-1. Since these are well depicted and self explanatory they are not explained again.

1.4.3 O&M Procedure

The O&M procedure with details depicted in Fig. X/1.4.2-1 needs to be developed. The procedure shall be developed taking into account various factors from requirements and considerations for O&M in light of the main objective. Major issues toward developments of procedure are elaborated in Fig. X/1.4.2-1. Documentation is an important part of the procedure. These documentations are subject to statistical analysis, which at a later date will provide the means for updating the procedure, as shown in Fig. X/1.4.2-1.

1.4.4 Outcome

Appropriate recording and documentation of the procedure is as follows and is briefly shown in Fig. X/1.4.2-1:

- Maintenance of functional safety in normal and abnormal conditions of operation of the facility
- Actions to be undertaken to prevent unsafe condition and how to reduce consequences of hazards
- Handling of system failure and demand rates
- SIS audits and test results
- Maintenance procedure during failure situations
- Proper calibration of test equipment

For details on the O&M phase technical requirements, considerations, etc., Fig. X/1.4.2-1 may be referred to. With this discussion on SIS implementation concluded (excluding reference to management of change), the next clause will focus on SIL certification and proof testing.

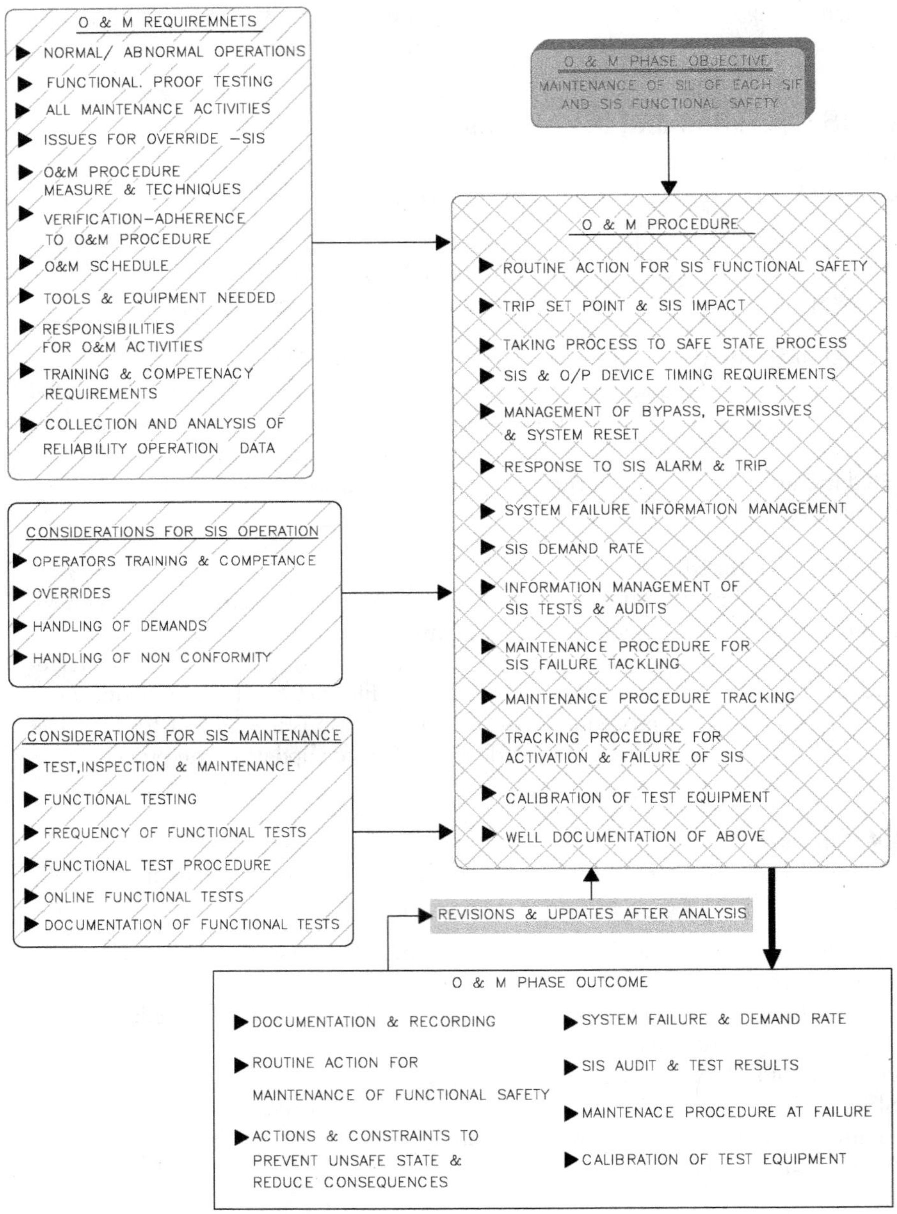

FIGURE X/1.4.2-1

Safety instrumentation system (SIS) life cycle operation and maintenance (O&M) phase details. *SIF*, safety instrument functions; *SIL*, safety integrity level.

1.5 Modification and Management of Change

Management of change (MOC) is part of the operation section of the life cycle and the last phase in the active safety life cycle. At first modification is proposed (due to any requirement), then it is followed by MOC. A brief MOC process is elaborated in Fig. X/1.5-1. On account of (say) modification in legislature or development in a software program there is a need for change in SIS design. This phase deals with how to handle these changes. From Chapter VI it is clear that all such changes require detailed analysis as well as authorization of the entire process.

1.5.1 *Objective*

The basic objective of this phase mainly includes the following:

- Modification of any SIS must be accompanied with proper planning, analysis, and authorization prior to incorporation of any such changes.

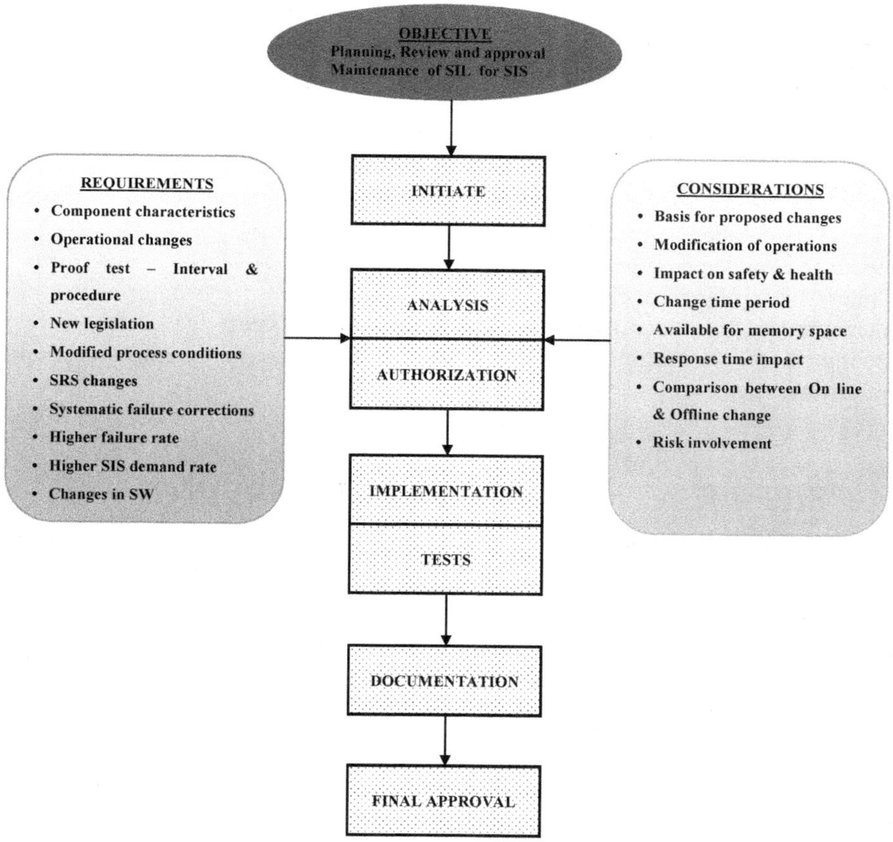

FIGURE X/1.5-1

Management of change (MOC) process. *SIL*, safety integrity level; *SIS*, safety instrumentation system; *SRS*, safety-related system.

- There must be an assurance that required safety integrity of the SIS is maintained even after the changes.

1.5.2 *Requirements and Considerations*

Based on the requirements for changes there are a few considerations for analysis, along with the main requirements of changes. Salient points are well elaborated in Fig. X/1.5-1 for MOC. The following issues need to be addressed prior to proposed MOC:

- Description of changes
- Reasons for changes
- Preliminary hazard identification
- Configuration history and audit trails

  In addition, testing is necessary for proper implementation of changes and impact, etc. This is stated under MOC.

  Overall, the MOC process is mainly concerned with:
- Changes
- Impact of changes
- Required testing
- Date
- Author
- Approving authority

These are detailed in Fig. X/1.5-1, which is self-explanatory.

With this implementation (actual application of Phases 3–7), the active part of the safety life cycle is concluded. Now, the focus will be on the other aspect of implementation of various requirements pertinent to the safety life cycle. In the following clause, concentration will be given to issues related to certification and proof testing.

2.0 CERTIFICATION AND PROOF TESTING

Certification and proof testing are two important steps in the safety life cycle process. In the first part, discussions will be on various issues related to certification. In the other part, the discussion will be on proof testing methods and proof test intervals.

2.1 Certification (By Third Party)

Certification is mainly concerned with the reliability of functional safety and associated testing and certification of the product for SIL. Normative standards establish minimum requirements for the design, testing, and performance of these safety devices necessary to maintain protection of personnel, property, and the environment already discussed. These standards provide specifications to address safety and reliability based on means and methods to reduce the risks to as low as reasonably practicable as well as optimizing the process for risks. These are applicable to instrumentation as well as FGS.

2.1.1 Basics Issues Related to Certification

Functional safety demands continue to increase. SIS designers (for end users to comply to IEC 61511) for electrical/electronics/programmable electronics (E/E/PE) prefer to integrate safety certified products (sensors, logic solver, and final element) into their systems and require manufacturers to provide certificates/assessments in regard to IEC 61508. At present almost all markets demand compliance to both functional safety testing and international standards for safety, especially compliance to IEC 61508 and IEC 61511. The IEC 61508 standard establishes the requirements for manufacturers to follow during product development to ensure that developed products offer a high level of resistance to random hardware and "systematic" design failures. Therefore this is extremely important for businesses seeking to expand internationally. Products and systems must possess high standards of safety, reliability, and quality. If products fail to meet functional and other safety requirements on account of faulty design/technical defects/faulty sequence of events/human failure, then the environment, people, and property could be endangered. In the process of third party certification, manufacturers are offered expert opinions to combat such eventualities.

With functional safety testing and certification one can expect to receive a number of financial benefits, for example:

- Compliance with domestic and international standards
- Reduced risk liability with documented safety standards
- Compliance from fulfilling requirements in the standard, hence better technical quality of the product
- The majority of certifying authorities have a global network, so it is possible to use these experts to integrate safety-related products produced at different locations into SIS and applications.
- Benefit from fast audit times and utilization of expert opinion especially in planning, design and testing of the products
- Possibility to challenge competitors on account of authentic third party certification

  Compliance is evaluated by a certifying authority, which assesses and certifies that a product has been designed and developed in accordance with the standard. Similar to third party certification and approval, a certifying authority provides benefits [15,16] like:

  - Assurances that safety-related products are built with sufficient functional safety to meet the intended SIL.
  - Proof that the product meets the requirements of the relevant product standards.
  - Assurance that the product can be used for the specified range of operations.

SIL is a qualitative measure of *safety* and a quantitative measure of the *safety level*. It is not just the probability of failure. Because of the complexity of today's devices, tests should be carried out parallel to product development. Streamlining development efforts are an important advantage for manufacturers.

2.1.2 Coverage and Documentation for Certification

The following are the major issues covered and documentation needed [15,16]:

- Total proposed facility documentation to be presented for documentation assessment. If recommended by a certifying authority, necessary modification/augmentation should be incorporated.
- Product development stages to be assessed
- Assessment of product design and testing process against IEC 61508
- Assessment of business process
- Relevant lessons learned assessment
- Operational safety assessment:
  - Technical assessment of product design to ensure requirements of standard are met.
  - Depending on applicability, assessment for "proven in use"
  - Assessment of failure modes, effects, and diagnostic analysis (FMEDA)
  - Assessment of probability of failure on demand (PFD)/probability of failure per hour calculations
- Assessment of validation planning and associated activities
- Assessment of verification activities
- Documentation of complete assessment results
- Final auditing of entire activities
- On successful final audit report, issuance of certificate
  Many of the certifying authorities use their own software tools (e.g., Exida's Safety Case tool) for the entire system.

2.1.3 Discussions on Various Points on SIL Certification

International standards IEC 61508 and 61511 do not mention specific requirements for acquiring/issuing SIL certification. Naturally, there remains scope for a variety of interpretations. From the foregoing it is clear that such certifications are issued for SIL ratings. It is therefore suggested to look at the following issues carefully:

- Why is SIL important? From Chapter VIII it is clear that SIL rating corresponds to the PFD for SIF, hence SIS. Also it is clear that the higher the SIL target, the lower the PFD and better the performance of SIS.
- What does SIL rating mean? SIL rating signifies that the entire system, not just an individual product, will reduce the risk to a specific level. Also from the discussions in Chapter VIII, it is clear that SIL rating of the system is dictated by the minimum SIL rating, hence SIL suitability of the individual device is immaterial when SIL rating of the entire system is inadequate.

- Certification: Both products and processes can receive such a certification. Certification for the former is most common and is issued by an independent agency to show that the appropriate SIL calculations have been performed and analysis has been completed on a product. Self-certification, though *not* common, is also possible. Such certifications are used to signify that it is compatible for use within a system up to the certified SIL. As discussed earlier, FMEDA is normally used to determine the safe/unsafe and detected/undetected failure modes of a product. FMEDA is useful for calculations of safety failure fraction and PFD. Although not common, full certification of IEC 61508 is also possible for manufacturer's design and quality processes.
- Who can issue certificates? According to the standards there are no preferred agencies or companies that have the authority to issue SIL certifications. However, Clauses 8.2.15—8.2.18 along with Tables 4 and 5 of IEC 61508-1: 2010 do specify minimum levels of independence of those carrying out a functional safety assessment. It is always better to have a very close look at this part of the standard to understand authority for issuance of the certificates. A variety of consulting firms and agencies provide SIL-related services and in some cases generate certificates. Here the main approach is complying with the functional safety standards. After the standards have been in place for so many years, it is always better to have the product duly certified by independent third party established certifying agencies. During certification, experts will look carefully inside a product's design, safety, and reliability. There will be a thorough audit and after passing the audit, SIL certification will be issued with FMEDA (hardware probabilistic failure analysis based on a huge collection of data) numbers. All these will help in the development of the product. For the self-certified method, it will be prudent to ask for and check a "detailed assessment report," a document that describes the steps completed to show compliance to IEC 61508. In this regard the blog http://www.exida.com/Blog/Who-certifies-the-certification-agency may be referred to. According to Dr. William Goble "I have yet to see a valid self-certification." Having said this, one needs to remember and understand that using certified products alone cannot ensure safety of the entire system forever. It is not something that can be bought and applied. Certification can only help in selecting appropriate safe and reliable products for the application. System performance depends on a multitude of factors such as accurate placement, installation, correct maintenance, proof testing, etc. [14]. Now, focus will be on another important issue of proof testing and proof testing interval.

2.2 Proof Testing (With Proof Testing Interval)

With the help of the proof test phase and inspection, results are analyzed to verify that actual performance matches expectations. Demand mode functions need periodic tests to validate their operation and availability. The acceptance criteria for the proof test establish the minimum threshold for successful operation of the SIF. Passing the proof test directly validates the SIF at the time of the test. It is important to remember that some degree of inspection and maintenance is also expected, concurrent with the proof test [19].

The proof test is defined in Clause 3.2.58 of IEC 61511:2003 (see Appendix I). Proof testing is a very important issue pertinent to SIS. The main objective of proof testing is to manipulate the process variable in a simulating mode under various operating conditions, without actually driving the system to the demand condition. This means that the aim is to achieve a full end-to-end test of the system. However, this is a theoretical requirement and practicability is very much **dependent** on the nature of the process, the process materials, associated risks, and on the tolerable upsets to the process and to production. The effectiveness of SIS proof tests confirms correct operation of sensors and final elements. IEC 61511:2003 Clause 16.3 discusses proof testing and testing intervals. Also in proof testing, good practice guidelines are available in IEC 61508:2010. The major issues covered under proof testing shall include but not be limited to the following:

- Proof testing principles
- Proof testing types and procedures
- Format of proof testing procedures
- Proof test records
- Planning and scheduling
- Competence
- MOC

A typical flow chart describing the proof testing process is shown in Fig. X/2.2-1. Let the discussions begin with proof testing principles followed by types of tests and procedure.

2.2.1 Basic Proof Testing Principles

The basic principles of proof testing are:

- The proof test of an SIS should reflect real operating conditions as accurately as possible. If possible, the SIS will be activated by manipulating process variables without driving the dangerous situation. If at all the process is driven in the demand state, followed by a proper risk *assessment, accompanied* by additional controls.
- When process variables cannot be safely or reasonably practicably manipulated, then the correct operation of sensors should be attained by other means, such as comparison with other measurements.
- The inherent difficulties associated with testing valves and inline flow meters are addressed during the *design* phase of SIS with additional means, for example, additional corroborative measurements.
- Proof tests should address the necessary functional safety requirements of SIS, such as response time and valve leakage class.
  According to Clause 16.2.8 of IEC 61511-1:2003, the written proof test procedures shall describe every step that is to be performed and shall include:

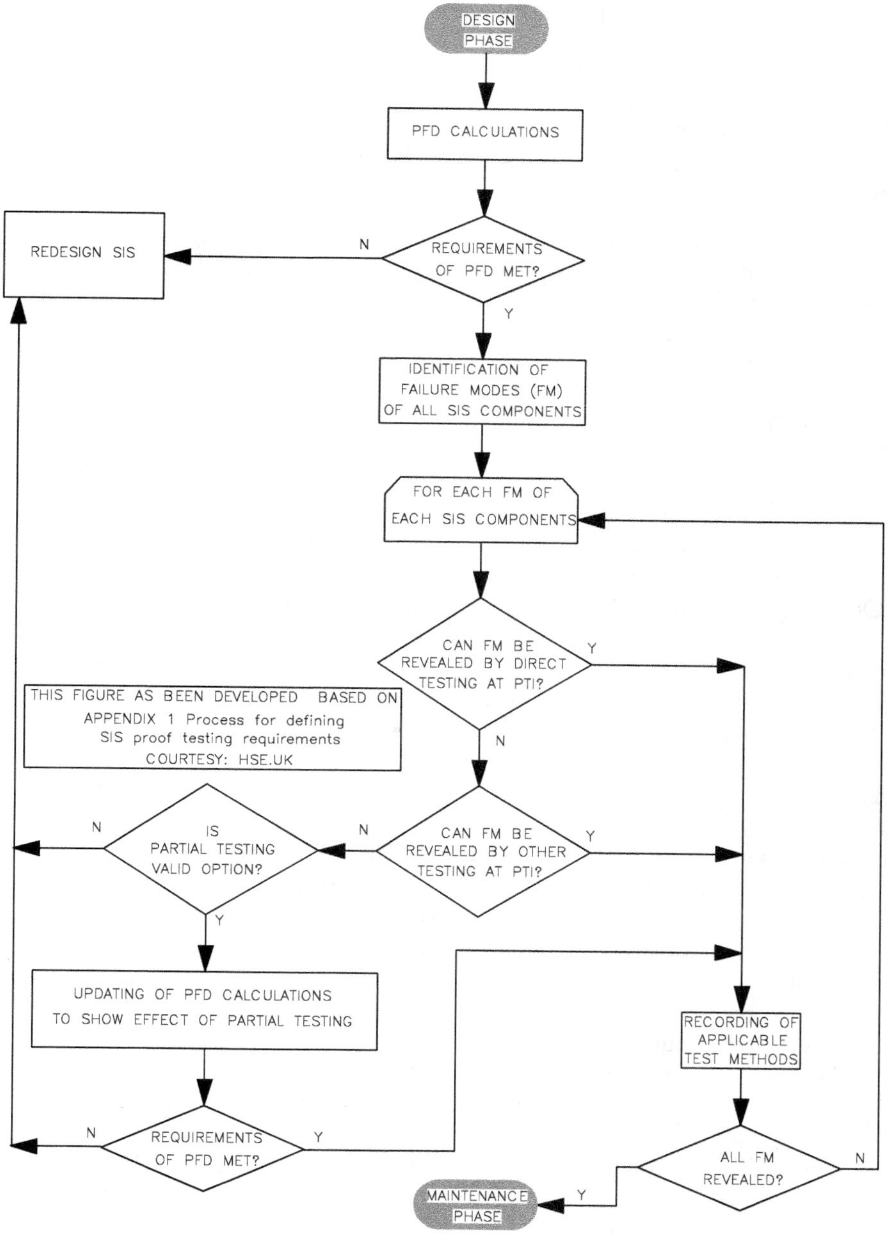

FIGURE X/2.2-1

Safety instrumentation system (SIS) proof testing process. *PFD*, probability of failure on demand; *PTI*, proof test interval.

- Correct operation of each sensor and final element
- Correct logic action
- Correct alarms and indications

2.2.2 *Types of Proof Testing*

There are three methods for proof testing, namely, direct testing, other methods, and partial proof testing.

- Direct testing: In direct testing, test equipment is normally used to create specific conditions and then the response of the component is observed. A typically direct test simulation method includes:
  - Specific process conditions
  - Calibration
  - Valve stroking
  - Trip condition simulation
  - Simulation of electrical signals, etc.

    This means that direct testing methods attempt to simulate a situation to reflect real operating conditions as accurately as possible without driving the system to a demand situation. The direct test methods are conducted at the *proof test interval* assumed in the PFD calculation.
- Other methods: These may be used when direct testing is not possible. Typical other methods include: cross-checking sensor history data against other instruments with the same range coverage, periodic replacement and sampled testing of single-use components, valve overhauls to ensure tight shutoff, etc. These other methods are conducted at the proof test interval assumed in the PFD calculation.
- Partial proof testing: The following are the major issues related to partial proof testing:
  - Good practice as per IEC 61511-1:2003 (Clause 11.2.5) requires that testing requirements are considered during the design stage. Therefore for new facilities, testing should be built into the design to facilitate online testing.
  - Judgment by a *competent* person is necessary for implementation of *partial* proof testing. Such decisions are taken on the basis of the component types, relative costs of periodic maintenance, replacement and overhaul, and the access to the equipment.
  - At times it is not possible to reveal ***all*** of the undetected failure modes at the proof test interval assumed in the PFD calculation, for example, on account of operational constraints. In such cases, a partial proof testing can be applied as long as the overall integrity requirements of the SIS are met.
  - A partial proof testing strategy requires the definition of two or more proof test intervals, as shown in Fig. X/2.2.2-1 [10,23]. Normally, this is defined as a more frequent partial test followed by a less frequent full test (typically to fit in with normal plant shutdown periods).

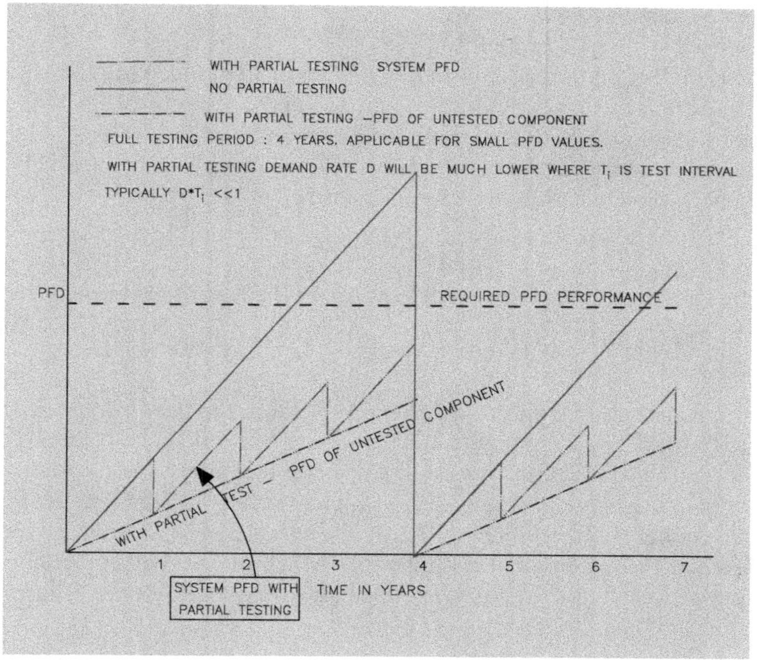

FIGURE X/2.2.2-1

Partial proof test to improve probability of failure on demand (PFD).

- For each of the defined proof test intervals, it is necessary to define which failure modes of the components will be tested, and which will not be tested [17].
- The partial proof test details need to be recorded in the proof test procedure. This **also** is applicable for direct testing and/or other methods.
- Each component has different requirements, so for each component, it is necessary to determine which failure modes are revealed at each of the defined proof test intervals and marry this with the PFD calculation. Therefore partial proof testing requires detailed **understanding** of the components' failure modes and their relative failure rates. When the result of the PFD calculation meets the required PFD, then the partial proof testing strategy is valid, otherwise there will be calls for a redesign. All undetected dangerous failure modes must be revealed at some point, otherwise SIS will have a lower than required PFD [17].
- Many modern component manufacturers design and provide test and built-in diagnostic facilities in their devices so that the unrevealed failure modes are insignificant in the expected lifetime. However, a partial test can reveal the reality, especially when the devices are operated beyond their lifetime or there are changes in actual operating conditions (operational constraints).

Normally, manufacturers' safety manuals furnish details about such issues. These need to be read carefully.

- The significance of partial proof testing will be clear from Fig. X/2.2.2-1. From Appendix B2 of IEC 61508-6:2010 one can access the formula for PFD calculations. Assuming MTTR is much smaller than proof test interval, dangerous failure rate λ_D and proof test interval T_i are directly related with PFD_{av} as per the following equation in its simplest form:

$$PFD_{av} = (\lambda_D * T_i)/2 \quad (X/2.2.2\text{-}1)$$

So, for the loop comprising sensor (S), logic solver (LS), and final elements (FEs):

$$PFD_{av} = \{(\lambda_{DS} * T_{iS})/2\} + \{(\lambda_{DLS} * T_{iLS})/2\} + \{(\lambda_{DFE} * T_{iFE})/2\} \quad (X/2.2.2\text{-}2)$$

From these equations, it is clear that lower the T_i value, the proportionately lower is the PFD_{av}. This is detailed in Fig. X/2.2.2-1 [courtesy Appendix 3 [20] of the Health and Safety Executive (HSE)]. In this case a full proof test interval is assumed to be 4 years. Also from the figure, it is seen that with proof testing in between, the PFD_{av} value can be kept below the required PFD_{av}. It also shows that in addition to the full proof test interval there are partial test intervals, which means that in partial proof tests there may be more proof test intervals based on devices (for different failures and failure rates).

2.2.3 *Basic Proof Test Procedure*

The following are major issues related to proof testing:

- Standard: As per IEC 61511-1:2003 standard (Clauses 16.3.1.2 and 16.3.1.4) all components in SIS shall be subject to a proof test and any deficiencies found shall be repaired in a safe and timely manner.
- Purpose: Proof tests are meant to reveal all undetected failure modes that would prevent the SIS operating as per SRS, even if this is completed at different intervals (e.g., partial testing). Proof testing should be designed and developed in a way to reveal any reasonably foreseeable undetected failure mode, that is, unrevealed (dangerous) fault conditions in each of the components (e.g., sensors, logic solvers, and final elements) in each SIF, and in the means of connecting the SIS to the process.
- Human error: The proof test procedures should be designed in such a manner that it would minimize human error. For this, a checklist including tick boxes and cross-checking for critical steps could be helpful. Many structure the tests in a way so that errors that could be introduced are revealed by subsequent test steps [17].
- Overrides: All SIS overrides are subject to **strict** controls to ensure safe application and timely removal.
- Calibration: For SIS proof testing, all test equipment must be calibrated as per national standards.

- Requirements for partial proof testing: As discussed earlier, for partial testing of SIS, the impact on process operation, functional safety, and overall test coverage must be established with the help of proper *assessment* and as necessary *additional* controls need to be deployed *without* disturbing SIS installation.
- Installations: SIS installations are tested on an as-found basis.
- Inspection: As per IEC 61511-1:2003 (Clause 16.3.2) as part of proof testing each SIS shall be subject to periodic visual inspection, including all components. This is to detect and ensure that there are no *unauthorized* modifications or *deteriorations*, etc., for example, missing bolts or instrument covers, rusted brackets, open wires, etc. The level of inspection depends greatly on the site conditions and access controls.
- Responsibility: All SIS components should be subject to appropriate maintenance, usually described by the component manufacturer [17].
- Procedure documents and timings: In an actual case the proof test procedure may not follow a single procedure document, but may be made up of different documents, for example, reference to other procedures. Also it may include procedures for testing different parts of SIS at different times. However, in all such cases, measures should be taken to ensure no failure modes or components are missed. It may include partial test procedures and full test procedures.
- Proof testing process description: For understanding the process, Fig. X/2.2-1 may be referred to. The process starts after calculating the PFD for SIS to meet the requirements of PFDs; if not, then it is to be redesigned. When the PFD calculations meet the requirements then all the failure modes are identified. For each of *failure modes* pertinent to each *SIF* in SIS, proof testing and inspections are carried out. As seen in Clause 2.2.2, there are three ways in which proof testing can be carried out. First, a check is made to see if direct testing is possible, which is then recorded. After recording, the test looks for another set of failure modes. When direct testing is not possible, then other means or partial testing are considered. If none of them are possible or valid for partial testing, then it is necessary to redesign the system. Whenever any failure mode, other method, or partial testing is valid and done, then, as discussed earlier, it is recorded and another failure mode is sought. In this way when all failure modes are tested the process is complete and proceeds to the maintenance phase.

2.2.4 Documentations

This is very important because proper record keeping could help in future analysis and resolving future maintenance problems. Also traceable data keeping is helpful in dealing with regulatory authorities. As per IEC 61511-1:2003 Clause 16.3.3 "The user shall maintain records that certify that proof tests and inspections were completed as required." Minimum data recording shall include but not be limited to:

- Description of the tests and inspections performed including the following:
  - Dates
  - Name of the person(s)
- Serial number or other unique identifier of the system tested

- Results of the tests and inspection

  Based on the foregoing the following are the details of basic data to be recorded:
- Steps undertaken during testing
- Results of steps
- Details of all faults
- Details of corrective actions including:
  - Recording of time to repair
  - Effectiveness of remedial action
- Test duration (start and end time)
- Details of tester and approving authority:
  - Signature
  - Date

2.2.5 *Planning and Scheduling*

Proof test planning is an **integral part** of safety management planning and it is dictated by the need to maintain the functional safety of the system and should **never** be compromised with other aspects such as operation/production needs or business policies. As this is very much an *object-oriented* program, it is justifiable that suitable explanations will be made prior to its adoption.

2.2.6 *Competence*

All persons concerned with the proof test must be competent to carry out the job. Also the responsibility of each of these persons must be well defined. The approval process of verification and validation cannot be overestimated. So, those who will verify the documented proof test must be experienced as well as experts in this field, so that they can check and verify the adequacy of the tests. Also they must be well aware of what is *demonstrated* and what the *object* is.

2.2.7 *Discussions*

Here, a few pertinent points will be discussed:

- Influencing factors: The following are influencing factors for proof test results:
  - Scope and coverage
  - Partial testing
  - Trip initiations
  - Overrides
  - Calibration
  - Diagnostics
  - Diagnostic coverage
- Proof test approaches: Proof tests of "wetted parts" are not always the same. In some cases pressure injection to the main impulse line with isolation and vent valves may be possible, but in a case where there are chances of leakage of energy or fluids, such an approach may not be feasible. So, based on this some of the major approaches are:
  - Manipulation of process variables

- Pressure injection
- Current/voltage/resistance simulation
- Use of smart instrument features
- Comparison with alternative measurements
- Valve closure
- Inspection and overhaul

There are major testing problems with valves and inline flow meters, for example, the sealing capabilities of valves under operating conditions. Also some processes pose high risks during *shutdown* and *startup* and there is a real need for SIS. When proof testing is likely to prove difficult, then the design phase of SIS should seek to provide the *specified functions and integrity* in a manner that accommodates these difficulties [18]. Use of proven capabilities of smart positioners could be a solution. Inline flow meters present very little scope to apply proof tests. The use of an alternative approach or portable measurements to corroborate inline flow measurements could be considered!

Another important aspect in safety is area classification of the plant based on hazardous situations. We will now focus on this in a separate section.

SECTION 2: EXPLOSION PROTECTION

3.0 HAZARDOUS AREA CLASSIFICATION AND ELECTRICAL SAFETY

Use of electricity in households as well as in industry has been available since the 19th century. Electricity has also been in use in coal and other mining applications. Using electricity in potentially explosive areas has been a concern from very early days. This has prompted developments of methods to use electricity in hazardous areas. After a number of initial accidents, developmental work has been undertaken to take care of safety. In this modern era, the number of accidents caused by electrical ignition sources is low. So, expenditure on development, manufacturing, and statutory regulations has proven to be successful. Today, many codes and standards toward safe use of electricity in hazardous areas have been developed. Like process hazards, it is also the duty of the designers to follow codes and standards during engineering processes so that quantified risks are well maintained within acceptable limits. This means the risks associated with use of electrical energy in hazardous/flammable areas are first assessed and quantified. Based on this, designers need to follow the recommendations of codes and standards to develop their design and engineering documentation. Also O&M personnel need to operate and maintain their plants accordingly. Obviously, the starting point of the discussions will be identification of the sources of release of flammable gas or vapor and/or accumulation of dusts. Such situations may arise from constant activity; from time to time in normal operation; or as the result of some unplanned event [22]. Also in a process, inside the equipment may be a source of hazard where both flammable materials and air are present, without actual release. Naturally, it is important to understand the properties of the materials and their impact on an explosion. In brief it is necessary to have a clear understanding of the principles of explosion and how people, property, and the environment can be protected. To have a standardized approach there have been a number of standards developed. Most internationally recognized standards have been developed by the IEC generally in line with CENELEC. After issuance of Directive 1999/92/EC, it has now become a precondition for setting up and operating a potentially explosive facility in Europe. These standards made it possible to select systems, equipment, and components with regard to explosion protection and to install, operate, maintain, and eventually repair them as required by the standards. *At this point it is worth noting that loss of containment of a vessel, that is, vessel rupture or line rupture, is not part of area classification discussions and these should be treated as part of PHA discussed in previous chapters*. In this clause, discussions shall cover mainly the following topics:

- Explosion discussions
- Electrical area classification (EAC)
- A few relevant standards
- Combustible gas detection
- Explosion protection
- Enclosure class
- Intrinsically safe circuits

During the discussions the main topic will be instrumentation and associated electronic parts, because these are mainly related to safe instrumentation.

Starting with explosion principles, now is the time to look at the details of the system.

3.1 Explosion Discussions

Flammable gases, vapors, mists, dusts, and fibers escape during production, processing, transportation, and storage of flammable substances in many industries. These flammable gases, vapors, mists, and dusts form an explosive atmosphere with oxygen or air. When this explosive atmosphere comes in contact with energy such as ignitions, sparks, hot surfaces, etc., explosions may take place and cause severe damage/harm to people, property, and the environment.

3.1.1 *Explosion of Flammable Substances*

According to the HSE.UK "Explosive atmospheres can be caused by flammable gases, mists or vapors or by combustible dusts. If there is enough of a substance, mixed with air, then all it needs is a source of ignition to cause an explosion." This means chemical explosions involve an oxidation and reduction reaction. However, all oxidation and reduction reactions do not cause explosions, for example, the rusting of iron. An explosion is a sudden, violent change of potential energy, which transfers to its surroundings in the form of a rapidly moving rise in pressure. Therefore combining the two, it is seen that an explosion is a sudden reaction involving a rapid physical or chemical oxidation reaction or decay, generating an increase in temperature or pressure or both simultaneously. So, in an explosion caused by a chemical reaction of a flammable substance with oxygen, there will be a simultaneous release of high energy. The following are major issues involved in explosions:

- Triangle: Similar to a fire triangle, there is also a triangular requirement for flammable materials that cause an explosion, as shown in Fig. X/3.1.1-1.

  So, for explosions to happen in atmospheric air, all three of the following factors must be present [23]:
  - Flammable material (in ignitable quantities)
  - Oxygen (in the air)
  - Ignition source/source of energy

In workplaces, hazardous areas can develop wherever the first and third preconditions for an explosion are fulfilled. In a workplace there has to be air and oxygen. Typical hazardous areas are created by flammable gases and vapors from chemical/petrochemical factories, refineries, offshore plants, tank facilities, loading areas for flammable materials, and/or boiler areas of natural gas-based power plants, and also dusts in milling plants, pulverizers in power plants, etc. This means that emphasis will be on *flammable* materials and *sources* of energy for ignition because oxygen cannot be dispensed with. Joint efforts from manufacturers, users, and established standards offer protection against such explosions as shown in the figure, with special emphasis on excluding the ignition source.

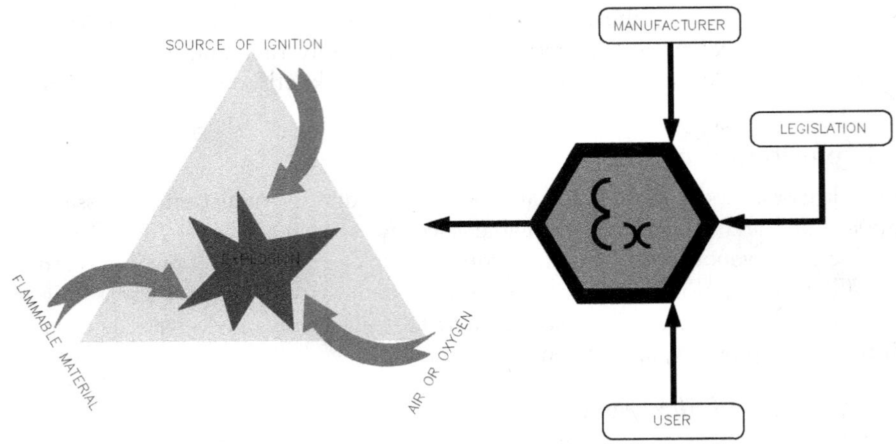

FIGURE X/3.1.1-1

Explosion and protection.

- Flash point, flammable liquid classifications: As indicated earlier, it is important to understand various properties of the flammable materials for the study of explosions involving flammable materials. Out of various characteristic properties of flammable materials, flash point is a very important safety consideration. According to the Occupational Safety and Health Administration (OSHA): "Flash point is the minimum temperature at which a liquid gives off vapor within a test vessel in sufficient concentration to form an ignitable mixture with air near the surface of the liquid. The flash point is normally an indication of susceptibility to ignition." When the flash point of a flammable liquid is well above the maximum operating temperature or hazardous atmosphere temperature, then an explosive atmosphere cannot be formed. The flash point of a mixture of various liquids may be lower than that of the individual components. As per OSHA, combustible liquids are defined as "any liquid having a flash point at or above 100°F (37.8°C)." Different nomenclatures for flammable and combustible liquids have been presented in Table X/3.1.1-1.

Table X/3.1.1-1 Flammable Liquid Classification (US Based)

Standards/Flash Point (Below)	OSHA	NFPA 30	ANSI
Flash point <−7°C	Flammable	Class I	Extremely flammable
Flash point −7°C–38°C	Flammable	Class I	Flammable
Flash point 38°C–60°C	Combustible	Class II	Flammable
Flash point 60°C–66°C	Combustible	Class III	Combustible
Flash point 66°C–93°C	Combustible	Class III	Combustible

NFPA, *National Fire Protection Association.*

Similarly, as per Council Directive 98/24/EC, flammable liquid can be divided as per Table X/3.1.1-2. Boiling point is one of the parameters to be considered here.

Table X/3.1.1-2 ***Flammable Liquid Classification as per Directive 98/24/EC***

Designation of the Flammable Liquid	Flash Point and Boiling Point (°C)
Highly flammable	Flash point <0°C and boiling point <35°C
Easily flammable	Flash point <0°C and boiling point >35°C or 0°C < flash point < 21°C
Flammable	21°C < flash point < 55°C

3.1.2 Range of Explosion

To form an explosive atmosphere, the flammable substance must be present in a certain concentration, as shown in Fig. X/3.1.2-1. Therefore lower explosive limit (LEL) and upper explosive limit (UEL) are two extreme limits and play an important role in determining the cause of an explosion. As seen in the drawing, if the concentration of flammable material is too high or too low (beyond the limits), then no explosion will occur on account of lack of oxygen and lack of flammable materials, respectively. Instead, there is just a steady-state combustion reaction or none at all.

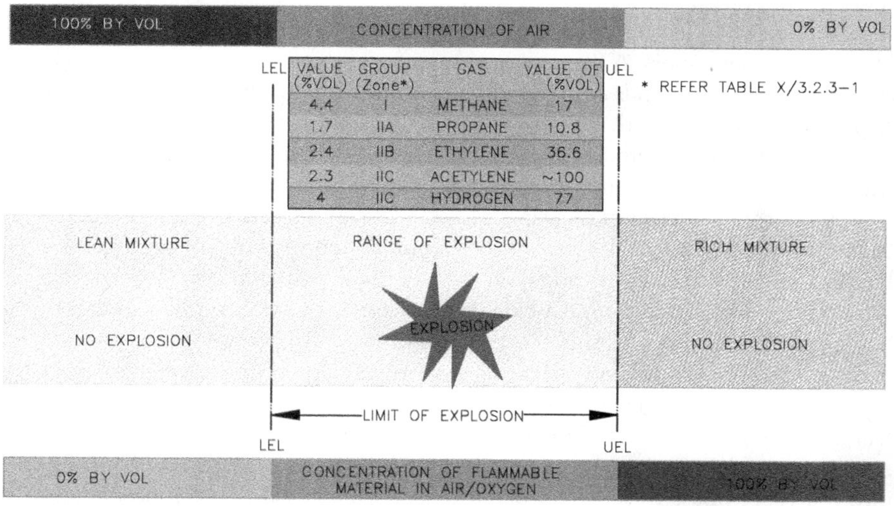

FIGURE X/3.1.2-1

Range and limit of explosion. *LEL*, lower explosive limit; *UEL*, upper explosive limit.

It is generally assumed that a volume of 10 L of an explosive mixture in an enclosed space can cause damage—particularly to people. For this reason, any area in which such a volume of an explosive mixture can collect is described as a potentially explosive atmosphere. It is only in the range between the LEL and UEL that the mixture reacts explosively when ignited. So, from a safety point of view, it is better to keep this concentration around 20% below LEL. Also see Clause 3.4.1 for further explanations. LEL and UEL are defined in Fig. X/3.1.2-2. The LEL and UEL of a few important gases are also shown in Fig. X/3.1.2-1 with associated gas group as per zone system (and NEC 505).

Lower Explosion limit (LEL): ***The lowest concentration (normally in %) of flammable gas or vapor in air capable of producing a flash of fire or explosion in the presence of an ignition sources is LEL. A concentration below is "too lean" to burn. This is also called the lower flammable limit (LFL).***

Upper Explosion limit (LEL): ***The highest concentration (normally in %) of flammable gas or vapor in air capable of producing a flash of fire or explosion in the presence of an ignition sources is UEL. A concentration above is "too rich" to burn. This is also called the upper flammable limit (UFL).***

The range between LEL and UEL is called the explosion limit or flammable range.

FIGURE X/3.1.2-2

Definition of lower explosive limit (LEL), upper explosive limit (UEL), and range of explosions.

3.1.3 *Flammable Substances*

There are three kinds of flammable materials. These are gases/vapors, liquids, and solids in the form of dust or flying.

- Flammable gases/vapors: A flammable gas may be in the form of an element such as H_2 or often compounds of carbon and hydrogen (hydrocarbon—e.g., natural gas), which can react with oxygen with very little additional energy. Also oxidation reaction of carbon (in complete combustion) like CO can cause explosion. A vapor is the proportion of a liquid that has evaporated into the surrounding air as a result of the vapor pressure above the surface of the liquid, for example, around a jet or droplets of the liquid. When spraying a flammable liquid, a mist can form consisting of very small droplets with a very large overall surface area. These mists can explode and in cases of fine mists a flash point is formed, as discussed in Clause 3.1.1. Flash points for mists are not that important because mists are behaviorally nearer a vapor than a liquid. So, it is necessary to consider mists as highly flammable material [21].
- Flammable liquids: Flammable liquids are mostly hydrocarbon compounds, for example, petroleum spirit. At room temperature some of these can even change into the vapor phase in sufficient quantities to form an explosive atmosphere near the surface. Many other liquids can form such an atmosphere near the surface at increased temperatures. So, the flash point discussed in Clause 3.1.1 is

important. From that discussion, it is clear that flammable liquids with a high flash point are less dangerous than those with a flash point at room temperature or below. Fine mists are formed from a flammable liquid and this is included (covered) above because of its behavior toward explosion.

- For solids, particle fineness, surface area, and chemical properties are important. Generally, solids require more energy than do gases and vapors to activate the explosion in air. However, once combustion starts, the energy released by the reaction can produce high temperatures and pressures. Dust reacts very differently, depending on whether it is in a deposited layer or whether it is in a suspended dust cloud [23]. Normally, smoldering first starts in dust layers on hot surfaces (secondary combustion in a hot furnace is an example), whereas in a dust cloud when ignited locally or through contact with a hot surface, it can explode immediately. Dust explosions could be a consequence of smoldering dust layers. When such layers are stirred up (say, as an extinguishing effort) they can lead to a dust explosion. A gas or vapor/air explosion can also stir up the dust. In such cases, first is the gas explosion, followed by the dust explosion. In deep coal mines, methane explosions often cause coal dust explosions.

3.1.4 Air/Oxygen

For any chemical reaction to take place it is known that definite quantities of constituents are necessary. The quantity of oxygen available in the air near flammable material can only oxidize/burn a certain quantity of the flammable material. A commonly known term, "stoichiometric" ratio, is responsible for determining the quantity of oxygen necessary to react with available flammable materials. Naturally, when the quantity of the flammable material and the available atmospheric oxygen are near to the stoichiometric ratio, the reaction will be near completion and cause an explosion with increase in temperature and pressure. The explosion will be violent. When the quantity of flammable material is too small, combustion cannot spread and may cease. The situation is similar when the quantity of flammable material is too large, because the lack of the required quantity of O_2 means that the reaction cannot proceed further. As indicated in an earlier clause (Fig. X/3.1.2-1), all flammable materials have their explosion ranges and limits: LEL and UEL. It may be possible to dilute flammable materials in excess air, but it is very difficult to create a situation where there is a dearth of oxygen because of the work force, hence this is only applicable inside equipment.

3.1.5 Sources of Ignition

With reference to Fig. X/3.1.1-1, we now need to look in depth at the third requirement for explosion—the sources of ignitions. There are many sources from where energies may come to ignite an explosive atmosphere. In this connection it is better to refer to EN 1127-1:2007. It is also important to consider the ignition properties of the explosive atmosphere such as listed here (see Clause 4.3 of the aforesaid standard):

- Minimum ignition energy requirement
- Ignition temperature of explosive atmosphere

- Minimum temperature for ignition of dust layer
  Also classification of ignition sources is useful. Ignition sources can be classified as follows (Clause 5.3.1 of the said standard):
- Likelihood of occurrences:
  - Continuously/frequently
  - Rare situation
  - Very rare situation
- Operation/equipment protection:
  - During normal operation
  - Result of malfunction
  - Result of rare malfunction
    The basic features of these sources are itemized in the following. The reference indicated in parentheses against each source corresponds to the clause in EN 1127-1:2007.
- Hot surface (5.3.2): Heated surface (e.g., heater, metal cuttings), dust layers, etc. can cause ignition when in contact with an explosive mixture. The capability of a heated surface causing an ignition is dependent on the type and concentration of the particular substance in the explosive mixture. Also the capability of causing an ignition is *dependent* on the temperature of the surface, surface area size, and shape of the heated body as well as how long it is in contact with the heated surface. In the event of a malfunction, for example, with overloading or tight bearings, etc., there will be a higher temperature to cause the ignition. Technical equipment, for example, tight bearing, resistor etc., must always be assessed.
- Flames and hot gases including hot particles (5.3.3): Even very small flames are among the most effective sources of ignition. Flames, their hot reaction products, or otherwise highly heated gases can ignite an explosive atmosphere. Therefore it is necessary to prevent flame propagation, for example, welding beads with a very large surface area during welding or cutting.
- Mechanically generated sparks (5.3.4): These could be generated as a result of friction, impact, or abrasion processes, for example, in grinding particles are hot because of the energy required for separation. These particles may be oxidized to reach higher temperatures and can be the cause of sparks to ignite flammable materials. In deposited dust, smoldering can be caused by the sparks and this can be a source of ignition for an explosive atmosphere. Cracks in rotating parts and sliding materials one over the other without lubrication can cause similar sparks, for example, a rusty hammer in contact with other tools.
- Electrical apparatus (5.3.5). These are always considered as a sufficient source of ignition. Electric sparks can be generated:
  - When electric circuits are opened and closed
  - On account of loose connections
  - Because of stray currents
    It is to be noted that low voltages (<50 V) are created against personal protection not against explosion. However, lower voltages can still produce sufficient energy to ignite an explosive atmosphere. For this reason, intrinsic

safety (IS) circuits are used in low power instrumentation. Very low energy sparks with minimal energy, say microwatt seconds, may be regarded as too weak to start an explosion.

- Stray electric current cathodic corrosion protection (5.3.6): Stray currents can flow in electrically conductive systems or parts of systems:
  - Return currents: In the vicinity of electric railways and large welding systems, conductive electrical system components such as rails and cable sheathing laid underground lower the resistance of this return current path—could be a source.
  - Result of short circuit or short circuit to earth because of faults in electrical installations
  - Result of magnetic induction

  A highly conductive connection to all the electrically conductive parts of the equipment should be provided to reduce potential difference to a safe level. An equipotential bonding (discussed later) must always be provided to meet the situation.
- Static electricity (5.3.7): The discharge of charged, insulated conductive parts can easily lead to incendive sparks, that is, electrical sparks can be caused by static electricity. This means that the stored energy can be released in the form of sparks that can act as an ignition source. Cone discharges from bulk material and cloud discharges can also occur. Friction is one of the possible causes of the generation of static electricity. Brush discharges can ignite almost all explosive gas and vapor atmospheres. The ignition of explosive dust/air atmospheres with extremely low minimum ignition energy by brush discharges is also a possibility. Sparks, propagating brush discharges, cone discharges, and cloud discharges can ignite all types of explosive atmospheres, depending on their discharge energy. Static electricity must be prevented from becoming an ignition source by taking appropriate measures, for example, synthetic enclosure.
- Lightning (5.3.8): When lightning strikes in an explosive atmosphere, ignition will always occur. Also there can be ignition of an explosive mixture because of high temperature caused by lightning. Large currents flow from where the lightning strikes and these currents can produce sparks in the vicinity of the point of impact. Even in the absence of lightning strikes, thunderstorms can cause high induced voltages in equipment, protective systems, and components.
- Radiofrequency electromagnetic waves from 10^4 Hz to 3×10^{12} Hz (5.3.9): Systems that generate and use radiofrequency electrical energy, such as radio transmitters or radiofrequency generators for heating, drying, hardening, welding, and cutting, emit radiofrequency waves. All conductive materials in this radiofrequency field functions as receiving *aerials*. When the field is powerful, the receiving energy is sufficiently large. So, these conductive parts can cause ignition in explosive atmospheres. The quantum of energy received is a *function* of the distance between transmitters and receivers and the dimension of the receiver. Suitable protective measures need to be taken against this.
- Radiofrequency waves from 10^{11} Hz to 3×10^{15} Hz (5.3.10): Radiation in this spectral range can—especially when focused—become a source of ignition through absorption by explosive atmospheres or solid surfaces.

- Ionizing radiation (5.3.11): This is generated, for example, by X-ray tubes, and radioactive substances can ignite explosive atmospheres (especially explosive atmospheres with dust particles) as a result of energy absorption.
- Ultrasonic (5.3.12): In the use of ultrasonic sound waves, a large proportion of the energy emitted by the electroacoustic transducer is absorbed by solid or liquid substances. As a result, the substance exposed to ultrasonic sound waves warms up so that, in extreme cases, ignition may be induced.
- Two other types are adiabatic compression and shock waves (5.3.13) and exothermic reactions (5.3.14) including self-ignition of dusts.
  With this brief knowledge about explosions, we will now focus on area classifications, which are the basis for explosion protection.

3.2 Electrical Area Classification (EAC)

Basically, area classification is for hazardous area classification. Since the main focus is on instrumentation systems, it has been designated as EAC to indicate that attention will be given to selection of instruments and associated enclosures, etc. The EAC process is quite rigorous and critical. The main objective of this classification is to divide the entire plant area according to the degree of presence of explosive environments [21]. Here degree mainly refers to the degree of severity. There are various codes and standards available to provide suitable guidance for selection, building up, and installation of electrical equipment in a particular area. Here, two important points are worth noting; first, that such classifications have been made in two ways:

- Class and division method: Mainly followed in North America and
- Zone division method: Mainly followed by Europe and other parts of the world. However, with introduction of IEC standards this is somewhat similar to the second type and is mainly followed globally; even the NEC has included zone divisions.

Second, the discussions will be focused mainly on instrumentation and control parts. The specifics concerning other electrical equipment such as motors, etc. will not be covered unless called for as part of instrumentation discussions. In general, the guiding factors for EAC, hence hazardous area classification, shall include but not be limited to the following:

- Type of hazards
- Type of flammable material present
- Flammable material properties and concentration
- Likelihood of the presence of flammable material
- Flammable materials present as normal or because of malfunction
- Area/location temperature
- Autoignition temperature of flammable material
- Sources and types of ignition energy

From NEC 497/499 and IEC 60079-10 it may be inferred that hazardous (classified) areas are those where fire and explosion hazards exist because of the presence of flammable materials (gas, vapor, liquid, dust). Area classification is a method of analyzing and classifying the environment where an explosive atmosphere may occur so as to facilitate the proper selection and installation of apparatus to be used safely in the environment, taking into account various groups and temperature classes. In practice, in places where there exist flammable materials, it is very difficult to ensure that (1) an explosive atmosphere will never exist or (2) apparatus will never give rise to a source of ignition. Therefore reliance is placed on using apparatus with low likelihood of creation of a source of ignition. All cases will not provide the same solution, however. In some cases, a suitable enclosure, say an explosion-proof enclosure (see Clause 3.6), will suffice, whereas in some cases suitable means should be chosen to limit the energy release, say an IS circuit (see Clause 3.7). Two kinds of area classification are shown in Fig. X/3.2-1.

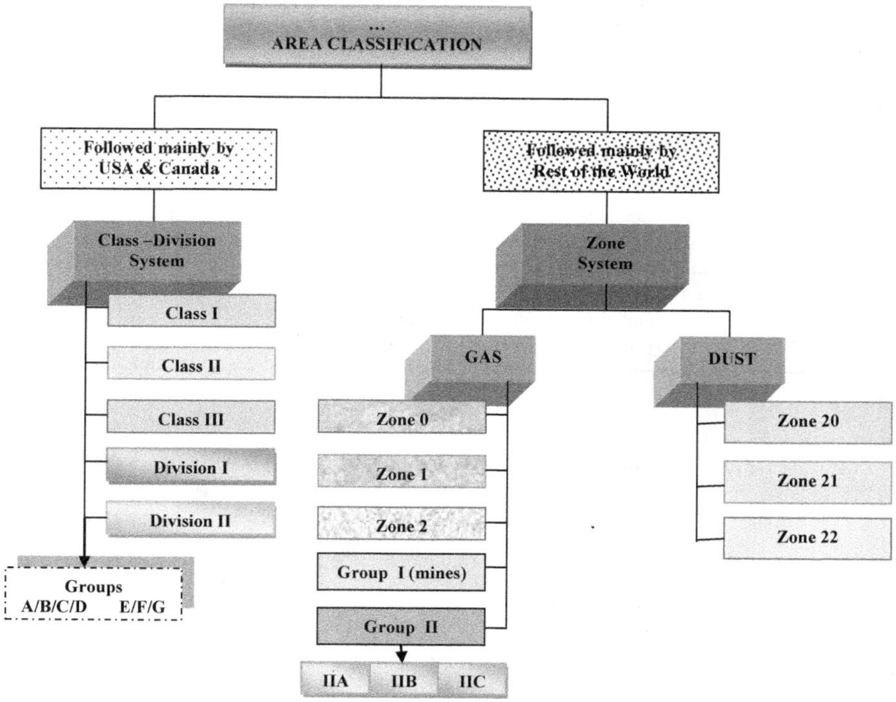

FIGURE X/3.2-1

Hazardous area classification chart.

3.2.1 *Class Division System*

According to this system, as shown in Fig. X/3.2-1, the entire hazardous area is divided into three classes based on the general nature or properties of the flammable materials such as gas/vapor, dusts, fibers, etc. Based on the likelihood of the presence of flammable materials in the atmosphere, the areas are further classified into divisions. Grouping of flammable materials is done based on type of hazardous materials present. Also there is categorization as per temperature.

- Class: Classes are presented in Table X/3.2.1-1.

Table X/3.2.1-1 Class—General Nature of Hazardous Materials

Class	Area
I	Area/location in which flammable gases or vapors[1] may (or may not[2]) be present in sufficient quantities to produce explosive or ignitable mixtures. *1: important is gas/vapor* *2: may not: qualifies as sufficient to indicate chances of explosion.*
II	Area/location in which combustible or conductive dusts[1] (in suspension intermittently, or periodic) may (or may not[2]) be present in sufficient quantities to produce explosive or ignitable mixtures. *1: important is combustible or conductive dusts* *2: may not: qualifies as sufficient to indicate chances of explosion.*
III	Area/location in which ignitable fibers[1], not likely to be in suspension, may (or may not[2]) be present in sufficient quantities to produce explosive or ignitable mixtures. *For this class, grouping does not have any relevance.* *1: examples: wood chips, cotton, nylon, etc.* *2: may not: qualifies as sufficient to indicate chances of explosion.*

Table X/3.2.1-2 Division—Likelihood of Hazardous Material Present

Division	Presence of Hazardous Materials Present
I	The substance referred to by class is present *continuously*, *intermittently*, or *periodically* or present *because of normal operation*, and has high probability of producing explosive or ignitable mixtures.
II	The substance referred to by class is present only in abnormal conditions (such as a container failure or system breakdown) or for a short duration and has low probability of producing explosive or ignitable mixtures.

Table X/3.2.1-3 Grouping of Hazardous Materials

Group	Hazardous Materials	Applicable
A	Acetylene.	Gases/vapors Class I
B	Hydrogen, fuel, and combustible process gases containing more than 30% hydrogen by volume or gases of equivalent hazard such as butadiene, ethylene, oxide, propylene oxide, and acrolein.	
C	Cyclopropane, CO, ether, hydrogen sulfide, morpholine, ethyl ether, and ethylene or gases of equivalent hazard.	
D	Acetone, ammonia, benzene, butane, cyclopropane, ethanol, gasoline, hexane, methanol, methane, vinyl chloride, natural gas, naphtha, propane, or gases of equivalent hazard.	
E	Combustible metal dusts: aluminum, magnesium, and their commercial alloys or other combustible dusts whose particle size, abrasiveness, and conductivity present similar hazards in connection with electrical equipment.	Dusts and flying Class II
F	Carbonaceous dusts, carbon black, coal black, charcoal, coal/coke dusts (>8% total entrapped volatiles).	
G	Dusts not included in E and F such as flour dust, grain dust, flour, starch, sugar, wood, plastic, and chemicals.	

- Division: Area classification based on likelihood is presented in Table X/3.2.1-2.
- Group: Hazardous materials in the surrounding areas are grouped. Their grouping and applicability are given in Table X/3.2.1-3.
- Temperature class: As discussed in Clause 3.1.5, flammable materials especially gases may ignite by coming in contact with a hot surface. Many factors, such as size, shape, type and surface quality, gas concentration, etc. have an influence on the ignition temperature. In all standards (say, IEC 60079-20-1), gases and vapors are divided into temperature classes. Approved equipment receives a

temperature code indicating the maximum surface temperature limit of the equipment. This means that surface temperatures in explosion protected equipment and other technological objects are controlled in such a way that ignition by the surface is not possible [23]. Table X/3.2.1-4 presents the ignition temperatures as per NEFA 497 article 500 and ATEX (atmosphere explosives). There are subdivisions in T2, T3, and T4. Permissible temperatures for electrical apparatus are also given in the table.

As shown in Fig. X/3.2-1, there is another way of classifying hazardous area in terms of zones and gas group. However, prior to this it is necessary to have some knowledge of ATEX directives, which is discussed in the following clause. Later, the zone type of classification is addressed.

3.2.2 ATEX Directives

The "Directive on Equipment and Protective Systems Intended for use in Potentially Explosive Atmospheres" (94/9/EC) came into force on March 1, 1996. All

Table X/3.2.1-4 Surface Temperature Division and Electrical Apparatus Temperature Limit

Temperature Code	Ignition Temperature[a] in °C (°F)	Ignition Temperature[b] Range in °C	Permissible Temperature Electrical Apparatus in °C[c]
T1	450 (842)	>450	450
T2	300 (572)	>300 to ≤450	300
T2A	280 (536)		
T2B	260 (500)		
T2C	230 (446)		
T2D	215 (419)		
T3	200 (392)	>200 to ≤300	200
T3A	180 (356)		
T3B	165 (329)		
T3C	160 (320)		
T4	135 (275)	>135 to ≤200	135
T4A	120 (248)		
T5	100 (212)	>100 to ≤135	100
T6	85 (185)	>85 to ≤100	85

[a] *As per NFPA 497 article 500.*
[b] *As per ATEX.*
[c] *Based on ATEX data.*

manufacturers of mechanical/electrical equipment, as well as protective systems, intended for safe operation in and around potentially explosive atmospheres in the European Union (others also follow) need to follow this directive. It is popularly known as ATEX 95. This directive is meant for equipment (electrical in relation to this text) to be used in potentially explosive atmospheres. This should not be confused with workplace area classification, which is covered by a separate directive known as ATEX 137. In this directive there are two distinct groups: Group I meant for mines and Group II meant for surface installations with potential explosive atmospheres. Essential classification as per ATEX Directive 95 is elaborated in Table X/3.2.2-1.

- Group I: Meant for mining applications, has the following categories as per the table:
  - Category M1: Equipment in this category should remain functional in an explosive atmosphere (in mining). Required protection level "Very High."
  - Category M2: Equipment required to be deenergized in an explosive atmosphere (in mining). Required protection level "High."
- Group II: Meant for equipment to be used in locations (other than mine use) endangered by potential explosive atmospheres. Here both gas and dust systems are considered and the markings are distinguished by "G" or "D" to indicate gas and dust, respectively. This group has the following categories as per Table X/3.2.2-1:
  - Category 1: Equipment in this category is intended for use in areas in which explosive atmospheres caused by mixtures of air and gases/vapors/mists or by air/dust mixtures are present continuously, for long periods or frequently. Equipment or protective systems need to guarantee a "very high level" of protection and are duly certified by a notified body (e.g., CESI/IMQ/TUV). See Table X/3.2.2-1.
  - Category 2: Equipment in this category is intended for use in areas in which explosive atmospheres caused by mixtures of air and gases/vapors/

Table X/3.2.2-1 ***Area Classification for Equipment in Explosive Atmospheres (ATEX 95)***

Group/Category	Group I: Equipment in Mines Below or Above Ground	Group II: Equipment in Other Locations, Endangered by Explosive Atmospheres
Category 1	Category M1 (Cat M1)	Category 1 (Cat 1)
Category 2	Category M2 (Cat M2)	Category 2 (Cat 2)
Category 3	Not applicable	Category 3 (Cat 3)

mists or by air/dust mixtures are likely to occur. Equipment or protective systems need to guarantee a "high level" of protection with due certification from a notified body. See Table X/3.2.2-1.

- Category 3: Equipment in this category is intended for use in areas in which explosive atmospheres caused by gases, vapors, mists, or air/dust mixtures are unlikely to occur or, if they do occur, are likely to do so only infrequently and for a short period only. Equipment or protective systems need to guarantee "a normal level" of protection. Self-certification (internal control) is allowed in this category if manufacturers have sufficient testing facilities. See Table X/3.2.2-1.

Product certification methods: In line with ATEX Directive 95, Fig. X/3.2.2-1 briefly describes the acceptability criteria for products. In all categories, the manufacturer

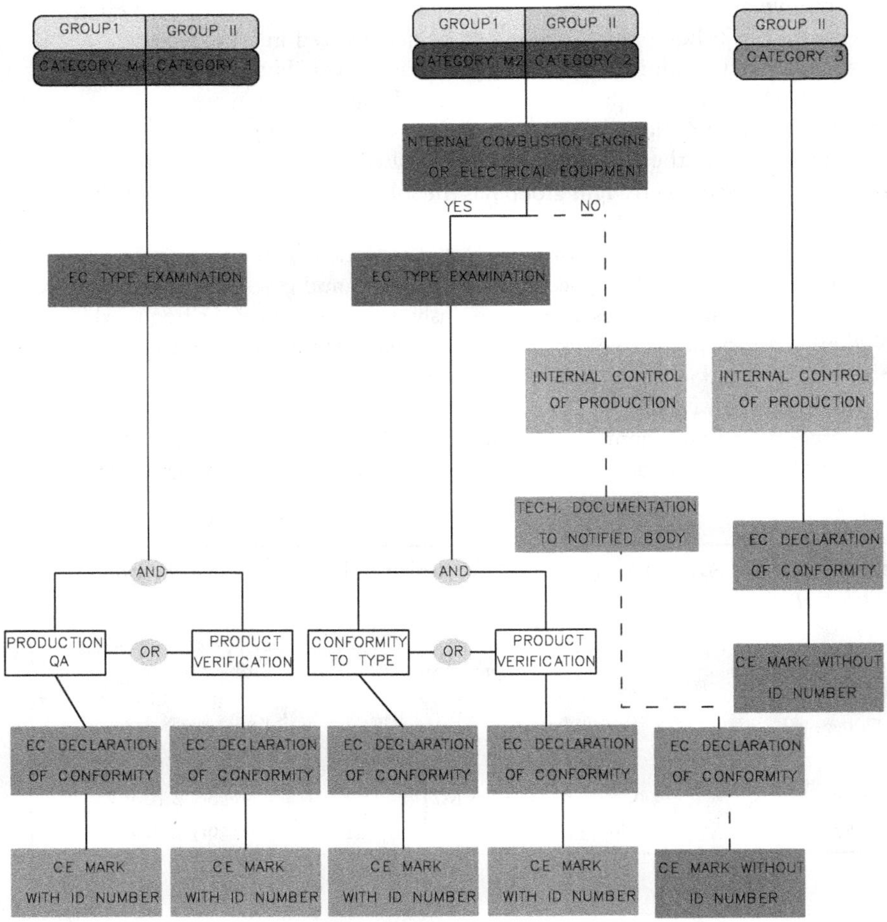

FIGURE X/3.2.2-1

Product acceptability criteria for ATEX 95.

needs to prepare a full technical file for the product(s) that demonstrates their conformity with the requirements of the ATEX Directive 94/9/EC.

The meaning of various terms in this regard may be noted:

- EC Type Examination: Examination by a notified body for compliance of documentation and a sample of the product to be certified
- Production Quality Assurance: The manufacturer's quality control system for production, inspection, and testing shall be approved and assessed regularly by a notified body.
- Product Verification: Each unit produced shall be assessed by a notified body.
- Conformity to Type: The manufacturer shall take all necessary steps to ensure that products are manufactured in compliance with the directive and to the same design and quality as the sample type tested. Each product manufactured will be subjected to testing approved by a notified body.
- Internal Control of Production: Manufacturers shall prepare technical documentations that allow the conformity of a product to be assessed. In some cases a copy of these documentations shall be presented to a notified body for review. Here it is important to note that there are two categories of CE certifications, namely, CE marking with or without identification number, as shown in Fig. X/3.2.2-1. Surface temperature details are available in Table X/3.2.1-4. There is another ATEX directive (99/92/EC), which is known as the "ATEX workplace directive" or ATEX 137. This is meant for the duty of employees to minimize the risk from explosive atmospheres. See also Clause 3.2.3.

3.2.3 Zone System

In line with the directive, employers need to classify into zones the areas in which explosive atmospheres may be present, and to ensure that the directive is observed. In the zone system there are clear divisions between gas and dust. For gas systems, zones are classified as zone 0, 1, and 2, whereas for dust they are designated as zone 20, 21, and 22, as detailed in Table X/3.2.3-1.

In the zone system the gases are further divided into Groups I and II. Again Group II is further subdivided as Groups IIA, IIB, and IIC. Classification criteria are the maximum experimental safe gap (MESG) and the minimum ignition current (MIC). See Fig. X/3.2.3-1 for MESG and MIC definitions. Also there are symbolic representations of these zones, typical symbols for gas have been depicted in Fig. X/3.2.3-2.

Grouping details are found in Table X/3.2.3-2 where one may note that ignitability *increases* from top to bottom.

For surface temperature, details in connection with ATEX and the zone system, including apparatus temperature limits, Table X/3.2.1-4 may be referred to.

The discussion on classification will be concluded by trying to marry both directives of ATEX, namely, ATEX 95 and 137, and linking them with IEC 60079-0 in Table X/3.2.3-3.

Table X/3.2.3-1 *Zone Classification (Gas and Dust)*

Material	Zone	Explanation
Gas	0	A location in which an explosive atmosphere consisting of a mixture with air of flammable substances in the form of gas, vapor, or mist is present continuously or for long periods or frequently.
Gas	1	A location in which an explosive atmosphere consisting of a mixture with air of flammable substances in the form of gas, vapor, or mist is likely to occur in normal operation occasionally.
Gas	2	A location in which an explosive atmosphere consisting of a mixture with air of flammable substances in the form of gas, vapor, or mist is not likely to occur in normal operation but, if it does occur, will persist for a short period only.
Dust	20	A place in which an explosive atmosphere in the form of a cloud of combustible dust in air is present continuously, or for long periods or frequently.
Dust	21	A place in which an explosive atmosphere in the form of a cloud of combustible dust in air is likely to occur in normal operation occasionally.
Dust	22	A place in which an explosive atmosphere in the form of a cloud of combustible dust in air is not likely to occur in normal operation but, if it does occur, will persist for a short period only.

MESG: **The maximum experimental safe gap is the gap of the test apparatus with a width of flame proof joint of 25mm at which internal ignition of an explosive mixture is not propagated to exterior (IEC 60079-1-1) [IIA: >0.9, IIB: 0.9-0.5, IIC: <0.5]**

MIC : **The ratio of the minimum current required from an inductive spark discharge to ignite the most easily ignitable mixture of a gas or vapor, divided by the minimum current required from an inductive spark discharge to ignite methane under the same test conditions. Ref: (IEC 60079-1-1) [IIA: >0.8, IIB: 0.45-0.8, IIC: <0.45]** ***Refer the said standard for further details.***

FIGURE X/3.2.3-1

MESG and MIC definitions.

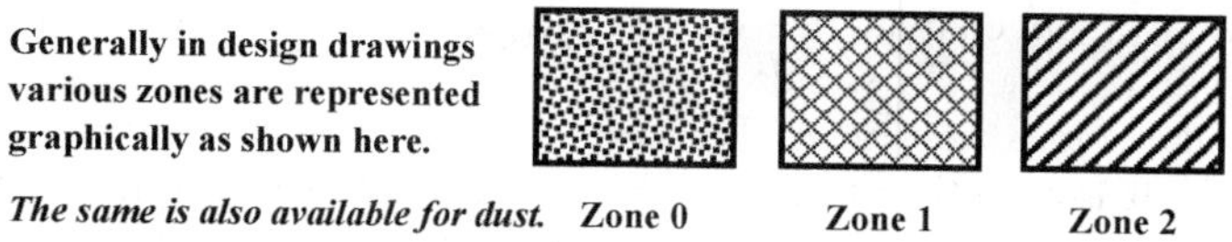

FIGURE X/3.2.3-2

Graphical symbols for zones (gas).

Table X/3.2.3-2 Gas Groups in Zone Systems

Group	Gas
I	Used in mines with methane gas or gas/vapor of equivalent hazard
IIA	Atmosphere with propane gas or gas/vapor of equivalent hazard
IIB	Atmosphere with ethylene gas or gas/vapor of equivalent hazard
IIC	Atmosphere with H_2/acetylene gas or gas/vapor of equivalent hazard

Table X/3.2.3-3 Relation Between ATEX 95,137, and IEC 60079-0

Material Type	ATEX 137 Zone	ATEX 95 Group	IEC 60079-0 Group (Equivalent)	Equipment Category (ATEX 95)	EPL IEC 60079-0
Gas/vapor	Zone 0	II	II	1G	Ga
Gas/vapor	Zone 1	II	II	2G, 1G	Gb, Ga
Gas/vapor	Zone 2	II	II	3G, 2G, 1G	Gc, Gb, Ga
Dust	Zone 20	II	III	1D	Da
Dust	Zone 21	II	III	2D, 1D	Db, Da
Dust	Zone 22	II	III	3D, 2D, 1D	Dc, Db, Da
CH_4/ C-dust (mine)	–	I	I	M1	Ma
CH_4/ C-dust (mine)	–	II	II	M2, M1	Mb, Ma

C-dust, *carbon dust (e.g., coal dust)*; EPL, *equipment protection level.*

In IEC there is an equivalence of category and it is termed the equipment protection level (EPL). EPL stands for the level of protection assigned to the equipment. Like categories in ATEX they have different designations. From Table X/3.2.3-3 it is seen that in the ATEX category of protection, IEC-EPLs are designated by 1, 2, 3 and a, b, c, respectively. In both cases, gases are designated by "G" and dusts by "D" for Group II and all Group I items in both cases by M. In the ATEX category, designation is followed by G/D/M, whereas in IEC it is the reverse. To have a better understanding it is important to draw a comparison with US (NEC 506)/Canadian (CEC Annex J) systems of area (and material) classifications with respect to the foregoing (see Tables X/3.2.1-1 and X/3.2.3-3).

- Zone 0 and/or zone 1 in gas/vapor is equivalent to division 1 class I.
- Zone 0 and/or zone 1 in dust is equivalent to division 1 class II.
- Zone 0 and/or zone 1 in fiber (not separately shown) is equivalent to division 1 class III.
- Similarly, zone 2 gas/dust in the table is equivalent to division 2 classes I and II, respectively. Also there is division 2 class III for ignitable fiber.

3.3 Discussions on Miscellaneous Standards

A short discussion is presented here on various standards so that different approaches can be understood. Major standards in these connections shall include but not be limited to what has been listed below. It is worth noting that the standards more related to electrical and instrumentation systems have been listed. There could be other standards such as National Fire Protection Association (NFPA) 50A, 59, etc.

- NFPA:
  - NFPA 30: Flammable and Combustible Liquids Code (cabling installations)
  - NFPA 70: National Electrical Code (Chapter 5 article 500 series)
  - NFPA 497: Classification of Flammable Liquids, Gases, or Vapors
  - NFPA 499: Classification of Combustible Dusts
- ANSI C2: National Electrical Safety Code: section 127
- Factory mutual
- Underwriter's laboratory standard
- OSHA
- IEC standards
  - IEC 60079-0 Explosive atmosphere
  - IEC 60079-1: Gas—Flameproof enclosures "d"
  - IEC 60079-2: Gas—Pressurized enclosures "p"
  - IEC 60079-5: Gas—Powder filling "q"
  - IEC 60079-6: Gas—Oil immersion "o"
  - IEC 60079-7: Gas—Increased safety "e"
  - IEC 60079-10: Electrical apparatus for explosive gas atmosphere
  - IEC 60079-10-1: Explosive atmosphere (classification—explosive gas atmosphere)

  - IEC 60079-10-1 Explosive atmosphere (classification—explosive dust atmosphere)
  - IEC 60079-11: Gas—Intrinsic safety "i"
  - IEC 60079-15: Gas—Type of protection "n"
  - IEC 60079-18: Gas and dust—Encapsulation "m"
  - IEC 61241-1: Dust—Protection by enclosures "tD"
  - IEC 61241-2 (IEC 61241-4): Dust—Pressurization "pD"
  - IEC 61241-11: Dust—Intrinsic safety "iD"
  - IEC 60079-14 for electrical installations
  - IEC 60079-17 for electrical apparatus maintenance
- ATEX Directives
  - ATEX 95
  - ATEX 137

Major features of these standards will be outlined as guidelines only. It is extremely important to note that for application of any of these standards, it is necessary to consult the latest version of the relevant standard so that it may be interpreted correctly before its implementation.

3.3.1 *Area Classification Principles/Procedures*

Very briefly, the gist of classification procedures in various standards is enumerated in the following. It is necessary that the applicable code be consulted:

- IEC (60079-10): Usually, this process is undertaken after basic documents such as a piping and instrumentation diagram (P&ID), layout, and other related documents for pertinent properties of flammable materials are available. The following are the major steps toward this:
  - Identification of source of release and its grade. It is better to consider all the applicable equipment (individually) caused by loss of containment.
  - Selection of type of zones mainly based on the likelihood of the presence of gas, which is very much dependent on ventilation and grade of release.
  - Decision about extent of zone, that is, distance from source (in all directions) up to which an explosive atmosphere exists before the material is dispersed. Rate of release (velocity, size, and geometry of sources), ventilation, and relative density of flammable material (mainly gas) are important considerations.
  - Other influencing factors are topology of the area including barriers, etc. and climatic conditions.
- NFPA: This is also basically the same as what was discussed previously. The major steps are:
  - Assembling all pertinent information such as P&ID, layout, flammable materials, operating description safety procedure, etc.
  - Listing all flammable materials and pertinent properties such as ignition temperature, surface temperature, etc. Identification of all release sources as mentioned earlier (for further guidance see NFPA 497 and 499).

    - Determination of class based on the properties of the flammable materials present.
    - Based on likelihood of occurrence, divisions 1and 2 are determined.
    - Extent of hazardous area is determined including boundary walls, etc.
    - Documentation of EAC is determined.
- Documentation during classification procedure: The general documentations during area classification shall include but not be limited to the following. These are mainly based on ATEX directives but are generally applicable to others:
    - Identification of person responsible
    - Structural and geographical details about plant site
    - Description of process and operational details
    - Material data
    - Risk assessment (risk analysis as in ATEX 137)
    - Protection concept and principles
    - Organizational measures, for example, training

3.3.2 *Discussions on Zone Systems*

Zone systems are followed in area classification by ATEX and IEC, and even NFPA has included zone systems of classification in their standard. The zone classification system probably has the edge over the class/division system because many consider the zone classification system to be more flexible and safer in hazardous locations. However, zone systems do not take care of the consequence of release. In case the consequences are important, as per the advice of the HSE.UK, it is better to go for a more **conservative** equipment selection. Although not officially accepted yet, various sources have tried to place time limits on these zones. The most commonly used values are as shown in Table X/3.3.2-1 based on probability and duration of explosive atmosphere.

Table X/3.3.2-1 Probability and Time Considerations Commonly Used for the Zone System

Zone	Likelihood of Explosive Atmosphere	Overall Duration of Explosive Atmosphere
Zone 0	$P > 10^{-1}$	More than 1000 h/year
Zone 1	$10^{-3} < P < 10^{-1}$	Between 10 and 1000 h/year
Zone 2	$10^{-5} < P < 10^{-3}$	<10 h/year

3.3.3 *ATEX Directives—Explanations*

As discussed earlier, the European Union has adopted two harmonized directives on health and safety, known as ATEX 94/9/EC (also ATEX 95) and ATEX 99/92/EC (also ATEX 137). ATEX Directive 94/9/EC focuses on the essential safety requirements (SRSs) for products and protective systems for use in potentially explosive atmospheres and the respective conformity assessment procedures. ATEX Directive

99/92/EC, on the other hand, looks after the minimum health and safety requirements for workplaces with a potentially explosive atmosphere. In ATEX 95, entire systems have been divided into groups and categories as already discussed. In Directive 137, workplace area has been divided into zones. Use of various categories of equipment in various zones is elaborated in Table X/3.2.3-3. The basic purposes of each of these directives are elaborated in Fig. X/3.3.3-1. This figure also shows the correlation between the two directives so readers can have better understanding of the directives with their distinctive features.

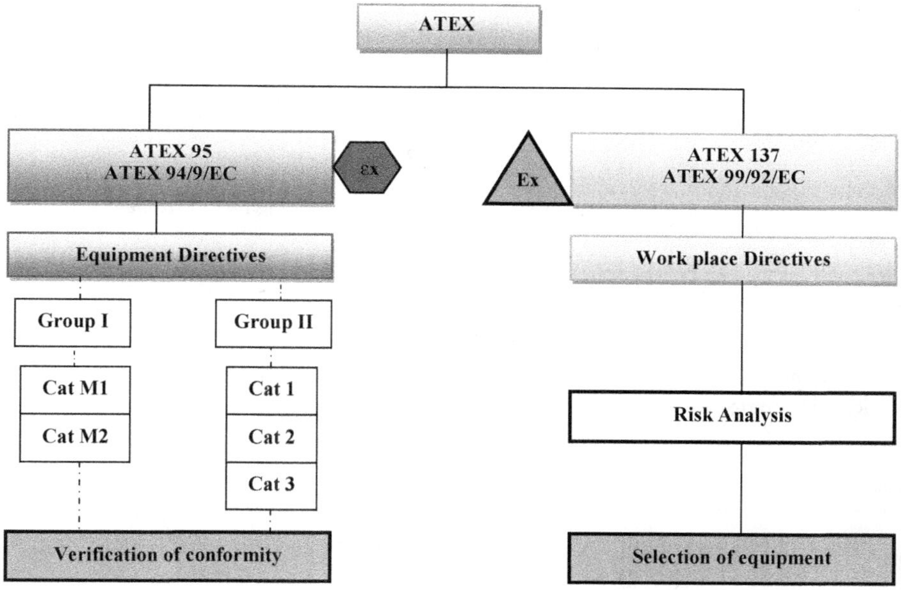

FIGURE X/3.3.3-1

ATEX directives explanation.

3.3.4 Protection Selection

Protection selections in accordance with various standards shall be discussed here to help in electrical equipment selections.

- ATEX: The ATEX concept of protection is elaborated in Table X/3.3.4-1. This may be compared with Table X/3.3.4-2 for IEC.
- EPL IEC: So far as IEC is concerned the various protection types and associated IEC standard references have been elaborated at the beginning of Clause 3.3 (under IEC standard). The EPL for gas with applicable zone and protection principles is elaborated in Table X/3.3.4-2 to help in the selection process. For EPL symbols see Clause 3.2.3. Ingress protections (IPs) mentioned in the table are covered in Clause 3.6 of this chapter.

Table X/3.3.4-1 ATEX Protection Concept

Concept	Code	Zone	Category	Principle of Protection
Increased safety	Ex e	1, 2	2, 3	No arc, spark, or hot surface
Nonsparking	Ex nA	2	3	No arc, spark, or hot surface
Flame proof	Ex d	1, 2	2, 3	Contains the explosion. Quenches the flame
Enclosed break	Ex nW	2	3	
Quartz/sand filled	Ex q	1, 2	2, 3	Contains the explosion. Quenches the flame
Intrinsic safety	Ex ia	0, 1, 2	1, 2, 3	Limits both the energy of spark and the temperature
Intrinsic safety	Ex ib	1, 2	2, 3	
Energy limitation	Ex nL	2	3	
Pressurize	Ex p	1, 2	2, 3	Keeps the flammable gas from hot surface and ignition-capable equipment
Simplified pressurization	Ex nP	2	3	
Encapsulation	Ex m	1, 2	2, 3	Keeps the flammable gas from hot surface and ignition-capable equipment
Oil immersion	Ex o	1, 2	2, 3	Keeps the flammable gas from hot surface and ignition-capable equipment
Restricted breathing	Ex nR	2	3	
Special	Ex s	0, 1, 2	1, 2, 3	Any proven method

The EPL for dust with applicable zone and protection principles is elaborated in Table X/3.3.4-3 to help in the selection process.

- Protection concept NFPA: NFPA 70 (NEC) article 500 (500-7) 2012 specifies various explosion equipment for electrical apparatus as listed in Table X/3.3.4-4, where sequential details as per NFPA 70 article 500-7 A through L are given.

The discussion on hazardous area standards is concluded with the recommendation that the applicable standard part must be read thoroughly before starting any equipment selection. Since area classification is mainly carried out by mechanical and safety engineers, good coordination is helpful. Gas detection systems for hazardous locations is another area of concern for instrumentation and control engineers so brief discussions on combustible gas detection systems will be dealt with in the next clause.

3.4 Combustible/Flammable Gas Detection

Combustible liquid and flammable liquids are actually not same although the two terms are frequently used interchangeably. Some of the differences between combustible liquid and flammable liquids are given in Table X/3.4-1.

Table X/3.3.4-2 IEC EPL Protection Techniques (Gas) (See Also Appendix II)

Method	Code Ex	EPL	Zone	IEC[a]	Principle of Protection
Increased safety	e	Gb	1, 2	7	No arc, spark or hot surface. IP54 or better
Type n (nonsparking)	nA	Gc	2	15	
Flame proof	d	Gb	1, 2	1	Contains the explosion.
Type n (enclosed break)	nC	Gc	2	15	Quenches the flame
Quartz/sand filled	q	Gb	1, 2	5	Quenches the flame
Intrinsic safety	ia	Ga	0, 1, 2	11	Limits energy of spark and surface temperature
Intrinsic safety	ib	Gb	1, 2	11	
Intrinsic safety	ic	Gc	2	11	
Type n (energy limiting)	nL	Gc	2	15	
Pressurized	p, px	Gb	1, 2	2	Keeps the flame gas out
	py	Gb	1, 2	2	
	pz	Gc	2	2	
Type n (hermetic sealing)	nC	Gc	2	15	
Type n (restricted breathing)	nR	Gc	2	15	
Type n (simple pressurized)	nZ	Gc	2	15	
Encapsulation	ma	Ga	0, 1, 2	18	Keeps the flame gas out
Encapsulation	mb	Gb	1, 2	18	
Oil immersion	o	Gb	1, 2	6	Keeps the flame gas out

EPL, *equipment protection level.*
[a]*IEC column: IEC 60079 Part, e.g., 7 in IEC column = IEC 60079-7.*

Table X/3.3.4-3 IEC EPL Protection Techniques (Dust) (See Also Appendix II)

Method	Code Ex	EPL	Zone	IEC[a]	Principle of Protection
Enclosure	t	Da, Db, Dc	20	31	Standard protection for dust-tight enclosure
Intrinsic safety	i	Da, Db, Dc	21	11	Same as above, some relaxation for intrinsic circuit
Encapsulation	m	Da, Db, Dc	22	18	Protection by encapsulation of incendive part
Pressurized	p	Db, Dc	21, 22	2	Protection by pressurization of enclosure

EPL, *equipment protection level.*
[a]*IEC column: IEC 60079 Part, e.g., 7 in IEC column = IEC 60079-7.*

Table X/3.3.4-4 ***Protection Concepts Class Division System***

Method	Class	Division	Remarks
Explosion proof	I	1, 2	
Dust ignition proof	II	1, 2	
Dust tight	II	2	
	III	1, 2	
Purged and pressurized	—	—	Classified area for which it is identified
Intrinsic safety	I	1, 2	
	II	1, 2	
	III	1, 2	
Nonincendive circuit	I	2	
	II	2	
	III	1, 2	
Nonincendive equipment		As above	
Nonincendive component		As above	
Oil immersion	I	2	
Hermetically sealed	I	2	
	II	2	
	III	1, 2	
Combustible gas detector		Means of protection in industry with restricted public access elec	
Inadequate ventilation		Classified as class 1 division 1 for inadequate ventilation. Electrical apparatus class 1 division 1	

To develop an appropriate strategy to combat the hazards of combustible/flammable vapor or gas it is essential to have a better understanding of the basic properties of combustible materials. So, discussions on combustible/flammable gas detection systems start with the general properties necessary for determining a

Table X/3.4-1 ***Differences Between Combustible and Flammable Liquids***

Characteristics	Combustible (Vapor) Liquid	Flammable (Vapor) Liquid
Flash point	Flash point >37.8°C	Flash point <37°C Vapor pressure <276 kPa
Typical nature	When heated shows many similar characteristic of flammable gas	Volatile vapor heavier than air
Nature of vapor	Remains near source till atmospheric temperature > flash point	Flows down in a slope and is collected at low point

material's hazardous potential, then attention will turn to detection systems and their location, etc.

3.4.1 General Properties of a Combustible/Flammable Gas System

To gather knowledge about the properties of combustible/flammable gases it is necessary to address a few terms associated with creating combustible/flammable atmospheres:

- Flash point: Discussions on flash point were covered in Clause 3.1.1. Many hazardous liquids have flash points at or below room temperature and are very dangerous. These liquids are covered by a layer of flammable vapors that will ignite immediately if exposed to an ignition source. *Open cup* and *closed cup* are testing methods for flash points of liquids. The former system is less accurate but useful. Acetylene, ammonia, hydrogen, and LPG are examples of common explosive gases. However, there are many other explosive gases encountered in industries.
- Autoignition temperature: Also known as spontaneous ignition temperature, autoignition temperature is defined as the minimum temperature for self-sustained combustion of a substance, independent of heating. See also Clause 3.2.1 and Table X/3.2.1-4.
- Explosion range, LEL and UEL: See Clause 3.1.2 for detailed discussions and explanations. A few characteristics regarding LEL and UEL are important:
  - LEL: Actual LEL for different gases varies widely. LEL is measured as a percentage by volume in air and typical values are 1.4–5% by volume. When the temperature is increased, less energy is necessary, hence the LEL percentage by volume decreases.
  - UEL: In the case of an oxygen-enriched condition, there will be a rise in UEL of gas and its rate and propagation intensity will also increase. A high level of O_2 increases the flammability of any material.
- Vapor density: The vapor density of gas is the density of the gas in comparison to air = 1. So vapor density can be defined as the *weight ratio of a volume of flammable vapor* compared to an *equal volume of air*. [Vapor density <1 (air) will rise, and vapor density >1 will fall.] This is important in the placement of detectors. Major flammable vapors are heavier than air. CH_4 (0.55), CO (0.6), and NH_3 (0.6) are lighter than air, whereas petroleum vapor (3.0), H_2S (1.45), and propane (2.48) are heavier than air.
- The relation between vapor pressure and boiling point is noted in the following:
  - Vapor pressure: Some of the molecules at the surface of a liquid have enough kinetic energy to escape to the surrounding atmosphere. These molecules naturally exert pressure at the wall of the closed container. Depending on temperature, this pressure of escaped molecules will increase up to a certain threshold. When this threshold is reached, the space is considered to be saturated. At saturation, rate of molecules escape equals they return. Vapor pressure is the pressure exerted by these escaping molecules when the molecules leave the surface at the same rate at which

they return. Naturally, when the intermolecular forces in the liquid are *less*, then the molecules can easily escape and the liquid will have a higher vapor pressure. The reverse is noted when vapor pressure is low. Low vapor pressure means there are fewer molecules of the substance to ignite, so there is generally less of a hazard present. However, from an instrumentation point of view, this is more challenging, and a highly sensitive instrument will be required as there are fewer molecules present to be detected. If the vapor pressure is higher, then there will be more molecules to ignite, hence the hazard will be greater. Vapor pressure is temperature dependent. The higher the temperature, the higher will be the vapor pressure. Also the nature of dependency on temperature varies with the gas types.

- Boiling point: The boiling point of a liquid may be defined as the temperature at which it changes state from a liquid to a gas throughout the bulk of the liquid. Also the boiling point of a liquid may be defined as the *temperature* at which the vapor pressure of the liquid equals the external environmental *pressure* surrounding the liquid. This is because at that point, bubbles of vapor that form below the surface of the liquid contain vapor at a pressure that matches the external pressure. Therefore the bubbles are not crushed by the surrounding liquid and their buoyancy causes them to rise through to the surface of the liquid and give the familiar appearance of a boiling liquid.
- Interrelation: If there are *high* intermolecular forces, the molecules will be strongly attracted to each other, so fewer will enter the gas phase, and the vapor pressure will be *low*. This means that more heat will be required to separate the molecules; hence the boiling point will be *higher*. On the contrary, if the liquid has a high vapor pressure, little heat will have to be added to separate the molecules and the boiling point will be low. Therefore it may be concluded that the higher the vapor pressure of a liquid at a *given temperature*, the lower the normal boiling point of the liquid.

Now let us explore combustible/flammable gas detection systems and their associated technology.

3.4.2 *Combustible/Flammable Gas Detector*

Combustible gas detectors are required whenever there is a possibility of a hazard to life or property caused by the accumulation of combustible gases. There are mainly two types of combustible/flammable gas detectors that are often used in most industries. These are the low-cost catalytic bead sensor and infrared sensor [26,27].

- Catalytic bead-type detector: This is a low-cost detection system used to detect combustible gases. The system uses a catalytic bead to oxidize combustible gases. On account of combustion there will be a change in temperature, which is detected by resistance with the help of a Wheatstone bridge. Typical sensor and measuring principles are shown in Fig. X/3.4.2-1.

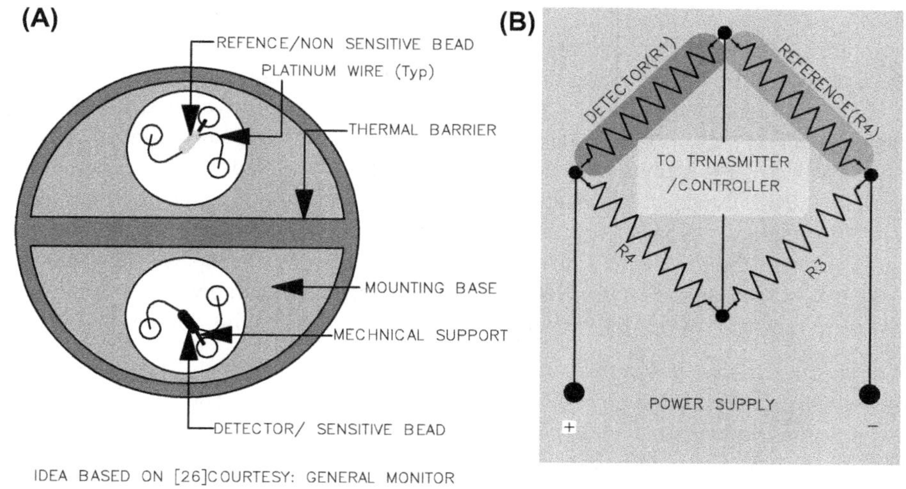

FIGURE X/3.4.2-1

Catalytic bead flame detector. (A) Catalytic bead sensor details, (B) catalytic bead sensing operation.

The system consists of a very small sensing element popularly known as a "bead." Beads are made of platinum wire coil, covered first with a ceramic base like alumina and then with a final outer coating of palladium or rhodium catalyst (Fig. X/3.4.2-1A). When a combustible gas/air mixture passes over the hot catalyst surface, combustion occurs and the heat evolved increases the temperature of the "bead." This in turn alters the resistance of the platinum coil. The change is measured by a Wheatstone bridge stated earlier. The resistance change is then directly related to the gas concentration in the surrounding atmosphere. As is typical in a Wheatstone, imbalance current can be sent to the transmitter or controller for further control and display. The necessary requirements of design safety mean that the catalytic type of sensor has to be mounted in a strong metal housing behind a flame arrestor to prevent propagation of flame outside. This slightly slows down the response. Typical response time is around 20—30 s. To ensure temperature stability under varying ambient conditions, the best catalytic sensors use thermally matched beads, as shown in the figure. The instrument sensitivity is affected by poisoning and blockage of the flame arrester. These are calibrated in terms of % LEL. For calibration and its implications refer to EN 60079-29-2.

- Infrared sensors: Although these are costlier with respect to catalytic bead sensors they offer a number of advantages such as faster response (~ 10 s), low maintenance, self-checking with the help of microprocessor circuitry, and measurement does not depend on O_2 to detect the gas. It may be designed in such a way to be unaffected by poisonous gases. This system can be developed with a single or dual beam, as shown in Fig. X/3.4.2-2.

The technique operates on the principle of dual wavelength infrared absorption. Light passes through the sample mixture at two wavelengths, one of which is set at the absorption peak of the gas to be detected, while the other is not. The two light sources are pulsed alternatively and guided along a common optical path to emerge via a flameproof "window" and then through the sample gas. The beams are subsequently reflected again by a retroreflector, returning once more through the sample and into the unit. Here a detector compares the signal strengths of sample and reference beams and by subtraction can give a measure of the gas concentration [28]. Internal compensation eliminates the chances of drift.

On account of gas leaks, the cloud created could be a relatively stationary or readily dissipated type. This depends on factors like wind, rate of leak, density of leaked gas, and the structural environment around the leak. Gas leak clouds normally have the following characteristics:

- The highest gas concentration is at the source and decreases toward the edge.
- The shape of the gas cloud is dependent on air current.
- Outdoor environment gas clouds dissipate faster and hence have low gas concentrations.

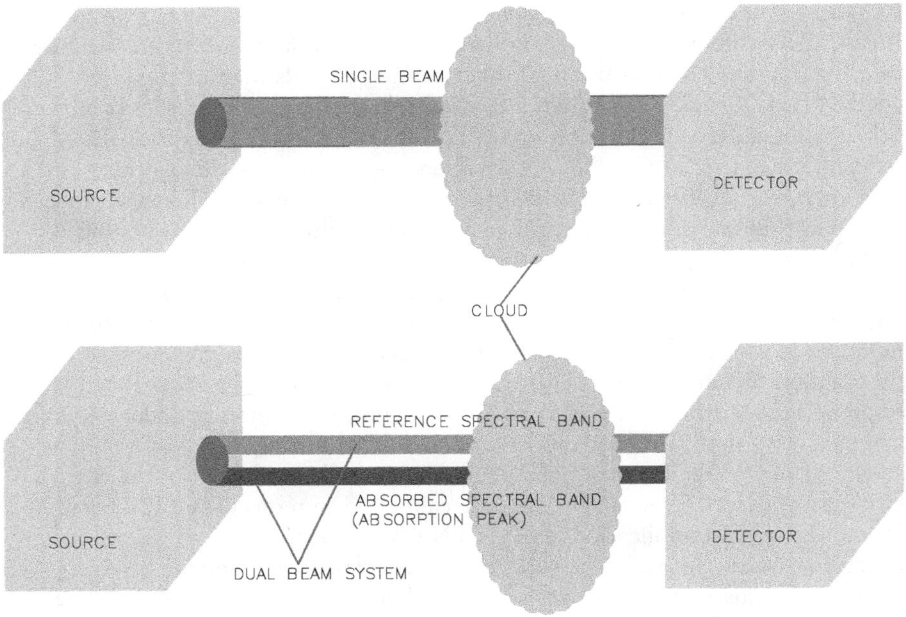

FIGURE X/3.4.2-2

Infrared flame detector.

Open path detection is very important in applications like floating production storage and offloading vessels, loading/unloading terminals, pipelines, perimeter monitoring, offshore platforms, and liquid natural gas storage areas. The conventional method of gas leak detection is *point type detection*. A number of individual sensors are used to cover an area or perimeter. Over time, however, instruments have become available that make use of infrared and laser technology in the form of a broad beam (or open path), which can cover a distance of several hundred meters [28]. Open path applications are affected by fog and rain. In today's instruments, necessary means have been incorporated to compensate for these effects. A brief comparison between both types of instruments is presented in Table X/3.4.2-1.

So, it is necessary to assess the site risk and prime objective, then to identify the gas to be detected, and finally, based on requirement, choose the detector taking into account the financial aspect.

Table X/3.4.2-1 Comparison Between Catalytic Bead and Infrared Instruments

Points of Comparison	Catalytic Bead Sensing	Infrared Sensing
Inert atmosphere	Requires O_2 (combustion)	Yes
Susceptibility to poison	No	No but less affected
Diatomic gas, e.g., H_2	Yes	No
Dust and drift immunity	Yes with proper protection	Yes with proper protection
O_2-enriched condition	Yes	Yes
Speed of response	20–30 s (typical)	<10 s (typical)
Cost	Less costly	Higher capital investment
Maintenance requirement	High	Low

3.4.3 Combustible Detector Placements

The placement of detectors should be determined following the advice of experts having sound knowledge of both gas dispersion and process/equipment and safety. All the agreed locations must be recorded. Obviously, detectors shall be located where the gas is most likely to be present. In this instance, Clause 3.6 of Chapter VIII may be referred to. Major issues in this regard shall include but not be limited to the following:

- To detect gases that are lighter than air (e.g., NH_3, CH_4), detectors should be mounted at high level and preferably use a collecting cone, whereas gases heavier than air (e.g., SO_2) should be mounted at a low level as discussed.
- To study the behavior of escaping gas against natural or forced air currents, detectors may be mounted in ventilation ducts if appropriate.
- For locating detectors, consideration must be given to natural calamities such as flood, rain, storm, etc. Outdoor detectors should use weatherproof enclosures as applicable. It is preferable to use sunshade and avoidance of direct sunlight.

- Process condition is an important consideration. Also it is better to locate the detector away from high pressure parts to avoid the formation of clouds.
- Ease of access for functional testing and servicing is recommended.
- To avoid dust, it is better to locate the detector at a suitable location with the detector pointing downward.
- When installing open path infrared devices it is important to ensure that there is no permanent obstruction or blocking of the infrared beam. Short-term blockage from vehicles, site personnel, birds, etc. can be accommodated [28].
- Structures that open path devices are mounted to should be sturdy and not susceptible to vibration [28].

We will now concentrate on explosion prevention/protection principles.

3.5 Explosion Protection

The main aim of explosion protection in instrumentation is to prevent ignition sources or the coming together of such sources with potentially explosive atmospheres. The other two factors of explosion (Fig. X/3.1.1-1)—the oxygen in the air and often the flammable substance—cannot be reliably and permanently ruled out in workplaces. Therefore the only way left is to isolate the source of ignition from the explosive atmosphere. In primary explosion protection, it is best to substitute something for the flammable substances or the atmospheric oxygen, or reduce their quantities to the point where there is no danger of an explosive mixture forming. Increased air circulation is another way of achieving this. These are achieved by structural measures, for example, the open layout of filling stations. Removal/replacement of atmospheric oxygen cannot be an option for areas where people work. For such locations the options available shall include:

- Avoidance or restriction of flammable substances
- Avoidance or restriction of release of flammable substances by:
  - Limiting their concentration
  - Using enclosures filled with an inert substance—natural or artificial ventilation
  - Concentration monitoring by means of a gas detection system.

3.5.1 Basic Protection Principles (Fig. X/3.5.1-1)

Protection principles focus on the ways and means to exclude equipment and components as ignition sources. Ignition sources are prevented in explosion protected equipment by selecting appropriate materials and by constructive measures. Another way is to limit the energy level from the equipment to prevent explosion. All these must be verified and confirmed by the appropriate tests.

Generally, four protection principles as detailed in Fig. X/3.5.1-1 are used to prevent equipment from any explosion by protecting it and limiting its energy so that it cannot become a source of ignition. In all the protection principles depicted in Fig. X/3.5.1-1, it is necessary to ensure that an explosive atmosphere must not be able to reach nonpermitted temperatures with respect to the ignition temperature of substances present in the surrounding atmosphere. The alternative is that all

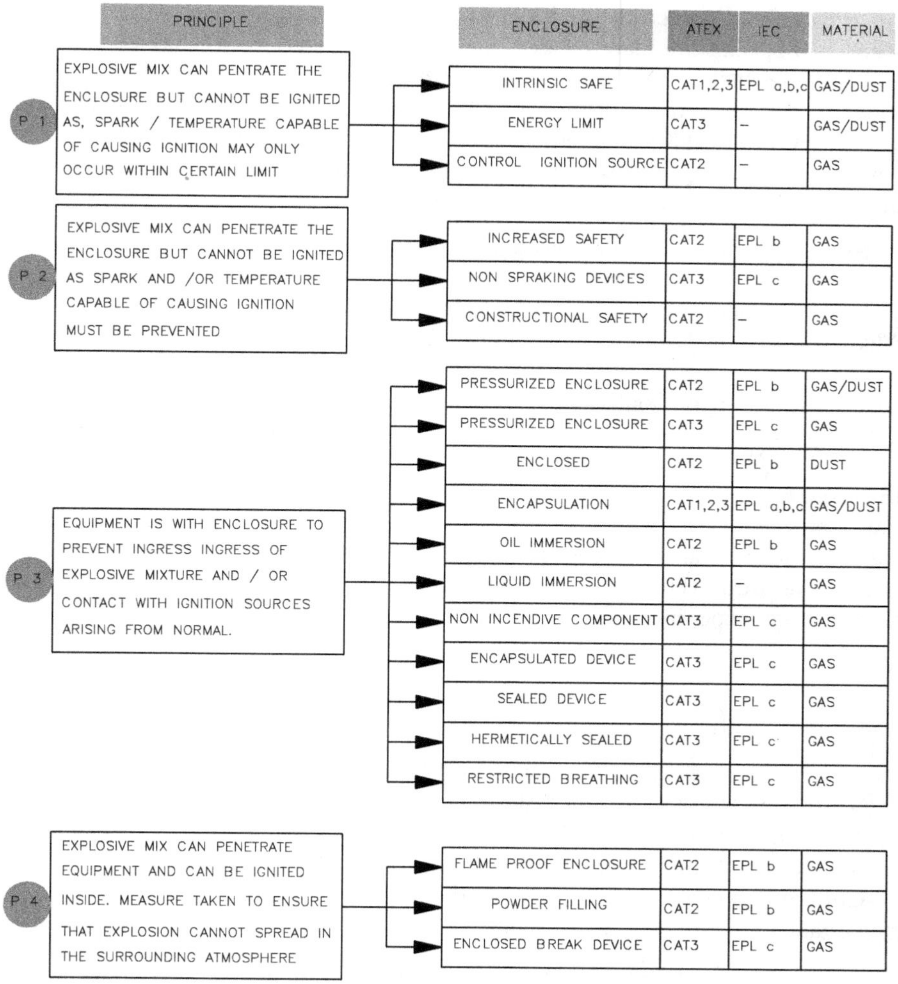

FIGURE X/3.5.1-1

Explosion protection principles.

propositions may not hold good. This also shows the relevance of the ignition temperature in all protection principles. Universal uniform requirements have been formulated by international standardization committees to combat hazards from flammable material emerging because of physical/chemical processes. Manufacturers and operators are required to adhere to these, and where there are increased protection requirements, they are monitored by notified bodies and authorities [23]. For wiring methods and cable standards in explosive atmospheres, guidelines from IEC session 13 [35] may be referred to. Explosion protection principles are applicable to fieldbus systems also, and this is where our attention will now turn.

3.5.2 *Explosion Protection for Fieldbus*

Conventional systems are mainly based on two aspects: each device has a separate barrier and they shall be wired separately from other field circuits. Probably the first definition of the required parameter for intrinsic application in a fieldbus system was done in the FOUNDATION fieldbus. Fieldbus devices are low-power devices, so IS is very much suitable for fieldbus installations in hazardous areas. The following are distinct features found in fieldbus intrinsic applications:

- There can be several *devices* connected through a single barrier; similarly, several *barriers* may be used for a single device.
- The same segment may have IS as well as other devices.
- Alike, conventional system, a fieldbus has the same philosophy of limiting energy.
- Fieldbus IS has more flexibility and is cost effective.

There are several approaches for fieldbus explosion protection:

- IS barrier for fieldbus
- Entity model
- Fieldbus intrinsic safety concept (FISCO)
- High-power trunk (HPT) concept
- Dynamic arc recognition and termination (DART)
- Fieldbus nonincendive concept (FNICO)

Only brief conceptual discussions on these are presented because of limited space. For more detail, standard literature on fieldbus systems may be consulted.

- IS (multiple) barriers in fieldbus system: A typical fieldbus IS connection is shown in Fig. X/3.5.2-1A. The major points are:
  - Avoidance of single point failure
  - Integration of shielding and grounding
  - Current and power limitations to field device
  - Distance limitation of cable length

A trunk and spur topology is used. In a typical fieldbus installation, 10—12 fieldbus devices are connected via one cable with length up to 1900 m. This single cable needs to support power and communications. It is recommended to use short-circuit protection and energy limitation to isolate a fault condition from negatively affecting device communications on that segment. As seen in the figure, there are a few fieldbus barriers being fed by one power supply. Each of the *spurs* connected to one field *barrier* is Ex i. From the ignition curve described in Fig. X/3.5.2-3 it is clear that based on voltage there will be a limit in the maximum current drawn from each of the field barriers. Typically, 110 mA are allowed in gas groups A and B and 235 mA in groups C and D [31]. Again various devices (at 24 V power supply) like pressure/differential pressure transmitters (20 mA), temperature transmitters (16 mA), valves (25 mA), and flow meters (10 mA) have different current consumptions (typical current consumption indicated in parentheses). So, while

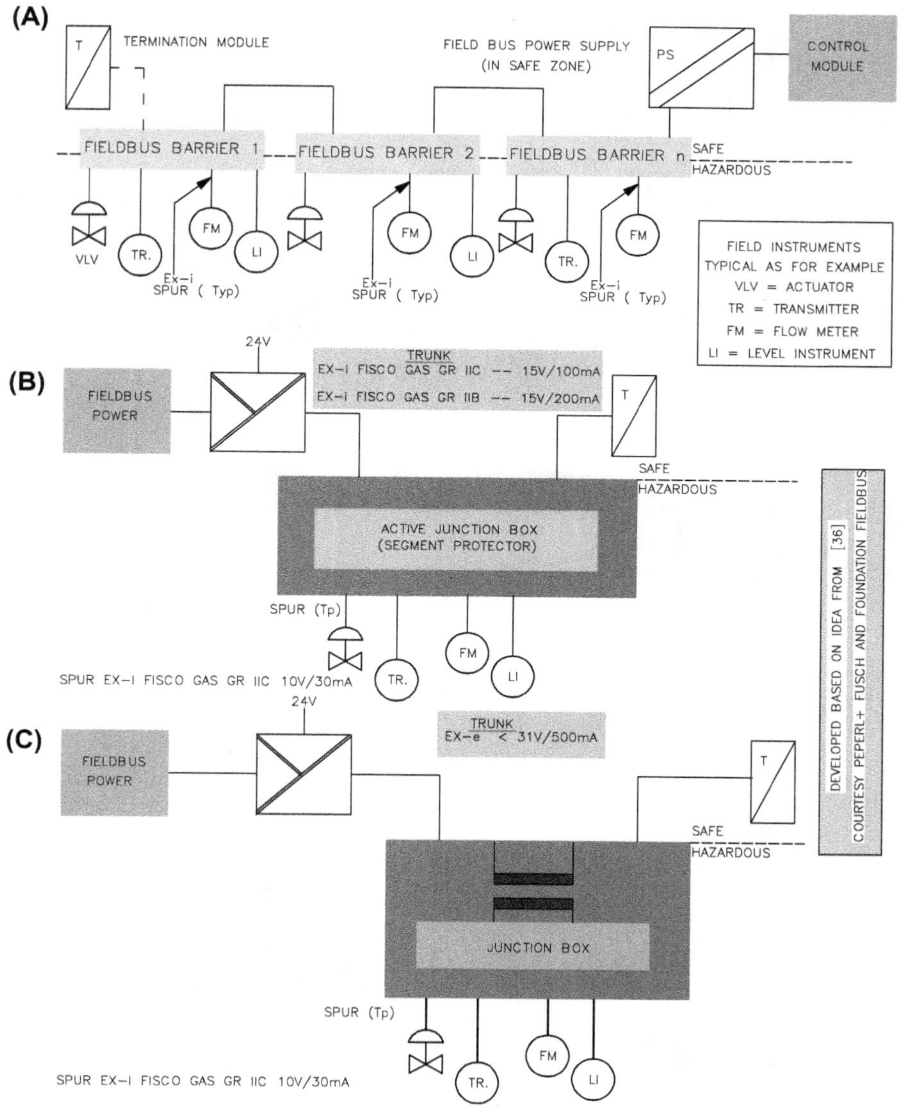

FIGURE X/3.5.2-1

Fieldbus intrinsic safety. (A) Multiple barriers fieldbus connection (typical), (B) FISCO concept, (C) HPT concept. *HPT*, high-power trunk.

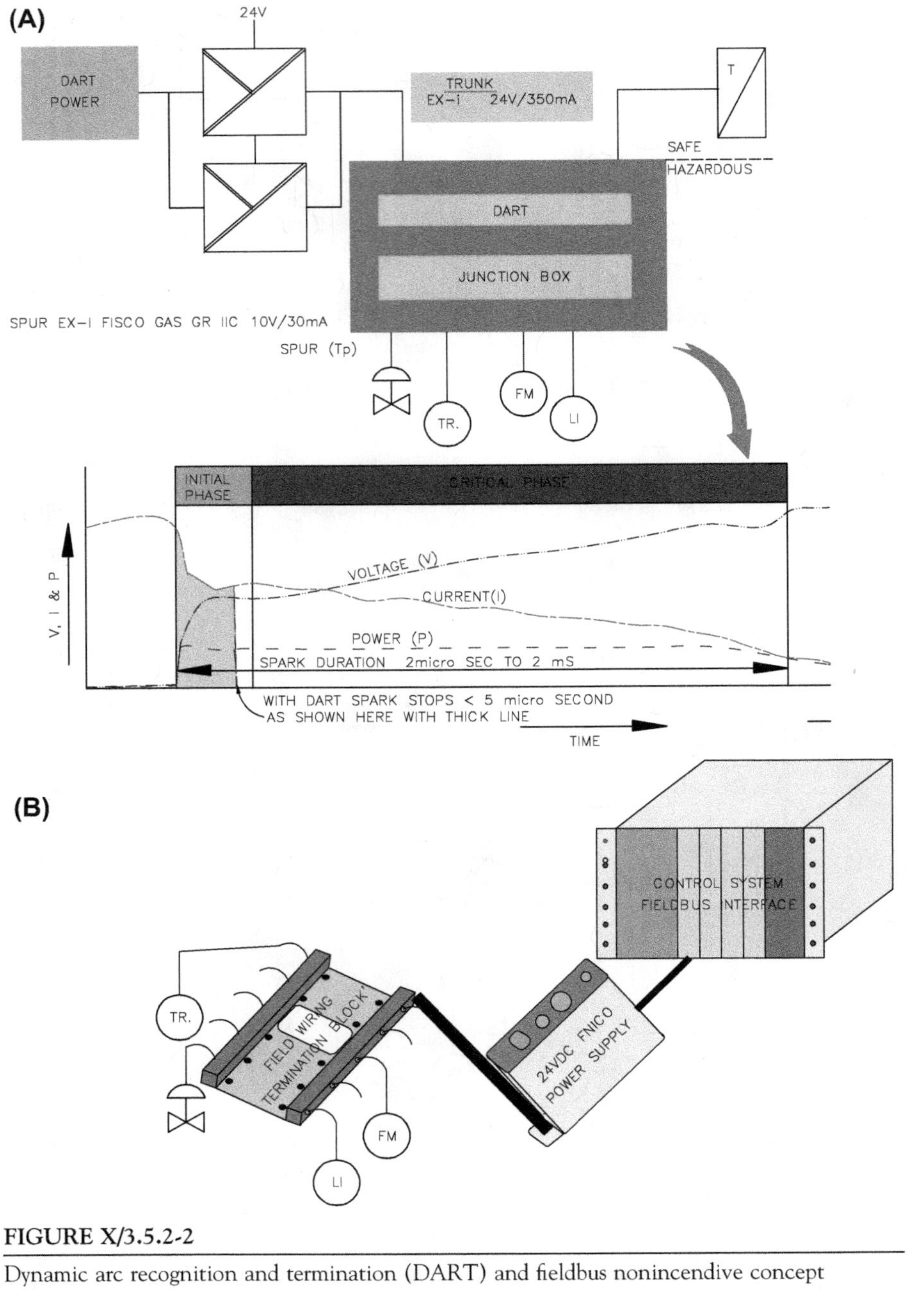

FIGURE X/3.5.2-2

Dynamic arc recognition and termination (DART) and fieldbus nonincendive concept (FNICO) for fieldbus protection. (A) Dart with charactistic curves, (B) FNICO connection system (for zone 2).

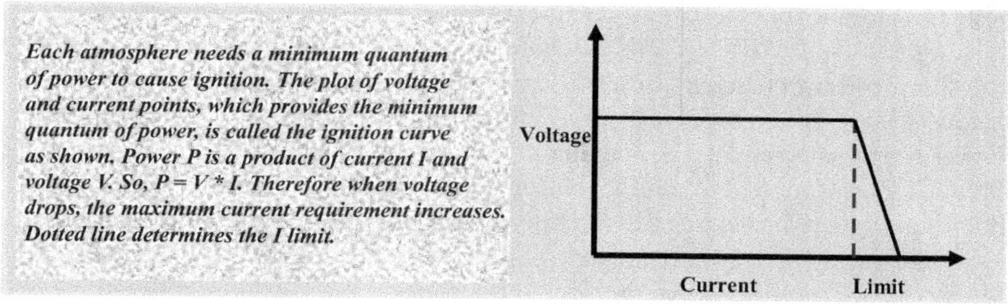

FIGURE X/3.5.2-3

Ignition curve.

designing a system with a number of these devices in use, it is necessary to combine them under each barrier so that the maximum allowed current is not exceeded. Maximum cable length, from a voltage drop point of view, is also an important consideration. This discussion applies to typical IS in a fieldbus.

- Entity model: The entity model defined in IEC 60079-1 is meant to use intrinsically safe parameters for validation of IS of a fieldbus. The entity model considers that all electrical parameters of the wire are concentrated at the point of a fault and electrical wire is a source of energy. In this method, 83 mA maximum current and 18.4 V maximum voltage are allowed. This model does not enjoy much popularity. Only a few power supply-conforming entity models are available. At best, these power supplies support two to three devices [32].
- FISCO: A typical FISCO model is depicted in Fig. X/3.5.2-1B. In this, various field devices are connected to an active junction box with Ex i spurs and Ex i FISCO trunk cable; also the topology provides isolation for intrinsically safe area equipment, as shown in the figure. In the topology of FISCO, there shall be isolation between safe and hazardous areas. *Active* junction boxes have *segment protectors* (current limiting devices). There are active junction boxes with foldback short-circuit protection. Suitable selection is important. These isolate a single short from affecting the entire network. Limitation is 200 mA @ 17 V for group IIB.
- High-power trunk concept: In the foregoing, FISCO power supply redundancy is not possible, whereas in the case of a high-power trunk redundancy is possible. In this method, Ex e has been considered for the trunk connection cable. This topology provides the highest possible cable length and at the same time supports the largest number of field devices per segment. However, spur cables are the same as FISCO. Typical current limitation is higher at 500 mA @ 31 V.
- DART: This is a newer concept in IS of the fieldbus system. To understand the system it is better to look at the curve showing how an arc is developed, as shown in Fig. X/3.5.2-2A. According to IEC 60079-11 a circuit is considered intrinsically safe when "electrical energy within the apparatus and of interconnecting wiring exposed to the potentially explosive atmosphere is

restricted to a level below that which can cause ignition by either sparking or heating effects." DART detects the characteristic behavior of a spark, especially the sharp current change di/dt in the initial phase and extinguishes the spark before it becomes incendive. It uses Ex i in both trunk and spur. Redundant power supply is possible. The maximum current limit is 35 mA @ 24 V.

- FNICO: This is applied in zone 2/division 2 hazardous areas, where the explosion hazard is expected to exist only in abnormal circumstances. At these locations, application of IS is not a compulsion. The type of protection "*n*" (see Clause 3.6) for zone 2 and nonincendive protection for division 2 are well-established techniques. The financial benefits of a zone 2 approach for instrumentation are seen in reduced capital expenditure and lower operating costs during the life of the plant [30]. Here, Ex nL is used and is applicable for zone 2 applications. A typical comparison of Ex i and Ex nL is presented in Table X/3.5.2-1.

Table X/3.5.2-1 Comparison of Fieldbus Protection by Ex i and Ex nL

Protection	Code	Fault Tolerance	Applicable Zone
Energy limited (see Clause 3.6)	Ex e	0	Zone 2
Intrinsic safety ib (see Clause 3.7)	Ex ib	1	Zones 1 and 2
Intrinsic safety ia (see Clause 3.7)	Ex ia	2	Zones 0, 1, and 2

Typical FNICO is shown in Fig. X/3.5.2-2B. The major components are FNICO power supply, field cable, and field cable termination devices. Power supply connections to the host control system are made at its "safe area" terminals, and those to the field trunk at its "hazardous area" terminals. It also includes the necessary functions for a reliable fieldbus with those of an energy-limited interface. This incorporates a repeater function, connections for 24 VDC supply input, and a switchable terminator [30]. Field terminators take the form of a DIN rail-mounted terminal module. In an FNICO system, the requirements are simple: the wiring hub and its enclosure must be certified for zone 2 or division 2 as appropriate, and be suitable for the environment.

- Other important issues: From these discussions it is clear that each of the systems has some limitations. Such limitations are compared in Table X/3.5.2-2.

The parameters in the table are standard values; however, on account of power supplies available in the market, in reality the values are far less than standard values. Since the foregoing discussions are mainly based on IS and energy limiting devices it is better to look in detail at the various enclosures to cover the various protection principles described (in Fig. X/3.5.1-1) in Clause 3.6 and then look at the details of IS circuits in Clause 3.7.

Table X/3.5.2-2 Comparison of Various Fieldbus Protection Concepts [29]

Fieldbus Parameter	IS Entity	FISCO		HPT	FNICO	
		Gr. IIB	Gr. IIC		Gr. IIB	Gr. IIC
Maximum current (mA)	80	265	120	500	320	180
Maximum devices	4	13	6	25	16	9
Maximum trunk length (m[a])	1900	1900	1000	1900	1900	1000
Maximum spur length (m[a])	120	60	60	120	60	60
Hazard zone	0 and 1	1	1	0 and 1	2	2

FISCO, *fieldbus intrinsically safe concept;* FNICO, *fieldbus nonincendive concept;* HPT, *high-power trunk;* IS, *intrinsic safety.*
[a]*Length in meters (m)*

3.6 Enclosure Class

From Fig. X/3.5.1-1 it is seen that the majority of explosion protections are undertaken with the help of various kinds of enclosures. In the figure there is another method of protection, namely, IS, which is extremely important for instrumentation, and this will be dealt with greater detail in the following Clause 3.7. While discussing enclosure class, it is obvious that ingress protection (IP) is very important. So, short discussions on this have been included also. During discussions, necessary references will be touched upon. Also it is necessary to note that nonelectrical protections are out of the scope of this discussion. Within electrical protection, the main focus will be on instrumentation. For each of the enclosures, applicable standards and broad areas of application are mentioned in the corresponding figures.

3.6.1 *Flameproof Enclosure*

See Fig. X/3.6.1-1 for a flameproof enclosure. It is marked as Ex d.

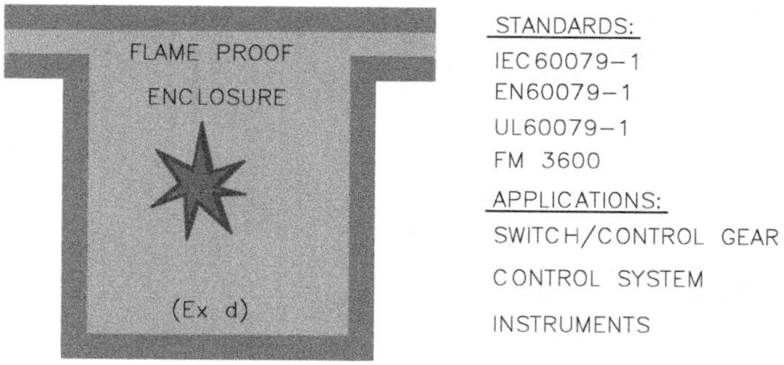

FIGURE X/3.6.1-1

Flameproof enclosure.

- Standards and application: Refer to the *figure*.
- Principles: Parts that can ignite a potentially explosive atmosphere are located inside an enclosure, which can withstand the pressure of an explosive mixture inside. Also enclosure prevents the transmission of the explosion to the atmosphere surrounding the enclosure. So, the major issue is related to the gap during manufacturing so that these are either sealed or explosion propagation is decelerated.
- Design considerations: The following points may be considered:
  - Mechanical strength in line with the safety factor considered for internal explosion pressure: Depending on applicability this could 1.5 times.
  - Gap: When a gap is not sealed after manufacture, then it shall (if at all) be very narrow and long to retard the propagation of explosion to the outside atmosphere. Gap parameters vary with gas groups and are stringent for group IIC.

3.6.2 *Increased Safety*

See Fig. X/3.6.2-1 for increased safety. It is marked as Ex e.

- Standards and application: Refer to the *figure*.
- Principles: Additional measures are applied to increase the level of safety protection to ensure prevention of unacceptably high temperatures and sparks or electrical arcs, both on the internals of the enclosure and on exposed parts of electrical apparatus, where such ignition sources would not occur in normal service.
- Design considerations: The following points may be considered:
  - Protective requirements for the uninsulated, live parts
  - Wider air gap and special requirements for IP (discussed later)
  - Minimum cross-sections for winding wires

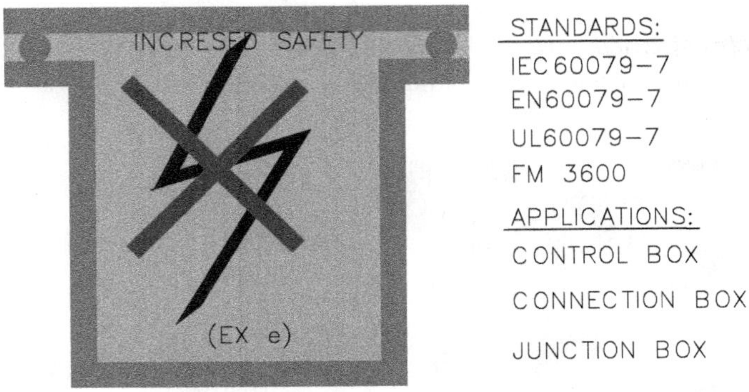

FIGURE X/3.6.2-1

Increased safety.

3.6.3 *Pressurized Enclosure*

See Fig. X/3.6.3-1 for pressurized enclosure. It is marked as Ex p (Ex pD for dust application).

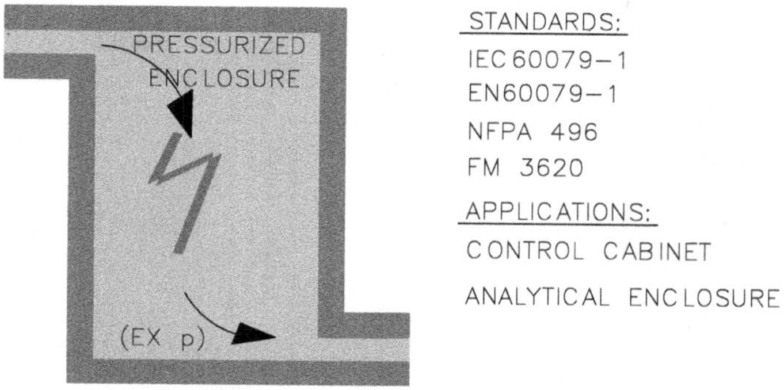

FIGURE X/3.6.3-1

Pressurized enclosure.

- Standards and application: Refer to the *figure*.
- Principles: In relation to the surrounding atmosphere, positive internal pressure of protective gas (air, inert, or a different suitable gas) is maintained inside the enclosure to prevent formation of a potentially explosive atmosphere. As necessary, a flow of protective gas, which acts to dilute any combustible mixtures, is supplied.
- Design considerations: The following points may be considered.
  - Strength to withstand at least 1.5 times overpressure
  - Alert when flushing gas *fails*

3.6.4 *Oil Immersed*

See Fig. X/3.6.4-1 for oil immersed. It is marked as Ex o.

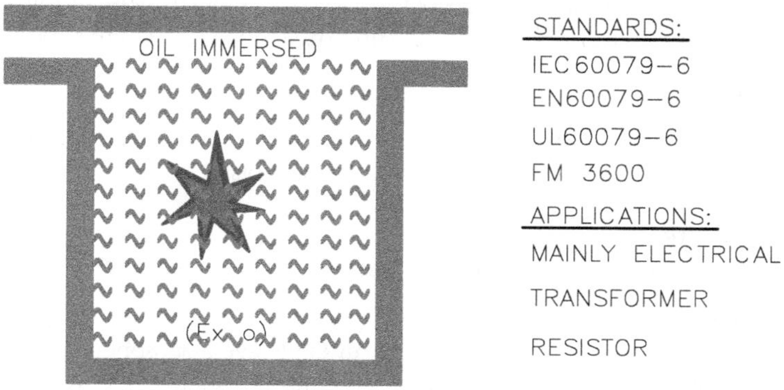

FIGURE X/3.6.4-1

Oil immersed.

- Standards and application: Refer to the *figure*.
- Principles: Electrical apparatus or parts of electrical apparatus, which might ignite an explosive atmosphere, are immersed in oil or other nonflammable, insulating liquid so that a potentially explosive atmosphere above the oil level and/or outside the enclosure cannot be ignited by electric arcs or sparks generated.
- Design considerations: The following points may be considered:
  - *Protection* of oil/liquid from contamination and moisture
  - Means of monitoring oil *level*
  - Not applicable for portable apparatus

3.6.5 *Powder Filling*

See Fig. X/3.6.5-1 for powder filling. It is marked as Ex q.

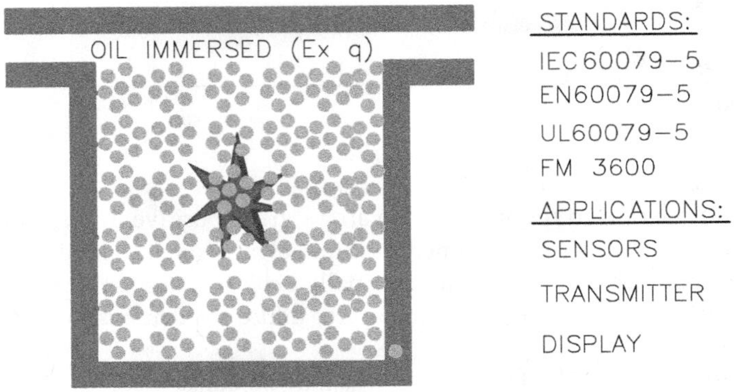

FIGURE X/3.6.5-1

Powder filling.

- Standards and application: Refer to the *figure*.
- Principle: By filling the enclosure with a finely grained powder, any arc within the enclosure under normal conditions is unable (with correct use) to ignite a potentially explosive atmosphere surrounding the enclosure. There should not be any risk of ignition by flames, nor by increased temperatures at the surface of the enclosure.
- Design considerations: Filling is normally done with sand or glass balls, which should not *leave* the enclosure under any circumstances.

3.6.6 *Encapsulated*

See Fig. X/3.6.6-1 for encapsulated. It is marked as Ex m for gas and mD for dust.

- Standards and application: Refer to the *figure*.
- Principle: Parts that are capable of igniting an explosive atmosphere by either sparking or heating are encapsulated in a compound in such a way as to avoid

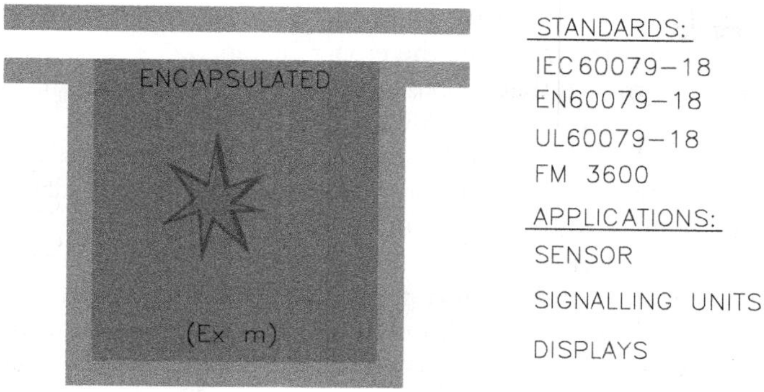

FIGURE X/3.6.6-1

Encapsulated.

ignition of a dust layer or cloud. This compound is normally *resistant* to physical—especially electrical, thermal, and mechanical—and chemical influences.

- Design considerations: The following points may be considered:
  - High *breakdown* strength
  - Low water *absorption*
  - *Resistant* to chemical influence
  - Casting compound penetrated only by the cable entries

3.6.7 *Type of Protection "n-"*

See Fig. X/3.6.7-1 for type of protection "n-." This consists of a number of types of protection, as listed in the figure.

- Types, standards, and application: Refer to the *figure*.
- Principles: In this case the electrical apparatus cannot ignite potentially explosive atmospheres surrounding them in normal and defined abnormal situations. These are applied for zone 2 applications:
  - nA: Unacceptably high temperatures and sparks or electrical arcs, on both the internal and external parts of electrical equipment are prevented by *construction*. Also normal operation does not call for high temperatures,

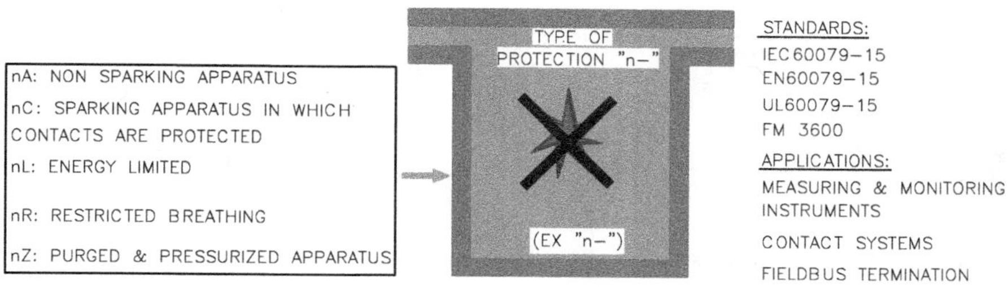

FIGURE X/3.6.7-1

Type of protection "n-."

sparks, etc. These are achieved by applying special *protective layers* on uninsulated live parts, specifying the proper air gap.

- nC: It is so constructed that an explosive atmosphere cannot enter. It is *hermetically* sealed—sealing by welding/solder, glass fusion, etc. Contacts of some process switches offer this kind of sealing.
- nL: Energy limiting apparatus: These are very much in use in fieldbus systems in zone 2 discussed earlier (FNICO). These are circuits in which there is no spark or thermal effect occurring under normal operation and certain fault conditions stipulated in the standard. However, they can ignite the explosive atmosphere of subgroups IIA, IIB, and IIC or of an air/dust mixture only when energy level exceeds stipulated value. The permissible currents or voltages exceed those stipulated for the IS type of protection. Also the requirements to be fulfilled by the circuit and the loads on the components are lower than those for the IS type of protection.
- nR: Restricted breathing: The construction does not permit ingress of explosive gases.
- nZ: Purging pressurization: It is protected by purging and pressurization.

3.6.8 *Ingress Protection*

Enclosure types: Mainly two standards, namely, IEC 60529 and National Electrical Manufacturer Association (NEMA), are used in industries for enclosures pertinent to electrical systems. Both these standards are covered briefly during the discussions. Discussions will be concluded by providing a comparison between them (Fig. X/3.6.8-2):

- IEC: According to IEC 60529 definitions for degrees of protection provided by enclosures of electrical equipment as regards protection cover:
- Access by persons to hazardous parts inside the enclosure
- Ingress of solid foreign objects
- Harmful effects caused by the ingress of water

  The standard also puts distinct designations for these degrees of protection and requirements for each designation. The tests to be performed to verify that the enclosure meets the requirements of this standard are also prescribed in the standard. The standard recommends the necessary measures to be taken to protect the enclosure and equipment inside against external influences like:

  - Mechanical impact
  - Corrosion
  - Corrosion solvent
  - Fungus
  - Vermin
  - Solar radiations
  - Moisture
  - Icing
  - Explosive atmospheres

Various IP code designations are depicted in Fig. X/3.6.8-1—developed based on IS/IEC 61529:2001.

SUPPLEMENTARY LETTER (OPTIONAL)	SUPPLEMENTARY INFORMATION SPECIFIC TO	CLAUSE NO IEC 60529
H	HIGH VOLTAGE	8
M	MOTION DURING TEST	8
S	STATIONARY IN WATER TEST	8
W	WEATHER CONDITIONS	8

ADDITIONAL LETTER (OPTIONAL)	PROTECTION AGAINST HAZARDOUS PART WITH	CLAUSE NO IEC 60529
A	BACK OF HAND	7
B	FINGER	7
C	TOOL	7
D	WIRE	7

SECOND CHARACTER	PROTECTION AGAINST WATER– HARMFUL EFFECT	CLAUSE NO IEC 60529
0	NO PROTECTION	6
1	VERTICALLY DRIPPING	6
2	DRIPPING AT 15 Degree	6
3	SPRAYING	6
4	SPLASHING	6
5	JETTING	6
6	POWERFUL JETTING	6
7	TEMPORARY IMMERSION	6
8	CONTINUOUS IMMERSION	6

FIRST CHARACTER	PROTECTION AGAINST SOLID FOREIGN OBJECT	PROTECTION AGAINST PERSON	CLAUSE NO IEC 60529
0	NO PROTECTION	NO PROTECTION	5
1	>= 50 mm DIA	BACK OF HAND	5
2	>= 12.5 mm DIA	TOOL	5
3	>= 2.5 mm DIA	FINGER	5
4	>= 1 mm DIA	WIRE	5
5	DUST PROTECTED	WIRE	5
6	DUST TIGHT	WIRE	5

FIGURE X/3.6.8-1

Ingress protection code.

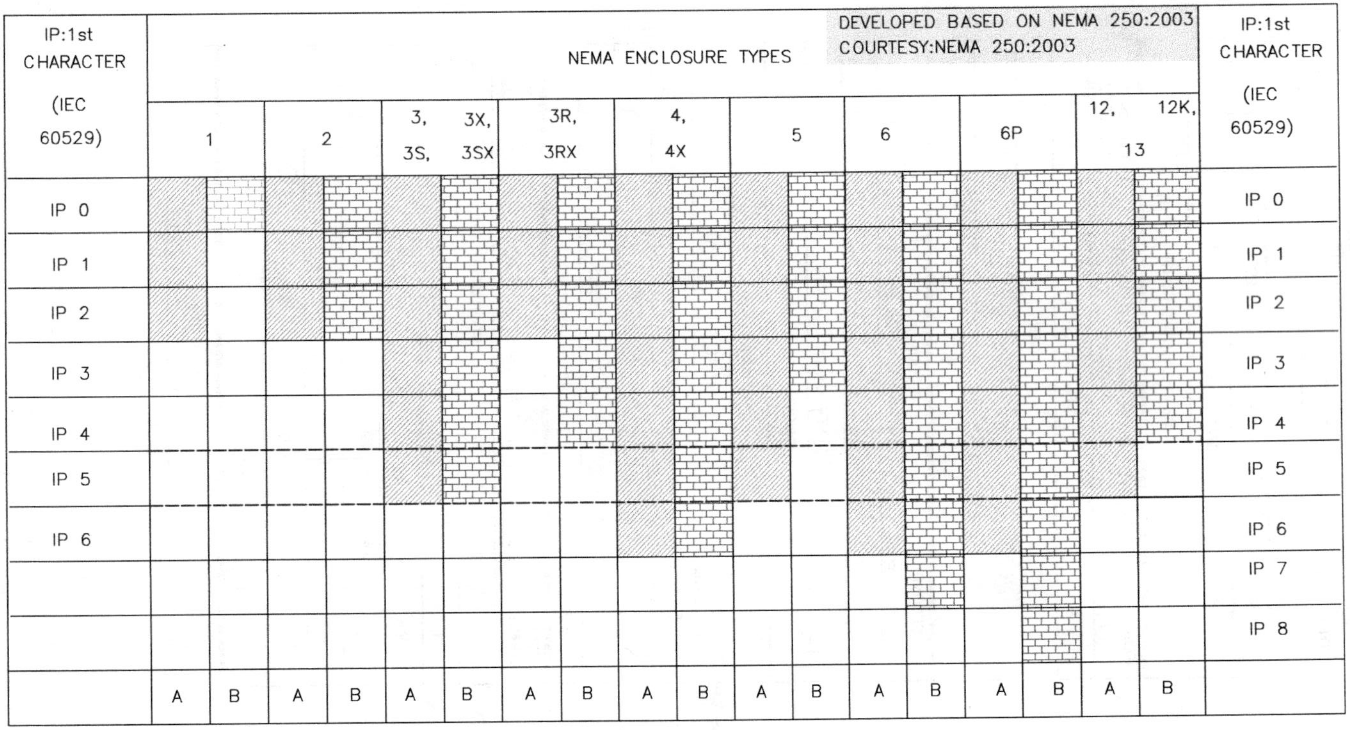

A REPRESENTS NEMA ENCLOSURES TYPES EXCEEDS THE REQUIREMENTS FOR IEC 60529 IP FIRST CHARACTER SHOWN BY CYAN HATCH

B REPRESENTS NEMA ENCLOSURES TYPES EXCEEDS THE REQUIREMENTS FOR IEC 60529 IP SECOND CHARACTER SHOWN BY RED BRICKS

SO, NEMA 4X = IP66 FROM ABOVE CHART

EXAMPLE: CONVERSION OF IP 45: FIRST 4 MET BY 3,3X,3S,3SX/4,4X/5/6/6P/12,1K,13 BUT FOR SECOND CHARACTER 5 ONLY 3,3X,3S,3SX/4,4X/6/6P QUALIFY SO, THEY ARE EQUIVALENT TO IP45. HOWEVER MANY LIKE TYPE 3 EXCEEDS IP 45 REQUIREMENT ON ACCOUNT OF CORROSION GASKET AGING TESTING.

FIGURE X/3.6.8-2

Conversion of NEMA to ingress protection (IP) rating.

Thus from the figure it is clear that the *first* letter code indicates protection against the back of hand, finger, tool, and wire. Also it indicates protection of equipment from foreign bodies of different sizes and dust. Detailed explanations of protections are important and it is recommended to consult the standard prior to application. For protection against people and foreign bodies, Tables 1 and 2 of the standard may be referred to. Similarly, Table 3 of the standard defines protection against water. The harmful effect of water and protection against the same is represented by the second letter. In all the cases, relevant test method references are also indicated. So, an enclosure of IP 66 stands for protection against wire, and equipment in the enclosure is protected against dust and powerful jets of water. It also stands for protection against corrosion.

NEMA: This organization has a standard for enclosures (standard 250—latest edition 2014). Similar to IEC the individual rating is identified against various conditions, as shown in Table X/3.6.8-1. One important point is that NEMA specifies indoor and outdoor separately, and for outdoor a corrosion test is done.

A chart showing conversion of NEMA to IP and vice versa is presented in Fig. X/3.6.8-2 (based on NEMA 250:2003).

3.6.9 *Markings*

Typical markings in line with IEC 60079 are depicted in Fig. X/3.6.9-1. The typical meaning of each of the markings is indicated in the figure.

Normally, the following information is marked on the plate:

- Manufacturer's name details and model and serial number, etc.
- Conformity mark and ID number
- Designation for identification
- Application zone including:
  - Group, vapor/dust/mine
- Categories of approval for specific zones
- Type(s) of protection the equipment fulfills
  - Explosion group and subgroup
- Temperature class
- Ambient specification
- Test laboratory where the test certificate was issued
- Standard with versions for certification

  Some examples:
  - Equipment group II other areas
  - Category 2 suitable for zone 1, 21
  - Gases, vapors marking with prefix G
  - Dusts marking with prefix D
  - Enclosure protections are important for instrumentation, so further details are given in Appendix II (Fig. APII/2.3-1). IS is very important from an instrumentation point of view, and will now be the focus of the next clause.

Table X/3.6.8-1 NEMA Enclosure Rating With Interpretation

Type	I	O	Protection Against: Personnel Access	Protection (Ingress): Foreign Solid Object	Protection (Ingress): Water Harmful Effect
1	X		To hazardous part	Falling dirt	
2	X		To hazardous part	Falling dirt	Dripping/light splashing
3	X	X	To hazardous part	Falling dirt, windblown dust	Rain, sleet, snow, and external ice formation
3R	X	X	To hazardous part	Falling dirt	Rain, sleet, snow, and external ice formation
3S	X	X	To hazardous part	Falling dirt, windblown dust	Rain, sleet, snow. Operable when ice laden
3X	X	X	To hazardous part	Falling dirt, windblown dust	Rain, sleet, snow, and external ice formation and additional corrosion protection
3RX	X	X	To hazardous part	Falling dirt	Rain, sleet, snow, external ice formation, and additional corrosion protection
3SX	X	X	To hazardous part	Falling dirt, windblown dust	Rain, sleet, snow, external ice formation, and additional level of protection for corrosion. Operable when ice laden
4	X	X	To hazardous part	Falling dirt, windblown dust	Rain, sleet, snow. Splashing and hose-directed water. Undamaged because of external ice formation on enclosure
4X	X	X	To hazardous part	Windblown dust	Rain, sleet, snow. Splashing and hose-directed water. Additional level of protection for corrosion and undamaged because of external ice formation on enclosure

Table X/3.6.8-1 NEMA Enclosure Rating With Interpretation **Continued**

Type	I	O	Protection Against: Personnel Access	Protection (Ingress): Foreign Solid Object	Protection (Ingress): Water Harmful Effect
5	X	-	To hazardous part	Falling dirt and settling airborne dust, lint, fibers and flying	Dripping and light splashing
6	X	X	To hazardous part	Falling dirt	Hose-directed water, entry of water because of occasional temporary submersion at a limited depth, and undamaged because of external ice formation on enclosure
6P	X	X	To hazardous part	Falling dirt	Hose-directed water, entry of water during prolonged submersion at a limited depth. Additional level of protection for corrosion and undamaged because of external ice formation on enclosure
12/12K[a]	X		To hazardous part	Falling dirt, circulating dust, lint, fibers, flyings	Dripping/light splashing
13	X		To hazardous part	Falling dirt, circulating dust, lint, fibers, flyings	Dripping/light splashing. Also protection against spraying, splashing, and seepage of oil and noncorrosive coolants

I, *indoor application;* O, *outdoor application;* X, *applicability.*
[a]*12K with knockout whereas 12 without knockout.*

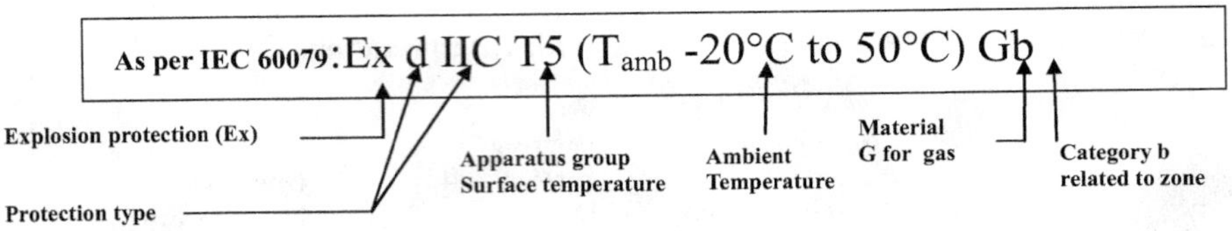

FIGURE X/3.6.9-1

Typical markings (IEC 60079).

3.7 Intrinsic Safety

IS circuits: IS in instrumentation and controls for hazardous areas is very important and a common means of explosion protection. Distribution of use of IS circuits among various types of instrumentation items is depicted in Fig. X/3.7-1 based on data from Ref. [34].

From the foregoing it is clear that the IS technique is used in cases of low-energy signaling. For instrumentation and control and communication (fieldbus), low-energy level signals are utilized in an effective way. IS circuits utilize the low-energy signaling technique, which prevents an explosion occurring from that energy, that is, to ensure that the available energy level is below the energy level required to cause ignition. This prompts the question: if instrumentation uses low-energy levels then why IS? There are a few reasons for this but one simple answer is that if there is a fault in the system, then a high-level signal may flow! IS prevents this from happening. This is well understood from Fig. X/3.7.2-1. Spark, arc, and hot surface are the major reasons for such chances of ignition. This is also clear from IS definition.

- IS equipment definitions:
  - IEC definition of IS: As per IEC 60079-11, IS can be conceived as a "type of protection based on the restriction of electrical energy within apparatus and of interconnecting wiring exposed to the potentially explosive atmosphere to a level below that which can cause ignition by either sparking or heating effects."

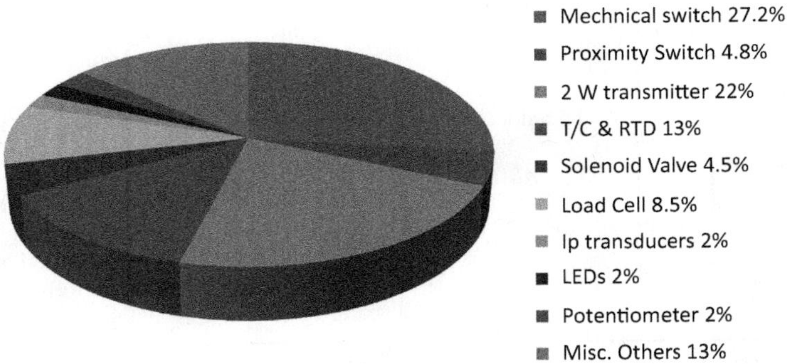

FIGURE X/3.7-1

Intrinsic safety circuit use (in %). *IP*, ingress protection; *RTD*, resistance temperature detector; *T/C*, thermocouple.

    - ISA R12-6 defines IS as "A type of protection in which a portion of the electrical system contains only intrinsically safe equipment (apparatus, circuits, and wiring) that is incapable of causing ignition in the surrounding atmosphere" (ref: Clause 3.12 of ANSI/ISA-RP12.06.01-2003)
- Categories of IS: There are three categories of IS circuit, namely, ia, ib, and ic. The category distinctions are as follows:
    - "ia": "ia" tops the category offering the highest level of protection. It is generally considered as being adequately safe for use in the most hazardous locations (zone 0), can offer fault tolerance up to two "faults," and has a factor of safety of 1.5.
    - "ib": "ib" is in the middle of the category. "ib" apparatus is adequately safe with a single fault with a factor of safety of 1.5. It is considered for use in hazardous areas—zone 1.
    - "ic": "ic" apparatus has a factor of safety of unity and is considered for use in hazardous zone 2. The "ic" concept (2005) is relatively new and replaces the "energy-limited" (nL) of the type "n" standard IEC 60079-15 and possibly the "nonincendive" concept of North American standards [33]. "ic" is considered equivalent to Cat3 of ATEX, and self-certification by manufacturers with a suitable testing facility is accepted for "ic."
- Comparison: Comparison and usage of three categories are shown Table X/3.7-1.
- Advantages of IS: IS enjoys a number of advantages over other types of explosion protection. The most prominent ones are as follows:
    - The IS technique is a well-accepted technique internationally for both gas and dust. Relevant standards provide detailed guidance, which is not available in other techniques.
    - IS, as a technique, is applicable for all the zones based on its level, as shown in Table X/3.7-1. Incidentally, "ia" is the only technique that gives the most satisfactory results in zone 0.
    - Intrinsically safe apparatus and systems are usually allocated a group IIC gas classification, which ensures that the equipment is compatible with all gas/air mixtures. Occasionally, IIB systems are used, as this permits a higher power level to be used [33].
    - A temperature classification of T4 (135°C) is achievable to meet the requirements of most of the industrial gases.
    - IS circuits can be interfaced with almost all types of instrument types and I/Os, namely, resistance temperature detector (RTD), Thermocouple, mV, mA loop, Digital I/Os, and Fieldbus systems.

***Table X/3.7-1** Comparison and Usage of Intrinsically Safe (IS) Categories*

IS Level	Fault Tolerance	Factor of Safety	ATEX Category	IEC EPL	Zone
ia	2	1.5	1	0	0
ib	1	1.5	2	1	1
ic	0	1	3	2	2

EPL, *equipment protection level.*

- Live maintenance is *possible* in IS circuits without a gas clearance certificate. However, suitable care should be taken. In all installations, IS circuits are properly *marked*. IS apparatus is well documented to avoid any mistake.
- Normal cables *can* be used in IS circuit applications only with special care to be taken for long cable for inductance and capacitance value especially in zone 0, 1 and group C gas. Normally, **blue-colored** cables are used for IS circuits to distinguish them from others. IS cable installations need special attention; in this connection associated standards (e.g., NORSOK) may be referred to.

Generally, each apparatus in a system is individually certified. From standard IEC 60079-25, for certification of the entire system necessary documentation is to be developed utilizing the information from each device/apparatus including cable details. Starting with general discussions, let us look a little deeper into the system with special reference to the design aspects.

3.7.1 *General Discussions*

A circuit in which any spark or thermal effect is incapable of causing ignition of a mixture of flammable or combustible material in air, under prescribed test conditions, is an IS circuit. Intrinsically safe circuits have three components: *field* device, referred to as the intrinsically **safe apparatus**; *barrier* (**associated apparatus**), the energy-limiting device, and the associated *wiring*. From this it is clear that intrinsically safe apparatus is interconnected by wiring and the safety of each device is highly dependent on the performance of the other pieces of apparatus in the circuit. So, this safety technique is a system concept and relies on accurate design of the entire system. In an intrinsically safe circuit, design and analysis starts with the field device, which will determine the type of barrier to be used for proper and safe functioning of the system under normal operating conditions as well as under fault conditions. The numerous products available in the market and various classes of calculation make things increasingly confusing.

- *Circuit* parameters: Intrinsically safe apparatus is classified as *simple apparatus* or *nonsimple apparatus*.

  There are a number of publications on simple apparatus worldwide, defined in similar terms. A few of them are EN 50020:1995 (P: 3.11, 5.4), EN 60079-14: 1997 (P: 3.21), IEC 60079-11—appendix, and ANSI/ISA RP12.6-1987 (P: 3.12). As per the ATEX directive (original), "simple apparatus is not capable of causing an explosion through its *own* potential source of ignition." Also "Simple apparatus is considered *not* to require certification from a notified body." This means that self-certification is acceptable. Electrical junction boxes, RTDs, thermocouples, contact passive devices, and semiconductors are examples of simple apparatus. As per ISA a simple apparatus cannot generate or store energy limited by a few parameters, as given in Table X/3.7.1-1.

Table X/3.7.1-1 Parameter Limits for Simple Apparatus (ISA)

Voltage	Current	Power/Energy
1.2–1.5 V	0.1 A	25 mW/20 μJ

Naturally, those devices that *cross* the limit given in Table X/3.7.1-1 are considered as nonsimple devices. Transmitters, relays, solenoid valves, etc. are examples of nonsimple devices and require certification by a notified body. One thing is clear, however; all these are mainly connected with energy and/or storage of energy. In an *entity approval* for nonsimple apparatus, each apparatus is examined separately and *assigned* a set of entity parameters. For nonsimple devices these parameters are: maximum voltage (V_{max}) allowed, maximum current allowed (I_{max}), internal capacitance (C_i), and internal inductance (L_i). The first two parameters indicate that when either of these are exceeded, then intrinsically safe apparatus (a field device) can heat up or spark in the hazardous area. L_i and C_i indicate the ability of the device to *store* energy. For designing the system it is necessary to obtain matching parameter data for associated apparatus. These are:

V_{oc}: Maximum open-circuit voltage that is allowed to appear *across* the intrinsically safe connections of the associated apparatus under fault conditions.

I_{sc}: Maximum allowable short-circuit current that can be drawn *from* the intrinsically safe connections of the associated apparatus under fault conditions.

C_a: *Allowed* capacitance that can safely be connected to the associated apparatus.

L_a: *Allowed* inductance that can safely be connected to the associated apparatus. The relationship between the two sets of parameters is as indicated in Table X/3.7.1-2.

- Zener diode: A diode is a solid-state device; when it is forward biased it allows current flow in one direction only. A Zener diode is a special diode that blocks current flow in reverse bias until a critical voltage is reached. At critical voltage on account of avalanche it is able to conduct current in this reverse bias mode without damaging itself. (This current cannot be controlled until the source is controlled.) Any standard book on solid-state devices may be referred to recapitulate Zener diode.

Table X/3.7.1-2 Relation Between Parameters of Barrier and Field Devices

Barrier Parameter	Relations	Field Device Parameter
Open-circuit voltage: V_{oc}	$\leq$	Maximum voltage: V_{max}
Short-circuit current: I_{sc}	$\leq$	Maximum current: I_{max}
Allowed capacitance: C_a	$\geq$	Internal capacitance: C_i[a]
Allowed inductance: L_a	$\geq$	Internal inductance: L_i[a]

[a] *Actually* $C_a \geq C_i + C_{cable}$ *and* $L_a \geq L_i + L_{cable}$.

- Generalized circuit: A typical barrier circuit (DC) is shown in Fig. X/3.7.1-1. Under normal conditions, the device really has no function except to allow a field device to function properly. Under fault conditions, the barrier has to protect the field circuit by preventing excess voltage and current from reaching the hazardous area. Three components in the barrier that limit current and voltage are: a resistor, a set of Zener diodes (three Zener diodes for "ia" and two for "ib"), and a fuse. The resistor limits the current to $\mathbf{I_{sc}}$. The set of Zener diodes limits the voltage to $\mathbf{V_{oc}}$. The fuse will blow when the set of Zener diodes conducts. This interrupts the circuit, which prevents the diode from burning and allowing excess voltage to reach the hazardous area. There are always at least two Zener diodes in parallel so that each can fall back in case of failure of the other. Because the set of Zener diodes is in *single* configuration, operation is in one direction only, hence it can be considered as DC. Cables have *distributed* inductance and capacitance, and hence possess energy storage capabilities, so they can affect system safety. As a result the system design imposes restrictions on the amount of each of these parameters. These are shown as a bulk separately in Fig. X/3.7.1-1. In this connection, Table A.2 in IEC 60079-11 for cable parameters may be referred to. Special calculations and considerations are applied for system analysis because of cable faults—especially when multicore cables (see IEC 60079-14 for type A and B cables in connection with multicore cables) are used in the installations [33]. Other important issues here are that: (1) voltage drop across the resistor under normal operating conditions has to be taken into account in the *design* so that there is sufficient voltage at the field device and (2) Zener barriers are properly *earthed* as shown because without an IS earth they are not safe. To obtain independence from

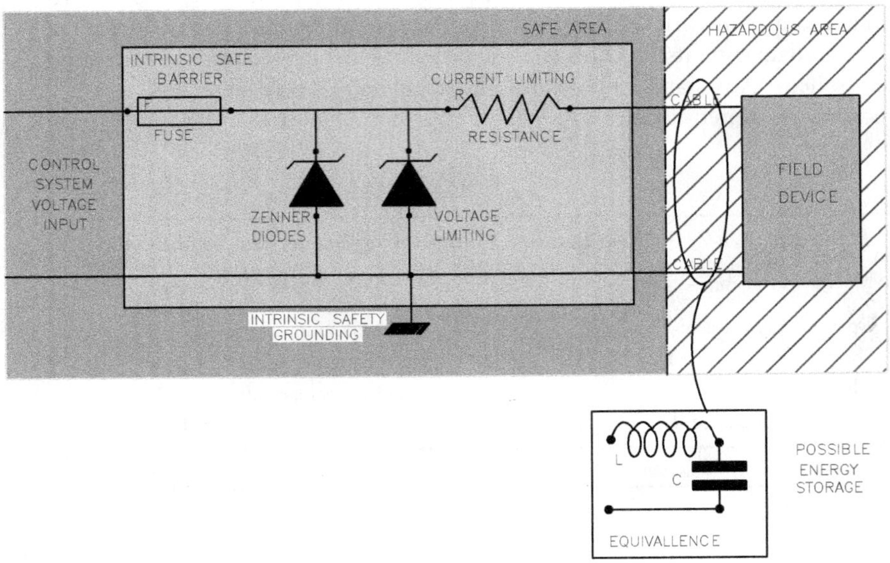

FIGURE X/3.7.1-1

Intrinsic safety (IS) barrier in a hazardous area.

earthing, galvanic isolators could be deployed. In the case of a galvanic isolator (another type of IS module) grounding is not done and hazardous side connection is isolated from the safe side connection and ground loop. In the case of galvanic isolators (Fig. X/3.7.1-4) longer cables and higher voltages can be used.

- Safety energy level: As already discussed, there are different levels of energy needed to cause ignitions in different groups of gases. Ignition curves for various groups of gases are shown in Fig. X/3.7.1-2. It is worth noting that the curves shown here are not to true scale and are replicas of actual curves. They are used for illustration

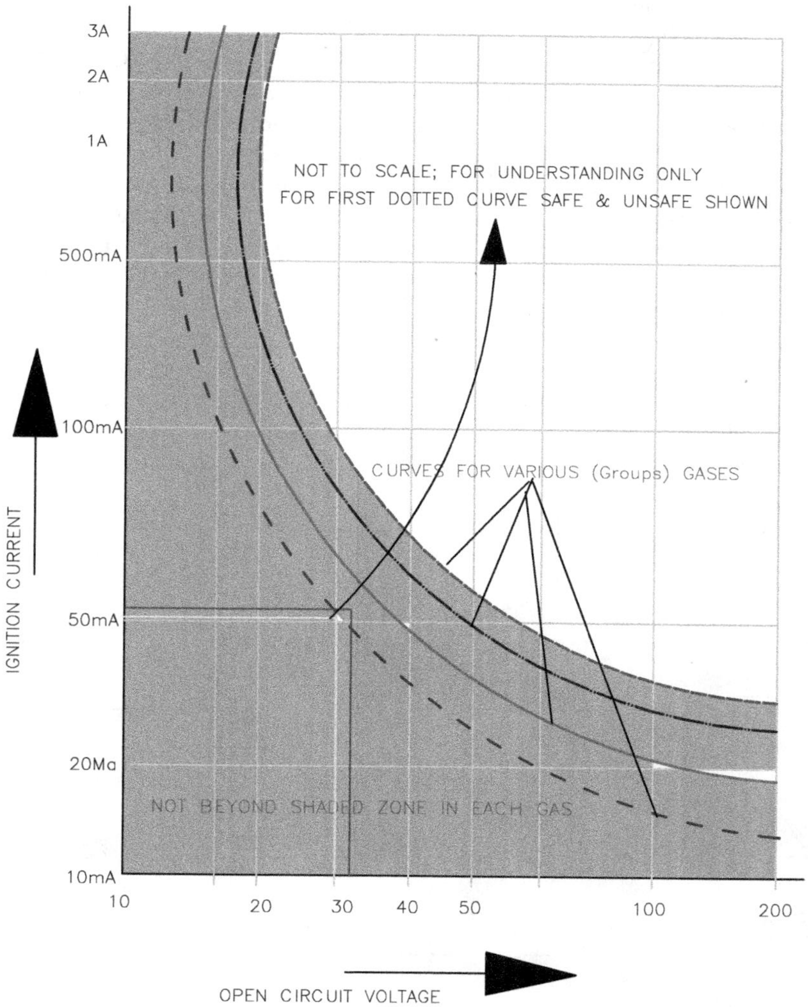

FIGURE X/3.7.1-2

Ignition curve—resistance for gases.

purposes only. Actual curves are available in any standard handbook on IS systems. In this log—log graph with current on the vertical axis and voltage on the horizontal axis, ignition energy required for various gas groups has been depicted. The portion to the left and below the group curve is the safe zone and shown by shading. Also for a better understanding of the gas group—shown by a dotted line—safe and unsafe operating points are also marked in the figure. A safety factor of 1.5 is applied for calculating the allowable energy.

These curves reflect the worst-case scenario. Because of the low power required by today's electronics, most manufacturers of intrinsically safe equipment find it safer (and easier) to start by designing the equipment for the worst-case specifications [24].

- Safety barrier types: There are two types of safety barrier: **DC** and **AC** type. The DC type is shown in Figs. X/3.7.1-1 and X/3.7.3-1A where it is seen that Zener diodes are in one configuration only. In this type, polarity is an important factor and if it is connected in reverse it will not function. On the other hand, to overcome the polarity problem, the AC type (double) is used, especially for thermocouples and RTDs. An AC double-type barrier is shown in Fig. X/3.7.1-3. Here it is seen that two sets of Zener diodes are in back-to-back connection to accept any polarity of connections. Because in this connection both fault current directions are accepted, it is an AC-*type barrier*.
- Galvanic isolator: A typical galvanic isolator IS module is shown in Fig. X/3.7.1-4.
  As seen in the figure the module consists of the following components:
  - Energy limiting circuit (in hazard side connection)
  - Hazard area circuit
  - Safe area circuit
  - Both are powered by two different secondary coils of an approved transformer

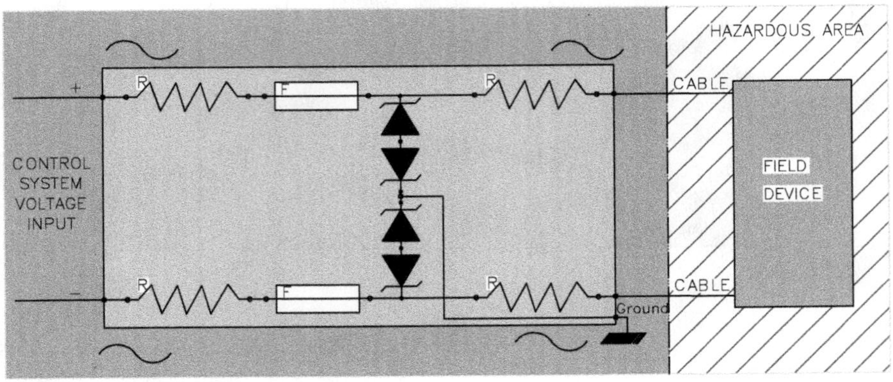

FIGURE X/3.7.1-3

AC safety barrier—double.

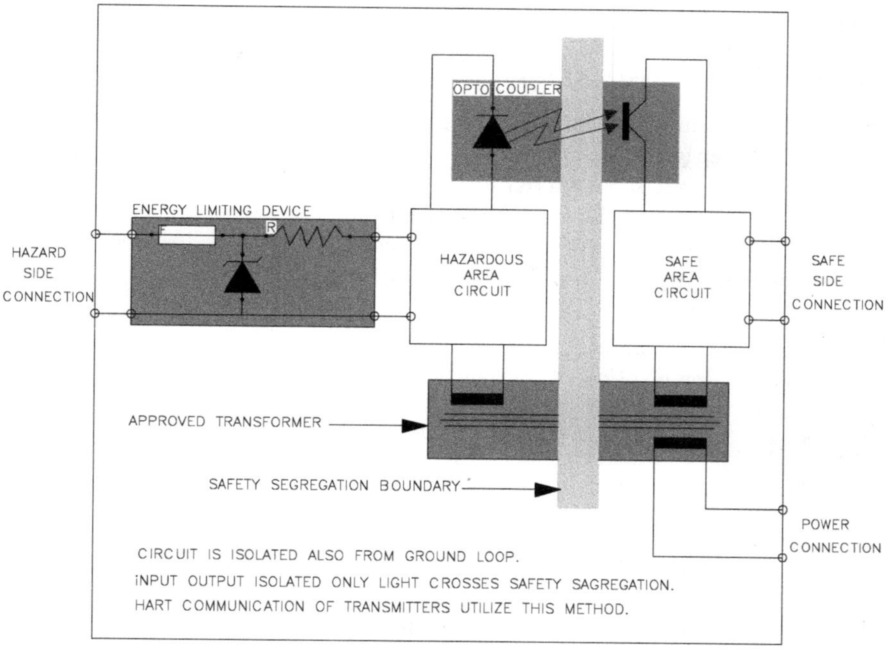

FIGURE X/3.7.1-4

Galvanic isolator intrinsic safety (IS) module.

- Power supply is also galvanically isolated through the aforesaid transformer primary coil.
- Optocoupler/approved relay

A few relevant issues shall include but are not limited to the following:

- Transformer-isolated barriers do not require a high integrity IS ground on account of isolation of two areas by the transformer.
- The system will be floating unlike grounding need for IS barrier.
- Transformer-isolated barriers are repairable IS barriers, which are encapsulated.
- Transformer-isolated barriers are mainly used when grounding is impractical and when regulated power supply is too costly. Some application examples are transmitters/thermocouples/switching inputs.
- These are used where noise caused by the *ground* loop is undesirable and when total *resistance* of the loop exceeds the specification for the desired instrument application.

There is approved safety segregation; only light from the optocoupler can cross approved safety segregation. There is no grounding, so the hazard side is isolated from the safe side and ground loop. Highway Addressable Remote Transducer communication in a transmitter is an example of its application.

- IS approaches: Fundamentally there are two approaches for IS circuits:
  - Grounded IS barrier
  - Grounded/ungrounded repeaters (galvanic isolator)

Each of these systems has some advantages and disadvantages, which are listed in Table X/3.7.1-3. Repeaters require certain components for which approval from notified bodies is required, such as a transformer in ungrounded repeaters. Usually in case of repeaters, a single product can be used for different applications. However, these are costly and require more space. In the case of a grounded safety barrier, precise engineering and selection of IS products are necessary.

Table X/3.7.1-3 ***Pros and Cons of Different Approaches to an Intrinsic Safety (IS) Circuit***

Pros and Cons	Grounded IS Barrier	Grounded Repeater	Galvanic Isolator
Advantage	Less costly and smaller size Precise signal response	Single product use Usable in transmitters with higher power supply	Single product use No need for ground Usable in transmitters with higher power supply
Disadvantage	Precise engineering and grounding are necessary	More expensive, and more power and space needed	More expensive, and more space needed Radiofrequency interference

3.7.2 *Thermocouple/Resistance Temperature Detector (RTD) Input*

Discussions begin with thermocouple/RTD inputs. Since both belong to the simple apparatus category they do not need certification by a notified body. Polarity, rated nominal voltage, and internal resistance are major design parameters to be considered. Both have similar requirements for these parameters.

- Thermocouple input: A thermocouple is a simple device, so it is incapable of creating or storing enough energy to ignite any mixture of volatile gases. When the energy level of a typical thermocouple circuit is seen against the ignition curve (like one shown in Fig. X/3.7.1-2) of the gas group it is seen that these are very much on the left and lower part of group A.

  Now, if a fault occurs on the secondary device, say in the logic solver, then it could cause excess energy to reach the hazardous area, as seen in Fig. X/3.7.2-1. To make sure that the circuit remains intrinsically safe, that is, the fault does not reach a hazardous area, a barrier limiting the energy is required as shown in the bottom part of Fig. X/3.7.2-1.

  Design issues: A few design issues are discussed:
  - Polarity: A thermocouple has two wires with positive and negative polarity. Two single-channel barriers, each with the proper polarity, could be used but

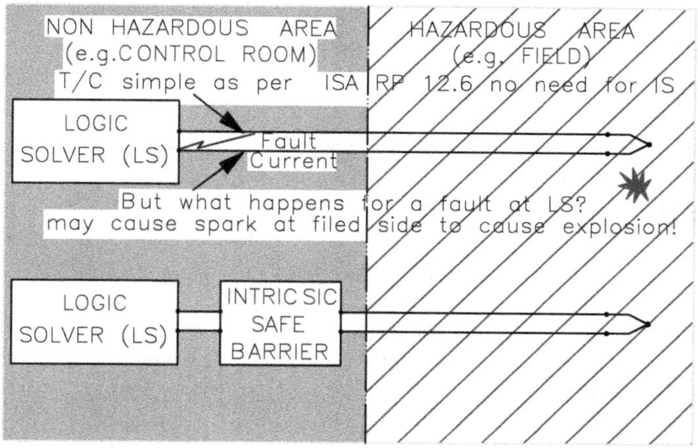

FIGURE X/3.7.2-1

Intrinsic safety (IS) circuit for a thermocouple. *LS*, logic solver; *T/C*, thermocouple.

the problem comes when polarities are reversed by mistake. To avoid polarity problems on the terminals, a double AC barrier could be used.

- Rated voltage (V_n): A thermocouple produces a very small voltage. Since the thermocouple produces such a small voltage, it makes sense to choose a double AC barrier with a higher rated nominal voltage ($V_n > 1$ V) [34].
- Internal resistance (R_i): Because the mV signal has a very small current and is going to a high-impedance voltmeter, the resistance of the barrier will not affect functioning of circuit. However, it is wise to select a barrier with a low resistance ($<110\ \Omega$) [34].

- Resistance temperature detector (RTD) input: Like the thermocouple, in the case of an RTD, IS barriers prevent excess energy from possible faults on the safe side, say from the logic solver, from reaching the hazardous area. A typical RTD input IS circuit is depicted in Fig. X/3.7.2-2.

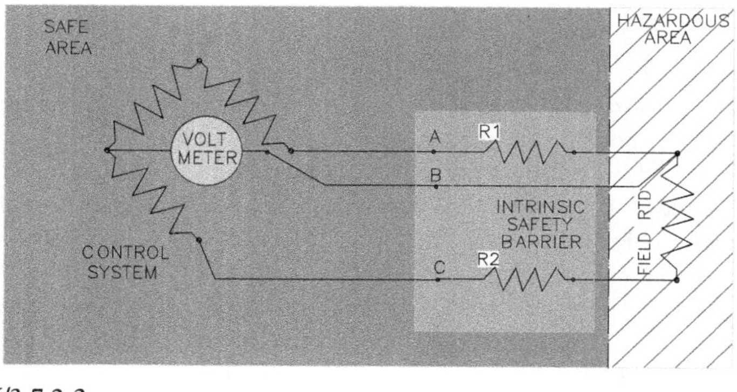

FIGURE X/3.7.2-2

Intrinsic safety (IS) barrier for resistance temperature detector (RTD) input.

In most cases, three-wire RTDs are used in the industry, so these will be considered here. Mainly, bridge circuits using a modified Wheatstone bridge are used. In these circuits, measured output voltage is a function of the RTD resistance. Requirements for RTDs are similar to thermocouples. Use of a double-channel AC barrier with parameters discussed later will be a cost-effective solution.

- Polarity: The current loop of the RTD has a positive and negative polarity. Hence *one each* of a standard DC barrier, one standard DC barrier (+ve and −ve), one double AC barrier are to be chosen. The last one is a better choice to avoid polarity problems.
- Rated voltage (V_n): The constant current sent to the RTD is very small at the 10^{-6} level. RTDs are recommended for use in the range 600–700°C. Even if a very high temperature is considered, the maximum resistance of the RTD Pt100 is 390 Ω at 1560°C [34]. So, the voltage drop across the RTD will be very low in the order of mV. Naturally, the V_n of the RTD loop is similar to the thermocouple, that is, $V_n > 1$ V.
- Internal resistance (R_i): Any constant current source will have a rated maximum load that it can drive. Considering a minimum standard load of 500 Ω and RTD value at high temperature as 390 Ω, the R_i to be chosen is less than 110 (500−390) Ω.

3.7.3 *Transmitter Input/Analog Input*

Transmitters are not simple apparatus and there is the possibility for storing energy. So, they need certification by a third party for use in IS applications. Generally, high-level analog (4–20 mADC) inputs also interface in a similar manner. As stated in Clause 3.7.1 and in Table X/3.7.1-3, there are three approaches to IS circuits and all are applicable here. These are detailed in Fig. X/3.7.3-1.

Various approaches shown in Fig. X/3.7.3-1 are not only applicable for transmitters but are also applicable to other applications discussed later. In transmitters, physical parameters are converted to an electrical signal (generally 4–20 mADC) for transmitting to DCS over a long distance. Usually, this mA signal is converted to 1–5 V with the help of conditioning resistance (250 Ω), so that it can be used for analog-to-digital conversion. Here, design issues are type of safety barrier, voltage input (V_n), and internal resistance (R_i).

- Barrier type: The type of safety barrier is largely determined by the placement of the conversion resistor. When a conditioning resistor is placed on the supply line, a simple DC positive barrier can be used, as shown in Fig. X/3.7.3-1A. However, in the majority of cases conditioning transmitters are placed in the return path, where a double-channel supply and return barrier are used, as shown in Fig. X/3.7.3-1B. The supply line positive DC barrier prevents a fault on the safe side from transferring excess energy to the transmitter. When looking at the return line, it is seen that there are two diodes in series to ensure that the signal flows only in one direction, that is, return to the DCS, and prevent any excess fault energy from being transferred to the transmitter. Both

(A)

(B)

(C)

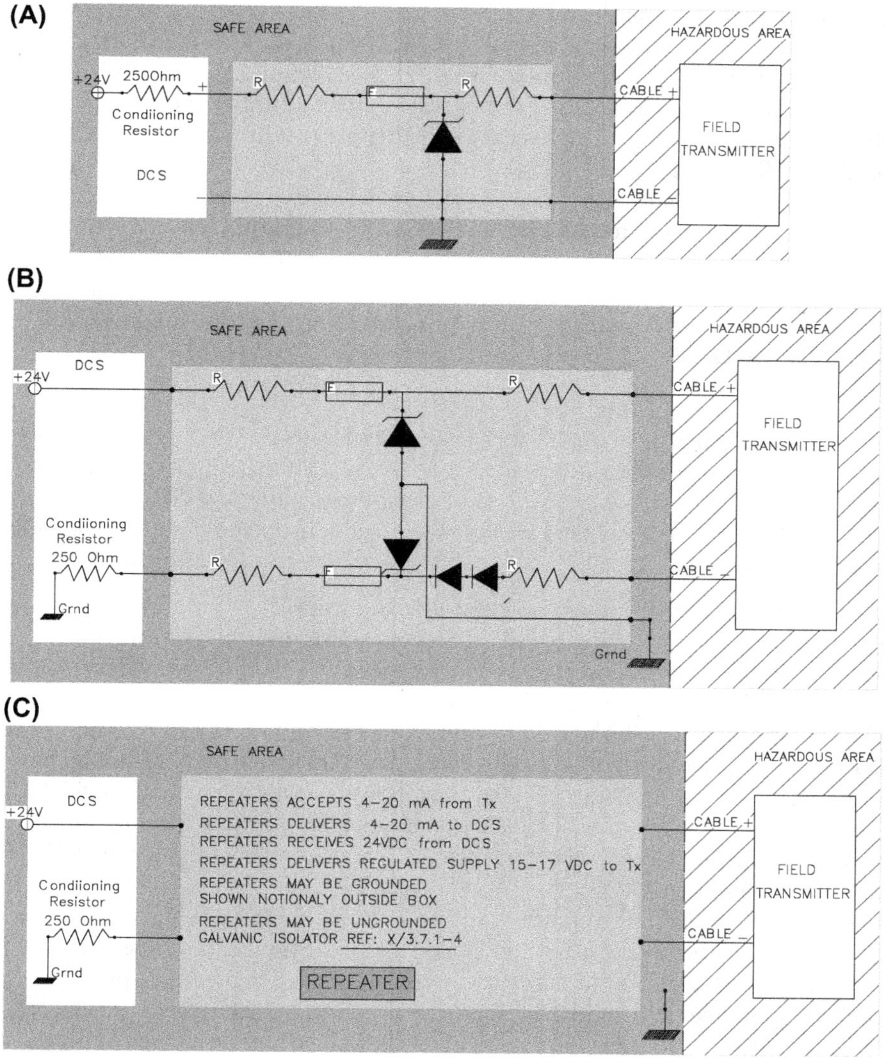

FIGURE X/3.7.3-1

Safety barrier and repeater for a transmitter (20 mA). (A) DC safety barrier (+ve), (B) both side safety barrier, (C) safety repeater for transmitter. *DCS*, distributed control system.

these approaches are grounded safety barriers. However, there could be situations where there is no proper ground or transmitters have higher operating voltages. In such situations, repeaters—grounded and ungrounded (see Clause 3.7.1 and Table X/3.7.1-3)—could also be deployed, that is, transmitters, with a loop indicator or communicator. Repeaters supply a regulated power supply of

15–17 V to the transmitters to drive a conversion resistor load of 750–1000 Ω [34]. A typical repeater circuit is shown in Fig. X/3.7.3-1C. The repeaters could be grounded or ungrounded galvanic isolators, as shown in Fig. X/3.7.1-4.

- Rated voltage: A regulated supply with optimum tolerance that does not exceed the barrier rating should to be selected. Keeping parity with normal industrial usage, 24 VDC ± 1% supply is a good choice.
- Internal resistance: Normal transmitters at 20 mA require at least a 12 VDC drop across them. When the total loop is considered it is seen that for a 24 VDC power supply major drops are transmitter drop and conditioning resistance drop, which in the highest case of 24 VDC will be 17 (12 + 5) VDC. So, 7 V are left at best for the IS barrier and cable drop. Applying Ohm's law one gets $R_{total} = 7/20\ mA = 350\ \Omega$. Taking 10 Ω for the cable it should be <340 Ω.

3.7.4 *Digital/Binary Inputs*

There are a good number of digital inputs interfaced with the DCS through safety systems. Mostly, these inputs come from mechanical or reed contacts, transistors, limit switches, pushbuttons, etc. All these are simple apparatus and do not need any certification by a third party. There are two ways digital inputs for IS applications are handled, either a switching amplifier (NAMUR sensor) or by grounded IS barrier.

- Switching amplifier (IS): Switching through an intrinsically safe relay or switching amplifier is quite popular in IS applications when power supply and space are available in the associated control panel of the DCS. A switch amplifier is an intrinsically safe relay. A NAMUR-style proximity switch is a *two-wire DC sensor* that operates at 8.2 V with switch points operating between 1.2 and 2.1 mA [34]. These switches require a remote amplifier for operation. Most of the available switching amplifiers have ratings: an intrinsically safe voltage of 8.2 V 8 mA current (through the contacts in hazardous areas). Switch amplifiers have two outputs: relays for *slow* switching application and optocouplers for the DCS and *faster* switching.
- Grounded IS barrier: When power or space is not available, or field switches are powered from other places, then a grounded safety barrier may be chosen. There are two types of grounded IS barriers: current source type and current sink type.
  - Current source type: Here this IS type used because it is the same as the supply and return type IS shown in Fig. X/3.7.3-1B. Blue IS cables are used between field sensor and IS barrier.
  - Current sink type: For current sink switching, a single-channel DC barrier as seen in Fig. X/3.7.3-1A can be utilized. In this type of binary input, the DCS sees high voltage when the contact is open whereas when the contact is closed the DCS sees lower voltage as is normal with sink type sensing. Suitable care must be taken for proper grounding and to avoid unnecessary group loop formation to avoid noise and influence actual signal measurements.

3.7.5 *Analog Output*

An example of an analog output is the output from the DCS to converters. When a current to a pneumatic (input) converter in the field in a hazardous area signal is to be sent from the DCS, a grounded barrier may be used. These converters need entity approval. The DC positive polarity grounded IS barrier as shown in Fig. X/3.7.3-1A may be utilized. Voltage rating is determined by type of converter chosen. Since converters have resistance to the tune of 150 Ω, it should be maximum *driving load* of analog output of the DCS minus converter load. So, if the driving maximum load is 1000 Ω, then 850 Ω will be R_i (including cable). A typical grounded IS barrier is shown in Fig. X/3.7.5-1.

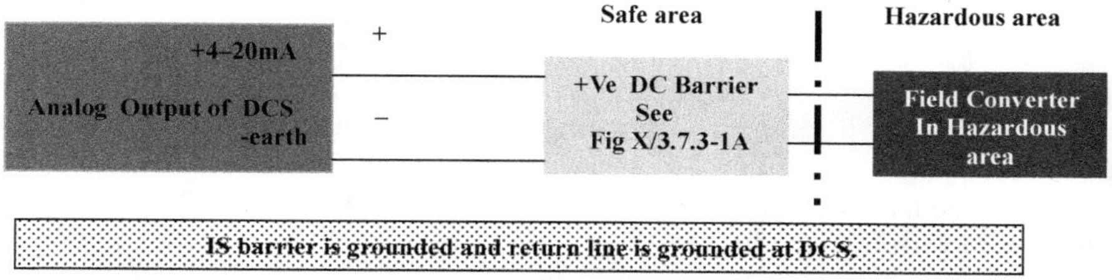

FIGURE X/3.7.5-1

Grounded intrinsic safety (IS) barrier for analog output.

3.7.6 *Digital Output*

Digital outputs are frequently used for actuation of valves, LEDs, etc. An LED being a simple apparatus does not require entity approval whereas a valve actuator such as a solenoid valve needs entity approval. Positive DC IS barriers are used. In most cases these barriers are rated for 24 VDC. R_i is calculated from the selected device. Typical values for LEDs and solenoid valves can be 480 and 350 Ω [34]. System configuration is similar to what has been shown in Fig. X/3.7.5-1, only +4—20 mA will be replaced by +24 VDC via the contact. The return line will be grounded at the DCS as shown. The IS barrier will be grounded. With discussions on IS barrier for various I/Os concluded we will now move on to wiring practices.

3.7.7 *Wiring and Installation*

Short discussions on installation practices are as follows:

- IS cable: There is not much difference between the specification of "IS" cables and other instrumentation cables. IS cables in the field are recognized by their light blue-colored (similar to RAL 5015) outer sheathe and proper labeling. A short description of IS cables in their general form is presented in Table X/3.7.7-1.
- Wiring: Two major differences between IS wiring and other unclassified wiring are separation and identification. The intrinsically safe conductors are separated from non-IS wiring by using separate conduits or by a *separation* of 2 inches of air

Table X/3.7.7-1 Short Intrinsic Safety (IS) Cable Specification (General)

Item	Specification	Item	Specification
Conductor	Finely stranded tinned Cu	Sheath color	Light blue (RAL 5015)
Insulation	Fluoropolymer/PVC	Standards	IEC 62228/VDE, etc.
Cores/pairs	1, 2, 6, etc. as required	Flame resistance	Flame retardant
Screening	Individual/overall	Capacitance[a]	100–120 pF/m
Screen	Braided wire tinned Cu	Inductance	0.5 μH/m
Outer sheathe	PVC chemical resistant	Voltage rating	500 V

[a]*Conductor to conductor.*

space. In an enclosure the conductors can be *separated* by a grounded metal or insulated partition. Intrinsically safe wiring is light blue, and other conductors must not have the same color. The raceways, trays, open wiring, and terminal boxes must be labeled for intrinsically safe *wiring* to prevent unintentional interference with the circuits. Usually, 25 × 40 mm size labels are used, and these are placed at a distance of not more than 7.5 m apart. Fig. X/3.7.7-1 depicts the safe distance between wiring in a cabinet.

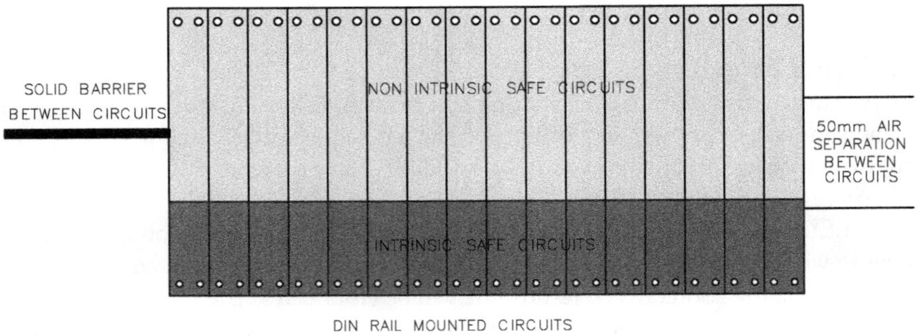

FIGURE X/3.7.7-1

Typical intrinsic safety (IS) wiring in a cabinet.

- Barrier enclosure and placement: Normally, barriers are located in a safe area in a NEMA 4/12 enclosure depending on indoor or outdoor installation because dust and moisture can reduce the air segregation distance. Another criterion is that they should be as close as possible to the hazardous area to minimize cable runs so that there is not much of an increase in capacitance of the circuit because of the cable. In such cases, the enclosure to be chosen is the one that is most suitable for the hazardous area. In the case of an explosion-proof enclosure, suitable sealing shall be used, as shown in Fig. X/3.7.7-2.

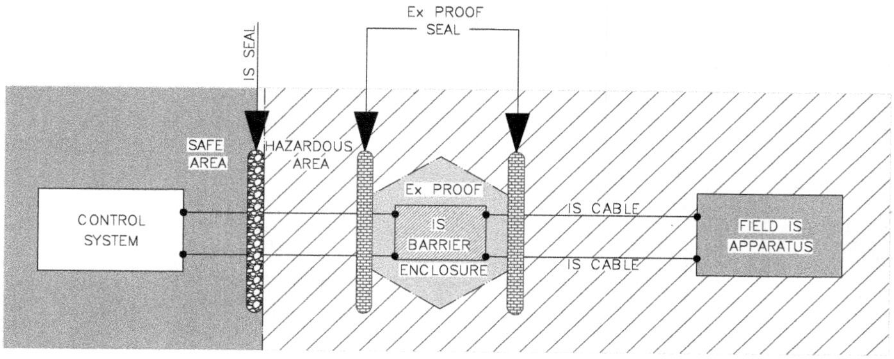

FIGURE X/3.7.7-2

Intrinsic safety (IS) barrier in a hazardous area.

Developed based on an idea from Intrinsic safety circuit design; P.S. Babiarz; Omega US; http://www.omega.com/temperature/Z/pdf/z131-148.pdf. Courtesy: OMEGA USA.

- Barrier grounding: As discussed earlier, IS can be grounded or ungrounded. In the case of a grounded IS barrier the ground is important as it is the path for releasing excess energy. Major grounding rules are:
    - Permanent, secured, separate isolated grounding path to be used.
    - Grounding path must be visible and subject to regular inspection.
    - Grounding path resistance is normally less than 1 Ω and a minimum 12 AWG must be used.
- Replacement and inspection: When fuses blow it is necessary to replace the module. This is done by first disconnecting the module in the following sequence: nonhazardous terminal (isolation of source), then hazardous terminal, and finally ground terminal. Another important aspect of inspection is that, as per IEC 60079-17, "a technical person with executive function" should be identified and should be responsible for inspection.

  With this, the discussion on electrical classification is concluded. However, some supplementary information is also presented in Appendix II.

LIST OF ABBREVIATIONS

AI/O	Analog input/output
ALARP	As low as reasonably practicable
ATEX	Atmosphere explosibles'
BMS	Burner management system (boiler)
BPCS	Basic plant control system
C&I	Control and instrumentation (engineering)

Continued

CCF	Common cause failure
CFSE	Certified functional safety experts
DART	Dynamic arc recognition & termination (fieldbus)
DCS	Distributed control system
DI/O	Digital input/output
DNV	Det Norske Veritas (Norway)
EAC	Electrical area classification
EC	Examination certificate
E/E/PE	Electrical/electronics/programmable electronics
E&I	Electrical and instrumentation
EHSR	Essential health and safety requirements
EMI	Electromagnetic interference
EPL	Equipment protection level
ESD	Emergency shutdown system *or* electro static disturbance
EUC	Equipment under control
FAT	Factory acceptance test
FDT	Field device tool
FEED	Frontend engineering and design
FF (SIF)	Foundation fieldbus (SIF)
FISCO	Fieldbus intrinsically safe concept
FNICO	Fieldbus non-incendive concept
FP	Flash point
FPL	Fixed program language
FSSS	Furnace safeguard and supervisory system (boiler)
FVL	Full variability language
HC	Hydrocarbon
HFT	Hardware fault tolerance
HMI	Human machine interface
HPT	High power trunk (fieldbus)
HW	Hardware
I&C	Instrumentation and control
IEC	International Electrotechnical Commission
I/O	Input/output
I/P or O/P	Input or output
IP	Ingress protection
IPL	Independent protection layer
IR	Infrared (sensing)
IS	Intrinsic safety
IT	Information technology
LD	Ladder diagram
LEL	Lower explosive limit
LHS	Left hand side
LOPA	Layer of protection analysis

LS	Logic solver
LVL	Limited variability language
LVM	Limit value monitor
MESG	Maximum experimental safe gap
MIC	Minimum ignition current
MOC	Management of change
NC	Normally close (contact)
NEMA	National electrical manufacturer association
NO	Normally open (contact)
O&M	Operation and maintenance
OREDA	Offshore Reliability Equipment Database Association
PFD	Probability of failure on demand or doc: process flow diagram
PFH	Probability of failure per hour
P&ID	Piping and instrumentation diagram
PLC	Programmable logic controller
PST	Partial stroke testing
QA	Quality assurance
RFI	Radio frequency disturbance
RHS	Right-hand side
RRF	Risk reduction factor
SFF	Safety failure fraction (IEC)
SIF	Safety instrument functions
SIL	Safety integrity level
SIS	Safety instrumentation system/supervisory information system (in case of DCS)[25]
SRS	Safety related system OR safety requirement specification
SW	Software
T/C	Thermocouple
TMR	Triple modular redundancy
UEL	Upper explosive limit
WRT	With respect to

References

[1] Safety Instrumented System Engineering Handbook, Kenexis Consulting Corporation, Columbus, OH, 2010. http://www.kenexis.com.

[2] S. Basu, A.K. Debnath, Power Plant Instrumentation and Control Handbook, Elsevier, November 2014. http://store.elsevier.com/Power-Plant-Instrumentation-and-Control-Handbook/Swapan-Basu/isbn-9780128011737/.

[3] Alarms Trips: The Ups and Downs, Moor Industries World Wide, June 2006. Write up.

[4] Safety Lifecycle Workbook; For the Process Industry Sector, Emerson Process Management, 2010. https://scadahacker.com/library/Documents/ICS_Basics/Emerson%20-%20Safety%20Lifecycle%20Workbook.pdf.

[5] Process safe book 1, Functional safety in the process industry principles, standards and implementation, Rockwell Automation.

[6] Technical Report; ISA-TR84.00.02-2002-Part 3, Safety Instrumented Functions(SIF)-Safety Integrity Level (SIL) Evaluation Techniques Part 3: Determining the SIL of a SIF via Fault Tree Analysis, June 2002.

[7] SIMATIC Safety, Matrix the Management Tool for all Phases of the Safety Lifecycle, Siemens AG, Brochure, September 2010.

[8] Safety Requirements Specification; General fact sheet, Process Safety, ABB Limited, 2012.

[9] A brief discussion over safety costs in new enterprises; A. Esparza, M. L. Hochleitner; EXIDA LLC; http://www.exida.com/images/uploads/CCPS_LA_2010_SIS_Esparza Hochleitner.pdf.

[10] T.V. Capelle, M.J.M. Houtermans, Functional Safety: A Practical Approach for End-Users and System Integrators, HIMA Paul Hildebrandt GmbH Co KG, June 2006. White Paper.

[11] E. Tran, Requirements & Specifications, Spring, Carnegie Mellon University, 1999, http://users.ece.cmu.edu/~koopman/des_s99/requirements_specs/.

[12] Safety Requirements Specification Guideline, Safe Prod; Process Industry, October 2005. IEC 61511; Version 1.

[13] Recommended Guidelines for Application of IEC 61508 and IEC 61511 in the Petroleum Activities on the Norwegian Continental Shelf, The Norwegian Oil Industry Association, 2001. No 70.

[14] SIL Information; How Can IEC 61508 Approval Aid Users in the Selection and Application of SIL Suitable Fire and Gas Detectors? General Monitors; http://www.gmsystemsgroup.com/sil/sil_info_105.html.

[15] What is the importance of third party certification and SIL rating of SISdevices? L.Duran; Triconex; http://www.iceweb.com.au/sis/Truth%2010%20What%20is%20the%20importance%20of%203rd%20party%20certification%20Paper.pdf.

[16] Certification; Exida; http://exida.com/Certification/Functional-Safety.

[17] Appendix 1: Process for Defining SIS Proof Testing Requirements; Health and safety executive; http://www.hse.gov.uk/foi/internalops/og/og-00054-appendix1.pdf.

[18] S. Nunns, Principles for Proof Testing of Safety Instrumented Systems in the Chemical Industry, Prepared by ABB Ltd. for the Health and Safety Executive; Contract research report 428/2002, 2002, http://www.hse.gov.uk/research/crr_pdf/2002/crr02428.pdf.

[19] W.H. Hearn, P. Skweres, A.D. Arnold, A.E. Summers, SIF Proof Testing Yields Process Sector Reliability Data, SIS-TECH Solutions, LP, October 2009, in: http://sis-tech.com/paper/sif-proof-testing-yields-process-sector-reliability-data.

[20] Appendix 3 Partial Proof Testing Example; Health and safety executive; http://www.hse.gov.uk/foi/internalops/og/og-00054-appendix3.pdf.

[21] Product bulletin; 9.2:001, Hazardous Area Classification and Protection, Fisher; Emerson Process Management, June 2012 http://www.documentation.emersonprocess.com/groups/public/documents/bulletins/d103222x012.pdf.

[22] Hazardous classifications and control of ignition sources; Health and safety executive; http://www.hse.gov.uk/comah/sragtech/techmeasareaclas.htm.

[23] BARTEC Brochure, Basic Concepts for Explosion Protection, 11 revision, 2012.

[24] B. Svacina, B. Larson, Understanding Hazardous Area Sensing; Intrinsic Sensing Handbook; http://www.parrinst.com/wp-content/uploads/downloads/2011/06/Svacina-Larson_Understanding-Hazardous-Area-Sensing_Intrinsic-Safety.pdf.

[25] ATEX Guide, Basic concepts for protection in explosive atmospheres, SCAME Electrical Solutions Brochure.
[26] Fundamentals of Combustible Gas Detection; General monitors; A white paper; http://s7d9.scene7.com/is/content/minesafetyappliances/Combustible%20Gas%20Detection%20White%20Paper.
[27] Gas Detection Handbook, Fifth ed., MSA; http://www.gilsoneng.com/reference/gasdetectionhandbook.pdf.
[28] Honeywell Gas book, Honeywell gas detection, Honeywell; http://www.honeywellanalytics.com/~/media/honeywell-analytics/documents/english/11296_gas-book_v5_0413_lr_en.pdf?la=en.
[29] H.G. Mistry, Foundation Fieldbus Engineering Consultant Perspective, Aker Powergas Pvt. Ltd., Fieldbus Foundation – India Marketing Committee; Automation 2008, Mumbai, 2008.
[30] P. Saward, MTL Instruments, Fieldbus Non-Incendive Concept takes FISCO into Zone 2 and Division 2 hazardous areas, Foundation Fieldbus End Users Council Australia Inc.; http://www.iceweb.com.au/Instrument/Fieldbus/FNICO-PSf.pdf.
[31] Fieldbus 203 Intrinsic safety, Emerson Process management, Plantweb University, 2002, in: http://www2.emersonprocess.com/siteadmincenter/PM%20Central%20Web%20Documents/Eng%20Sch%20-%20Fieldbus%20203.pdf.
[32] A. Beck, A. Hennecke, Intrinsically Safe Fieldbus in Hazardous Area, Pepperl + Fuchs, Technical white paper, 2008.
[33] AN9003-A Users Guide to Intrinsic Safety, Reasons for selecting the intrinsically safe concept; Cooper Crouse-Hinds, MTL.
[34] Intrinsic safety circuit design; P.S. Babiarz; Omega US; http://www.omega.com/temperature/Z/pdf/z131-148.pdf.
[35] Session 13 – Wiring Methods & Cable Standards; IEC. http://www.sourceiec.com/Catalogs/Chapter%2013%20EX%20Wiring%20Methods.pdf.
[36] I.G. Kegel, Foundation Fieldbus topologies for intrinsic safety (IS) installations, in: Foundation Fieldbus End User Council Meeting; Middle East, Pepperl+Fuchs GmbH, Maltaqa, 2007.

Further Reading

Application Note AN 9026, FISCO intrinsically safe fieldbus systems; Technical brochure, MTIL.

ATEX Directive 94/9/EC, Technology International Group; http://www.techintl.com/atex.cfm.

ATEX Reference Guide, Hoffman; http://www.hoffmanonline.com/stream_document.aspx?rRID=119786&pRID=119783.

Basics of Explosion Protection, STAHL Brocheure; http://www.rstahl.com/fileadmin/Dateien/tgus/Documents/ExProtection_Global-America-Basics.pdf.

Flame and Gas Detection Handbook; Ampac Industries; http://www.iceweb.com.au/Technical/flame_and_gas.htm.

IS & Ex; G. Friend, How to Calculate an Intrinsically Safe Loop Approval, South African Instrumentation and Control, October 2012 http://www.instrumentation.co.za/7571a.

Junction box selection guide for fieldbus infrastructure; Pepperl +Fuchs; Technical catalog.

NEMA Enclosure Types, National Electrical Manufacturers Association; Approved by NEMA Enclosures Section, November 2005. https://www.nema.org/Products/Documents/nema-enclosure-types.pdf.

911, X-NI FNICO Power Supplies, MTIL documents; http://www.mtl-inst.com/images/uploads/datasheets/fieldbus/Fnico_Overview.pdf.

PROFIBUS Installation Guideline for Planning, PI International, August 2009. Version 1.0.

Real-time Industrial Networks: Fieldbus network design: H1 design cookbook; T. Ozkul; Google book; Internet.

R.K. Saini, Practical Classification for Determining Electrical Classification, Powermag 3 (March 2007). http://www.powermag.com/practical-guidelines-for-determining-electrical-area-classification/.

Session 12 IEC Protection Techniques, Source: IEC; http://www.sourceiec.com/Catalogs/Chapter%2012%20EX%20Protection%20Techniques.pdf.

SIL Certification for dfunctional safety systems regarding IEC 61508; Certifying SIL component safety; Bureau Veritas; http://www.bureauveritas.com/services+sheet/sil+certification+for+functional+safety+systems.

SIS Project Execution Services; SIS Installation Services; Emerson Process management; Service Data Sheet, December 2010, in: www.EmersonProcess.com/ProcessAutomationServices.

CHAPTER

Fault Tolerance, Protection Layer, and System Security XI

It is needless to argue that the common goal for everybody today is to have a reliable system that can tolerate or accept single or multiple failures as long as the system is not degraded or the required operation is not disturbed. Reliability is therefore a necessary goal of a system and is set by designers and users depending on the type of system in question. Different applications have different requirements for reliability, availability, recovery time, data protection, and maintainability. So, there exist different fault tolerant techniques, layers of protection, and security arrangements to increase system safety reliability and availability. During the discussions on safety integrity level (SIL) it has been seen that different SILs demand different fault tolerances. Apparently, it may appear that by increasing redundancy the fault tolerance can be increased. This notion is not always true. Also an increase in redundancy will call for higher cost. Therefore one needs to concentrate on how, by designing fault tolerant control and computing systems, system availability as well as reliability and safety can be increased. The fault tolerant design of control and computing systems is a vast issue and is dealt with in this chapter only briefly. Layers of protection have a direct impact on the SIL of the system. So, additional aspects of SIL will be covered in this chapter. As stated earlier, today's control systems are based on real-time computing systems that allow networking and integrating of different systems, which make system design and control more economic and easier to handle. There is a flip side to this success: the system will become vulnerable to cyber attacks. "Open-based standards have made it easier for the industry to integrate various diverse systems together, it has also increased the risks of less technical personnel gaining access and control of these industrial networks" (courtesy National Communication Systems Bulletin) [2]. So, network security demands special attention. In this chapter, brief discussions on network security and firewall arrangements are also covered. Let the discussions begin with fault tolerant design.

1.0 FAULT TOLERANCE

Let us first examine the two words "fault tolerance." One can define fault as an incorrect step, process, or data function, etc., meaning that it is the malfunction or deviation from an expected result or behavior. On the other hand, tolerance stands for endurance, in this case, continuance of operation even after a fault has occurred. So, the two words together stand for the ability of the system to function even after a fault(s). To start with it is better to define the basic terms that will be used in subsequent discussions.

1.0.1 *Definition of Terms*

The following are definitions of a few commonly used terms:

- Error: *Error* can be defined as the *deviation* from correctness or accuracy. Since it is associated with certain values, it can be alternatively defined as the difference in the result of a computation and correct result.
- Failure: *Failure* is the *nonperformance* of some action that is due or expected. The result of a fault could be failure. Error leads to failure or failure may be an effect of error.
- Fault: Fault may be defined in various ways:
  - Fault: *Fault* in a system is defined as some *deviation* from the expected behavior of the system—a malfunction.
  - Fault can be defined as an incorrect step, process, or data definition in general terms.
  - In terms of computing fault can be defined as the deviation of one or more logic variables in computer hardware from a specified value.
- Fault tolerance: *Fault tolerance* is the property by which a system *continues* to operate properly in the event of the failure of (or one or more faults within) some of its components.
- Fault tolerant design: A system that is designed with fault tolerance philosophy is basically a fault tolerant design. So, a fault tolerant design enables a system to continue its intended operation uninterruptedly and/or at a reduced level, rather than failing completely, when some part(s) of the system or component(s) fails. Also in a fault tolerant design, if the operating quality decreases at all, then the decrease is normally *proportional* to the severity of the failure; this is in contrast to complete failure of the system for a nonfault tolerant design. The most common example to understand this is the running of a car. When the tire of a car is punctured, the car can run but at a lower speed (of course there is the possibility of the wheel rim being damaged) for some distance. If the puncture is slow the speed reduction may not be appreciable (i.e., proportional to fault). There are fault tolerant designs where designers provide redundant components, and there are cases where redundant computers are used. Such designs are called fault tolerant architecture. In cases of information and communication systems there is another popular term called fault resilient design (which often considered as economic version of fault tolerant design, where only critical parts are duplicated). The main purpose of a fault tolerant system is to develop a system that is dependable. Fault tolerance is a means of dependability (see Clause 1.0.2).
- Fault tolerant unit (FTU): FTU is a part of fault tolerant design. A device (maybe a controller) continues to operate even in the presence of faults. This is achieved primarily by using replication of hardware, software information, and

time. In control applications there are two types of FTU, *node oriented* and *application oriented.*

- Node-oriented FTU: In computer architecture nodes normally operate independently for their associated application. In a node-oriented FTU, there is replication of the node(s). In case of failure of the primary node, the *designated* replicas of the primary nodes will take over with a necessary *mechanism* to allow just one replicated unit to take over to avoid collisions or contention [9]. There are a number of boiler controls in a power plant, namely, a closed loop control such as a superheater temperature control, furnace safeguard and supervisory system, high pressure bypass, etc., each with different apparently independent applications. In node-based architecture, there may be FTUs for each of these control systems acting as one node. In a particular node, when one of the main units fails any one of the standby controllers takes over irrespective of the application. Here the major concern is the *size* and *capacity*, and also the *changeover* mechanism.
- Application-oriented FTU: The units of replication in an application-oriented FTU are also nodes belonging to a common pool of redundant nodes defined for a specific application. When the primary node fails, only nodes that are *designated* replicas of the application to which the failed node belongs will take over [9]. There are separate controllers for boilers and turbines. Similarly, each of them may have separate subsystems for application processing. Now, if any of these application processors fails, one application-oriented FTU from a common application processing subsystem can take over the function so that the entire system continues to operate uninterrupted.

- Graceful failure/shutdown: A system that is designed to be fail safe or fail *gracefully* means that it functions at a reduced level or fails completely, and does so in such a way that it protects people, property, or data from injury, damage, intrusion, or disclosure. Normally, a distributed control system (DCS) is designed in such a way that even it fails for some reason, data are not lost, so that it can be started after it is functional or part repaired. It is common that in the case of power failure, with a trailing edge below a certain value, computing systems save all necessary data so that they can be started when power is available.
- Failed deadly (see http://en.wikipedia.org/wiki/Fail-deadly): Fail deadly is the opposite strategy of graceful failure, which can be used in weapons.

1.0.2 Dependability

Dependability is the ability of a system to deliver its intended level of service to its users [5]. Also it can be conceived as the reliance on a system for the quality of services it provides during an *extended* interval of time. There are attributes, measures, means, and impairments pertinent to dependability. Fault tolerance is one of the

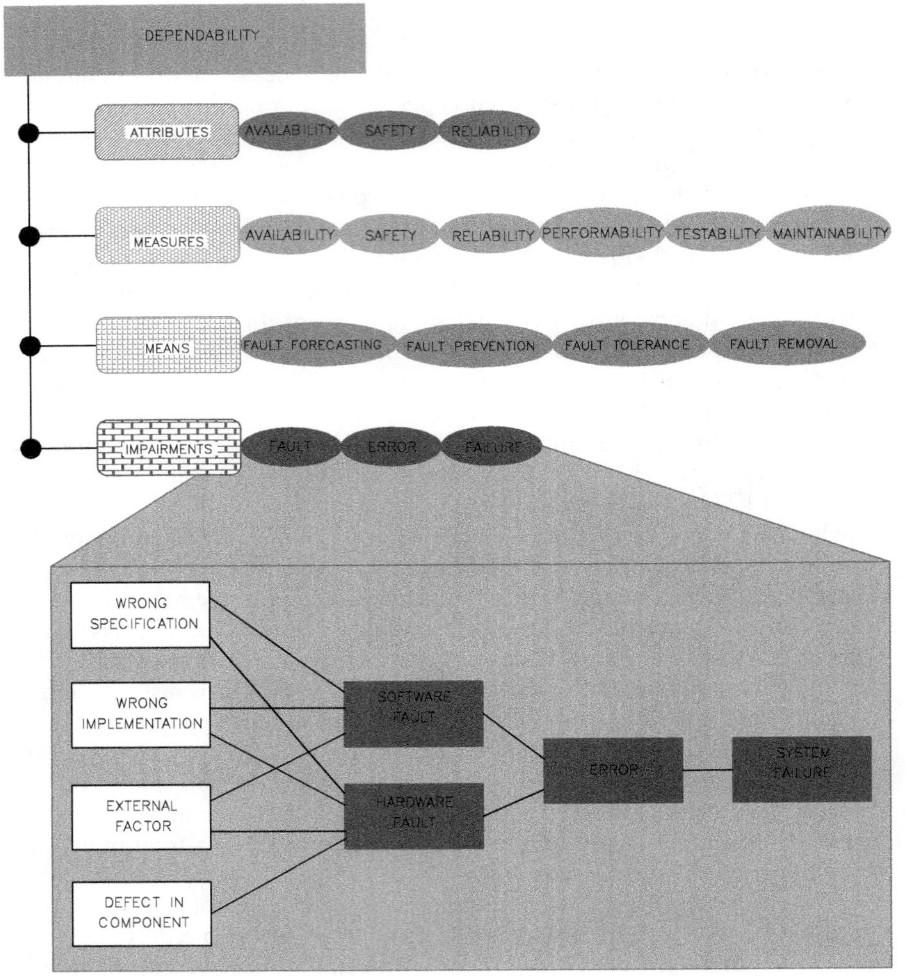

FIGURE XI/1.0.2-1

Dependability and fault tolerance.

means of dependability, as shown in Fig. XI/1.0.2-1. Various contributing factors as shown are elaborated in the following:

- Attributes: There are three major attributes of dependability to signify the properties expected of the system. These are briefly discussed as follows:
  - Availability: Availability $A(t)$ of a system at time t is the probability that the system is functioning correctly at the instant of time t, where $A(t)$ stands for instantaneous availability. The interval of availability is given by:

$$A(T) = \frac{1}{T}\int_0^T A(t)dt \qquad \text{(XI/1.0.2-1)}$$

At steady state it shall be:

$$A(\infty) = \lim_{T\to\infty} \frac{1}{T}\int_0^T A(t)dt \qquad \text{(XI/1.0.2-2)}$$

For further details see Clause 1.1.4. In this connection two other terms are also important: "probability of failure" and "mean time to repair" (see Chapter VII).

- Safety: Safety $S(t)$ of a system at time t is the probability that the system either performs its function correctly or not in a fail safe manner in the interval $[0, t]$, given that the system was operating correctly at time 0. The issue here is *fail safe operation* or not.
- Reliability: As already discussed in earlier chapters, given that the system was performing correctly at time 0, reliability $R(t)$ of a system at time t is the probability that the system operates without a failure in the interval $[0, t]$. Reliability is a measure of the continuous delivery of correct service.

- Measure: All these attributes need to be suitably measured to get an idea of the dependability of the system. Major issues like availability, safety, and reliability were discussed earlier and hence are not repeated here. Apart from these there shall be a few other measurable issues such as performability, testability (self-explanatory), and maintainability.
  - Performability signifies how the system performs with respect to dependability.
  - Maintainability: This is the probability of a system/subsystem being repaired. Major factors affecting maintainability shall include but are not limited to the following:
    - Troubleshooting and troubleshooting tools
    - Fault diagnosis and isolation
    - Fault alarms
    - Training of personnel
    - Accessibility
    - Addition and removal of components

In various intelligent systems, maintainability has reached such a position that it is possible to perform online addition/removal of cards, partial editing of programs, as well as diagnostics via fieldbus systems/Highway Addressable Remote Transducer (HART) [6].

- Means: There are several means for dependability. Fault tolerance is one of them and was defined earlier. The other means are:
  - Fault forecasting is a set of techniques for estimating the number of faults, possibilities of occurrence in the future, as well as their consequences. This evaluation can be qualitative or quantitative. The former technique only ranks them. The following are the issues here:
    - Estimate faults:
      - Present number
      - Future number
      - Consequences

    - Qualitatively:
      - Causes of faults
    - Quantitatively:
      - Failure rate
      - Time to failure
      - Time between failures
  - Fault prevention techniques: These are aimed at preventing the introduction or occurrence of faults in the system. Fault prevention is achieved by quality control techniques during the specification, implementation, and fabrication stages of the design process. Design reviews, component screening, and testing (burn-in, power supply failure, etc.) used for hardware are also types of prevention technique. For software, structural programming, modularization, and formal verification techniques are used.
  - Fault tolerance: As defined earlier, fault tolerant designs are aimed at development of systems that could function correctly in the presence of faults. This is primarily achieved by some kind of redundancy to detect or mask a fault. Masking/detections are followed by fault location, containment, and recovery.
  - Fault removal: Fault removal techniques are used for reduction of the quantity of faults during the developmental phase as well as during the operational life of a system:
    - Developmental stage: Verification, diagnosis, and correction.
    - Operational stage: Corrective and preventive maintenance.
- Impairment: This consists of fault, error, and failure, as already defined. The relation between them is detailed in Fig. XI/1.0.2-1. These are dealt with in detail in subsequent subclauses.

1.1 General Discussions on Fault Tolerance

Here, general characteristics and features of fault tolerant designs are discussed.

1.1.1 Characteristics

The following are general characteristics found in fault tolerant systems:

- High availability, no single point of failure: When a system experiences a failure, it should continue operation without interruption during repair, that is, failure recovery.
- Fault isolation of the failing component: The system should be able to isolate the failure and there should be a number of dedicated failure detection mechanisms for fault isolation. Recovery from a fault condition requires classification of the fault or failing component.
- Fault containment and prevention of failure propagation: The system needs to have a mechanism for fault containment to prevent fault propagation. Firewalls or other mechanisms are examples.
- Availability of reversion modes
- Planned and unplanned service outage

1.1.2 Fault and Failure Types

The most important characteristic of fault is duration. So, fault can be classified utilizing this characteristic feature. Based on duration, the fault can be classified as a "permanent/solid/hard" fault and a "transient/intermittent/soft" fault. In most systems the majority of faults are transient faults (*above* 80%). Transient faults, or intermittent faults, can be defined as random failures that prevent the proper operation of a unit for only a short period of time—not long enough to be tested and diagnosed as a permanent failure.

- Permanent faults: *Permanent* faults are caused by failures of components. This type fault is persistent and continues to exist until the faulty component is repaired or replaced. These faults are easier to diagnose but normally require more rigorous correction than transient faults. A computer disk crash is an example.
- Transient/intermittent faults: *Transient* faults occur once and then disappear. The majority of DCS or computer system faults (80–90% [4]) are the transient type, for example, a message is sent but does not reach the recipient, but when it is resent, it reaches the recipient. There is a peculiar characteristic of transient faults. This is termed intermittent fault, which occurs and vanishes in a cyclic manner. A simple loose connection in a component can cause such a situation. Transient and intermittent faults are mainly caused by random failure of components and the faults stay for short while and are then diagnosed or tested. It is quite possible that transient faults become permanent with further deterioration of the equipment.

From discussions in Chapter VII it is seen that there are mainly two types of failure: "random" and "systematic" [common cause failure (CCF) can be either of these]. Here, the faults discussed are caused by random failure. In an electronic system, especially in a computing system, two special types of failures are encountered:

- Fail silent failure: When the system either stops producing output or produces output that clearly indicates that the component has failed.
- Byzantine failure: This is derived from the Byzantine general rule/problem (Fig. XI/1.1.2-1). Here, the faulty unit continues to run but produces incorrect results. Byzantine fault is difficult to handle.

As discussed earlier, the aim of system reliability is to forecast/prevent/recover from faults. So, to achieve a high reliability, it is essential that components be highly

> Byzantine General: **Given three processes, if one fails, consensus is impossible. Similarly given N processes, if F processes fail, consensus is impossible if $N \leq 3F$**

FIGURE XI/1.1.2-1

Byzantine general (rule).

reliable and there is high immunity to software fault induction into the system. Therefore fault tolerant computers need to use proper selection of redundancies, as discussed in the next subclause. There can be other faults like hardware/software design fault, operator error, physical damage, etc. However, most fault tolerant design deals with random hardware faults.

1.1.3 *Redundancy*

Before starting discussions it should be made clear that fault tolerance may be aided by redundancy, but redundancy and fault tolerance are not the same. Redundancy does not ensure continuance of operation of a system. In the case of the previous example of the car (Clause 1.0.1), if the car tire bursts, then one needs to stop the car and change the tire before continuing. Here, there is a redundant tire, which can be used, but car has to stop before the operation can continue. On the other hand, if fault tolerant design operation is uninterrupted, for example, in a programmable logic controller an uninterrupted power supply (UPS) is used if the main supply fails, output of the UPS is unaffected or a hot standby processor can continue the operation of the PLC. Most of the instrumentation and control systems currently in use are intelligent types comprising hardware and software. So, to achieve a fault tolerant design both can be utilized. There two types of redundancy, namely, hardware redundancy and software redundancy. When retrial of instructions in a computer system is done it is time redundancy [4].

- Hardware redundancy: In hardware redundancy a set of hardware components needs to be introduced into the system to provide fault tolerance with respect to operational faults. These components are a duplication or triplication of the original set of components; hence when there is *no fault* these are superfluous or redundant. Obviously, removal of the components in the *absence of faults* has no real impact on system performance. As discussed earlier, for hardware fault tolerance, the most reliable components available should be used. However, increased component reliability has hardly any impact on fault tolerance if the fault is an operational one. Hardware redundancy is more important in *recovery* than in prevention.
- Software redundancy: Software redundancy refers to all additional software installed in a system to handle a fault exclusively. An intelligent system consists of hardware and software and for such a system, when there is no hardware design fault, then one of the reasons for the fault in the system can be because of a software design fault. To handle a software design fault, software redundancy is applied since it is hardly possible to handle a software design fault with hardware. In intelligent systems, software redundancy is very important, because it *can* be used even to handle hardware malfunctions. Based on the requirements of recovery needed, the *level* of software redundancy is decided. The recovery design is dependent on the *type* of error or malfunction. In some cases only hardware *or* software redundancy is sufficient for the recovery, while in most cases a mix of both redundancies is called for. When hardware fails there is the possibility that

there will be a loss of data retrieval. In such cases for critical and important programs designers use redundant storage as a preventive measure for duplication of all necessary programs and data storage. This is necessary for retrieval. Also critical programs, for example, error recovery programs, are placed in nonvolatile storage with critical data in nondestructive readout memories.

Redundancy is very much related to availability, dealt with in the next subclause. However, it should be borne in mind that to attain technoeconomical benefit, redundancy must be deployed judiciously. Another issue that is often overlooked is voting pattern, which is also important for achieving better results, for example, a conventional dual redundant system can provide either availability when the voter is set to 1 out of 2, or integrity when the voter is set to 2 out of 2. Not both [1]!

- Redundancy and voting: Fig. XI/1.1.3-1 shows how a redundancy and voting system increases the fault tolerance. In the first case in Fig. XI/1.1.3-1A, output

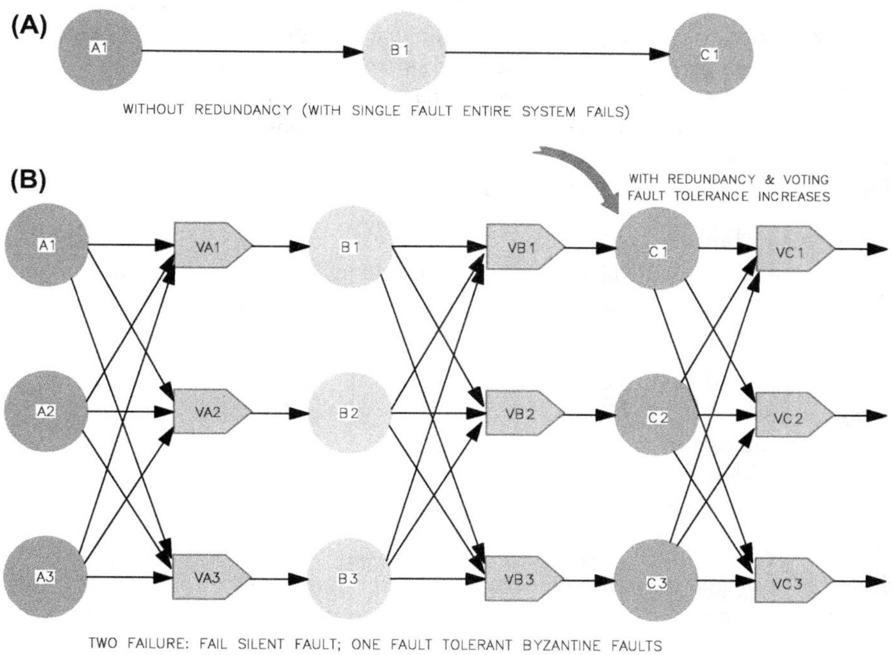

FIGURE XI/1.1.3-1

Redundancy and voting. (A) Without redundancy, (B) TMR redundancy.

of A goes to B and then C to produce the final output. If one fails, then there will be failure of the entire system. On the other hand, in Fig. XI/1.1.3-1B, each item has three-way redundancy (e.g., for A it is A1, A2, and A3). In each case good ones are selected via voting circuits. This is a typical triple modular redundancy

(**TMR**). If in place of three there are *N* numbers of such selection, then voting is *N*-modular redundancy (**NMR**). In this case of TMR, in the case of a double failure (item or voting circuit) output will fail "silent fault", whereas in the case of a single fault (in each item or voting circuit) it is a single fault tolerant Byzantine fault.

1.1.4 *Availability*

As discussed earlier, availability can be defined as the probability that a system is operating successfully when needed. To help make sure that products meet customer expectations, reliability can be designed using techniques such as component derating and design through Six Sigma [6]. For the Six Sigma technique Fig. XI/1.1.4-1 may be referred to. It is a common notion that the availability of a system increases with redundancy. This is *partially* true (think of systematic/CCF for duplicate and triplicate redundant hardware). However, when redundancy in terms of duplication and triplication is done, it obviously costs more money and chances of failure are increased. So, redundancy must be chosen judiciously along with *voting logic*. Also redundancy alone is not responsible for higher availability. Since availability is related to failure of devices and components there are other factors that are also responsible for higher availability. For this reason in addition to mathematical expression of availability defined in Eq. (XI/1.0.2-1), however in terms of reliability, availability can be expressed in terms of mean time between failure (MTBF), mean downtime (MDT), and mean time to repair (MTTR) as follows:

$$\text{Availability } A = (\text{MTBF})/(\text{MTBF} + \text{MDT}) = (\text{MTBF})/(\text{MTBF} + \text{MTTR}) \quad (\text{XI}/1.1.4\text{-}1)$$

(Assuming MDT and MTTR are the same, though there is some difference.). As MTBF and MTTR depend on many other factors like quick fault detection, etc. So, availability depends on the following factors:

- Redundancy and fault tolerance
- Diagnostics:
  - Easy troubleshooting
  - Easy modification and repair
  - Easier maintainability

Modern field devices are intelligent with field communication, so they provide a high degree of diagnostics. Also modern intelligent control systems provide a means of fault forecasting and diagnostics and maintenance management. Therefore, with

Six Sigma is a set of techniques and tools developed by Motorola in 1986 with special emphasis for setting extremely high objectives, collecting data and analyzing results to a fine degree to reduce DEFECTS in products and services. It can be conceived as a disciplined, data-driven approach and methodology for elimination of defects in products and services. this technique mainly uses empirical and statistical methods and data for analysis.

FIGURE XI/1.1.4-1

Six Sigma technique.

these modern intelligent devices having high degree of diagnostics, along with judicious redundancy and fault tolerant design, it is possible to ensure higher availability, which is of prime importance, especially for higher SIL systems. Concentration will now be on the original issue of fault tolerance and discussions will be focused on various techniques used for achieving fault tolerant design and recovery.

1.1.5 Fault Tolerant Techniques (Computing System)

The fault tolerant design discussed here mainly pertains to computing systems and intelligent systems for real-time computer systems such as DCS/PLC and/or associated intelligent devices. Here, the discussion is on the basics of hardware and software fault tolerant principles in computing systems, whereas that applicable to control systems is covered in Clause 1.2. Two ways in which fault tolerant designs can be developed are *hardware* technique and *software* technique.

- Hardware technique: Hardware fault tolerant designs are aimed at developing computing systems that can automatically recover from random faults in hardware components. In this type of technique, generally the computing system is *portioned* as modules that act as fault containment regions. Each module is backed up with protective redundancy so that, if the module fails, others can assume its function [7]. Here, suitable mechanisms are deployed for error detection and recovery. There are mainly two variations in hardware fault tolerant design, namely, the masking (static) technique and dynamic recovery.
  - Masking: Fault masking makes use of redundant modules that completely mask faults within them. A number of identical modules perform the same functions, and their outputs are voted to remove errors created by a faulty module. The masking (often called static) technique, which is automatic and instantaneous, can handle both transient and permanent failures. This error correction method is transparent to the user and often to the software. Redundant components effectively mask the hardware failure effect from other components. Depending on the degree of reliability and cost, various techniques can be used. The simplest one is shown in Fig. XI/1.1.5-1. Single fault in either left or right side pair or even single failure in both pairs output will be there.

  TMR is a commonly used form in fault masking. Here, the circuit uses triplicate identical main and voting circuits, for example, A1, A2, A3, as shown in Fig. XI/1.1.3-1. A TMR system fails whenever two modules in redundant

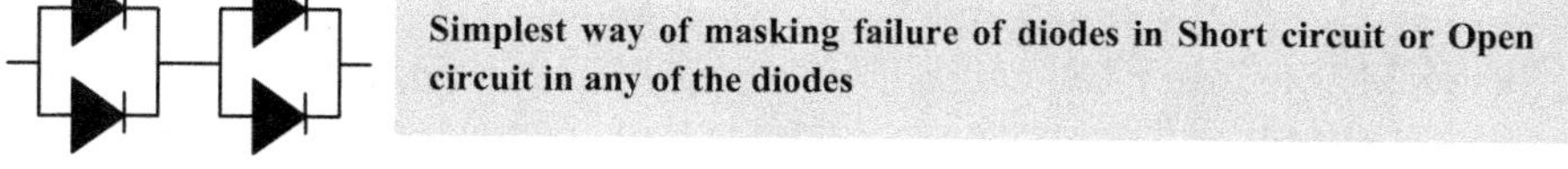

FIGURE XI/1.1.5-1

Example of the simplest way of hardware masking.

modules create errors so that the vote is no longer valid. When in place of three numbers N is used, it is known as NMR.

- Dynamic recovery: Dynamic recovery is generally more hardware *efficient* than voted systems and can be deployed where there is hardware constraint. This advantage can be seen in conjunction with disadvantages like *delay* in fault recovery, *lower* fault coverage, and involvement of *special* mechanisms. Dynamic recovery is applied when a computation is running and *self-repairing* is required. However, *higher* levels of fault tolerance can be achieved with the help of dynamic redundancy. Dynamic recovery involves both fault detection and recovery. Like fault masking, spare protective redundant modules are also used. In this method, special mechanisms are used to *switch* out a faulty module, switch in a *spare*, and instigate those software actions (rollback, initialization, retry, restart) necessary to restore and continue the computation [7]. Error detecting and error correcting codes can be used for messaging also. Coding may be used to detect faults in the modules. In a single computing system this involves *additional* hardware and software, while in multiple computing systems it can be managed by *other* processors. When there is involvement of software and hardware for achieving fault tolerance, then it is called *hybrid redundancy*. Hybrid redundancy is an extension of TMR. In this method, triplicate modules are backed up with additional spares, which are used as a replacement of faulty modules to tolerate more faults. Voting systems require more than three times as much hardware than do nonredundant systems, but they have the advantage that computations can continue without *interruption* when a fault occurs, allowing existing operating systems to be used [7].

- Software fault tolerance: For software, fault tolerant design redundancies are required to mask residual design faults. Some of the issues related to this shall include but are not limited to:
  - Planning software fault tolerance strategy
  - Defensive programming, namely, "to do the same in many ways," "review and test," and "verify," etc.

  Because software fault tolerance is based on hardware fault tolerance, it is a bigger challenge. Additional software is used in computing systems for fault handling and for fault-free computation. A few major software fault tolerance techniques somewhat similar to their hardware counter parts have been enumerated below.
  - N-version programming: This is similar to static redundancy. Here, independently written programs (versions) for the same functions are executed parallel and their outputs are voted at special checkpoints. Naturally, the voted data may not be exactly the same. So, criteria must be used to identify and reject faulty versions and to determine all good versions for use. See Fig. XI/1.1.5-2A.

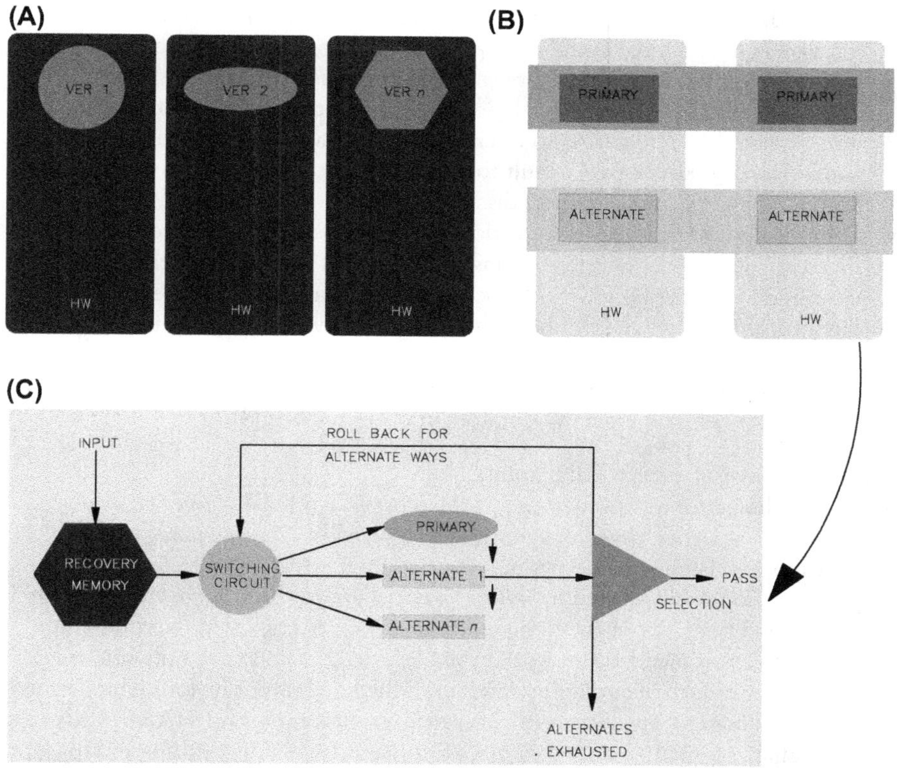

FIGURE XI/1.1.5-2

Software fault tolerance methods. *FT*, fault tolerance. (A) Software fault *n* version programming, (B) software fault - recovery block, and (C) software fault - recovery block schematic.

- Recovery block scheme: There are two types of recovery block scheme, namely, *forward rollback* (produces the correct result by repetitive and continuous processing—it is application dependent) and *backward rollback* (redundant information is recorded in the process of execution and rolls back the interrupted process to the point where correct information is available, e.g., checkpoint, retry, etc.). See Fig. XI/1.1.5-2B and C.
- Dynamic approach: In this approach, like the hardware dynamic approach, programs are partitioned into blocks and acceptance tests are executed after each block. If case of failure of an acceptance test, a redundant code block is executed.

- Design diversity: This approach is rather costly. It combines hardware and software fault tolerance in different sets of computing channels. Each channel is developed in different hardware *and* software in redundant mode to provide the same function. This method is deployed to identify deviation of a channel from the others. The goal is to tolerate both hardware and software design faults [7]. After developing a fault tolerant design it is necessary to validate it from a reliability point of view, discussed later.
- Other issues: With software detection, it is not possible to localize the error sources. So, diagnostic test programs are frequently used identify the module concerned. Almost all DCS have some form of diagnostic routines to pinpoint the faults as best as possible. The accuracy of such pinpointing may vary from system to system. In a fault tolerant system, the system itself initiates these tests and interprets their results, as opposed to the outside insertion of test programs by operators in other systems [4].

1.1.6 *Validation of Fault Tolerance*

Fault tolerant design for reliability is one of the most difficult tasks to verify, evaluate, and validate. It is either time-consuming or very costly. This requires creating a number of models. Fault injection is an effective method to validate fault tolerant mechanisms. Also an amount of modeling is necessary for error/fault environment and structure and behavior of the design, etc. It is then necessary to determine how well the fault tolerant mechanisms work by analytic studies and fault simulations [7]. The results from these models after analyses shall include but not be limited to: error rate, fault rate, latency, etc. Some of the better known tools are: HARP—hybrid automated reliability predictor (Duke), SAVE—system availability estimator (IBM), and SHARPE—symbolic hierarchical automated reliability and performance evaluator (Duke).

The design of fault tolerance is different from other fault tolerant computing systems. So discussions will now be on fault tolerance in control systems.

1.2 Fault Tolerance in Control Systems

Defects or faults in any component of the loop can develop into malfunctions. Faults are not always visible to the operator immediately, but may appear in such a way that they give rise to complete loop failure. In safety-critical applications, no failure can be tolerated [3]. Redundancies in hardware and software facilitate fault recovery. So, for increased dependability fault tolerant control (FTC) is an ideal solution. In critical controls it may be disastrous to tolerate any failure of control systems. In FTC the system continues to operate with single failure in components and/or subsystems. Also in cases of critical controls, FTC will make a controlled shutdown to a safe state in a critical situation. FTC systems use the help of redundancies in hardware and software, discussed earlier, and fault diagnostics and intelligent software to monitor health and behavior of components and function blocks and take remedial action. With these tools the faults are isolated and suitable action is initiated to prevent faults from propagating, which may cause a critical failure in the system. Before going further it is best to clarify the meanings of a few terms that will be used in FTCs.

1.2.1 *Frequently Used Terms in Fault Tolerant Control*

Following are a few important terms frequently used in fault tolerant control:

- Fail operational: A system that can continues to function without any change in objectives or performance after a single failure.
- Fail safe: A system that fails to a state that is considered safe in the particular context.
- Fault accommodation: This is a common approach to achieve fault tolerance. Fault accommodation is limited to internal controller changes. In this method there will be changes in controller parameters or structure to avoid fault consequences. However, the input—output (I/O) between controller and plant remains *unchanged*, that is, the loop is not completely restructured.
- Reconfiguration: Control reconfiguration is an approach to achieve FTC for dynamic systems. In contrast to fault accommodation, in this system there will be change in I/O between the controller and plant, because there will be changes in controller structure and parameters. In reconfiguration, the interconnections among modules are altered. The original control objective is achieved although performance may degrade [3].

1.2.2 *Ways and Means for Fault Tolerance*

In FTC systems certain measures are taken to combat common failures such as:

- Power failure
- Control signal failure/error
- Address failure
- Module internal failure
- Module data transfer errors
- Mistiming/failure
- Communication channel failure
- Reconfiguration faults

There has been much advancement in system design for fault tolerance, for example, the use of dual and triple redundancy in the processing unit. These (dual and triple redundancy) are quite age-old systems but quite effective in many applications.

- In the duplex system, there are two identical main processing units (controllers) operating in parallel. In this parallel mode one may be active and the other may be standby or hot standby. In the case of hot standby configuration, at all times both processors are alive, in the sense that they are supplied by parallel inputs and other functions. Only the output of hot standby is inhibited, as long as the other processor is in action. In case of failure of the active processor, hot standby immediately takes over. Since the hot standby processor is always updated, transfer will usually be smooth. Each duplicated system of the processors will have internal communication with the help of parallel data and control buses, which are used for high-speed communication and interrupt handling. In the dual/duplex system there will be comparison of data for fault detection with rollback and recovery to handle transient errors [4]. To overcome the problem of comparison between two processors and resource management, there are architectures where a separate processor is used for diagnostic purposes to identify faulty processor. See Fig. XI/1.2.2-1 for typical dual redundancy processor configurations.

- Triple redundant system: In a triple redundant system, there will be three processors operating synchronously. In addition to error detection and

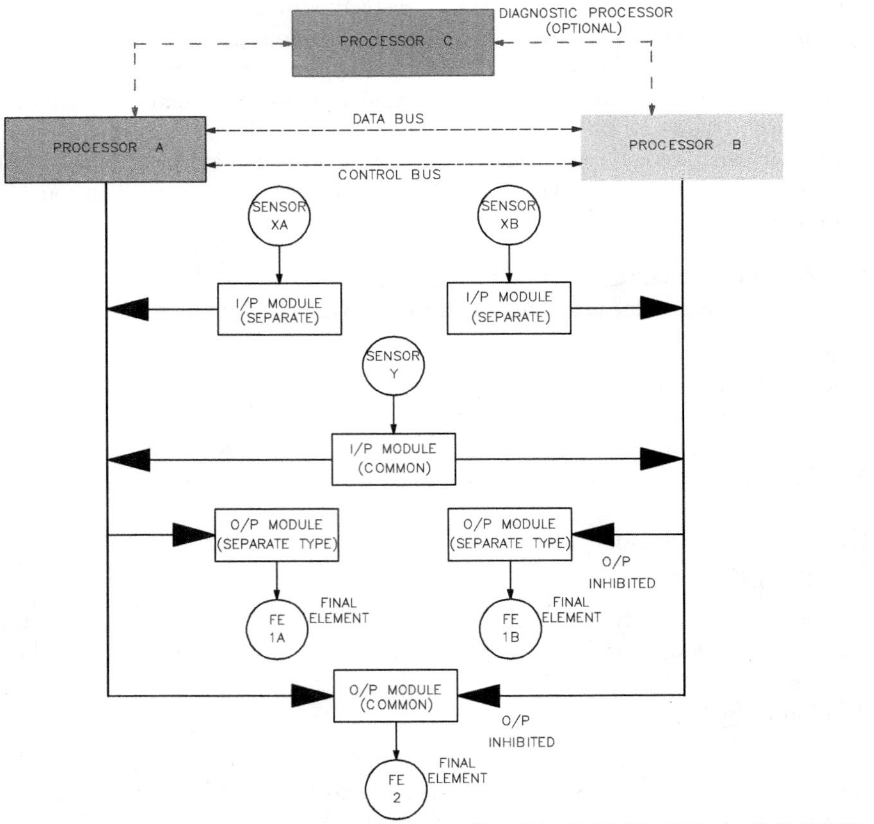

FIGURE XI/1.2.2-1

Dual processor redundancy.

correction capabilities, there will be fault tolerant features in software for both systems. Triple redundant systems have a software voting system for transient fault recovery. Naturally, there is no issue related to rollback. All computer elements will communicate with higher-level systems via a full-duplex synchronous serial bit bus, a bus that will permit simultaneous message transfer in both directions through the protocol microprocessor [4]. Fig. XI/1.2.2-2 depicts typical triple processor redundancy.

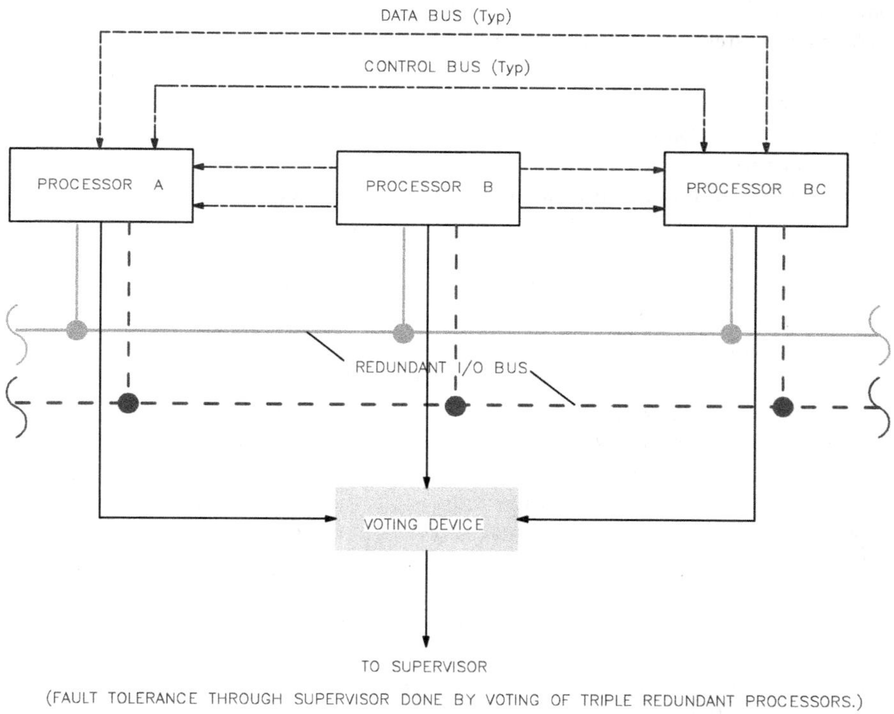

FIGURE XI/1.2.2-2

Triple processor redundancy. *I/O*, input/output.

1.2.3 Practical Application

Under this clause, short discussions are provided to see how practical application in a system is realized.

- Confession and declaration: To describe the actual application of fault tolerance, internationally reputable manufacturers' technical literature has been used and two FTCs from reputable manufacturers have been described so that readers can get a feel for the actual systems. Since the data have been taken from only two manufacturers the description may not match all systems and efforts have been made to make the data as **generalized** as possible but keeping the functionality intact.

- Redundant host network: A fast Ethernet network may be used as a redundant host network. This facilitates the communication medium for application development, operator interface, and data exchange among nodes. Additionally, there could be peer-to-peer communication for direct transfer of data between nodes over the high-speed Ethernet link. To handle high data traffic, only changes in data are exchanged for getting the best efficiency in peer-to-peer communications.
- Node: Many systems use separate processor modules to achieve system flexibility and performance. The node processors are responsible for all system functions pertinent to the node. The I/O or chassis processors are responsible for I/O scanning, floating point conversion, and results voting. This is a part of resource management to relieve the load of the node processor as well as to minimize the data corruption and to maximize system response required for critical controls. In this connection, processor connections pertinent to safe PLC, discussed in Chapter IX, may be referred to:
  - Fault tolerance: The highest level of fault tolerant operation can be configured with redundant node processors, redundant I/O, and dual power supply chassis.
  - Data integrity: Each node processor performs the system function(s) and sends the resultant outputs to the associated I/O chassis. Then the I/O processor performs data validation checking and independent voting for its associated outputs. If the system's outputs are located in multiple I/O chassis, the voting is distributed over these chassis. Distributed voting, closed to the process, enhances integrity and makes the loop a much more responsive, high-speed system. In this connection, I/Os pertinent to safe PLC discussed in Chapter IX, may be referred to.
  - High availability: In case of a hardware/network fault, the operators could be notified and the node processor could be taken offline. However, the application may continue running unaffected on the remaining node processor(s). When the error condition is cleared through online maintenance activities, nonintrusive reinitialization of the restored node processor may take place [8].
  - Signal validation: Signal validation routines are supported in most of the DCS/PLC family. Signal validation routines allow for the control of one logical input produced from up to four redundant inputs (three hardware and one additional logic) [8].
  - High-integrity validation: Communications are validated with the help of cyclic redundancy check (CRC) routines for main processors and the redundant I/O networks. System information is validated for remote I/O to ensure hardware availability, error -free performance. Error checking on the data transfers diagnoses data corruption. Field wiring is supervised to ensure error-free output data, I/O card diagnostics, even calibration checks on the

analog/digital converters. For TMR systems, three processors may be installed in three physically separate chassis for better reliability.
- Flexible modular redundancy: This is another feature available in many DCS, for example, Siemens "Simatic PCS7." Depending on the automation task and the associated safety requirements, the degree of redundancy may be defined separately for the controller, fieldbus, and distributed I/O level, and coordinated with the field instrumentation [10]. The great advantage of this type of redundancy is that redundancy is applied only when it is necessary. This uses the help of different types of FTUs discussed earlier.

1.3 Redundancy and Voting in Field Instrumentation

Fault tolerance in field instrumentation is mainly concerned with redundancies that in case of a basic plant control system (BPCS) ensure continuance of operation. In the case of a safety instrumentation system (SIS), they additionally minimize nuisance or spurious interventions and alarms. Redundancy in BPCS also improves safety. With properly selected and installed transmitters, improved performance can be achieved by measuring the same variable with more than a single field device. When field devices are more than 2, voting circuits are necessary. In this way "m"measurements out of a total of "n"number of signals are made so that $m > n/2$, for example, $m = 2$, $n = 3$, 2oo3 is the selection by voting. There are standardization of redundancy and voting techniques. Some of these are presented next. These are applied for both BPCS and SIS.

1.3.1 *Field Instrument Redundancy Selection Details*

In this clause, some of the typical redundancy schemes for field instruments are presented. The selection and voting circuit may be implemented in the I/O section of DCS/PLC or could be hardware. A basic selection scheme is important. For details see [2].

Some typical schemes of 1oo2 and 2oo3 are presented in Fig. XI/1.3.1-1.

High and low limit checks for transmitters have been in practice for quite some time, even when discrete instrumentations were in use. Now with the DCS it is very easy to monitor out of limits for the transmitter and open-circuit and short-circuit tests for sensors like resistance temperature detectors and thermocouples. Most of the transmitters are monitored for out of span (e.g., <4 or >20 mA). Also since smart transmitters have a diagnostic system, they also can detect faults and isolate them, that is, the output of a faulty transmitter could be inhibited generating an alarm. The transmitter is connected via HART/Profibus/fieldbus, and such detections are more explicit and well reported in the system. Also there exists a facility for the operator to select any transmitter manually.

- 1oo2 selection: In this mode, as shown in Fig. XI/1.3.1-1C, two transmitter signals are fed to an averaging circuit (soft average selection in the signal processing part of the DCS). The output from the average unit is taken through one selection switch. In auto mode, normally the average output is selected, but

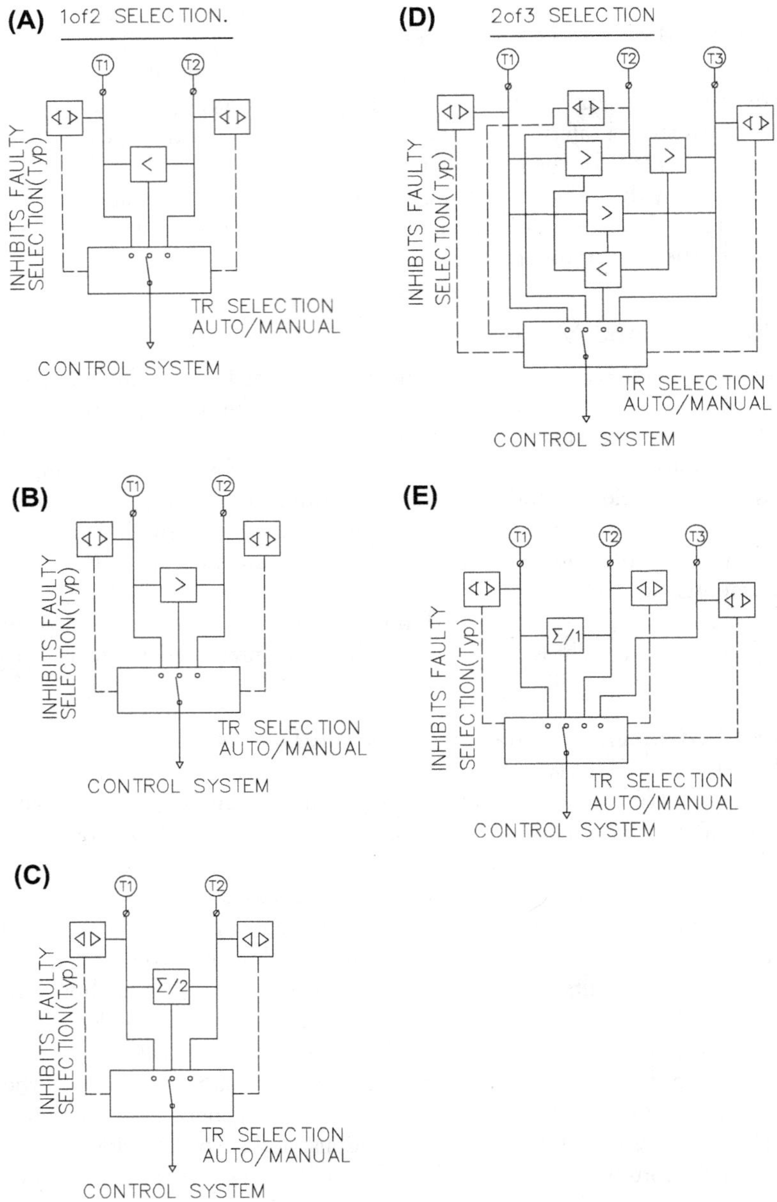

FIGURE XI/1.3.1-1

Redundant instrument selection methods. (A) 1of2 selection by Lo select, (B) 1of2 selection by Hi select, (C) 1of2 selection with average, (D) 2of3 voting selection (digital also), and (E) 2of3 selection with average. *TR*, transmitter.

From S. Basu, A.K. Debnath, Power Plant Instrumentation and Control Handbook, Elsevier, November 2014; http://store.elsevier.com/Power-Plant-Instrumentation-and-Control-Handbook/Swapan-Basu/isbn-9780128011737/. Courtesy Elsevier.

if out of two transmitters (sensors) one is detected faulty (by transmitter diagnostics or by an out-of-range detector), then it will be inhibited, so other transmitters will be selected. However, any one of the two transmitters or average output can be selected manually.

- 2oo3 selection: There are two ways this can be selected, either by average or by voting.
  - Selection with average: In this mode, as shown in Fig. XI/1.3.1-1E, three transmitter signals are fed to an averaging circuit (soft average selection in the signal processing part of the DCS). The output from the average unit is taken through one selection switch. In auto mode, normally average output is selected, but if one transmitter (sensor) is detected faulty (by transmitter diagnostics or by an out-of-range detector), then it will be inhibited, so the average output will be from the other two transmitters. In manual mode, any one of the three transmitters or average output can be selected. If there is a fault in any transmitter, it will be alarmed and healthy transmitter(s) will be selected in auto mode (i.e., if output is not manually selected).
  - Median Selection: In this mode, as shown in Fig. XI/1.3.1-1D, three transmitter signals are initially voted through high selection between two transmitters (soft selection in the signal processing part of the DCS). The output of these three high selections is fed to the low selection for final voting as shown. Each of the transmitters, like other systems, is checked for health. The faulty transmitter is automatically voted out. The voted transmitter is selected in auto mode. In manual mode it is possible to select any of the three transmitters or the voted transmitter, but in no case is the faulty transmitter selected. *This method is applicable for process switch selections.*

1.3.2 Input Redundancy Interface at Intelligent Control

From IEC 61508 Part 6, a definition of commonly used architectures in safety instrumented systems is available. The elements used in a single or multiple configuration can be either sensors or final elements—mainly for input sensors, and only a few for the final element on account of cost [4]. Typical interfaces of these with an intelligent control (DCS/PLC) system are shown in Fig. XI/1.3.2-1. The configuration may be 1oo1, which is quite vulnerable because single instrument failure will make the loop unavailable.

- Dual field devices: With dual transmitters there are three possibilities: 1oo2 (not in standard), 1oo2*D*, and 2oo2. The first two cases are such that if anyone gives the signal, action will be taken. "*D*" at the end stands for diagnostics resident mainly in the control system. 1oo2 gives higher availability of the system as if anyone is true, and action is taken, but integrity may suffer because of nuisance trip. On the contrary, in the case of the 2oo2 system, availability may suffer because action will be initiated only when both are agreeing, but system integrity will be higher on account of no nuisance trip, etc. So, with a dual field device either of availability or integrity of the system will be better catered to not both. This may be compared with 1oo2 discussed in Clause 1.3.1 also.

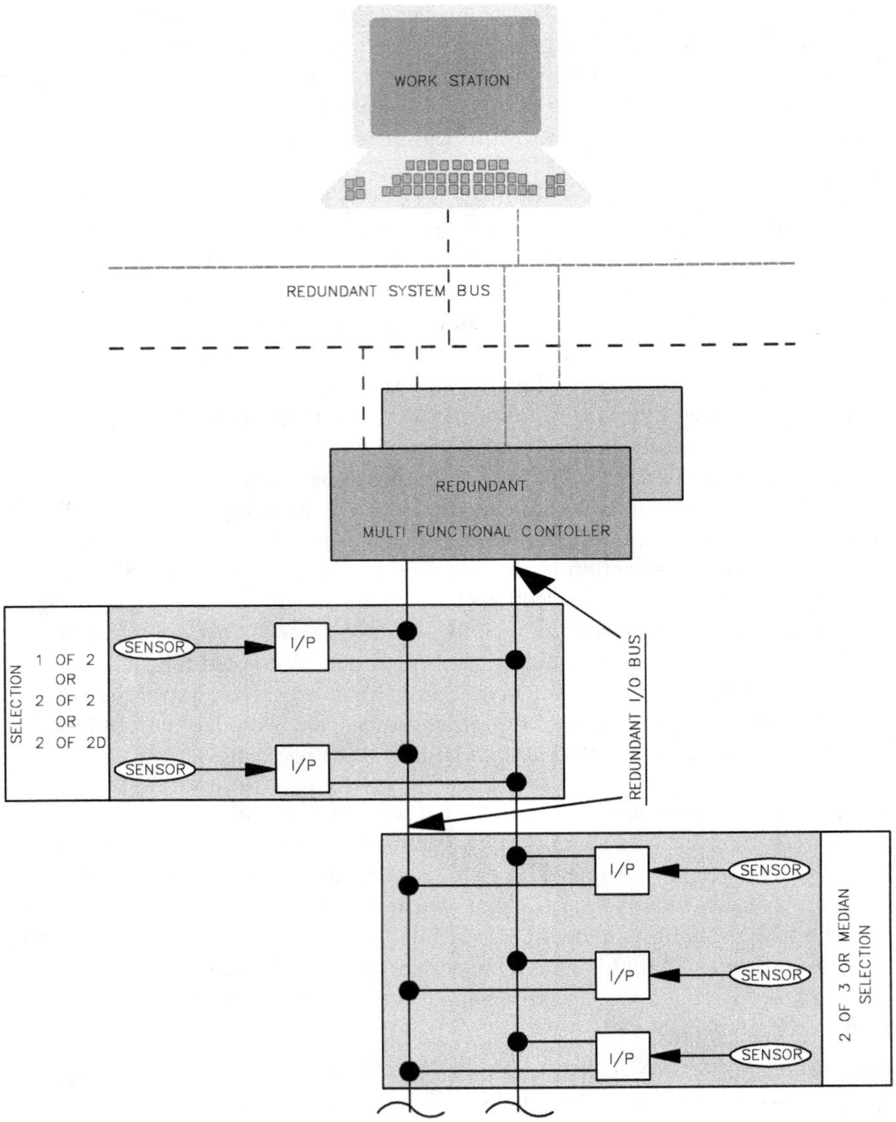

FIGURE XI/1.3.2-1

Input redundancy in an intelligent control system. I/O, input/output.

- Triple field devices: As discussed earlier (Clause 1.3.1) in the case of 2oo3 there are two possibilities. In the case of digital inputs, voting is done (median selection). When such voting is done three times it becomes TMR, as discussed in Fig. XI/1.1.3-1.
- Diagnostics and allied discussions: Usually, the diagnostic coverage (DC) in the BPCS is much less than in the SIS. This is more so when discrete controllers are used in the BPCS. However, in the case of *integrated* DCS/PLC different diagnostic capability may not be that *wide*. This is because DCS/PLC have enough power to accept various types of signals and compute the difference between them to detect the fault. When an inconsistency is detected, the DCS is capable of signaling the abnormal situation and can continue to run the control system uninterrupted with the correct field device(s). Smart devices also have the capability to detect faults.
  - 1oo1*D*: The diagnostic coverage can be partly integral to the transmitter and/or external in the control system (rate of change alarms, over range alarms detecting the individual fault) [4].
  - 1oo2*D*: This was discussed in Clause 1.3.1; hence it will not be repeated here. Normally, within the valid range, the difference between the two transmitters should be within 3% (*typical value depends on measurement*, e.g., for pressurized vessel level measurement, e.g., drum level); if this is exceeded, an alarm is issued to the operator. In such a case the average signal may not be acceptable. One of the possibilities could be that, the control system stays put at the last good value and control may be forced to manual by the operator with an alarm.
- 2oo3: Here also differences between transmitters are computed, as just discussed. Naturally, there will be three such differences, namely, $x-y$, $y-z$, $z-x$. As long as these differences are within the preset limit and/or one difference is beyond the preset limit, transmitters may be correct and the median/average may be taken, but an alarm may sound for the operator to check the reason why one difference exceeded the limit. If two differences exceed the preset limit, the value of the transmitter involved in *both* the excessive differences is discarded, an alarm is issued to the operator, and the average value of the remaining two is used as process value [4]. When there are three differences that exceed the preset limit, this shows that sources are unreliable. Usually, the control system stays put at the last good value and control may be forced to manual by the operator via an alarm. As indicated before, the preset value is set based on measurement types.

1.3.3 *Final Element Redundancy*

Final control elements: In rare instances the final control elements can be duplicated, in cases when the erosive/corrosive or sticking characteristics of the fluid could cause unacceptable downtime or in cases of critical controls (viz, boiler drum level control with control valves in medium-sized power plants). The major cases are as follows:

- 1oo1/1oo1*D*: In typical control loops, a single control valve is used. A valve malfunction (e.g., sticking) could be detected, with some time delay, because of a drift in process variables caused by the incorrect positioning of the trim [4]. Use of a positioner or a remote position indicator is a good solution to overcome this. However, with the use of an electropneumatic positioner it is possible to check (and correct) the valve's actual position against the required one and verify that the dynamic response of the valve has not changed over time. An intelligent electropneumatic positioner provides feedback to the DCS on valve behavior for the DCS to generate an alarm, and a loop may be transferred to manual. In the majority of control systems there could be a bypass manual inching (modulating) valve (less costly) to the control valve so that control can be maintained manually, for example, a bypass valve for the main condensate valve in a power plant. For on—off control valves, to prevent trim from sticking the diagnostic functions can *occasionally* command the valve to move from the current condition only shortly and slightly, performing a partial travel. Such a movement command is given based on process characteristics and these movements are monitored.

- 1oo2/1oo2*D*: Two control valves with diagnostic coverage are used in cold standby mode, as discussed. In certain cases, similar 1oo2 are achieved in a separate way also, for example, use of two of three fans or pumps with speed controls as in cases of induced draft (ID) fans and boiler feed pumps (BFPs). In such cases, in case of failure of the final control element, say a hydraulic coupling scoop tube, a standby fan/pump with a scoop control is started. In such cases the scoop (speed) control of the standby follows the running fan/pump scoop position. Diagnostics referred to earlier help to switch from main to standby. On account of criticality of application, dosing pumps (in large boiler plants) used to have a cold standby.

1.4 Fault Tolerant Network

Based on the application, there are variations of type of computer or computing system needed. Spacecraft controls must have long-life, maintenance-free computers. Typically, an application calls for computers to operate correctly without maintenance for 5—10 years. On the other hand, applications such as aircraft, mass transportation systems, and nuclear power plants demand computers for which an error or delay can prove to be catastrophic. In these cases TMR processors and duplicated memories, etc. can be used. So far various requirements for computing systems, control systems, and field instruments have been discussed. *But what about communication fault tolerance?* In modern control systems where controls are highly distributed, communication between the nodes is becoming a *critical part* of the system architecture. In this clause a short discussion on this and on network fault tolerance will be covered. In certain cases a diverse redundancy scheme is employed, for example, redundant media (copper cable and fiber optic cable) are employed for highway communication, but this is effective only if they are routed through two different paths. This will prevent not only electromagnetic interference but also cables being cut. Media redundancy is an important issue.

1.4.1 *Media Redundancy*

Media redundancy is the formation of a backup path when part of the network is unavailable. IEEE 802.1D *Spanning Tree Protocol* (STP) supports redundant configurations of any type such as meshes, rings, or a combination of these and thereby avoids looping problems in Ethernet connections. However, it has one major limitation: lower convergence speed (30–40 s). When fast fault recovery is necessary this is not suitable. Another standard, IEEE 802.1w Rapid Spanning Tree Protocol (RSTP), has been created for faster recovery time (1 s) from topology changes. RSTP provides faster recovery by *monitoring* link status of each port and then generating a topology change after a link status change. RSTP also improves recovery time by *adding* a new port designation, which is used as a backup to the root port.

1.4.2 *Network Node Redundancy*

Another aspect is the failure of electronics. Switches are used for critical devices to set dual network paths. To keep the system running when a network fails, critical devices support two Ethernet interfaces to connect to both redundant switches.

1.4.3 *Communication Diagnostics*

There shall be *diagnostic* information available in the device about network communication status, node communication status, and diagnostic information for the single node concerned.

1.4.4 *Fault Tolerant Ethernet*

Fault tolerant ethernet (FTE) is quite a good solution. This has been developed by Honeywell. The FTE connects a group of nodes typically associated with communication paths between them, so the network can tolerate all single faults and many multiple faults. FTE can rapidly detect faults and, in case of communication failure,

Table XI/1.4.4-1 **Comparison Between Dual Ethernet and Fault Tolerant Ethernet (FTE)**

Connectivity	Supporting Figure Reference	Dual Ethernet	FTE
Number of networks	Fig. XI/1.4.4-1A	2	1
Dual connected nodes (DCNs) to DCN	Fig. XI/1.4.4-1B[a]	2	4
DCN to single connected nodes (SCN)	Fig. XI/1.4.4-1D[a]	1	2
SCN to SCN in same tree	Similar to Fig. XI/1.4.4-1C	1	1
SCN to SCN in different tree	Fig. XI/1.4.4-1C[a]	0	1

[a]*Each possible path(s) is shown by* colored lines *in the figure.*
From S. Basu, A.K. Debnath, Power Plant Instrumentation and Control Handbook, Elsevier, November 2014; http://store.elsevier.com/Power-Plant-Instrumentation-and-Control-Handbook/Swapan-Basu/isbn-9780128011737/. Courtesy Elsevier.

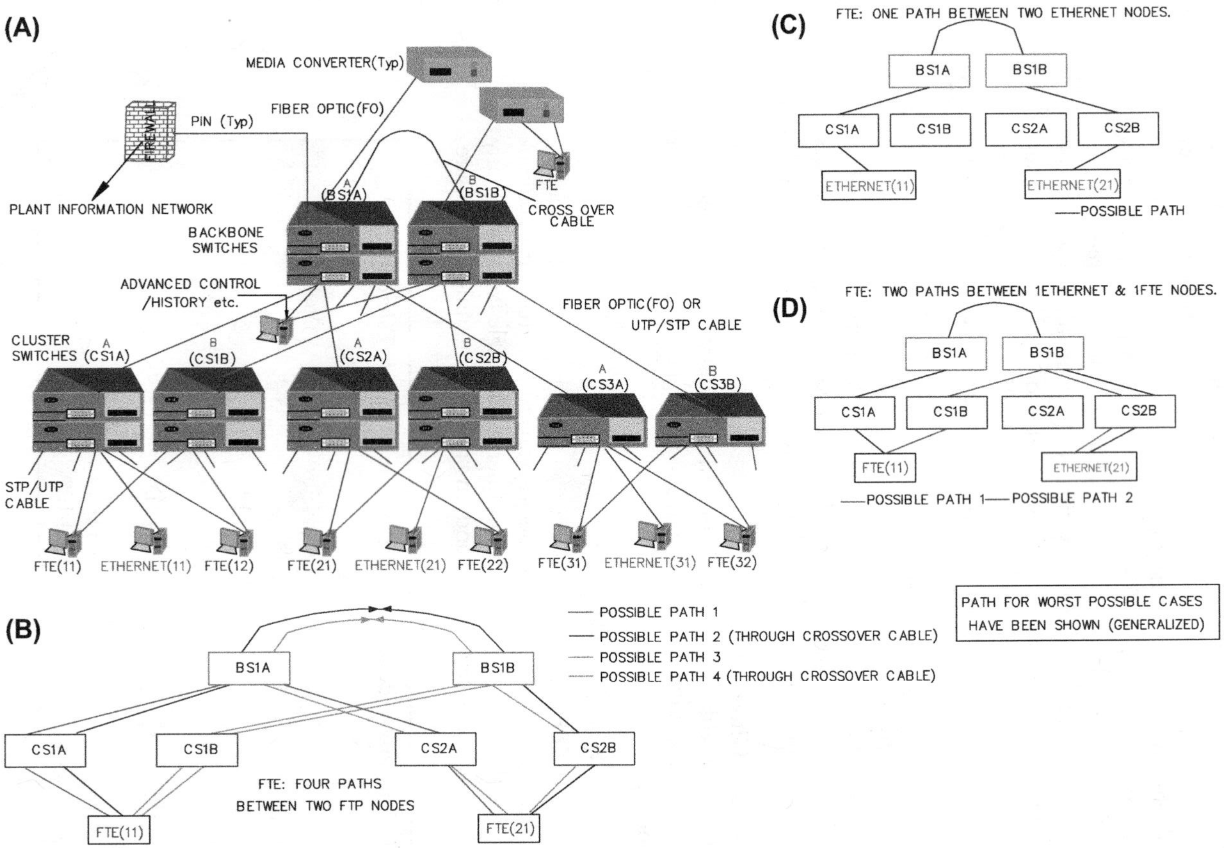

FIGURE XI/1.4.4-1

Fault tolerant Ethernet network. (A) Fault tolerant Ethernet network, (B) FTE network communication; FTE to FTE node communication; four possible communication paths, (C) FTE network communication; common between 2 Ethernet nodes; single possible communication path, (D) FTE network communication; common between FTE & Ethernet nodes; two possible communication path. *FTE*, fault tolerant Ethernet; *UTP/STP*, unshielded twisted pair/spanning tree protocol.

The drawing is based on FTE network of honeywell (Courtesy: Honeywell); (A and B) From S. Basu, A.K. Debnath, Power Plant Instrumentation and Control Handbook, Elsevier, November 2014; http://store.elsevier.com/Power-Plant-Instrumentation-and-Control-Handbook/Swapan-Basu/isbn-9780128011737/. Courtesy Elsevier.

the switchover time is around 1 s. FTE uses commercial off-the-shelf (COTS) equipment but with increased system availability.

- Some benefits (author's book [2] courtesy Elsevier):
  - Rapid response: In conventional Ethernet, there are two separate networks with each node (server) connected to both networks. The switchover time, in case of communication failure, is 30 s. FTE employs a single network and does not require a server, so changeover time is less.
  - Possible communication path: FTE provides more communication path possibilities than the dual Ethernet networks, as is clear from Table XI/1.4.4-1.
  - Full redundancy in a single network: A conventional Ethernet network with redundancy usually has two independent Ethernets, and naturally there will be a difference in performance and configuration between the two. However, in an FTE single Ethernet there is no such problem and at the same time it provides multipath capabilities in its unique topology.
- Network topology: A typical network topology based on Honeywell (courtesy Honeywell), FTE is shown in Fig. XI/1.4.4-1 [2]. Two parallel trees of switches and cabling "A" and "B" are linked at the top to form one fault tolerant network. Each FTE node has two ports that connect to a switch in each tree. In contrast, Ethernet nodes can connect to either if the switches are A or B. There may be one or more levels of switches and there can be multiple pairs of switches in each level. These have been designated as "cluster" and "backbone" switches in Fig. XI/1.4.4-1A. FTE to FTE communication paths and possible connections are shown in Fig. XI/1.4.4-1B—D. These are self-explanatory. However, for *further detail* the book [2] may be referred to. Having gathered some knowledge on fault tolerance, its time to focus on IPL and operator actions, from control systems point of view in next clause.

The discussion on fault tolerance is now concluded.

2.0 PROTECTION LAYERS

A Center for Chemical Process Safety (CCPS) publication gives the following definition: "An IPL is a device, system or action which is capable of preventing a scenario from proceeding to its undesired consequence independent of the initiating event or the action of any other layer of protection associated with the scenario. The effectiveness and independence of an IPL must be auditable" [12]. Discussions on layer of protection analysis (LOPA) were covered in Chapter V, so they are not repeated here. However, a few characteristic features of protection layers are presented. Similar to fault tolerance and security, this is also important so that the control system is always safe. As per IEC 61511 standard the core idea for integrated safety and security is "defense-in-depth" with independent layers of protection to reduce process risk. The strategy behind this is that the BPCS, critical alarms, operator actions, SIS, fire and gas (F&G) systems, and any other system intended to reduce risk in the processes are capable of acting **independently** from each other. The major reasons for the basic requirement are to avoid common cause faults, minimize systematic errors, and provide security against unintentional access. The nature of all layers of protection is not the same. Some of them may be preventive in nature such as emergency shutdown (ESD); some may be mitigating in nature, for example, F&G(!) (which mitigates after it has happened). Other layers may be deterring in nature.

2.1 IPL Characteristics

In Clause 4.0.2 of Chapter V, the necessary characteristics of independent protection layers (IPLs) were discussed. Definitions are given here again to elaborate further an understanding of the importance of assigning IPLs. The following are major issues:

2.1.1 Specificity

An IPL is designed solely to prevent or to mitigate the consequences of one potentially hazardous event (IEC 61511-3:2003). Multiple causes may lead to the same hazardous event; the action of one IPL is necessary.

2.1.2 Independence

The performance of a protection layer is not degraded or affected by the initiating event nor is it influenced by the failure of other protection layers. This is mainly for common cause error.

2.1.3 Functionality

The protection layer must be responsive to the targeted hazardous event, meaning that it is applicable for the event so that the concerned protection layer operates in response to a hazardous event.

2.1.4 Integrity

This function is related to risk reduction, which can be reasonably expected of the protection layer in question with suitable design and management.

2.1.5 Dependability (/Reliability)

This is the probability that a protection layer will operate accurately toward the intended event under stated conditions for a specified time period.

2.1.6 Auditability

The IPL must be designed to permit validation of function and probability of failure on demand (PFD) (including drill for human error), in a regular periodic manner, that is, the ability to inspect information, documents, procedures, etc. to demonstrate the adequacy of protection and adherence to the requirements.

2.1.7 Access Security
This encompasses administrative and physical controls to prevent unauthorized access for making any change.

2.1.8 Management of Change
This is the formal prior process of reviewing, documenting, and approving any modification proposals before implementation.

2.2 Impact and PFD Guidelines
A few guidelines put forward by CCPS are summarized here:

2.2.1 Initiating Event Validation
All initiating events and IPLs should be properly maintained and validated to provide current initiating event frequency and PFD for an initiating event and IPL, respectively.

2.2.2 Human Error
Human error and other systemic errors found during maintenance and testing and restoration need to be considered in assigning the PFD (especially for IPLs with a PFD < 0.1). This is particularly important for process *safety* valves with *block* valves.

2.2.3 Human Failure During Fabrication
Human failure during fabrication of equipment can affect the failure rate of the equipment. This is important for sensitive equipment like pressure vessels from alloy.

2.2.4 Advanced LOPA
Advanced LOPA or LOPA integrated with quantitative risk analysis requires a greater degree of expertise, knowledge, and judgment. The major issues here are:

- Evaluation of common mode failure in LOPA
- Use of multiple failures in a BPCS in LOPA
- High demand rate for IPLs
- Complex mitigating controls in LOPA
- Human reliability analysis

2.3 Protection Layer Effectiveness
In line with the requirements of IEC 61511-3:2003 the standard protection layers are shown in Fig. XI/2.3-1.

The effectiveness of each of these layers is quantified in terms of PFD, that is, the probability that the IPL will fail to perform a specified function on demand. IEC 61511-3:2003 specifies typical PFDs expected from different protection layers. Typical values are indicated in Table XI/2.3-1.

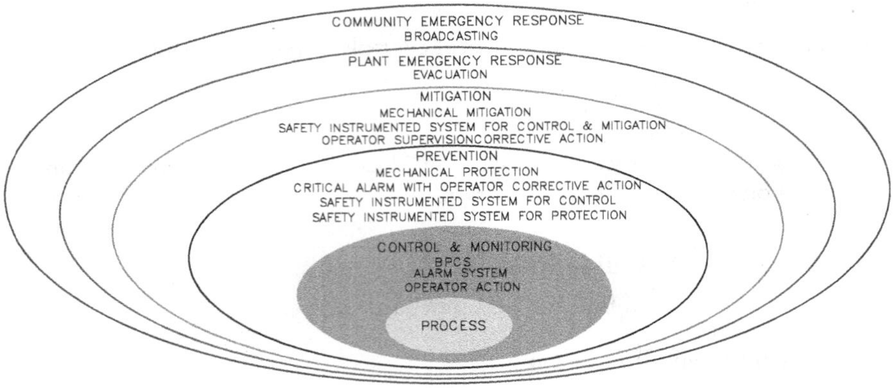

FIGURE XI/2.3-1

Standard protection layers. *BPCS*, basic plant control system.

Table XI/2.3-1 *Typical Protection Layer Probability of Failures on Demand (PFDs)*

Protection Layer	PFD
Control loop	1×10^{-1}
Human performance (trained, no stress)	1×10^{-2} to 1×10^{-4}
Human performance (under stress)	0.5–1.0
Operator response to alarm	1×10^{-1}
Vessel pressure rating above maximum challenge from internal and external pressure sources	10^{-4} or better when vessel integrity is maintained

As indicated in the table the IPL is quantified by PFD of the layers, so if there are n independent layers, then the mitigated consequence frequency is given by:

$$f_c = f_i \times \text{PFD}_1 \times \text{PFD}_2 \times \ldots \text{PFD}_n \qquad \text{(XI/2.3-1)}$$

where, f_i is the frequency of the initiating cause; PFD_n is the probability of failure on demand of the nth independent protection layer; and f_c is the mitigated frequency of the consequence. The main condition is that each protection layer is *independent*.

2.4 Operator Action: Protection Layer and Risk Reduction

From Clause 2.2 it is seen that operator action plays an important role both in protection layer and in risk reduction. Hence this has direct influence on PFD and therefore SIL. In this clause this will be briefly discussed.

2.4.1 Operator Action in Protection Layers

From the discussions in Chapter II it is clear that in any facility risk is a function of the frequency of a hazardous event, and the severity or consequence of the event. Also, depending on the facility function, location, design, hazardous materials, etc. and the risk tolerance limit, each facility sets its risk criteria function. From IEC 61511-3:2003 a generalized protection layer is presented in Fig. XI/2.3-1. Here it is clear that there are three active protection layers where there is scope of operator action in response to process parameters that exceed safety limits. The first is in the BPCS for the alarm system (independently considered in BPCS). The second is when the operator action is an integral part of an SIS both in preventing and mitigating an event. The third place is emergency response of the plant. This is not shown so explicitly, but it is known that the operator activates a facility emergency response system for evacuation, that is, action is mainly to initiate an evacuation process. In either the first or second case an operator may respond to an alarm/indication in the control room and initiate an action. The distinctive part in the second case is that the SIS has a PFD associated with it so operator action may alter it and thereby change the credit to the particular SIS layer.

2.4.2 Operator Action in BPCS

In BPCS, operator actions in response to process conditions are not part of a safety system (see IEC 61511 -1:2003 Clause 9.4.2) if risk reduction is less than 10. It is needless to say that BPCS should be designed as per ISA standard for alarm systems. Also the design of a BPCS operator interface should incorporate human factors engineering principles to ensure adequate response of the operator to displays and alarms. It is extremely important that operator response during both normal and abnormal conditions in the facility should not unduly violate process safety limits and norms and put the facility in an unsafe or undesirable mode or condition.

2.4.3 Operator Action in SIS

As per IEC 61511-3:2003 Clause 9.4.3, operator action as part of safety instrument functions (SIFs) can be credited with a level of risk reduction greater than 10 when the system from the sensor to the final element can be designed and evaluated as an SIS per the requirements of IEC 61511. A typical automated SIS, popularly known as an "industrial automation and control system (IACS)," from the sensor to the final element can be conceived, as shown in Fig. VIII/1.4-1 or Fig. VII/1.3-1 where the main constituents are sensor, logic solver, and final element. When an operator action such as through the display/alarm is necessary this needs to be as shown in Fig. XI/2.4.3-1.

The key point here is to recognize the additional factors that affect the PFD. The two main factors that affect the SIL of SIS with operator action are human errors and support system reliability. Human error essentially is the failure of the operator to respond correctly to the alarm/display and to take the corrective action(s) necessary to return the process/facility to a safe state. As already discussed in previous chapters

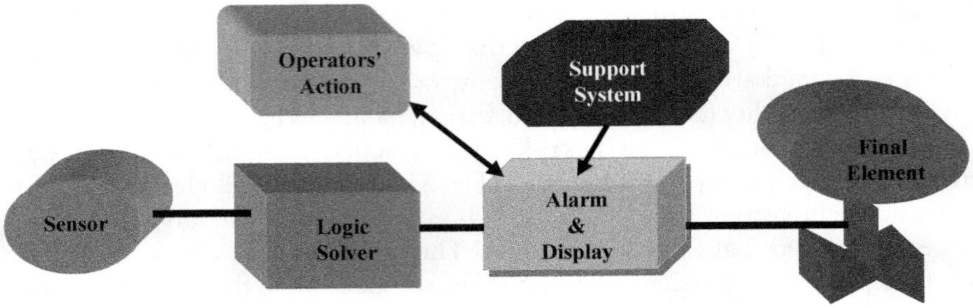

FIGURE XI/2.4.3-1

Operator action in an SIS block diagram (IACS).

in connection with alarms, the human response can be broken down into four functions:

- Identification and recognition of unsafe condition
- Proper analysis of the condition
- Initiation of the required safety action
- Observation of the response of the process to the safety function

There are a number of methods for evaluation of the probability of human error, for example, the technique for human error rate prediction, discussed earlier (Clause 6.2.1 of Chapter V). The best source for determining the human error rate would be company/facility-specific historical data, but in most organizations this is not available [11]. So, other means need to be explored. The reliability of support systems necessary for an operator's action is also an important issue that can influence risk reduction. The majority of SIS systems are designed as deenergize to actuate. The calculation of PFD for these SIS systems does not generally have to take into consideration any system outside of the SIS. See also Clause 3.2.2.

With this discussion on protection layers now concluded we will now look into network security. IACS, discussed earlier, has security problems for which there are specific standards such as IEC 62443. In subsequent clauses issues related to security in networking will be discussed briefly. This is especially important for integrated network systems. However, prior to looking at security issues it is important to understand why segregation between BPCS and SIS has been discussed in IEC standards. In the following clause the discussions have been presented on segregation between BPCS and SIS from an architectural point of view, so that the discussions on security issues pertinent to networks will be meaningful.

3.0 BPCS AND SIS INTEGRATION: ARCHITECTURAL ISSUES

From a commercial and economic point of view the majority of IACS manufacturers and system integrators are developing systems with integration of both BPCS and SIS.

Such demands are so high they could hardly be ignored. In Clause 2.2 of Chapter VII some aspects were already discussed. Here the discussions will be on architecture of the integrated systems: due consideration will be given to IEC standards. It is advisable that Clause 7.4.2.3 of IEC 61508-2:2010 and Clause 11.2.2/11.2.3/11.2.4/ 11.2.9/11.2.10 of IEC 61511-1:2004 are referred to.

3.1 Major Issues Behind Separate Systems

Following are the main reasons as per IEC 61511 [15]:

3.1.1 *Impact*

Impact on SIS because of common cause and mode and systematic failure of BPCS

3.1.2 *Flexibility*

Retention of flexibility of changes, maintenance, testing, and documentation for BPCS

3.3.3 *Facilitation*

Facilitation of functional safety assessment and validation of SIS

3.1.4 *Analysis Time*

Analysis time reduction to ensure requirements for safety

3.1.5 *Supports*

Support for access security and enhancement of cyber security for SIS so that revisions in BPCS do not affect SIS

3.2 BPCS and SIS Architectures

The following are several ways the two systems can be conceived.

Completely separate (air gap): BPCS and SIS are completely separate with no physical connections between them, for example, PLC-based BPCS and hardware SIS.

Interfaced: Two separate systems with a link, for example, RS 232 and MODBUS. This may be for data exchange for display/monitoring.

Integrated: Separate BPCS and SIS (separate sensor, logic solver, and final element) but connected through a common network. Here there may be chances of commonality of hardware/software; hence a CCF issue! On the contrary, if there are different suppliers, the chances of CCF systematic failure may be less. Further integration is possible when there is a common engineering station or separate I/Os but fallback of BPCS and SIS controller. In these cases, obviously the suppliers may be the same; hence the chances of CCF, etc. will be higher but with the use of different technology the issue may be circumvented.

It is seen from the foregoing that as the standards were developed for completely separate BPCS and SIS, naturally not only is third party certification necessary but many considerations must be taken into account at all stages so that the main philosophy behind the standards is not diluted. Again it is also a fact that most of the systems available in the market are integrated systems.

3.2.1 *Integration Approach*

Right from the planning stage due consideration must be given to integrating BPCS and SIS. A few relevant points in this regard shall include but not be limited to the following:

- Adherence to duty holder philosophy at all stages such as specification, design, engineering verification validation, etc. separately for BPCS and SIS
- Evidence of confidence of all stakeholders (see IEC 61511) or suppliers and duty holders
- Adherence to local and international regulations including IEC
- Compliance with IEC 61508-1:2010 and 61511-1:2004 for functional safety management

It is known that in SIS, separate IPLs are to be considered. Therefore it is necessary that such independence must be demonstrated properly. Also if there are any credits to be claimed for BPCS, then Clause 9.4.3 of IEC 61511-1:2003 should be followed (see Chapter VI).

3.2.2 *Integration Guidelines*

There have been separate guidelines for this and these must be followed [16]. The following are the major issues to be addressed and documented:

- Training and culture
- Competence
- Safety and security
- Location
- Access control
- Manufacturer's guidelines
- Procedure
- Human interface
- Separation
- Segregation
- Redundancy
- Diversity

3.2.3 *Salient Issues*

In view of the foregoing, some relevant and important issues need to be considered for integrated BPCS and SIS and shall include but not be limited to:

- **Diversity** in hardware and software between BPCS and SIS to avoid CCF, etc. So, it is necessary to address these *explicitly* and they should be certified by a *third* party.
- IEC 61511 must be followed for validation and functional safety assessment and documentation for SIS.
- Electrical and logical separation of processing units between two systems
- Specific I/Os and processing units for SIS, preferably with fault tolerant design

Black channel communication: In black channel communication a new safety functionality is built on the top of the existing protocol without the safety protocol knowing it, that is, it is in darkness. *A fault tolerant wireless "black channel," realized with Parallel Redundancy Protocol according to IEC 62439-3, is achieved that is able to fulfill soft real-time availability plus providing redundancy. Wireless technologies resembling the method used in fieldbuses are also spreading to safety-related applications. Both Profisafe as well as the new FOUNDATION fieldbus utilize black channel communication SIS applications.*

FIGURE XI/3.2.3-1

Black channel communication.

- Specific and separate development, engineering and systematic capability in line with the standards for SIS
- Separate SIS zone and use of a firewall to combat risk from cyber security in an integrated system. IEC 61511and 62443 standards need to be followed.
- Use of the black channel technique as per IEEE design for communication (Fig. XI/3.2.3-1)

The subject of integration of BPCS and SIS is now concluded and we can move on to issues related to security in SIS.

4.0 SECURITY ISSUES IN SIS

Security issue is one of the most important aspects in the current design of SIS. With the help of an open interface like Open Platform Communications (OPCs) (Fig. XI/4.0-1) it is possible to integrate not only BPCS and SIS but the entire enterprise network. Common and open communications protocol architecture standards are replacing the diverse and disparate proprietary systems of industrial control systems. This migration empowers users to access new and more efficient methods of communication as well as more robust data, quicker time to market, and

***What is OPC?* This is another tool for system integration. OPC is open connectivity in industrial automation for interoperability supported by the creation and maintenance of open standards and specifications. OPC is a standardized interface for accessing process data. Object linking and embedding (OLE), component object model (COM)/distributed component object model (DCOM), was developed by Microsoft. When this is applied to the process control, OPC (OLE for process control) is developed. OPC is based on the Microsoft COM/DCOM standard and has been expanded according to the manufacturer's requirements.**

FIGURE XI/4.0-1

OPC open interface.

From S. Basu, A.K. Debnath, Power Plant Instrumentation and Control Handbook, Elsevier, November 2014; http://store.elsevier.com/Power-Plant-Instrumentation-and-Control-Handbook/Swapan-Basu/isbn-9780128011737/. Courtesy Elsevier.

interoperability. Integrated systems developed by system integrators offer communications and security solutions that are flexible enough to collaborate with a variety of third party DCSs and easy enough to deploy. However, all these advantages are coupled with new cyber-related vulnerabilities and risks. "Open-based standards have made it easier for the industry to integrate various diverse systems together, it has also increased the risks of less technical personnel gaining access and control of these industrial networks"(courtesy National Communication Systems Bulletin). Some of the issues could be: use a denial of service (DoS) shutdown, delete system file (downtime), modify logging (data loss), and plant a Trojan and gain control [2]. In a modern SIS, which is a digital system and often connected to a network, there is a real concern that a targeted cyber attack can disable or affect its performance. Cyber security is increasingly critical for maintaining control and safety integrity and for ensuring both communications security and integrity. Security risks are increased in the case of a totally integrated system. SIS demand integrator skills significantly more advanced than those required for the usual PLC project. Most safety systems need to have their communications functions integrated into the PLC/DCS communications infrastructure safely and securely.

As stated there are open standards, for example, OPC, that make it possible for integrators to work with a standard protocol that gives them greater flexibility and economy but with the probability of higher risks because of security. Another important issue is that SIS functions are partitioned appropriately from the PLC/DCS functions so that a loss of communications or integrity will not prevent the SIS from performing its function, and keep the system in a safe state. Now it is time to look at the issues closely.

4.1 Security Issues: General Discussions

According to M. Barzilay of ISACA: "Cyber security is the sum of efforts invested in addressing cyber risk…" From an ISA point of view, security issue refers to the prevention of illegal or unwanted penetration, intentional or unintentional interference with the proper and intended operation, or inappropriate access to confidential information in industrial automation and control systems [17]. Cyber security therefore is mainly concerned with protection against unauthorized access (intentional or unintentional) to save data and information systems from theft or damage to prevent the system from any disruption of operation and unwanted functioning of the system. IEC/ISA 62443 (formerly ISA 99) is the relevant standard. Many propose to treat cyber risks as physical risks, that is, to check and assess vulnerability, frequency of occurrence, consequences, etc.

4.1.1 Vulnerability Check

Vulnerability check is a very important assessment. Vulnerability assessment requires checks at entry points, architecture, and current protective measures. Introduction of information technology (IT) components into industrial controls further complicates the situation. Deployment of wireless and remote control technologies coupled with trends to integrate the process system into the business networks opens the gates for cyber attacks. The usefulness of such technologies cannot be undermind. Therefore the balance between security and operational functionality needs to be maintained with a clear mission to detect vulnerability at the earliest possible opportunity so that it can be treated accordingly to prevent the system from serious damage.

- Major causes of cyber attacks: Vulnerability of business networks from cyber attacks make people more concerned about cyber security. Major causes that make the systems vulnerable for cyber attacks may be as follows:
  - High dependency on automation in secured external network connections and lack of qualified cyber security business in industrial controls
  - Trends for moving toward COTS operating systems and designs to integrate BPCS, SIS, and business networking and access through standard open system protocols
  - Adaption of open protocols such as internet protocol to access IACS, that is, common protocol vulnerability
  - Use of joint venture, alliance with other partners, meaning number of organizations accessing the same network parts—database attack
  - Added complexity of disgruntled staff, amateur attackers, criminals and terrorists, and availability of automated tools to access networks
  - Upward trend of malicious code attacks on business networks year by year to collect information in a dishonest way and attacks through the backdoor or a hole in the network perimeter and/or man-in-the-middle or insecure coding
  - Chances of attack through field devices
  - Other types of attacks such as communication hijack
- Major effects of cyber attacks: A few major effects of cyber attacks shall include but not be limited to the following:
  - DoS:Network flooded with spurious data, denying access to legitimate users
  - Penetration: Gaining control to disrupt the network, develop backdoor entry, and steal confidential information, for example, imported tools, viruses, etc.
  - Social engineering: System can be vulnerable if a member of staff unsuspectingly divulges sensitive information [18]. Attacker entry through legitimate user.

4.1.2 Probable Checklist to Prevent Cyber Attacks

The following is a typical checklist:

- Secure components (locking if necessary) so that settings cannot be changed easily
- Implement and use robust passwords for hardware and software
- Check vulnerabilities of "smart" devices and secure them as far as possible
- Check vulnerability of wireless transmission and secure using proper encryptions
- Disable unused ports, file transfer protocol (FTP) connections as far as possible, and minimize keep-alive and other settings that hold a disconnected port open; also reassess site strategy for security situations and measures taken [18]

- As far as possible disable USB and related common connections
- Use hardware and software firewalls and antimalware protection
- Create a site-specific strategy for software (especially open software) updates
- Create multilayer access points for information and enforce login/logout. Some use constantly changing numbers (generated in small dongles given to authorized persons) to be added after the desired password
- Report any employee leaving an organization who has had access to a dongle or anyone who loses a dongle for the generation of arbitrary numbers; access must be disabled immediately by IT staff
- Implement a change management policy to control access and track changes to configurations and programs [18]
- Implement recovery by only authorized persons with backup files
- Train all staff regarding the dangers of cyber attacks and social engineering impacts

4.1.3 *Architectural Aspects*

In modern transmission control protocol (TCP)/IP-based computing environments, implement a single integrated corporate network for the business that drives operations in a control system. The main concern of IACS as part of larger conjoined architectures is providing *security* procedures that cover the control system domain as well. Unique vendor-specific protocols and inherent legacy system security may not be adequate to protect mission-critical systems against *modern* cyber attacks [19]. Because there are several key differences between traditional IT environments and control system environments, it will be unwise to simply applying IT security technologies into a control system. Major differences are given in Table XI/4.1.3-1.

Also the requirements of three basic parameters such as availability, confidentiality, and integrity have different priorities, namely, for control systems like SIS the order of priority will be: integrity, availability, and confidentiality, whereas in the case of IT, confidentiality will be of the highest priority.

"Homeland Security Defense in Depth" [19] gives good guidelines for divisions in zones, etc. The salient features are:

- To implement multiple layers of defenses
- Entire integrated network to be divided into functional zone and interconnections made between them, for example, in offshore mud handling, pipe handling, etc. process control zones may be divided into area zone, for example, low-pressure mud handling, high-pressure mud handling/choke kill section, etc. as per requirements of the facility.
- Rules of data exchange between zones through conduits
- Deployment of firewalls for hardware and software (better to choose from different vendors)
- All external data or access to be routed through DMZ to act as buffer

For a better understanding the foregoing is depicted in Fig. XI/4.1.3-1. In the figure demilitarized zone (DMZ) is a buffer between external traffic and a trusted

Table XI/4.1.3-1 Differences in Security Handling in Information Technology (IT) and Industrial Automation and Control Systems (IACSs) [19]

Security Issue	IT	IACS
Antivirus mobile code	Easy to implement and update	Impact on control system difficult to implement
Time criticality	Delay may be allowed	No delay—real time
Security awareness	Moderate	Not much developed except physical means
Patch management	Easily defined and automatic	Original equipment manufacturer-specific (long time to manage)
Technology support	2–3 years multivendor	10–20 years one vendor
Test and audit	Easy modern method	Modern method not suitable
Change management	Regular scheduling	Strategic scheduling for impact on control system
Incident response	Easily developed and deployed	Uncommon beyond system resumption activities
Physical and environmental security	Poor to excellent	Excellent, but remote places may be unmanned
Security system development	Integral part of system development	Not an integral part of development
Compliance	Limited regulatory	Specific regulatory

internal network. There is no direct connection conduit between external traffic and internal network. DMZ acts as a buffer to permit allowable data exchange between external system and internal network without a direct conduit.

4.1.4 Major Cyber Attacks

The following are major issues related to cyber attacks:

- Backdoor and network perimeter hole: "Backdoor" is used to gain unauthorized, but not necessarily physical, access. A single point of compromise in an integrated network may provide extended access because of preexisting trust established among interconnected resources. Often the security perimeter of a control system is *compromised* for better connectivity without *precautions* being taken. In wireless communications, the residual effects of default installations and attackers, once having discovered wireless communications points, can use this situation to their advantage. Significant service-based vulnerabilities in the 802.15.4 protocol could lead to jamming and DoS [19]. Modern architecture with *remote* accesses for controls has the security perimeter *relocated* to the remote access making the system vulnerable. To allow robust information to be provided via external services, such as a web or FTP server, communication must be made from the web server to the internal databases or historians, and this connection is made via the firewall.

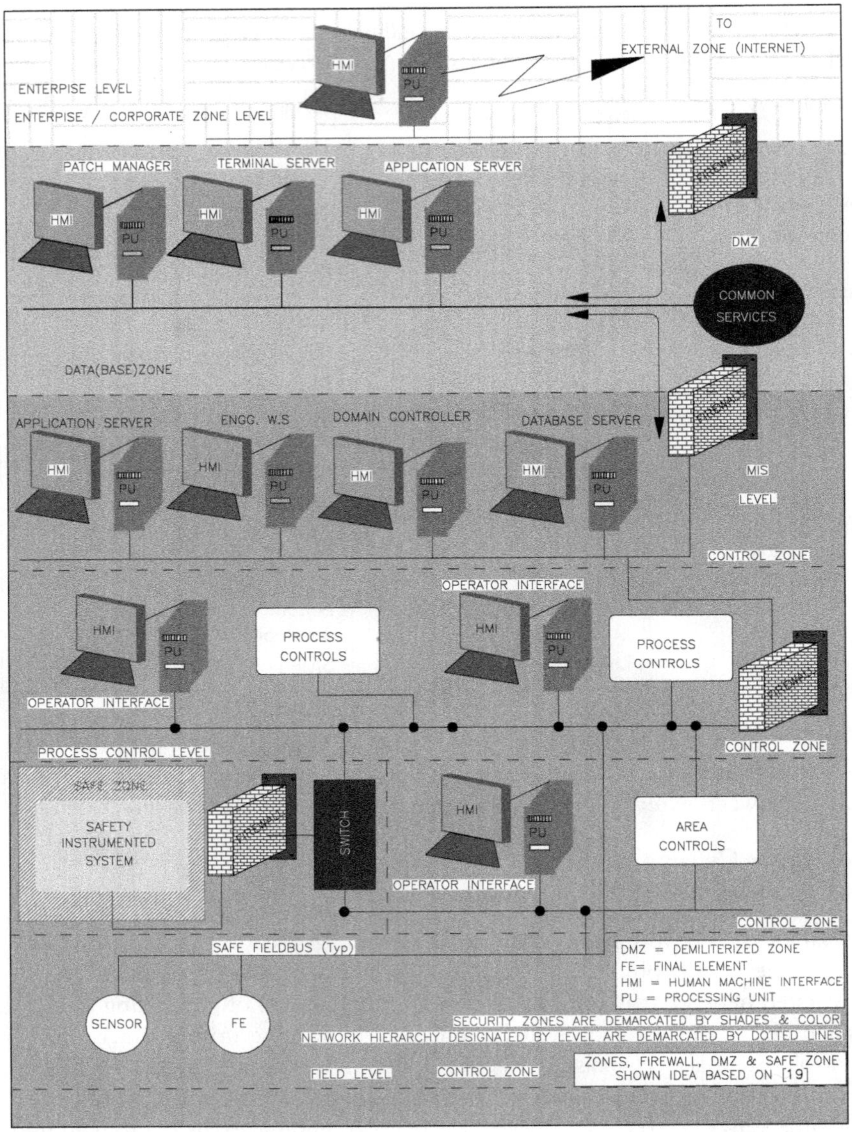

FIGURE XI/4.1.3-1

Zones and levels with firewall and demilitarized zone (DMZ) in an integrated network. *FE*, final element; *HMI*, human–machine interface; *MIS*, management information system; *PU*, processing unit.

- Common protocol attacks: Although a wide variety of security implications and vulnerabilities have been identified with OPC services and standards, OPC standards and application programming interfaces are common in control system environments. OPC vulnerabilities could be simple system enumeration and password vulnerability or they could be more complex like remote registry tampering and buffer overflow flaws, etc. These could result in threats to IACS. The installation of undetected malware, DoS attacks, escalated privileges on a host, and/or even the accidental shutdown of IACS are vulnerabilities for which solutions are *available* but these are not **always** very successful.
- Field devices: Remote accessed field devices and IT compatible instruments are quite vulnerable. For remote access, to provide for the collection of operational and maintenance data, some modern equipment has embedded file servers and web servers to facilitate robust communications [19]. In a trusted network this issue is not that serious but for an *integrated* network this could be **dangerous**. In addition to a dedicated communication channel, engineers often use publicly accessible telephone networks or dedicated lines for modem access. When **left** unsecured, an attacker can connect remotely with little effort, and the remote connection may be difficult to detect. Again field devices are part of an internal and trusted domain, so access to these devices can provide an attacker with access to the control systems. Attackers often add these field devices to their list of targets. If a device is compromised, an attacker can leverage control over the device and cause unauthorized actions without being easily recognized.
- Man-in-the-middle attack: Control system environments with air gapping exchanges are often less secured. Major issues in this type of attack are:
  - The ability for an attacker to reroute data
  - The ability to capture, stop the system functioning, and analyze critical traffic
  - The ability to reverse engineer control protocols to gain control over communications

Address resolution protocol (ARP) helps to maintain routing. Poisoning (manipulation) of the ARP tables is the key to this attack. With ARP table poisoning, an attacker can force system traffic to be rerouted through the computer the attacker has compromised, as shown in Fig. XI/4.1.4-1.

- Structured query language (SQL) data injection: IT databases are maintained by SQL. In an integrated network of IT and IACS, an attacker may use this to gain control of data from IACS by exploiting the communication channel (between the enterprise network and the control network) and disregarding the protection mechanism of IACS.
- Insecure coding: Some industrial control systems have very old custom-built or unsupported (by vendor) *programming* code. The programming code can suffer from insecurities such as:
  - Control environments have been built with no security *training*.
  - Custom applications have not gone through a proper cycle of security *testing* [19].

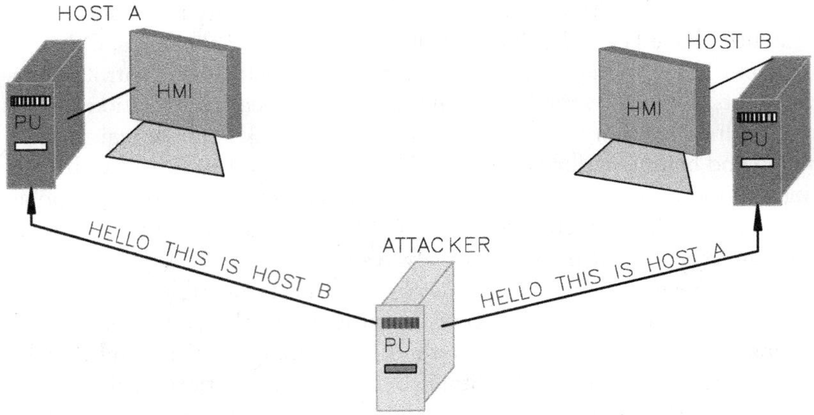

FIGURE XI/4.1.4-1

Man-in-the-middle attack. *HMI*, human–machine interface; *PU*, processing unit.

Many control system codes do not have any authentication or encryption to avoid slow response caused by code encryption. As a result the systems become insecure.

- Improper procedure: On account of large network integration and remote access there has been wide use of modems within the industrial control system environment. When modems are improperly managed from a security perspective, and they are always left on without any type of authentication, there is every *possibility* of a cyber attack. Even with a detailed procedure to access through a remote capability, many control system devices have poor *logging* capabilities and have not been properly turned on for *auditing* purposes. In such cases, guidelines in standards like NERC-CIP 002-009 and NIST SP 800-53 may be helpful for reference.

There are many other types of cyber attacks such as database attack or data injection, etc.

4.1.5 *Cyber Attack Defense Mechanisms*

Cyber security is not just the application of specific technologies to combat risks. From Aberdeen Research in November 2011 (Ismail, 2011) it can be inferred that the best-in-class companies must establish a formalized risk management strategy and ingrain safety as part of the culture. In a similar way, effective security programs for an organization depend on the willingness of the company to accept security as a constant *constraint* on all cyber activities. Accordingly, the necessary *strategic* framework involving personnel, operational philosophy is to be developed. This will help to acquire in-depth knowledge of security programs so that they can be handled effectively to the overall *gain* of the company. Major issues are:

- Know the *security risks* that an organization faces
- Identify qualitative and quantitative *risks* involved

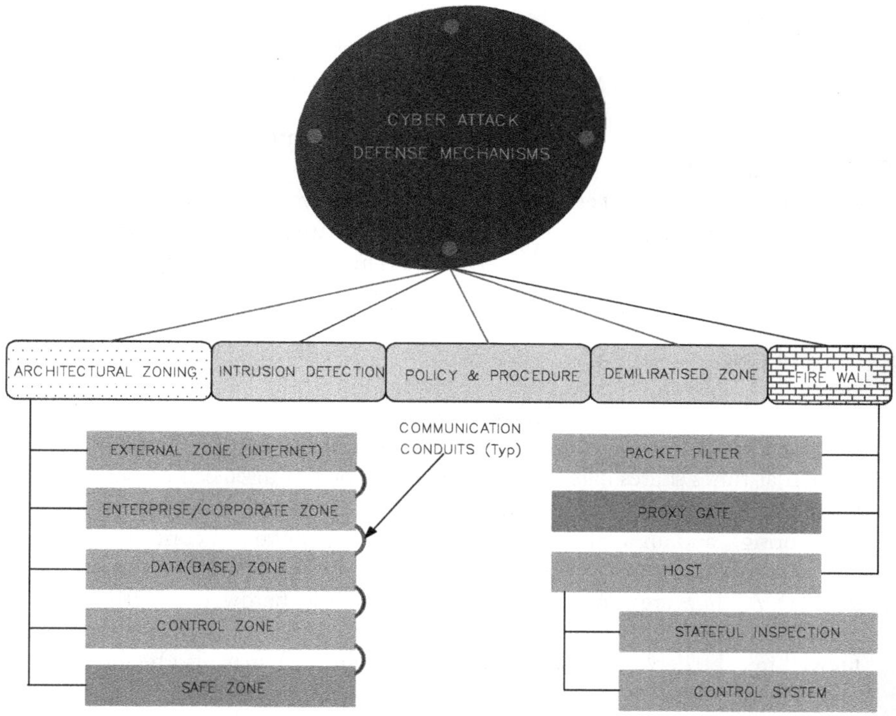

FIGURE XI/4.1.5-1

Defense against cyber attack.

- Use resources suitable to *mitigate* the risks
- Follow existing or emerging security *standards* at different phases as applicable
- Create and *customize* specific controls most suitable for the organization

As shown in Fig. XI/4.1.5-1 there are five major defense mechanisms to combat cyber attack:

- Architectural zone (conduit)
- Intrusion detection
- Miscellaneous policies and procedures
- DMZ
- Firewall
- Architectural zones: It is important to understand how the entire integrated network is connected and how it could be segmented. Based on functional relationships and data flow, large integrated enterprise/corporate networks including control systems can be divided into five basic zones:
  - External zone: The main connection is through the internet to the external world for global networking. With respect to IACS it has the least priority but the highest risk as it is not trusted, and has neither a DMZ nor any protection.

  - Corporate zone: This is the business area such as email servers, domain name servers (DNS), etc. On account of connectivity (Fig. XI/4.1.3-1) to the external zone it has a lot of risks and less priority (however, much higher than the external zone).
  - Data(base) zone: The majority of monitoring and control takes place in this zone. It is a critical area for continuity and management of a control network. Operational support and engineering management systems are located in this zone. Naturally, huge data handling is involved with the help of data acquisition servers and historians. This zone is in the middle of corporate zone and control zone. Since there is a direct connection with the corporate zone it is a DMZ with firewalls on either side, as shown in Fig. XI/4.1.3-1.
  - Control zone: This is the area of connectivity to control systems such as controllers (PLCs), human—machine interfaces (HMIs), and basic I/O devices such as actuators and sensors. Basically, there are three sections shown, namely, a management information system for plant management (which at times shares data with the database zone as discussed earlier), a main process control (namely, closed/open loop control system and data monitoring), and an applicable area control (e.g., choke kill control in offshore drilling, or offsite control like a coal handling plant in a power station). All I/Os are connected to this zone either by hardware directly or by a fieldbus system. This zone has very high priority and firewalls like a control firewall may be deployed. Additional external firewalls may also be used.
  - Safe zone: This usually has the highest priority because these devices have the ability to automatically control the safety level of an end device such as SIS. Typically, the risk is lower in this zone because these devices are only connected to the end devices.
- Commonly used tools: The following are the major tools that could be used to segment the integrated network, keeping control environment and SIS in mind:
  - Firewalls (single, multihomed, dual, cascading)
  - Routers with access control lists
  - Configured switches
  - Static routes and routing tables
  - Dedicated communications media
- Intrusion detection: Intrusion detection is a type of security management system for networks. The system gathers and analyzes information from various areas within a computer or a network to identify possible security breaches. There could be two types of breaches: intrusions (outside) and misuse (within the organization). Intrusion detection uses vulnerability assessment (also referred to as *scanning*), which is a technology developed to assess the security of a network. Intrusion detection is a **comprehensive** set of tools and processes providing network *monitoring* that can give an administrator a complete picture of how the network is being used. Implementing a variety of these tools helps to create a *defense-in-depth* architecture that can be more effective in identifying attacker activities, and using them in a manner that can be preventive. Each organization

must assess its particular situation, identify the *criticality* of the impacted devices, and develop a prioritized course of action. Unfortunately, a simple and prescriptive *remedy* applicable for all situations does not exist.

- Policy and procedure: A well-documented and detailed policy and procedure that is *specific* to the industrial control systems environment is essential to combat cyber attacks. Regular *periodic* review and maintaining policies and procedures will be very useful.
  - Security policy: Just incorporating the security policy of the IT sector in IACS will not do. Control system security must be *practical* and *enforceable*, and it must be possible to comply with the policy. The policy should not have an *adverse* effect on productivity and cost. It should be well supported and could be better developed by involving personnel from management and system administrator during policy development.
  - Log and event management: Commonly used security incident event management *technologies* can be deployed for centralized log and event management. Central security data consoles give security personnel a complete view of security tools, for example, ID system logs, firewall logs, and other **logs** that can be generated from any number of devices.
  - Security training: Security training and robust security *awareness* programs that are specific to the IACS domain are **critical** to the security of industrial control systems as well as the safety of those involved with any automated processes.
  - Patch management: This is common in the IT sector. A good *patch* management plan and procedure is also required in the IACS environment to create a layer of *defense* against published vulnerabilities.
  - Incident response and forensics: In the event a security-related incident in the control system domain, activities to *recognize*, *respond*, *mitigate*, and *resume* need to be established. An incident response procedure will instruct employees on the steps to take if a computer on the network has been compromised.
- DMZ: DMZ stands for demilitarized zone; actually it is a demarcation zone. As discussed earlier in Clause 4.1.3, DMZ is created as a buffer between external zones and trusted internal zones by isolating the host. This does not allow a direct conduit between the two. Multiple DMZs are created for separate *functionalities* and access *privileges*. Usually, in a good network all connections to IACS LAN are routed through the firewall. *Multiple* DMZs have proved to be very *effective* in protecting large integrated architectures. Network administrators need to keep an accurate network diagram of their system and its connections to other protected subnets, DMZs, and the outside. As is seen in Figs. XI/4.1.3-1 and XI/4.4-2, enterprise/business servers having access from outside have DMZs through firewalls. Basically, a DMZ is the *outward* facing level of an application. It is a subnetwork that resides between a known/trusted internal network and an external network, providing services to the outside without allowing direct access. The following are some system components of a DMZ [2]:

  - Public-facing server
  - Public-facing FTP server
  - Email gateway
  - Public-facing DNS
  - Traffic management and security server
  - Streaming video, etc.
- Firewall: Firewalls provide additional levels of defense. Types and functional aspects of firewalls are extremely important for IACS. This is discussed separately in the next clause.

4.1.6 Operational Issues

Once the defense mechanisms are set it is necessary to implement and maintain them. A few common points regarding these are elaborated here:

- Use a strong password as protection
- Restrict physical and electrical access to a needs-only basis. Also restrict access to subcontractors, etc.
- Implement separate and distinct authentication mechanisms, especially for control system access
- Form a security team with trained security personnel
- Revoke access for dismissed or resigned employees
- Enforce a consciousness program for employees on security issues so that they keep access through a public line for long periods (even it is left open after a short duration, new authentication may be required, if the application allows this)

A firewall is an important defense and demands separate discussions.

4.2 Firewall

A firewall protects a computer network from unauthorized access. Firewalls may be hardware, software, or a combination of both. The first firewall in an external and internal trusted network is a proxy server acting intermediately by receiving and selectively blocking data at the boundary. It also helps in *hiding* the LAN addresses from outside (to avoid ARP poisoning) [2]. Functional details of a firewall are presented in Fig. XI/4.2-1.

4.2.1 Category and Classification of a Firewall

Depending on the hardware/software, firewalls can be categorized and the classifications are based on function.

- Category: There are three categories of firewall, namely, hardware firewall, software firewall, and a combination of the two. Advantages and disadvantages of hardware and software firewalls are enumerated in Table XI/4.2.1-1.
- Classification of firewalls: Apart from being categorized as hardware/software or a combination of both, the broad classification of firewalls is depicted

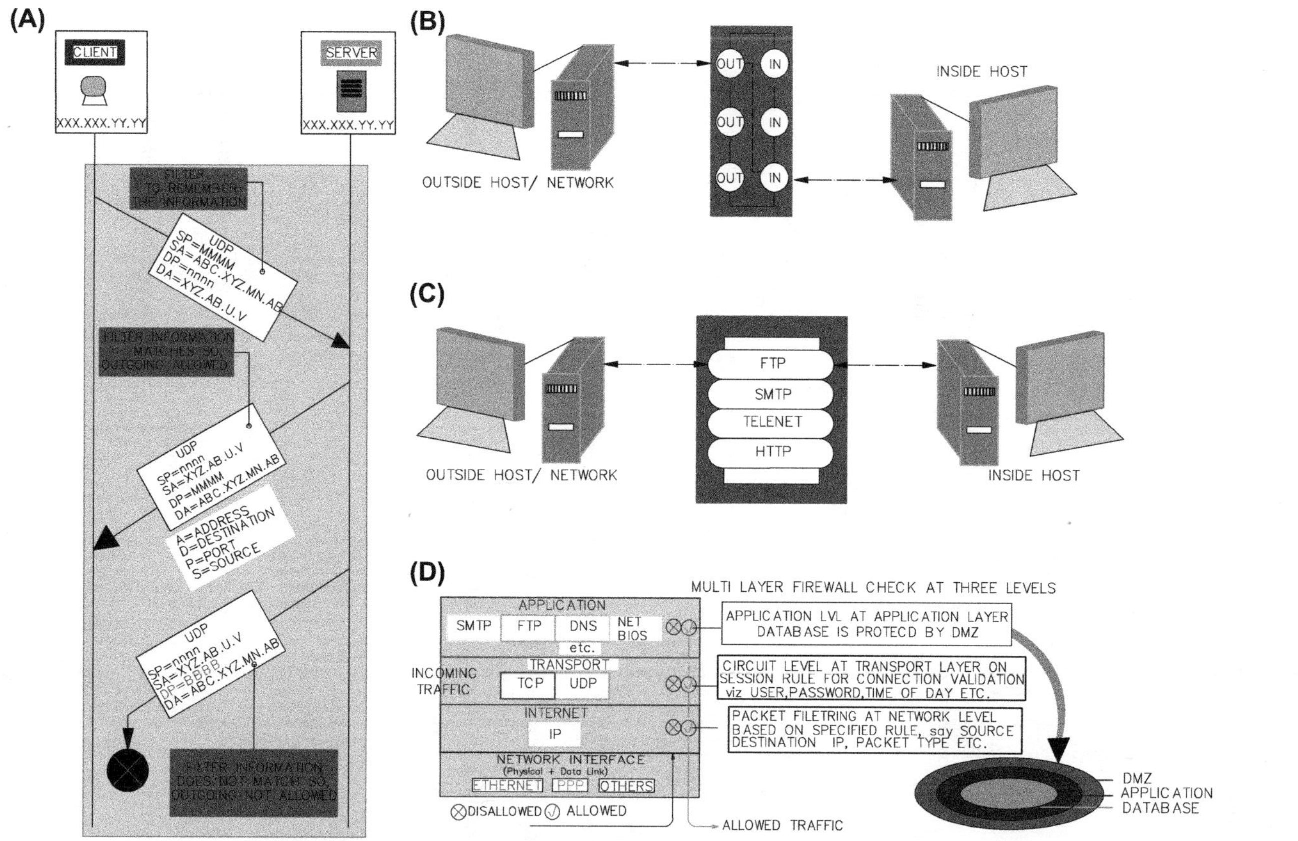

FIGURE XI/4.2-1

Firewall functional details. (A) Firewall used as Packet filter, (B) firewall as circuit level relay, (C) firewall as application gateway, and (D) multi layer firewall structure. *DMZ*, demilitarized zone.

From S. Basu, A.K. Debnath, Power Plant Instrumentation and Control Handbook, Elsevier, November 2014; http://store.elsevier.com/Power-Plant-Instrumentation-and-Control-Handbook/Swapan-Basu/isbn-9780128011737/. Courtesy Elsevier.

Table XI/4.2.1-1 Hardware Versus Software Firewall [2]

Comparison	Hardware Firewall	Software Firewall
Advantages	• Independent of operating system so immune to viruses generally • Very fast so better performance • Very effective because of single duty	• Less expensive, some free software possible • Can be implemented in existing network • Low administrative cost
Disadvantages	• On a single failure system may collapse • Proprietary nature prior knowledge may be necessary • High cost for maintenance as well as for installation	• Vulnerable to malicious attack • Operating system dependent, may be affected by viruses, etc. • Requires additional host resources like CPU/memory, etc. • Low performance

in Fig. XI/4.1.5-1. Based on TCP/IP layers, firewalls are classified as network layer firewall, application layer firewall, etc. For control systems, firewalls are as follows:

- Packet filtering firewall
- Proxy server firewall
- Host firewall

Short description of each of them has been presented in next clause.

4.2.2 *Short Discussions on Various Firewalls*

- Packet filtering firewall: These look after and analyze the packets going out and coming in to allow or deny access based on certain rules. Types of packet filters are shown in Table XI/4.2.2-1.

Table XI/4.2.2-1 Types of Packet Filter

Types	Details
Stateless packet	This is also known as a static IP filtering firewall. It does not remember the information about the previously passed packet. This is not a smart filter and can be fooled quickly. This type is vulnerable to user datagram protocol-type packets. It has a very high throughput but is costly. It is included with router configuration software or with most open source operating systems. It is highly vulnerable for security.
IP packet	IP packet filtering firewall: Every packet is handled on an individual basis. Previously forwarded packets belonging to a connection have no bearing on the filter's decision to forward or drop the packet.
Stateful packet	This is a pure packet filtering environment. It is known as smart firewall or dynamic-type packet firewall, because it remembers the information about previously passed packets.

- Proxy server firewall: These firewalls are critical in hiding the networks they are protecting and are used as primary gateways to proxy the connection initiated by a protected resource. They offer more security but at the cost of *speed*. Here, traffic does not flow through the proxy server gateway firewall because it acts as a buffer and has a direct *effect* on network performance. This type of firewall is suitable to *segregate* business networks from control networks. A firewall proxy server essentially turns a two-party session into a four-party session, with the firewall emulating the two real hosts, as shown in Fig. XI/4.2.2-1. They filter at the application layer of the open system interconnectivity model so they are often called application layer firewalls. As a gateway they require users to direct their connection.

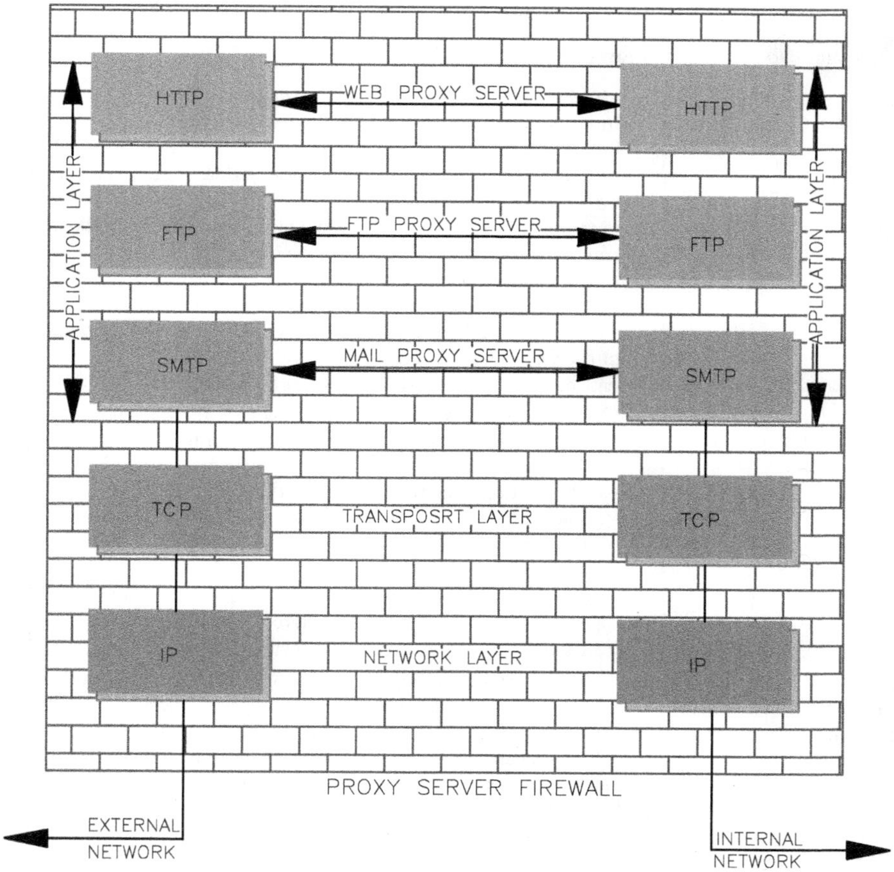

FIGURE XI/4.2.2-1

Proxy server firewall.

- Host firewalls: This software firewall protects ports and services specifically for the device on which it is installed. There are also third party software packages and these are host resident for servers, workstations, laptops, etc. There are various types of host firewalls such as control firewall and stateful inspection firewall, as shown in Fig. XI/4.1.5-1. Stateful inspection firewall filters at the network layer to determine the legitimacy of the sessions, and evaluate contents of the packets at the application layer. Control firewalls are hardware based directly on the controller to regulate traffic.

4.2.3 Firewall Functionality

The functionality of various firewalls is detailed in Table XI/4.2.3-1. The table should be seen in conjunction with Fig. XI/4.2-1.

Table XI/4.2.3-1 ***Firewall Functionality***

Type Name	Feature
Packet filter (see Fig. XI/4.2-1A and D)	• First line of defense (Fig. XI/4.2-1A). • Internet and other digital network data travel in packets of limited sizes. Consists of the Data, Ack, Request or Command, Protocol Information, Source and Destination IP Address, Port Error Checking Code, etc. • Filtering consists of examining incoming and outgoing packets compared with a *set of rules* for allowing and disallowing transmission or acceptance. • Rather fast because it really does not check any data in the packet *except* IP header. Works in the network layer (internet) of the Open Systems Interconnection (OSI) model. • Fast but not *foolproof*. IP address can be spoofed.

Circuit relay/gateway (see Fig. XI/4.2-1B and D)	• One step above the packet filter and commonly known as "stateful packet Inspection" to check the legitimacy or validation of the connection between two ends (in addition to packet filtering operation) based on the following:
	  ◦ Source destination IP address/port number
	  ◦ Time of day
	  ◦ Protocol
	  ◦ Username and password
	• Operates on the transport layer. Stateful inspection makes the decision about connection based on the data stated above.

Application gateway (application proxies) (see Fig. XI/4.2-1C and D)	• Acts as proxy for the application at the application layer of the OSI. See Fig. XI/4.2.2-1 also. • Authorizes each packet for each protocol differently. • Follows specific rules and may allow some commands to a server but not others, OR limits access to certain types based on authenticated user. • Setup is quite complex; every client program needs to be set up. Also each protocol must have a proxy in it. • True proxy is much safer.

Table XI/4.2.3-1 Firewall Functionality Continued

Type Name	Feature
NAT/PAT	• Firewalls using network address translation (NAT) or port address translation (PAT) completely hide the network. • In NAT there will be a single IP address used for the entire network. • Has the disadvantage that it cannot properly pass protocols containing IP addresses in the data portion.

From S. Basu, A.K. Debnath, Power Plant Instrumentation and Control Handbook, Elsevier, November 2014; http://store.elsevier.com/Power-Plant-Instrumentation-and-Control-Handbook/Swapan-Basu/isbn-9780128011737/. Courtesy Elsevier.

4.2.4 General Discussions

A firewall policy is extremely important from a cyber security point of view. To obtain effective results it is essential that there is a well-planned policy for a firewall and its management. It should be properly maintained and periodically audited. Any weakness in policy and failure to implement the policy will result in failure of the firewall. If one is considering an IP virtual private network (VPN) one needs to consider the *placement* of the VPN with respect to the firewall. The following are some of the guidelines for a firewall policy:

- Internal and external access and their extent
- Remote user access
- Virus protection and avoidance
- Encryption requirement if any and permitted
- Program usage

With this the discussions on firewalls are concluded. Next we will explore the standards regarding cyber security, especially for SIS.

4.3 Cyber Security Standards

From the discussions in Clause 4.1 it is quite clear that there are many holes or vulnerabilities in cyber security. Such vulnerabilities come in various forms such as improper authentication, improper input validation, etc. The percentage share of each of them varies. To appreciate the percentage share of each category of such vulnerabilities, Fig. XI/4.3-1 may be referred to. This is based on ICS-CERT vulnerability disclosure [20].

Since most of the control systems are intelligent and there is a clear tendency to opt for integrated systems with COTS and open system protocol, the vulnerability of the network from cyber attack is always in the positive gradient of growth. Also those with malicious intentions will always try to access the network. So, all these have accelerated the rate of attacks. Naturally, some countermeasures to save the business are necessary. Initially, ISA 99 standards, created by the International Society for Automation (ISA), were released as the American

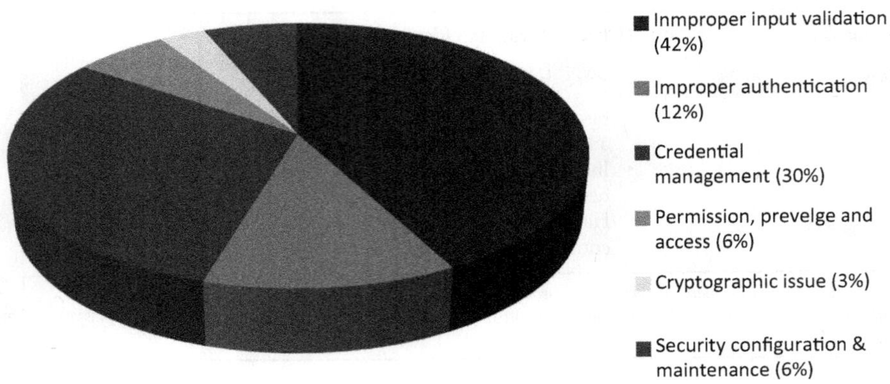

FIGURE XI/4.3-1

Vulnerability percentages.

National Standards Institute (ANSI) documents. These were the original standards for cyber security in IACS. Later, in 2010, to align these standards with the international standards organization International Electrotechnical Commission (IEC), the standards were renumbered and IEC 62443 emerged. *Generally, IEC and ISA standards are functionally identical.* These standards cater for the end user, system integrators, product manufacturers, etc. The standard is numbered IEC 62443-*x*-*y*; where *x* stands for categories 1, 2, 3, and 4. Each category has divisions (indicated by *y*), for example, category 1 has four divisions— ISA/IEC 62443-1-1/2/3/4. Similarly, other categories are: ISA/IEC 62443-2-1/2/3/4; ISA/IEC 62443-3-1/2/3; ISA/IEC 62443-4-1/2. These are further elaborated as follows.

4.3.1 Categories and Divisions in Standards

There are four general categories:

- Category 1: General: This has *four* divisions mainly concerned with general things like concepts, models, and terminology. Also included are security metrics and security *life cycles* for IACS.
- Category 2: Policy and procedure: This has *four* divisions, and coverage includes various aspects of *creating* and *maintaining* an effective IACS security program. Targets mainly the asset owner.
- Category 3: System: This has *three* divisions. This category mainly focuses on security technology, risk assessment, and security requirements including *zone conduit* concepts.
- Category 4: Components: This has *two* divisions. These include work products that describe the specific product *development* and technical requirements of *control* system products. This is primarily intended for control product vendors, but can be used by integrators and asset owners.

Table XI/4.3.2-1 62443 Series Standards (Standard Number IEC/ISA 62433-x-y)

IEC	ISA	62443-x-y	Title	Status
IEC/ TS	ISA	62443-1-1	Terminology, concepts and models	P, UR
IEC/ TR	ISA/ TR	62443-1-2	Master glossary of terms and abbreviations	D
IEC	ISA	62443-1-3	System security compliance metrics	D
IEC/ TR	ISA/ TR	62443-1-4	IACS security life cycle and use case	PL
IEC	ISA	62443-2-1	IACS security management system requirements	P, UR
IEC	ISA	62443-2-2	IACS security management system implementation guidance	PL
IEC/ TR	ISA/ TR	62443-2-3	Patch management in the IACS environment	P
IEC	ISA	62443-2-4	Requirements for IACS solution suppliers	P
IEC/ TR	ISA/ TR	62443-3-1	Security technologies for IACS	P, UR
IEC	ISA	62443-3-2	Security assurance levels for zones and conduits	V
IEC	ISA	62443-3-3	System security requirements and security assurance levels	P
IEC	ISA	62443-4-1	Product development requirements	V
IEC	ISA/ TR	62443-4-4	Technical security requirements for IACS components	V

D, *development;* P, *published;* PL, *planned;* UR, *under revision;* V, *voting.*

4.3.2 *ISA/IEC 62443 Series Standards and Technical Reports*

It is clear from the foregoing that ISA/IEC 62443 is series of standards not a single standard. Some of them have been published and some are in development stages, etc. See http://isa99.isa.org for the latest status. The title and status of these are elaborated in Table XI/4.3.2-1.

4.3.3 *Objective and a Few Definitions of Terms*

The following are the objective of the standard and definitions of a few relevant terms:

- Objective and mission of the standard: It is needless to state that critical requirements of IACS are to ensure that the system should never have the potential to cause **impacts** to *essential* services and functions, including emergency procedures. This is quite different from the requirements for IT security, which puts more emphasis on information—integrity, availability, and confidentiality. Accordingly, the mission for the standard will be to develop

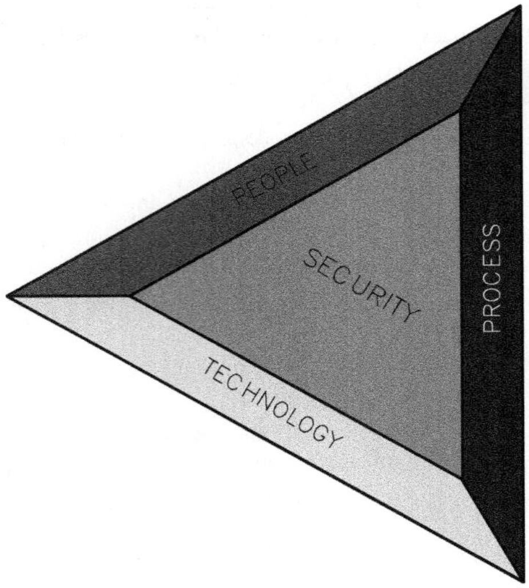

FIGURE XI/4.3.3-1

Security element groups.

Developed based on draft ISA 62443-1-1. Courtesy: ISA.99.

Table XI/4.3.3-1 ***Connection Between Various Elements (Based on ISA 62443-1-1)***

Element	Connections
People	Role: Asset owner/operator, system integrator, product supplier, service provider, compliance authority.
People	*Beyond the scope of the standard* but are connected to security indirectly: Resourcing, relationship, intent, support decision, awareness.
Process	Security policy, organization of security, asset management, human resource security, physical and environmental security, access control, communication/operation/business management, incident management, system acquisition, and maintenance management.
Technology	Use control, system integrity, data confidentiality, restriction of data flow, response to event in time, availability of resource.

a set of well-engineered specifications and processes for the testing, validation, and certification of IACS products. Standards are to be developed so that uniform methods and rules will be developed to reduce the time, cost, and risk of developing, acquiring, and deploying control systems. Such a standard is developed by establishing a collaborative industry-based program among asset owners, suppliers, and other stakeholders.

- Security elements: According to the standard, three elements, namely, people, technology, and process, constitute the security standard, as shown in Fig. XI/4.3.3-1.

Table XI/4.3.3-2 Security Maturity Phases (Based on ISA 62443-1-1 Draft)	
Phase	**Step**
Concept	Identification concept
Functional analysis	Definition
Implementation	Functional design, detailed design, and construction
Operation	Operation and compliance monitoring
Recycle and disposal	Disposal and dissolution

Security connections between these elements are listed in Table XI/4.3.3-1.

- Security level: Assets that make up the system under consideration shall be assigned a security level in accordance with standard ISA 62443-2-2.
- The security life cycle: The life cycle is focused on the security level of the system over time. A *change* in asset will trigger *changes* in security level or a change of vulnerability or an asset may trigger changes in a physical asset. There are a few basic steps for the security development life cycle assessment (SDLA), which are detailed in Clause 4.3.5.
- Zone and conduit: This is a process of segmenting or dividing a system under consideration for the purpose of assigning a security level and associated measures; it is an essential step in the development of the program.
- Security maturity program: A security maturity program *integrates* all aspects of cyber security incorporating desktop and business computing systems with IACS. The development program shall recognize that there are steps and milestones in achieving this maturity. Security maturity phases are shown in Table XI/4.3.3-2.
- Defense-in-depth: In a complex system it is impossible to achieve the security objective with a single countermeasure. In such cases, superior approaches involving application of multiple countermeasures in a layered or stepwise manner are used, for example, firewall and intrusion detection.

4.3.4 *Conformity Assessment as Per IEC*

There are two kinds of violations: casual or coincidental violation and intentional violation. Therefore conformity requirements call for protection against:

- Casual or coincidental violation
- Intentional violation using:
  - Low resources, generic skill, and low motivation
  - Advanced ways to moderate resources, motivating IACS-specific skills
  - Higher resources and motivation coupled with IACS-specific skills

4.3.5 *Security Development Life Cycle Assessment*

The basic steps as per IEC activity for SDLA are as follows [21]:

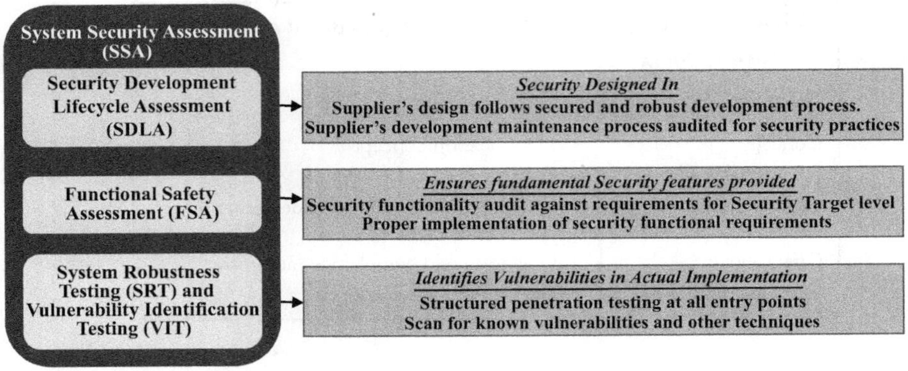

FIGURE XI/4.3.6-1

System security assessment program.

Developed based on an idea from IEC.

- Security management process
- Security requirements specification
- Security architecture design
- Security risk assessment (*threat* model)
- Detailed software design
- Document security guidelines
- Module implementation and verification
- Security integration testing
- Security process verification
- Security response planning
- Security validation testing
- Security response execution

4.3.6 *System Security Assessment*

The major issues involved in conformity assessment are:

- The system has been designed and developed as per a robust security process and norms.
- Supplier's design and maintenance process has been thoroughly audited to ensure robust security practices as per stipulation in the standard.
- Each system has a target security level, so it is necessary to audit security functionality of the system against the stipulated requirements.
- Systems have been audited for proper implementation of all security requirements.
- *Vulnerability* of the implemented system has been *identified*:
  - Structured penetration testing at all entry points
  - Scan for known vulnerabilities

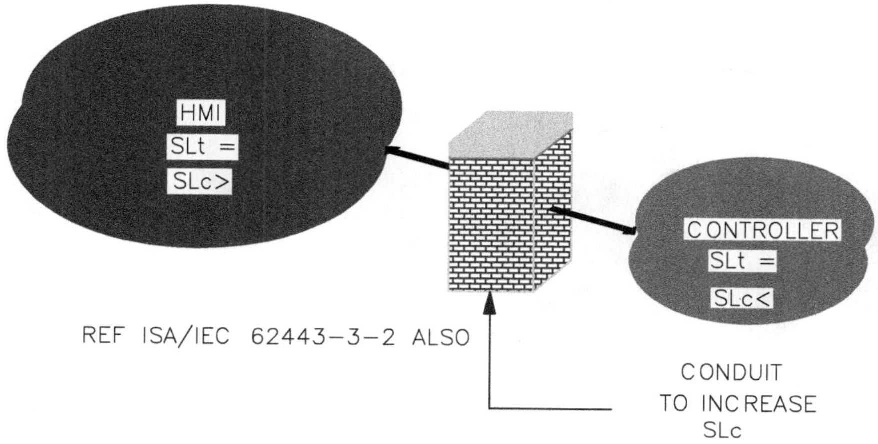

FIGURE XI/4.4-1

Zone conduit concept.

The system security assessment program as per the foregoing details is depicted in Fig. XI/4.3.6-1. It is worth noting that in many places, many parts of IEC/ISA 62443 have been mentioned as standard, while in reality these may be only in draft form. One may refer to http://isa99.isa.org for the latest status.

This partially concludes the discussion on the IEC/ISA standard for cyber security. We now move on to the separate and specific issues of zone and conduit, which are quite important in the context of SIS security, especially when the architecture is a large integrated network.

4.4 Zone and Conduit Concept

Earlier it was shown that there is a difference in the requirements of security in IT and control systems. The security level is not uniform in a network especially when it is complex, large, integrated, and includes an enterprise network. Discussions on zones and conduits are mainly in line with the international standard discussed in the previous clause. To understand the zone conduit concept Fig. XI/4.4-1 may be referred to, in conjunction with Clause 4.4.3. Prior to starting discussions, a short explanation on the relevant terms is presented:

- Zone: Difference in security level at various parts of the network can be addressed by dividing the network into zones, **defined** as logical and physical informational, physical and application assets. To get an idea, Fig. XI/4.4-2 may be referred to. In fact, Fig. XI/4.1.3-1 also has zone division in a similar manner. One point worth noting is that SIS in both cases is segregated from the process control system via firewall protection so that it is in a safe condition and in a safe zone. In fact, this is a zone within a zone acting as a safety layer and an example of defense-in-depth. There can be multiple zones and separate zones.

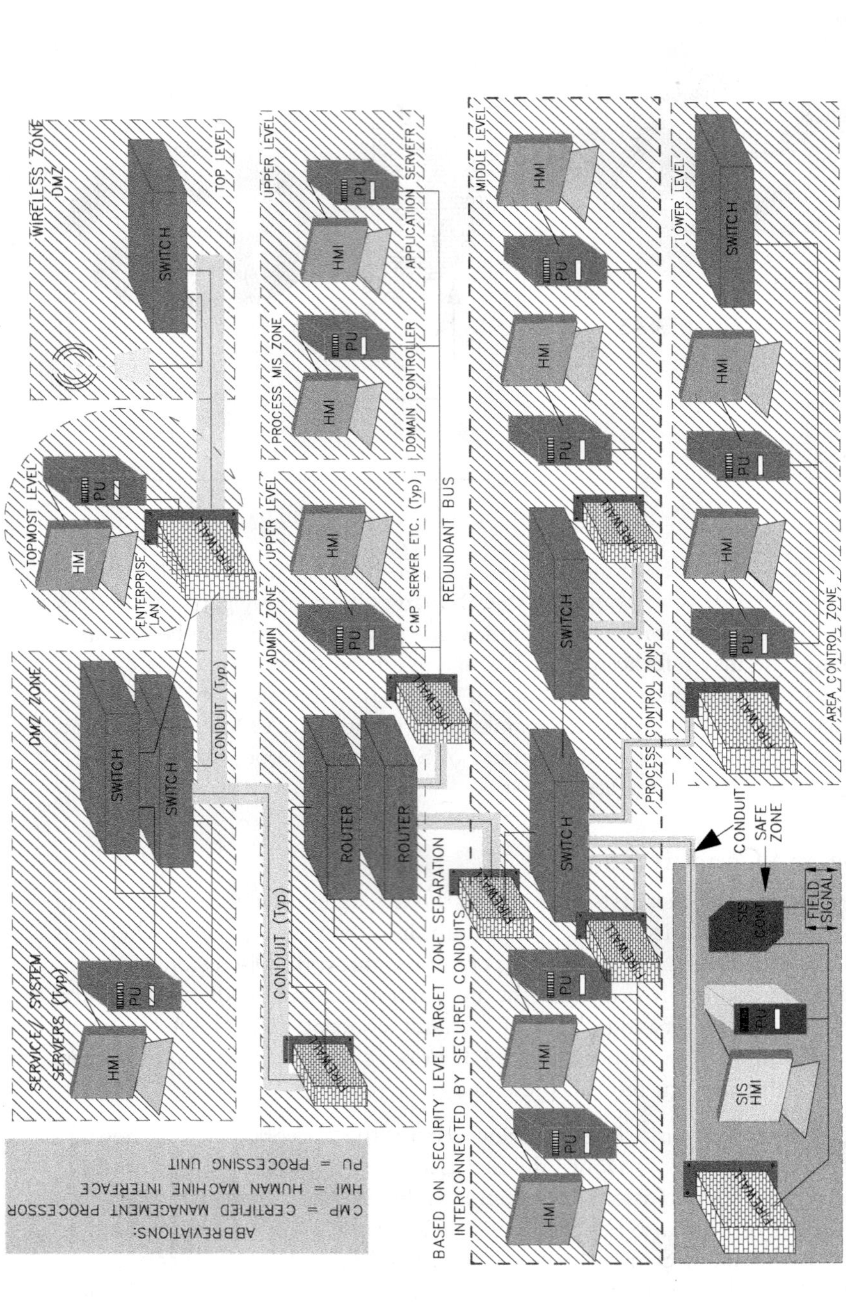

FIGURE XI/4.4-2

Zone conduit-based integrated architecture. *CMP*, certified management processor; *DMZ*, demilitarized zone; *HMI*, human–machine interface; *MIS*, management information system; *PU*, processing unit; *SIS*, safety instrumentation system.

- Conduit: In a network, information flows in and out or within zones. For secured communication, ISA/IEC 62443 series standard defines a special type of security zone communication as a *communication conduit*. Conduit types are:
  - Single—single Ethernet
  - Multiple data carrier, multiple network cables, or direct physical access
  - Trusted: Never crosses the zone boundary—if it does, then it must be secured at both ends
  - Untrusted: Conduits are not at the same security level at the zone end points Conduits can *connect* different zones or various entities within a zone.
- Channel: Channels possess the security properties of conduits and are used as communication media within the conduit. Like a conduit, a channel can be trusted or untrusted. Trusted channels allow *communication* between security zones or may extend the *virtual* zone to the entity outside the physical security zone. Untrusted channels are the *communication* path to another zone having a different security level.
- Security requirement: Before going for zone structure, one first need assess the security goal/requirements, then to place the asset in zone of the zone structure of the network. A group of assets within a security border should have a link to communicate with the outside zone. Such a link may be *physical movement and/or electronic communications*. Electronic communications may be two kinds: remote access and local communication:
  - Remote access: This takes place when the entity is not in close proximity and mainly refers to out-of-zone communication.
  - Local communication: This refers to communication between entities *within* a zone. Assets within the borders are protected to give the *same* security level.
- *Reference model*: Reference model refers to the framework for understanding the relationship among the entities for a particular (network) environment and it is helpful for development of standards and specifications to support the environment. The major constituents are:
  - Governance
  - Risk program
  - Vulnerability management
  - Awareness and training
  - Identity access management
  - Information and knowledge management
  - Document control
  - Monitoring and reporting
  - Internal and external audit

4.4.1 *Discussions on Zone*

There are a few issues that need to be addressed:

- Definition explanation: Zone is an important tool for security program success. Therefore it is important for zone to be correctly defined. For this, a reference model architecture and asset model must be used to develop the security zone and security level for achieving the security goal. When a device is to cater for *different* security levels, then it is better to create a separate zone blending various zones or mapping

Table XI/4.4.1-1 *Zone Characteristics*

Characteristic Issue	Related Issues
Security attributes	Zone: Scope and risk; security: Level, target, strategy and policy; permitted activity and communication
Asset inventory	Hardware: External, computer, and development; Access: Authentication and authorization; spares, monitoring and control; simulation, training and reference manual
Access	Access and control requirements of the zone
Threat and vulnerability	Identification and evaluation of vulnerability for risk assessment and necessary documentation. Suggestion of suitable countermeasure for vulnerabilities in the zone
Consequence: Security breach	As a part of risk assessment the consequences shall be analyzed to suggest necessary countermeasures
Authorized technology	IACS technologies are evolving to meet the business requirements in better and more efficient ways, but there are a number of vulnerabilities. Naturally, proper selection is important to minimize security vulnerability along with an efficient system to meet the challenges
Change management	Formal process to maintain the accuracy of the zone and how to change the security policy to meet the security goal without any compromise

devices for more stringent requirements. A historian, which is needed to connect to both management information system and control system, is an example of such a device. For these kinds of devices, a suitable logical boundary can be created. Only those with privilege to that level and application can access them.

- Zone characteristics: Zone characteristics as per the standard draft are shown in Table XI/4.4.1-1.

4.4.2 *Conduit Details*

The following are important issues related to conduits:

- Definition explanation: Conduits are used for protection of communication assets, that is, applicable for communication processes. Conduits are responsible for physical and logical grouping of communication assets. As the name suggests it protects the security of channels comprising physical connection data, etc. Like a pipe it connects various zones and assets. In IACS, conduits are like network elements such as switches, routers, etc. Conduits can group two dissimilar network technologies. Conduits analyze communication threats and vulnerabilities.
- Characteristics: Conduit characteristics are enumerated in Table XI/4.4.2-1.

4.4.3 *Security Level*

There is some similarity and commonality between safety and security. Safety integrity is represented in terms of SIL. Similarly, security is represented in terms of **security level**. However, security systems have a large set of consequences and circumstances, which lead to a large number of events. Initially, when sufficient data

Table XI/4.4.2-1 *Conduit Characteristics*

Characteristic Issue	Related Issues
Security attributes	Conduit: Scope and risk; security: Level, target, strategy and policy; permitted channels; documentation
Asset inventory	Similar to zone; accurate lists of communication channels
Access	Access to limited sets of entities and access and control requirements
Threat and vulnerability	Identification and evaluation of vulnerability for risk assessment of assets within conduits that fail to meet business requirements; necessary documentation. Suggestion of suitable countermeasure for vulnerabilities in the zone
Consequence: Security breach	As a part of risk assessment the consequences shall be analyzed to suggest necessary countermeasures
Authorized technology	IACS technologies are evolving to meet the business requirements in better and more efficient ways, but there are a number of vulnerabilities to conduits. Naturally, proper selection is important to minimize security vulnerability to conduits along with an efficient system to meet the challenges
Change management	Formal process to maintain the accuracy of the conduit's policy and how to change the security policy to meet the security goal without any compromise
Connected zones	Description in terms of the zone to which it is connected

are not available, security level is *measured* qualitatively like low/medium/high for the purpose of *comparison* within an organization. However, in the long term, it is necessary to adopt a *quantitative* approach using mathematical equations. Basically, security level is a quantitative representation of various security zones that need to be developed with security targets in mind. There are *three* types of security level: security level target (**SL-T**), security level capability (**SL-C**), and security level achieved (**SL-A**).

- SL-T: This defines the *desired* level of security of a particular system—each security zone and/or entire system comprising security zones. This is normally determined only after performing risk assessment of the system and ascertaining a particular level of security.
- SL-C: This is the security level that components/systems can provide when properly implemented. This shows the *capability* of the component/system without the use of additional compensating device(s).
- SL-A: This represents the actual level of security of a particular system. This is *measured* with available data, when the system design is complete, or the system is in place to establish that the security level target is achieved.

4.4.4 Integrated Network With Zones and Conduits

Having some idea of the various relevant terms it is time to apply these to the network normally encountered in IACS. Fig. XI/4.4-1 is a good example to show how zones and conduits are helpful in meeting the same security level target with different security level capabilities.

HMI designed with a standard Windows-based operating system may have inherent higher security level capability than that achievable in a standard controller. Now, if both systems are set to have the same security level target in the network, then additional device(s) may be necessary to bring the controller security level capability to the same available in HMI. This is a very costly proposition; instead conduits may be applied that effectively can increase the security level capability of a controller and bring both to the same security level. In Clause 4.1.3 the necessity for zone divisions was discussed and is shown in Fig. XI/4.1.3-1. Now, similar network conduits are shown in Fig. XI/4.4-2.

In this architecture, starting from the enterprise top level, various levels are shown. Such level differentiations are mainly based on functional use of information. Here it is worth noting that both IT and IACS levels are shown. The topmost level may be connected to the internet (not shown here). As is seen, different zones have been created. It is worth noting that each has different functions, which demand different security levels and also different communication devices such as switches and routers as shown. Top-level wireless zones and service zones have been connected with the help of a *firewall* and separate *conduits*. Admin zone and process control MIS zone have separate security zones with different security *targets*. Here it is worth noting that an application processor (related with historian discussed in Clause 4.4.1) is a device that is connected to both process control system and process MIS with the help of a conduit to be connected to two separate security zones. Also switches and routers are used to maintain such a connection. Area controls are basically an extension of process control. These may have the same or separate security zone and security level targets with different vendors/networks, namely, choke and kill or cementing controls are examples of area controls. So, these are shown separately. SIS controls, for example, SIS, F&G, ESD, always demand separate treatment and have different security zone and security level targets. SIS has been shown separately and connected via a different conduit to cater to different security level targets. It is better to focus on the connections of SIS and BPCS and MIS parts. Also there could be other variations regarding configurations for SIS and BPCS integration, as elaborated later and as shown in Fig. XI/4.4.4-1 [22]. This is to be seen in conjunction with Fig. XI/4.4-2.

- SIS controllers and HMIs in the same bus of BPCS as shown in Fig. XI/4.4.4-1A. In such configurations there is a high probability of DoS attack on account of both network flood or malicious traffic and open bus malware attacks.
- In another configuration as shown in Fig. XI/4.4.4-1B, HMI of SIS may be on the same bus with SIS controllers in a separate bus with peer-to-peer communication. The controller communicates with HMI via an interface module. Here SIS controllers may be susceptible to fewer DoS attacks because the interface between SIS controllers and HMI will have a similar attack as discussed in the first case.

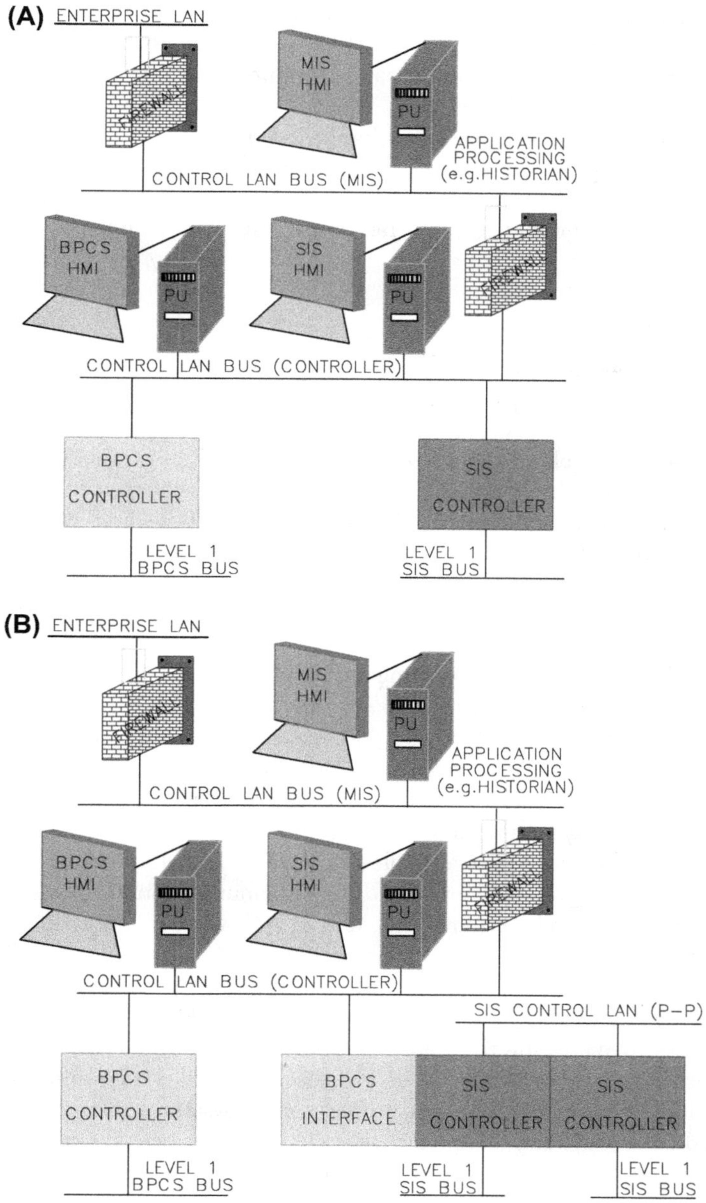

FIGURE XI/4.4.4-1

Basic plant control system (BPCS) and safety instrumentation system (SIS) integration in a common bus. (A) Open bus integration, (B) common bus integration. *HMI*, human–machine interface; *LAN*, local area network; *MIS*, management information system; *PU*, processing unit.

Developed based idea from Z.D. Tudor, Cyber Security Implications of SIS Integration with Control Networks; The LOGIIC SIS Project; ISA automation week; https://logiic.automationfederation.org/public/Shared%20Documents/LOGIIC%20SIS%20AW11%20Final%20PPT.pdf.

- The third configuration is when there is a separate bus as shown in Fig. XI/4.4-2; the chances of cyber attack may be much less, but it is also an integrated system. The same may be achieved without real integration, by a simple RS link with Modbus protocol also for data polling.

  It is worth noting that SIS may exchange information with a process control system but may not use the same data bus to ensure separate security level capability economically. With this, the discussions on zone conduit are concluded. We will now look into other possibilities to conclude the discussions on security issues.

4.5 Discussions on Security

There are a few other means to meet the security requirements of a network. OPC is one such choice. In many systems there are OPC servers and clients for security analysis. Access to an OPC server is restricted to the persons with higher level privilege, whereas OPC client is allowed to others. Some SIS systems also self-police communications access. In one case, Invensys Operations Management (www.iom.invensys.com) collaborated with Byres Security (www.tofinosecurity.com), a cyber security firm, to add an OPC firewall to its Tricon Communications Modules (TCM). The firewall enabled a layer of defense-in-depth that lets systems integrators enjoy the flexibility and integration benefit of OPC Classic without worrying about security systems that have in the past been associated with distributed component object model (DCOM)-based systems. According to Byres Security "A reliable OPC firewall means that in addition to blocking hackers and viruses from accessing the safety system, integrators can deliver dynamic port management and built-in traffic-rate controls to prevent many basic network problems from spreading throughout a plant." Trinity Systems, UK developed a remote viewer that takes advantage of the communications security features of the Triconex TCM and Triconex Firewall. The viewer allows the end user to have a simple window into the SIS from the business or primary control network. The Firewall and the Communication Module's on-board User Access Security Model ensures that it is a *read-only* window that can never *impact* the safety functionality. This combination of OPC-based accessibility with true defense-in-depth security provides cost-effective and secure access. Joe Scalia, portfolio architect, Invensys Operations Management, said "An OPC firewall mitigates those risks by managing the traffic to and from the communications module, providing further assurance that a cyber incursion will not compromise integrated communications between the safety and critical control systems and supervisory HMI or distributed control systems." Implementing the HMI portions of a safety system is critical to securing communications between the SIS and the DCS. Communications integrity, including cyber security, must be ensured so that safety-based actions such as reads from the HMI to the safety system can be executed securely and without interruption. The new module from MTL Instruments and Byres Security is said to provide a safe and secure means of locating what is on control system networks. The new module from Tofino listens for traffic and then uses special characterization techniques to determine the types of control devices on the network. When it discovers a new device, it prompts the system administrator to either accept its deductions and insert the new device into the network inventory diagram, or flag the device as a potential intruder. This way, an up-to-the-minute network map is always available to the control engineer. The module also guides the user while creating appropriate firewall rules to allow or block messages, based on what it has learned about the network traffic. Technical complexities such as IP addressing and TCP/UDP port numbers are managed behind the scenes, making firewall configuration easier for a controls professional.

As an example, Yokogawa Electric Corporation has the ProSafe-RS safety instrumented system, which has obtained the ISASecure Embedded Device Security Assurance (EDSA) certification. The ISASecure program has been developed by the ISA Security Compliance Institute with the goal of accelerating the industry-wide improvement of cyber security for IACS. It achieves this goal by offering a common industry-recognized set of device and process requirements that drive device security, simplifying procurement for asset owners and device assurance for equipment vendors. The ISASecure EDSA certification has three elements: communication robustness testing, functional security assessment, and software development security assessment, and is based on the IEC 62443-4 standard [13–15].

With this, the discussions on cyber security are concluded with a hope that an even better development in this area would make SIS more secure and easier to handle. For further details, interested readers may consult reference books and watch out for the release of the international standard ISA/IEC 62443 series.

LIST OF ABBREVIATIONS

ARP	Address resolution protocol
BPCS	Basic plant control system
CCF	Common cause failure
CCPS	Center for chemical process safety
COTS	Commercial off the shelf
DCOM	Distributed component object model
DCS	Distributed control system
DMZ	Demilitarized zone
DoS	Denial of service
ESDA	Embedded device security assurance
ESD	Emergency shutdown system
F&G	Fire and gas
FTC	Fault tolerant control
FTE	Fault tolerant Ethernet
FTP	File transfer protocol
FTU	Fault tolerant unit

Continued

HART	Highway addressable remote transducer
HMI	Human—machine interface
IACS	Industrial automation and control systems
IEC	International Electrotechnical Commission
I/O	Input/output
IP	Internet protocol
IPL	Independent protection layer
IT	Information technology
LOPA	Layer of protection analysis
MDT	Mean downtime
MIS	Management information system
MTBF	Mean time between failure
MTTR	Mean time to repair
NMR	*N*-modular redundancy
OPC	Open platform communications
OSI	Open systems interconnection
PFD	Probability of failure on demand
PLC	Programmable logic controller
SDLA	Security development life cycle assessment
RSTP	Rapid spanning tree protocol
SIF	Safety instrument functions
SIL	Safety integrity level
SIS	Safety instrumentation system/supervisory information system (in case of DCS) [16]
SQL	Structured query language
STP	Spanning tree protocol
TCM	Tricon communications modules
TCP	Transmission control protocol
TMR	Triple modular redundancy
VPN	Virtual private network

References

[1] R. Prew, Why the Architecture of Safety Systems Doesn't Matter, Document ID: 3BNP100416, ABB, UK, December 2008.

[2] S. Basu, A.K. Debnath, Power Plant Instrumentation and Control Handbook, Elsevier, November 2014. http://store.elsevier.com/Power-Plant-Instrumentation-and-Control-Handbook/Swapan-Basu/isbn-9780128011737/.

[3] M. Blanke, M. Staroswiecki, N.E. Wu, Concepts and methods in fault-tolerant control, in: Tutorial at American Control Conference, June 2001. http://www.iau.dtu.dk/secretary/pdf/ACCmb30a4.pdf.

[4] B.G. Liptak, Instrument Engineers' Handbook, Process Measurement and Analysis; Chapter 1.10: Redundant and Voting System, vol. 1.

[5] E. Dubrova, Fundamentals of dependability. Chapter 2. Fault-Tolerant Design, Springer Science+Business Media, New York, 2013.

[6] Designing a Control System for High Availability; A. Pietrzyk, B. Root; Rockwell Automation; P. Gruhn; ICS Triplex; https://www.isa.org/FileStore/Intech/.../High%20Availability.doc; and https://www.controlglobal.com/assets/14WPpdf/140324-ISA-ControlSystemsHighAvailability.pdf.

[7] D.A. Rennels, Article98; Fault- Tolerant computing; http://web.cs.ucla.edu/~rennels/article98.pdf.

[8] Fault-Tolerant Scalable DCS; 2300 family; RTP Corporation; Technical literature; www.rtpcorp.com.

[9] Dependability of Distributed Control System Fault Tolerant Units; J.R. Pimentel, Kettering University; M. Salazar, Universidad de los Andes; http://paws.kettering.edu//~jpimente/flexcan/pmentel_iecon2.pdf.

[10] Simatic PCS7; Siemens; Technical Brochure, April 2008; www.siemens.com/simatic-pcs7.

[11] Operator Action Within a Safety Instrumented Function; L.T. Suttinger, Westinghouse Savannah River Company; C.L. Sossman, Westinghouse Safety Management Solutions.

[12] A. Franks, Lines of Defence/Layers of Protection Analysis in the COMAH Context.

[13] Yokogawa Obtains ISA Secure® EDSA Certification for ProSafe®-RS Safety Instrumented System. Technical Write Up; Yokogawa Electric Corporation, January 2014.

[14] Cyber security: firewall device creates its own rules; Control Engineering; Control Engineering staff; July 05, 2008; http://www.controleng.com/index.php?id=483&tx_ttnews[tt_news]=13757&cHash=062910.

[15] Protecting Safety Instrumented Systems from Cyber Attack Using a Zone and Conduit Design; E. Byres, Byres Security; K. Staggs, Honeywell Process Solutions; Process Control Systems Industry Conference; https://ics-cert.us-cert.gov/sites/default/files/pcsf-arc/protecting_safety_instrumented_systems-byres.pdf.

[16] Combined BPCS and SIS Demonstration of Independence; The 61508 Association; www.61508.org; Version 2, December 2015; http://61508.org/wp-content/uploads/2015/11/2015.12.11-Guide-Combined-BPCS-SIS-V21.pdf.

[17] The Rocky Relationship between Safety and Security, Control Engineering, November 2012.

[18] Cyber Security Working Group; The 61508 Association; http://61508.org/wp-content/uploads/2015/11/61508-cyber-security-initial-concepts-V0.5-110612-The-original11.pdf; Internet document.

[19] Recommended Practice: Improving Industrial Control Systems Cyber Security with Defense-in-depth Strategies, Homeland Security; Control System Security Program; National Cyber Security Division, October 2009. http://www.us-cert.gov/control_systems/practices/documents/Defense_in_Depth_Oct09.pdf.

[20] Common Cyber Security Vulnerabilities in Industrial Control Systems, Homeland Security; Control System Security Program; National Cyber Security Division, May 2011.

[21] IEC Cyber Security Activity, Cyber Security in Industrial Automation – IEC Activities, September 2015. https://www.iec.co.il/EN/Documents/IECs_presentation.ppt.

[22] Z.D. Tudor, Cyber Security Implications of SIS Integration with Control Networks; The LOGIIC SIS Project; ISA automation week; https://logiic.automationfederation.org/public/Shared%20Documents/LOGIIC%20SIS%20AW11%20Final%20PPT.pdf.

Further Reading

A simple definition of cyber security; M. Barzilay, ISACA; http://www.isaca.org/Knowledge-Center/Blog/Lists/Posts/Post.aspx?ID=296.

M. Blanke, R. Izadi-Zamanabadi, S.A. Bøgh, C.P. Lunau, Fault-tolerant control systems — a holistic view, Control Engineering Practice 5 (1997).

Black channel constraint; Profinet.com/technology; Internet document.

Fault tolerance and resilience: meanings, measures and assessment, in: L. Strigini, K. Wolter, et al. (Eds.), Resilience Assessment and Evaluation of Computing System, Springer-Verlag, Berlin, Heidelberg, 2012.

Functional Safety Standards: achieving an effective compliance strategy; Honeywell Process Solutions; White Paper.

P. Gruhn, Things to consider when selecting a safety system, in: ISA Safety Division Symposium, 2009.

Impact of Guidelines for Independent Protection Layers and Initiating Events; Developed by: CCPS - Center for Chemical Process Safety.

Integrated or separate. The 61508 Association; www.61508.org; and http://61508.org/wpcontent/uploads/2015/11/2015.12.11.Intro-Combined-BPCS-SIS-V22.pdf.

ISA Cyber Security Brochure. Safe and secure: multiple challenges one solution; https://www.isa.org/cybersecuritybrochure/.

Is It Really an Independent Protection Layer? A. M. (Art) Dowell III, PE, 2011.

H.H. Kollmeier, Reconfiguration for Fault Tolerance and Performance Analysis (Technical reports), Department of Computer & Information Science; University of Pennsylvania, November 1987.

C.A. Lassen, Layer of Protection Analysis (LOPA) for Determination of Safety Integrity Level (SIL), Aker Solutions; NTNU, June 2008. http://frigg.ivt.ntnu.no/ross/reports/stud/ lassen.pdf.

Providing independent layers of protection with integrated safety systems; L.M. Duran, Ron Johnson; ABB Technical write up; https://library.e.abb.com/public/f3038ff4f251587585257658006fc0e7/1730%20Layers%20of%20protection%20VP.pdf.

M. Rentschler, P. Laukemann, Towards a Reliable Parallel Redundant WLAN Black Channel; http://wwwlehre.dhbw-stuttgart.de/wrentschler/Publications/WFCS_2012_Paper_8_Rentschler_WLAN_PRP_Black_Channel.pdf.

V.M. Sivagami, K.S. Easwara Kumar, Survey on fault tolerance techniques in cloud computing environment, International Journal of Scientific Engineering and Applied Science 1 (9) (December 2015). http://ijseas.com/volume1/v1i9/ijseas20150952.pdf.

The ABCs of Spanning Tree Protocol: Contemporary Controls, February 2006.

(NT technical report), Validation of Safety-related Wireless Machine Control Systems, Nordic Invitation Center-NORDEN, 2007. http://www.nordicinnovation.org/Global/Publications/Reports/2007/Validation%20of%20safety-related%20wireless%20machine%20control%20systems.pdf.

White Paper, Ethernet Fault Tolerance and Redundancy, Emerson Process Management, March 2007.

CHAPTER

XII Discussions on the Industrial Application of SIS

From earlier discussions we can gather knowledge about various methods for hazard analysis systems, and implication of these methods from international standards. Also details about safety instrumentation system (SIS) and safety integrity level (SIL) have been covered. From previous discussions it is quite clear that the essence of a safety system can be described by Ref. [2]:

- A business that operates any form of hazardous process needs safety systems.
- Safety systems do not work without good management.

Though SIS is only one of the parts of overall risk reduction measures that facilities need to deploy to combat the hazards associated with the facility, it is the most important issue at the current time. For successful implementation of SIS and other safety systems, the project depends not only on instrumentation and control engineering discipline but also on the support and knowledge of other disciplines. However, company culture and full commitment from company management structures are also very important factors for proper implementation and success of SIS and other safety systems. There should be a well-defined safety management system and team within the company. This safety management system must enjoy full support from corporate management. Without proper support structures and understanding by all involved in safety management, SIS as well as safety systems as a whole will not be in a position to deliver good results and the levels of safety that are expected of the systems. Keeping in view the current international standard IEC/ISA, the major components of safety management with special reference to SIS will include:

- Systematic methods to identify and record all hazards and risks present in the facility
- Reduction of all unacceptable risks to an acceptably low level (ALARP—as low as reasonably practicable) by well-recognized and controllable means sustainable in the life cycle of the plant (see IEC standard)
- Monitoring and reviewing the system to look at the implementation and performance of SIS
- Participation of personnel from all relevant departments who should be well aware of their individual responsibilities

- Response to local and international laws and regulations for adequate safeguards against people, property, and the environment
- Maintenance of a risk register and a safety case report
- Development of specification design for a safety system as a control measure to combat hazards
- Validation and verification that safety measures are in place and are being maintained at all times throughout the life cycle.

With these in mind it is highly beneficial to see how SIS, etc. are applied to various plants. Application of SIS in various plants should be studied to complete the circle of discussions about SIS. However, before this, a short discussion on recapitulation of the entire system is presented to make the discussion meaningful.

1.0 RECAPITULATION OF SIS

Initially, all plants are provided with a basic control system and protection interlock for day-to-day operation of the plant. However, this may not be sufficient to combat various risks persisting in the plant. Therefore plants handling dangerous materials and/or energy, that is, any hazardous process, need safety systems. To satisfy the need for safety requirement it is first necessary to identify not only the hazards but also their frequency of occurrence and consequences. These are the determining factors for risk ranking. During the initial phase, risks may be at an unacceptable limit. One then needs to consider different independent protection layers (IPLs) so that the risk levels could be brought to a level acceptable for the system. This is ALARP and is shown in Fig. XII/1.0-1.

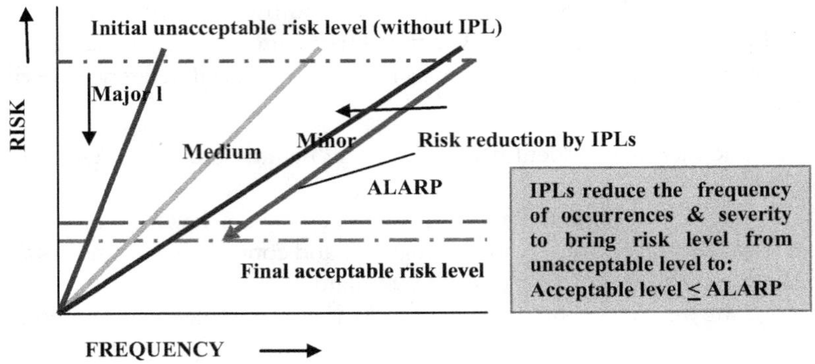

FIGURE XII/1.0-1

Risk reduction by independent protection layer (IPL). *ALARP*, as low as reasonably practicable.

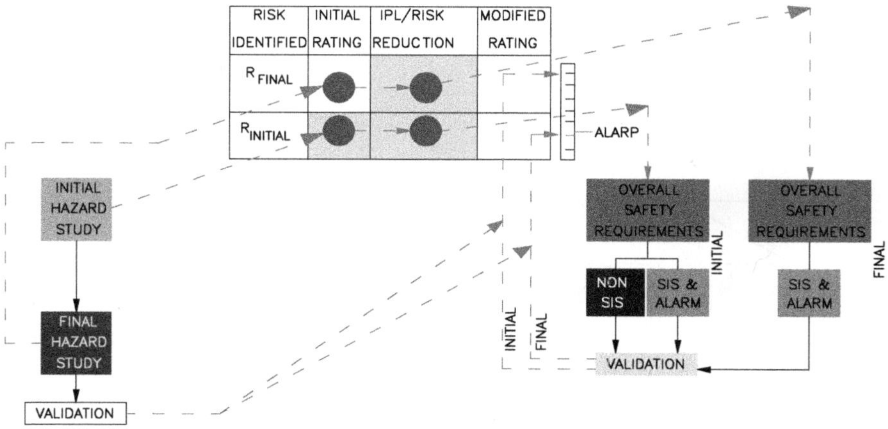

FIGURE XII/1.0-2

Overall safety instrumentation system (SIS) implementation. *ALARP*, as low as reasonably practicable; *IPL*, independent protection layer.

IPLs comprising SIS and non-SIS systems reduce the unacceptable risk levels to acceptable risk levels, which should be equal to or below the ALARP level.

Various probable IPLs are shown in Fig. V/4.0-1. Each IPL has a risk reduction factor (RRF), which is an effective index for safety integrity. Higher SILs have higher RRFs. Our concentration will be on SIS. The overall process for SIS determination is shown in Fig. XII/1.0-2.

With initial assessment, all risks are listed and suitable IPLs are deployed. After this the entire system is reassessed. Because we are interested in SIS, in the final assessment only SIS is shown. After each assessment the IPLs and SIS are validated. After final assessment of overall safety requirements it is ensured that the risk level is at an acceptable level, as shown in the figure. This shall be a level equal to or below the ALARP level. In principle, what has been discussed here is more or less the same as discussed earlier, the only difference is that here the basic implementation process is shown. Similarly, SIS in the design phase is shown in Fig. XII/1.0-3. The reason for showing the figure is to recapitulate further details about SIS design.

With this the recapitulation of SIS is concluded. We will now look at the application of SIS in various plants to see how in each case the requirements change. Boilers, heaters, and furnaces are considered critical equipment in most operating facilities, posing challenges to safe and reliable operation. Similarly, a turbine as rotating equipment is commonly used in many hazardous plants. In fossil fuel plants both of these are used as main equipment, so it will be better to start the discussion on the application of SIS in industrial plants with fossil fuel power plants. Also the use of SIS in fossil fuel plants is quite a recent development.

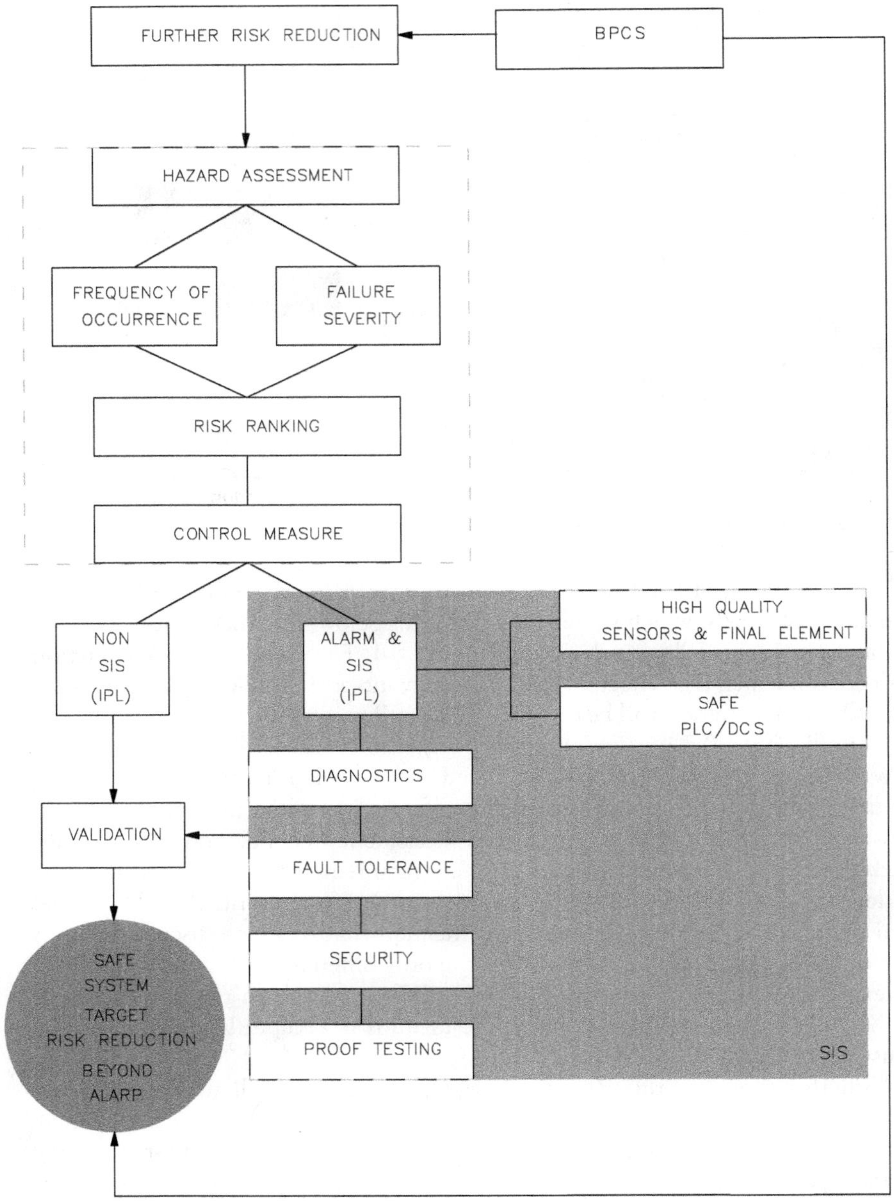

FIGURE XII/1.0-3

Design steps for a safe system. *ALARP*, as low as reasonably practicable; *BPCS*, basic plant control system; *DCS*, distributed control system; *IPL*, independent protection layer; *PLC*, programmable logic controller; *SIS*, safety instrumentation system.

2.0 FOSSIL FUEL POWER PLANTS

The rise in scarcity of and demand for energy all over the world, coupled with the requirements to reduce emission levels, gave birth to newer technologies such as supercritical/ultrasupercritical plants and integrated gasification and combined cycle (IGCC) plants in the area of fossil fuel power plants. People are increasingly interested in obtaining plants with higher and higher efficiencies. This has resulted in the so-called "megawatt race" among the utility companies. These newer technologies are not benign; they also give rise to a number of complexities. It is not that there were no safety considerations in power plant designs. Safety issues were always an embedded part of system design based mainly on mutual understanding between owners and suppliers founded on a number of codes and standards such as NFPA, ASME, TUV, etc. However, in cases of newer technologies offering higher efficiency, embedded safety, and protection may not be sufficient on account of complexities involved in the operation of these newer plants. Naturally separate safety systems are called. SILs are not new in plants like chemical/petrochemical plants. Now in power plant operations are also adapting SILs in their plants. Another helping factor could be that, in most cases rotating equipment like the turbine and steam generator are in use also in hazardous plants where it is necessary to follow safety standards like those in IEC 61508/61511. This would also help in adding safety standards in modern power plants. As such, in power plants there are many hazardous chemicals such as NO_x, SO_x, fly ash, coal dust, CO, CO_2, and Si to name a few and these could cause health and environmental hazards. In the case of a gas turbine there will be the hazard of explosion because of natural gas handling. Therefore adapting safety standards in power plants will be helpful and is recommended especially for large size plants. In this clause discussions are presented in three parts: SIS for both steam generators and turbine generators in normal fossil fuel power plants and gas turbines in combined cycle/cogeneration plants.

2.1 SIS in Boiler Controls

ISA BMS technical report (ISA TR 84.00.05) focuses on the use of SIS in burner management system (BMS) applications. This is informative and not mandatory. However, other relevant standards also support this.

2.1.1 Various Standards and SIS

Standard API 556 (2011) is meant for instrumentation, control, and protective systems for fired heaters and steam generators. It has been found that it brings the concepts from the ISA BMS technical report. It also covers instrumentation and control (I&C) for gas-fired heaters and steam generators in refinery, chemical, and petrochemical plants. However, it includes neither heat recovery steam generators nor any guidelines for SIL. Other major standards relevant to boiler controls are the latest edition (2015) of NFPA with different application areas:

- NFPA 85—Boiler and Combustion Systems Hazards Code
- NFPA 86—Standards for Ovens and Furnaces
- NFPA 87—Recommended Practice for Fluid Heaters

Out of many requirements of **NFPA 85 (2015)**, a few are as follows (NFPA clause ref indicated in the bracket in applicable cases):

- "The burner management system (BMS) shall be provided with independent logic, independent logic solving hardware, independent input/output systems, and independent power supplies and shall be functionally and physically[1] separate device from other logic systems" (Ref. Clause 4.11.8).
- BMS designers shall evaluate failure modes (Clause 4.11.6).
- "Diagnostics shall be included in the BMS design (Clause 4.11.7) to monitor processor logic function."

These are listed here to indicate that according to the foregoing there is some scope left to include SIS in BMS. NFPA 87 allows use of five types of logic solver [3]:

- Hardwired system
- Listed safety relays
- Listed programmable logic controllers (PLCs)—none exist in the marketplace at this time
- Nonlisted PLCs
- Safety PLC implemented per ISA S84

There are a number of requirements specified for nonlisted PLCs. According to A8.3 (P37) of NFPA 87 (2015) *Controls that meet the performance-based requirements of standards such as ANSI/ISA 84.00.01 Application of Safety Instrumented Systems for the Process Industries, can be considered equivalent. The determination of equivalency involves complete conformance to the safety lifecycle including risk analysis, safety integrity level selection, and safety integrity level verification, which should be submitted to the authority having jurisdiction.*

All these are discussed here to show that even various other standards also recognize the implementation of IEC 61508/61511* (especially) in BMS applications[2]. Typically, BMS shall include at least one SIL2 SIF. Therefore the ***logic solver (PLC/DCS—distributed control system) should be selected accordingly***.

2.1.2 *Major Hazards in Boilers*

Major hazards associated with boilers are listed in Table XII/2.1.2-1.

2.1.3 *BMS and SIS*

BMSs have traditionally been designed and implemented according to prescriptive and other good engineering practices. These prescriptive practices do not require evaluation of the risk reduction capability of the BMS, whereas IEC 61511 (ISA 84.00.01) is a performance-based standard. NFPA standards reference ANSI/ISA

---

[1]Note that in view of discussions in previous chapters about physical separation of SIS from BPCS, here physical separation is not dealt with, because it is integrated in many systems meaning that there may not be any physical separation in the *true* sense. Here it means that a separate logical system may be integrated into a network.

[2]As already discussed in Chapter VI that ISA 84.00.01 is basically based on IEC 61511.

Table XII/2.1.2-1 Major Hazards in Boilers

Hazard	Probable Cause	Consequence
High/low furnace pressure	Loss of draft control	High pressure: Loss of containment, fire and gas escapes causing damage to property, seal, and personnel, explosion Low pressure: Furnace implosion, damage to seal, excess draft, combustion
Unburnt coal, flammable gas buildup	Loss of air, higher fuel/air ratio, ignition failure during startup, loss of flame, starting without purging (prohibited)/purge failure	Chances of later explosion, injury to people
Soot buildup	Poor combustion, nonuniform heat transfer	Soot fire, tube failure
NO_x/SO_x	High temperature at burner for thermal NO_x Higher content of sulfur in fuel or flue gas desulfurizing plant failure	Environmental impact, acid rain, pollution
High dust	Failure of electrostatic precipitator/dust control equipment	Dust and smoke pollution
Very high pressure steam	Loss of combustion control, excess firing, sudden loss of electrical load	Rupture, injury because of high pressure steam
High steam temperature	Failure of steam temperature control	Damage to tube, tube leakage
Low feed water flow or low drum level (drum boiler)	Loss of feed water flow control, tube leakage	Burning of circulating tube, boiler damage caused by starvation
High feed water flow or high drum level (drum boiler)	Sudden loss of load or feed water control failure	Carry over to turbine, turbine overspeed, thermal shock
Thermally induced stress	Rapid variations in temperature in pressure parts, uneven heating	Tube leakage, material failure

84.00.01-2004 as a performance-based standard that can be followed to determine the hazard and the appropriate risk mitigation strategy. IEC 61511 (ISA 84.00.00-2004) safety life cycle addresses SISs used to prevent unacceptable hazardous events. The life cycle is supported by a management system that focuses on reducing the potential for SIS failure through effective SIS design and management. Therefore BMS developed based on this will offer these additional benefits. Technical report ISA-TR84.00.05 provides guidance for assessment and implementation *safety logic* in BMS. It also helps to *identify* the deficiencies in its integrity and/or reliability, resulting in the need to upgrade it to achieve the required probability of failure on demand (PFD_{avg}) and spurious trip rate [6]. Burners, furnaces, and boilers are very critical and complex systems. So, most of the major manufacturers as well as end users (*by specifying either SIS BMS or conformance to API 556/FM 7605, which in turn calls for SIS BMS in line with IEC 61511/61508*) that opt for SIS BMS to comply with standards (IEC/NFPA), or to meet certain insurance requirements, will have to **classify** BMS as SIS [5]. As per IEC 61511 definition, it would need certification by a third party agency and this will ensure *higher safety* for BMS. For SIS BMS, all associated sensors, a logic solver, and a final control element shall also be designed according to IEC/61508/61511. In SIS PLCs all critical BMS functions shall be managed via IEC 61508 compliant components to ensure that the relevant safety metrics are met. With this it is possible to achieve up to SIL3 integrity. Also it is not that these are unavailable in the market. PLCs from Siemens/Rockwell, Emerson etc., are available for the same purposes. All that is required is for the end user to specify SIS BMS. Major advantages shall include but not be limited to the following [4]:

- Reduced complexity
- Improved operations and maintenance
- Increased safety, reliability and availability

Already readers are aware of requirements for IEC 61511, so it is better to examine how API 556 conforms to IEC 61511 and how all these standards are consistent with the requirements. API RP puts forward the following suggestions/recommendations for the need of [7]:

- SIL assignment
- No nuisance trip
- Diagnostics and online testing
- Redundancies
- Layer of protection analysis
- Reducing PFD
- Common mode assessment
- O&M procedure

Table XII/2.1.3-1 elaborates the compliance requirements for various standards. For further details [1] may be referred to.

Table XII/2.1.3-1 Typical Major Compliance Requirements for Various Standards

Conditions NFPA 85	NFPA 86	API 556	ISA TR 84.05
Fuel block valve closed	YES	YES	YES
No flame	YES	YES	YES
Air flow in prepurge condition	YES	YES	YES
Prepurge timer complete	YES	YES	YES
Fuel pressure adequate		YES	YES
Air proved at low fire rate		YES	YES
Postpurge air flow OK			
Purge complete			
Pilot flame detected in time	YES		YES
Main fuel set at minimum position	YES		YES
Main flame detected in time	YES		YES
Loss of flame	YES	YES	YES
Loss of combustion air	YES	YES	YES
Furnace pressure high/low	YES	YES	
High/low fuel pressure	YES		YES
Loss of atomizing medium	YES		YES
Power failure/loss of actuating energy	YES	YES	YES
Emergency shutdown	YES	YES	

- SIS and SIL for BMS: A master fuel trip required by design codes demands multiple actions. The verification results shall confirm that the required risk reduction is achieved. However, the validation can be compromised when an SIF is not defined properly and its functional requirements are poorly specified or when all actions for total shutdown are included in the same functional requirements of the same SIF [8]. From discussions in previous chapters it is clear that the safety life cycle model not only helps with necessary ways and means to avoid systematic failures, but also helps to ensure the required integrity level to prevent random failures. The safety standards (IEC 61508/61511) required to identify a set of parameters and factors for PFD_{avg} calculations are:
  - Diagnostic (also in NFPA 85)
  - Requirements for O&M and testing
  - Requirements for reliability

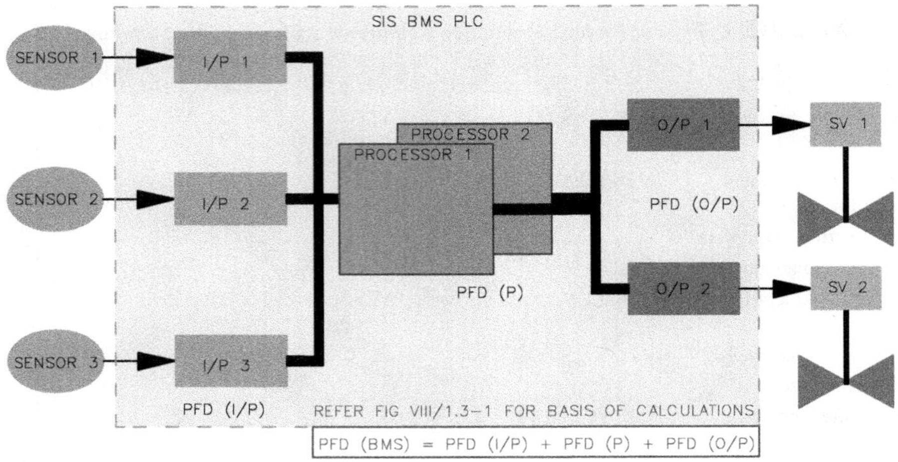

FIGURE XII/2.1.3-1

Burner management system safety integrity level (BMS SIL) calculation. *BMS*, burner management system; *I/P*, input; *O/P*, output; *PFD*, probability of failure on demand; *PLC*, programmable logic controller; *SIS*, safety instrumentation system; *SV*, solenoid valve / shutdown valve.

For SIL calculation it is better to recall the discussions presented in Clause 1.3 of Chapter VIII. Fig. XII/2.1.3-1 shows the calculation principle of the BMS SIL level. With reference to Figs. VIII/1.3-1 and XII/2.1.3-1, if the input stage has PFD_{avg} (input) $= 1.5 \times 10^{-5}$, and the same for the processor and output are 2×10^{-4} and 1.5×10^{-5}, respectively, then the PFD_{avg} of BMS will be $(1.5 \times 10^{-5} + 2 \times 10^{-4} + 1.5 \times 10^{-5})$ 2.3×10^{-4}, meaning SIL3. There are a number of vendors who can offer SIL BMS, for example, Triconex, Invensys, and Siemens. With this the discussion on SIS in boiler controls is concluded. For further details refer to Ref. [1].

2.2 Steam Turbine Controls

When discussing steam turbines one needs to keep in mind that these are applicable for nuclear plants also. Refer to Fig. XII/3.0-1 to see that the turbine generator part in both cases is the same. In nuclear applications, naturally reliability and availability are of immense importance. Similarly, in fossil fuel power stations, if turbine safety is compromised, then the high-cost turbine cannot survive and would incur huge financial loss. There following points need to be considered for turbine control and safety systems:

- Highly reliable system with high availability
- No single point of failure to interrupt continuous plant operation
- Online diagnostics features
- Online repair as applicable
- Proper easy method of maintenance

- Flexibility for modification, upgrade, and expansion
- Well-designed field interface
- Interface with other control systems
- Easily implementable control strategies and operational procedure
- Automatic startup and warmup until synchronization
- Automatic testing facility for critical items
- Well-interfaced/dedicated turbine protection
- Detection and safe shutdown of the system when unsafe plant conditions arise [9]
- Electrohydraulic governing system with hydraulic governing *backup*
- Automatic *switching* between
  —Governing system
  —Full/partial arc system
- Speed droop controls/load controls
- Triple modular redundancy (TMR) logic for critical controls measurements
- Supports IEC programming languages and data type
- Well-designed advanced human—machine interface (HMI) and software support

In this connection it is worth noting that in many large-sized turbines, like those from (say) Siemens-KWU, it has been found that even though there is some integration of turbine control with protection, there is a separate entity for protection systems. Even for a few selected important protections like overspeed and condenser vacuum they also use mechanical trips as backup. Major equipment/devices used for protection (for some manufacturers) have facilities for automatic testing (automatic turbine testing). For important trips TMR logics are also deployed. *These experiences are elaborated to indicate the huge importance of protection systems for turbines and to emphasize that use of SIS is beneficial for the end user. For further details on turbine controls [1] may be referred to.*

With this in mind one needs to see how general safety procedures discussed in IEC 61511/61508 could be applied to turbine safety. Traditionally, turbine safety systems are embedded with turbine control systems. However, there is a trend to separate them and to apply general safety systems for application of IEC 61508/61511. Readers' attention is drawn to Clauses 7.4.2.3—7.2.4.5 of IEC 61508-2:2010 and Clause 11.2.10 of IEC 61511 (in this connection Clauses 4.2 and 5.0 of Chapter VI may be referred to). In these clauses, integration of safety functions with nonsafety functions has been dealt with. To look deeper into the system it is necessary to understand not only electronics/programmable systems but one must also understand the process. It is important to understand "how these are used" and "how they fail and what the impact of such failure is." Also to meet the requirements of safety standards it is necessary to validate the system to different degrees based on the turbine control system in question. As discussed earlier, failure modes, effects, and diagnostic analysis could be deployed to verify that the control logic does not affect the performance of SIS pertinent to turbine safety. A few major issues in this regard are as follows:

- Added modules do not affect SIS integrity level.
- Central processing unit power remains unaffected.

- Operating system prioritizes safety over control logic.
- Sufficient security protection at HMI level is to be shown.

When turbine control modules are integrated with SIS, as per IEC 61511, all control modules should have a certified SIL level the same as that required for SIS function. Such certification is meant for software and firmware also.

API 670 is a good guideline standard not only for turbines but also for many other rotating machines. So, readers are advised to familiarize themselves with these devices. API 670 edition 5 provides good engineering practices and guidelines for integration architectures. There have been a number of changes in the 5th edition with respect to previous revisions, such as incorporation of redundant power supplies, TMR logic, redundant communications, etc.

2.3 Gas Turbine

In view of frequent accidents, industries are opting for and preferring to use functional safety standards in their plants, keeping in mind the long-term benefit that could balance the additional cost involvement. Gas turbines are no exception to this. The major functionality of SIS is to ensure that all associated equipment are in a fail safe condition based on preselected inputs that are continuously interrogated by the logic solver [10]. As discussed earlier, SIS ensures higher reliability and availability of safety systems to ensure safety to personnel, property, and the environment in an emergency situation. In the case of gas turbines the main objective is to ensure optimum efficiency with higher availability. The major benefits expected shall include but not be limited to the following:

- Scalable SIL in line with requirements of IEC 61511
- Optimization of availability
- Necessary redundancy to ensure no single point failure
- Flexible allocation of safety channels and controls [10]
- Complete life cycle validation and verification
- Common operations, maintenance, and troubleshooting tools [10]
- Multivendor options for owner with marginal increase in cost

From the GE fact sheet: "The Mark VIe Turbine Control System with SIL capable protection meets the requirements of controller separation, but also provides additional benefits not offered by other stand-alone systems." The major tasks involved are:

- Gas turbine works at optimum efficiency.
- Control and protection systems need to offer high availability.
- Protection strategies are needed against overtemperature and overspeed, etc.

In an integrated control and protection system any fault or deterioration in the system could not only result in a reduction of plant efficiency but also failure of a protection function, and therefore result in damage to the engine, which is safety critical to the plant. Major SIF involved in a gas turbine protection system shall include but not be limited to the following:

- Overspeed
- Overtemperature
- Loss of cooling
- Emergency stop (local/remote)
- Flame detection
- Exhaust purge timer
- Excess firing fuel flow
- Absolute vibration
- Fire detection
- Hazardous gas detection
- False start drain temperature

So, specifying SIL as per IEC 62061 (manufacturing industry) under the umbrella of IEC 61508 is a good way for achieving this.

3.0 NUCLEAR PLANT SAFETY DISCUSSIONS

It is meaningless to continue describing advantages and promises from digital instrumentation. Even though I&C systems are the nerve system of a nuclear power plant, nuclear plants are comparatively more conservative than other plants in implementing digital instrumentation rather than programmable systems in I&C safety applications. One of the major issues related to this could be software logic errors. Software is used for physical representation of the actual physical condition, so, if there is any deviation from this, then there will not be logical error. This is applicable to all cases. However, as safety in nuclear plants is of prime importance and a justified priority, naturally, special care needs to be taken. Another important issue is that the discussions made here are mainly for nuclear plants with fission technology as shown in Fig. XII/3.0-1. On account of requirements for very high temperature fusion, the technology is mainly in the research stage [see the International Thermonuclear Experimental Reactor (ITER) project discussed later].

3.0.1 *Generic Approach for Safety*

Safety guide NS-G-1.3, International Atomic Energy Agency (IAEA) Safety Standards series provides very good guidelines toward a safety approach in nuclear plants.

A systematic approach is followed to identify the systems, structures, and components that are required to fulfill these safety functions. There is an extended set of functions necessary to avoid or prevent accident conditions as well as functions necessary to mitigate the consequences of accident conditions. Some of these are as follows:

- Safety functions for the control of reactivity:
  - Normal reactivity control within safe limits
  - Prevention of unacceptable reactivity transients

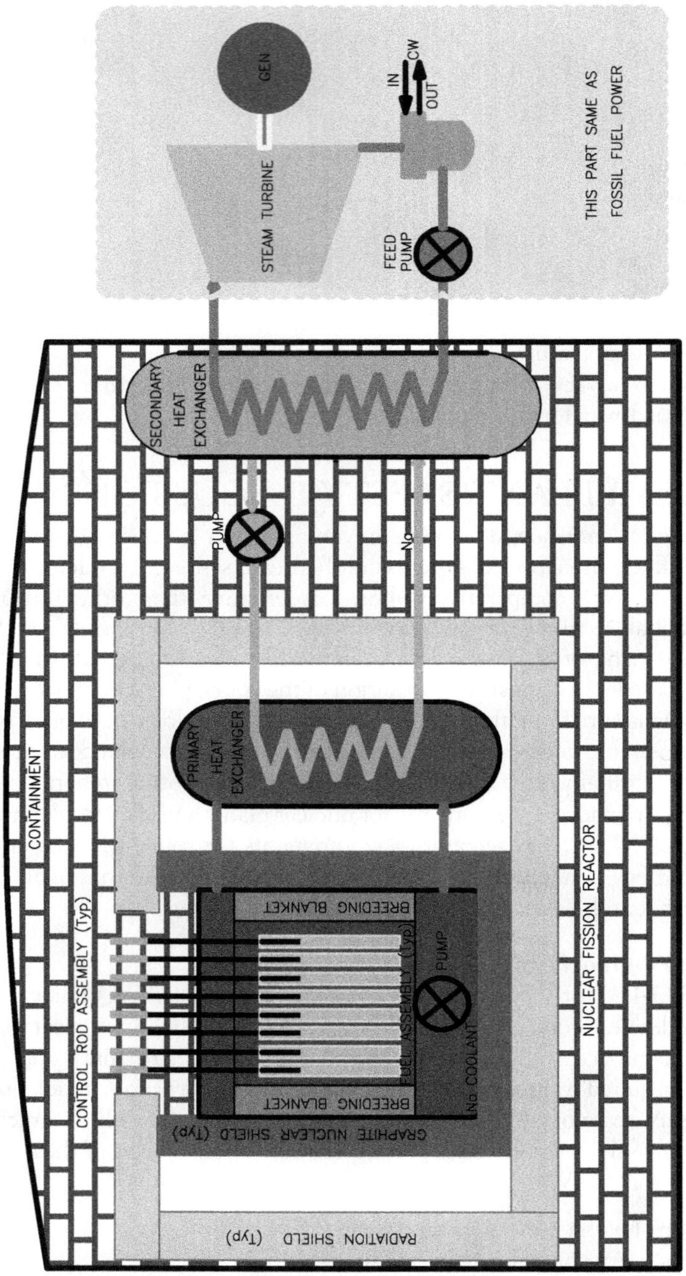

FIGURE XII/3.0-1

Nuclear reactor and power generation.

  - Shutdown of the reactor when anticipated maloperational occurrences lead to accident conditions
  - Shutdown of the reactor to mitigate the consequences of accident conditions
  - Maintenance of the reactor in a safe shutdown condition after all shutdown actions
- Safety functions for the removal of heat from the core:
  - Remove heat from the core during power operations
  - Remove residual heat in appropriate operational states and design basis accident conditions
  - Maintain sufficient coolant inventory for core cooling in normal operation
  - Remove heat from the core after a failure of the reactor coolant pressure boundary to limit fuel damage
  - Transfer heat to the ultimate heat sink from intermediate heat sinks

So, for I&C systems, typical primary functions that are important to safety include [15]:

- Protection functions
- Control functions
- Monitoring and display functions
- Testing functions

In addition to these service functions, utilities such as power supply, pneumatic/hydraulic power, data communications, etc. are important, because these support the systems that perform the primary functions.

- Protection functions: These are the most critical of the safety functions. These relate directly to nuclear safety and provide a line of defense against serious failure in other plant systems. The following are major issues:
  - Automatically initiate the operation of appropriate systems
  - Detect design basis accidents and initiate the operation of necessary systems
  - Limit the consequences of such accidents within the design basis
  - Be capable of overriding unsafe actions of the control system [15]
- Control functions are meant to ensure that the plant is controlled and kept within its operating envelope under all conditions. At times control functions take certain mitigating actions to lower the demand for protection functions. Interlock is a part of the control system. Interlock systems prevent unsafe conditions or operations, protect personnel, and prevent hazards. Interlocks prevent actions that could lead to or increase danger or damage to the plant, and do not normally initiate actions to correct conditions. Interlocks could be mechanical, administrative, or electrical. The electrical interlock is of interest to this discussion and is part of IAEA guidelines.
- Monitoring and display functions are the interface between the plant and personnel.
- Testing functions provide assurance of the availability and effectiveness of other functions.

3.0.2 Risk Reduction and System Theoretic Process Analysis

Risk reduction systems encompass all equipment and components provided specifically to reduce the probability of core damage in the event of a multiple failure sequence, as well as to prevent the initiator [15]. The way in which risk reduction is implemented into the system varies with countries also.

Hardware failure and software failure are two kinds of failures encountered in programmable systems, as already discussed. In cases of hardware failure, fault tolerant designs such as redundancy could be applied. Software failure, as discussed, has to overcome certain procedures, but certain failures (design failure) could include behaviors that can be unsafe. A new technique known as system theoretic process analysis is applied in nuclear installations. This is required to identify the control requirements and then check conditions caused by inadequate control actions such as:

- Required control action is not followed.
- Incorrect control action is followed.
- Safe control action is applied too early or too late (timing).
- Control action is stopped too quickly.

3.1 Digital I&C

Because of higher reliability, accuracy of measurement, and diagnostic features, digital programmable systems are slowly(!) but steadily replacing analog instrumentation. Even nuclear safety systems are not far behind. About 40% of the world's operating reactors have been modernized to include at least some digital I&C systems. In Japan, the first fully digital I&C system was integrated into the Kashiwazaki-Kariwa-6 advanced boiling water reactor in 1996 [11]. When looked at closely one can understand how I&C systems are critical to nuclear installations. There are more than 10,000 sensors with vast quantities of cabling and huge networks. Unlike fossil fuel power stations where there is one main control room, in the case of nuclear power installations there are *primary* and *secondary* control rooms. Similarly, there are various classes of safety aspects in nuclear power plants such as critical safety (e.g., related to reactor shutdown). Apart from this critical control there are other safety functions that influence plant safety to varying degrees. The Digital Instrumentation & Control Task Force (DICTF) of the Reactor Design Evaluation and Licensing Working Group (CORDEL WG) was set up in 2013. The key issues in digital I&C were the main focus for them. On the basis of an internal survey, CORDEL DICTF has identified four main issues [13]:

- Safety classification for I&C systems in nuclear power plants
- In-depth defense and diversity
- Field-programmable gate arrays (see Appendix V)
- Reliability predictions

As discussed earlier, the nuclear industry has an inherently conservative approach to safety. Therefore substantial effort is required to provide the necessary evidence

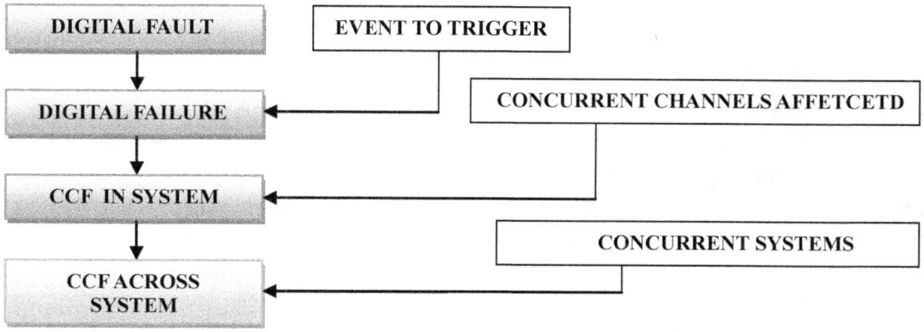

FIGURE XII/3.1.1-1

Common cause failure (CCF) in software.

and analysis to ensure that digital I&C systems can be used in safety-critical and safety-related applications [11].

3.1.1 *Common Cause Failure*

One of the most important issues connected with digital I&C is common cause failure (CCF). There exists the probability that a latent fault can exist in the system. This fault can be triggered and may propagate throughout the system as shown in Fig. XII/3.1.1-1. In most software one subprogram/subroutine may be used many times in a full program. Naturally, if there is a fault, then it can be repetitive, and when used even with redundant channels the same mistake may recur. Naturally extra care must be taken to ensure with thorough validation and verification.

3.1.2 *Security*

Various network attacks by external agencies are quite common and have been covered at length in the previous chapter. Nuclear plants are always the major targets. So, nuclear power plant I&C systems are generally kept isolated from external communication systems. Also the network deployed for safety and safety-related systems must be very well secured utilizing various techniques discussed in the previous chapter, such as zone conduit methods. Possible major attacks could be from:

- Unauthorized access to information
- Loss of integrity by interception and change of information
- Loss of system availability by blocking data transmission
- Unauthorized intrusion

For further details, Clause 4.0 of Chapter XI may be referred to. We will now look into the various classifications of I&C safety in nuclear plants.

3.2 Classification of I&C Systems

On the basis of the importance of safety pertinent to functions and I&C systems support, I&C systems are classified. In this graded approach, greater attention is given to systems and equipment that are important to safety than to systems that have less or no safety impact. Systems having higher safety importance should have higher quality and be more tolerant of failures and more resistant to hazards both inside and outside the plant. Thus the safety class of an I&C system and its assigned defense-in-depth level have a direct impact on the requirements on qualification, quality assurance, fault tolerance, system architecture, physical layout within the plant, and the extent of engineering documentation [14]. In this connection, Annex B of IEC 61513 (Categorization of functions and classification of systems) may be referred to. This will be clearer from an example. For reactivity control, control rods are either withdrawn or inserted in a typical pressurized water reactor. There are two modes of operation: *normal mode*, in which speed of movement of the rod is the parameter for reactivity control and withdrawal and insertion of the control rods is done manually by the operator or automatically, and *emergency mode* when rods can be released manually by the operator or automatically, so that they drop by gravity to shut down the reactor.

Each of these functions has a different degree of safety significance. Safety classification of I&C systems and components is an outcome of combined effects of process functions and I&C system architectural design. As per IEC 61513, the design of I&C architecture shall divide the entire system to meet the requirements of Ref. [13]:

- Independence of the functions in different lines of defense
- Adequate separation of the systems of different classes
- Fulfillment of the constraints on the physical separation and electrical isolation arising from the environmental and layout constraints, hazard analysis, and constraints from startup activities, testing, maintenance, and operation

3.2.1 *Safety Class Selection Criteria*

Major factors for classifying safety shall include:

- The safety function(s) to be performed
- The consequences caused by failure of a safety function
- The frequency at which the safety function is called upon
- The time period for which the safety function will be called upon
- The design shall be developed in such a way to ensure that there is no interference between items responsible for safety, and the propagation of failure from lower level to higher level is prevented.
- When equipment has multiple functions, then the equipment shall have a safety class consistent with the *most* important function it performs.

3.2.2 *Generic Classification of Safety*

The basis of the generic classification of safety is shown in Fig. XII/3.2.2-1.

Here it is important to note that at the initial stage the equipment not related to safety is set aside. In safety there are two categories: one is direct safety and the other is

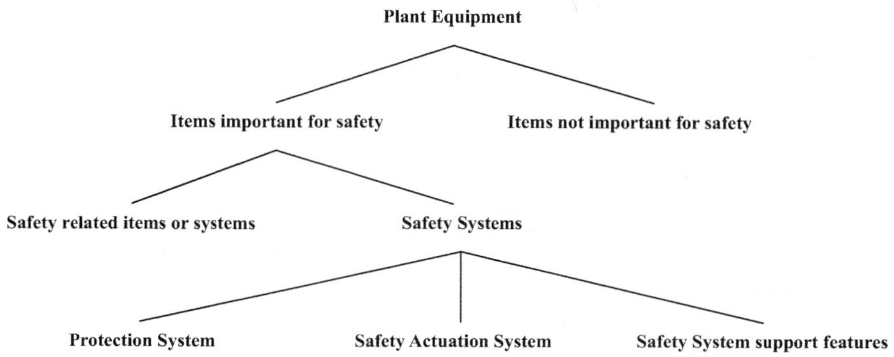

FIGURE XII/3.2.2-1

Generic system of instrumentation and control (I&C) safety classification.

Developed based on details and idea from Safety Classification for I&C Systems in Nuclear Power Plants — Current Status & Difficulties; CORDEL digital instrumentation & control task force; World Nuclear Association.

safety related. Main safety has three parts: protection, safety actuation, which may be in a different system, and safety support system. There are a number of standards in this regard and each has different divisions as discussed next.

3.2.3 *Comparison of Safety Classifications in Various International Standards*

There are various international and local regulations based on each country's safety system classifications, for example, India has IA, IB, IC, and NINS (nonsafety), and France and the European Union have F1A, F1B, F2, and NS (nonsafety for the European Union and nonclassified for France). In these classifications generally IAEA is kept at the top. For a detailed elaboration and understanding it is recommended that designers should refer to the associated standards, namely, IEC 61513 and 61226. Fig. XII/3.2.3-1 is a chart showing classifications of system safety as per various international standards.

As stated earlier, each country has local regulations and these pose diverse classifications of system safety, which also make it difficult to select criteria for diverse backup systems. Major difficulties in this regard are as follows:

- Inconsistency between international standards and local regulations
- Ambiguous requirements for safety classification
- Incomplete rules for categorization of "other I&C functions"
- Criteria for diverse backup systems

However, people are working together to bring uniformity and a consistent approach to all international standards for system safety classification.

<table>
<tr><td colspan="2">INTERNATIONAL STANDARDS/ ORGANIZATION</td><td colspan="4">SAFETY CLASSIFICATION OF I & C FUNCTIONS & SYSTEMS IN NUCLEAR PLANTS</td></tr>
<tr><td colspan="2" rowspan="2">IAEA NS-G-1.3</td><td colspan="3">SYSTEMS IMPORTANT TO SAFETY</td><td rowspan="4">SYSTEMS
NOT IMPORTANT
TO SAFETY</td></tr>
<tr><td>SAFETY</td><td colspan="2">SAFETY RELATED</td></tr>
<tr><td rowspan="2">IAEA SSG -30</td><td>FUNCTION</td><td>SAFETY CATEGORY 1</td><td>SAFETY CATEGORY 2</td><td>SAFETY CATEGORY 3</td></tr>
<tr><td>SYSTEM</td><td>SAFETY CLASS 1</td><td>SAFETY CLASS 2</td><td>SAFETY CLASS 3</td></tr>
<tr><td colspan="2"></td><td colspan="3">SYSTEMS IMPORTANT TO SAFETY</td><td rowspan="3">NON CLASSFIED</td></tr>
<tr><td rowspan="2">IEC 61226</td><td>I&C FUNCTION</td><td>CATEGORY A</td><td>CATEGORY B</td><td>CATEGORY C</td></tr>
<tr><td>I&C SYSTEM</td><td>CLASS 1</td><td>CLASS 2</td><td>CLASS 3</td></tr>
<tr><td colspan="2" rowspan="2">IEEE</td><td colspan="3">SYSTEMS IMPORTANT TO SAFETY</td><td rowspan="2">NOT SAFETY RELATED</td></tr>
<tr><td>SAFETY RELATED</td><td colspan="2">NO NAME BUT
NOT CLASSIFIED 'SAFETY RELATED'</td></tr>
</table>

EURO REVISED TO FOLLOW SSG 30'
COUNTRYWISE FURTHER DETAILS FOLLOW [13] WHICH WAS THE BASIS FOR THIS DEVELOPMENT

FIGURE XII/3.2.3-1

System safety classification. *I&C*, instrumentation and control.

3.3 Fusion Technology

Discussions will not be complete if fusion technology is not touched upon, especially in view of ITER project aims to produce electricity in fusion power stations. The ITER develops 500 MW of power utilizing fusion technology. This project is located in France and is funded by many countries in the European Union, as well as India, Japan, China, Russia, South Korea, and the United States. There is also research into reactors for which there are a number of guidelines for safety in fusion technology. The IAEA has produced guidelines for these.

3.3.1 *Safety Guidelines*

The fusion safety philosophy now includes the following concepts:

- Passive systems and inherent safety features
- Fail safe design
- Reliability
- Redundancy of components:
  - Diversity
  - Independence
  - Simplicity
  - Surveillance
- Consideration of human factors
- Remote maintenance capability
- Safety culture in worker attitudes
- Quality assurance:
  - Codes and standards
  - Verification
  - Validation
  - Safety analysis

Table XII/3.3.1-1 Relationship Between Safety Instrumentation System (SIS) and Category Classes

Structure System Component (ITER)	IEC 61226 Safety Category	IEC 61513 Safety Class
SIC1	A	1
SIC2	B	1
SIC2	C	3
SR	C	3
SR	Nonsafety	Conventional I&C

On account of cost class 2 is not used in ITER.

ITER, *international thermonuclear experimental reactor.*

- Operational controls:
  - Fault detection
  - Automatic corrective response
- Safety systems:
  - Accident preparedness
  - Management, to preserve confinement integrity
  - Emergency planning to mitigate effects of radioactive release

  Many IAEA activities in the area of radiation safety are relevant to fusion safety. For safety analysis, event tree analysis may be adapted.

From previous discussions it is seen that there are safety categories and classes as defined by IEC standards. IEC 61226 defines safety categories for nuclear safety I&C functions (category A, B, C). IEC 61513 also defines three safety classes for I&C systems (Class 1, Class 2, Class 3). A "nonsafety" class (NS) can be added to this classification. Based on the safety analysis, all safety I&C subsystems will be assigned to a safety class on a case-by-case basis. Safety importance class (SIC) describes a classification scheme for structures, systems, and components that perform a safety function and contribute to the general safety objectives at ITER during incident/accident situations [12]. In the case of fusion technology, these classifications are termed SIC1, 2, and SR. The relationship between these is shown in Table XII/3.3.1-1.

3.3.2 I&C Design Philosophy

I&C design philosophy, especially safety philosophy, is depicted in Fig. XII/3.3.2-1.

Single failure criterion must be taken into account for the design of the SIC1 and SIC2 safety I&C systems by using adequate solutions from all of the following:

- Redundancy
- Physical separation

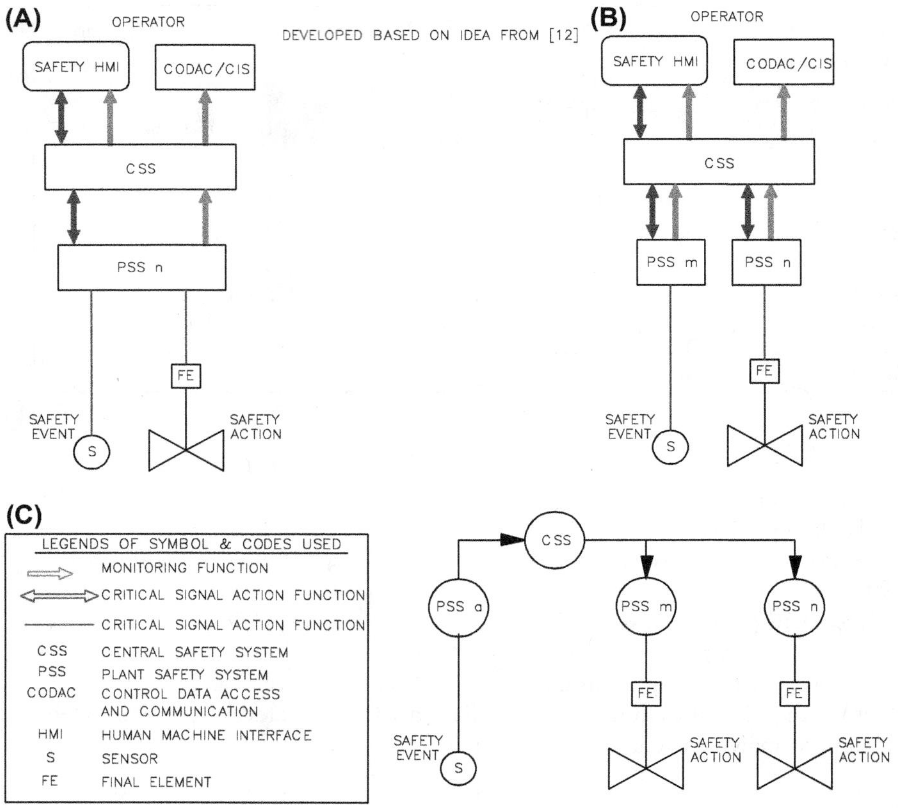

FIGURE XII/3.3.2-1

Design philosophy for a thermonuclear plant. (A) Local function, (B) central function, and (C) typical example of central safety system.

- Independence
- Electrical isolation

For detailed guidelines the ITER plant handbook [12] and IAEA standard SSG-37 [16] may be referred to. With this, discussions on nuclear plants are concluded.

4.0 SAFETY INSTRUMENTATION IN OIL AND GAS FACILITIES

The major portion of safety instrumented facilities is applicable in the case of oil and gas facilities. It is important to note that this is a very broad area, starting with exploration right through to petrochemical plants. Each of these facilities has different problems and issues, and it is not possible to put all these facilities in one basket. Prior to starting discussions it is better to have some idea of oil and gas facilities. In the limited space available it is not possible to describe each of these facilities; instead it is better to refer to Fig. XII/4.0-1 to get an idea of the various facilities and processes that have been developed around the natural resources of oil and gas.

As seen from the figure, the oil and gas system starts with oil exploration. Oil exploration can be onshore or offshore. There are millions of wells in operation worldwide. In oil-rich areas there can be huge productions of 1 million barrels of oil per day. Offshore production is quite different to onshore and different types of structures are used based on size and depth of water. Gravity-based, tension-leg platforms, semisubmersible platforms/ships, and floating production storage and offloading (FPSO) are a few examples for offshore installations. One needs to treat petrochemical exploration, production, and refining separately. However, prior to this it is better to have a look at standardization efforts through ISA 106. *What is ISA 106?* ISA 106 is not yet a standard. In 2013 a technical report, TR106.00.01-2013, *Procedure Automation for Continuous Process Operations— Models and Terminology* was published. A second report in draft form is also available. This technical report (nonnormative) is not a standard (normative) [18]. ISA 88 already exists for batch process (discussed in Clause 3.0 of Chapter VI); however, from point of view of procedure, continuous process hardly anything to be adapted from batch process

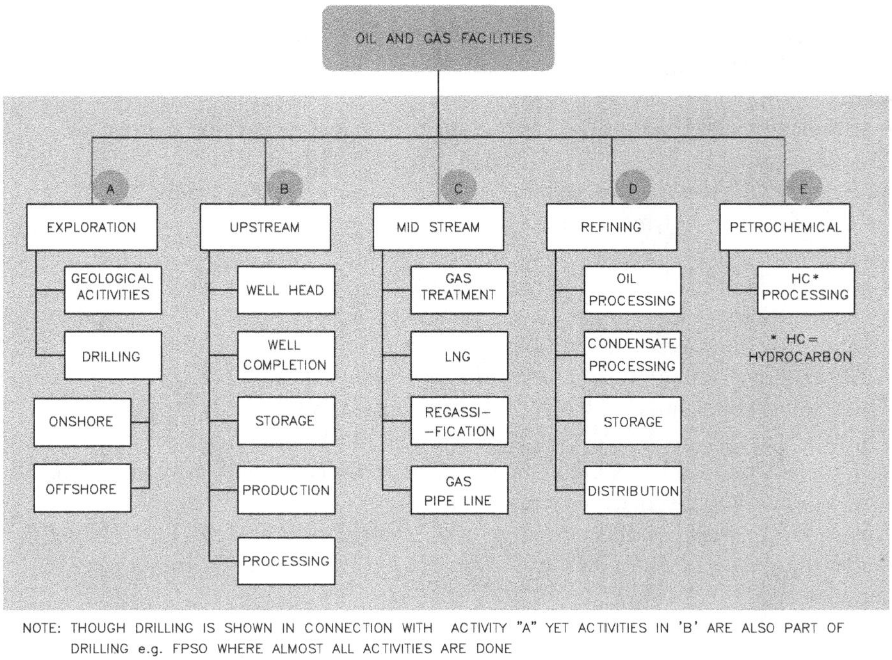

FIGURE XII/4.0-1

Oil and gas facilities. *FPSO*, floating storage production storage and offloading.

so all these efforts have been undertaken to increase safety procedure in continuous process also.

4.0.1 ISA 106 Discussions

ISA 106 was mainly developed for various oil and gas, chemical plant, paper plant furnace facilities, etc. *Why is this?*

- Major expected benefits for using ISA 106: These shall include but not be limited to:
  - Improved safety performance
  - Improved reliability
  - Improved training and operator effectiveness
  - Improved response to disturbance (reduced loss)
  - Consistent operation
  - Improved process insight
  - Reduced loss for operational/human error
  - Reduced cost of enterprise adaption
  - Better transition and improved quality
  - Increased production
  - Better startup and shutdown—greater production
  - More efficient change control
  - Common terminology/definitions
  - Higher retention and better dissemination of knowledge
- Target automation in oil and gas: In oil and gas, refining and petrochemical plants are the major targets:
  - Petrochemical: Startup/shutdown, transition management line switchover and cleaning
  - Refinery: Transition management crude switchover, regeneration
  - Others plants: Polymer, furnace, and paper plants
- Major themes behind the technical report are:
  - Models: Physical, procedural requirements, procedure implementation, and automation
  - Sharing of best practices
  - Procedure automation structure using process states

  This is somewhat similar to batch process, discussed in Clause 3.0 of Chapter VI. The model approach of ISA 106 is shown in Fig. XII/4.0.1-1. Here each procedure and implementation has three basic execution steps: command (trigger), perform (action), and verify (check for success/failure).
- Automation style: Organized structure of implementation modules utilizes three modes of operations as follows. It also includes HMI philosophy and degrees of automation:
  - Manual: Operator responsible for commands and work
  - Semiautomatic: Computer and operator share responsibilities for commands and control performances
  - Automatic: Computer responsible for commands (major) and control performance

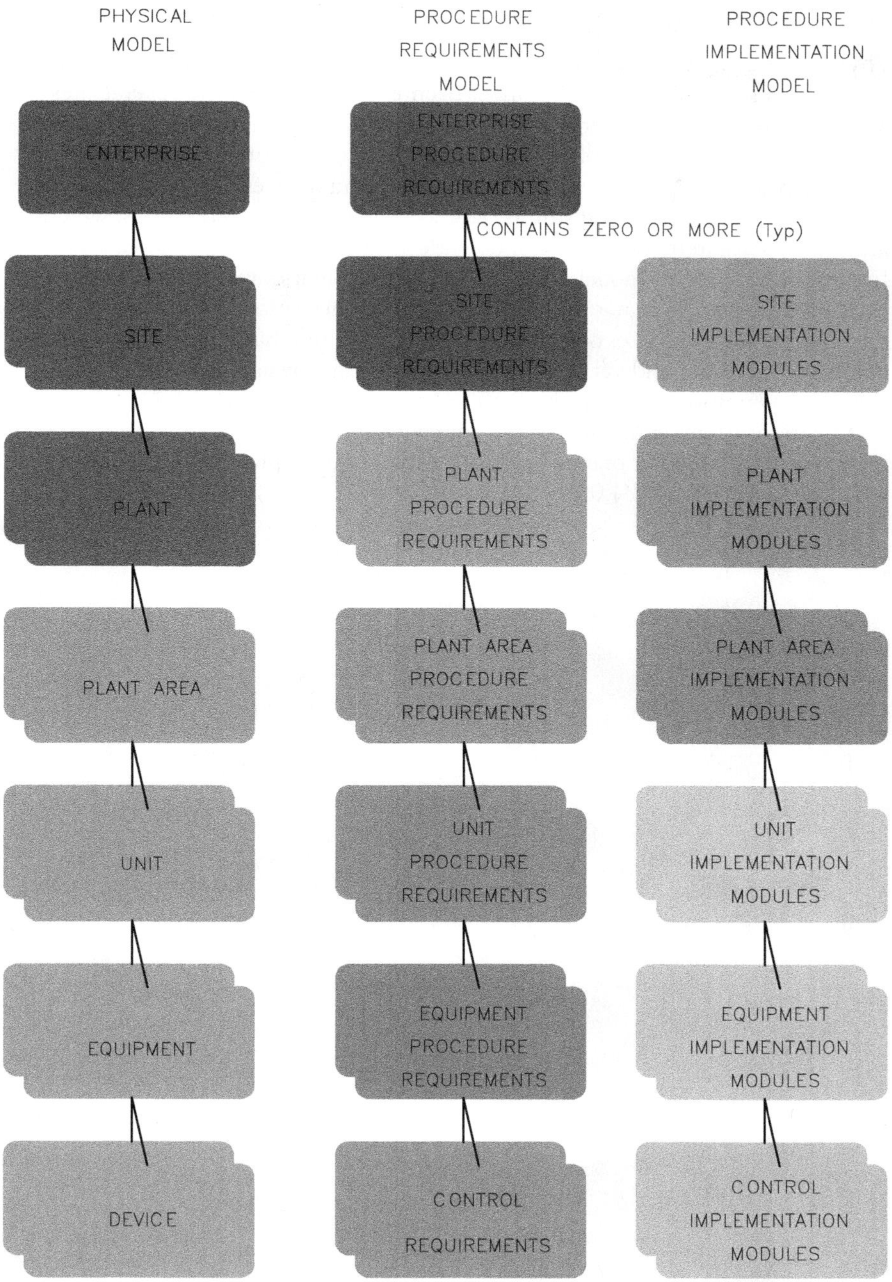

BASED ON TECHNICAL REPORT ISA 106

FIGURE XII/4.0.1-1

Model approach.

As is seen in the Fig. XII/4.0.1-1, there are three classes of models:

- Physical model: This is developed based on dividing the entire enterprise into smaller and smaller independent units. Each enterprise may have several sites and each site may have several plants, each of which may comprise various units viz. distillation column, hydrotreater, gas processing, or isomerization plant or unit. Each unit, of course, will have equipment and devices.
- Procedure: Each of the plant's units, etc. will follow certain procedures, which are shown separately for systematic operation.
- Implementation: This model is responsible for systematic implementation of each of the procedures mentioned above each of these procedures or its implementation comprises one or more task(s). Each task has three definite parts: command, perform, and verify. This is shown specifically against a single task in the implementation block in the highlighted part of Fig. XII/4.0.1-2.
- Automation: As per automation type, command, perform, and verify are done by an operator or computer or a mix of both. A typical example is shown for a state-based system in Fig. XII/4.0.1-2.

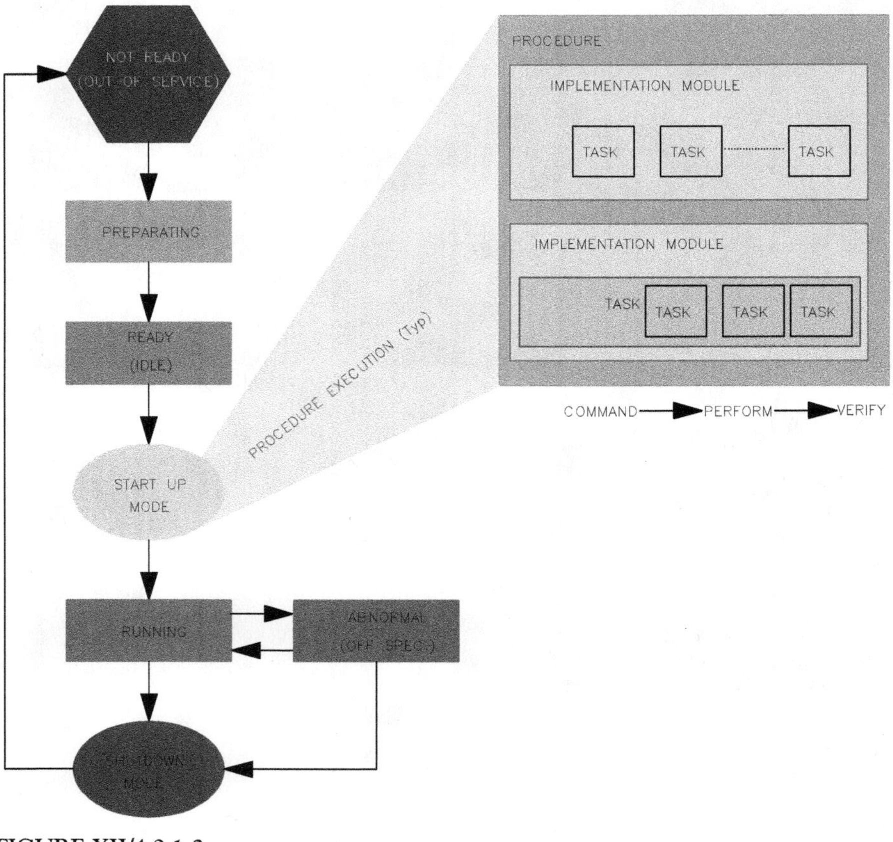

FIGURE XII/4.0.1-2

State-based control and procedure execution.

4.0.2 *Advantages With Integration*

In oil and gas industries there are a few distinct advantages of using integrated control and SIS systems:

- Higher system availability and flexibility
- Better fault tolerant design
- Fewer problems caused by interface
- Improved engineering quality
- Less expense for engineering and commissioning
- Overall cost savings

There are reservations against such integration also, as already discussed. Even certain standards do not allow direct integration on account of CCF problems. However, there are ways and means to circumvent such difficulties in new technologies, discussed later.

4.0.3 *Loss of Containment*

This is an important issue in connection with oil and gas SIS. API RP 754 plays an important role here. Important definitions of the standard are elaborated in Appendix I. Process safety indicator in line with the said standard is presented in Fig. XII/4.0.3-1.

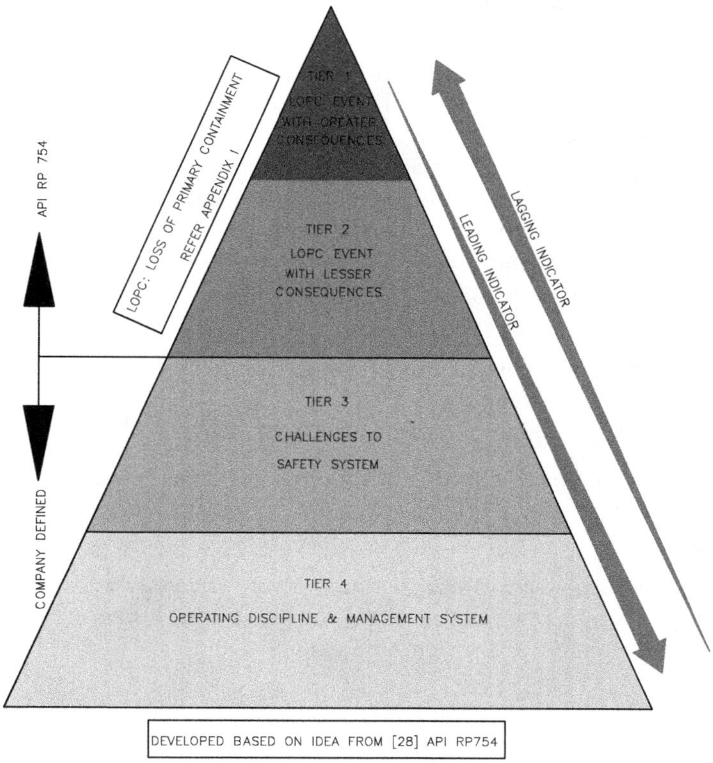

FIGURE XII/4.0.3-1

Process safety indicator pyramid.

Here it is to be noted that the first two tiers are defined in API RP 754, which deals with loss of containment, etc. The bottom two tiers are as per company definitions. It is recommended that readers refer to the said standard.

4.0.4 Other Issues

There are a few other issues such as HMI, alarm guidelines, and abnormal situation management (ASM) Guidelines on Effective Procedure Automation. HMI requirements and alarm details have already been discussed in previous chapters and there is no need to repeat them here. ASM guidelines cover:

- Effective use of procedures
- Content and format
- Development
- Deployment
- Maintenance
- Training

With this, general discussions on oil and gas systems are concluded and we now move on to specific issues regarding offshore facilities.

4.1 Drilling (Upstream)—Offshore

In this clause, the focus will be on drilling, that is, upstream production including FPSO. Productivity from traditional land-based installations has declined, and naturally more attention is now given to other unconventional sources, such as heavy crude oil, oil sands, oil shale, and offshore operations. When compared with onshore, offshore operations are quite complex. The complexity of any production facility increases exponentially with production capacity in limited spaces, weight capacity, and harsh environments along with a few other factors. FPSO could be considered as an example of offshore facilities for broader coverage. FPSO could also be used for deep sea explorations (~2000 m). A typical FPSO has three main parts:

- Vessel for oil storage, accommodation, and vessel services
- Turret for the mooring system
- Topsides for oil, gas, and water processing facilities

Offshore productions have driven innovations including the need to provide very large production facilities. There are a few major challenges for deep sea operations:

- Extremely precise engineering and coordination are necessary for the installation of instrumentation and wiring. Smart control systems coupled with communication systems are extremely critical for O&M of the facilities.
- Safety factors greatly depend on company culture, safety attitudes, and maintenance philosophy. There are huge numbers of equipment from several

original equipment manufacturers (OEMs) installed on the seabed. Also there will be a great deal of complexity involved in their connections, operations, and communications. Therefore obtaining, consolidating, and analyzing proper information is a mammoth task. Naturally, one needs to depend on equipment status and performance.

- There are a number of challenges, namely, maintaining topside and subsea assets over their lifespan, obtaining optimum profitability, minimizing risks, and maintaining reusability of assets decommissioning. To meet these targets, valve activities, human activities, and information management must be well coordinated and controlled.
- The cost of abnormal situations and safety responses to abnormal conditions is critical, especially in deep sea installations.
- Operational risk increases with increased manpower and weight on the topside. On the other hand, if more equipment is placed on the seabed to reduce weight there will be complexity in accessibility and increased dependence on communications. Also the cost of subsea repair will be enormous in terms of lost production, monetary value, and environmental impact.

In addition to standard BPCS and SIS, the FPSO needs an increased need for an efficient operations environment on board and accessibility from a remote operation center onshore. The system is operated from the central control room with the help of an intelligent control system and HMI. With modern systems the same information is available to remote locations such as an onshore corporate operations support center [17]. Main control systems provide the operator with:

- Access alarm and events
- Ability to assess initiating events from sequence of event data, SIS, and historian
- Seamless reaction to diagnostics
- Personalized workspaces for operational response and commands

4.1.1 Integration Needs in Offshore

As topic already discussed in previous chapters, North Sea exploration standards do not permit such integration. However, it is a question of interpretation of the standard. It is more important to demonstrate that the two protection layers are independent. It is also a fact that integrated systems are becoming slowly accepted in offshore operating environments. These may be on account of.

- Technological development with higher performance standards
- Significant operational and maintenance benefits
- Pressure from OEMs

However, there is another need, especially for subsea operations. In the case of subsea operation there are distinctly two separate sections: topside and subsea side. For a few items in subsea controls a *specialist* OEM is necessary. When each of these sides is *integrated* seamlessly then there will be a smooth flow of information, that is, interface problems could be avoided. Such a design must be operator oriented to avoid operator error. Consistent HMI design is critical for safety and shutdown systems so that the operator is relieved of various alternative shutdown procedures. For this, guidelines from ISA 106/ISA 18.2 and EEMUA 191 may be followed.

4.1.2 *Key Features for Subsea System Architecture*

Some of the special features pertinent to subsea operation are as follows:

- In spite of many similarities between land-based counterparts, subsea control systems have additional complexities that need to be addressed.
- The subsea master control station needs to interface directly with topside control systems, which could be from multiple vendors and protocols for communication.
- Various valve commands and interlocks, emergency shutdown (ESD), choke and kill controls and commands, etc. receive and monitor subsea controls and provide HMI for the operator. So, seamless information flow is very important. Following open system architecture it may be possible to integrate the systems. In fact, as long as the two control logics are separate (subsea items normally supplied by specialized OEM), NORSOK requirements are not really violated.
- From the foregoing it is clear that the better the standardization of hardware and software, the lesser will be the interface issue.
- Standard smart-tested field input/output (I/O) and skid controllers need to be used.
- Master control station (MCS) should have simple but effective integration capability.

4.1.3 *Shutdown and FGS*

Major safety issues related to offshore: ESD and process shutdown (PSD) systems need to maintain set points for process values to provide process precondition alarm, to indicate that the process is going out of range and malfunctions in the system might have set in. Fire and gas systems (FGSs) are designed to be self-contained. However, process interlocks such as partial closing of the ventilation fire damper are necessary. So, FGS needs to interface with process systems.

4.1.4 *Additional Offshore Safety Considerations*

In addition to a few conventional safety considerations (independence of SIS from BPCS, use of redundancy and fault tolerant design, etc., discussed at length in previous chapters) the following points are worth considering as safety issues related to offshore. Risks associated with typical BOP has been presented in Fig. XII/4.1.4-1. Depending on applicability, reader to decide the associated standard for reader's application.

- Use of self-diagnostics at systems and device level to reduce mean time to repair and overall SIF PFD. Use of deenergize to trip logic.
- Collision detect: In drilling and pipe handling, collision detect is important and could be a part of the safety system [19].

Risks associated with BOP: *Readers may think and decide if risks associated with BOP in offshore are designed as per API 14C alone or if IEC 61508/61511 should be considered along with the API standard. There are diverse expert opinions of so the choice is left to the reader!*

FIGURE XII/4.1.4-1

Risks associated with blowout preventer (BOP).

- Subsea PSD system: The purpose of the subsea PSD system is to control abnormal operating conditions by stopping hydrocarbon flow by shutting down processes. Two such processes are:
  - High-pressure trip: High integrity pressure protection system—an active and preventive protection layer.
  - Low-pressure trip: This is used for probable major subsea leakages indicated by sudden pressure drop.
- Subsea ESD system: This is used to prevent escalation of abnormal conditions and to prevent unwanted consequences from occurring.
- Isolation of riser: This is interlocked with ESD for closing of the ESD valve pertinent to the riser.
- Isolation of subsea well: Wells and Christmas trees are like subsea IPLs. Actions like chemical injections, production master valve operation, etc. are issues here.

4.1.5 Common Cause Issues

The traditional approach for reducing common cause was to use totally different systems for the BPCS and SIS, using different hardware and software to reduce CCFs. One way was to use dedicated control systems from different vendors. However, this could pose problems in seamless integration, advocated earlier. With the help of better technologies a new degree of integration required to fulfill the production targets in offshore facilities and to meet the challenges of CCF and the requirements of demonstrating that the SIS is not subject to CCFs with the BPCS have been developed. Also it is required to demonstrate that although they are integrated, both systems can provide independent protection layers. So, the same vendor may be in a position to develop an integrated system using diverse design engineering and programming teams provided with different software architecture specifications and guided by an overall concept for diversity from the start of the detailed design specifications. In these systems it is possible to provide very high diagnostic coverage for dangerous failures. Such integrated systems can be developed: "by combining two different technologies [such as micro processor (MPA) or micro controllers and field programmable gate arrays (FPGA)] to perform the same functionality in parallel to each other, the design achieves a truly redundant and diverse implementation with a minimum of possible common cause failures" [17]. The 800xA integrated control and safety system of ABB is an example of this. In these cases not only requirements of safety standards are fulfilled but also various benefits of integrated systems for offshore facilities are fully satisfied.

4.1.6 *Safety and Security Issues*

Safety and security issues are so bonded together it is not possible to separate them. This is more so when one advocates the use of integrated control and safety systems in drilling and production processes. A memory system in an intelligent system can be partitioned and each partition can utilize different execution processes. So, memory management units of intelligent control systems can provide better results for independence of SIS. Security and testing are not isolated activities. In the previous chapter, security issues have been discussed at length. Another issue is third party certification, which needs to be adapted. In this connection, the following points may be considered:

- Full vulnerability assessment including threat modeling and testing of subsystems and interfaces
- Identification and adaption of the best security mechanism
- Maintenance of the security mechanisms throughout the system life cycle
  With this, discussions on upstream systems are concluded.

4.2 Midstream Processing

On account of shale playing a great role, there has been tremendous growth in midstream processing in oil and gas systems. Because of the abundance of shale formation in parts of the United States there has been an increase in the production of natural gas and associated liquefied natural gas (LNG—primarily ethane). The price difference between natural gas and LNG determines how gas processing will be done. Midstream mainly consists of gas processing, compressors, block valve station, and pipelines. Associated instrumentation and control systems shall include mainly sensing and actuating field devices, production control, safety management, supervisory control and data acquisition (SCADA), integrated CCTV, and asset and production management. Major processes here are:

- Pipelines comprising pumps and valves
- Intermediate storage
- Compressor monitoring and control
- LNG liquefaction
- LNG regasification
- Gas processing
- Gas metering
- Networking and SCADA
- Gas terminal distribution

Almost all major and reputable I&C manufacturers can provide dedicated optimum solutions for these control and safety systems. Control and safety solutions need to provide improved monitoring and management of hydrocarbon transportation through pipelines, tankers, and terminals, ensuring reliable operations, functional safety, system availability, and compliance with environmental requirements. Relational database management for data integrity should be reliable, real time and object oriented [e.g., use of structured query language (SQL) compliant database management]. Another important issue here is energy management to monitor and control energy consumption as well as to see how the energy is used and identify areas of waste and specific areas for improvement. On account of major energy consumers such as gas compressors, oil pumps along pipelines, etc. are important equipment for midstream systems.

4.2.1 *Pipeline SCADA*

For pipeline control, SCADA is extremely important.

As shown in Fig. XII/4.2.1-1, SCADA mainly consists of:

- Field devices and instrumentation
- Control system: PLCs/remote terminal units (RTUs)
- Telecommunications
- SCADA host
- Applications

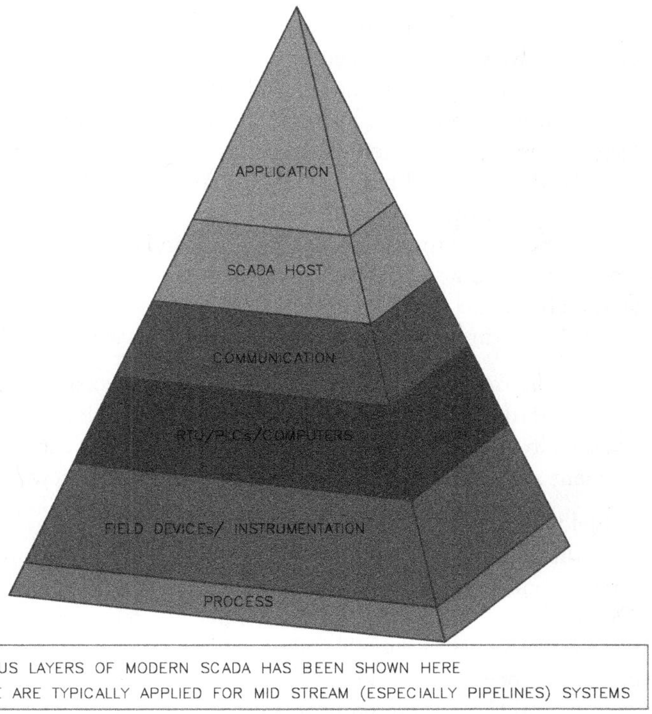

FIGURE XII/4.2.1-1

Modern supervisory control and data acquisition (SCADA) in midstream. *PLCs*, programmable logic controllers; *RTU*, remote terminal unit.

The essence of SCADA is built on a reliable, high-performance, real-time relational database management system. A dedicated SCADA system is installed for effective and efficient monitoring and control of the entire pipeline from the MCS. In an SCADA system there may be one MCS and several (local) station control centers (SCCs) at attended stations, namely, pump stations/delivery stations/pump-cum-delivery stations. Major reasons for pipeline SACDA installation shall include:

- Updated field conditions available automatically for monitoring
- Consolidation of information and management information system
- Sharing of information for better management within an enterprise
- Automatic alerting and alarm management of abnormal field circumstances
- Remote control of abnormal field circumstances
- Maintenance of consistent, accessible archival records

For data integrity, technologies like SQL may be deployed. Present-day SCADA is designed with open connectivity to external corporate databases, improving compatibility and performance. However, the security issues discussed in Chapter XI need special attention while developing an SCADA system. Live data, databases, and HMI configuration are broadcast in real time through one of two modes:

- Hierarchical (local, regional, and corporate levels)
- Peer-to-peer (offsite backup control center for immediate disaster recovery)

As stated earlier, SCADA system integrates RTUs and PLCs/hybrid controllers and provides local control, monitoring, and specific dedicated functionalities, such as ESD, fire and gas, and asset management. Replication of functionality configurations can be centralized or decentralized based on the owner's choice.

- MCS: The following are the major features of a standard MCS:
  - Effective and efficient monitoring, safe operation, and control of SCCs/RTUs
  - Configured with redundant computer systems
  - Distributed architecture using intelligent units
  - All interconnected through a redundant local area network

A typical modern networked SCADA architecture is shown in Fig. XII/4.2.1-2. There may be some variation in actual implementation, but functionally architectures similar to this are used in almost all SCADA. Not shown in the figure are historians for database management and management information systems separately.

- RTU/PLC—Field interface device: RTUs are connected to the field devices, like sensors/transmitters, local switches, and actuators. RTUs also interface to convert the data from supervisory systems to necessary electronic signals for the actuators. On account of versatility, flexibility, and configurability, PLCs are also used as a field device interface like RTUs. Obviously, PLCs incorporate programmed intelligence in the form of logical procedures to be executed in the event of certain conditions. Communication modules connected with PLCs allow it to communicate with remote devices [1].
- Media for data transfer: The following are generally used as media:
  - Leased lines
  - Satellite communication

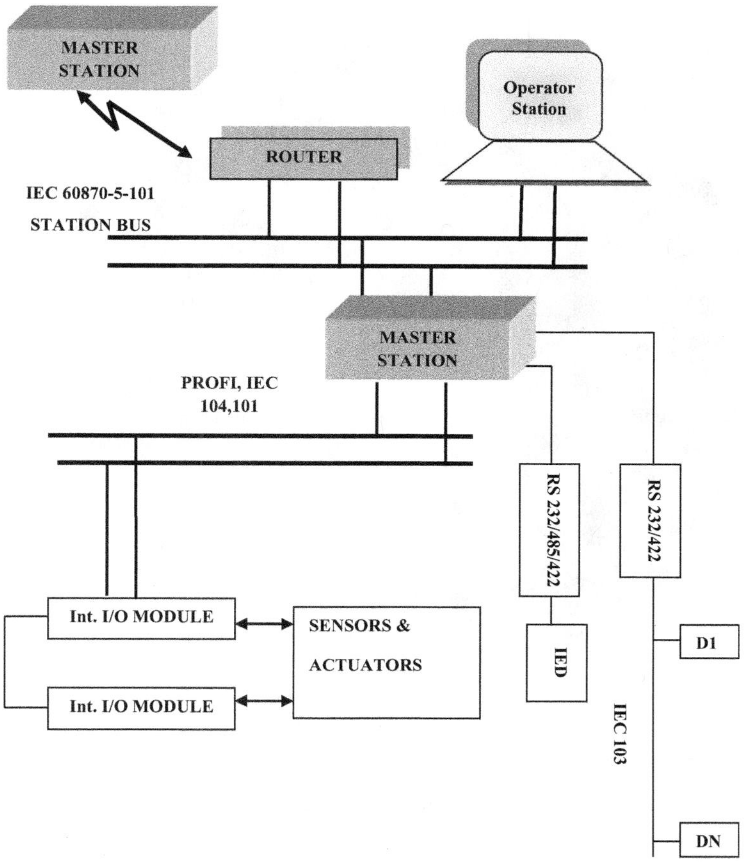

FIGURE XII/4.2.1-2

Modern networked supervisory control and data acquisition (SCADA). *D1*, device 1; *DN*, device *N*; *IED*, intelligent electronic device; *Int.*, intelligent; *I/O*, input/output.

Adapted from author's book S. Basu, A.K. Debnath, Power Plant Instrumentation and Control Handbook, Elsevier, November 2014; http://store.elsevier.com/Power-Plant-Instrumentation-and-Control-Handbook/Swapan-Basu/isbn-9780128011737/. Courtesy Elsevier.

 - Digital microwave communication
 - Fiber optics
- Effective operator interface: The following are the major issues:
 - Equipment-based configuration
 - Automatic generation of displays
 - State-of-the-art presentation, pan and zoom
 - Dashboard-type presentation
 - Task-based filter for operator's focus and pipeline controller attention

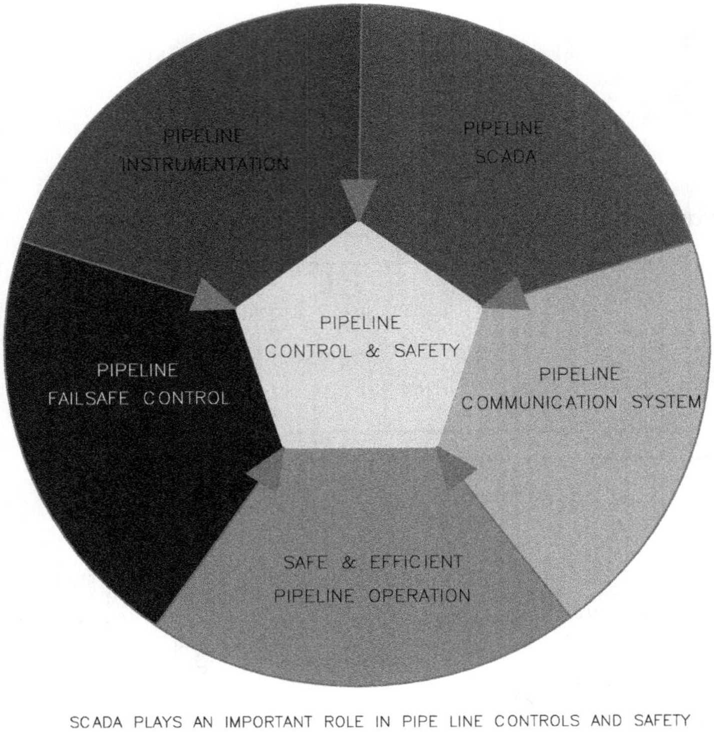

FIGURE XII/4.2.1-3

Typical pipeline controls and safety. *SCADA*, supervisory control and data acquisition.

- SCADA software: Software plays a very important role in SCADA, especially with MMI packages, which contribute significantly in cost also. For a better performed SCADA system the software needs to be well defined, designed, checked, and tested [1]. Major software components in SCADA shall include the following:
  - PLC/RTU Software: Local automation software
  - Server(s) software: Typically, operating system, application software, user interface (GUI) for historian, trending and mimic functions, etc.
  - MMI software: Similar to the above and able to access servers in a network.
  - Communication software: Suitable communication protocol drivers resident in a server as well as at PLC/RTU. It is the responsibility of protocol drivers to prepare data for communication.
  - Security: Security issues in SCADA are very important. Security issues have been discussed at length in Chapter XI so will not be repeated here. For in depth discussions on the same, Clause 2.4 of Chapter VII of [1] may be referred to. Normally, SCADA is integrated with various applications like

pipeline management, fatigue management, alarm management, change management, operational experience integration, etc.

Functional divisions in typical pipeline controls and safety are depicted in Fig. XII/4.2.1-3. This is self-explanatory. One thing to note is that in pipeline controls, the SCADA system plays a major role. Also it is used in terminal automation.

4.2.2 *Terminal Automation*

A terminal automation system (TAS) has a wide application in marine and inland terminal management. This is mainly responsible for handling various operations including receipt, storage, and dispatch of the complete product portfolio. Since these terminals receive products from different sources such as pipelines, rail lines, and ship offloading facilities, TAS needs to be equipped to cope with all these requirements. TAS also includes various control and instrumentation functions like tank gauging, inventory management, SCADA, safety, and regulatory and transactional business.

4.3 Refinery Safety Issues

In oil and gas systems, the refinery is a major industrial setup for obtaining various desired/specified petroleum products from crudes. Naturally, safety issues in a refinery are of prime importance. At present with the introduction of digital technology in instrumentations there have been a number of changes in refineries. Now refineries are smart refineries. The following basic objectives remain the same:

- Maintaining safe operations
- Enhancing environmental stewardship
- Sustaining high equipment availability and reliability
- Maximizing plant and product value
- Achieving higher efficiency and optimized operation
- Attaining smaller footprint and easier manageability of I&C

Modern digital technology provides a better return on investment. Benefits of smart instrumentation are achieved during the project with commissioning, calibration, and startup completed more quickly, more accurately, and more efficiently [24].

In earlier days there were several control rooms to control the process. However, in the modern era control rooms are consolidated to have a common control room. Similarly in safety issues, in some cases preventive measures were fully operational while in others plans were made to take action after an incident happens. According to M. Opheim, Director of Marketing for Oil and Gas, Honeywell Process Solutions: "…To focus on only one of these aspects is the wrong approach." So, effective safety and security demands clear understanding of the process and associated best practices. This will not only help in preventing incidents before they happen, but also help to respond properly when they occur [21]. Another important issue is monitoring and analyzing parameters. Measurements of temperature and level parameters are extremely important in oil and gas industries. Also it is extremely important to have a thorough and continuous check of these parameters to avoid explosion and fire incidents, like what happened in a Texas refinery in 2005, which killed at least 15 people and resulted in a huge loss of property [20].

4.3.1 General Technical Issues

Major challenges in a typical refinery are presented in Fig. XII/4.3.1-1.

Major challenges in a modern refinery shall include but not be limited to the following:

- Safer plant design philosophies and equipment
- Operator and plant safety hazards
- Centralized focused equipment/process monitoring

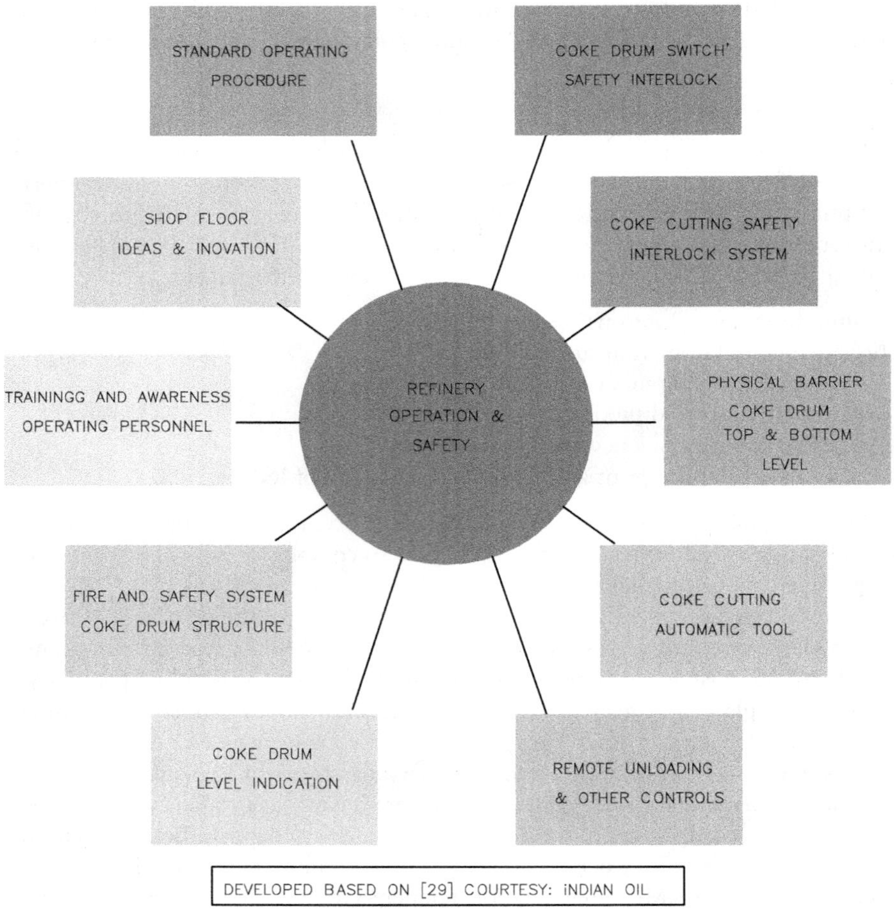

FIGURE XII/4.3.1-1

Plant operation and safety for a typical refinery.

- Synchronized startup/operation of all the auxiliary facilities
- SIL-rated emergency shutdown system

4.3.2 Centralized Approach

As indicated earlier, in a modern refinery a consolidated control room approach and central monitoring of equipment and process is preferred. For proper implementation of this, in some cases a supervisory DCS is used and is placed at a higher level than the process control DCS. The main function of this supervisory DCS is to effectively carry out the coordination between various DCS units and assist central monitoring of equipment and process. The major advantages associated with such centralized monitoring and supervision shall include:

- High efficiency
- Central place for overall monitoring and operation
- Refinery-wide coverage
- Utilization of cost efficient advanced controls and SIS

A typical architecture [22] is presented in Fig. XII/4.3.2-1.

To supervise the refinery operation a supervisory DCS has been kept at a higher level, which collects the data refinery-wide plant bus. There are a few other control and safety systems connected at this level. These are ESD system and advanced control system for overall coordination among the unit systems. ESD is a diverse independent system integrated with the system for better coordination and ease of operation. These also help in having a consolidated database management and management reporting. The supervisory DCS also has a refinery-wide control function such as utility balance, and plant manipulations will be carried out through local DCS HMI. These supervisory DCS may play the role of set point control when seen from unit DCS. Each unit DCS in turn has its own alarm management system and SIS, which could be a diverse system integrated to the system as discussed earlier for efficient operation. Each of these unit DCS may be meant for (say) landing, utility, and crude oil tanker control or may be for blenders and secondary processing units as per actual configuration and overall layout and functional requirements of the plant [22]. The redundancy and fault tolerant designs are not detailed but systems are developed based on these. Also here, SIS has been shown in the same bus for simplicity; in reality it may not be so, as discussed in Chapter XI, to take care of IEC requirements. There will be different hardware and software (may be different technologies as discussed earlier) for SIS and BPCS.

4.3.3 SIS Functions

Integrating equipment diagnostics and status information within the safety logic solvers permits improvements in availability and reliability of SIS. In supervisory DCS, the consolidated SIS could be deployed in addition to unit SIS to enhance the safety of the entire refinery [22]. As stated earlier, use of diverse technology is important to overcome CCF. ESD shall have SIL certification by a third party. It is possible to integrate the digital SIS with DCS, and ESD triggers and events can be seen on HMI of DCS, for example, as in Prosafe [22].

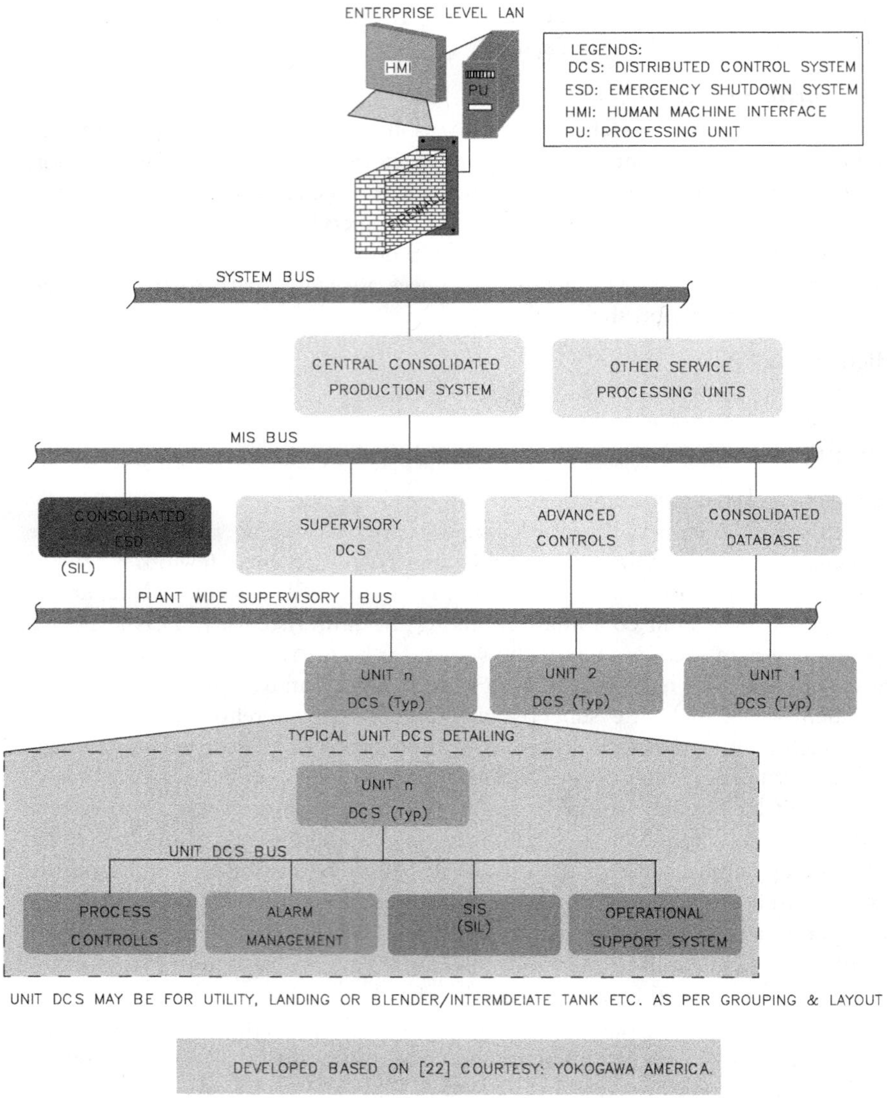

FIGURE XII/4.3.2-1

Centralized control approach. *DCS*, distributed control system; *ESD*, emergency shutdown system; *HMI*, human–machine interface; *MIS*, management information system; *PU*, processing unit; *SIL*, safety integrity level; *SIS*, safety instrumentation system.

For better operational safety the following features may be incorporated into ESD:

- Failure detection diagnostics of the entire path
- Diagnosis of field devices and wiring from ESD
- Partial stroke test of ESD valves

Modern SIS provides an audit trail to ease change management with automated testing documentation to demonstrate regulatory compliance. Also these SIS solutions provide considerable assistance in complying with regulations by completely integrating change management and generating—on demand—detailed documentation from several aspects, including the device audit trail, calibration history, control and/or safety configuration audit trail, process history, and event history [24].

4.3.4 Coker Safety Switch

Coker drum safety switch interlocks and controls are important. A few issues related these systems shall include but not be limited to the following:

- Permissive for coker drum "feed isolation valve" operation
- Permissive for "switch valve" operation
- Permissive for coker drum "drain valve" operation
- Permissive for coker drum "heading/unheading"
- Physical barrier

4.3.5 Coker Drum Safety

The coker drum safety system (CDSS) logic solver needs to automatically inhibit the opening of valves under unsafe process conditions. This will prevent the release of hydrocarbons and/or hydrogen sulfide to the atmosphere, or introduction of hydrocarbons to an open drum. According to the Operational Safety and Health Administration (OSHA), in hazards of delayed coker unit (DCU) operations, "The batch portion of DCU drum switching and coke cutting operations creates unique hazards, resulting in frequent and serious accidents." Each DCU has at least two coker drums, and an operator is responsible for manually lining up the coker drums and valves. Failure to open or close valves in the correct sequence can result in the release of hot hydrocarbons and/or hydrogen sulfide gas, and can also cause a fire. CDSS is an SIS independent of the coker interlock system pertinent to TAS and other plant control systems. The CDSS monitors process conditions to prevent operations from opening the wrong valve or performing operations on the wrong drum.

Coker drum deheading is a challenge to the safe and reliable operation of the coker unit. Safety issues are generally related to release of hydrocarbons to the atmosphere by opening the top or bottom head in a service drum [23]. Independent safety systems are also available in the market, namely, Diamond SIS of SIS Tech.

4.3.6 Other Safety and Monitoring Issues

There are a few other safety and monitoring issues:

- Shot coke: Bottom nozzle plugging, uneven cooling
- Coker drum: Vibration, heating cooling rate

- Heaters: Skin temperatures, input temperature, pressure, box temperature, etc.

With this, the discussions on refinery are concluded. Next we will discuss petrochemicals, the last major phase under oil and gas systems.

4.4 Petrochemical Plant

In petrochemical plants, various chemicals are made from petroleum or natural gas. However, petrochemical products are not really the end chemicals but act as primary sources of so many products and chemicals we use in our daily lives. Based on the structure, these are separated into three major divisions:

- Olefins: Ethylene, propylene, and butadiene are major products under this category. The first two are major sources of various industrial chemicals and plastics, etc. used in our daily lives.
- Aromatic: Benzene, toluene, and xylene are major products under this category. These products are the sources of detergents, explosives, dyes, etc.
- Synthesis gas: This is a mixture of CO and hydrogen used for manufacturing NH_3 and methanol.

Naphtha cracking is mostly used as the main process.

4.4.1 Major Process Hazard Issues

There are a number of hazardous situations in petrochemical plants and a few of them are enumerated here:

- Firebox explosions (lighting of burners)
- Overtemperature in furnace convection and radiation section
- Overtemperature protection in transfer line exchangers/steam superheat coils
- Firing/overheating caused by backup fuel switching (use of Wobbe meter)
- Fire from coil outlet caused by thermo-well erosion
- Runway reaction caused by excess hydrogenation and polymerization reaction
- Cold box under slight positive nitrogen pressure

4.4.2 Major Process Issues

Major safety issues related to the process shall include but not be limited to the following:

- Remote isolation of flammable inventories in emergency situations
- Depressurization of equipment from remote interlock and operation
- Use of class 6 isolation valves
- Overpressure and thermal protection of heat exchangers
- Continuous monitoring of pressure and temperature on this equipment
- Provision of safety interlocks in fault tolerant mode
- SIL3 interlock system for flare load mitigation
- Continuous monitoring and safety interlock to protect process equipment from temperature excursions beyond design limit
- Separate cold dry and wet flare system

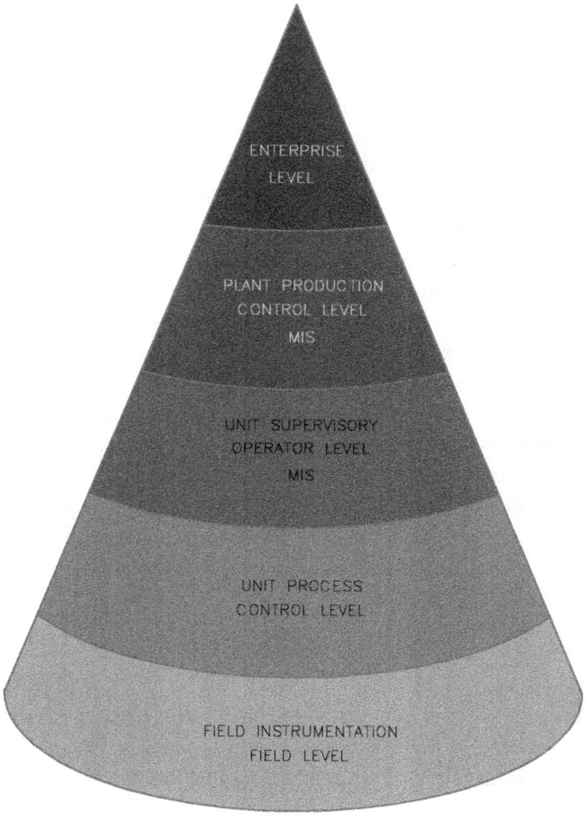

FIGURE XII/4.4.3-1

Automation structure of a typical petrochemical plant. *MIS*, management information system.

4.4.3 *Automation Structure and Safety Issues*

A typical automation structure in a petrochemical unit is depicted in Fig. XII/4.4.3-1, which is similar in structure to that of refineries discussed earlier. In petrochemical plants there are a few local controls that are integrated with the main system.

Safety is of paramount importance in the petrochemical industry. Working with flammable and hazardous substances under severe process conditions is always challenging. Naturally, all up-to-date information of the plant must be available at suitable places. There shall be ESDs/PSDs to cope up with emergency situations. In most interlock and safety-related instrumentation, fault tolerant circuits with 2oo3/1oo2 or TMR voting are deployed. SIS SIL is maintained in most cases at SIL3 in these plants.

4.4.4 *Fired Heaters and Automation*

Process fired heaters, which are extensively in use in refineries and petrochemical plants, present significant safety risks. These are used for heating, vaporization, and thermal cracking of various process fluids. Besides maintaining temperature and charge rate these heaters are required to control the system to maintain efficient combustion of fuel and safe operation. From a safety point of view fuel must not be allowed to accumulate in the firebox as subsequent introduction of an ignition source could be catastrophic. Also fired heaters present risks associated with the process (unlike boilers, where the process stream is water); the process stream for most fired heaters is highly flammable hydrocarbons [26]. Every fired heater must have some type of safety system in place. In cases of larger units these are fully automated with special emphasis on SIS. BMSs like those in boilers are also not uncommon. BMS offers great opportunity to go beyond the minimum requirements and can simultaneously meet safety targets and provide operational benefits [26]. As discussed in connection with boilers, fully automated BMSs utilize sequence logic designed for startup, normal running, protection trips, and emergency situations in conjunction with other systems. Consistent with the standards and recommended practices, a BMS can be treated as an SIS, as already discussed in this chapter. Naturally, a suitable SIL shall be determined based on fault tolerance selected for the logic solver. Use of a BMS also reduces startup time, and makes the system safe from human error. Diagnostic data can also play a role in an automated BMS. The use of smart devices with a fieldbus system also helps in fault identification and reduction of cost because of cabling. Also these help in system integrations.

4.4.5 *Automation of Naphtha Cracker Decoking*

During the pyrolysis reaction, coke forms inside the (naphtha cracking unit) NCU coils, resulting in decreased heat transfer and a decline in reaction efficiency; the coke could even crack or rupture the coils (as with soot formation in a boiler). To prevent this from happening, the plant operators need to perform a decoking process regularly or based on temperature. Also dryers adjacent to separators need regeneration when saturated with moisture. In modern control systems such procedures are automatically carried out by software programs so that operators at each step are aware of the progress of the procedure. Exapilot of Yokogawa is an example [25].

5.0 SHORT DISCUSSION ON SIS APPLICATION IN OTHER INDUSTRIES

IEC 61508 and IEC 61511 are applicable to electrical/electronics/programmable electronics and based on their use they are applicable for all industries. In this connection, OSHA guidelines may be referred to. These standards as well as SIF and SIS are not industry specific, but mainly depend on application. This will be clear from the following example. The importance of cooling water in steel industries for converters and furnaces cannot be overestimated. Failure of water supply could be catastrophic. Naturally, instruments needed for flow measurements in cooling lines must be reliable and should have a very good "PFD" requirement. In many such applications, even though there may not be any requirement for SIL rating, one would always prefer an instrument manufactured as per IEC 61508. A few such applications are elaborated next to throw some light on the application of SIS in a few other industries.

SIS in steel production: Production of direct reduced iron involves reduction of iron ores in pallet or lump form in the presence of reducing gas comprising CO and H_2. SIS plays an important role in this process, which is a chemical process rather than a normal iron extraction process. The shaft furnace used in this process operates at 6 kg/cm^2 and 1000°C. The greatest hazard is associated with the failure of the O_2 injection process, which may cause backflow of H_2 into the O_2 injection system. So, normally, three block valves are used in the system. A suitable safe trip has been suggested, which first opens the N_2 injection valve and closes the O_2 block valves. The third one is kept open for the N_2 purging operation, receiving the closing of block valves 1 and 2. This is another hazard when CO_2, CO, and H_2 gases emerge from the furnace. This could happen if the water used to cool the pipe fails. So, a high-temperature trip could be used for the furnace. For this kind of hot iron production, IEC 61508 for life cycle analysis is necessary. Usually, SIL2/3 SIS are used. In cases of important measurements, 2oo3 are deployed. In most of the cases, integrated SIS and BPCS are used and IEC 61511 is applied to demonstrate that SIS is independent.

In the case of steel production in normal ways also find SIS applications in the integrated process, for example, sinter plant and ore processing plants.

6.0 FINAL DISCUSSIONS

It is now time to conclude discussions on SIS. This means the basic philosophy of SIS and what we have learnt in previous chapters can be summarized as follows.

All plant/facility owners at present are looking for SIS manufacturers/suppliers that are capable of offering a set of products and services to provide complete support of their installed systems for the entire life cycle. On account of steep competition, no plant/facility owner can afford to accept downtime because of system failure/property loss/penalty caused by violation of regulations/deterioration of their product quality; at the same time the products shall be competitively priced also. The best way to reduce risks in a plant/facility is to design inherently safe processes. However, this is neither possible nor can it ensure that the system will be completely safe, especially in cases where there are inherent risks (say) because of handling of hazardous or toxic materials. Industry culture is very much responsible and related to safety of the system. Apart from the safety culture of the company there are a few other issues responsible for increased risks that at any cost should be avoided as far as possible. The major reasons for increased risks are:

- Operation of plant and machinery close to their limits
- Safety culture missing
- Deployment of untrained personnel
- Human factors, for example, stress

- Use of hazardous/toxic materials without proper care
- Transient operations (namely, startup/shutdown/trip condition)

In contrast, it is also true that people are now much more aware of:

- Environment impacts
- Pressure because of local/international regulations
- The need not to tarnish the company image
- Safety standards and associated requirements

So, most plant/facility owners look for safety system suppliers that possess the following qualitative requirements, so that safety of the plant is taken care of:

- Supplied control and/or safety system has predictive maintenance [27].
- Smart field safety devices are provided with necessary fault diagnostics.
- Logic solver shall have:
    - Flexibility
    - Fault tolerant architecture
    - Scalability to meet specific requirements
- Integrated control and safety systems can provide:
    - Controllers of diverse technology and software
    - Variable configuration of controllers
    - Scalability on an application basis
- Tools and support are needed for life cycle management.
- Comprehensive system support is necessary for a full life cycle.
- Well-defined built-in security system is needed:
    - To avoid and prevent undesired intruders
    - To isolate the system with firewalls
    - To develop zone and conduit

It has been stated many times that there are still two schools of thought regarding integrating control and safety, that is, BPCS and SIS. With advancement of technology it is possible to take the advantage of integration yet eliminate the problems of systematic failure and CCFs. Based on the foregoing the following recommendations are suggested.

6.1 Recommendations (Suggestive)

The following are a few recommendations suggested on a generalized basis:

- Implement a well-thought-out, intelligent SIS and safety strategy that can offer:
    - Safety and protection for people, property, and the environment
    - Reduced downtime
    - Increased productivity
    - Improved machine efficiency
    - An edge over competitors
    - Lowering of overall cost of ownership

- Carry out standards-based hazard and risk analysis to decide on the right level of protection
- Implement IEC 61508/61511 or ISA 84 as applicable
- For SIS evaluation, choose a certified system to meet the highest applicable SIL requirement, based on the fault tolerant system chosen
- Depending on applicability, carry out integrations for existing systems and new systems as far as feasible
- Implement total life cycle management for new and/or integrated systems
- While integrating, apply diversification in technology and software to control systems and SIS
- Apply the huge potential in fieldbus systems to the system to ensure a safety fieldbus (FDI, refer Chapter IX fieldbus discussions) system for future augmentations.
- Always use a certified product from manufacturers/suppliers (preferably the same supplier for sensor, final element, and logic solver to avoid interface issues, if any) that can offer a comprehensive set of products with diverse technology and services and comprehensive support for your installed systems [27].

With this, the main discussions on plant hazard analysis and SIS are concluded. To supplement the discussions presented in the main chapters, additional pertinent information is provided in the appendices.

I shall be more than happy to share my experiences with you properly. This will bring fruitful results ***only*** *when utilized in association with some of the issues people are facing day to day in their plants. Any good suggestions are always welcome.*

LIST OF ABBREVIATIONS

AI/O	Analog input/output
BMS	Burner management system
BPCS	Basic plant Chapter IX system
C&I	Control and instrumentation (engineering)
CCF	Common cause failure
COTS	Commercial off the shelf
DCS	Distributed control system
DI/O	Digital input/output
E/E/PE	Electrical/electronics/programmable electronics
EMI	Electromagnetic interference
EPL	Equipment protection level
ESD	Emergency shutdown system ***or*** electro static disturbance
EUC	Equipment under control
FAT	Factory acceptance test
FPSO	Floating storage production storage and offloading
FTC	Fault-tolerant control

Continued

HC	Hydrocarbon
HIPPS	High integrity pressure protection system
HMI	Human—machine interface
HW	Hardware
IACS	Industrial automation and control systems
IAEA	International atomic energy agency
I&C	Instrumentation and control
IEC	International Electrotechnical Commission
IGCC	Integrated gasification and combined cycle
I/O	Input/output
I/P or O/P	Input or output
IP	Ingress protection or internet protocol
IPL	Independent protection layer
IT	Information technology
LHS	Left hand side
LOPA	Layer of protection analysis
LS	Logic solver
LVM	Limit value monitor
MIS	Management information system
MOC	Management of change
MTBF	Mean time between failure
MTTR	Mean time to repair
O&M	Operation and maintenance
OREDA	Offshore reliability equipment database association
PFD	Probability of failure on demand or doc: process flow diagram
PFH	Probability of failure per hour
P&ID	Piping and instrumentation diagram
PLC	Programmable logic controller
QA	Quality assurance
RFI	Radio frequency disturbance
RHS	Right hand side
RRF	Risk reduction factor
SCADA	Supervisory control and data acquisition
SIF	Safety instrument functions
SIL	Safety integrity level
SIS	Safety instrumentation system/supervisory information system (in case of DCS)
SOE	Sequence of event
SW	Software
T/C	Thermocouple
TMR	Triple modular redundancy
WRT	With respect to

References

[1] S. Basu, A.K. Debnath, Power Plant Instrumentation and Control Handbook, Elsevier, November 2014. http://store.elsevier.com/Power-Plant-Instrumentation-and-Control-Handbook/Swapan-Basu/isbn-9780128011737/.

[2] Overview of Safety Instrumented Systems; IDC Technologies; http://www.idc-online.com/downloads/ALL_BOOKBOON_SS_PDF_REV1.pdf.

[3] Burner Management System Codes and Standards Update; M. Scott; A.E. Process solutions; https://www.isa.org/pdfs/microsites327/website-akisaapi556nfpa87/.

[4] Safety Controls and Burner Management Systems (BMS) on Direct-Fired Multiple Burner Heaters; http://www.onquest.com/docs/BurnerManagementSystems_Pres.pdf.

[5] D. Sheppard, Burner Management — A Straightforward Approach Using DeltaV SIS for Typical Systems; Plantweb; Emerson process management; Emerson global user exchange, 2010.

[6] Burner Management Systems (BMS); SIS technical application; SIS-TECH Applications, LP; www.sis-tech.com and http://sis-tech.com/wp-content/uploads/2011/05/DIAMOND-SIS-BMS.pdf.

[7] M. Schmidt, C. Miller, What now? More Standards for Safety and Regulatory Compliance, Emerson global users exchange; Emerson process management; http://www.slideshare.net/EmersonExchange/what-now-more-standards-for-safety-and-regulatory-compliance.

[8] J. Sanchez, SIL application in burner management systems - a case study: thermal burner, in: Safety Control & Instrumentation Systems Conference, IDC Technologies, 2009.

[9] The Tricon Turbine Control System; White paper; Invensys; Triconex; N. Desai; Invensys.

[10] Mark* VIe for Gas Turbines with SIL Capable Protection; Fact sheet; GE Energy; https://www.gemeasurement.com/sites/gemc.dev/files/mark_vie_with_sil_capable_protection_fact_sheet_english_1.pdf.

[11] Instrumentation and Control (I&C) Systems in Nuclear Power Plants: A Time of Transition; https://www.iaea.org/About/Policy/GC/GC52/GC52InfDocuments/English/gc52inf-3-att5_en.pdf.

[12] Technical Specifications (In-Cash Procurement): Plant Control Design Handbook for Nuclear control systems; IDM UID 2YNEFU; F. Robles C, January 2013, Version 3; http://static.iter.org/codac/pcdh7/Folder%201/17-Plant_Control_Design_Handbook_for_Nuclea_2YNEFU_v3_0.pdf.

[13] Safety Classification for I&C Systems in Nuclear Power Plants — Current Status & Difficulties; CORDEL digital instrumentation & control task force; World Nuclear Association.

[14] O. Bäckström, J.-E. Holmberg, Use of IEC 61508 in nuclear applications regarding software reliability, in: 11th International Probabilistic Safety Assessment and Management; Conference & the Annual European Safety and Reliability Conference, 2012 (Digital Open Access Repository of VTT).

[15] Instrumentation and Control Systems Important to Safety in Nuclear Power Plants; Safety guide No. NS-G-1.3; IAEA Safety Standards Series; International Atomic Energy Agency, Vienna.

[16] Instrumentation and Control Systems and Software Important to safety for Research Reactors; IAEA Safety Standards; Specific Safety Guide; No. SSG-37.

[17] Integrated and Safe Operations in the Offshore Oil & Gas Industry; ABB; https://www.controlglobal.com/assets/knowledge_centers/abb/assets/Integrated-and-Safe-Operations-in-Offshore-OG-d1.pdf.

[18] D.E. Reed, An Update on Process Automation Standard ISA 106; Rockwell automation.

[19] Drill-floor Machinery and Tubular-handling Safety; Health and Safety Executive, UK; http://www.hse.gov.uk/offshore/infosheets/is2-2013.pdf.

[20] Overlooked Instrumentation can Lead to Safety Incidents; wika – Instrument Guru, August 2013; http://blog.wika.us/blog/2013/08/16/overlooked-instrumentation-lead-safety-incidents/.

[21] Safety, security concerns growing challenges for refineries and contractors, Oil & Gas Journal (March 2007). http://www.ogj.com/articles/print/volume-105/issue-12/supplement-to-oil-gas-journal/technology-forum-refining-equipment-services/safety-security-concerns-growing-challenge-for-refiners-contractors.html.

[22] H. Isobe, Consolidation of Control Rooms for Achieving Safe, Stable and Efficient Operation by Refinery Operation Modernization Services, White Paper, Yokogawa Corporation of America; http://www.yokogawa.com/us/technical-library/resources/white-papers/consolidation-of-control-rooms-for-achieving-safe-stable-and-efficient-operation-by-refinery-operation-modernization-service/.

[23] SIS-TECH Introduces Coker Drum Safety System for Crude Oil Refineries, September 2010. http://www.automation.com/product-showcase/sis-tech-introduces-coker-drum-safety-system-for-crude-oil-refineries.

[24] S. Sonnenberg, Smart Refinery, Emerson Process Management, 2010, in: http://www2.emersonprocess.com/siteadmincenter/PM%20Articles/SmartRefinerySupplement_Sept2010_Final.pdf.

[25] Exapilot Automates Naphtha Cracker Decoking and Dryer Regeneration Processes at Large Ethylene Plant; Yokogawa America; Library Document; http://www.yokogawa.com/us/technical/resources/references/exapilotautomatesnaphthacrackerdecokinganddryerregenerationprocessesatlarg.

[26] N. Bishop, D. Sheeppar, Role of fired heater safety systems, Emerson Process Management; http://www2.emersonprocess.com/siteadmincenter/PM%20Articles/PTQ2013.pdf.

[27] ARC White Paper, Siemens Offers a Comprehensive Product Portfolio for Safety Lifecycle Management, ARC Advisory Group, May 2009, http://w3.siemens.com/topics/mea/en/safety-integrated/Documents/white-paper-process-safety-2009.pdf.

[28] K. Keim, ANSI/API RP-754; Process Safety Performance Indicators for the Refining & Petrochemical Industries.

[29] A.S. Shaney, Challenges in Commissioning & Operation of Delayed Coker at IOCL Gujarat Refinery, Indian Oil Corporation, April 2012.

Further Reading

A New Edition of API 670 -25 Years and Still Going Strong; S. Sabin; Bently Nevada Corporation; http://www.ge-mcs.com/download/orbit-archives/2001-2005/1st_quarter_2001/1q01sabin2.pdf.

A. Bobbio, S. Bologna, E. Ciancamerla, P. Incalcaterra, C. Kropp, M. Minichino, et al., Advanced Techniques for Safety Analysis Applied to the Gas Turbine Control System of ICARO Co-generation, Università del Piemonte Orientale, ENEA CR Casaccia, Università dell'Aquila; TESEC, Genova, Italy, 2001.

Burner Management System SIMATIC BMS400F Brochure; Process Automation System, April 2010. http://www.industry.usa.siemens.com/verticals/us/en/chemical/Documents/bmsbrochure.pdf.

Burner Management Systems Solutions Increase the Safety and Availability of Your Combustion Process; Invensys; Triconex; Schneider electric; Technical write up.

Coker Drum De-Heading; SIS Tech Application; Catalog; http://sis-tech.com/applications/diamond-sis/coker-drum-de-heading/.

H. Devold, Oil and Gas Production Handbook An Introduction to Oil and Gas Production, Transport, Refining and Petrochemical Industry, ABB, ISBN:978-82-997886-3-2.

Drilling deeper for offshore oil and gas production, Yokogawa Corporation of America; Media Publication; Oil and Gas; http://yokogawa.com/us/technical-library/resources/ media-publications/drilling-deeper-for-offshore-oil-and-gas-production/.

D. Emerson, Automating Manual Procedures in Continuous Process Applications Using the Concepts of ISA-106, Vigilant Pant; Yokogawa User Conference and Exhibition, 2014. http://cdn1.us.yokogawa.com/document_12008.pdf.

Gas processing; Yokogawa India; http://www.yokogawa.com/in/industries/oil-gasmidstream/gas-processing/.

Interlock for drill floor machinery; HSE, 2006 Interlock for drill floor machinery; HSE.GOV.UK; safety notice December 2006; http://www.hse.gov.uk/offshore/notices/sn_02_06.htm.

ISA-106 Procedure Automation for Continuous Process Operations, ARC Industry Forum, Minimizing Operational Risk through Procedural Automation, 2014, in: http://www.arcweb.com/events/arc-industry-forum-orlando/arcindustryforumorlando2014presentations/ISA-106%20Procedure%20Automation%20for%20Continuous%20Process%20Operations.pdf.

R. Kamphaus, Comparison of Integrated versus Dedicated Turbine Safety System,Woodward Inc., USA; http://pennwell.websds.net/2013/vienna/pge/papers/t5s5o4-paper.pdf.

G.G. Lopez, E. Noriega, Hot Iron; Iron Reduction Technology Keeps Plant Shutdown Safe and Free, Emerson Process Management; Intech e ISA, December 2008.

E.M. Marszal, The State of Safety Instrumented System in Offshore Production, Kenxis; Offshore Magazine, January 2011. http://www.offshore-mag.com/articles/print/volume-71/issue-6/departments/offshore-automation-solutions/the-state-of-safety-instrumentedsystems-in-offshore-production.html.

E.M. Marszal, Preparing for subsea safety instrumented system,Kenxis; Offshore Magazine; http://www.offshore-mag.com/articles/print/volume-72/issue-6/departments/offshoreautomation-solutions/preparing-for-subsea-safety-instrumented-systems.html.

C. Miller, Can performance- based safety process prevent offshore oil spills? Emerson process management; World oil, January 2012.

Optimization of plant performance. Automation Solutions for the Oil & Gas Industry, Totally Integrated Automation, Siemens; https://w5.siemens.com/france/web/fr/ad/secteur/OG/Documents/Brochure_TIA_OG_EN.pdf.

Power plant in a new safety perspective, "Engineerlive" Magazine (February 2013). http://www.engineerlive.com/content/24229.

Pipeline SCADA-Automation and Safety Infrastructure for Pipelines; Honeywell Process Automation; https://www.honeywellprocess.com/en-US/explore/solutions/industrysolutions/oil-and-gas/pipelines/Pages/automation-and-safety-infrastructure.aspx.

Safety instrumented system for the over pressure protection of pipeline risers; HSE.UK; Spec/tech/OSD/31; http://www.hse.gov.uk/foi/internalops/hid_circs/technical_osd/spc_tech_osd_31.htm.

L. Scibile, J.Y. Journeaux, W. Klotz, I. Yonekawa, A. Wallander, An overview of the ITER interlock and safety systems, in: Proceedings of ICALEPCS, ITER Organization, Saint-Paul-lez-Durance, France, 2009. Kobe, Japan; WEC005.

Specification of Safety Integrated Burner Management Systems; Automation World and Design World; Totally Integrated Automation, March 2012. http://www.totallyintegratedautomation.com/2012/03/specification-of-safety-integrated-burnermanagement-systems/#comment-92039.

TelecommunicationInstrumentation & SCADA Systems In, Telecommunication, Instrumentation & SCADA Systems In Indian Oil Pipelines; K. Kumar; Indian Oil Corporation; petrofed.winwinhosting.net/upload/15e18June11/8.ppt.

J. Thomas, F. Luiz de Lemos, N. Leveson, Evaluating the Safety of Digital Instrumentation and Control Systems in Nuclear Power Plants, November 2012. Research Report: NRCHQ-11-6-04-0060, http://sunnyday.mit.edu/papers/MIT-Research-Report-NRC-7-28.pdf.

T10-Standards Review e Burner Management and Combustion Control; Rockwell Automation; https://www.rockwellautomation.com/resources/downloads/rockwellautomation/pdf/events/raotm/sessions/tech/T10StandardsReviewBurnerManagementandCombustionControl.pdf.

Popular Terms and Their Definitions

The definitions given as follows have references to various standards, but are slightly modified to suit the requirements of the book.

Term	Definitions
Accident	An unplanned event or sequence of events that results in undesirable consequences. An incident with specific safety consequences or impact.
Accommodation area	Space used for cabins, offices, lavatories, corridors, hospitals, cinemas, public spaces, etc. Service spaces and control stations may be a part it.
Administrative control	A procedural requirement for directing and/or checking engineered systems or human performance associated with plant operations.
Application software	Software specific to the user application is application software such as: logic sequences, permissive, calculations, and decisions necessary to meet the safety instrumented function requirements.
Architecture	It represents the arrangement of hardware and/or software elements in a system. Normally this is related to programmable electronic devices, e.g. internal structure of a safety instrumented system subsystem.
Bonding	Bonding refers to the interconnection of two pieces of conducting material so as to prevent a potential difference between them, that is, equipotential.
BPCS	Basic plant control system, or BPCS, is the control system responsible to operate the facility in the desired manner. It may have protection system but the safety integrity level will be ≤ 1. This also includes manual control (see CHI and IEC61511:1-2003).
Cause consequence analysis	A method for illustrating the possible outcomes arising from the logical combination of selected input events or states. It could be a combination of event and fault tree.
Chronic hazard	A group of hazards which arise from continuous conditions (e.g., large-scale toxic contamination), which accumulate over time.
Code/coding	See programming.
Common cause failure	The occurrence of two or more channel/device failure in redundant configuration, leading to system failure. This results from one or more events or circumstances. Normally this is common in cases of the same redundant hardware/devices (see Chapter VII).

Continued

—cont'd

Term	Definitions
Common mode failure	Failure of two or more channels/devices in a redundant configuration, on account of the same failure mode, causing the same erroneous result.
Continuous mode SIF	The cases where in the event of a dangerous failure of the safety instrumented function (SIF), a potential hazard will occur without further failure unless action is taken to prevent it.
Cut set	Group of events that if all occur would cause occurrence of top event (fault tree analysis).
Dangerous failure	This represents those failures which have the potential to put safety instrumented system in a hazardous or fail-to-function state.
Demand mode SIF	Demand mode safety instrumented function (SIF) where a specified action is taken in response to process conditions or other demands. In the event of a dangerous failure of SIF, a potential hazard only occurs in the event of a failure in the process or the basic plant control system.
Diagnostic coverage (DC)	It stands for the ratio of the detected failure rate to the total failure rate of the component or subsystem as detected by diagnostic tests but excluding any fault detection during proof tests (see Chapter VII).
Diversity	Diversity represents the existence of different means performing a required function. This could be physical or design approach also.
Drilling area and utility area	This includes the derrick, drill floor, blowout preventer (BOP) area, and the area containing shale shakers and degassers. There is also a drilling utility area for drilling utilities, viz. mud mixing, pumping, bulk storage, and cementing. Utility area generally refers to power area.
Earthing	Earthing is a way of a specific return path for fault currents in order to operate protective devices in a very short time (see grounding also).
E/E/PE	It represents the control/protection/monitoring system and/or combination of them, based on one or more electrical/electronic/ programmable electronic (E/E/PE) devices including all elements of the system such as power supply, sensor, input/output (I/O), control system, final elements, etc. The terms encompasses hardware, software, and I/O.
Embedded software	Software that is part of the system supplied by the manufacturer, not accessible by the end-user. It is also referred to as firmware or system software.
Emergency response	Action to safeguard the health and safety of persons on or near the unit or installation.
Engineered control	A specific hardware or software system designed to maintain a process within safe operating limits, to safely shut it down in the event of a process upset, or to reduce human exposure to the effects of an upset.
Error	It is the discrepancy or difference (control system) between a computed, observed, or measured value or condition and the true, specified, or theoretically correct value or condition.

—*cont'd*

Term	Definitions
EUC	IEC 61508 refers to EUC (equipment under control), while IEC 61511 refers to process. As per IEC 61508-4; 2010, EUC stands for equipment, machinery, apparatus, or plant used for manufacturing, process, transportation, medical, or other activities.
EUC control	The system that responds to input signal from process/facility and/or from an operator and generates output signal causing equipment under control (EUC) to operate in the desired manner.
Event	An occurrence related to equipment performance or human action, or an occurrence external issue to the system to cause system upset or an accident. An event could be the cause or a contributor to the accident.
Event sequence	A specific series of events composed of an initiating event and intermediate events that may lead to an incident.
Failure	It stands to mean the end or termination of the ability of a functional unit to perform a required function. (see Chapters VII and XI for difference with fault).
Failure mode	A condition or fashion in which hardware fails or it could be a symptom of hardware failure. A failure mode could be: loss of function, non timely function, out-of-tolerance condition, etc.
Fault	Condition that may cause a reduction in, or loss of, the capability of a *functional unit (entity of hardware, software, or both)* to perform a required function.
Fault avoidance	The techniques and procedures which aim to avoid the introduction of faults during any phase of the safety life cycle of a safety instrumented system.
Fault tolerance	It is the capability of a *functional unit (entity of hardware, software, or both)* to continue to perform a required function in the presence of faults or errors.
Final element	This represents that part of a safety instrumented system which is responsible for implementation of the physical action necessary to achieve a safe state.
Fixed program language (FPL)	This is a type of language; the user has limited adjustment of a few parameters (viz. alarm level).
Full variability language (FVL)	This type of language is designed to be comprehensible to computer programmers and provides the capability to implement a wide variety of functions and applications.
Functional safety	As per IEC 61508:4:2010, "*it is part of overall safety relating to EUC and EUC control system that depends on functioning of E/E/PE safety related system and other risk reduction measures*." From IEC 61511, one gets that, "*it is that part of the overall safety related to the process and the BPCS which depends on the correct functioning of the S1S and other protection layers (not only SIS)*." The difference is due to the fact that IEC 61511 is related to process.
Functional safety assessment	It represents evidence-based investigation to judge the functional safety achieved by one or more protection layers (including safety instrumented system).

Continued

—cont'd

Term	Definitions
Ground and grounding	*Ground*: Usually ground refers to earth, or conducting system connecting a part to earth. *Grounding*: With reference to National Electrical Code (NEC) and bonding, it requires bonding of metal raceways, cable trays, cable armor, cable sheath, enclosures, frames, fittings, and other metal non-current-carrying parts that are to serve as grounding conductors, even if a separate grounding conductor is used. For grounding, low resistance conductors are used.
Hazard event	Hazard event refers to parameters which are the cause of an accident.
Human error	Any human action (or lacking) that goes beyond limit of acceptability set by the system. This may be cause of an accident.
Human factor	A discipline concerned with designing machines, operations, and work environments to match human capabilities, limitations, and needs.
Initiating event	The first event in an event sequence to cause an accident unless engineered protection systems or human actions intervene to prevent or mitigate the accident.
Intrinsic safety	It is a type of protection in which a portion of the electrical system contains only intrinsically safe equipment (apparatus, circuits, and wiring), which is not capable of causing ignition in the surrounding atmosphere. Generally, no single device or wiring is intrinsically safe by itself, but is intrinsically safe only when employed in a properly designed intrinsically safe system [ISA].
Intrinsic apparatus	Apparatus in which all the circuits are intrinsically safe.
Intrinsic safe barrier	This is a network designed to limit the energy available to the protected circuit in the hazardous location, under specified fault conditions.
Intrinsic safe circuit	A circuit in which any spark or thermal effect is incapable of causing ignition of a mixture of flammable or combustible material in air under prescribed test conditions.
Intrinsic safety ground system	This is a grounding system with a dedicated conductor isolated from the power system, except at one point, so that ground currents will not normally flow, and that is reliably connected to a grounding electrode.
Life cycle	Time interval between a product's or project's conception and its disposal (including decommissioning).
Life cycle cost	Cumulative cost of product or project over life cycle (fixed and variable).
Lightning protection level	Based on minimum and maximum lightning current parameter for designing surge protection devices, it has been divided in to four levels. This is lightning protection level. This is based on 10/350 μs wave form. Levels and corresponding maximum/minimum current values in KA are: *LPL I* (200/3) *LPL II* (150/5) *LPL III* (100/10) *LPL IV* (100/16)[refer BS EN/IEC62305-1].

—cont'd

Term	Definitions
Lightning protection system	As per UL 96A, lightning system is defined as a complete system of strike termination devices, conductors, grounding electrodes, interconnecting conductors, surge protective devices, and connectors or fittings.
Lightning protection zone	Lightning protection zone is a concept developed to protect equipment from lightning, especially meant for hazardous and potentially explosive areas. The concept allows planning, implementation, and monitoring protection measures. Based on these zones, protection measures can be determined in particular lightning and surge protection devices and components. There are mainly LPZ0A, LPZ0B, LPZ 1, and LPZ2 lightning protection zones.
Limited variability language (LVL)	This type of language is designed for comprehensible use in process sector. This provides the capability to combine predefine application specific, library functions to implement the safety requirements specifications. An LVL provides a close functional correspondence with the functions required to achieve the application.
Logistic support	The materials and services required to operate, maintain, and repair a system/process.
Loss of primary containment	An unplanned or uncontrolled release of material from primary containment, including non-toxic and non-flammable materials (viz. steam, hot condensate, etc.) – API.
Mitigation	Action that reduces the consequences of a hazardous event.
Mitigation system	Equipment and/or procedures designed to interfere with incident propagation and/or reduce incident consequences.
Mode of operation	The way in which a safety instrumented function operates is mode of operation.
Performance level (PL)	Performance level is a measure of the reliability of a safety function. PL is divided into five levels (a–e).
Primary containment	A tank, vessel, pipe, truck, rail car, or equipment intended to serve as the primary container or used for processing or transfer of material – API (*see loss of containment*).
Prior use	See Proven-in-use.
Process audit	An inspection of a plant or process unit, drawings, procedures, emergency plans, and/or management systems, etc., usually by an independent, impartial team.
Product storage area	Part of a unit or installation which contains: The storage spaces, the pump rooms and/or cofferdams adjacent to product storage tanks, and includes deck areas.
Programming	Process of designing, writing, and testing a set of instructions for solving a problem or processing data.
Proof test	Test performed to reveal undetected faults in safety instrumented functions so that, if necessary, the system can be restored to its designed functionality.

Continued

—cont'd

Term	Definitions
Protection layer	Any independent mechanism to reduce risk by control, prevention, or mitigation.
Protective system	Devices or systems that prevent the occurrence of, or mitigate the effects of an accident.
Proven-in-use	*"When a documented assessment has shown that there is appropriate evidence, based on the previous use of the component, that the component is suitable for use in a safety instrumented system."*—IEC 61511.
Recovery factors	Feedback factors that limit or prevent the undesirable consequences.
Riser area	Area containing import and/or exports risers, and includes the isolation valve on the riser.
Risk avoidance	Based on risk evaluation, the decision not to be involved in, or to withdraw from, an activity based on the level of risk.
Risk criteria	Risk criteria are based on internal/external context/standard, law/policy. These are regularly reviewed to ensure continued relevance.
Safe failure	Type of failure which does not have the potential to put a safety instrumented system in a hazardous or fail-to-function state.
Safe failure fraction	Fraction of the overall random hardware failure rate of a device pertinent to safe failure or a detected dangerous failure (see Chapters VI and VII).
Safe state	State of the process when safety is achieved.
Safety function	Function implemented by a safety instrumented system or other safety related technological system for reduction risk of the facilities, i.e. to achieve or maintain a safe state for the process, with respect to a specific hazardous event.
Safety instrumented function (SIF)	The requirements that the system is safely functioning with a specified safety integrity level to achieve functional safety. This could be a safety instrumented control/protection function.
Safety instrumented system (SIS)	The instrumentation function used to implement one or more safety instrumented functions. SIS is comprised of a combination of sensors, logic solvers, and final elements.
Safety integrity	It is the average probability of safety instrumented systems satisfactorily performing the required safety instrumented functions under all the stated conditions within a stated period of time.
Safety integrity level (SIL)	It represents the discrete level (1−4) for specifying the safety integrity requirements of the safety instrumented functions to be allocated to the safety instrumented system.
Safety life cycle	All required activities involved in the implementation of safety instrumented functions occurring during a period of time that starts at the concept phase of a project and finishes with decommissioning including management of change.
Safety requirements specification	It stands for specification that contains all the necessary requirements of the safety instrumented functions to be performed by the safety instrumented systems.
Safety review	Scrutiny or inspection of a facility or part thereof, drawings, procedures, emergency plans, and/or management systems, etc. This is usually by a team for the purpose of solving the issues.

—cont'd

Term	Definitions
Safety system	Equipment/devices/systems and/or procedures designed for mitigating the accident and its consequences.
Secondary containment	Secondary containment systems include tank, dikes, curbing around process equipment, drainage collection systems, the outer wall of double walled tanks, etc. — API (*see Loss of containment*).
Sensor	This represents the device or combination of devices, which measure the process condition (viz. transducer/transmitter).
Stakeholder	Any person or organization that can affect, be affected by, or perceive themselves to be affected by a decision or activity.
Systematic failure	Failure related in a deterministic way to a certain cause, and can only be eliminated by a modification of the design or of the manufacturing/operational process.
Tolerable risk	Risk which is accepted in a given context based on the current values of society.
Turret area	Area containing mooring equipment, which enables the unit to rotate relative to fixed facilities or pipelines on the seabed. Import and export risers are usually located within the turret area.
Utility software	Software tools for the creation, modification, and documentation of application programs. These software tools are not required for the operation of the safety instrumented system.
Validation	Activity for demonstrating that safety instrumented functions and safety instrumented systems under consideration after installation meets in all respects the safety requirements specification.
Verification	Activity of demonstrating for each phase of the relevant safety life cycle by analysis and/or tests, for specific inputs, the outputs meet in all respects the objectives and requirements set for the specific phase.
Watch dog	Combination of diagnostics and an output devices meant to monitor the correct operation of the programmable electronic (PE) devices and taking action upon detection of an in correct operation.

Electrical Protection and Enclosure Class

1.0 GENERAL

The details discussed here shall be read in conjunction with Chapter X. In fact, engineering details for electrical enclosure class and area classification has been discussed in details in Chapter X. Here, supplementary information has been included to complete the discussion on the subject. Explosion proof and flame proof enclosures are the two most commonly used hazardous area protection techniques.

Really they are same, or they have some differences?

1.1 Explosion Proof versus Flame Proof Enclosures

There are many similarities between the two in the sense that flame proof and explosion proof enclosures have the same objectives:

- To keep the insides isolated from the outer hazardous atmosphere.
- In case of an explosion inside the box, the products of the explosion remain inside, i.e. they prevent the transmission of the explosion to the explosive atmosphere surrounding the enclosure.
- Unavoidable gaps are so long and narrow that these gases will not ignite the volatile gases in a hazardous area.

However there are some differences between the two as elaborated in Table APII/1.1-1.

Different standards followed by different countries for explosion proof and flame proof enclosures are listed in Table APII/1.1-2.

1.2 ATEX Directive

The discussions presented as follows are supplementary to what has been discussed in Clauses 3.2.2 and 3.3.3 of Chapter X.

1.2.1 ATEX Core Idea

The core idea behind ATEX is as follows:

- No references to product standards.
- Covers basic health and safety requirements.
- Applicable to electrical and non-electrical equipment.
- Applicable to potentially explosive environments due to gas and dust.

Plant Hazard Analysis and Safety Instrumentation Systems.

- Applicable to mines and surfaces.
- Accounts for all potential hazards from the equipment.

Table APII/1.1-1 Differences Between Flame Proof and Explosion Proof Enclosures

Comparing Point	Explosion Proof Enclosure	Flame Proof Enclosure
Pressure testing	4 times maximum possible pressure	1.5 times maximum possible pressure
Testing	Generally individual	Generally individual
Country	USA/Canada	As per Europe and international – IEC
Drilling	Field drilling allowed	Field drilling not allowed
Joints	Bolted or joints	Bolted or joints,[a] but require special tool to open

[a]*Bolts/joints are: flanged with bolt, threaded, or spigot joint (spigot cotter and socket).*

Table APII/1.1-2 Standards Used for Explosion Proof and Flame Proof Enclosures

Canada	USA	Europe	International–IEC
CSA 60079-0	FM3600	EN 60079-0	IEC 60079-0
CSA 60079-1	ISA 60079-0	EN 60079-1	IEC 60079-1
C22.2-30	ISA 60079-1	EN 13463-1	IEC 13463-1
	UL 1203	EN 13463-2	IEC 13463-2
		EN 13463-3	IEC 13463-3

1.2.2 ATEX Exemptions

The following cases are exempted from ATEX directive:

- Medical devices in a medical environment.
- Equipment where the explosion hazards result exclusively from the presence of explosive substances or unstable chemical substances.
- Equipment in domestic and non-commercial environments.
- Personal protective equipment, ref: 89/686/EEC.
- Seagoing vessels and mobile offshore covered by the International Maritime Organization (IMO) convention and other units (not exempted: fixed platforms, floating production storage and offloading (FPSO), etc. intended for navigation on high seas).
- Means of transport not intended for use in a potentially explosive atmosphere.

1.3 Duties of Various Agencies

The most popular symbol of Ex in a hexagon stands to represent the duty sharing and coordination among various concerned agencies as shown in Fig. APII/1.3-1.

In order to achieve the best possible safety in potentially explosive areas, it is necessary that all the concerned parties work hand-in-hand. So, close cooperation

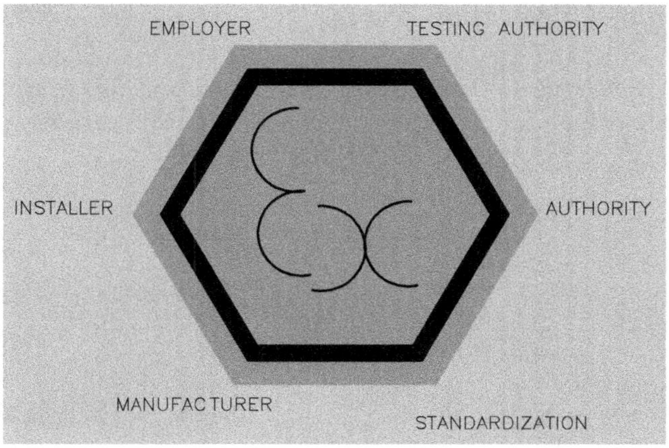

FIGURE APII/1.3-1

Symbol of cooperation.

Developed based on idea from basics of explosion protection: STAHL. Courtesy: STAHL.

among them is essential. The hexagon associated with Ex symbol stands to represent the same [3].

1.3.1 *Employer*

The employer is responsible for:

- The safety of entire installations.
- To check and verify area classification of facility is completed properly.
- Thorough inspection of all installations.
- To ensure proper installations as per regulation applicable.
- Periodic inspection and maintenance.

1.3.2 *Installer*

The installer is required to make sure that all installation requirements as per norm and manufacturer's recommendation are carried out, i.e. selection of correct electric apparatus and proper installation.

1.3.3 *Manufacturer*

Certification, routine testing, and documentation are major responsibilities of manufacturers of explosion protected apparatus. Design, manufacturing, and compliance with approved design are also responsibility of the manufacturer.

2.0 ENCLOSURE TYPES, APPLICATION NOTES, AND STANDARDS

Various enclosure classes normally used in various applications have been discussed in Clause 3.6 of Chapter X. Here, the supplementary information related to that will be covered.

2.1 Selection of Protection Types

It is better to try to find the type of enclosure best suited for different applications [2]. From Table APII/2.1-1, various enclosure types and intrinsic safety can be selected based on application. This is a limited guide table with electrical apparatuses only.

Table APII/2.1-1 Enclosure Type as per Application

Application	Prot. type.	Gas/Vapor	Dust	Cat	EPL
Explosive can enter enclosure to cause explosion inside, but cannot spread surrounding atmosphere.	Ex d/ Ex q/ Ex nc*	YES		2	b
The item/equipment with enclosure to prevent ingress of explosive mixture/ignition source.	Ex p/ Ex nR/ Ex m/ Ex o/ Ex nc	YES YES YES YES YES	YES YES	2/3 3 1–3 2 3	b/c c a–c b c
Explosive can enter enclosure but unable to cause explosion. Spark and temperature is restricted.	Ex e	YES		2	b
Explosive can enter enclosure but unable to cause explosion. Sparks and temperature are able to cause ignition but energy is limited.	Ex i Ex nL	YES	YES	1–3	a–c

Cat, *category*; EPL, *equipment protection level*; Prot, *protection*.

2.2 Protection Concepts and Comparison of Standards

It is not unlikely that there may be some confusion in the reader's mind, after going through so many ways of area classifications, again, each class further divided according to material types. Also, these are not uniform across standards. For example, in ATEX for dust in mines, and other than mines, have different representations and ignitable fibers mentioned in the United States standard is not separately treated in other standard. In order to circumvent the situation, a small comparative study has been presented after, so the reader can understand the same properly. In order to supplement Table X/3.2.3-3, at this point some detailed comparisons of various standards have been presented in Table APII/2.2-1 and Fig. APII/2.2-1 (mainly for enclosure class) to get a better grasp on the issue about applicability (as per location/country).

Table APII/2.2-1 General Requirements of Various Protection Standards

Explosive Atmosphere	Geographical Location	Standards	Code
Class I Division 1 and 2	USA	FM3600	–
Class I Division 1 and 2	Canada	CSA C22.2–0	–
Class I Division 1 and 2	USA	ISA 60079-0	AEx
Class I Division 1 and 2	Canada	CSA C22.2-60079-0	Ex
Category 1G/2G/3G	Europe union	EN 60079-0	Ex
EPL Ga/Gb/Gc	International IEC	IEC 60079-0	Ex

See Fig. APII/2.2-1 also.

PROTECTION TYPE	PROTECTION PRINCIPLE	EXPLOSIVE ATMOSPHERE	GEOGRPHICAL LOCATION	STANDARD	CODE
EXPLOSION PROOF	CONTAINS THE EXPLOSION AND EXTINGUISH THE FLAME	CLASS I DIVISION1	USA	FM3615	XP
		CLASS I DIVISION1	CANADA	CSA C22.2-30	XP
FLAME PROOF		CATEGORY 1G/2G/3G	EUROPE UNION	EN 60079-1	EX da/db/dc
		EPL Ga/Gb/Gc	International- IEC	IEC 60079-1	
		CLASS I ZONE 1	USA	ISA 60079-1	AEx d
		CLASS I ZONE 1	CANADA	CSA C22.2-60079-1	Ex d
POWDER- FILLED		CATEGORY 2G	EUROPE UNION	EN 60079-5	Ex q
		EPL Gb	International- IEC	IEC 60079-5	Ex q
		CLASS I ZONE 1	USA	ISA 60079-5	AEx q
		CLASS I ZONE 1	CANADA	CSA C22.2-60079-5	Ex q
ENCLOSED BREAK		CATEGORY 3G	EUROPE UNION	EN 60079-15	Ex nC
		EPL Gc	International- IEC	IEC 60079-15	Ex nC
		CLASS I ZONE 2	USA	ISA 60079-15	AEx nC
		CLASS I ZONE 2	CANADA	CSAC22.2-60079-15	Ex nC
INTRINSIC SAFETY	LIMITED ENERGY OF SPARK AND SURFACE TEMPERAURE	CLASS I DIVISION1	USA	FM3610	IS
		CLASS I DIVISION1	CANADA	CSA C22.2-157	IS
		CATEGORY 1G/2G/3G	EUROPE UNION	EN 60079-11	EX ia/b/c
		EPL Ga/Gb/Gc	International- IEC	IEC 60079-11	
		CLASS I ZONE 1	CANADA	CSA C22.2-60079-11	-DO-
		CLASS I ZONE 1	USA	ISA 60079-11	AEX ia-c
LIMITED ENERGY		EPL Gc	International- IEC	IEC 60079-15	EX nL
		CLASS I ZONE 2	USA	ISA 60079-15	AEX nC
		CLASS I ZONE 2	CANADA	CSAC22.2-60079-15	EX nL
PRESSURIZED	KEEP FLAMMABLE GAS OUT	CLASS I DIVISION1	USA	FM 3620(NFPA496)	TYPE X/Y
		CLASS I DIVISION1	CANADA	NFPA496	
		CLASS I DIVISION2	USA	FM 3620(NFPA496)	TYPE Z
		CLASS I DIVISION2	CANADA	NFPA496	
		CATEGORY 2G	EUROPE UNION	EN 60079-2	Ex px/py
		EPL Gb	International- IEC	IEC 60079-2	
		CLASS I ZONE 1	CANADA	CSA C22.2-60079-2	
		CLASS I ZONE 1	USA	ISA 60079-2	AEx px/py
		CATEGORY 3G	EUROPE UNION	EN 60079-2	Ex pz
		EPL Gc	International- IEC	IEC 60079-2	
		CLASS I ZONE 2	CANADA	CSA C22.2-60079-2	
		CLASS I ZONE 2	USA	ISA 60079-2	AEx pz
RESTRICTED BREATHING		CATEGORY 3G	EUROPE UNION	EN 60079-15	Ex nR
		EPL Gc	International- IEC	IEC 60079-15	Ex nR
		CLASS I ZONE 2	USA	ISA 60079-15	AEx nR
		CLASS I ZONE 2	CANADA	CSAC22.2-60079-15	Ex nR
ENCAPSULATED		CATEGORY 1G/2G/3G	EUROPE UNION	EN 60079-18	Ex ma/mb/mc
		EPL Ga/Gb/Gc	International- IEC	IEC 60079-18	
		CLASS I ZONE 0/1/2	CANADA	CSAC22.2-60079-18	
		CLASS I ZONE 0/1/2	USA	ISA 60079-18	AEx ma/m or mb/mc
OIL IMMERSED		CATEGORY 2G/3G	EUROPE UNION	EN 60079-6	Ex o ob/oc
		EPL Gb/Gc	International- IEC	IEC 60079-6	
		CLASS I ZONE 1	CANADA	CSAC22.2-60079-6	Ex o
		CLASS I ZONE 1	USA	ISA 60079-6	AEx o

DEVELOPED BASED ON FM APPROVAL POSTER.

FIGURE APII/2.2-1

Comparison of standards and protection concept.

2.3 Ex Marking

All enclosures have Ex marking stamped on the enclosure. Here are a few as per various standards. Fig. APII/2.3-1 depicts various markings according different standards.

In addition to what has been shown in Fig. APII/2.3-1, there are some electrical parameters such as voltage, current rating, wattage, etc. related to the equipment that

(A)

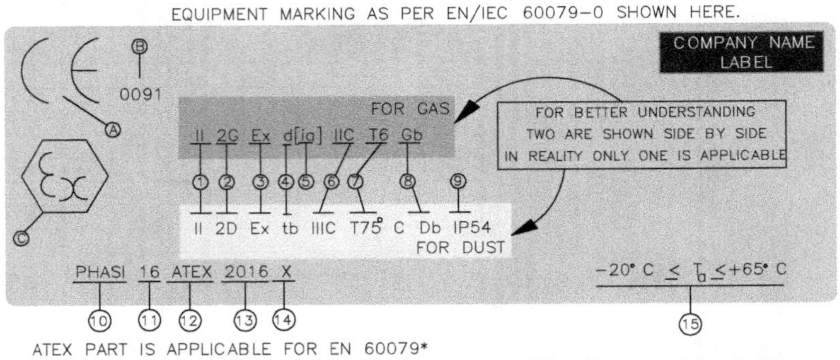

ID	EXPLANATION	REMARKS
A	CONFIRMATORY MARK	
B	ID. NO. OF NOTIFIED BODY	CERTIFYING AUTHORITY; BY QA/PRODUCT TEST
C	Ex MARKING	
1	EQUIPMENT GROUP	NON MINING HERE
2	EQUIPMENT CATEGORY (2*)	*G GAS/D DUST
3	EXPLOSION PROECTION	EN /IEC 60079–
4	PROTECTION TYPE	GAS; FLAME PROOF DUST: BY ENCLOSURE
5	INTRINSIC SAFE OUTPUT	SHOWN ONLY WHERE APPLICABLE.
6	EXPLOSION GROUP	GAS: IIC HERE DUST: IIIC HERE(Cond)
7	SURFACE TEMPERATURE CLASS	GAS: CLASS; t6 HERE DUST Surfcae 75 C
8	EQUIPMENT PROTECTION LEVEL	Gb/Db FOR GAS & DUST (Cat:2G/2D–EN*)
9	INGRESS PROTECTION (VIZ.IP54)	FOR DUST APPLICTION
10	CERTIFYING BODY	
11	CERTIFIED – FIRST YEAR	2016 IN THIS CASE.
12	ATEX GENERATION	FOR ATEX ONLY
13	ID DETAILS OF NOTIFIED BODY	ATEX ONLY
14	SPECIAL CONDITION IF ANY	
15	AMBIENT TEMPERATURE	TEMPERATURE RANGE

(B)

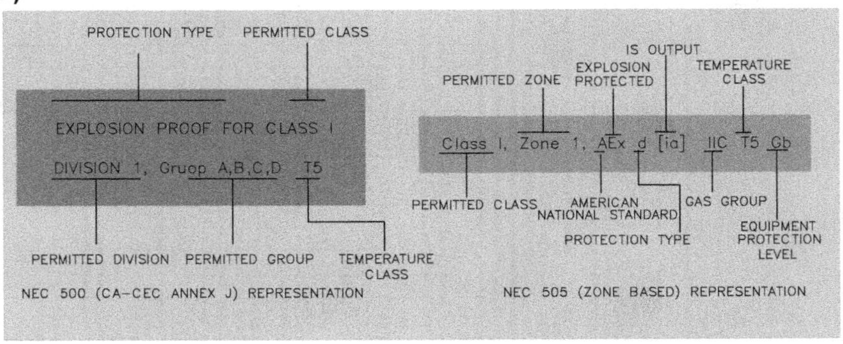

FIGURE APII/2.3-1

Equipment marking. (A) Equipment marking—EU/IEC, (B) equipment marking—NEC based.

are also marked in the equipment marking plate. For intrinsic safety [e.g. ib], they are also marked and instrinsic safety (IS) cables are used.

2.4 External Factors to be Considered for Enclosures

The following external factors need attention while selecting an enclosure for a hazardous area:

- Thermal effects
- Expected ambient temperature
- Chemical effects
- Mechanical effects
- Effects of movement and vibration
- Electrical effects
- Moisture
- Ingress of material
- Corrosion
- Spark avoidance
  - Use of lighter material
  - Potential equalization

3.0 LIGHTNING AND SURGE PROTECTION OF INTRINSIC SAFETY CIRCUIT

Surge voltages, caused by switching operations, fuse tripping, or lightning strikes, are an important issue to be considered to retain the electrical systems functionality and availability. In chemical or oil and gas industries, potentially explosive areas are developed quite frequently during plant operation and/or storage, etc. Therefore, it is essential to keep any ignition/energy source apart. Atmospheric discharge such as lightning is a potential source of such energy. Naturally such there shall be protective measures to save the plant and keep the same operating safely. As stated earlier, there could be energy surge due to other reasons also. Therefore protection shall be there for lightning and other surges. For low voltage applications, such as instrumentation and/or telecommunication devices located indoors, surge protective devices will be sufficient. However, instrumentation and/or telecommunication devices for located outside and in the boundary need separate requirements. As per requirements of the standards, coordination between surge protection devices (SPDs) in different lightning protection zones and coordination of SPDs with terminal equipment are important. Therefore, it is important to know *what a lightning protection zone is*.

3.1 Lightning Protection Issues and Zoning

A lightning protection zone (LPZ) is a concept developed to protect equipment from lightning, especially meant for hazardous and potentially explosive areas. When assessing the potential risk related to lightning following issues as per EN 1127-1, the following should be considered as ignition/energy sources:

- Melting at the point of strike.
- Heating of discharge paths.
- Uncontrolled flashover for insufficient separation distance.
- Induced voltages in cables.
- Lightning strikes into metallic cables entering hazardous areas.

According to IEC 62305-2 [1], all equipment, protective systems, and components of all categories must be protected by adequate lightning and surge protection measures. One needs to note that apart from direct damage due to lightning, there could be damage also from the electromagnetic effect of the lightning surge current on the electrical system present in potentially explosive atmospheres. The very high surge of the lightning current in a short time would cause a high change velocity of the magnetic field, to induce a high voltage (ignitable energy) in various cables (external conductive parts) that form the safety loop. Thus, when an explosive atmosphere is present, this ignition energy may cause fire or explosion.

3.1.1 *Damage and Losses*

IEC/BS EN 62305 identifies the following sources of damages and associated losses:

- S 1: Flashes to the structure
- S 2: Flashes near to the structure
- S 3: Flashes to the services
- S 4: Flashes near to the services

Each of these will cause one or more damages listed below:

- D 1: Injury to living being due to step and touch voltages
- D 2: Physical damage (fire/explosion, chemical, etc.) due to lightning current effect including spark
- D 3: Failure to internal systems due to lightning electromagnetic impulse (LEMP).

Also there may be losses, which are categorized as:

- L 1: Loss of life
- L 2: Loss of services to public
- L 3: Loss of cultural heritage
- L 4: Loss of economic value

The preceding issues have been addressed in IEC 60079-14, which may be consulted for details.

3.1.2 *Application of IEC 60079-14*

IEC 60079-14 deals with lightning protections in line with requirements specified in IEC 62305. Also IEC 60079-14 aims to reduce the effect of lightning strikes to a safe level. This is in line with ATEX directives. The basic guidelines towards the same shall include but are not limited to the following [4]:

- A risk analysis to determine the necessity of lightning protection and the required lightning protection measures.
- Measures to protect structures from material damage and damage to persons caused by direct lightning strikes by installing a lightning protection system (LPS) consisting of air-termination systems, down conductors, earth-termination system, lightning equipotential bonding, and separation distance.
- Protection measures against the effects of LEMP by LEMP protection system (LPMS) for structures with electrical and electronic systems. An individual combination of protection measures are earthing and equipotential bonding, spatial shielding, cable routing and screening, coordinated protection by SPDs.

3.1.3 *Lightning Protection Zone*

From BS EN/IEC 62305, one gets that the concept of LPZ was introduced to assist in determining the protection measures required to counter LEMP within a structure. It is based on general principle that the equipment requiring protection should be located in an LPZ whose electromagnetic characteristics should be compatible with the equipment stress withstand or immunity capability. Here higher the number of LPZ, the lower the expected electromagnetic effect will be. So, an electronic control system in a control room would be LPZ2 and that in the field device (say a sensor on a tank) may be LPZ1. Table APII/3.1.3-1 details out such LPZ numbers with area class.

Basic LPZ concept has been depicted in Fig. APII/3.1.3-1 for better understanding.

Table APII/3.1.3-1 Lightning Protection Zone

LPZ	Area	App. Note.
$LPZ0_A$	Zone where the threat is due to the direct lightning flash and the full lightning electromagnetic field. The internal systems may be subjected to full or partial lighting surge current.	External zone; subject to direct lightning strike.
$LPZ0_B$	Zone protected against direct lightning flashes but where the threat is the full lightning electromagnetic field. The internal systems may be subjected to partial lightning surge currents.	External zone; side wall structure.
LPZ1	Zone where the surge current is limited by current sharing and by SPDs at the boundary. Spatial shielding may attenuate the lightning electromagnetic field.	Internal area; partial lightning strike.
LPZ2	Zone where the surge current may be further limited by current sharing and by additional SPDs at the boundary. Additional spatial shielding may be used to further attenuate the lightning electromagnetic field.	Further interior structure, e.g. control room.

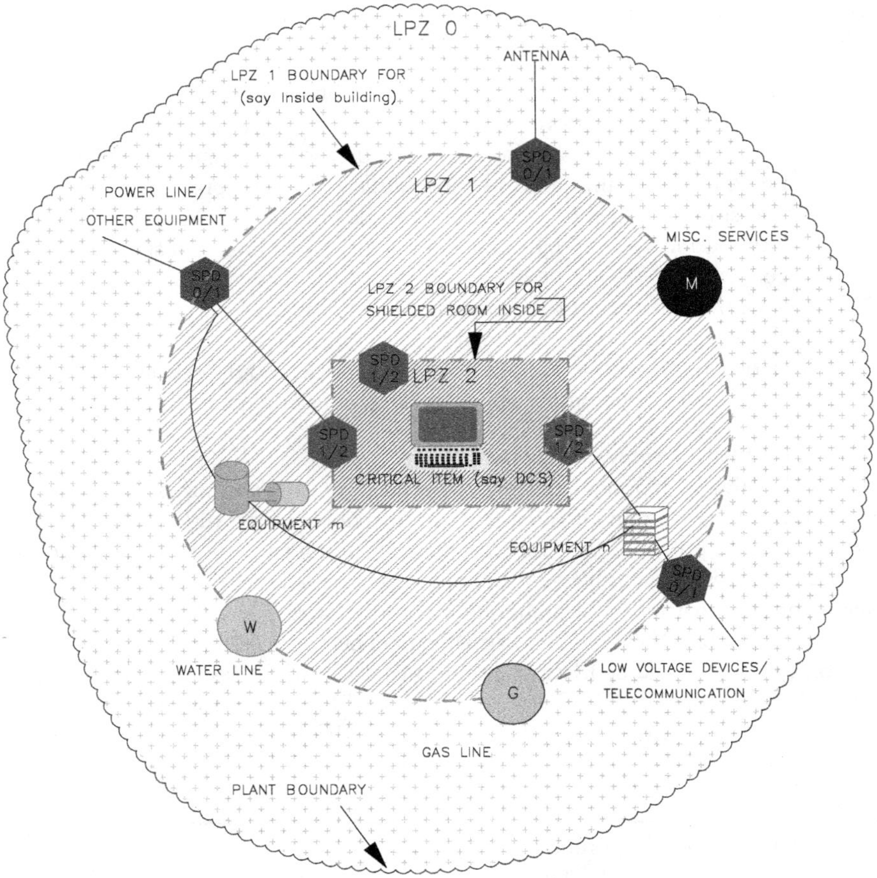

FIGURE APII/3.1.3-1

Lightning protection zone concept.

3.1.4 Equipotential Bonding

Equipotential bonding is an electrical connection maintaining various exposed conductive parts and extraneous conductive parts at substantially the same potential. So, equipotential bonding is the process of making an electrical connection between the grounding electrode and any metal conductor—pipes, plumbing, flues, etc.—that may be exposed to a lightning strike and become a conductive path for lightning energy. In all areas where there is potential for explosion, it is necessary maintain consistent equipotential bonding. All structures, pipelines, containers, etc. should be a part of the same so as to eliminate any fear from voltage difference as shown in Fig. APII/3.1.4-1. IEC 60364-4/6 may be referenced.

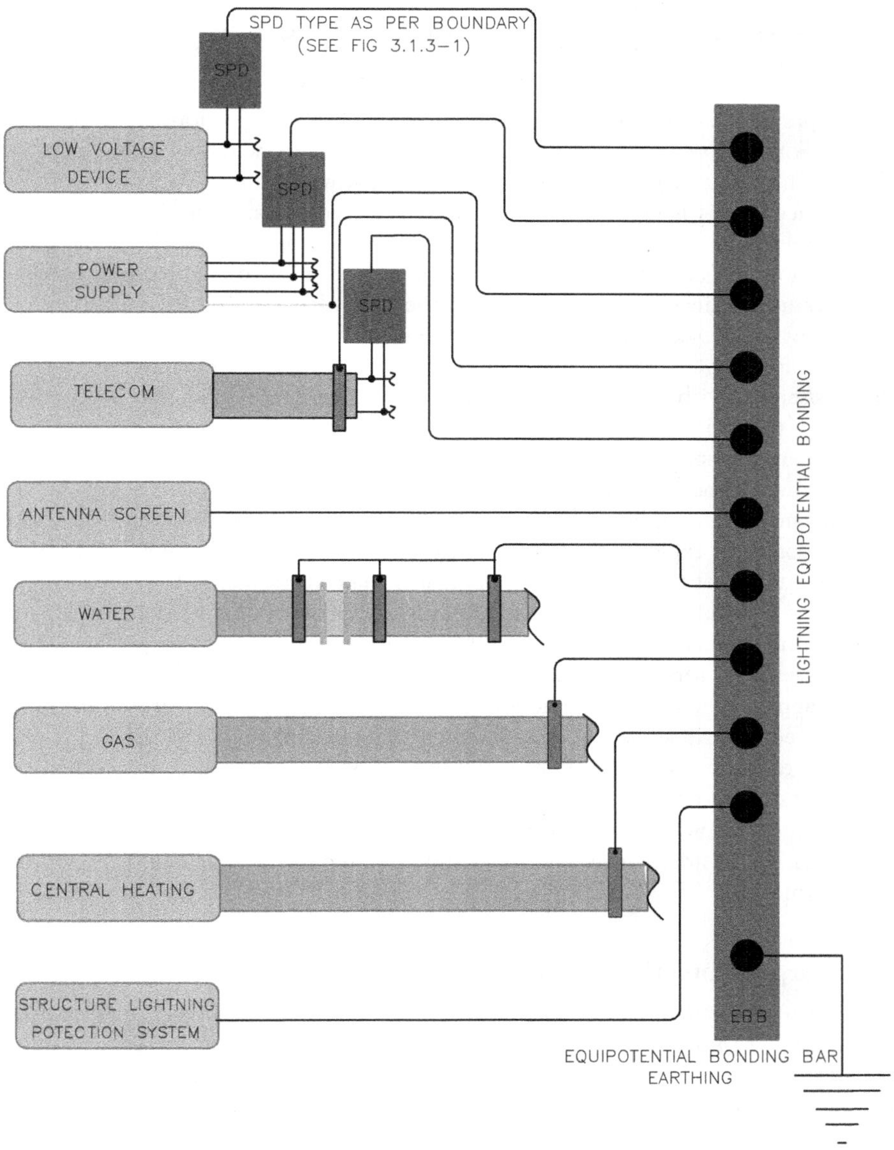

FIGURE APII/3.1.4-1

Equipotential bonding (bar).

4.0 SURGE PROTECTION DETAILS FOR INTRINSIC SAFETY CIRCUITS

In the previous clause, details regarding lightning with protections have been discussed. For any surge types, be it lightning or surge due to switching or fuse tripping, there will be surge voltage, or rather surge energy, which needs to be dissipated to safe devices. This is especially critical in case of potential explosive atmosphere. Day by day, on account of market competitions people are looking for faster, smaller, low-power consuming, and cheaper versions of instruments. Now these qualities are achieved by the devices at the cost of their ability to dissipate any significant impressed energy making the devices more *vulnerable* to induced over voltages and surge currents. These devices and systems are connected to the field devices/sensors with the help of cables (where optical fiber connection is not used), which are potential entry routes for surges and transients. SPDs are meant to improve the resilience of such systems to induced transients including lightning. In this part, discussions will be on SPDs pertinent to IS circuits. In other circuits, similar discussions may be good, but those will not be so stringent, as in the case of IS, i.e. in case of potential explosive atmosphere the vulnerability is much more. SPDs in IS circuits must meet the *same standards of design and construction* as the IS equipment (discussed in Chapter X), i.e. either be "simple apparatus" or must be certified as being within the safety parameters of the intended application. It is important that in many cases inductors are used to reduce the resistance value so SPDs are not always simple apparatus and nor suitable for intrinsic application. However, *certified* SPDs can be used in intrinsic safe circuits. Before going into further detailed discussions, it is better to clarify one ambiguity regarding the surge protection provided by the manufacturers as an option with most of the process transmitters. These optional surge protectors with process transmitters are basically surge *diodes* between I/P terminals. These are *not effective* SPDs, and are often either removed or augmented by a normal SPD [5].

4.1 Surge Protection Device Details

In order to prevent the damages of I&C devices, from excess energy on account of any surges, SPDs are used. The basic purpose of SPDs is to divert the excess voltage and current from transient or surge into grounding wire and prevents it from flowing through the electronic I&C devices — but at the same time allowing the normal voltage to continue along its path. In this section, a short discussion will be presented on SPD functions.

4.1.1 Functions of SPDs

The two major functions of the surge protector are:

- Providing low impedance path for conducting surge current to avoid excess voltage.

- SPD absorbs and diverts the extra current to ground to protect electronic instrumentation and control (I&C) devices from surge or transient.
  It is worth noting that actually same principles are applicable for other electronic devices as well as electrical appliances/devices. Since we are interested about I&C devices, we mention here I&C devices.

4.1.2 Surge Protection Types

Surge protectors are classified into two types:

- Filter: It is some sort of low pass filter, to act as barrier to high frequency signal, and allows the low frequency normal signal/power current to pass through unaffected.
- Transient diverter: This device presents a very low impedance path to ground, whenever voltage across the device exceeds certain value, but reduces voltage that could be presented to the sensitive equipment.
  Out of these two, commonly transient diverter types are often used in I&C devices.

4.1.3 Main Elements of Transient Diverters

Short details of various components used in a typical transient diverter type SPD has been discussed as follows:

- Gas discharge tube (GDT): A GDT diverts the excess current from line to ground with the help of inert gas as a conductor from hot line to ground line. In normal conditions, the inert gas is a poor conductor, but when voltage is above a specified level, the inert gas is ionized, and becomes a good conductor to pass on current to ground until voltage returns to a normal level. The GDT will start conducting at a voltage (>ionization voltage) and able to conduct more current for their size than other components, hence GDT is a better choice in many cases. A GDT has a finite life expectancy, and can take a few very large transients or a greater number of smaller transients [6].
- Transient voltage suppressor (TVS): "TVS" is basically a silicon avalanche diode (SAD) to clamp the incoming signal as shown (functioning shown in the bubble shown separately) in Fig. APII/4.1.3-1. In fact, in a circuit, the "TVS" is "invisible" until a transient appears. So, a TVS becomes active when a transient appears and provides the limiting/clamping action for the input signal in order to protect the load devices/components. When voltage increases above a preset level for TVS, it will tolerate avalanche breakdown resulting voltage is conducted to ground. Surge ratings for silicon TVS families are normally specified in kilowatts of peak pulse power (PPP) during a given waveform. These are available to meet 8/20 (8 μs to rise to peak and 20 μs decay time) are test waveforms. Now, available power ratings for TVS could range from 5 kW for 10/1000 μs, down to 400 W for 8/20 μs.
- Other components: Resistors, inductors, and capacitors are other passive components used in various circuit as typically shown in Fig. APII/4.1.3-1.

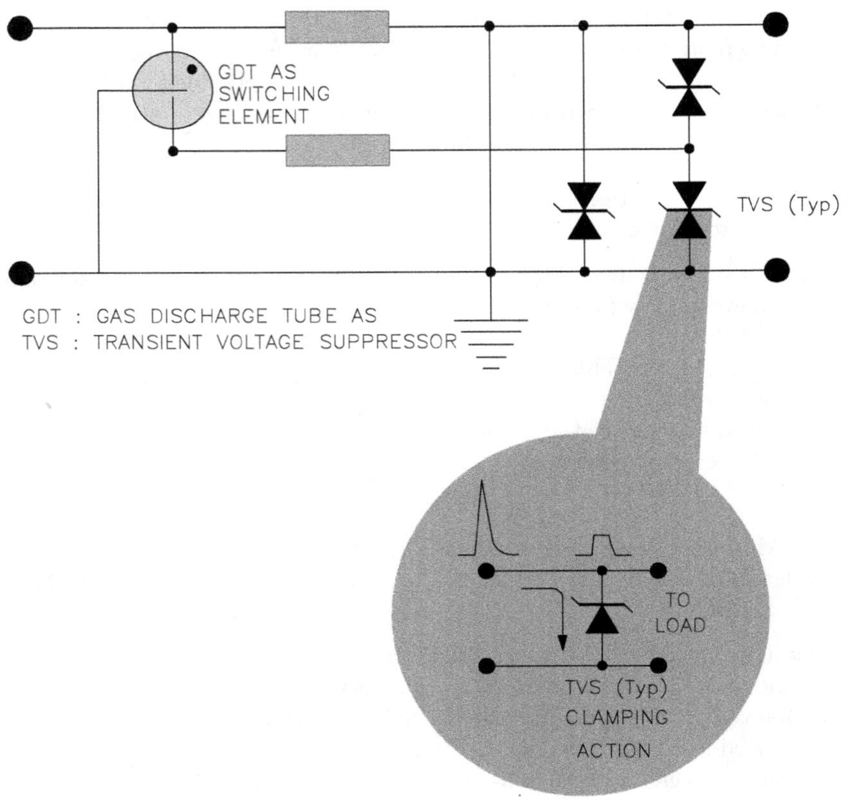

FIGURE APII/ 4.1.3-1

Surge protection device operating principle.

4.1.4 Typical Circuit Explanation

Fig. APII/4.1.3-1 shows the typical configuration for SPDs. As seen in the figure, there is protection both in differential mode (line to line) and common mode (line to earth). GDT, discussed previously, provides the first switching action to divert high voltage and current. Second elements comprising TVS provides the second line of defense especially for slower responding surges. Inductors are used to decouple the two protection stages. Basic purpose and functioning of various components have already been discussed in previous clause.

4.2 Surge Protection Device Application Issues

In this part of discussions, concentration will be on requirements as per standards and installation details.

4.2.1 *Requirements of EN 60079-14*

Following are the general requirements:

- SPDs need to have a minimum discharge capacity of 10 impulses with 10 KA (8/20 μs)
- SPDs shall be mounted inside metallic-shielded enclosures, properly earthed with at least 0.4 mm^2 Cu cables.
- Cables between the SPD and equipment shall be done by shielded cable (maximum length 1m) properly earthed at the ends or the ordinary cable can be used when the same run through properly earthed (at both ends) metallic conduit.

  During design, the hazardous area and lightning protection zone shall be properly harmonized.

4.2.2 *Use of SPD in the Circuits*

It is possible to insert SPD at any place in an IS loop between the field device and the IS barrier. It is quite common to locate them at the back of the panel or it can be used as I/O marshalling to accept direct field terminations. The SPDs and IS interfaces should be mounted close to each other but on separate Deutsches Institut für Normung (DIN) rails in order to maintain the required 50 mm clearance between safe area and hazardous area terminals [7]. In hazardous areas, SPDs are mounted in IP 54 (weather proof enclosure). Typical mounting of SPDs in hazardous locations have been detailed in Fig. APII/4.2.2-1. In Fig. APII/4.2.2-1A for safe distances, SPD is in hazardous location and IS barrier is in safe area.

In Fig. APII/4.2.2-1B also, SPD is in hazardous location and there is another SPD is in safe area to show that on account of two LPZ locations SPDs are different as distinctly shown. In this case, one SPD is safe zone for connecting cable, but SPD and IS barrier need to maintain a safe distance of 50 mm minimum. With this, discussions on surge protection have been completed and now discussion will concentrate on grounding.

4.3 Surge Protection Standards for Intrinsic Safety Circuits

Short discussions on IS wiring and grounding have been covered in Chapter X. Here, some supplementary *information* on the same shall be discussed. NEC 504 and 505 of NFPA 70 cover quite good details regarding the same. In article 504.10, details about installation and 504.20, details about wiring practices are presented. 504.30 has three distinct parts; (1) distance between conductors of IS circuit and non-IS circuits; (2) distance between IS conductors; and (3) from grounded metal. 504.50 deals with grounding, 504.60 deals with bonding, and 504.70 deals with sealing.

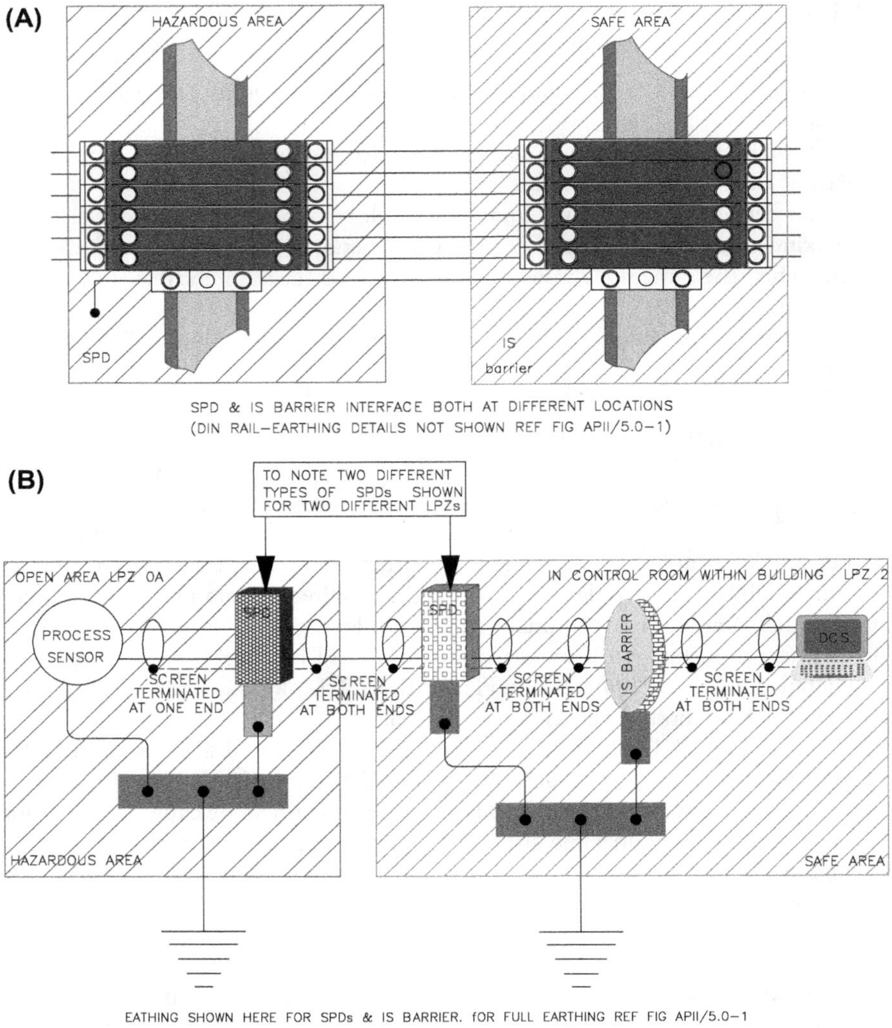

FIGURE APII/4.2.2-1

Surge protection in intrinsic safety application. (A) IS barrier interface with SPD (single SPD), (B) IS barrier interface with SPD (two different SPDs).

(A) Developed based on idea from MTL Surge Technologies — surge protection for intrinsically safe systems; MTL Technical Brochure; Application Note TAN 1004. Courtesy: MTL.

4.4 Distance of Separation Between Conductors and Non-Intrinsic Safety Conductors

4.4.1 Separation

Conductors of IS conductors shall not be placed in any raceway, cable trays, etc. carrying non-IS conductors. However, there are a few exceptions such as:

- IS conductors are secured and separated by 50 mm, or separated by grounded metal/approved insulating partition.
- All conductors of IS and non-IS conductors are within grounded metal sheathed or metal clad cables and the associated metal sheathed or metal clad cables are capable of carrying the fault current.
- Where IS conductor is in or passing through (for supply to zone 0/1 or division 1), Zone 2/division 2 shall be permitted to be in raceway along with non-IS conductor when installation is done as per NEC 504.30B.

4.4.2 Within Enclosure

The conductor of intrinsic safety circuit is secured, and even if the conductor comes loose, it is unlikely to come into contact with another conductor. However:

- Separation distance of 50 mm is to be maintained between conductors of IS and non-IS.
- Separation with a non-IS conductor by use of metal partition of thickness $\geq$0.91 mm.
- Separation with a non-IS conductor by use of approved insulating partition extended within 1.5 mm within enclosure.

4.4.3 Distance Between Conductors of Two IS Circuits

- Clearance between two terminals terminating two conductors from different IS circuits shall be 6 mm.
- Conductor of each circuit within grounded metal shield.
- Conductor each circuit has insulation of 0.25 mm thick.

Clearance between un-insulated part of field wiring connected to terminal and grounded metal and other conducting part shall be 3 mm.

5.0 EARTHING FOR INTRINSIC SAFETY CIRCUITS

A short discussion on earthing/grounding systems for IS has been covered in Chapter X. Here, the systems will be discussed in greater depth. In most of the cases, reference voltage point of computer and I&C systems and the barrier bus bar are linked, making the earth returns are combined. Combined earthing helps to avoid interference and ensures intrinsic safety. When isolators are used, the barrier bus bar may be omitted and the screens of the field wiring are connected to the 0 V rail of the system. Safety barriers shall be connected to the main electrical system earth or equipotential boding with conducting thick bars or with a dedicated conductor of at least 4 mm^2 cross-sectional area, so as to ensure that a total connection resistance does not exceed 1 Ω. Similarly SPDs also need effective earthing. A typical earthing system with SPDs (two different classes for two LPZs) and one IS barrier has been depicted in Fig. APII/5.0-1.

In this figure, both SPDs located at two zones have earth bus systems, which are connected; alternatively each of them could be connected separately by earthing bus bar or conductor to a common earthing bus (as optional shown). In case of conductor connection it shall be at least 8 AWG to offer minimum resistance as

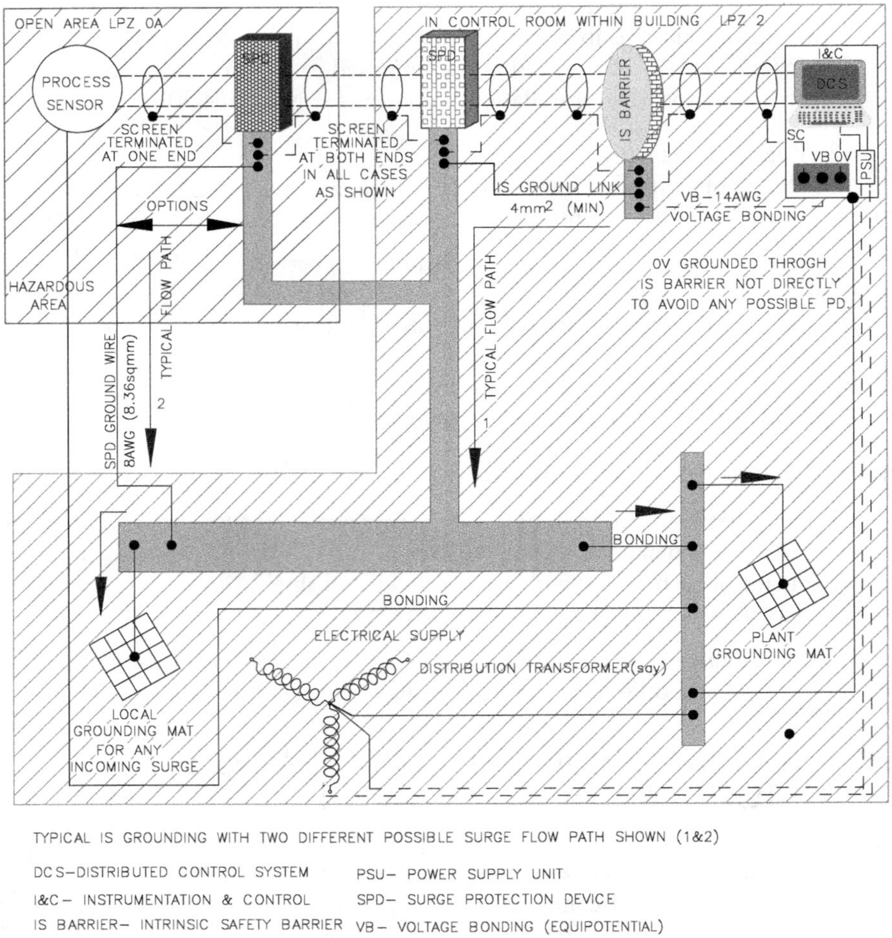

FIGURE APII/5.0-1

Complete earthing system for intrinsic safety barrier and surge protection device.

shown. IS barrier has earthing route through SPD with the help of voltage bonding (especially shown to illustrate possible options) with a 14 AWG conductor. Earthing of surge at IS barrier has been illustrated with (blue) line with arrow to show that it is earthed SPD (in safe area) to earth bus then to earth mat of the plant. Here one thing to be noted that 0 V point is connected to earth via SPD earth bus and direct earth not directly to earth so that 0 V and power supply unit (PSU) are maintained at IS earth potential, and no potential difference between them. Also in such connections, there is possibility of *SPD surge current finding the same as parallel path and instrumentation system might be affected.* Instrument body and control system body earthing is connected to plant earth mat. Also, a local earthing mat helps in dissipate surge energy at SPD (in hazardous area) as shown. Earthing paths for SPDs have also been shown in the figure. Typically these are earthed with <0.5 Ω i.e. nearer earthing point with (say) 8 AWG conductor. Earth wiring should be ***labeled*** distinctly to deter unauthorized removal and should be made with reliable connectors [5]. Isolator type barriers do not normally need a high-integrity earth connection — these may follow SPD manufacturer's recommendations.

From IEC, one gets that circuits, which cross boundaries between Zone 0 and Zone 1, and which may be subjected to significant voltage transients, are bonded at the crossing points to eliminate the possibility of Zone 0 being invaded by lightning or stray fault currents [5]. So, while designing the system, the same must be considered. With this discussion, Appendix II is concluded. For further details, ANSI/ISA-RP12.06.01-2003 may be referenced.

LIST OF ABBREVIATIONS

DCS	Distributed control system
EBB	Equipotential bonding bar
EPL	Equipment protection level
IEC	International Electrotechnical Commission
GDT	Gas discharge tube
I/O	Input/output
I/P or O/P	Input or output
IP	Ingress protection
LEMP	Lightning electromagnetic impulse
LPMS	LEMP protection system
LPS	Lightning protection system
LPZ	Lightning protection zone
PPP	Peak pulse power
SAD	Silicon avalanche diode
SPD	Surge protection device
SW	Software
TVS	Transient voltage suppressor

References

[1] R.K. Saini, Practical Classification for Determining Electrical Classification, Powermag, March 3, 2007. http://www.powermag.com/practical-guidelines-for-determining-electrical-area-classification/.

[2] Basic Concepts for Explosion Protection; BARTEC Brochure, eleventh revision ed., 2012.

[3] Basics of Explosion Protection; STAHL Brocheure; http://www.rstahl.com/fileadmin/Dateien/tgus/Documents/ExProtection_Global-America-Basics.pdf.

[4] Lightning and Surge Protection for Intrinsically Safe Circuits; M. Kienlein; Stahl; Ex Magazine 2011; https://www.r-stahl.com/fileadmin/Dateien/ex-zeitschrift/2011/en/10_Lightning_and_surge_protection_for_intrinsically_safe_circuits_Ex-Magazine_2011_low-10.pdf.

[5] MTL Surge Technologies — surge protection for intrinsically safe systems; MTL Technical Brochure; Application Note TAN 1004.

[6] Basics of Surge Protector, TVSS, Transient Voltage Surge Suppressor; Leonics; Technical Catalog; http://www.leonics.com/support/article2_7j/articles2_7j_en.php.

[7] SD Series; Technical Data Sheet; MTL Instruments; http://www.mtl-inst.com/images/uploads/datasheets/surge/data_and_sig/901-107%20SD.pdf.

Further Reading

ATEX Directive 94/9/EC; Technology International Group; http://www.techintl.com/atex.cfm.

ATEX guide; Basic concepts for protection in explosive atmospheres; SCAME Electrical Solutions Brochure.

Flame and Gas Detection Handbook; Ampac Industries; http://www.iceweb.com.au/Technical/flame_and_gas.htm.

Gas detection handbook; fifth ed.; MSA; http://www.gilsoneng.com/reference/gasdetectionhandbook.pdf.

Guide to BS EN/IEC 62305; Furse; Technical Write Up/Catalog; www.furse.com.

Hazardous Classifications and Control of Ignition Sources; Health and Safety Executive; http://www.hse.gov.uk/comah/sragtech/techmeasareaclas.htm.

Hoffman, ATEX Reference Guide; http://www.hoffmanonline.com/stream_document.aspx?rRID¼119786&pRID¼119783.

Honeywell Gas Book; Honeywell Gas Detection; Honeywell; http://www.honeywellanalytics.com/w/media/honeywell-analytics/documents/english/11296_gasbook_v5_0413_lr_en.pdf?la¼en.

How to calculate an intrinsically safe loop approval IS & Ex; G. Friend; South African Instrumentation and Control, October 2012; http://www.instrumentation.co.za/7571a.

https://www.fmglobal.com/assets/pdf/FM_EXGas_HazardPoster.pdf.

Intrinsic safety circuit design; P.S. Babiarz; Omega US; http://www.omega.com/temperature/Z/pdf/z131-148.pdf.

Protection of data and signal lines in intrinsically safe circuits; Application note AN013 data and signal line protector ESP SL15X and ESP SL30X; Thomas & Bett; Furse; www.tnb-europe.com.

Recommended Practice for Wiring Methods for Hazardous (Classified) Locations Instrumentation. Part 1: Intrinsic Safety; ANSI/ISA-RP12.06.01-2003.

B. Svacina, B. Larson. Understanding Hazardous Area Sensing; Intrinsic Sensing; Handbook; http://www.parrinst.com/wp-content/uploads/downloads/2011/06/Svacina-Larson_Understanding-Hazardous-Area-Sensing_Intrinsic-Safety.pdf.

P. Thurnherr, Electrical installations design, selection, erection and inspection, in: IECEx International Conference; Kuala Lumpur, Malaysia, 2014.

L.C. Towle, A Definitive Guide to Earthing and Bonding in Hazardous Areas, The MTL Instruments Group Plc, May 1997. TP1121−1.

Simplified Mathematical Equations

1.0 GENERAL

Basic mathematical and statistical formulas have been discussed in Clause 5.0 in Chapter I. Also, the Weibull distribution and some mathematical treatments of reliability and have been covered in chapter clause numbers 1.2.3, 1.3.1, and Table VII/1.3-2 of Chapter VII. In this appendix, these basics will be discussed further with a little more detail and with application approaches in safety assessments so that the reader will be in a position to apply the same in future. All endeavors have been put to eliminate detailed theoretical part and other mathematical detailing. As the focus is on application part, so in this limited space only basic mathematical structures have been touched upon. For further reading, standard books on statistics and reliability will be helpful.

1.1 Basics of Probability Recapitulations: Some Recapitulations Begin the Discussion

- Discrete and continuous variables and probability distributions: From Clause 5.3.3 of Chapter I, we get the probability mass function and cumulative distribution functions. For a single dimension, discrete random variable X, the discrete probability function is defined by $f(x_i)$, such that $f(x_i) \geq 0$, for all $xi \in R$ (range space), and $\sum_{xi<x} f(xi) = F(x)$ where $F(x)$ is known as cumulative distribution function. Similarly for a continuous variable X, there may exist a function $f(x)$, such that, for $a \leq b$ $\int_a^b f(x)\mathrm{d}x = P(a \leq X \ll b)$. It satisfies the conditions: (a) $f(x) \geq 0$; (b) $\int_{-\infty}^{\infty} f(x)\mathrm{d}x = 1$.
  This function is called probability density function.
- Various other distributions: There are a few numbers of probability distribution functions which are frequently used. The functional details for the same have been presented in Table I/5.3.5-1, "important probability distributions and their characteristics." These are not repeated.
- Conditional probability: $P(A/B) = \{P(B/A) \cdot P(A)\}/P(B)$, where events are A, B, and $P(A)$ $P(B)$ are independent probabilities, and $P(A/B)$ or $P(B/A)$ are conditional probabilities. A typical example is, what is the probability of drawing "A" from a pack of cards after one "A" has been drawn (see Clause 5.3 of Chapter I).

  With this, the recapitulation part of basic probability ends to pave the way for Bayesian posterior predictive distribution.

Plant Hazard Analysis and Safety Instrumentation Systems.

1.2 Bayesian Posterior Predictive Distribution

Frankly speaking, Bayes' rule involves the manipulation of conditional probabilities. However, there exists a sharp difference in classical and Bayesian approaches. The major difference between the Bayesian and classical approaches to statistical inference is in the definition of probability. The classical approach asserts that the probability of a fair coin tossed, landing heads is ½ after repetitive tests. In contrast to this, as per Bayesian approach, will say that the probability a coin lands heads is ½ expressing a ***degree of belief***, *and would argue* that based on the symmetry of the coin, there is no reason to think that one side is more likely to come up than the other side. This definition of probability is usually termed ***subjective*** probability [1]. The classical approach uses probability to express the frequency of certain types of data to occur over repeated trials. The Bayesian approach uses probability to express *belief* in a statement about unknown quantities. This will be clearer from the difference between the two approaches which have divided statistical approaches in two clear divisions, i.e. now in statistics there are two schools of thoughts:

- Classical approach: In classical approach, the null hypothesis is that no activation occurred, and would reject null hypothesis if data is sufficiently unlikely. This approach gives the likelihood of getting the data, given no activation and cannot accept null hypothesis.
- Bayesian approach: This approach uses the likelihood and prior distributions to construct the posterior distribution, and would compute probability that activation exceeds some threshold directly from the posterior distribution. This approach can determine activation/no activation and can compare models of the data as well as gives the probability distribution of the activation given the data.
- Basic steps for Bayesian distribution: A typical Bayesian analysis consists of:
  - Formulate a probability model for the data.
  - Decide on a *prior distribution*, to quantify the uncertainty values of the unknown model parameters *before* the data are observed.
  - Observing data to construct the *likelihood* function based on the data and the probability model formulated at the beginning.
  - Next the likelihood is combined with the prior distribution to determine the *posterior distribution*.
  - *Posterior distribution* quantifies the uncertainty in the values of the unknown modelparameters *after* the data are observed.
  - Finally calculate quantities of interest based on the posterior distribution.
- Mathematical approach: From conditional probability (see Chapter I), one gets that joint probability of A and B, i.e. probability of A and B occurring simultaneously:

$$P(A, B) = P(A|B) * P(B) = P(B|A) * P(A) => P(A|B) = \{P(B|A) * P(A)\}/P(B)$$

$$P(A/B) = \frac{P(B/A) * P(A)}{P(B)} \qquad \text{(APIII/1.2-1)}$$

This Eq. (APIII/1.2-1) is so-called Bayes' rule.

In Bayesian approach, one of these "events" is the hypothesis, H or parameter (θ) and the other is data, D. The main aim is to find the probability of event or parameter for a given data, i.e.:

$$P(H/D) = \frac{P(D/H) * P(H)}{P(D)} \quad \text{(APIII/1.2-2)}$$

The term $P(D|H)$ represents the *likelihood* function and provides the probability of the observed data arising from the hypothesis. Normally, this is known because it expresses one's knowledge of how to look at the data, given that the hypothesis is true. The term $P(H)$ or $P(\theta)$ is called the *prior distribution*, as it reflects one's prior knowledge before the data are considered. The advantages of Bayesian probability theory is that one's assumptions are made up front, and any element of subjectivity in the reasoning process is directly exposed [2].
Now repeating Eq. (APIII/1.2-2), one considers parameter as (θ):

$$P(\theta/\text{data}) = \{P(\text{data}/\theta) * P(\theta)\}/P(\text{data}) \quad \text{(APIII/1.2-3)}$$

where, $P(\theta/\text{data})$ stands for posterior distribution.; $P(\text{data}/\theta)$ stands for likelihood; $P(\theta)$ stands for prior distribution (*i.e. parameter/prior information*). Take away from the preceding, it could be summarized as:

- Bayesian distribution estimates the probability of parameter values as conditional on the data, may be needed for a decision-making.
- Bayesian distribution is proportional to the likelihood.
- Bayesian distribution combines two independent sources of information, the likelihood and the prior distribution.
- Normal distribution is conjugate to itself for μ when σ^2 (see: Table I/5.3.5-1) is known. When a normal *prior* distribution is combined with the normal *likelihood* function, it would lead to a normal posterior distribution.
- An advantage of Bayesian methods is that additional observations can be used to update the output. Once a joint probability distribution for *all observable and unobservable quantities* has been chosen, posterior distributions and Bayesian posterior predictive distributions can be calculated.
  The Bayesian posterior predictive distribution can be obtained by integrating Eq. (APIII/1.2-3), so, for a future observation $\hat{y}$ is given by:

$$\begin{aligned} p(\hat{y}/\text{data}) &= \int_\theta p(\hat{y}/\theta, \text{data})p(\theta/\text{data})d\theta \\ &= \int_\theta p(\hat{y}/\theta)p(\theta/\text{data})d\theta \end{aligned} \quad \text{(APIII/1.2-4)}$$

[since $\hat{y}$ is independent of sample data].

This clearly shows how new data to behave/vary is predicted. If the data observed follow closely then prediction is correct.

The posterior predictive distribution can also be used to predict for any number of future observations. In Fig. APIII/1.2-1 adopted from one real time

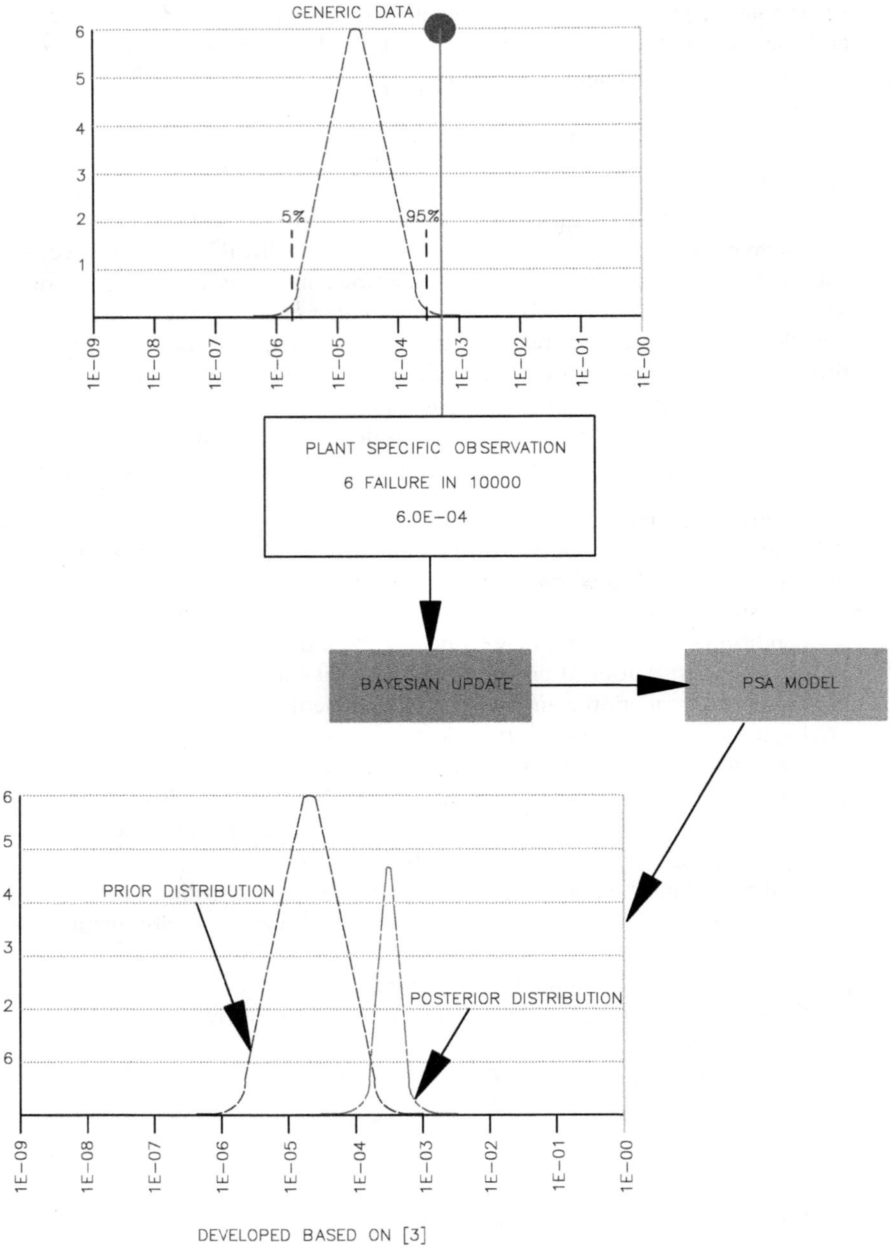

FIGURE APIII/1.2-1

Bayesian distribution in probabilistic safety assessment (PSA).

Introduction to Probabilistic Safety Assessments (PSA); O. Nusbaumer; Leibstadt NPP; http://nusbaumer.tripod.com/resources/publications/nusbaumer_introduction_to_probabilistic_safety_assessments.pdf.

probabilistic safety assessment (PSA) [3], here one may note:
As prior distribution is a normal distribution. Posterior distribution is also normal.
As stated above, prior distribution has been updated with actual data in PSA model to get posterior distribution as shown in lower graph side by side with prior distribution.
Bayesian distribution is extensively used in PSA. Now concentration will be on reliability engineering and the associated mathematical approach.

2.0 SIMPLE MATHEMATICS FOR RELIABILITY

Reliability is better understood with the help of the load strength curve relation. Supposing a person has strength to lift a load of (say) 60 kg; when he has to lift 40 kg, a person can do the same as person has still some margin. Now if person has to lift, (say) 70 kg, he will try best but may fail. In this connection, Fig. APIII/2.0-1 may be referenced.

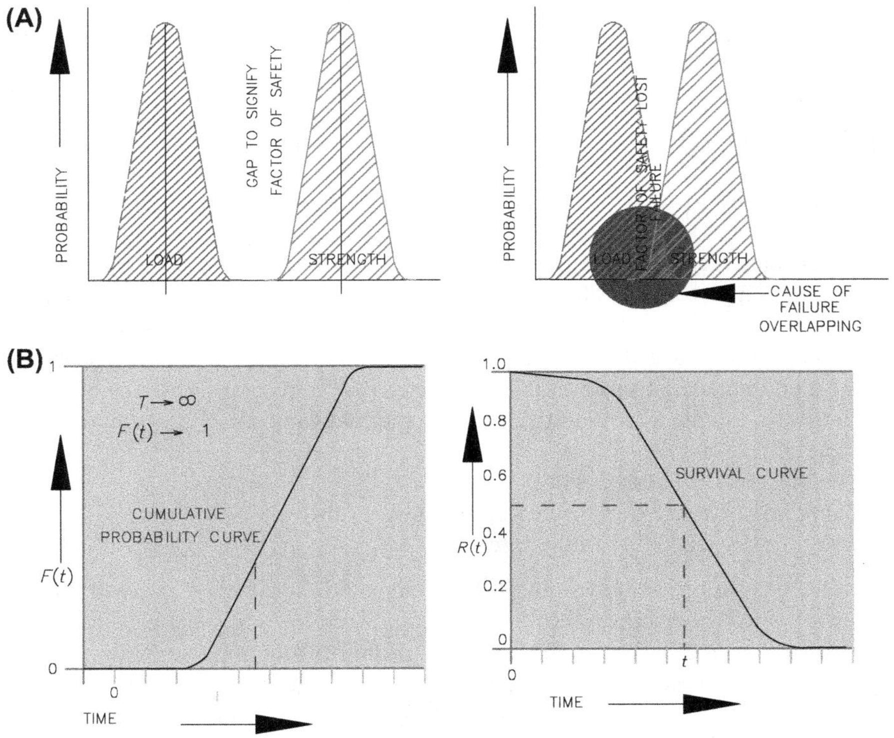

FIGURE APIII/2.0-1

Reliability in safety assessment. (A) Significance of relibility, (B) reliability curves.

Previous knowledge on random variables, the probability density function, and the cumulative probability distribution function will be utilized to study reliability mathematical equations where random variable time (t) is the major issue:

$$P(t_1 \leq t \leq t_1 + \Delta t) = f(t_1)\Delta t = F(t_1 + \Delta t) - F(t_1) \qquad \text{(APIII/2.0-1)}$$

where: $f(t_1)$, value of the probability density function at point t_1; $F(t_1)$, value of the cumulative probability distribution function at t_1 so, from Eq. (APIII/2.0-1) one gets:

$$f(t) = dF(t)/dt \qquad \text{(APIII/2.0-2)}$$

So,

$$F(t) = \int_0^t f(x)dx \qquad \text{(APIII/2.0-3)}$$

Thus if probability of failure by time t_1, $P_f(t_1)$, as the probability that the time to failure occur in the interval $0 < t < t_1$. So, from Eq. (APIII/2.0-2), one can argue:

$$P_f(t) = F(t) = \int_0^t f(x)dx \qquad \text{(APIII/2.0-4)}$$

Now if probability of success is $P_s(t)$, then:

$$P_s(t) + P_f(t) = 1 \qquad \text{(APIII/2.0-5)}$$

Hence, reliability is:

$$R(t) = P_s(t) = 1 - P_f(t) = 1 - \int_0^t f(x)dx \qquad \text{(APIII/2.0-6)}$$

Failure data are generally obtained mainly from the failure times of various items in a population placed on a life test, or repair reports or from similar plant data. Since such data are sequential discrete data, whereas probability density functions considered as continuous, it is necessary to define piecewise-continuous failure density and hazard-rate functions in terms of the data.

Generally, people are interested to know the following data:

- Time of failure
- Time interval between failures
- Cumulative failures experienced up to a given time
- Failures experienced in a time interval

Now our mission will be to find how these could be calculated from the above curves. This will be clarified with the help of a few numerical examples. From previous discussion, it is clear that $f(t)$ is the failure density function over a time interval (t_1 and t_2) (from piecewise linear concept). Therefore, it the ratio of the number of failures occurring in the interval $\{n(t_2) - n(t_1)\}$ to the size of the original population (N), divided by the length of the time interval (say) t_2-t_1:

$$f(t) = \frac{\left\{\frac{n(t_2)-n(t_1)}{N}\right\}}{(t_2 - t_1)} \qquad \text{(APIII/2.0-7)}$$

If $Z(t)$ the failure rate, then it can be defined as the probability that a failure occurs in some interval $[t_1, t_2]$, given that the system has survived up to time t. It is the ratio of the number of failures occurring in the time interval to the number of survivors at the beginning of the time interval. Now, when it is divided by time interval it gives the rate of failure. So:

$$Z(t) = \frac{\left\{\frac{n(t_2)-n(t_1)}{n(t_1)}\right\}}{(t_2 - t_1)} \quad \text{(APIII/2.0-8)}$$

meaning that if there are two failures up to t_2 and one failure in t_1 then $n(t_2) - n(t_1) = 1$, and if there are 10 failures total over a given period, then number of survivors will be 9 at t_1. The survivor rate curve has been shown in Fig. APIII/2.0-1B.

The cumulative distribution function or CDF, $F(t)$, gives the probability that a measured value will fall between $-\infty$ and t $(0 - t)$, i.e. theoretically by definition it is:

$$F(t) = \int_{-\infty}^{t} f(t)\mathrm{d}t \ \ \text{(basically from 0 to } t\text{)} \quad \text{(APIII/2.0-9)}$$

This has been illustrated in Fig. APIII/2.0-1A.

So:

$$R(t) = 1 - F(T) = 1 - \int_{-\infty}^{t} f(t)\mathrm{d}t \quad \text{(APIII/2.0-10)}$$

Another important function is the hazard function, $h(t)$; it is the conditional probability of failure in the interval t to $(t + \mathrm{d}t)$, given that no failure has occurred by t:

$$h(t) = 1 - \frac{f(t)}{F(t)} \quad \text{(APIII/2.0-11)}$$

Now concentration will be on various types of reliability distribution. Refer to Fig. VII/1.2-1 and compare with Eq. (APIII/2.0-8). Reliability distributions can be many types as presented in Table VII/1.3-2. One important survival function is the Weibull distribution shown below:

$$R(t) = \mathrm{e}^{-\left(\frac{t}{\eta}\right)\beta} \quad \text{(Ref: Table VII/1.3} - 2) \quad \text{(APIII/2.0-12)}$$

Mean time between failure (MTBF) and mean time to failure (MTTF) have been discussed in Clause 1.3.1 of Chapter VII. Here, in this connection, it is important to note that normally for MTBF/MTTF people are interested to know the interval and point estimates for them. If there are failures at 10, 20, 30, 45, and 55 h, then point estimation will be: $(10 + 20 + 30 + 45 + 55)/5 = 32$ h which is between 30 and 40 h so, interval is between 30 and 40 h. This part should be read in conjunction with Chapter VII.

With this, Appendix III discussions have been concluded to look into the issue related to little process language important for hazard and safety analysis.

LIST OF ABBREVIATIONS

AI/O	Analog input/output
DCS	Distributed control system
DI/O	Digital input/output
E/E/PE	Electrical/electronics/programmable electronics
ESD	Emergency shutdown system ***or*** electro static disturbance
IEC	International Electrotechnical Commission
I/O	Input/output
I/P or O/P	Input or output
IP	Ingress protection
MTBF	Mean time between failure
MTTF	Mean time to failure
MTTR	Mean time to repair
P&ID	Piping and instrumentation diagram
PSA	Probabilistic safety assessment
SW	Software

References

[1] M.E. Glickman, D.A. van Dyk, Basic Bayesian Methods; http://www.glicko.net/research/glickman-vandyk.pdf.

[2] B.A. Olshausen, Bayesian Probability Theory, March 2004. http://redwood.berkeley.edu/bruno/npb163/bayes.pdf.

[3] Introduction to Probabilistic Safety Assessments (PSA); O. Nusbaumer; Leibstadt NPP; http://nusbaumer.tripod.com/resources/publications/nusbaumer_introduction_to_probabilistic_safety_assessments.pdf.

Further Reading

Bayes for Beginners; Shuman ji & Nick Todd; https://www.google.co.in/url?sa=t&rct=j&q=&esrc=s&source=web&cd=5&ved=0ahUKEwi16s3eusfLAhUWBo4KHbEzBkAQFgg2MAQ&url=http%3A%2F%2Fwww.fil.ion.ucl.ac.uk%2F~chris%2Fmfd_archive%2F2013%2Fpage1%2Fbayes.pptx&usg=AFQjCNGl-4Fc0PNS9ya8JLQjDfVd2qsZLA&sig2=OKOBFI3x5k-D1ndVyA77Vw&cad=rja.

V. Montgomery, New statistical methods in risk assessment by probability bounds, in: A Thesis Presented for the Degree of Doctor of Philosophy, Department of Mathematical Sciences; Durham University, UK, February 2009. http://maths.dur.ac.uk/stats/people/fc/thesis-VM.pdf.

Little JIL Process Language

1.0 GENERAL

In the area of plant hazard analysis and safety instrumented system, business process definitions are tremendously helpful in process analysis and system automation. It is obvious that in order to contemplate how best to support the clear understanding of risk assessment processes, and how to use this understanding to support an orderly discipline of engineering of these processes, a clear and effective engineering of processes should provide strong support for better engineering of the systems that support the processes. All these needs are the major reasons behind development of process definition languages. Also, process definition languages have good role in coordination. Therefore, process definition languages are also useful in process efficiency improvement, and automation. Most real world business processes are large and complex, so good coordination for capturing, analyzing, and automating data requires huge coordination among these factors. Naturally, a good process definition language needs to have central focus on coordination. A few distinct advantages of process definition languages have been listed as follows:

1.0.1 Better Coordination and Communication

It is needless to mention that all members of a team coordinate their efforts effectively to get a fruitful result from the effort. In order to do this, it is important for each team member to understand their roles in the overall process, and to perform that role in a manner that is consistent with the overall process. Realization of all these demands a strong focus on the value of clear descriptions of the processes to be coordinated for all human and automation efforts.

1.0.2 Better System Analysis

There is every possibility that team members will commit mistakes to reduce the system efficiency. One way to circumvent the situation is to deploy a skilled person and improve control system. This is not always possible. Similar results can be achieved by modifying the process itself. This is more so in cases of large and complex processes. A simple example may shed some light. After thoroughly scrutinizing a process, it may be possible to find some activities that could be done parallel, hence improving the efficiency of the system (e.g. in order to start a boiler it is necessary to start ID fan prior to FD fan. So, first start the control oil and lube oil pumps of ID fan, then start ID fan and PARALLELLY start the control oil and lube oil of FD fan so, for starting FD fan some time can be saved. An important issue is exception handling, which when recognized earlier in analysis phase, then with proper planning many inefficiencies can be circumvented through proper planning and taking precautions beforehand. This is important for safety assessment systems.

1.0.3 *Continuation of Process Improvement*

As a part of good engineering practice, it is natural to invite a constant flow of new ideas to improve the functioning of the team and team efforts. Obviously, detailed, precise, and complete process description seems most useful as the focus of these activities. Also, such process description becomes a key to judge the pros and cons of any proposed modification, phasing and batching of proposed modifications, as well as to identify if there be any conflicting issues related to the proposed modification.

1.0.4 *Improvement of System Performance and Efficiency*

Deployment of more people and inappropriate automation at times can lead to inefficiencies like "too many cooks spoil the broth." Practitioners in many domains have recognized that extra people can actually reduce efficiency and productivity. Naturally, the efforts shall be there to improve the performance of the system by utilization of superior processes which can help elucidate how project work should be phased and sequenced, where parallel activities should be expected to help, and what the various members of teams should be doing at various times [3]. Detailed but precise, clear process descriptions help to identify the bottlenecks and inefficiencies. Therefore, process descriptions are always major tools to improve system performance and efficiencies.

1.0.5 *Supports Development and Reinvention*

Any incremental improvements are always welcome. However in many cases there are issues related to radical changes for system improvement. Now, prior to implementation of such changes it is necessary to thoroughly evaluate the system. Similarly, it is essential to have detailed but precise and clear definition of the process so that modified system can be fit in.

1.0.6 *Efficient Application of Automation in All Respect*

Augmentation of automation is always a natural tendency to improve efficiency and productivity. However, such augmentation must be proper. Otherwise inappropriate application or wrong automation will be not only be ineffective, but could be counterproductive also., Automation can provide improvements when automation activities closely match tasks and activities in existing processes. Thus, in all such cases, detailed but precise and accurate description of the process cannot be overestimated.

1.0.7 *Support and Training*

A clear, precise, and complete description of the team's processes is absolutely essential as a part of training for a newcomer joining the team. Such descriptions should specify what each participant is expected to do, what each can expect from the others, and the overall nature of the team activity [3]. Also, process descriptions at times are used as simulation unit during training to have a feel about actual working environment. This also requires detailed but precise and clear description of process.

It is better to take a survey about the various requirements for a process definition language before taking a call on Little Julia input language (*JIL*) process language.

1.1 Requirements of Process Definition Language

There are a number of personnel belonging to different disciplines, involved in large processes, so analysis of process descriptions can help each of the concerned personnel to know respective jobs better and can interact with others in more efficient way. Also it is possible to identify the system bottlenecks, inefficiencies, process defects, and faulty coordination. Well-defined business processes also lay the foundation for potential automation and e-Business exchanges by way of IT systems across organizations [2]. Apparently this approach is simple and straightforward but in reality it is not so. Major challenges comes from actual definition of such large and complex processes with required details. Proper coordination is also a big issue. Major issues related to process definition languages which are meant to be used in computer shall include but not limited to the following:

1.1.1 *Clear and Understandable*

Rigorous efforts are put to make software languages clear. Similarly, efforts are also necessary to make the process definition language clear and well understood.

1.1.2 *Detailing*

Computer languages have the specification of arbitrary levels of detailing. In most cases, iterative hierarchical decomposition is used for doing this. Thus, similar approach and specification is called for in process language also.

1.1.3 *Resource Management*

Control flow and activities have been given major thrust in most of the process languages. This is possible only when equal and more importance have been allotted to various resources such as personnel, systems, and other physical entities. Proper specification resource capability, policies for resource allocations are major semantic issues in actual process definition language.

1.1.4 *Exception Handling*

When looking at the issue, as the name suggests, may seem that it is a rare issue but in reality it is not so. In many cases, exceptions may dominate business scenarios. This could be well understood from the following example: Fire from an explosive atmosphere or leakage in nuclear installations is rare, because during design process a lot of importance are put handling such exceptions. Therefore, process definitions must put proper importance to the specification of exception management and exception handling strategies.

1.1.5 *Rigorous*

Process languages will be used in computers, which executes programs in coding languages with rigorous semantics. Therefore, such requirements of language must be considered as bases for the specification of processes as well.

1.1.6 *Process Variations*

On account of different reasons, there could be variations in a processes. So, the process definitions language must have flexibility to cater to such process variations. In reality, addressing this issue requires large-scale process reuse through both planned variation points and property observing mechanisms [2].

1.1.7 *Control and Data Flow*

Data flows and data-driven information systems are main enveloping information and vital for process definitions. The definition of the activity aspects of processes cannot be isolated from process data. Control and data flow improves process understanding, enables process analysis, and guides unambiguous process implementation and deployment, so reconciliation of these are important.

There are a few numbers of such process definition languages such as business process modeling notation (BPMN) or extensible markup language (XML). Little JIL is one of the most promising process definition languages, which shall be discussed in this part of the discussion.

1.1.8 *Systematically Broad*

In computer software languages, there are a few familiar features such as artifacts, scoping, looping, if-then-else, and sequencing, etc. So, it seems reasonable to have such software language approach in process definition languages also.

1.2 What Is Little JIL Process Definition Language

Little JIL is a process definition language to coordinate the activities of multiple people and software tools operating in a distributed environment and sharing data and resources (all those discussed previously). It is different from usual programming language normally used for computations. Various aspects of Little JIL language shall include mainly:

- Coordination diagram: The graphical representation of process steps. It also serves as the skeleton to which various annotations about other aspects of the process can be attached.
- Artifact space/model: This is the collection of data types (using Java), which are used by the objects during the process execution.
- Resource model: The resources shared by the interfacing agents during execution of the process.
- Agents: Java programs.

1.3 Juliette

It is an interpreter. Little JIL operating environment consists of:

- Editor: This is visual-JIL used to create the process in the language.
- JSim: It is the simulator for Little JIL.
- Little JIL checker: Used for syntax and semantic checking.
- Jul: This is interpreter for process execution.

Now it is time to explore to see why it is popular as process definition language.

2.0 LITTLE JIL PROCESS DEFINITION LANGUAGE

From the preceding discussions, it is clear that any process definition language has some requirements to be fulfilled so that the same can be used as process definition language. Little JIL is one of such process definition language which is powerful. So, in this part, language shall be looked into with a little depth for proper understanding of the same.

2.0.1 Little JIL Explanation

Little JIL could be conceived as a programming language for coordination in processes. This high-level language possesses formal graphical syntax and rigorously defined operational semantics. The focus of Little JIL processes is coordination of autonomous agents, shared resources, and information flow. All such coordinations are done with the help of graphical notation. The steps and sub-steps are the focal points for coordination, with scoping mechanism for control, data, and exception flow. The steps also provide interface for agent and resource assignment. Steps are organized usually in a hierarchy, but can have the possibility of recursion and concurrency. Agents are: autonomous entities, human, and software program. Assignment of tasks to agents is a part of Little JIL, but how agents perform the *task* is *not* part of Little JIL. Also, resources are external entities but these are monitored and controlled, as the resource needs are specified in little JIL. Information/data, (information/data – type, characteristics and manipulations are external) are propagated between agents to complete their jobs.

2.0.2 Basis of Little JIL

Little JIL is based on two main hypotheses:

- *Coordination structure* is separable from other *process language issues*. Although it offers rich control structures, it relies on separate systems for resource, artifact, and agenda management.
- *Agents*, who execute the process, *know how to perform their tasks* with benefit from coordination support. Therefore, each step has an execution agent responsible for performing the work of the step. This approach has proven effective in supporting the clear and concise expression of agent coordination.

2.0.3 *Little JIL as Computer Language*

As discussed above, there are a few characteristics a process definition language when used in computer needs to meet. These are rigorous, detailed, broad semantic, and clear understanding. These characteristics are met in the following ways by Little JIL:

- Rigorous: Little JIL's semantics are defined in terms of finite state machines [3].
- Detailed: Little JIL uses hierarchical decomposition as the primary vehicle for supporting the iterative, incremental elaboration of details [3].
- Semantically broad: A Little JIL process definition consists of three components, viz. artifact (see Fig. APIV/2.0.3-1), resource, and coordination structure.
- Clear and understandable: Diagrammatical representation to support clarification of the coordination structure.

2.0.4 *Approaches [3,4]*

Little JIL has been developed in a factored approach that allows the core coordination language to be simpler and easier to understand, develop, and use. This kind of approach also allows a few aspects of process definition to be developed and evolved in independent ways. The design of Little JIL coordination features was guided by following primary principles:

- Simplicity: Concerted efforts have been made to keep the language simple and to get clarity, ease-of-use, and understandability.
- Expressiveness: Software and workflow processes are semantically rich domains, and a process language, one tightly focused on coordination [4]. The language is highly expressive to reflect a variety of semantics. The language allows users to address entire the range of concerns relevant to a process and to express their goals in a clear manner.
- Precision: The language semantics are precisely defined for automatic execution of process programs. Such precision supports the analysis of process programs. Analysis is a key to assuring that process programs indeed have properties that are desirable for process safety, correctness, reliability, and predictability [4].

An artifact in connection with software stands for one kind of tangible (document) or intangible (reading from an instrument) by-product produced during software development. It includes many things such as test plan, documentation, class diagram etc. There can be analog (flicker, scratches etc) or digital (developed when digital camera is zoomed or distortion in larger display) artifact. A process works by creating or modifying artifacts. In little –JIL artifact and artifact flow have great role to perform to make Little-JIL successful.

FIGURE APIV/2.0.3-1

Artifacts.

- Flexibility: Little JIL is designed to give process programmers the power to choose what level of flexibility to give to the process's agents [4]. This is important for the cases where more than one sequencing is acceptable, or when a different set of steps may be acceptable.
- Separability: Little JIL relies on separate systems for certain functionality that is not deemed central to the expression of coordination structures. This makes the core coordination language simpler and easier to understand, develop, and use [3].

2.1 The Language

As stated earlier, Little JIL language has hierarchical structure and is a tree of step types, each of which can be multiply instantiated at runtime. Each leaf represents the smallest specified units of work and the tree's structure represents the way in which this work will be coordinated. As processes execute, steps go through several states. Typically, a step can be posted only when it is *assigned* to an agent who starts the step. Any of these said step is either successfully completed or may be terminated due to an exception.

2.1.1 *Representation*

Typical Little JIL language representation has been shown in Fig. APIV/2.1.1-1. As shown in the figure, there are a number of badges that make up a step, as well one step's possible connections to other steps.

- Interface badge: To the circle at the top of the step to this interface badge, an edge from the parent may be attached. The circle is *filled* if there are local declarations such as parameters and resources associated with the step. Otherwise it will be empty. Filled in blue circle to indicate some attributes.
- Step name: Just below the interface badge.
- Prerequisite badge: The triangle left side of step is a called the prerequisite badge. The badge appears filled if the step has a prerequisite *step/condition* and an edge may be shown that connects this step to its prerequisite.
- Post requisite badge: Similar as previous and shown in the right side.
- Badges in box: There is a box below the step name. Within the box there are two badges. As shown, from left to right are: the control flow badge, the exception handler.
- Control badge: This indicates the order in which sub-steps are to be executed, and to which the sub-steps are attached.
- Exception handler: Red Cross (X) to which exception handlers are attached.

The edges that come from these badges can be annotated with parameters (passed to and from sub-steps). This is also shown in the figure under reference. In case, when there is no sub-step/exception handler, respectively, then the badges previously described (control flow or exception handler) are omitted (refer to *blank boxes* at the bottom side). It is possible for an exception to have a null handler, in which case the continuation badge alone determines how execution proceeds.

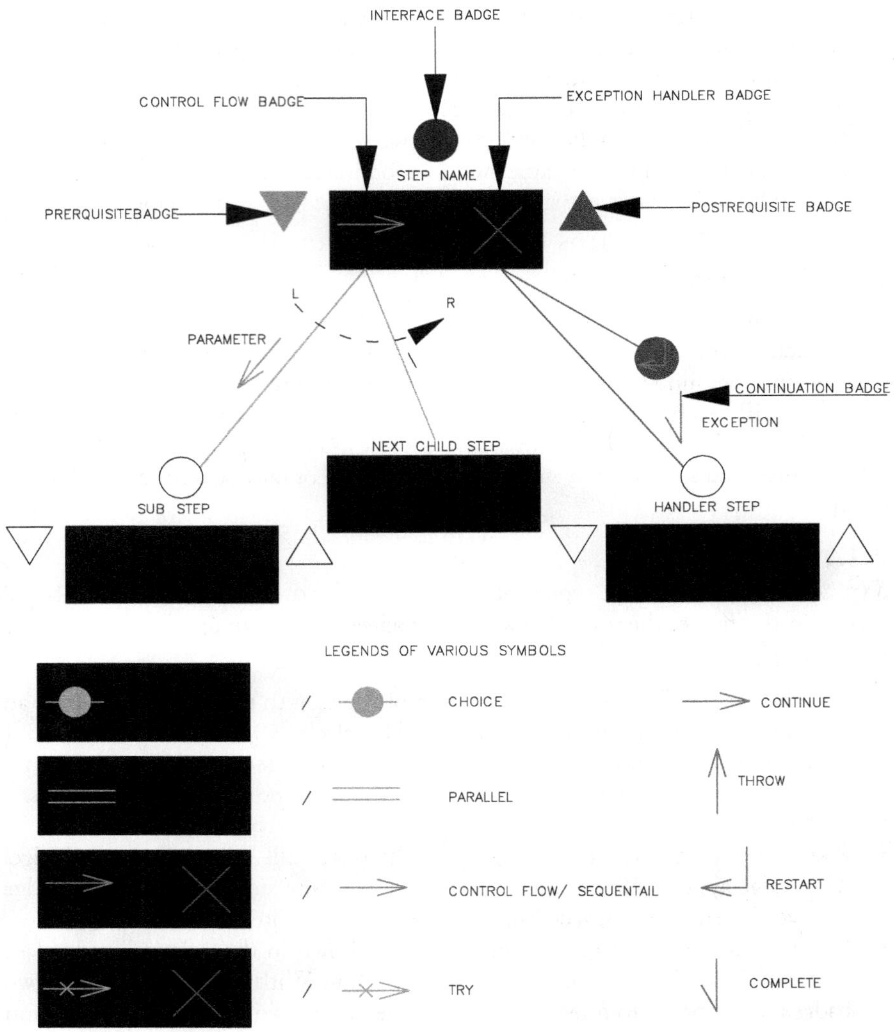

FIGURE APIV/2.1.1-1

Little JIL language representation.

Figure courtesy of Professor L.J. Osterweil—University of Massachusetts. A.G. Cass, B.S. Lerner, E.K. McCall, L.J. Osterweil, S.M. Sutton Jr., A. Wise, Little-JIL/Juliette: A Process Definition Language and Interpreter; http://laser.cs.umass.edu/techreports/00-66.pdf.

Non-leaf step: There are four non-leaf step kinds provide control flow. These are:

- Sequential: It is similar in representation to control flow go sequentially (*analogy*: For starting a large fan/pump, one needs to carry out a number of steps sequentially such as starting of control oil supply, lube oil supply to close inlet or outlet damper, etc. before giving starting command to the fan).

- Parallel: To indicate parallel action (*analogy*: starting of ID fan and starting of control and lube oil supply to FD fan).
- Try: This may be stage-by-stage protection in a system. (*analogy*: first to start an auto means to combat a fault, if fails, try operator alarm then, say, trip the machine).
- Choice: Selection of various paths to meet the target (*analogy*: if a pump fails automatically start any of the standby pump).

It is important to note that the language offers through "the parallel" and "choice" steps to the human user to *exercise* their judgment and to make choices regarding subsequent steps.

2.1.2 Little JIL — The Concept

From the previous discussions it is clear that Little JIL as a coordination language rests on four major components. The concept of Little JIL language has been as shown schematically in Fig. APIV/2.1.2-1. Major components are:

- Agent — actor (taking action)
- Step — the activity
- Artifacts
- Resources

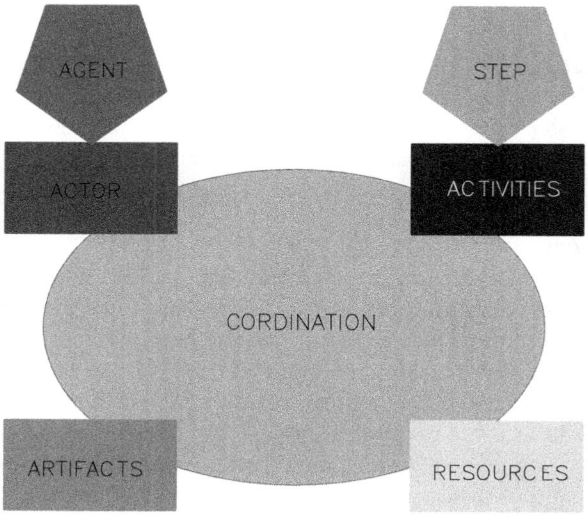

FIGURE APIV/2.1.2-1

Little JIL concept.

Figure courtesy of Professor L.J. Osterweil—University of Massachusetts. L.J. Osterweil, The Little-JIL Process Definition Language, Laboratory for Advanced SE Research; University of Massachusetts, USA, March 2008.

2.2 Important Issues of Little JIL

There are a few important issues related to Little JIL process definition, which have been enumerated as follows:

2.2.1 *Requisites*

These are mechanisms for checking before and after a step is executed to ensure that all the conditions needed before a step are satisfied and also the step has been executed "correctly." A prerequisite is a step that must be completed, prior to starting of a step, to which it is attached. Similarly, a post-requisite must be completed after completing the activities of the step, to which it is attached. This will be clear from an analogy: starting fuel fire in a boiler prerequisite is to check purge is complete. So, prerequisite is purge completion (and no master fuel relay) and after starting of firing to ensure flame is on, so "flame OK" is post-requisite. . The need for pre- and post-requisites appears to be common in process programs, and requisite step semantics seem different enough from other kinds of sequential steps that a special notation was introduced [1].

2.2.2 *Exceptions and Handlers*

As the name suggests, exceptions and handlers are used to indicate and fix exceptional conditions or errors during program execution and provide a degree of reaction control and management to handle the condition. The exception mechanisms in Little JIL are designed to be simple yet expressive. It is based on the use of steps to define the scope of exceptions and handlers. Exceptions are passed up the step *decomposition* tree (call stack) until a matching handler is found [1].

2.2.3 *Parameters*

Parameters are basically information passed from one step to the other, necessary for execution of the step to get step execution results. The parameters types, such as type definition, equality, and others, are not related to coordination. These are kept out of Little JIL.

2.2.4 *Resources*

Resources are entities which are specified and required during every step execution by Little JIL process definition. Resources needed by a step are specified as types only, offering flexibility for the execution of the process. This enables the use of scheduling technologies to make superior use of the resources available. As only the type of resource is specified by the step, so *either* human interface or automation could be used as long as the resource has the ability to meet the need defined in the resource specification. Resources may include: the agent, permissions to use tools, a database, software tools, and various physical artifacts. When software is specified as agent, then that context can also be viewed as a useful basis for defining the requirements that the software system is to meet [3]. However resource management is *not* part of Little JIL. Little JIL tries to minimize the requirements placed on the resource specification.

2.2.5 *Artifacts*

A process generally executes its functions by creating and modifying artifacts described in Fig. APIV/2.0.3-1, normally utilizing other artifacts which could be internally generated ones or inputs to the process or produced by a step. There are unique notations for the artifacts as these are required for process to identify them. Step definitions incorporate types of artifacts, so these are entities. In process applications, artifacts could be simple single ones, or could be an aggregate of artifacts [3].

2.2.6 *Discussions*

Little JIL does not specify parameters' data type, resources, expressions, or imperative commands. Little JIL relies on *agents* to know how the tasks represented by leaf steps are performed. It is meant for specifying step coordination and NOT step *execution* (i.e., what steps or actions to be executed as these are process dependent). This is the precise reason in previous discussions the word "analogy" has been used in place of "example," because in all the example cases these are executive *commands*.

2.3 Coordination Structure

The Little JIL coordination structure is typically represented visually as a hierarchical structure of steps as detailed in Fig. APIV/2.1.1-1. Some useful features and characteristics are as follows:

- Step name: Each step has a name.
- The badge: The badges represent step features, e.g. control flow among the step's sub-steps or the step's interface, including parameters.
- The exceptions are the step handles.
- Leaf step: A step with no sub-steps (activity to be performed by an agent, without the process).

Short details about these have already been covered in Clause 2.1.1; hence are not repeated.

2.3.1 *Sub Step Decomposition*

There are two sub-steps types, viz. Ordinary sub-steps and exception handlers (Ref: Clause 2.3.2). The ordinary sub-steps: this is the step execution and connection to parent step by edges with possible annotation (artifacts specifications for artifact flow) and cardinality specifications (define the number of times the sub-step is to be instantiated).

2.3.2 *Exception Handler*

These define the way exceptions are to be handled. The edge connection with parent has annotation with the type of the exception being handled and an indication of how to continue after exception handling.

2.3.3 *Step Sequencing*
Every non-leaf step has a sequencing badge as shown in bottom part of Fig. APIV/2.1.1-1.

- Arrow: The badge defines the order in which its sub-steps execute, e.g. left to right arrow indicates that sub-steps are to be executed sequentially from left to right, and is only considered completed after all of its sub-steps have been completed.
- Parallel: A parallel step indicates that parallel execution is possible. Here also it is considered completed after its entire sub-steps have completed.
- Choice: A choice step indicates agent's choice of step.
- Sub-steps: This is important for human agent and automation, for choices, try step, etc.

2.3.4 *Interface Specification*
Following are the features of interface specification:

- Circular icon located, top of the step bar.
- Parameter specification for the step exchanges information with its parent, and required resources for the step execution.
- Artifacts and artifact flows: An artifact is an entity discussed in Clause 2.2.5. Artifacts flow through steps can be two types:
  - Hierarchical: This type is as the flow of artifacts between parent and child steps indicated by attaching to the edges between parent and child identification of the artifacts as well as arrows indicating the direction of flow of each artifact.
  - Through channels: The flow of artifacts similar as above laterally or between sub-steps.
- Resources and agents: Already discussed in Clause 2.2.4—hence not repeated.
- Channels: Channels are *direct named* entities. Data channels are like queue. A Little JIL channel is defined as a first in first out queue. The channel construct can be used as a vehicle to coordinate and synchronize steps executing in parallel. Channels can specify steps for additional artifact of same type.

2.3.5 *Requisites*
Refer to Clause 2.2.1.

2.3.6 *Exception Handling*
A step to indicate exceptional conditions when some aspects of the step's execution fail. This will trigger the execution of a matching exception handler, associated with the parent of the step that throws the exception in this connection Clause 2.2.2 may be referred to.

2.3.7 *Scope and Recursion*
All steps represent a scope, specifying what artifacts are considered local to that scope. It also supports recursive execution of steps, which specifies the iterative application of a process step to specified inputs.

2.3.8 Abstraction

A step can be referred to in more than one location in a process definition. One such reference includes required details, e.g. the step's parameters and resource, and other information about the step's exception handling facilities, etc. This reference thus serves to define the step as an abstraction [3].

With this idea in mind, it is better to explore the application side of the language in plant hazard analysis automation.

3.0 APPLICATION OF LITTLE JIL IN HAZARD ANALYSIS AUTOMATION

During the discussions in previous chapters on plant hazard analysis, several methods were discussed for automating plant hazard systems. In fact, in clause number 2.6.5 of Chapter IV, automated failure mode and effect analysis (FMEA) with Little JIL has been touched upon. When the entire system or process is modeled with sufficient detail, then it is possible to automate FMEA with the help of Little JIL process definition language, which has precise semantic definitions for all its language constructs. In semiconductor manufacturing and assembly line, supposing there are *n* numbers of subassemblies and there are *n* numbers of workers working together to put some subassemblies at *n* working stations, which are then finally assembled at a place. Now in this case, the entire system can be divided into steps and each step into substeps. By applying Little JIL, the entire process can be defined with the help of several artifacts (say address of any subassembly). Each artifact and step has to be precise to get a good batch of production. Now, if one worker develops power supply unit by combining unit A and B, whereas the next worker develops RF stage by putting together C and D unit. The output of these two operators is finally assembled by another operator. If there is a fault at just one operator (say first operator made a mistake in specifying C in place of A or robot who is working for workers made similar mistake), when final assembly comes out with this failure then the entire batch may have to be rejected, meaning that it has severe adverse effect on production. In such cases Little JIL can be applied for automating the process. Here, a point to be noted is that for *single example of failure* there could be two types of faults such as:

- Type 1: Artifact (wrong address item), step S is wrong, i.e. worker specified C in place of A.
- Type 2: Artifact (wrong address) from step S is wrong, i.e. robot by mistake brought C in place of A.

The preceding is just an example to show how Little JIL can be applied in a manufacturing process. In the example, it is only one failure mode and type has been depicted. There could be a large number of failure modes/artifact-related failure modes or can easily be *turned* into artifact-related failure modes. Also, if a step is done incorrectly, one would expect that failure to be evident in one or more of the out artifacts associated with that step. For detailed and effective analysis, an artifact flow graph (AFG) may be developed from by traversing the process tree with necessary algorithm. Derived FMEA information is obtained from AFG. When the system is large, one may take the help of fault tree analysis (FTA) to determine one failure mode in a complex system. Little JIL process definition can help in deriving automated FTA. Here, there are two issues:

- How to automatically extract fault tree events that could lead to a given event.
- How to use appropriate gates for arriving at the event.

For this, each type discussed in the preceding needs to be addressed, maybe separately, by Little JIL. Therefore Little JIL language can be applied in automating FMEA/FTA for finding failure modes in large and complex systems.

The main aim of this appendix is to bring readers attention to this powerful process language, which is very useful and helpful in automating hazard analysis and failure mode identifications. With this, the discussions on Little JIL is completed to look into another important issues of embedded controls which are now used in process control application including nuclear installations.

LIST OF ABBREVIATIONS

AFG	Artifact flow graph
AI/O	Analog input/output
BPMN	Business process modeling notation
DCS	Distributed control system
DI/O	Digital input/output
E/E/PE	Electrical/electronics/programmable electronics
FD (fan)	Forced draft (fan)
FIFO	First in first out
ID (fan)	Induced draft (fan)
IEC	International Electrotechnical Commission
I/O	Input/output
I/P or O/P	Input or output
IT	Information technology
JIL	Julia input language
MFR	Master fuel relay
P&ID	Piping and instrumentation diagram
PSA	Probabilistic safety assessment
SW	Software
XML	Extensible markup language

References

[1] A.G. Cass, B.S. Lerner, E.K. McCall, L.J. Osterweil, S.M. Sutton Jr., A. Wise, Little-JIL/Juliette: A Process Definition Language and Interpreter; http://laser.cs.umass.edu/techreports/00-66.pdf.

[2] L. Zhu, L.J. Osterweil, M. Staples, U. Kannengiesser, Challenges Observed in the Definition of Reference Business Processes; www.google.co.in/url?sa=t&rctj=&q=&esrc=s&source=web&cd=1&ved=0ahUKEwiUlJjPmc_LAhWSHY4KHdvZBwA-QFggbMAA&url=http%3A%2F%2Fwww.nicta.com.au%2Fpub%3Fdoc%3D486&usg=AFQjCNHH_ynpor1YxkULCvztCLOX1BwZgA&sig2=wUU_ZBz2wNyEnC61yeu21-w&bvm=bv.117218890">https://www.google.co.in/url?sa=t&rct=j&q=&esrc=s&source=web&cd=1&ved=0ahUKEwiUlJjPmc_LAhWSHY4KHdvZBwAQFggbMAA&url=htt-p%3A%2F%2Fwww.nicta.com.au%2Fpub%3Fdoc%3D486&usg=AFQjCNHH_ynpor1 YxkULCvztCLOX1BwZgA&sig2=wUU_ZBz2wNyEnC61yeu21w&bvm=bv.117218890, d.c2E&cad=rja.

[3] Osterweil, L.J., December 2008. Formalisms to Support the Definition of Processes. Department of Computer Science; University of Massachusetts, USA. http://www.ccf.org.cn/web/resource/files/tuijianwenzhang/08-4114.pdf.

[4] A. Wise, B.S. Lerner, E.K. McCall, L.J. Osterweil, S.M. Sutton Jr., Specifying Process Coordination Using Little-JIL; Lederle Graduate Research Center; University of Massachusetts, USA; https://www.mtholyoke.edu/~blerner/papers/99-71.pdf.

[5] Osterweil, L.J., March 2008. The Little-JIL Process Definition Language. Laboratory for Advanced SE Research; University of Massachusetts, USA.

Further Reading

A.G. Cass, B.S. Lerner, E.K. McCall, L.J. Osterweil, A. Wise, Logically Central, Physically Distributed Control in a Process Runtime Environment; University of Massachusetts, USA; https://www.mtholyoke.edu/wblerner/papers/99-65.pdf.

B. Chen, G.S. Avrunin, L.A. Clarke, L.J. Osterweil, Automatic Fault Tree Derivation from Little-JIL Process Definitions; Department of Computer Science; University of Massachusetts, USA; http://www.umass.edu/eei/EEI%20Website%20Articles/Automatic%20Fault%20Tree%20Derivation%20from%20Little-JIL%20Process%20Definitions.pdf.

Lerner, B., May 2010. Getting Started with Little-JIL Case Study: Measuring Stream Discharge. Mount Holyoke College. In: https://www.mtholyoke.edu/wblerner/process/GettingStarted.pdf.

D. Wang, J. Pan, G.S. Avrunin, L.A. Clarke, B. Chen, An Automatic Failure Mode and Effect Analysis Technique for Processes Defined in the Little-JIL Process Definition Language; http://www.umass.edu/eei/EEI%20Website%20Articles/An%20Automatic%20Failure%20Mode%20and%20Effect%20Analysis%20Technique%20for%20Processes%20Defined%20in%20the%20Little-JIL%20Process%20Definition%20Language.pdf.

A. Wise, A.G. Cass, B.S. Lerner, E.K. McCall, L.J. Osterweil, S.M. Sutton Jr., Using Little- JIL to Coordinate Agents in Software Engineering; http://laser.cs.umass.edu/techreports/00-45.pdf.

Embedded Control

1.0 MICROPROCESSOR, MICROCONTROLLER, AND FIELD PROGRAMMABLE GATE ARRAY

In instrumentation and control applications, these three types of devices have a large contribution. In this part through short discussions, the application side of these devices in instrumentation and control (I&C) will be established.

1.1 General

In previous chapters, it has been made clear that any safety instrumented system (SIS) consists of three basic parts, viz. sensors, logic solver (LS) and final element (FE). Usually LSs are implemented with the help of programmable logic controller (PLC) or distributed control systems (DCS). Both PLC and DCS normally are developed and meant for generalized applications; *naturally* these are developed with microprocessor-based systems. This is especially true when these have to handle large number of input/output (I/O) data. However, this may not be true for sensors, FEs, or various peripherals connected to PLC/DCS. All the examples (sensors, FEs, and peripherals) illustrated are application-specific. Microcontrollers and/or field programmable gate array (FPGA) find their applications in these areas. In this appendix, a short discussion will be presented on microcontrollers and FPGA (also comparatively newer in application), so that readers have a fair idea on basics and application side of them so that it will be easier to handle them in real life.

The terms "microprocessor" and "microcontroller" (or microcontroller unit—MCU) can be confusing. Both find their uses in real time applications and share many common features, yet these two are neither the same, nor are they really used for the same type of applications. Currently, another solid state device, FPGA, came into the real time application field to further increase these apparent confusions. However, once somebody looks into all these a little more closely, then it becomes clear that each of these have specific areas of application in real time. Prior to going into each topic separately, the confusions will be eliminated. In order to do so, the differences between these devices will be discussed first.

1.2 Differences Between Microprocessor and Microcontroller

In spite of so many commonalities, there are a few differences between them, which have been illustrated in Table APV/1.2-1.

Table APV/1.2-1 Differences Between Microprocessor and Microcontroller

Points of Difference	Microprocessor	Microcontroller
Term	Powerful brain or heart of computer.	Computer on chip. Considered as heart of embedded system.
Device	It is just a processor[a] with external I/O, RAM/ROM memory.	As complete computer on chip,[a] it consists of processor with memory and peripherals.
Connection size and power consumption	On account of external connections, circuits sizes are large and have higher power consumption.	Smaller in size in view of internal memory and I/O. Normally CMOS is used (with immunization of power spikes) for manufacturing of MCU. All these make MCU power consumption less.
Programmable pins	A few only.	Most of the pins are programmable.
Registers	Possesses less registers and operation is memory oriented.	Possess more register so easier to write program.
Boolean functions	Not capable.	Capable of handling.
Instruction Type	Instructions are mostly external.	Majority instructions are internal because of configuration.
Instruction execution	Microprocessor performs the instruction cycle (fetch, decode, and execute) repeatedly.	Microcontroller performs normal and environment-based instruction cycle as well.
Bit handling	Few bit handling instruction.	Many bit handling instruction.
Interrupt latency	More emphasis on instruction throughput than interrupt latency.	Optimizes interrupt latency.
Bit manipulation	A few bit manipulations instructions are supported.	Powerful bit manipulation and used widely.
Application	Microprocessors find general purpose wide range of applications, viz. SW development, PLC, website development.	These are application-oriented, and designed for specific tasks with defined I/O relationship, for example, specific instrument, digi-cam, remote control.
Resource	On account of general purpose application it has higher requirements for RAM/ROM, I/O port, etc. resources.	On account of specific application it has much lesser requirements for RAM/ROM, I/O port etc. resources.
Clock speed	It has much higher clock speed requirements to the tune of X GHz.	Microcontrollers need lesser clock speed to the tune of XY MHz.
Compactness	On account of external connections not compact.	Compact design possible.
Computational power	Computational power is much more, hence used for generally complex problems.	Computational power is less, used for specific cases with defined simple tasks.

Table APV/1.2-1 Differences Between Microprocessor and Microcontroller—cont'd		
Points of Difference	**Microprocessor**	**Microcontroller**
Co-processor	They have math co-processor to support floating point complex mathematical problems.	No math co-processor hence takes the help of SW for math solutions, hence slower.
Power saving	May or may not have.	It has power saving feature.
Cost consideration	Much costlier than microcontroller but cannot be used in place of microcontroller cost will be higher.	Much cheaper than its counterpart. However, it should not be used to replace microprocessor then cost will go up.
Example	Intel 8086	Intel 8051

[a]*Ref Fig. APV/2.1-1A and B.*

1.3 Relation Between Application Specific Integrated Circuit and Field Programmable Gate Array

FPGA can be conceived as one application specific integrated circuit (ASIC) in a prototype stage. Two technologies have their own pros and cons, hence have their market share based on application. Major differences between ASIC and FPGA come from costs, tool availability, performance, and design flexibility. The major distinct features have been compared in Table APV/1.3-1.

Table APV/1.3-1 Comparison of Features for Application Specific Integrated Circuit (ASIC) and Field Programmable Gate Array (FPGA)

Points of Comparison	ASIC	FPGA
Cost and optimization	Higher fixed cost, neutralized with higher volume of production—finally cost-effective. Cost of tools is higher when compared with FPGA. It is optimized for application/use.	Comparatively lesser fixed cost, but overall recurring cost may be higher. Lower volume of production may not be so cost-effective. It is not optimized for application.
Configuration and application	Generally fixed configuration and design, after design and manufacture.	Reconfigurable to achieve the goal as many times as one wishes but not optimized.
Production	Higher volume speed, and complexity in design and production.	Lower volume speed, and simpler design and production.
Marketing	Comparatively slower to market on account of design and production process.	Faster to market for no masking layout and production steps.
Power off	Static with power off configuration is not lost.	Dynamic and with power off the current state as well as configurations are lost. Now flash chip / MCUs are used to circumvent the situation.

Continued

Table APV/1.3-1 Comparison of Features for Application Specific Integrated Circuit (ASIC) and Field Programmable Gate Array (FPGA)—cont'd

Points of Comparison	ASIC	FPGA
Hardware Description Language	Utilized.	Utilized.
Pros and cons	High port density applications, full custom capability (as these are manufactured as per design) are plus points of ASIC over FPGA. Costlier design tool, higher time to design, and marketing leave FPGA ahead of ASIC.	Short time to market, simpler design cycle, field programmability, reusability, and more predictable project cycle are clear advantage over ASIC. Higher power consumption and low scale production are disadvantages of FPGA.

1.4 Comparison of Feature of Microcontroller Unit and Field Programmable Gate Array

When one compares the characteristic features of MCU and FPGA from the preceding discussions, one may notice that both of them are specific application oriented. In that case, a natural question may come to one's mind then, *why two different technologies?* Comparison between the two technologies in Table APV/1.4-1 would reveal requirements of both of them separately.

Table APV/1.4-1 Comparison of FPGA and Microcontroller Unit (MCU)

Points of Comparison	MCU	FPGA
Device functionality	Microcontroller has its own circuitry and instruction set, that is, MCU will come with a pre-designed set of features. So, MCU can be conceived as predesigned logic system, where codes are written to clarify what work it has to perform.	FPGA is fully configurable to user's choice. FPGA can be electrically configured to perform a task. It can be conceived as a programmable logic where VHDL, verilog help to describe hardware. MCU can be written in FPGA.
Computation	It is a computing system with a processor and governing rule for I/O regulation.	It is just array of gates to be used as per user desire and hence no computing power as such in true sense.
Use and speed	Application specific may be complex operation. It has lower speed than FPGA.	Used for comparatively simpler operation with higher speed.
Power consumption	Typically lower than FPGA.	Typically higher than MCU.
Functionality	Always sequential. Addition/deletion function not possible.	Concurrent sequential is also possible. Addition /deletion function possible.
Cost	Cheaper.	Costlier.
Application	Because of lower cost, finds applications where large volumes are required.	Suitable for larger, complex problems with lower volume of use.

VHDL stands for VHSIC hardware description language, where VHSIC stands for high speed integrated circuit used by US military in 1980s. VHDL is one of the commonly used HDL. VHDL is a non proprietary, international standard regulated by IEEE. It basically describes the behavior and structure of electronic systems especially for ASICS and FPGA and other related digital circuits. It is not purely an information model/database/ simulator/tool set or a methodology. Even though methodology and tool sets (simulation and synthesis) are part and parcel of VHDL for its effective use.

FIGURE APV/1.4-1

Very high speed integrated circuit hardware description language (VHDL).

With the help of various tables presented, readers could gather some knowledge about the nature of the devices very commonly used in embedded circuits which at present enjoy tremendous advantages over microprocessor based systems (e.g., PLC) in application specific instrumentation devices such as sensors, intelligent positioners, etc.

In fact these embedded systems also find applications in LS, viz. ABB 800XA DCS. So, it is necessary to gather some knowledge on MCU and FPGA in subsequent clauses to supplement the discussions on LS presented in previous chapters.

As indicated in Fig. APV/1.4-1, VHDL stands for VHSIC hardware description language, where VHSIC stands for very high speed integrated circuit, as used by US military in 1980s. VHDL is one of the commonly used HDLs. VHDL is a non-proprietary, international standard regulated by (Institute of Electrical and Electronics Engineers) IEEE. It basically describes the behavior and structure of electronic systems especially for ASIC and FPGA, and other related digital circuits. It is not purely an information model/database/ simulator/tool set or a methodology; although methodology and tool sets (simulation and synthesis) are part and parcel of VHDL for its effective use. See Fig. APV/1.4-1.

2.0 MICROCONTROLLER UNIT: *WHAT IS MICROCONTROLLER?*

A microcontroller is a self-contained computing system with peripherals, memory, and central processing unit. Mostly the system is embedded into any products/systems for which its used. For this reason, it is also referred to as "embedded controller." The largest single use for microcontrollers is in the automobile industry, but it finds its application in almost all day-to-day use devices such as ovens, toasters, and clock systems. Also it has a number of applications in medical units viz. an artificial kidney or heart. Even in sophisticated spacecraft, microcontrollers are used. In instrumentation, there has been extensive use of microcontrollers, viz. sensor, field controller, positions, etc. Naturally some knowledge about the same is essential for studying SIS.

2.1 Functional Details

At this point, discussions on MCU architecture and various subsystems that constitute MCU will be presented. This is a short description of MCU. Further details are available in any *standard book on microcontroller*. Following are the major parts that constitute MCU.

- Central processing unit (CPU)
- Random access memory (RAM)

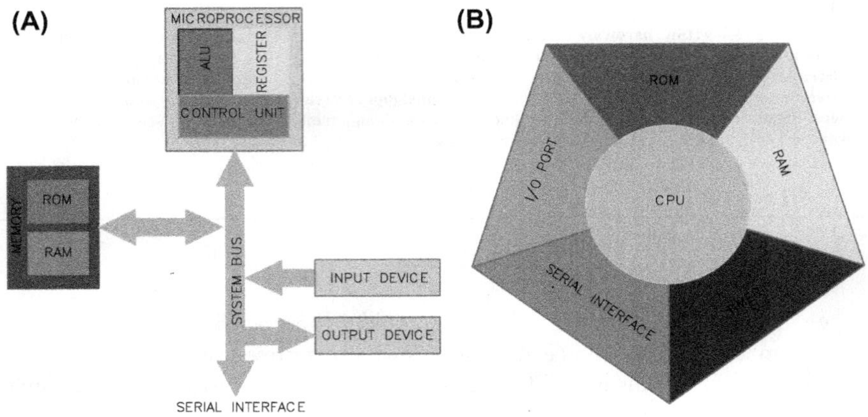

FIGURE APV/2.1-1

Microprocessor and MCU block (comparison). (A) Microprocessor with basic components (for comparing with micro controller), (B) micro controller self contained unit (for comparing with microprocessor).

- Read only memory (ROM)
- Input/output ports
- Timers and counters
- Interrupt controls
- Analog to digital converters
- Digital analog converters
- Serial interfacing ports
- Oscillatory circuits

Various part of MCU has been shown in Fig. APV/2.1-1B as a compact unit around the central CPU for comparing MCU with typical microprocessor configuration shown in Fig. APV/2.1-1A. In this diagram, interrupt control, analog to digital converter (ADC)/digital to analog converter (DAC), and timing control, etc. are not detailed out, only basic blocks have been depicted.

2.1.1 Architectural and Functional Description

An architectural block diagram of a typical MCU (mainly based on 8051) has been presented in Fig. APV/2.1.1-1. Like any other computing device, the CPU is the brain of MCU. It meticulously scrutinizes and manages all processing functions carried out in MCU. CPU interprets program in storage space (ROM) to carry out all the duties assigned. CPU has to manage general purpose register (GPR) and special function register (SFR) to discharge its duty.

As seen in the figure, the first box to encounter is oscillator and timing circuit directly connected to the CPU. MCU, being a digital circuit, needs timer for its functioning. This is carried out by a chip oscillator, which acts as a time source to synchronize all functions. For functioning of an MCU, it needs a program in form of command sets. An interrupt program enlightens the MCU to perform precise tasks. Interrupt is a subroutine call which assists an MCU to perform some program of higher importance. All other units are connected to the CPU with the help of a bus system, which is basically a wire

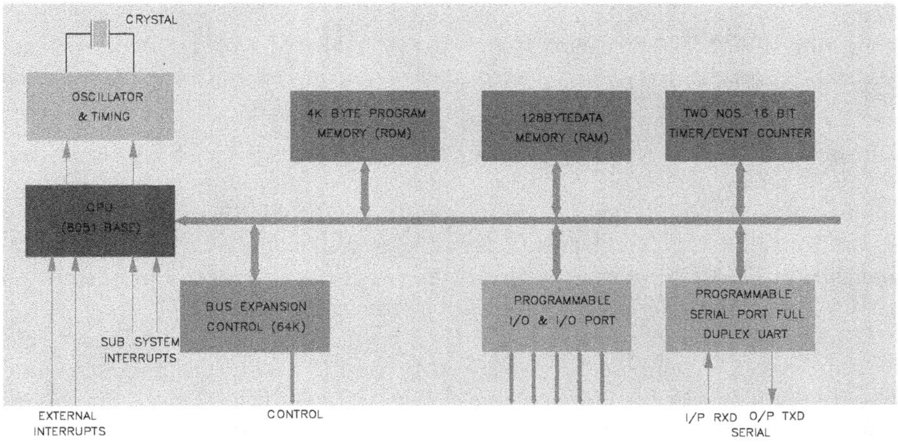

FIGURE APV/2.1.1-1

Architectural block diagram of typical MCU.

connection that acts as a communication channel. It could be 8/16 wires. There are two types of bus systems, viz. address bus and data bus. An address bus may be a 16 bit (8051) bus to bring the address of the memory position (also transmission from CPU). Also, a data bus could be 8 (8051)/16 (I/O port) bits. Now it is time to look into the functionality of different subsystems.

2.1.2 Central Processing Unit (CPU)

As a brain of the device, its primary function is to fetch instructions and decode them. It connects every part of a microcontroller into a single system. This is in addition to regulation of registers discussed earlier. It consists of following major sub-units which has similar functionalities of a microprocessor:

- Program counter: As an internal register, it stores the memory address of next (subsequent) instruction for CPU. So as CPU executes instruction it is incremented by one (0000 to 0001).
- Instruction decoder: The decoder decodes each instruction read from memory to drive the MCU through arithmetical and operational circuitry. It keeps encoded instruction to original form.
- Arithmetic and logic unit (ALU): Primarily it performs the arithmetic and logical operations. It is regulated by an instruction decoder.
- Internal registers: As the name suggests, these are transient storage units. There are two types of registers:
  - GPR is used for storing results arithmetic and logical operations. 128 byte visible or user accessible RAM is available in this form, hence 128 byte RAM is shown in Fig. APV/2.1.1-1
  - SFR: Used for specific information storage such as flag (intermediate result of logical operation), carry flag, etc. This is additional 128 bytes in the form of special register as shown in Fig. APV/2.1.1-1, for example, ports /timer counter, etc.

2.1.3 *Memory*

Memory consisting of ROM, flash memory (source code), and RAM is used to store data and programs which needs storage space for the CPU to interpret and act upon a specific task. There are two types of memories:

- ROM: It stands for the memory to accumulate the program of MCU. This is also called code memory. ROM retains its content even when power is off, so it is used to store the startup program for reading only. Many MCUs use flash memory in place of ROM. Microcontroller 8051 contains 4 kB ROM. Many MCUs have electrically erasable programmable ROM (EEPROM) (field programmable by EEPROM programmer). At times basic configurations can be stored in EEPROM so that after power off the same can be retrieved.
- RAM: This memory loses its contents when power is off, but it can be freely rewritten. In 8051 there is 128 byte general purpose (shown as RAM) and 128 byte special purpose memory, for example, timer, counter, and SFR. Details are elaborated in Fig. APV/2.1.3-1. So, this memory is mainly used to store program variables/data. Two varieties of RAM are:
    - SRAM: Static RAM, or SRAM, which has complex internal circuitry, supports faster access and does not require periodic refreshment. However, it is not suitable for large quantities of storage.
    - DRAM: Dynamic RAM, or DRAM, has simple circuit structure to store large quantities storage in small spaces. DRAM sizes are much bigger than typical SRAM sizes but are not suitable for high-speed logic. This is the reason why it is not popular in MCU use except peripheral interface controller (PIC) use. 8051 has storage space of 128 byte RAM (general purpose memory).

2.1.4 *Parallel Port*

These are used to drive various devices such as LED/LCD/ printer connected with the MCU.

2.1.5 *Serial Port*

This is also used for the same purpose discussed previously; the only difference is that these are used for serial interface.

2.1.6 *Timer & Counter*

A microcontroller may have more than one timer and counters (two sets for 8051), to provide all timing and counting functions in the microcontroller. The major operations involve clock functions, modulations, pulse generations, frequency measuring, making oscillations, etc. This also can be used for counting external pulses [1].

2.1.7 *A/D and D/A Converter*

As the name implies, ADC is used for converting analog signal (sensor output) to digital form. Similarly, DAC is used for converting digital signals of MCU to analog for to be used (say) for final elements. These are at times subject to noise, or acts as a source of noise. For eliminating noise, researchers are now trying to use advanced *chaos theory* (see Fig. APV/2.1.7-1).

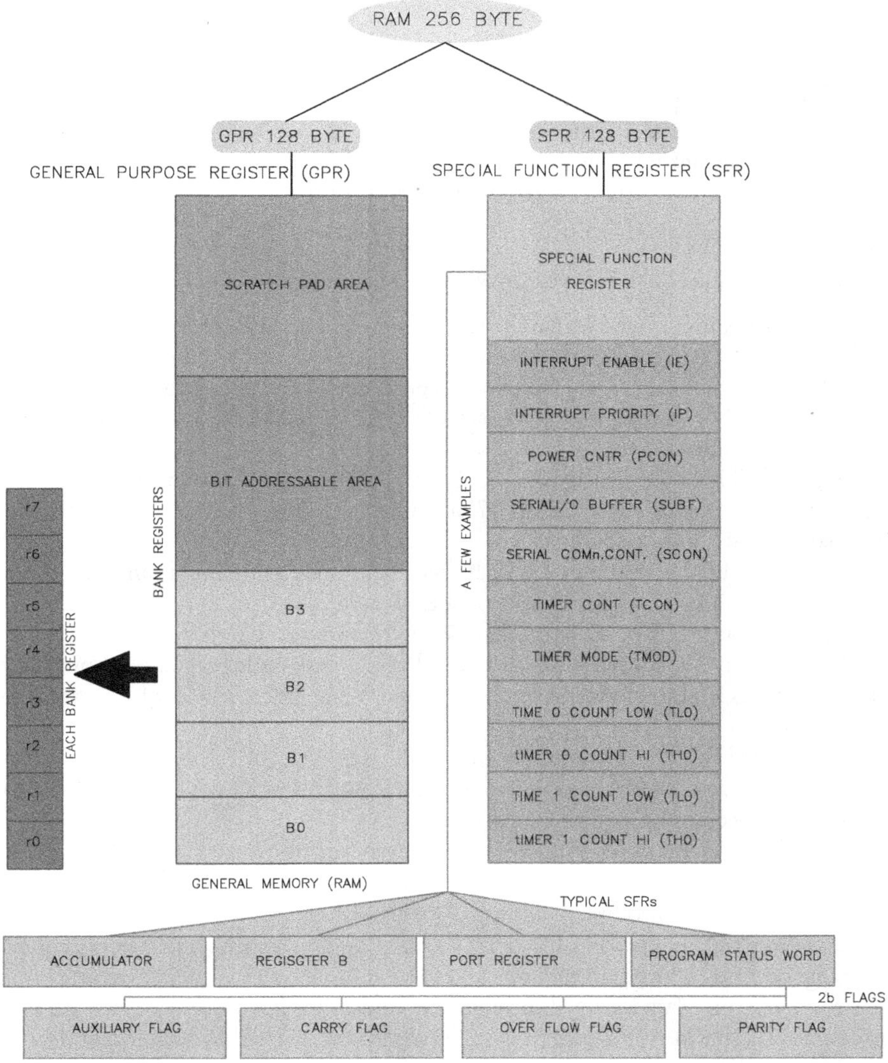

FIGURE APV/2.1.3-1

Registers of MCU (typical).

Chaos theory: Chaos theory is a mathematical sub discipline i.e. branch of mathematics to deal with complex systems whose behavior is highly sensitive to very small changes in existing condition and can result great consequences at the end (may be due to accumulative effect). This finds its application in wide varieties of subjects including those in embedded controls *e.g. Noise controls in ADC and signal processing of MCU.*

FIGURE APV/2.1.7-1

Chaos theory.

2.1.8 *Interpret Control*

The characteristic of 8051 interrupt is extremely constructive to meet the emergencies. Interrupt provides a method to postpone or delay the current process, carry out a sub-routine task, and then all over again restart standard program implementation. When a sub-routine task is finished, then the implementation of core program initiates automatically as usual. The interrupt may be external (activated by using interrupt pin), or internal. There are 5 interrupts in 8051 microcontroller (2 peripheral interrupts, 2 timer interrupts, and 1 serial port interrupt).

2.1.9 *Discussions*

A few important issues related to microcontroller discussed below in brief:

- GPR: As discussed earlier, there are two kinds of registers, viz. GPRs and SFRs. GPRs are normally Byte-addressable, but there are cases for bit addressing also. GPR is often loosely referred to as RAM, as these are used for general purpose. Typical MCU registers (major) are shown in Fig. APV/2.1.3-1. There are mainly three parts of GPR, viz. bank register, bit-addressable area, and scratch pad area, as shown.
  - Bank registers: There are four banks, viz. B0, B1, B2, and B3. Each of them has eight general-purpose registers ranging from 'r0' to 'r7' as shown. Bank operations are selected by PSW discussed later. By default, B0 is selected. B0 (00h to 07h), B1 (08h to 0Fh), B2 (10h to17h), and B3 (18h to 1Fh) so naturally r1 of B0 will have address 01h and same for r7 will be 07h.
  - Bit addressable: 20h to 2Fh is bit-addressable area RAM.
  - Scratch pad area: In GPR, 30h to 7Fh is scratch pad area.
- SFR: SFRs are mainly Bit-addressable. However, PSW is byte-addressable. As shown in Fig. APV/2.1.3-1, major SFRs shall include but are not limited to the following (for bit byte addresses refer to Clause 2.2.1):
  - Accumulator (SFR): The accumulator (also known as ACC or A), is used to hold the results of most arithmetic and logical operations. A is a bit as well as a byte-addressable register by an address of the accumulator [2].
  - B Register (SFR): The major purpose of this register is in executing multiplication and division. Register B is also byte-addressable and bit-addressable. To access bit o or to access all 8 bits (as a single byte), physical address F0 is used.
  - Port register (SFR): There are 4 I/O ports for 8051, and these are named as P0, P1, P2, and P3. Data must be written into port registers first to send it out to any other external device through ports. Similarly, any data received through ports must be read from port registers for performing any operation [2]. Pin configuration very important for a microcontroller that depends on the logic states of the registers
  - Stack pointer: Stack pointer is also known as SP. It is byte-addressable. It is an 8 bit register, and its content is the last stored location.
  - Processor status word (PSW-SFR): PSW is an important SFR to show the status of operation being carried out in MCU processor. Out of 8 bits, here 6 are used and 2 are left for flags as typically shown in Fig. APV/2.1.3-1.
  - Other SFRs: Other byte-addressable SFRs, as shown typically in Fig. APV/2.1.3-1, include power management register, timer counter, data pointer, etc.

- Special I/O: There are a number of special I/Os found in modern MCUs. Major such I/Os shall include:
    - Hardware interrupt: Special input to get the attention of MCU and run special sub-programs.
    - Mixed signals: These may include universal I/Os normally encountered in instrumentation. Many MCUs are designed to handle both in the same chip.
    - External reset: Input to reset an MCU.
    - I/P pull up/down: These registers are necessary to handle various special control electronics.
    - Signal debouncer: This feature cleans any signal transition by mechanical means.

2.2 Special Function Register Address and Microcontroller Unit Pin Details

Like GPRs, it is necessary to have some knowledge about SFRs also. Similarly, pin allocation of MCU is necessary to handle. In this part, these two issues will be addressed very briefly.

2.2.1 SFR Addresses

As there are many kinds of SFRs, it better to describe them with the help of Fig. APV/2.2.1-1.

2.2.2 Pin Allocation

All the pins of MCU have been allocated. The majority are for ports p0-p3. Typical pin allocation with associated functional details has been enumerated in Figs. APV/2.2.1-1 and APV /2.2.2-1, so the same is not repeated in write up.

DIRECT ADDRESS	BIT ADDRESS								
F04	F7	F6	F5	F4	F3	F2	F1	F1	A REG
E04	E7	E6	E5	E4	E3	E2	E1	E1	B REG
D04	D7	D6	D5	D4	D3	D2	D1	D1	PSW
B84	BF	BE	BD	BC	BB	BA	B9	B8	IP
B04	B7	B6	B5	B4	B3	B2	B1	B0	P3 PORT
A84	AF	AE	AD	AC	AB	AA	A9	A8	IE
A04	A7	A6	A5	A4	A3	A2	A1	A0	P2 PORT
984	9F	9E	9D	9C	9B	9A	99	98	SCON
904	97	96	95	94	93	92	91	90	P1 PORT
884	8F	8E	8D	8C	8B	8A	89	88	TCON
804	87	86	85	84	83	82	81	80	P0 PORT

TO BE READ IN CONJUNCTION WITH FIG APV/2.1.3-1 & CLAUSE 2.1.9

BYTE ADDRESS OF BYTE ADDRESSABLE SFRs

REGISTER	ADDRESS
SP	81H
DPTR	DTATA POINTER (2B)
DPL	LOW BYTE 82H
DPH	83H
TMOD	89H
TL0	8AH
TH0	8CH
TH1	80H
TL1	86H
PCON	87H
SBUF	99H

TO BE READ IN CONJUNCTION WITH FIG APV/2.1.3-1 & CLAUSE 2.1.9

FIGURE APV/2.2.1-1

Special function register address.

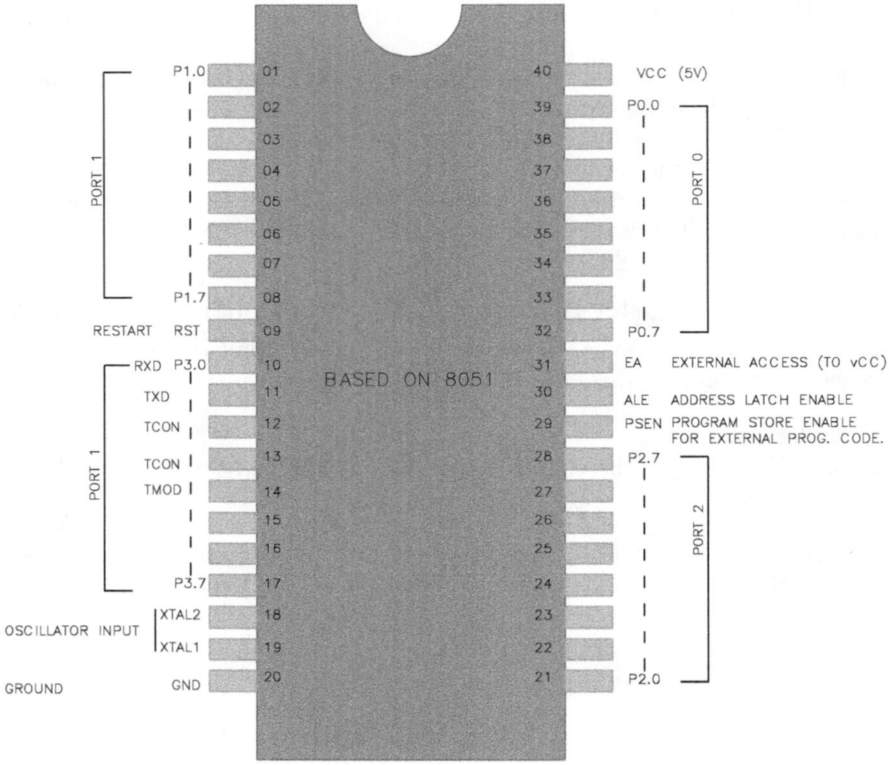

FIGURE APV/2.2.2-1

Pin allocation of MCU (typical).

It is worth noting that MCUs are often graded according to type of memory used especially for MCUs in peripheral interface controllers. These are designated in their series numbers with "C" for EPROM and "F" for flash memory versions. With this, it is time to conclude the discussions on the MCU and move on to another important device, FPGA, in the next clause.

3.0 FIELD PROGRAMMABLE GATE ARRAY (FPGA)

FPGA is another important milestone in embedded technology, which has quite a good presence in specialized SISs (with special reference to nuclear installations) as well as in local controls of sensors and final elements (even in safe PLC in specialized area). Prior to going into details about the system, it is necessary to put forward a few definitions of terms which will be used quite frequently.

3.0.1 *Commonly Used Terms*

- *What is FPGA?* FPGAs are programmable logic devices (PLDs). Internally, FPGAs contain gate arrays of pre-manufactured programmable logic elements called cells. These are specialized chips that are programmed to perform very specific functions in hardware. When a circuit board is manufactured containing FPGA as part of the circuit, then the program during the manufacturing process, can later be *re-programmed* to reflect any changes. So, many think of this as "***soft hardware***," as it can be freely reprogrammed. Another important issue here is that not all the logic in an FPGA is designed to be directly used by the customer. Therefore, FPGAs are not always optimized. All these advantages are accompanied by disadvantages, like that these are slower in operation when compared with ASIC (see Clause 1.4).

- CLB: CLB stands for configurable logic block. Some call this a logic array block or LAB. CLBs are the programmable logic blocks of FPGA. The basic cell structure for FPGA is somewhat more complicated than the basic cell structure of standard gate array. These are constituted by these CLBs and programmable interconnects which are analogous to wires for electrical connections. In case of FPGA, these are connected through interconnect blocks (switches).
- IOB: IOB stands for input/output block. IOBs provide a programmable interface between the internals, such as array of logic blocks and the device's external package pins.
- Net and Netlist: A net stands for a conductor that interconnects two or more component terminals. Netlist is a description of the connectivity of an electronic circuit. In its simplest form, a Netlist consists of a list of the terminals of the electronic components in a circuit and a list of interconnecting electrical conductors. As a programming file of the FPGA contains the Netlist information.
- RISC: Reduced instruction set computer (RISC) is a type of microprocessor architecture that utilizes a small, highly-optimized set of instructions in one cycle execution time. So, RISC processors have clock per cycle instruction. A few characteristic features shall include but are not limited to:
  - Load/store Architecture: This architecture is also called register-to-register (RR). In this architecture, operands and results are indirectly fetched from main memory through a lot of scalar registers.
  - Instructions: Instructions are simple, small, and limited in number.
  - Fixed length instruction: These are easier to decode; faster and inexpensive memory could be used.
  - Fused and compound instructions: These types of instructions can be heavily optimized for commonly used functions.
  - Pipelining: This technique allows for simultaneous execution of parts, or stages, of instructions more efficiently, for example, assembling, painting, and fitting of a device in *one go*.
  - Utilization: Mainly utilized for hardware control units. RISC is used in Xilinx FPGA also.
  - Registers: Large numbers of registers, hence less need for memory access.
  - Power consumption and performance: Less power consumption and high performance.
- ARM (advanced RISC machines) processor: An ARM processor is one of a family of CPUs based on the RISC developed by ARM. ARM makes 32-bit and 64-bit RISC multi-core processors. These are used in smart instrumentation, smart phones, and tablets to name only a few. The salient features of ARM processors shall include:
  - Load/store architecture.
  - 64- and 32-bit execution states for scalable high performance.
  - Enhanced power-saving mode.

- MIPS (originally an acronym for microprocessor without interlocked pipeline stages) is a RISC instruction set architecture (ISA) developed by MIPS Technologies (formerly MIPS Computer Systems, Inc.) (*Definition from Wikipedia*). MPIS in FPGA has been covered in *Clause 3.2*.

3.0.2 FPGA Internals and Designing Process

As indicated previously, the internal parts mainly consists of three elements; CLB, IOB, and programmable interconnect or routing channels. The key word related to the performance of FPGA is its flexibility of operation. Therefore, it is essential that all these elements operate successfully so that FPGA can perform well. The actual implementation depends upon the manufacturer the series of FPGA being used [4]. The major variations come from, the number of I/Os, the general complexity of the logic block. All these will directly impact on the size of the chip and type. The connections /routing within the FPGA comprise wires which are connected via electrically configurable switches—interconnect block. In view of complexity of FPGA, software is used. HDL/ VHDL languages or Verilog are used by the designers. Upon completion of this stage, Netlist, which describes the connectivity required within FPGA, is prepared. The Netlist can then be fitted to the actual FPGA architecture using a process called place-and-route. This is done with the help of *proprietary* software of the company [4].

3.0.3 FPGA and Microprocessors

In order to ease system design, now by developing circuits by combining FPGA with embedded microprocessor and related peripherals. Fig. APV/3.1-1C has been developed to show the same. These are at times referred to as "system on a programmable chip." Zynq-7000 of Xilinx, Inc. uses an ARM high-end processor. Another way is to use hard-macro processors with soft processor cores that are implemented within the FPGA logic; Nios II, MicroBlaze, and Mico32 are examples.

3.1 Field Programmable Gate Array Architecture

Typical architecture of FPGA has been presented in Fig. APV/3.1-1. There are three distinct figures here for better understanding.

Fig. APV/3.1-1A shows various mostly used components and their dispositions. Fig. APV/3.1-1B depicts one simplified connection of logic gates. Fig. APV/3.1-1C has been used to show the types of devices those could be embedded, like processor, logic, and memory.

3.1.1 General Layout

The most common FPGA architecture consists of an array of CLB/LABs (e.g., 32 × 2 bit RAM for combinational logic), I/O pads (comprising I/Os—direct or registered), and programmable interconnect or routing channels (which are normally of same width, i.e., number of wires). Multiple I/O pads may fit into the height of one row or the width of one column in the array. Application circuits are mapped in to FPGA with adequate resources. While numbers of CLBs or LABs and I/O requirements are easily determined from the design but number of programmable interconnect /routing channels may vary with design and type of I/O block switch, for example, crossbar requires more routing channels. Also, the CLB contains RAM memory cells and can be programmed to realize functions (e.g., SRAM cells are used to store the function values). The functions are stored in the truth table form, so the number of

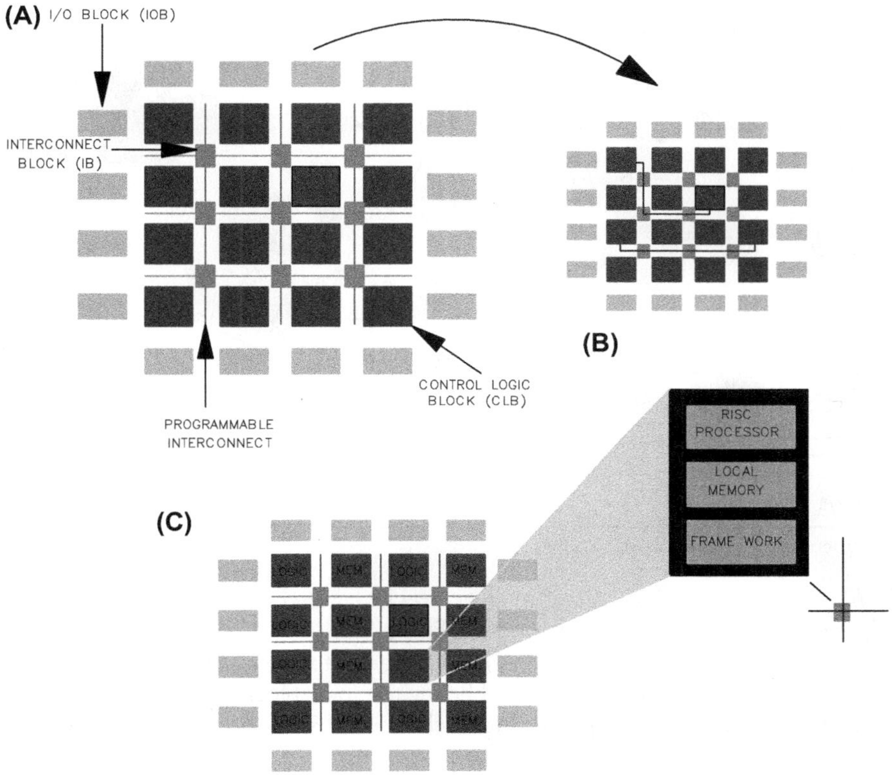

FIGURE APV/3.1-1

Structural details of FPGA. (A) Structure and parts of FPGA, (B) typical connection details through interconnect block, (C) FPGA with logic, memory & processor.

Developed based on idea from the epiphany multicore solutions.

gates required to realize the functions is not so important. FPGA manufacturers try to provide just enough tracks so that most designs that will fit in terms of lookup tables (LUTs) and I/Os can be routed.

3.1.2 Programmable Interconnect/ Routing Channel

The programmable interconnections between the configurable logic blocks and I/O blocks can be made in several ways, for example, general purpose, direct and vertical, and horizontal long line interconnect.

3.1.3 Pin Allocation and Routing Connections

Before starting the discussions, it is better to refer to Fig. APV/3.1.3-1.

In order to understand the pin allocation system it is better to take an example, where there is a logic block comprising 4 I/P LUT and D flip flop, shown in Fig. APV/3.1.3-1. There are four I/Ps and

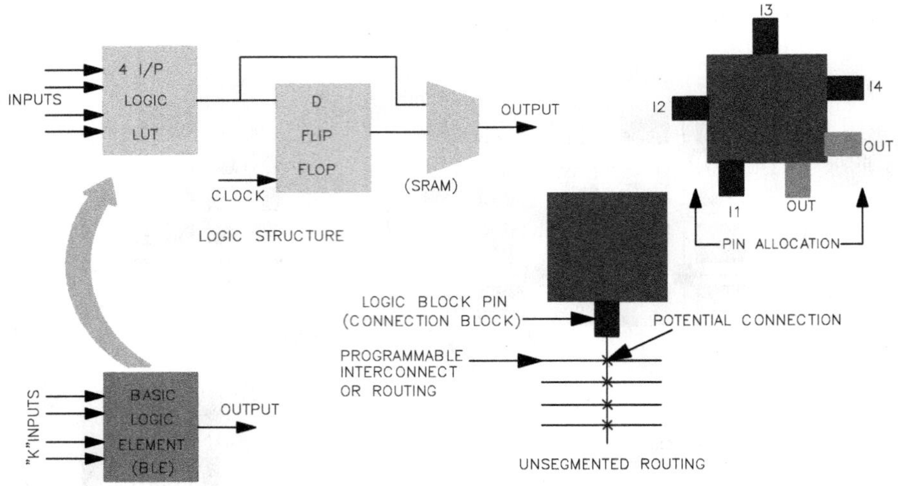

FIGURE APV/3.1.3-1

Logic block structure pin allocation and connection.

one clock I/P. Routing network pertinent to clock may be ignored, since it is normally connected via a special purpose dedicated routing connection in FPGAs. Each input is accessible from one side of the logic block (from left side—serially numbered), while the output pin can connect to routing wires in both the channel to the right and the channel below the logic block. Each logic block input pin can connect to any one of the wiring segments in the channel adjacent to it. Each logic block output pin can connect to any of the wiring segments in the channels adjacent to it. Based on this philosophy, pin allocations have been shown in Fig. APV/3.1.3-1. Similarly, an I/O pad can connect to any one of the wiring segments in the channel adjacent to it. The programmable routing *within* each CLB consists of interconnection matrix. The programmable routing between the CLBs consists of mainly switch blocks, connection blocks, and the programmable switches. A switch block appears at each intersection between horizontal and vertical programmable interconnect or routing channels. The connections in the switch blocks and connection blocks (for logic block) are made by programmable switches and part of such *programming* lies within each CLB. The FPGA routing is un-segmented. That is, each wiring segment spans only one logic block before it terminates in a switch box which can be turned on to construct a longer path also.

3.1.4 Switching Block

A programmable switch or switching matrix are mainly following types:

- SRAM cell for SRAM-based FPGA
- An anti-fuse for anti-fuse FPGAs
- A non-volatile memory cell for floating gate FPGAs

As transistors in the switch block add capacitance and resistance loading to the each track in a channel, and hence the *switch block has direct impact* on the speed of each routable connection and thus a major impact on the speed of the FPGA as a whole.

3.1.5 *Design Issues*

The computer aided design (CAD) software can be used to convert the given digital circuit (schematic or description in VHDL or Verilog) into a stream of bits, which is then used to program the FPGA. However, when dynamic RAM is used in FPGA, it is not possible to remove it without losing the configurations. So, it is better to load FPGA on power up.

Discussions on basic architecture of FPGA are concluded to start discussions on use of processors in FPGA.

3.2 Processor in Field Programmable Gate Array

FPGA becomes versatile when it has embedded processors. First, these types with advantages shall be explored. Hard and soft processors:

3.2.1 *Hard and Soft Processors*

Embedded processor in FPGA can be of two types these are:

- Hard processor: A processor built from dedicated silicon is referred to as hard processor, for example, ARM922T of Altera and PowerPC405 of Xilinx.
- A soft processor is built using the FPGA general-purpose logic and is described in HDL. In both cases, local memory, peripherals, peripheral controllers, and memory controllers are built from FPGA.

3.2.2 *Advantages of FPGA Embedded Processor*

FPGA embedded processor system offers the following major advantages over microprocessors:

- Component and cost reduction: It is evident from earlier discussions that a single FPGA can replace a number of components. Even a custom *co-processing engine* can be designed in the FPGA [5]. This means that there will be reduction of components and wiring hence lower board size and finally reduction in cost.
- Customization: In embedded processor of FPGA, designer has total flexibility for selection of any combination of peripherals and controllers. Even it is possible to develop new, unique peripherals for connecting to the processor bus.
- Obsolescence mitigation: Component obsolescence mitigation is a difficult issue [5]. This is possible by soft processor developed with HDL and HDL source codes can be purchased for longer life span.
- Hardware and software trade off: An important issue related to use of embedded processor in FPGA is its ability to make *tradeoff between hardware and software to maximize efficiency and performance*.

3.2.3 *Peripherals and Memory Controllers*

Major manufacturers of FPGA offer extensive library for peripherals and memory controllers. These shall include, but are not limited to:

- Peripheral controllers:
  - General purpose I/O
  - UART (universal asynchronous receiver/transmitter)

- Timer
- Debug
- DMA (direct memory access) controller
- Ethernet interface
- Memory controllers:
  - SRAM
  - SDRAM
  - DDR SDRAM
  - Flash
  - Compact Flash

3.2.4 MIPS Processor and FPGA

In FPGA applications, MIPS has been utilized to improve performance of the system. MIPS has RISC (see Clause 3.0.1) microprocessor architecture and defines large numbers of 32-bit GPRs. The CPU uses byte addressing for word and a byte boundary divisible by four. There are three types of instructions listed as follows, and each instruction has different format as shown in Fig. APV/3.2.4-1A.

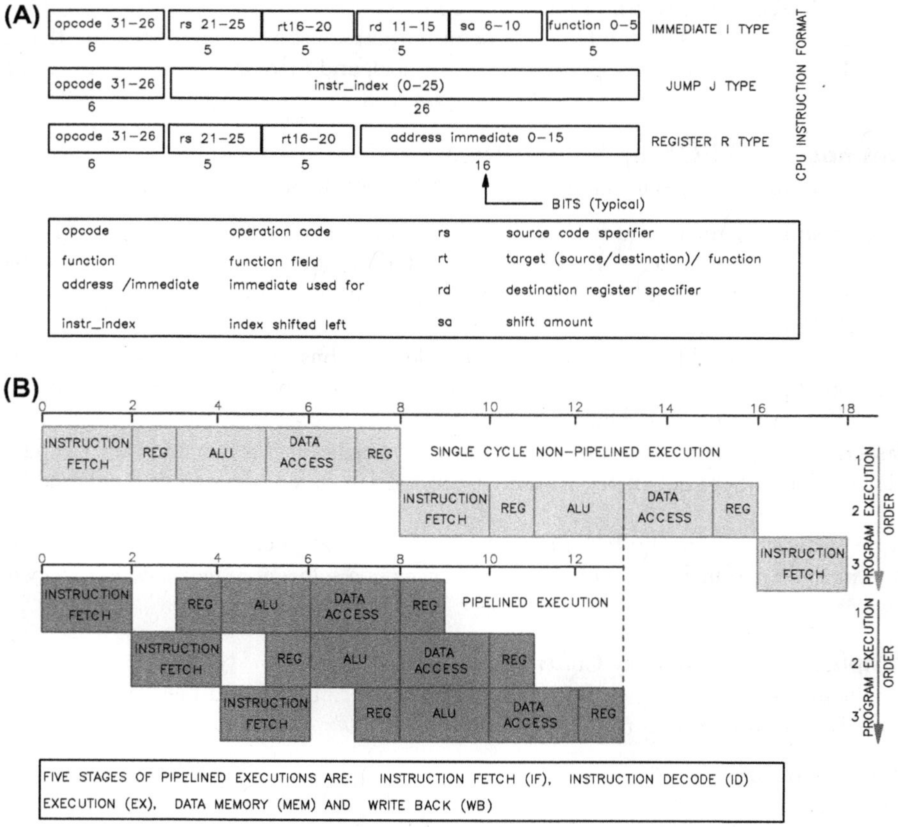

FIGURE APV/3.2.4-1

Instruction and execution type. (A) MPIS instruction types with format, (B) execution types.

- I-type-load and stores instructions: As MIPS is based on RISC processor; it is load/store type architecture, meaning that all operations are performed on operands in the processor registers. The main memory is only accessed through the load/store instructions which are used to load a value in to register from memory or store a value from register to memory respectively.
- R-type-arithmetic instructions: ALU Immediate, three-operand (e.g., add, and, slt), and shift instructions.
- J-type-jump instructions: These are jump and branch instructions.

Like any other processors, the MIPS implementation also involves two types of logic elements: combinational and sequential elements. Combinational elements are elements that operate on data values, while sequential elements are elements that contain and hold a state (hence two I/Ps, one being clock and one O/P). State elements in the MIPS implementation include the register file, instruction memory, and data memory [3]. There are two types of executions one is single cycle non-pipelined and pipelined execution. Pipelining, a standard feature in RISC processors, is used to improve both clock speed and overall performance. Fig. APV/3.2.4-1B shows how pipelined execution allows a processor to work on different steps of the instruction at the same time. So, more instruction can be executed in a shorter period of time. It is worth noting that both implementations use the same hardware components. There are five stages in MIPS pipelined single cycle execution as depicted in Fig. APV/3.2.4-1. The beauty of pipelined single cycle is the application of pipeline registers, which are used to separate data path into five stages; IF, ID, EX, MEM, and WB (shown in Fig. APV/3.2.4-1). With this brief outline on salient features of MIPS processor in FPGA, discussions on processors in FPGA is concluded. The discussions may not be complete without brief focus on HDL/VHDL etc. So, now the focus will be on the same.

4.0 CONCEPTUAL DISCUSSIONS ON HARDWARE DESCRIPTION LANGUAGE (HDL)

In this part, a short discussion on HDL shall be put forward so that reader can develop concepts about the requirements, objectives, and application of HDL. HDL is essential for electronic system/device design and development, especially in embedded controls. There are several abstraction levels involved in digital circuits for design developments in embedded systems, for example, it may be necessary to describe the *layout* of the wires, resistors, and transistors on an integrated circuit (IC) chip for designing an embedded digital circuit. Also it may be necessary to describe the *characteristics* of the circuits. Designers of electronic hardware describe the *behavior and structure* of system and *circuit designs* using HDLs. HDL has been developed specifically for this purpose. HDL is specialized programming language to include *propagation time*, *signal strength*, etc., hence is not like traditional programming language, viz. JAVA or C. Two major HDLs used currently are: **VHDL** (see Fig. APV/1.4-1) and **Verilog** (there is another called system Verilog). In spite of similarities in capability, scope, and industrial supporting software tools, they have different Syntax and "appearance." It is better to compare the two languages.

4.0.1 *Comparison of VHDL and Verilog*

Table APV/4.0.1-1 has been presented to show the similarities and differences between the two types of HDLs.

Table APV/4.0.1-1 Comparison of VHDL and Verilog

Comparing Point	VHDL	Verilog
Usage	General purpose digital design languages. Additionally it is easily available, support many methodology (bottom up/top down), and independent of technology. Large scale design and reuse.	General purpose digital design languages.
Style and Similarity	More like PASCAL[a] OR ADA[a] PL.	Similarity with C language.[a]
Modeling	Effective for hardware structural and functional modeling and simulation.	Effective for hardware structural and functional modeling and simulation. However it is harder to model with abstract modeling.
Language characteristic	VHDL is a rich and strongly typed language, deterministic with strong verbose and self documenting.	Verbose is not that strong and it is not considered self documenting.
Learning	Engineers need to do extra coding to convert from one data type to another.	Easier to learn.
Error detection	More capability.	Less capability.
Modeling spectrum	Covers higher level in behavioral abstraction.	Covers lower level in behavioral abstraction.
EDA tool	Unambiguous semantic and portability of tools are allowed.	Portability is missing.
Compilation	Multiple design-units in same file may be separately be compiled if desired.	Still rooted to original interpretation mode.
Data type	Huge language and user data type.	Simple language but data types in models are based on Verilog language, and not user defined.
Package	Procedure and functions are packaged hence reusable.	No concept of package.
Library	Supported for storage.	No such concept.
Procedure and task	Concurrent procedure calls are allowed.	Concurrent procedure calls are not allowed.
Structural replication	Generation of statement for replication of instances.	No such generation of statement.

[a]*These are different programming language with usage in instrumentation.*

4.0.2 The Concept (HDL)

The concept of HDL will be clear from the study of Gajski and Kuhn chart shown in Fig. APV/4.0.2-1A. Basic characteristic features of HDL concept shall include mainly the following issues [6]:

- Timing
- Concurrency
- Hardware Simulation: Analysis, elaboration, and simulation
  - Signal update
  - Selective re-simulation

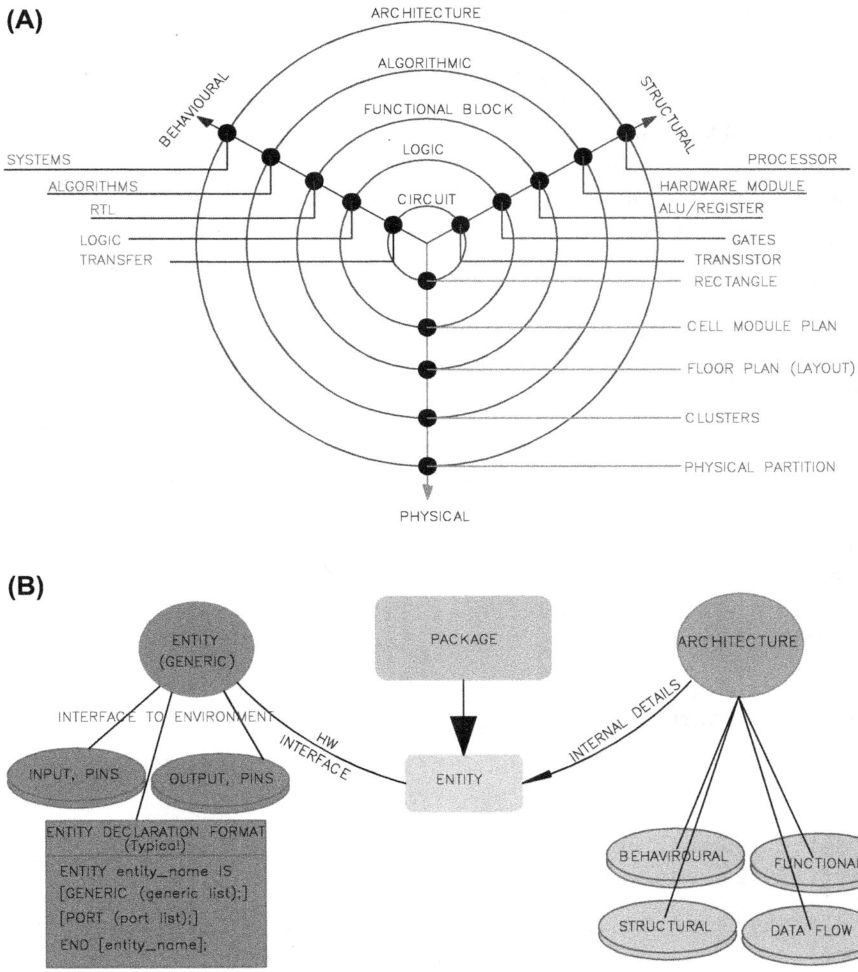

FIGURE APV/4.0.2-1

Abstraction level and VHDL model. (A) Gajski and Kuhn's chart (in VHDL), (B) VHDL conceptual model.

The very basic concept of HDL lies with the following philosophy designers need to adopt:

- Hierarchical design.
- Modular architecture.
- Various abstraction levels.
- Text-based, rather than pictorial descriptions.
- Re-use of existing resources.

These abstraction levels here refer to functional, structural, or physical views of the design as shown in Fig. APV/4.0.2-1A. When looking at any abstraction level, say structural abstraction, it is found that at higher level functional block, (say) processor at lower level, (say) register, and further lower level gates and circuits. In basic files, there are generic entity and architectural details as shown in Fig. APV/4.0.2-1B.

Entity declaration defines an entity and the interface between entity and the external environment. It is defined in a specific format as shown in Fig. APV/4.0.2-1B.

An HDL must be able to simulate a system whose components have been designed to different levels of detail. It is advantageous to do as much work at higher level. With this concept in mind, it is time to look at the basics of HDL.

4.0.3 Application Side

HDLs are used for:

- Design management
- Hardware generation
- Design entity
- Description of
    - Interfaces
    - Behavior
    - Structure
- Synthesis
- Test benches
- Verification
- Documentation

4.1 Basics of Hardware Description Language

During discussions on this concept, it has been stated that HDL is in hierarchical structure. In that, there are several approaches:

4.1.1 Hierarchy of Design

There two types such as: bottom—up and top—down approaches.

- Bottom—up Design: In this traditional approach of electronic design, design starts with transistors and moving to a higher level of gates, and finally, the system. With the increase in design complexity, this approach has to give way to new structural, hierarchical design methods.
- Top—down design: Top—down design allows early testing, fabrication technology independence, a structured system design, etc. So, in HDL, this approach is adopted. In reality, some mix of both the approaches is used in place of pure top—down approach.

4.1.2 Abstraction Domain

The abstraction hierarchy has been shown in Fig. APV/4.0.2-1A. Various levels in generalized form has been elaborated as follows:

- Behavioral or algorithmic Level: This level describes a system by concurrent algorithms. Only the functional description of the individual blocks is described and not the intricacies of the system.
- Register—transfer level (RTL): At RTL, the characteristics of a circuit by operations and the transfer of data between the registers, etc. are defined. An explicit clock is used. RTL design

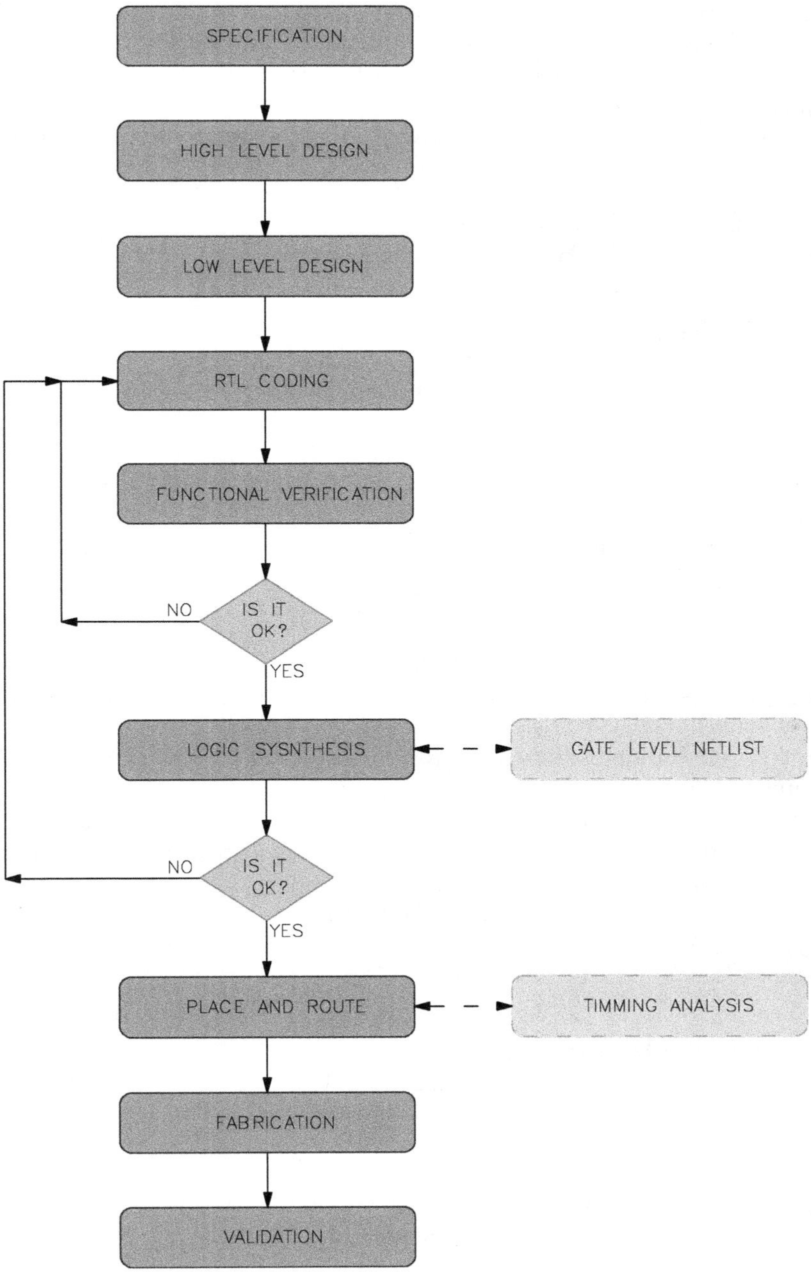

FIGURE APV/4.2-1

Typical design flow chart for hardware description language.

contains exact timing possibility; operations are scheduled to occur at certain times. The modern definition of a RTL code is "any code that is synthesizable is called RTL code" [7].
- Gate level: At this level, the characteristics of a system are described by logical links and their timing properties.
- Physical development: This is the time to develop doing the layout place and route, and then fabrication.

4.1.3 *Important Characteristic Features*

With reference to the concept of HDL discussed above, a few characteristic features of HDL shall include, but are not limited to the following:

- The design process has to be hierarchical.
- Hardware components are concurrent.
- HDL must be able to describe and simulate a system (with components designed at different levels of detailing) at
  - Behavioral level
  - Structural level
  - Mixed level
- HDL shall have ability to synthesize (structure from behavior).

4.2 Design Flow

Design flow in HDL is very important. Design flow broadly covers five major procedures such as: functional design, synthesis, place and route, system integration, and fabrication. Typical HDL design flow has been elaborated in Fig. APV/4.2-1.

The design flow is not a waterfall diagram, as it has a number of iterative processes in which the designer can return to or redo any step until the proper functionality is achieved. Within each step, with the exception of the fabrication process, there is a complete generate—simulate cycle (not shown explicitly), in which the design components are developed and then simulated to ensure correct functionality before moving on to the next step.

With this, the discussion of HDL as well as embedded control is concluded. Also, short discussions on embedded control comes to an end. To see the impact of cost for life cycle, see the next appendix, that is, , Appendix VI, cost impact discussions.

LIST OF ABBREVIATIONS

ADC	Analog to digital converter
AI/O	Analog input/output
ARM	Advanced RISC (*see below*) machines
ASIC	Application specific integrated circuits
BLE	Basic logic element
CPU	Central processing unit
DAC	Digital to analog converter
DCS	Distributed control system

DI/O	Digital input/output
EDA	Electronic design automation
E/E/PE	Electrical/electronics /programmable electronics
FE	Final element
FPGA	Field programmable gate array
GPR	General purpose register
HDL	Hardware description language
HW	Hardware
IEC	International Electrotechnical Commission
I/O	Input/output
I/P or O/P	Input or output
IT	Information technology
LS	Logic solver
MCU	Microcontroller unit
MIPS	Microprocessor without interlocked pipeline stages
P&ID	Piping and instrumentation diagram
PL	Programming language
PLC	Programmable control system
PLD	Programmable logic device
PSW	Processor status word
RAM	Random access memory
RISC	Reduced instruction set computer
ROM	Read only memory
SFR	Special function register
SIS	Safety instrumented system
SW	Software
UART	Universal asynchronous receiver/transmitter
VHDL	VHSIC hardware description language
VHSIC	Very high speed integrated circuit

References

[1] Basics of Microcontrollers; Circuit Today, January 2011; http://www.circuitstoday.com/basics-of-microcontrollers.

[2] T. Agarwal, Different Types of Registers Used in 8051 Microcontroller; https://www.elprocus.com/know-about-types-of-registers-in-8051-microcontroller/.

[3] V.P. Rubio, J. Cook. A FPGA Implementation of a MIPS RISC Processor for Computer Architecture Education; New Mexico State University, July 2004. http://www.ece.nmsu.edu/~jecook/thesis/Victor_thesis.pdf.

[4] FPGA Tutorial; FGPA Radi-Electronics; http://www.radio-electronics.com/info/data/semicond/fpga/fpga.php.

[5] H. Fletcher, FPGA Embedded Processors Revealing True System Performance; Memec USA; Embedded Systems Conference San Francisco, 2005; http://www.xilinx.com/products/design_resources/proc_central/resource/ETP-367paper.pdf.

[6] D. Sharma, Hardware Description Languages Basic Concept; Microelectronics Group, EE Department; IIT Bombay, May 2006. https://www.ee.iitb.ac.in/~smdp/DKStutorials/hdl.pdf.

[7] Design of Embedded Processors; Version 2; EE IIT, Kharagpur; http://www.nptel.ac.in/courses/108105057/Pdf/Lesson-21.pdf.

Further Reading

ASIC-System on Chip VLSI Design: What is the Difference Between FPGA and ASIC; Digital Chip Design Article; http://asic-soc.blogspot.in/2007/11/what-is-difference-between-fpga-and_06.html.

S. Brown, J. Rose, Architecture of FPGAs and CPLDs: A Tutorial; Department of Electrical and Computer Engineering; University of Toronto; http://www.eecg.toronto.edu/wjayar/pubs/brown/survey.pdf.

G. Budzy_n, Programmable Logic Design; http://www.ue.pwr.wroc.pl/pld/pld_12.pdf.

P. Chu, Hardware Description Language; RTL Hardware Design; http://academic.csuohio.edu/chu_p/rtl/chu_rtL_book/silde/chap02_1.pdf.

Difference Between Microprocessor and Microcontroller; Z Series Innovations; http://www.zseries.in/embedded%20lab/8051%20microcontroller/difference%20between%20microprocessor%20and%20microcontroller.php#.VvPpMeJ97IU.

Difference Between Microprocessor and Microcontroller; Electronics Hub; Administrator, May 2015; http://www.electronicshub.org/difference-between-microprocessor-and-microcontroller/.

Difference Between FPGA and Microcontroller; Internet Document; http://www.differencebetween.net/technology/difference-between-fpga-and-microcontroller/.

FPGA or ASIC? Pro's & Con's of Each Technology; Brocade Community Forums: Brocade Blogs: Industry: Service Providers; Blog; http://community.brocade.com/t5/Service-Providers/FPGA-or-ASIC-Pro-s-amp-Con-s-of-Each-Technology/ba-p/709.

FPGA vs. ASIC; What is the Difference Between a FPGA and an ASIC? Xilinx; http://www.xilinx.com/fpga/asic.htm.

FPGA Architecture for the Challenge; Internet Document; http://www.eecg.toronto.edu/wvaughn/challenge/fpga_arch.html.

R. Kamal, PIC Family Microcontroller Pearson Education, 2011. http://www.dauniv.ac.in/downloads/MController_PPTs/MicroC2_eCh13L01PICMicrocontrollerExamples.pdf.

K. Leyba, Verilog VS VHDL; http://electronicdesign.com/what-s-difference-between/what-s-differencebetween-vhdl-verilog-and-systemverilog.

Microcontrollers MCU Basic Structure/Operation; Renesas Engineer School; http://am.renesas.com/edge_ol/engineer/07/index.jsp.

Robot Bonanza Application Notes; http://www.robotoid.com/appnotes/programming-microcontrollerspecial-functions.html.

F. Salewski, S. Kowalewski. Hardware Platform Design Decisions in Embedded Systems e A Systematic Teaching ApproachRWTH Achen University; Embedded Software Laboratory, 2006. http://www.artistembedded.org/docs/Events/2006/WESE06/WESE2006salewski.pdf.

8051 Special Function Registers and Ports; Circuit Today; http://www.circuitstoday.com/8051-specialfunction-registers-and-ports.

What are Microcontrollers and What are They Used for? Architecture and Programming of 8051 MCU;http://learn.mikroe.com/ebooks/8051programming/chapter/what-are-microcontrollers-and-what-are-theyused-for/.

Cost Impact Discussion

1.0 GENERAL

So far during the discussions, various technical aspects have been covered with main focus on safety instrumented systems (SISs) and life cycle (LC). Naturally, it is clear that there is no substitute for LC studies and SIS to face hazardous conditions. However, nothing in the world is available free of cost. Therefore, LC and SIS efforts will also have some cost impact. In this part, short discussions will be put forward on cost impact on account of LC and SIS. In line with guidelines from HSE. UK cost benefit analysis will be very helpful in assessing whether risk reductions are reasonably practicable. One has to weigh cost and benefit prior to applying risk reduction measure. When cost divided by benefit is a very high value, that is, disproportionate factor is much greater than 1, it may be prohibitive. However, cost benefit analysis (CBA) has certain limitations like:

- A CBA cannot be used to argue against *implementation* of relevant *good practice* or any *statutory* duties.
- CBA of its own does *not* constitute as low as reasonably practicable (ALARP).

On the other hand, in order to run a unit, it has to be below ALARP. So certain risk reduction measures and SIS are inevitable. Naturally there will be cost impact on account of the same. So, the discussion that follows is based on the fact that LC and SIS are required and justified. Having accepted the above basis cost analysis, is still essential to know real impact as well as what could be the optimum one.

1.1 Safety Instrumented System Justification

From the management point of view, it is essential that the cost impact on account of buying, maintaining, and operating a safety system (i.e., SIS and associated LC). So, a CBA toward SIS will be helpful for all especially to convince the top management with reasoning. In view of IEC 61508 and 61511, LC is closely associated with SIS, so all the times two are not used separately, meaning that SIS will always have LC with it—hence associated cost for the same. SIS may be justified may be on account of a few reasons listed as follows:

- Essential for safety and no alternative methods exist
- Statutory requirement/international standard/code demands
- Lowest cost option for safety (for this CBA will be helpful)
- Prevention of environmental harm/violation of emission standard limits
- Protection against asset losses due to plant damage/lost production

- Protection against human safety
- Protection against loss of company image for not following good engineering practices, which have following major influencing factors:
    - Reputation
    - Share price
    - Customer base
    - Market share

1.2 Cost Impact on Safety Integrity Level

From the discussions in Chapters VII and VIII, it is clear that SIS and safety integrity level (SIL) are very closely coupled. Also, in order to achieve higher SIL, at times people may have to go for higher redundancies and/or fault tolerant design, which by itself will add cost to SIS. The higher the SIL value, the higher will be the cost of the system as is clear from Fig. APVI/1.2-1.

The figure clearly shows that costs escalate significantly with increasing SIL targets. This is because of the fact that on account of higher SIL, one may have to go for redundancies at various levels, as well as a few other factors such fault tolerance issues, safe programmable logic controller, and/or higher maintenance requirements (e.g., as per IEC 61511, SIS will be proof tested at an interval directed by the SIL) will

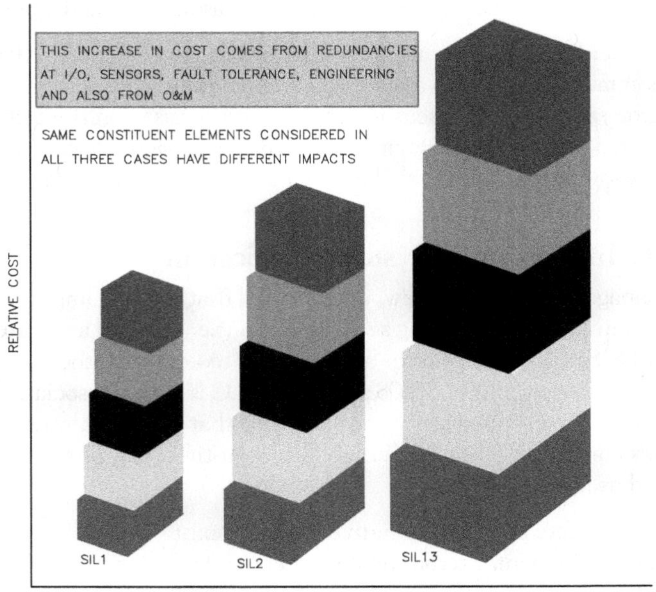

FIGURE APVI/1.2-1

Cost impact on safety integrity level.

increase the cost. With the same in mind, it is better to look at details into the system, by exploring life cycle costing process.

2.0 LIFE CYCLE COSTING PROCESS

Life cycle costing (LCC) shall be done in such a way that it is clear to the investor. Appropriately developed LCC, along with good engineering judgment, provides a rich set of information for making cost-effective, long-term decisions in a disciplined manner. A typical LC structure for SIS has been depicted in Fig. APVI/2.0-1. As is seen from this simplified figure, there are two clear parts in the cost structure.

2.1 Initial Capital Cost

The total SIS cost comprises two parts; one is initial capital cost, and the other is recurring cost. Fixed costs come from:

- Design engineering cost
- Supply charges for:
  - Sensors
  - Logic solver
  - Final element
- Training charges
- Other service related charges (including supplies), for:
  - Erection/installation
  - Testing
  - Commissioning (including start up)

These costs are mainly incurred at the initial stage of the project up to start up.

2.2 Recurring Charges

Recurring charges are incurred during main running conditions of the plant in terms of the operation and maintenance support charges. Again, there are two parts; one is fixed and the other is variable. These shall include mainly:

- Fixed charges:
  - Staff salary
  - Service charges (annual maintenance contract)
  - Training
- Maintenance charges in the form of
  - Spare inventory
  - Repair charges
  - Software maintenance
  - Testing (e.g., proof testing)
- Variable charges (incidental charges):
  - Hazardous events
  - Spurious trips

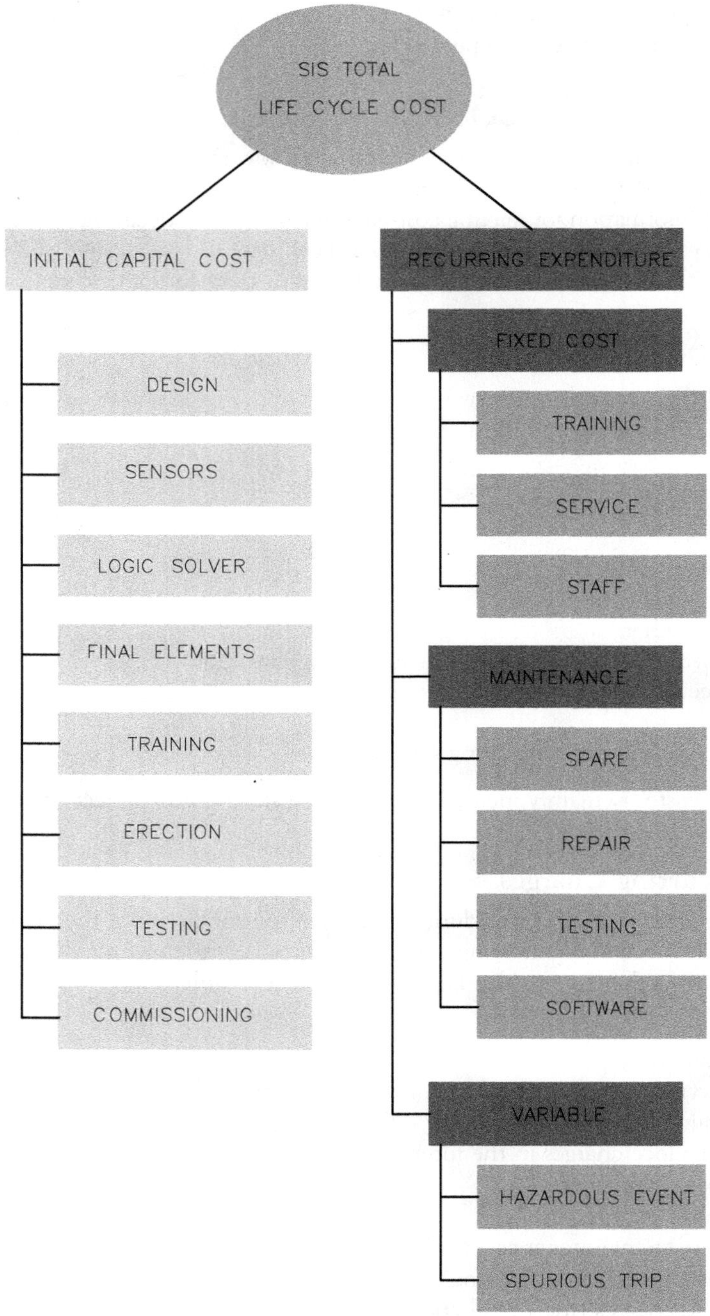

FIGURE APVI/2.0-1

Life cycle cost structure for safety instrumented systems (typical).

2.3 Variation of Costing Due to Different Reliability Model

In Clause 1.2, it has been shown that on account of variations in SIL requirements, there are variations in overall costing. Now from a reliability point of view, there will be variations in costing also, even if there is no change in SIL number. HSE.UK [3] gave a good example to show that in two cases for same SIL number (same overall PFD_{avg}), the number of spurious trips are different due to different redundancies chosen at sensor and at LS levels. Since there is variation in numbers of spurious trips, hence there will be a change in overall costing. These are already discussed in Chapter VIII. For details, HSE.UK [3] may be referenced also. With this preliminary knowledge in mind, it is better to explore the LCC analysis part, which is also very interesting.

3.0 LIFE CYCLE COST ANALYSIS

In a project/process/product, there is involvement of a number of disciplines. Each of these disciplines has different ways of looking at LCC. A person from project engineering would like to reduce the capital expenditure. A maintenance engineer is interested in ensuring that LCC be done in such a way that good maintenance strategy is adopted to reduce repair hours. A reliability engineer will like to have low failure rate, while production personnel will like to have higher operating hours. Accounting persons and share holders would like to see increase in net present values (NPVs) and stock value, respectively. Why all these are discussed? This is because LCC analysis is a joint effort requiring input from various disciplines. So good coordination is essential. With this in mind, let us look into the details of the analysis.

Net present value (NPV) is an important economic measure and concept used in projects to present actual situations taking into account discount factors, cash flow, and time. In LCC, analysis NPV is used. In order to make any major important decisions, project engineers/managers highly depend on life cycle costs. This calls for various considerations including how and when sustaining costs occur during the LC of the equipment or project. Adding expected equipment failure rates and renewals from a statistical viewpoint makes analysis about economics smarter and gets the rational decisions closer to real world conditions. In this connection, guidelines from international standard IEC 60300-3-3:2005 may be referenced. This standard is the basis for LCC analysis discussions. Engineers must supply facts (not *opinions*) for LCC calculations [2]. From Fig. APVI/2.0-1, some idea about the acquisition cost and sustaining cost (2—20 times acquisition cost) could be gathered. One needs to keep in mind that this evaluation process is not really a one-time process, but iterative to get the best alternative at that point of time. Before moving on to any other decision-making issue, it is better to have look how such process is carried out.

3.1 Life Cycle Cost Analysis Process

The process discussed here is in generic term as with different facilities there may be some variations. However, the procedure discussed here is more appropriate for process industries such as chemical and oil and gas plant LC cost analysis for SIS. Fig. APVI/3.1-1 gives a general overview of the process. As this is an iterative process, a feedback line from evaluation to problem definition has been shown. This is used to get best possible alternative by iteration.

There are the following seven main steps into which the whole system can be divided:

- Problems definition
- Cost elements definition
- System modeling

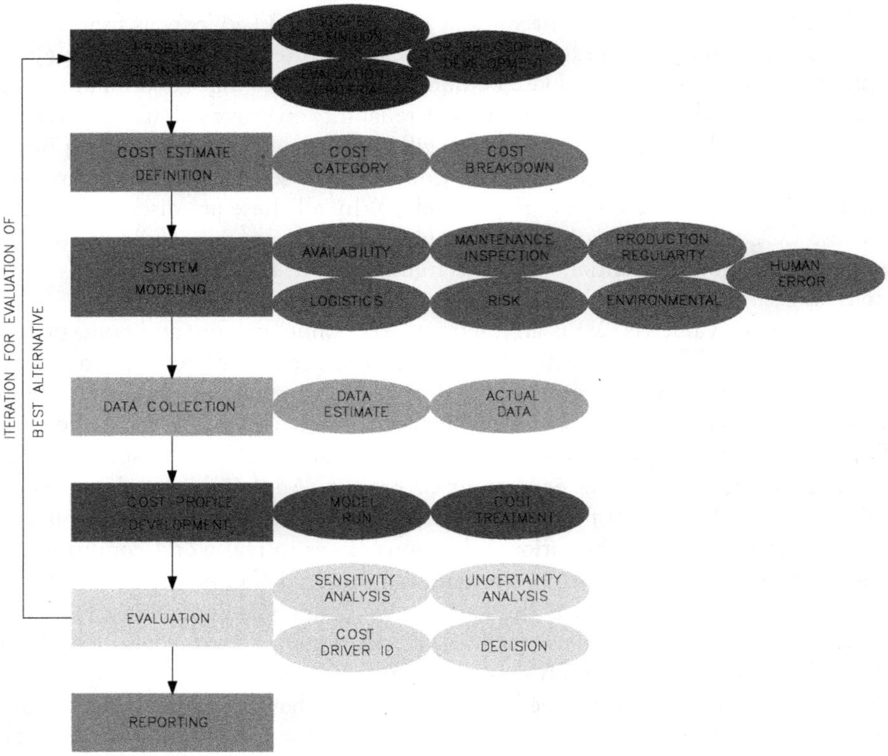

FIGURE APVI/3.1-1

Life cycle cost evaluation for safety instrumented systems.

This figure has been developed based on idea from Y. Kawauchi, M. Rausand, Life Cycle Cost (LCC) Analysis in Oil and Chemical Process Industries, June 1999; http://frigg.ivt.ntnu.no/ross/reports/lcc.pdf.

- Data collection
- Cost profile development
- Evaluation
- Reporting

Each of these sub-steps has subdivisions also, which are shown by the ellipse associated with each step. Now, short discussions on each of them will be taken up in brief so that concepts about the same are well understood.

3.1.1 *Problem Definition*

There are three sub-steps in problem definitions. These are:

- Scope definition: As with any other issues, at the starting point it is necessary that the problem, scope, and boundary limit are well-defined so that there is no ambiguity in targeting the issue. Here the aspects, boundary limits of program phases, and the equipment and activities to be covered in modeling are defined to get a clear definition of the cost elements.
- Evaluation criteria: The criteria based on which evaluation is to be carried out are defined at the beginning as part of problem definition. The criteria normally cover the total cost, system performance, and effectiveness. The system performance characteristics (like availability, maintainability, and SIL of shutdown, etc.) and the effectiveness (like production capacity, product quality, etc.) shall be covered. In many cases regulation, codes, and standards specification play a great role at this point.
- Operational philosophy: This specifies modes and requirements of operation, maintenance strategies (predictive maintenance, proof test requirements, etc.). This is also important in case of problem definitions for LCC analysis. This is also somewhat dependent on an owner's prerogatives.

3.1.2 *Cost Element Definition*

There are two major issues here; one is cost breakdown structure (CBS), and the other is cost category. As mentioned earlier basic guidelines for the same is available from IEC 60300-3-3:2005 (with special reference to dependability).

- Cost breakdown structure (CBS): It is needless to tell that most vital work is to identify all cost items/cost elements that have considerable influence on the total LCC of the system. Fig. APVI/2.0-1 gives broad structure for costing. Naturally, the same is to be defined in a systematic manner. Also, development of CBS is required as per IEC 60300-3-3. CBS may be developed by defining items along three independent axes, which are, "life cycle phase," "product/work breakdown structure," and "cost categories."
- Cost categories: It is difficult to define generalized cost elements that are applicable for every LCC analysis. This is because of the fact that LCC analysis may be applied to various types of systems each with different requirements. It is recommended that CBS and cost categories should be tailored for each application area for LCC analysis [1]. However, in line with

IEC 60300-3-3, "acquisition costs" and "ownership costs" are fairly applicable in most cases. Based on this assumption, Fig. APVI/2.0-1 has been developed. At any time, it is possible to expand the cost categories on the highest level depending on the system to be analyzed. In many cases, like oil and gas or power generation, cost of deferred production (may be generally quantified based on the unavailability performance of the production system, and a unit cost of the product) are considered as it has a lot of impact on LCC. There are other cost categories to include costs like hazard cost, spurious trip cost, etc. as shown in Fig. APVI/2.0-1.

3.1.3 System Modeling

It is necessary to make a model taking in to consideration various factors such as:

- Availability
- Maintenance and inspection
- Logistics
- Risk involved
- Production regularity
- Environmental effect
- Human factor

These are already shown in Fig. APVI/3.1-1. These are influencing factors, so suitable care must be taken to model them to get realistic cost element.

- Availability: As already discussed in previous chapters, availability, maintainability, and human factor (depending on applicability) in SIS are very important and have tremendous impact on cost element.
- Maintenance and inspection: Two kinds of maintenance, viz. "corrective maintenance" and "preventive maintenance" have direct impact on cost element. The frequency of maintenance or inspection directly affect "availability" and the "operating cost" in terms of cost toward man-hours spent, spare part consumption, etc. Turnaround time, mean time to repair (MTTR), etc., are measures for maintenance costs for modeling.
- Logistics: Logistic support in the form of the following may be considered for modeling purposes:
  - Maintenance personnel
  - Training and training support
  - Supply support
  - Spare inventory
  - Support equipment
  - Computer resources
  - Packing, handling, storage, and transportation
  - Maintenance facilities
  - Technical data and information systems
- Risk: The potential risk related to a system is not only useful information, but it also needs to be considered in modeling. In this connection, IEC-60300-3-3 may

be referenced (it recommends considering liability costs from risk analysis to be considered LCC analysis). It is also recommends to include warranty costs in the CBS of the LCC analysis [1].

- Production regularity is a term used to describe how a system is capable of meeting demand for deliveries or performance. It depends on system availability, production availability, and deliverability.
- Environmental: After the Convention on Climate Change, in the Kyoto Meeting in 1997, globally there is great concern over impact of production on the surrounding environment. Now almost all countries have their own laws or international laws and standards which need to be met by all concerned. Therefore, plant owners need to spend money toward pollution prevention, viz. SO_2, CO_2, and particulate emission from power plants; owners need to take necessary measures to limit such emissions within limits. Naturally, these will have impact on LCC.
- In the actual operation, the effect of human error cannot be overestimated. All such effects may have tremendous impact on hazardous situations, etc. So the contribution of the human error is not negligible in many cases, especially in cases where there are manual interventions frequently. There are broadly three categories of human error, viz.:
  - Omission error
  - Action error
  - Extraction error

Similarly, there many techniques to face and combat human error such as technique for human error rate prediction (THERP), human error assessment and reduction (HEART), etc. already discussed in previous chapters.

3.1.4 Data Collection

It is quite obvious that LCC analysis is done based on data, naturally the more accurate the data the more realistic the LCC will be. These data are input data necessary to carry out the analysis. There are two types of data; one is estimated data and the other is actual data. If latter one is available, then it can be directly applied to CBS. Otherwise one has to depend on estimated data based on expert judgments.

- Estimated data: When actual data is not available, the value of data may be estimated. The following types of methods are used for data estimations:
  - Stochastic: Specialized statistical method.
  - Parametric: Another statistical method used on historical data for estimation of cost factor and/or cost estimation, etc.
  - Analogous: Used to establish relationship between current data and previous data duly judged by experts.
- Actual data: There are many data sources available for various reliability, etc. data, but it is difficult to get actual cost data. So, one has to depend on operational and cost data from the database of the *operating companies* [1].

3.1.5 Cost Profile Development

One factor of LCC analysis is an affordability analysis with due considerations for long-term financial planning. Therefore, it is necessary to draw a cost profile over entire life. It is obviously noticed that the cost profile of each design case should be compared on a common basis or reference point when making financial judgments [1]. There are two aspects here:

- Model run: Cost profile is developed by running cost models developed in an LCC analysis with input data. Computer tools can be used to run a model or it could be manual calculations in a spreadsheet.
- Cost treatment: In this, various aspects like effect of inflation, interest rates, exchange rates and taxation, etc. are considered for financial judgment. Many times, the cost profiles are made on the basis of "constant prices." This is adopted for those cases where it is difficult to accurately predict inflation and exchange rate, etc. In such cases, it is necessary to compare the alternatives on a common baseline.

3.1.6 Evaluation

It is the aim of this analysis is to find out the most desirable alternative configuration. In order to do the same, one needs to check if the baseline system meets the criteria defined in the problem definition. If not, the baseline system should be modified as an alternative system, and the LCC of the alternative system should be evaluated. There are several issues and checkpoints; major issues are listed as follows:

- Sensitivity analysis: The main task of sensitivity analysis is to find the impact of changes in input parameters on the result. This is done by making variations in the input parameters over a range to see if the impact on cost can help highlight the major factors affecting costs. There are several methods available for sensitivity analysis. Mainly "deterministic" and "stochastic" approach are used.
- Uncertainty analysis: Uncertainty analysis is done to consider possible ranges of the estimate and their effect on decisions. Three categories of uncertainties are:
  - Parameter
  - Modeling
  - Completeness

  This will give *confidence* to decision-makers to make financial decisions.
- Cost driver: Identification of cost drivers in LCC analysis is one of the major issues. Cost drivers have major impact on the total LCC. Once a cost driver is identified, it is important to establish cause-and-effect relationships so that system design may be modified to effectively reduce the causes of cost drivers, which in effect will reduce the total LCC.

3.1.7 Reporting

Documentation of the entire process is extremely important and these reports could be used in the future as database.

3.2 Cost Analysis Timing

Theoretically, LCC analysis can be done any time in any phase of the project. In fact, it should be a continuous process. However, earlier identification of acquisition and ownership costs give the investor a better chance to balance reliability, performance, and maintenance.

Fig. APVI/3.2-1 provides curves of commitment and expenditure trends. There is another interesting curve, which is the dotted curve which shows how cost reduction chances vary with time during the LC of the plant/project/process/

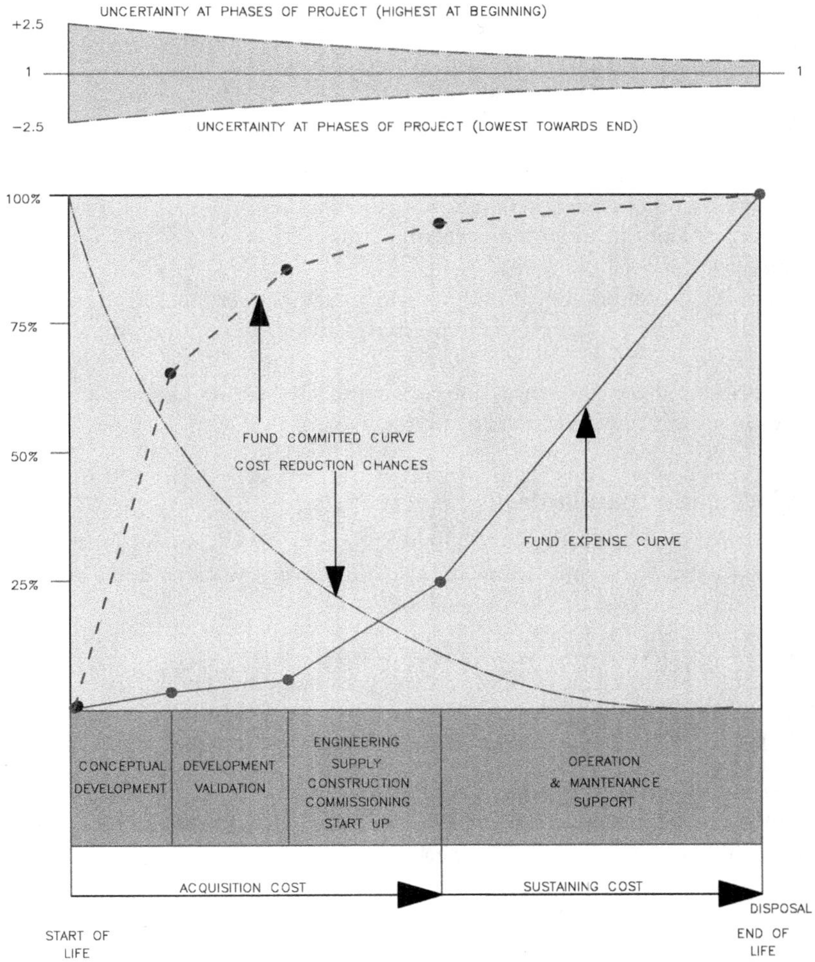

FIGURE APVI/3.2-1

Life cycle cost decision at phases of project.

product. Little LCC opportunity exists with start after construction when it is not possible to significantly change LCC. An interesting feature to note here, is that about 95% of commitment has to be done prior to the end of acquisition cost period, but only 50% is spent at this time. Major expenditures will be done during the sustaining period. Another issue is uncertainty. An uncertainty curve within the range of LC period has been shown outside. Obviously, as the LC is toward the end of life, uncertainty asymptotically approaches to one, which is quite obvious. In view of the discussions, it is believed that tradeoff between uncertainty curve and commitment curve will decide the best timing for taking up LCC analysis.

3.3 Life Cycle Costing Analysis Application

Application of LCC analysis shall include, but is not limited to the following:

- Assessing economic viability of projects/products.
- Evaluation and comparison of alternative design.
- Cost driver identification and improvement.
- Long term financial planning.
- Optimization of fund allocation for various activities and facilities.
- Evaluation and comparison of alternative strategies in different areas (e.g., maintenance).
- Evaluation and comparison of different approaches for renovation, etc.
- Assessment of product assurance criteria.

3.4 Codes and Standards

As indicated at the beginning, there will be variations of LCC analysis with industries and applications. There are various standards to cover the same; a few are noted as follows to conclude cost impact discussions.

- IEC 60300-3-1/23/9/11:2005 Life cycle costing (3-3).
- ISO 15663 Life cycle costing within the petroleum and natural gas industries.
- SAE ARP-4293: Life cycle cost - Techniques and applications.
- API RP 580/581: Risk based inspection.

Hope you enjoyed reading the book and look forward to your feedback in the form of review. Any feedback, comments (good or bad), or suggestions from you is very much valuable and is always welcome. — Author.

LIST OF ABBREVIATIONS

AI/O	Analog input/output
CBA	Coast benefit analysis
CBS	Cost breakdown structure
DI/O	Digital input/output
E/E/PE	Electrical/electronics/programmable electronics

FE	Final element
FPGA	Field programmable gate array
HEART	Human error assessment and reduction
HW	Hardware
IEC	International Electrotechnical Commission
I/O	Input/output
I/P or O/P	Input or output
IT	Information technology
LC	Life cycle
LS	Logic solver
MTTR	Mean time to repair
NPV	Net present value
P&ID	Piping and instrumentation diagram
SIS	Safety instrumented system
SW	Software
THERP	Technique for human error rate prediction

References

[1] Y. Kawauchi, M. Rausand, Life Cycle Cost (LCC) Analysis in Oil and Chemical Process Industries, June 1999. http://frigg.ivt.ntnu.no/ross/reports/lcc.pdf.

[2] P. Barringer, Life Cycle Cost Analysis—Who Does What? Barringer & Associates, Inc.; American Society of Mechanical Engineers, and PennWell Publishing; http://www.barringer1.com/pdf/Barringer-NPRA-5-27-04.pdf.

[3] Cost Benefit Analysis (CBA) Checklist; http://www.hse.gov.uk/risk/theory/alarpcheck.htm.

Further Reading

H.P. Barringer, Life Cycle Cost & Reliability for Process Equipment; Barringer & Associates, Inc., Humble, TX; http://www.barringer1.com/pdf/lcc_rel_for_pe.pdf.

D. Macdonald, Practical industrial safety, risk assessment and shutdown system; Newness.

Planning Your Safety Instrumented System; White Paper; Honeywell.

Overview of Safety Instrumented Systems; IDC Technologies; http://www.idc-online.com/downloads/ALL_BOOKBOON_SS_PDF_REV1.pdf.

Index

'*Note*: Page numbers followed by "f" indicate figures and "t" indicate tables.'

K

L

M

R

S

T

U

V

W